PIPING DESIGN HANDBOOK

EDITED BY

JOHN J. McKETTA

The University of Texas at Austin
Austin, Texas

MARCEL DEKKER, INC. NEW YORK • BASEL • HONG KONG

Library of Congress Cataloging-in-Publication Data

Piping design handbook / edited by John J. McKetta.
p. cm.
"The contents of this volume were originally published in Encyclopedia of chemical processing and design, edited by J.J. McKetta and W.A. Cunningham"--T.p. verso.
Includes index.
ISBN 0-8247-8570-3
1. Pipe lines--Design and construction--Handbooks, manuals, etc.
I. McKetta, John J. II. Encyclopedia of chemical processing and design.
TJ930.P578 1992
660'.283--dc20 91-43768
CIP

This book is printed on acid-free paper.

MARCEL DEKKER, INC.
270 Madison Avenue, New York, New York 10016

Current printing (last digit):
10 9 8 7 6 5 4 3

PRINTED IN THE UNITED STATES OF AMERICA

Preface

This book represents an encyclopedia of piping design methods. Almost every phase of piping design is covered and presented in an easy-to-read and easy-to-use format. Each article is prepared by a world expert in that particular area. For simplicity, *Piping Design Handbook* is arranged in 10 general sections to present the elements of fluid flow, fluidics, piping design, pipeline design, buried pipelines, pipeline supports, pipeline shortcut methods, operation and maintenance, pipeline failure, and the economics and costs of pipeline systems including a compendium of worldwide activity.

Every practicing professional engineer involved in piping design should have a copy of this extremely helpful handbook on his or her working shelf.

The basic principles are extremely well covered by experts in the field such as Professor Neil S. Berman. Another important section is the pipeline shortcut methods, which includes 22 articles presenting expanded rules of thumb for the piping design engineer.

I am grateful for all the help I received from the great number of experts who have contributed to this book.

JOHN J. McKETTA

Contents

Contributors

Charles H. Artus, P.E. Retired, The Gates Corporation, Denver, Colorado

T. N. Baker Retired, Pipeline Authority, Canberra City, Australia

John A. Barnette, P.E. Owner, John Barnette & Associates, Brenham, Texas

Neil S. Berman, Ph.D. Professor, Department of Chemical, Bio and Materials Engineering, Arizona State University, Tempe, Arizona

W. Wayne Blackwell Principal Process Engineer, S.I.P. Engineering Inc., Houston, Texas

Michael L. Bradford, P.E. Central Region Environmental Division Manager, Jacobs Engineering Group, Inc., Houston, Texas

Arthur W. Brooks, P.E. TERCHMAR Engineering, Inc., Plainsboro, New Jersey

Robert K. Broyles, P.E. Manager of Engineering, Pathway Bellows Inc., Oak Ridge, Tennessee

Michael V. Calogero, P.E. GESTRA, Inc., Hackensack, New Jersey

F. Caplan, P.E. Retired, Oakland, California

A. J. Carleton Senior Principal Research Officer, Warren Spring Laboratory, Stevenage, United Kingdom

Arnold J. Carrico Pipeline Engineer, Bechtel Inc., San Francisco, California

Lance W. Chantos

C. V. Char Consulting Civil Engineer, Baton Rouge, Louisiana

David C.-H. Cheng, Ph.D., C.Eng., F.I. Chem. E. Consultant in Applied Rheology, Warren Spring Laboratory, Stevenage, United Kingdom

Richard D. Chirillo L. D. Chirillo Associates, Bellevue, Washington

John D. Constance, P.E.[†] Consulting Engineer, Cliffside Park, New Jersey

Paolo Conte Technical Manager, IDECO SRL, Rome, Italy

Ralph A. Crozier, Jr., P.E. Project Engineer, E. I. du Pont de Nemours, Wilmington, Delaware

D. W. Culler Maintenance Manager, Shell Oil Company, Wood River, Illinois

[†]Deceased

John D. Dawson

R. H. Derammelaere Vice President, ETSI Pipeline Project, San Francisco, California

Joseph S. Dorsey, P.E. Consultant, Escondido, California

Robert J. Eiber Director, Transmission Pipeline Programs, Battelle Memorial Institute, Columbus, Ohio

Harry F. Fabisch, P.E. Supervising Control Systems Engineer, Fluor Engineers Inc., Irvine, California

I. H. Farina Process Manager, IPAKO S.A., Buenos Aires, Argentina

Maurizio Felici Piping Senior Engineer, TPL (Tecnologie Progetti Lavori) S.p.A., Rome, Italy

V. Ganapathy, Ph.D. Abco Industries, Abilene, Texas

Edward Gordon, Ph.D. Consultant, Laguna Hills, California

Nigel I. Heywood, Ph.D., C. Eng., F.I. Chem. E. Principal Research Chemical Engineer, Warren Spring Laboratory, Stevenage, United Kingdom

P. D. Hills

William B. Hooper, P.E. Senior Fellow, Engineering, Monsanto Chemical Company, St. Louis, Missouri

P. Mark Hope, M.Sc., D.I.C. Senior Reservoir Engineer, Carless Exploration Limited, London, England

Ronald E. Ingham Engineering Manager, Bechtel Inc., San Francisco, California

K. Jayaraman Senior Product Engineer, Mills, Bharat Heavy Electricals Limited, Tiruchirapalli, India

Dana J. Jones Researcher, Engineering Mechanics, Battelle Memorial Institute, Columbus, Ohio

John F. Kenefick President, JFK, Inc., Indialantic, Florida

Robert Kern Hoffman-La Roche, Inc., Nutley, New Jersey

John F. Kiefner, Ph.D. President, Kiefner and Associates, Inc., Worthington, Ohio

Graeme G. King Chief Engineer, Tensor Engineering Ltd., Calgary, Alberta, Canada

Peter Kipin President, World Pipe Services and Kipin Industries, Inc., Aliquippa, Pennsylvania

Joseph Koslov Retired Chemical Engineer, Great Neck, New York

Gregory S. Kramer Principal Research Scientist, Product and Machine Development, Battelle Memorial Institute, Columbus, Ohio

John A. Kremers, P.E. Manager of Application Engineering, Armstrong Machine Works, Three Rivers, Michigan

C. B. Lester C. B. Lester & Associates, Inc., Longview, Texas

Thomas C. Lippitt Senior Propellant Engineer, National Aeronautics and Space Administration, Kennedy Space Center, Florida

Joseph T. Lonsdale Manager, Sales and Marketing, Chemelex Division, Raychem Corporation, Redwood City, California

Peter A. Marcucci, P.E. Manager, Electrical Inspection, Ontario Hydro, Toronto, Canada

Willard A. Maxey Senior Research Scientist, Battelle Memorial Institute, Columbus, Ohio

G. R. Mayer Bechtel Inc., San Francisco, California

John J. McKetta, Ph.D., P.E. The Joe C. Walter Professor of Chemical Engineering, The University of Texas at Austin, Austin, Texas

Mileta Mikasinovic, P.E. Senior Design Engineer, Ontario Hydro, Toronto, Canada

David Mills, Ph.D. Chief Executive, Bulk Solids Handling Department, Glasgow Polytechnic, Glasgow, Scotland

Jerry E. Mundy Vice President, Business Development, Procon International, Des Plaines, Illinois

C. E. Myrick Bechtel Inc., San Francisco, California

R. G. Nelson Conoco North Sea Inc., London, England

Robert R. Olson Fluor Inc., Irvine, California

R. N. Parkins The William Cochrane Professor and Head of Department, Department of Metallurgy and Engineering Materials, University of Newcastle-upon-Tyne, Newcastle-upon-Tyne, United Kingdom

Ram Prasad Lecturer, Department of Chemical Engineering, Harcourt Butler Technological Institute, Kanpur, India

H. H. Rachford, Jr., Ph.D. President, DREM, Inc., Houston, Texas

Paul Rietjens N. V. Nederlandse Gasunie, Groningen, The Netherlands

James B. Riggs, Ph.D. Professor, Department of Chemical Engineering, Texas Tech University, Lubbock, Texas

G. G. Rochfort Technical General Manager, The Pipeline Authority, Canberra A.C.T., Australia

Richard P. Ruskin

Edwin P. Russo, Ph.D., P.E. Professor, University of New Orleans, New Orleans, Louisiana

L. G. Sala Espiell Process Engineer, IPAKO S.A., Buenos Aires, Argentina

Monte L. Schwartz Manager of Project Development, Lotepro Corporation, Valhalla, New York

Kenneth O. Simpson, Ph.D. Fluor Inc., Irvine, California

Bill Smith, P.E. Manager, Pipeline Engineering, Crest Engineering, Inc., Tulsa, Oklahoma

Leigh A. Smith Professor, University of New Orleans, New Orleans, Louisiana

Rafik Soliman, P.E. Principal Process Engineer, Fluor Engineers, Inc., Houston, Texas

Joseph W. Stanecki, P.E. Process Design Manager, Lummus Crest Inc., Bloomfield, New Jersey

A. A. Sultan, Ph.D. Technical Director, ECG Engineering Consultants Group, Cairo, Egypt

Warren R. True Pipeline/Gas Processing Editor, *Oil & Gas Journal*, Houston, Texas

Robert J. Tsal, Ph.D. Fluor Inc., Irvine, California

Arthur H. Tuthill Materials and Corrosion Consultant, Riverside, Connecticut

H. J. Van Dyke Bechtel Inc., San Francisco, California

W. R. Wade Senior Systems Engineer, DREM, Inc., Houston, Texas

E. J. Wasp ETSI Pipeline Project, San Francisco, California

C. E. Yamashiro Process Engineer, IPAKO S.A., Buenos Aires, Argentina

Adam Zanker, Ch.E., M.Sc. Senior Research Engineer, Oil Refineries, Ltd., Haifa, Israel

Conversion to SI Units

To convert from	To	Multiply by
acre	square meter (m^2)	4.046×10^3
angstrom	meter (m)	1.0×10^{-10}
are	square meter (m^2)	1.0×10^2
atmosphere	newton/square meter (N/m^2)	1.013×10^5
bar	newton/square meter (N/m^2)	1.0×10^5
barrel (42 gallon)	cubic meter (m^3)	0.159
Btu (International Steam Table)	joule (J)	1.055×10^3
Btu (mean)	joule (J)	1.056×10^3
Btu (thermochemical)	joule (J)	1.054×10^3
bushel	cubic meter (m^3)	3.52×10^{-2}
calorie (International Steam Table)	joule (J)	4.187
calorie (mean)	joule (J)	4.190
calorie (thermochemical)	joule (J)	4.184
centimeter of mercury	newton/square meter (N/m^2)	1.333×10^3
centimeter of water	newton/square meter (N/m^2)	98.06
cubit	meter (m)	0.457
degree (angle)	radian (rad)	1.745×10^{-2}
denier (international)	kilogram/meter (kg/m)	1.0×10^{-7}
dram (avoirdupois)	kilogram (kg)	1.772×10^{-3}
dram (troy)	kilogram (kg)	3.888×10^{-3}
dram (U.S. fluid)	cubic meter (m^3)	3.697×10^{-6}
dyne	newton (N)	1.0×10^{-5}
electron volt	joule (J)	1.60×10^{-19}
erg	joule (J)	1.0×10^{-7}
fluid ounce (U.S.)	cubic meter (m^3)	2.96×10^{-5}
foot	meter (m)	0.305
furlong	meter (m)	2.01×10^2
gallon (U.S. dry)	cubic meter (m^3)	4.404×10^{-3}
gallon (U.S. liquid)	cubic meter (m^3)	3.785×10^{-3}
gill (U.S.)	cubic meter (m^3)	1.183×10^{-4}
grain	kilogram (kg)	6.48×10^{-5}
gram	kilogram (kg)	1.0×10^{-3}
horsepower	watt (W)	7.457×10^2
horsepower (boiler)	watt (W)	9.81×10^3
horsepower (electric)	watt (W)	7.46×10^2
hundred weight (long)	kilogram (kg)	50.80
hundred weight (short)	kilogram (kg)	45.36
inch	meter (m)	2.54×10^{-2}
inch mercury	newton/square meter (N/m^2)	3.386×10^3
inch water	newton/square meter (N/m^2)	2.49×10^2
kilogram force	newton (N)	9.806

To convert from	To	Multiply by
kip	newton (N)	4.45×10^{3}
knot (international)	meter/second (m/s)	0.5144
league (British nautical)	meter (m)	5.559×10^{3}
league (statute)	meter (m)	4.83×10^{3}
light year	meter (m)	9.46×10^{15}
liter	cubic meter (m^3)	0.001
micron	meter (m)	1.0×10^{-6}
mil	meter (m)	2.54×10^{-6}
mile (U.S. nautical)	meter (m)	1.852×10^{3}
mile (U.S. statute)	meter (m)	1.609×10^{3}
millibar	newton/square meter (N/m^2)	100.0
millimeter mercury	newton/square meter (N/m^2)	1.333×10^{2}
oersted	ampere/meter (A/m)	79.58
ounce force (avoirdupois)	newton (N)	0.278
ounce mass (avoirdupois)	kilogram (kg)	2.835×10^{-2}
ounce mass (troy)	kilogram (kg)	3.11×10^{-2}
ounce (U.S. fluid)	cubic meter (m^3)	2.96×10^{-5}
pascal	newton/square meter (N/m^2)	1.0
peck (U.S.)	cubic meter (m^3)	8.81×10^{-3}
pennyweight	kilogram (kg)	1.555×10^{-3}
pint (U.S. dry)	cubic meter (m^3)	5.506×10^{-4}
pint (U.S. liquid)	cubic meter (m^3)	4.732×10^{-4}
poise	newton second/square meter ($N \cdot s/m^2$)	0.10
pound force (avoirdupois)	newton (N)	4.448
pound mass (avoirdupois)	kilogram (kg)	0.4536
pound mass (troy)	kilogram (kg)	0.373
poundal	newton (N)	0.138
quart (U.S. dry)	cubic meter (m^3)	1.10×10^{-3}
quart (U.S. liquid)	cubic meter (m^3)	9.46×10^{-4}
rod	meter (m)	5.03
roentgen	coulomb/kilogram (c/kg)	2.579×10^{-4}
second (angle)	radian (rad)	4.85×10^{-6}
section	square meter (m^2)	2.59×10^{6}
slug	kilogram (kg)	14.59
span	meter (m)	0.229
stoke	square meter/second (m^2/s)	1.0×10^{-4}
ton (long)	kilogram (kg)	1.016×10^{3}
ton (metric)	kilogram (kg)	1.0×10^{3}
ton (short, 2000 pounds)	kilogram (kg)	9.072×10^{2}
torr	newton/square meter (N/m^2)	1.333×10^{2}
yard	meter (m)	0.914

Bringing Costs up to Date

Cost escalation via inflation bears critically on estimates of plant costs. Historical costs of process plants are updated by means of an escalation factor. Several published cost indexes are widely used in the chemical process industries:

Nelson Cost Indexes (*Oil and Gas J.*), quarterly
Marshall and Swift (M&S) Equipment Cost Index, updated monthly
CE Plant Cost Index (*Chemical Engineering*), updated monthly
ENR Construction Cost Index (*Engineering News-Record*), updated weekly

All of these indexes were developed with various elements such as material availability and labor productivity taken into account. However, the proportion allotted to each element differs with each index. The differences in overall results of each index are due to uneven price changes for each element. In other words, the total escalation derived by each index will vary because different bases are used. The engineer should become familiar with each index and its limitations before using it.

Table 1 compares the CE Plant Index with the M&S Equipment Cost

TABLE 1 *Chemical Engineering* and Marshall and Swift Plant and Equipment Cost Indexes since 1950

Year	CE Index	M&S Index	Year	CE Index	M&S Index
1950	73.9	167.9	1971	132.3	321.3
1951	80.4	180.3	1972	137.2	332.0
1952	81.3	180.5	1973	144.1	344.1
1953	84.7	182.5	1974	165.4	398.4
1954	86.1	184.6	1975	182.4	444.3
1955	88.3	190.6	1976	192.1	472.1
1956	93.9	208.8	1977	204.1	505.4
1957	98.5	225.1	1978	218.8	545.3
1958	99.7	229.2	1979	238.7	599.4
1959	101.8	234.5	1980	261.2	659.6
1960	102.0	237.7	1981	297.0	721.3
1961	101.5	237.2	1982	314.0	745.6
1962	102.0	238.5	1983	316.9	760.8
1963	102.4	239.2	1984	322.7	780.4
1964	103.3	241.8	1985	325.3	789.6
1965	104.2	244.9	1986	318.4	797.6
1966	107.2	252.5	1987	323.8	813.6
1967	109.7	262.9	1988	342.5	852.0
1968	113.6	273.1	1989	355.4	895.1
1969	119.0	285.0	1990	357.6	915.1
1970	125.7	303.3			

TABLE 2 Nelson Inflation Refinery Construction Indexes since 1946 (1946 = 100)

Date	Materials Component	Labor Component	Miscellaneous Equipment	Nelson Inflation Index
1946	100.0	100.0	100.0	100.0
1947	122.4	113.5	114.2	117.0
1948	139.5	128.0	122.1	132.5
1949	143.6	137.1	121.6	139.7
1950	149.5	144.0	126.2	146.2
1951	164.0	152.5	145.0	157.2
1952	164.3	163.1	153.1	163.6
1953	172.4	174.2	158.8	173.5
1954	174.6	183.3	160.7	179.8
1955	176.1	189.6	161.5	184.2
1956	190.4	198.2	180.5	195.3
1957	201.9	208.6	192.1	205.9
1958	204.1	220.4	192.4	213.9
1959	207.8	231.6	196.1	222.1
1960	207.6	241.9	200.0	228.1
1961	207.7	249.4	199.5	232.7
1962	205.9	258.8	198.8	237.6
1963	206.3	268.4	201.4	243.6
1964	209.6	280.5	206.8	252.1
1965	212.0	294.4	211.6	261.4
1966	216.2	310.9	220.9	273.0
1967	219.7	331.3	226.1	286.7
1968	224.1	357.4	228.8	304.1
1969	234.9	391.8	239.3	329.0
1970	250.5	441.1	254.3	364.9
1971	265.2	499.9	268.7	406.0
1972	277.8	545.6	278.0	438.5
1973	292.3	585.2	291.4	468.0
1974	373.3	623.6	361.8	522.7
1975	421.0	678.5	415.9	575.5
1976	445.2	729.4	423.8	615.7
1977	471.3	774.1	438.2	653.0
1978	516.7	824.1	474.1	701.1
1979	573.1	879.0	515.4	756.6
1980	629.2	951.9	578.1	822.8
1981	693.2	1044.2	647.9	903.8
1982	707.6	1154.2	622.8	976.9
1983	712.4	1234.8	656.8	1025.8
1984	735.3	1278.1	665.6	1061.0
1985	739.6	1297.6	673.4	1074.4
1986	730.0	1330.0	684.4	1089.9
1987	748.9	1370.0	703.1	1121.5
1988	802.8	1405.6	732.5	1164.5
1989	829.2	1440.4	769.9	1195.9
1990	832.8	1487.7	795.5	1225.7

Index. Table 2 shows the Nelson Inflation Petroleum Refinery Construction Indexes since 1946. It is recommeded that the CE Index be used for updating total plant costs, and the M&S Index or Nelson Index for updating equipment costs. The Nelson Indexes are better suited for petroleum refinery materials, labor, equipment, and general refinery inflation.

Since

$$C_B = C_A(B/A)^n \quad (1)$$

Here, A = the size of units for which the cost is known, expressed in terms of capacity, throughput, or volume; B = the size of unit for which a cost is required, expressed in the units of A; $n = 0.6$ (i.e., the six-tenths exponent); C_A = actual cost of unit A; and C_B = the cost for B being sought for the same time period as cost C_A.

To approximate a current cost, multiply the old cost by the ratio of the current index value to the index at the date of the old cost:

$$C_B = C_A I_B / I_A \quad (2)$$

Here, C_A = old cost; I_B = current index value; and I_A = index value at the date of old cost.

Combining Eqs. (1) and (2):

$$C_B = C_A(B/A)^n(I_B/I_A) \quad (3)$$

For example, if the total investment cost of Plant A was $25,000,000 for 200-million-lb/yr capacity in 1974, find the cost of Plant B at a throughput of 300 million lb/yr on the same basis for 1986. Let the sizing exponent, n, be equal to 0.6.

From Table 1, the CE Index for 1986 was 318.4, and for 1974 it was 165.4.

Via Eq. (3):

$$\begin{aligned} C_B &= C_A(B/A)^n(I_B/I_A) \\ &= 25.0(300/200)^{0.6}(318.4/165.4) \\ &= \$61{,}200{,}000 \end{aligned}$$

JOHN J. MCKETTA

1

Fluid Flow

Fluid Flow

I. Introduction

The usual definition of a fluid is a substance that can sustain no static shear stress. Thus a fluid takes the shape of its container, but a shear stress will develop if the fluid flows. We know that some fluids flow more readily than others and a quantitative property of a fluid that describes the ease of flow is the viscosity. Let us put a thin layer of a fluid between two parallel plates, move one plate at a velocity U, and keep the other plated fixed. The width of the plates are to be large compared to the space between them. The force to move the plate divided by the area is the shear stress. The viscosity of the fluid is the ratio of this force to the shear rate (the shear rate here is the velocity divided by the plate spacing). Fluids with a low viscosity require less force to keep the plate in motion at the constant speed U than fluids with a high viscosity. This experiment can be carried out at many different speeds or plate spacings so that a complete curve of shear rate vs shear stress can be found. If this curve is a straight line which passes through the origin, the fluid is called Newtonian. Otherwise we say the fluid is non-Newtonian.

For Newtonian fluids the viscosity is a function of temperature. The viscosity of a gas will increase with temperature at low pressures, but the viscosity of liquids decreases as the temperature is increased. Although the viscosity of gases can be estimated at low pressures using equations derived from the kinetic theory of gases, the viscosity of liquids must be measured.

The solution to problems involving flows of fluids can be started by examination of the equations of change. These consist of conservation of mass, conservation of linear and angular momentum, and conservation of energy. The differential forms of these equations are obtained from a balance about a small element of fluid in the form

$$\{\text{rate of accumulation}\} = \{\text{net rate in by molecular motion}\} + \{\text{net rate in by fluid motion}\} + \text{generation} + \text{other net inputs}$$

Here we wish to consider only flows in pipes and not the entire field of fluid mechanics. Solutions to the equations of change for an isothermal flow with constant density are functions of the Reynolds number, $\rho UL/\mu$, and ratios of other characteristic length parameters. In the following sections we consider incompressible flow, compressible flow, and special flow problems including drag reduction and the flow of heavy crudes.

II. Incompressible Pipe Flow

A. Laminar

For an incompressible fluid in steady-state laminar flow in a straight horizontal circular pipe at constant temperature and neglecting any entrance or exit effects, the velocity is a function of the radial coordinate only. Then

$$\frac{1}{r}\frac{d}{dr}r\tau = -\frac{(P_0 - P_L)}{L} \tag{1}$$

where τ is the momentum flux, P_0 is the upstream pressure, and P_L is the downstream pressure at a distance L. The momentum flux must be zero at the pipe center or we will have an accumulation of momentum at the center. Thus

$$\tau = -\mu\frac{dv}{dr} = -\frac{(P_0 - P_L)}{2L}r \tag{2}$$

and at the wall $\tau_w = -(P_0 - P_L)R/2L$, where R is the pipe radius. A second integration with the condition that $v = 0$ at the wall gives

$$v = \frac{\tau_w R}{2\mu}\left[1 - \left(\frac{r}{R}\right)^2\right] \tag{3}$$

Then $v_{max} = \tau_w R/2\mu$, and when Eq. (3) is averaged across the pipe, $v_{ave} = \frac{1}{2}v_{max}$. We can also define a friction factor, f, so that the force on the pipe wall equals the product of the wall surface area, the friction factor, and a characteristic energy $\rho v_{ave}^2/2$. Then

$$f = \frac{2\tau_w}{\rho v_{ave}^2} = \frac{16\mu}{\rho v_{ave} D} \tag{4}$$

where D is the pipe diameter. The Reynolds number is $\rho v_{ave} D/\mu$ for pipe flow and

$$f = 16/N_{Re} \tag{5}$$

for laminar Newtonian flow. In many sources an alternate definition of friction factor is used which is a factor of 4 greater than the Fanning friction factor in Eq. (4).

The assumptions used to derive the laminar flow velocity profile hold when P_0 is measured far enough from the entrance for the parabolic profile to become fully developed. If we start with a flat profile, this distance from the entrance is

$$L_0 = 0.028DN_{\mathrm{Re}} \tag{6}$$

With great care to prevent disturbances in the flow, laminar flow in a smooth circular tube has been maintained up to Reynolds numbers approaching 10^5 in the laboratory. However, in industrial pipes disturbances begin to grow at Reynolds numbers near 2000, and fully developed turbulent flow is established for Reynolds numbers over 4000.

B. Turbulent

In turbulent pipe flow for steady-state, constant temperature and no end effects, Eq. (1) is valid but the momentum flux consists of the molecular flux as in laminar flow plus a contribution due to turbulent fluctuations. This turbulent momentum flux can be expressed in terms of the average over time of the product of the fluctuations in the radial and axial directions, $\overline{u'v'}$. So

$$\tau = -\mu\frac{dv}{dr} + \overline{\rho u'v'} = -\frac{(P_0 - P_L)}{2L}r = \frac{\tau_w r}{R} \tag{7}$$

where the term $\overline{\rho u'v'}$ is called the Reynolds stress. Very close to the pipe wall the molecular momentum transfer mechanism dominates and if $y = R - r$, we find

$$dv/dy = \tau_w$$

or (8)

$$v = \tau_w y/\mu$$

It is convenient to define a shear velocity $u^* = \sqrt{\tau_w/\rho}$ so that the dimensionless velocity U^+ and distance from the wall y^+ can be defined as

$$U^+ = v/u^* \quad \text{and} \quad y^+ = u^*\rho y/\mu \tag{9}$$

Then Eq. (8) becomes $U^+ = y^+$. Experimental measurements show that Eq. (8) holds for $y^+ < 5$. Clearly, for slightly larger values of y^+ we remain very close to the wall so

$$\mu\frac{dv}{dy} + \overline{\rho u'v'} = \tau_w = \rho u^{*2} \tag{10}$$

or

$$\frac{du^+}{dy^+} + \frac{u'v'}{u^{*2}} = 1 \tag{11}$$

The velocity U^+ will be a function of y^+, or the solution to Eq. (11) is

$$U^+ = f(y^+) \tag{12}$$

Further away from the wall $\overline{u'v'}/u^{*2} >> dv^+/dy^+$, and we must use a different scaling for the dimensionless distance from the wall. A balance on turbulent kinetic energy will give an equation in terms of $dU^+/d(y/R)$ where the radius R is now the characteristic length. Thus

$$U^+ = g(y/R) \tag{13}$$

If we match Eq. (12) as $y^+ \longrightarrow \infty$ and Eq. (13) as $(y/R) \longrightarrow 0$, we are examining

$$\lim_{y^+ \to \infty} \frac{dv}{dy} = \frac{u^{*2}}{\upsilon}\frac{df}{dy^+}$$

$$\lim_{y/R \to 0} \frac{dv}{dy} = \frac{u^*}{R}\frac{dg}{d(y/R)} \tag{14}$$

So

$$\frac{u^{*2}}{\upsilon}\frac{df}{dy^+} = \frac{u^*}{R}\frac{dg}{d(y/R)}$$

Then multiplying equation (14) by y/u^* gives

$$y^+\frac{df}{dy^+} = \frac{y}{R}\frac{dg}{d(y/R)} = \frac{1}{K} \tag{15}$$

where K is a constant. Therefore

$$f(y^+) = \frac{1}{K}\ln y^+ + B \tag{16}$$

and

$$g(y/R) = \frac{1}{K}\ln (y/R) + D \tag{17}$$

For smooth walled tubes Eq. (16) represents experiments very well for $y^+ > 30$, and for large diameters Eq. (17) holds. These equations are also valid in boundary layers.

The constants K and B are found from experiment to be $K = 0.4$ and $B = 5.5$. To get the average velocity we can assume that Eq. (17) applies across the entire pipe and integrate to get

$$v_{ave}/u^* = 1.75 + 2.5 \ln (R\rho u^*/\mu) \tag{18}$$

or

$$f^{-1/2} = 4 \log_{10} N_{Re} f^{1/2} - 0.4 \tag{19}$$

An empirical fit to this equation for N_{Re} between 4000 and 10^5 is the Blasius equation

$$f = 0.079 N_{Re}^{-1/4} \tag{20}$$

which corresponds to a velocity profile

$$v = v_{max}(1 - r/R)^{1/7} \tag{21}$$

These equations apply for smooth walls, but roughness will change the flow near the wall if the height of the wall irregularities extend beyond $y^+ = 5$. Then there is an additional dimensionless group, the ratio of roughness height to diameter, ϵ/D, to consider. Commercial pipe friction factors can be calculated from the empirical Colebrook formula which includes the roughness:

$$f^{-1/2} = -4.0 \log_{10}\left[\frac{\epsilon}{3.7D} + \frac{1.26}{f^{1/2} N_{Re}}\right] \tag{22}$$

The values of ϵ, the surface roughness, are given in handbooks or by manufacturers of pipes. The design value of ϵ for commercial steel pipe is 0.0002 ft (0.06 mm), and for copper tube ϵ is 5×10^{-6} ft (1.5×10^{-3} mm). A formula to calculate the friction factor explicitly has been proposed by Churchill:

$$f = 2[(8/N_{Re})^{12} + 1/(A + B)^{1.5}]^{1/12}$$
$$A = \{-2.457 \ln [7/N_{Re})^{0.9} + 0.27\epsilon/D]\}^{16}$$
$$B = (37530/N_{Re})^{16} \tag{23}$$

This formula includes the laminar and turbulent regimes. It is also possible to solve Eq. (22) for f using two iterations starting with an initial guess. If the initial guess is $f = .0075$,

$$f^{-1/2} = -4.0 \log\left[\frac{\epsilon}{3.7D} - \frac{5.02}{N_{\text{Re}}}\log\left(\frac{\epsilon}{3.7D} + \frac{14.5}{N_{\text{Re}}}\right)\right] \tag{24}$$

Equation (24) is adequate for rapid computer calculations and can be used to replace Reynolds number vs friction factor charts.

Some information regarding the optimum size of pipe can be obtained from the friction factor analysis. A more detailed analysis can be found in Perry's *Handbook*. We assume the cost of pipe per year including capitol, depreciation, interest, and maintenance can be expressed as $K_a D^a L$ and the power cost can be expressed as $K_b Q|\Delta_p|$, where Q is the flow rate $\frac{\pi}{4}D^2 v_{\text{ave}}$. The total cost C_T is then

$$\begin{aligned} C_T &= K_a D^a L + K_b Q|\Delta p| \\ &= K_a D^a L + 2f\rho v_{\text{ave}}^2 QL/D \\ &= K_a D^a L + 32f\rho Q^3 L/D^{5\pi 2} \end{aligned} \tag{25}$$

The optimum pipe diameter holding Q, f, and L constant is found when $dC_T/dD = 0$:

$$D_{\text{opt}} = kQ^{3/(5+a)} \tag{26}$$

where, for liquids, we can absorb ρ into the constant k. For a completely rough pipe, f is constant, but even for a smooth pipe the variation of f with diameter is small compared with D^5. Typically, a is between 1 and 2 so the optimum pipe diameter is roughly proportional to $Q^{1/2}$. Then substituting for Q in terms of $D^2 v_{\text{ave}}$, we find the velocity at the optimum diameter is a constant. This appears to be true in practice where average velocities of 3–8 ft/s (1–2.5 m/s) are typical for pipe flows of liquids.

For noncircular cross sections, the previous results can be used for turbulent flow if an equivalent hydraulic diameter is defined,

$$\begin{aligned} D_H &= \frac{4(\text{volume})}{\text{wetted surface}} \\ &= \frac{4(\text{cross-sectional area})}{\text{wetted perimeter}} \end{aligned} \tag{27}$$

The straight horizontal pipe itself may be only a small part of a complete piping problem. In addition, gases cannot always be considered incompressible. It is necessary to develop macroscopic momentum and energy balances to treat these problems.

III. Momentum and Energy Balances

The differential balances are developed in most elementary textbooks on fluid mechanics or transport phenomena. In a rectangular coordinate system, the balance of linear momentum in the x direction for a Newtonian fluid is

$$\underbrace{\frac{\partial v_x}{\partial t}}_{1} + \underbrace{\rho v_x \frac{\partial v_x}{\partial x} + \rho v_y \frac{\partial v_x}{\partial y} + \rho v_z \frac{\partial \rho_x}{\partial z}}_{2} =$$

$$\underbrace{\mu\left[\frac{\partial^2 v_x}{\partial x^2} + \frac{\partial^2 v_x}{\partial y^2} + \frac{\partial^2 v_x}{\partial z^2}\right]}_{3} + \underbrace{\rho g_x}_{4} + \underbrace{F_x}_{5} - \underbrace{\frac{\partial P}{\partial x}}_{6} \tag{28}$$

The terms in Eq. (28) are:

1. The rate of accumulation of linear momentum per unit volume in the x direction
2. The linear momentum in the x direction convected with the flow
3. The input of linear momentum due to molecular motions
4. The gravitational force (g_x is the x component)
5. The component of all other body forces in the x direction, F_x
6. The pressure force

A differential energy equation may be obtained in a similar manner for

$$E = \hat{U} + \tfrac{1}{2}v^2 \tag{29}$$

where E is the sum of internal energy per unit mass U, and the kinetic energy per unit mass, $\frac{1}{2}v^2$. Then

$$\underbrace{\rho\frac{\partial E}{\partial t}}_{1} + \underbrace{\rho v_x \frac{\partial E}{\partial x} + \rho v_y \frac{\partial E}{\partial y} + \rho v_z \frac{\partial E}{\partial z}}_{2} =$$

$$\underbrace{k\left[\frac{\partial^2 T}{\partial x^2} + \frac{\partial^2 T}{\partial y^2} + \frac{\partial^2 T}{\partial z^2}\right]}_{3} + \underbrace{\rho(\mathbf{v}\cdot\mathbf{g})}_{4} - \underbrace{\frac{\partial P v_x}{\partial x} - \frac{\partial P v_y}{\partial y} - \frac{\partial P v_z}{\partial z}}_{5} \tag{30}$$

$$+ \text{ work done by viscous forces}$$

The terms here are:

1. The rate of accumulation of energy per unit volume
2. The energy convected with the flow
3. The energy input due to molecular motion
4. The gravitational work
5. The pressure work

An equation for the conservation of kinetic energy can be found from the scalar product of the velocity vector and the vector form of Eq. (28):

$$\rho \frac{\partial(\frac{1}{2}v^2)}{\partial t} + \rho v_x \frac{\partial(\frac{1}{2}v^2)}{\partial X} + \rho v_y \frac{\partial(\frac{1}{2}v^2)}{\partial y} + \rho v_z \frac{\partial(\frac{1}{2}v^2)}{\partial z} =$$
$$-v_x \frac{\partial P}{\partial x} - v_y \frac{\partial P}{\partial y} - v_z \frac{\partial P}{\partial z} + \rho(\mathbf{v}\cdot\mathbf{g}) + \text{viscous work} + \text{viscous dissipation} \quad (31)$$

Equation (31) is often combined with Eq. (30) and simplified for many special cases of engineering interest to form the differential energy equation which is used for nonisothermal problems.

In pipe flow problems it is common to take a macroscopic balance over a much larger volume than a small differential element. This balance can be obtained from the previous equations by integrating over the volume. Many assumptions can be used to simplify the result. These include:

a. Steady-state
b. The fluid properties do not vary across the cross sections of the inlet or outlet pipes
c. The molecular transfer of momentum and energy can be neglected at the inlets and outlets
d. The mean velocity is parallel to the pipe walls

Then the macroscopic continuity equation is

$$\sum_{\text{inputs}} \rho_i\langle v_i\rangle S_i = \sum_{\text{outputs}} \rho_i\langle v_i\rangle S_i \quad (32)$$

where S_i is the cross-sectional area of pipe i and the average $\langle v\rangle$ is

$$\langle v\rangle = \frac{1}{S}\int\int_s v\,ds = \frac{2}{R}\int_0^R vr\,dr \quad (33)$$

for a circular pipe. The macroscopic momentum equation is

$$\sum_{\text{inputs}} \{\rho_i\langle v_i^2\rangle S_i\boldsymbol{n}_i + P_iS_i\boldsymbol{n}_i\} + \Sigma\boldsymbol{F} = \sum_{\text{outputs}} \{\rho_i\langle v_i^2\rangle S_i\boldsymbol{n}_i + P_iS_i\boldsymbol{n}_i\} + \boldsymbol{F}_{\text{drag}} - m_t\boldsymbol{g} \quad (34)$$

where $\boldsymbol{n}_i$ is the unit outward normal vector from the input or output surface. The sum of the external forces on the fluid $\mathbf{F}$, the net drag force $\boldsymbol{F}_{\text{drag}}$, and the weight of the fluid (the total mass m_t multiplied by the gravitational acceleration $\mathbf{g}$) also appear in this equation.

The macroscopic energy balance is

$$\sum_{\text{inputs}} \left\{\hat{U} + \frac{1}{2}\frac{\langle v^3\rangle}{\langle v\rangle} + \phi + \frac{P}{\rho}\right\}\rho\langle v\rangle S + \dot{Q} - \dot{W} =$$
$$\sum_{\text{outputs}} \left\{\hat{U} + \frac{1}{2}\frac{\langle v^3\rangle}{\langle v\rangle} + \phi + \frac{P}{\rho}\right\}\rho\langle v\rangle S \quad (35)$$

where ϕ is the potential energy per unit mass, $\dot{Q}$ is the rate of heat transfer through the walls, and $\dot{W}$ is the rate of work done by the system by pressure and viscous forces.

An additional equation can be obtained by integration of the mechanical energy balance, a special form of the energy equation. The result for a system with one inlet and one outlet and without chemical reaction is called the engineering Bernoulli equation:

$$\Delta\frac{1}{2}\frac{\langle v^3\rangle}{\langle v\rangle} + \Delta gz + \int_1^2 \frac{1}{\rho}dp + \hat{W} + \hat{E}_v = 0 \tag{36}$$

where g is the gravitational acceleration, z is the height above a reference, $\hat{W}$ is the work per unit mass, Δ represents out minus in, and $\hat{E}_v$ is the energy loss by viscous dissipation. This energy loss appears as an increase in internal energy in Eq. (35). The integral must be evaluated over some representative streamline in the system considering all of the thermodynamic states between the input and output. For flow in piping systems there are three limiting cases of interest: incompressible flow, isothermal flow, and adiabatic flow.

IV. Incompressible Flow

The engineering Bernoulli equation becomes

$$\Delta\frac{1}{2}\frac{\langle v^3\rangle}{\langle v\rangle} + \Delta gz + \frac{(P_2 - P_1)}{\rho} + \hat{W} + \hat{E}_v = 0 \tag{37}$$

In the straight pipe considered previously

$$\frac{(P_1 - P_2)}{\rho} = \frac{2\langle v\rangle^2 Lf}{D} = \hat{E}_v \tag{38}$$

Another example that can be solved exactly is a sudden expansion from a smaller pipe with cross-sectional area A_1 to a larger pipe with area A_2. The continuity, momentum, and Bernoulli equations are as follows:

$$S_1\langle v_1\rangle = S_2\langle v_2\rangle \tag{39}$$

$$P_1S_1 + \rho\langle v_1^2\rangle S_1 + F = \rho\langle v_2^2\rangle S_2 + P_2S_2 + F_d \tag{40}$$

$$\frac{1}{2}\frac{\langle v_2^3\rangle}{\langle v_2\rangle} - \frac{1}{2}\frac{\langle v_1^3\rangle}{\langle v_1\rangle} + \frac{P_2 - P_1}{\rho} + \hat{E}_v = 0 \tag{41}$$

Here $F = -P_1(S_1 - S_2)$ when we neglect the frictional drag and consider only the force of the fluid pressure on the expansion surface. Then let

$$\beta_i = \langle v_i^2\rangle/\langle v_i\rangle^2, \quad \alpha_i = \langle v_i^3\rangle/\langle v_i\rangle^3$$

so from Eq. (40)

$$\frac{(P_1 - P_2)}{\rho} = \beta_2\langle v_2\rangle^2\left[1 - \frac{\beta_1 S_2}{\beta_2 S_1}\right]$$

and from Eq. (41)

$$\hat{E}_v = \frac{(P_1 - P_2)}{\rho} + \frac{1}{2}(\alpha_1\langle v_1\rangle^2 - \alpha_2\langle v_2\rangle^2)$$

When $\alpha_i = \beta_1 = 1$, we find

$$E_v = \frac{\langle v_2\rangle^2}{2}\left(\frac{S_2}{S_1} - 1\right)^2 \tag{42}$$

The result is in the form $\hat{E}_v = K\frac{1}{2}v^2$. Extension to all forms of fittings and valves leads to the total friction loss:

$$\hat{E}_v = \sum_{\text{pipe}} \frac{2\langle v_i\rangle^2 L_i f_i}{D_i} + \sum_{\text{fittings}} \frac{1}{2}\langle v_i\rangle^2 K_i \tag{43}$$

V. Compressible Pipe Flow

When the downstream pressure P_2 is less than 90% of the upstream pressure for the flow of gases in pipelines, compressible flow calculations are recommended. The continuity and Bernoulli equations are needed in differential form to develop the design equations. Therefore we place Planes 1 and 2 very close together to obtain for one input and one output with $\alpha = \beta = 1$:

$$\frac{dp}{\rho} + \frac{d\langle v\rangle}{\langle v\rangle} = 0 \tag{44}$$

$$\langle v\rangle d\langle v\rangle + \frac{dp}{\rho} + d\hat{E}_v = 0 \tag{45}$$

Since $\hat{E}_v = f\langle v\rangle^2 z/2D$ in terms of the differential length dz,

$$\hat{E}_v = \frac{f\langle v\rangle^2 dz}{2D} \tag{46}$$

A. Isothermal Ideal Gas

Equation (45) can be integrated when $G = \rho\langle v\rangle$ is constant and $\rho = Mp/RT$ to give

$$\frac{M}{2RT}(P_2^2 - P_1^2) - G^2 \ln \frac{P_2}{P_1} + \frac{2G^2 fL}{D} = 0 \tag{47}$$

where R is the gas constant, M is the molecular weight, and T is the absolute temperature. This equation has a maximum mass flow rate which can be found by equating the derivative of G^2 with respect to $(P_2/P_1)^2$ to zero. This maximum corresponds to the exit velocity $V_{2\,\max} = (P_2/\rho_2)^{1/2}$, the speed of sound in an isothermal ideal gas. The pressure at the exit cannot fall below the value corresponding to $V_{2\,\max}$, and the velocity cannot exceed this value in a straight pipe of constant diameter.

For pipe flow in long small pipes or where the velocities are less than 30% of the speed of sound at the exit, the isothermal equation is valid. The adiabatic flow results are closely approximated by the isothermal results under these conditions also.

Usually the second term containing the $\ln P_2/P_1$ is small compared to the other terms, and Eq. (47) without this term can be solved directly for G.

Equation (47) applies to the flow of a compressible gas from Point 1 to Point 2 in a horizontal pipe. If there is also a flow through a nozzle from a reservoir at pressure P_0 preceeding P_1, the velocity at 1 is related to the pressure ratio by

$$N_{\mathrm{Ma},1}^2 = 2 \ln (P_0/P_1)$$

where

$$N_{\mathrm{Ma},1}^2 = v_1^2 M/RT$$

is the square of the isothermal Mach number, and v_1 is the average velocity at 1.

B. Adiabatic Flow

There is a maximum velocity for adiabatic flow, the sonic velocity a, where

$$a = \left(\frac{\gamma P}{\rho}\right)^{1/2} \tag{48}$$

for an ideal gas with isentropic path $P\rho^{-\gamma} =$ constant. Here γ is the ratio of specific heats C_p/C_v. We define the Mach number N_{Ma} as the ratio of the velocity to the speed of sound from Eq. (48). Then integration of Eq. (45) for an ideal gas with adiabatic flow from Point 1 to Point 2 in a horizontal pipe gives

$$\frac{4fL}{D} = \frac{1}{\gamma}\left(\frac{1}{N_{\mathrm{Ma},1}^2} - \frac{1}{N_{\mathrm{Ma},2}^2} - \frac{\gamma + 1}{2} \ln A\right) \tag{49}$$

where

$$A = \frac{N_{\mathrm{Ma},2}^2\left(1 + \frac{\gamma - 1}{2} N_{\mathrm{Ma},1}^2\right)}{N_{\mathrm{Ma},1}^2\left(1 + \frac{\gamma - 1}{2} N_{\mathrm{Ma},2}^2\right)}$$

and f is an average friction factor $(f_1 + f_2)/2$. The average friction factor can be used because the variation of viscosity with temperature for a gas is small.

The use of Eq. (49) to solve for the flow in a pipe is trial and error. An additional factor is the limiting condition of sonic velocity at the exit. This limits the mass flow rate for a given length of pipe or the length for a given mass flow rate. The actual pressure at the exit may be greater than the pressure of the downstream reservoir. This comes from the limiting pressure corresponding to sonic velocity. When the upstream reservoir is at pressure P_0, adiabatic flow into the pipe at P_1 for a rounded entrance gives

$$\frac{P_0}{P_1} = \left[1 + \frac{G^2}{2}\frac{(\gamma - 1)}{\gamma}\frac{RT_1}{MP_1^2}\right]^{\gamma/(\gamma - 1)} \tag{50}$$

$$\frac{T_1}{T_0} = \left(\frac{P_1}{P_0}\right)^{(\gamma - 1)/\gamma} \tag{51}$$

Similar relations between 1 and 2 in the pipe are

$$\frac{P_1}{P_2} = \frac{N_{\mathrm{Ma},2}}{N_{\mathrm{Ma},1}}\sqrt{\frac{1 + [(\gamma - 1)/2]N_{\mathrm{Ma},2}^2}{1 + [(\gamma - 1)/2]N_{\mathrm{Ma},1}^2}} \tag{52}$$

and

$$\frac{T_1}{T_2} = \frac{1 + [(\gamma - 1)/2]N_{\mathrm{Ma},2}^2}{1 + [(\gamma - 1)/2]N_{\mathrm{Ma},1}^2} \tag{53}$$

The relationship between the maximum length and the flow rate can be found from Eq. (49) when $N_{\mathrm{Ma},2} = 1$. Lapple's method of solution using charts for three different valves of γ has been reproduced in Perry's *Handbook*. Present-day computers and hand calculators make an iterative solution of the equations about as easy when P_1 does not differ significantly from P_0. Both Lapple's graphs and the equations presented are limited to one pipe diameter, ideal gases, and constant specific heat ratio.

In industrial designs these assumptions may be relaxed when the following approximations are used.

1. More than one pipe diameter—The most accurate calculations are made by computing each portion separately. An approximation can be made by setting $(f_b L_b/D_b) = (f_a L_a/D_a)(D_b/D_a)^4$.
2. Inclusion of fittings—The equivalent L/D of the fittings is used unless the equivalent L/D for any fitting is greater than 100.
3. Nonideal gases—When the gas does not condense to form two phases, the temperature T is replaced by zT; that is, the gas equation becomes $Mp = z\rho RT$. If z and γ change with pressure, calculations over incremen-

tal lengths of pipe can be used to increase accuracy. Similarly, when two phases are present, it would be appropriate to divide the total length into several segments.

4. Exit pressure—There is usually a pressure recovery at the pipe exit. About 10% of the kinetic energy is converted to pressure energy at an exit to atmospheric pressure. Except for a limited set of conditions, the exit pressure can be assumed to be atmospheric. These conditions are $P_2/P_1 < 0.75$ and $4fL/D < 2$. Then use

$$P_2 = P_{atm} - (3 \times 10^{-7})W^2 T_2 z_2/(MP_2 d^4) \tag{54}$$

which is solved by trial and error for P_2. Here W is the mass flow rate in lb/h when T is in °R, d is the pipe diameter in inches, and all pressures are in lb/in.^{2}abs. When W is in units of kg/h, T in °K, and d in meters, the constant factor in Eq. (54) becomes 5.2×10^{-5} for pressures in Pascals.

VI. Special Topics

A. Drag Reduction

High molecular weight polymers when added to liquids in parts per million by weight proportions can often cause large reductions in pressure drop at constant flow rate. Although these additives are expensive, there can be significant savings in capital costs if the equipment can be scaled down to take advantage of drag reduction. Some examples of the use of drag-reducing additives are in the 48-in. Trans-Alaska Pipeline as a replacement for unconstructed pump stations, and in an old storm sewer system as a substitute for an enlarged system.

Unfortunately, there is not a sufficient understanding at present to predict the effect of adding these long-chain molecules to a liquid from laboratory data in small pipes. For the Alaska Pipeline a scale-up was made from a 14-in. experimental study to the full-sized 48-in. pipeline by Burger, Chou, and Perkins of Arco Oil and Gas Co. Their method, based on the technique originally developed by Savins and Seyer, will be described here. Additional references on drag reduction theory are listed in the Bibliography.

Let

$$\sigma = \Delta P_P/\Delta P_0 \tag{55}$$

where ΔP_P is the pressure drop for the solution containing the polymer additive and ΔP_0 is the pressure drop for the pure solvent and both pressure drops are for the same flow rate. The fractional drag reduction is usually defined as $1 - \sigma$. Additional parameters in the model are

$$\alpha = \frac{u^*}{\langle v \rangle} = \left(\frac{f}{2}\right)^{1/2}$$

$$\theta_P = a\phi^b(\gamma_0\sigma)^c$$

and

$$\gamma_0 = \rho u^{*2}/\mu$$

where $u^* = (-\Delta P_0 D/4\rho L)^{1/2}$, ϕ is the polymer concentration, and a, b, and c are constants. All of the quantities are in terms of the solvent system. Then

$$\sigma[1 + \alpha(2^{3/2} \log \sigma + 1.454\gamma_0\sigma\theta_P - 0.8809)]^2 = 1 \tag{56}$$

From experiments on small pipes for different concentrations ϕ, in parts per million by weight, the constants a, b, and c can be determined. These constants are unique for a polymer–solvent system and must be determined from experiments. Once these constants are found, Eq. (56) can be used to find the expected drag reduction in the larger pipes.

For example, data from 1, 2, 14, and 48 in. pipes for a Conoco drag reducing additive in Sadlerochit crude gave $a = 0.0516$, $b = 0.489$, and $c = -0.579$. To find the expected pressure drop in a 48-in. pipe with 20 ppm of this polymer in Sadlerochit crude at a Reynolds number of 3×10^5 and a temperature giving a kinematic viscosity of 9 cSt, we proceed as follows.

The friction factor is 0.0037 at the Reynolds number of 3×10^5 and the velocity is 7.42 ft/s. Then

$$\alpha = (f/2)^{1/2} = 0.043$$
$$\gamma_0 = f\langle v\rangle^2/2\upsilon = 1.05 \times 10^3 \ s^{-1}$$
$$\theta = (0.0516)(20)^{0.489}(\gamma_0\sigma)^{-0.579} = 3.98 \times 10^{-3}\sigma^{-0.579}$$

and Eq. (56) becomes

$$\sigma[0.9621 + 0.1216 \log \sigma + 0.2613\sigma^{0.421}]^2 = 1$$

Solving by trial and error, σ is 0.725, or with the 20 ppm polymer additive the pressure drop would be 27.5% less than with the crude oil alone. This is equivalent to the elimination of one of every four pumps.

Some polymer additives degrade readily in pipelines and must be continuously added along the pipeline. Scale-up from laboratory pipe data with inside diameters of less than 1 in. is not recommended.

B. Flow of Heavy Crudes

Another example of an industrial fluid flow problem where extensive laboratory investigation is necessary to design the facility is the flow of waxy crudes. For these crudes the pour point is between 60 and 115°F and the fluids must often be pumped at temperatures below the pour point. At temperatures below approximately 20°F above the pour point, the crudes have non-Newtonian viscosities. The wax can crystallize as the crude is cooled to form a gel or a partial gel. Under static conditions a rigid gel is formed, but if the crude is cooled while in motion, the apparent viscosity will increase but the material remains fluid. Therefore, the rheological properties are functions of temperature, shear rate, shear stress, and past history. Problems in pumping these crudes will occur if the temperature drops and the fluid becomes non-Newtonian and if gel formation occurs after a shutdown. The pipeline facility must be designed to recover from these problems or prevent them.

First of all, laboratory work is required to find the properties of the crude. Measurements must include pour point, density, specific heat, wax content, distillation, nonorganic content, and experimental flow data. The flow data are necessary to find the apparent viscosity vs rate of shear. Several other flow tests are desirable including yield stress under static and dynamic conditions of cooldown.

Operation of a heavy crude pipeline involves start-up, shutdown, and continuous operation. The start-up process will require pump pressures to overcome the yield stress of the gel formed under static conditions. This start-up pressure may be higher than the maximum pressure at the maximum flow rate for continuous operation. From the experimental yield stress Y, the restart pressure is

$$P = YLC/A$$

where L is the pipe length, C is the circumference, and A is the cross-sectional area.

When the fluid cools and the flow rate is reduced, the pressure is at first reduced. Lowering the flow rate will lower the required pump pressure until a minimum pressure occurs. Further lowering of the flow rate will generate a pressure increase as the fluid becomes non-Newtonian. Therefore, high pressure at low flows may be found in the cool-down process.

For normal operation the waxy crudes are Newtonian, but temperature gradients may lead to a buildup of wax on the walls.

Some design procedures to avoid problems in the pumping of heavy crudes have been suggested by Smith. All possibilities of avoiding gel formation as a result of an unforeseen shutdown cannot be built into a facility design, so standby pumps, heaters, and injection equipment should be available.

Example 1: Laminar Flow. A 1-mm diameter capillary tube is to be used to monitor small flows to a laboratory reactor. If pressure taps are 2 m apart and the maximum pressure drop is to be 7×10^4 Pa, find the maximum flow rate when $\rho = 1000$ kg/m^3 and $\mu = 1$ cP.

$$q = v_{\text{ave}}A = \frac{(-\Delta P)\pi D^4}{128\mu L}$$

$$= \frac{(7 \times 10^4)\pi(10^{-3})^4}{(128)(10^{-3})^2} = 8.59 \times 10^{-7}\ \text{m}^3/\text{s}$$

$$= 51.54\ \text{cm}^3/\text{min}$$

Check to see if the flow is laminar:

$$\text{Re} = \frac{q\rho D}{A\mu} = \frac{(8.59 \times 10^{-7})(10^3)(10^{-3})}{(\pi/4)(10^{-3})^2(10^{-3})} = 1094$$

Example 2: Turbulent Flow. A fluid is to pump 200 km through a 0.7366 m i.d. cast iron pipeline at an average velocity of 1 m/s. Find the pressure drop when $\mu = 2$ cP, $\rho = 10^3$ kg/m^3, and $\epsilon = 0.46$ mm. Churchill's equation is appropriate here, but only the term involving A is different from zero.

$$N_{\text{Re}} = (0.7366)(1)(10^3)/(2 \times 10^{-3}) = 3.683 \times 10^5$$

$$f = 2[1/A^{1.5}]^{1/12}$$

$$A = [-2.457 \ln([(7/3.683 \times 10^5) + (0.27)(0.46)(736.6)]]^{16}$$

$$= (20.6376)^{16}$$

$$f = 2/(20.6376)^2 = 0.00469$$

$$f = \frac{(-\Delta P)D}{2L\rho v_{\text{ave}}^2}$$

$$-\Delta P = 2.55 \times 10^6 \text{ Pa } (368 \text{ lb/in.}^2)$$

Example 3. What flow rate can be obtained in a 4-in. schedule 40 pipe (0.10226 m i.d.) for a pressure drop of 2×10^5 Pa. Let $\rho = 10^3$, $\mu = 2$ cP, and $\epsilon = 0.06$ mm, and the length is 1000 m. When pressure drop, diameter, and length are known, the right-hand side of Eq. (22) is all known.

$$f^{1/2}N_{\text{Re}} = \left[\frac{-\Delta PD}{4\rho L}\right]^{1/2}\frac{D\rho}{\mu} = 3.6558 \times 10^3$$

$$f^{-1/2} = -4.0 \log\left[\frac{0.06 \times 10^{-3}}{(3.7)(0.10226)} + \frac{1.26}{3.6558 \times 10^3}\right]$$

$$f = 0.005745 = \frac{-\Delta PD}{2L\rho v_{\text{ave}}^2}$$

Solving for the one unknown,

$$v_{\text{ave}} = 1.334 \text{ m/s}$$

Example 4. A hydrocarbon product must be pumped from a chemical plant to the loading dock. An existing 3 in. schedule 40 line (77.93 mm i.d.) and a pump capable of producing a 100 lb/in.^{2} (6.925×10^5 Pa) pressure difference between inlet and outlet are available for this project. If the fluid density is 860 kg/m^3 and the viscosity is 3 cP, find the maximum flow rate that can be supplied. The line is 1000 m long and has eight 90° elbows and 4 gate valves. Here

$$N_{\text{Re}}f^{1/2} = [(-\Delta P\rho D^3)/(2L\mu^2)]^{1/2}$$

does not contain the unknown average velocity, and the friction factor can be found directly from Eq. (22). The result is $f = 0.0062$. Then Eqs. (38) and (43) can be used to find $\langle v \rangle$. A 90° elbow has an L/D equivalent of 30 and an open gate valve of 15. The total L/D is

$$(L/D)_{\text{total}} = 8(30) + 4(15) + (0.1000)/0.07793$$
$$= 13{,}130$$

then

$$v^2 = \frac{(-\Delta P)}{2\rho f (L/D)_{\text{total}}}$$

and

$$v = 2.23 \text{ m/s}$$

A check is necessary to confirm turbulent flow

$$N_{\text{Re}} = (2.23)(0.07793)(860)/0.003$$
$$\cong 50{,}000$$

Example 5: Compressible Isothermal Flow. A 2-in. schedule 40 (52.5 mm i.d.) air line runs 2000 m from a tank at a pressure of 10^6 Pa to another tank at a pressure of 5×10^5 Pa. The temperature is 300 K and the process is assumed to be isothermal. Find the flow rate.

We assume $1/\sqrt{f} = -4 \log(\epsilon/3.7D)$ to get $f = 5.07 \times 10^{-3}$. Then we neglect the ln term in Eq. (47).

$$\frac{2G^2(5.07 \times 10^{-3})(2000)}{(0.0525)} = \frac{29[10^{12} - 0.25 \times 10^{12}]}{2(8.314 \times 10^3)(300)}$$
$$G = 106.24 \text{ kg/m}^2 \cdot \text{s}$$

The density is $\rho_1 = PM/RT = 11.63$ kg/m^3 and $v_1 = 9.14$ m/s. Thus $v_2 = 18.27$ m/s. A check of the assumptions shows that less than a 1% change in v_2 is obtained if a correction to f is made and the ln term is included. We have also assumed that $P_o = P_1$. This can be checked:

$$V_{\max} = \sqrt{\frac{RT}{M}} = 293.3 \text{ m/s}$$
$$M_1 = 9.14/293.3 = 0.0312$$
$$P_1 = P_0 e^{-M_1^2/2} = 0.9995 P_0 \approx P_0$$

Example 6. Air ($\gamma = 1.4$) is exhausted from a vessel at a pressure of 7×10^5 Pa to the atmosphere through 120 m of 2 in. schedule L/D steel pipe (52.5 mm i.d.). The pipe contains fittings with an equivalent L/D of 300. The temperature in the vessel is

310 K. Find the flow rate. (None of the individual fittings has a L/D greater than 100.)

First we must estimate the friction factor by assuming only the roughness is important.

$$f = [-1/(4 \log \epsilon/3.7D)]^2$$
$$= [-1/4 \log (0.06)/(3.7)(52.5)]^2 = 0.005$$

Then the total $4fL/D$ is

$4fL/D$ pipe	45.7
$4fL/D$ fittings	6
For entrance add 0.5	0.5
	$4fL/D = 52.2$

We assume $P_0 = P_1$ and solve for $N_{Ma.1}$ from Eq. (49) when $N_{Ma.2} = 1$. Let $N_{Ma.1} = x$.

$$52.2 = 0.714\left(\frac{1}{x^2} - 1 - 1.2 \ln \frac{1 + 0.2x^2}{1.2x^2}\right)$$
$$x \neq 0.112$$

From Eq. (52) we now find $N_{Ma.2}$ for $N_{Ma.1} = 0.112$. Let $y = N_{Ma.2}$.

$$\frac{7 \times 10^5}{1.018 \times 10^5} = \frac{y}{0.112}\left[\frac{1 + 0.2y^2}{1.0025}\right]^{1/2}$$
$$y = 0.733$$

We go back to Eq. (49) to get an improved value of x until the two unknowns x and y converge. Here $x = 0.112$ does not change upon another iteration. To find the mass flow rate

$$G = N_{Ma}\left[\frac{\gamma P^2 M}{RT}\right]^{1/2}$$

and at Point 1:

$$G = 0.112\left[\frac{(1.4)(7 \times 10^5)^2(29)}{(8.314 \times 10^3)(310)}\right]^{1/2} = 311.2 \text{ kg/m}^2 \cdot \text{s}$$
$$W = 311.2(3600)\left(\frac{\pi}{4}\right)(0.0525)^2 = 2425 \text{ kg/h}$$

Several checks need to be made here. First, from Eq. (50)

$$\frac{P_0}{P_1} = \left[1 + \frac{\gamma - 1}{2} N_{Ma.1}^2\right]^{\gamma/(\gamma - 1)} = 1.0088$$

and the assumption that $P_0 = P_1$ is verified. The final temperature $T_2 = 0.905T_1$ from Eq. (53). The friction factor at Position 1 is $f = 0.0054$; and at Position 2, $f = 0.005$. The average $f = 0.0052$ is slightly greater than the initial assumption. Only a small change in G is expected from this change. Finally, P_2 within the pipe is actually slightly less than atmospheric pressure. The estimate from Eq. (54) gives

$$P_2 = P_{\text{atm}} - \frac{(5.2 \times 10^{-5})(2425)^2(0.905)(310)}{(29)(P_2)0.0525)^4}$$

$$= 9.78 \times 10^4 \text{ Pa (96\% of atmospheric)}$$

Again the change is small. For this problem Lapple's charts are easy to use. $P_2/P_0 \doteq 0.145$ and $4f/LD = 52.2$. The ratio $G/G_{0_i} = 0.22$ and the temperature ratio $T_2/T_0 = 0.9$ are read directly from the chart.

$$G_{0_i} = P_0[M/2.718RT_0]^{1/2}$$

then

$$G = 313 \text{ kg/s} \cdot \text{m}^2$$

which is the same answer using the equations.

Symbols

A_i	area at location i
a	sonic velocity
C_T	total cost
C	circumference
D	pipe diameter
D_H	hydraulic diameter
E	energy
$\hat{E}_v$	lost energy by viscous dissipation
f	Fanning friction factor
g_i	gravitational acceleration component
F_i	i component of force
G	mass velocity, $\rho\langle v\rangle$
k	thermal conductivity (Eq. 30)
L_0	entrance length
L	characteristic length or pipe length
M	molecular weight
M_t	total mass
N_{Re}	Reynolds number
N_{Ma}	Mach number
n_i	unit outward normal vector
P_i	pressure at location i
P_0	pressure at upstream reservoir
P_L	pressure at downstream location
Q	flow rate
$\dot{Q}$	heat transfer rate
r	radial coordinate

R	pipe radius or gas constant
S_i	cross-sectional area at location i
T	temperature
u^*	shear velocity
$u'v'$	turbulent stress term
U	velocity
U^+	dimensionless velocity
v	local velocity
v_{ave}	average velocity
v_{max}	maximum velocity
v_i	i component of velocity
$\hat{U}$	internal energy per unit mass
W	mass flow rate
$\hat{W}$	work per unit mass
$\dot{W}$	rate of work
Y	yield stress
y	distance from the wall
y^+	dimensionless distance from wall
z	vertical coordinate or compressibility factor
γ	ratio of specific heats
ϵ	roughness height
υ	kinematic viscosity, μ/ρ
μ	viscosity
ρ	density
τ	shear stress
τ_w	wall shear stress
ϕ	potential energy per unit mass

Bibliography

The derivation of the laminar flow equations can be found in most unit operations or fluid mechanics texts. For example:

Bird, R. B., Stewart, W. E., and Lightfoot, E. N., *Transport Phenomena*, Wiley, New York, 1960.

Denn, M. M., *Process Fluid Mechanics*, Prentice-Hall, Englewood Cliffs, New Jersey, 1980.

The derivation of the turbulent flow profile is found in

Millikan, C. B., "A Critical Discussion of Turbulent Flows in Channels and Circular Tubes," in *Proceedings of the Fifth International Congress on Applied Mechanics*, Wiley, New York, 1939, p. 386.

and also

Tennekes, H., and Lumley, J. L., *A First Course in Turbulence*, MIT Press, Cambridge, Massachusetts, 1972.

The working formulas are based on the following additional sources:

Churchill, S. W., "Friction Factor Equation Spans All Fluid-Flow Regimes," *Chem. Eng.*, *84*, 91 (November 7, 1977).
Colebrook, C. F., and White, D., "Turbulent Flow in Pipes with Particular Reference to the Transition Region between Smooth and Rough Pipe Laws," *J. Inst. Civil Eng.*, *II*, 133 (1938–1939).
Shacham, M., "Comment on 'An Explicit Equation for Friction Factor in Pipe,'" *Ind. Eng. Chem., Fundam.*, *19*, 228 (1980).

The optimum pipe size analysis is presented in

Perry, R. H., and Chilton, C. H. (eds.), *Chemical Engineers Handbook*, 5th ed., McGraw-Hill, New York, 1973.
Peters, M. S., and Timmerhaus, K. D., *Plant Design and Economics for Chemical Engineers*, 3rd ed., McGraw-Hill, New York, 1980.

The derivations in Sections III, IV, and V are based on the equations in

Bird, Stewart, and Lightfoot and McCabe, W. L., and Smith, J. C., *Unit Operations of Chemical Engineering*, 3rd ed., McGraw-Hill, New York, 1976.

Additional sources on drag reduction are

Berman, N. S., "Drag Reduction by Polymers," *Annu. Rev. Fluid Mech.*, *10*, 47 (1978).
Burger, E. D., Chorn, L. G., and Perkins, T. K., "Studies of Drag Reduction Conducted over a Broad Range of Pipeline Conditions When Flowing Prudhow Bay Crude Oil," *J. Rheol.*, *24*, 603 (1980).
Virk, P. S., "Drag Reduction Fundamentals," *AIChE J.*, *21*, 625 (1975).

Discussions on pumping heavy crudes are given in

Smith, B., "Pumping Heavy Crudes," *Oil Gas J.*, p. 111 (May 28, 1979); p. 148 (June 4, 1979); p. 110 (June 18, 1979); p. 105 (July 2, 1979); p. 69 (July 16, 1979).
Withers, V. R., and Mowill, R. T. L., "How to Predict Flow of Viscous Crudes," *Pipe Line Ind.*, p. 45 (July 1982).

NEIL S. BERMAN

Crude Oils

Many pipelines around the world are now successfully transporting waxy crudes under conditions where the ambient temperature is lower than the pour point of the liquid.

In each case the most economical method had to be determined for transporting the particular crude.

An economic evaluation must consider the energy required to pump the waxy crude. Prime considerations are the pipeline diameter and the temperature of the crude. These factors determine the pump horsepower and govern the selection of the pump.

Flow characteristics must be determined in the early stages of pump selection and design. Determining the flow characteristics requires a look at the complete system and at alternate methods of transporting high-pour-point crude.

There are many design alternates which are "industry-accepted" methods in pipelining high-pour point oil. Some of these are:

1. Preheating the crude to a higher inlet temperature to allow it to reach the destination of intermediate station before cooling to below its pour point. The pipeline may or may not be insulated.
2. Pumping the crude at a temperature below the pour point.
3. Adding a hydrocarbon diluent such as a less-waxy crude or light distillate.
4. Injecting water to form a layer between the pipe wall and the crude.
5. Mixing water with the crude to form an emulsion.
6. Processing the crude before pipelining to change the wax-crystal structure and reduce the pour point and viscosity.
7. Heating both the crude and the pipeline by some method such as steam tracing or electrical heating.
8. Injection of paraffin inhibitors.
9. Combinations of these methods.
   Before deciding which design alternate to use, several design parameters must be investigated: Physical properties of crude heat transfer, restart after shutdown, facilities design.

High-Pour-Point Crude Properties

High-pour-point crudes require higher than normal temperatures before their pour points are reached (normally between 60 and 115°F). Additionally, high-pour-point crudes exhibit non-Newtonian viscosity behavior at temperatures below about 20°F above the pour point.

This means the effective viscosity is not a function of temperature alone, but is also a function of the effective rate of shear in the pipeline. Shear stress and rate of shear must be determined to predict the pressure required to deliver specified production volumes.

With each waxy crude discovery, extensive laboratory tests should be made to determine its exact behavior under temperature variation. These tests determine the crude's rheology.

In determining the rheology of liquids, any one of five basic behavior patterns (fluid types) may be found upon agitation of the liquid at constant temperature (Table 1).

TABLE 1 Basic Fluids

Fluid Type	Fluid Characteristics
Newtonian	Unaffected by magnitude and kind of motion to which they are subjected
Dilatant	Viscosity will increase as agitation is increased
Plastic (Bingham)	Have a definite yield value which must be exceeded before flow starts. After flow starts, viscosity decreases with increase in agitation
Pseudoplastic	Do not have a yield value, but do have decreasing viscosity with increase in agitation
Thixotropic	Viscosity will normally decrease upon increased agitation, but this depends upon duration of agitation and viscosity of fluid and rate of motion before agitation

The crude may exhibit pseudoplastic or thixotropic behavior and/or act as a Bingham plastic in the transition phase between the onset of wax crystallation and the fully gelled state.

With pseudoplastic behavior the fluid displays increasing viscosity and decreasing shear rate. This means there is a nonlinear relationship between shear stress and shear rate. Also, pseudoplastics are time dependent.

A thixotropic liquid's viscosity varies as a function of time and shear rate.

A waxy crude may exhibit Bingham plastic characteristics after gelling. The behavior of this type of waxy crude varies from that of Newtonian fluids only in that its linear relationship between shear stress and shear rate does not go through the origin. A finite shear stress is required to initiate flow. Further, the viscosity or pumpability of a Bingham plastic is rate-of-shear and time dependent.

When waxy crude is allowed to cool below its pour point under static conditions in a pipeline (no flow in the pipeline), the paraffins will crystallize, causing the entire mass of crude oil to gel. Because of this gelling effect upon cool-down, most operators' initial reaction is to specify that the line be operated at temperatures above the crude's pour point.

While this may or may not be a valid operating criterion, there are no particular problems in pumping waxy crude below its pour point, provided the fluid is kept in motion. More pressure is required to pump in the non-Newtonian range, but there is no sudden change in fluid characteristics at the pour point.

It must be kept in mind that, if a waxy-crude pipeline being pumped below its pour point is shut down for any reason, the resulting gelled state will require, upon restart, substantially more pressure to put it in motion.

However, this additional restart pressure will be substantially less than if the crude oil had been pumped above the pour point, then allowed to cool down statically.

Heat Transfer

In order to determine the required pumping equipment in a pipeline system, pressure-loss calculations must be made. With conventional crude oils, this determination is made using conventional hydraulic equations.

These calculations become more complex with non-Newtonian crude since the viscosity is both temperature and shear-rate dependent. Pressure-loss calculations for non-Newtonian liquids usually are made by dividing the pipeline into segments and analyzing each segment separately. This is best achieved by computers.

To analyze a segment effectively, its inlet temperature and the temperature loss must be calculated. This involves an examination of the heat transfer from the oil to the surrounding environment.

Defined as the transmission of energy from one region to another as a result of a temperature difference between them, heat transfer has three distinct modes of heat transmission:

1. Conduction is the process by which heat flows from a region of higher temperature to a region of lower temperature within a medium (solid, liquid, or gas) or between different mediums in direct physical contact.
2. Radiation is the process by which heat flows from a high-temperature body to a lower temperature one when the bodies are separated in space.
3. Convection is the process by which heat flows by fluid motion between regions of unequal density caused by nonuniform heating. Convection is the mechanism of energy transfer between a solid and a liquid or a gas.

A pipeline's surrounding environment is critical in determining its potential heat transfer. Table 2 lists possible environments. The resulting range of overall heat-transfer coefficients (U) is shown to indicate the critical nature of this variable.

Once a U value has been ascertained, calculation of heat loss in the pipeline can be determined by the Eq. (1) which takes into account heat lost to the environment as well as heat gained from friction between the oil and pipe wall:

$$Q = UA(T_0 - T) - (1.0381)PM \tag{1}$$

where Q = heat flow (Btu/h/ft)
U = overall heat-transfer coefficient (Btu/h/ ft^2/°F)
A = outside surface area (ft^2/ft)
T = outside temperature of air ground, or water (°F)
T_0 = bulk oil temperature (°F)
P = pressure loss due to friction in the length being considered (lb/in.^{2}/ft)
M = flow rate (bbl/h)
1.0381 = conversion factor for dimensional consistency.

TABLE 2 Environment vs U Value[a]

Environment	U value, (Btu/h · °F · ft²)
Above-ground pipeline, exposed to atmosphere:	
Uninsulated	1.5–0.7
2-in. thick insulation	0.1–0.21
Buried on shore pipeline dry soil (desert 2-ft cover):	
Uninsulated	0.15–0.65
2-in. thick insulation	0.05–0.15
Buried onshore pipeline, moist to wet soil (2-ft cover):	
Uninsulated	0.3–0.8
2-in. thick insulation	0.1–0.2
Offshore, unburied pipeline lying on bottom exposed to water currents:	
Uninsulated	8–12
2-in. thick insulation	0.1–0.2
Offshore uninsulated pipeline, suspended, unburied, exposed to water currents	15–100
Offshore buried pipeline:	
Uninsulated	0.5–0.7
Insulated	0.1–0.2

[a]Overall heat-transfer coefficient U is based on pipe-surface area in square feet. Values are from various sources and experimental data.

If the pipeline is insulated or coated, the heat loss can be described by

$$Q = \frac{T_0 - T}{\dfrac{1}{2\pi r_1 h_i} + \dfrac{\ln r_2/r_1}{2\pi K_1} + \dfrac{\ln r_3/r_2}{2\pi K_2} + \dfrac{1}{2\pi r_3 h_o}} \tag{2}$$

where In = logarithm base e
r_1 = pipe inside radius (ft)
r_2 = pipe outside radius (ft)
r_3 = pipe center to insulation surface (ft)
h_i = inside heat transfer (film) coefficient (Btu/h/ft²/°F)
h_o = outer surface heat transfer coefficient (Btu/h/ft²/°F)
k_1 = thermal conductivity of pipe (Btu/h/ft/°F)
k_2 = thermal conductivity of insulation (Btu/h/ft/°F)

A further effect that must be considered is that a film of wax or coating formed on the inside of the pipeline that shields the hot oil from direct contact with the pipe wall. While the net effect on the overall heat-transfer coefficient U is not extreme, it should be considered. Wax deposits from the crude oil would have an insulating effect similar to that of an internal lining.

Since it cannot be assumed that wax will automatically form on pipe walls just because wax is present in the crude, laboratory tests should be performed to determine at what conditions such formations could start. Wax will not deposit at high temperatures, particularly in turbulent flow, so part of a pipeline may have a thin wax film and part may be clean.

When a hot pipeline is buried, it heats the soil surrounding it. Heat moves from the pipe in a parabolic shape, with the widest area of the parabola at the ground surface. The term "shape factor" (S_f) expresses the amount of heat flow through this parabolic-shaped field. The magnitude of the shape factor depends primarily on the burial depth and pipe diameter, as shown by

$$S_f = \frac{2\pi}{\ln\left[\dfrac{2D_n + (4D_n^2 - D_i^2)^{1/2}}{D_i}\right]} \tag{3}$$

or

$$S_f = \frac{2\pi}{\cosh^{-1} \dfrac{2D_n}{D_i}} \tag{4}$$

Heat flow from the pipeline then can be calculated by

$$Q = S_f K_s (T_i - T_s) \tag{5}$$

where S_f = shape factor
Q = soil thermal conductivity (Btu/h/ft/°F)
K_s = soil thermal conductivity (Btu/h/ft/°F)
T_i = pipe surface temperature (°F)
T_s = ambient temperature (soil) (°F)
D_n = depth of burial, soil surface to pipe center (ft)
D_i = pipe outside diameter (ft)

To determine heat flow (Q) by the above equation, the soil thermal conductivity must be measured. Methods of direct measurement of onshore soil thermal conductivity are available and well documented. However, these methods are not yet adapted to direct measurements offshore. The same basic concepts could be used, but equipment modifications would be required.

Restart after Shutdown

As waxy oil congeals under stationary conditions, a finite pressure is required to initiate flow. This starting pressure gradient is related to the rheological property known as yield stress.

Yield stress is the shear stress that must be developed in the oil to initiate flow. Yield stress for waxy crude is an inverse function of temperature, the stress increasing with decreasing of temperature. It is also a function of the rate of shear when cool-down occurs and depends upon whether or not the cool-down occurs during static or dynamic conditions.

Since start-up pressure depends to a large extent on whether the oil is cooled under static or dynamic conditions, rheological studies should be made for static and dynamic cool-down rates so that the exact start-up pressures can be determined for use in designing pumping equipment.

Experiments have shown that restart pressures can be 5 to 10 times higher for a statically cooled pipeline (the fluid was above the pour point while flowing and allowed to cool while shutdown) than for one that has been dynamically cooled (the fluid was already below the pour-point temperature before shutdown).

Facilities Design

In addition to restart problems and their effect on operation and pumping equipment, consideration also must be given to the requirements of facilities appurtenant to the pipeline, such as storage tanks, scraper traps, test equipment, auxiliary lines, and spare parts for equipment repair.

When dealing with waxy crudes, pump selection for the mainline units requires a thorough understanding of the pipeline system, as well as the crude characteristics. However, once the design parameters are determined, the same approach as for Newtonian fluids is used.

As in the normal pump design, a flow vs head requirement must be calculated. Figure 1 shows a typical waxy crude flow vs pressure curve.

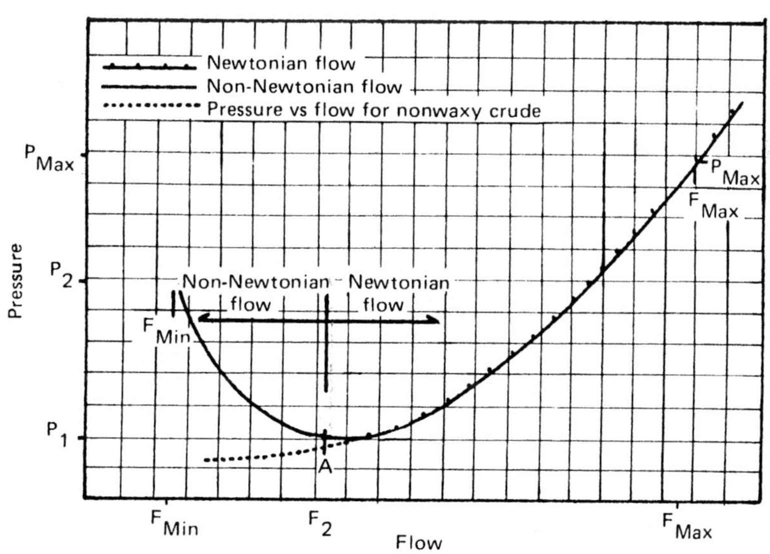

FIG. 1. Flow vs pressure for waxy crude.

It can be shown that, as flow decreases from some optimum point A, the pressure requirement will increase; because the oil is in the line longer, it cools more, and its viscosity increases. Additionally, as the flow decreases, the velocity decreases, which decreases the rate of shear.

As the rate of shear decreases in non-Newtonian flow, the viscosity increases (Fig. 2). When the flow starts to increase past point A, Newtonian characteristics take effect and pressure again increases—this time because increased pressure is required to offset increased friction losses.

With increased flow, resulting incoming temperatures will start to increase after the pipeline and its surrounding soil have adjusted to the new temperature profile.

In choosing a main pump for the system curve outlined in Fig. 1, the operating point P_{max}, F_{max} must be met. If the point P_2, F_{min} cannot be met by the mainline pumps because of temperature restrictions at the low flow rate,

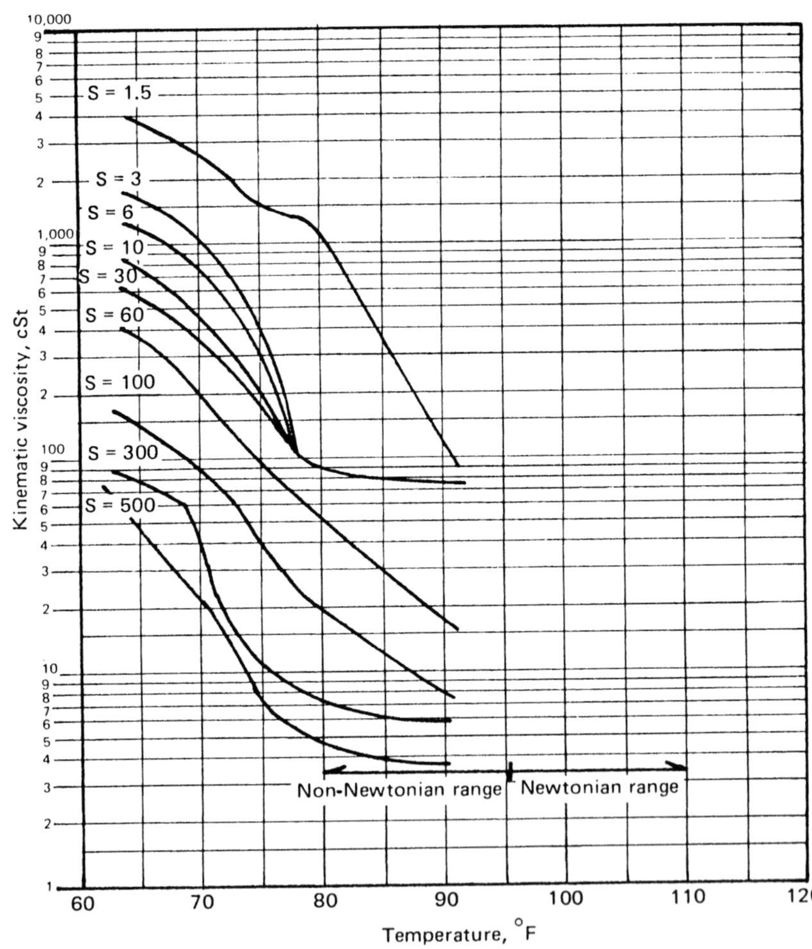

FIG. 2. Temperature vs viscosity (experimental data) for various shear rates S.

or because P_2 might be greater than P_{max}, either special start-up/restart pumps, such as positive-displacement (high pressure, low flow) pumps, or the other possible solutions for restart presented elsewhere might be considered.

An example of transporting high-pour-point crude oil is a field producing oil with a pour point of 80°F. After comprehensive rheology tests were performed in the laboratory, heat-loss and hydraulic calculations resulted in the following recommended design premises and operating philosophy:

1. Design pumping facilities to handle maximum expected volumes. Pumps should be selected to allow a parallel/series arrangement which could transport early production volumes at slower rates and higher pressures when necessary. The piping could be manifolded so that parallel arrangement would be accommodated by repositioning of valves to handle higher flow rates.
2. Use pour-point depressants.
3. Design production separators for higher than normal outlet pressures to allow as much gas and light hydrocarbons as possible to remain in solution.
4. Schedule pipeline start-up during period of warmest ambient temperatures.
5. Establish minimum flow rate to be maintained during initial start-up and for at least 2 weeks thereafter.
6. Include redundant provisions for emergency and planned shutdowns. Systems to consider would include crossover connection with a parallel gas line, standby pumps for displacing the crude oil in the pipeline with water, and pour-point-depressant injection facilities.

Once the pipeline system is defined in terms of fluid rheology, hydraulic head requirement, and operating philosophy, pump selection may begin.

Part A: How to Determine Properties of Crude Oil

Several design parameters must be investigated before a determination is made of how to pipeline a high-pour-point crude. These parameters are:

1. Physical properties of crude
2. Heat transfer
3. Restart after shutdown
4. Facilities design

This section will describe the steps needed to determine the first parameter.

To determine the physical properties of a high-pour-point crude, samples of the crude must be tested in the laboratory.

As an example of this complicated process, consider a project which involved the feasibility of transporting a waxy crude, designated R-82, from an oil field in North Africa to a coastal sales terminal. Since the production from the R field was only 10,000 bbl/d, installation of a separate pipeline was precluded.

The principal question to be answered was whether it would be technically feasible to inject the R-82 into an existing pipeline handling C-65 crude.

The work involved collection of field oil samples, examination of the subject crude oils through laboratory analysis, evaluation of existing C-65 operations, prediction of the total system response to blended oil transport, estimation of new and/or additional facility requirements, and consideration of the effect of blended oil on saleability.

Before conclusions could be drawn, detailed laboratory work was needed.

Sample Information

The C-65 crude oil samples were taken at a sampling station in the oil field. During the sampling period, the pipeline pressure was 30.24 kg/cm^2 (430 lb/in.^{2}gauge). The temperature was 57.8°C (136°F) and the production rate was 165,000 barrels of oil per day (bo/d).

The R-82 crude-oil samples were taken downstream of the producing separator at a new well location remote to the producing oil field. During sampling the separator pressure was 15 kg/cm^2 (213.3 lb/in.^{2} gauge) and the temperature was 55°C (131°F).

At the laboratory, the following tests were performed on the samples: Pour point, density, specific heat, wax content, and water and sediment. Also, tests were run to determine salt content, distillation, rate of shear, and yield stress.

Pour Point

The test for the pour point of R-82 crude and a blend of the C-65 and R-82 crudes (80%/20% by weight, respectively) was conducted according to ASTM Method D97-66. This is accomplished by heating the sample, then cooling it at a specified rate and examining it for flow characteristics at each interval of 3°C (5°F) in temperature drop.

The lowest temperature at which movement of the oil is observed is recorded as the pour point. The results of three separate experiments were:

Crude oil R-82: Mean pour point 47°C (117°F).
Crude oil C/65/R-82 (80%/20%): Mean pour point 33°C (91°F).

Density

The densities of R-82 and the blend C65/R82 crudes were determined with a DMA 50 digital precision densitometer. The determination is based on the measurement of the vibration frequency of a thermostated U-shaped tube filled with a sample. The precision of the measurements with this apparatus is $\pm 1 \times 10^{-4}$ g/mL.

Two series of duplicate measurements were performed. The first series started at 55°C (131°F) and ended at 10°C (50°F). The second series started at 10°C and ended at 55°C. The heating and cooling rate was about 2°C/min. The equilibration time at the measuring temperature was about 10 min. The results of the two series are given in Table 3.

The observed differences probably were caused by slow crystallization effects.

Specific Heats

The specific-heat measurements were made with an adiabatic calorimeter. The sample energy in the measuring vessel is dissipated by means of a calibrated electrical heating coil. The temperature of the sample is recorded as a function of the energy input.

The energy input in the sample and measuring vessel amounted to 6 cal/min. Of this, about 50% was used for the oil and 50% for the vessel with accessories. Heat input into the oil was about 0.12 cal/g/min.

In order to avoid slow crystallization effects, the measurements were performed at rising temperature after equilibration at about 0°C. The mean specific heat was calculated in intervals of 5°C. The temperatures given in Table 4 are the mean temperatures at the chosen intervals.

Wax Content

The wax contents of crudes R-82 and C-65 were determined by crystallizing the wax in methylene chloride at −25°F (−32°C) according to BP Method

TABLE 3 Density Test Results

Temperature (°C)	Crude R82 Density (g/mL)		Crude Blend R82/C65 Density (g/mL)	
	Cooled	Heated	Cooled	Heated
10	0.8514	0.8514	0.8543	0.8543
15	0.8458	0.8461	0.8496	0.8496
20	0.8397	0.8403	0.8444	0.8444
25	0.8330	0.8338	0.8390	0.8391
30	0.8255	0.8266	0.8337	0.8340
35	0.8175	0.8186	0.8389	0.8298
40	0.8099	0.8124	0.8250	0.8255
45	0.8057	0.8077	0.8212	0.8216
50	0.8013	0.8022	0.8175	0.8177
55	0.7972	0.7977	0.8140	0.8141

TABLE 4 Specific Heats

Temperature (°C)	Specific Heat (cal/g°C)	
	Crude R82	Crude Blend R82/C65
5	0.68	0.60
10	0.71	0.63
15	0.75	0.63
20	0.81	0.66
25	0.90	0.68
30	0.89	0.60
35	0.80	0.54
40	0.87	0.57
45	0.79	0.57
50	0.73	0.55
55	0.63	0.54

237/55. Asphaltic materials were removed by first swirling a solution of the oil in petroleum ether with concentrated sulfuric acid. The solution was then centrifuged and filtered through a layer of a filler aid (Celite).

After this, the wax was crystallized from methylene chloride and isolated by filtration. The wax was dissolved again in petroleum ether and filtered through a glass filter. The results by weight were:

	Crude (%)	
	R-82	*C-65*
Experiment 1	39	17.6
Experiment 2	38.2	18.2
Experiment 3	38.2	17.0
Mean	38.5	17.6

Water and Sediment

The water and sediment contents of crudes R-82 and C-65 were determined according to ASTM Method D96-63. The samples were diluted with an equal volume of "water-saturated" toluene and centrifuged until no droplets or particles were observed microscopically in the oil solution. The volume of sediment was then measured. The results were:

	Crude (vol. %)	
	C-65	*R-82*
Experiment 1	0.0	0.2
Experiment 2	0.0	0.2

The amount of water condensed was 0.1–0.2% for the C-65 oil and negligible for R-82 oil.

Salt Content

The salt content was determined by careful extraction of a solution of 100 mL of crude oil in petroleum ether with three 50-mL portions of warm distilled water. The chloride content of the extract was determined by adding 0.1*N* of silver nitrate.

The obtained turbidity was compared with the turbidity of standard solutions of salt with the same amount of silver nitrate. The results were:

Crude	
C-65	*R-82*
<1 ppm NaCl	<1 ppm NaCl

Distillation

About 350 g of crude oils C-65 and R-82 were distilled, using a distillation still, according to ASTM Method D1160. Briefly, the sample was distilled at some predetermined and accurately controlled pressure between 1 mmHg,

TABLE 5 Distillation Results

Fraction No.	Weight %	Pressure (mmHg)	Distillation range (°C)	Density (g/mL)	Molecular Weight
		Crude Oil C-65			
1a	7.4	760	<112	—	—
1b	12.2	760	112–180	0.7577 (25°C)	135
1c	12.8	760	>180	0.7963 (25°C)	195
2	7.2	10	110–165	0.8174 (25°C)	223
3	7.2	10	165–197	0.8264 (25°C)	267
4	6.9	2	115–166	0.8406 (25°C)	358
5	6.3	2	166–185	0.8424 (40°C)	361
6	6.6	2	185–220	0.8489 (50°C)	331
7	6.9	<1	220–250	0.8650 (50°C)	435
Residue	26.5	—	—	0.9385 (70°C)	1,020
		Crude Oil R-82			
1a	7.7	760	<112	—	—
1b	7.1	760	112–180	0.7583 (25°C)	141
1c	11.7	760	>180	0.7885 (25°C)	201
2	6.0	25	160–188	0.8041 (25°C)	220
3	6.6	10	140–184	0.8106 (25°C)	251
4	6.6	2.5	140–170	0.8088 (40°C)	300
5	6.2	1	145–180	0.8010 (55°C)	359
6	6.7	1	180-200	0.8013 (60°C)	341
7	6.2	0.5–1	180–215	0.8070 (60°C)	401
8	6.4	0.5–1	175–230	0.8153 (60°C)	464
9	5.3	0.5–1	230–255	0.8177 (70°C)	373
Residue	23.5	—	—	0.8551 (70°C)	882

absolute and atmospheric, under conditions which provided about one theoretical plate fractionation.

From the data obtained, a distillation curve, relating volume distilled and boiling point at the controlled pressure, was prepared. The results are given in Table 5.

Finally, a pipeline model was used to establish the following relationships:

1. Viscosity–temperature–rate of shear.
2. Yield stress vs temperature for static and dynamic cool down.

The pipeline model consisted of four trains. A hydraulic system, inert-gas return system, and recycle heating and cooling system were shared by all the trains. Each train contained two membrane separators, a test length of steel tubing, a hydraulic activator, and instrumentation.

The velocity was controlled by adjusting the pressure drop across the valve that supplies hydraulic oil to the membrane separator. The sample oil was transferred through a 12.19-m (40-ft) straight length of 0.9525-cm (0.375-in.) tubing at a preset velocity for various temperature settings.

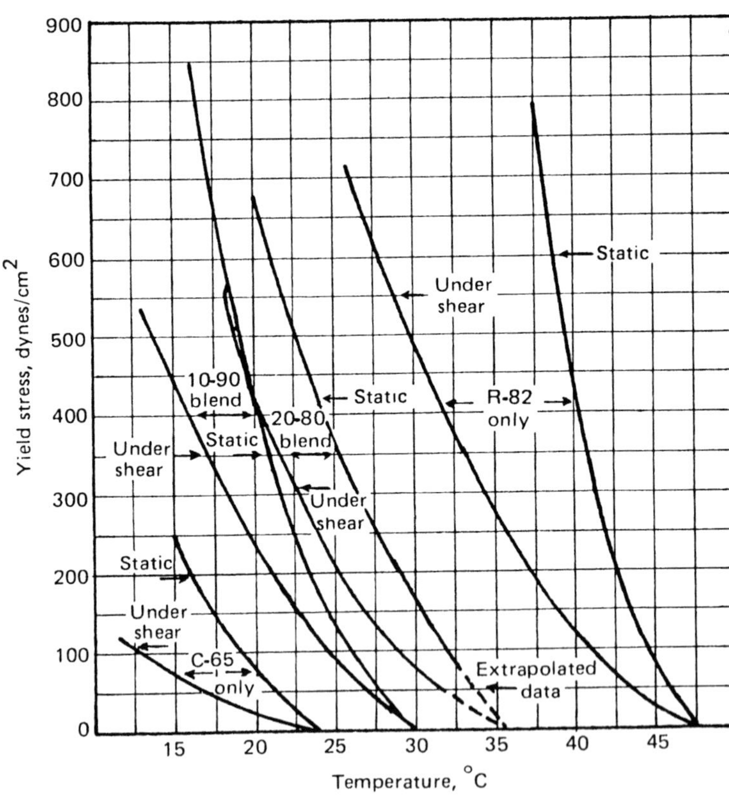

FIG. 3. Yield stress vs temperature of various C-65/R-82 blends.

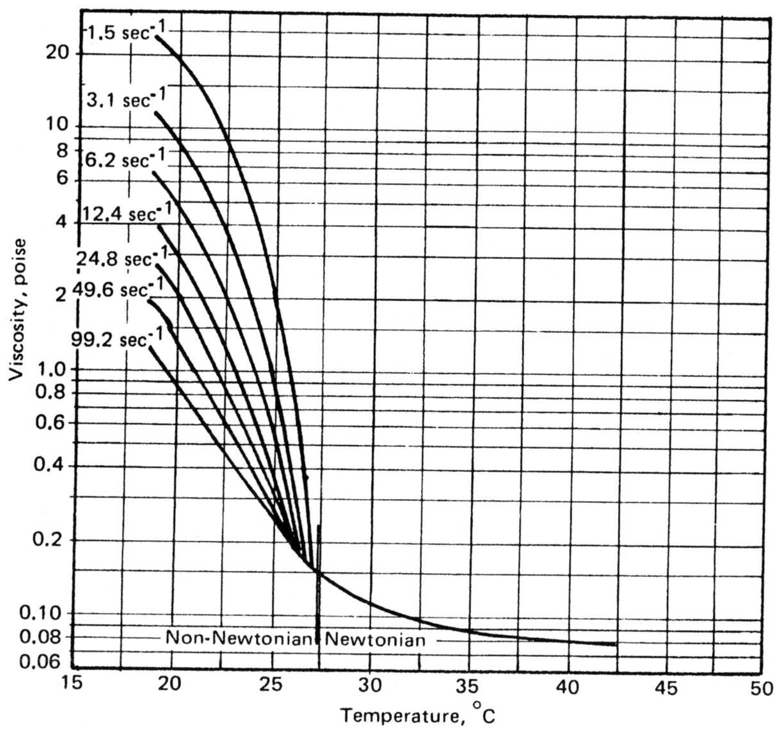

FIG. 4. Viscosity vs temperature at various shear rates for C-65 crude oil.

For each velocity a differential pressure was developed across the test length of tubing. Start and stop times were recorded and the velocity computed. Figures 3 and 4 describe the results obtained.

Once the physical properties of the crude were determined, it was concluded that injecting up to 10% blends of R-82 into C-65 under certain specified conditions was technically feasible.

This conclusion could not have been made without extensive laboratory work. Such a determination of physical properties of waxy crude is fundamental to the safe and complete design of a pipeline system.

Part B: Heat Transfer

Before the required pumping equipment in a pipeline system can be specified, pressure-loss calculations must be made. With conventional crude oils, these calculations are straightforward. But they become more complex with non-Newtonian crude, because the viscosities are both temperature and shear-rate dependent.

To determine the amount of heat loss in a pipeline, Eq. (1) is used.

Many designers prefer to assume an overall heat-transfer coefficient (U) based on previous measurements. Heat flow also can be calculated by obtaining the soil thermal conductivity K (Btu/h/ft/°F) which is inserted in Eq. (5).

Heat Conduction in Soils

Heat transfer in very dry soils takes place primarily through small but finite areas of contact between soil particles. The amount of heat transferred depends on the mineral composition of the soil and the number and magnitude of the soil contact points.

Heat will follow the path of least resistance. Because of this, most of the heat will be transferred through the soil particles, which have a thermal conductivity of about 0.2 Btu/ft/h/°F, rather than through the air in the voids between the particles, which has a thermal conductivity of about 0.015 Btu/ft/h/°F.

The amount of heat transferred increases when relatively small amounts of moisture (7–20%) are added to the soil. This is because the liquid is held by surface tension at the soil contact points, effectively increasing the contact area.

As more water is added, and the void spaces become completely filled, heat transfer takes place by conduction through both the soil particles and the water. The heat now can follow parallel paths rather than the limited-point contact paths.

The net result is a significantly greater thermal conductivity of the soil–water mixture. Experimental results based on physical measurements indicate that the apparent thermal conductivity of the soil–water mixture is higher than that of either of its constituents.

While dry soil conductivity is 0.2 to 0.4 and water thermal conductivity is 0.3 to 0.4, depending on temperature, measured conductivities of the soil–water mixture sometimes exceed 1.0. Wet or saturated soils have measured values above 0.6.

It appears that some mechanism other than pure conduction is at work, since the direct addition of soil conductivity and water conductivity is 0.5 to 0.8 Btu/h/ft/°F. Some of the explanations offered for this phenomenon are structural change or density increase with the increasing water content of soils.

It is possible that, as soils become saturated, the mixture becomes essentially soil particles floating in water. The K_s will probably decrease to about the thermal conductivity of water (0.35–0.4) when this happens. Convective movement, however, is more probable in this state.

Assigning finite values to variables, such as the number of soil-particle contacts and the increase in contact area for varying water content, is virtually impossible. Researchers, therefore, have resorted to field measurements to arrive at soil thermal conductivities.

The most significant research work indicates that the following relationships should be taken into consideration when selecting K_s values for design.

At any constant moisture content, including absolutely dry, increased soil density results in increased conductivity. The reason: more points of intergranular contact.

At constant density, an increase in moisture content causes an increase in conductivity. This relationship is true up to the point of saturation, where the possibility of convection must be considered.

Thermal conductivity generally varies with soil texture. At a given density and moisture content, the conductivity is relatively high in coarse-textured soils such as gravel or sand, somewhat lower in sandy loam soils, and lowest in fine textured soils such as silt loam or clay. Thermal conductivity of a soil also depends upon its mineral composition. Sands with a high quartz content have greater conductivities than sands with high contents of such minerals as plagioclase, feldspar, and pyroxene. Soils with relatively high contents of kaolinite (clay) have relatively low conductivities.

Soil Thermal-Conductivity Calculation

Equations for calculating the soil thermal-conductivity coefficient (K_s) for varying moisture contents have been published by several investigators. These equations are based on a correlation of measured thermal conductivity with known soil properties and moisture contents.

These values were derived for the particular soils available to the investigators.

In situ field measurements of soil thermal conductivity should be made for onshore pipelines. This is because thermal conductivity depends on soil texture. The presence of different soil factions or organic matter may significantly affect a soil's actual K_s. Offshore *in situ* measurements are more difficult, but thermal conductivity measurements can be made on recovered samples.

The following is a brief summary of what appear to be the most reliable formulas proposed by various investigators. (All the techniques discussed assume the moisture content of the soil remains stationary.)

Kersten [1] established experimentally the following for silts and clays:

$$K_s = (0.9 \log w - 0.2)10^{0.01} r_d$$

For sandy soils:

$$K_s = (0.7 \log w + 0.4)10^{0.01} r_d$$

where log = log base 10
K_s = thermal conductivity (Btu/ft^2/in./°F)
w = moisture content (percent of dry soil weight)
r_d = dry density (16/ft^3)

It is suggested by others [2] that these values could be in error by up to 25%.

Based on additional experimental measurements of soil in the London area, Makowski and Mochlunski [3] revised Kersten's original general formula to the following:

$$K_s = (A \log w + B)10^c$$

where K_s = thermal conductivity of soil, $(w/m - °K)$
$A = 0.1424 - 0.000465S_c$
$°K = (°F + 460)/1.8$
$B = 0.0419 - 0.000311S_c$
$c = 6.24 \times 10^{-4} r_d$
S_c = weight percent of clay (less than 0.002 mm particle size) referred to the total weight of dry soil.
w = water content of the soil (%)
r_d = dry density of soil (kg/m^3)

McGaw [4] suggests this expression for thermal conductivity of a saturated granular material:

$$K = (n - N_c)\, K_f + (v + N_c)\, \epsilon\, K_s(v + N_c)/(V + \sigma N_c)$$

where K_f = bulk conductivity of a continuous liquid phase
K_s = bulk conductivity of a dispersed granular phase
$\sigma = K_s/K_f$
ϵ = interfacial efficiency
n = porosity
$v = (1 - n)$ = volume fraction of solids
N_c = volume of fluid

Convection Heat Transfer

The mechanism of heat transfer which involves the physical movement of heat-laden particles away from the heat source is called convection. The heat-laden particles in a marine-pipeline installation are most often water.

In a soft slurry, however, which often backfills an offshore pipeline, it is possible that a light slurry could migrate away from the pipeline with increasing temperature.

When water is transporting the heat away, the magnitude of the heat loss is significant because water has an extremely high specific heat compared to other liquids and materials.

Two types of convective flows are possible, forced and natural. A forced convection could occur when a pipe is exposed to ocean currents or if an onshore pipeline is buried in sloping ground.

In buried, offshore pipelines, natural convection due to heating of the water is of primary concern.

The rate of heat transfer by convection depends on the velocity with which the water can move away. The velocity is controlled by resistance of the soil to flow of moisture. This resistance is described by D'Arcy's law, $V = Ki$, where V is the velocity, i is the hydraulic gradient, and K is the coefficient of permeability. The value of K depends on the properties of both soil and water and is, therefore, not a constant.

In general, K values fall into the following ranges:

Soil types	*Permeability coefficient (m/s)*
Gravels	100 to 10^{-1}
Sand	10^{-1} to 10^{-4}
Silts	10^{-3} to 10^{-7}
Clays	10^{-5} to 10^{-10}

Due to the density differentials of the hot and cool water, hydraulic gradients will depend primarily on buoyant forces. For most temperature-differential ranges in offshore buried pipelines, the hydraulic gradient is about 1 in 100. Once the velocity is calculated from the above equation, an estimate of the heat transfer can be made.

Many factors make exact measurement of the permeabilities of offshore soils impractical. Undisturbed samples are difficult to obtain. In addition, when offshore pipelines are buried, they usually are allowed to backfill by normal sediment transport and trench caving.

It would be futile to try to predict accurately the eventual permeability of this mixed backfill.

To compensate, most designers assign a high K_s value.

More detailed discussion on problems encountered in heat transfer of pipelines handling waxy crudes are available in the technical articles and books referenced in this article.

Part C: Restart after Shutdown

Under stationary conditions, waxy oil congeals when it cools below the pour point. If this cooling takes place, a certain pressure is required to initiate flow of the gelled waxy crude.

As stated earlier, the starting-pressure gradient is related to the rheological property known as yield stress. Yield stress is the shear stress that must be developed in the oil to initiate flow. This relationship is illustrated in a rheogram of a non-Newtonion liquid (Fig. 5).

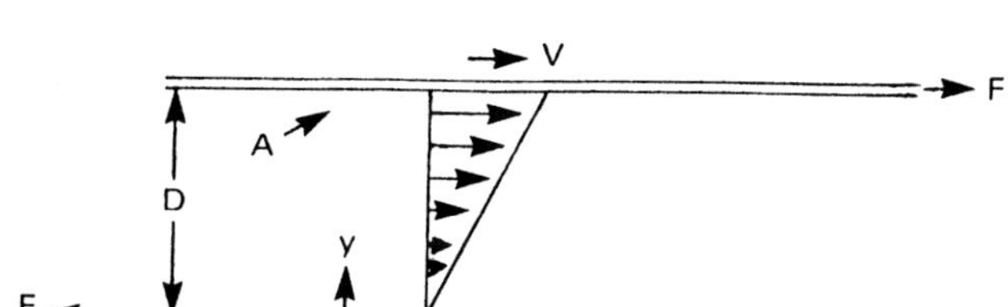

FIG. 5. Rheogram of non-Newtonian liquid

From the values in Fig. 5, the following relationships are determined:

$$\text{Shear stress} = F/A \ (\text{lb/in.}^2)$$

$$\text{Rate of shear} = V/d = dv/dy \ (\text{s}^{-1})$$

$$\text{Viscosity } \eta = \frac{\text{shear stress}}{\text{rate of shear}} = (F/A)/(V/d) \ (\text{lb} \cdot \text{s/in.}^2)$$

$$F = \eta A(V/d) = \eta A(dv/dy) \ (\text{lb})$$

Yield stress for waxy crude is an inverse function of temperature and increases with decreasing temperature. With a crude oil that is below its pour point, the structures in the oil start to break down as flow begins. This breakdown depends on both time and rate of shear.

Almost all characteristics of waxy crudes, such as the ratio of shear stress to rate of shear (called apparent viscosity or pumping viscosity), pour point, yield stress, and breakdown in structure, also depend on the history, thermal, and other pretreatment of the oil.

Start-up pressure, for instance, depends to a large extent on whether the oil is cooled under static or dynamic conditions. Rheology studies of static and dynamic cool-down rates for the particular crude under consideration are necessary to determine the exact shear-stress values to use in design.

The maximum pressure buildup when a gelled pipeline is restarted occurs when the gel is sheared at the wall. To establish whether restart is possible for a given shutdown period, the available pumps must be evaluated for their ability to provide the required shear stress.

This can be determined by Eq. (6). If S, the available shear stress, is equal to or greater than the maximum force needed to overcome the shear stress at the inner surface of the pipe, then flow will be initiated. This available shear stress is defined as

$$S = \frac{(P_1 - P_2)d}{4L} \tag{6}$$

where S = available shear stress (lb/in.^{2})
P_1 = available pressure at the pump (lb/in.^{2})
P_2 = Static pressure at end of line segment (lb/in.^{2})
L = length of line segment (ft)
d = diameter (ft)

By increasing the initial pressure (P_1) and/or reducing the section length (L), the available shear stress (S) is increased.

A general equation for calculating restart pressure requirements for a cooled line is

$$P = \frac{YLC}{A}$$

where P = pressure required (lb/in^2)

Y = yield strength of gelled crude (this value must be determined in laboratory tests and is different for static and dynamic cooling) (lb/in.^{2})
L = line length (ft)
C = circumference of inside pipe wall (ft)
A = cross-sectional area of pipe (ft^2)

Restart problems must be anticipated in the design phase of the project. Here are design questions that must be answered:

What is the length of anticipated shutdowns? Will they be static or dynamic?
What will be the resulting start-up pressures?
Can any precautions be taken to prevent gelling in case of a planned shutdown?
What facilities must be designed into the system to permit restart under the various anticipated shutdown periods?

Lengths of planned shutdown can range up to 10 for land pipelines and much longer for offshore lines. The exact time period depends on how well the operator is prepared for the shutdown.

The following list can be used to envision the cause and length of shutdown a pipeline might have:

Normal operational shutdowns
Power failures
Malfunction of ancillary equipment
Line damage or leaks
Authority regulations (period proof testing)
Force majeure

Lengths of downtime for these reasons can vary. Therefore, the designer must decide what would be the longest foreseeable shutdown and ensure that, should a shutdown occur, the pipeline flow could be resumed. Because start-up pressure will be higher than normal, with low initial flows, minimum flow conditions of the pumps must be investigated.

If the initial flow-rate conditions cannot be met with normal-operation pumps, special restart pumps with appropriate characteristics will have to be used. Under emergency conditions, it might be necessary to install pumps along the line at intermediate locations.

Several alternates are available to help solve or reduce restart problems:

1. Displacement with water or lighter hydrocarbon liquids.
2. Crossover connections at frequent intervals with a parallel gas line to allow gas displacement of the crude.
3. Side traps at frequent intervals to allow short sections to be started separately.

4. Chemical injection prior to shutdown to prevent or retard wax crystallization.
5. Separation at higher pressures to allow more light hydrocarbons and gas to remain in the crude.
6. Circulation of the crude oil through heaters to maintain the temperature above the pour point.
7. Using a "gut line"—a smaller line installed inside a larger line to allow either circulation or continuous pumping of hot fluid through the smaller line.
8. Reverse pumping to create a back-and-forth pumping sequence which prohibits static cool-down (storage and pumping facilities are required on each end of the line).

Part D: Design of Heavy Crude Facilities

Of primary concern in facilities design is pump selection. Once the properties of the crude, its heat-transfer characteristics, and the restart requirements are established, mainline-pump selection will not be much different for high-pour-point crudes than it would for Newtonian crudes.

As in normal pump design, a flow vs head requirement is calculated. Fig. 1 shows what the results might be for a typical waxy crude.

It can be shown, as noted earlier, that, as flow decreases from some optimum point A, the pressure requirement will increase, since the longer the oil is in the line, the more it cools and the more its viscosity increases. Additionally, as the flow decreases, so does the velocity, thus decreasing the rate of shear.

As the rate of shear decreases in non-Newtonian flow, the viscosity increases. When the flow starts to increase past point A, Newtonian characteristics control, and pressure again increases because, as expected, increased pressure is required to offset the increased friction losses caused by the additional flow.

In choosing a main pump for the system curve outlined in Fig. 1 the operating point P_{max}, F_{max} must be met. If P_2, F_{min} cannot be met by the mainline pumps because of temperature restrictions at the low flow rate, or because P_2 might be greater than P_{max}, special start-up/restart pumps, such as positive-displacement (high pressure, low flow) pumps, might be considered.

Pump Selection

Types

Generally, pumps are of two types—dynamic (centrifugal) and positive displacement. Dynamic pumps impart velocity energy to the oil being pumped by means of a rotating impeller on a shaft. This velocity energy is converted to pressure energy as the oil leaves the impeller and goes through a stationary volute or diffuser casing.

Positive-displacement pumps, on the other hand, impart energy to the oil in a fixed displacement volume. Examples are casings or cylinders, vanes or screws, reciprocating pistons or plungers, or the rotary motion of gears.

Generally, high-volume low-to-moderate pressure requirements can be met by centrifugal pumps. Positive-displacement pumps handle low-volume, high-pressure requirements.

Flows between F_{min} and F_2, if it were found that P_2 was greater than P_{max}, likely would be handled by positive-displacement start/restart pumps. If proper recirculation is provided and P_2 is less than P_{max}, centrifugal application might be possible.

Specific Speed

Further pump information can be obtained by calculating the pump's specific speed (N_s). This dimensionless number is defined as

$$N_s = \frac{N(Q)^{1/2}}{H^{3/4}}$$

where N_s = pump specific speed
N = rotative speed (r/min)
Q = flow at or near peak efficiency (gal/min)
H = head (ft/stage)

Centrifugal pumps have N_s values that run from 400 to over 10,000. Reciprocating pumps have much lower values

Specific speeds are also useful in determining the impeller type for centrifugal-pump application. In general, pump impellers are classified as radial when the specific speed is 500 to 1500, mixed flow from 2000 to 6000, and axial from 7000 to over 10,000.

For services requiring intermediate to high heads, radial-flow pumps are used. Mixed-flow pumps are used when an intermediate head is needed. Axial-flow pumps are used for low head requirements.

Net Positive Suction Head

A further design consideration is NPSH (net positive suction head). All pumps require this to permit the oil to flow into the pump casing.

This value is influenced by the speed of rotation, the inlet area of the impeller in a centrifugal pump, and the number and type of vanes in the impeller. The NPSH of a reciprocating pump is based mainly on speed and valve design. Providing the required NPSH is important to eliminate any chance of cavitation during operation.

Temperature Considerations

Since in most design applications, high-pour-point crude is heated to such a temperature as to avoid operating problems caused by downstream cooling, several modifications are required to conventional centrifugal pumps so they can withstand the severe conditions imposed by the heated oil.

Bearing frames should be cooled because the impeller drive shaft conducts heat from the hot oil directly to the bearing housing. While water cooling may be optional to extend bearing life in temperature ranges of 200 to 350°F, it must be provided when oil temperatures are above 350°F.

Thrust bearings also must be cooled. This protects them from excessive loads generated when differential temperatures distort the pump frame, changing shaft dimensions.

Mechanical seals, which should be used instead of packing in a hot oil system, also should be cooled. Without cooling, the seal life will be shortened considerably.

Centerline-supported pumps are sometimes used to provide for equal expansion above and below the pump centerline. This is essential in maintaining alignment.

Operating Problems

The operation of a pipeline system is most economical and efficient when a continuous constant flow rate is maintained. This minimizes storage volumes, energy consumption, and investment in plant facilities.

Pipelines that transport high-pour-point crude oil especially should be designed for continuous operation because of restart problems after shutdown. Since it is not always practical or possible to operate the entire pipeline system at a temperature above that at which the crude begins to form wax crystals, system components must be designed carefully and operating procedures selected to allow more flexibility.

Obviously, it is essential that the pipeline system be tested to the design requirements prior to start-up. Since restart pressures may approach the yield strength of the pipe, each component of the pipeline system will be required to perform to design capacity.

Start-up

During initial start-up, soil temperatures will be low and the rate of cooldown for the oil will be rapid if shutdown occurs. Several days to several months will be required to reach an equilibrium temperature. Therefore, provisions should be made to prevent a shutdown while the line is being filled and until temperature equilibrium has been reached.

In the start-up period, the arrival temperature of the oil will be substantially less than operating design predictions, which are based on equilibrium conditions.

Operation

Wax buildup on the pipeline wall also is to be expected. This problem will be minimized by running scrapers frequently. As wax buildup occurs, throughput will decrease. Flow, temperatures, and pressures should be monitored closely for several weeks after start-up.

Varying frequencies for scraper running should be analyzed to determine the effects on flow and temperature. This data can then be used to prepare an operating plan that will incorporate the most economical scraper-running frequency.

Most scrapers used in waxy crude lines are of the partial-bypass type. This means they have holes in the scraper body to allow some of the crude behind the scraper to bypass and cause a jetting action in front of the scraper.

This jetting action keeps the dislodged wax in suspension and agitated. Otherwise, the scraper may push a long cylinder of packed wax into the receiving trap, requiring complicated disposal procedures.

Due to unforeseen circumstances, heating facilities may break down. When this occurs, full tanks of crude can gel, lines can plug, and operation becomes impossible. In anticipation of such emergencies, standby provisions should be made, such as the installation of heating coils near the storage-tank effluent line and the use of temporary pumps connected at gravity lines.

Plans

For all pipelines there should be detailed operating and emergency plans. A complete description of all phases of operation at both ends of the system should be contained in the operating plan.

If the crude oil is to be displaced with water in an emergency, the plan should indicate where the line-fill crude is to be stored and how the fill water will be disposed of when operations are resumed.

In addition, if pour-point depressants are used, the plan should specify what special handling will be required at the terminal because of changed crude-oil characteristics.

Emergency plans must include notification procedures for shutdown or leaks. Emergency plans also must list type, location, and specified use of all equipment needed in leak repair and/or oil-spill clean-up operations. For waxy-crude-oil pipelines, the procedures must be keyed to the length of the shutdown period.

Restart procedures also must be detailed in the emergency plan. For short-term shutdowns, it may be possible to shift pumps from a parallel operation to a series operation and restart with installed pumps. If existing positive-displacement pumps are not available, it may be necessary to bring in standby pumps.

Items which should be investigated and defined during preparation of the emergency plan are:

Subdivision of the pipeline into sections. Maximum available shear stress should be calculated for each definable line segment. Provision should be made for handling the discharged oil during the start.

Flexibility of pumping arrangement. It is necessary to clarify ways in which the pipeline pumps can be used. For instance, pumping in a closed circle with installed pumps may allow small pumping rates to be maintained at maximum pressure to prevent static cool-down of the oil.

Availability of standby pumps. Items to consider are: (1) which kind of transportable pumps, owned or rented, are to be installed at the different locations; (2) how standby pumps will be transported to the site; (3) piping arrangements to connect pumps in the pipeline system; (4) personnel to operate the pumps; and (5) if owned pumps are not available, agreements with a contractor for the renting of pumps.

Determination of whether or not an emergency power supply is necessary at the various locations.

In case of anticipated line drainage at certain locations, a comparison of available tank capacity with the content of the line section which is to be drained. If required, make an agreement with contractors to provide moveable tankage, select transportation routes, and select temporary storage locations for the drained oil.

This material appeared in *The Oil and Gas Journal*, pp. 111–114, May 28, 1979; pp. 150–152, June 4, 1979; pp. 110–111, June 18, 1979; pp. 105–106, July 2, 1979; and pp. 69–70, July 16, 1979; and is used with the permission of the editor.

References

1. M. S. Kersten, *Thermal Properties of Soils*, University of Minnesota Institute of Technology Bulletin No. 28, 1949, Minneapolis, Minnesota.
2. H. F. Winterkorn and H. Y. Fang, *Foundation Engineering Handbook*, Van Nostrand-Reinhold, Princeton, New Jersey, 1975, pp. 105–110.
3. M. W. Makowski and K. Mochlunski, "An Evaluation of Two Rapid Methods of Assessing the Thermal Resistivity of Soil," *Proc. Inst. Electr. Eng.*, *103A*, 453 (October 1956).
4. R. McGaw, "Heat Conduction in Saturated Granular Materials," in *Proceedings of the International Conference on Effects of Temperature and Heat on Engineering Behavior of Soils*, Highway Research Board Special Report 103, Washington, D.C., 1969, pp. 114–131.
5. Davenport and Conti, "Heat Transfer Problems Encountered in the Handling of Waxy Crude Oils in Large Pipelines," *J. Inst. Pet.*, *57*(555), (1971).

BILL SMITH

Gas Flow Pressure Drop

In a gas-duct network, the dimensions and gas velocity vary for each section. Computing the Reynolds number, friction factor, and head loss for each section is tedious. This nomograph (Fig. 1) presents a single sheet of curves so that head loss can be obtained directly.

The nomograph is based on the following formulas. The pressure loss due to flow through ducts is represented by Darcy's equation:

$$\Delta h = 4flV^2/(2g_cd) \tag{1}$$

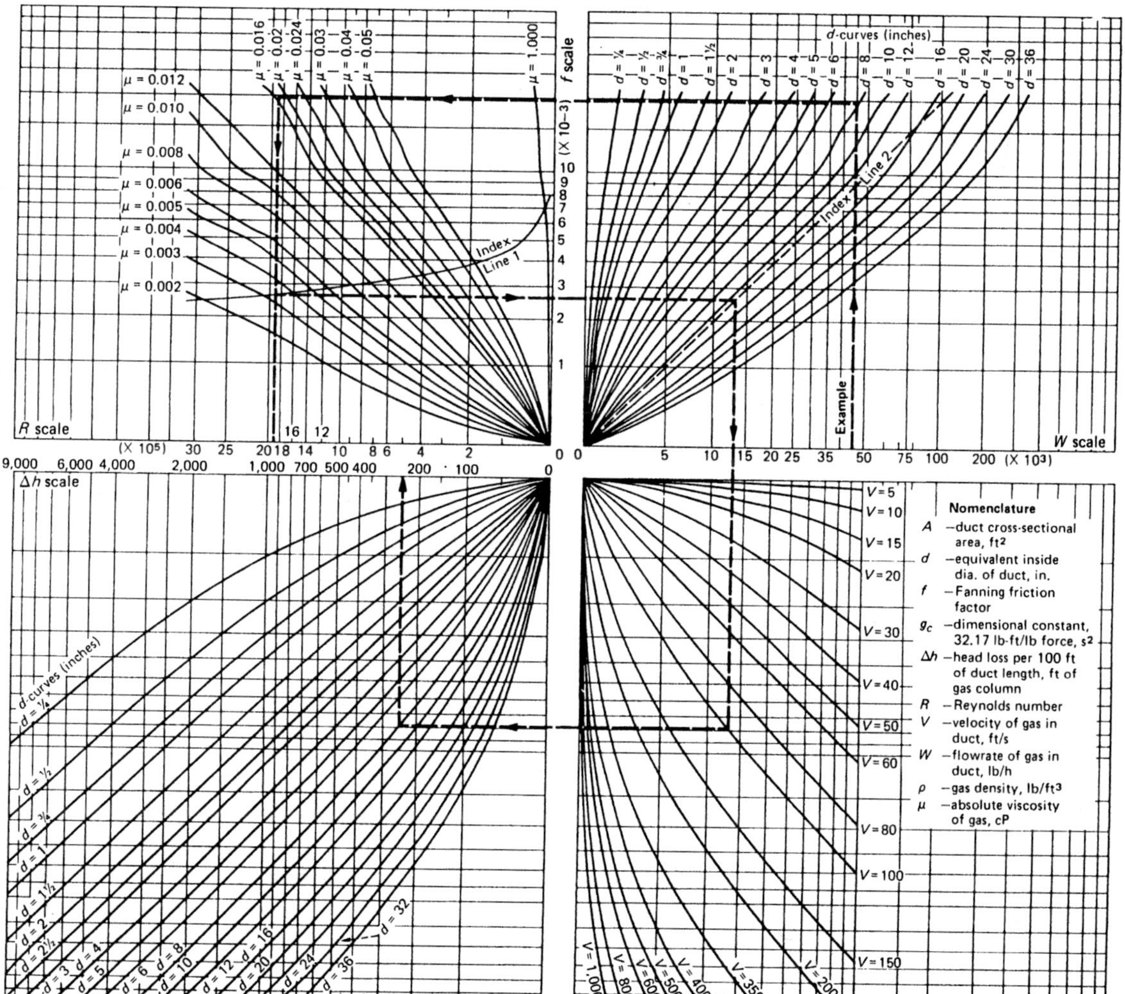

FIG. 1

For $l = 100$ ft of duct, and for d in inches:

$$\Delta h = 2400fV^2/(g_c d) \tag{2}$$

Also:

$$V = W/(3600A) \tag{3}$$

For turbulent flow (where $R > 2100$), f and R are interrelated by:

$$f = 0.0791R^{-0.25} \quad (\text{when } 0.021 \times 10^5 < R < 10^5) \tag{4}$$

and

$$4f = 0.0032 + 0.221R^{-0.237} \quad (\text{when } 0.2 \times 10^5 < R < 32.4 \times 10^5) \tag{5}$$

Now:

$$R = Vd\rho/\mu \tag{6}$$

Rewriting V and ρ in terms of W, and with consistent units:

$$R = 6.32W/(d\mu) \tag{7}$$

Note that Darcy's equation is for incompressible flow. It is a good approximation for gases only if the pressure drop across the duct is less than 15% of the initial pressure.

Also, note that the equivalent diameter d is four times the ratio of duct cross-section to wetted perimeter. Equivalent diameters for some of the commonly used duct types are listed below for ready reference:

Duct cross-section	*Equivalent diameter*
Circle of diameter d	d
Annulus of outer diameter d_o and inner diameter d_i	$d_o - d_i$
Square of side L	L
Rectangle of sides A and B	$2AB/(A + B)$

Example. What is the head loss in an 8-in.-inside-diameter duct for 45,000 lb/h of gas, with a viscosity of 0.0183 and a velocity of 100 ft/s?

Starting with the known value of airflow, find 45,000 on the W-scale and draw a vertical line to intersect the $d = 8$" curve. Draw a horizontal line to intersect the $\mu = 0.0183$ curve and draw a vertical line to intersect Index Line 1. From this point, proceed vertically to read the Reynolds number (R-scale) and horizontally to read the friction factor (f-scale). Continuing horizontally, intersect Index Line 2. From this point, draw a vertical line to intersect the $V = 100$ curve. Draw a horizontal line to intersect the $d = 8$ curve. Now, Draw a vertical line to the Δh-scale and read that the head loss is 240 ft of gas column.

K. JAYARAMAN

Measurement

The term flow measurements can refer to many different types of measuring devices and to the flow of solids, liquids, and gases. This article will be limited to the measurement of the flow of liquids and gases and to the most common devices. We will consider measurements of flow rate usually in a pipeline and measurements of velocity at a point as separate topics.

Flow Rate Measurements

Obstruction or Head Meters

The most common way to measure the mean velocity or flow rate in a pipeline is to use an obstruction in the flow and find the pressure drop resulting. We consider only the obstruction here and not the method to measure the pressure. The basic principle is that when a fluid flows through a contraction, the fluid must accelerate so the pressure must decrease.

Figure 1 shows an orifice plate in a pipe with Plane 1 located upstream of

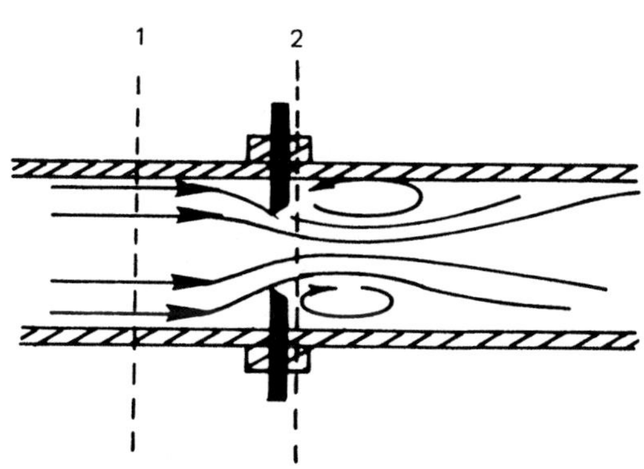

FIG. 1. Schematic diagram of orifice plate and flow pattern

the constriction and Plane 2 located somewhat downstream of the constriction. For a sharp-edged opening the narrowest part of the flow stream occurs downstream from the orifice plate. This vena contracta is also the minimum pressure location. Pressure measurements are to be made at the two positions 1 and 2. For steady-state, assuming velocities parallel to the pipe wall, the macroscopic mass and energy balances give

$$A_1\langle v_1\rangle = A_2\langle v_2\rangle \tag{1}$$

$$\frac{1}{2}\frac{\langle v_2^3\rangle}{\langle v_2\rangle} = \frac{1}{2}\frac{\langle v_1^3\rangle}{\langle v_1^1\rangle} - \int_{P_1}^{P_2}\frac{1}{\rho}dp - 1_v \tag{2}$$

where A_1 is the pipe area, A_2 is the orifice area, $\langle v_i\rangle$ is the space average, P is the pressure, ρ is the density, and 1_v is the friction loss. We have also assumed that properties are constant across the pipe at Planes 1 and 2 and that there is no change in elevation between the planes. In addition, we assume constant density and let

$$1_v = \frac{1}{2}K\langle v_2\rangle^2 \tag{3}$$

$$\alpha_i = \langle v_i\rangle^3/\langle v_i^3\rangle \tag{4}$$

then

$$(P_1 - P_2)/\rho = \frac{1}{2}\langle v_2\rangle^2\left\{\frac{1}{\alpha_2} + K - \frac{A_2^2}{A_1^2\alpha_1}\right\} \tag{5}$$

or

$$\langle v_2\rangle = C_d\left\{\frac{2(P_1 - P_2)}{\rho\left[1 - \left(\frac{A_2}{A_1}\right)^2\right]}\right\}^{1/2} \tag{6}$$

where C_d, the discharge coefficient, is a function of Reynolds number, area ratio, and location of pressure taps. For incompressible flow, $\langle v_2\rangle A_2$ is the flow rate. Under certain conditions the discharge coefficient is a constant. If the taps are located one pipe diameter upstream of the orifice and one-half pipe diameter downstream, $0.5 < A_2/A_1 < 0.7$ and $Re_2 > 3 \times 10^4$, then $C_d = 0.62$. Other possibilities exist for orifice geometry and tap locations. These are standardized, and complete information on discharge coefficients are given in other publications. The sharp-edged orifice plate is typically used with the following tap locations:

Flange Taps—1 in. upstream and 1 in. downstream of the orifice
Vena Contracta Taps—One pipe diameter upstream and at the minimum pressure location downstream
Radius Taps—One pipe diameter upstream and one-half pipe diameter downstream

Pipe Taps—2.5 pipe diameters upstream and 8 pipe diameters downstream

An orifice meter has the following advantages:

1. Construction is simple with no moving parts and the cost is low
2. Standards for construction and calibration are readily available
3. The method can be used with most fluids

and disadvantages:

1. The accuracy of the orifice and pressure measurement instrumentation is usually ±1% with calibration and less if uncalibrated
2. A large proportion (50–80%) of the pressure drop is not recovered
3. A long, straight run of pipe is necessary to avoid effects of upstream conditions
4. The pressure drop flow rate relationship is not linear so that the range is limited to a 4 : 1 ratio between maximum flow and minimum flow
5. Highly viscous fluids and fluids containing particulates are not suitable

If pressure taps are used, there are some limitations on their size so that errors are minimized. A rule of thumb is to have the length of a pressure tap at least 2.5 times the diameter, and the diameter should be less than 1/8 of the pipe diameter. The taps should not be placed on the bottom of the pipe where dirt can collect and cause plugging.

In order to minimize the permanent pressure loss across the meter, constrictions which avoid separated flow on the downstream side of the constriction are available. However, the cost is considerably higher than the simple orifice plate. The Venturi meter has a conical entry with a cone angle about 21° and an exit cone angle of about 7°. Taps are located at the throat and upstream of the entrace cone. For pipe Reynolds numbers greater than 2×10^5, the discharge coefficient is 0.984 and the permanent pressure loss is less than 15% of the pressure differential, $P_1 - P_2$. Another common constriction is the flow nozzle. It has a flared approach section leading to a short cylinder. Typical permanent pressure losses are 60% of the pressure drop.

The differential pressure meters can also be used for compressible flow. We need to integrate the differential form of Eq. (2), assuming an adiabatic expansion from P_1 to P_2 and subsonic flow. If $\alpha_1 = \alpha_2 = 1$, we have

$$\frac{1}{\rho}dp + \frac{1}{A}dA + \frac{1}{\langle v \rangle}d\langle v \rangle = 0 \tag{7}$$

and

$$\frac{1}{\rho}dp + d\left(\frac{\langle v \rangle^2}{2}\right) + d(1_v) = 0 \tag{8}$$

For Venturi meters, flow nozzles, and orifice plates, we can obtain a working equation for the average velocity in the form of Eq. (6) with an additional factor **Y** on the right to account for expansion. In terms of the mass flow rate which is constant, $\rho_1\langle v_1\rangle A_1 = \rho_2\langle v_2\rangle A_2 = \dot{m}$,

$$\dot{m} = C_d \mathbf{Y} A_2 \left\{ \frac{2(P_1 - P_2)\rho_1}{\left[1 - \left(\frac{A_2}{A_1}\right)^2\right]} \right\}^{1/2} \tag{9}$$

When $r = P_2/P_1$, $\gamma = C_p/C_v$, and $\beta = D_2/D_1$, for a Venturi or flow nozzle

$$\mathbf{Y} = \left\{ r^{2/\gamma}\left(\frac{\gamma}{\gamma - 1}\right)\left(\frac{1 - r^{\frac{\gamma - 1}{\gamma}}}{1 - r}\right)\left(\frac{1 - \beta^4}{1 - \beta^4 r^{2/\gamma}}\right) \right\}^{1/2} \tag{10}$$

For orifice plates, empirical fits to experimental data are used for flange taps, vena contracta taps, or radius taps.

$$\mathbf{Y} = 1 - \frac{1 - r}{\gamma}(0.41 + 0.35\beta^4) \tag{11}$$

and for pipe taps

$$\mathbf{Y} = 1 - \frac{1 - r}{\gamma}[0.333 + 1.145(\beta^2 + 0.7\beta^5 + 12\beta^{13})] \tag{12}$$

For a Venturi or flow nozzle the maximum flow through the constriction is limited to sonic velocity at the minimum area. For a sharp-edged orifice the area of the vena contracta increases as downstream pressure decreases. All cases can be treated similarly if the differences are taken into account by changing the discharge coefficient.

For $r_c = P_2/P_1$ at the point of maximum velocity and assuming a thermodynamically reversible adiabatic process in an ideal gas,

$$r_c^{(1-\gamma)/\gamma} + \frac{\gamma - 1}{2}\beta^4 r_c^{2/\gamma} - \frac{\gamma + 1}{2} = 0 \tag{13}$$

When $\beta \leq 0.2$, the second term can be neglected and

$$\dot{m}_{\text{max}} = C_d A_2 \left\{ \gamma P_1 \rho_1 \left(\frac{2}{\gamma + 1}\right)^{\frac{\gamma + 1}{\gamma - 1}} \right\}^{1/2} \tag{14}$$

Flow nozzles are often used as gas meters when operated at the maximum mass flow rate.

Instead of varying the pressure differential across a fixed sized constriction as a function of flow rate, a meter can be constructed to vary the flow area of the constriction while the pressure drop is constant. The rotameter shown in Fig. 2 is an example of such an area meter. A float is free to move up

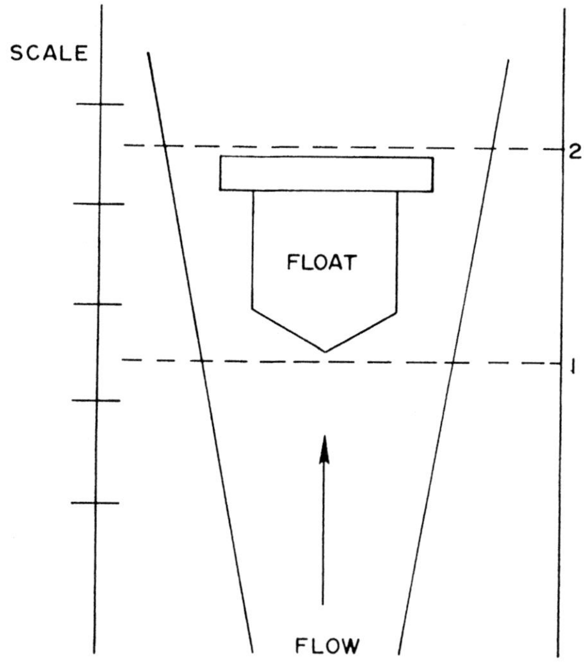

FIG. 2. Schematic diagram of a rotameter.

or down in a tapered tube which has a linear scale inscribed on it. The tubes are usually glass or plastic so the float can be seen. Metal tubes with a magnetic readout can also be used. The meter operates by letting fluid (liquid or gas) enter the bottom of the tube and the flow through the annulus between the float and the tube leads to a pressure drop which balances the other forces on the float.

The theoretical analysis of the rotameter must use a force balance in addition to the mass and energy balances. We place Plane 1 at the bottom and Plane 2 at the top of the float and assume that the tube area A_T is constant between the two planes. Thus for an incompressible fluid $\langle v_1 \rangle = \langle v_2 \rangle = q/A_T$, where q is the flow rate. At Plane 1 the velocity is considered uniform across the plane and at Plane 2 let

$$\beta_2 = \frac{\langle v_2^2 \rangle}{\langle v_2 \rangle^2} = \frac{A_T}{A_T - A_F} \tag{15}$$

$$\alpha_2 = \frac{\langle v_2^3 \rangle}{\langle v_2 \rangle^3} = \left(\frac{A_T}{A_T - A_F} \right)^2 \tag{16}$$

then the balances become

$$\frac{1}{2}\langle v_1 \rangle^2 - \alpha_2 \langle v_2 \rangle^2 + \frac{P_1 - P_2}{\rho} + g(h_1 - h_2) - 1_v = 0 \tag{17}$$

$$\dot{m}[\langle v_1 \rangle - \beta_2 < v_2 \rangle] + (P_1 - P_2)A_T - \rho g[(h_2 - h_1)A_T - V_F] - \rho_F g V_F = 0 \tag{18}$$

where V_F is the float volume and ρ_F is the float density. These equations can be combined to eliminate the pressure, and if

$$1_v = \frac{1}{2}K\left(\frac{q}{A_T - A_F}\right)^2$$

we will get

$$q = (A_T - A_F)\left[\frac{A_T/A_F}{1 + K(A_T/A_F)^2}\right]^{1/2}\left[\frac{2gV_F}{A_F}\left(\frac{\rho_F}{\rho} - 1\right)\right]^{1/2} \quad (19)$$

If $\sqrt{(A_T/A_F)\,[1 + K(A_T/A_F)^2]^{-1}}$ is constant, q is linear in A_T and the useful range of the rotameter can be a factor of 10 in flow rate. Additional increases in range can be made by including two floats, each with a different density.

The equation was put in the same form as the orifice equation by Shoenborn and Colburn:

$$q = C_R(A_T - A_F)\left[\frac{2gV_F}{A_F}\left(\frac{\rho_F}{\rho} - 1\right)\right]^{1/2} \quad (20)$$

The rotameter coefficient C_R was found to correlate with the Reynolds number $D_{eq}\langle v_2\rangle\rho/\mu$ which is also

$$\mathrm{Re} = \frac{2}{\sqrt{\pi}}\frac{\rho q}{\mu(\sqrt{A_T} + \sqrt{A_F})} \quad (21)$$

Denn prefers to correlate K with the same Reynolds number, although commercial rotameters usually are purchased with calibration curves relating to linear scales on the meter.

The rotameter is a simple, cheap instrument, and a wide range of velocities can be measured with the same tube. It is not necessary to locate the meter far from values or fittings. Disadvantages are the head loss is high, the tube must be vertical so extra piping is necessary, the accuracy is only 2–3% unless an expensive high precision rotameter is manufactured as a special item, and rotameters are not suitable for high capacities.

When the flow rate is small, the orifice discharge coefficient varies a lot with flow rate. Laminar flow through a capillary or a bundle of capillaries can be used as a flow measurement technique. A bundle of capillaries can be manufactured so uniformly that calibration is not required for accuracies of 2–3%.

In the laminar flow method the pressure drop is related to the flow rate by

$$\Delta P = \frac{q}{\pi R^4}(8\mu L + k\rho q) \quad (22)$$

where R is the capillary radius, L is the total length, and k is a factor to account for entrance and exit losses. If $k\rho q$ can be neglected, then the equation is linear and a wider range of flow rates can be accommodated than with an orifice meter. Within the capillary the usual assumptions for laminar flow must hold.

Turbine Flow Meter

The turbine flow meter was developed for the aviation industry during World War II, and after the war uses were rapidly found in other industries. The rotor is usually mounted to spin freely on bearings. Flow straighteners are positioned upstream of the rotor. A fluid flowing through the blades will strike the blades at a small angle to balance the forces on the rotor, and the device rotates at an angular speed proportional to the flow rate. Usually a magnetic method is used to detect the rotation frequency f. Either the propeller blades are magnetic or magnetic material is embedded in the tips. A wall-mounted pick-up is used to detect the total counts (number of revolutions) or the frequency. Calibration curves to give $q = f/K$, where K is the calibration constant, are supplied with the meters. For liquid flows the effect of viscosity can be included in the calibration to give $q = f/\nu K$ or a different curve is supplied for each fluid.

The turbine flow meter is sufficiently accurate so that it is used as a calibration standard. The range of flow rates is 10 : 1 and the response is typically better than 10 ms. The instrument is compact and of the same diameter as the pipe, and the unrecoverable pressure drop is moderate. The disadvantages include high cost, problems with bearings, and the dependence of calibration upon viscosity.

Electromagnetic Flow Meter

When a conductor moves across a magnetic field, a voltage is induced in the conductor which is proportional to the speed of the conductor. Thus we can put a magnetic field around a pipe with an insulating wall and measure the voltage across electrodes mounted in the pipe wall. When the velocity profile is symmetrical and the magnetic flux density B is uniform, the velocity is

$$\langle v \rangle = E/BD$$

where E is the voltage and D is the pipe diameter.

There is no obstruction in the flow so there is no head loss. The range of velocity or flow rate can be increased beyond 10 : 1 by increasing the sensitivity for the voltage detection up to the limit of extraneous electronic noise. The measurement is also independent of density and viscosity although the fluid must be an electrical conductor. For an industrial meter an electrical conductivity greater than 10 μmho/cm is adequate. Other disadvantages are the low accuracy and the expense of small meters. Such small meters are also bulky.

Ultrasonic Flow Meters

Ultrasonic flow meters use a transducer to send an ultrasonic beam along a path in the flow and a receiver to detect either along the same path or at some angle to the path. There are two different principles that are usually used to form the basis for the operation of these meters.

One method uses the principle that the resultant velocity of sound waves in a moving fluid is the vector sum of the fluid velocity and the sound velocity in the motionless fluid. Thus detection of the time of flight of a pulse along a path through the fluid is related to the average velocity along the path. This average velocity is not the flow rate divided by the pipe cross-sectional area, and a correction or the use of several paths is necessary to get the flow rate.

The second method is based upon the Doppler effect. The sound waves that are scattered from moving particles in the fluid are shifted in frequency when observed at an angle to the incident path. Again some averaging or calibration technique must be used to relate the measured frequency shift to the true average velocity.

Ultrasonic flow meters can be inexpensive, do not present any obstruction to the flow, and do not require a conductive fluid.

Vortex-Shedding Flow Meters

When a cylinder or other bluff body is placed across a flowing stream, vortices are regularly shed alternately from the top and bottom surfaces. A flow meter can be manufactured using this principle along with a suitable detector of the frequency of the vortex shedding. Usually the detector element is incorporated in the bluff body.

A vortex-shedding flow meter covering a 360:1 range of Reynolds numbers above 10^4 has been reported. These flow meters compete with the orifice meter and have the same limitations on upstream conditions. In some dirty fluid conditions the vortex-shedding flow meters have shown better life then orifice meters.

Other Methods

Almost any physical measurement technique can be adapted to the measurement of flow rate. These methods include nuclear magnetic resonance, dye dilution, time of flight, induced oscillatory motions in the fluid, and positive displacement meters.

Local Velocity Measurements

Detailed measurements of mean velocities and velocity fluctuations are often required in research, especially when the flow is accompanied by heat transfer and chemical reaction. In addition, there are measurements of flow rates that can only be made by doing a complete traverse of point measurements of velocity. Examples are in two-phase flows and in short stacks where the measurement must be made a short distance from a bend.

Two general classes of local velocity measurements can be defined, probe methods and tracer techniques. Some of these are smaller size versions of the overall flow-rate techniques that have been previously mentioned. All require some "art" in their use, and although theoretical analyses guide the use of these local velocity measurement techniques, frequent calibration in use must be done.

Usually a choice of techniques is available and the measurement problem is often a unique one, so cost is a secondary concern. Instruments should be compared with respect to the following requirements:

1. The measured velocity should be as close as possible to the velocity in the absence of the detecting element or tracer. This implies that the detector in the flow does not disturb the flow, that the velocity variation across the detector is small, and that in the case of a tracer, the tracer follows the flow.
2. The relationship between the measured variable and the velocity should be direct and preferably linear.
3. The technique should be easy to calibrate and there should be no change in the calibration over a reasonable period.
4. The method must be sufficiently sensitive to detect necessary small differences.
5. The instrument must not be damaged by the flow.
6. If the fluctuations in velocity about the mean must be measured, the method must respond to the most rapid of these.

Probe Techniques

Head Probes

The most common total head probe is the Pitot tube. Although an open tube inserted in the flow with a second pressure tap in the wall will do, many commercial Pitot tubes are based on the Prandtl design as shown in Fig. 3.

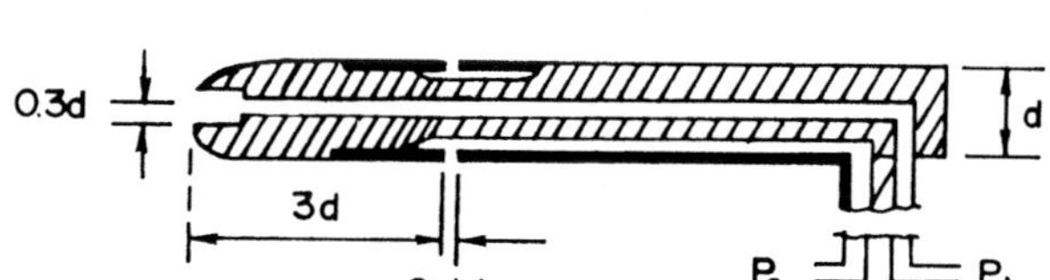

FIG. 3. Prandtl design of a Pitot-static tube.

When Bernoulli's equation is applied to the Pitot tube, the fluid velocity v is

$$v = [2(P_t - P_s)/\rho]^{1/2}$$

where P_t is the pressure at the nose of the Pitot tube and P_s is the static pressure in the fluid measured at the wall or away from the nose along the wall of the tube. This equation is applicable when the assumptions in Bernoulli's equation are valid. At low Reynolds numbers based on the Pitot tube diameter, the viscosity effects cannot be neglected. A calibration is necessary for Reynolds numbers below 100 to account for viscous effects.

The Pitot tube is simple and inexpensive so the method is widely used. However, the probe size is large and the response is very slow so that Pitot tubes are not suitable for measurements of rapidly fluctuating velocities. In some cases such as in viscoelastic fluids, the presence of the probe disturbs the flow enough to give erroneous measurements. Small probes with piezoelectric transducers mounted directly on the probe have been used to improve the response. However, the minimum outside diameter of the tube of about 1 mm is still larger than other flow velocity measurement probes.

The Hot Wire Anemometer

The principle of the hot wire anemometer is to detect the change in the heat transfer rate (which is proportional to the fluid velocity) for a wire heated electrically in a flowing fluid. The heat transfer rate depends upon the flow velocity, the temperature difference between the wire and the fluid, the dimensions and physical properties of the wire, and the physical properties of the fluid. Typical hot wire anemometers consist of a cylindrical wire several microns in diameter and 1 to 2 mm long mounted normal to the flow.

If the heat generation by electrical current in the wire is set equal to the heat loss by convection to the flowing fluid,

$$I^2R = hA_w(t_w - t) \tag{23}$$

where I is the current, R is the wire resistance, h is the heat transfer coefficient, A_w is the surface area of the wire, t_w is the wire temperature, and t is the fluid temperature. Heat loses due to radiation, natural convection, and conduction along the wire to the supports have been neglected. The heat transfer coefficient can be expressed in the form

$$h = A + Bv^{1/n} \tag{24}$$

where A, B, and n are constants, and n is approximately 2. The hot wire anemometer can be operated as a constant current or as a constant temperature device. There are some advantages to constant temperature operation because the constants A and B are functions of the temperature. Although the relationship between the measured voltage drop across the wire and the velocity is nonlinear, an electronic linearizer can be used with the constant temperature system.

Hot wire anemometers can be used to measure temperature, thermal conductivity, pressure, mass fraction, heat transfer coefficient, and mass flow in addition to velocity. Because many variables are involved and because the wire is easily contaminated in the flow, it is necessary to frequently clean and calibrate. Mean flow measurements in highly turbulent flows may require averaging over a long time, and the hot wire anemometer must be carefully calibrated. Small diameter hot wires may have a response time of less than 0.1 ms, and these are particularly suited to measurements of instantaneous velocities. Such fine wires are easily broken and are not suitable for liquids or high velocity systems. Hot film sensors which have a conducting film on a ceramic substrate are much larger and can be used in these severe environments. The principles of operation are the same as for the hot wire.

Electrochemical Probes

Velocity measurements based on mass transfer to a film or wire are also possible, but commercial instruments are not available. The principle of operation is similar to the hot wire analysis but the mass transfer coefficient is substituted for the heat transfer coefficient.

Tracer Techniques

Laser Light Scattering

Small particles in a flowing fluid will follow the flow and can be detected by observing the scattered light from a focused laser beam. In liquids, natural seeds can be used so the probe does not disturb the flow. In gases, some method of particle seeding is necessary. The probe volume is formed by the most intense part of the focused beam. Laser beams have a Gaussian intensity profile across the beam and the size is usually given by the diameter to the points where the intensity has dropped to e^{-2} times the maximum. Using such e^{-2} points of the probe volume at the focus of a typical laser beam, the size is 50 μm by 1 mm long. Laser light-scattering techniques have advantages over hot wires or hot films in compressible flow, for very low velocities where natural convection limits the hot cone, and where a probe that does not disturb the flow is necessary. Laser methods are generally more expensive than hot wires, but no calibration is involved.

There are many commercial laser velocimeters available. Although all involve light scattering from particles, there are many methods to extract the velocity information from the scattered light. One possibility is to measure the intensity vs time as a particle travels through the scattering volume. The velocity is found from the time between the e^{-2} intensity points. The most common method of detection is to measure the Doppler shift of the scattered light. The frequency of scattered light from a moving particle at an angle to the incident light is shifted in frequency. This Doppler effect is the same as the familiar change in frequency of a train whistle as the train comes toward or recedes from an observer. The frequency change of scattered light can be measured by combining the scattered light and a second light beam at a photodetector. Several different optical systems use different combinations of beams, but the resulting equations are all the same:

$$\Delta \nu = \frac{2n\nu}{\lambda} \sin \frac{\theta}{2} \tag{25}$$

where $\Delta \nu$ is the frequency shift, n is the index of refraction, λ is the wavelength of the light, and θ is the scattering angle or the angle between two intersecting focused beams.

Methods of detection of the Doppler signal at the photodetector are numerous. Commercial instruments use two different forms of electronic signal processing to follow a time-varying velocity. One method, frequency tracking, requires the presence of a particle in the scattering volume at all times. The other electronic technique, the counter, gives the frequency shift for a single particle in the scattering volume. The response of the signal processors is very fast so that the only limits to the response are effects due to the finite measuring volumes, random distribution of particles, and electronic noise. Attempts to account for these limitations or to devise ingenious optical and electronic ways around them can be found in the literature.

A third single spot method measures the distribution of particle arrivals. These methods can be extended to more than one spot to measure more than one velocity component. In addition, two spot time of flight and two spot particle arrivals can yield mean velocities and velocity fluctuations. Other laser optical methods include holography and fluorescence.

Flow Visualization

Three typical flow visualization techniques are hydrogen bubbles, dye tracers, and particle tracers. If we put two electrodes in water and apply a potential difference between the electrodes, then hydrogen bubbles will appear on the cathode. For flow visualization the cathode is a fine platinum wire and the voltage is pulsed to create bubble stripes in the flow. The local velocity can be found from photographs of the flow field.

The hydrogen bubble method needs a platinum wire which is inserted into the fluid. To avoid this defect, some photochemical reactions are used to make the flow visible. Some years ago the dye tracer method meant that a dye was injected from a small hole into a flow system. More recently many other dye tracer techniques have been uncovered.

Thymol blue changes color with pH. When a pulsating electrical reaction occurs on the surface of a platinum wire, the resulting line of color can be followed just like the hydrogen bubbles. Other studies have used dyes sensitive to ultraviolet light. A large pulsed ruby laser is used to irradiate a small spot in the fluid. The colored strip that is produced can be followed photographically.

If a particle is selected to move in just the same way as a fluid element in a flowing system, we can follow the movement of this particle with a suitable detecting method. Two problems arise in this technique. The first one is how to select the particle, and the second one is how to detect the particle movement.

To follow a rapidly fluctuating flow as, for example, in highly turbulent flow, we need a particle with a diameter smaller than the smallest turbulence scale and a density the same as the fluid density. In practice it is often impossible to meet either of these criteria, so we settle on the smallest particles. Typical techniques to detect the moving particles are photographic (streak stroboscope, and high-speed movies) and tracking with multiphoto-detectors or with digital techniques and TV cameras. In each case a large amount of data must be processed to get a representation of the velocity field.

Example 1: Sizing a Liquid Orifice Meter. An orifice meter is to be used to measure the flow rate of a liquid hydrocarbon side stream for a distillation tower. The maximum flow rate $\dot{M}$ is 887 kg/min (1951.4 lb/min) and the maximum differential pressure is to be 4×10^4 Pa (5.776 lb/in.^{2}). The hydrocarbon has a specific gravity of 0.9 and a viscosity of 3 cP. The pipe size is a 4 in. schedule 40 (102.26 mm i.d.). Find the orifice diameter.

$$\text{Re} = \frac{4\dot{M}}{\pi D \mu} = \frac{(4)(887\ \text{kg/min})}{\pi(60\ \text{s/min})(0.10226\ \text{m})(0.003\ \text{kg/m}\cdot\text{s})} = 61{,}400$$

Let

$$x = A_2/A_1 \quad \text{and} \quad \langle v_2\rangle = (1/x)\langle v_1\rangle$$

$$\langle v_1\rangle = r/\rho A_1 = 2.0\ \text{m/s}$$

Substitution in Eq. (6) gives

$$(2/x)^2 = (0.62)^2 \left[\frac{2(4\times 10^4)}{(900)[1-x^2]}\right]$$

$$x = 0.324$$

$$D_2 = x^2 D_1 = 0.569 D_1 = 58.18\ \text{mm}\ (2.291\ \text{in.})$$

Example 2: Converting a Pressure Measurement into Flow Rate. The orifice meter of Example 1 is installed. What is the flow rate when the pressure drop is 4 lb/in.^{2} (2.77×10^4 Pa). The equation for the flow rate is obtained from Eq. (6).

$$\dot{M} = \frac{900\ \text{kg}}{\text{m}^3}(0.62)\left(\frac{\pi}{4}\right)(0.05818)^2\ \text{m}^2 \left[\frac{2(2.77\times 10^4\ \text{kg/m}\cdot\text{s}^2}{(900\ \text{kg/m}^3[1-0.105]}\right]^{1/2}$$

$$= 12{,}303\ \text{kg/s}\ (1624\ \text{lb/min})$$

For this meter

$$\dot{M} = (7.39186 \times 10^{-2})(-\Delta P)^{1/2}$$

where ΔP is in Pa, and Re > 3×10^4 ($\dot{M}$ > 433 kg/min or $-\Delta P$ > 9550 Pa).

Example 3. A natural gas flows in a 3 in. schedule 40 (77.927 mm i.d.) pipe with a maximum flow rate of 1000 m^3/h. The upstream pressure is 40 lb/in.^{2}abs (4.072×10^6 Pa) and the temperature is 300 K. The maximum pressure differential is to be 2 lb/in.^{2}abs (1.385×10^4 Pa) at the maximum flow rate. Find the orifice diameter when flange taps are to be used. The natural gas properties are $\gamma = 1.33$, $\mu = 0.011$ cP, $\rho = 0.7674$ kg/m^3.

If **Y** is 1.0 and $C_d = 0.62$, the orifice diameter is found as in Example 1. From Eq. (9), we let $x = A_2/A_1$.

$$q = \dot{M}/\rho = (0.62)x\left(\frac{\pi}{4}\right)(0.077927)^2\left\{\frac{2(1.385\times 10^4)(0.7674)}{1-x^2}\right\}^{1/2} = 1000/3600$$

Then

$$x = 0.542 \quad \text{or} \quad D_2/D_1 = \beta = 0.736$$

and

$$\mathbf{Y} = 1 - \frac{0.05}{1.33}[0.41 + 0.35(0.736)^4] = 0.981$$

and the corrected $D_2/D_1 = 0.741$. Therefore

$$D_2 = 57.74 \text{ mm}$$

Symbols

A_i	area at cross-section i
A_T	tube are in a rotameter
A_F	float area projected perpendicular to the flow in rotameter
C_d	discharge coefficient
C_R	rotameter coefficient
D	diameter
g	gravitational acceleration
h	heat transfer coefficient
h_i	height at cross-section i
I	electrical current
K	proportionality factor in various equations
k	factor in Eq. (22)
l_v	friction loss
$\dot{M}$	mass flow rate
n	index of refraction, Eq. (25)
P_t	pressure at nose of Pitot tube
P_s	static pressure
P	pressure
q	volumetric flow rate
R	electrical resistance
r	P_2/P_1
r_c	P_2/P_1 at maximum velocity

Re	Reynolds number $D<v>\rho/\mu$
t_w	hot wire temperature
t	fluid temperature
v_i	velocity at point i
$<v_i>$	space average velocity
V_F	float volume in rotameter
Y	compressibility factor in Eq. (9)
α	$<v_i^3>/<v_i>^2$
β	D_2/D_1
β_i	$<v_i^2/<v_i>^2$
γ	C_p/C_v
θ	angle
λ	light wavelength
μ	viscosity
ν	light frequency
ρ	density
ρ_F	float density in rotameter

Bibliography

A handbook with detailed design procedures for all flow meters is

Miller, R. W., *Flow Measurement Engineering Handbook*, McGraw-Hill, New York, 1983.

Additional equations and charts for orifices and Venturi meters are available in ASME Research Committee on Fluid Meters, *Fluid Meters—Their Theory and Applications*, 6th ed., 1971.

Other recent books on the general topic are

Cheremisinoff, N. P., *Applied Fluid Flow Measurement*, Dekker, New York, 1979.
Hayward, A. T. J., *Flowmeters*, Wiley, New York, 1979.
Wendt, R. E. (ed.), *Flow—Its Measurement and Control in Science and Industry. Part Two. Flow Measuring Devices*, Instrument Society of America, 1974.

A section on flow measurement is in Perry's *Handbook*:

Perry, R. H., and Chilton, C. H., (eds.), *Chemical Engineer's Handbook*, 5th ed., McGraw-Hill, New York, 1973.

Most textbooks on fluid mechanics or unit operations contain derivations of the general equations for orifice meters, Venturi meters, and rotameters. For example:

Denn, M. M., *Process Fluid Mechanics*, Prentice-Hall, Englewood Cliffs, New Jersey, 1980.
McCabe, W. L., and Smith, J. S., *Unit Operations of Chemical Engineering*, 3rd ed., McGraw-Hill, New York, 1976.

NEIL S. BERMAN

Natural Gas

Introduction

In spite of the great amount of time and effort which has been devoted to two-phase flow problems over the past 30 years, present design methods remain imprecise. A multitude of different equations and correlations are available for calculating the expected pressure drop in a two-phase pipeline. In deciding which method to use, the pipeline designer will find a bewildering number of diverse opinions expressed in the published literature.

A further complication arises when designing wet-gas pipelines: How much condensate can be carried in the gas stream before single-phase gas-flow equations become invalid? Published data from large diameter hydrocarbon-carrying pipelines are scarce and so it is often difficult to test predictive methods with real data.

This article presents an analysis of operating data which was carried out in order to determine the flowing capacity of an offshore wet-gas pipeline. The actual capacity of the pipeline was calculated to be 978 MMSCF/d (27.7×10^6 m^3/d) compared to an original design capacity of 921 MMSCF/d (26.1×10^6 m^3/d). The analysis is based upon equations developed for dry-gas flow. The significance of the results is discussed with reference to both single-phase and two-phase design methods.

The pipeline transports gas from a North Sea gas field to an onshore terminal on the east coast of England. The shore terminal supplies gas directly to the UK gas grid. The line is 86 mi (138 km) long and has an outside diameter of 28 in. (711 mm). The gas field produces relatively dry gas: the quantity of liquids in the line is approximately 5 bbl/MMSCF (28×10^{-6} m^3/m^3).

The data analyzed in this article were provided by two extensive pipeline pressure surveys which were conducted in the winters of 1973/4 and 1974/5. Tables A1 and A2 in the Appendix contain a list of the measured flow data.

Gas Flow Equations

The general horizontal gas flow equation may be written as

$$Q = C\left[\left(\frac{T_0}{P_0}\right)\left(\frac{1}{f}\right)\right]^{1/2} \frac{(P_1^2 - P_2^2)^{0.5}D^{2.5}}{(LZTG)} \tag{1}$$

Several simplifying assumptions are made in order to arrive at Eq. (1) from the mechanical energy equation:

1. Flow is assumed to be isothermal.
2. Compressibility of the gas is assumed to be constant.
3. Kinetic energy losses are assumed to be negligible.
4. Flowing velocity is assumed to be accurately characterized by the apparent bulk average velocity ($v = 4Q/\pi D^2$, where Q is the volume flow rate at pipeline conditions).
5. Friction coefficient is assumed to be constant along the length of the pipeline.

If these assumptions are deemed valid, the only major problem which remains is that of estimating the friction factor.

Many friction factor correlations have been published: Table 1 contains some of those which are still commonly used. In describing the various friction factors it is useful to tabulate transmission factors ($\sqrt{1/f}$) since pipeline flow-rate is directly proportional to the transmission factor. Table 1 includes equivalent Panhandle transmission factors: these values are implied by the Panhandle equations which are discussed in the following section.

The Blasius equation and the smooth pipe law of Nikuradse both apply to turbulent flow in smooth pipes. Since commercial pipes are not usually smooth, these equations merely provide a lower limit on friction factor (i.e., upper limit on transmission factor).

An upper limit on the friction factor is described by Nikuradse's rough pipe law. This equation was developed by Nikuradse using data collected from a large number of experiments with artificially roughened pipes. The pipes were roughened by cementing sand particles to the inside surface. The roughness factor (ϵ) represents the mean height of uniform sand grains from the pipe wall.

In the case of artificially roughened pipes, the friction factor is usually independent of Reynolds number, being solely dependent upon the relative roughness (ϵ/D) of the pipe.

Typical commercial pipes are considerably smoother than the sand-

TABLE 1 Dry-Gas Flow Transmission Factors

Title	Transmission Factor ($\sqrt{1/f}$)	Ref.
Weymouth	$11.2D^{0.167}$	12
Blasius	$3.56\text{Re}^{0.125}$	4
Panhandle A	$6.87\text{Re}^{0.073}$	
Modified Panhandle	$16.5\text{Re}^{0.0196}$	
Smooth pipe law (Nikuradse)	$4 \log (\text{Re}\sqrt{f}) - 0.4$	9
Rough pipe law (Nikuradse)	$4 \log \frac{(D)}{(2\epsilon)} + 3.48$	9
Colebrook	$4 \log \frac{D}{2\epsilon} + 3.48 - 4 \log (1 + 9.35\frac{D}{2\text{Re}\sqrt{f}})$	5

coated pipes used by Nikuradse, and the friction factor may be dependent upon both Reynolds number and relative roughness. At low values of Reynolds number the friction factor follows the smooth pipe law. (Flow behavior in this region is loosely termed "partially turbulent.") At high Reynolds numbers the friction factor is constant and is limited by the rough pipe law. ("Fully turbulent" flow.) At some intermediate point (or region) a transition between the two flow types occurs.

The most widely used equation for turbulent flow in naturally rough pipes is probably the Colebrook equation [5]. The Colebrook equation suggests a broad transition between partially and fully turbulent flow.

Subsequent research by the United States Bureau of Mines [10] indicates an abrupt transition between the two flow types, and this approach is used in the American Gas Association's dry-gas design manual [11].

Many other methods of gas pipeline design derive from early practical experience. Three equations which still survive are the Weymouth equation and the Panhandle equations (Panhandle A and Modified Panhandle).

The Weymouth equation defines a friction factor which is simply dependent upon diameter (and thus could only be expected to be accurate in fully turbulent flow situations). Both Panhandle equations, which are discussed in some detail in the following section, contain Reynolds number-dependent friction factors.

It is generally agreed that the most accurate design of dry-gas pipelines can be accomplished by following the procedure recommended in the American Gas Association design manual *Steady Flow in Gas Pipelines* [11].

This method uses the general flow equation, Eq. (1). A transmission factor is calculated with regard to Reynolds number, a different relationship applying for partially turbulent as opposed to fully turbulent flow. A separate drag factor is incorporated to account for bends and fittings in the pipe.

The Panhandle Equations

The original Panhandle equation (Panhandle A) was developed in the early 1930s from operating data taken from the Texas Panhandle Field–Chicago gas pipeline (Natural Gas Pipeline Company of America). This pipeline was constructed with bolted-flange connections and was operated at a pressure in the region of 900 lb/in.^{2} (6.2 MPa) which at the time was considered a very high operating pressure. The pipeline was operated at flow rates toward the upper end of the partially turbulent region, and the Panhandle A equation quite closely represents flow behavior in this region.

The Modified Panhandle (or Panhandle–Eastern Pipeline Company) equation was developed from operating data taken from another Panhandle pipeline which was constructed in the early 1950s. This pipeline operated at flow rates high enough to encounter fully turbulent flow, and the modified Panhandle equation reflects this by being less dependent upon Reynolds number and including, implicitly, a fixed value of pipe roughness for each diameter to which it is applied.

The Panhandle equations appear in the literature in several forms. The most usual forms are

Panhandle A:

$$Q = 435.9E(T_0/P_0)^{1.07881}[(P_1^2 - P_2^2)/TLG^{0.8539}]^{0.5394}D^{2.6182} \quad (2)$$

Modified Panhandle:

$$Q = 737.2E(T_0/P_0)^{1.02}[(P_1^2\,(1 + SP_1) - P_2^2(1 + SP_2))/TLG^{0.961}]^{0.51}(D^{2.53}) \quad (3)$$

where S is a compressibility correction.

The term E, which appears in both Panhandle equations, is called, unfortunately, the "efficiency" factor. This name is very misleading since E does not represent an efficiency in any remotely scientific meaning of the word. E is merely a factor which is used to shift a calculated value of flow rate (using the Panhandle equation) onto a real, measured point.

Many authors have attempted to define those factors upon which E is dependent (such as age and cleanliness of pipe, bends and valves in pipe, and material of which pipe is constructed). Although such a definition is often useful, it tends to obscure the fact that E is a fudge factor whose value will vary simply due to the inadequacy of the Panhandle relationships.

The differences between the Panhandle equations and various other gas flow equations are discussed later (Results and Discussion). For the present, suffice to say that the Panhandle equations are of very limited application. They may be useful in providing first estimates of flow rates in gas pipelines operating in partially turbulent flow (Panhandle A) or near to the partially/fully turbulent flow transition (Modified Panhandle). But most recently constructed offshore gas lines are operated at pressures between 1000 lb/in.^{2} (6.9 MPa) and 2000 lb/in.^{2} (13.8 MPa). These pipelines are usually operating in the fully developed turbulent flow regime and neither Panhandle equation adequately represents these conditions.

Further confusion arises when the Panhandle A equation is used to design, or assist in the design, of wet-gas lines. Baker [1] suggested that the Panhandle efficiency could be varied to account for the presence of liquids in gas lines and presented a correlation of efficiency versus liquid flow rate. Flanigan [7], through independent work, reached a similar conclusion.

Although these correlations were helpful design tools at the time, the continuing use of Panhandle equations to attempt to predict wet-gas line performance may be ill-advised. The inherent limitations of the Panhandle relationships only serve to cloud a problem which is already extremely complex.

The analysis of the data which follow will illustrate this point.

Computational Method

The initial objectives of the pipeline pressure surveys were (1) to determine the pipeline's maximum operating capacity and (2) to construct a diagram of flow rate versus input pressure for various stipulated output pressure in order to facilitate operational control of the pipeline.

In the initial analysis of the data, it was assumed that the Modified Panhandle equation would apply and that efficiency would be constant at flow rates above 500 MMSCF/d (14.2×10^6 m^3/d). Flow rate was then correlated with the pressure term in the Modified Panhandle equation using an expression of the form: $Q = mX + b$, where m and b are constants and X is the pressure term in the Modified Panhandle equation.

The equation $Q = 0.5387X + 40.1$ was found to be the best fit for the data. Figures 1 and 2 illustrate the results of this initial study.

At flow rates above 500 MMSCF/d, Fig. 1 is a useful and accurate aid to pipeline control. However, from a theoretical standpoint, the straight line fit was not completely satisfactory. The fact that a positive flow-rate is indicated when zero pressure differential is acting over the pipeline suggests that the Modified Panhandle equation is not accurately describing pipeline performance.

A further study was carried out to determine which gas-flow equation most accurately represented the operating data. Three short computer programs were written to facilitate this analysis.

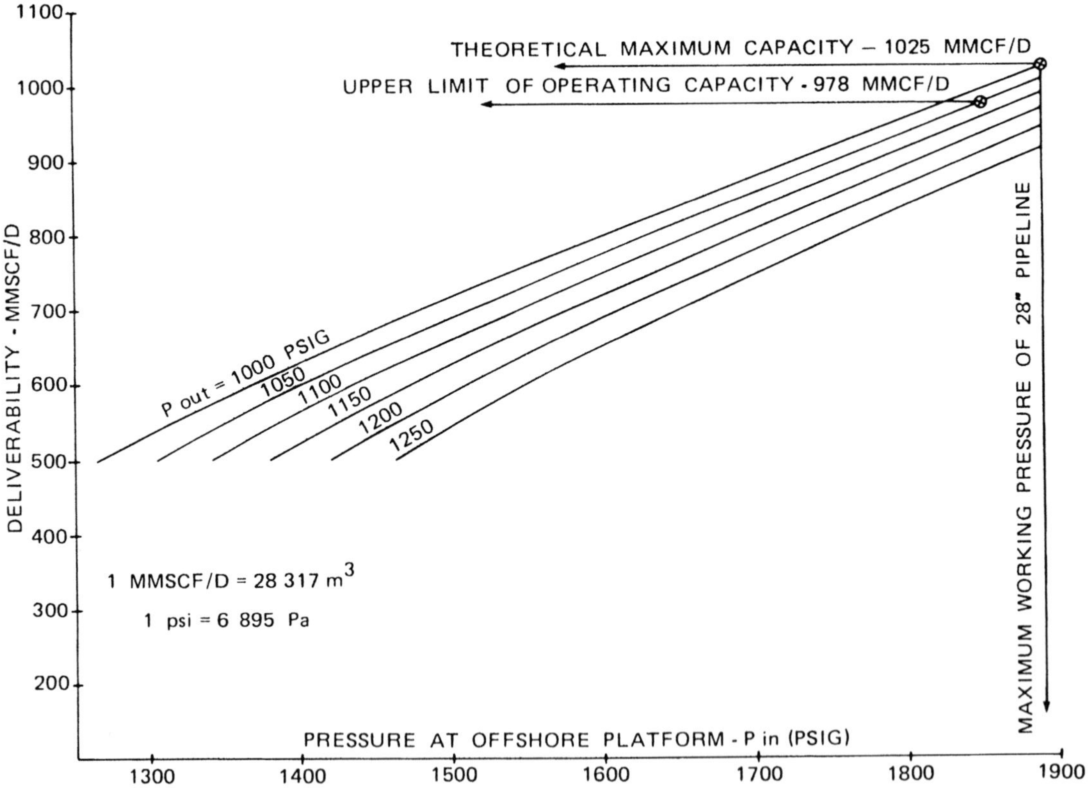

FIG. 1. Determination of gas pipeline capacity. Offshore platform to shore terminal.

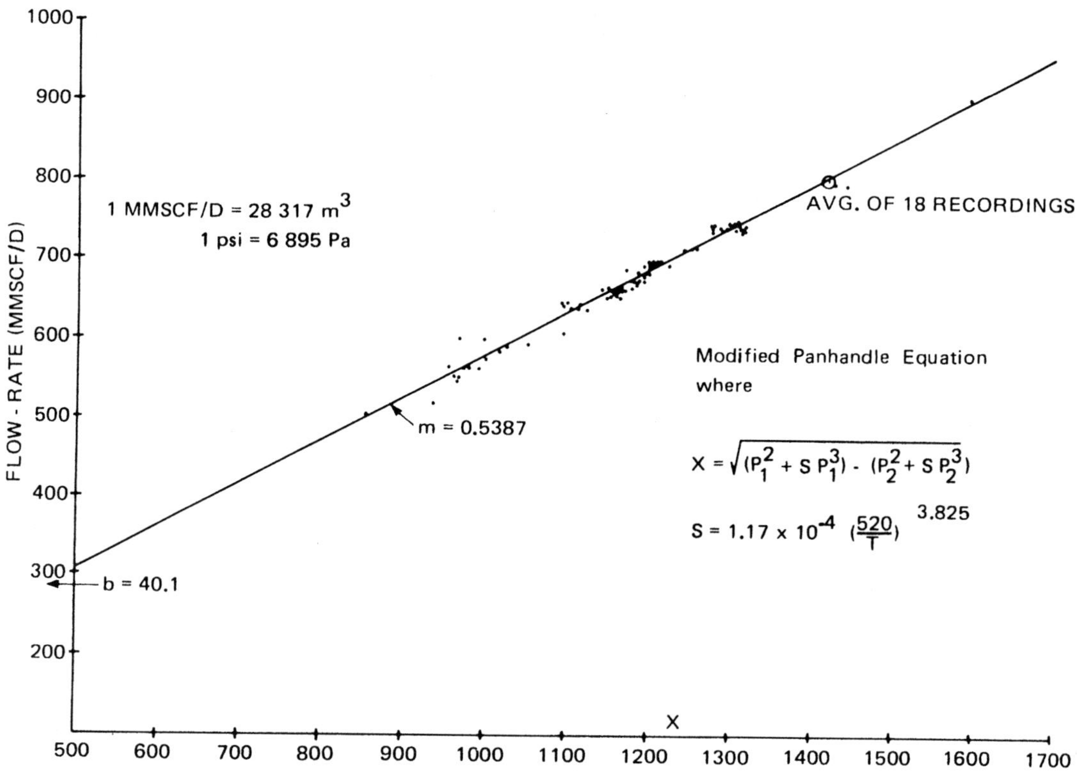

FIG. 2. Determination of gas pipeline capacity. Offshore platform to shore terminal.

Program 1 uses the usual form of the Panhandle equations (eqs. 2 and 3) to calculate a flow rate for each pair of measured pressures (inlet pressure and outlet pressure). This calculated flow rate is then compared with the measured flow rate and an efficiency factor is calculated (Figs. 3 and 4).

Program 2 uses the general flow equation (Eq. 1) to calculate a transmission factor from the measured values of Q, P_{in}, and P_{out} (Fig. 6).

Program 3 uses the general flow equation (Eq. 1) plus the Panhandle transmission factors (given in Table 1). This enables the Panhandle efficiency factors to be compared directly (Fig. 5).

Use of Eqs. (2) and (3) does *not* facilitate this comparison because the transmission factor definition is *not* the only difference between the Panhandle equations and the general flow equation. For example, in Panhandle A the compressibility factor (Z) is omitted. Some authors specifically state that Z is implicitly included in the efficiency factor but they do not usually suggest a normal value of E other than 0.92. It is possible that Z is included in the efficiency term by default rather than by design.

The three programs were run with the data which are contained in the Appendix.

Results and Discussion

The results of this study are illustrated by Figs. 3 through 7.

Figure 3 shows a generalized plot of Panhandle A efficiency (Eq. 2) versus flow rate.

Figure 4 shows a similar plot for Modified Panhandle (Eq. 3).

The actual data points are shown on these diagrams to illustrate a simple but important point: presentation of a single, generalized curve of efficiency versus flow rate obscures the fact that in *all* two-phase lines a *spread* of points is to be expected even where so-called "steady-rate" conditions are thought to exist.

Figure 5 shows both Panhandle efficiency plots using Eq. (1) plus the Panhandle transmission factors (Table 1). Ignoring, for the moment, the shape of the curves, comparison of Fig. 5 with Figs. 3 and 4 displays the wide variation in "efficiency" which results from the omission or estimation of Z.

Figure 6 shows the plot of pipeline transmission factor versus flow rate. This plot enables a direct comparison to be made between the measured operating data and the various gas-flow equations. Figure 7 illustrates this comparison. This diagram displays the inherent character of the various gas-flow equations for the 28-in. pipeline. The American Gas Association (AGA)

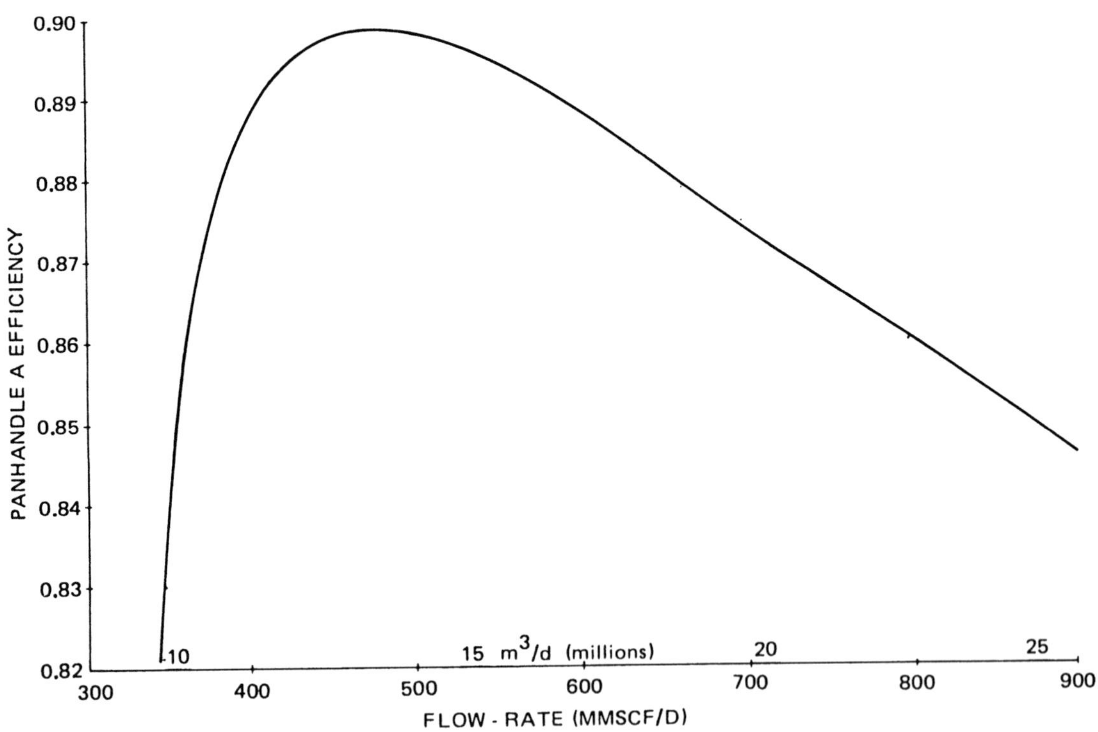

FIG. 3. Generalized plot of Panhandle A efficiency versus flow rate.

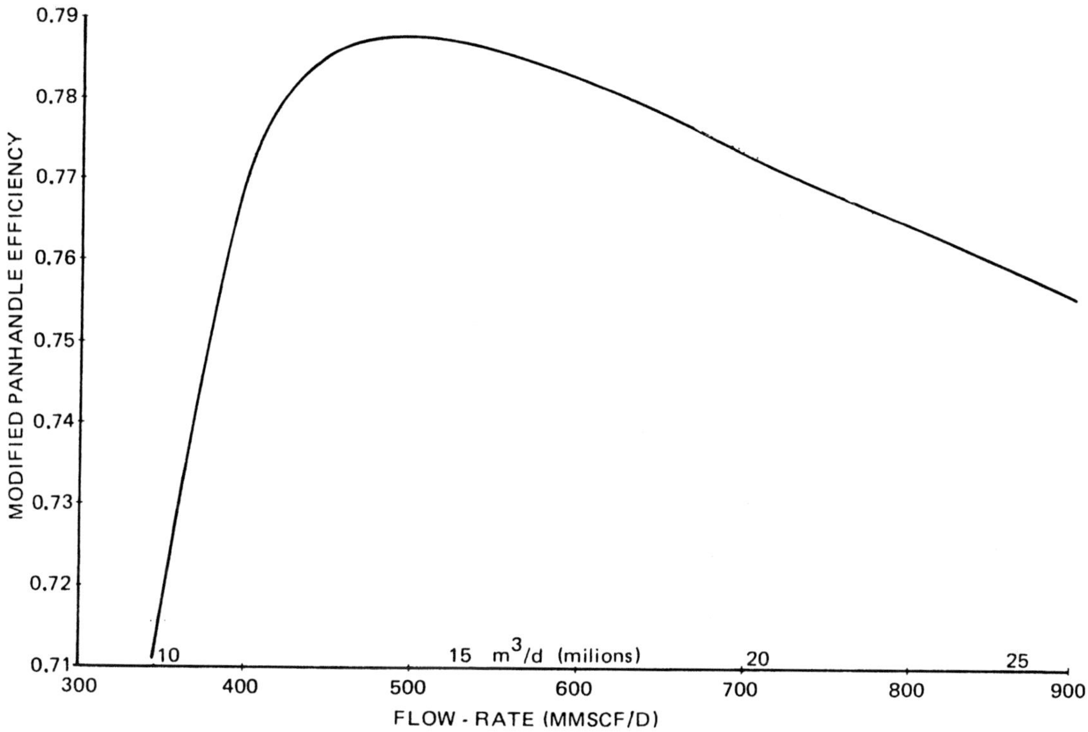

FIG. 4. Generalized plot of Modified Panhandle efficiency versus flow rate.

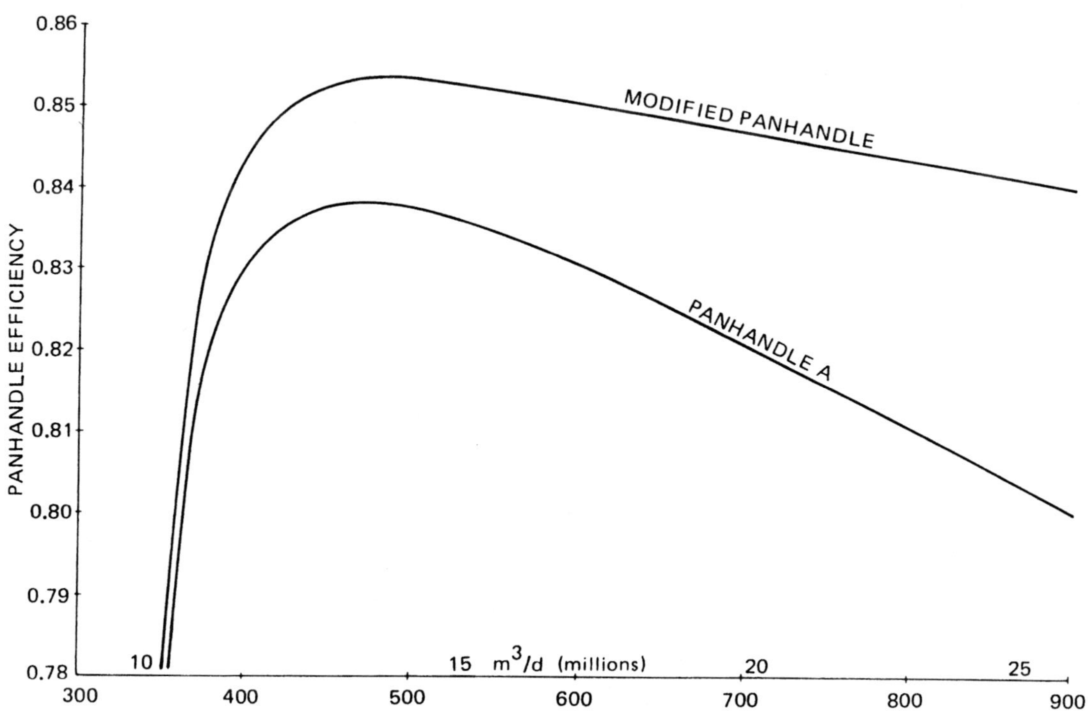

FIG. 5. Panhandle plots of Eq. (1) plus transmission factors.

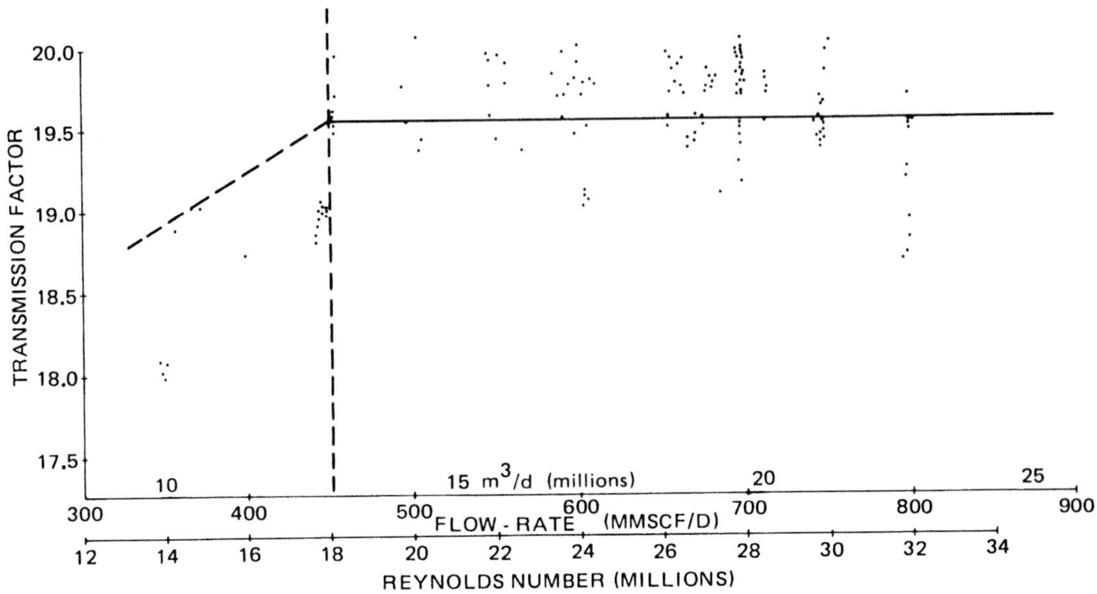

FIG. 6. Calculated transmission factor versus Reynolds number.

plot should be regarded as the most reliable indication of dry-gas behavior.

It can be seen that use of "efficiencies" in the region of 90% brings the Panhandle plots into reasonable agreement with the AGA plot (assuming the use of that Panhandle equation which yields the lowest transmission factor for any given Reynolds number).

As expected, the Weymouth equation has no relevance to partially turbulent flow. In the fully turbulent flow region, Weymouth's equation is consistently conservative by about $5\frac{1}{2}$% relative to the AGA plot.

Comparison of the operating data plot with the various gas equations reveals several interesting points.

It appears that the North Sea gas pipeline operates in the fully turbulent flow region at rates above about 450 MMSCF/d (12.7×10^6 m^3/d). In fully turbulent flow, the presence of condensates in the gas stream effectively reduces the transmission factor (predicted by AGA) by about $4\frac{1}{2}$%. Hence the Weymouth equation closely follows the operating data at flow rates above 450 MMSCF/d.

The overall behavior of the pipeline bears little resemblance to either Panhandle equation. The Panhandle "efficiency" factor could be varied to enable the Panhandle equations to characterize the pipeline's performances but it is more logical to compare the line's performance with that predicted by the AGA (dry-gas) design method.

Comparison of the operating data plot to the AGA plot is interesting for two principal reasons.

First, the apparent fully turbulent transmission factor is approximately 5% below that suggested by the AGA dry-gas method (assuming a pipe

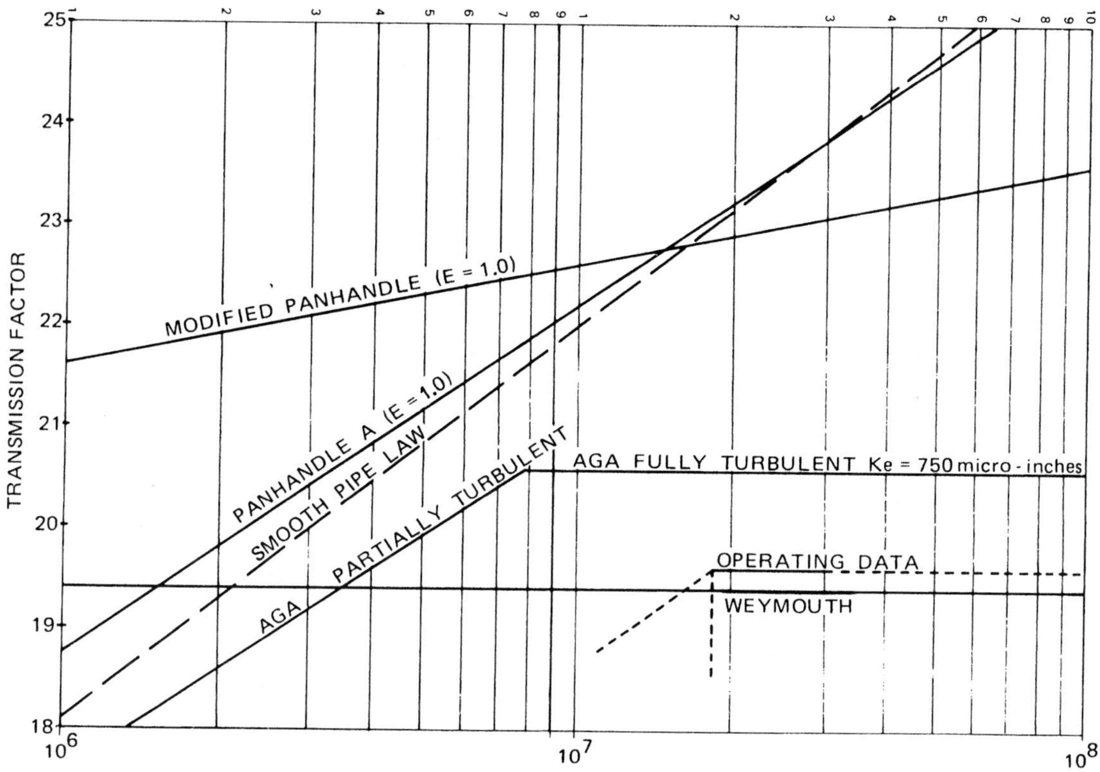

FIG. 7. Transmission factors plus operating data.

roughness of 750 micro-inches (19 μm)). It is reasonable to attribute this apparent increase in fraction factor to the fact that the gas is transporting liquids.

Three reasons are usually proposed as being responsible for the increase in pressure drop due to small quantities of liquids in gas lines.

1. The presence of the liquids reduces the diameter of the pipe available for gas (and flow rate is proportional to diameter to the 2.5 power).
2. The flowing mixture density is slightly higher than that of gas alone.
3. A two-phase head term is present in inclined pipelines (i.e., in *all* offshore gas lines which extend to shore).

Whatever the precise reason for the increased pressure drop, the presence of liquid has caused a direct shift of the fully turbulent flow plot. Thus, if the concept of "efficiency" must be used, the efficiency should be made to define displacements from the fully turbulent flow plot (i.e., *parallel* displacements) rather than displacements from a smooth pipe law-type plot. The Weymouth equation represents such a parallel displacement.

Second, the real data indicate that the transition from partially turbulent to fully turbulent flow has not occurred until a relatively high Reynolds number has been reached. The AGA equation indicates a transition to fully turbulent flow at a Reynolds number of approximately 8 million (for pipe roughness of 750 micro-inches). In practice, this transition appears to occur at a Reynolds number of approximately 18 million (Fig. 6).

It would appear that, in the case of the North Sea gas line, the transition between fully turbulent and partially turbulent flow occurs at the same point at which liquids start to accumulate in the pipeline due to the gas velocity being insufficient to maintain mist flow. (The partially/fully turbulent transition occurs at a similar flow rate to that rate at which it becomes necessary to sphere the pipeline to prevent random liquid slugs occurring at the shore terminal.)

This coincidence may simply be due to those particular conditions which happen to prevail in this particular pipeline. However, it seems probable that the presence of liquid in the gas stream modifies the partially/fully turbulent transition.

If more data from other operating wet-gas pipelines were available, it might be possible to show that the presence of a separate, segregated liquid phase in a two-phase pipeline *prevents* the development of fully turbulent flow conditions. Once pipeline conditions enable all the liquid to be carried as a dispersion in the gas (mist flow), fully turbulent flow may develop.

The previous statement is of critical importance in relation to two-phase flow design methods. Three commonly used two-phase correlations were looked at in some detail: AGA (Baker et al. [2], Dukler [6], and Beggs-Brill [3]. All of these methods include friction factors which are dependent upon Reynolds number (i.e., basically smooth pipe law-type factors). No specific allowance is made for fully turbulent gas flow. Thus these methods could not be expected to predict the performance of the North Sea gas line at flow rates above about 450 MMSCF/d (12.7×10^6 m^3/d) with any accuracy.

Furthermore, the widely held belief that pipeline efficiency does not vary with flow rate (in horizontal pipelines) is fallacious whenever fully turbulent gas flow occurs. As illustrated by Fig. 7, a smooth pipe law-type plot is only relevant to partially turbulent flow situations.

The danger of overlooking this fact is well illustrated by considering the diagram of the Federal Power Commission (FPC) which was presented by Gould and Ramsey [8] (Fig. 8). The curve is supposed to represent the efficiency of a 15-in. (381 mm) wet-gas pipeline for liquid loadings from 0 to 100 bbl/MMSCF/d (0 to 0.00056 m^3/m^3). As Gould and Ramsey correctly point out, a separate curve should be presented for each different liquid loading. However, a more fundamental failing of this diagram is overlooked: the dry-gas line (dashed) should *not* be a straight line. Dry-gas efficiency may be approximately constant at low flow rates but at rates above about 175 MMSCF/d (5×10^6 m^3/d) in this example, fully turbulent flow will exist and the Panhandle A equation will progressively overestimate flow rate (i.e., the efficiency will decrease). Thus the effect of liquids in the line is not to present a curved plot as opposed to a straight line plot, but rather to present one curve as opposed to a different curve.

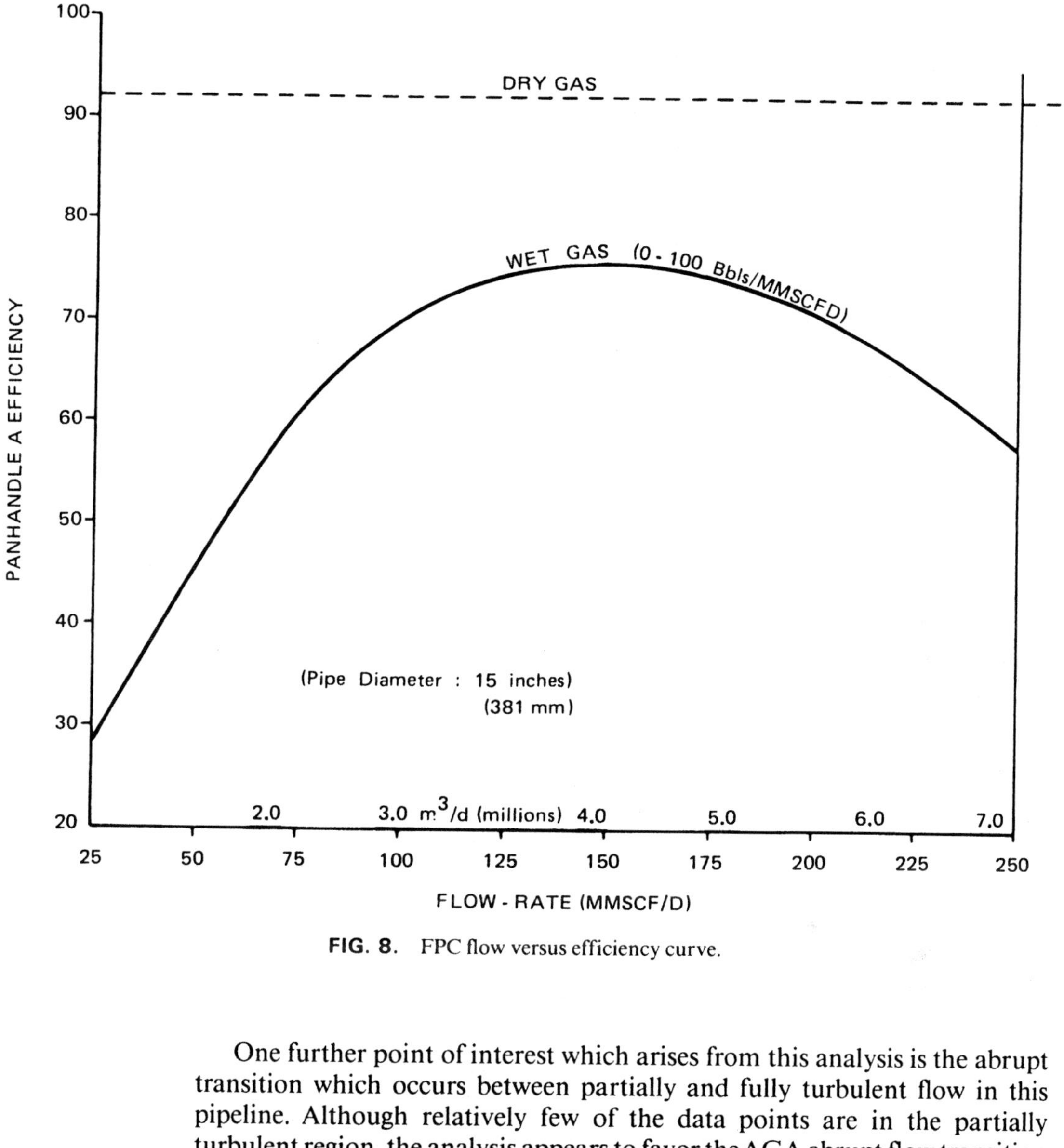

FIG. 8. FPC flow versus efficiency curve.

One further point of interest which arises from this analysis is the abrupt transition which occurs between partially and fully turbulent flow in this pipeline. Although relatively few of the data points are in the partially turbulent region, the analysis appears to favor the AGA abrupt flow transition rather than the broad Colebrook transition.

Transient Behavior and Sphering

Regarding the data analysis presented here, it is noteworthy that "steady-state" conditions are unlikely to be attained in this pipeline due to the frequent changes in flow rate dictated by varying demands for gas by the purchaser.

Data points were eliminated from the pipeline pressure surveys if the measured flow rate had altered by more than 10 MMSCF/d (280,000 m^3/d) in the previous 2 hours. This does not eliminate transients since a large jump in flow rate may cause pressure oscillation for several hours.

One small and predictable source of transients is provided by the practice of sphering wet-gas lines. This practice is very common since the use of spheres can both increase the "efficiency" of the line and prevent large random liquid slugs appearing at the pipeline outlet.

At the time the pipeline pressure surveys were carried out, spheres were launched at about 6-h intervals when the flow rate was below 600 MMSCF/d (17.0×10^6 m^3/d). Sphering was unnecessary at flow rates above 600 MMSCF/d.

This sphering policy had been formed from operating experience with the pipeline. When the line was not sphered, large slugs occasionally appeared at the shore terminal at rates below about 550 MMSCF/d (15.6×10^6 m^3/d). The slugs caused handling problems at the terminal and large pressure drops in the pipeline.

It appears, therefore, that the practice of sphering this pipeline has the effect of extending the range of flow rates for which fully turbulent flow behavior occurs. Figure 6 indicates a transition at about 450 MMSCF/d (12.7×10^6 m^3/d) with sphering as opposed to 550 MMSCF/d (15.6×10^6 m^3/d) without sphering.

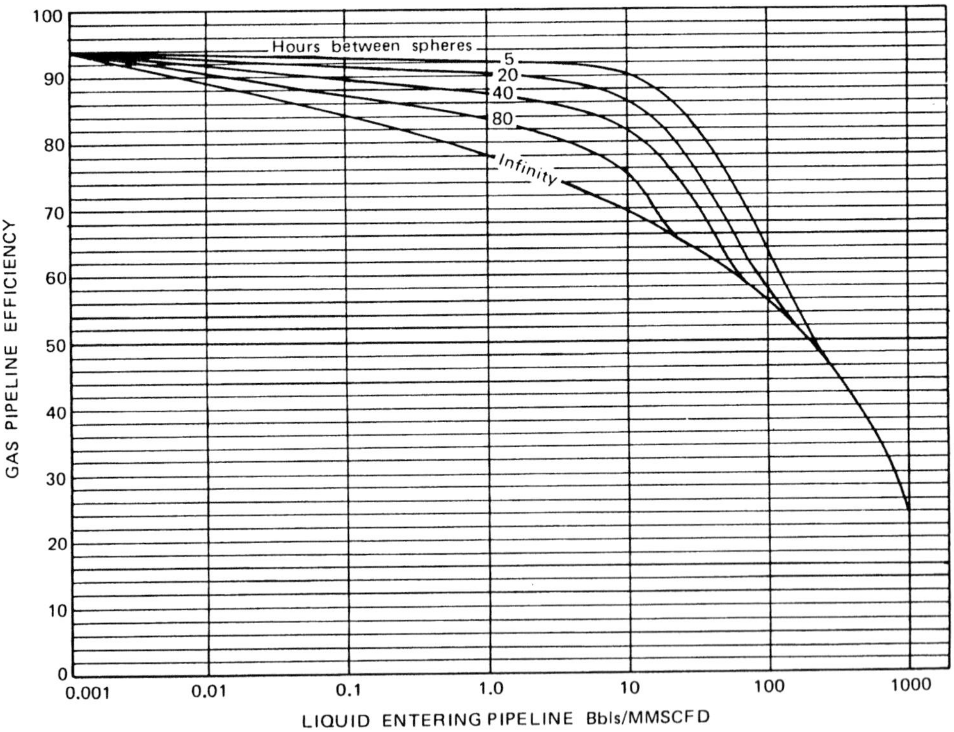

FIG. 9. Effect of spheres in multiphase pipelines on gas pipeline efficiency. Reproduced from Ref. 2.

One of the few published diagrams indicating the effect of spheres on gas pipeline efficiency is reproduced in Fig. 9 (from Baker et al. [2]). It is apparent from the previous discussion that Fig. 9 is not very helpful unless the pipeline diameter and flow rate (hence gas velocity) are given. If gas flow rates are high enough to maintain mist flow, no benefit will be gained by sphering.

Wet-Gas Pipeline Design

The foregoing discussion has indicated that currently published two-phase design methods may not accurately predict the performance of wet-gas pipelines with low liquid loading. The use of Panhandle-type equations with adjusted "efficiencies" is an arbitrary method which tends to cloud an already difficult problem.

It has been suggested that the AGA dry-gas design procedure may represent a sounder basis from which to study gas pipelines carrying small quantities of condensate.

The present analysis illustrates the behavior of an offshore pipeline carrying about 5 bbl/MMSCF. Pipelines carrying greater quantities of liquids can be expected to exhibit a smaller operating range in the turbulent flow region. Above some critical liquid loading, whose exact value will vary with gas and liquid composition, physical conditions, and pipeline profile, the whole operating range of the pipeline will exhibit partially turbulent behavior. At liquid loadings above this critical value, smooth pipe law-type friction factors will be applicable.

Figure 10 illustrates this approach. This diagram indicates the situation

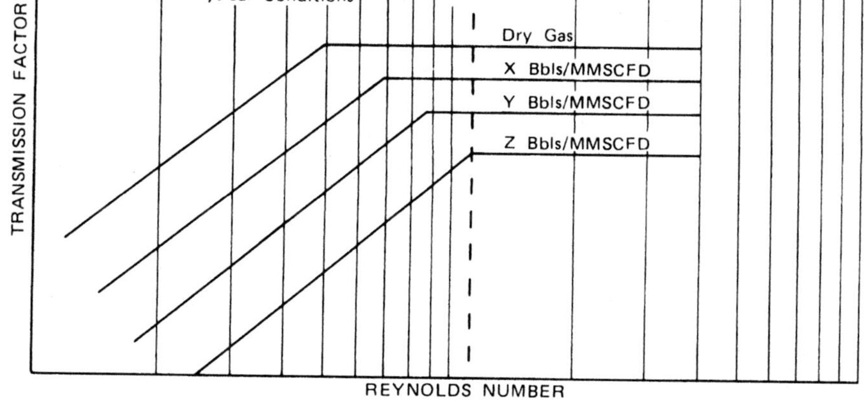

FIG. 10. Proposed approach to wet-gas pipeline design.

which might be expected for a gas pipeline of a certain diameter carrying various quantities of condensate (where $Z > Y > X$). The vertical broken line represents the Reynolds number which will occur in the pipeline at maximum flow rate. The maximum flow rate will be dictated by the maximum operating pressure of the pipeline and hence by the pipe wall thickness.

In the situation illustrated in Fig. 10, the pipeline will operate in fully turbulent flow (at maximum flow rate) if there is less than Z bbl/MMSCF condensate in the line. At all liquid loadings greater than Z bbl/MMSCF, fully turbulent flow will not occur within the operating range of the pipeline.

It is apparent from Fig. 10 that a change of pipe diameter not only affects the ranges of Reynolds number and transmission factor which may occur, but also determines the flow behavior in the pipeline.

At the present time insufficient pipeline data have been published to enable the presentation of a complete design method. However, the authors believe that the approach outlined earlier provides a sound basis from which to study wet-gas pipeline performance. If pressure surveys from other pipelines, operating under different conditions, were made available, the future design of wet-gas pipelines could be greatly improved.

Conclusions

This analysis suggests the following general conclusions for high-pressure gas pipelines carrying small quantities of condensate.

1. At flow rates above a certain value, fully turbulent flow occurs (i.e., friction factor is independent of Reynolds number). Thus two-phase flow correlations intended for use in mist flow conditions should contain friction-factor relationships which distinguish between partially turbulent and fully turbulent flow.
2. The Panhandle equations should only be used for "first estimate" design calculations. Even in this case, a Reynolds number calculation *must* be carried out to ascertain the appropriate flow regime. In fully turbulent flow situations, Weymouth's equation is superior to the Modified Panhandle equation.
3. The smooth pipe law and the rough pipe law provide a much sounder basis for gas pipeline design than either Panhandle equation. Hence the AGA dry gas design method is superior to the other methods of gas pipeline design mentioned in this article.
4. The AGA dry-gas design method also provides a sounder basis from which to study the behavior of gas lines carrying small quantities of liquids.
5. Results of this study indicate, albeit tentatively, that the transition from partially to fully turbulent flow can be abrupt in a wet-gas line. Thus models based upon the AGA dry-gas design method may prove to be marginally superior to Colebrook-type models in predicting two-phase pressure drops in the mist flow regime.

With the data published herein, the authors would like to challenge and encourage other sources to publish actual operating data from other two-phase pipelines. A critical and comprehensive evaluation of existing design methods cannot be achieved until a wide variety of operating data are generally available for analysis.

Appendix: Data Used in Analysis

The data analyzed in this article were taken from two extensive pipepline pressure surveys conducted in the winters of 1973/4 and 1974/5.

The pressure surveys consist of hourly readings of flow rate (measured by orifice meters at the pipeline outlet), inlet pressure, and outlet pressure (measured with deadweight testers).

General pipeline data are presented in Table A1. The data tabulated in Tables A2 and A3 are the most steady flow-rate points which occurred during the pressure surveys.

TABLE A1 General Pipeline Data

Length	86 mi	138 km
Internal diameter	26.624 in	676.25 mm
Sea temperature	42°F	5.6°C
Gas gravity	0.63	0.63
Condensate gravity	56°API	0.76
Average Z	0.82	
Base pressure	14.73 lb/in.^{2}	0.102 MPa
Base temperature	60°F	15.6°C
Maximum pipe depth	120 ft ss	36.6 m ss

TABLE A2 Measured Operating Data—Winter 1973/74[a]

Q	P_{in}	P_{out}	Q	P_{in}	P_{out}
955	1820	1069	745	1623	1145
920	1790	1084	743	1628	1140
			743	1621	1142
797	1667	1115	743	1626	1140
797	1683	1119	741	1622	1140
797	1665	1126	741	1630	1145
797	1667	1114	740	1628	1142
797	1671	1107	738	1630	1145
795	1667	1115	736	1621	1151
795	1685	1102	736	1621	1151
795	1692	1103			
793	1696	1107	719	1606	1164
787	1693	1113	717	1603	1166
			716	1610	1166
752	1623	1157	716	1605	1164
749	1620	1151	699	1537	1097
749	1627	1147	699	1581	1164
749	1608	1110	699	1580	1167
749	1616	1110	699	1582	1167
747	1605	1114	697	1592	1187
745	1625	1137	695	1591	1184
745	1625	1146	695	1530	1113

TABLE A2 Measured Operating Data—Winter 1973/74[a]

Q	P_{in}	P_{out}	Q	P_{in}	P_{out}
695	1548	1093	694	1591	1179
695	1573	1129	694	1592	1181
694	1591	1184	692	1588	1176
694	1543	1114	690	1566	1137
694	1543	1113	684	1583	1164
694	1542	1113	684	1591	1167
694	1531	1108			
694	1528	1101	600	1458	1137
694	1565	1131	594	1460	1139
694	1566	1134	592	1462	1142
694	1555	1130	590	1461	1150
694	1567	1124	590	1461	1142
694	1558	1134	590	1460	1138
694	1558	1134	590	1460	1139
694	1552	1132	589	1461	1143
694	1587	1176	587	1461	1150

[a]Units: Q = MMSCF/d, P_{in} and P_{out} = lb/in.^{2}gauge. 1 MMSCF = 28,316.8 m^3. 1 lb/in.^{2} = 6894.76 Pa. Pressures tabulated are gauge pressures.

TABLE A3 Measured Operating Data—Winter 1974/75[a]

Q	P_{in}	P_{out}	Q	P_{in}	P_{out}
686	1558	1149	609	1527	1182
683	1558	1152	609	1505	1177
677	1570	1169	607	1515	1193
677	1563	1164	607	1515	1191
675	1570	1167	605	1518	1176
675	1571	1170	605	1510	1167
675	1548	1147	603	1506	1176
675	1546	1146	602	1523	1182
668	1571	1170	600	1520	1208
668	1568	1167	594	1503	1182
668	1558	1158	565	1456	1167
668	1558	1166	563	1481	1203
666	1566	1164	561	1488	1197
664	1548	1162	555	1458	1184
662	1545	1167	550	1446	1160
662	1546	1166	548	1448	1180
662	1547	1167	548	1446	1173
662	1546	1167	546	1452	1185
660	1558	1173	546	1443	1168
660	1560	1173	509	1437	1194
659	1542	1161	509	1433	1188
659	1542	1162	508	1425	1200
659	1542	1159	498	1431	1202
657	1540	1150	495	1431	1209
655	1533	1164			
			462	1401	1203
611	1507	1177	462	1401	1200

TABLE A3 Measured Operating Data—Winter 1974/75[a]

Q	P_{in}	P_{out}	Q	P_{in}	P_{out}
462	1398	1205	439	1396	1197
460	1397	1200	439	1395	1199
458	1400	1206			
449	1415	1217	399	1382	1218
449	1408	1208	371	1335	1194
447	1418	1224	366	1333	1197
447	1411	1218	357	1338	1204
447	1408	1210	351	1366	1230
445	1391	1194	349	1373	1236
443	1411	1215	346	1373	1242
441	1385	1191	346	1361	1226

[a]Units: Q = MMSCF/d, P_{in} and P_{out} = lb/in.^{2}gauge. 1 MMSCF = 28,316.8 m^3. 1 lb/in.^{2} = 6894.76 Pa. Pressures tabulated are gauge pressures.

Symbols

C	numerical constant
D	diameter (in.)
E	"efficiency" factor
ϵ	roughness factor
f	friction factor (Fanning)
G	gas specific gravity
L	length (mi)
P	pressure (lb/in.^{2}abs)
P_0	base pressure (14.73 lb/in.^{2}abs)
Q	flow rate (scf/d)
Re	Reynolds number
S	compressibility correction term
T	temperature (°R)
T_0	base temperature (520 R)
Z	gas deviation factor

This article was originally presented at the Energy Conference of the Petroleum Division of the American Society of Mechanical Engineers, Houston, Texas, 1977.

References

1. Baker, O., *Oil Gas J.*, *55*, 45 (1957).
2. Baker, O., et al., "Gas/Liquid Flow in Pipelines, 11" *Design Manual*, AGA-API, October 1970.
3. Beggs, H. D., and Brill, J. P., *J. Pet. Technol.*, May 1973.
4. Blasius, H., *Mitt. Forschungsarb., No. 131* (1913).
5. Colebrook, C. F., *J. Inst. Civ. Eng.*, *11*, 133 (1939).
6. Dukler, A. E., "Gas/Liquid Flow in Pipelines, 1," *Research Results*, AGA-API, May 1969.

7. Flanigan, O., *Oil Gas J.*, March 10, 1958.
8. Gould, T. L., and Ramsey, E. L., *J. Pet. Technol.*, March 1975.
9. Nikuradse, J., *Forschungshegt, No. 356,* V.O.I. Verlag, Berlin, 1932.
10. Smith, R. V., et al., United States Bureau of Mines, *Monograph 9*, 1956.
11. Uhl, A. E., et al., *Project NB-13*, AGA, New York, 1965.
12. Weymouth, T. R., *Trans. ASME, 34*, 203 (1912).

P. MARK HOPE
R. G. NELSON

Slurry Systems, Nomograph

In the minerals industries, insoluble mixtures of ore and liquids (slurries) are often used for transport and for processing. Examples are movement of coal–water mixtures in pipe lines and flotation–separation of iron ore concentrates.

Slurry calculations require knowledge of the mixture average specific gravity, percent dry solids by volume, and the amount of liquids required. Let P_S and P_L = weights of dry solids and liquids, respectively; D_S, D_L, D_M and D_W = densities of solids, liquids, mixture, and water, respectively; and S, L, and M = specific gravities of solids, liquid, and mixture and water, respectively. The density of an insoluble mixture is the total weight of all components divided by the total volume:

$$D_M = \frac{P_S + P_L}{\dfrac{P_S}{D_S} + \dfrac{P_L}{D_L}} \tag{1}$$

Mixture specific gravity, $M = D_M/D_W$, and $S = D_S/D_W$ and $L = D_L/D_W$ Dividing both sides of Eq. (1) by D_W:

$$M = \frac{D_M}{D_W} = \frac{P_S + P_L}{P_S\left(\dfrac{D_W}{D_S}\right) + P_L\left(\dfrac{D_W}{D_L}\right)} = \frac{P_S + P_L}{\dfrac{P_S}{S} + \dfrac{P_L}{L}} \tag{2}$$

Let W = percent dry solids by weight:

$$W = 100\left[\frac{P_S}{P_S + P_L}\right] \tag{3}$$

and

$$P_L = P_S\left[\frac{100 - W}{W}\right] \tag{3a}$$

Substituting (3a) in (2) and solving for M:

$$\frac{1}{M} = \frac{1}{L} + \frac{W}{100}\left[\frac{1}{S} - \frac{1}{L}\right] \tag{4}$$

Let V = percent dry solids by volume:

$$V = \frac{\left[\dfrac{P_S}{D_W S}\right]}{\left[\dfrac{P_S + P_L}{D_W M}\right]} = 100\left[\frac{P_S}{P_S + P_L}\right]\frac{M}{S} = \frac{WM}{S} \tag{5}$$

By manipulating Eqs. (4) and (5):

$$M = \frac{VS}{W} = \frac{L(100 - V)}{(100 - W)} = L + \frac{V}{100}(S - L) = \frac{100}{\dfrac{W}{S} + \dfrac{(100 - W)}{L}} \tag{6}$$

$$V = \frac{WM}{S} = \frac{100}{1 + \dfrac{S}{L}\left(\dfrac{100 - W}{W}\right)} = 100 - \frac{M}{L}(100 - W) = 100\frac{(M - L)}{S - L} \tag{7}$$

$$W = \frac{VS}{M} = \frac{VS}{L + 0.01V(S - L)} = 100 - \frac{L}{M}(100 - V) = 100\left(\frac{M - L}{M}\right)\left(\frac{S}{S - L}\right) \tag{8}$$

$$S = \frac{WM}{V} = \frac{WM}{100 - \dfrac{M}{L}(100 - W)} = L + \frac{100}{V}(M - L) = L\left(\frac{100 - V}{V}\right)\left(\frac{W}{100 - W}\right) \tag{9}$$

In the usual case, the liquid is water (L = 1.00), and the equations become as shown in the nomographs.

Water at 60°F weighs 8.333 lb/gal, so for any solid of specific gravity (SG),

$$\text{gpm} = \frac{[\text{tons/hr}][2000\ \text{lb/short ton}]}{[60\ \text{min/h}][8.333 \times (\text{SG})\ \text{lb/gal}]} = \frac{4(\text{tph})}{(\text{SG})} \tag{10}$$

$$W = 100\left[\frac{(\text{tph})_S}{(\text{tph})_S + (\text{tph})_W}\right] = \frac{100(\text{tph})_S}{[\text{tons/h slurry} = (\text{tph})_{SL}]} \tag{11}$$

or

$$\frac{(\text{tph})_W}{(\text{tph})_S} = \left(\frac{100 - W}{W}\right) \tag{12}$$

and

$$\frac{(\text{gpm})_W}{(\text{tph})_S} = 4\left(\frac{100 - W}{W}\right) \tag{13}$$

$$\frac{(\text{tph})_{SL}}{(\text{tph})_S} = \frac{100}{W} \tag{14}$$

$$\frac{(\text{gpm})_{SL}}{(\text{tph})_S} = \frac{400}{WM} \tag{15}$$

and since

$$\frac{1}{M} = \frac{(100 - W) + W/S}{100} \tag{6}$$

$$\frac{(\text{gpm})_{SL}}{(\text{tph})_S} = 4\left[\left(\frac{100 - W}{W}\right) + \frac{1}{S}\right] \tag{16}$$

where gpm = gallons per minute.

Figure 1 permits a rapid, simultaneous solution of these equations for water slurries. If any two of the variables are set, the others are determined. Example: What are the other quantities if the dry solids specific gravity, S, = 2.5 and the percent dry solids by volume, V, = 40.0? Align S and V and read mixture average specific gravity, M = 1.60; percent dry solids by weight W = 62.5; gpm of slurry mixture per tph dry solid = 4.00; and that 2.4 gpm of water is required per tph dry solids.

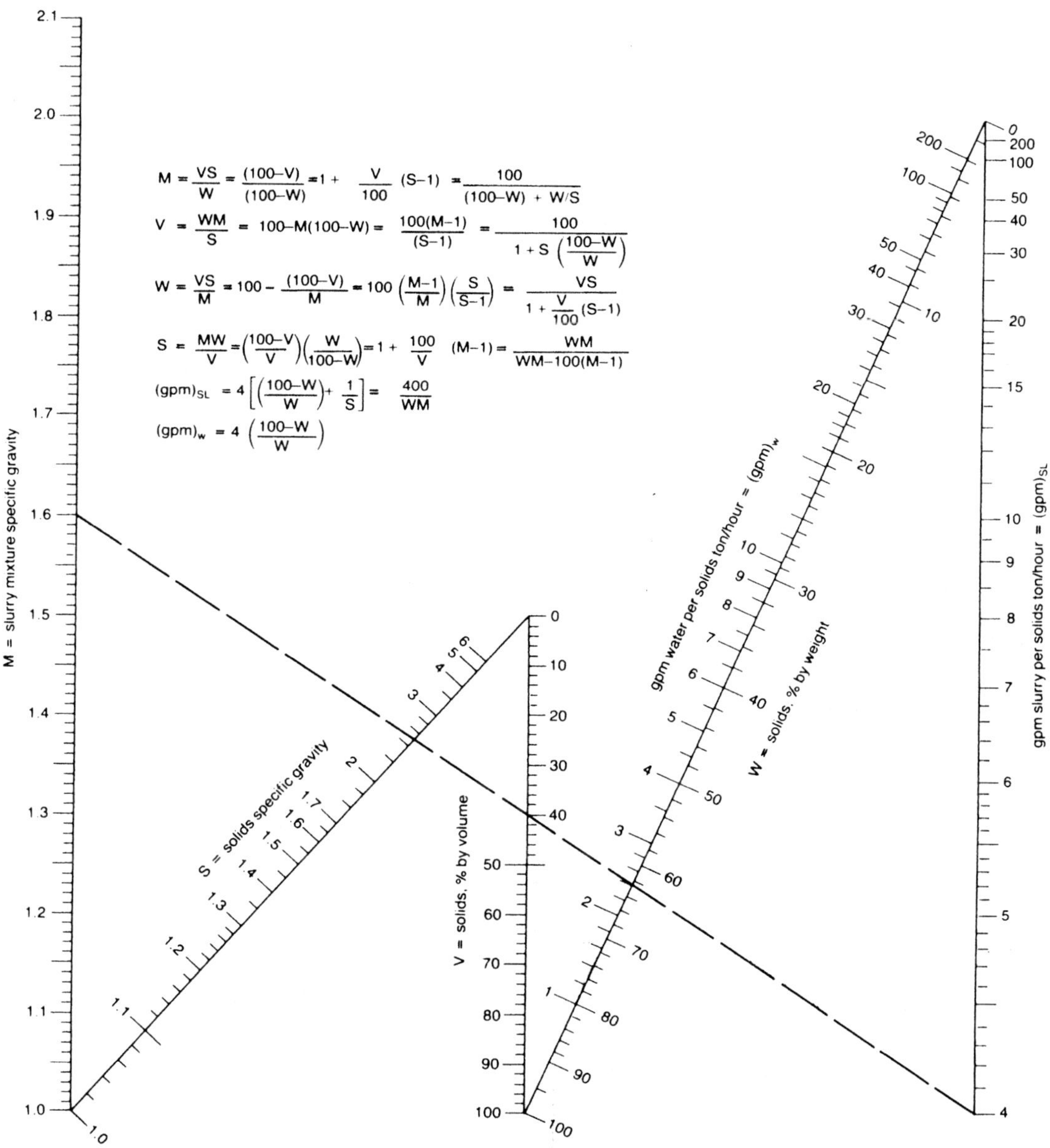

FIG. 1. Water slurry properties.

$$V = \frac{100}{1 + \left(\frac{100-W}{W}\right)\left(\frac{S_S}{S_L}\right)}$$

$$S_s\left(\frac{100-W}{W}\right) = S_L\left(\frac{100-V}{V}\right)$$

KEY:
S_s – W – (1)
(1) – V – S_L

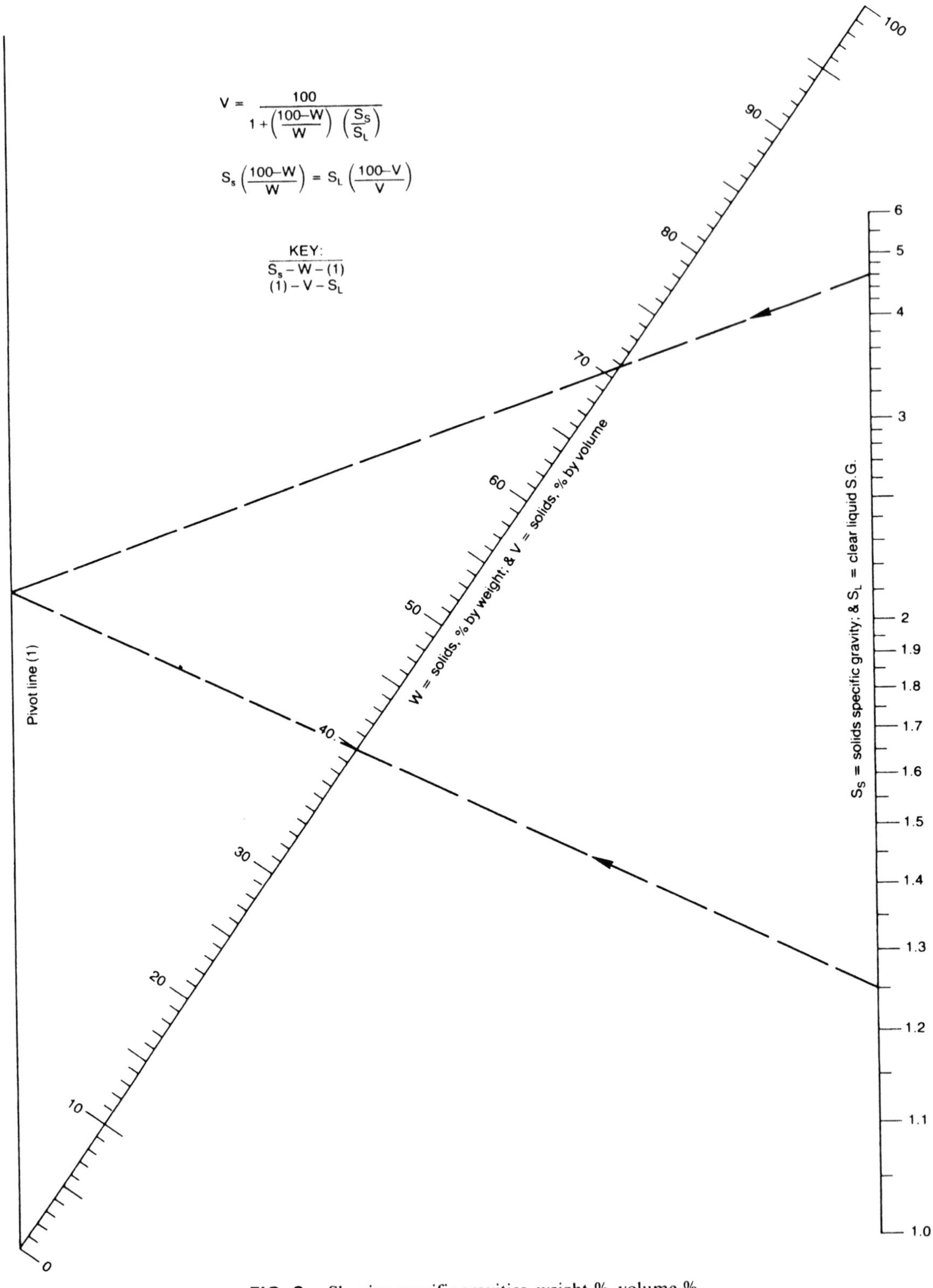

FIG. 2. Slurries: specific gravities, weight-%, volume-%.

Figure 2 permits a fast determination of volume or weight percent for any specific gravity liquid. Example: what is V if $S_S = 4.6$, $S_L = 1.25$, and $W = 71\%$? Align $S_S = 4.6$ with $W = 71$ and mark intersection on Pivot Line (1); align (1) with $S_L = 1.25$ and read $V = 40\%$.

This article appeared in *Pipe Line Industry*, June 1981, and is used with the permission of the editor and author.

F. CAPLAN

Slurry Systems and Pipelines

Slurry pipeline systems are an efficient and reliable transportation mode. They originated in the mining industry, where wet grinding of minerals made slurry transportation an inherent part of the process. In those early applications, distance seldom exceeded several hundred feet except in tailings lines. The latter could transport *gangue* material from the concentrator to a tailings pond no more than 5 to 10 km away.

Innovative slurry system design now permits cross-country transportation over hundreds of kilometers. Particle-size consist, which is an important slurry transport parameter, can be made compatible with other process requirements. The pipeline can be buried using conventional cross-country construction techniques.

These systems had their modern beginnings in the 1950s with the 175-km, 0.254-m-diameter Consolidation Coal pipeline and the 116-km, 0.152-m-diameter American Gilsonite pipeline. Considerable technical and operating knowledge has been gained which, coupled with broad experience with short-distance slurry pipelines and slurry handling, has advanced the design and building of long-distance systems from an art to a maturing technology.

The slurry transportation concept remained relatively obscure to the general public until the oil embargo and rising oil prices forced attention to the critical energy situation in the latter half of the 1970s. Coal is being recognized as one of the United States' major short-term solutions to the energy crisis.

The economics of coal is heavily dependent on transportation costs. Until recently, the railroads have had a virtual monopoly on coal transportation, but being labor-intensive and using diesel-powered locomotives, they suffer from heavily escalating operating costs. Long-distance coal slurry pipelines are emerging as a strongly competitive transportation mode with many environmental advantages.

On August 14, 1970, coal began flowing in an underground slurry pipeline 440 km across the state of Arizona to a 1500-MW power plant near Davis

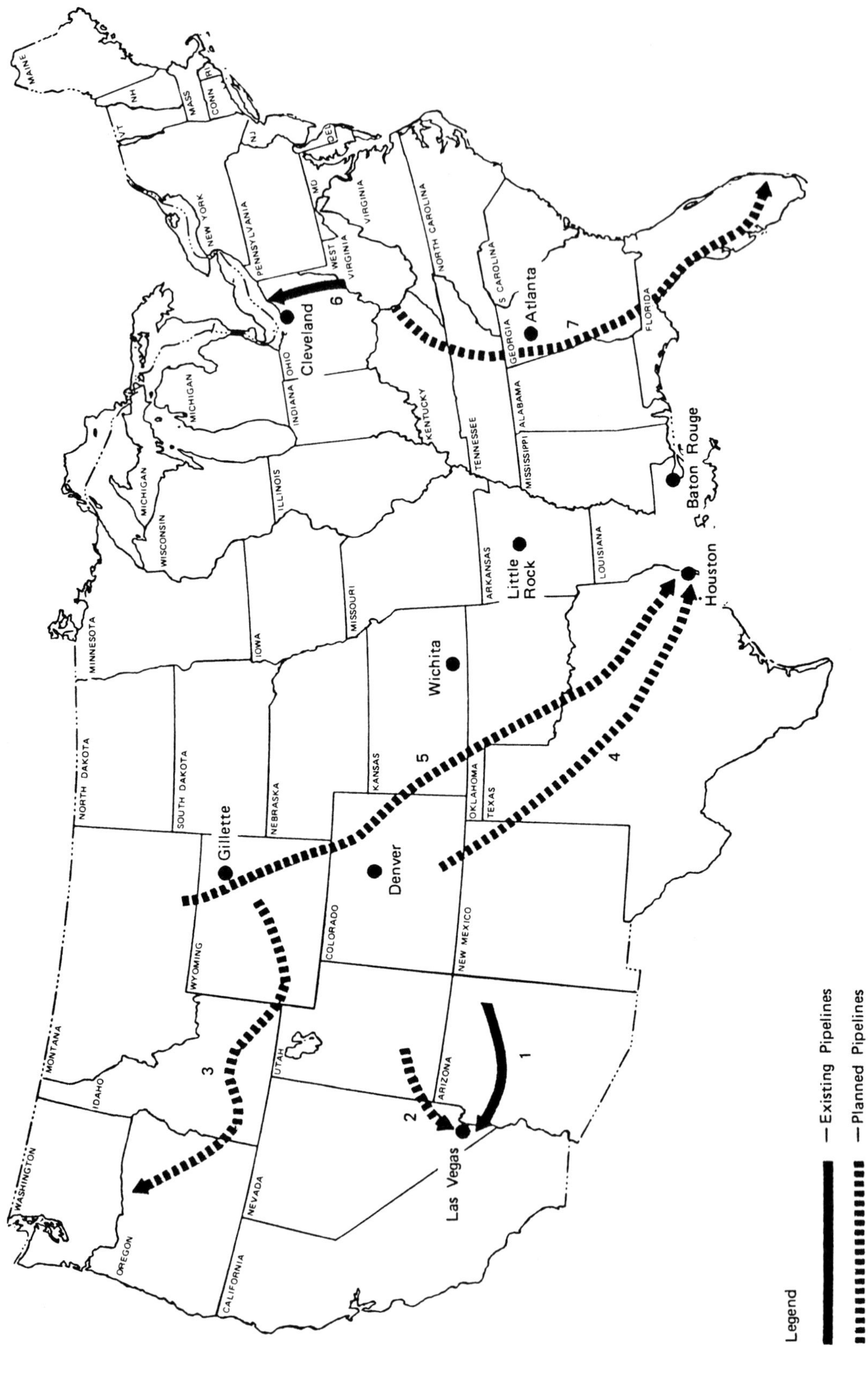
Cleveland
Atlanta
Baton Rouge
Houston
Little Rock
Wichita
Gillette
Denver
Las Vegas
1
2
3
4
5
6
7
Legend
—Existing Pipelines
—Planned Pipelines

FIG. 1. Coal slurry pipeline systems. Coal has steadily grown in importance as a fuel for generating electricity, particularly in states formerly dependent on natural gas. Several slurry pipelines are under development to help transport increased coal production. Source: Energy Transportation Systems, Inc., San Francisco, California.

Pipeline System	*Length (miles)*	*Annual Capacity (tons)*
1. Black Mesa Pipeline	273	4,800,000
2. Alton Pipeline	183	11,600,000
3. Gulf Interstate–Northwest Pipeline	1100	10,000,000
4. San Marco Pipeline	900	15,000,000
5. ETSI Pipeline	1378	25,000,000
6. Ohio Pipeline	108	1,300,000
7. Florida Pipeline	1500	15–45,000,000

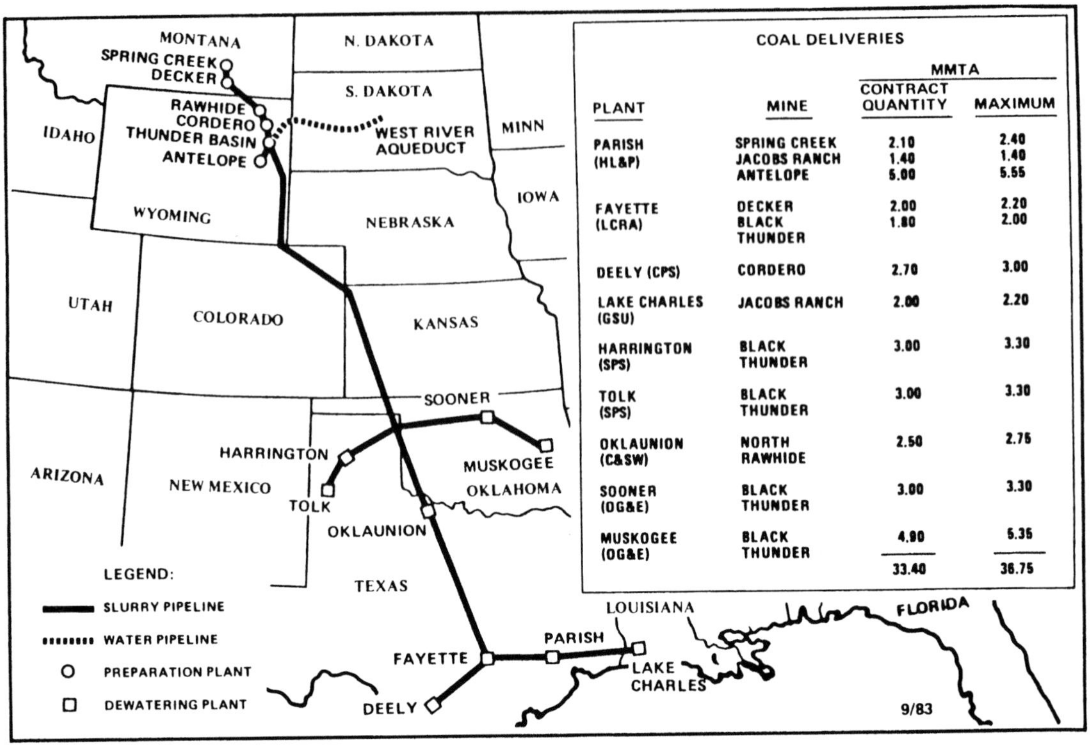

COAL DELIVERIES

PLANT	MINE	MMTA CONTRACT QUANTITY	MMTA MAXIMUM
PARISH (HL&P)	SPRING CREEK JACOBS RANCH ANTELOPE	2.10 1.40 5.00	2.40 1.40 5.55
FAYETTE (LCRA)	DECKER BLACK THUNDER	2.00 1.80	2.20 2.00
DEELY (CPS)	CORDERO	2.70	3.00
LAKE CHARLES (GSU)	JACOBS RANCH	2.00	2.20
HARRINGTON (SPS)	BLACK THUNDER	3.00	3.30
TOLK (SPS)	BLACK THUNDER	3.00	3.30
OKLAUNION (C&SW)	NORTH RAWHIDE	2.50	2.75
SOONER (OG&E)	BLACK THUNDER	3.00	3.30
MUSKOGEE (OG&E)	BLACK THUNDER	4.90	5.35
		33.40	36.75

FIG. 2. The ETSI pipeline route. Originating in the Powder River Basin, the ETSI line will run underground 1378 mi to major power plants along its route and ultimately in Texas. Coal can be delivered en route or transferred to barges for further shipment on the waterway system.

Dam in Nevada. This installation is the 0.457-m-diameter Black Mesa coal slurry pipeline, which is equivalent in capacity to two 100-car railroad trains per day and is the longest slurry pipeline in the world to date [14]. The relative length and geographic relationship of this system to others now in the planning stage is shown in Fig. 1. The magnitude of anticipated expansion is quite evident. The most advanced of the planned facilities is ETSI (Energy Transportation Systems, Inc.). This is a multibillion-dollar pipeline system intended to transport coal from the Powder River Basin in Wyoming to utilities in Oklahoma, Arkansas, and Louisiana. ETSI's pipeline capacity would be 20 to 25 million metric tons of subbituminous coal, enough to generate approximately 7000 MW. It is shown in Fig. 2.

Slurry Pipeline Design

Slurry pipeline design is a complex procedure that cannot be completely covered here. The methods of hydraulic analysis, adapted to materials that are neither solid nor liquid but something in between, are essential to the design process. Two reference textbooks deserve special attention from those interested in in-depth study of slurry characteristics and flow. *The Flow of Complex Mixtures in Pipes* by Govier and Azia [8] provides an excellent theoretical background for behavior and flow of non-Newtonian fluids and other complex mixtures. *Slurry Pipeline Transportation* by Wasp, Kenny, and Gandhi [26] is a current practical design manual for the slurry engineer.

For a given material to be moved, the design process aims at the selection of an appropriate pipe type and size and pumping pressure. Abrasion, corrosion, usable velocities, friction losses, amount and gradient of rise and fall, and ambient and generated temperatures are but some of the more important parameters to be considered.

Seven basic design-calculation steps are required:

1. Classify the slurry as being either homogeneous or heterogeneous.
2. Establish the slurry concentration.
3. Select a pipe size, based on the system's throughput requirement.
4. Calculate the critical velocity.
5. Check that the design velocity is at least 1 ft/s above, but not excessively above, the critical velocity. It may be necessary to select another trial pipe size and repeat the calculations until an acceptable relationship between critical velocity and design velocity is achieved.
6. Calculate the design friction losses (distinguishing between horizontal and vertical pipe for heterogeneous slurries).
7. Calculate the system pressure gradient and pump discharge pressure.

Slurry Characteristics

Flow of mixtures of solids and liquids (i.e., slurries) in pipes differs from flow of homogeneous liquids in several important ways. With liquids, the complete range of velocities is possible, and the nature of the flow (laminar, transitional, or turbulent) is defined by the physical properties of the fluid and system. With slurries, two additional distinct flow regimes and several more physical properties are superimposed on the liquid system. The two regimes of slurry flow involve the following.

Homogeneous Slurries

Here, the solid particles are homogeneously distributed in the liquid media, and the slurries are characterized by high solids concentrations and fine particle sizes. Such slurries often exhibit non-Newtonian rheology (i.e., the effective viscosity is not constant, but varies with the applied rate of shearing strain). Some examples are sewage sludge, clay slurries, and cement-kiln-feed slurry.

Heterogeneous Slurries

Here, concentration gradients exist along the vertical axis of a horizontal pipe even at high flow rates; i.e., the fluid phase and the solid phase retain their separate identities. Heterogeneous slurries tend to be of lower solids concentration and have larger particle sizes than homogeneous slurries. Florida phosphate rock is a good example.

Many slurries encountered commercially are of mixed character; the finer particle-size fractions join with the liquid media to form a homogeneous vehicle, while the coarser sizes act heterogeneously. Pipeline coal slurry is a prime example of the mixed characteristic.

Thus, the designer will often be faced with defining the dominant characteristic of a slurry. Evaluation of the complex mixed regime is outside the scope of this article. This analysis will, however, allow the designer of short systems to bracket the answer and usually achieve a practical solution.

Critical Velocity

The two types of slurries, homogeneous and heterogeneous, have entirely different critical-velocity characteristics, as seen in Fig. 3. On the chart, Curve A illustrates the deposition critical velocity typical of heterogeneous slurries. The characteristic hook in the friction-loss-vs-velocity curve results from deposition of solids on the bottom of a horizontal pipe, as shown in the sketch. (In vertical pipes, the solids that in a horizontal pipe would be deposited are easily transported because their settling velocity is usually much lower than normal flow velocities.)

The deposition critical velocity is directly related to the settling velocity of the coarser particles in a heterogeneous slurry and the degree of turbulence in the pipe; it therefore increases with increasing particle size or specific gravity, and with increasing slurry concentration or viscosity. Deposition velocity generally exhibits an increase proportional to the square root of pipe diameter.

Curve B shows viscous-transition critical velocity, which is characteristic of homogeneous slurries. While design of a system for operation below the transition critical velocity is acceptable for truly homogeneous slurries, no turbulent forces exist to suspend even trace amounts of heterogeneous particles.

The transition velocity and laminar friction losses are very sensitive to the rheological properties of a homogeneous slurry. Transition velocity tends to increase with viscosity and therefore increase with solids (volume) concentration, greater quantity of fines and lower solids gravity. Transition velocity for slurries with a yield stress (Bingham plastic*) is very little affected by pipe diameter, whereas it is directly proportional to diameter for slurries with Newtonian rheological properties.

*Although Bingham-plastic fluids are similar to Newtonian fluids in that there is a linear relationship between shear stress and shear rate, Bingham plastics require a finite shear stress to initiate flow and overcome the "rigidity" of the fluid (as discussed later).

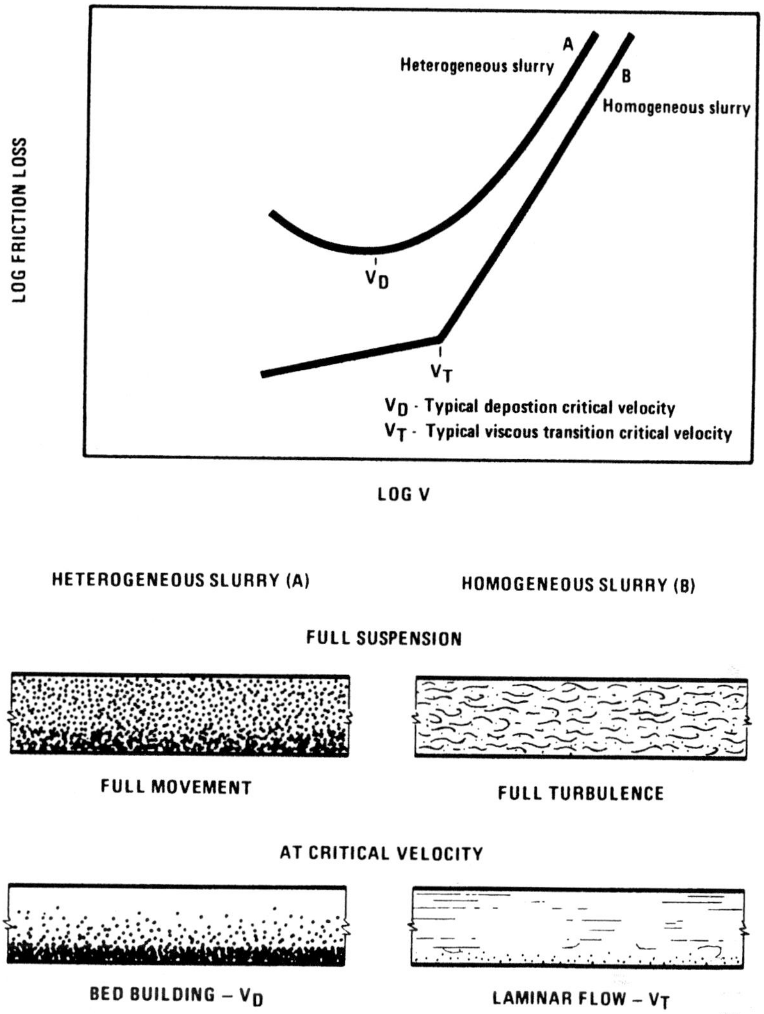

FIG. 3. Critical velocity characteristics depend on whether slurry is heterogeneous or homogeneous.

Design Parameters

Before making design calculations, the engineer must measure or estimate the slurry concentration, rheology, and flow regime, and must select a "trial" pipe diameter.

Solids Concentration

The concentration of solids in a slurry will often be controlled by the upstream or downstream process. In this case the dependent variables of slurry viscosity (or rheology) and specific gravity are also fixed. When addition of liquid before transportation is required, the amount of dilution may be an important factor in the downstream process, as well as in the slurry transportation.

Solids concentrations up to about 40% by volume are readily handled when the solids surface area is low (coarse slurries); however, fine materials such as clay or sewage sludge may not be fluid at even a 10% solids concentration. As a general rule, solids that are less than 25% finer than 325 mesh (0.44 μm) will be quite fluid when slurried to a 40% solids volume. For thicker and finer slurries, a knowledge of the relationship between slurry rheology and concentration may be required to select a pumpable concentration.

A good estimate of a comfortable pumping concentration can be made by reducing the static settled slurry-concentration 10 to 15 percentage points in volume concentration. Slurries approaching static settled concentration can be pumped, providing the high friction losses are tolerable (as, for instance, in

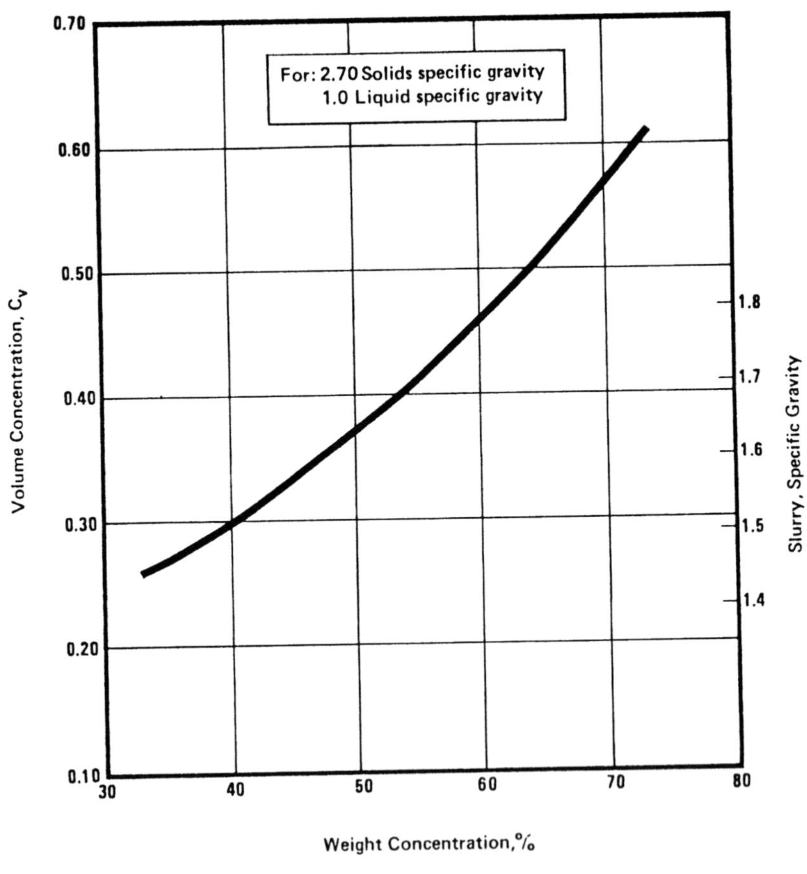

FIG. 4. Slurry weight concentration bears this relationship to volume concentration and specific gravity.

short systems). The static settled concentration is determined by allowing a slurry of known concentration to settle under static conditions in a graduated cylinder until the slurry/water interface reaches equilibrium. The settled concentration is then calculated from the starting concentration and the beginning and ending volume. If the slurry does not settle, it is already too thick.

The slurry volume concentration and specific gravity are directly related, depending only on the solids specific gravity and the liquid specific gravity. Figure 4 relates these properties for solids having a specific gravity of 2.7 and water. It is helpful to prepare a chart such as this for the slurry being evaluated.

Slurry Rheology, Viscosity

Much of the above discussion of solids concentration is based on consideration of the slurry rheology (or viscosity). The rheological properties of a slurry, for the purposes of this article, determine the "viscosity" used for friction loss calculation and for the transition critical velocity of fine, thick slurries. The following two rheological cases adequately cover most commercial slurries for short pipeline design.

Newtonian slurries are described by the simple rheological property of viscosity (Fig. 5), and can be treated as true fluids, provided the flowing velocity is high enough to suspend the solids. The slurries are characteristically composed of graded materials, with few fines at moderate concentrations. Figure 6 shows reduced viscosity as a function of slurry concentration for spheres, which is the minimum viscosity case at a given concentration [23].

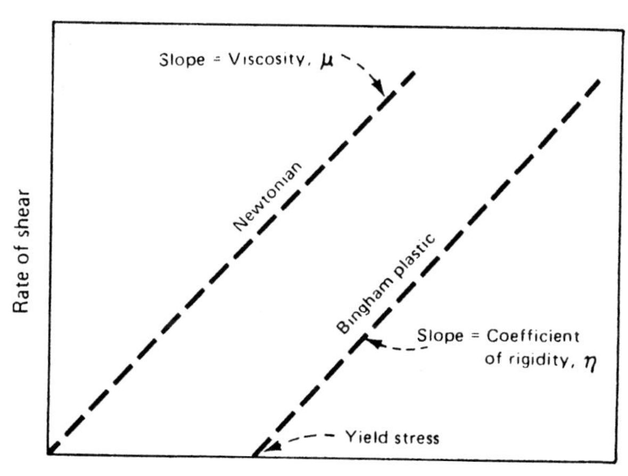

FIG. 5. Typical rheogram for two types of slurries.

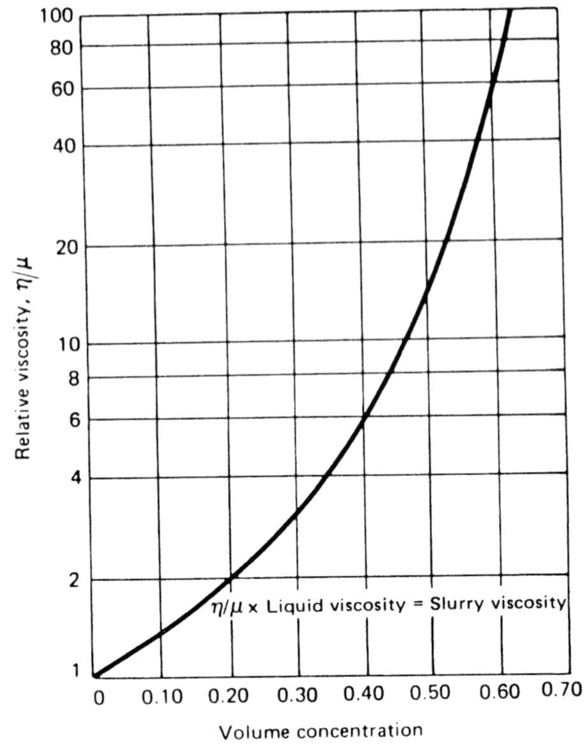

FIG. 6. Viscosity for spherical-particle slurry is a function of slurry concentration [14].

Bingham-plastic slurries require a knowledge of shear stress as a function of shear rate (a rheogram, as in Fig. 5) to determine the parameters of coefficient of rigidity (η) and yield stress (τ_0) at any concentration. These data must be obtained in the laboratory with a rheometer such as the Contraves or Fann; the simple Stormer or Brookfield viscometers can be used to get a conservative estimate of these properties. Bingham-plastic slurries are characteristically composed of fine solids at higher concentrations. See Kenny [13] for information on the analysis of rheometer data, and on other rheological aspects of these slurries.

Design Characteristics

The seven basic design-calculation steps were listed in the section entitled "Slurry Pipeline Design."

The first two steps were discussed in the design-parameters section; let us examine the remaining ones, first for homogeneous slurries, then for heterogeneous ones.

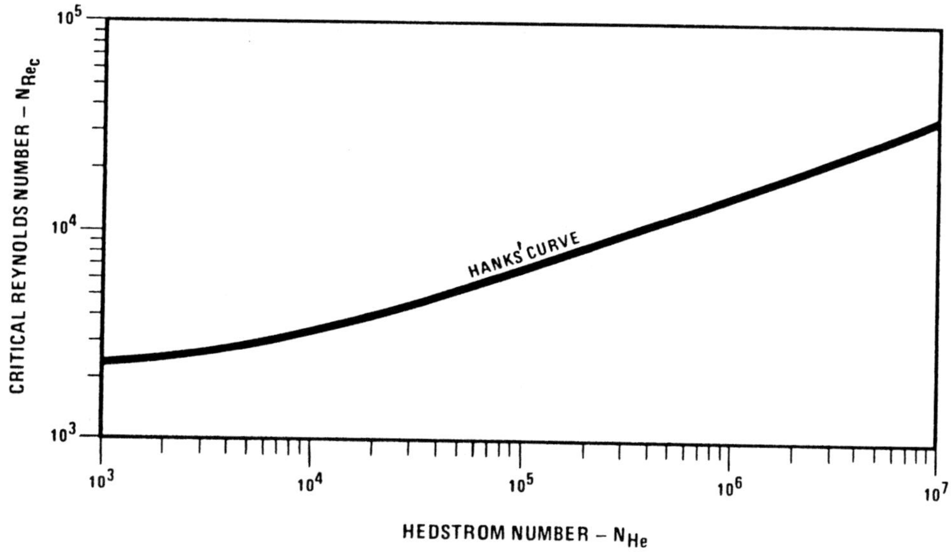

FIG. 7. Hanks' curve relates critical Reynolds number to Hedstrom number for Bingham flow in pipes.

Homogeneous Slurries: General Procedure

The viscous transition from turbulent to laminar flow has been defined as the critical velocity for homogeneous slurries. As discussed earlier, the absence of turbulent energy in laminar flow allows coarser particles to settle and accumulate. Design in laminar flow for very fine (nonsettling) slurries requires an accurate knowledge of the slurry rheology, because the laminar (or viscous) friction losses are quite sensitive to that property.

Having established the design parameters and classified the slurry as homogeneous, the viscous-transition critical velocity can be calculated by using an analysis of transition velocity for Bingham-plastic materials developed from Hedstrom [10] and Hanks [9]. Hanks' curve (Fig. 7) is used to correlate the Hedstrom number (based on system and slurry properties) to the critical Reynolds number. This critical Reynolds number becomes the familiar 2100 for slurries with Newtonian (true fluid) rheology; i.e., with no yield stress.

Once satisfied that the selected pipe size results in a design velocity above critical transition velocity, friction losses are calculated using the normal relationship between the Fanning friction factor and Reynolds number. For Bingham-plastic material, the coefficient or rigidity (η) is substituted for viscosity in Reynolds number calculations for turbulent flow—i.e., flow above the transition velocity.

Flow Regime

As discussed earlier, slurries are broadly classified as being in the homogeneous or heterogeneous regime, each regime having a distinctive character, particularly as to critical velocity. Figure 8 provides a means of classifying a slurry as homogeneous or heterogeneous based on the top particle size (largest 5%) and the specific gravity of the solids. The middle area between fully homogeneous and fully heterogeneous slurries is controlled by a compounding of the properties of each type. Analysis of the compound regime is beyond the scope of this paper; however, Wasp has covered the subject in some detail [24, 25].

Design Velocity and Pipe Diameter

Because the volumetric flow is fixed by solids throughput and concentration in the slurry, and because the design velocity and pipe diameter are directly related, we can concentrate on velocity. The design velocity should, of course, be above critical velocity, which in turn depends on pipe diameter to a greater or lesser extent, depending on the slurry regime.

A velocity/diameter combination must be selected to enter the calculations described below. A velocity in the range of 4 to 7 ft/s is usually practical and economical. Velocities below 4 ft/s are seldom desirable, whereas, at the other end of the range, velocities above 7 ft/s may be necessary for strongly heterogeneous slurries. Pipe abrasion is a consideration above 8 to 10 ft/s and can be serious at higher velocities.

These calculations are described in detail below; also, an example is worked out later in this article.

Viscous Transition Velocity

In Fig. 7 the critical Reynolds number is related to the Hedstrom number. The relationship between the design number and the Hedstrom is:

$$N_{\mathrm{He}} = N_{\mathrm{Re}} \times P_L \tag{1}$$

where

$$N_{\mathrm{Re}} = DV_{\rho/\eta} \tag{2}$$

and

$$P_L = \frac{\tau_0/\eta}{V/D} \tag{3}$$

Note that the velocity (V) actually cancels out of the Hedstrom number; however, engineers will probably find it more convenient to work with the dimensionless ratios shown. This can be done by using the system properties of velocity (V) and pipe inside diameter (D), and the slurry properties of density (ρ), coefficient of rigidity (η), and yield stress (τ_0).

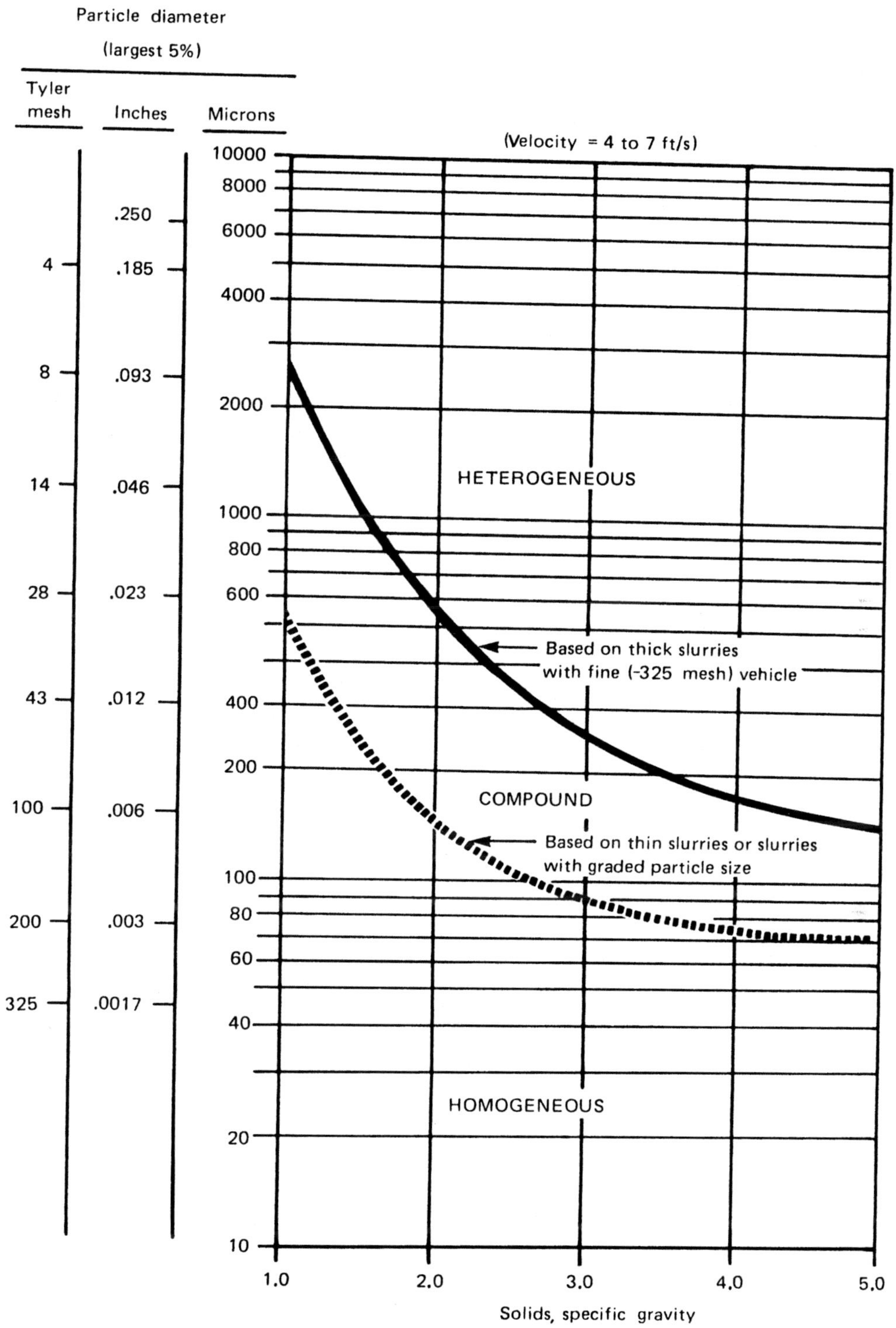

FIG. 8. Slurry flow regime (heterogeneous, homogeneous) is a function of solids size and specific gravity.

The critical Reynolds number is directly proportional to the viscous-transition velocity, as follows:

$$(N_{Re})_c = (DV_{tp})/\eta \tag{4}$$

Therefore, we can use the simple ratio

$$\frac{V_t}{V} = \frac{(N_{Re})_c}{N_{Re}} \quad \text{or} \quad V_t = V \times \frac{(N_{Re})_c}{N_{Re}} \tag{5}$$

If the design velocity (V) is not at least 1 ft/s greater than the transition velocity (V_t), select a smaller pipe size and recalculate the transition velocity.

Friction Losses

The Fanning friction factor (f) is found by using the Reynolds number calculated with Eq. (2) and the normal f vs N_{Re} chart. Friction losses can then be found via

$$\Delta P = \frac{2f_p V^2 L}{gD} \tag{6}$$

It is usually simpler to use some unit length, like 100 ft, for L, then extend to the system length in another step.

Selection of the friction factor for Eq. (8) requires an assumption of the pipe roughness that will exist in the operational slurry system, just as for pipes carrying liquids. This can be a major factor in longer slurry pipelines, so

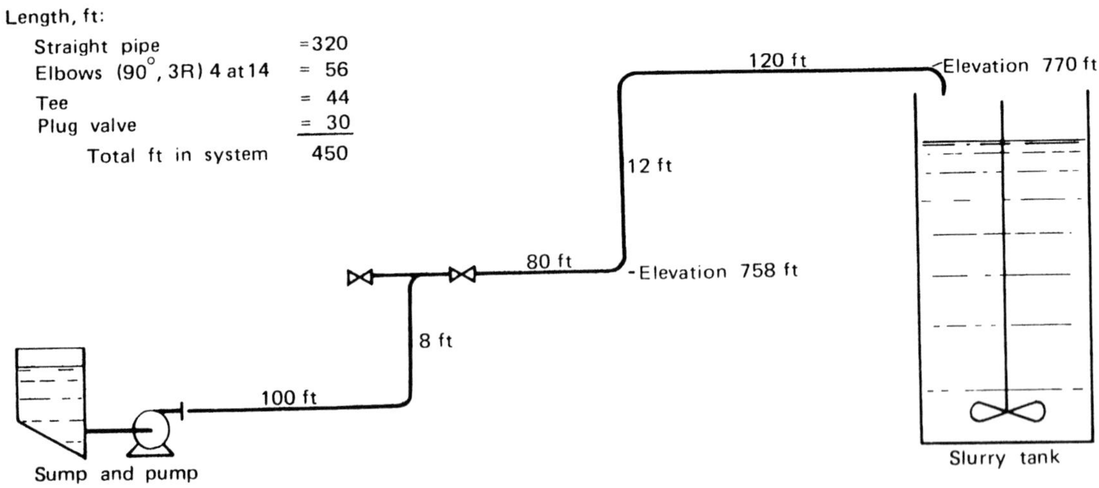

FIG. 9. Flow systems for Examples 1 and 2.

provision may have to be made for sending cleaning "pigs" through the pipe, and for using corrosion inhibitors during operation. In short, designing for rough pipe is usually the better solution. The often quoted Hazen Williams "*C*" factor of 100 will result in a pressure-loss capability about 100% greater than would be provided for new steel-pipe conditions.

See Fig. 9 for the flow system of Examples 1 and 2.

Example 1: Homogeneous Slurries: Calculations for an Inplant System

Basis for Problem

Slurry type	Fly ash in water
Solids content	60%, by weight
Solids specific gravity	2.7
Slurry volume concentration	0.356
Slurry specific gravity	1.61
Solids density	100 lb/ft^3
Slurry rheology	at 68°F
Solids coefficient or rigidity	60 cP (100 cP = 1 dyn/cm^2 · s)
Slurry yield stress	507dyn/cm^2
System data (see Fig. 9):	
Flow	800 gal/min
Pipe diameter, trial basis	8.625-in. o.d.; 8.125-in. i.d.
Velocity	5 ft/s
Length (including fittings)	450 ft
Inlet elevation	750 ft
Discharge elevation	770 ft (to atmosphere)

Calculating Friction Losses

$$f = 0.0082$$

For $N_{\text{Re}} = 8400$,and commercial steel-pipe roughness:

$$\Delta P = \frac{2 f_p V^2 L}{gD}$$

$$= 2 \times 0.0082 \times \frac{100 \text{ lb}}{\text{ft}^3} \times \frac{5^2 \text{ ft}^2}{s^2} \times 100 \text{ ft} \times \frac{s^2}{32.2 \text{ ft} \times 8.125 \text{ in.}} \times \frac{1 \text{ ft}}{12 \text{ in.}}$$

$$\Delta P = 1.3 \text{ lb/in.}^2 \text{ per } 100 \text{ ft pipe}$$

$$\Delta h = 1.3 \text{ lb/in.}^2 (0.433 \times 1.61 \text{ ft})$$

$$= 1.88 \text{ ft (slurry) per } 100 \text{ ft pipe}$$

Calculating the Critical Velocity

$$N_{He} = N_{Re} \times P_L = \frac{DV_\rho}{\eta} \times \frac{\tau_0/\eta}{V/D}$$

$$N_{Re} = \frac{8.125 \text{ in.}}{12 \text{ in./ft}} \times \frac{5 \text{ ft}}{\text{s}} \times \frac{100 \text{ lb}}{\text{ft}^3} \times \frac{1{,}488 \text{ cP/(lb/ft} \cdot \text{s)}}{60 \text{ cP}} = 8{,}400$$

$$P_L = \frac{50 \text{ dyn}}{\text{cm}^2} \times \frac{100}{60 \text{ cP}} \times \frac{8.125 \text{ in}}{5 \text{ ft/s}} \times \frac{\text{ft}}{12 \text{ in.}} = 11.2$$

$$N_{He} = 8400 \times 11.2 = 95{,}000$$

$$(N_{Re})_c = 6500 \text{ (see Fig. 7)} = DV_{t\rho/\eta}$$

Transition velocity:

$$\frac{V_t}{V} = \frac{(N_{Re})_c}{N_{Re}}$$

$$V_t = \frac{(N_{Re})_c}{N_{Re}} \times V = \frac{6500}{8400} \times 5 \text{ ft/s} = 3.8 \text{ ft/s}$$

Calculating Pump Discharge Pressure

Static pressure: 770 ft − 750 ft	= 20 ft	= 13.9 lb/in.^{2}
Friction pressure: 1.88 ft × 450 ft/100 ft	= 8.5 ft	= 5.9 lb/in.^{2}
Total pressure:	= 28.5 ft	= 19.8 lb/in.^{2}

Heterogeneous Slurries: General Procedure

The velocity of deposition of the coarsest size fraction on the bottom of a horizontal pipe is the critical velocity for heterogeneous slurries. Of course, this deposition will not occur, and is not a concern, in purely vertical systems. Also, friction pressure losses are usually lower in vertical pipes for heterogeneous slurries because there can be no concentration gradient across a vertical pipe.

Calculations for both the vertical and horizontal case are described below, and a numerical example is worked out.

Deposition Velocity

For heterogeneous slurries, for our purposes defined by Fig. 8, the volume concentration of solids in the slurry, and the solids density, are related to a critical Froude number as shown in Fig. 10 [24]:

$$(N_{Fr})_c = V_d/\sqrt{gD} \tag{7}$$

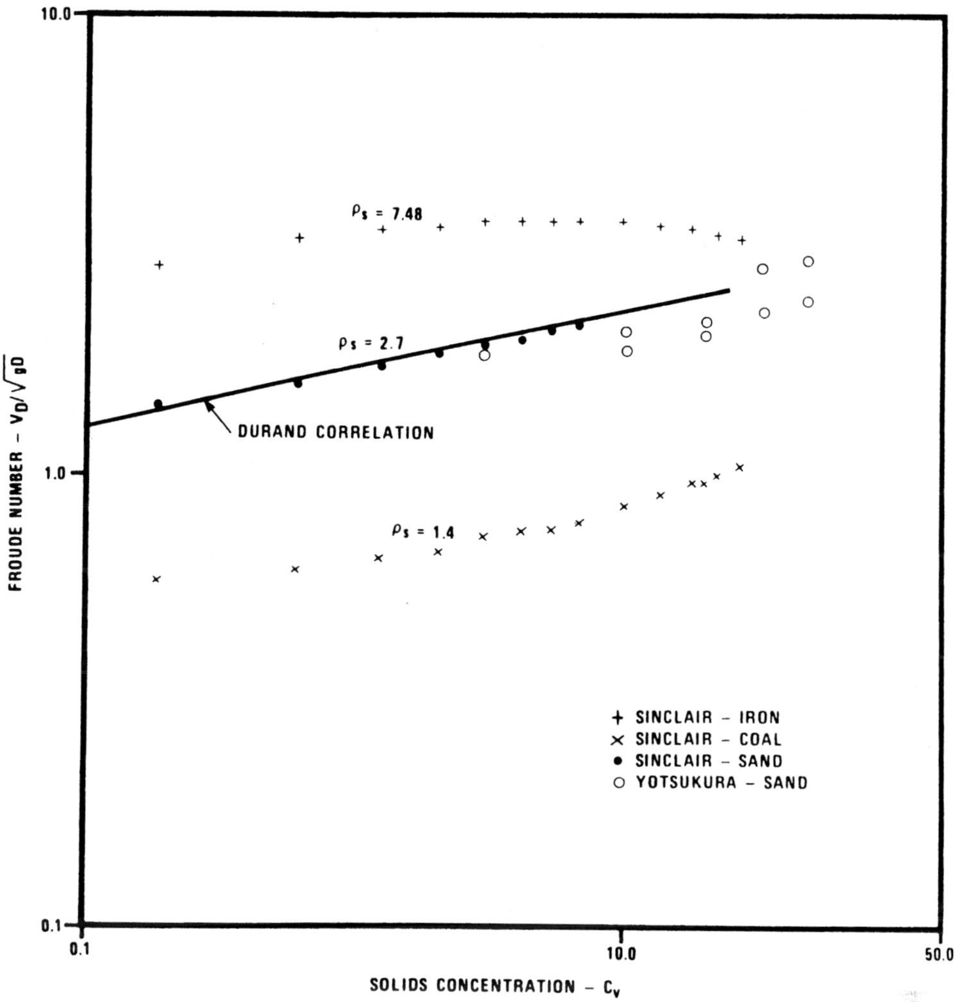

FIG. 10. Froude number at deposition, for various suspensions, shows this relationship to concentration [24].

therefore,

$$V_d = (N_{\text{Fr}})_c\sqrt{gD} \tag{8}$$

The effect of particle size on deposition velocity in this approach was covered in the regime selection of Fig. 8. Although the velocity of first deposition is not affected by particle size (i.e., above a certain size), the sensitivity of the system and horizontal friction losses are greater for coarser slurries.

Friction Losses

Having confirmed that the selected pipe diameter results in a velocity above critical, turbulent friction losses are calculated using the familiar f vs N_{Re} relationship, with appropriate overpressure corrections for heterogeneous solids distribution in horizontal pipes.

Heterogeneous slurries flowing in horizontal pipes have higher friction losses due to nonuniform distribution of solids across the pipe. Durand's empirical relationship corrects the later friction factor for the solids loading as follows:

$$f = f_w\left[1 + 82\left(\frac{gD}{V^2}\frac{\rho - \rho_w}{\rho_w}\right)^{3/2}\frac{C_V}{C_d^{\,3/4}}\right] \tag{9}$$

It has been found that the overpressure correction,($f - {}_w$), should be found for each size fraction for materials with wide granularity to completely account for the relationship of drag coefficient to particle size [25]. The volume concentration of each size fraction is used in that calculation. A quick, conservative estimate can be made by assuming the minimum value of the drag coefficient (0.44) for all sizes. See Perry's *Chemical Engineers' Handbook* for the drag-coefficient calculation [20].

The sum of the overpressure corrections ($f - f_w$) for the several size or solids–gravity fractions, times the water friction factor (f_w), gives the friction factor that is used in Eq. (8) to calculate horizontal-pipe, heterogeneous-slurry friction losses.

The friction losses of many slurries are known to be dampened below the values calculated from the Fanning friction factor; however, this phenomenon is not completely understood and therefore cannot be generalized. In any event, neglecting this possible dampening leads to a conservative prediction.

In a vertical pipe, the solids are uniformly distributed, so the pressure losses are calculated as described earlier for homogeneous slurry. This takes advantage of the lower pressure losses encountered in vertical pipes for heterogeneous slurries.

Example 2: Heterogeneous Slurries: Calculations for an Inplant System

Basis for Problem

Slurry type	Graded sand in water
Solids content	35% by weight
Solids specific gravity	2.7
Solids size	60% is 8 × 14 mesh 40% is 14 × 28 mesh
Slurry volume concentration	0.166
Slurry specific gravity	1.28
Slurry density	80 lb/ft^3
Slurry viscosity	1.8 cP (Fig. 5)

System data (configuration is the same as in Example 1):

Flow	1940 gal/min
Pipe diameter, trial basis	8.625-in. o.d.; 8.125-in. i.d.
Velocity	12 ft/s
Length (including fittings)	430 ft horizontal; 20 ft vertical
Inlet elevation	750 ft
Discharge elevation	770 ft (to atmosphere)

Slurry Friction Losses: Horizontal Pipes

$$N_{\text{Re}}\,(\text{water}) = DV_{\rho w/\mu}$$

$$= \frac{8.125 \text{ in.}}{12 \text{ in./ft}} \times \frac{12 \text{ ft}}{\text{s}} \times \frac{62.3 \text{ lb}}{\text{ft}^3} \times \frac{1{,}488 \text{ cP/(lb/ft} \cdot \text{s)}}{1 \text{ cP}}$$

$$= 755{,}000$$

$$f_w = 0.0037$$

$$f = \text{See Eq. (9) for heterogeneous slurries}$$

$$f - f_w = 82\left(\frac{gD}{V^2} \times \frac{\rho - \rho_\omega}{\rho_w}\right)^{3/2} \frac{C_V f_w}{C_d^{3/4}}$$

The friction factor increment (f - f_w) is found for each size and then added to f_w:

Size (mesh)	*Wt.%*	C_V	C_d
8 × 14	60	0.10	0.5975
14 × 20	40	0.066	1.05
		0.166	

In this example a single calculation for f can be made, due to the single value of C_d:

$$f = 0.0037\left[1 + 82\left(\frac{32.2 \text{ ft/s}^2}{12^2 \text{ ft}^2/\text{s}^2} \times \frac{8.125 \text{ in.}}{12 \text{ in./ft}} \times \frac{2.7 \times 62.3 - 62.3}{62.3}\right)^{3/2} \times \left(\frac{0.1}{0.5975^{3/4}} + \frac{0.066}{1.05^{3/4}}\right)\right]$$

$$= 0.0037\,(1.23) = 0.00455$$

$$\Delta P = \frac{2f\rho\, V^2 L}{gD}$$

$$= \frac{2 \times 80 \text{ lb/ft}^3 \times 12^2 \text{ ft}^2/\text{s}^2 \times 100 \text{ ft}}{32.2 \text{ ft/s}^2 \times 8.125 \text{ in.} \times 12 \text{ in./ft}}$$

$$= 11.8 \text{ lb/in.}^2 \text{ per 100 ft of horizontal pipe}$$

$$h = 11.8 \text{ lb/in.}^2\,(0.433 \times 1.28 \text{ lb/in.}^2/\text{ft})$$

$$= 21.3 \text{ ft (slurry) per 100 ft of horizontal pipe}$$

Slurry Friction Losses: Vertical Pipe

$$N_{Re}\,(\text{slurry}) = DV_{\rho/\eta}$$

$$= \frac{8.125\ \text{in.}}{12\ \text{in./ft}} \times \frac{12\ \text{ft}}{\text{s}} \times 80\,\frac{\text{lb}}{\text{ft}^3} \times \frac{1{,}488\ \text{cP/(lb/ft}\cdot\text{s)}}{1.8\ \text{cP}}$$

$$= 537{,}000$$

$$f = 0.0038$$

$$\Delta P = \frac{2f\rho V^2 L}{gD}$$

For commercial steel pipe,

$$\Delta P = 2 \times 0.0038 \times 80\,\frac{\text{lb}}{\text{ft}^3} \times 12^2\,\frac{\text{ft}^2}{\text{s}^2} \times 100\ \text{ft} \times \frac{\text{s}^2}{32.2\ \text{ft} \times 8.125\ \text{in.}} \times \frac{1\ \text{ft}}{12\ \text{in.}}$$

$$\Delta P = 2.8\ \text{lb/in.}^2\ \text{per 100 ft of pipe}$$

$$\Delta h = 2.8\ \text{lb/in.}^2 / 0.433 \times 1.28\ \text{lb/in.}^2\text{/ft}$$

$$= 5.0\ \text{ft (slurry) per 100 ft of vertical pipe}$$

Calculating the Critical Velocity

$$(N_{Fr})_c = 2.3 = V_d/\sqrt{gD} \qquad \text{(Fig. 10)}$$

$$V_d = 2.3\sqrt{gD} = 2.0\sqrt{\frac{32.2\ \text{ft}}{\text{s}^2} \times \frac{8.125\ \text{in}}{12\ \text{in./ft}}}$$

$$= 10.7\ \text{ft/s}$$

Calculating the Pump Discharge Pressure

Static pressure, 20-ft rise	= 11.0 lb/in.^{2}
Friction pressure, horizontal:	
21.3 ft/100 ft × 430 ft = 91.6	= 50.8 lb/in.^{2}
Friction pressure, vertical:	
5.0 ft/100 ft × 20 ft = 1.0 ft	= 0.6 lb/in.^{2}
Total pressure 112.6	= 62.4 lb/in.^{2}

Comparing the Two Cases

Pump Discharge Pressure

In both examples the calculations are carried to a pump discharge pressure for the assumed system, to illustrate the effect of static elevation change on system pressure. Note that the threefold difference in friction losses causes only a one-third increase in system pressure, as a result of the moderate rise (20 ft in 450 ft) in the system.

The effect of fittings on slurry friction losses is adequately compensated for by including the usual equivalent length for fittings and valves in the system length.

Having developed the pump discharge pressure, and knowing the system flow and configuration, we can proceed to the mechanical design, and to the economic evaluation of alternates. Pumping horsepower is directly related to pumping pressure and flow:

$$\text{Horsepower} = \frac{(\text{pressure, lb/in.}^2)(\text{flow, gal/min})}{1.714e}$$

where e is the pump and drive train efficiency. The pumping pressure and system configuration define the operating and static pressures that must be contained at any point in the system.

In many cases a preliminary design and an estimate of capital and operating costs are required before a final pipe diameter can be selected.

Operability

By operability, we mean sensitivity to operating upset or shutdown. Here again, the two slurry regimes have distinctive characteristics.

Homogeneous slurries tend to be slow or nonsettling under static conditions, and the settled slurry is usually soft and fluffy. Shutdown and restart of the system will probably not be a problem, even if vertical piping is involved. On the other hand, homogeneous slurries can be very sensitive to moderate concentration increases above the design concentration. It is not difficult to conceive of plugging a system by pumping enough high-concentration (high yield-stress) slurry to exceed the system's pressure capabilities.

Heterogeneous slurries settle rapidly and form a sandy or hard deposit under static conditions. The settled solids may close and plug a vertical or inclined section, but usually can be washed out of a horizontal pipe. Vibrations from other equipment will greatly increase the possibility of plugging where the pipe cross section is closed by settled solids. Frequent mechanical joints are provided in coarse slurry systems to facilitate unplugging and replacement for wear.

Slurry Processing

Processing may be broadly classified into slurry preparation and/or slurry utilization. Figure 11 indicates the various processing steps that may be required. The purpose of this section is to indicate the processing steps that are associated with the emerging slurry-transport systems, to discuss types of equipment presently used, and to note various key design parameters that will be useful to the design engineer.

Slurry Preparation

This is the physical and chemical processing necessary to give the slurry characteristics required for hydraulic transport and utilization. Preparation normally involves both size reduction (crushing and grinding) and slurrification or addition of the liquid phase. Chemical treatment may also be part of slurry preparation for corrosion inhibition, thinning, and improving the characteristics of the final product.

Some systems require special slurry-preparation facilities (see Table 1). Coal preparation for slurry pipeline transportation is an example. In this case, a particle size specifically suited for slurry transportation must be produced.

In all slurry systems a balance has to be made between pumpability and dewatering characteristics. If sizing is too fine, pumpability may be good but the slurry may be difficult to dewater at the pipeline terminal. If the size is too coarse, the slurry is heterogeneous and must be pumped at higher flow rates to maintain suspension. The cost of pumping then goes up. The choice of particle size for a slurry depends on (1) overall cost of preparation, pumping, and use of the slurry; and (2) operability of the slurry, including shut-down/start-up characteristics and critical velocity.

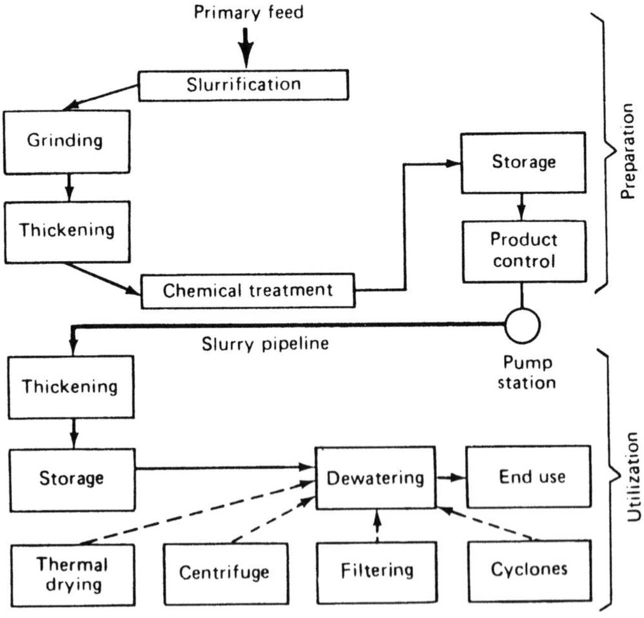

FIG. 11. Typical processing for slurry system.

TABLE 1 Commercial Slurries for Pipelines

Material	Specific Gravity	Maximum Particle Diameter (mesh)	Special Preparation Required?	Size Reduction Equipment Commonly Used
Gilsonite	1.05	4	Yes	High-pressure water jets, crushers
Coal	1.40	8	Yes	Impactors, cage mills, rod mills
Limestone	2.70	48	No	Impactors, ball mills
Copper	4.30	65	No	Crushers, autogenous mills, ball mills
Magnetite	4.90	150	No	Autogenous mills, ball mills

Grinding Solids

Mineral extraction generally requires very fine grinding of ore; 70 to 80% of particles passing 325 mesh (44 μm) is common. Size reduction of the ore normally involves a wet process and the resulting slurries are readily handled hydraulically. As a result, the minerals industry is ideally suited for hydraulic transportation of solids.

Where grinding is done specifically to prepare a material for pipelining, the step will normally involve conventional milling equipment. In some cases the equipment may be a new application to grinding the specific material [2]. In the Black Mesa coal-preparation plant, shown in Fig. 12, rod mills (typical of those found in ore-dressing plants) are used for the final coal-grinding stage. It was necessary to apply the known milling technology of metal-ore dressing to the grinding of coal [3]. Preparing coal for slurry transport involves screening, crushing, grinding, and storage in mixing tanks, as depicted in Fig. 12.

In slurry preparation, two variables are quite important which may have only minor significance in other crushing processes. These are the slurry density and the product top size.

Control of slurry density is necessary to produce a consistent material that fits the hydraulic design of the pipeline. Slurry concentration plays an important role in both the friction loss in the pipeline and the critical velocity of the slurry. The preparation step is usually performed at a solids concentration slightly higher than that required for the pipeline, with dilution control instrumentation provided downstream in the line.

Top-size control is extremely important for any slurry transported over a long distance. Large quantities of coarse, fast-settling particles can cause pipeline plugs. To prevent this, safety screens are installed to prevent coarse particles from entering the pipeline. In systems where a minimum amount of

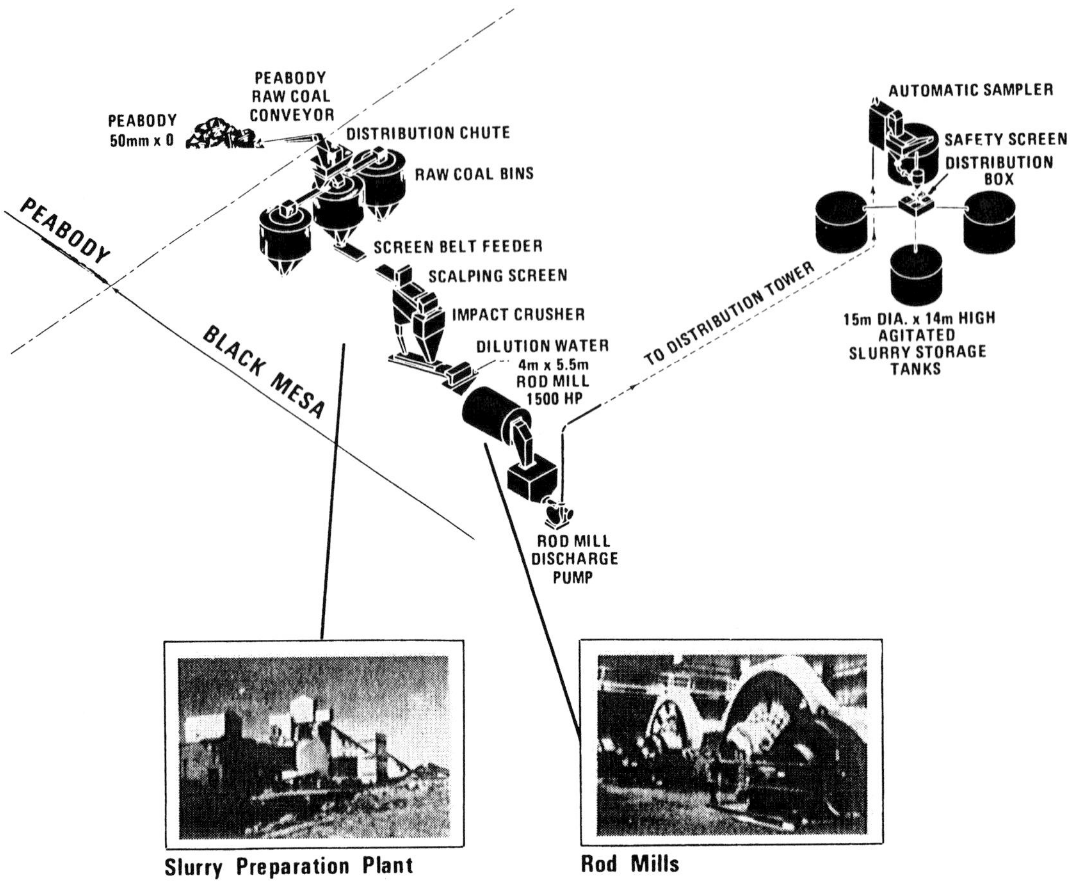

FIG. 12. Preparing coal for slurry transport involves screening, crushing, grinding, and storage in mixing tanks.

crushing is done and where the slurry is relatively high in density and slow settling, conventional classification devices such as cyclones and screen are not satisfactory for top-size control unless the slurry can be diluted and later thickened prior to transportation.

Available Equipment

There are many types of mills available for grinding ores. The most common include autogenous, impact cage, rod, and ball mills. Selection of the most suitable mill depends on the characteristics of the ore, particularly hardness, and the required fineness of the final product.

Slurrification: Mixing Water and Solids

Slurrification normally takes the form of mixing water and solids together in a gravity-feed chute, where the water assists the travel of solids through the chute. This is common practice in a wet-grinding process. The water and coarse solids feed directly into the grinding mill, which discharges a uniform slurry. A notable exception to this is the cutter-head suction dredge. A rotating cutter head provides the shearing action to break down the bed of sand or gravel into particles that can be transported hydraulically in slurry form. The solids in the sand are sheared downward, thus are dragged down toward the dredge suction, where they are picked up in the suction flow. The maximum suction lift of a slurry pump limits the density of a slurry that can be recovered by a dredge.

Slurrification is also often achieved by using hopper-shaped sumps (Fig. 13). Solids disperse over the surface of the liquid in the sump, usually by gravity flow through a screen. Water is sprayed onto the surface of the screen to assist in screening, to break up conglomerates, and to distribute the particles. Liquid in the sump is maintained at a constant level independent of solids addition. In the sump the particles settle at their terminal settling velocity. As the cross-sectional area of the sump narrows (an inverted cone or pyramid shape), slurry concentration increases.

Another slurrification method requires using energy of a high-velocity stream of water to break up the solids and form a slurry. The high-velocity water stream is formed by a high-pressure nozzle. Hydraulic mining and the

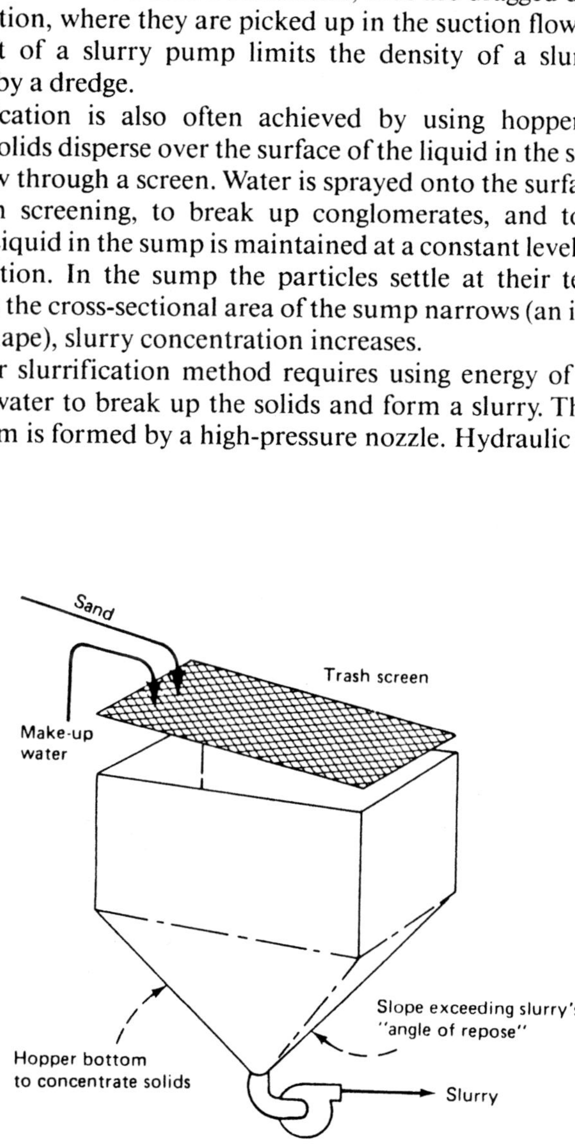

FIG. 13. Screen system for reslurrying coarse sand.

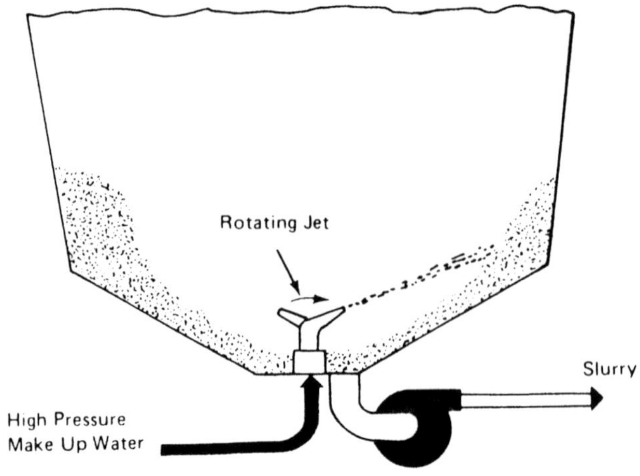

FIG. 14. High-pressure jet can reslurry caked solids.

Marconaflo jet system, slowly rotating jets at about 300 lb/in.^{2}, decompact and reslurry caked concentrate [17]. The reslurried concentrate then flows by gravity through grates to sumps and sump pumps (Fig. 14).

Chemical Treatments

Corrosion in a pipeline transporting a water–solids slurry is controlled by adding corrosion inhibitors [21]. Laboratory-scale tests determine the most suitable type and dosage of inhibitor to be added to a slurry. Various inhibitors may be used, depending on the corrosion mechanism taking place. For example, sodium dichromate will form a film on the pipe wall and protect the pipe; severe attack by dissolved oxygen can be controlled by adding an oxygen scavenger such as sodium sulfite [1]. In other systems, a high-slurry pH controls corrosion.

Normally, a solution containing inhibitor is added to the slurry by a metering pump. A corrosion meter monitors internal pipe corrosion, and slurry pH can be controlled by addition of slurried lime.

Shear-thinning materials, or dispersants, such as lignin compounds and polyphosphates, are examples of materials that are effective in breaking up agglomerates. Such chemicals are used in the cement industry. These materials reduce the electric charges on the edge of the particles so that they are dispersed in the fluid trapped in the agglomerates. This fluid is released and becomes part of the free fluid. The effect is to increase the dilution of the slurry. A surprising reduction in pressure losses is possible with low dosage additions [27].

Drag reducers suppress turbulence in the pipe. These materials are usually macromolecules of very large molecular weight (large-chain polymers). The exact mechanics of how they reduce turbulence is not understood. They are quite effective in astonishingly small quantities in nonslurry systems, but they are less effective in slurry systems because considerable suppression of turbulent energy dissipation already has taken place. In addition, most slurry systems operate near the laminar-turbulent transition point, where turbulent dissipation is not very high.

Controlling Product Quality

As a final control step to ensure that only specification slurry is committed to the main pipeline, a "test loop" should be included in the system. This test loop, made of the same-size-diameter pipe as used in the main slurry pipeline, should have a length 400 to 500 times the pipe diameter. All slurry must pass through this section before going into the main pipeline. The test loop on the Savage River system (iron concentrates transportation in Tasmania, Australia) is located between the feed tank and the mainline pumps.

Since the test loop is simply a short in-line version of the main pipeline, it is useful in monitoring changes in slurry hydraulics. Pressure drop (usually by differential pressure cell) is measured in the loop together with flow, temperature, and density. Any increase in the test-loop pressure drop above specified limits signifies a deleterious change in slurry properties and probably indicates that coarse, fast-settling particles are forming a bed in the loop. In such an event, the off-specification slurry is either recycled or sent to the dump pond.

Slurry Piping

The designer of slurry systems is always faced with the selection of the most economical pipe that will have an expected life consistent with the particular application. Pipe choice depends on pressure, temperature limitations, corrosiveness, and abrasiveness of the slurry. Available pipe includes:

Conventional unlined carbon steel
Rubber-lined carbon steel
Concrete-lined carbon steel
Special abrasion-resistant steel
Aluminum

The last four types are more expensive than conventional unlined carbon steel and would only be used with a very abrasive or corrosive slurry. As noted, general pipe abrasion becomes a consideration at velocities of 4 m/s. Also, pipe wear increases exponentially with velocity [15] above a certain threshold velocity. This exponential value has been reported to range from 2.1 to 2.9.

The amount of annual wear on a pipe is a function not only of the velocity but, of course, the abrasive character of the slurry. This must be evaluated on a case-by-case basis. However, some general relative values of abrasivity as measured by the Miller Number [15] on a scale of 0 to 1000 are given in Table 2.

Corrosiveness must also be evaluated on a case-by-case basis. If conventional unlined pipe is used, extra wall thickness must be added to compensate for the metal that will corrode away during the life of the system. As discussed earlier, it may also be desirable to add corrosion inhibitors.

After selecting the pipe, the designer must specify the pipe minimum yield strength and the wall thickness required to contain the expected pressures. This is done using standard formulas.

In areas subjected to prolonged subfreezing temperatures, slurry pipelines require the same degree of protection as any water-supply piping. Although a flowing solids–water slurry will usually generate sufficient heat from wall friction to prevent freezing, a prolonged system shutdown in subzero temperatures may produce frozen pipes. To minimize this risk, exposed piping in the immediate area of the plant can be protected by external insulation or heat tracing. For long-distance slurry pipelines, burying the pipe below the frost line is the accepted method of protecting against water–solids slurry freezing.

Before utilization, there is virtually always a storage step involved, because it is rarely practical to close-couple the pipeline transport system with the processing plant. Various storage facilities, such as tanks and ponds and associated agitators, and recovery dredges, must be included in the system. The utilization process may include facilities to change the concentration of the slurry—normally to increase the solids content via thickening, decanting, cycloning, and screening. More expensive dewatering methods may be needed, depending on the slurry and its ultimate use, such as vacuum filtration, centrifugation, or thermal drying. Further, the filtrate or effluent might need treatment before it returns to the environment. Additional processing steps can be used to enhance final product characteristics (e.g., heating to improve centrifugation or chemical additions to enhance thickening).

Slurry Utilization

Storage

Size and type of storage depend on the throughput of the preparation plant and slurry pipeline, the operating factor of the preparation plant, and the type of slurry. The storage facilities provide a buffer capacity between the preparation plant and pipeline and between the pipeline and terminal facilities.

For most long-distance slurry pipelines it is good practice, as a rule of thumb, to have a minimum storage capacity equivalent to 6 h of plant throughput. The time between receiving the slurry into storage and committing it to the pipeline is necessary for laboratory analyses, such as slurry percent solids and size distribution. If the slurry is outside the specification for pumping, the retention time in storage can be used to carry out corrective measures without shutting down the system.

TABLE 2

Material	Miller Number
Detergent	0
Coal A	11
Coal B	28
Fine magnetite	64
Hematite	260
Carborundum	1000

Agitated Storage. Slurries may be stored under agitation to maintain desired characteristics. The engineer must specify carefully what is needed in the way of performance. Two comments on this:

1. The variation of concentration of the slurry upon withdrawal from the tank must be held within specified limits. However, there is usually little interest in the actual concentration gradient in the tank. Situations have been observed where a water layer was present at the top of the tank when full, and upon emptying, the concentration remained relatively uniform.
2. The design should avoid large quantities of solids being deposited at the edge of the bank due to insufficient turbulence.

The Mohave coal tanks, with 22.7 m^3 in useful capacity per tank, are the largest agitated slurry tanks in existence, being 38 m in diameter by 26 m high. The agitator-drive horsepower is 500, the upper larger agitator blade is 10 m in diameter, and the lower agitator blade is 9 m in diameter. These blades rotate at a speed of 7 r/min.

The calculation of mixing power requirements is explained by Holland and Chapman [12].

Static Storage. For the bulk storing of large quantities of slurry, either nonagitated tanks or ponds are usually the most economical method. Usually, large-volume storage is necessary only in processes that operate very intermittently or in industries that require large, on-site feed stockpiles. The bulk storing of iron-ore concentrate at marine terminals for shipment using the Marconaflo concept is an example of nonagitated storage. Another example is the storing of a coal slurry in nonagitated, emergency storage ponds for use in the event of long-term stoppage of the pipeline process.

In operation, the slurry is discharged directly into the tank or pond, and the solids settle to their terminal concentrations. Depending on particle size of the solids, in-flow slurry concentration, and specific gravity of the solids, the settled slurry will remain semifluid for extended periods (provided that it is not compacted by induced vibration). It is possible to reclaim this settled slurry via suitably located drawoff points. Fast-settling, high-specific-gravity slurries settle into a compact bed, and shearing action is required for reslurrification. One method now being adopted is to use water jets to reslurry the solids to a pumpable consistency. Another method involves dredge recovery.

Thickening

The thickening of a slurry prior to pipeline pumping is necessary with some slurry pipelines. This is done for two reasons: (1) to raise the slurry density to a controllable level for pumping, and (2) to limit the amount of unnecessary water pumped into the slurry. But if the density is too high, the slurry will be too thick for economic pumping. Therefore, the economic trade-off must be analyzed. At the terminal end of the pipeline, the slurry is often thickened as the first step in the dewatering process, to reduce the size of the more expensive filtering equipment.

A *thickener* is a tank or basin, usually circular, in which solids settle by gravity. The primary objective of a thickener is to recover settled solids as a concentration slurry (clarifiers produce a clear liquor from a dilute suspension). Sedimentation in a thickener is by zone or line settling. The particles in the feed slurry cohere in a floc structure, and the solids settle as a consolidated mass, leaving a sharp boundary between the settling suspension and the supernatant liquor. Thickeners usually have a raking mechanism to convey the settled solids to the center-cone discharge points.

The degree to which a slurry can be thickened depends on the settling characteristics of the solids, the physical size of the thickener (area and depth), the feed rate, and the underflow concentration.

Bench-scale batch testing of the settling characteristics of various concentrations of slurry as measured in graduates are useful for sizing thickeners as large as 300 ft (90 m) in diameter. In these tests the subsidence of the slurry supernatant interface is measured as a function of time [6].

Decanters. Thickening by decanting is helpful in dewatering a slow-settling, low-specific-gravity slurry such as coal. The procedure is simply to fill a nonagitated tank or pond with slurry through a distribution pipe with evenly spaced spigots to ensure an even, uniform deposit of slurry and allow time for the solids to settle. If necessary, coagulating chemicals that are not harmful to the final process can be added to assist in clarification of the excess liquors. The clear liquid above the settled solids is then decanted.

Cyclones. Cyclones have been very successfully used in dewatering of heavy beach-sand minerals, such as rutile, which is a uniformly graded, round-shaped material. For very little expenditure of energy (only 10 to 15 lb/in.^{2} of head loss), a 25% slurry can be dewatered to a moist sand of 80 to 90% solids by weight.

Screens. Dewatering screens also thicken certain selected slurries. Solids, retained in the slurry, form the screen oversize and water is the screen undersize. This is especially effective on slurries that have a gapped size distribution.

Dewatering

The principal aim in dewatering a slurry is to recover the solids with a minimum moisture content. Clarification of the liquid phase of the slurry is usually of subsidiary importance. Table 3 shows a cross section of available dewatering equipment for solids recovery. In most cases a centrifuge or a filter is necessary.

Vacuum Filters. Continuous vacuum filters are popular devices for dewatering slurries. In most of these situations, primary thickening precedes the filter.

A typical vacuum filter incorporates a filtering cloth or surface attached to a moving frame. The frame moves through the thickened slurry, and the

TABLE 3 Classifying Dewatering Equipment

Major Function	Operation	General Equipment Classification	Equipment Subclassification
		Solid-bowl centrifuge	Cylinder-conical bowl (vertical, horizontal); solid/screen bowl combination
	Continuous	Centrifugal filter	Conical screen (helix conveyor, oscillator); cylinder screen (pusher, conveyor)
		Vacuum filter	Rotary belt/drum, horizontal belt, horizontal pan
		Other	Various wet screens, cyclones, special filters and centrifuges, settling tanks
		Centrifugal filter	Vertical perforated basket, constant speed; horizontal basket, variable speed
Recover solids	Batch automatic	Screen-basket centrifuge	Vertical basket; constant speed; horizontal basket, variable speed
	Batch	Pressure leaf filter	Plate and frame; pressure leaf, vertical/horizontal leaf
		Settling tank	

solids are caked onto the filter through the action of a vacuum. The frame continues its journey to a point where the cake is removed by mechanical and pneumatic methods. This cake goes onto a conveyor and the filtrate flows to a clariflocculator. The filter frame then moves back into the thickened slurry to repeat the cycle.

Flood et al. provide a practical guide on equipment selection and a good explanation of the filtration [7].

Centrifuges. Continuous centrifuges are often used for dewatering slurries. The solid-bowl centrifuge consists of a rotating bowl and a rotating-screw-conveyor section that revolves concentrically within the bowl. The bowl "conveyor" conforms to the inside of the bowl. Slurry enters via a stationary feedpipe concentrically located within the hollow conveyor shaft. This feedpipe extends inward to a rotating chamber located within the screw conveyor. Here the slurry is accelerated and moves radially outward to the bowl. The liquid portion of the feed tends to form a cylinder of revolution within the bowl. Liquid depth is controlled by adjustable weir plates at the outboard end of the cylindrical section of the bowl. A rotating-screw conveyor keeps the solids moving up the conical slope. At the discharge ports, the solids fall into a discharge hopper.

Tests have shown that screen-bowl centrifuges (Fig. 15) can remove more moisture than solid-bowl centrifuges for pipeline slurry. Surface moistures can be as low as 12 to 15% with screen bowls. The effluent of the screen bowl after being treated in a clariflocculator can be burned directly in the furnace, as was done at Mohave, or can be further treated in filter presses or solid-bowl centrifuges before being blended with the screen-bowl product. This product

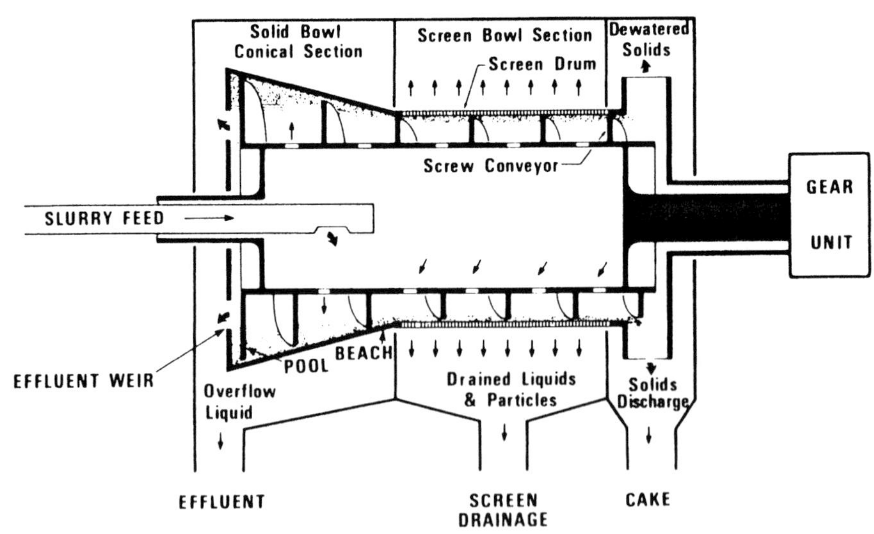

FIG. 15. Principles of screen-bowl centrifuge.

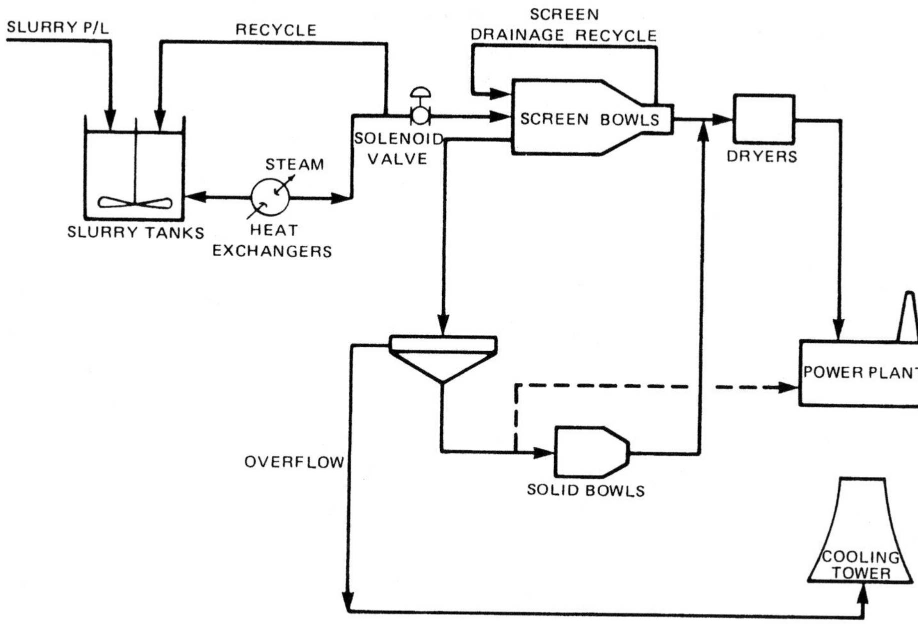

FIG. 16. Typical system for dewatering coal slurry.

can be further dried in fluid bed dryers, rotary drum dryers, flash dryers, or other devices. Figure 16 represents a typical dewatering flow sheet. Heating of the slurry prior to centrifuging has been found to improve moisture removal.

Mechanical Considerations

Slurries are abrasive by their very nature. They "sandblast" on impingement at high velocity; they are a "grinding compound" between moving mechanical parts; they are a "cutting tool" when throttled through a restriction; and they are "sandpaper" when dragged along the bottom of a pipe. The abrasive nature of slurries is a major consideration in design and selection of equipment for the slurry system. Selection of major mechanical components of the system—pipe, valves and fittings, and instrumentation—is discussed below.

Slurry Pumps

Slurry pumping is roughly split between centrifugal and positive-displacement pumps, depending on the system pressure requirements. Various lock-hopper and surge-leg pumping systems have been constructed to isolate the primary pump from the slurry.

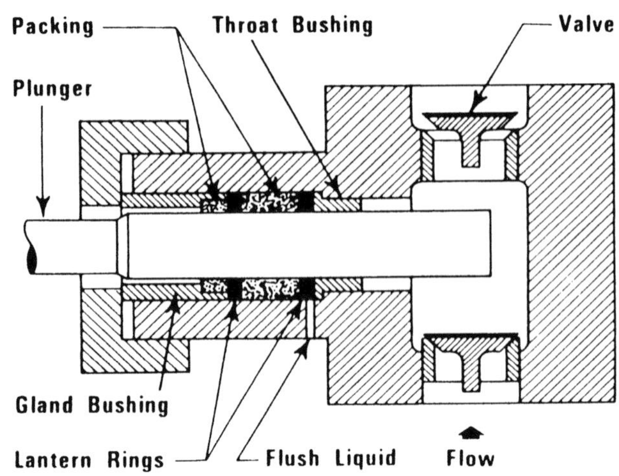

FIG. 17. Plunger pumps flushing arrangements are useful for very abrasive slurries.

Centrifugal Pumps

These are limited in casing pressure and efficiency, owing to the nature of the material they are designed to handle. Their casing-pressure capacity is limited by the vertical-split-casing design necessary for easy replacement of the impeller and of the wear linings of the casing. Impeller tip speed is generally limited to 1350 m/s to minimize wear of the volute. Multiple pumps in series can develop final-stage discharge pressures up to about 600 lb/in.^{2}.

Positive-Displacement Pumps. These are used where pumping pressures above 600 lb/in.^{2} are required. For very abrasive slurries, the plunger-type pump is used, with a flushing arrangement injecting clear liquid to keep solids away from the plunger packing. This is illustrated in the fluid-end schematic diagram of Fig. 17.

Figure 18 shows a schematic diagram for a double-acting piston pump. Piston pumps have application in less-abrasive service, such as for coal slurry; they have the advantage of displacing slurry on both the in and out strokes, but depend on sealing the full differential pressure across the piston. Positive-displacement pumps up to 1700 hp are now in service at flows about 2000 gal/min. Units in excess of 3000 hp are being considered, but large-volume and high-pressure systems will require multiple units in parallel.

A pulsation-dampening system is required when positive-displacement pumps are used. The primary elements of this system should be proper piping configuration and restraint, and gas-filled pulsation dampeners. Short, straight, suction lines are preferable; longer ones call for a centrifugal pump to charge the positive-displacement pumps. Pulsation dampeners should be fitted as close to the pump as possible on both the suction and discharge side.

Multiple-pump installations will require detailed vibration analysis of the piping system. Instrument connections to positive-displacement-pump piping, such as pressure gauges, are quite susceptible to vibration damage to the instrument, or failure at the line connection.

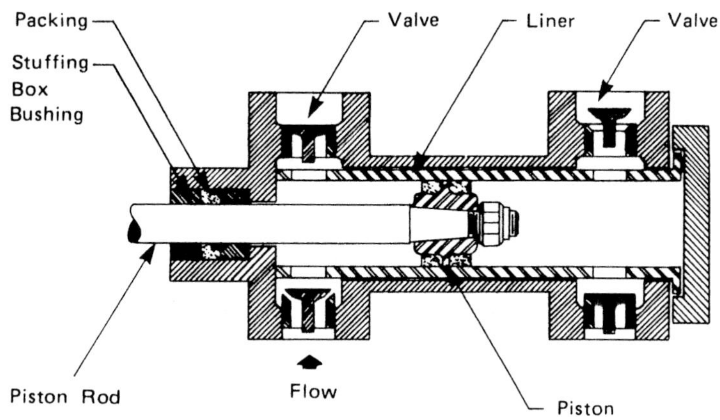

FIG. 18. Piston pumps are useful for moderately abrasive slurries such as coal.

Other Pumps. Various pump designs have been developed to isolate the abrasive slurry from the pumping equipment.

1. The "lockhopper" system shown in Fig. 19 is designed to allow the use of conventional, multistage water pumps, and to let them develop high heads while the slurry is switched in and out of the lockhopper by sequenced valves. This system is now used in hydraulic hoisting from

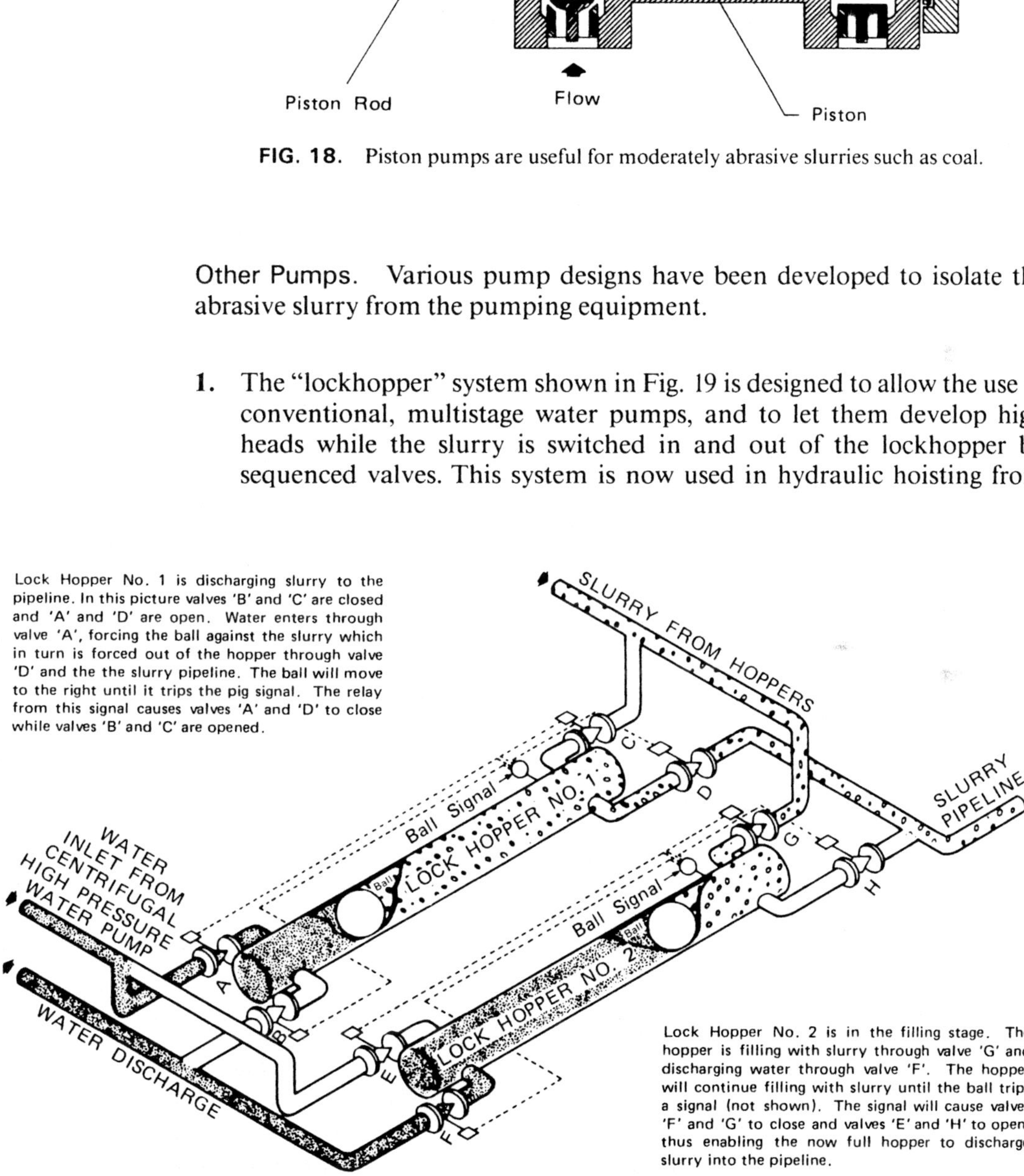

FIG. 19. Lockhopper system is designed to allow use of centrifugal water pumps by isolating the abrasive slurry.

underground mines, where the pump is mounted on the surface and the lockhopper at the working level.

2. The "surge leg" pump is a special case of the lockhopper concept, where a positive-displacement pump is fitted with a chamber (surge leg) full of clear liquid between the piston and pump valves. The isolating fluid may be oil or water, and makeup fluid must be periodically added. This type of pump has been installed in some very abrasive services.
3. Diaphragm pumps have been used in low-volume, low-head systems; however, they are emerging as competition to the plungers pumps. They can now deliver high volumes at high discharge heads and have superior expendable parts life.
4. The "advancing cavity," Moyno pump is ideal for slurries at moderate flow and pressures where steady delivery is required. It is well suited to very thick slurries.

Piping System

When laying out the slurry piping system, the designer must consider:

Flushing or draining the piping on normal or emergency shutdown
Replacement of wear points: near pump discharge, sharp bends, at and downstream of restrictions
Rotation of straight horizontal sections (very coarse slurries)
Access for unplugging
Elimination of dead spaces at tees and tappings

These considerations apply much more strongly to heterogeneous slurries.

Corrosion may be a consideration in pipe-schedule selection, but this must be evaluated case by case. The designer should realize that the internal pipe-wall protection often gained from corrosion products may be eroded away in a slurry system, resulting in a much higher metal loss than would be expected in a liquid pumping system having the same chemical properties.

The abrasive slurry effects discussed under slurry pumps above are also operating in the pipeline system, although to a lesser degree (Fig. 20). General pipe abrasion becomes a consideration above about 2 m/s and is a major consideration at velocities about 4 m/s. Rubber or concrete lining may be desirable for these high velocities. Extra-long-radius elbows are often used to minimize wear at changes of direction.

Another consideration in piping design is the slope limitation. Upon shutdown with slurry in the pipeline, solids will settle and, where slopes are too steep, will slough to the valleys, thereby plugging the pipeline (Fig. 21).

Valves, like pumps, must be designed for abrasive service and with consideration to sedimentation and plugging. They preferably provide a full opening, should not depend on machined-metal surfaces for closure, and do not have dead pockets that can fill with solids and restrict operation. Valves

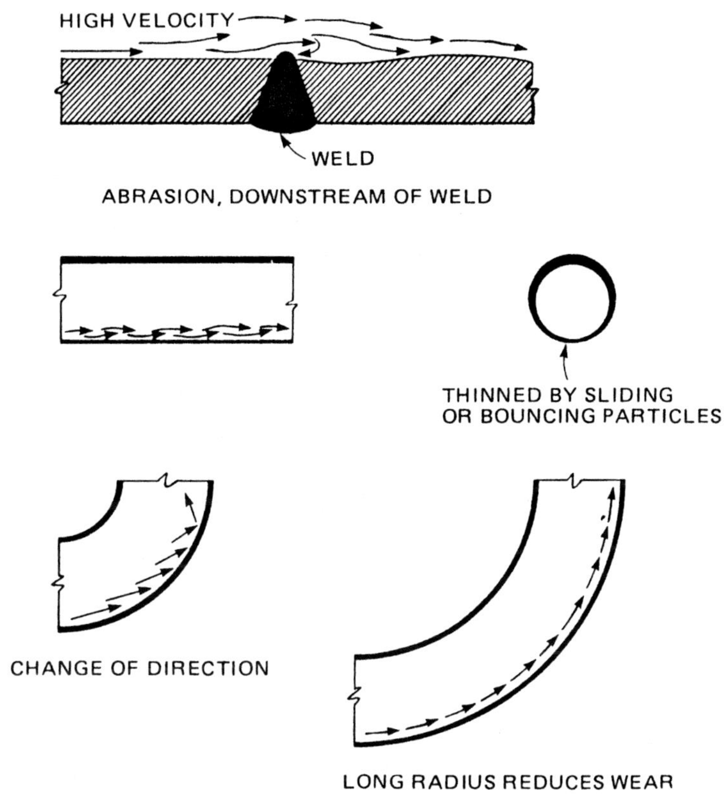

FIG. 20. Abrasive wear can have various causes.

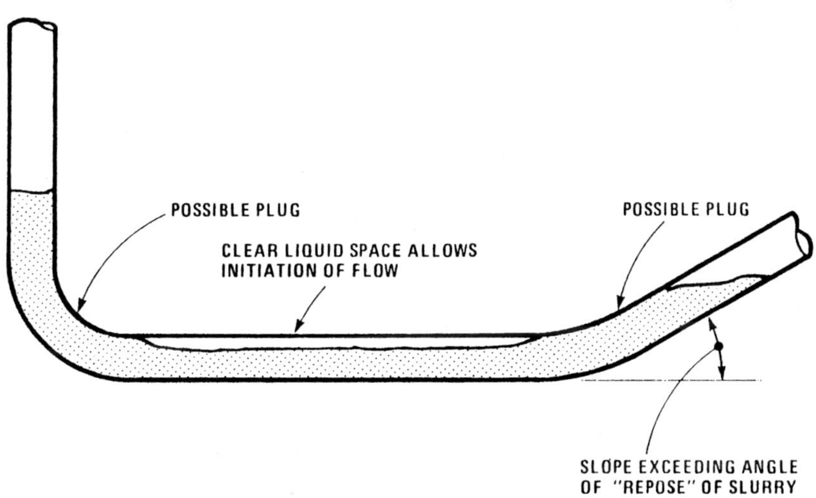

FIG. 21. Vertical or inclined sections may undergo plugging when shutting down the system.

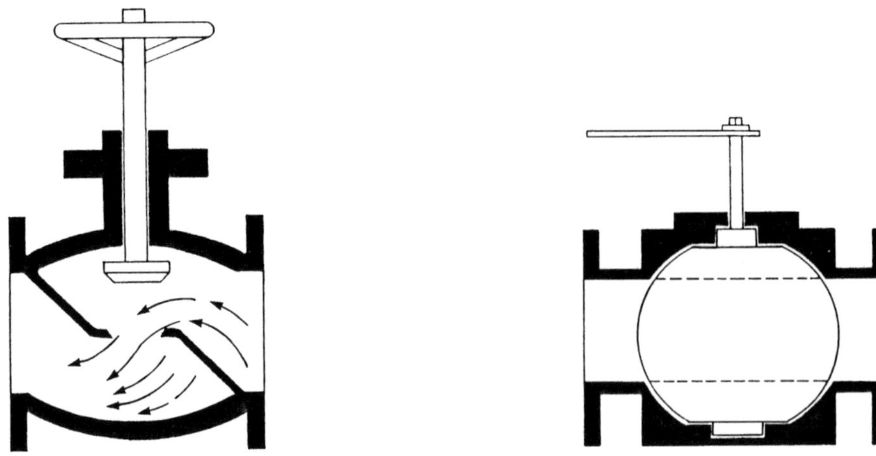

FIG. 22. Valves: impingement on seat and walls will cause excessive wear (left); the full-opening flow arrangement (right) minimizes wear.

with restricted or circuitous openings create abrasion downstream (Fig. 22). Many suitable low-pressure valves are available, often with rubber-to-rubber or metal-to-rubber sealing. To date, high-pressure/large-diameter applications have used lubricated plug or ball valves, neither of which is ideally suited for slurry service.

In some cases plug valves are specified with Stellite facing on the plug and internal body to reduce wear. Most ball valves must be fitted with flush and drain connections so that solids can be removed from the body after operation.

Instrumentation

The presence of solids in a slurry system complicates measurement of the system variables, because many conventional primary measuring elements will be worn or plugged by the solids. Wear may affect the measuring elements themselves (such as orifices or turbine blades) or the system (i.e., due to turbulence created by the measuring element).

The need to avoid plugging of the sensing element or impulse lines is a major consideration in selecting measuring elements for slurry systems. It is a particularly troublesome problem at pressure tappings, and can disable any element with small clearances or static zoens.

Segregation of a slurry in the measuring element must also be considered. A radiation density meter or magnetic flow meter will not read correctly in a horizontal pipe unless the suspension is ideally homogeneous. Fast-settling slurries will foil instruments, requiring a side stream if flows are not maintained above critical velocity.

An excellent review of instrumentation for slurry systems is given by Liptak [16]. The major sensing elements are discussed briefly below:

1. Magnetic flow meters have proved to be adequate, although expensive, primary flow-measuring elements; their chief advantage is that they create no flow restriction.
2. Where positive-displacement pumps are used, their speed provides an accurate flow-rate indication.
3. Radiation-density meters provide a slurry-concentration indication, again without flow restriction, but they require frequent calibration. Continuous weighing devices such as the Halliburton densometer have been used, although they present the problem of taking a side stream and returning it to the main stream.
4. The difficulty in pressure measurement is in keeping the pressure taps from plugging. This is best overcome by using a diaphragm close-mounted to the pipe to separate the slurry from the pressure sensor. In some cases a continuous backflow of pressure-sensing lines may be required.
5. Transmitters or gauges should not be mounted on the piping around positive-displacement pumps, because they will often quickly fail from the vibration.
6. Pressure devices can be supported separately with a capillary connecting element and diaphragm.

Economic Aspects

Having followed the general design procedures, the design engineer is usually faced with several choices that can only be decided on the basis of economics. Often, after considering critical velocity limitations, it will be found that more than one pipe diameter is technically feasible. To select the correct diameter, the alternatives must be compared both on the basis of first cost and annual operating costs. Because operating costs are incurred over the life of the project, they must be discounted in order to realistically weigh them against capital costs. For very short distance lines, however, operating costs (power and supplies) may be quite low, which would allow the diameter decision to be made on the basis of capital costs alone.

For quick evaluations of alternatives, some very simple procedures can be used. These should be considered as initial screening techniques only. If the economic impact of the decision is large, such as would be the case in longer-distance pipelines, more sophisticated techniques must be employed.

The concept of capital charges is necessary for the analysis. Capital charges are simply the average annual payments necessary to support a capital investment. They include depreciation, interest, property taxes, profit, and income taxes. As a percent of initial investment, they usually range from

15% for a utility to 25% or more for more speculative ventures. A good percentage for preliminary evaluations is 20%.

The significant operating costs are usually power (e.g., electricity or fuel) and supplies such as pump parts. Normally, the operating labor for the various alternatives would be the same and therefore could be neglected. Capital expenses are largely comprised of pipeline-steel and pump costs (including installation labor). For preliminary analysis of alternatives, these two cost components will suffice; instrumentation costs usually do not have to be estimated until later.

In addition to the pipe-diameter decision, the type of pump selected often must also be based on economics. For example, in a particular application, either a series of centrifugal pumps or positive-displacement pumps in parallel might be technically feasible. The fact that positive-displacement pumps have a higher efficiency than centrifugal pumps should not be overlooked (85 versus 60% is typical). In this case the higher initial cost of positive-displacement pumps should be weighed against higher power costs for centrifugal ones. Maintenance costs should also be included in the evaluation.

In certain very abrasive slurries there may be a choice of installing a more expensive, abrasion-resistant or lined pipe versus a lower-cost conventional pipe that would have to be replaced during the life of the project. Assuming that replacement is tolerable, the familiar "present-worth" analysis must be made at several appropriate interest rates to select the economic alternative.

Slurry Pipeline Applications

Pipeline applications may be considered in two distinct classifications: short-distance in-plant transport and long-distance transport. The latter, for purposes of this discussion, will involve systems whose major purpose is to move the slurry to another geographical area that is closer to the point of end use, usually distances from 20 to several hundred kilometers.

Short-distance in-plant or intraplant systems, on the other hand, are primarily associated with moving slurry from one point in a process to another, and usually are in the range of a few meters to several hundred meters. Pipelines for tailing and dredging materials will also be considered as short-distance, although in certain cases they can be many miles in length.

Short-Distance, In-Plant Applications

Short-distance applications are as varied and numerous as types of chemical processes. The types of slurry pumped number in the hundreds [18] and include various foodstuffs, chemicals, chemical wastes, industrial wastes, sewage, and mine and quarry products.

One of the most interesting specialized applications is the transport of a thorium–uranium slurry for nuclear power generation; a great deal of basic theoretical work in slurry transport was done at Oak Ridge for this application.

Virtually every mineral-processing installation has a tailings pipeline to transport gangue and waste from the mineral-disposal site. On the Mesabi range in Minnesota, for example, 40,000 tons of tailings are carried daily through nine 305-mm-diameter pipelines for distances up to 12,200 m. On the island of Bougainville in the Solomons, Bougainville Copper Pty. Ltd.'s new copper facilities will include an extendable 1.22 to 4.57-m-diameter tailing pipeline to move more than 6800 m^3/h.

Design of tailing pipelines involves nearly all the considerations of longer pipelines, and is further complicated by the variability in screen analysis, concentration, and tonnage throughput. The tailings line must be able to handle all these variables. Since the processing plants are usually modular with several lines, consideration must be given to the operating mode when one or more process lines are interrupted. Owing to critical velocity considerations, the answer in many instances is a multiple-line system.

Dredging pipelines are normally associated with the deepening of harbor channels, but could also involve mining. Probably the best known hydraulic mining operation is in the Florida phosphate fields, where phosphate-rock systems pumping up to 5700 m^3/h are in use. Pipelines with diameters up to 508 mm and lengths up to 10 km exist.

Long-Distance Applications

Although materials amenable to long-distance slurry pipeline transport are fewer than for short-distance in-plant applications, the list is still quite large. Materials presently transported by long-distance slurry pipelines include limestone and other cement raw materials, iron concentrate, coal, gilsonite, salt in brine, phosphate rock, kaolin, copper concentrate, sewage sludge, uranium-bearing slimes, sugar cane, and wood pulp. Other materials for which the technical feasibility of long-distance slurry transport has been established include potash, lead-zinc concentrates, sulfur, laterite (nickel ore), pyrite, coke wood chips, and solid wastes. Table 4 summarizes some selected commercial slurries in operation or planned. Several of these are described briefly below.

Black Mesa Coal Pipeline

The longest slurry pipeline in operation to date is the 440-km Black Mesa coal pipeline, which traverses the state of Arizona and was commissioned in the fall of 1970. This 457-mm-diameter pipeline can transport over 5½ million tons of coal annually from the mine site in the Navajo–Hopi Indian reservation to the Mohave power plant site on the Nevada side of the Colorado River, south of Davis Dam. The coal will supply all the fuel requirements for the two 750-MW generating units at Mohave.

TABLE 4 Selected Commercial Slurry Pipelines

	Length (mi)	Pipe Size (in.)	Capacity (tons/yr × 10^6)	Operation (year)
Coal:				
Consolidation	108	10	1.3	1957
Black Mesa	273	18	4.8	1970
ETSI	1378	38	25	198-
Alton	180	24	10	198-
Iron concentrate:				
Savage River	53	9	2.25	1967
Waipipi (iron sands)	6	8 and 12	1.0	1971
Peña Colorada	28	8	1.8	1974
Las Truchas	17	10	1.5	1976
Sierra Grande	20	8	2.1	1978
Samarco	253	20	12	1977
Copper concentrate:				
Bougainville	17	6	1.0	1972
West Irian	69	4	0.3	1972
Pinto Valley	11	4	0.4	1974
Limestone:				
Rugby	57	10	1.7	1964
Calaveras	17	7	1.5	1971
Phosphate concentrate:				
Valep	80	8	2.0	1979

The >50-mm coal provided at the mine site by Peabody Coal Co. is reduced to 19 mm by impactors and then ground to pass 14 mesh in three parallel rod-mill lines. From the rod mills the slurry is stored in three agitated slurry tanks that feed the initial pump station. Three additional pump stations are required along the route. Pumps are electric-powered 1700-hp Wilson-Snyder double-acting duplex piston pumps, the largest of this type in existence. Each station has three of these pumps in parallel, except Station Two, which has four. The pipeline system is owned and operated by Black Mesa Pipeline Inc., a subsidiary of the Southern Pacific Transportation Company.

At the power plant the coal discharges into one of three 24-h holding tanks and is then pumped to centrifuges for dewatering. There are 20 centrifuges for each 750-MW generating unit. The dewatered coal, at about 25% total moisture, then goes through a mill and is blown into the boiler with heated air [22]. Coffey et al. provide a more extensive description of the system [4].

Savage River Pipeline: Mountains and Valleys

The first long-distance iron-concentrate pipeline in the world is the 85-km 244-mm-diameter slurry pipeline operated by Pickands Mather & Co. This

operate by batching for the immediate future, as it has more capacity than the concentrator. The profile shown in Fig. 23 reflects the rugged terrain the pipeline has to cross.

The pipeline was designed in conformance with ANSI Code B31.4. The pipe is API 5L × 60 with wall thicknesses of 8.7 to 21 mm and averaging over 13 mm. Most o f the pipeline is 508 mm in diameter; however, a 40-km sector of 457 mm was installed on the downslope leading to the terminal to help dissipate some of the potential energy.

The most challenging of the many engineering feats on the SAMARCO slurry pipeline was resolution of the slack flow problem [5]. As with many other concentrate slurry pipelines, the product must be transported from the mine to a terminal on the coast for export; the difference in elevation between the mine site and the terminal causes release of energies exceeding those required to overcome friction losses in normal "packed-line" conditions. With no restriction in the line, the excess energy dissipates itself through the formation of slack or open-channel flow in which velocities are an order of magnitude higher than normal packed-line flow velocities. If slack flow conditions were allowed to continue with an abrasive slurry, the pipeline would rapidly wear out, because wear is in an exponential relation with flow velocity. An orifice station is used to apply variable backpressure, required to eliminate slack flow.

Symbols

C_d	drag coefficient (dimensionless)
C_V	volume concentration of solids
D	pipe inside diameter (in.)
f, f_w	Fanning friction factor
g	gravitational constant (ft/s^2)
Δh	frictional head loss (ft of fluid flowing/ 100 ft)
L	length (ft)
$(N_{Fr})_c$	critical Froude number (dimensionless)
N_{He}	Hedstrom number (dimensionless)
N_{Re}	Reynolds number (dimensionless)
$(N_{Re})_c$	critical Reynolds number, (dimensionless)
ΔP	frictional head loss (lb/in.^{2}/100 ft)
P_L	plasticity number (dimensionless)
V	velocity (ft/s)
V_d	deposition critical velocity (ft/s)
V_t	viscous-transition critical velocity (ft/s)
η, μ	coefficient of rigidity, or viscosity for Newtonian slurries (cP)
ρ, ρ_w	density (lb/ft^3)
τ_0	yield stress (dyn/cm^2)

A portion of this article appeared in *Chemical Engineering Magazine*, June 28, 1971, and is reprinted with the permission of the editor.

References for Further Reading

Ellis, J. L., and Bacchetti, P., *Pipeline Transport of Liquid Coal*, Lignite Symposium, Bismarck, North Dakota, May 1971.

Hanks, R. W., and Pratt, D.R., "On the Flow of Bingham Plastic Slurries in Pipes and between Parallel Plates," *Soc. Pet. Eng. J.*, pp. 342–345 (December 1967).

Hedstrom, B. O. A., "Flow of Plastic Materials in Pipes," *Ind. Eng. Chem.*, *44*, 651 (1952).

Job, A. L., *Transport of Solids in Pipeline, with Special Reference to Mineral Ores, Concentrates, and Unconsolidated Deposits*, Literature Survey, Department of Energy, Mines and Resources of Canada, Information Circular 230, October 1969.

Kenny, J. P., et al., *FWPCA Waste Management Study*, Vol. 3, *Technical Aspects of Pipelining of Waste Materials*, Bechtel Corp., September 1969.

Perry, J. H., *Chemical Engineers' Handbook*, 4th ed., McGraw-Hill, New York, 1963, p. 5-59.

Rotary Grinding Mills—Selection and Capacity Determination, Allis-Chalmers, PM3.2, August 30, 1963.

Thomas, D. G., "Non-Newtonian Suspension, Part I," *Ind. Eng. Chem.*, *55*(11), 18–29 (1963).

Thomas, D. G., "Transport Characteristics of Suspension. VIII. A Note on the Viscosity of Newtonian Suspensions of Uniform Spherical Particles," *J. Colloid Sci.*, *20*(3), 267–277 (1965).

Walker, G. H., and Wasp, E. J., *Experience and Prospects in Economic Transportation of Coal in Pipelines*, 6th World Power Conference, Melbourne, Australia, October 1962.

Wasp, E. J., et al., *Depositions Velocities, Transition Velocities and Spatial Distribution of Solids in Slurry Pipelines*, First International Conference on Hydraulic Transport of Solids in Pipe, Coventry, England, September 1970.

Wasp, E. J., et al., *Hetero-homogeneous Solids–Liquid Flow in the Turbulent Regime*, ASCE, International Symposium on Solid–Liquid Flow, University of Pennsylvania, Philadelphia, March 1968.

Wasp, E. J., Thompson, T. L., and Aude, T. C., *Slurry Pipeline Economics and Application*, First International Conference on Hydraulic Transport of Solids in Pipes, Coventry, England, September 1970.

References

1. Bomberger, D. R., "Hexavalent Chromium Reduces Corrosion in Coal-Water Slurry Pipeline," *Mater. Prot.*, January 1965.
2. Bond, F. C., *Crushing and Grinding Calculations, Parts I and II*, Allis-Chalmers (rev. January 1961); Reprint from *British Chemical Engineering*.
3. Bond, F. C., *Metal Wear in Crushing and Grinding*, 56th Annual Meeting, AIChE, Houston, Texas, December 1–5, 1963.
4. Coffey, R. C., Lyons, H. G., and Oakes, A. C., *Mohave Generating Station, Design Features*, American Power Conference, Chicago, April 1969.
5. Derammelaere, R. H., and Chapman, J. P., "Slack Flow in the World's Largest Iron Concentrate Slurry Pipeline," in *Proc. 4th Int. Tech. Conf. Slurry Transp.*

6. Fitch, B., "Batch Tests Predict Thickener Performance," *Chem. Eng.*, August 23, 1971.
7. Flood, J. E., Porter, H. F., and Rennie, F. W., "Filtration Practice Today, Centrifugation Equipment," *Chem. Eng.*, June 20, 1966.
8. Govier, G. W., and Azia, K., *The Flow of Complex Mixtures in Pipe*, Van Nostrand Reinhold, New York, 1972.
9. Hanks, R. W., and Pratt, D. R., "On the Flow of Bingham Plastic Slurries in Pipe and between Parallel Plates," *Soc. Pet. Eng. J.*, pp. 342–345 (December 1967).
10. Hedstrom, B. O. A., "Flow of Plastic Materials in Pipes," *Ind. Eng. Chem.*, *44*, 651–656 (1952).
11. Hill, R. A., Jennings, M. E., and Derammelaere, R. H., "SAMARCO Iron Ore Slurry Pipeline," in *Proc. 3rd Int. Tech. Conf. Slurry Transp.*
12. Holland, F., and Chapman, F., *Liquid Mixing and Processing in Stirred Tanks*, Reinhold, New York, 1966.
13. Kenny, J. P., et al., *FWPCA Waste Management Study*, Vol. 3 *Technical Aspects of Pipelining of Waste Materials*, Bechtel Corp., September, 1969.
14. Levene, H. D., "The Longest, Largest Coal Slurry Pipeline Ever Built," *Coal Min. Process.*, February 1971.
15. Link, J. M., and Tuason, C. E., *Pipe Wear in Hydraulic Transport of Solids*, American Mining Congress, Las Vegas, Nevada, October 14, 1971.
16. Liptak, B. G., "Instrumentation for Slurries and Viscous Materials," *Chem. Eng.*, p. 133 (January 30, 1967); p. 151 (February 13, 1967).
17. "Macronaflo—The System and the Concept," *Eng. Min. J.*, May 1970.
18. Maurer, G. W. "Pipelining Can Transport Your Bulk Solids," *Mater. Handl. Eng.*, p. 56 (March 1966).
19. McDermott, W. F., et al., *Savage River Mines—The World's First Long Distance Iron Ore Slurry Pipeline*, Society of Mining Engineers Fall Meeting, Preprint 68-b-364, September 1968.
20. Perry, J. H., *Chemical Engineers' Handbook*, 4th ed., McGraw-Hill, New York, 1963, p. 3-59
21. Swan, J. P., et al., "Corrosion Control Achieved in Coal Slurry Pipeline," *Mater. Prot.*, September 1963.
22. Taylor, D. M., "Liquefied Coal Piped Directly into Boiler," *Pipe Line Ind.*, December 1961.
23. Thomas, D. G., "Transport Characteristics of Suspension. VIII. A Note on the Viscosity of Newtonian Suspensions of Uniform Spherical Particles," *J. Colloid Sci.*, *20*, (3), 267–277 (1965).
23a. Thomas, D. G.' "Non-Newtonian Suspension, Part I," *Ind. Eng. Chem.*, *55*(11), 18–29 (November 1963).
24. Wasp, E. J., et al., *Deposition Velocities, Transition Velocities and Spatial Distribution of Solids in Slurry Pipelines*, First International Conference on the Hydraulic Transport of Solids in Pipe, Coventry, England, September 1970.
25. Wasp, E. J., et al., *Hetero-Homogeneous Solids–Liquid Flow in the Turbulent Regime*, ASCE, International Symposium on Solid–Liquid Flow, University of Pennsylvania, March 1968.
26. Wasp, E. J., Kenny, J. P., and Gandhi, R. L., *Slurry Pipeline Transportation*, TransTech Publication, 1977.
27. Zandi, I., "Decreased Head Losses in Raw Water Conducts," *J. Am. Water Works Assoc.*, February 1967.

R. H. DERAMMELAERE
E. J. WASP

Two-Phase Design

Here is the simplest and least frustrating method for sizing pipes when two-phase flow exists in a pipeline. Also presented are useful tips that exploit the inherent flexibility in the distribution of pressure losses in a piping system so that the designer can obtain reasonable sizes for pipes and components.

Two-phase-flow theories and experiments have a threefold significance for the process-piping designer. It has been shown that:

1. If the vapor content of a liquid line increases, the friction loss is greater than the single-phase liquid pressure loss, and is greater than the pressure loss calculated with the average density.
2. For a given vapor–liquid ratio and associated physical properties, a characteristic flow pattern develops.
3. Between the various flow patterns, unit pressure losses can differ when comparing borderline cases.

Piping design for two-phase flow has been investigated by a great number of researchers through rational and empirical steps [1, 2]. And limitations, generalizations and simplifications have been introduced for providing practical methods of design.

We will assume here that two-phase flow is isothermal, turbulent in both the liquid and vapor phases, and steady (liquid and vapor move with the same velocity), and that pressure loss is not more than 10% of the absolute downstream pressure.

An often-asked question is: How accurate are two-phase-flow calculations? If actual pipelines closely resemble the experimental conditions, deviations are small. The application of correlations for two-phase flow to process-piping design is arbitrary. Experiments are usually done with small-diameter, straight, and relatively short pieces of horizontal or vertical pipe. Under laboratory conditions, flow patterns are kept constant and flow conditions consistent. However, most process piping probably has changing flow patterns in various segments of the line because of the three-dimensional pipe configurations in which one finds horizontal and vertical runs, elevation changes, offsets, branch connections, manifolds, pipe components, reducers, and other restrictions. Sizable deviations can be expected in prediction of friction loss compared to actually measured values. Because of this, pipe runs for two-phase flow should be short and simple.

We present here the simplest, and in its practical application the least frustrating, method of design from among the many available ones. Let us compute the resistance for two-phase flow in two main steps. These are:

Select a possible flow pattern by calculating the coordinates of a flow-region chart.

Determine unit pressure losses by calculating only the vapor-phase unit loss, corrected by an applicable correlation for two-phase flow.

Two-Phase-Flow Regions: Baker Parameters

We select two-phase-flow patterns [3] from Fig. 1. The borders of the various flow patterns in Fig. 1 are shown as lines. In reality, these borders have rather broad transition zones [4].

We can establish a particular flow region from the Baker parameters, B_x

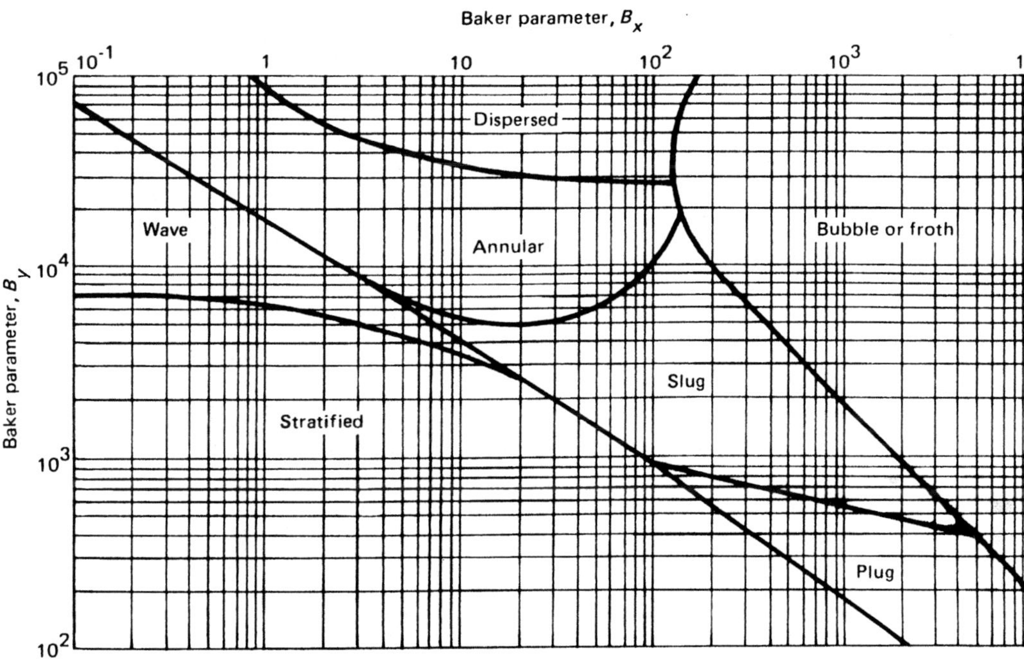

Two-Phase Flow Correlations						
Dispersed	Bubble	Slug	Stratified	Wave	Plug	Annular
Use Fig. 3 and Eq. (3)	$\phi = \dfrac{14.2\, X^{0.75}}{(W_l/A)^{0.1}}$	$\phi = \dfrac{1{,}190\, X^{0.815}}{(W_l/A)^{0.-5}}$ Avoid slug flow	$\phi = \dfrac{15{,}400\, X}{(W_l/A)^{0.3}}$ Horizontal pipe	Use Fig. 5 and Eq. (9) and (10) Horizontal pipe	$\phi = \dfrac{27{,}315\, X^{0.855}}{(W_l/A)^{0.17}}$	$\phi = aX^b$ $a = 4.8 - 0.3125\,d$ $b = 0.343 - 0.021\,d$ d = I.D. of pipe, in For pipe 12-in and over, use $d = 10$.

Courtesy: Mr. Ovid Baker and The Oil and Gas Journal.

FIG. 1. Baker parameters determine the type of two-phase flow and the appropriate two-phase-flow correlation sets unit loss.

and B_y. From data supplied or usually available to the piping designer, the Baker parameters can be expressed as:

$$B_y = 2.16 W_v / A\sqrt{\rho_l \rho_v} \tag{1}$$

B_y depends on the vapor-phase flow rate, vapor and liquid densities, and pipe size. The practical significance of pipe size is that by changing pipe diameters, the type of flow might also be changed, which in turn changes friction losses in the pipe.

$$B_x = 531\left(\frac{W_i}{W_v}\right)\left(\frac{\sqrt{\rho_l \rho_v}}{\rho_l^{2/3}}\right)\left(\frac{\mu_l^{1/3}}{\sigma_l}\right) \tag{2}$$

In Eq. (2), we can substitute the ratio of percent liquid to percent vapor for W_l/W_v, and

$$\sqrt{\rho_l \rho_v}/\rho_l^{2/3} = (\rho_v)^{0.5}/(\rho_l)^{0.166}$$

B_x depends on the weight–flow ratio and the physical properties of the liquid and vapor phases. Once calculated, B_x does not change with alternative pipe diameters. The position of the B_x line in Fig. 1 changes only if the liquid–vapor proportion changes and, to a much lesser extent, if the physical properties of the concurrently flowing liquid and vapor change. This can occur in long pipelines where relatively high friction losses reduce the pressure. Consequently, the vapor content of the mixture increases with a corresponding decrease in vapor density. The B_x line will shift somewhat to the left.

In Eq. (2), σ_l is the liquid-phase surface tension [7, 8]. For the surface tension of water at various temperatures, see the chart by Yaws and Setty [9]. For the surface tension of paraffinic hydrocarbons and mixtures, consult the *Engineering Data Book* [10].

The intersection of B_x and B_y in Fig. 1 determines the flow region for the calculated liquid–vapor ratio and the physical properties of the liquid and vapor. With increasing vapor content, the point of intersection moves up and to the left.

Unit Losses for Two-Phase Flow

The calculations of unit losses for vapor–liquid mixtures are based on the method of Lockhart and Martinelli [5]. Only the essential relationships are repeated here. We will use these with the customary data for practical piping design. The general equation is:

$$\Delta p_{100(\text{two-phase})} = \Delta p_{100(\text{vapor})}\phi^2 \tag{3}$$

We calculate the pressure drop of the vapor phase by assuming that only vapor flows in the pipeline. We then correct the calculated vapor-phase unit loss with the correlations shown in Fig. 1. Most of these correlations result from experiments with large-scale, horizontal industrial piping [4].

The forms of the correlations (in Fig. 1) are identical:

$$\phi = aX^b \tag{4}$$

where a includes the vapor-phase flow rate and the pipe's cross-section, and b is a constant (in annular flow, only pipe diameters appear as variants in a and b), and X is the Lockhart-Martinelli, two-phase-flow modulus:

$$X^2 = \Delta p_{100(\text{liquid})}/\Delta p_{100(\text{vapor})} \tag{5}$$

In Eq. (5), $\Delta p_{100(\text{liquid})}$ is calculated by assuming only liquid flows in the pipe, and $\Delta p_{100(\text{vapor})}$ by assuming only vapor flows in the same size pipe. The modulus, X^2, remains constant for one set of flow conditions and is independent of pipe size within two to three sequential diameters.

After inserting Darcy's equation in the numerator and denominator of Eq. (5) and simplifying, the two-phase-flow modulus becomes:

$$X^2 = (W_l/W_v)^2(\rho_v/\rho_l)(f_l/f_v) \tag{6}$$

where f_l is the liquid-phase and f_v the vapor-phase friction factor. The modulus can be obtained directly by calculating the liquid-phase and vapor-phase Reynolds numbers and using Fig. 2. (Alternatively, friction factors can be determined from Figs. 5 and 6, *Chem. Eng.*, p. 65 (December 23, 1974).)

Reynolds numbers are calculated separately for the vapor and liquid phases by using the same pipe diameter, corresponding flow rates, and viscosities from:

$$N_{\text{Re}} = 6.31\,W/d\mu$$

A convenient form of Darcy's equation for unit pressure-loss calculations in pipelines for liquid or vapor is restated here:

$$\Delta p_{100} = 0.000336(fW^2)/d^5\rho \tag{7}$$

In Eq. (7), we use the same diameter for the liquid-phase and vapor-phase calculations, and the corresponding phase flow rate, density, and friction factor. Also, in Eq. (7), we must use the Moody friction factors from Fig. 2.

A generalized form suggested by Blazius for the Lockhart-Martinelli relation expresses the friction factors for turbulent flow as:

$$f_l = 0.046/(N_{\text{Re}})_l^{0.2}$$

$$f_v = 0.046/(N_{\text{Re}})_v^{0.2}$$

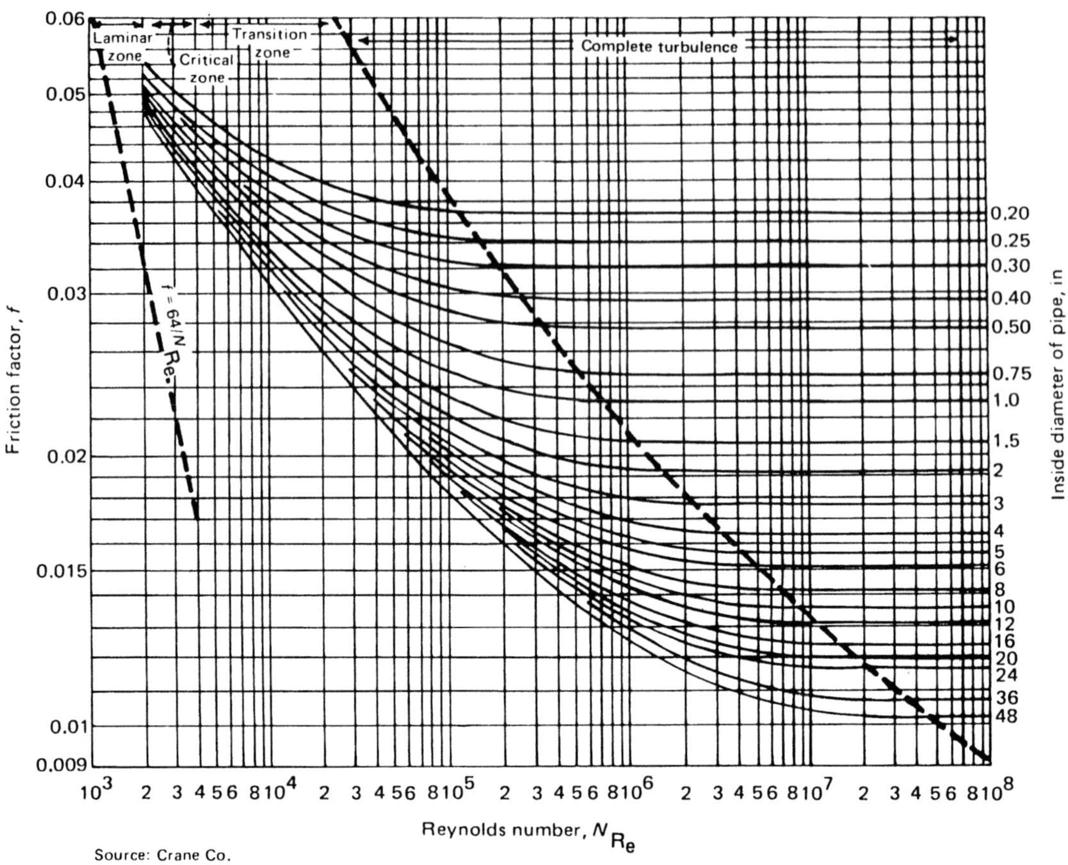

FIG. 2. Friction factors in new commercial-steel pipe for vapor-phase or liquid-phase flow.

Inserting these values and the equivalent relation for the Reynolds numbers into Eq. (6) and simplifying yields

$$X^2 = (W_l/W_v)^{1.8}(\rho_v/\rho_l)(\mu_l/\mu_v)^{0.2} \tag{8}$$

The Blazius friction-factor expression does not take into account the relative roughness of the pipe wall and the consequent relative reduction of the friction factor as the pipe diameter increases.

The widely used Moody friction factor (Fig. 2) will give somewhat smaller values for f_l/f_v, and consequently smaller unit losses than the Blazius form of f_l/f_v, depending on pipe size and liquid–vapor proportion.

As with all line-sizing procedures, pipe sizes must first be estimated. After selecting a pipe size, the flow-region coordinates can be calculated and the flow type determined from Fig. 1. After finding the vapor-phase unit loss and the applicable two-phase-flow correlation (Fig. 1), unit losses for two-phase flow can be calculated from Eq. (3).

Overall friction loss in the pipeline between two points will be

$$\Delta p_{(\text{two-phase})} = \Delta p_{100(\text{two-phase})}(L/100)$$

where L is the equivalent length of pipe and fittings.

Dispersed Flow

When nearly all of the liquid is entrained as spray by the gas, the flow is dispersed. This type of flow can be expected when the vapor content is more than roughly 30% of the total weight flow rate. Some overhead-condenser and reboiler-return lines have dispersed flow.

The curve in Fig. 3 represents the Lockhart-Martinelli correlation for all pipe sizes and for turbulent-liquid and turbulent-vapor phases. Turbulence for this correlation is defined as $(N_{\text{Re}})_l \geqq 2{,}000$, and $(N_{\text{Re}})_v \geqq 2{,}000$.

Let us find the unit loss in a 12-in. Schedule 40 line (d = 11.938 in., d^5 = 242,470 in.^{5}, D = 0.9965 ft, A = 0.7773 ft^2) for the following flow data:

	Liquid	*Vapor*
Flow, W, lb/h	300,000	350,000
Molecular weight, M	78.8	75.7
Density, ρ, lb/ft^3	33.5	2
Viscosity, μ, cP	0.1	0.01
Surface tension, σ, dyn/cm	5.7	—

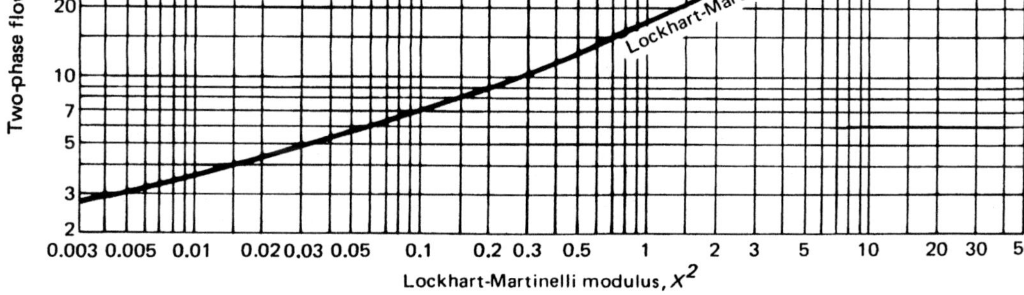

FIG. 3. Lockhart-Martinelli correlation relates vapor and liquid properties to establish two-phase flow modulus.

We first calculate the Baker parameters from Eqs. (1) and (2):

$$B_x = 531\left(\frac{300{,}000}{350{,}000}\right)\left(\frac{\sqrt{33.5(2)}}{33.5^{0.67}}\right)\left(\frac{0.1^{0.33}}{5.7}\right) = 29.3$$

$$B_y = \frac{2.16(350{,}000)}{0.7773\sqrt{33.52(2)}} = 118{,}800$$

We enter Fig. 1 with these coordinates, and find they intersect in the dispersed-flow region. Next, we calculate the modulus X^2 from Eq. (8):

$$X^2 = \left(\frac{300{,}000}{350{,}000}\right)^{1.8}\left(\frac{2}{33.5}\right)\left(\frac{0.1}{0.01}\right)^{0.2} = 0.072$$

From Fig. 3, we find $\phi^2 = 6.3$ for $X^2 = 0.072$.

We determine the vapor-phase unit loss from Eq. (7). To do so, we must find $(N_{\text{Re}})_v$ and f_v, as follows:

$$(N_{\text{Re}})_v = \frac{6.31(350{,}000)}{11.938(0.01)} = 18.5 \times 10^6$$

With this Reynolds number we find $f_v = 0.013$ from Fig. 2. By substituting the appropriate values into Eq. (7) we find the vapor-phase unit loss:

$$\Delta p_{100(\text{vapor})} = \frac{0.000336(0.013)(350{,}000)^2}{242{,}470(2)}$$

$$\Delta p_{100(\text{vapor})} = 1.1 \text{ psi/100 ft}$$

And, from Eq. (3), we determine the dispersed-flow unit loss as:

$$\Delta p_{100(\text{dispersed})} = 1.1 \times 6.3 = 6.93 \text{ psi/100 ft}$$

Bubble Flow

Bubble flow develops when vapor or gas bubbles, distributed in the liquid, move along at about the same velocity as the liquid. This type of flow can be expected when the vapor content is less than about 30% of the total weight flow rate. The sizing procedure is similar to that for dispersed flow. The correlation for bubble flow in Fig. 1 should be used.

Slug Flow

Slug flow develops when waves of liquid are picked up periodically in the form of frothy slugs that move at a greater velocity than the average liquid velocity. Typically, this phenomenon may occur in a pocketed line between an overhead condenser at grade and an elevated reflux drum. At low velocities, liquid collects at the low point and periodically can become a slug in the line. With a sufficiently high velocity, the liquid phase can be carried through without developing slug flow.

Unit losses for slug flow in process piping are generally not calculated. In borderline cases, unit losses for slug flow are smaller than those for bubble or annular flow.

Slug flow causes pressure fluctuations in piping, which can upset process conditions and cause inconsistent instrument sensing. Slug flow can be avoided in process piping by:

Reducing line sizes to the minimum permitted by available pressure differentials.
Designing for parallel pipe runs that will increase flow capacity without increasing overall friction loss.
Using valved auxiliary pipe runs to regulate alternative flow rates and thus avoid slug flows.
Using a low-point effluent drain or bypass.
Arranging the pipe configurations to protect against slug flow. For example, in a pocketed line where liquid can collect, slug flow might develop.

Slug flow will not occur in a gravity-flow line. A hard-tee connection (i.e., flow through the branch) at a low point can provide sufficient turbulence for more effective liquid carryover.

A diameter adjustment, coupled with gas injections, can also alter a slug-flow pattern to bubble or dispersed flow. Gas addition (used solely to avoid slug flows) can be expensive.

Let us determine a reasonable pipe size, downstream of the control valve, for the piping configuration sketched in Fig. 4. The available pressure difference, ΔP, is 10 lb/in.^{2}, including that for the control valve. The two-phase flow data in the line after the control valve are:

	Liquid	*Vapor*
Flow, W, lb/h	59,033	9,336
Molecular weight, M	79.47	77.2
Density, ρ, lb/ft^3	31.2	1.85
Viscosity, μ, cP	0.11	0.0105
Surface tension, σ, dyn/cm	5.07	—

From Eq. (1) and (2) we calculate the Baker coordinates for several sizes of

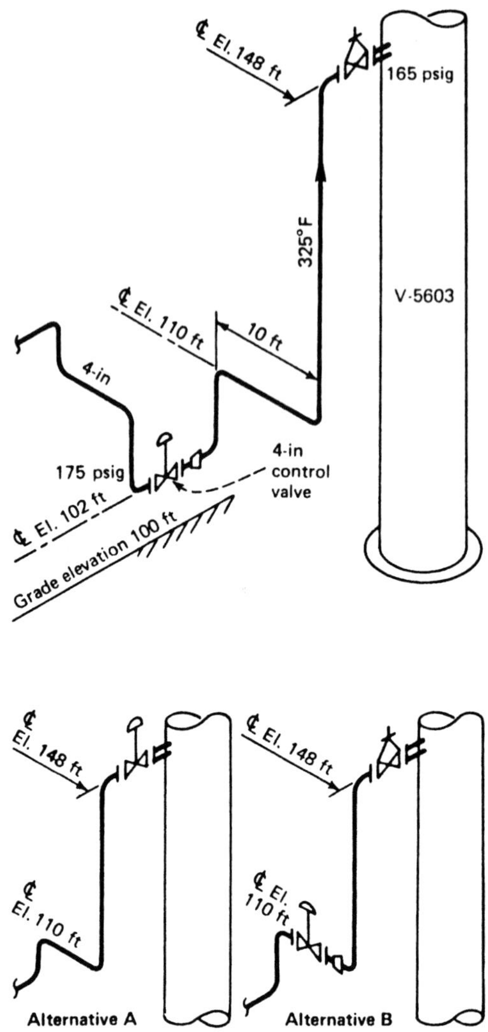

FIG. 4. Configurations of piping for sample problem.

Schedule 40 pipe, including and larger than 4 in. Using the Baker coordinates, we establish the flow region from Fig. 1. The results are:

Pipe size (in.)	*Baker coordinates* B_x	B_y	*Flow region*
8	245	7,650	Bubble/slug
6	245	13,250	Bubble
4	245	30,000	Bubble

Flow in the 8-in. line falls on the borderline of slug and bubble flow. In the 6-in. line, we will assume both bubble and slug flow.

Using the correlations for the appropriate two-phase flow (see Fig. 1), we compute the unit losses and the overall friction losses, as summarized here:

Line size (in.)	*Two-phase Δp_{100} (lb/in.^{2})*	*Equivalent length (ft)*	*Total Δp (lb/ in.^{2})*	*Flow region*
8	0.13	204	0.27	Slug
6	0.3	168	0.48	Slug
6	0.96	168	1.6	Bubble
4	6.9	132	9.1	Bubble

Reviewing the data in the preceding table, we find that a 6-in. line is the optimum size. The 8-in. line falls in the slug-flow region, which should be avoided. The total loss in the 4-in. line of 9.1 lb/in.^{2} plus the static-head backpressure of 3.15 lb/in.^{2} (computed on the basis of two-phase density) is greater than the available ΔP of 10 lb/in.^{2}. Consequently, the 4-in. line cannot be used.

There is always an uncertainty in making reasonably accurate predictions of pressure loss due to two possible types of flow. Vertical upflow and flashing in the line are affected proportionally by the decreasing static head and friction-loss gradients. Overall friction can be assumed to range from 0.5 to 2 lb/in.^{2}. In this example, if we add the static-head backpressure to 3.15 lb/in.^{2} to the overall friction loss, we obtain a range of 4.85 to 6.35 lb/in.^{2} for the control valve, i.e.:

$$\Delta P_{cv} = 10 - (2 + 3.15) \text{ to } 10 - (0.5 + 3.15)$$

A 4-in. double-seated conventional control valve can accommodate this range within acceptable operating points.

Alternative locations for the control valve are also shown in Fig. 4. For alternative A, slug flow cannot develop. For alternative B, a shortened and self-draining pipeline improves the pipe configuration but at the expense of convenient access to the control valve. In all cases there will be considerable turbulence after the control valve, which helps to provide slugfree liquid-phase carryover.

This example emphasizes two points. First, the majority of process-piping systems have an inherent flexibility in the pressure-loss distribution. Second, a control valve is able to operate within wide, available, pressure differentials. The process-piping designer can exploit these advantages for reasonable piping and components sizing.

There are a number of ways to adjust the pressure-loss distribution in a pipe system. Pipe sizes can be changed. A section of pipeline can be designed with increased or decreased pipe diameter. The amount of energy that can be lost at the terminating point of the pipeline can be varied. The static head of elevated vessels can be adjusted. Valve and orifice restrictions can be changed to consume more or less pressure differential. Within design limits, a pump can work against changing pipe resistances.

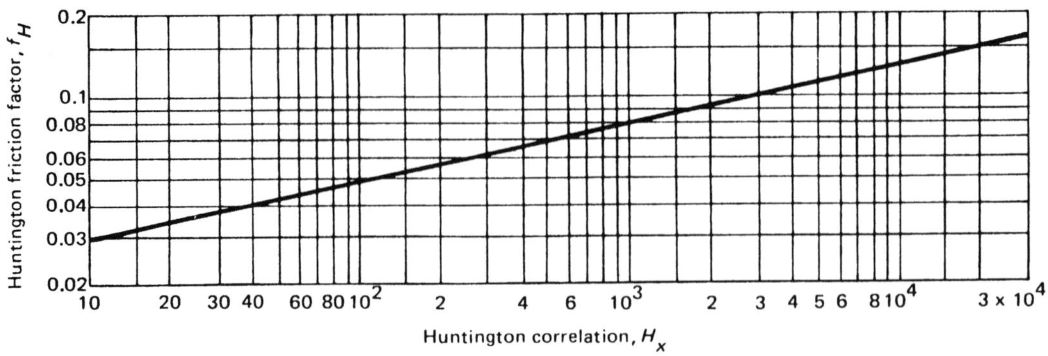

FIG. 5. Unit losses that occur during wave flow in pipelines for use with the Huntington correlation.

Stratified Flow

In horizontal lines, liquid flows along the bottom of the pipe, and the gas flows over a smooth gas–liquid interface. For stratified flow, a correlation is given in Fig. 1.

Wave flow is similar to stratified flow except the interface has waves moving in the direction of flow. A useful method for calculating the unit loss for wave flow is the Schneider-White-Huntington [6] correlation, H_x:

$$H_x = (W_l/W_v)/(\mu_l/\mu_v) \tag{9}$$

After computing H_x, we find the friction factor, f_H, from Fig. 5 and use it in the Darcy relation, Eq. (7), to determine the unit loss:

$$\Delta p_{100(\text{wave})} = 0.000336 f_H (W_v)^2/d^5 p_v \tag{10}$$

Plug flow develops when the interface is high in the pipe's cross-section. The waves congregate and touch the high point of the pipe wall. Here, plugs of liquid alternating with pockets of gas in the liquid close to the upper wall move along the horizontal line.

Gravity-flow lines have such types of flow. The initial, minimum pipe sizes can be selected with velocities typical of those for pump-suction lines.

Annular Flow

Annular flow develops when liquid forms a film or ring around the inside wall of the pipe, and gas flows at a higher velocity as a central core.

The annular-flow relation in Fig. 1 correlates well with pipe sizes up to 10 in. For pipes larger than 10 in., use $d = 10$.

Let us find the unit loss in a 6-in., Schedule 40 line handling the following flows:

	Liquid	*Vapor*
Flow, W, lb/h	6,150	21,500
Density, p, lb/ft^3	52	1.92
Viscosity, μ, cP	0.1	0.01
Surface tension, σ, dyn/cm	6.25	—

Pipe data are: $d = 6.065$ in., $d^5 = 8{,}206$ in.^{5}, and $A = 0.2$ ft^2.

Using Eqs. (1) and (2), we calculate the Baker parameters as

$$B_x = \left(\frac{6{,}150}{21{,}500}\right)\left(\frac{\sqrt{52(1.92)}}{52^{0.67}}\right)\left(\frac{0.1^{0.33}}{6.25}\right) = 8$$

$$B_y = \frac{2.16(21{,}500)}{0.2\sqrt{52(1.92)}} = 23{,}200$$

With these coordinates we find that the point of intersection in Fig. 1 falls in the annular-flow region.

From Eq. (8) we calculate the two-phase flow modulus, X^2:

$$X^2 = \left(\frac{6{,}150}{21{,}500}\right)^{1.8}\left(\frac{1.92}{52}\right)\left(\frac{0.1}{0.01}\right)^{0.2} = 0.0062$$

$$X = 0.0787$$

From Fig. 1 we find that the two-phase flow correlation for annular flow is $\phi = aX^b$, where

$$a = 4.8 - (0.3125 \times 6.065) = 2.9$$

$$b = 0.343 - (0.021 \times 6.065) = 0.216$$

Consequently:

$$\phi = 2.9(0.0787)^{0.216} = 1.67$$

$$\phi^2 = 2.80$$

To calculate the vapor-phase unit loss, we must find the Reynolds number:

$$(N_{\mathrm{Re}})_v = \frac{6.31(21{,}500)}{6.065(0.01)} = 2.24 \times 10^6$$

With this Reynolds number we find $f_v = 0.015$ from Fig. 2. Using this value of the friction factor in Eq. (7),

$$\Delta p_{100(\text{vapor})} = \frac{0.000336(0.015)(21{,}500)^2}{8{,}206(1.92)}$$

$$= 0.15 \text{ psi/100 ft}$$

By substituting into Eq. (3) we now find the unit loss:

$$\Delta p_{100(\text{annular})} = 0.15 \times 2.8 = 0.42 \text{ psi/100 ft}$$

Symbols

A	internal cross-section of pipe (ft^2)
D	internal diameter of pipe (ft)
d	internal diameter of pipe (in.)
L	equivalent length of pipe and fittings (ft)
M	molecular weight
W	flow rate (lb/h)
Δp	pressure loss (lb/in.^{2})
ΔP	available pressure difference (lb/in.^{2})
μ	viscosity (cP)
ρ	density (lb/ft^3)
σ	surface tension (dyn/cm)

Dimensionless Quantities

B_x	Baker parameter for Eq. (2)
B_y	Baker parameter for Eq. (1)
f	friction factor for Eq. (7)
f_H	Huntington friction factor (Fig. 5)
H_x	Huntington correlation in Eq. (9)
N_{Re}	Reynolds number
X	Lockhart-Martinelli modulus
ϕ	two-phase-flow modulus

Subscripts

l	refers to liquid
v	refers to vapor
100	refers to 100 ft of pipe

References

1. A. E. DeGance and R. W. Atherton, "Chemical Engineering Aspects of Two-Phase Flow," *Chem. Eng.*, pp. 135–139 (March 23, 1970); pp. 151–158 (April 30, 1970); pp. 113–120 (May 4, 1940); pp. 95–103 (July 13, 1970); pp. 119–126 (August 10, 1970).
2. O. Baker, H. W. Brainerd, C. L. Coldren, O. Flanigan, and J. K. Welchen, *Gas–Liquid Flow in Pipelines: Design Manual*, American Gas Association, Arlington, Viginia, October 1970.
3. G. E. Alves, *Co-Current Liquid–Gas Flow in Pipe Line Contactor*, Presented at the AIChE National Meeting, San Francisco, September 14, 1953.
4. O. Baker, "Simultaneous Flow of Gas and Oil," *Oil Gas J.*, July 26, 1954.
5. R. W. Lockhart and R. C. Martinelli, "Proposed Correlation of Data for Isothermal Two-Phase, Two-Component Flow in Pipe," *Chem. Eng. Prog.*, January 1949.
6. F. N. Schneider, P. D. White, and R. L. Huntington, "Correlation for Two-Phase Wave Flow," *Pipe Line Ind.*, October 1954.
7. *A.P.I. Technical Data Book—Petroleum Refining*, American Petroleum Institute, Washington, D.C., Chap. 10.
8. O. Baker and W. Swerdloff, "Calculations of Surface Tension," *Oil Gas J.*, November 21, 1955; December 5, 12, and 19, 1955; January 2 and 9, 1956.
9. C. L. Yaws and H. S. N. Setty, "Water and Hydrogen Peroxide," *Chem. Eng.*, p. 69 (December 23, 1974), Fig. 7-2.
10. *Engineering Data Book*, 9th ed., Natural Gas Processors Association, Tulsa, Oklahoma, 1972, pp. 16-27 to 16-33 and p. 16-39.

ROBERT KERN

Two-Phase Pressure Drop Computation

A computer program for quick calculation of two-phase and single-phase flow pressure drops per 100 ft of pipe is presented. It is written in BASIC language for a computer with a printer, but can easily be modified for use without a printer.

Methodology

The Darcy equation with a friction factor for old, steel pipes [1] is used to calculate the pressure drop. For two-phase flow, the Baker [2] and Lockhart-

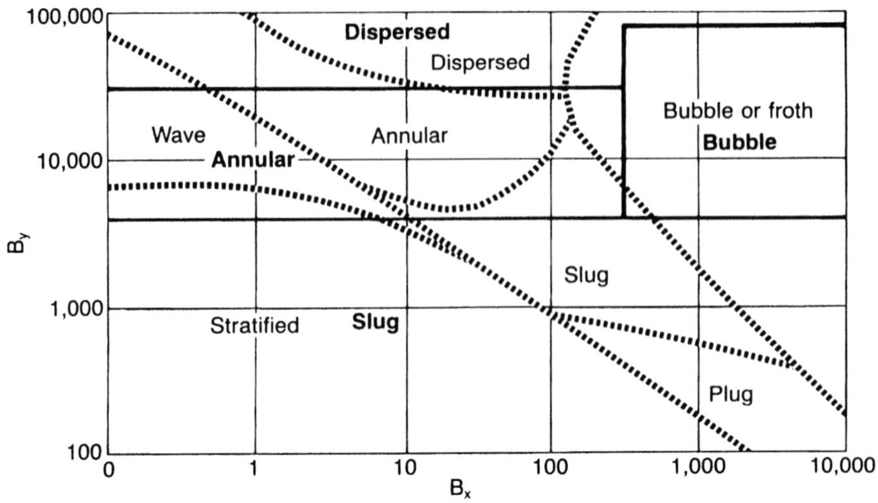

FIG. 1. The seven conventional flow regions are grouped into only four regions to simplify the computer program.

Martinelli [3] correlations were used as outlined by Kern [4]. The flow region is determined by the program using a simplified Baker graph (B_y vs. B_x as shown in Fig. 1.

The Baker graph in Fig. 1 shows the original flow region boundaries and the computer program's simplified boundaries. The wave and stratified regions are rare and only occur in long horizontal pipes; otherwise, the flow is annular. Thus, the program assumes the flow region is annular. Plug flow is also rare and thus, omitted from the program. The B_y and B_x coordinates are printed by the program so that the engineer can check the flow region.

Table 1 lists the flow regions and the corresponding two-phase flow modules used by the program.

TABLE 1

Flow regions	ϕ Calculation technique
1. Dispersed	The Lockhart-Martinelli correlation curve, ϕ^2 vs X^2, was approximated by 3 straight line segments, with ϕ^2 determined from the equation for the appropriate line.
2. Annular	$\phi = ax^b$ $a = 4.8 - 0.3125d$ $b = 0.343 - 0.021d$ d = pipe inside diameter (in.). 10 in. used for 12 in. and larger sizes.
3. Bubble	$\phi = (14.2)^{0.75} \div (W_L/\pi(d/24)^2)^{0.1}$
4. Slug	$\phi = (1190)^{0.815} \div (W_L/\pi(d/24)^2)^{0.5}$

Equations Used

$$\Delta P_{100,\text{ single phase}} = 0.129 f V^2 \rho / d \tag{1}$$

$$\Delta P_{100,\text{ two-phase}} = (\Delta P_{100,\text{ vapor as single phase}})\phi^2 \tag{2}$$

$$f = [(8/N_{\text{Re}})^{12} + 1/(A + B)^{3/2}]^{1/12} \tag{3}$$

$$A = \left[2.457 \ln \frac{1}{(7/N_{\text{Re}})^{09} + (0.27\epsilon/d)}\right]^{16} \tag{4}$$

$$B = (37{,}530/N_{\text{Re}})^{16} \tag{5}$$

$$N_{\text{Re}} = 123.9 d V \rho_{\text{av}} / \mu_L \tag{6}$$

$$B_x = 531(W_L/W_V)(\mu_L{}^{0.33}/\sigma_L)(\rho_V{}^{0.5}/\rho_L{}^{0.166}) \tag{7}$$

$$B_y = 2.16 W_V/\pi(d/24)^2(\rho_L \rho_V)^{0.5} \tag{8}$$

$$X^2 = \Delta P_{100,\text{ as liquid}}/\Delta P_{100,\text{ as vapor}} \tag{9}$$

$$\rho_{\text{av}} = [W_L + W_V]/[(W_L/\rho_L) + (W_V/\rho_V)] \tag{10}$$

$$V = [W_L + W_V]/[3600\rho_{\text{av}}\pi(d/24)^2] \tag{11}$$

$$\phi = \text{calculated according to Table 1} \tag{12}$$

Using the Program

The program is listed in Table 2. The subroutines are identified in Table 3. The program is written in BASIC for wide application among several brands of computers. Although the program is written for use with a Radio Shack TRS-80, PC-1 with printer, it can be easily modified for use without a printer. To run the program, enter RUN and then enter the following values:

For single-phase flow, enter:

1. d, in.
2. ρ, lb/ft^3
3. μ, cP
4. W, lb/h
5. 0
6. 0
7. 0
8. 0
9. 0

TABLE 2 BASIC Program Calculates Two-Phase and Single-Phase Flow

```
10:INPUT I,D,V,
   Q,G,L,F,J,S
11:PRINT I,D
12:PRINT V,Q
13:PRINT G,L
14:PRINT F,J
15:PRINT S
18:PRINT "ID=",
   I
19:E=((I/24)^2)
   *π
20:IF 0=FGOTO 4
   8
30:GOSUB "BYX"
40:C=M*T
41:PRINT "PRES
   DROP=",C
42:PRINT "PHI S
   Q=",M
43:PRINT "VEL="
   ,U
46:END
48:GOSUB "SIN"
50:PRINT "VEL="
   ,U
51:PRINT "PRES
   DROP=",P
55:END
60:"SIN"
80:U=Q/(3600*D*
   E)
90:R=123.9*I*U*
   D/V
100:A=(((LN (1/(
    ((7/R)^.9)+(
    .0005/I))))*
    2.45)^16)
110:B=(37530/R)^
    16
120:W=8*((((8/R)
    ^12)+(1/((A+
    B)^1.5)))^.0
    8)
130:P=.129*W*D*(
    U^2)/I
140:RETURN
150:"BYX"
170:Y=2.16*G/(E*
    (√(F*J)))
180:X=531*(L/G)*
    ((V^.33)/S)*
    (√J/(F^.166)
    )
182:PRINT "BX=",
    X
183:PRINT "BY=",
    Y
190:IF 80000<=Y
    GOSUB "DIS"
200:IF 4000>=Y
    GOTO 525
210:IF 300<=X
    GOSUB "BUB"
220:IF Y< =30000
    GOSUB "ANR"
221:IF X>=2GOSUB
    "DIS"
222:GOSUB "ANR"
230:RETURN
240:"DIS"
250:GOSUB "XXV"
251:K=1
252:PRINT "DISPE
    RSED"
260:IF .1>=Z
    GOSUB "C"
270:IF .5>=Z
    GOSUB "D"
280:M=((6.71*(-.
    5))+12.8)
281:GOTO 40
290:RETURN
300:"C"
310:M=((36.67*(Z
    -.01))+3.6)
320:GOTO 40
330:RETURN
340:"D"
350:M=((14.75*(Z
    -.1))+6.9)
360:GOTO 40
370:RETURN
380:"ANR"
390:GOSUB "XXV"
400:IF 10>=I
    GOSUB "E"
410:M=((1.68*((√
    Z)^.13))^2)
420:K=2
421:PRINT "ANNUL
    AR"
422:GOTO 40
430:RETURN
440:"E"
450:M=(((4.8-(.3
    1*I))*((√Z)^
    (.34-(.02*I)
    )))^2)
460:K=2
461:PRINT "ANNUL
    AR"
462:GOTO 40
470:RETURN
480:"BUB"
490:GOSUB "XXV"
500:M=((14.2*((√
    Z)^.75)/((L/
    E)^.1))^2)
510:K=3
511:PRINT "BUBBL
    E"
512:GOTO 40
520:RETURN
525:IF 20>=X
    GOSUB "ANR"
530"SLG"
540:GOSUB "XXV"
550:M=((((1190*((
    √Z)^.815))/(
    √(L/E)))^2)
560:K=4
561:PRINT "SLUG"
562:GOTO 40
570:RETURN
580:"XXV"
590:D=F,Q=L
600:GOSUB "SIN"
601:H=P
610:D=J,Q=G,V=.0
    1
620:GOSUB "SIN"
621:T=P
630:Z=H/T
640:D=((G+L)/((L
    /F)+(G/J)))
650:Q=G+L,U=Q/(3
    600*D*E)
660:RETURN
```

TABLE 3 Subroutine Labels

Label	Line	Label	Line
ANR	380	DIS	240
BUB	480	E	440
BYX	150	SIN	60
C	300	SLG	530
D	340	XXV	580

Example 1: What is the pressure drop in the two-phase flow in an 8-in., Schedule 40 line:

W_L = 59,033 lb/h $\quad$ W_V = 9,336 lb/h
ρ_L = 31.2 lb/ft^3 $\quad$ ρ_V = 1.85 lb/ft^3
σ_L = 5.07 $\quad$ μ_L = 0.11 cP

Example 2: What is the unit loss in a 6-in., Schedule 40 line with the following flow data:

Liquid: W_L = 572 lb/h $\quad$ ρ_L = 52 lb/ft^3
μ_L = 0.1 cP $\quad$ σ_L = 6.25 dyn/cm
Vapor: W_V = 2,000 lb/h $\quad$ ρ_V = 192 lb/ft^3
μ_V = 0.01 cP

Example 3: What is the unit loss in a 6-in., Schedule 40 line with the following flow data:

Liquid: W_L = 6,150 lb/h $\quad$ ρ_L = 52 lb/ft^3
μ_L = 0.1 cP $\quad$ σ_L = 6.25 dyn/cm
Vapor: W_V = 21,500 lb/h $\quad$ ρ_V = 1.92 lb/ft^3
μ_V = 0.01 cP

Example 4: What is the unit loss in a 4-in., Schedule 40 line with the following flow data:

Liquid: W_L = 59,033 lb/h $\quad$ MW = 79.47
ρ_L = 31.2 lb/ft^3 $\quad$ μ_L = 0.11 cP
σ_L = 5.07 dyn/cm
Vapor: W_V = 9,336 lb/h $\quad$ MW = 77.2
ρ_V = 1.85 lb/ft^3

Example 5: What is the unit loss in an 18-in., Schedule 40 line with the following flow data:

Liquid: W_L = 607,769 lb/h $\quad$ MW = 78.8
ρ_L = 33.5 lb/ft^3 $\quad$ μ_L = 0.1 cP
σ_L = 5.7 dyn/cm
Vapor: W_V = 718,094 lb/h $\quad$ MW = 75.7
ρ_V = 2 lb/ft^3 $\quad$ μ_V = 0.01 cP

(continued)

TABLE 3 *(continued)*

Example 1.	Example 3.	Example 5. (Two phase)
7.981	6.065	16.874
0.	0.	0.
0.11	0.1	0.1
0.	0.	0.
9336.	21500.	718094.
59033.	6150.	607769.
31.2	52.	33.5
1.85	1.92	2.
5.07	6.25	5.7
ID=	ID=	ID=
7.981	6.065	16.874
BX=	BX=	BX=
245.6045231	8.174199236	29.11618511
BY=	BY=	BY=
7640.28259	23165.99324	122020.8845
ANNULAR	ANNULAR	DISPERSED
PRES DROP=	PRES DROP=	PRES DROP=
6.15251E-02	4.99281E-01	4.929414405
PHI SQ=	PHI SQ=	PHI SQ=
6.366754472	2.558124489	4.8845524
VEL=	VEL=	VEL=
5.547859463	15.66781997	67.46731884

Example 2.	Example 4.	Example 5. (Single phase)
6.065	4.026	6.065
0.	0.	62.4
0.1	0.11	1.
0.	0.	325000.
2000.	9336.	0.
572.	59033.	0.
52.	31.2	0.
1.92	1.85	0.
6.25	5.07	0.
ID=	ID=	ID=
6.065	4.026	6.065
BX=	BX=	VEL=
8.172870098	245.6045231	7.211195519
BY=	BY=	PRES DROP=
2154.976115	30024.54969	1.502004025
ANNULAR	DISPERSED	
PRES DROP=	PRES DROP=	
5.39312E-03	3.050334689	
PHI SQ=	PHI SQ=	
2.786387811	9.445	
VEL=	VEL=	
1.457469149	21.80180905	

An echo of the input will be printed followed by the inside diameter (ID, in.), the velocity (VEL, ft/s), and the pressure drop per 100 feet (psi/100 ft).

For two-phase flow, enter:

1. d, in.
2. 0
3. μ_L, cP
4. 0
5. W_V, lb/h
6. W_L, lb/h
7. ρ_L, lb/ft^3
8. ρ_V, lb/ft^3
9. σ_L, dyn/cm

An echo of the input will be printed followed by the inside diameter (in.), the B_x and B_y Baker parameters, the flow region, the pressure drop per 100 feet for two phase flow (psi/100 ft), the square of the two phase flow modulus, and the velocity (ft/s).

Some Examples

For comparison the accompanying examples are the same as Kern's examples [4]. The computer results are very close to Kern's graphical solutions.

Pipe Sizing Rules

In using the correlations presented in this article, the following general pipe sizing rules [4] are suggested:

1. *Dispersed Flow*. Apply $\Delta P_{100,\text{ two-phase}}$ throughout three-dimensional pipe (horizontal, up, and downflow sections.) Use dispersed flow correlation for pipe smaller than 2½ in. for all flow regions.
2. *Annular and Bubble Flow*. Apply $\Delta P_{100,\text{ two-phase}}$ for 3-in. and larger pipe throughout the process line.
3. *Slug Flow*. Slug flow unit losses in process piping are generally not calculated. In borderline applications, slug flow unit losses are much smaller than either bubble or annular flow unit losses. Slug flow should be avoided in process piping. It causes pressure fluctuations, which can upset process conditions and cause inconsistent instrument reading and recording.

Slug flow can be avoided in several ways:

By reducing line sizes to a minimum permitted by available pressure differentials
By designing parallel pipe runs that will increase flow capacity without increasing the overall friction loss
By using valved auxiliary pipe runs to regulate alternative flow rates and avoid slug flows
By using a low point effluent drain or bypass or other solutions
By arranging the pipe configurations to protect against slug flow

In a pocketed line where liquid can collect, slug flow will develop. Slug flow will not occur in a gravity flow line. Hard tee connections can provide sufficient turbulence for a more effective liquid carryover. A diameter adjustment coupled with gas injections can also alter a slug flow pattern to bubble or dispersed flow.

Gas addition used only to avoid slug flow can be expensive. Gas flow quantities to get the fluid flow out of the slug flow region can be estimated by choosing a B_y value falling well into the annular or bubble flow region. Expressing the vapor-phase flow rate from the equation, the total vapor content of two-phase flow will be

$$W_{V,\,\text{total}} = B_y \pi (d/24)^2 (\rho_L \rho_V)^{0.5} \qquad (8')$$

Summary

This program is accurate and will save the user valuable time. Since Baker's flow regime boundaries were approximated, care should be given to the regime printed by the computer. B_x and B_y printed by the computer can be used with the original-flow regime boundaries. If the regime is different than the original, the accuracy should still be reasonable. If it is desired to hand-calculate the two-phase pressure drop, use the printed ϕ^2. Single phase is accurate unless the limitation on Darcy is exceeded. In this case, break the line to smaller segments.

Symbols

Text	Program	Definition
A	A	factor defined by Eq. (4)
B	B	factor defined by Eq. (5)
B_x	X	Baker parameter defined by Eq. (7)

B_y	Y	Baker parameter defined by Eq. (8)
d	I	pipe inner diameter (printout shows "ID") (in.)
—	E	pipe inner diameter cross-sectional area, $E = \pi(d/24)^2(\text{ft}^2)$
f	—	friction factor (dimensionless)
–	W	$W = 8f$, used in program line 120
ln	LN	natural logarithm
MW	—	molecular weight
N_{Re}	R	Reynolds number defined by Eq. (6) (dimensionless)
ΔP_{100}	C, P	pressure drop per 100 ft of pipe (psi/100 ft)
—	H	pressure drop per 100 ft if fluid were a liquid (psi/100)
—	T	pressure drop per 100 ft if fluid were a vapor (psi/100)
V	U	average fluid velocity (printout shows "VEL") (ft/s)
W_v	G	vapor flow rate (lb/h)
W_L	L	liquid flow rate (lb/h)
—	Q	total fluid rate, $W_V + W_L$ or $G + L$ (lb/h)
X	—	Lockhart-Martinelli modulus (dimensionless)
—	Z	$Z = X^2$ or H/T (dimensionless)
ϵ	—	effective pipe roughness, assumed 0.00015 for old, steel pipe, $0.27 \times 12 \times \epsilon = 0.0005$ used in program line 100
μ_L	V	liquid viscosity, $\mu_L = 0.01$ used in program line 610 (cP)
ρ_{av}	D	two-phase average fluid density (lb/ft^3)
ρ_V	J	vapor density (lb/ft^3)
ρ_L	F	liquid density (lb/ft^3)
σ_L	S	liquid surface tension (dyn/cm)
ϕ	—	two-phase flow modulus
ϕ^2	M	printout shows "PHI SQ"
—	K	flow region identifier

This material originally appeared in *Hydrocarbon Processing*, pp. 155–157, April 1984, and is used with the permission of the editor and the author.

References

1. S. W. Churchill, *Chem. Eng.*, *84*(24), 91–92 (November 7, 1977).
2. O. Baker, *Oil Gas J.*, July 26, 1954.
3. R. W. Lockhart and R. C. Martinelli, *Chem. Eng. Prog.*, *45*, 39–48 (1949).
4. R. Kern, *Hydrocarbon Process.*, *48*(10), 105–116 (1969).

RAFIK SOLIMAN

Water Hammer

With a better understanding of the nature and severity of the water hammer problem, we can avoid its destructive forces. This greater understanding should also help with the introduction of more preventive measures into system designs and installations, which will help provide maximum safety for personnel, lower maintenance cost, and reduce system downtime.

Where Water Hammer Occurs

Water hammer can occur in any water supply line, hot or cold. Its effects can be even more pronounced in heterogeneous or biphase systems. Biphase systems carry water in two states, as a liquid and as a gas. Such a condition exists in a steam system where condensate coexists with live or flash steam: in heat exchangers, tracer lines, steam mains, condensate return lines and, in some cases, pump discharge lines.

Effects of Water Hammer

Water hammer has a tremendous and dangerous force that can collapse floats and thermostatic elements, overstress gauges, bend mechanisms, crack trap bodies, rupture fittings and heat exchange equipment, and even expand piping. Over a period of time, this repeated stress on the pipe will weaken it to the point of rupture.

Water hammer is not always accompanied by noise. Some types of water hammer, resulting from localized abrupt pressure drops, are never heard. The consequences, however, may be just as severe.

Conditions Causing Water Hammer

Three conditions have been identified that can cause the violent reactions known as water hammer. These conditions are hydraulic shock, thermal shock, and differential shock.

Hydraulic Shock

Visualize what happens when a valve is open. A solid shaft of water is moving through the pipes from the point where it enters the house to the valve. This could be 100 lb of water moving at 10 ft/s, about 7 mi/h.

When the valve is shut suddenly, it is like a 100-lb hammer coming to a stop. There is a "bang." This shock wave is similar to a hammer hitting a piece of steel. The shock pressure wave of about 600 lb/in.^{2} is reflected back and forth from end to end until the energy is dissipated. Similar action can take place in the suction or discharge piping of a pump when the pump starts and stops if check valves are in the line. Slow closure of the valve and slow-closing check valves along with water hammer arrestors are solutions to these problems. If the column of water is slowed before it is stopped, its momentum is reduced gradually and, therefore, damaging water hammer will not be produced.

Water hammer arrestors, if correctly sized, placed, and maintained, will reduce water hammer by providing a controlled expansion chamber in the system. As the forward motion of the water column in the pipe is stopped by the valve, a portion of the reversing column is forced into the water hammer arrestor. The water chamber of the arrestor expands at a rate controlled by the pressure chamber and gradually slows the column, preventing hydraulic shock.

If a check valve is used in a system without an arrestor, excessive pressure may be exerted on the system when the reversing water column is violently stopped by the check valve. If a float-type air vent is located between the check valve and the closing valve, the float could easily be ruptured.

Thermal Shock

In biphase systems, steam bubbles may become trapped in pools of condensate in a flooded main, branch, or tracer line, as well as in heat exchanger tubing and pumped condensate lines. Since condensate temperature is almost always below saturation, the steam will immediately collapse.

One pound of steam at 0 lb/in.^{2} gauge occupies 1600 times the volume of a pound of water at atmospheric conditions. This ratio drops proportionately as the pressure increases. When the steam collapses, water is accelerated into the resulting vacuum from all directions. This happens when a steam trap discharges relatively high-pressure flashing condensate into a pump discharge line.

Another cause of water hammer is lack of proper drainage ahead of a steam control valve. When the valve opens, a slug of condensate will enter the equipment at a high velocity, producing water hammer when it impinges on the interior walls. In addition to this, the mixing of the steam that follows with the relatively cool condensate will produce water hammer from thermal shock. This condition can be corrected by dripping the supply riser as shown in Fig. 1.

Water hammer can also occur in steam mains, condensate return lines,

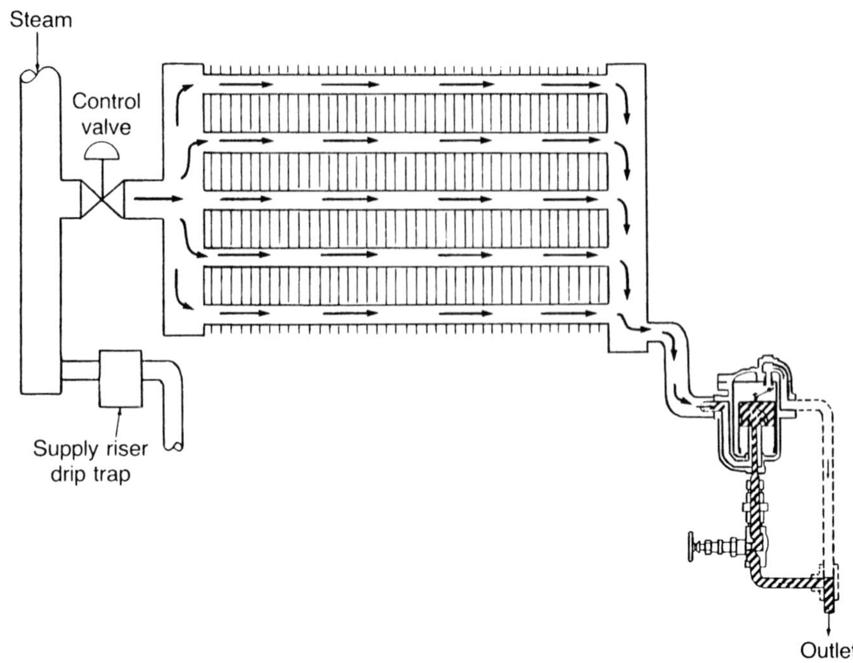

FIG. 1. Condensate controller maintains a positive pressure differential across all the tubes.

and heat exchange equipment where steam entrapment can take place (Fig. 1). A coil constructed and installed as shown here, except with just a steam trap at the outlet, permits steam from the control valve to be directed through the center tube(s) first. Steam then gets into the return header before the top and bottom tubes are filled with steam. Consequently, these top and bottom tubes are fed with steam from both ends. Waves of condensate are moved toward each other from both ends, and steam can be trapped between the waves.

Water hammer results from the collapse of this trapped steam. The localized sudden reduction in pressure caused by the collapse of the steam bubbles has a tendency to chip out pipe and tube interiors. Oxide layers that otherwise would resist further corrosion are removed, resulting in accelerated corrosion.

One means of overcoming this problem is to install a condensate controller, which maintains a positive pressure differential across all the tubes (Fig. 1). A condensate controller provides a specialized purge line, which assures a positive flow through the coil at all times.

Differential Shock

Differential shock, like thermal shock, occurs in biphase systems. It can occur whenever steam and condensate flow in the same line, but at different velocities, such as in condensate return lines.

In biphase systems, the velocity of the steam is often 10 times the velocity of the liquid. If condensate waves rise and fill a pipe, a seal is formed with the pressure of the steam behind it (Fig. 2). Since the steam cannot flow through the condensate seal, pressure drops on the downstream side. The condensate seal now becomes a "piston" accelerated downstream by this pressure differential. As it is driven downstream it picks up more liquid, which adds to the existing mass of the slug, and the velocity increases.

If this slug of condensate gains high enough momentum and is then required to change direction, for example at a tee, elbow, or valve, great damage can result.

Since a biphase mixture is possible in most condensate return lines, their correct sizing becomes essential.

Condensate normally flows at the bottom of a return line. It flows because of the pitch in the pipe and also because of the higher velocity flash steam above it, dragging it along. The flash steam moves at a higher velocity because it moves by differential pressure.

Flash steam occurring in return lines, due to the discharge of steam traps, creates a pressure in the return line. This pressure pushes the flash steam at relatively high velocities toward the condensate receiver, where it is vented. Condensing of some of the flash steam, due to heat loss, contributes to this pressure difference and amplifies the velocity. Because the flash steam moves faster than the condensate, it makes waves. As long as these waves are not high enough to touch the top of the pipe and so do not close off the flash steam's passageway, all is well.

In our glass pipe demonstrator, cold water is used to simulate condensate and air pressure is applied to simulate the flash steam in the top portion of the pipe (Fig. 2).

Damaging water hammer similar to that just described is experienced also when elevated heat exchange equipment is drained with a long vertical drop to a trap.

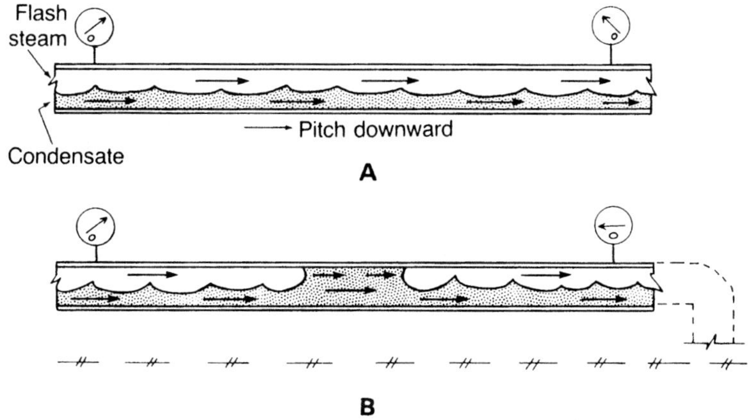

FIG. 2. Formation of a condensate seal.

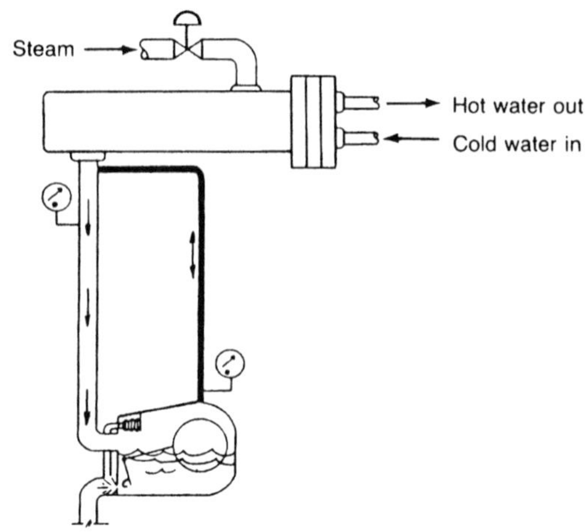

FIG. 3. Back-venting a trap on elevated heat exchangers.

Condensing of steam downstream of the slug produces a drop in pressure and, therefore, a pressure differential across the slug. This pressure differential, together with gravity, accelerates the slug downward. It does not take much of this strong force to collapse a ball float in a trap. Back-venting the trap to the top of the vertical drop can correct this problem (Fig. 3).

Since condensation is what produces the acceleration, uninsulated pipes and their appurtenances would be expected to suffer greater damage than those with insulation.

To control differential shock, the condensate seal must be prevented from forming in a biphase system. Steam mains must be properly pitched, condensate lines must be sized and pitched correctly, and long vertical drops to traps must be back-vented. The length of lines to traps should be minimized, and pipes may have to be insulated to prevent water hammer.

This material originally appeared in *Hydrocarbon Processing*, pp. 67–68, March 1983, and is used with the permission of the editor and the author.

JOHN A. KREMERS

2

Fluidics

Fluidics

Modern fluidics dates from 1957 when physicist Billy M. Horton of the Harry Diamond Laboratories in Washington had difficulties with his home fireplace. Smoke flowed into his living room rather than up the flue. Horton found that with a small bellow he could deflect, or switch, the flue gases up the chimney.

Like so many other "new" technologies, fluidics uses physical phenomena long ago recognized but not put to practical use. The devices originated by Horton and others are based on the Coanda effect, named after the Rumanian engineer, Henri Coanda, who discovered it in 1926. Coanda found that a fluid jet attaches itself to one wall of a flat nozzle and remains there because of a stable, dynamically formed and sustained pressure gradient across the fluid stream. Coanda did not find an application for his discovery.

In 1962, Ray Auger filed a patent application for a fluid logic element, now called a turbulence amplifier. However, a 1922 patent issued to R. E. Hall for a device called a relay is very much like the Auger device. The same situation exists for the vortex amplifier. Fluidic diodes can be found as early as 1916. Only momentum exchange devices date to the early 1960s.

The National Fluid Power Association (NFPA) recommends the following definition or fluidics: "The technology associated with sensing, control, information processing, and/or actuation functions performed solely through utilization of fluid dynamic phenomena." The emphasis is on control.

Control and logic functions, both digital and analog, normally performed by electronic or electromechanical devices, can be advantageously performed with fluidics. In most commercial devices the fluid used is compressed air. However, there are pure fluidic devices based on the interaction of air and liquid streams.

At the inception of the technology some 20 years ago, many companies and engineers embraced fluidics enthusiastically, but fluidics failed to live up to early expectation and successes. The effects of internal contamination of the fluid medium apparently had not been sufficiently investigated. Some systems were improperly engineered. In addition, the rapid rise of miniaturized electronics and intrinsically safe electrical systems doomed the general use envisioned for fluidics. Despite these setbacks, fluidics is still a viable technology, with a definite place in the chemical process industries for many actual and potential applications.

Fluidic systems still offer certain unique advantages. Foremost, they are absolutely safe in hazardous areas. Second, they are immune to damage or malfunction under severe environmental stress such as mechanical shock, vibration, nuclear radiation, or temperature extremes. In the last two attributes, fluidic systems are superior to electromechanical or electronic systems.

The use of electrical power, even at the lowest energy levels, involves a finite risk. There is always some possibility of malfunction or unforeseen phenomena. Fluidic systems completely eliminate this risk. For this reason alone, fluidic controls should be considered for hydrogen compressors, for example. Because of their simplicity and lack of moving parts, fluidic systems are not affected by mechanical shock or vibration. Sensors and logic elements for these systems are available in metallic or ceramic construction for resistance to high temperature and corrosive environments. Fluidic devices are also unaffected by nuclear radiation, magnetic or electric fields, and stray noise. The environmental reliability of fluidics was studied as early as 1962, when researchers reported that failures induced by temperature change were reduced five times when fluidic devices were substituted for equivalent electronic units. Nuclear radiation failures were reduced by a factor of six.

The dynamic phenomena in all true fluidic devices are based upon the interaction of two jets of fluid or upon the reaction of a fluid jet to a confining wall. The description of the various devices that follows will illustrate how jet interaction and the geometry of the devices are used to accomplish sensing, control, amplification, and logic functions.

Sensing Devices

The most common and useful sensing devices will be described here. Perhaps the most widely used is the proximity sensor or fluidic gauge. In the simplest air actuated gauge, when an object is brought very close to a jet of air issuing from a nozzle, a backpressure is developed within the nozzle. This type of gauge has generally been used with fluidic-logic devices to initiate corrective or protective action when the sensing pattern is abnormal. It has also been used as a sensor for fluidic counters and shaft speed sensors.

An acoustic sensor (called the fluidic ear) can detect the presence or absence of an object over a span, between transmitter and receiver, of up to 15 ft. It consists of an acoustic wave generator or whistle (not audible) and a receiver containing a sound sensitive amplifier—actually a turbulence amplifier (to be described later). This acoustic sensor can be used to generate a sensing curtain that may serve as an intrusion alarm. It is the fluidic equivalent of the photocell.

Note that the nozzle–baffle system in most pneumatic process-control instruments uses the principle of the fluidic gauge.

Fluidic temperature sensors have been developed for use by the U.S. Air Force for operation at 2000°F. The transient response of the fluidic sensor greatly exceeds that of thermocouples. A fluidic oscillator changes frequency as the density of the gas changes with temperature. The same principle can also be used to detect changes in gas densities or sense the ratio of two gases at a fixed temperature.

Two fluidic devices for flow measurement, both having unique capabilities, are commercially available. One is simply a fluidic proportional amplifier (also known as a jet deflection amplifier) and is placed in the fluid stream to be measured. The proportional amplifier is shown schematically in Fig. 1.

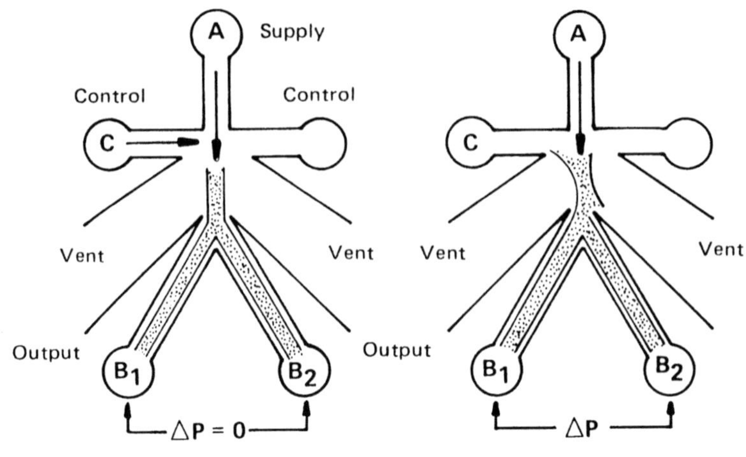

FIG. 1. A fluidic device is a proportions amplifier.

The jet of fluid issuing from nozzle A is directed at the two receiver nozzles, B_1 and B_2, with the control jet, C. If C is off, the jet from A is split equally between B_1 and B_2 and no differential pressure exists between the outlet parts. By applying pressure at the control nozzle, C_1, the jet from A is diverted away from B_1 toward B_2, and a pressure differential is developed between the two nozzles. If an open passage exists between A and B_1 and B_2, a fluid stream through this passage will divert the jet, and the differential pressure between B_1 and B_2 will be proportional to the velocity of the fluid stream (within limits). This is the principle of the flow element shown in Fig. 2. Output nozzles B_1 and B_2 may be purged when particulate contamination is present.

This flow element is designed for insertion in a pipe or duct for measuring point velocity. It can be used on both gases and liquids. The element may be located at the point of average velocity, or several units may be manifolded and the output signals averaged in order to meter the total flow rate. The chief advantages of this device over other flow elements are sensitivity to very low velocities and the relatively high differential developed at low flow rates. It is customary to input this element to a differential-pressure transmitter for interfacing with a standard process-control system. This principle is also the basis for a sensitive wind-velocity sensor. This type of element is well suited for use in exhaust and flare stacks. A gas densitometer operates on the same principles as the flow element, see Fig. 2A.

The other fluidic flow element is based upon the principle of the fluidic oscillator. In this device the supply jet is the flowing stream to be metered. Hence, total flow is monitored and the device is a line size fitting. This meter depends for its operation upon the Coanda effect, illustrated in Fig. 3.

The operation of the flow element based on this principle can be explained by examining Fig. 4. As the fluid starts to flow through the meter body (commercial models are suitable for liquids only), the stream attaches itself at random to one or the other of the internal deflectors. The geometry of the

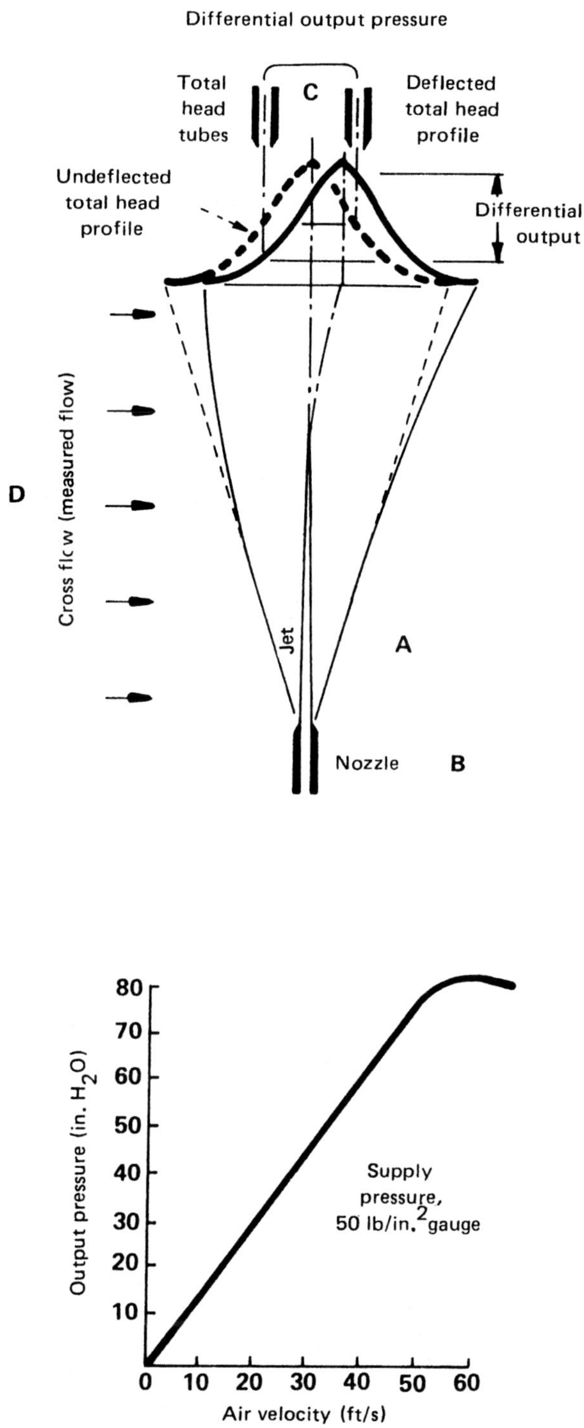

FIG. 2. A fluidic element for gas and liquid flow.

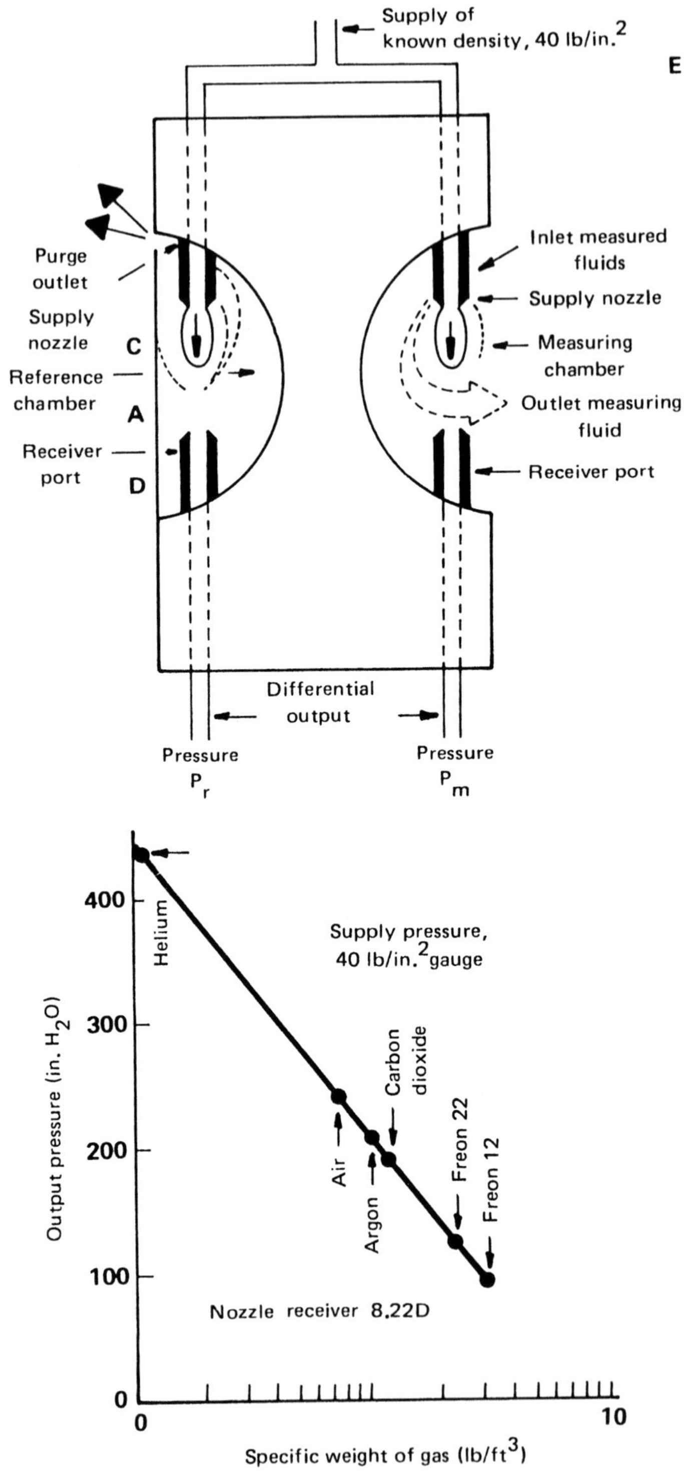

FIG. 2A. Gas densitometer.

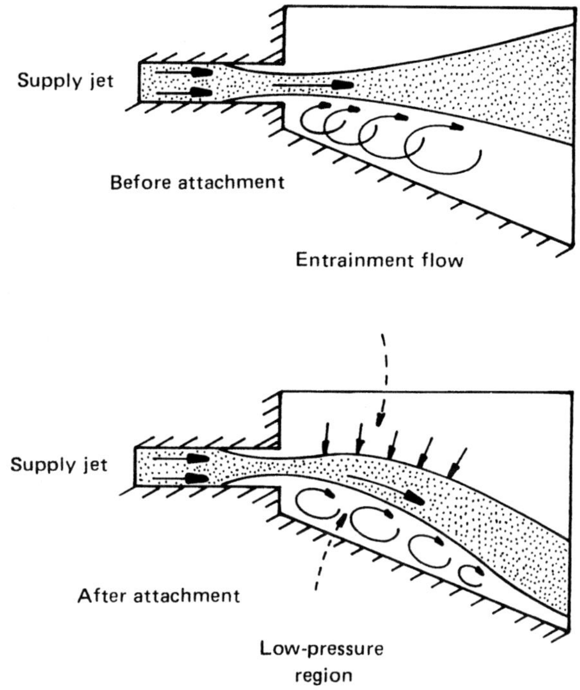

FIG. 3. Coanda effect in a fluidic oscillator.

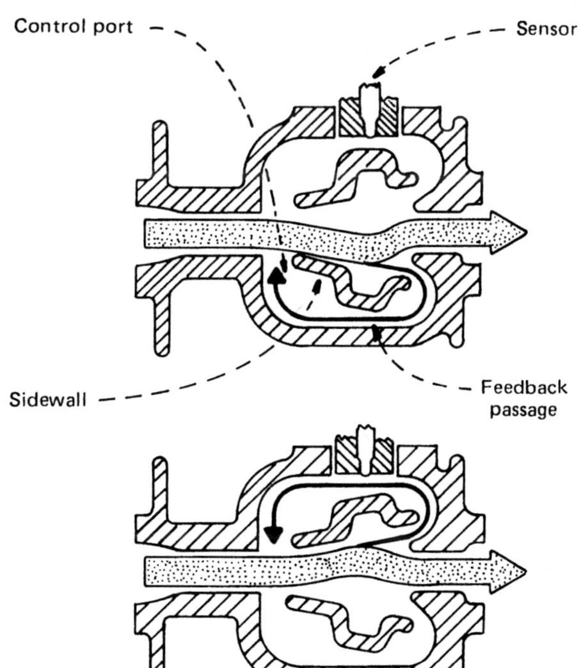

FIG. 4. Flow meter acts like a fluidic oscillator.

meter body is such that a small feedback stream is caused to flow to the upstream side of the deflector, where it creates a slight pressure increase and thereby causes the jet to switch over to the other deflector. Here again, the feedback stream causes the jet to switch back. The jet oscillates back and forth between the deflectors, and the rate of oscillation is proportional to the flow velocity.

A thermistor in a bridge circuit on one side of the meter body detects each oscillation, so the output of the meter is digital. Below a certain velocity the fluid jet does not oscillate. This minimum velocity depends on the kinematic viscosity of the liquid. For example, the minimum measurable flow of water through a 1-in. meter is one times the viscosity in centistokes, or approximately 1 gal/min at 70°F.

The advantages of this flow meter are digital output and no moving parts. Accuracy and linearity are comparable to the turbine meter. However, the meter generates some pulsations in the flow stream and is not recommended for slurries.

For level sensing the conventional dip tube is readily interfaced with fluidic devices. Backpressure on the dip tube may be used as the control press of a proportional amplifier for an analog output. Figure 5 shows a fluidic level control system using a dip tube and a proportional fluidic amplifier. A booster relay on the output of the amplifier actuates a diaphragm valve. If the backpressure on the dip tube is used to control a digital fluidic amplifier (also known as a flip–flop), the output is digital (i.e., on–off).

Several fluidic speed sensors have been developed. These are designed to detect speed of rotation and are used to actuate fluidic governors. They are all based upon the principle of interrupting a jet with a notched wheel, slotted disk, or wobble plate which is driven off the shaft of the rotating machine or attached directly to it. The pulsing pressure signal so generated is compared to

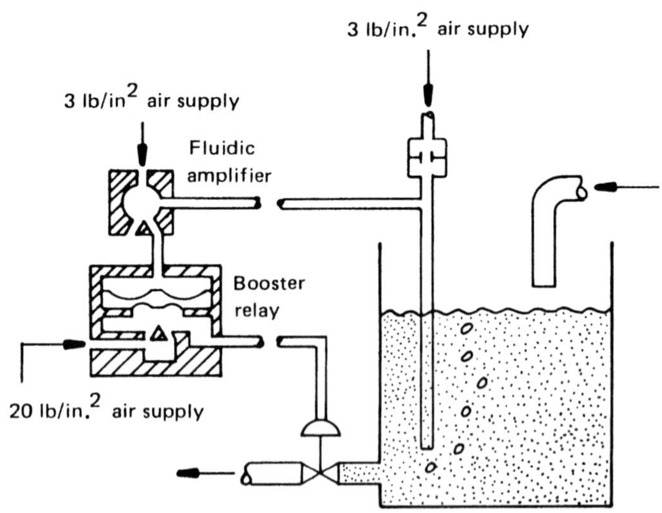

FIG. 5. Fluidic amplifier controls the level in a tank.

a reference signal, and a pressure output that is proportional to phase shift is obtained.

The fluidic sensors just described are representative. It is important to note that any sensor with a pneumatic output, such as a pneumatic transmitter, can be interfaced with fluidic circuits.

Final Control Element

The principle of the bistable flip–flop is used in the two-position fluidic diverter valve, a fluidic device for which successful applications have been found in the chemical process industries. The value is shown schematically in Fig. 6. Its operation depends on the Coanda effect. The flow may be diverted from either outlet to the other by means of the pilot valves. The process fluid serves as the supply jet.

When flow is first started, the jet will attach itself to one wall or the other, since the valve body is symmetrical. If the pilot valve on the side to which the jet has attached is opened, ambient air will be drawn in, and the vacuum created between the wall and the jet on that side will be relieved. The jet will now be directed to the other side. It can be diverted back again by reversing the pilot valve positions. Note that the geometry of the valve is such that the flow is diverted to the outlet port opposite the side to which the jet is attached. Obviously, the flow rate must be sufficient to create a jet, or the valve will not work properly.

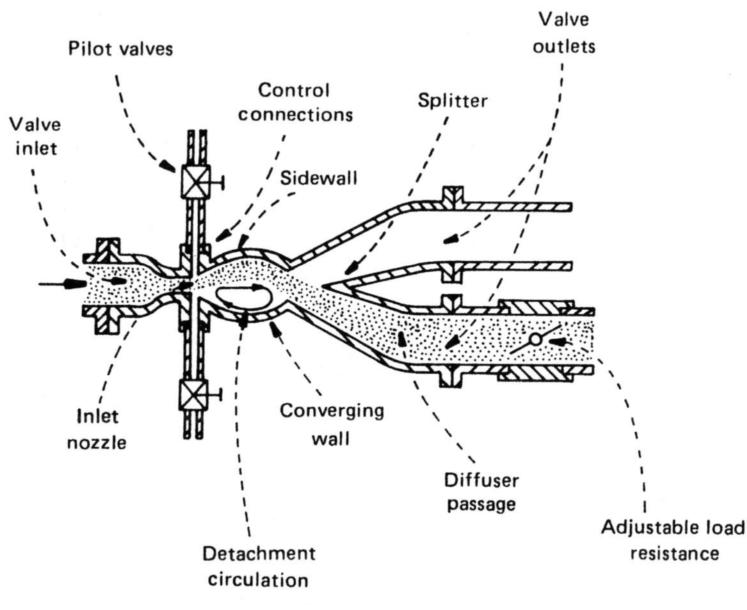

FIG. 6. Diverter valve depends on the Coanda effect.

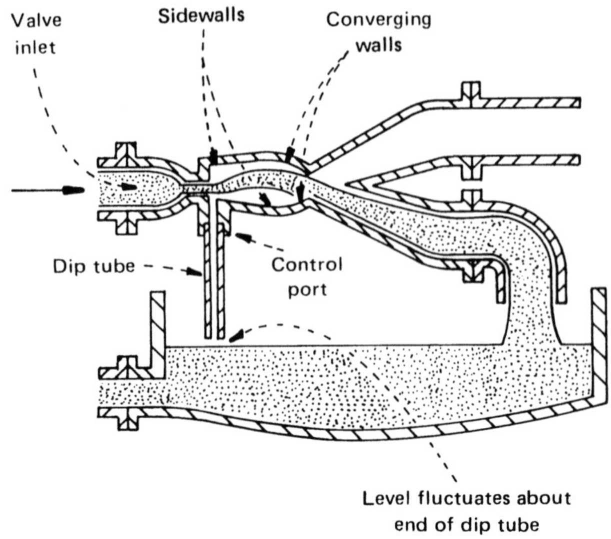

FIG. 7. Diverter valve sets the level in a receiver.

The diverter valve can be modified to a self-contained level control valve, as shown in Fig. 7. This valve is based on the principle of the monostable flip–flop. There is only one control port, and the valve body is asymmetrical. When there is no pressure at the control port (dip tube closed on high level), the jet discharges through the upper outlet port. As the level drops and the dip tube is uncovered, the jet breaks away from the lower sidewall, attaches to the upper sidewall, and discharges through the lower outlet, refilling the vessel.

Proposed applications of the diverter valve must be carefully studied and properly engineered to assure satisfactory valve operation. Flow through the valve depends upon the pressure drop between the inlet and the splitter. The pressure at the splitter is the same as that of the non-flowing outlet. The pressure at the flowing outlet is greater than that at the splitter and the outlet port. When the pressure at the splitter is atmospheric, the pressure at the flowing outlet is approximately 60% of the inlet pressure. If the pressure at the flowing outlet is less than this, air will be aspirated from the nonflowing outlet into the fluid stream. If pressure is more than this, leakage of fluid into the nonflowing outlet will occur.

The two-position diverter valve should only be used where the flow required through either port is the same, and upstream conditions are constant. Any fluid may be used for control. If the control ports plug due to entrained solids or from cooling or evaporation from the ingested air, a purging fluid may be used for control. If dilution is a problem, a pulse may be used to divert the flow.

By modifying the design of the valve, proportional diversion can be achieved. In this case, the valve becomes a proportional amplifier. An oscillating type is also possible.

Fluidic Logic

Fluidic logic systems may incorporate devices having nonmoving and moving parts. True fluidic logic devices have no moving parts. Logic performed with devices having moving parts (moving parts logic or MPL) might better be referred to as pneumatic logic control. Pneumatic control is a long-established technology that makes use of diaphragm operated pilot valves, check valves, poppets, etc. Hybrid systems are frequently indicated where the output elements, because of their greater power handling ability, are moving parts devices. Moving parts devices often contain fluidic logic and are often called fluidic devices but, strictly speaking, this is a misnomer.

The difference between fluidic and air control devices is analogous to the difference between electronic and electromechanical devices. Fluidic and electronic devices have no moving parts, operate at lower power levels than their air control and electromechanical counter-parts, are smaller, and are generally less expensive. Both fluidics and electronics, to varying degrees, lend themselves to physical integration.

Perhaps the most common fluidic logic element is the bistable flip–flop shown in Fig. 8. Application of control pressure at either of the control parts

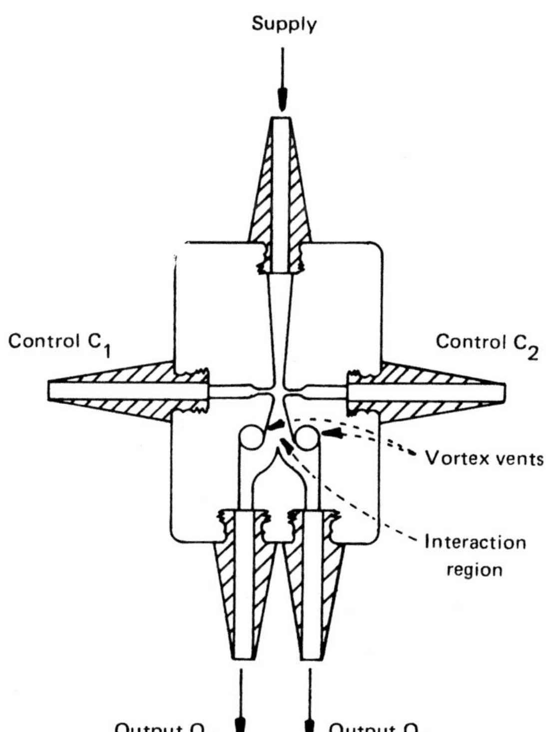

FIG. 8. Bistable flip–flop is under dual control.

will divert the supply jet to the opposite output port, where it remains after the control pressure is removed. If a flowing outlet is blocked, the jet will be diverted to the other outlet, but it returns to the original outlet when the blockage is removed. This characteristic of fluidic devices is referred to as memory.

By changing the internal geometry and eliminating one of the control ports, the flip–flop becomes monostable. The jet discharges through one of the outlets and is diverted to the other only when pressure is applied at the control port. The flip–flop in Fig. 8 has connections for flexible tubing for use in a breadboard design configuration (i.e., an experimental working model).

Connections for rigid tubing can be supplied, or the flip–flop may be incorporated in an integrated circuit module containing other fluidic elements. The latter may consist of a block of plastic, formed from plates stacked together, with many passages and cavities creating a series of interconnected fluidic elements. The gain of a flip–flop as a digital amplifier is typically 3 to 4. Switching time is typically 1 ms.

With some changes in configuration, the flip–flop becomes an AND or OR-NOR gate as shown in Fig. 9. The AND gate has three output ports and two supply jets. There is a signal at Output 2 only when there is a signal present at both supply ports. The OR-NOR gate is a monostable flip–flop with two control ports on the same side. With no signal at either control port, the jet discharges from the left-hand output port. When a signal is present at either one of the control ports, the jet is diverted to the right-hand output port.

The turbulence amplifier, Fig. 10, may also be used for logic functions. This device is based upon fluid-dynamic phenomena first described by Lord Rayleigh in the nineteenth century. A jet of fluid flowing in the laminar mode becomes turbulent when interrupted by a transverse control jet. A single turbulence amplifier is the counterpart of the monostable flip–flop, and functions as an OR-NOR gate. Three NOR gates may be combined to provide the AND function. Various combinations provide other logic func-

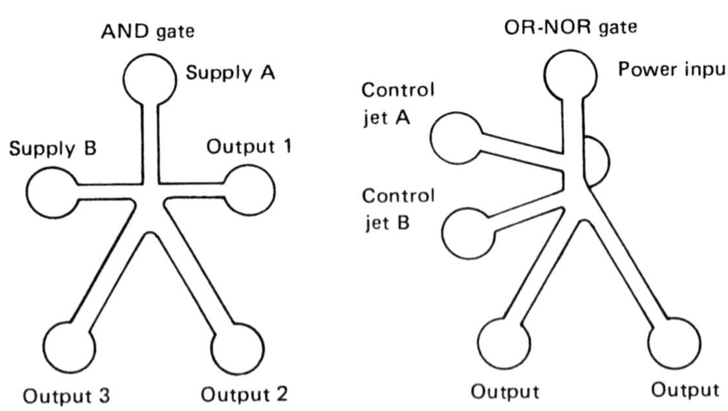

FIG. 9. Fluidic devices for AND and OR-NOR logic.

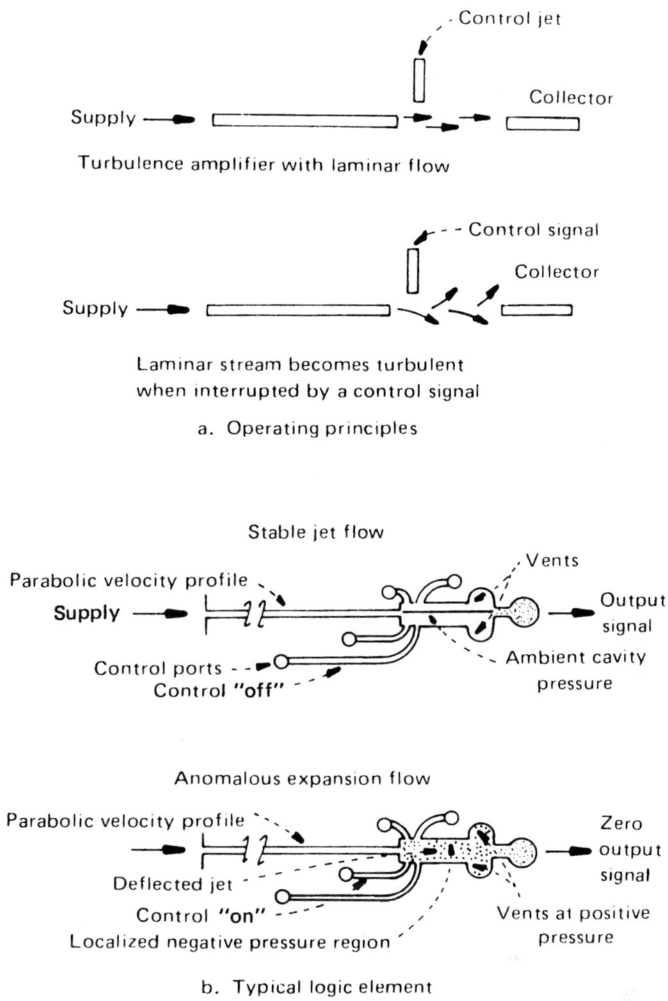

FIG. 10. Turbulence amplifier for fluidic.

tions. Like wall attachment logic element, turbulence amplifiers may be integrated in the form of plastic (or other material) blocks, either with internal connections for a specific operation or with external connections so that the user can build his own logic system.

Another type of proportional fluidic device is a vortex amplifier, shown in Fig. 11. The essential parts of this amplifier are a cylindrical vortex chamber, a supply port, one or more control ports, and an outlet port. The primary flow is admitted through the supply port, and the secondary or control flow is admitted through the tangential control ports. The orifice size of the outlet port determines the level of flow which the amplifier can handle. The vortex chamber contains the flow passing through the device so that a fluid vortex can be generated.

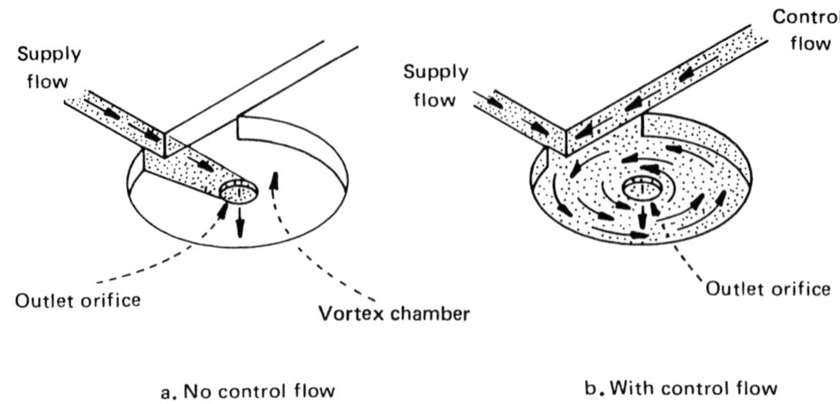

FIG. 11. Vortex valve has no moving parts.

When control flow is injected into the amplifier, a vortex is generated which reduces the outlet flow rate. This flow reduction is the result of the radial steeper pressure gradient and therefore less outlet flow. When the control flow is discontinued, the vortex disappears and the power flow returns to its original state.

A vortex amplifier is essentially a variable restrictor which achieves resistance to flow by means of the constrained vortex. Vortex amplifiers will modulate flow from full output down to 10 to 20% of full output. This is a negative gain device.

The impact amplifier, shown in Fig. 12, is another concept for proportional amplifiers. It uses two axially opposed power jets, $S1$ and $S2$, to provide a planer "impact region." The input signal is applied at port C, and the amplifier output is at port O. When the input flow is increased, the impact region is moved to the right. This results in increased flow from port O. The amplifier is a single-side arrangement. The outstanding feature of the impact

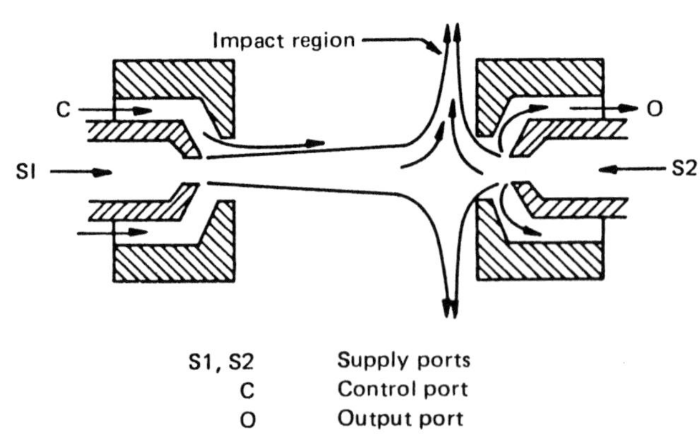

FIG. 12. Impact amplifier.

amplifier is its high gain with reported values up to 200. This amplifier can also be operated as a digital device.

Passive circuit elements, in addition to basic fluidic elements, are usually needed to assemble a practical circuit. The most generally used passive elements are:

Restrictors to provide fluid *resistance*.
Volume chambers to produce fluid *capacitance* (for gases only)
Long, small cross-sectional area line to provide fluid *inductance*

Fluidic display systems consist basically of pistons having one end coated with a colored reflective surface. One large piston may correspond to a bull's-eye light, or many small pistons may be arranged in a matrix for displaying alphanumeric characters. A fluidic signal causes the pistons to extend out and become visible. The chief advantage of these displays is that they are 100% explosionproof. if desired, these systems can be interfaced with electrical display systems.

Applications of Fluidic Systems

Most commercial applications of fluidics have been in the machine-tool industry and assembly line operations, since air gauging and air sensing are well established in these areas.

Fluidic devices are the most versatile and simplest way of sensing the presence or absence of objects on a conveyor. They are so precise that they can detect the presence or absence of a paper label on an object. Fluidic logic is readily adapted for sequencing operations and is generally used with fluidic sensors. However, fluidic sensors can readily be interfaced with electronic or electromechanical logic.

Some examples of fluidics in the machine-tool industry are protection of the dies on stamping presses, safety controls on presses, and verification and checking in machining operations. The presence or absence of a part in a die is sensed very simply by drilling small orifices in the die and detecting backpressure.

A modification of the air gauge is used to detect changes in diameter of fiber or wire filaments, rope, rods, etc., and the density of filter bundles. Fluidic turbine governors are available. Exhaust flow and temperature of gas turbines is being measured with fluidics.

In general, fluidics has not penetrated the chemical process industries, except for very specialized applications. Fluidic devices have not replaced conventional pneumatic and electronic instrumentation. This is not meant to imply that there is no place for fluidics in these industries. As successful applications increase in number, variety, and ingenuity, fluidics will find a place in the CPI.

Some successful applications of fluidics that should be considered are:

1. Position sensors for sensing valve or damper position, or excess expansion of a rotary-dryer shell. The position sensor when combined with the proper logic circuit may be used as a motion detector or speed sensor if placed near the teeth of a rotating gear or other moving device that will provide a continually interrupted signal.
2. The acoustic sensor might be used as a high level or low level detector for dry, granular material in a bin. The 50-kHz signal used in this device will penetrate porous material such as foam rubber. Hence, such material with backup could be used to protect the transmitter and receiver from contamination.
3. Sequence control and timing is readily accomplished with fluidics by using fluidic resistance/capacitance circuits. A fluidic sequence control system has been developed for programming a process chromatograph. For long cycle times, a hybrid system using electromechanical timers may be used.
4. Level control with a diverter valve has already been noted. On–off control is easily achieved with fluidics. Commercial units are available for single or multilevel sensing. These use a dip tube and Schmitt triggers (also known as differential pressure sensors). These systems can be used on closed tanks if the pressure does not exceed 10 lb/in.^{2}.
5. Fluidic flow sensors have already been described. Their use in flare stack flow measurement is increasing as environmental controls are tightened.
6. Gas densitometers are good for high temperatures and pressures. A commercial unit is available with high sensitivity and speed of response.
7. Temperature sensors have been successfully used in high temperature service, and in oxidizing and reducing atmospheres.
8. Any logic function can be performed with fluidics, whether or not fluidic sensors are used. In general, fluidic logic should be seriously considered wherever their unique characteristics can be employed to advantage, or where environmental factors may jeopardize the functioning of electrical or electronic devices.

Advantages of Fluidics

The following advantages of fluidic systems should be taken into account.

Simplicity. Fluidics is based upon a few basic operating principles. Once these are understood, troubleshooting and modifying existing circuits, or designing new ones, becomes relatively simple. A specialized technical background is not required. There are no moving parts and no mechanical adjustments to be made as with standard pneumatic instrumentation. Troubleshooting generally consists of checking pressures with a manometer at various points in the circuitry while the system is operating. Routine maintenance is not required except the replacement of air filters.

Compactness. Fluidic systems are smaller and lighter in weight than comparable electromechanical or standard pneumatic systems. In this respect, however, fluidics cannot compete with microelectronics.

Speed. Fluidic systems operate at speeds sufficiently high (typically 1 ms per logic function) for most industrial operations.

Reliability. Fluidic systems are highly reliable provided care is taken in the design of the system. For practical purposes there are two causes of failure of fluidic systems: system interaction and dirt. In the early days of fluidics, some systems were designed that failed due to system interaction (interference and feedback between signals and components). Fluidic system designers have now become more sophisticated. Hence, system interaction can be ruled out as a cause of failure. The other cause of failure of fluidic systems has been dirt—primarily from the air supply. Better filters and regular filter changes have eliminated this problem. Reliability of fluidic systems in high temperature service is five times that of equivalent electronic systems. Fluidic systems are also immune to nuclear radiation as well as electromagnetic radiation.

Nonhazardous. Fluidic systems are 100% safe in hazardous areas.

Noise-free. Fluidic systems are not affected by vibration, electrical, or electromechanical noise.

Disadvantages of Fluidics

Energy and Speed Limitations

Although much of fluidics is based on simple and well-known scientific principles, a significant portion of the phenomena are not clearly understood, such as, for example, the Coanda effect. Fluidic devices suffer from relatively low speeds. Also, only low pressure is used and thus only low power output levels can be attained. Direct actuation of powered equipment is, in most cases, not possible, which requires amplification. This is expensive, and may negate to some extent the reliability and simplicity of fluidic devices in complete systems.

The reasons for the low speeds and pressure lie in the fact that almost all these devices use air for power. A supply of 5 lb/in.^{2} air is adequate for 0.01 std. ft^3/min for a single logic element to 0.5 std. ft^3/min for a proximity sensor.

The velocity of gases going through ordinary nozzles is practically limited to the speed of sound in the particular gas. In air, sonic choking of a nozzle occurs at approximately twice the vent pressure. And since almost all fluidic devices are vented to atmosphere, the upper pressure limit is 35 lb/in.^{2}. It may be possible to go to supersonic flow and thus increase both velocity and pressure. However, considerable basic research is required to understand shock wave phenomena and to design suitable equipment.

Relatively incompressible fluids, such as hydraulic oil, have been attempted in fluidic devices, but with minor success. At low pressures, liquids are not economical, while at higher pressures, even only 100 lb/in.^{2}, cavitation occurs with resulting nozzle erosion.

System Limitations

Highly complex systems involving a great number of individual functions are not practical for fluidics because of physical size and cost limitations. A fluidic minicomputer would be completely impractical. A total of 1000 logic functions is an extreme limit for a fluidic system.

A corollary limitation is speed. A relatively simple fluidic system may run at speeds more than adequate for the operation that is monitored or controlled. As complexity is added, the system may become too slow. Long transmission lines should be avoided.

Practical Considerations

To ensure a successful fluidic installation containing logic, the process engineer should do the following:

1. Seek experienced help. The vendors of fluidic systems have capable engineers to help one determine whether fluidics is right for the application and to help design the system.
2. Provide a reliable, clean, and dry air supply. Most manufacturers of fluidic devices recommend a 1-μm filter and mist coalescer on each air supply line.
3. Keep the equipment clean before installation and make sure no dirt gets in during installation. Factory assembled parts and modules are generally put together in clean rooms to avoid contamination.
4. Provide filters for vents on fluidic devices that operate under slight vacuum and aspirate ambient air.
5. Keep the equipment clean during shutdowns. If the equipment that is monitored is shot down, the fluidic sensors should be kept in operation or covered up to prevent dirt from getting in and finding its way to the logic elements. Isolators are available that will isolate the sensors from the logic elements, and block the passage of dirt. The sensors are self-purging and less affected by dirt.
6. Provide commonsense mechanical protection. Sensors should not be installed where they may be used as handholds or footsteps. Plastic tubing should be protected from heat and installed in trays, ducts, or conduits, if necessary.

Comparative Costs

How does the cost of fluidic systems compare with equivalent electro-mechanical and electronic systems? Relay logic, using low power reed relays, is somewhat less expensive than fluidics. Electronic logic using integrated circuits is much less expensive than fluidics—typically 10% of the cost of equivalent fluidic systems. However, installed costs of fluidic systems compare favorably with electro-mechanical and electronic systems.

Bibliography

American National Standard Glossary of Terms for Fluid Power, ANSI/B93.2, American National Standards Institute, New York, 1971.

"Fluidic Oscillator Measures Flow," *Control Eng.*, p.56 (September 1973).

What You Should Know about Fluidics, Technical Manual, National Fluid Power Association, Thieusville, Wisconsin, 1972.

Fluidic Control (When to Use It), General Electric Co., Schenectady, New York, 1970.

Fluidics, General Electric Co., Schenectady, New York, 1970.

"A High Performance Reactor Shutdown System with Fluidic Controls," *ISA Trans.*, pp. 124–130 (December 1973).

Product Bulletin, FluiDynamic Services Ltd., Mississauga, Ontario, Canada, 1971.

Iron Age, p. 55 (July 1, 1965).

NASA Contributions to Fluidic Systems, NASA SP-5112, National Aeronautics and Space Administration, Washington, D.C., 1972.

Engineer, p. 14 (January/February 1969).

Chemical Engineering, pp. 117–124 (December 9, 1974).

HARRY F. FABISCH

3

Piping Design

Piping Design

The piping designer needs three essential source documents: engineering flow diagrams, nomenclature, and equipment elevations. These documents are usually furnished by the piping analyst.

Engineering Flow Diagrams

Engineering flow diagrams contain a schematic flow of the process connecting the various items of equipment such as columns, exchangers, pumps, etc. They include all the information essential for the operation of the process.

Figure 1 shows part of a typical engineering flow diagram. In addition to showing schematically the piping required, the engineering flow diagram also includes all the valves and control devices that will be required.

Engineering flow diagrams do not present all the essential information for the piping designer. They do not include equipment elevations and insulation specifications, which may be obtained from the *equipment elevation summary* and from the *nomenclature list.*

Every line on the engineering flow diagram is identified by a special number. This same number is used in the nomenclature list and on the equipment elevation list. In addition to information from the process flow diagram, the engineering flow diagram contains start-up conditions required by the unit, normal operating conditions, and shutdown requirements. It also shows the valving and piping required for spare pump arrangements. The number of heat exchangers and the shell-and-tube inlet and outlet connections are also shown.

On the average-size process plant engineering flow diagram, there are approximately 1500 separate pipe lines. Information on each of these lines is absolutely necessary for the piping designer. The information that must be shown on each line includes the size, the line specification showing materials of construction, the wall thickness, the flowing fluid, and the equipment or line number at both terminals of the line. The routing of the line is shown by the use of arrows. Normal and emergency temperatures and pressures are given for each line. Field testing requirements are shown. The type of tests and testing pressure are specified in the nomenclature.

Symbols

The symbols on the engineering flow diagram are key tools for piping design. Every engineering contractor or operating company uses a standard set of

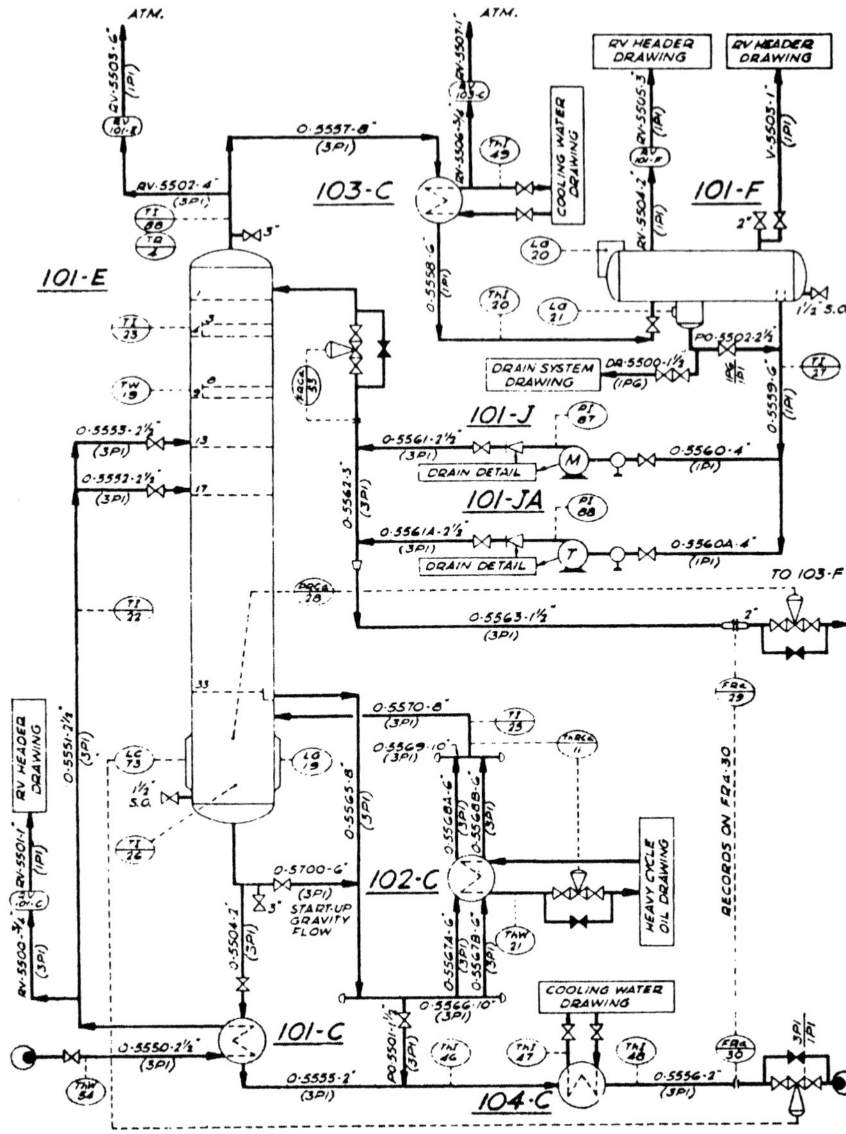

FIG. 1 Portion of typical engineering flow diagram.

piping symbols; all are based on the ANSI standard symbols for process piping. The only fitting symbols which usually appear on engineering flow diagrams are symbols for reducers, which indicate a change in line size, and the cap symbol, which indicates a header.

Symbols for instrumentation are usually based on the standard reference *Basic Instrumentation Symbols RP5* developed by ISA. Instrumentation symbols fall into three categories: temperature, pressure and flow indicators.

Process Flow Diagram

The process engineering department furnishes the piping analyst a process flow diagram (see Fig. 2). The flow diagram shows how the various items of process equipment are interconnected. It indicates the flow rates, specific gravity, molecular weights, general flowing materials, expansion factors, temperatures, and pressure. It also shows the process requirements for fixed points in the process.

Process data sheets supplement the process flow diagram. These sheets give physical data required of the process equipment, such as vapor–liquid

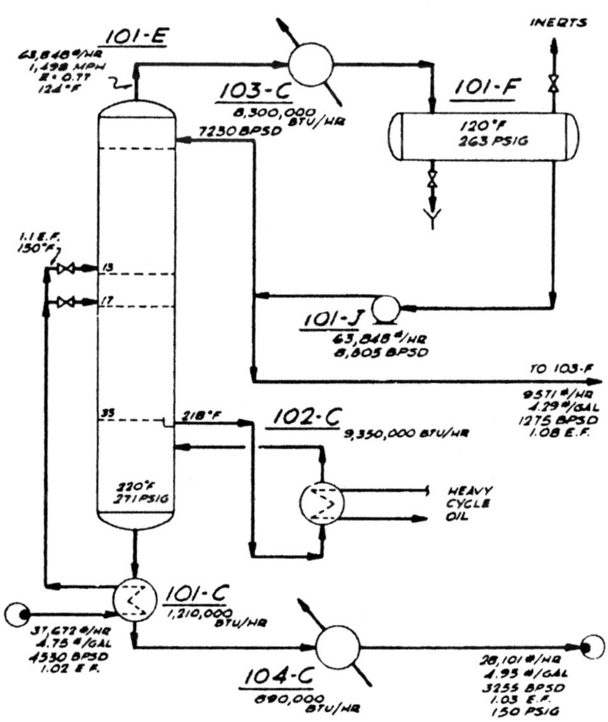

FIG. 2 Portion of typical process flow diagram.

proportions of tower trays, physical data, safety factors for pumps, detailed furnace flow conditions, etc.

Process Control Diagram

This diagram, supplied by the Instrument Department, shows the entire instrumentation of the plant (see Fig. 3).

Specifications

The specification group provides the piping analyst with detailed design requirements for the piping, valves, and other piping-related equipment.

The specifications give the minimum pipe wall thickness, the pipe schedule, and insulation requirements.

The specifications include a print of the process flow diagram showing special piping materials such as glass-lined pipe, alloy pipe, high-pressure pipe, corrosion allowances, etc.

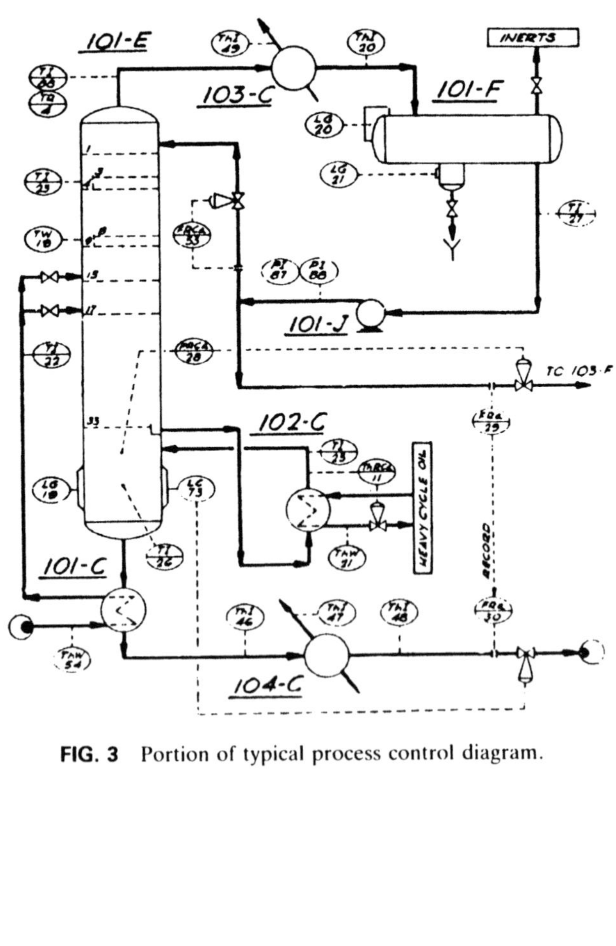

FIG. 3 Portion of typical process control diagram.

Engineering Design Data

The project engineering department furnishes the piping analyst special requirements for all utility and auxiliary systems which are not shown on the process flow diagram.

The auxiliary and utility flow diagrams are laid out according to a plot plan arrangement. The schematic form is not used for these diagrams.

Equipment Elevations

The piping analyst will furnish the piping designer elevations for the main equipment items and for the critical circuits where gravity flow is necessary, for example. The gravity flow requirement can easily change the location on the final drawings made by the piping designer. The piping analyst does not want to restrict the piping designer but must provide him with critical elevation points and special elevation requirements.

Several major elevation requirements are: Pump NPSH which sets the fractionator tower elevations based on the bottoms line condition going to the pump.

Reboiler circuits may set tower elevations. A thermosyphon reboiler requires enough liquid static head to provide a driving force so that the reboiler will work properly. This head determines the circulation ratio and the amount of vapor returned to the tower, thereby setting the entire tower gradient. Reboiler circuits must be considered with pump NPSH considerations as they set the tower elevation.

A flashing liquid must have enough static head to offset the line friction loss and the total loss through an orifice. When the process includes flashing liquids, the vessel from which the liquid is being drawn and the orifice flange itself must be given a definite elevation or a relative elevation.

Gravity flow determines the relative elevations of related equipment and probably will determine the exact elevation of the related equipment itself.

The Piping Analyst

Using information supplied by the process, instrument, specification, and project engineering groups, the piping analyst can review the process design and expand this design on the engineering flow diagram. In addition, he can design the utility and auxiliary systems required to support the process flow.

To do this, the piping analyst decides on the valves required to meet the process and specification requirements including all the instruments shown on the process control diagrams. He then sizes all the lines to be sure of a realistic pressure drop. He indicates on every line the material specifications and the points at which the specification changes. He schematically draws on the engineering flow diagram the actual number of lines required which may have been a single line on the process flow sheet.

The Piping Designer

From information furnished by the piping analyst, the piping designer sketches the most economical piping arrangement. He starts with the larger size lines and most expensive alloys and follows down to the least expensive, smaller lines. He completes an overall layout and draws individual isometrics of each line. On a summary sheet, he shows whether the pipe is shop or field fabricated. For the shop-fabricated piping, he shows the break points so that the shop will provide flanges for field assembly.

Piping Layout

Main Process Flow Lines

Main process flow lines represent the main process flow. Such streams pass through furnaces, reactors, and dryers. They continue as tower bottom and feed inlet to the next tower, often with exchangers and pumps between them. These lines will be shortest if towers are arranged in process flow sequence as close to each other as equipment sizes and access space permits. With smaller interconnecting lines, towers can be located farther apart with little increase in piping cost. For example, cooling water lines can be shortened if condensers can be grouped between towers.

A common steam line can be designed for grouped reboilers. Grouping condensers and reflux drums allows a common supporting structure. Figure 4 shows an example of an alternative tower arrangement. Many configurations are possible and justified if shortening of these process lines is the ultimate result.

Process flow is not always a simple straight through-flow but can split into two or three streams, as is often done with a number of distillation columns. Subsidiary circuits to process flow, such as the refrigeration circuits in ammonia or ethylene units, must also be considered.

Equipment Interconnecting Lines

The spacing of towers depends on the number and size of other equipment connected to them. This leads to the second group of lines on a process flow diagram, lines which interconnect closely related equipment. For example, this group includes the pipe lines interconnecting towers with reboilers and condensers. These are generally large-diameter lines and should have preference over the first group which are usually smaller process lines.

Feed and Product Lines

Feed lines and the usually small-diameter product lines can be minimized if they start at equipment close to that bettery limit where feed and product lines terminate.

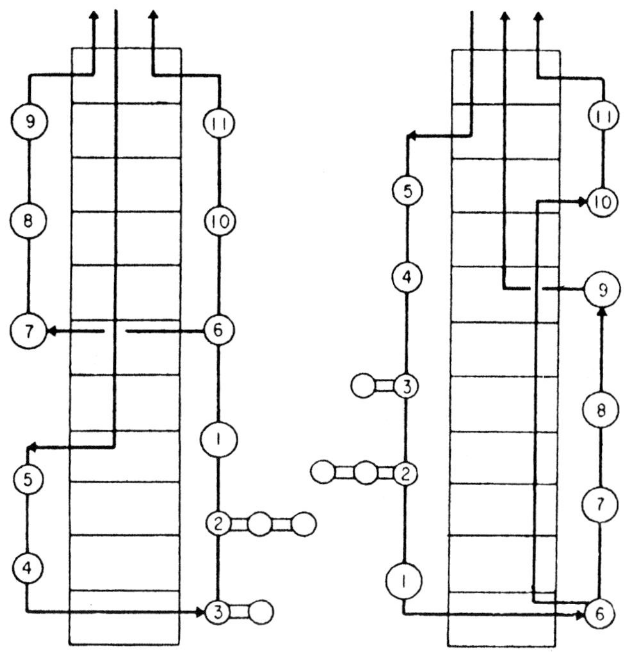

FIG. 4 Alternative tower arrangements can shorten main process flow lines.

For pipe sizing, process equipment sizes are required. These include types and sizes of vessels, heat exchangers, pumps, compressors, intercoolers, scrubbers, silencers, floor space of lube and seal-oil console, furnace and heater details, control, and switch house arrangements.

Location for Minimum Pipe Runs

For plant layout, in addition to tower sequence, every equipment item has an optimum location for minimum pipe runs. For exchanger, drum, and pump locations, the following general classifications can be made.

Exchangers

Thermosyphon reboilers and condensers use short pipe runs. Short reboiler and overhead lines are essential for both economy and reliable operation.

Some exchangers should be close to other process equipment; for example, exchangers in closed pump circuits such as some reflux circuits. For a bottom-draw-off exchanger pump, the flow exchangers should be close to the tower or drum to give short suction lines.

The preferred location of exchangers between two distant process equipment items is in the yard piping where the two streams meet. These exchangers have process lines connected to both the shell and tub side. Figure 5 shows the extra piping required for the exchanger located in the process equipment area instead of either preferred location in the yard piping.

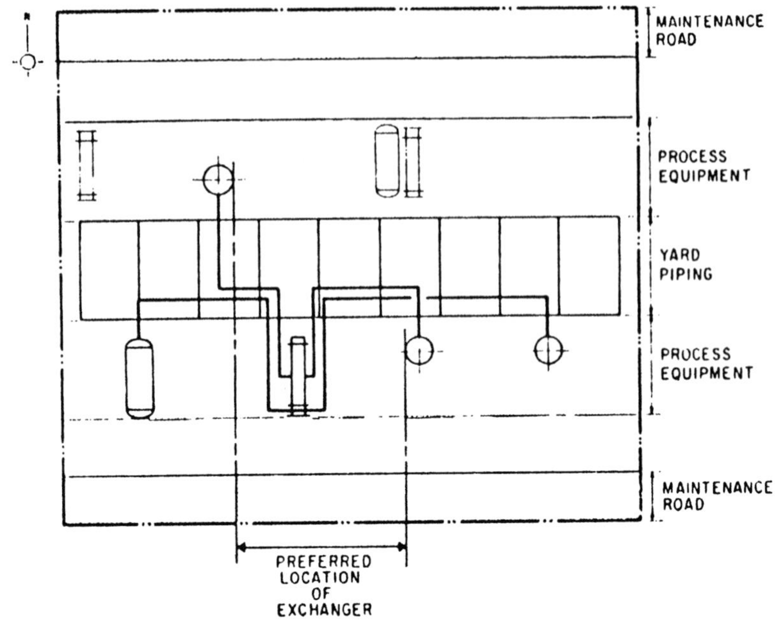

FIG. 5 Exchanger location for minimum pipe runs.

Exchangers located between process equipment and the unit limit can be located at one end of the plant. Product coolers are an example.

Drums

When a tower bottom flows by gravity into a collecting drum, the drum should be under or next to the tower. A reflux drum should be next to the condenser. Compressor suction drums and knock-out drums should be close to the compressor.

Most process and utility drums serve as separators, surge, and reflux drums. These drums should be arranged in process flow sequence.

Storage drums or tanks, located within a unit, usually are given secondary consideration and are located as space permits, mostly at the outer limits of the unit.

Construction

Expense can rise rapidly with poor access to equipment or difficult-to-erect piping. Available crane and construction clearances, access width and location requirements, and difficult construction points must be checked for each plant arrangement. In some arrangements, increased structural and piping cost for simpler construction is more than offset by the saving in construction cost.

Operation and Maintenance

Road access is essential for exchanger bundle removal, for tower tray removal, to pumps, for catalyst loading and removal, for crane access to compressors, etc. It is advisable to study the piping layout from this standpoint. Remember that roads and access space spread the unit apart and add to the length and cost of piping.

Convenient and adequate access to points of operation and instrument adjustment is essential. Grouped, lined-up manifolds, and functionally located control valves help maintain economical operation.

The Plot Plan

Figure 6 shows a plan of the feed gas compressor area for a 200,000 tons per year ethylene unit. Main pipe runs are also shown. It is an in-line-layout with equipment in process flow sequence. The large diameter gas lines directly interconnect process equipment. On the complete plot plan, equipment (including compressors) is arranged on both sides of a central yard in process flow sequence. Pumps are located at their point of suction and are

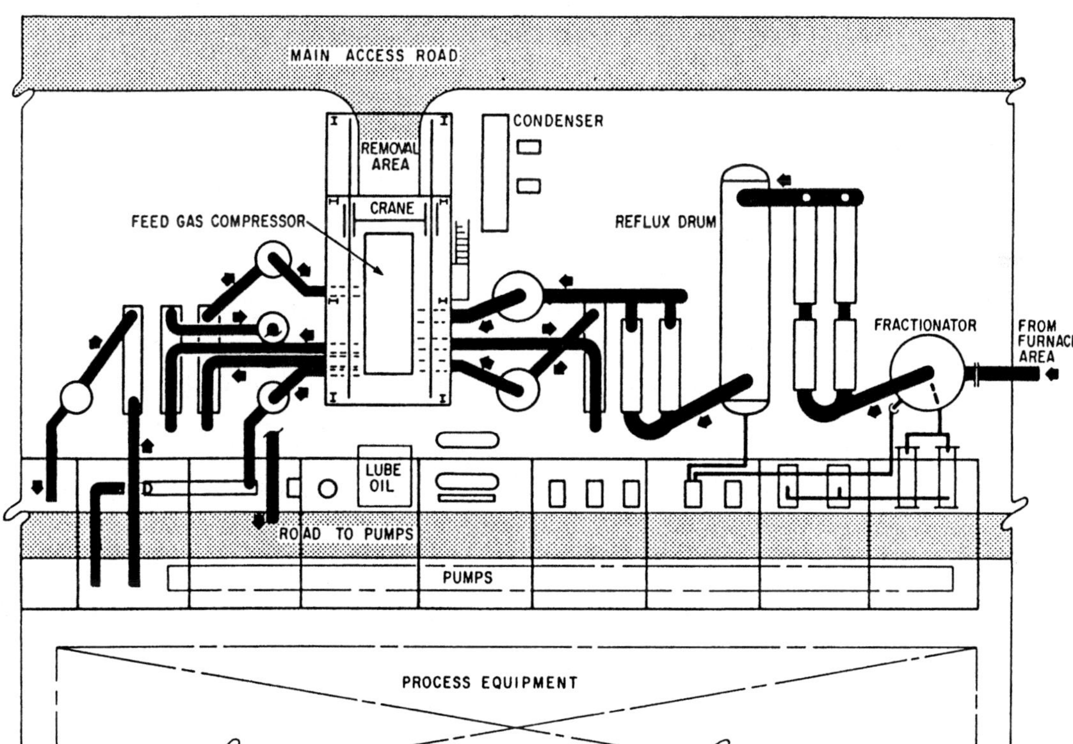

FIG. 6 Part of ethylene unit plot plan showing direct routing of piping.

lined up under the yard. A parallel road is arranged to every line of equipment for convenient construction and maintenance access.

Many equipment groupings can be used for economical plot arrangements. Two obvious groupings are furnaces and reactors. Small furnaces, however, are often placed in several locations as required by the process flow. For safety and economy, these furnaces should be located at the outer limits of the process unit.

Another often-used equipment grouping is of housed compressors. Economy is achieved here by common building and maintenance facilities and by the operation of the grouped compressors.

In Fig. 6 the feed gas-compressor has been separated from the refrigeration compressors. The savings in piping and construction cost justified building two compressor houses.

Some layout systems use similar equipment groupings more extensively. Several towers can be lined up fairly close to each other on one side of the yard, providing common interconnecting platforms. Piping economy is usually sacrificed for convenient access to manholes, valving, and instruments on the towers. Exchangers can also be grouped on the other side of the yard and a common gantry crane provided for convenient maintenance. In such arrangements, tower, overhead, and other process lines to exchangers cross the yard, increasing pipe length and number of fittings.

When piled foundations are used, regrouping equipment can often save a number of piles, more than paying for increased piping cost.

Yard Piping Economy

The main arterial system of a plant is yard piping. It is here that long process lines are located interconnecting distant equipment with lines entering and leaving the unit. Also located in the yard are utility headers supplying steam, air, gas, and water to process equipment. Here are located all relief and blow-down headers. Often instrument lines and electrical supply conduits are also supported on the yard steel.

Figure 7 shows those critical dimensions which influence yard piping costs.

Dimension A is the total length of the yard. It is governed by the amount and size of equipment, structures, and buildings arranged along both sides of the yard. If the same amount and size of equipment can be arranged on a shorter yard length, yard piping can be reduced considerably. Equipment in pairs, stacked exchangers, exchangers under elevated drums, drums or exchangers supported on towers, two vessels combined into one, closely located towers with common platforms, drums supported on exchangers, and process equipment located under the yard are only a few examples of ways to help shorten yard length. These arrangements, of course, not only shorten process lines—either directly interconnecting equipment or in the yard—but also shorten lines passing through this area and utility headers serving this area.

Equipment arranged along but not associated with the yard increase yard piping cost unnecessarily. A control house located along the yard, for ex-

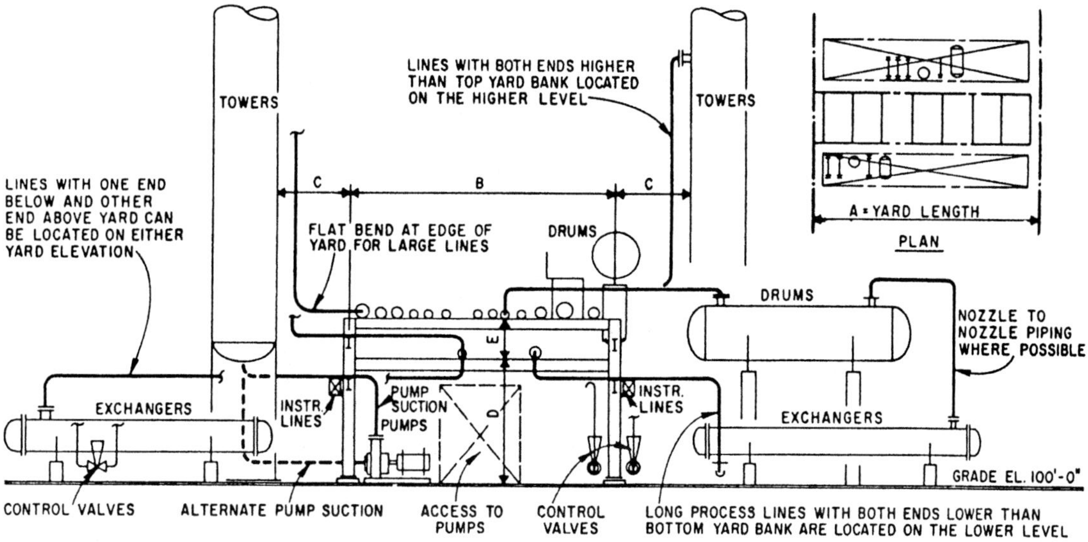

FIG. 7 Typical cross-section of yard piping showing general pipe runs.

ample, will increase yard piping cost because all lines must pass by without really being associated with the relatively long control house.

The careful selection of Dimensions B and C (Fig. 7) can minimize the pipe length between the yard and process equipment and the pipe length interconnecting equipment on opposite sides of the yard. Keeping yard height no higher than necessary (Dimension D and E) will minimize vertical pipe runs.

When changing direction, change elevation is an old rule in piping design. It applies to all lines connecting to yard piping. However, some large diameter lines can make a flat turn when entering at the edge of the yard.

Cost and Equipment Elevations

Towers, drums, and exchangers can be elevated for the following reasons.

Pump Net Positive Suction Head

A pump with a high NPSH requirement pushes equipment to a higher elevation, increasing the cost of piping and supports. On the other hand, a pump with a lower NPSH may cost more. The difference in support and piping costs must be considered in the selection of a more expensive pump with a lower NPSH.

With turbine surface condensers, it often pays to choose a pump with a lower NPSH, especially when the condenser is directly below the turbine. The compressor house floor can be lowered. With the lower elevated condenser, operation and maintenance access is improved. Vertical pumps are usually specified, with a minimum height from grade to equipment because their suction inlet nozzle is below grade.

If for some reason equipment is elevated above the required NPSH, a reduction in line size and pump differential is often possible.

Thermosyphon Reboiler Circuits

The driving force in a reboiler circuit is the static head difference between the head of the liquid draw-off line and that of the vapor–liquid mixture in the return line minus friction loss. For horizontal reboilers at grade, an increase in driving force requires a greater elevation of the tower or drum. Line sizes can be reduced because a higher friction loss can be allowed. Decreasing the vertical legs of reboiler piping will also decrease driving force. Consequently, the line size of the system must be increased to provide lower friction losses.

Liquid Flow Measurement

The requirement of accurate liquid flow measurement can also elevate process equipment (see Fig. 8). If liquid is near the boiling point, a static head is required in the front of the control valve to overcome pipe friction losses and avoid flashing in the line. Minimum equipment elevation, orifice range, and minimum line size can be used if the orifice is as close to the equipment as possible and the piping has only one elbow up to the control valve.

Grade Location

The most economical and common location of process equipment is at grade. Supporting structures and platforms are not required. Construction is easy. Most valves and instruments can be made accessible from grade. Operation and maintenance are convenient. Elevated equipment with its structures, platforms, handling beams, etc. means cost increases in several design areas.

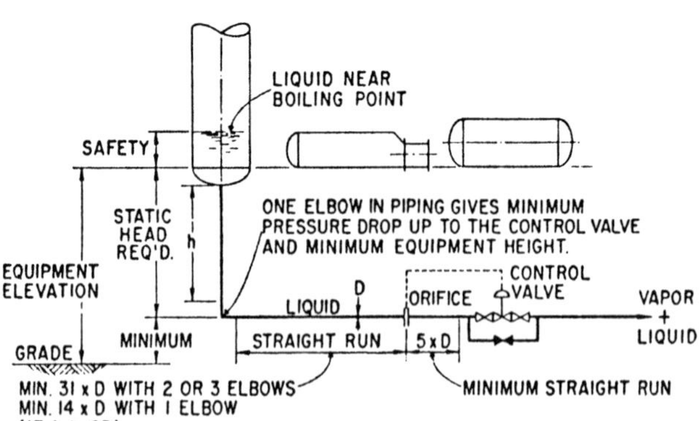

FIG. 8 Put control valve close to process equipment for economy and reliability of operation.

In layout and design, the first attempt should be to eliminate structures, extra supporting columns, and extra platforms. Smaller equipment can be supported on towers, on yard columns, or structures for larger equipment. The second attempt can be to combine two or three structures supporting elevated equipment. Some equipment regrouping will increase the overall plant layout, but savings in structural costs can be greater than the increase in piping costs.

Piping Design Economy

Line sizes give a readily available basis for comparison. However, accurate cost estimation depends on weight, type of material, insulation, and construction. Consequently, pipe lines for economical comparison are better represented with an in-place cost figure per unit length than with line size, schedule, and material alone. Special attention should be given to alloy lines, high pressure piping, and large diameter carbon steel piping. As a rough comparison, in-place piping cost is about double material cost for carbon steel piping.

At an early stage of plant layout, line sizes are not available. Two items of process data from the process flow diagram give a good feel for rough line size comparison—flowing quantities and pressure differences between two vessels. Obviously, smaller flow quantities or higher available pressure differences for friction losses will allow smaller diameter lines. For suction and discharge to pumps, only quantities should be compared for the feel of line size.

For line size calculations, several companies use economical pressure drops. This is the most direct approach because available pressure differences are readily accessible from the process flow diagram. A hydraulic slide rule has been developed for fast and convenient liquid flow calculations.

Line sizes based on velocity limitations are calculated only in special cases where corrosion, erosion, or deposits on the pipe wall have to be accounted for or where critical flow conditions exist.

Valves are generally line size. The maximum control valve size is line size. In most applications, control valves are one size smaller than line size. When a larger pressure drop is available, control valves can be two or three sizes smaller than line size.

Sometimes it is feasible to compare piping costs and the cost of control valve assemblies. Increased line size will give a higher proportion or pressure drop of the total line loss for controlling. Perhaps a smaller control valve can be chosen. Money spent to increase line size should be less than the savings for a smaller control valve assembly.

Orifice Runs

Because of metering accuracy, orifice runs are often increased for the necessary straight runs. With very large lines, a cost comparison can be made between Pitot tube cost plus the necessary straight runs of piping or the more expensive orifice and flanges and shorter straight runs of piping. It should be remembered that there is a much higher friction loss through an orifice than when a Pitot tube is used. But an orifice needs a shorter straight run of pipe with less friction loss than a Pitot tube.

Safety devices should be piped with a minimum of inlet pipe runs because excess pressure drop before a safety valve will affect its operation. For a minimum of discharge piping, relief valves are located high on a tower in open relief systems. In closed systems, relief valves are located just above the relief header.

Reboiler Piping

Two types of thermosyphon reboilers are used: vertical and horizontal.

A vertical reboiler has very little piping and its length determines the height of the tower skirt. Supports at grade are saved but supports on the tower must be added.

Many towers have a bottom draw-off pump. NPSH requirements usually elevate the tower above the reboiler's minimum height, thereby increasing the static heads in the reboiler's vertical legs and the driving force in the circuit. With increased tower height it is worthwhile to check the reboiler circuit to see if liquid and return line sizes can be reduced.

A symmetrical piping arrangement between the draw-off and reboiler inlet nozzles and between the reboiler outlet and return connection on the tower is preferred for equal flow in the reboiler circuit. Nonsymmetrical arrangements may also be accepted for a more economical or more flexible piping design.

Overhead Lines

Several variations exist for overhead reflux circuits. A condenser can be elevated above the reflux drum or the reflux drum can be elevated with the condenser at grade. These arrangements can be adjacent to or somewhat remote from the tower. The simplest overhead line is shown on Fig. 9, Sketch A.

Usually, little pressure drop is available in overhead lines; and longer overhead lines with more elbows quickly involve increased line size (see Sketch B).

Piping configuration is also important. In the piping layout of the total condensing overhead circuit (see Fig. 9, Sketch D), a smaller line size may be used. The static head backpressure (Dimension X) has been reduced by raising the condenser and entering the reflux drum from the bottom with a standpipe. Even if line size cannot be reduced, Sketch D provides a better working arrangement than Sketch C.

Compressor Piping

Pipes in a compressor circuit should connect directly point-to-point. Bends instead of elbows cause less friction loss and less vibration. Angular branch

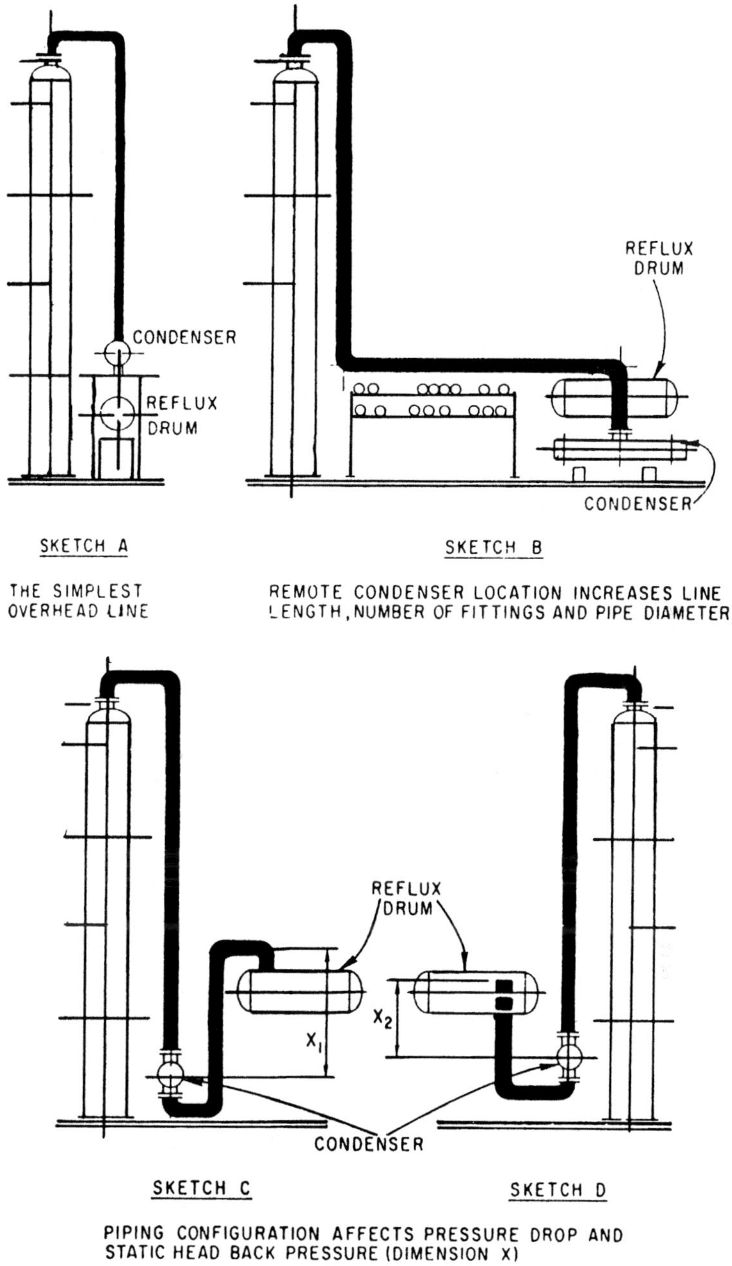

FIG. 9 Typical overhead piping arrangements.

connections eliminate hard tees and give a smoother flow. Double offsets for a directional change should be avoided. Intercoolers closely integrated with the machine minimize piping. Pulsation dampeners should be located on the cylinders without any interconnecting pipe. Knockout drums should be adjacent to the machine. Several aftercoolers or exchangers in the circuit should be stacked as much as possible for a direct gas flow. Equipment in the circuit should be in process flow sequence.

Because of the vibration problems ever present with reciprocating compressors, pipe supports have a very important role in piping design. Supports independent of any other foundation or structure are almost mandatory. Pipe systems firmly fastened close to grade are much preferred.

The selection of an optimum pipe size is also involved. The cost of piping increases with increased line sizes, but pressure drop and utility costs decrease. Here is how the yearly utility cost per unit pressure drop can be calculated. This cost multiplied by amortization time (number of years) gives the total cost of compression for the calculated period and process conditions.

The example in Table 1 is a tabulation for comparing line sizes, pressure drops, alloy piping cost, and utility cost for a portion of a centrifugal compressor circuit. This example shows that for a 2-year payout time, a 12-in. line is the most economical. For 5 years, any line from 12 to 16 in. is economical, and a 16-in. line should be selected. For 10 years, a 16-in. line will be the most economical. For maintaining these calculated economies, line sizes should be calculated, at least, with a good preliminary layout.

Optimum pressure drops and sizes can be established for all equipment groups in the compressor circuit. Table 1 assumes that the compressor works well within its capacity and pressure range. In borderline applications, the cost difference between the price of a smaller or larger compressor will also enter an overall cost comparison.

Piping and overall economy for very large pumps can be similarly evaluated as described for compressors. Time-consuming calculations might not be justified with smaller than average pumps.

Pumps

Very small pumps, in-line or vertical pumps, are usually adjacent to their suction vessel. With many pumps taking suction from the same vessel (a

TABLE 1 Alloy Pipe Size Selection for Various Payout Times[a]

			Payout Time with Yearly Utility Cost of 1 lb/in.^{2}, Δp = \$2,340					
			2 Years		5 Years		10 Years	
1	2	3	4	5	6	7	8	9
Line Size (in.)	Δp (lb/in.^{2})	Alloy Pipe Cost (\$)	Utility Cost (\$)	Total Cost, Columns 3 and 4 (\$)	Utility Cost (\$)	Total Cost, Columns 3 and 6 (\$)	Utility Cost (\$)	Total Cost, Columns 3 and 8 (\$)
8	9.75	10,800	45,630	56,430	114,075	124,875	228,150	238,950
10	3.0	15,240	14,040	29,280	35,100	50,340	70,200	85,440
12	1.28	20,850	5,990	26,840	14,976	35,826	29,952	50,802
14	0.84	25,250	3,931	29,131	9,828	35,028	19,656	44,856
16	0.39	28,500	1,825	30,325	4,563	33,063	9,126	37,626

[a]The cost data should be updated from these 1980 figures with current cost indices.

crude fractionator, for example), adjacent pump location is possible with only four or six pumps. For a great number of medium or large-sized pumps, road access must be provided. It is advantageous and economical to have a common road to all the pumps in the plant for convenient operation and maintenance. Too many dead-end access roads between process equipment lengthen the yard bank.

Suction piping should be designed without loops or pockets. The suction line is generally one or two sizes larger than the pump suction nozzle. Table 2 gives unit pressure drops at various flow rates, which will give economical pump discharge header sizes. Leads to the header are one size smaller—but not smaller than the pump nozzle.

Reactor-Furnace Piping is usually the most expensive alloy piping in a process unit (because of high temperatures and pressure), and it is often part of a compressor circuit.

Figure 10 illustrates that despite its high cost, a considerable length of piping and fittings had to be included in this arrangement. Engineering and space limitations restricted the layout. Allowance had to be made for pipe expansion. Furnace and reactor design determined the physical size and location of nozzles. Under such circumstances, piping layout economy depends on the ingenuity of the designer, who can critically examine his layout and eliminate every unnecessary fitting, flange, and field weld, determine optimum equipment locations and interconnect a piping system with a minimum of piping and fittings.

Figure 10 shows some extensive valving in reactor piping and gives an idea of how much piping and valving cost can be saved with line size reduction. It pays to recalculate and check line pressure drops with an exact piping layout. It also pays to investigate the pressure drop distribution in the entire compressor circuit. Decreased pressure drop in other equipment groups (exchangers, furnaces, reactors) can help in decreasing alloy line and associated valve sizes and still hold the overall pressure differential constant.

Line Pressure Drop

A nomograph may be used for sizing lines and estimating pressure drop. It is easy to use and gives rapid results of sufficient accuracy for ordinary line sizing problems.

TABLE 2 Economical Unit Pressure Drops for Pump Discharge Line Sizing

	Flow Rate at Pumping Temperature (gal/min)	Δp lb/in.^{2} per 100 ft	
		Steel Pipe	Alloy Pipe
Optimum friction losses for pump discharge header	0–250	2.5–10	6–15
	251–700	1.5–7	4–11
	701 and over	0.9–4	2–7
Optimum friction losses for extended payout time	0–150	1.5–6	4–10
	151–500	1.0–4	2.5–7
	501 and over	0.5–2	1.5–4

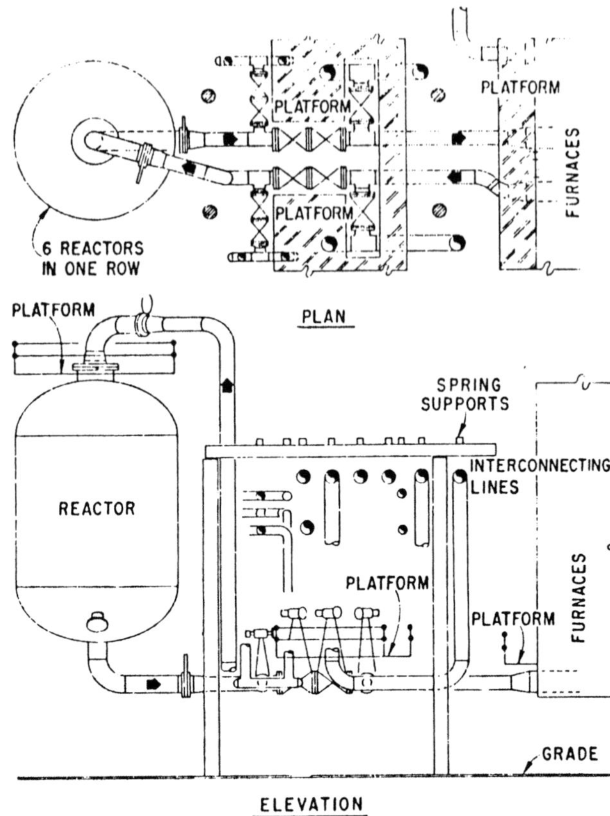

FIG. 10 Piping and valving between reactor and furnaces.

Using this nomograph (Fig. 11) you can find the following:

1. The pressure drop in a given line
2. The line size giving a desired pressure drop
3. The line size giving the most economical pressure drop (for gases only)
4. Correction necessary for a different friction factor
5. The pressure drop in a new line size
6. The allowable flow in a given line

This nomograph solves the Fanning equation expressed in the form

$$\Delta p = \frac{fW^2}{74{,}000pd^5}$$

where Δp = pressure drop, lb/in.^{2} per 100 ft of pipe
W = flow, lb/h
p = fluid density, lb/ft^3
d = inside diameter, in.

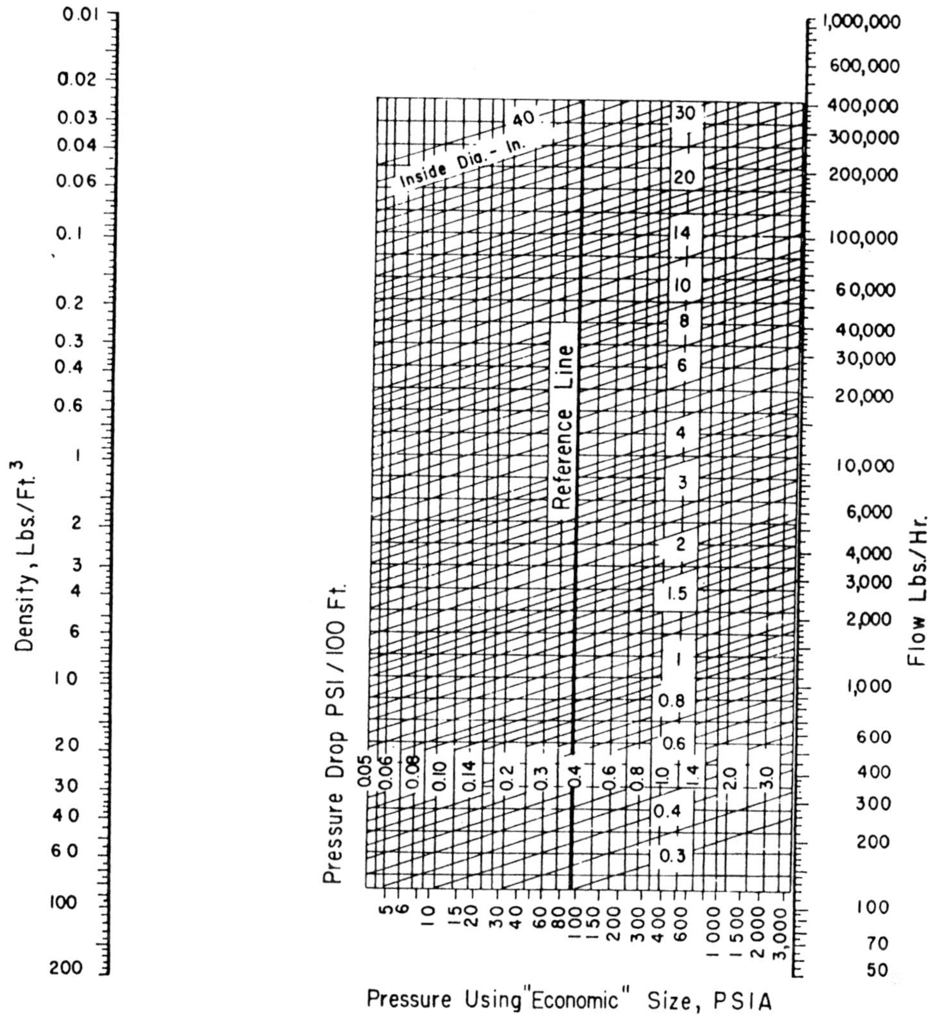

Instructions—1. Locate a working point on the reference line by laying a straight edge from the density scale to the flow scale. 2. Move horizontally to the desired pressure drop or the desired line size, or to the operating pressure to find the "economic" size. Note: Chart is based on a friction factor of 0.004.

FIG. 11 Nomograph for sizing lines and estimating pressure drop.

f = friction factor, assumed equal to 0.004. The chart can be corrected for other friction factors as described below

To find the pressure drop in a given line:

1. Locate a working point on the reference line by laying a straight edge from the density scale to the flow scale.
2. Move horizontally from the working point until the diagonal line showing the correct line size is intersected.
3. Read the pressure drop.

To find the line size giving a desired pressure drop:

1. Find a working point as in Step 1 above.
2. Move horizontally from the working point until the vertical line showing the desired pressure drop is intersected.
3. Read the line size on the diagonal scale.

To find the line size for the most economical pressure drop (gases only):

1. Find a working point (as in Step 1) under *pressure drop* above.
2. Move horizontally from the working point until a vertical line extended up from the pressure scale is intersected.
3. Read the line size on the diagonal scale. Note: The pressure scale does not necessarily show actual pressure except at the most economical size.

To correct for a different friction factor:

1. Find a working point as in Step 1 under *pressure drop* above.
2. Move diagonally to 100 times the new friction factor as shown on the pressure drop scale. Note: This is the only time a diagonal move should be made.
3. Move horizontally back to the reference line to find the corrected working point.

Continue with Steps 2 and 3 as in the above paragraphs. For example, suppose you have 1000 lb/h of material with a density of 0.12 lb/ft^3, and preliminary work shows that a 2.7-in. diameter line is required to give the desired pressure drop of 0.3 lb/in.^{2} per 100 ft. You decided to use a 3-in. line. A careful check later shows that the actual friction factor is 0.006 instead of 0.004. Make the correction as follows:

1. Start where the 2.7-in. line intersects the 0.3 lb/in.^{2} line.
2. Move horizontally back to the reference line to find the old working point, then move diagonally to the 0.6 lb/in.^{2} vertical line.
3. Move horizontally back to the reference line to find the corrected working point.
4. Move horizontally back to the 0.3 lb/in.^{2} vertical line. The new required line size is 2.9 in. instead of 2.7 in. as estimated originally.
5. Instead of stopping at the 0.3 lb/in.^{2} vertical line, you could continue moving horizontally past the 0.3 lb/in.^{2} line to the 3-in. diagonal line. Here you can read that the actual pressure drop will be 0.27 lb/in.^{2} per 100 ft in a 3-in. line.

To find the pressure drop in a new line size:

1. Find the point on the chart where the old line size intersects the existing pressure drop.
2. Move horizontally to the new line size and read the new pressure drop.

To find the allowable flow in a given line:

1. Use the given line size and the allowable pressure drop to find a point on the chart.
2. Move horizontally to the reference line to find a working point.
3. Lay a straight edge from the density scale through the working point to find the allowable flow.

Graphical Piping Analysis

Precise analytical methods and computer analysis should be limited to critical and hazardous lines. The term *critical* applies to lines with high or low operating temperatures and pressures. It may, either as well or instead, refer to the type of equipment to which they are connected. *Hazardous* refers to the nature of the fluid being conveyed by the lines—highly inflammable, etc. Designers should segregate all critical and hazardous lines which demand the attention of a piping stress analyst. For simpler, noncritical lines an approximate calculation method is permitted by the Code for Pressure Piping USASI B31.3.

The total deflection in a piping system usually is known. For example, if the pipe length of a symmetrical loop is 50 ft (U length between anchor points), the thermal expansion is 4 in. per 100 ft of linear length. Then the total deflection = 50/100(4) = 2 in. The height and width of the loop are generally determined by the space available. Once the shape of the loop is decided by the piping designer, forces, moments, and stresses can easily be found by using graphs.

Allowable Stress

It is recommended that the stress computed by this method should be compared by the code allowable stress range S_A:

$$S_A = f(1.25S_c + 0.25S_h)$$

where S_c = allowable stress (S value) in the cold condition
S_h = allowable stress (S value) in the hot condition
S_c and S_h are to be taken from tables in the applicable sections of the code

f = stress-range reduction factor for cyclic conditions. Use a value of 1.0 for one cycle per day or less. Consult ANSI B31.3

Weight and other sustained external loadings shall not exceed S_h.

Pipe supports and restraints are not considered in the flexibility calculation. It is assumed that the supports which have not been considered in the analysis should be located and designed so as not to interfere with the flexibility of the system. The reactions computed by this method shall not exceed the limits which the attached equipment can safely sustain. Application of such equipment as pumps, turbines, and similar strain-sensitive machines should receive the manufacturer's approval; and the piping system design should be flexible enough to comply with their recommendations.

Sample Problems

The following data apply to all sample problems, unless otherwise stated.

Pipe size = 3 in.
Schedule = 40
Operating temperature = 860°F
E = Young's modulus (cold) = 27.9 (10^6) lb/in.^{2}
Thermal expansion = 7.37 in. per 100 ft
I = moment of inertia = 3.02 in.^{4}
Z = section modulus = 1.714 in.^{3}
i = stress intensification factor = 1.78
Material = carbon steel
Code = USASI B31.3
S_A = allowable stress = 16,800 lb/in.^{2}

Solution: Use Figs. 12a, 12b, and 13.

Sample Problem 1. For a simple U-type expansion loop as shown in Fig. 14, check maximum stress in the loop, and if the calculated stress is much less than the allowable, suggest a loop which will produce a maximum stress equal to or near S_A. Space does not permit a change in H, but G can be changed.

Given: 3-in. carbon steel pipe, Sch. 40, thermal expansion 2 in. per 100 ft, 325°F.

Data: I = 3.02 in.^{4}; Z = 1.714 in.^{3}; i = 1.78. Allowable stress S_A = 18,000 lb/in.^{2}.

Solution:

$$\beta = W/H = 10/10 = 1$$

$$\alpha = G/H = 10/10 = 1$$

$$\delta_0 = \text{total thermal expansion} = \frac{2}{100}(100) = 2 \text{ in.}$$

Note: β and α and other nomenclature have no connections with similar symbols used in piping design books. These are used here just as symbols.

Step 1. Determine Force F_A using Fig. 12a. Enter scale β with $\beta = 1$, then move vertically upward to the curve $\alpha = 1$, and then move horizontally to the right to the line $\delta_0 = 2$ and then vertically move down to the line $H = 10$ ft and horizontally from this point to scale F_A/I and read the value $F_A/I = 38$. The force $F_A = 38(I) = 38(3.02) = 114.76$; say $F_A = 115$ lb.

Step 2. Determine the maximum moment in the bend using Fig. 12b. Enter β scale with $\beta = 1$. Move to curve $\alpha = 1$ and then horizontally on the right to $H = 10$ ft and vertically down to line $F_A = 115$ lb. From this point move horizontally to the moment scale to read BM (bending moment) = 6300 lb-in.

Step 3. Calculate the maximum stress in the bend.

$$S_E = \text{expansion stress} = \frac{BM}{Z}(i)$$

$$= \frac{6300}{1724}(1.78) = 6520 \text{ lb/in.}^2$$

$$S_A = 8000 \text{ lb/in.}^2, \; S_E < S_A$$

The system is acceptable. The calculated stress is less than the allowable stress.

Step 4. Determine the flexible loop which will produce a stress of 8000 lb/in.^{2}. Use Fig. 13.

$$\text{Stress ratio} = \frac{S_A}{S_E} = \frac{18{,}000}{6520} = 2.76$$

Find $\gamma = 0.475$ from Fig. 13 when $\alpha = 1$ and $\beta = 1$.

The new $\gamma = (0.475)$ (stress ratio) $= 0.475(2.76) = 1.312$.

Enter Fig. 13 with the new $\gamma = 1.312$, move to $\beta = 1$, then move vertically downward to read a new $\alpha = 0.25$.

Since $\alpha = G/H = 0.25$,

$$G = 0.25(H) = 0.25(10) = 2.5 \text{ ft}$$

$$\beta = W/H = 1, \; W = 10, \; H = 10$$

The suggested loop will have the following dimensions: $H = 10$ ft, $W = 10$ ft, $G = 2.5$ ft, $U = 100$ ft. This new loop will produce a stress = 18,000 lb/in.^{2}.

Check stress:

$$\text{Calculated } BM = 6300 \text{ lb/in.}$$

$$\text{Stress ratio} = 2.76$$

$$(BM)(\text{stress ratio}) = 6300(2.76) = 17{,}400 \text{ lb/in.}$$

$$\text{New stress, } S_E = \frac{BM}{Z}(i) = \frac{17{,}400}{1.724}(1.78)$$

$$= 18{,}000 \text{ lb/in.}^2$$

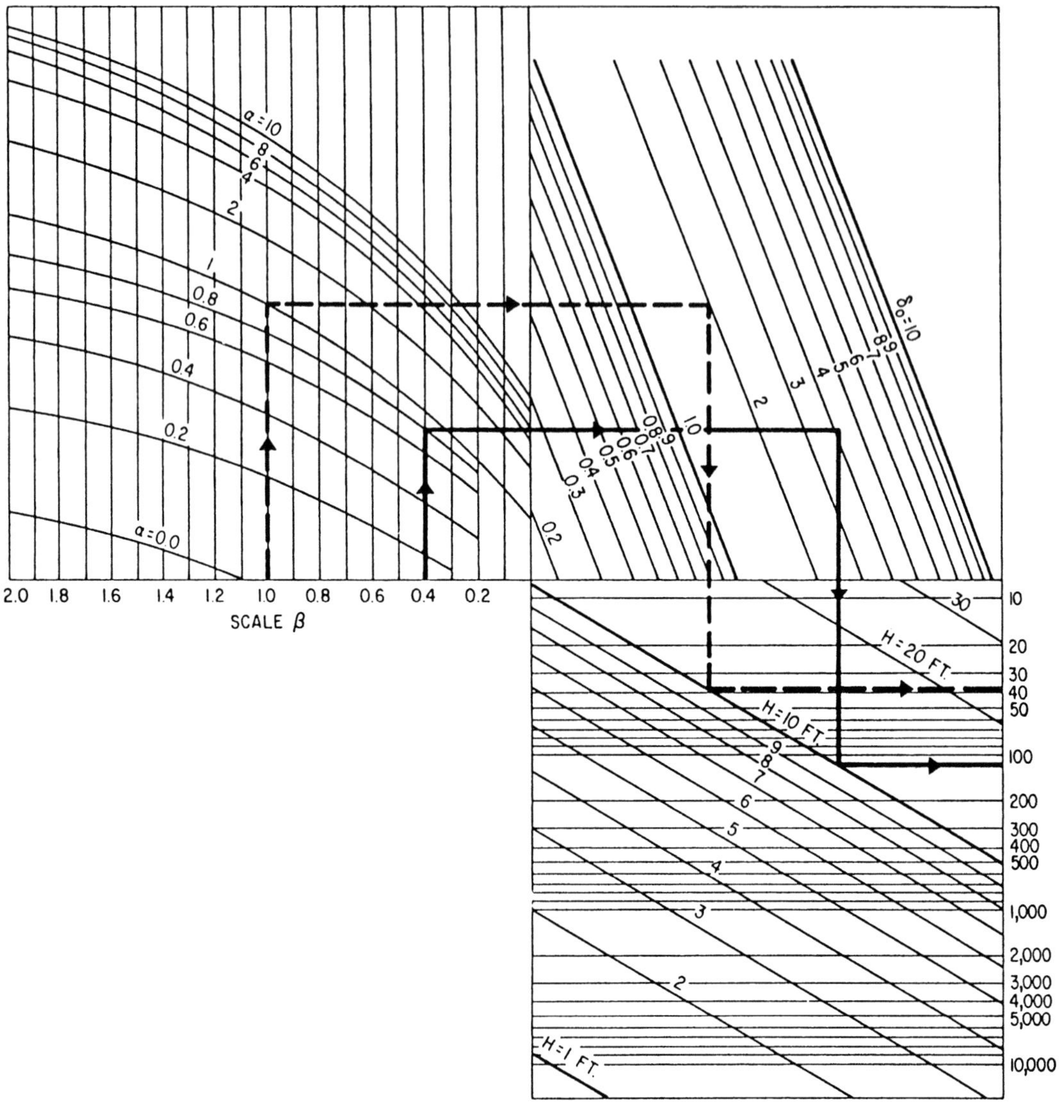

FIG. 12a Force graph for U-type pipe expansion loops.

Sample Problem 2. Calculation of forces at the nozzles, see Fig. 15.

$$\beta = W/H = 4/10 = 0.4$$

$$\alpha = G/H = 8/10 = 0.8$$

$$\begin{aligned}\delta_0 &= \frac{7.37}{100}(55) \\ &= 4.05 \text{ in. (thermal expansion of } U)\end{aligned}$$

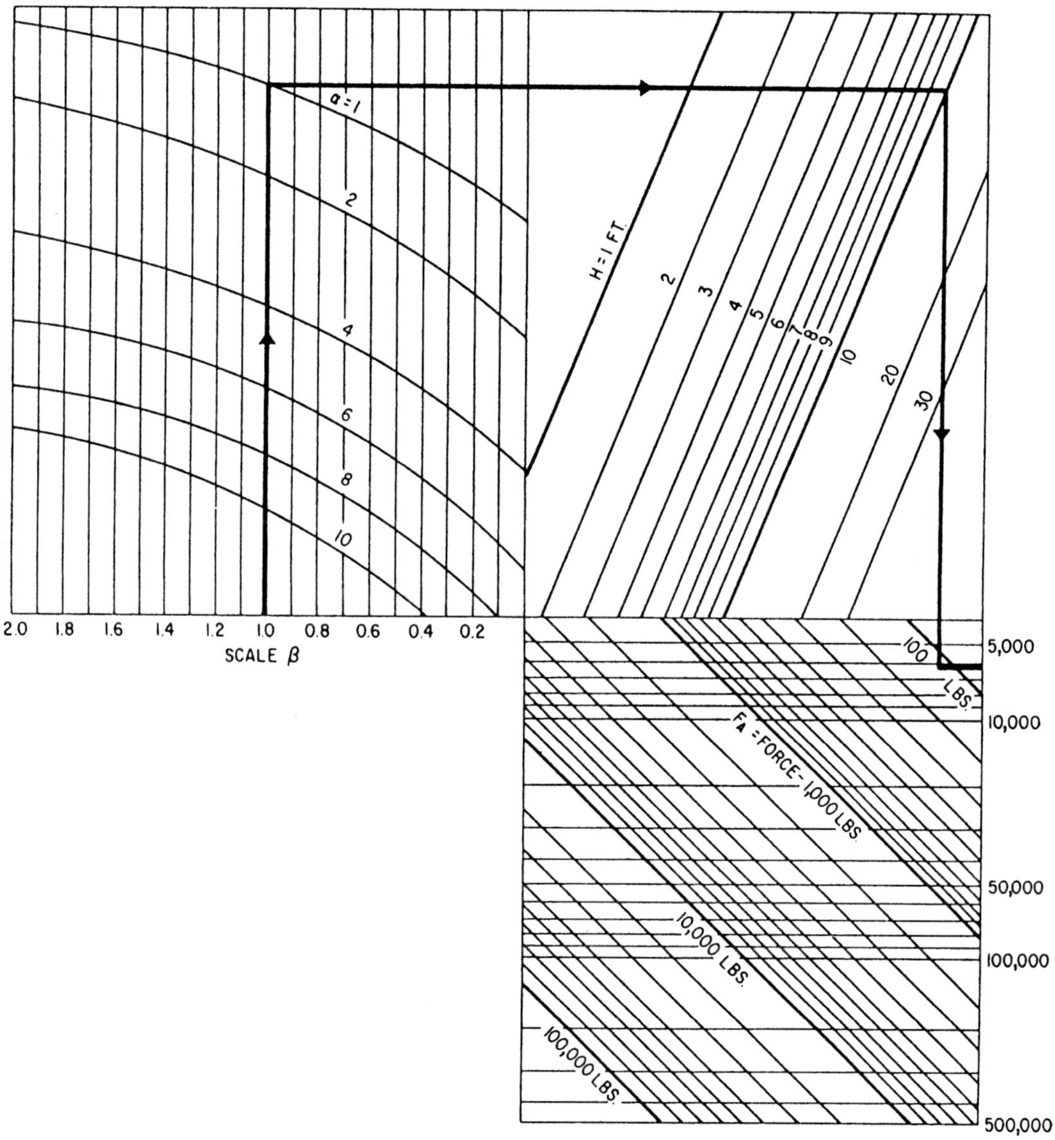

FIG. 12b Moment graph for U-type pipe expansion loops.

From Fig. 12a find $F_A/I = 120$.

$$\therefore F_A = 120(I) = 120(3.02) = 362 \text{ lb}$$

$$\text{Computer result} = 252 \text{ lb}$$

$$\text{Kellogg graphical} = 368 \text{ lb}$$

F_A = force of pipe on anchor or nozzle, caused by thermal expansion

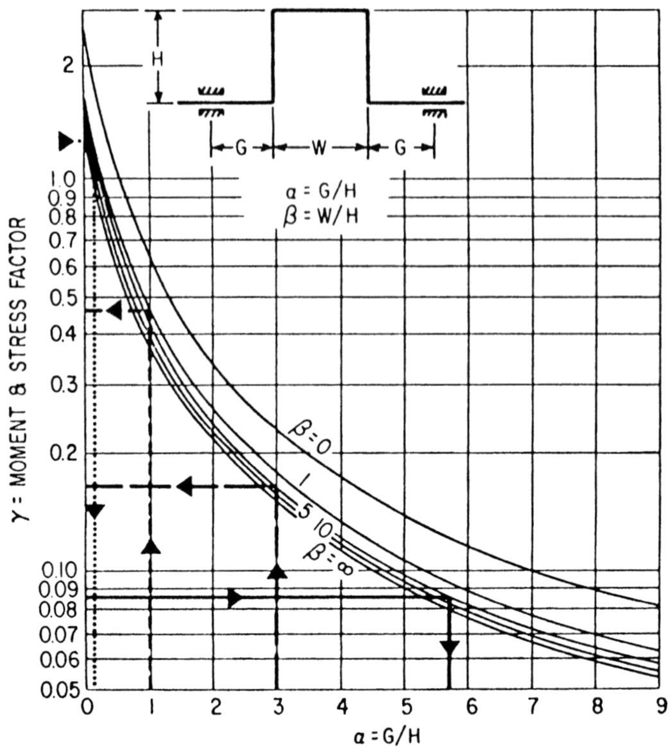

FIG. 13 Variation of bending moment and stress with β and α.

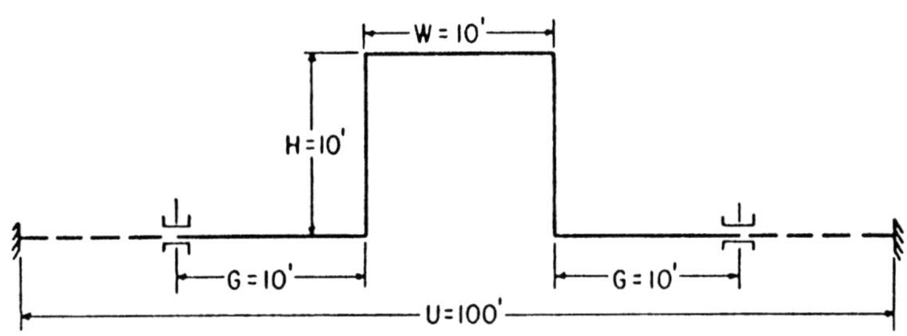

FIG. 14 Simple U-type expansion loop for Example 1.

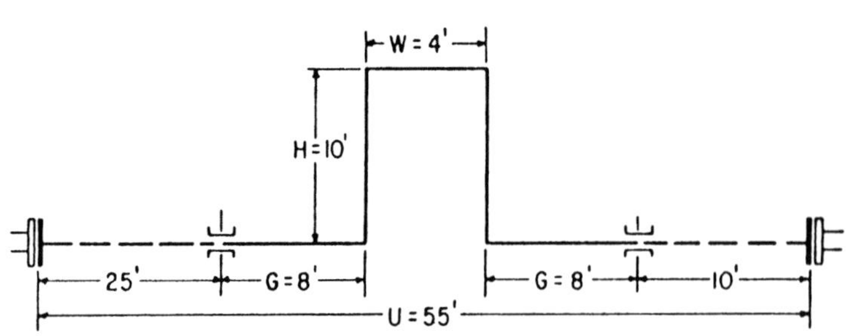

FIG. 15 Calculations of forces on nozzle. Sample Problem 2.

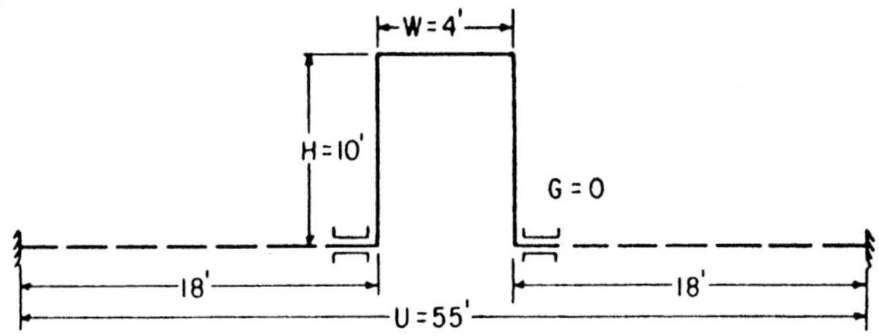

FIG. 16 Figure for Sample Problem 3.

Sample Problem 3. See Fig. 16.

$$\beta = W/H = 4/10 = 0.4$$

$$\alpha = G/H = 0/10 = 0$$

$$\delta_0 = \frac{7.37}{100}(55) = 4.05 \text{ in.}$$

From Figure 12a, by interpolation, $F_A/I \doteq 270$. Therefore, $F_A = 270(I) = 270(3.02) = 815$ lb. By computer $F_A = 642$ lb.

Sample Problem 4. See Fig. 17.

$$\beta = W/H = 4/10 = 0.4$$

$$\alpha = G/H = 0/10 = 0$$

$$\delta_0 = \frac{7.37}{100}(4) = 0.295 \text{ in.}$$

Find force from Fig. 12a, $F_A/I = 20$. Therefore $F_A = 20(I) = 20(3.02) = 60.40$ lb. By computer $F_A = 46.0$ lb.

From the above three examples, it is obvious that the results using this method compared with computer analysis indicate safe and reasonably accurate values.

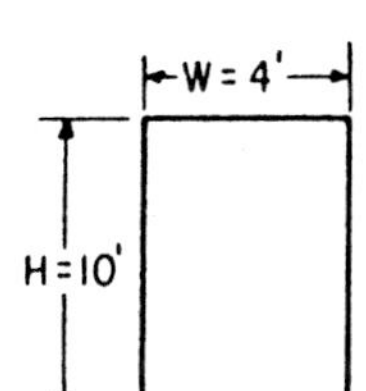

FIG. 17 Figure for Sample Problem 4.

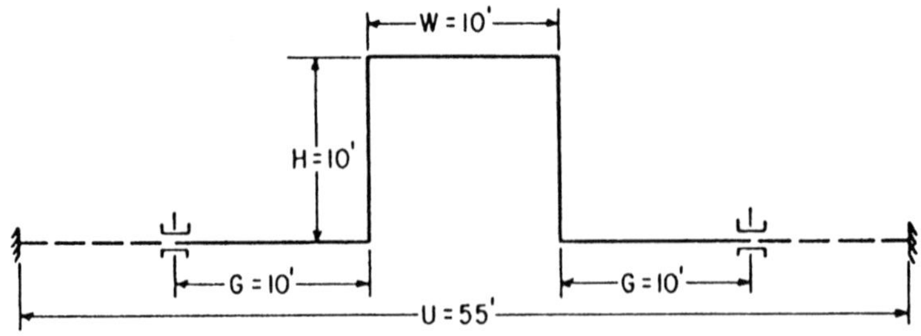

FIG. 18 Figure for Sample Problem 5.

Sample Problem 5. See Fig. 18.

$$\beta = W/H = 10/10 = 1$$

$$\alpha = G/H = 10/10 = 1$$

$$\delta_0 = \frac{7.37}{100}(55) = 4.05 \text{ in.}$$

Find from Fig. 12a, $F_A/I = 78$.

$$F_A = 78(I) = 78(3.02) = 236 \text{ lb}$$

$$M_A = 12{,}000 \text{ lb/in.}$$

Stress in the bend,

$$S_E \frac{M_A}{Z}(i) = \frac{12{,}000}{1.724}(1.78) = 12{,}400 \text{ lb/in.}^2$$

$$S_A = 16{,}800 \text{ lb/in.}^2$$

$$S_E < S_A$$

The reactions and stresses are within the allowable limits. The system is, therefore, adequately flexible.

Economic Loop Design

If the anchor forces are not the factors which dictate the design, then the loop can be quickly determined by the use of this method which will produce maximum moments and stresses equal to its allowable limits. No trial and error method is needed. By using Fig. 13 a great deal of labor can be saved. This also eliminates a large number of mathematical computations and reduces the chances of errors to negligble. The graph is self-explanatory and the results are sufficiently accurate for most engineering purposes.

If the calculation, based on either this method or an analytical method, indicates that the moments and stresses exceed the allowable limits, then Fig. 13 can be used to predict the guide distance G which will allow the shape to become flexible enough and to yield moments and stresses equal to the allowable limits.

Sample Problem 6. A loop which has the ratio $\alpha = G/H = 3$ and $\beta = W/H = 5$, and the solution indicates the bending moment = 50,000 lb/in. and the expansion stress = 30,000 lb/in.², the loop exceeds the allowable limits.

Allowable BM = 25,000 lb/in. Allowable S_A = 15,000 lb/in.². Design a U-type symmetrical expansion loop to yield the maximum moment and stress equal to the allowable limits.

Solution: Step 1. Determine from Fig. 13 when $\alpha = 3$ and $\beta = 5$. Follow the arrows and read the value of $\gamma = 0.17$ which is the moment and stress factor on the left-hand vertical scale.

Since the allowable stress S_A = 15,000 lb/in.² and the calculated expansion stress S_E = 30,000 lb/in.², the ratio = 15,000/30,000 = 0.5.

Therefore, the corrected $\gamma = 0.17(0.5) = 0.085$.

It is assumed that $\beta = W/H = 5$ remains unchanged.

Step 2. Determine the new α. Enter in Fig. 13 the corrected $\gamma = 0.085$ on the vertical scale from the left-hand side of the graph, and move horizontally to the right to the curve $\beta = 5$, then move vertically down to the α scale and read the new $\alpha = 5.7$.

Now summarize the new values as follows: $\alpha = 5.7$, $\beta = 5$.

Expansion stress S_E = 15,000 lb/in.².

Bending moment, BM = 50,000(0.5) = 25,000, equals the allowable moment.

Step 3. Determine the distance of the guide G. The new $\alpha = 5.7 = G/H$, when H remains as before.

Therefore, the distance of the guide $G = 5.7(H)$. If the G is increased to $5.7(H)$, then the new shape will yield a stress of 15,000 lb/in.², which is equal to the allowable stress.

The following conditions apply:

The entire system lies in one plane.
The system is treated with square corner intersections.
The system is composed of straight pipe elements of uniform size and thickness.
The thermal expansion of a given element is absorbed by the elements oriented perpendicular to the direction of deflection.
The effects of dead weight, wind, etc. are neglected.
The clearance between guides and pipe is nil.
The compressive stresses within the element are neglected
The flexibility of the elbow allowed by an oval shape is neglected.

The three deflections in the horizontal plane are:

$$\delta_1 + \delta_2 + \delta_3 = \delta_0 = \text{total deflection}$$

Force, Moment, and Stress

The following units apply in the formula below:

Modulus of elasticity = E (lb/in.²)
Resisting force = F_A (lb)

Moment of inertia of the pipe $= I$ (in.^{4})
Total thermal expansion between anchor points is in inches.
Height of the loop $= H$ (ft)
Width of the loop $= W$ (ft)
Distance of first guide $= G$ (ft)
Section modulus of pipe $= Z$ (in.^{3})
Stress intensification factor for the bend $= i = 0.9/h$ where $h = tR/r^2$

See code for pressure piping ANSI B31.3.
Then the force is

$$F_A = \frac{EI\delta_0}{2(12H)^3}\left[\frac{1}{\frac{1}{3} + \frac{\beta}{2} - \frac{1 + 2\beta + \beta^2}{4\alpha + 4 + 2\beta}}\right]$$

Loop Restraints and Supports

Design engineers should make certain that the loop between the two guides is made to function without any environmental obstructions or restraints. Also the system must be fully supported and no branch connections made within the flexible portion of the loop. It is not recommended that any external loading be placed on the loop. The designer should avoid locating rigid sections, such as large valves, etc., within the loop. A good practice is to locate valves and other rigid sections near the guide or between the anchor and the guide.

Line Size Limits

This method provides engineers with reactions and stresses that are reasonably accurate for pipe up to 6 in. in diameter. Lines above 6 in. in diameter can be safely analyzed by this method. However, the results obtained will be on the conservative side. Therefore, where large diameter piping systems are required, a precise analysis should be made.

Round Corners

Loops having round corners can be solved as follows:

1. If the square corner solution causes reactions and stresses which exceed acceptable limits
2. If the radius of the bend is more than 1.5 pipe diameters
3. If the line exceeds 6 in. in diameter

Refer to Figs. 19 and 20. Use the equations below to modify the width (W), height (H), and guide distance (G), respectively.

Use Figs. 12a, 12b, and 13 to determine forces and moments.

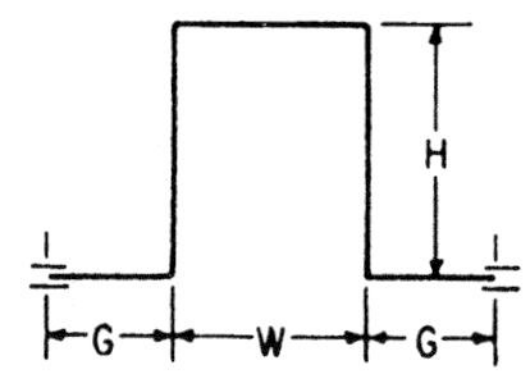

FIG. 19 For round corners, the width, height, and guide distances must be modified.

Assumption: R = radius of all the bends which must be the same; H = height of the loop which must be equal on both sides; G = distance of the guides which must be equal on both sides.

$$W = w + 1.57R(K/3)$$

$$H = h + 1.57R(K/3)$$

$$H = h + 1.57R(K/6)$$

$$G = g + 1.57R(K/6)$$

$$K = 1.65/h = \text{flexibility factor}, K \geq 1$$

where $h = tR/r^2$
t = thickness of pipe
r = mean radius of pipe
R = radius of the bend

Note: For force calculation, use modified height (H) of the loop. For moment calculation, use original height (H) of the loop.

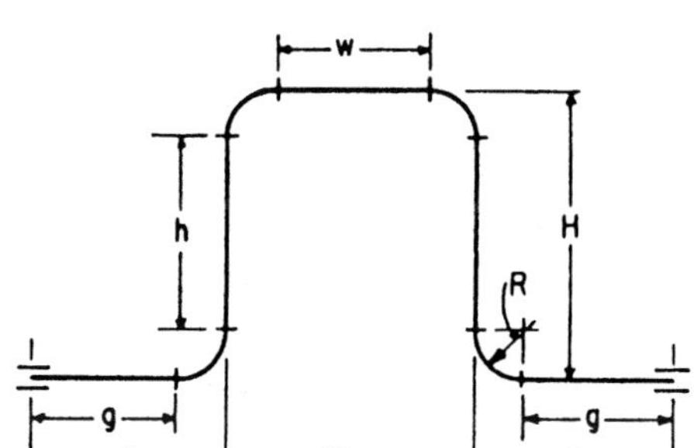

FIG. 20 Loop designations for round corners.

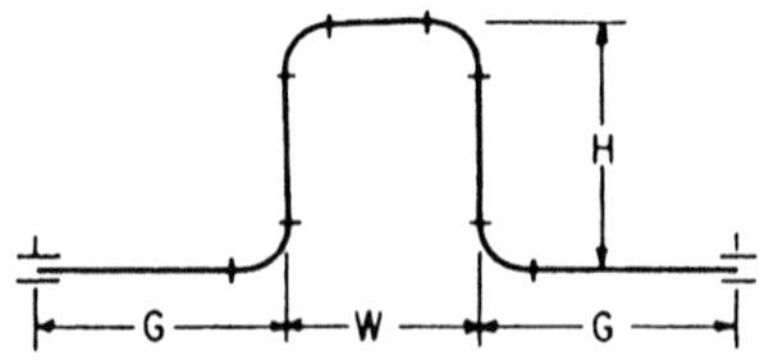

FIG. 21 Symmetrical loop with guide *G* distance from bend.

For Figs. 21–23:

$$w = W - 2R$$

$$h = H - 2R$$

$$g = G - R$$

Symmetrical Loop. Refer to Figs. 20 and 21. Modify *H* as follows:

$$H = h + 1.57R(K/3)$$

Symmetrical Loop. Refer to Fig. 22. Note that the guide is very near the bend. For *H*, use

$$H = h + 1.57R(K/3)$$

Since the guide is at or near the bend, $G = 0$.

Two-Plane Loop. Refer to Fig. 23. For *H*, use

$$H = h + 1.57R(K/3)$$

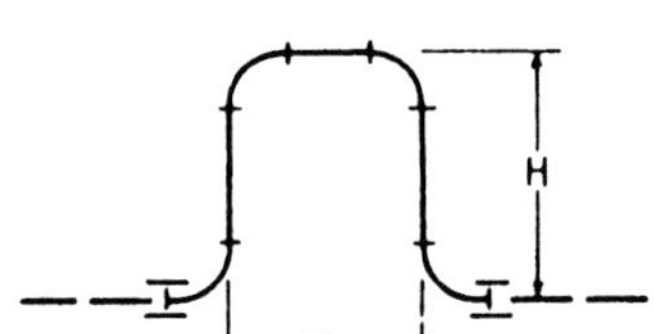

FIG. 22 Symmetrical loop with guide very near to the bend.

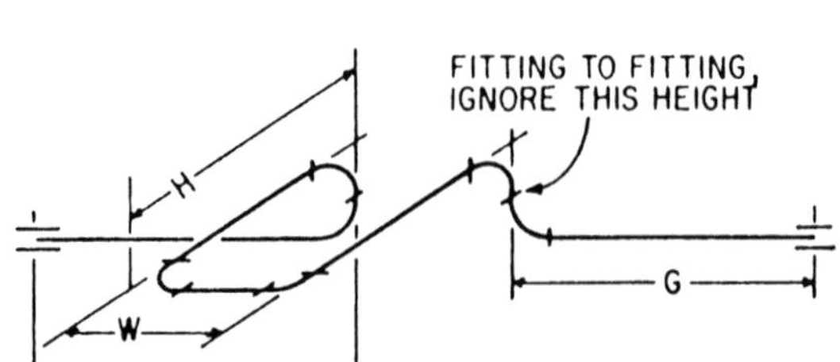

FIG. 23 A two-plane loop configuration.

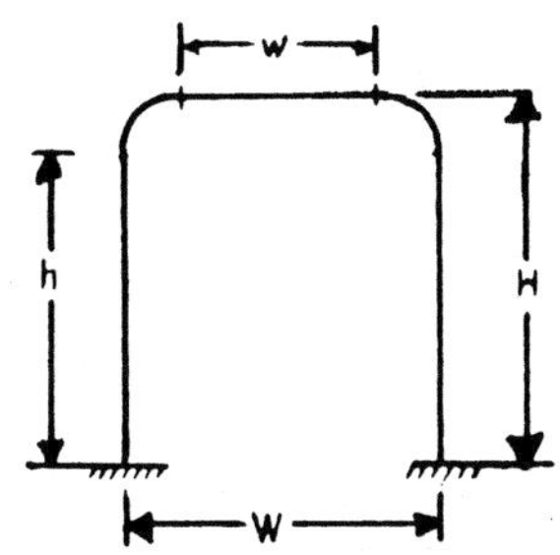

FIG. 24 A U-loop configuration with equal legs.

U-Loop with Equal Legs. Refer to Fig. 24.

$$w = W - 2R$$

$$h = H - R$$

For H, use

$$H = h + 1.57R(K/6)$$

$G = 0$.

Two-Plane Loop with Y-Leg Longer Than Fitting to Fitting. Refer to Fig. 25.

$$w = W - 2R$$

$$h_1 = H_1 - 2R$$

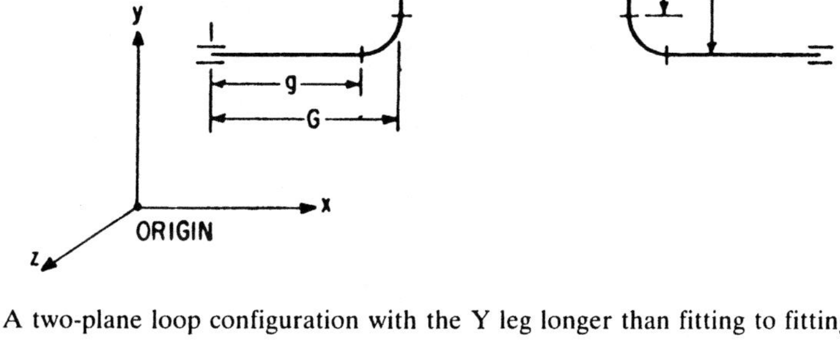

FIG. 25 A two-plane loop configuration with the Y leg longer than fitting to fitting.

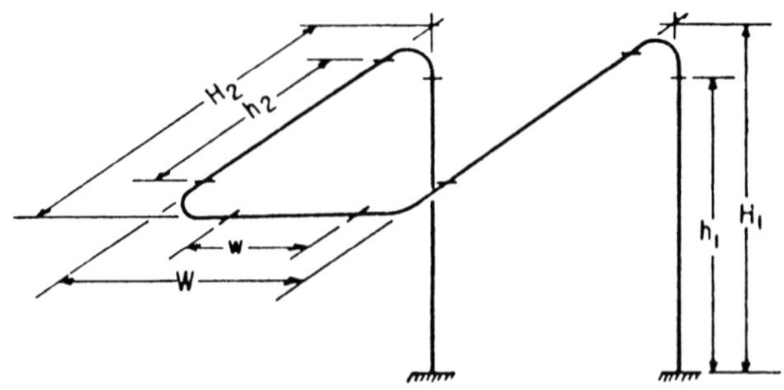

FIG. 26 A two-plane loop with equal legs.

$$h_2 = H_2 - 2R$$

$$g = G - R$$

$$H_1 = H_1 + H_2$$

Two-Plane Loop with Equal Legs. Refer to Fig. 26.

$$w = W - 2R$$

$$h_1 = H_1 - R$$

$$H_2 = H_2 - 2R$$

For H_1, use

$$H = h + 1.57R(K/6)$$

For H_2, use

$$H = h + 1.57R(K/3)$$

$$H = H_1 + H_2$$

$G = 0$

U-Loop with Equal Legs and Single Tangent. Refer to Fig. 27 and solve the same as Fig. 24.

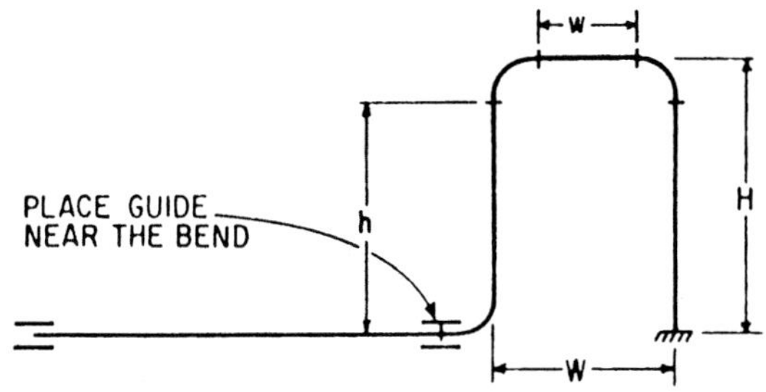

FIG. 27 A U-loop configuration with equal legs and single tangent.

Graphical vs Computer Solutions

The following examples are given to illustrate the results obtained graphically and by the computer.

Example 1. Given: 4-in. pipe, schedule 40; radius of bend = 0.5 ft; operating temperature = 350°F; material = ASTM 106 Gr. B; thermal expansion = 2.26 in./100 ft; allowable stress = 22,500 lb/in.^{2}; code = power piping; moment of inertia = I = 7.23 in.^{4}; section modulus = Z = 3.22 in.^{3}; stress intensification factor = i = 1.95. (See Figs. 28 and 29.)

$$\alpha = G/H = 12/10 = 1.2$$

$$\beta = W/H = 8/10 = 0.8$$

$$\delta_0 = (2.26/100)(180) = 4.06 \text{ in.}$$

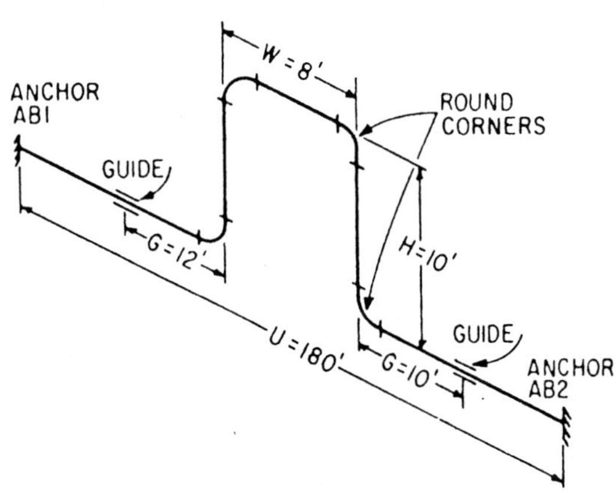

FIG. 28 An example configuration calculated by the graphical and computer methods.

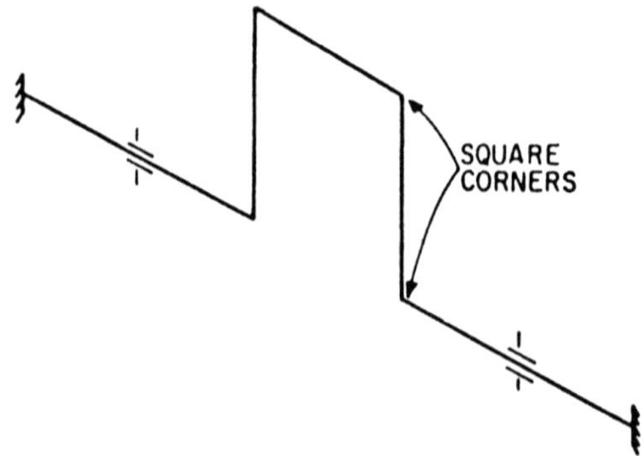

FIG. 29 A square corner configuration calculated by graphical and computer methods.

From Fig. 12a,

$$F/I = r = 86$$

$$\text{Force} = r(I) = 86(7.23) = 620 \text{ lb}$$

From Fig. 12b,

$$BM = 27{,}100 \text{ lb-in.}$$

$$\text{Stress} = \frac{27{,}100}{3.22}(1.95) = 16{,}450 \text{ lb/in.}^2$$

Method	*Force (lb)*	*Moment at Guide (lb-in.)*	*Maximum Stress at Bend (lb/in.²)*	*Remarks*
Graphical	620	27,100	16,450	Square corner solution
Computer	617	27,000	16,479	1.5 *D* elbow solution

Example 2. Given: 6-in. pipe, schedule 40. All other data as in example above. $\delta_0 = 4.06$ in.; $I = 28.1$; $Z = 8.50$; $i = 2.27$.
From Fig. 12a,

$$r = 86, \; r(I) = 86(28.1) = 2410 \text{ lb}$$

From Fig. 12b,

$$BM = 90{,}200 \text{ lb-in.}$$

$$\begin{aligned}\text{Stress} &= 90{,}200/8.5(2.27)\\ &= 24{,}550 \text{ lb/in.}^2\end{aligned}$$

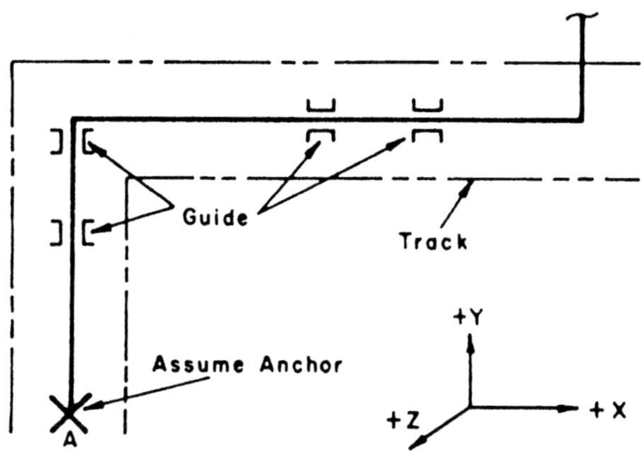

FIG. 30 Rack piping in a single plane.

Method	*Force (lb)*	*Moment (lb-in.)*	*Stress (lb/in.²)*	*Remarks*
Graphical	2410	90,200	24,550	Square corner solution
Computer	2179	90,500	20,122	Square corner solution

Single Plane Systems

To illustrate the use of the graphical method in a single plane shape, a portion of rack piping will be analyzed, as shown in Fig. 30. This shape should be guided as indicated.

Figure 31 shows the expansion of the pipe when heated, under guided deflection.

Results can be determined from the graph (Fig. 32) directly.

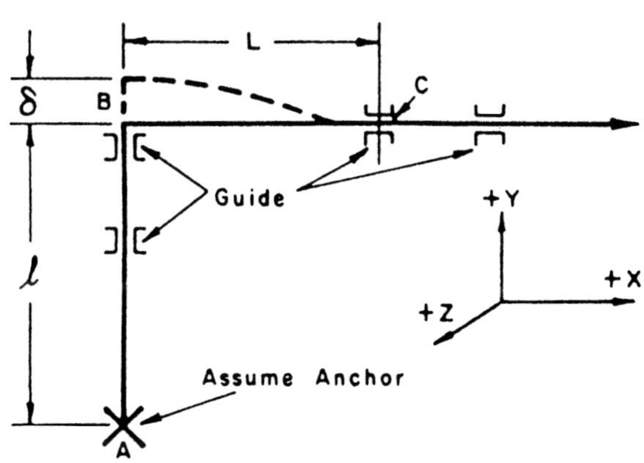

FIG. 31 Expansion of rack piping when heated.

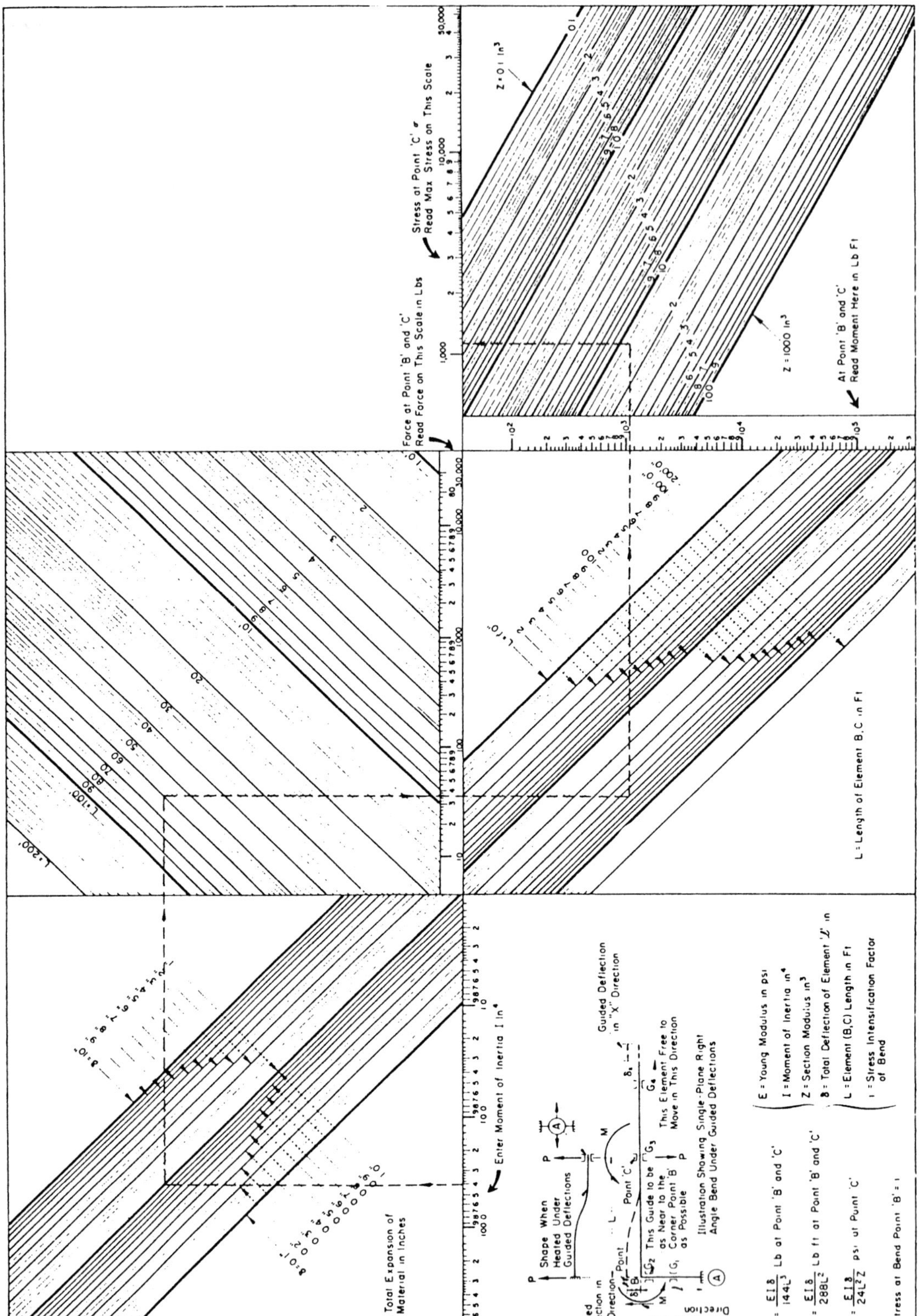

FIG. 32 Nomograph to solve single plane systems.

The maximum stress in the bend is found by multiplying the stress by the stress intensification factor β.

Expansion stress, S_E, must be less than or equal to the allowable stress, S_A. $S_E \leqslant S_A$.

Note: It is recommended that S_A should be taken from ANSI B31-3—Refinery Piping Code.

Assumptions: Clearance between the pipe and guide is negligible. The influence of axial force on the pipe is negligible. The pipe throughout its length, L, is of the same size. Young's modulus is constant ($E = 26.9 \times 10^6$ lb/in.^{2}). Flexibility of a *bend* at the Point A is neglected.

Example 1. Given a 6-in. pipe, schedule 80, carbon steel, operating temperature of 640°F, $e = 5.01$ in./100 ft, $I = 40.5$ in.^{4}, $Z = 12.23$ in.^{3}, $\delta = 5.01(20)/100 = 1.0$ in.

Read from Fig. 32, $F = 35.0$ lb, $M = 1050$ lb-ft, $\sigma = 1120$ lb/in.^{2}.

The maximum stress will occur at the bend point $B\sigma_{max} = 1120\ (1.6) = 1790$ lb/in.^{2}.

Example 2. Given: Same configuration as Example 1, carbon steel pipe operating at 640°F; pipe size, $D = 12$-in.; pipe schedule = 80; moment of inertia, $I = 475$ in.^{4}; section modulus, $Z = 74.5$ in.^{3}; effective length, $L = 30.0$ ft; expansion length, $l = 15.0$ ft; stress intensification factor, $i = 1.85$; thermal expansion, $e = 5.01$ in./100 ft; l_1 and $l_2 = 2.0$ ft.

Solution: Determine $\delta = el = (5.01)(15)/10 = 0.75$ in.

Use Fig. 32 as in Example 1 and find:

$$F = 2{,}400 \text{ lb}$$

$$M = 34{,}000 \text{ lb-ft}$$

$$\sigma = 6{,}000 \text{ lb/in.}^2 \text{ at Point B}$$

$$\begin{aligned}\sigma_{max} &= \sigma i \\ &= (6000)1.85 \\ &= 11{,}100 \text{ lb/in.}^2\end{aligned}$$

The force acting on Guides $G1$ and $G2$ caused by the bending moment is

$$F_{G1} = M/l_1 = 34{,}000/2 = 17{,}000 \text{ lb}$$

$$F_{G2} = M/l_2 = 34{,}000/2 = 17{,}000 \text{ lb}$$

The force acting on Guide $G3$ is

$$\begin{aligned}F_{G3} &= -F - (M/l_2) \\ &= 2400 - (34{,}000/2) \\ &= -19{,}400 \text{ lb}\end{aligned}$$

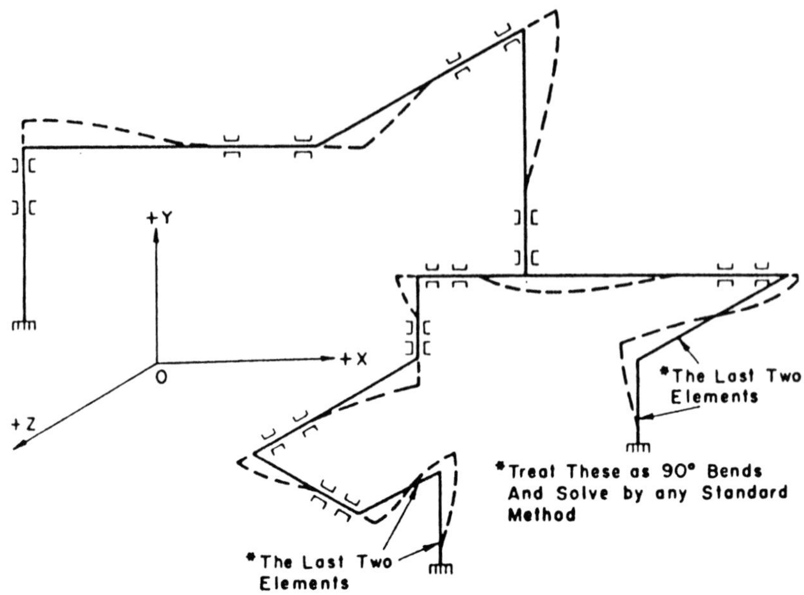

FIG. 33 Multiplane, multianchor problem.

The force acting on Guide $G4$ is

$$M/l^2 = 17{,}000 \text{ lb}$$

Multiplane, Multianchor Problems

The Guided Deflection Method can be used to analyze multiplane, multianchor problems as shown in Figs. 33 and 34. It should be noted that the

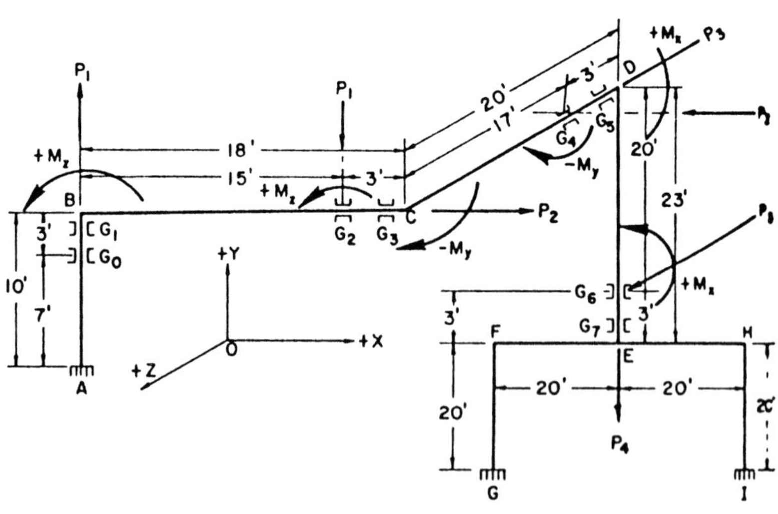

FIG. 34 Moments, forces, and dimensions added to Fig. 33.

last two elements forming a 90° bend must be solved by any other accepted method, such as the Elastic Center Method.

Example 3. The following data are given: 6-in., schedule 80 pipe; design temperature, 690°F; operating temperature, 640°F; expansion, e = 5 in./100 ft; moment of inertia, I = 40.5 in.^{4}; section modulus, Z = 12.23 in.^{3}:

$$\text{Element A–B} = \Delta_1 = (10/100)(5) = 0.5 \text{ in.}$$

$$\text{Element B–C} = \Delta_2 = (18/100)(5) = 0.9 \text{ in.}$$

$$\text{Element C–D} = \Delta_3 = (20/100)(5) = 1.0 \text{ in.}$$

$$\text{Element D–E} = \Delta_4 = (23/100)(5) = 1.15 \text{ in.}$$

$$\text{Force on Guide } G_0 \text{ along the } X \text{ axis} = -M_Z/l_0 = -8000/3 = -2660 \text{ lb}$$

$$\text{Force on Guide } G_1 \text{ along the } X \text{ axis} = -P_2 + M_Z/l_0 = -1400 + (8000/3) = 1260 \text{ lb}$$

$$\text{Force on Guide } G_2 \text{ along the } Y \text{ axis} = F_1 + M_Z/l_1 = 1100 + (8000/3) = 3766 \text{ lb}$$

$$\text{Force on Guide } G_2 \text{ along the } Z \text{ axis} = M_Y/l_1 = 11{,}500/3 = 3833 \text{ lb}$$

$$\text{Force on Guide } G_3 \text{ along the } Z \text{ axis} = -M_Y/l_1 + F_3 = (-11{,}500/3) + 980 = 4813 \text{ lb}$$

$$\text{Force on Guide } G_3 \text{ along the } Y \text{ axis} = -M_Z/l_1 = -8000/3 = -2660 \text{ lb}$$

$$\text{Force on Guide } G_4 \text{ along the } X \text{ axis} = F_2 + M_Y/l_2 = 1400 + (11{,}500/3) = 5230 \text{ lb}$$

$$\text{Force on Guide } G_4 \text{ along } Y \text{ axis} = M_X/l_2 = 9000/3 = 3000 \text{ lb}$$

$$\text{Force on Guide } G_5 \text{ along the } X \text{ axis} = -M_Y/l_2 = -11{,}500/3 = -3{,}830 \text{ lb}$$

$$\text{Force on Guide } G_5 \text{ along the } Y \text{ axis} = F_4 - M_X/l_2 = 2620 - 3000 = -380 \text{ lb}$$

$$\text{Force on Guide } G_6 \text{ along the } Z \text{ axis} = -F_3 - M_X/l_3 = -980 - (9000/3) = -3980 \text{ lb}$$

$$\text{Force on Guide } G_7 \text{ along the } Z \text{ axis} = M_X/l_3 = 9000/3 = 3000 \text{ lb}$$

This simple, effective, and time-saving method can be used to analyze any complex piping structure subjected to thermal expansion without any tiresome computations.

It is important to note that the method is *approximate.* The *accuracy increases* with:

Decrease of the *clearance* between pipe and guide
Decrease of the *width* of the guide
Increase of the *element L* (see Fig. 31)

Computer Program for Piping Analysis

A computer program in Fortran IV was written (Fig. 35) to solve expansion loop design based on the manual method described above. The program does not require manual interpolation techniques. It uses terms and measurements common to any piping drafting room and is written to allow the piping draftsman or designer to identify and solve expansion loop problems without depending upon consultants. If the loop input has too great an anchor force or too great a stress, the program provides the required flexibility by increasing the length of a designated member of the loop within user input limits. The locations of guides normally are not altered because guides usually are mounted on supports whose locations are subject to controls other than piping flexibility.

Input Data

The entries for an individual piping loop require a *single* card (Fig. 36) of 80 letters or digits maximum containing the following information:

1. Pipe loop identification name or number.
2. Outside diameter of the pipe and wall thickness (in inches). (There are no limitations on diameter or thickness.)

```
C      S AND B ENGINEERING SERVICES PIPE LOOP DESIGN PROGRAM
C      MAINLINE  PIPE LOOP DESIGN --SINGLE PLANE -2 TO 6 PIPE MEMBERS
C                 WITH 1 TO 5 ELBOWS OR BENDS OF SAME BEND RADIUS
C      WRITTEN BY W.W.SHULL AND G.N.BOGEL, HOUSTON TEXAS, MAY 1967.
C      PROGRAM USES FORMULAS AND TABLES PRESENTED IN ''PIPING DESIGN AND
C      ENGINEERING'' THE GRINNELL COMPANY,INC., PROVIDENCE, RHODE ISLAND,
C      USA, SECOND EDITION, 1963.
C      ALL SUBS FOLLOW PROCEDURE AS DETAILED ON PAGES 52,53 OF GRINNELL
C          INPUT STARTING WITH ANY MEMBER AND ENDING WITH ANY MEMBER.
C          EACH MEMBER HAS TWO JOINTS (WELDED) WHICH ARE ANALYZED FOR
C          STRESS.
C          THE LOWEST AND HIGHEST JOINT NUMBERS ARE CONSIDERED TO BE THE
C          LOCATION OF THE INNERMOST PAIR OF GUIDES OR THE LOCATON OF
C          ANCHORS IF THE INPUT LENGTH 'OUTSIDE OF GUIDES' IS ZERO.
C          THE EXCEPTION TO THIS RULE IS FOR TWO MEMBER (SINGLE ELBOW)
C          JOINTS FOR WHICH THE PROGRAM WILL INCREMENT ONE MEMBERS LENGTH
C          TO PROVIDE THE NECESSARY FLEXIBILITY.
C
      COMMON D(6),B(6),V(6),X(12),Y(12),XL(12),T(12),TX(12),TY(12),R(12)
     1,S(12),RM(12),ST(12),AA(20)
      COMMON SMP,PI,XLA,XK,SBETA,XBAR,YBAR,XLP,YLP,PXY,PIX,PIY,TMAX,CT,
     1FX,FY,CONSTX,CONSTY,DA,THK,FSTMAX,DEM,XLEN,ALWST
      COMMON TLNG,XORY,OLNG
      COMMON TRX(5),TRY(5),BX(5),BY(5),DX(5),BST(5),BR,STMAX,KBD
      COMMON KPRT,KST,KEND,KFST,KLOOP,J,K,N,KMAX
      DOUBLE PRECISION PIPEID
C     X(J) = DST TO CENTROID  X DIRECTION
C     Y(J) = DST TO CENTROID Y DIRECTION
C     XL(J)= LENGTH OF SEGMENT
C     R(J) = X DIST FROM CENTROID
C     S(J) =  Y DIST FROM CENTROID
C     PXY  =  PROD OF INTERIA
C     PIX  =  MOM OF INTERIA ABOUT X
C     PIY  =  MOM OF INTERIA ABOUT Y
C     CT   = EXPANSION AT TEMP
C     XLG,YLG = X AND Y DIST END POINT TO END POINT
C     PI   =  PIPE MOMENT OF INTERIA
C     SMP  =  PIPE SECTION MODULUS
C     FUNCTIONS
C     CENTROID OF STRAIGHT LINE IN PLANE OF PROJECTION - COLN
C     A = LENGTH  B= X OR Y DIST      EQN 1
C     COLN(A,B) = A*B
C     CENTROID OF BEND XK=FLEX.FACT.R=RADIUS OF BEND.X=DIST TO AXISEQN3
C     CDBND(XK,R,X)= 1.57*XK*X *R
C     PROD. OF INTERIA  LINE PARALLEL TO AN AXIS EQN6    XL=LENGTH
C     PLNXY(XL,X,Y)= XL * X * Y
C     PROD OF INTERIA BEND NEG WHEN AXIS X RADIAL OR ARC
C     PBNDN(XK,R,X,Y)=-XK *.137*R*R*R + XK * 1.57 * R * X * Y
C     PROD OF INTERIA BEND POS WHEN AXIS X RADIAL AND ARC
C     PBNDP(XK,R,X,Y)= XK*.137*R*R*R + XK*1.57*R*X*Y
C     MOMENT OF INTERIA ST LINE IN PLANE OF PROJ PARALLEL TO AXIS X
C     PLMILN(XL,Y) = XL *Y *Y
C     MOM OF INT ST LINE PERPENDICULAR TO AXIS Y
C     PPMILN(XL,X)  = XL*XL*XL/12.+ XL*X*X
C     MOM OF INT BEND FROM AXIS Y
C     PMIBN (XK,R,X) = XK *(0.149*R*R*R) + 1.57* XK *R *X *X
C     PIPE SECTION MODULUS   DA = ACTUAL DIA.  THK= THICKNESS WALL
      SM(DA,THK,SMX)=0.25/DA*SMX*(DA*DA-2.*(DA*THK)+2.*(THK*THK))
C     LAMBADA   XLAM      R = RADIUS OF PIPE  BEND
      XLAM(THK,R,DA)=THK*R/((DA-THK)*(DA-THK)*.25)
C     FLEX. FACTOR       XLA = LAMBADA
      FF(XLA)=1.65/XLA
C     STRESS INTENSIFICATION FACTOR, BETAS, FOR WELDED ELBOWS OR BENDS.
      BETAS(XLA)=0.9/XLA**0.6667
C     PIPE CROSS SECTIONAL METAL AREA
      PAREA(DA,THK)=3.1416*THK*(DA-THK)
C     PIPE MOM OF INT
      PMINT(DA,THK)=3.1416*(DA-THK)**03*THK/8.
C     XBAR DISTANCE
C     D IS DIST  V= VARIABLE  B=BEND RADIUS
C     KPRT--A NON-ZERO ENTRY IN COL 78 PRINTS VALUES IN COMMON
C     LPRT--NON-ZERO ENTRY IN COL. 79  SHOWS LAST OF PIPE LOOPS USING
C             HEADER CARD AND PROGRAM WILL EXPECT A NEW HEADER CARD.
C     NPRT--PRINTS STRESS AS PIPE MEMBER LENGTH IS VARIED FOR DESIGN
C            PROBLEMS. ENTER IN COL. 80 DIGIT 1 FOR STRESS EACH .5 FEET.
C                                       DIGIT 2 FOR STRESS EACH 1. FEET.
C                                       DIGIT 3 FOR STRESS EACH 1.5 FEET. ETC.
C      GUIDE INDICATES DISTANCE IN X OR Y DIRECTION OUTSIDE THE SECTION
C      DEFINED BY D1-D6 ,  0=X  1=Y DIRECTIONS  X=D1,D3,D5
C     TEST DATA FROM GRINNELL SECOND EDITION PAGE 52.  CT FROM PAGE 9
C A6  F5.0 F5.0 F5.0 F5.0 F5.0 F5.0 F5.0 F5.0 F5.0 F5.0 F5.0 I1F4.I1F4.I1F4. 3I1
C ID  DIA  THK  ALWST CT  BND D  D1 D2   D3   D4   D5   D6    GUIDEVARYLVARYA PRT
CP.52R10.75 .50 17675 996. 50. 40.  24.  12.   14.
CP.52D10.75 .50 17675 996. 50. 40.  10.  12.   4.             2 10.14000        2
CP.52RGRINNELL ANSWERS: ANCHORS X=2795,Y=1467LBS. END MOMENTS=24364,-23970 FT-LB
C     FOR PROBLEM P.52: MAX.BEND STRESS BET. JOINT 2 & 3 = 9030PSI, MOM.=29672
C     MANUAL-GRAPHICAL: GREATEST STRAIGHT PIPE IS AT JOINT 2 = 8714 PSI
C                       INERTIA OF PIPE=212. XBAR=34.63,YBAR=9.60FT.
C                       SECTION MODULUS=39.43  IXY=10457.  IX=9415. IY=21369.
C
C     COMPUTER RESULTS: MAX.BEND STRESS BET.JOINTS 2 & 3 = 9182 PSI
C     FOR PROBLEM P.52: GREATEST JOINT STRESS IS AT JOINT 2 = 8723 PSI
C                       INERTIA OF PIPE=212. XBAR=34.62,YBAR=9.59 FT.
C                       SECTION MODULUS=39.43  IXY=10446.  IX=9395. IY=21354.
C
      KONTRL=2
      WRITE (3,135)
101   READ (1,129) AA
      NPAGE=1
102   WRITE (3,130) NPAGE,AA
      DO 103 J=1,132
103   B(J)=0.0
      DO 104 J=153,211
104   B(J)=0.0
      READ (1,131) PIPEID,DA,THK,ALWST,CT,BRAD,D,KFST,FSTMAX,KSND,SNDMAX
     1,KDIR,OLNG,KPRT,LPRT,NPRT
      WRITE (3,132)
      WRITE (3,133) PIPEID,DA,THK,ALWST,CT,BRAD,D,KFST,FSTMAX,KSND,
     1SNDMAX
      IF (KFST) 106,106,105
105   FSTMAX=FSTMAX+D(KFST)
106   CONTINUE
      KN=NPRT
      SMX=PAREA(DA,THK)
      SMP=SM(DA,THK,SMX)
      PI=SMP*(DA*.5)
C     SET BENDS TO ZERO FOR SQ. CORNER ANAYLSIS
      IF(BRAD) 74,75,74
75    BRAD = 0.1E-30
      XK=0.1E-30
      XLA=0.1E-30
      GO TO 110
74    CONTINUE
      IF (BRAD-(DA-2.0*THK-0.12*THK)) 107,108,108
107   XLA=THK/((DA-THK)/2.)
      XK=1.52/XLA**0.8333
      GO TO 109
108   XLA=XLAM(THK,BRAD,DA)
      XK=FF(XLA)
109   SBETA=BETAS(XLA)
      IF (SBETA-1.0) 110,111,111
110   SBETA=1.0
C     TEST FOR BEG & END OF LOOP MEMBERS
111   KST=0
      KEND=0
      DO 115 J=1,6
      IF (D(J)) 115,115,112
112   IF (KST) 114,113,114
113   KST=J
114   KEND=J
115   CONTINUE
C     INSERT VALUES WHERE THERE ARE BENDS
      BR=BRAD/12.
      KEN=KEND-1
      DO 116 J=KST,KEN
116   B(J)=BR
C     TEST FOR KIND OF LOOP
      KLOOP=KEND-KST
      GO TO (121,117,118,119,120), KST
117   KLOOP=KLOOP+5
      GO TO 121
118   KLOOP=KLOOP+9
      GO TO 121
119   KLOOP=KLOOP+12
      GO TO 121
120   KLOOP=KLOOP+14
121   CONTINUE
C     SET LIMITS ON END POINTS
      CONSTX=D(1)+D(3)+D(5)
      CONSTY=D(2)-D(4)+D(6)
C     SUB P5CXY CALCULATES LINE & BEND CENTER OF GRAVITY
122   CALL P5CXY
C     SUB P5IXY CALC. PRODUCTS  OF INERTIA ABOUT X-AXIS AND Y-AXIS
      CALL P5IXY
C     SUB P5IXAY CALS. MOMENTS OF INERTIAS ABOUT THE TWO AXIS.
      CALL P5IXAY
      CALL P5DIST
C     SUB P5FORC CALC. ANCHOR FORCES AND ADJUSTS FOR LENGTH OUTSIDE GUIDES.
      CALL P5FORC (KDIR)
      CALL P5MARM
C     SUB P5MARM CALCS. MOMENT ARM DISTANCES TO EACH POINT OF INTEREST
C          AND THEN ADDS X & Y MOMENTS TO DETERMINE STRESS AND
C          MULTIPLIES BY STRESS INTENSIFICATION AS REQUIRED.
      IF (KSND-1) 150,140,140
  140 IF (SNDMAX) 150,150,141
  141 GO TO (142,143),KSND
  142 IF (ABS(FX)-SNDMAX) 150,150,124
  143 IF (ABS(FY)-SNDMAX) 150,150,124
  150 IF (ABS(STMAX)-ALWST) 123,123,124
C     SUB P5PRNT PRINTS RESULTS--CONTAINS FORMAT STATEMENTS.
123   CALL P5PRNT (PIPEID)
      NPAGE=NPAGE+1
      WRITE (3,135)
      IF (LPRT) 102,102,101
C     SUB P5CNTL INCREMENTS LOOP LENGTHS BASED ON CONTROL PARAMETERS.
124   CALL P5CNTL (KONTRL,KSND,SNDMAX)
      GO TO (125,123), KONTRL
125   IF (NPRT) 122,122,126
126   IF (KN/NPRT*NPRT-KN) 128,127,128
127   DV=D(KFST)+V(KFST)-.5
      WRITE (3,134) DV,STMAX,FX,FY
128   KN=KN+1
      GO TO 122
C
129   FORMAT (20A4)
130   FORMAT (T110,'PAGE NO.',I5,
     2   T51,' S AND B ENGINEERING SERVICES',/,
     3   T53, ' PIPE LOOP DESIGN PROGRAM'///,
     4   T26, 20A4,//)
131   FORMAT (A6,11F5.0,3(I1,F4.0),1X,3I1)
132   FORMAT (8X'PIPE LOOP',3X,'PIPE',4X,'WALL',3X,'ALLOW',2X,' FACTOR C
     1',2X,'RADIUS',2X,'LENGTHS OF MEMBERS (TO CENTER OF CORNERS)',2X,
     2'D_ WHICH  ANCHOR FORCE LIMIT',/,10X,'IDENT.    O.D.   THICK.  STR
     3ESS  @ DEG.F  OF BEND    D1      D2      D3      D4      D5      D6
     4MAY VARY  0=NO. 1=X 2=Y DIR.',/,20X,'INCHES INCHES    PSI   SEE P.
     5 9  INCHES    FEET   FEET   FEET   FEET   FEET   FEET  NO. FEET  C
     6ODE  POUNDS FORCE')
133   FORMAT (1H+, 7X,'_________   _____ _____  ______  ________  ____
     1___ ______ ______ ______ ______ ______ _____  ___ ____  ____  ___
     2__________', ///,1X,'INPUT',4X,A8,2F8.3,F8.0,F9.1,
     3F9.3,1X,6F7.2,I3,F7.2,I4,F10.1,/)
  134 FORMAT(1X,'WITH VARYING MEMBER LENGTH =',F7.1,' FEET, STRESS IS',
     1F9.0, ' LBS/SQ.IN., ANCHOR FORCES X-DIR.=',F8.0,', Y-DIR.=',F8.0)
135   FORMAT (1H1)
      END
```

FIG. 35 Pipe expansion loop program. Note: Subroutines are not included in the program listing. Solve step-by-step by using the manual method described, using the functions duplicated as comment statements in the mainline program listing. The comment statements reveal all the basic thoughts and methods.

3. An allowable stress range at bends and terminal joints based upon code or connected equipment limitation (lb/in.^{2}).
4. A value of c obtained from the following equation:

$$c = \frac{\text{expansion, in./100 ft}(E_c)}{1728(100)}$$

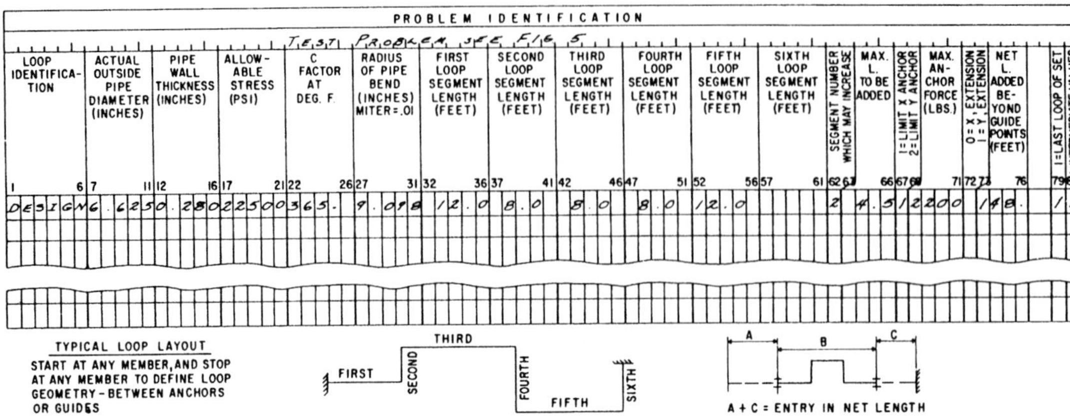

FIG. 36 Input card with data for the problem shown as Fig. 39.

or from Table 3. This value combines the lineal thermal expansion in inches from 70°F to the operating temperature with the tensile modulus of elasticity at 70°F (lb/in.).

5. Radii of bends (in inches)—a single bend radius is applied to all bends.
6. Lengths between corners of the loop members within the innermost pair of guides (in feet). [A *corner* is defined as the intersection of the centerline extensions of straight pipe members (however short) which connect to a 90° bend or square elbow. The input length must be not less than the sum of the bend radii (expressed in feet) included in the particular member.]
7. The number of a member whose length may be varied by the program and the limit of this variation (in feet). Parallel member lengths will be adjusted automatically, and the original distance between guides will be maintained except in certain configurations where options are made to hold anchors and move guides.
8. The maximum permitted anchor force (in pounds) and the X or Y direction of this limitation.
9. The piping length (in feet) which is outside the innermost pair of guides and a code to indicate the X or Y axis of this additional length between anchors which contributes force because of thermal expansion.

An additional header card precedes a group of pipe loop entries. This card contains any desired alphabetic or numeric descriptive data desired in the output printout (job title, user's name, accounting information, etc.).

What the Program Solves

The program is capable of expansion to handle many piping loop design possibilities but is used for the purposes of this example for the following

TABLE 3 Expansion Factor $c^{a,b}$

Temperature F (°F)	Carbon Steel, C ≤ 0.30%	Carbon Steel, C > 0.30%	C-Mo and Low Cr-Mo, C ≤ 3%	Cr-Mo, 5% ≤ Cr, Mo ≤ 9%	Austenitic Stainless Steels	Cr Stainless Steels: 12 Cr, 17 Cr, and 27 Cr	25 Cr-20 Ni	Wrought Iron
70	0	0	0	0	0	0	0	0
100	37	40	40	35	54	34	47	44
150	98	106	106	92	143	90	125	120
200	160	171	171	149	232	145	204	195
250	228	244	244	212	323	204	287	273
300	294	315	315	271	414	264	368	352
350	365	391	391	335	509	326	455	434
400	436	467	467	396	603	389	541	514
450	510	547	547	465	699	455	629	598
500	581	626	626	531	794	520	716	681
550	664	711	711	603	893	590	809	768
600	743	796	796	672	989	659	901	855
650	827	886	886	714	1089	730	995	946
700	909	974	974	815	1189	799	1088	1035
750	996	1068	1068	891	1292	874	1186	1125
775	1038	1113	1113	929	1344	909	1235	1171
800			1159	967	1395	946	1284	1216
825			1208	1005	1448	983	1335	
850			1256	1043	1500	1022	1384	
875			1303	1081	1552	1061	1435	
900			1351	1121	1605	1097	1484	
925			1398	1161	1659	1134	1533	
950			1445	1200	1713	1174	1585	
975			1492	1240	1766	1212	1634	
1000			1538	1278	1820	1250	1681	
1050			1639	1357	1928	1328	1781	
1100			1737	1435	2036	1404	1879	
1150				1511	2144	1480	1980	

[a] $c = \dfrac{\text{expansion in inches per 100 ft} \times E_c}{1728 \times 100}$

[b] Properties of pipe. The straight and curved pipe: D_n = nominal pipe size, D = outside diameter, t = wall thickness, d = inside diameter ($= D - 2t$). Inside area: $A_l = \pi d^2/4$. Metal area: $A_m = \pi t(D - t)$. Moment of inertia: $I_p = 0.0491(D^4 - d^4) = 0.0625A_m(D^2 + d^2)$. Section modulus: $S_m = 2I_p/D$.

frequently designed single-plane loop types which have no external loads or restraints between anchors except guides:

1. U-shaped loop with unequal or equal legs plus up to two tangent members of unequal or of equal length
2. U-bend expansion loops similar to above (use the term *legs* instead of *tangents*)
3. Simple two-member loop with one elbow
4. Z-shaped expansion loop
5. Hooked Z-shaped loops with up to one tangent member

Program Tests

The program has been tested with the problem shown as Fig. 37 and with the examples from the graphical method (Figs. 38 and 39). In general, answers given by the program agree with the problems and examples within ±2%. Extreme care was necessary to make sure the problems were the same and that the results were identifiable.

For the second example given in the preceding section under "Graphical vs Computer Solutions," the stress using long radius bends and computed intensification factor is in agreement, but the anchor force is lower than the

```
      PIPE LOOP   PIPE    WALL    ALLOW    FACTOR C   RADIUS   LENGTHS OF MEMBERS (TO CENTER OF CORNERS)  D_ WHICH  ANCHOR FORCE LIMIT
       IDENT.     O.D.    THICK.  STRESS   @ DEG.F    OF BEND    D1     D2     D3     D4     D5     D6    MAY VARY  0=NO. 1=X 2=Y DIR.
      ________   INCHES  INCHES   __PSI_   ________   INCHES_  _FEET_ _FEET_ _FEET_ _FEET_ _FEET_ _FEET   NO. FEET  CODE  POUNDS FORCE

INPUT   DESIGN   6.625   0.280   22500.    365.0      9.098    12.00   8.00   8.00   8.00  12.00   0.0   2  4.50    1      2200.0

WITH VARYING MEMBER LENGTH =    8.0 FEET. STRESS IS   32369. LBS/SQ.IN., ANCHOR FORCES X-DIR.=  2089., Y-DIR.=    -0.
WITH VARYING MEMBER LENGTH =    8.5 FEET. STRESS IS   30079. LBS/SQ.IN., ANCHOR FORCES X-DIR.=  1831., Y-DIR.=    -0.
WITH VARYING MEMBER LENGTH =    9.0 FEET. STRESS IS   28063. LBS/SQ.IN., ANCHOR FORCES X-DIR.=  1618., Y-DIR.=    -0.
WITH VARYING MEMBER LENGTH =    9.5 FEET. STRESS IS   26276. LBS/SQ.IN., ANCHOR FORCES X-DIR.=  1438., Y-DIR.=    -0.
WITH VARYING MEMBER LENGTH =   10.0 FEET. STRESS IS   24682. LBS/SQ.IN., ANCHOR FORCES X-DIR.=  1286., Y-DIR.=    -0.
WITH VARYING MEMBER LENGTH =   10.5 FEET. STRESS IS   23251. LBS/SQ.IN., ANCHOR FORCES X-DIR.=  1157., Y-DIR.=    -0.
RESULTS  DESIGN                  21960.                   12.00  11.00   8.00  11.00  12.00   0.0

STRESS AT JOINT :  J1      J2      J3      J4      J5      J6      J7      J8      J9      J10     J11     J12
                 6473.  14561.  12044.  19443.  21960.  21960.  19443.  12044.  14561.   6473.     0.      0.

     JOINT  6 HAS GREATEST STRESS        ANCHOR FORCE IN X-DIRECTION =   1045.       ANCHOR FORCE IN Y-DIRECTION=     0.

     LENGTH  OF   SUM OF MEMBERS  COORDINATE OFFSET FEET  MAXIMUM STRESS IS IN BEND  LENGTH OUTSIDE  AXIS DISTANCE
     FLUID PATH   D1-D6 LENGTHS   JOINT J 1 THRU J 10     BETWEEN JOINTS  6 AND  7   LOOP  GUIDES    ANCHOR-ANCHOR
       FEET          FEET          X-AXIS   Y-AXIS          IN POUNDS/SQ.INCH           FEET          FEET   AXIS
      200.70        54.00          32.00    0.0                 21960.                 148.00        180.00   X

     MOMENT AT JOINT  1 =                 DIAGRAM OF PIPE LOOP                     MOMENT AT JOINT 10 =
         815. FOOT-POUNDS                                                              815. FOOT-POUNDS

                                             8.000 FEET
                                     _J5_____MEMBER D3_____J6_
                                     I                        I
                                     J4                       J7
                                     I                        I
                                     I                        I
                                     I  11.000 FEET           I  11.000 FEET
                                  MEMBER D2                MEMBER D4                J12
                                     I                        I                      I
                                     I                        I                      I
              /                      I                        I                      I
              /      12.000 FEET     J3                       I                      I   0.0   FEET
     BEGINNING = J1_____MEMBER D1_____J2_I                     I                  MEMBER D6
     GUIDE    =                                               I                      I
              /                                               I                      I
              /                                               I                      I
                                                              J8     12.000 FEET    J11
                                                              I_J9_____MEMBER D5___J10_I

LOOP O. K. FROM STRESS STANDPOINT
```

FIG. 37 Test problem based on book input data.

```
       PIPE LOOP   PIPE     WALL    ALLOW    FACTOR C   RADIUS  LENGTHS OF MEMBERS (TO CENTER OF CORNERS)  D_ WHICH   ANCHOR FORCE LIMIT
        IDENT.     O.D.     THICK.  STRESS   @ DEG.F    OF BEND    D1     D2     D3     D4     D5     D6   MAY VARY   0=NO. 1=X 2=Y DIR.
       ________   INCHES   INCHES   __PSI_   _______    INCHES_  _FEET_ _FEET_ _FEET_ _FEET_ _FEET_ _FEET  NO. FEET   CODE  POUNDS FORCE

INPUT    RATING     6.625   0.280   22500.    365.0      9.098   12.00  10.00   8.00  10.00  12.00   0.0   0   0.0    0          0.0

RESULTS  RATING                     24682.                       12.00  10.00   8.00  10.00  12.00   0.0

STRESS AT JOINT :  J1      J2      J3      J4      J5      J6      J7      J8      J9      J10     J11     J12
                 7196.  16187.  13088.  21583.  24682.  24682.  21583.  13088.  16187.   7196.     0.      0.

       JOINT  5 HAS GREATEST STRESS         ANCHOR FORCE IN X-DIRECTION =   1286.       ANCHOR FORCE IN Y-DIRECTION=     0.

       LENGTH  OF  SUM OF MEMBERS  COORDINATE OFFSET FEET  MAXIMUM STRESS IS IN BEND  LENGTH OUTSIDE  AXIS DISTANCE
       FLUID PATH  D1-D6 LENGTHS   JOINT J 1 THRU J 10     BETWEEN JOINTS  6 AND  7   LOOP  GUIDES    ANCHOR-ANCHOR
          FEET         FEET          X-AXIS    Y-AXIS        IN POUNDS/SQ.INCH            FEET          FEET   AXIS
         198.70        52.00         32.00     0.0                24682.                143.00         180.00   X

       MOMENT AT JOINT  1 =                    DIAGRAM OF PIPE LOOP                   MOMENT AT JOINT 10 =
           906. FOOT-POUNDS                                                               906. FOOT-POUNDS

                                                  8.000 FEET
                                           _J5_____MEMBER D3_____J6_
                                            |                      |
                                            J4                     J7
                                            |                      |
                                            |                      |
                                            |  10.000 FEET         |  10.000 FEET
                                       MEMBER D2              MEMBER D4                J12
                                            |                      |                    |
                                            |                      |                    |
                 /                          |                      |                    |
                 /      12.000 FEET         J3                     |                    |  0.0    FEET
       BEGINNING_=_J1_____MEMBER D1_____J2_|                      |                 MEMBER D6
       GUIDE     =                                                 |                    |
                 /                                                 |                    |
                 /                                                 |                    |
                                                                   J8     12.000 FEET   J11
                                                                   |_J9_____MEMBER D5___J10_|

THIS LOOP OVERSTRESSED
```

FIG. 38 Test problem based on graphical method input data.

```
       PIPE LOOP   PIPE     WALL    ALLOW    FACTOR C   RADIUS  LENGTHS OF MEMBERS (TO CENTER OF CORNERS)  D_ WHICH   ANCHOR FORCE LIMIT
        IDENT.     O.D.     THICK.  STRESS   @ DEG.F    OF BEND    D1     D2     D3     D4     D5     D6   MAY VARY   0=NO. 1=X 2=Y DIR.
       ________   INCHES   INCHES   __PSI_   _______    INCHES_  _FEET_ _FEET_ _FEET_ _FEET_ _FEET_ _FEET  NO. FEET   CODE  POUNDS FORCE

INPUT    TEST      10.750   0.500   17675.    996.0     50.000   40.00  24.00  12.00  14.00   0.0    0.0   0   0.0    0          0.0

RESULTS  TEST                        9183.                       40.00  24.00  12.00  14.00   0.0    0.0

STRESS AT JOINT :  J1      J2      J3      J4      J5      J6      J7      J8      J9      J10     J11     J12
                 7307.   8723.   7033.   6330.   8020.   6380.    962.   7426.      0.      0.      0.      0.

       JOINT  2 HAS GREATEST STRESS         ANCHOR FORCE IN X-DIRECTION =   2803.       ANCHOR FORCE IN Y-DIRECTION=  1470.

       LENGTH  OF  SUM OF MEMBERS  COORDINATE OFFSET FEET  MAXIMUM STRESS IS IN BEND  LENGTH OUTSIDE  AXIS DISTANCE
       FLUID PATH  D1-D6 LENGTHS   JOINT J 1 THRU J  8     BETWEEN JOINTS  2 AND  3   LOOP  GUIDES    ANCHOR-ANCHOR
          FEET         FEET          X-AXIS    Y-AXIS        IN POUNDS/SQ.INCH            FEET          FEET   AXIS
          84.63        90.00         52.00    10.00                9183.                  0.0           52.00   X

       MOMENT AT JOINT  1 =                    DIAGRAM OF PIPE LOOP                   MOMENT AT JOINT  8 =
        -24011. FOOT-POUNDS                                                             24402. FOOT-POUNDS

                                                 12.000 FEET
                                           _J5_____MEMBER D3_____J6_
                                            |                      |
                                            J4                     J7
                                            |                      |
                                            |                      |
                                            |  24.000 FEET         |  14.000 FEET
                                       MEMBER D2              MEMBER D4                J12
                                            |                      |                    |
                                            |                      |                    |
                 /                          |                      |                    |
                 /      40.000 FEET         J3                     |                    |  0.0    FEET
       BEGINNING_=_J1_____MEMBER D1_____J2_|                      |                 MEMBER D6
       GUIDE     =                                                 |                    |
                 /                                                 |                    |
                 /                                                 |                    |
                                                                   J8      0.0   FEET   J11
                                                                   |_J9_____MEMBER D5___J10_|

LOOP O. K. FROM STRESS STANDPOINT
```

FIG. 39 Test problem using graphical method data.

square corner solution by a factor of 2. The anchor force for the computer square corner solution of the same problem agrees with the force from the computer program mentioned in the graphical method and is 9% lower than the graphical method.

Square Corner Technique

The use of the square corner technique with its conservative answers (sometimes by a factor of 2 or more) tends to indicate problems where there are none and may result in wasting investment and operating capital for unnecessary expansion loops.

Force and Moment at Anchors

Computers have taken the hand calculation work out of piping stress analysis. However, because of time and cost, the computer is not available for many simple two-anchor pipe loops.

Fast estimating procedures have been developed for less complicated configurations based on the cantilever principle. The procedure predicts excessive loop requirements which, in turn, increase piping cost estimates. An alternative to the cantilever principle is to rely on generalized charts for estimates, but very often they do not cover the range for the problem at hand.

Another method, though slightly more complex than the cantilever approach, has the advantage of greater accuracy and economy when compared with the cantilever method. The economy results from the need of less pipe for a required flexibility.

The improved accuracy of this method results from the use of *generalized parameters* obtained from several calculated pipe loops. No claim is made that the procedure will equal the accuracy of a detailed pipe loop calculation. However, it is expected that the results will deviate by a reasonable amount from the most probable values.

The following equations were determined from several calculated pipe loop arrangements:

$$\frac{1}{K_\theta} = \frac{50}{(L/D_\alpha)^{4.83}}$$

$$\frac{1}{K_\delta} = \frac{2900}{(\Sigma L/D_\alpha)^{8}}$$

With the values of K_θ and K_δ determined from these equations, the value of the resultant force and moment at an anchor, for a given pipe arrangement, may be found. The following equations may be used to solve for the desired resultant forces and moment:

$$F = \frac{EI\delta(1/K_\delta)}{D_\alpha^3(1728)}$$

$$M = K_\theta F D_\alpha$$

Refer to Fig. 40a for a diagrammatic representation of key variables.

Limiting Reactions

It is always necessary to limit the force and moment on a piece of machinery such as a pump, compressor, or turbine. When the manufacturer of equipment is asked for their limitations, some sort of negotiable approach is used, starting with zero allowable. Obviously, no piping can be made sufficiently flexible to result in zero thrust or torque. The following formulas yield values of force and moment which have been found acceptable to several equipment manufacturers. The limiting force:

$$F_L = 140 \log_e N/6$$

The limiting moment:

$$M_L = 1.72 F_L$$

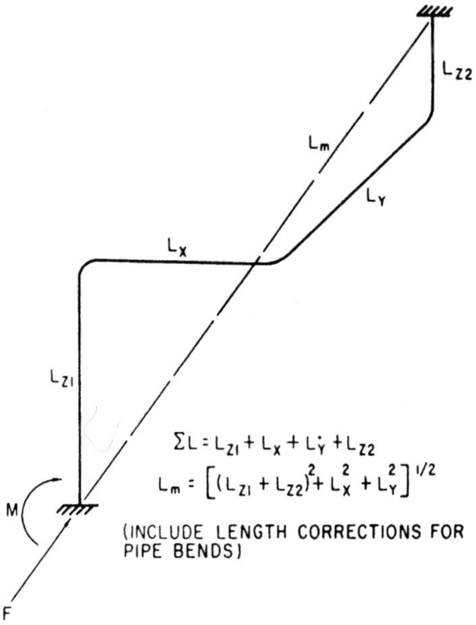

FIG. 40a Typical pipe arrangement.

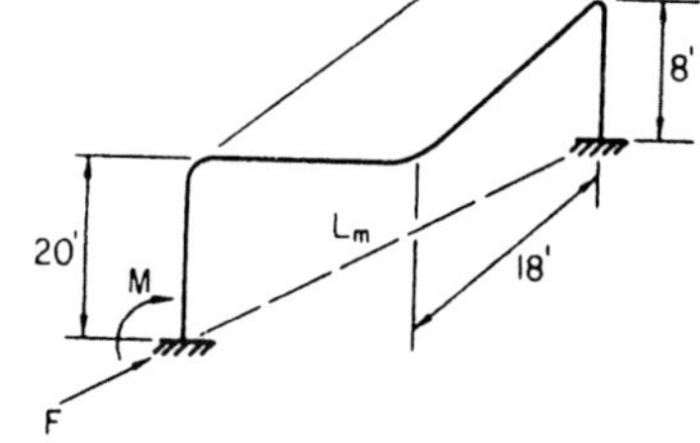

FIG. 40b Piping configuration for example given.

From these equations, a reasonable value for the limiting force and moment may be obtained. It is cautioned that the equipment manufacturer's approval should be obtained before proceeding with final piping design.

Example. Find the resultant force and moment for the arrangement shown in Fig. 40b. Given: 10-in., schedule 40, C.S. pipe, L.R. bends, $I = 160.7$ in.^{4}, $T = 750$°F, $E = 25(10^6)$ lb/in.^{2}, $e = 7.35(10^{-6})$ length/°F.

$$\Sigma L = 20 + 14 + 18 + 8 - 1.6 = 58.4 \text{ ft}$$

Note: 1.6 feet corrects the length for pipe bends.

$$D_\alpha = [14^2 + 18^2 + (20 - 8)^2]^{0.5} = 25.8 \text{ ft}$$

Note: See Fig. 40a.

$$\Delta T = 750 - 70 = 680°\text{F}$$

$$\delta = e(\Delta T)L_M 12$$
$$= (7.35/10^6)680(25.8)12 = 1.55 \text{ in.}$$

$$L/D_\alpha = 58.4/25.8 = 2.26$$

$$1/K_\theta = 50/2.26^{4.83} = 0.98$$

$$1/K_\delta = 2900/2.26^8 = 4.3$$

$$F = \frac{25(10^6)(160.7)1.55(4.3)}{(23.8)^3 1728} = 900 \text{ lb}$$

$$M = 1/0.98(900)25.8 = 23{,}700 \text{ lb-ft.}$$

Pipe Stress

After the thrust and moment at an anchor has been calculated, it is necessary to determine whether the selected pipe size and configuration have sufficient strength to withstand the combined stresses to which the pipe will be subjected. The contributing factors to pipe stress may be caused by any combination or all of the following:

Internal or external pressure
Bending in the straight pipe or at an elbow
Torsion which subjects the pipe section to shear
Direct stress by axial forces
Pipe support spacing affecting local bending

Expansion Loops

Thrust and stress (the main considerations in piping flexibility calculations) are controlled entirely by the piping configuration. Since the piping configuration is arbitrary, it often produces excessive stresses or thrust. When this happens the designer must start over or revise his piping configuration until another try produces results that are within the desired limitations.

A better approach is to select the maximum stress and desired thrust first, eliminating further concern about these two main considerations. The piping configuration is then selected so that it can be adjusted with little effort to give the desired total moment for the stress and thrust given.

How the Method Works

This method assumes that the moment of inertia is constant throughout the piping configuration, that the flattening of pipe bends during expansion or contraction will be negligible, and that all anchor points are 100% fixed.

It should be noted that systems having a combination of high expansion factors, low tensile strength, and extremely high temperatures should be examined more closely by some appropriately rigorous mathematical method which considers such things as flattening of bends, varying of wall thickness, and other such peculiarities of the system subjected to the above conditions. However, few systems require such an examination.

Example 1. Assume a piping configuration and conditions as shown in Fig. 41a. Find the most economical configuration to meet these conditions.

Solution: 1. Assume a maximum stress (12½% of tensile strength):

$$S_{max} = 60{,}000(0.125) = 7500 \text{ lb/in.}^2$$

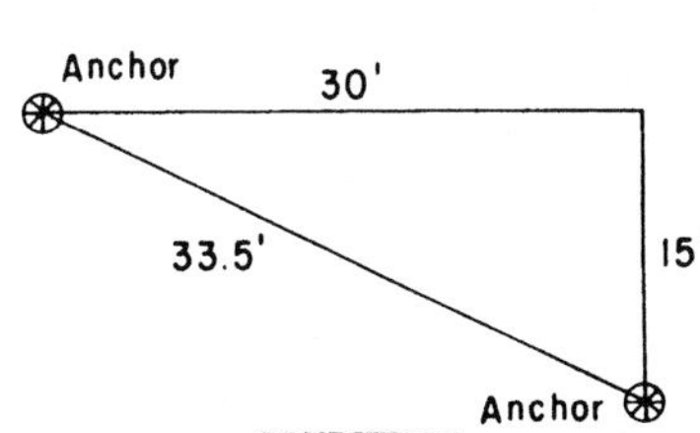

CONDITIONS

Pipe	6-in. X-Stg. pipe
Moment of Inertia	40.49 in.^{4}
Max. Temp.	800° F
Fab. Temp.	70° F
Δt (differential temp.)	730° F
Expansion per 100 ft. of pipe	6.9 in.
E at max. temp.	24,200,000 psi
Tensile Strength	60,000 psi
Dist between anchors	33.5 ft.

FIG. 41a Piping configuration and conditions for Example 1.

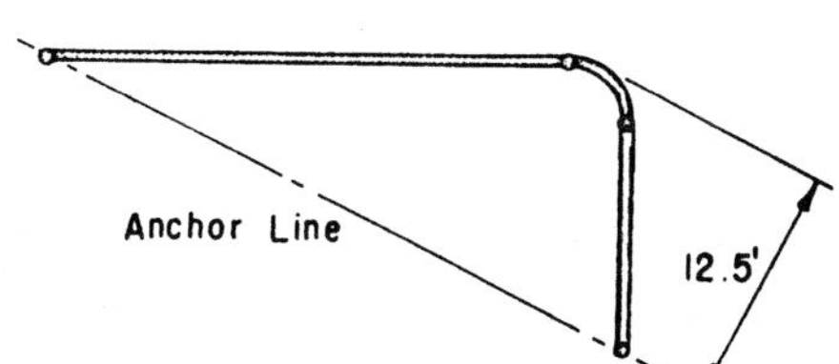

FIG. 41b The expansion loop can't be taken off the long side.

2. Assume a maximum thrust:

F_T = 750 lb (This figure may be any one the designer may choose, but preferably one that conforms to his standard anchoring system.)

3. Calculate b:

$$b = \frac{S_{max} I}{6F_T \text{ (o.d.)}} = \frac{7500(40.49)}{6(750)(6.625)} = 10.1 \text{ ft}$$

4. Compute Δc:

$$\Delta c = \frac{D_\alpha e}{100} = \frac{33.5(6.90)}{100} = 2.31 \text{ in.}$$

5. Compute M_2':

$$M_2' = \frac{(\Delta c)EI}{1728F_T}$$

$$= \frac{(2.31)(24{,}200{,}000)(40.49)}{(1728)(750)} = 1750 \text{ ft}^3$$

6. Draw an assumed piping configuration and calculate the neutral axis position. As shown in Fig. 41b, it is doubtful that the piping configuration can be extended enough to allow for proper flexibility without exceeding b if we take the expansion loop off the long side. Therefore, the short side is more desirable. See Fig. 42 for the proposed configuration and calculation of the neutral axis distance (η). Note: All lengths in Fig. 42 are true and all moment arms are to be perpendicular to a

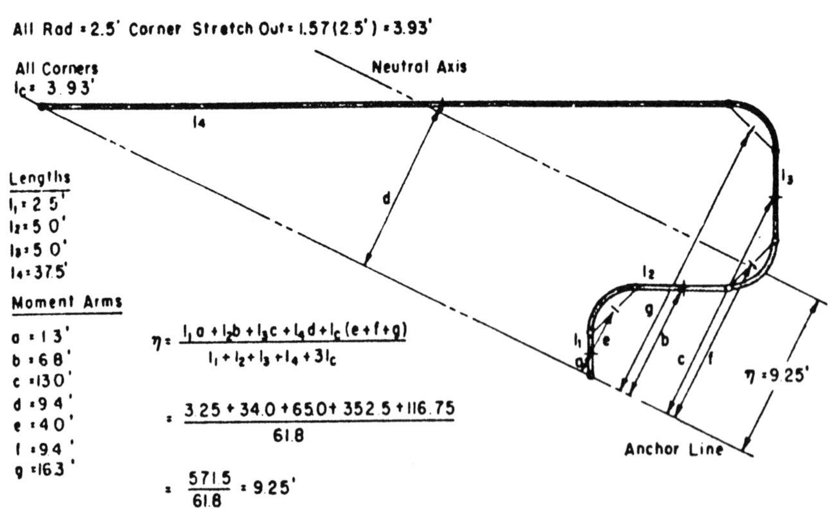

FIG. 42 Proposed configuration for Example 1 and calculation of neutral axis.

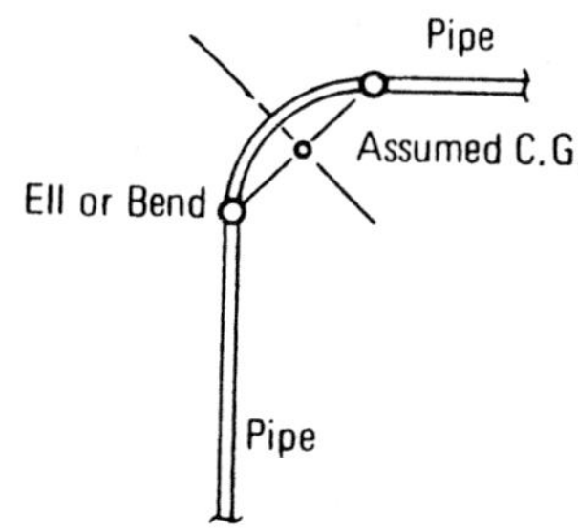

FIG. 43 Assumed center of gravity for elbows.

line drawn through the anchor points. For layout purposes the center of gravity of each bend may be assumed as shown in Fig. 43 except for extremely long radius bends.

7. Draw new moment arms ($\bar{l}$) (see Fig. 44) perpendicular to the neutral axis and extend them to the center of gravity of each of the components and the equivalent lengths (l') of each component. Note that new components were made wherever the neutral axis crossed any section, with the exception of bends. The difference in moment by considering a bend as two components is in general negligible. Measure each component (l), each moment arm ($\bar{l}$), and each equivalent length (l') from the layout, place in the appropriate column of the "moment calculation form" (Table 4) and calculate moment M_2.

$$L = l_1 + l_2 + l_3 + l_4 + l_5 + 3l_6 = 61.8$$

See Fig. 42 for l_c lengths.

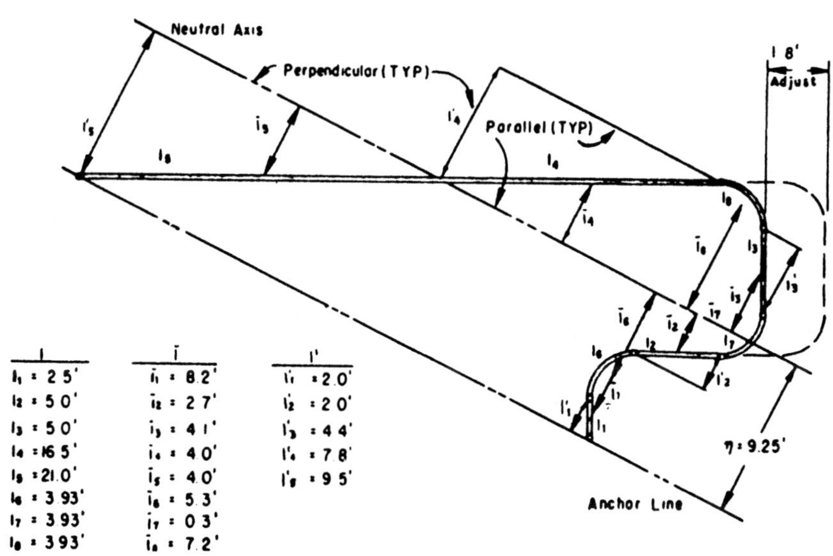

FIG. 44 New moment arms are drawn.

TABLE 4 Moment Calculation Form for Example 1

Section	Length l	Length l'	$(l')^2$	Length $\bar{l}$	$(\bar{l})^2$	$A = l(\bar{l})^2$	$B = \frac{l(\bar{l}')^2}{12}$	Ms per Section $= A + B$
1	2.5	2.0	4.0	8.2	67.2	168.0	0.8	168.8
2	5.0	2.0	4.0	2.7	7.3	36.5	1.7	38.2
3	5.0	4.4	19.4	4.1	16.8	84.0	8.1	92.1
4	16.5	7.8	60.9	4.0	16.0	254.0	83.7	337.7
5	21.0	9.5	90.2	4.0	16.0	336.0	157.8	493.8
Ms total								1030.6

Corner	Radius R	$l = 1.57R$	R^3	Length $\bar{l}$	$(\bar{l})^2$	$C = l(\bar{l})^2$	$D = 0.15R^3$	Mc per Corner $= C + D$
1	2.5	3.9	15.6	5.3	28.1	109.7	2.3	112.0
2	2.5	3.9	15.6	0.3	0.1	0.4	2.3	2.7
3	2.5	3.9	15.6	7.2	51.9	202.1	2.3	204.4
Mc total								319.1

$M = Ms + Mc = 1030.6 + 319.1 = 1349.7\ \text{ft}^3$

8. Adjust configuration: Calculate the radius of gyration of the configuration about the neutral axis based on M:

$$K = \sqrt{\frac{M_2}{L}} = \sqrt{\frac{1.350}{61.8}} = 4.7\ \text{ft}$$

Calculate the approximate radius of gyration based on M_2':

$$k_g'\ (\text{approx.}) = \sqrt{\frac{M_2'}{L}} = \sqrt{\frac{1.750}{61.8}} = 5.3\ \text{ft}$$

Calculate the approximate Δx (based on k_g and k_g' (approx.) above):

$$\Delta x\ (\text{approx.}) - [k_g'\ (\text{approx.}) - k_g] = (5.3 - 4.7) = 0.6\ \text{ft}$$

Compute Δl. Assume a Δx based on the above Δx (approx.); try $\Delta x = 0.4$ ft, then:

$$\Delta l = c(\Delta x) = (33.5/15)(0.4) = 0.9\ \text{ft}$$

Compute k_g' (actual):

$$k_g' = k_g + \Delta x = 4.7 + 0.4 = 5.1\ \text{ft}$$

Compute L':

$$L' = L + 2N_m(\Delta l) = 61.8 + 2(2)(0.9) = 65.4\ \text{ft}$$

Check adjusted M_2' to see how closely it agrees with the required M':

$$M_2'(\text{adj.}) = L'(K')^2 = 65.4(5.1)^2 = 1700 \text{ ft}^3$$

The Δx value of 0.4 ft checks. Note that, with respect to significant figures, this is as close to the desired M' as we can get with the figures used in this example.

9. Change the configuration by adding $2\Delta l$ to the length of the expansion loop and check to see that b is not exceeded by this change.

Example 2. Assume a piping configuration and conditions shown in Fig. 45. Find the most economical configuration to meet these conditions.

Solution: 1. Assume maximum stress:

$$S_{\max} = 60{,}000(0.125) = 7500 \text{ lb/in.}^2$$

2. Assume thrust:

$$F_T = 925 \text{ lb}$$

3. Calculate b:

$$b = \frac{(7500)(781.3)}{6(925)(12.75)} = 83.0 \text{ ft}$$

4. Compute Δ_c:

$$\Delta_c = \frac{6.4(51)}{100} = 3.26 \text{ in.}$$

5. Compute M_2':

$$M_2' = \frac{3.26(24{,}550{,}000)(781.3)}{1728(925)} = 39{,}100 \text{ ft}^3$$

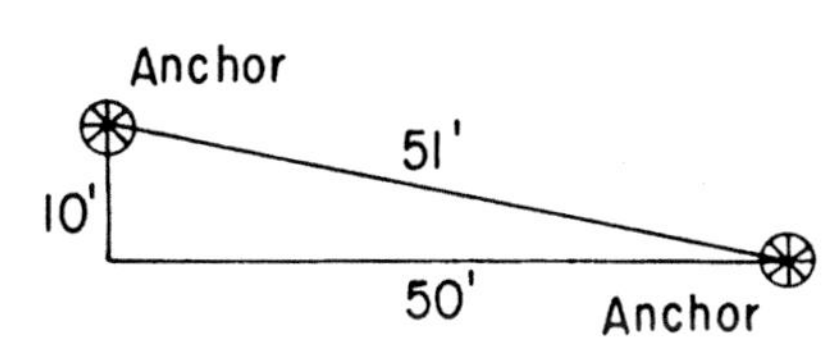

CONDITIONS

Pipe	12" sch. 160 pipe
Moment of Inertia	781.3 in.^{4}
Max. Temp.	750° F
Fab. Temp.	70° F
Δt (differential temp.)	680° F
Expansion per 100 ft. of pipe	6.4 in.
E at max. temp.	24,550,000 psi
Tensile Strength	60,000 psi
Dist. between anchors	51.0 ft.

FIG. 45 Piping configuration and conditions for Example 2.

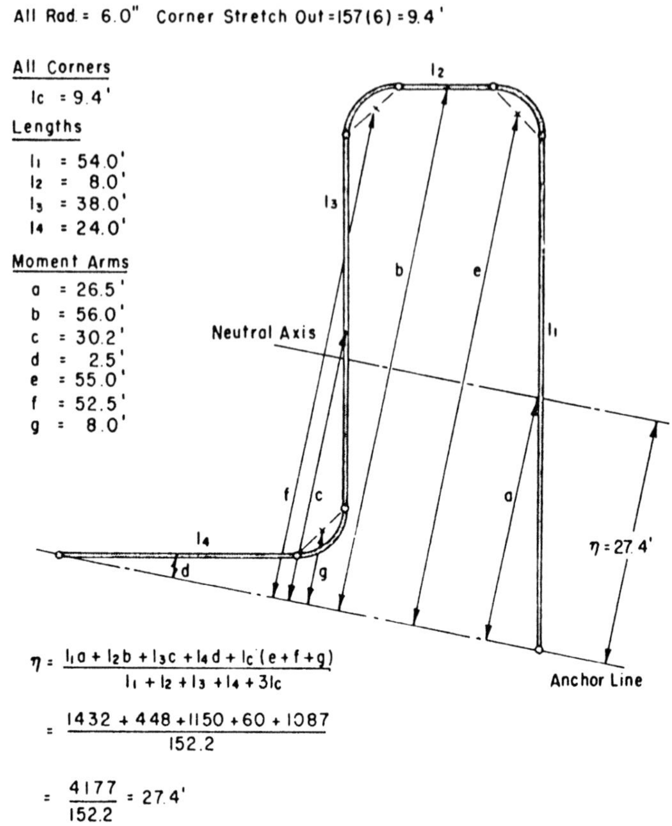

FIG. 46 Proposed configuration for Example 2 and location of neutral axis.

6. Draw an assumed configuration and locate the neutral axis (see Fig. 46).
7. Draw and measure l, l', and $\bar{l}$, and compute M_2. (See Fig. 47 and Table 5.)
8. Adjustment. Calculate k_g:

$$k_g = \sqrt{\frac{M_2}{L}} = \sqrt{\frac{53{,}051}{152.2}} = 18.7 \text{ ft}$$

Calculate k_g' (approx.):

$$k_g' \text{ (approx.)} = \sqrt{\frac{M_2'}{L}} = \sqrt{\frac{39{,}100}{152.2}} = 16.0 \text{ ft}$$

Calculate Δx (approx.):

$$\Delta x \text{ (approx.)} = (k_g' - k_g) = 16.0 - 18.7 = -2.7 \text{ ft}$$

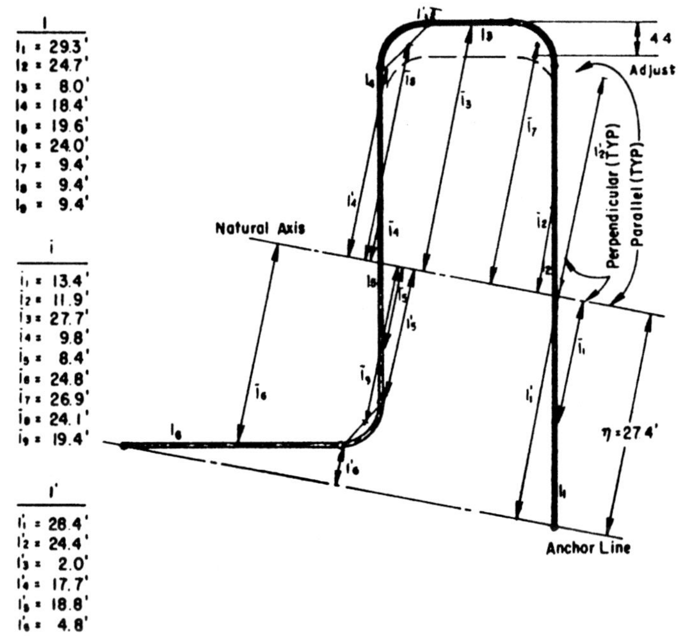

FIG. 47 New moment arms drawn for Example 2.

TABLE 5 Moment Calculation Form for Example 2

Section	Length l	Length l'	$(l')^2$	Length $\bar{l}$	$(\bar{l})^2$	$A = l(\bar{l})^2$	$B = \frac{l(\bar{l}')^2}{12}$	Ms per Section $= A + B$
1	29.3	28.4	807.0	13.4	179.8	5260.0	1970.0	7,230.0
2	24.7	24.4	596.0	11.9	141.8	3500.0	1260.0	4,760.0
3	8.0	2.0	4.0	27.7	768.0	6140.0	3.0	6,143.0
4	18.4	17.7	313.5	9.8	96.0	1766.0	480.0	2,246.0
5	19.6	18.8	354.0	8.4	70.5	1382.0	578.0	1,960.0
6	24.0	4.8	23.0	24.8	615.2	14770.0	46.0	14,816.0
Ms total								37,155.0

Corner	Radius R	$l = 1.57R$	R^3	Length $\bar{l}$	$(\bar{l})^2$	$C = l(\bar{l})^2$	$D = 0.15R^3$	Mc per Corner $C + D$
1	6.0	9.4	216.0	26.9	724.5	6800.0	32.0	6,832.0
2	6.0	9.4	216.0	24.1	581.0	5460.0	32.0	5,492.0
3	6.0	9.4	216.0	19.4	376.5	3540.0	32.0	3,572.0
Mc total								15,896.0

$M = Ms + Mc = 37,155 + 15,896 = 53,051 \text{ ft}^3$

Compute Δl. Assume Δx: try $\Delta x = -2.2$ ft. Then:

$$\Delta l = c(\Delta x) = (51/50)(-2.2) = -2.2 \text{ ft}$$

Compute k_g' (actual):

$$k_g' \text{ (actual)} = k_g + x = 18.7 - 2.2 = 16.5 \text{ ft}$$

Compute L':

$$L' = L + 2N_m(\Delta l) = 152.2 + 4(-2.2) = 143.4 \text{ ft}$$

Check adjusted M_2' to see how closely it agrees with the required M_2':

$$M_2' \text{ (adj.)} = L'(k_g')^2 = 143.4(16.5)^2 = 39{,}080 \text{ ft}^3$$

9. Change the configuration by *subtracting* $2\Delta l$ from the length of the expansion loop. In this case, b does not require a further check.

Example 3. Assume a piping configuration and conditions shown in Fig. 48. Find the most economical configuration to meet these conditions.

Solution: 1. Assume maximum stress:

$$S_{\text{max.}} = 60{,}000(0.125) = 7500 \text{ lb/in.}^2$$

2. Assume thrust:

$$F_T = 925 \text{ lb}$$

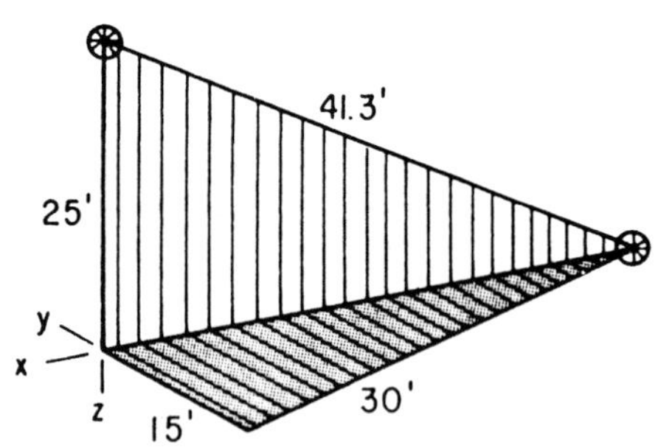

CONDITIONS

Pipe	12" X-Stg. pipe
Moment of Inertia	361.5 in.^{4}
Max. Temp.	750 °F
Fab. Temp.	70 °F
Δt	680 °F
Expansion per 100 ft. of pipe	6.4 in.
E at max. temp.	24,550,000 p.s.i.
Tensile Strength	60,000 p.s.i.
Dist. between anchors	41.3 ft.

FIG. 48 Piping configuration and conditions for Example 3.

3. Calculate b:

$$b = \frac{(7500)(361.5)}{6(925)(12.75)} = 38.3 \text{ ft}$$

4. Compute Δc:

$$\Delta c = \frac{6.4(41.3)}{100} = 2.64 \text{ in.}$$

5. Compute M_2':

$$M_2' = \frac{2.64(24{,}550{,}000)(361.5)}{1728(925)} = 14{,}650 \text{ ft}^3$$

6. Draw assumed piping configuration and calculate the neutral axis position (see Fig. 49).

This problem involves a three-plane system which complicates the relatively simple method that was used in the two-plane problems of Examples 1 and 2. It may

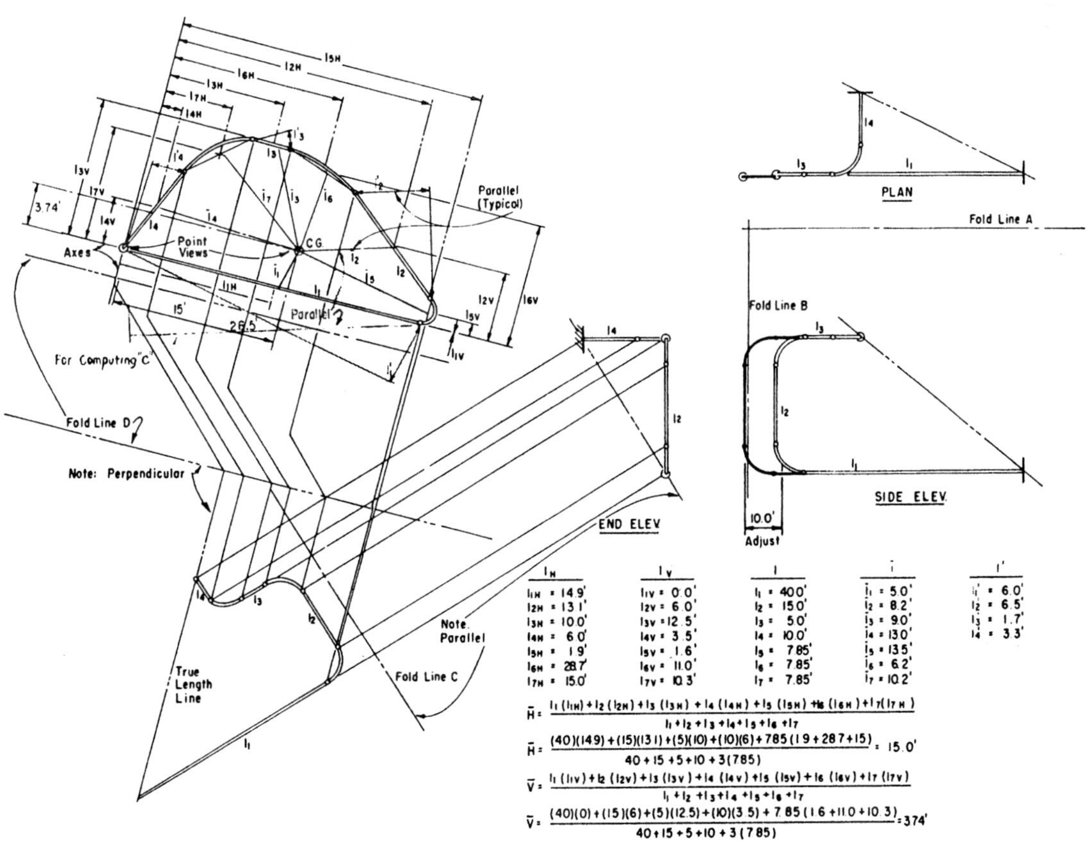

FIG. 49 Piping configuration and new axis calculations.

be greatly simplified by using a few of the more simple tools of descriptive geometry. They are: (1) the transfer of points, which should require no explanation here; (2) the true length of a line; and (3) the point view of a line.

Explanation. The true length of a line may be found by projecting a view of the system with the fold line parallel to the line drawn between the anchors or with the projection lines perpendicular to the line between the anchors.

The point view of a line may be found by projecting a view containing a true length line across a fold line that is drawn perpendicular to the true length line or with the projection lines parallel to the true length lines.

Project the true length line view (Fig. 49) and the point view of the line view. Erect a convenient pair of perpendicular coordinate axes and calculate the center of gravity, which will also be a point view of the neutral axis parallel to the point view of the neutral axis parallel to the point view of the line between the anchors.

Lay out moment arms (l) and equivalent lengths ($\bar{l}'$).

Measure component lengths (l), moment arms ($\bar{l}$), and equivalent lengths (l'). Place these lengths in the appropriate column of the moment calculation form and calculate the moment M_2 (see Table 6).

7. Adjust configuration. Calculate k_g:

$$k_g = \sqrt{\frac{M_2}{L}} = \sqrt{\frac{5970}{93.6}} = 8.0 \text{ ft}$$

Calculate k_g' (approx.):

$$k_g' \text{ (approx.)} = \sqrt{\frac{M_2}{L}} = \sqrt{\frac{14{,}650}{93.6}} = 12.5 \text{ ft}$$

TABLE 6 Moment Calculation Form for Example 3

Section	Length l	Length l'	$(l')^2$	Length $\bar{l}$	$(\bar{l})^2$	$A = l(\bar{l})^2$	$B = \frac{l(\bar{l}')^2}{12}$	*Ms* per Section $= A + B$
1	40.0	6.0	36.0	5.0	25.0	1000.0	120.0	1120.0
2	15.0	6.5	42.3	8.2	67.2	1010.0	53.0	1063.0
3	5.0	1.7	2.9	9.0	81.0	405.0	1.2	406.2
4	10.0	3.3	10.9	13.0	169.0	1690.0	9.1	1699.1
Ms total								4288.3

Corner	Radius R	$l = 1.57R$	R^3	Length $\bar{l}$	$(\bar{l})^2$	$C = l(\bar{l})^2$	$D = 0.15R^3$	*Mc* per Corner $= C + D$
1	5.0	7.85	125.0	13.5	182.2	911.0	18.8	929.8
2	do	do	do	6.2	38.4	192.0	do	210.8
3	do	do	do	10.2	104.4	522.0	do	540.8
Mc total								1681.4

$M = Ms + Mc = 4288.3 + 1681.4 = 5969.7 \text{ ft}^3$

Calculate Δx (approx.):

$$\Delta x \text{ (approx.)} = (k_g' - k_g) = 12.5 - 8.0 = 4.5 \text{ ft}$$

Compute Δl. Assume Δx: try $\Delta x = 3.3$ ft. Then:

$$\Delta l = c(\Delta x) = (40/26.5)(3.3) = 5 \text{ ft}$$

Compute k_g' (actual):

$$k_g' \text{ (actual)} = k_g + \Delta x = 8.0 + 3.3 = 11.3 \text{ ft}$$

Compute L':

$$L' = L + 2N_m(\Delta l) = 93.6 + 2(2)(5) = 113.6 \text{ ft}$$

Check adjusted M_2:

$$M_2 = L'(k_g')^2 = (113.6)(11.3)^2 = 14{,}500 \text{ ft}^3$$

8. Change configuration by *adding* $2(\Delta l)$ to the length of the expansion loop and recheck b.

Stress Analysis of Curved Pipelines

The transmission of high-pressure steam often presents special flexibility problems to the piping designer. Since the circular type transmission line shown in Fig. 50 is always symmetrical, the neutral axis passes through the

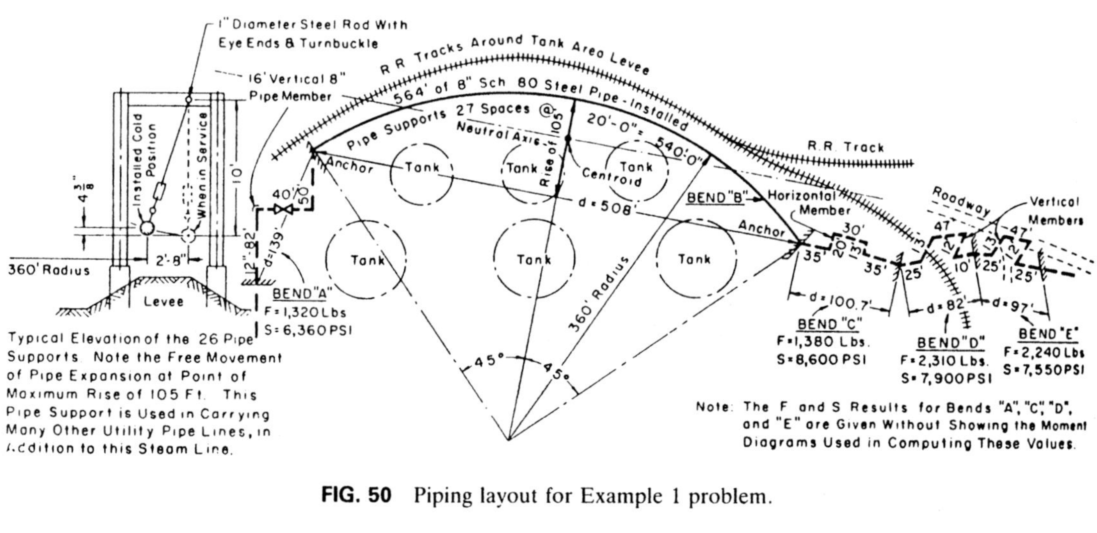

FIG. 50 Piping layout for Example 1 problem.

centroid and is also parallel to the straight line connecting the fixed anchor points. For convenience, the line connecting the anchor points is called the *d* distance.

In using circularly curved piping arrangements it is necessary that the piping be supported in a manner so as not to interfere with natural pipe movement when the pipe elongates because of thermal expansion. Using long swing eye supporting rods as shown in Fig. 50, suspended about 10 or more feet from the top of the steel frame and attached to overhead links, assured unrestricted pipe movement in this installation.

Example 1. The large curved pipe bend *B* shown in Fig. 50 may be stress-analyzed as follows.

The L/D_α ratio is 564/508 = 1.110. The ratio rise (ft)/distance (ft) = 105/508 = 0.207.

Using the next largest rise/D_α distance of 0.21, Table 7 gives an M_f factor of 0.0046 and an R_f factor of 0.138.

Total M_M value is determined by the equation, $M_M = d^3 M_f$. Thus

$$M_M = (508)^3 0.0046$$
$$= 603{,}044\ \text{ft}^3$$

The formula for the maximum moment arm is $R_M = D_\alpha R_f$.

$$R_M = 508(0.138)$$
$$= 70.1\ \text{ft}$$

In this example the high-pressure steam service piping for the curved pipe bend *B* is 8-in., schedule 80 carbon steel with a wall thickness of 0.50 in., a moment of inertia *I* for 105.7 in.^{4}, and an o.d. of 8.625 in. The service pressure is 680 lb/in.^{2}. The steam temperature is 650°F. The pipe expansion rate at this temperature is 5.584 in./100 ft. $\Delta = 5.584(508)/100 = 28.4$ in. The cold modulus of elasticity *E* of 29,000,000 lb/in.^{2} will be used. This puts the *F* and *S* values in the high stress range.

$$F = \frac{\Delta EI}{1728 M_M}$$
$$= \frac{28.4(29)(10^6)105.7}{1728(603{,}044)}$$
$$= 90\ \text{lb}$$

$$S_{\max} = \frac{6F(R_M)D}{I}$$
$$= \frac{6(90)(70.1)8.625}{105.7}$$
$$= 3100\ \text{lb/in.}^2$$

TABLE 7 Circularly Curved Piping M_f and R_f Factors

Use in Formula Total Mitchell Moment, ft^3, $M = d^3 \times M_f$ Factor		Use in Formula Maximum Mitchell Arm, ft, $R = d \times R_f$ Factor		Ratio of Total Pipe Length, L/d Distance	
Factor of rise/*d* Distance	M_f Factor	Factor of rise/*d* Distance	R_f Factor	Factor of rise/*d* Distance	*L*/*d*
.00	.00000	.00	.000	.00	0.000
.01	.00010	.01	.006	.01	1.002
.02	.00025	.02	.012	.02	1.004
.03	.00035	.03	.018	.03	1.006
.04	.00050	.04	.024	.04	1.008
.05	.00060	.05	.029	.05	1.010
.06	.00075	.06	.036	.06	1.014
.07	.00095	.07	.042	.07	1.016
.08	.00110	.08	.048	.08	1.019
.09	.00125	.09	.054	.09	1.022
.10	.00141	.10	.066	.10	1.026
.11	.00165	.11	.068	.11	1.031
.12	.00190	.12	.074	.12	1.040
.13	.00220	.13	.080	.13	1.045
.14	.00250	.14	.087	.14	1.054
.15	.00270	.15	.094	.15	1.060
.16	.0029	.16	.101	.16	1.070
.17	.0032	.17	.108	.17	1.076
.18	.0035	.18	.116	.18	1.086
.19	.00385	.19	.123	.19	1.095
.20	.00412	.20	.132	.20	1.103
.2083[a]	.0045[a]	.2083[a]	.136[a]	.2083[a]	1.110[a]
.21	.0046	.21	.138	.21	1.113
.22	.0051	.22	.146	.22	1.123
.23	.0056	.23	.153	.23	1.134
.24	.00625	.24	.160	.24	1.146
.25	.0070	.25	.167	.25	1.158
.26	.00775	.26	.174	.26	1.170
.27	.0088	.27	.181	.27	1.183
.28	.0096	.28	.187	.28	1.195
.29	.0105	.29	.194	.29	1.209
.30	.01153	.30	.196	.30	1.224
.31	.0126	.31	.205	.31	1.237
.32	.0137	.32	.210	.32	1.252
.33	.0148	.33	.216	.33	1.267
.34	.0158	.34	.221	.34	1.282
.35	.0171	.35	.227	.35	1.298

(*continued*)

TABLE 7 *(continued)*

Use in Formula Total Mitchell Moment, ft^3, $M = d^3 \times M_f$ Factor		Use in Formula Maximum Mitchell Arm, ft, $R = d \times R_f$ Factor		Ratio of Total Pipe Length, L/d Distance	
Factor of rise/d Distance	M_f Factor	Factor of rise/d Distance	R_f Factor	Factor of rise/d Distance	L/d
.36	.0183	.36	.232	.36	1.312
.37	.0196	.37	.237	.37	1.329
.38	.0209	.38	.242	.38	1.345
.39	.0223	.39	.248	.39	1.362
.40	.02373	.40	.252	.40	1.380
.41	.0254	.41	.256	.41	1.395
.42	.0270	.42	.262	.42	1.412
.43	.0286	.43	.268	.43	1.430
.44	.0302	.44	.273	.44	1.450
.45	.0319	.45	.279	.45	1.467
.46	.0337	.46	.285	.46	1.485
.47	.0356	.47	.292	.47	1.504
.48	.0375	.48	.298	.48	1.522
.49	.0396	.49	.305	.49	1.540
.50[b]	.0416[b]	.50[b]	.318[b]	.50[b]	1.570[b]

[a]Designates values for 90° arc.
[b]Designates values for semicircle (180°) pipe bend.

The longitudinal pipe stress is

$$S_L = \frac{\text{steam pressure, lb/in.}^2\ \text{(pipe i.d., in.)}}{4\text{(pipe wall thickness, in.)}}$$

$$= \frac{680(7.625)}{4(0.50)}$$

$$= 2600\ \text{lb/in.}^2$$

The allowable combined stress set for this piping is 14,500 lb/in.^{2}. Therefore, $S_{max} + S_L = 3100 + 2600 = 5700$ lb/in.^{2}, making this piping design below the allowable combined stress value of 14,500 lb/in.^{2} and on the safe side. The curvature in this application was determined by the contour of the earth levee along the railroad track.

Example 2. In the previous example the curvature of the levee and railroad track around the tank area determined the circular arc for the 680 lb/in.^{2}, 650°F steam line. This 8-in. schedule 80 carbon steel pipeline will be redesigned and recalculated with the idea of using the smallest possible rise with the maximum bending stress S_{max} and the resultant force F at the fixed anchor ends so they will be within the allowable stress range. A rise of 10 ft above the straight line 508 ft d distance will

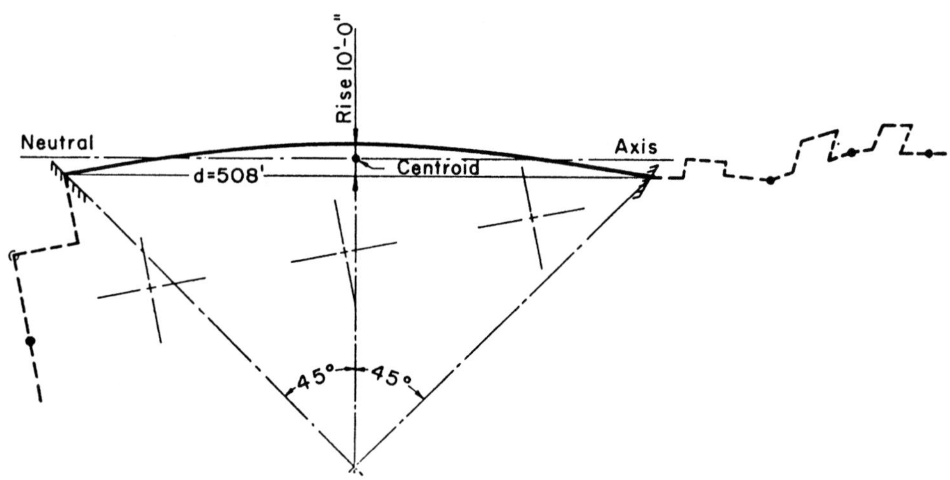

FIG. 51 Piping layout for Example 2 problem.

be tried first (see Fig. 51). The pipe will be suspended from free movement pipe supports similar to the elevation shown in Fig. 50.

The ratio of rise to d distance is 10/508 = 0.0197. From Table 7, the nearest rise to d distance ratio value is 0.02. For this value the factors are: M_f = 0.00025, R_f = 0.012, and the L/d is 1.004.

$$M_M = (D_\alpha)^3 M_f$$

$$= (508)^3(0.00025)$$

$$= 32{,}774 \text{ ft}^3$$

$$R_M = D_\alpha R_f$$

$$= 508(0.012)$$

$$= 7.1 \text{ ft}$$

$$F = \frac{\Delta EI}{1728 M_M}$$

$$= \frac{28.4(29)(10^6)(105.7)}{1728(32.774)}$$

$$= 1590 \text{ lb}$$

$$S_{\max} = \frac{6FRD}{I}$$

$$= \frac{6(1590)(7.1)8.625}{105.7}$$

$$= 5520 \text{ lb/in.}^2$$

$$S_L = \frac{\text{steam pressure, lb/in.}^2 \text{ (pipe i.d., in.)}}{4(\text{pipe wall thickness, in.})}$$

$$= \frac{680(7.625)}{4(0.50)}$$

$$= 2600 \text{ lb/in.}^2$$

Then

$$S_{max} + S_L = 5520 + 2600 = 8120 \text{ lb/in.}^2$$

If an allowable combined stress value of 14,500 lb/in.^{2} is used, the above piping arrangement will be satisfactory provided the pipe can be suspended to allow free movement and the end anchor point is able to support the end reactive thrust of 1590 lb.

Yard Piping

The main arterial system of a plant is yard piping. It is here that long process lines are located, interconnecting distant equipment and lines entering and leaving the units. Utility headers are located in the yard supplying steam, air, gas, and water to process equipment. All relief and blowdown headers are located here. Often instrument lines and electrical supply conduits are also supported on the yard steel.

Information Required

Specifications

Usually, only a few items included in the job specifications affect yard piping design. Such items are: minimum headroom over roads, overhead pipelines or steel beams; access, headroom, and handling requirements for equipment installed under the yard; ladder, catwalk, and platform requirements for valves, relief valves, orifice flanges, and instruments located in the yard; and details affecting piping and structures, operating, and safety requirements.

Project design data and site maps show required or existing conditions inside and outside the unit's perimeter. These are: required location of cooling water mains below or above grade; required location of furnaces, control, and switch house; location of utility and process lines entering and leaving unit limits; main pipe runs outside the unit; location of storage tanks relative to process units; blending, loading, filling station, and cooling tower location; and site grade level variations.

Plot Plan

The relationship between plant units, equipment, buildings, and yard piping is shown on the plot plan. Positions of incoming and outgoing lines can be seen. Major structures, location of buildings, and all equipment are shown. Roads crossing the yard or located under the yard steel are indicated.

Process flow diagrams show essential process lines interconnecting process equipment. Mechanical flow diagrams (developed from process flow diagrams) indicate the complete flow systems necessary for plant operation—also pipe sizes, valving, manifolds, all piping details, and instrumentation. Utility flow diagrams show the number and size headers for water, steam, condensate, gas, air, etc., as well as all equipment supplied by these headers with necessary valving and piping details.

Figure 52 shows a plot plan and process flow diagram. With these two drawings you can decide which portions of process lines will be located in the yard and which lines will interconnect directly to nozzles on adjacent or nearby process equipment. Heavy lines on the flow diagram indicate piping assumed to be located in the yard. These lines are also shown on the plot plan to give a visual idea of yard space requirement. A mechanical flow diagram is similarly interpreted. A greater number of lines can be drawn on the plot to give a more accurate estimate of required yard width.

In addition to process lines, utility flow diagrams show individual service lines and utility headers. (Utility mains generally run the whole length of the yard.) These lines should also be taken into account for estimating additional space requirement (see Tables 8 and 9).

Layout

The plant layout determines the main yard piping runs. Figure 53 shows yard piping layouts resulting from various plant arrangements. Smaller plants usually have the simplest yard piping as shown on Sketches A and B. On Sketch A, process and utility lines enter and leave the same end of the plot. Sketch B is a frequently adopted layout with utility lines entering at one end and process lines at the opposite end of the yard. Layout conditions sometimes result in an L-shaped yard as shown on Sketch C. Larger plants will have a more involved yard piping arrangement as shown in Sketches D, E, and F. Sketch G shows the yard piping arrangement of a very large plant. This layout can be considered a combination of several simpler yard piping arrangements.

Some of the pipelines arranged in the yard need special consideration. These lines are classified as follows.

Process lines (a) which interconnect nozzles on process equipment more than 20 ft apart (closer process equipment can be directly connected by pipelines); (b) product lines which run from vessels, exchangers, or more often from pumps to the unit limits, to storage or header arrangement outside the plant; (c) crude or other charge lines which enter the unit and usually run in the yard before connecting to exchangers, furnaces, or to other process equipment, e.g., holding drums or booster pumps.

Relief line headers, individual relief lines, blowdown lines, and flare lines should be self-draining from all relief valve outlets to the knockout drum, flare stack, or a point at the plant limit. A pocketed relief line system is more expensive because an extra condensate pot is usually required with

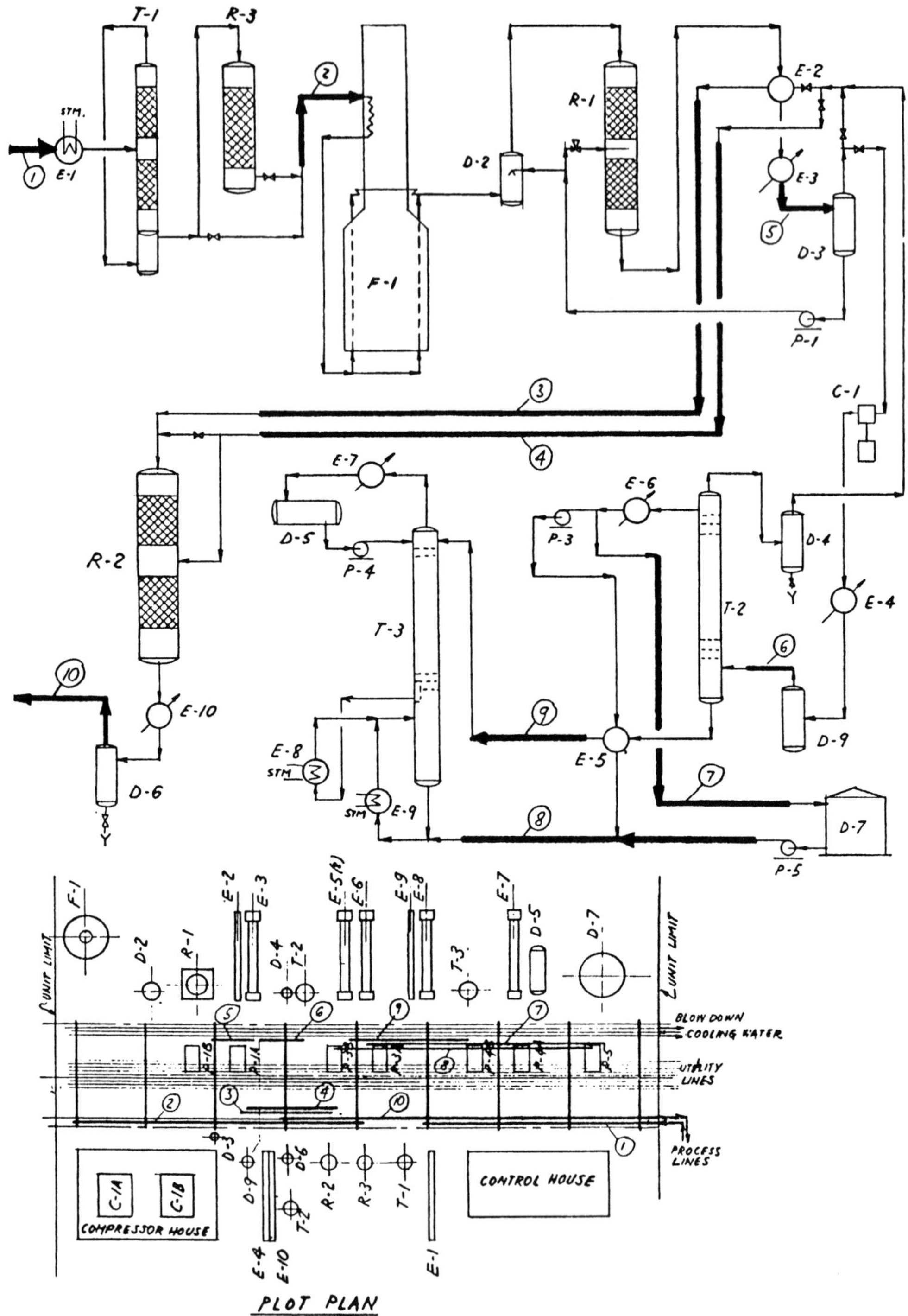

FIG. 52 The first step in yard piping design is careful study of plot plan and flow diagram. Notice that the heavy lines on the flow diagram have been located in the yard piping rack.

TABLE 8 Spacing of Yard Piping: Pipelines without Flanges

NOTE: 1. THIS TABULATION IS BASED ON THE FOLLOWING RELATIONSHIP:

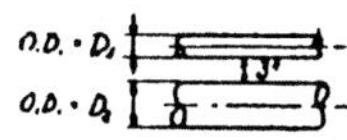

$A = \frac{D_1}{2} + \frac{D_2}{2} + 3''$ = TABULATED VALUES

2. FOR INSULATED LINES ADD INSULATION THICKNESSES TO TABULATED VALUES:

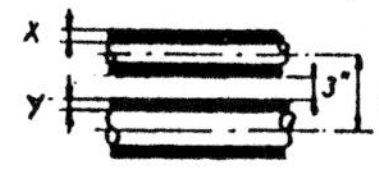

$B = A + X + Y$

3. REMEMBER THE 3" GAP BETWEEN LINES WHEN CHECKING LATERAL THERMAL MOVEMENTS OF ADJECENT PIPES. FOR EXCESSIVE LATERAL MOVEMENTS INCREASE GAP ACCORDINGLY.

PIPE SIZE	1	1½	2	2½	3	4	6	8	10	12	14	16	18	20	24		PIPE SIZE
24	15½	16	16	16½	17	17½	18½	19½	20½	21½	22	23	24	25	27	6	1
20	13½	14	14	14½	15	15½	16½	17½	18½	19½	20	21	22	23	7	7	1½
18	12½	13	13	13½	14	14½	15½	16½	17½	18½	19	20	21	7½	7	7	2
16	11½	12	12	12½	13	13½	14½	15½	16½	17½	18	19	8	8	7½	7½	2½
14	10½	11	11	11½	12	12½	13½	14½	15½	16½	17	9	8½	8½	8	8 7	3
12	10	10½	10½	11	11	11½	12½	13½	15	16	10	10	9½	9	9	8½ 7½	4
10	9	9½	9½	10	10	10½	11½	12½	14	12½	11½	11	10½	10½ 9½	10 9½	10 9	6
8	8	8	8½	9	9	9½	9½	9½	15	14	13	12	12 11	11½ 10½	11½ 10½	11 10	8
6	7	7	7½	8	8	8½	8½	17	16	15	14	13½ 12½	13 12	13 11½	12½ 11½	12½ 11	10
4	6	6	6½	6½	7	7½	19½	18½	17½	16½ 15½	15½ 14½	15 13½	14½ 13	14½ 12½	14 12½	14 12	12
3	5½	6	6	6	6½	21½	21	20 19	19 17½	18 16	17 15	16 14	16 14	15½ 13	15½ 13	15½ 12½	14
2½	5	5½	5½	6	24	23	22	21 20	20 18½	19 17	18 16	17½ 15	17 15	17 14	16½ 14	16½ 13½	16
2	5	5	5½	26	25	24	23½ 22½	22½ 21	21½ 19½	20½ 18	19 17	19 16	18½ 16	18 15	18 15	17½ 14½	18
1½	4½	5	28	27	26	25	24½ 23½	23½ 22	22½ 20½	21½ 19	20½ 18	20 17	19½ 17	19½ 16	19 16	19 15½	20
1	4½	33	31	30 29	29 28	28 26½	27½ 25½	26½ 24	25½ 22½	24½ 21	23 20	23 19	22½ 19	22 18	22 18	21½ 17½	24
PIPE SIZE		24	20	18	16	14	12	10	8	6	4	3	2½	2	1½	1	PIPE SIZE

IF FLANGES ARE IN LINES LISTED HERE USE TOP FIGURES

IF FLANGES ARE IN LINES LISTED HERE USE BOTTOM FIGURES

TABLE 9 Pipelines with Flanges or Line Size Valves (up to 400 lb rating)

NOTE: 1. THIS TABULATION IS BASED ON THE FOLLOWING RELATIONSHIP:

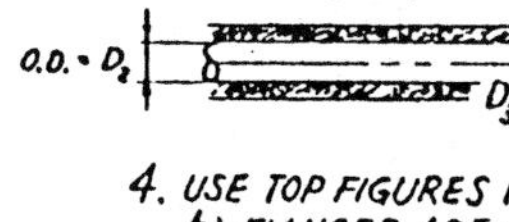

$C = \frac{D_1}{2} + \frac{D_3}{2} + 3''$ = TABULATED VALUES

2. IN CASE OF UNINSULATED D_1 LINE AND UNINSULATED D_3 FLANGE, BUT INSULATED D_2 LINE USE TABULATED VALUES:

3. FOR INSULATED LINES AND FLANGES ADD INSULATION THICKNESSES TO TABULATED VALUES:

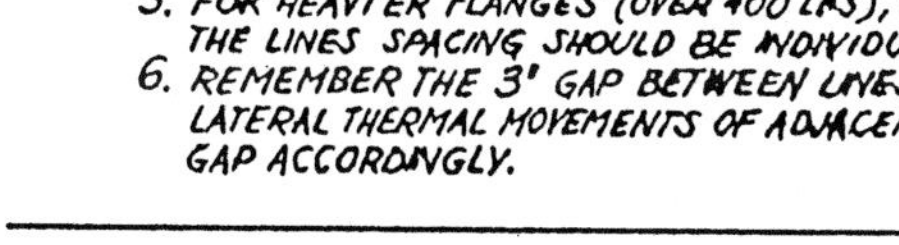

$D = C + X + Y = \frac{D_1}{2} + \frac{D_3}{2} + 3'' + X + Y, \quad D_3 \geq D_4$

4. USE TOP FIGURES IF a) FLANGES ARE IN LARGER DIAMETER LINE; USE BOTTOM FIGURES IF b) FLANGES ARE IN SMALLER DIAMETER LINE AND USE TOP FIGURES IF c) FLANGES ARE IN BOTH ADJACENT LINES. (FLANGES SHOULD BE STAGGERED).
5. FOR HEAVIER FLANGES (OVER 400 LBS), ORIFICE FLANGES, ANCHORS OR SPECIAL EQUIPMENT IN THE LINES SPACING SHOULD BE INDIVIDUALLY CALCULATED.
6. REMEMBER THE 3" GAP BETWEEN LINES OR FLANGES AND OUTSIDE PIPE WALL WHEN CHECKING LATERAL THERMAL MOVEMENTS OF ADJACENT PIPES. FOR EXCESSIVE LATERAL MOVEMENT INCREASE GAP ACCORDINGLY.

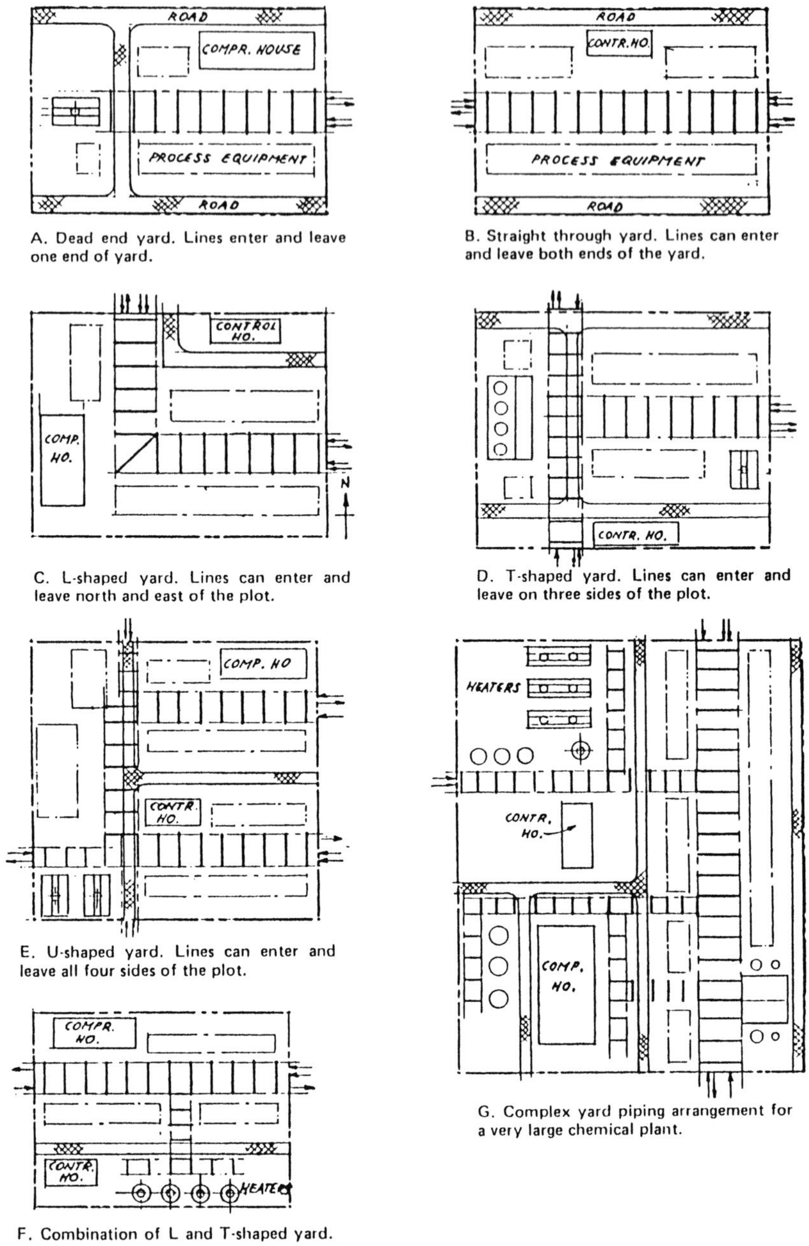

A. Dead end yard. Lines enter and leave one end of yard.

B. Straight through yard. Lines can enter and leave both ends of the yard.

C. L-shaped yard. Lines can enter and leave north and east of the plot.

D. T-shaped yard. Lines can enter and leave on three sides of the plot.

E. U-shaped yard. Lines can enter and leave all four sides of the plot.

F. Combination of L and T-shaped yard.

G. Complex yard piping arrangement for a very large chemical plant.

FIG. 53 In these typical yard piping arrangements, notice how a complex piping arrangement can be broken down into a combination of several of the more simple arrangements.

instruments, valves, and pumps. To eliminate pockets, some relief line headers must be placed at a higher elevation above the main yard, usually on a tee support on the extended yard column. However, on some noncondensing gas systems, self-drainage is not so essential.

Utility lines in the yard can be put in two groups. The first are utility headers serving equipment in the entire plant. Such lines are: low and high pressure steam lines, steam condensate, plant air, and instrument air lines. If requested, cooling water, as well as hot return, service, and fire water, can also be arranged in the yard. The second group contains utility lines individually serving one or two equipment items or a group of similar equipment (furnaces, compressors) in the plant. Such lines are: boiler feedwater, fire steam, compressor starting air, various fuel oil lines, lubricating oil, cooling oil, fuel gas, inert gas, and chemical treating lines.

Steam headers should drain to the steam separator for more effective condensate collection. Branch connections to steam headers usually connect to the top to avoid excessive condensate drainage to equipment.

Instrument lines and electrical cables are often supported in the yard, and extra space should be provided for them. The best instrument line arrangement eliminates almost all elevation changes between the plant and the control room. This can be easily achieved when instrument lines are supported outside the yard column at a suitable elevation.

Line Location

Figure 54 shows one level of yard piping. Regardless of the service, heavy lines (very large diameter lines, large lines full of liquid) are placed over or near the yard columns. (Centrally loaded column and reduced bending moment on the beam will result in a light structural design.) Next to these lines are placed all process lines and relief lines. Utility lines are in the center portion of the yard. A general sequence of utility lines is also shown on Fig. 54.

Underpositions in the yard depend on the number and size of the branch connections. If the majority of similar size branches connect to the header from the right, it is more economical to place it in the right half of the yard.

It is advantageous, from a support standpoint, to group hot lines requiring expansion loops as shown in Fig. 54. Elevating loops horizontally over the yard is the most common adopted solution with the hottest and largest diameter line outside. Usually line guides, line stops, and anchor points are also required along a hot line somewhere in the yard. Pipe expansion forces at some of these points will affect yard support design.

Those process lines which interconnect equipment on the same side of the yard should be near the edges of the yard bank. Lines which interconnect equipment located on both sides of the yard should be close to utility lines and can be placed on either side of the yard. The position of product lines is influenced by their routing after leaving the plant limit. Right- or left-turning lines should be on the right- or left-hand side of the yard. Utility lines individually serving one or two equipment items should be on the same side of the yard as the equipment to which they connect.

If, because of the large number of lines, two yard bank elevations are required, generally utility lines are placed in the top bank and process lines in the bottom bank. Obviously, exceptions always can be made to the elevation of individual utility or process lines. Line sequence arrangement will

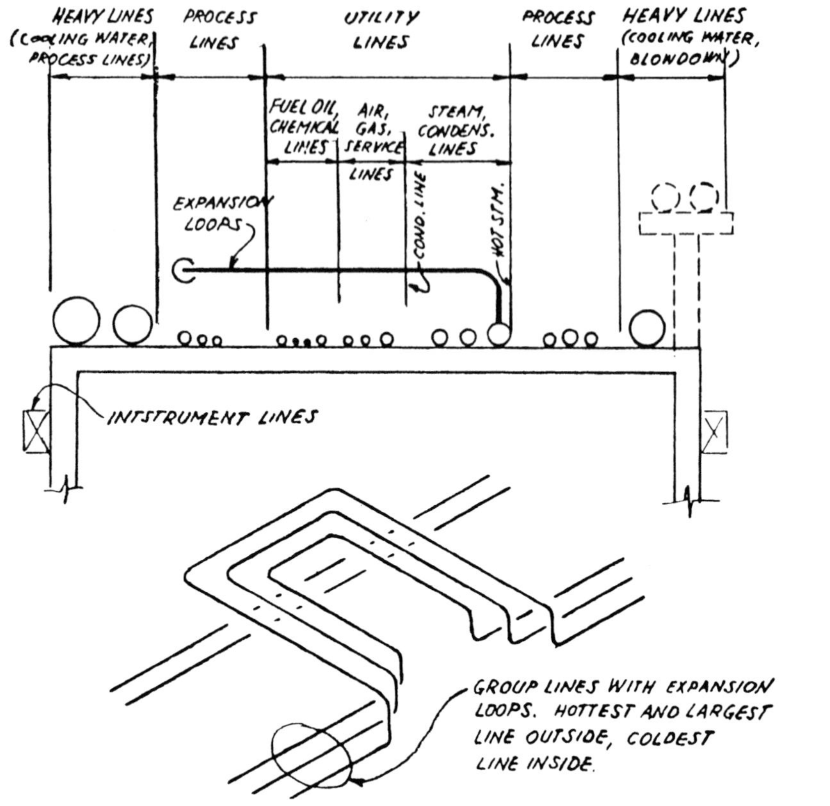

FIG. 54 Important points in this typical line position arrangement of yard piping are: heavy lines over columns, utility lines near the center, and horizontal expansion loops with hot lines on the outside of the loops.

be similar to the sequence already discussed for the one-level yard. Line spacing in the yard is shown and explained in Table 9.

Yard in Elevation

Figure 56 shows a typical yard section with main elevations. The elevation of yard piping is determined by the highest requirement of the following: (a) headroom over a main road, (b) headroom for access to equipment under the yard, and (c) headroom under lines interconnecting the yard and equipment outside the yard. The size of steel beams supporting yard piping should also be taken into account when considering headroom.

Generally, those process lines should be located in the top bank which interconnect two nozzles elevated higher than the top yard bank. Process lines with one end lower than the bottom yard elevation can run either in the top or the bottom bank. If both ends of a process line are lower than the bottom yard elevation, the line should be located in the bottom bank.

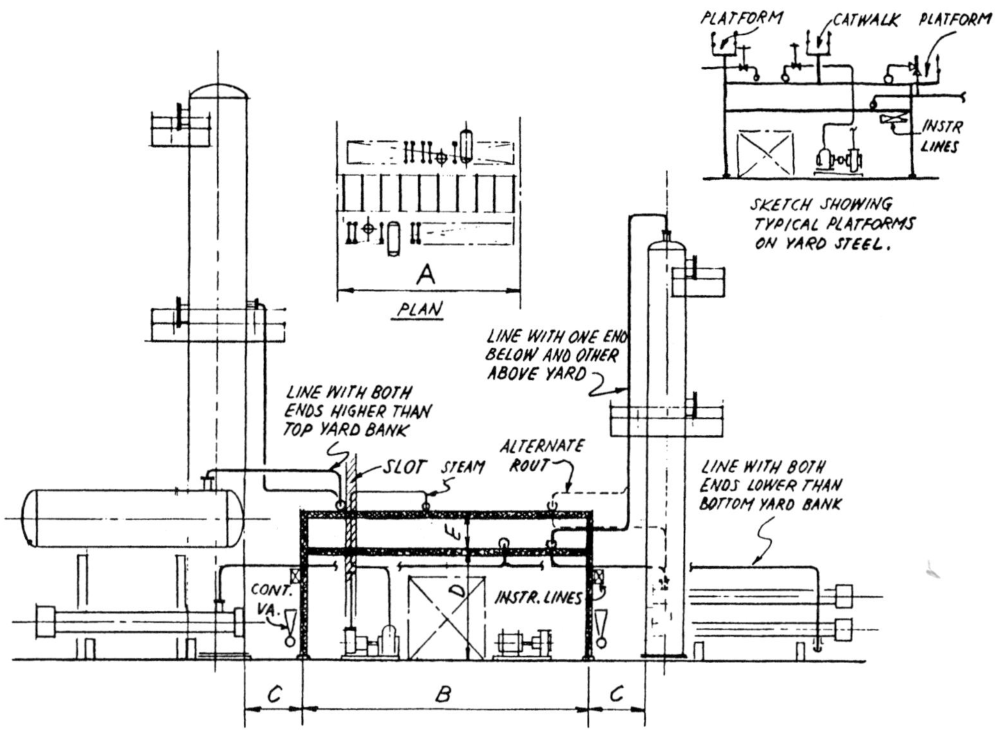

FIG. 56 Typical cross-section of yard piping showing critical dimensions which affect piping cost. Notice the arrangement for platforms for access to valves in the upper right-hand corner.

The elevation of a line can also be influenced by valves and instruments in the line. Often a more convenient access platform can be provided for valves arranged in the top yard bank. The preferred location of lines with orifice runs is near the edge of the yard with orifice flanges near a yard column, for more convenient portable ladder access.

The sketch in the upper right-hand corner of Fig. 56 shows platform and walkway arrangements to valves, relief valves, and instruments located in the yard.

When pumps are arranged under the yard, one or two slots are often required along the yard, usually over the pump discharge nozzle for process, steam, and other utility lines connecting from the yard to the pumps and driver.

Pipe Economy

Pipe economy depends primarily on the length of lines arranged in the yard. Fittings, valves, and instruments are relatively few compared to pipe length.

Figure 56 also shows those critical dimensions which will influence piping cost from a yard piping layout standpoint. These dimensions depend on the overall plant layout and should be carefully considered when the plot is arranged.

Dimension A is the total length of the yard and is governed by the number and size of equipment, structures, and buildings arranged along both sides of the yard. On an average, about 10 ft of yard length is required per process equipment item (exchanger, drum, tower, unhoused compressors, etc.). A control house located along the yard, for example, will increase the yard piping cost because all lines must pass by without really being associated with the relatively long control house.

If, with good layout practices, the same number and size of equipment can be arranged on a shorter yard length, yard piping cost can be reduced considerably. A 7 to 8-ft average length per process equipment item is not unusual in a well-arranged plant. Equipment in pairs, stacked exchangers, exchangers under elevated drums, drums or exchangers supported on towers, two vessels combined into one, closely located towers with common platforms, drums supported on exchangers, and process equipment located under the yard are only a few examples which help shorten the yard. These arrangements, of course, not only shorten process lines interconnecting equipment directly or in the yard, but also shorten those lines which pass through this area and utility headers serving this area.

The careful selection of dimensions B and C (Fig. 56) can minimize pipe length between the yard and process equipment and the pipe length interconnecting equipment on opposite sides of the yard. C is usually 6 to 7 ft. Not more than necessary yard height (dimensions D and E) will minimize vertical pipe lengths.

When changing direction, change elevation is an old rule in piping design. This happens with all lines connecting to yard piping. However, some large diameter lines can make a flat turn when entering the yard. Such lines should be placed at the edge of the yard. Any other spot will block excessive space in the yard.

Figure 57 shows commonly used elevations for main yard heights at a yard piping intersection. Note that the 14-ft elevation of the lateral yard

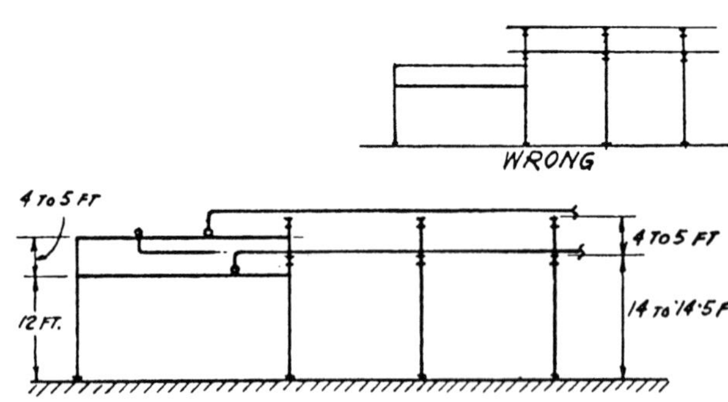

FIG. 57 This is a typical elevation for yard piping intersection. Notice that the 14-ft elevation of the lateral rack permits turning up or down at the intersection.

permits turning up or down at the intersection. It is important to elevate lateral pipe banks between the two elevations of the main yard.

The elevation difference between the main yard bank and laterally connecting pipelines is about 2 to 2½ ft. This gives an elevation difference of 4 to 5 ft between two main yard banks.

If a building (control house, pump house) is located under the main yard piping, elevations will be higher than without a building. Clearances in the building, pitching of the roof, steel structures, and pipeline clearances will affect the yard height.

An elevation difference is not required if a flat turn can be made within the yard. Line sequence in this instance must be identical before and after the turn as shown on Fig. 58, Sketch A. However, varying the line sequence in the two directions introduces an elevation difference and an additional elbow in each line, as shown on Fig. 58, Sketch B.

Piping Supports

The width of the yard is influenced by two conditions: (a) the number of lines, instruments, electrical lines, and space for future lines in the yard; and (b) space requirement for equipment arranged under the yard.

The number of lines can be estimated by marking up the yard on a print of the plot plan, with the help of flow diagrams, showing all lines located in the yard. Adding the number of lines (n) up to 18 in. in diameter in the most dense section of the yard, the total width (W_t, ft) will be

$$W_t = (fns) + A$$

where f = safety factor (f = 1.5 if the lines have been laid out on the plot with the help of process flow diagrams. f = 1.2 if the lines have been laid out with the help of fully detailed mechanical flow diagrams).

The estimated average spacing between lines is s, in feet, usually 1 ft. If lines in the yard are smaller than 10 in., the value of s = 0.75 ft.

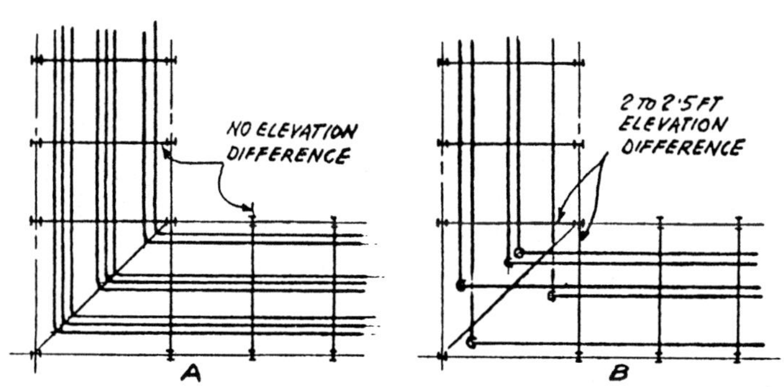

FIG. 58 In Sketch A, a flat turn is more economical if the line sequence can be kept the same in both directions. Sketch B shows the need for an elevation change when the line sequence changes after the turn.

A, in feet, is the additional width required: (a) for lines larger than 18 in., (b) for future lines, (c) for instrument lines (about 2 to 3 ft), and sometimes (d) for electrical cables (about 2 to 3 ft) if these are also supported on the yard steel, and (e) for one or two slots for pump discharge and driver utility lines (about 18 in. or 3 ft).

The total width of the yard W_t can be between 20 and 60 ft. If W_t is larger than 30 ft, usually 1½ or 2 yard banks will be required. The upper limit of yard steel span is 32 ft.

Space requirement for equipment plus access below the yard can also influence the yard width. For a single row of pumps and 8 to 10 ft access to the pumps, about 20 to 24 ft yard span is required depending on the length of the pumps. For a double row of pumps, a 28 to 32-ft span will be required.

Type of Supports

Figure 59 shows typical yard steel bents with dimensions. The total available yard width to each type of support is included in the table of dimensions. This tabulation can be used for selecting a type of yard support after the total required width has been estimated. The most commonly used yard piping supports are Types 2, 3, 4, and 5.

In almost all plants, spacing between yard support bents is about 16 to 20 ft. Neverless, consideration should be given to (a) line sizes (smaller lines have to be supported more frequently than large diameter lines); (b) liquid-filled lines require a shorter span than gas lines; (c) line temperature (very hot lines span shorter distances than cold ones of the same size and wall thickness); (d) insulation (heavily insulated small diameter lines with cold temperatures must be supported at relatively short intervals); and (e) space requirement for equipment at grade and under the yard can sometimes also influence the spacing between yard bents.

Figure 60 shows a yard piping junction with an adjacent exchanger supporting structure. The yard elevations have been governed by headroom requirement over the north-south access road. The top of the north-south yard also sets the height of the first platform of the exchanger supporting structure, because lines had to cross below the first platform to the top bank of the north-south yard.

All lines from the exchanger structure to the yard dropped along the east side of the structure. A slot has been left open in the yard adjacent to the exchanger structure for lines which connect to the lower north-south yard or to pumps below the yard. Process lines turning into the east-west yard from the exchanger structure have been arranged on the highest yard elevation.

A number of vertical reflex drums have been arranged on the first level of the exchanger structure. All suction lines to pumps turn horizontally below the lower north-south yard bank.

This, of course, is an unusual example but it illustrates the necessity of carefully choosing yard elevations and an overall system of design.

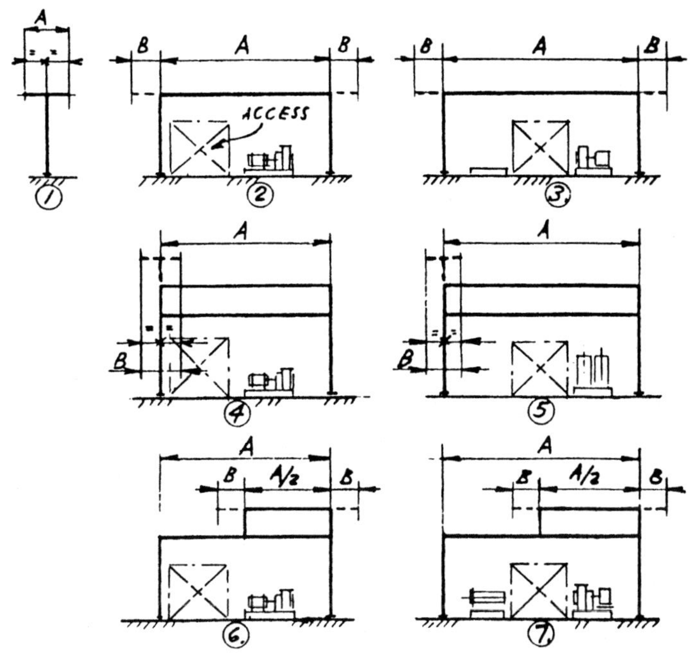

Total Available Width W_t (ft)		Dimensions (ft)		Number of	Sketch
Without Cantilever	With Cantilever	A	B	Elevation	No.
10	—	10	—	1	1
20–24	30–34	20–24	5	1	2
28–32	38–42	28–32	5	1	3
28–34	36–42	20–24	4 (5)	$1\frac{1}{2}$	6
39–47	45–53	20–24	3 (4)	2	4
40–44	48–52	28–32	4 (5)	$1\frac{1}{2}$	7
55–63	61–69	28–32	3 (4)	2	5

FIG. 59 From these data, total available width of these typical yard piping support bends can be determined.

Heat Exchanger Piping

Figure 61 shows frequently adopted nozzle arrangements on exchangers. Elbow nozzles permit lowering exchangers closer to grade. Stacked exchangers in parallel or dissimilar service can be arranged closer, with elbow nozzles between them. This provides better access to exchanger valves and instruments besides easier maintenance.

Angular connections can save one or two bends in the pipeline. These connections are more often applied to top shell nozzle or to the tube side. Excessive angular connection at the bottom can mean a separate drainage point on the shell or channel. The maximum angle from vertical centerline

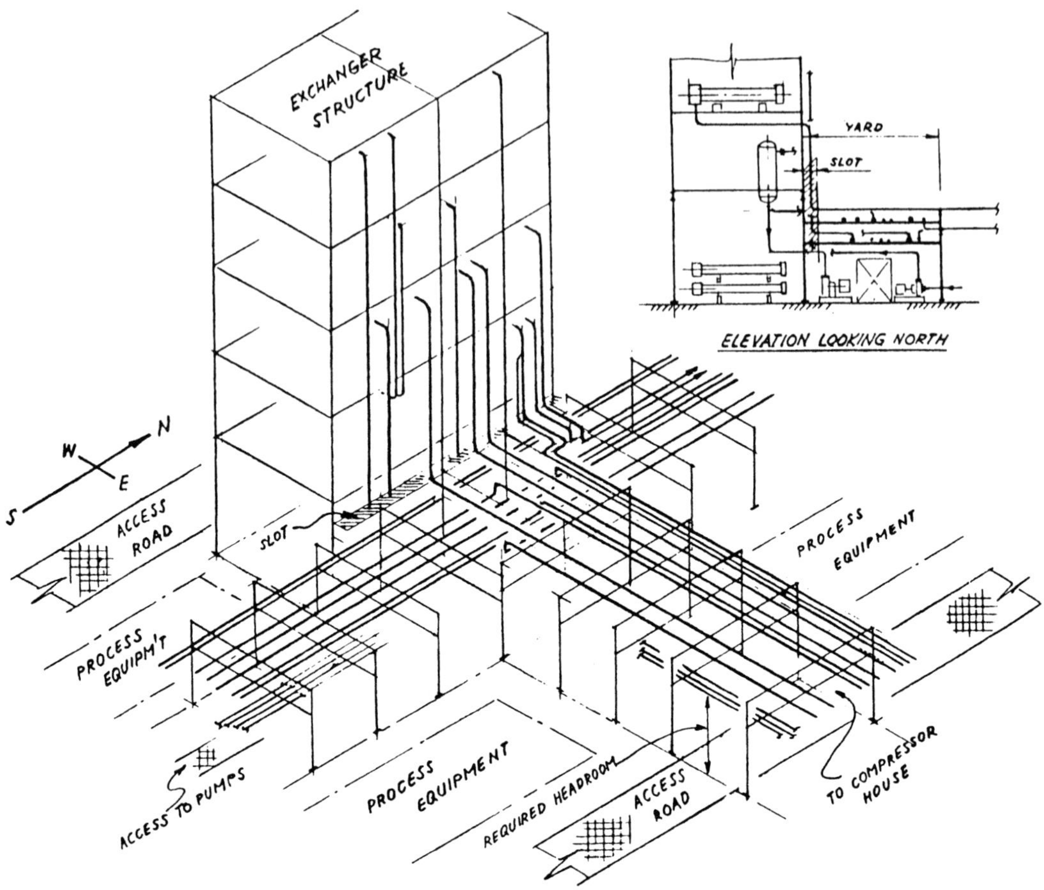

FIG. 60 In this example of yard piping layout, notice how the elevations have been governed by headroom requirements over he North-South road. The elevation also sets the height of the first platform of the exchanger supporting structure.

can be about 30°. This angle depends on nozzle and shell size, internals of the exchanger, baffle arrangement in the shell, and partitions in the channel.

Tangential connections are more expensive to make but can save fittings, making piping arrangement simpler, and improve access to valves.

If the piping designer blindly follows an exchanger designer's nozzle locations and flow requirements, he can end up with the piping arrangement shown in Fig. 62, Sketch A. It has a pocket and a loop in the line which means a longer line, more fittings, more vents and drains, and a longer pump suction with an undesired loop. Figure 62, Sketch B, was arranged by reversing the flow in the system. A loop, pocket, vent, and drain were eliminated. Friction losses through the piping system have been reduced and the pump suction shortened and simplified.

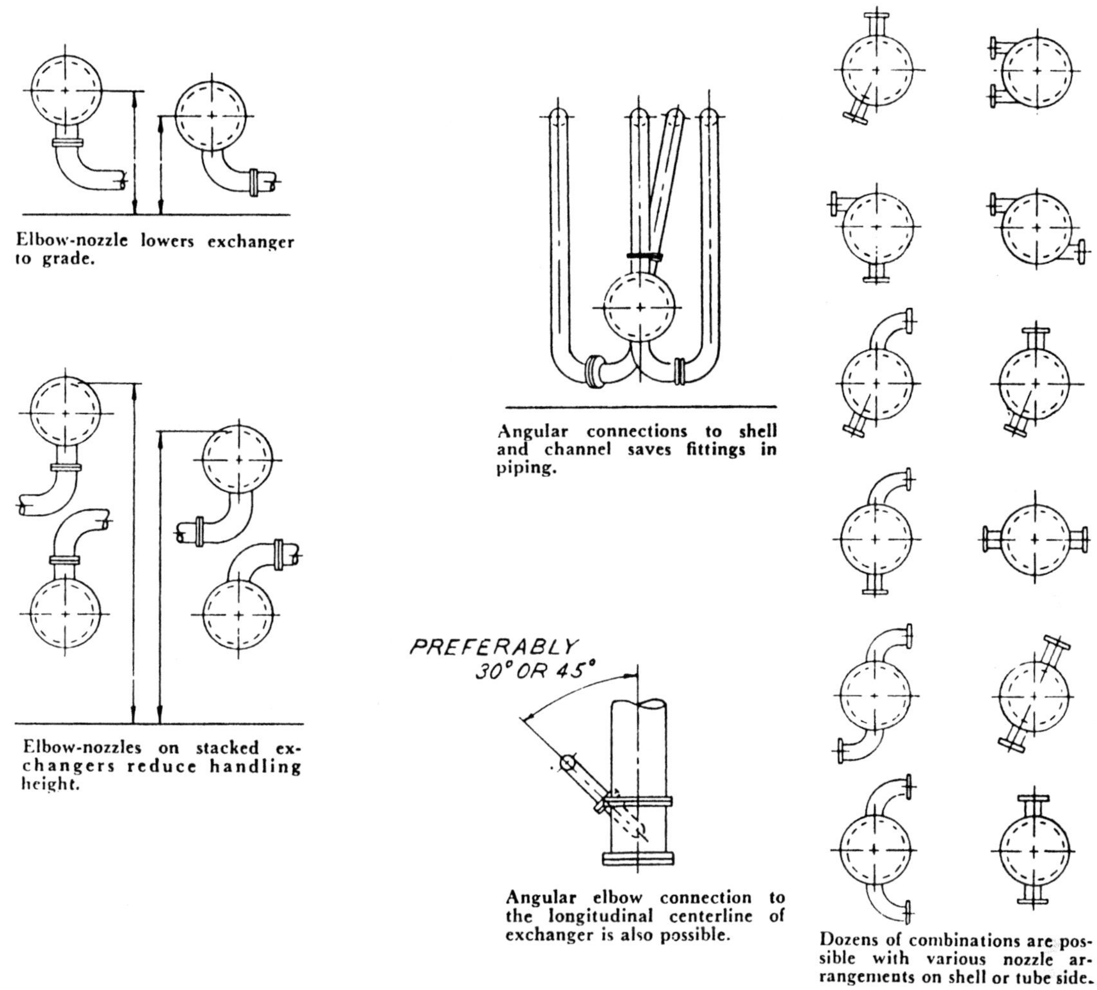

FIG. 61 Here are a number of variations for nozzle arrangement for better piping.

Another example is shown in Fig. 63. Relocating nozzles on the exchanger shell (Sketch B) results in a simpler piping arrangement than shown on Sketch A.

Exchanger drawings and data sheets give all the details of exchanger design: size, type, number of units, nozzle arrangements, weight of bundle, exchanger internals, tube and baffle arrangements.

Suggestions can be made for locations and alterations of exchanger footing, nozzles, and instrument connections. Similarly, possible changes in the direction of flow through the heat exchanger can be evaluated.

These suggested changes should be made at an early stage of design because they affect exchanger design, fabrication, and delivery schedules.

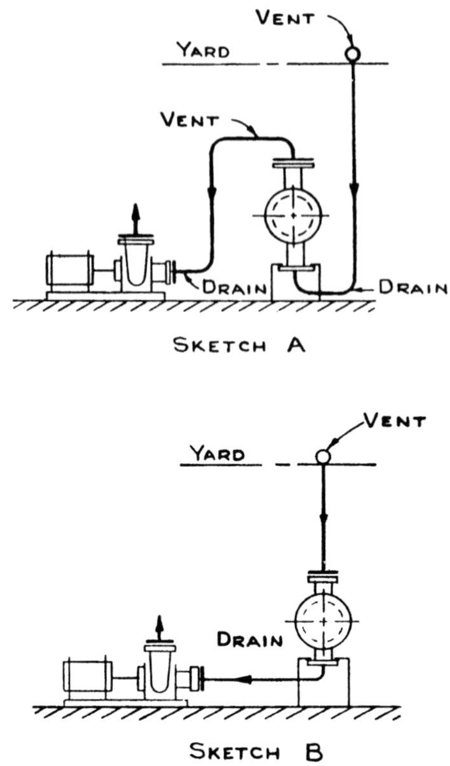

FIG. 62 Sketch A may work but Sketch B is better and is achieved by changing the flow direction through the exchanger.

Exchanger Elevation

Most exchangers are at grade, with centerline elevations of about 3.0 to 5.5 ft above ground level for exchangers about 1 to 3 ft in diameter. Exchangers at grade are the most economical arrangements. Most valves and instruments can be made accessible from grade, tubebundle handling is convenient, and maintenance is easy.

Some exchangers have a condensate or liquid holding pot after one of their outlets. These exchangers may have somewhat higher elevations than those at grade without a condensate pot.

Figure 64 shows an example. This arrangement is really a high capacity steam trap. In the layout, the top of the condensate pot should be at least in line with the bottom of the exchanger to avoid flooding the tubes with condensate and adversely affecting the exchanger heat transfer duty.

Figure 65 is another example. Here, holding of liquid level in an exchanger was required. The precise relationship between the exchanger and *control pot* is important and should be given by the flow system engineer for the physical layout.

Figure 66 shows a reboiler which has been elevated by the NPSH requirement of a centrifugal pump. Such an arrangement can elevate the

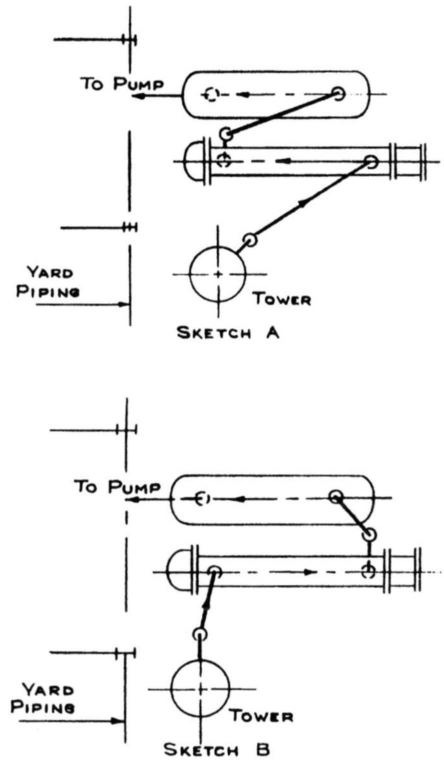

FIG. 63 Sketch A shows a zigzag flow pattern which should be avoided. Sketch B shows how relocating nozzles can provide a more functional flow pattern and shorter piping.

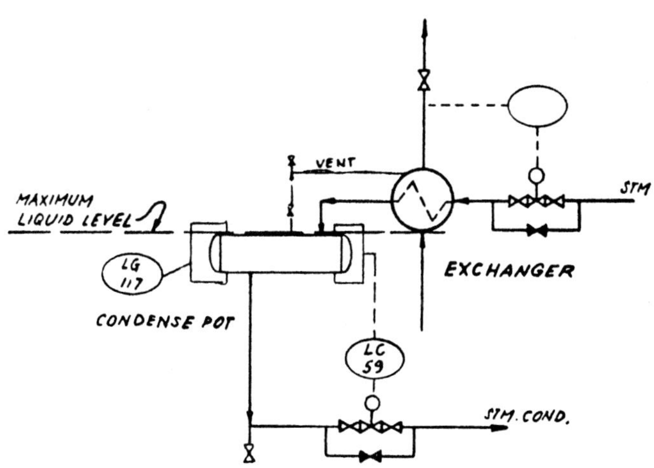

FIG. 64 With the exchanger elevated just above the condensate pot, flooding of tubes can be avoided.

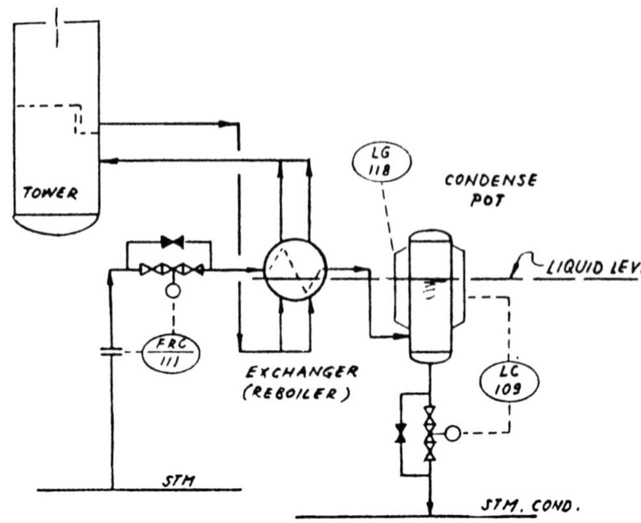

FIG. 65 When holding a liquid level in an exchanger, the precise relationship is provided by the process engineer.

exchanger from 5 to 18 ft. In most arrangements, access to valves, instruments, exchanger flanges, tube removal platforms, trolley, and trolley beam must be provided. (This elevated reboiler in turn elevates the tower, because the liquid level in the bottom of tower must be higher than the liquid level in the reboiler.)

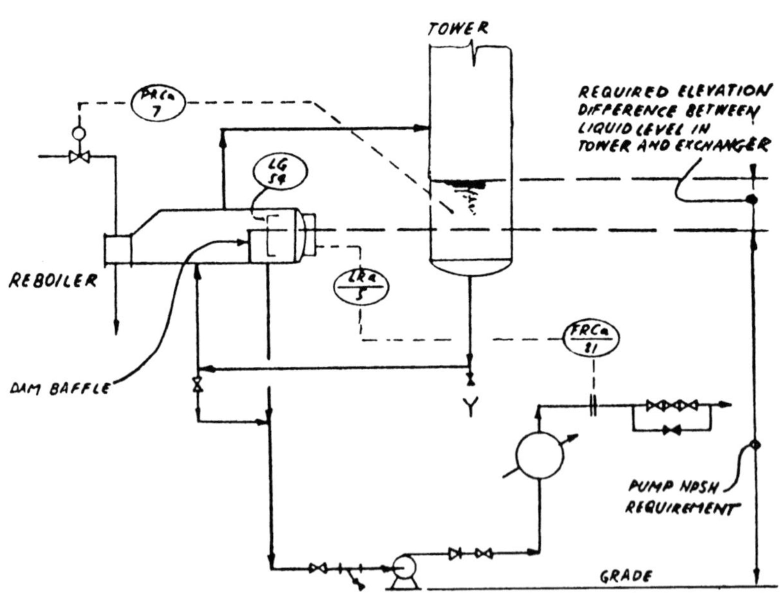

FIG. 66 When a reboiler is elevated because of pump NPSH, then the reboiler sets the tower elevation.

If gravity flow is required from the condenser outlet to a collecting drum, reflux drum, or separator, exchangers must be elevated. Examples are shown in Figs. 67 and 68.

At the condensing-reflux system in Fig. 67, the elevation of the condensers is influenced by several factors. First, the reflux drum must be elevated because of pump NPSH requirements, say 14 ft to the bottom of the drum (about 2–3 ft below this elevation, a platform is required). To this 14 ft must be added the drum diameter, estimated space for pipelines, depth of structural steel members, plus platform-to-exchanger-centerline dimension for establishing the exchanger elevation above grade.

In general, in vacuum service and in certain fractionating services where close pressure control is required, condensers are placed above the reflux drum. But most condensers are located at grade.

Figure 68 shows coolers in a gravity flow system. If the available pressure difference is small between the exchanger inlet and outlet, the exchangers must be placed above the respective tower inlet nozzles.

Figure 69 shows an exchanger in elevation with adjacent process equipment and single level yard piping. The main elevation for lines between the exchanger nozzles and yard piping is about 2 to 3 ft lower than the yard elevation. It is frequently preferred to have steam lines connecting to the top of the header to avoid condensate drainage toward the exchanger. (With steam traps at the low point, there is nothing wrong in having steam connections to the bottom of the header.) This elevation can be used for pump

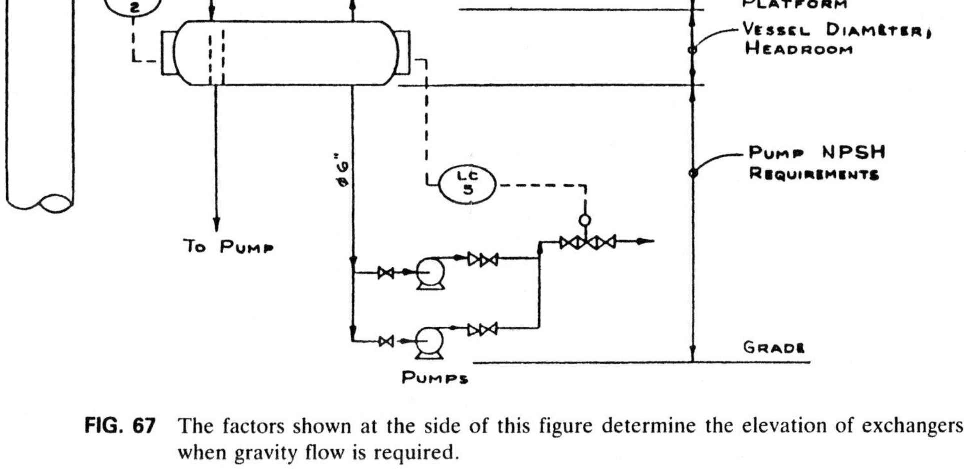

FIG. 67 The factors shown at the side of this figure determine the elevation of exchangers when gravity flow is required.

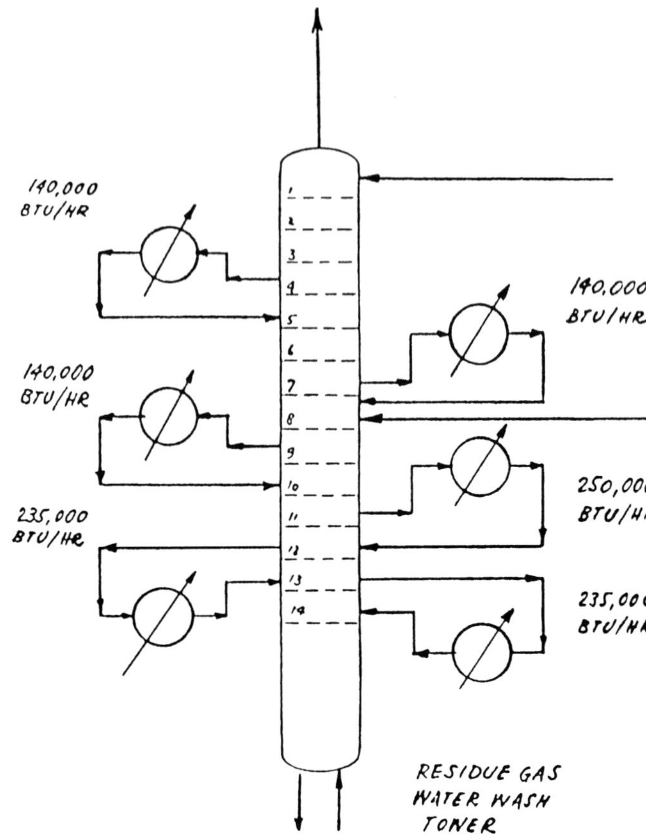

FIG. 68 When coolers in gravity flow are required, they should be located above the respective tower inlet.

discharge lines if the pump is under the yard piping and near the exchanger and for lines connecting to equipment arranged below the yard.

Pump suction lines from exchangers are just above grade. For nonvaporizing liquids under pressure, pump suction lines can be run overhead; however, loopless (and pocketless) suction lines are always preferred.

In a double-pipe yard, two elevations can be used for lines between yard and exchanger (see Fig. 69, dotted lines); but use only one supporting beam for all these lines at the lower or higher elevation depending on the number of lines supported. Hang or post-up one or two lines arranged on the other elevation.

Sometimes a flat turn at the edge of the yard piping can be arranged. Of course, here the exchanger yard interconnecting line will be raised to the elevation of the main yard piping.

Lines interconnecting exchangers with other process equipment can run just above the required headroom or about on the same level as the yard piping. Reboiler line elevations are determined by the draw-off and return line nozzles on the tower. Symmetrical reboiler piping arrangements—be-

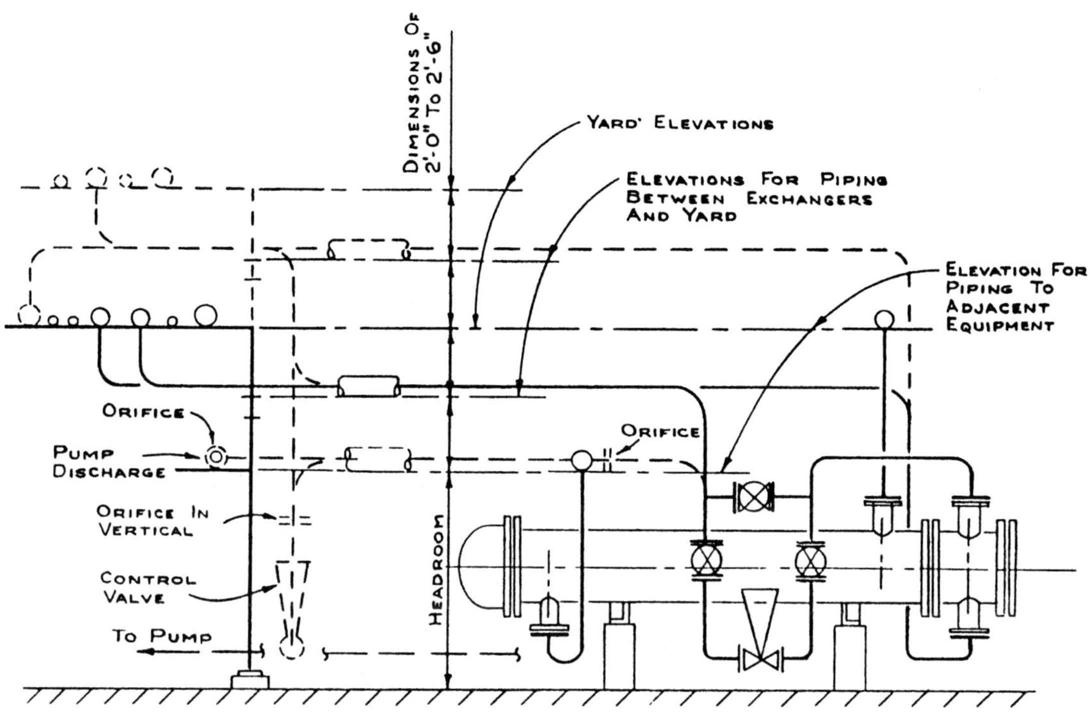

FIG. 69 An exchanger in elevation with adjacent process equipment and single-level yard piping.

tween the tower draw-off and reboiler inlet nozzles and between reboiler outlets and return connection on the tower—are preferred for equal flow in the reboiler circuit. Nonsymmetrical arrangements are also accepted for a more economical or more flexible piping design.

The overall plant layout already influences the main arrangement of exchanger piping and the necessary access (see Fig. 70). The channel ends of exchangers face the main plant access for convenient tube removal. The shell cover faces the yard and should be as close to the yard steel as practically possible (usually 6 to 7 ft from column center to outside of shell cover).

If piping is arranged on one elevation between only the exchanger and yard piping, one line will be located right over the exchanger centerline. Choose a top shell side nozzle for this location. The top tube side connection can be placed on a slight angle to miss the top shell side line without an offset in the line. Lines turning right in the yard should be right from the exchanger centerline and those turning left should approach the yard on the left-hand side of the exchanger centerline. Lines from bottom connections should also turn up on the right or left side of exchangers depending upon which way the line turns in the yard. Lines with valves should turn toward the access aisle with valves and control valves arranged close to the exchanger.

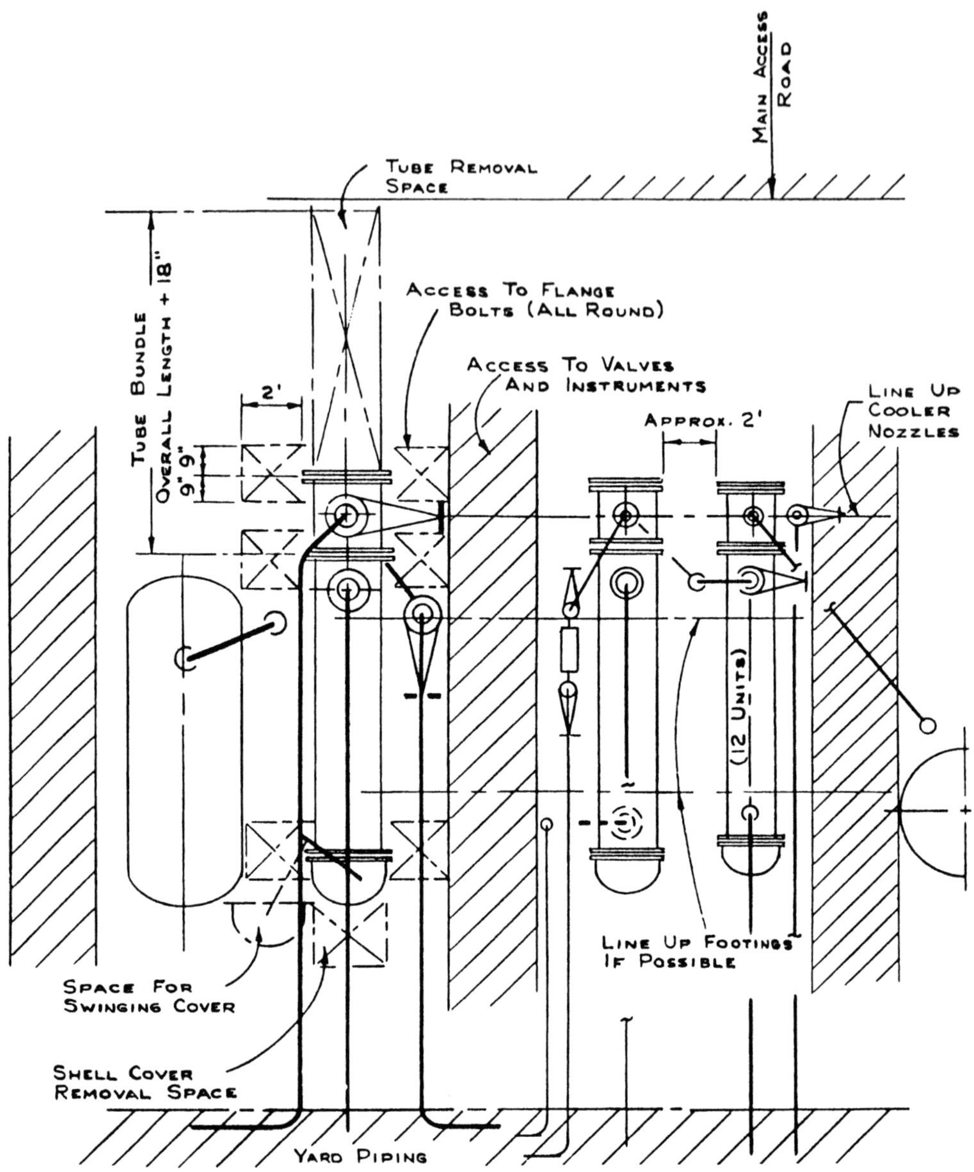

FIG. 70 This plan view shows provision for tube bundle removal with minimum removal of piping.

Utility lines connecting to a header in the yard can be arranged on any side of the exchanger centerline without increasing the pipe length.

Cooling water lines are generally below grade and should run right under the aligned channel nozzles of all coolers. The cooling water return header is usually adjacent to the cooling water line.

Access to valve headwheels and instruments will influence piping arrangement around heat exchangers. Valve handwheels should be accessible from grade and from a convenient access way. These access ways should be used for arranging manifolds, control valves, and instruments.

The piping arrangement should also provide access for tube removal. This usually means a spool piece or flanged elbow in the pipeline connecting to the top of the channel nozzle.

The designer should avoid unnecessary loops, pockets, and crossovers. He should investigate, nozzle-to-nozzle, the whole length of piping routed from the exchanger to some other equipment, aiming to provide not more than one high point and one low point, no matter how long the line. Very often a flat turn in the yard, an alternative position for control valves or manifold, changed nozzle location on the exchanger, etc. can accomplish this requirement.

Avoid excessive piping strains on exchanger nozzles from the actual weight of pipe and fittings and from forces of thermal expansion.

For valves and blinds, the best location is directly at the exchanger nozzle. An elbow nozzle on an exchanger should be checked to be sure that sufficient clearances are provided between the valve handwheel and the outside of the exchanger. Elevated valves are usually chain operated. The chain should hang freely at an accessible spot near the exchanger. Figure 71 shows sketches highlighting exchanger piping details.

Orifices

Orifice flanges in exchanger piping are usually in horizontal pipe runs. These lines should be located just above usual headroom and the orifice itself accessible with a mobile ladder. When convenient, lines with orifice and dp cell measuring elements can be at grade (grade to pipe centerline dimension should be about 2½ ft). Orifices in a liquid line and mercury-type measuring elements require more height. The long vertical measuring U-tube must be just below the orifice. At gas lines, the U-tube can be above the line with the orifice. Height here, consequently, is not critical. Lines with orifice flanges should have the necessary straight runs before and after the orifice flanges as required in specifications or standards.

Instruments

Locally mounted pressure and temperature indicators on exchanger nozzles or on the shell or process lines should be visible from the access aisles.

Design for Maintenance

In exchanger maintenance either the complete unit is removed, cleaned, and repaired, or only the tubebundle is removed for cleaning and repair—the shell is cleaned in place. If the complete unit is removed, all piping must

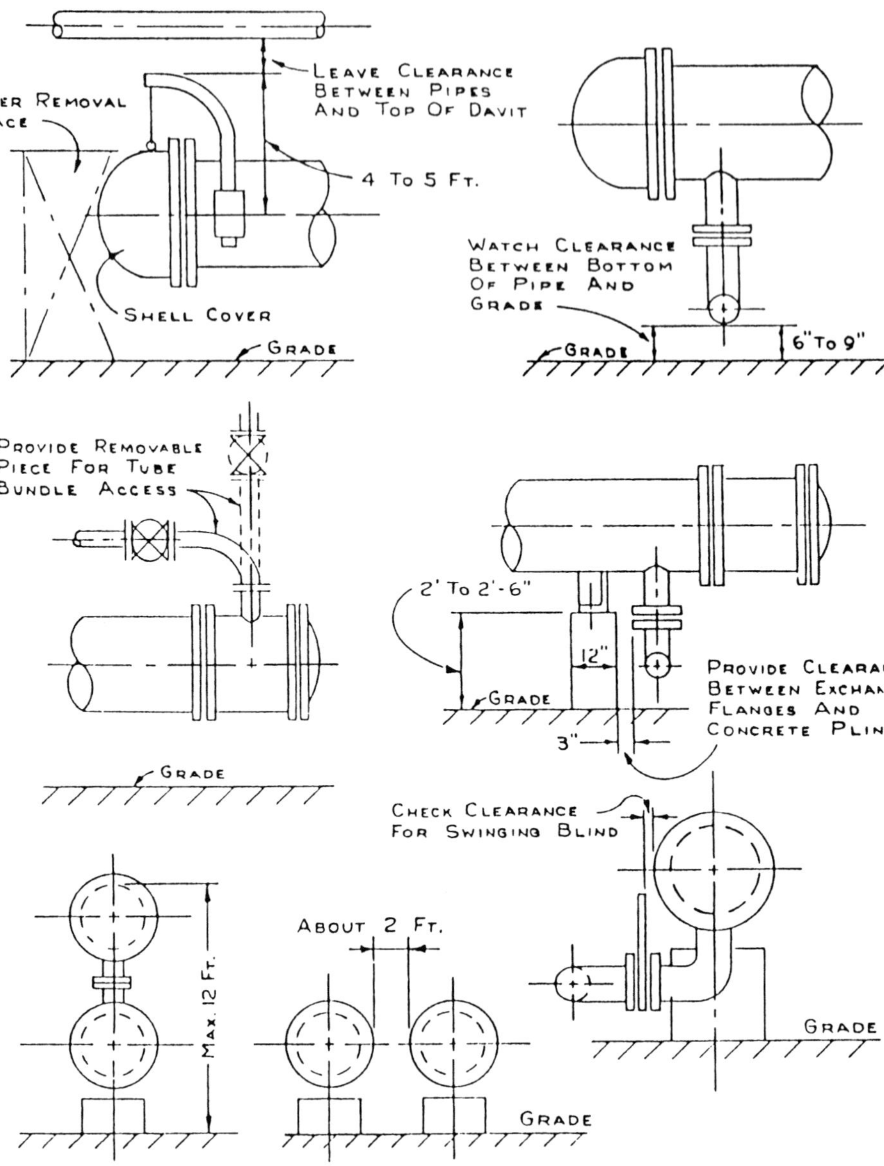

FIG. 71 By carefully following these sketches of piping arrangement, many problems can be avoided.

be disconnected. If only the tubebundle is removed, only the channel nozzle piping need be disconnected. The piping designer can help maintenance in three ways:

1. By designing and supporting piping so that no temporary support will be required when removing the channel and tubebundle, or at least temporary supports can easily be built

2. By providing easily removable spool pieces, flanged elbows, break flanges, or short pipe runs to provide adequate clearances for tube removal equipment
3. By leaving space and access around the exchanger as shown already in Fig. 70

Thermosyphon Reboiler Piping

Force of Circulation

Horizontal reboilers, with natural circulation, have a simple circulation system. Liquid flows from an elevated drum, tower bottom, or tower trapout boot through a downcomer pipe to the bottom of the exchanger shell. The liquid is heated and leaves the reboiler in the return piping as a vapor or vapor–liquid mixture and flows back to the tower or drum. There is no pressure difference between the inlet and outlet nozzles. The circulation is forced by the static head difference between the two liquid columns (see Fig. 72). Use the exchanger centerline as a reference line.

ρ_1 is the hot liquid density in the downcomer and $\rho_1 H_1/144 = P_1$ is the pressure at the level of the exchanger centerline. Similarly ρ_2 is the hot

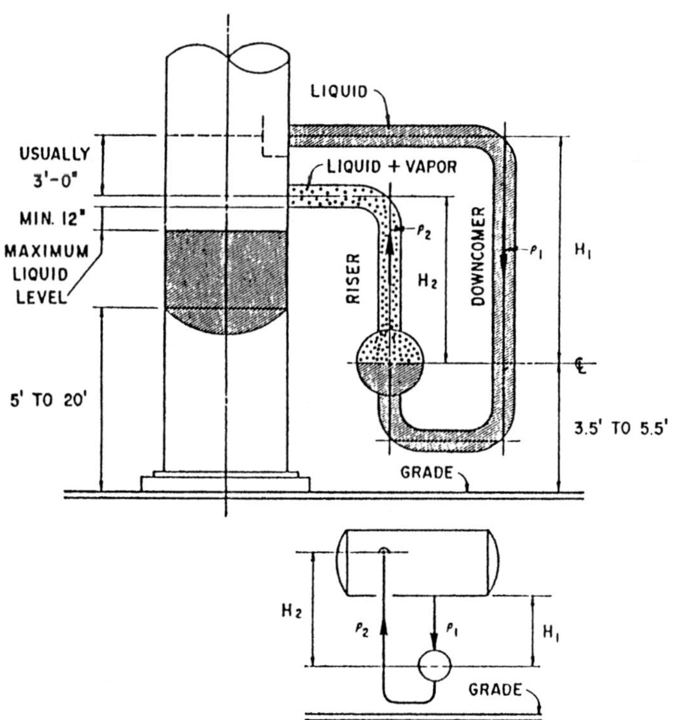

FIG. 72 Typical flow in thermosyphon reboiler piping.

mixture density and $\rho_2 H_2/144 = P_2$ is the vapor liquid column pressure at the same level. These two forces work against each other, and the difference between them is the driving force necessary for natural circulation. This is also the pressure difference available to overcome exchanger and piping friction losses:

$$P_1 - P_2 = P = (1/144)(\rho_1 H_1 - \rho_2 H_2)$$

If a safety factor of 2 is introduced, the available pressure difference for friction losses is halved:

$$\Delta P = (1/288)(\rho_1 H_1 - \rho_2 H_2)$$

$H_1 - H_2$ is usually 3 ft (see Fig. 72). Consequently, a minimum driving force of $\Delta P_{min} = 3/288$, $\rho_1 \cong 0.01$. ρ_1 is always available at horizontal exchangers.

Friction losses in reboilers are generally given as $\Delta\rho_e = 0.25$ to 0.5. It should be noted whether this figure includes entrance and exit losses. Generally, unit losses in downcomers and risers are in decimal fractions of 1 lb/in.^{2} per 100 ft.

To avoid trial-and-error calculations, a reboiler pipe size selection graph is presented in Fig. 73. It is based on the limiting velocities of 2 to 7 ft/s

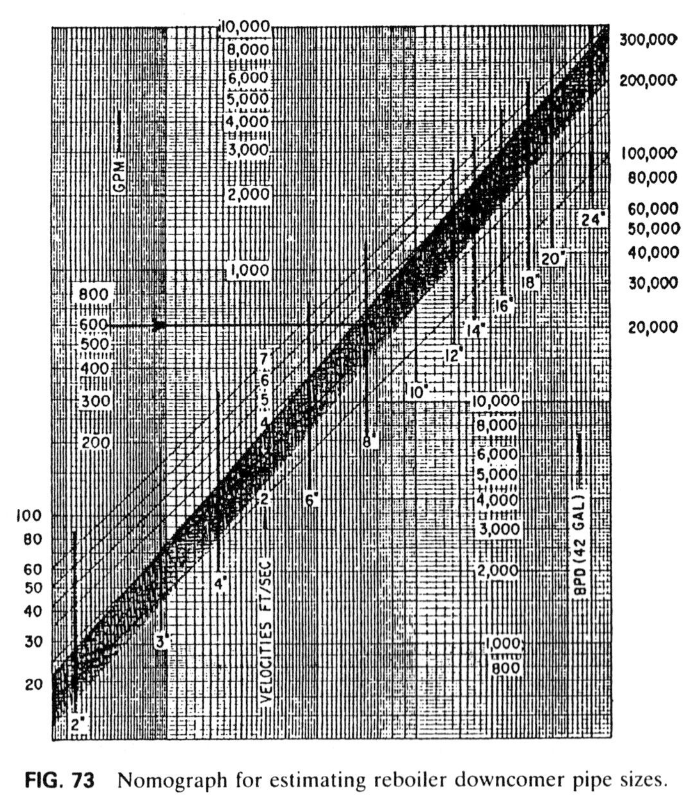

FIG. 73 Nomograph for estimating reboiler downcomer pipe sizes.

and pipe wall thicknesses of Schedule 40. Entering the graph with known liquid flow quantities, downcomer pipe sizes can be intersected in the shaded portion of the graph (corresponding velocities can also be obtained at the same time for calculating Reynolds numbers). The riser can be assumed as one or two sizes larger than the downcomer pipe size.

A kettle-type reboiler produces high evaporation rates. For this reboiler, a large diameter return line might be necessary. The process flow diagram, of course, should indicate flow rates and the physical properties of the flowing fluid.

(Figure 73 can also be used for estimating pipe diameters at gravity flow process piping.)

Draw-off Nozzle Elevation

Side Draw-off (see Fig. 72)

For the minimum downcomer nozzle elevation above the horizontal reboiler centerline, H_1 may be found from the equation for ΔP where $H_2 = H_1 - 3$ ft:

$$H_1 = \frac{288\Delta P - 3\rho_2}{\rho_1 - \rho_2}$$

The value of H_1 is useful when elevation adjustments are made to vessel heights during plant layout or when the vessel can be located at a minimum elevation. The coefficient for ρ_2 in the above equation is the elevation difference between the downcomer and riser nozzle. If this is other than 3 ft., the correct dimension, in feet, should be inserted.

The downcomer nozzle cannot be lower than H_1. Δf replaces ΔP in the above equation and is the sum of the downcomer pipe, riser, and exchanger friction losses:

$$\Delta f = \Delta f_d + \Delta f_r + \Delta f_e$$

Many towers have a bottom draw-off pump. NPSH requirements usually elevate the process vessel and the reboiler draw-off nozzle higher than that of the reboiler's minimum. This increases the static head in the vertical legs and also the driving force in the circuit. With the increased tower height, it is worthwhile to check the reboiler circuit for a possible reduction of liquid and return line sizes, especially where large diameter lines are required.

Bottom Draw-off

At horizontal reboilers, whether the drawoff nozzle is elevated or located in the bottom of the tower, the hydraulic conditions are the same. For available energy, the ΔP equation and Fig. 74 can be used.

For the minimum of downcomer length (H_1), the return line nozzle elevation (H_3) must be known. Inserting $H_2 = H_1 + H_3$ into the ΔP equation and solving for H_1:

$$H_1 = \frac{288\Delta f + \rho_2 H_3}{\rho_1 - \rho_2}$$

Δf is again the sum of the actual resistances of the reboiler plus the downcomer and riser pipe plus the inlet and outlet losses at the vessel and exchanger.

Vertical Reboiler

Figure 75 shows a vertical reboiler with the balancing static head dimensions and corresponding densities. A simplified, conservative, and convenient assumption is made that along the exchanger length, density varies in a straight line proportion. This mens that the fluid density in the reboiler will be an average of the liquid downcomer and return line densities:

$$\rho_3 = (\rho_1 + \rho_2)/2$$

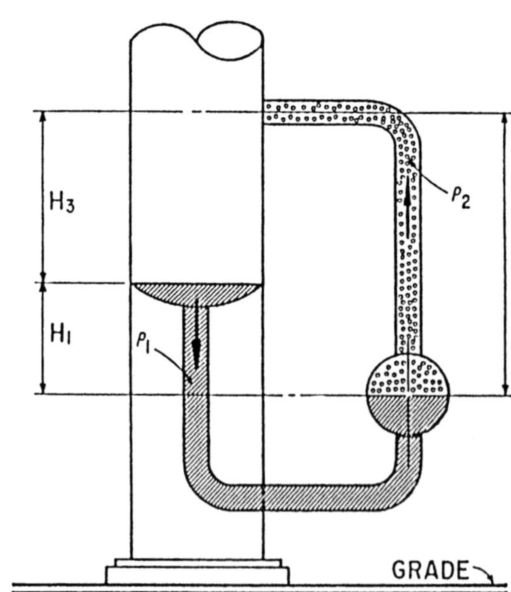

FIG. 74 Relative elevation between the tower bottom and a horizontal reboiler.

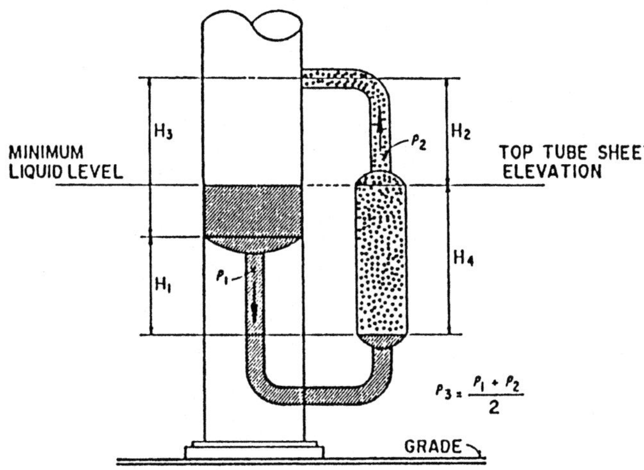

FIG. 75 Relative elevation between the tower bottom and the bottom of a vertical reboiler.

The sum of the static heads gives the following equation:

$$\Delta P = (1/288)(\rho_2 H_1' - \rho_2 H_2 - \rho_3 H_4)$$

This is the driving force for vertical reboilers. Friction losses should be smaller than this.

The relative position between the tower bottom tangent and the bottom of the reboiler must be a minimum of H_1' ft (see Fig. 75). Expressing H_1' from the above equation and inserting $H_2 = H_1' + H_3 - H_4$ from the dimensions of Fig. 75: $H_1' + H_3 = H_2 + H_4$. The relative elevation difference will be

$$H_1' = \frac{288\Delta f + \rho_2(H_4 - H_3) + \rho_3 H_4}{\rho_1 - \rho_2}$$

Δf is again the total pressure loss including exit and entrance losses at the tower and exchanger, H_4 is the exchanger length, and H_3 is the return nozzle location. ρ_1 and ρ_2 densities can be obtained from the process flow diagram or the exchanger data sheet. ρ_3 and H_1' can be calculated.

The vertical reboiler should be flooded. The maximum elevation of the top tubesheet should not be higher than the minimum liquid level in the tower.

Assuming that the tower bottom tangent line and the top tubesheet are on the same elevation, the above two equations can be simplified because $H_1' = H_4$ and $H_3 = H_2$. Consequently:

$$\Delta P = (1/288)[H_1'(\rho_1 - \rho_3) - \rho_2 H_3]$$

and

$$H_1' = \frac{288\Delta f + \rho_2 H_3}{\rho_1 - \rho_3}$$

In vertical reboiler circuits, reboiler losses are greater and pipe losses are smaller than in horizontal circuits. This often results in a difference of two pipe sizes between the downcomer and riser—also, in larger pipe sizes than in horizontal reboilers (assuming the same liquid, flow, and evaporation rates).

The H_2 dimension can be greater than H_3. The minimum liquid level static head above the bottom tangent line can also be taken into account as an additional driving force. Because of the predictable and very simple piping, a safety factor of much less than 2 can be allowed. [In the above equations, 288 = 144(safety factor).]

Figure 76 shows a vertical reboiler with a high draw-off nozzle. Using this figure, the driving force is

$$\Delta P = (1/288)(\rho_1 H_1'' - \rho_2 H_2 - \rho_3 H_3)$$

The draw-off nozzle elevation will be

$$H_1'' = \frac{288\Delta f - \rho_2(H_4 + 3) + \rho_3 H_4}{\rho_1 - \rho_2}$$

Once more, the total system loss (Δf), known densities, and exchanger length (H_4) will give the distance between the draw-off nozzle and the reboiler bottom tubesheet (H_1''). If the elevation difference between the draw-off and return nozzle is other than 3 ft, the correct dimension should be inserted in the H_1'' equation above.

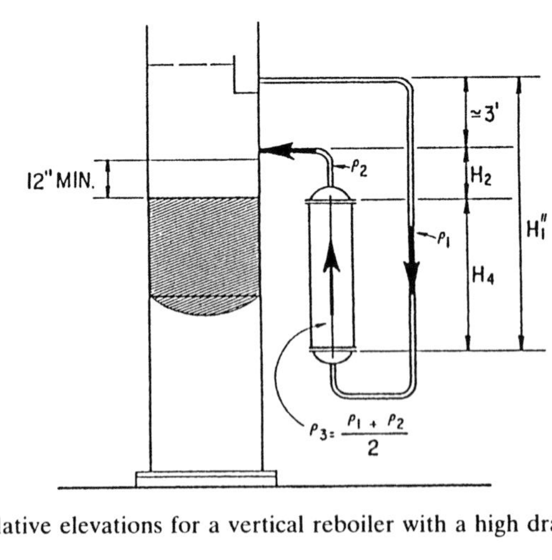

FIG. 76 Relative elevations for a vertical reboiler with a high draw-off nozzle.

Arrangements

The type of reboiler arrangements can be classified as follows:

Thermal circulation:
  Shell and tube
  Kettle type
  Vertical reboilers
Inserted reboilers:
  Closed coil helical
  U-type stub bundle
Pump circulated reboiler circuits

Vertical reboilers and the inserted-type reboilers have little or no piping. Larger diameter towers can have 1 to 4 U-tube stub bundles inserted directly in the liquid space through the tower nozzles, and extending across the tower diameter. Reboilers with small heat duties are most often designed using helical coils. Forced recycling is a pump piping design problem.

Piping to horizontal reboilers is designed to be as simple and direct as possible within the limitations of thermal expansion forces (see Fig. 72).

Symmetrical arrangements between the draw-off and reboiler inlet nozzles, as well as between the reboiler outlet and return connection on the tower, are preferred for equal flow in the reboiler circuit. A nonsymmetrical arrangement may also be accepted for a more economical or more flexible piping design.

When sizing and arranging nonsymmetrical piping, an attempt should be made to equalize the resistance through both legs of the piping. More resistance in one leg can mean less flow than in the other. Uneven heat distribution will occur in the reboiler—one riser will be hotter than the other.

A direct pipe connection to the trapout boot is much preferred (Fig. 77,

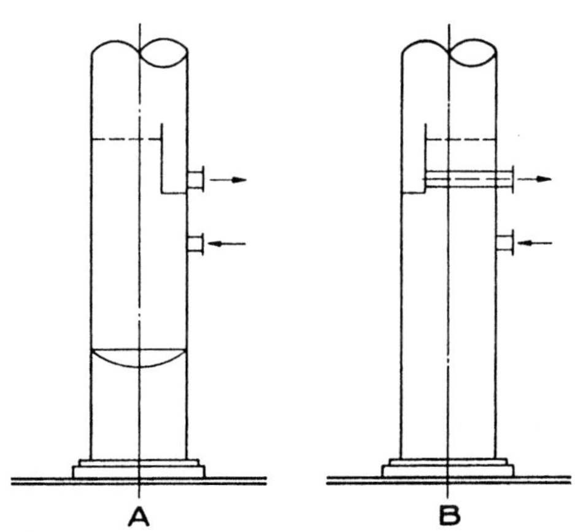

FIG. 77 Avoid an internal crossover from the downcomer nozzle to the trapout boot as shown in Sketch B.

Sketch A). Liquid can be heated in an internal crossover (Fig. 77, Sketch B) and the downcomer might have some vapor content. With even a small amount of evaporation, the liquid density rapidly decreases and the calculated positive liquid head will not be entirely available.

For startup, a 2 or 3-in. diameter, gravity-flow bypass is usually provided from the tower liquid space to a low point of the downcomer, at reboilers with high liquid draw-off nozzles.

Valves are rarely included in reboiler piping except, for example, when a standby reboiler is provided or when two or three reboilers are used and operated on an extremely wide heat capacity range. Some companies require line blinds at the tower nozzles for blanking off during shutdown, turnaround, and maintenance.

The heating mediums of steam or a hot process stream always connect to the tube side of horizontal reboilers. The inlet piping should have a control valve—with block valves and a bypass globe valve, if required. This should be arranged near the reboiler's tube side inlet.

Example. An example of reboiler piping design procedure is shown on Fig. 78 and Table 10. The necessary data for pipe sizing is given in Table 10. Entering the graph

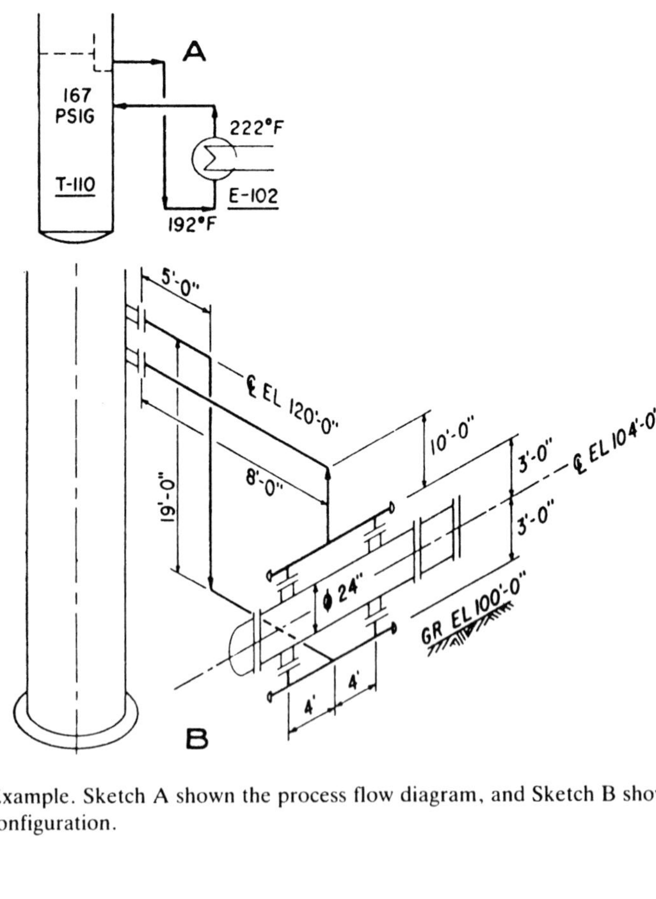

FIG. 78 Example. Sketch A shown the process flow diagram, and Sketch B shows the piping configuration.

TABLE 10 Reboiler Pipe Sizing

Downcomer: Liquid, 186,850 lb/h

$$\rho_l = 36.7 \text{ lb/ft}^3 \text{ (hot)}$$

$$Q = 633 \text{ gal/min (hot)}$$

Riser: Liquid, 130,750 lb/h

$$\rho_2 = 36 \text{ lb/ft}^3$$

Vapor, 56,100 lb/h (30% of total)

$$MW = 53$$

$$\rho_v = \frac{(53)181.7}{(10.73)682} = 1.32 \text{ lb/ft}^3$$

$$\rho_2 = \frac{100}{\dfrac{70}{36.7} + \dfrac{30}{1.32}} = 4.06 \text{ lb/ft}^3$$

Available Driving Force (safety factor = 2):

$$\begin{aligned}\Delta P &= (1/288)(\rho_1 H_1 - \rho_2 H_2) \\ &= (1/288)[(36.7)16 - (4.06)13] \\ &= 1.86 \text{ lb/in.}^2\end{aligned}$$

Friction Losses:

Downcomer Size	8 in. Δp	0.19 lb/in.^{2}	8 in. Δp	0.19 lb/in.^{2}
Exit + Entrance		0.096		0.096
Riser Size	8 in. Δp	1.13	10 in. Δp	0.433
Exit + Entrance		0.948		0.39
Exchanger Δp		0.35		0.35
Total Δp		2.714 lb/in.^{2}		1.459 lb/in.^{2}

Minimum Draw-off Nozzle Elevation:

$$\begin{aligned}H_1 &= \frac{288\Delta p - 3\rho_2}{\rho_1 - \rho_2} \\ &= \frac{(288)1.459 - (3)4.06}{36.7 - 4.06} \\ &\approx 13 \text{ ft}\end{aligned}$$

Actual Draw-off Nozzle Elevation: 16 ft
(from exchanger centerline)

on Fig. 73 at a flow rate of 633 gal/min, an 8-in. downcomer size can be selected. The riser can be 8 or 10 in. Figure 78 shows a piping configuration for estimating the equivalent length of the pipe and fittings. Calculations are shown in Table 10 for the available driving force, friction loss in the system, and minimum draw-off elevation. (Friction loss in pipes has been calculated with formulas and data published in Crane's Technical Paper No. 410, *Flow of Fluids*.)

The 8 and 10-in. lines selected in this example are the most economical sizes. Economy of reboiler piping depends on the simplest possible pipe configuration with the minimum number of pipe fittings and on choosing the minimum suitable pipe sizes. Uneconomical reboiler lines are just carelessly overdesigned or oversized.

Pressure Relief Piping

The most important design factor about pressure-relieving devices is the underlying principle of intrinsic safety. They must *fail safe* or not at all.

Hazardous Fluids

The arrangement shown in Fig. 79 should not be used for hazardous fluids unless the valve is at a considerably higher elevation than the surrounding equipment and the discharge can be directed away from such equipment. The possible need for a bird screen should be considered.

When hazardous fluids are discharged from valves installed as in Figs. 80a–e, the terminal point should be at least 10 ft above any walkways within a 25-ft radius (see also Table 11).

Drain Hole Plug

The safety valve drain hole plug (Fig. 80a) should be removed in those services where liquid could gather at the valve discharge. This includes

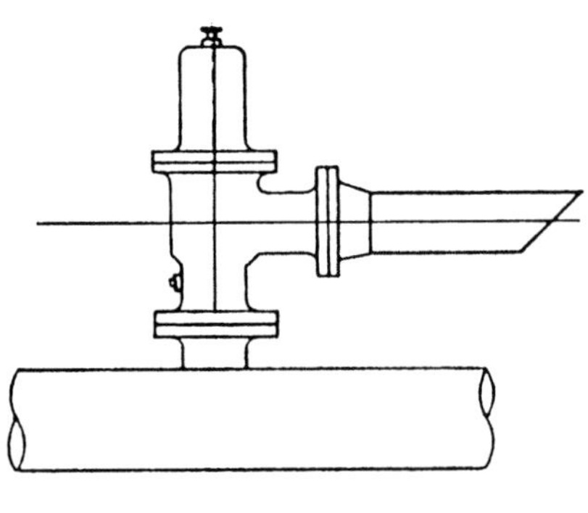

FIG. 79a Pressure relief piping for air or gas service.

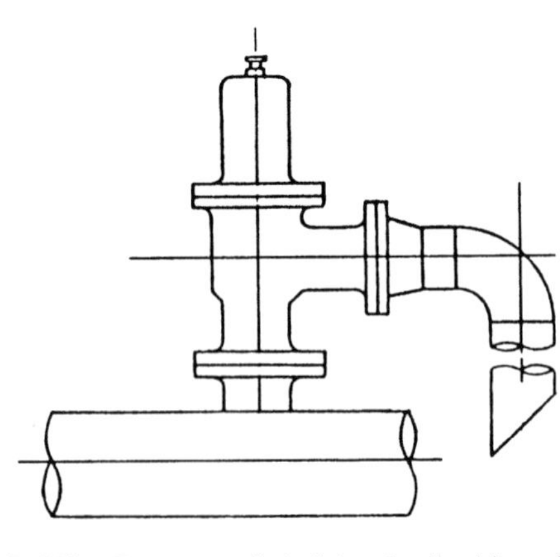

FIG. 79b Pressure relief piping for liquid service.

services where condensate may form or where rain or snow may enter the discharge pipe and collect as a liquid in the line. If the plug is removed, the drain hole must be piped for safe disposal if the fluid is hazardous or if the location of the hole is such that a sudden discharge through the opening might endanger personnel. If the pipe is not warm enough to melt any snow which may enter it, a cover must be provided. This may be either a lid, as shown in Fig. 81, or a light plastic bag fastened around the end of the pipe. Metal covers in some sizes are also available commercially.

'*Flashing liquids,* in which much of the liquid is vaporized on relieving, should not be emptied to a sewer or other ground location without proper protection for personnel. The vapor will propel the liquid at high velocity and may spatter passersby with the hot liquid.

Piping Supports

Supports for discharge piping should be designed to keep the load on the valve to a minimum. In high temperature service, high loads will cause permanent distortion of the valve because of creep in the metal. Even at low temperatures, valve distortion will cause the valve to leak at pressures lower than the set pressure and result in faulty operation. The discharge piping should be supported free of the valve and carefully aligned so that the forces acting on the valve will be at a minimum when the equipment is under normal operating conditions. Expansion joints or long-radius bends of proper design and cold spring should be provided to prevent excessive strain.

The major stresses to which the discharge pipe is subjected are usually caused by thermal expansion and discharge reaction forces. The sudden release of a compressible fluid into a multidirectional discharge pipe produces an impact load and Bourdon effect at each change of direction. The piping must be adequately anchored to prevent sway or vibration while the valve is discharging (Figs. 82 and 83).

Pressure loss in the discharge piping should be minimized by running the line as directly as possible. Use long-radius bends and avoid close-up fittings. In no application may the cross-sectional area of the discharge pipe be less than that of the valve outlet.

Outlet Piping

The size of outlet piping required for a safety disk is not necessarily the same as the disk receptacle size. Disks frequently are sized on the basis of pressure requirements rather than capacity requirements. In such instances it is possible for the outlet piping to be smaller than the pipe size of the disk receptacle. If an arrangement of this type is desirable, the pipe diameter must be calculated on the basis of the relief capacity requirements and the maximum allowable upstream pressure. The inlet piping, however, must have an area which is at least equal to that of the receptacle in order to comply with the ASME code.

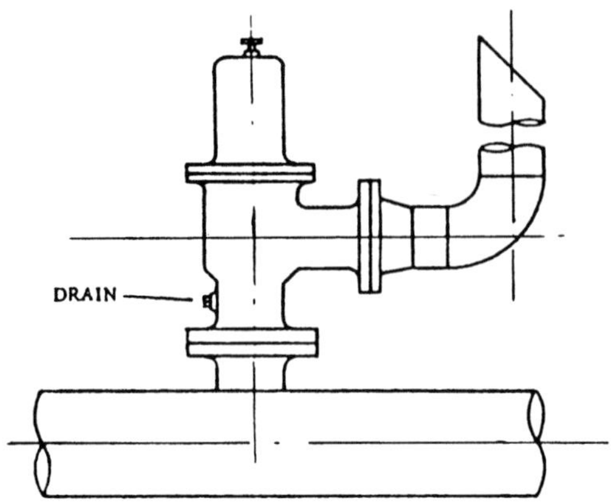

FIG. 80a Pressure relief piping for air, gas, or steam service.

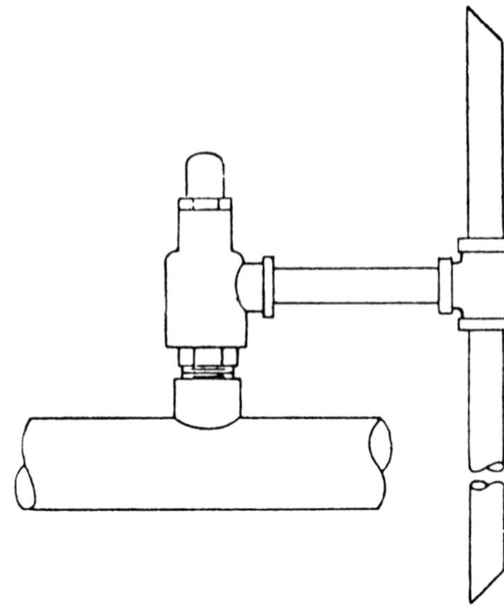

FIG. 80b Pressure relief piping for steam or vapor service.

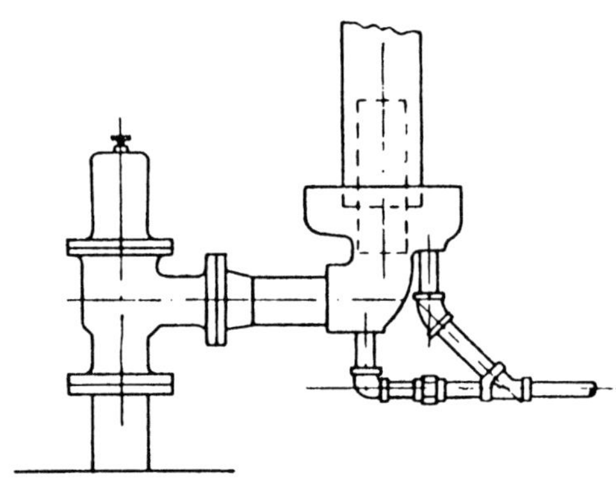

FIG. 80c Pressure relief piping for steam or vapor service to 3-in. pipe.

Pressure Drop

The pressure drop allowed through the inlet and discharge lines is unlimited as long as the capacity of the line is adequate for the relief requirements. That is, at the required flow rate the vessel pressure must not exceed the maximum allowable accumulated pressure. In sizing a safety disk, it is usually assumed that the entrance loss at the nozzle is the governing restriction insofar as capacity is concerned. Thus the effective orifice area is considerably larger than the effective orifice area of a safety valve of the same pipe size. Consequently, the pressure loss in the piping will be of greater importance in the disk application. It is advisable to check the effect of line

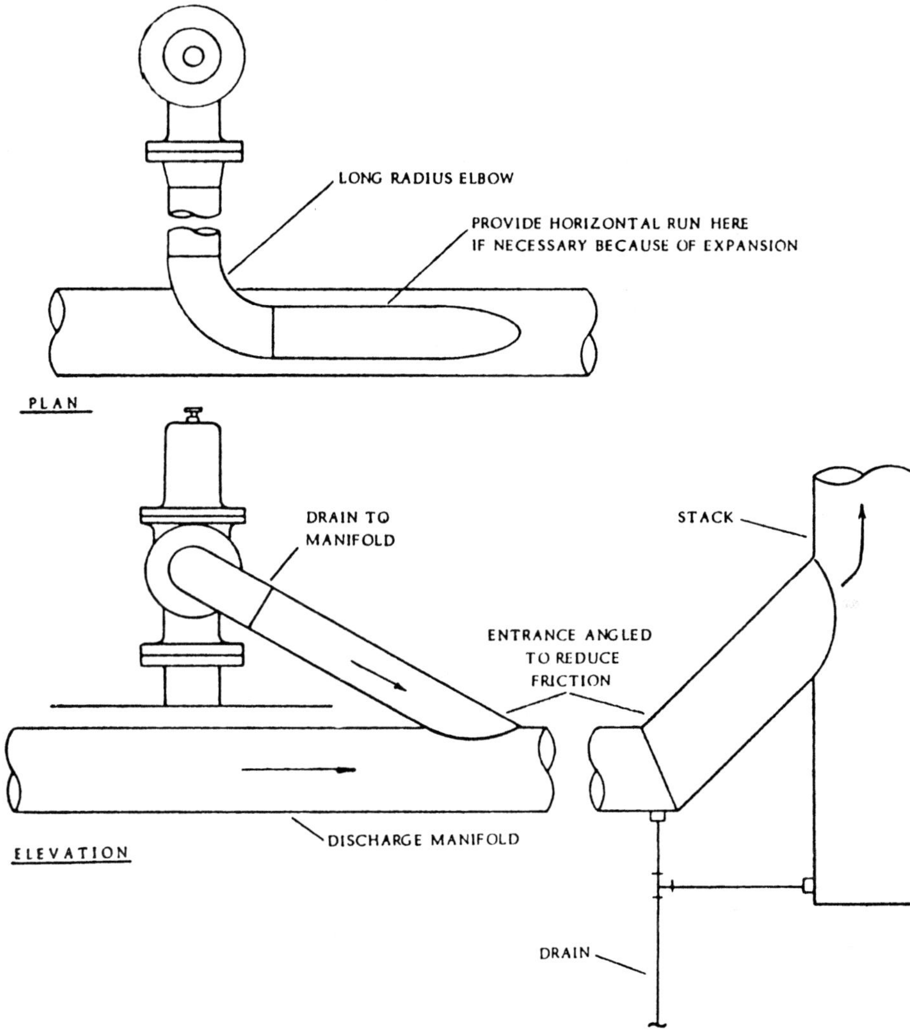

FIG. 80d Closed system for hazardous service.

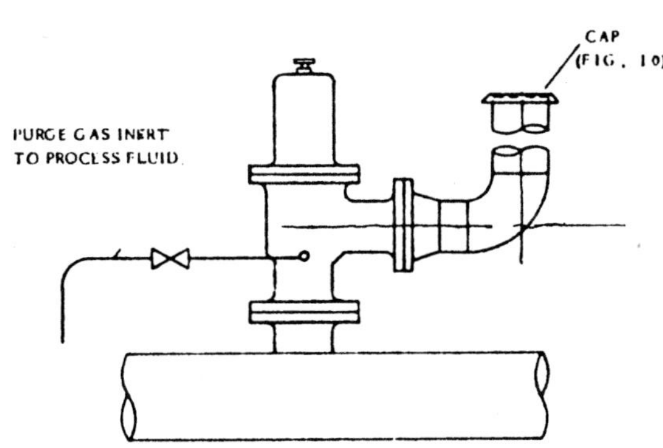

FIG. 80e Open system for pyrophoric gases.

TABLE 11 Design Guide for Discharge Piping

Service	Figure Number: Valve Indoors	Figure Number: Valve Outdoors
Nonhazardous service:[a]		
Air or gas	79a, 80a, 80c	79a, 80c
Liquid	79b	79b
Steam or vapor:[b]		
Discharge pipe size to 1 in.	80b	80b
Discharge pipe size to 1½–2½ in.	80a	80a
Discharge pipe size to 3 in. and over	80c	80a
Hazardous service:[a]		
Closed system (to vent stack, burning stack or scrubber)	80d	80d
Open system (to atmosphere):		
Gas[c]	79a	79a
Liquid[d]	79b	79b
Vapor[c,d]	79a, 80a, 80b	79a, 80a, 80b
Pyrophoric gases or vapor[e]	80e	80e

Low Temperature Service:

At or below ambient: Design discharge pipe so that snow or ice cannot accumulate at any point in the line where the temperature may be at or below freezing. Use Fig. 3, if possible. Where necessary Fig. 4 may be used with a cover.

Below 32°F: Locate safety valve to avoid need for discharge piping, if possible. Discharge opening and exposed spring must be protected from the weather. A housing or local heating may be required. The discharge, if properly designed, may be sealed with a low viscosity oil and covered with plastic to prevent the entrance of moisture.

[a]Flammable or toxic fluids are considered hazardous.
[b]Discharge pipe not required if outlet over 7 ft above walkway and/or directed away from personnel.
[c]Carry discharge outdoors to a safe elevation.
[d]Carry to an appropriate drain.
[e]Point of discharge must be safe for fire.

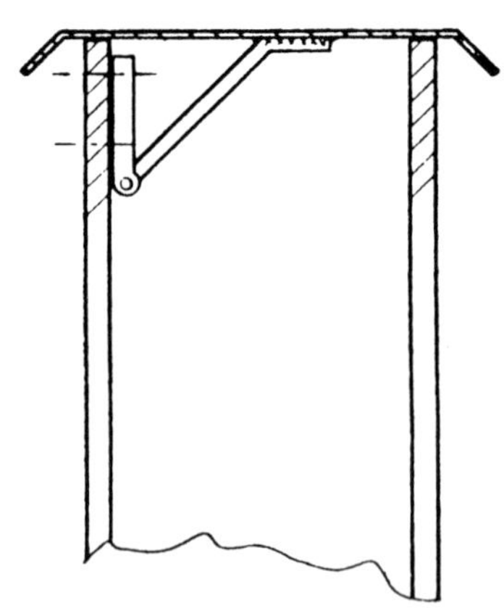

FIG. 81 A cap like this will protect a discharge pipe from being plugged with snow.

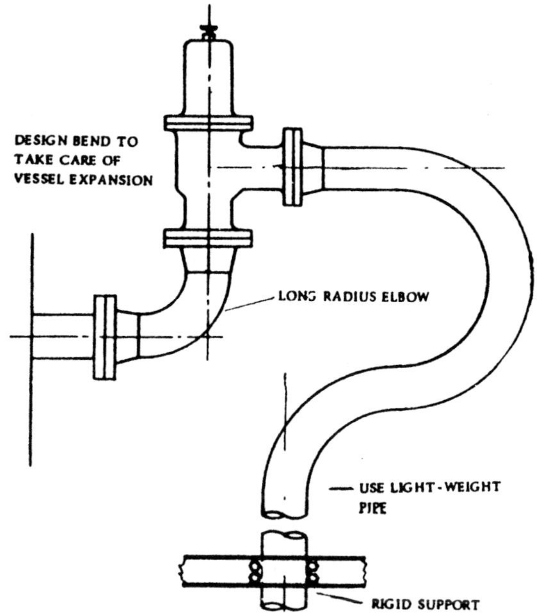

FIG. 82 Horizontal vessel nozzles, when used for safety valve mounting, can be connected this way.

pressure drop on any safety disk installation with a low rupture pressure or a long discharge line.

The allowable pressure loss through the discharge pipe, exclusive of the entrance loss, may be determined as follows:

$$P \leqslant P_a r_c$$

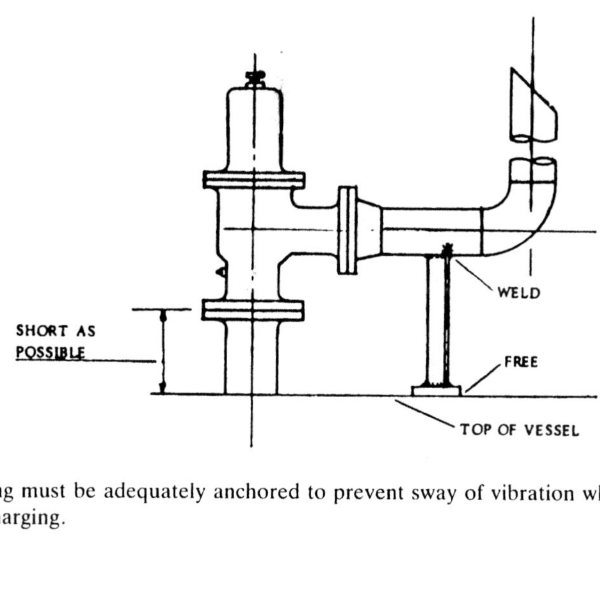

FIG. 83 Piping must be adequately anchored to prevent sway of vibration while the valve is discharging.

where P_a = accumulated relieving pressure, lb/in.^{2}abs
P = pressure inside relief pipe near the vessel, lb/in.^{2}abs
r_c = critical pressure ratio for sonic velocity

$$= \left(\frac{2}{k + 1}\right)^{k/(k-1)}$$

k = ratio of specific heats

If P exceeds $P_a r_c$, it becomes necessary to size the safety disk based on subsonic flow.

Steam Tracing Design

A typical steam tracing system is shown in Fig. 84. The keystone of this system is the steam trap. Correct selection and application of other components is to no avail if the steam trap is improperly chosen. Selection must be based on two choices: type and size of the trap.

Trap Types

The two types of traps normally used in steam tracing service are the bucket trap and the floating disk thermodynamic trap. Since the bucket trap is subject to freezing, the thermodynamic trap is generally used in outdoor

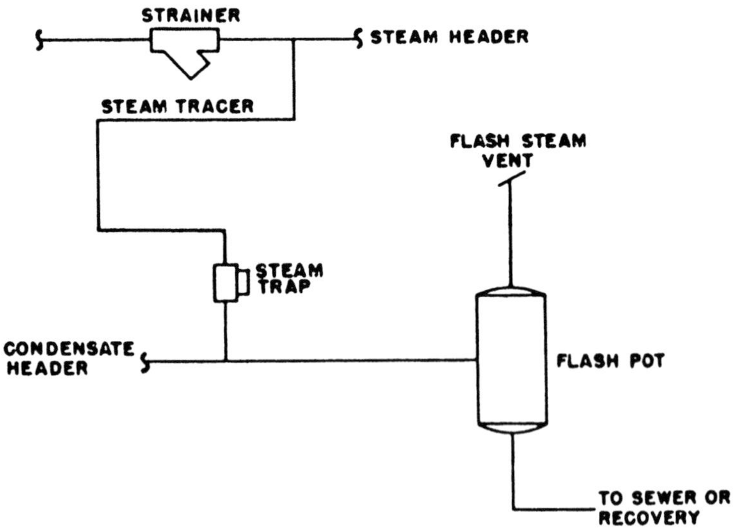

FIG. 84 Typical steam tracing system.

locations. Because of handling ease and reduced inventory, the thermodynamic trap is also used to a great extent in indoor locations. Other advantages of the thermodynamic trap are its small size and insensitivity to pressure variations (up to 600 lb/in.^{2}gauge).

Trap Sizing

The sizing of thermodynamic traps is more critical than for other types. Size a trap too small and it will back up condensate; too large and it will waste steam. Excessive oversizing will cause the trap to destroy itself. This occurs when the trap cannot get enough condensate to fulfill its energy requirements and, therefore, begins to cycle more rapidly, partly on live steam. This causes an accelerated wear rate, which causes an even more rapid cycling, which causes an even more rapid wear rate, and so on until the trap becomes useless.

Sizing to include potential startup loads leads to oversizing in thermodynamic traps. A thermodynamic trap will handle a great deal more cold condensate than hot condensate; and if a still greater rate is desired, the line can be manually blown down. For applications other than steam tracing, careful consideration should be given before introducing any startup allowance, especially if the addition requires an increase in trap size.

For steam tracing applications, a flow rate of 100 lb/h gives the best balance between trap sizing and traced length of pipe. The allowable lengths at various temperatures stays within the realm of possibility while the unavoidable short runs do not cause the flow rate to fall into the trap's inefficient range.

On this basis, the selection of a group of steam traps to handle all steam tracing situations is possible. Actually this consists of a ⅜-in. nominal size trap, but many steam trap manufacturers rate their traps by orifice size rather than by connection size. Suitable traps should have 150 lb/in.^{2}gauge steam condensate capacities from 350 to 600 lb/h at saturation temperature, and from 550 to 850 lb/h at 30°F below saturation temperature. Using 15 lb/in.^{2}gauge steam, the condensate capacity should range between 135 and 250 lb/h at saturation temperature and from 225 to 400 lb/h at 30°F below saturation temperature.

Field and laboratory tests have shown that the nominal ⅜-in. size trap will operate efficiently below 20 lb/h and has enough capacity to give safety factors of 2 or more, depending on the trap, based on a 100 lb/h normal load.

Low Loads and Cycling

The trap should be able to handle loads down to a very small percent of its rating which should be in the realm of 2 to 3% at operating temperature. Cycling, or condensate blowing, should occur at a rate of 3 to 4 cycles/min, with a blowing time of 5 to 10 s. With proper operation, the trap will not allow condensate buildup in the tracer and will blow down completely each time. This, of course, is a bare trap; insulation should not be applied to a thermodynamic steam trap.

Tracer Lengths

Recommended maximum traced lengths have been computed to give a steam flow of 100 lb/h and are shown in Table 12. For the majority of steam-tracing applications, the length of the tracer run is uncontrollably short. Only for long transfer line runs can the length be controlled for effective trap use. The usual length of a tracer run produces a condensate flow much below the capacity of the trap. Grouping of small systems is useful but limited, since combined systems rapidly become unwieldy. Basing traced lengths on a flow of 100 lb/h allows the use of a trap small enough to avoid oversizing and keeps the allowable runs reasonably long. In layouts where looping or pocketing of the tracer exists, the tracer should incorporate no more total pocket height than determined by the following API formula:

Sum of pocket heights = 2.31(10)% of inlet steam pressure, lb/in.^{2}gauge

TABLE 12 Maximum Tracer Lengths for ⅜-in. Trap

Based on the formula:

$$L = \frac{W(LH)}{Q(SF)}$$

where L = length of tracer per foot
W = steam flow, (100) lb/h
LH = latent heat of vaporization, Btu/lb
Q = heat loss, Btu/h-ft
SF = safety factor [2 (basic) 1.5–3.5 (actual)]

For 150 lb/in.^{2}gauge Steam:[a]

Line Size (in.)	Tracer Length (ft)	Insulation Thickness (in.)
2	400	1
3	375	1 to 325°, 1½ above
4	375	1 to 275°, 1½ above
6	275	1 to 225°, 1½ above
8	250	1 to 225°, 1½ to 325°, 2 above

[a]Chart is valid from 150 to 350°F fluid temperature. The tracer should incorporate no more total pocket height than computed by the following API formula:

Sum of pocket heights
= 2.31 × 10% of inlet steam pressure, lb/in.^{2}gauge

Insulation thicknesses are based on economic studies for the Philadelphia area.

This condition will occur in the majority of tracing applications within a refinery or petrochemical plant unit.

Strainers

Field tests have shown that 2 out of 14 traps in normal tracing service were malfunctioning because of dirt under the disk. Providing strainers for each and every tracer to prevent possible trouble during startup is undoubtedly false economy, especially when the frequency of such problems appears to be no more than 5%. A better solution is to install a strainer in each small header to feed a number of tracers, thereby covering all systems.

Integral-Strainer Traps

Integral-strainer traps are not suitable for large refinery or petrochemical units for the following reasons: their straining capacity is too small, especially for startup conditions; they cost considerably more than separate units; and for complete coverage, each tracer would have to be equipped with an integral-strainer trap.

Condensate Collection System

The remaining facet of tracing systems is the condensate collection system. For underground systems a small flash pot is often used which will handle a given number of traps. The tracing systems are broken down into groups, each having its own flash pot. Insufficient sizing of these, as well as any other open condensate collection system, will cause a steady rain of condensate in the area of the vent pipe, and the sewers in the area will emit flash steam. An underground system can cause maintenance problems when the coolers must be installed beneath concrete.

The alternative is to install an above-ground condensate collection system. From a maintenance viewpoint, a single centralized unit would be even better. Underground flash pots, being generally horizontal, do not give as good a separation as above-ground vertical units. The difficulty in drawing any specific conclusions on condensate collection systems is the nature of the application and the existence of local codes.

Thermowell Design

Flow Sheets and Temperature Points

The temperature points, with their index numbers, will be found on the process control diagram. These require thermowell installations in the piping. The temperature index numbers assigned to each point are designated by conventional symbols such as TI (temperature indicator), TH, TIC, TC, TRC, TR, TIA, TRCA, TT, and TW followed by the number assigned to each of the points.

The flow sheets will give the pipeline size and the piping specification reference to use. An identifying process reference line number may also be given to each line.

Layout

The piping must be run in the space provided, without interfering with structural steel platforms and other piping. Sufficient space must be provided for the valves, flanges, instrument connections, and for the installation of thermowells.

Possibly the shortest piece of pipe that may be shown on flow sheets into which thermowells are installed is a pipe reducer between two flanges. Such a connection may be used between two heat exchanger nozzles as shown in Fig. 85.

When two different process lines converge into a pipe tee and then flow into one line, it is necessary to get a good representative temperature. This pipeline must extend not less than 10 pipe diameters to obtain a good mix-

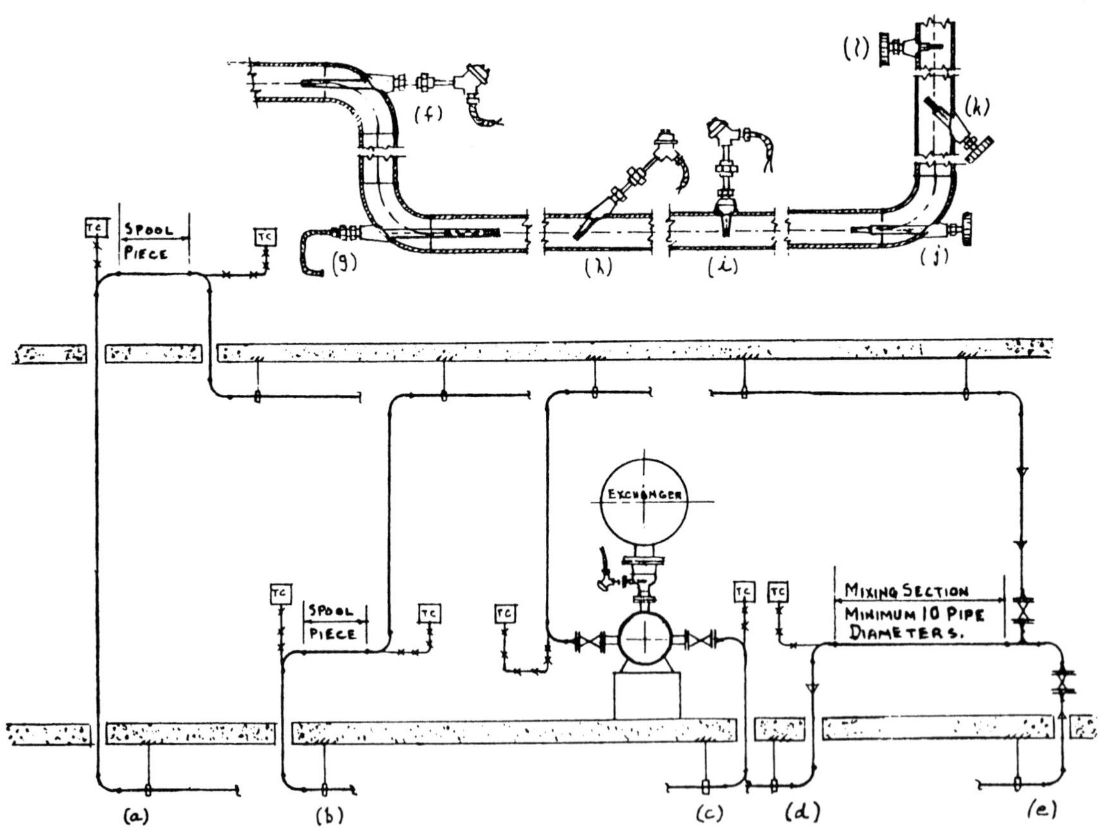

FIG. 85 Typical thermowell locations for process piping 3-inches and larger. Installation details: thermobulb (g), thermocouples (f, h, i), and dial thermometers (j, k, l).

ing before the flow reaches the thermowell. An example of this is shown in Fig. 85.

Thermowells for Large Size Pipe

Thermowells can be installed with ease in process pipelines 3 in. and larger. Every effort should be made to install the thermowell connection in a welding pipe elbow. The connection should be made by welding a 1-in., Type 6000 lb or other size forged steel elbow adapter in the heel of the pipe elbow (see Fig. 85f). If this is done, the thermowells with the longest required insertion length can be installed without running into interferences inside the piping. The pipe should, of course, be laid out so the thermowell will be accessible from grade or a platform. Since most thermowells must be removed for inspection during a plant shutdown, they must be located within the pipefitter's reach, preferably without resorting to the use of ladders. This will often require an additional process pipe elbow in the line.

Some of the typical piping layouts that can be used are shown in Figs. 85(a)–(e). Only the pipe elbow has been used in each of these details for installing thermowell connections. These are for dial thermometers, thermocouples, and thermobulbs. Such arrangements are necessary when the 3-in., 4-in., and larger process lines are used. It is often possible to install dial thermometers and thermocouple well assemblies on a 45° angle by using an elbow adapter as a lateral connection as shown in Figs. 85(h) and 85(k). Thermowell connections can be installed perpendicular to the pipe wall with a forged steel thread adapter. This arrangement can be applied best on 4 in. or larger pipe as shown in Figs. 85(i) and 85(l).

Thermowells for Small Pipelines

Thermowell installations in small-sized process lines (¾ to 2-in.) require special consideration. Because the lines are small, the thermowells cannot be installed directly into the process piping; this would restrict the flow in the line. The pipe is enlarged to overcome this or swaged up with a 2½-in. pipe elbow to accommodate the 1-in. elbow adapter connection for the thermowell. Thermowells with 6-in. insertion lengths or longer can thus be installed. Longer length thermowells required for long thermobulbs furnished with the instrument or process requirements can have the 2½-in. outlet of the pipe elbow extended with a 2½-in. pipe spool piece to accommodate these wells.

In Fig. 86, four typical piping arrangements are shown for installing thermowells. When the enlarged sections must be provided with a 2½-in. spool piece, the minimum length should be at least 12 in. Note that the Fig. 86 details are accessible so the thermowells can be taken out, inspected and reinstalled during plant shutdowns. When providing for an indicating dial thermometer, the swaged-up section should be located so that it may be seen from grade, a floor, or a platform. At times it may be necessary to install two thermowells together in the same pipe section for the same

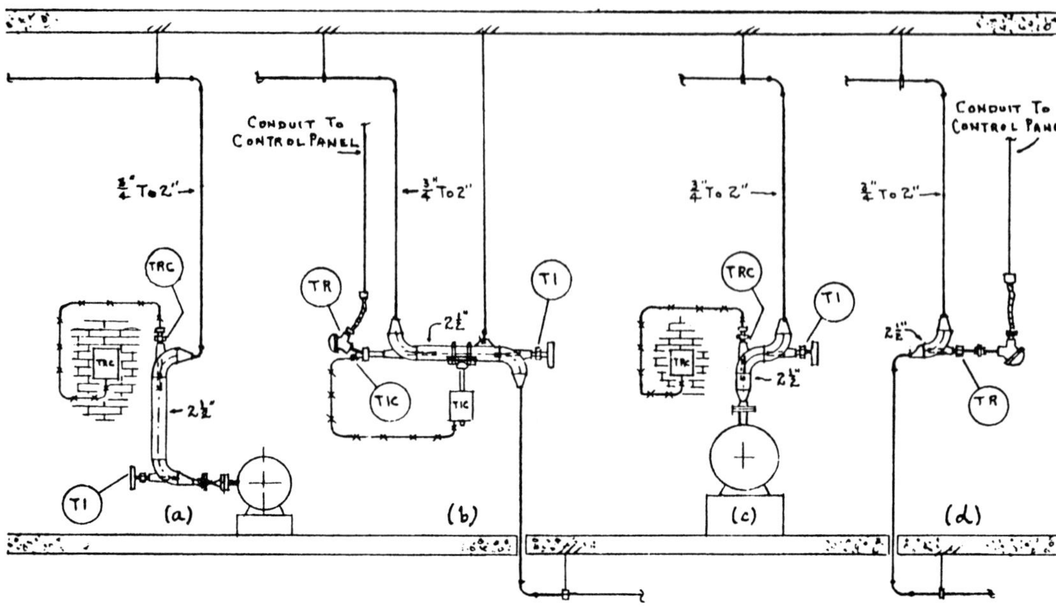

FIG. 86 Small process pipelines, $\frac{3}{4}$ to 2 in., swaged up to accommodate thermowells.

service. Such installations may require an indicating thermometer and another thermowell for a temperature transmitter thermobulb. The 2½-in. enlarged pipe elbow with the spool piece can provide for these duplex installations. These sections should be made up to enhance streamline flow. The fittings should be assembled to prevent the residue in the flowing product from accumulating inside the pipe and around the thermowell which would present a temperature lag to the operating instrument. The swaged up section could give the arrangement an awkward appearance, thus every effort should be made to blend this piping with the surrounding equipment.

In Fig. 86(a) this 2½-in. section is installed in the vertical, with a thermometer at the bottom and the thermobulb at the top. The vertically installed thermowell may be required if a thermal fluid is to be placed inside the well to increase the thermal transmission through the well to obtain a quicker temperature response to the instrument.

In Fig. 86(b) the 2½-in. section is installed with two 1-in. elbow adapters in the horizontal with one connection for the indicating thermometer and the other for a duplex thermowell containing a thermocouple and a ⅜-in. diameter thermobulb for a temperature transmitter. If vibration is likely to be transmitted in the piping, the transmitter should not be supported from the process piping but instead mounted on a building column, wall, or from a floor pedestal.

In Fig. 86(c), two 2½-in. pipe elbows are installed together with a pipe spool piece at the lower end to increase its length for the installation of a thermowell with a thermobulb. The other elbow adapter holds a dial thermometer in a 6-in. long thermowell. In Fig. 86(d) a single 2½-in. pipe elbow is installed with suitable pipe reducers to fit the small size process line. The 1-in. elbow adapter holds a 6-in.-long well with a thermocouple (T/C). The T/C head is connected with an electrical conduit that carries the circuit to the temperature recorder on the control panel.

Thermowell Pipe Elbow Connections

Most thermowells for process work will require a 1-in. pipe thread connection for its installation in process piping. In some engineering offices a 1-in. 3000 lb or 6000 lb forged steel screwed end (F.S.S.E.) pipe coupling is specified for the installation in a pipe welding elbow. This practice continues from the time when there was nothing better. The designer would find, if he checked, that the pipe coupling cannot be installed in pipe elbows as shown in Fig. 87(a). Designers new to the work are often under the erroneous impression that this makes a cheaper installation.

Piping fabricators are well aware of the problem, and will instead provide a steel boss 1½-in. in diameter by 4 in. long, drilled and tapped for a 1-in. pipe thread. This boss is welded into the pipe elbow as shown in Fig. 87(b). It is a more expensive operation than using an elbow adapter.

The best and cheapest installation for a 1-in. thermowell is to weld a 1-in. 6000 lb F.S.S.E. elbow adapter in the heel of a welding pipe elbow as shown in Fig. 87(c). The elbow adapter is shaped for welding and gives a streamline flow service and appearance. It will prevent unnecessary flow turbulence in comparison to a boss installation, mentioned above. During some refinery inspections it has been noticed that when bosses are installed in pipe elbows for mounting thermowells on hot oil services, considerable erosion has actually taken place inside the elbows in the welding area because of flow turbulence. It is therefore recommended that for all threaded-type thermowells installed in piping, whether ¾, 1, or 1¼ in., an elbow adapter rated at 6000 lb (F.S.S.E.) should be welded on all pipelines requiring a thermowell connection.

Fittings for Thermowell Connections

The piping specifications for a refinery or petrochemical plant will often give the allowable sizes permitted for screwed pipe branch connections. For usual piping services, the 2000-lb or 3000-lb F.S.S.E. fitting will be specified.

Thermowell requirements differ from the screwed pipe branch connection in that the threaded thermowell must be removed occasionally for inspection, especially during plant shutdown, and then replaced. This will require perfect pipe threads for the thermowell connections. Good threads can be assured if the ¾, 1, or 1¼-in. F.S.S.E.-type fitting rated at 6000 lb, such as a thread adapter, elbow adapter, or pipe coupling, is welded in the process piping or equipment.

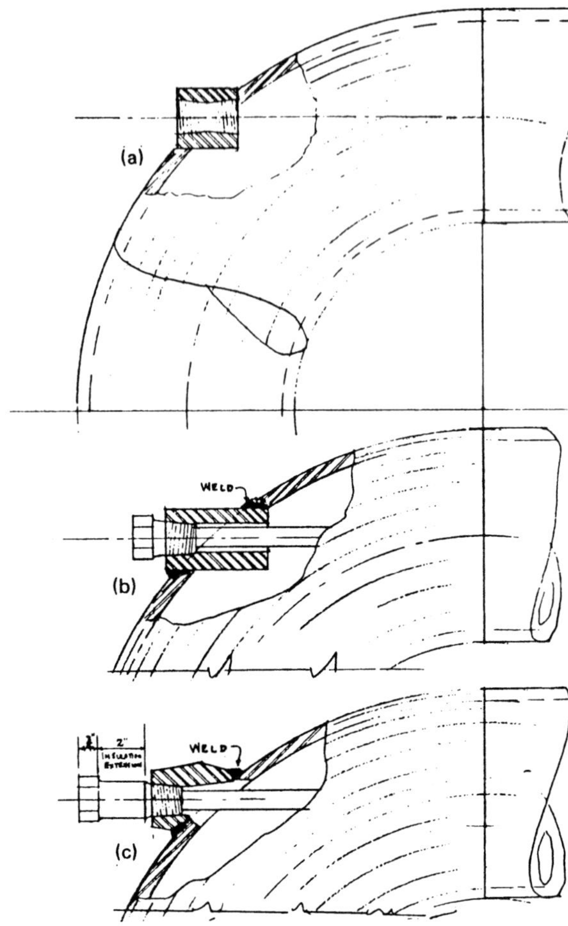

FIG. 87 Three types of 1-in. connections for thermowells installed on welding pipe elbows. (a) Regular pipe coupling that should not be used. (b) Improvised connection made with a boss. (c) 6000 lb forged steel elbow adapter provides the best thermowell connection at the lowest installed cost.

Pipe Thread Size for Thermowells

Thermowells can usually be obtained in pipe thread sizes of ½ in. as a special type, while ¾ and 1 in. are the usual available stock sizes and the 1¼ inch is a special size.

Thermowells in the ½-in. size are used where space on the equipment will only allow a ½-in. connection to be made for either a thermometer or thermocouple assembly.

The ¾-in. thermowell is generally selected where process operating service pressures and temperatures are in the low range. It is used for mounting a thermometer or thermobulb in a thermowell and for thermocouple assemblies. This size is a few dollars cheaper than the next larger one.

When a ¾-in. thermowell is to be used, the pipe designer is cautioned not to use the conventional ¾-in. full pipe coupling for mounting it. If the ¾-in. full coupling is welded on equipment, the narrowness of the inside of the coupling where the two pipe threads meet may prevent thermowell insertion. If, instead, a ¾-in. 6000-lb thread adapter or elbow adapter were welded on the piping, a better installation would result.

The 1-in. size thermowell is selected for most process services. A typical selection of thermowells is shown in Fig. 88. They should be used when the 1-in. pipe thread connection conforms with the piping specifications for the project. Sometimes they can be used on process services that operate at fairly high pressures and temperatures, as high as 750°F. The 1-in. connection for these thermowells should be made with an F.S.S.E.-type 6000-lb fitting. When process pressures are very high, approaching the superpressure range, the threaded connection should not be used. Specifications for very high pressures usually call for a flanged nozzle connection with a special type of ball ground joint for the thermowell. In some refineries the 1-in. threaded thermowell is used only on steam pressures 400°F and lower and on water and air services. For hydrocarbon services a ball ground joint thermowell is often used. Fits in a flanged nozzle made to fit the joint are often used. There are other special adapter-type thermowells used by refineries and petrochemical plants.

Petrochemical units have been designed by using piping specifications which allow for the installation of 1-in. pipe connections for all the 1-in. threaded-type thermowells used throughout the unit. This applies to the process piping and the 1-in. pipe thread connection that is also specified for use on vessels and drums for the threaded-type thermowell. Normally flange-type nozzle connections are specified for vessels including those used for thermowells. These processes usually operate in the low pressure and temperature range. The products are noncorrosive, so threaded thermowells are acceptable.

The 1¼-in. thermowell can, because of its large size thread, be provided with a thicker tapered-wall construction which starts at the bottom of the threads; it is shown in Fig. 88(a). It is used for higher pressure, temperature, and velocity steam services because of its sturdy construction.

Material for Thermowells

It is customary, when the piping and vessels are of carbon steel, to order the thermowells in stainless steel, AISI Type 304 or AISI Type 316, for greater protection. The AISI Type 316 is often specified for refinery and petrochemical plant services and preferred when process pressures and temperatures operate in the higher range.

The thermowell manufacturers usually provide a recommended-materials selection list or chart in their catalog. Such a list or chart can serve as a guide in selecting the right service material for various process fluids. Special material recommendations can be obtained from the thermowell manufacturer by specifying the service and fluid in which the thermowell will be

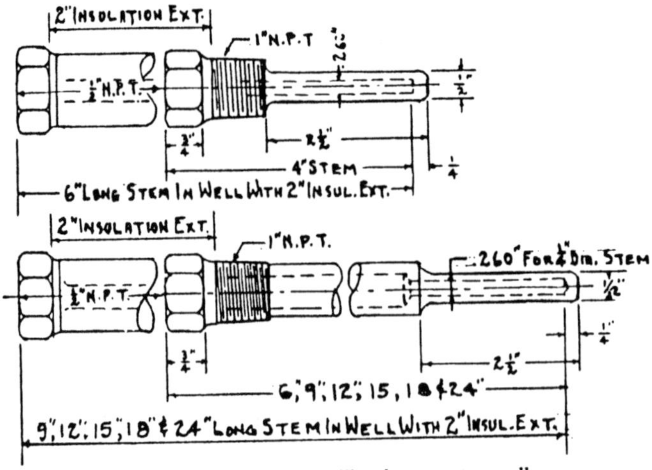

(a) Typical dial bimetallic thermometer well

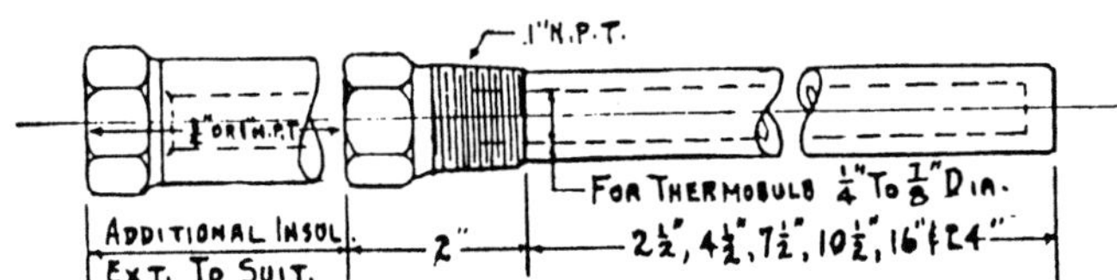

(b) Typical thermowells for thermobulbs with filled system

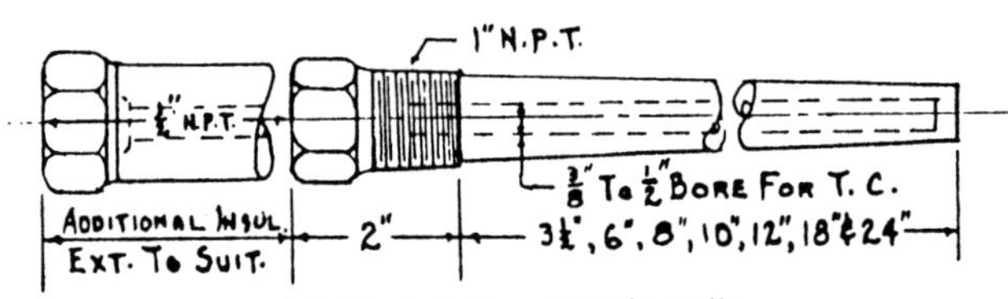

(c) Typical thermocouple wells

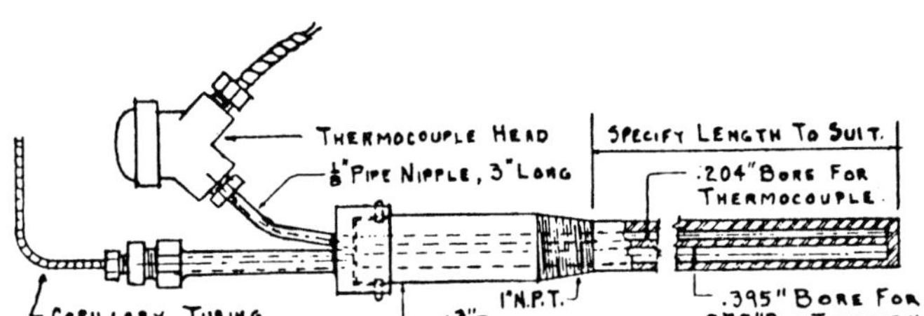

(d) Double bore well with bore for thermobulb and thermocouple including head

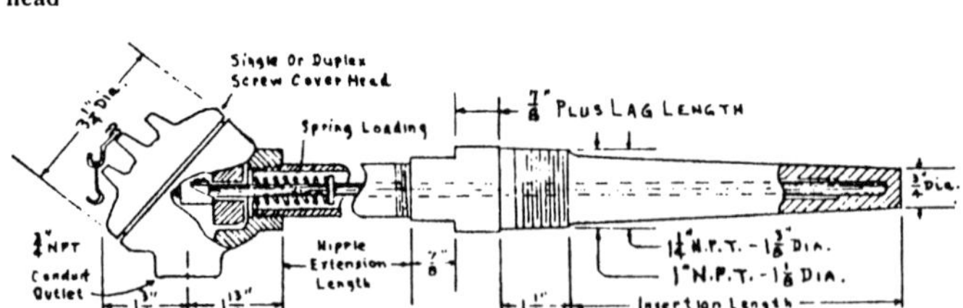

(e) Packed and spring loaded thermocouple in a threaded T/C thermowell, for high pressure, high velocity steam measurement

FIG. 88 Typical thermowell types for process piping.

submerged. It is not recommended that the threaded-type thermowell connection be made of the same material as the special alloy of the equipment. This could cause galling of the thermowell pipe threads. It is customary to provide a 1½-in. flange nozzle in the process piping and install a flange-type thermowell.

It is important when selecting the thermowell material that the composition will prevent an electrolytic action between the well, piping, or vessel.

Thermobulbs without Thermowells

The catalogs and service manuals for some of the instrument companies show recommendations for the installation of temperature transmitter thermobulbs being installed without thermowells similar to Fig. 89(a). A compression union fitting with gasket and packing gland permits the thermobulb to be inserted through this fitting and made pressure-tight at the capillary tubing. This type of installation should not be used as a general practice. It may sometimes be used where there is a noncorrosive process

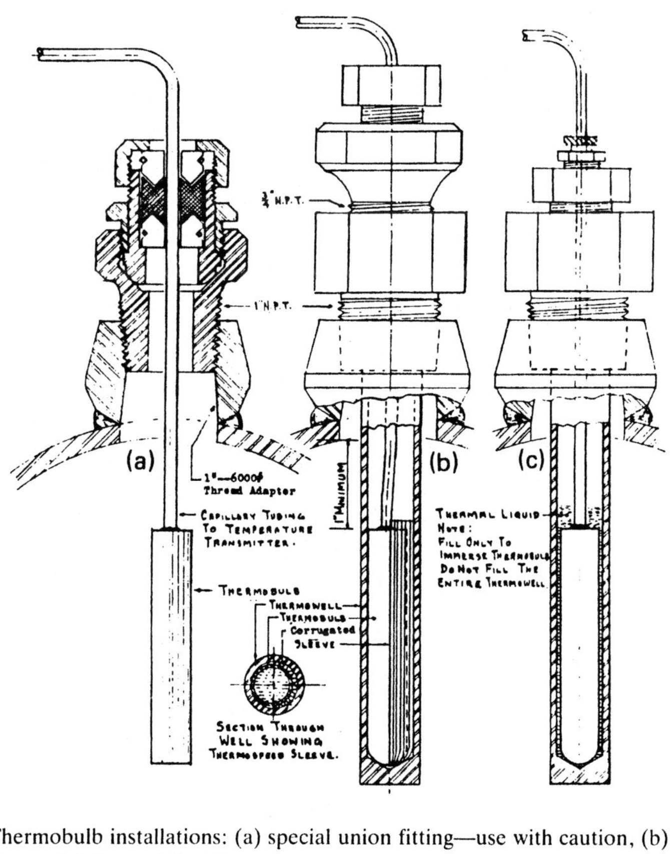

FIG. 89 Thermobulb installations: (a) special union fitting—use with caution, (b) corrugated sleeve, (c) thermal liquids.

service and low fluid flow to cause movement of the thermobulb suspended from the frail capillary tubing. The installation of thermobulbs in thermowells is shown in Figs. 89(b) and 89(c).

Reducing Lag

A thermowell is required to protect the thermal element of a thermometer or temperature transmitter thermobulb from corrosion and erosion, to give it adequate support, and to permit its removal without interrupting the process. The use of the thermowell will unavoidably introduce a temperature time lag to the changes in the process temperature and response to the temperature relayed to the instrument. This is caused by the transmission of heat through the thickness of the metal well and the inevitable dead air space between the well and bulb.

One manufacturer solved this problem by producing a corrugated sleeve shown on Fig. 89(b). This sleeve provides a metal-to-metal contact between the bulb and the thermowell by means of a very thin (0.005 in.) corrugated aluminum sleeve.

The sleeve forces the bulb to one side of the well, insuring a metal-to-metal contact throughout the length of the sensitive thermobulb. Two sleeves are sometimes used on both sides of the bulb. The corrugations in the sleeve itself provide a metallic path between the well and the bulb on the opposite side. The advantage of using a sleeve is that it can be applied to thermowells installed in a vertical, horizontal, or upside-down position.

The standard type thermowells manufactured and supplied are provided with the following bulb diameter-to-well bore relationship. The dimensions listed in catalogs are usually given as: a ¼-in. bulb fits into a 0.260-in. bore, a ⅜ in. into a 0.385 in., a ½ in. into a 0.510 in., a 9/16 in. into a 0.572 in., a ⅝ in. into a 0.635 in., an 11/16 in. into a 0.707 in., a ¾ in. into a 0.760 in., and a ⅞ in. into a 0.885 in.

When thermowells are installed in a vertical position in piping and equipment, various substances are used to increase the heat transmission and temperature response—one is mercury (see Fig. 89c). Although mercury is probably the best substance from a thermal consideration, it must not be used when the process operating temperature approaches its boiling point (375°C/674°F). Both the thermowell and bulb should be made of steel or ferrous alloy. A brass thermowell or bulb immersed in mercury would be destroyed by amalgamation.

A mixture of oil and graphite is better than pure graphite for use at higher operating temperatures. Other substances used in vertical thermowells are glycerine, napthalene, oils and various types of greases, and proprietary heat transfer fluids.

Thermowells in a horizontal position can be filled with graphite, carbon, metallic dust (such as copper, aluminum, or clean iron filings), or a corrugated aluminum sheath to reduce the insulating properties of the air gap. Solder is sometimes used between the thermobulb and the thermowell (tin, 420°F; lead, 600°F).

Pipe Flange Used for Thermocouple Well Assembly

Special types of thermocouple (T/C) wells can often be designed and installed in process piping to better advantage than the conventional cantilever-type thermowell. One such arrangement, shown in Fig. 90, uses an orifice flange for mounting the T/C well. This T/C well is ½ in. in diameter. Half of the extended length is a solid extension which fits into the opposite flange orifice opening that has been counterbored to 0.510-in. diameter. The solid end of the T/C well holds the entire well securely in place regardless of the flow velocity. The slight expansion that may exist between the parts will slide in the flange openings.

The active other half of the T/C well is drilled to hold either a ⅛ or ¼-in. o.d. pencil-type T/C assembly. The hot junction extends into the center of the process pipe flow area. This T/C assembly is shown in Fig. 90 and is connected to the T/C head. The entire well and assembly must be designed special for each flange and ordered with the drawing from the thermowell manufacturer.

If an orifice flange is not available for this purpose, the regular pipe welding neck flange can be drilled, bored, and taped as shown in Fig. 90, providing the flange is at least 1⅛-in. thick—preferably thicker.

Supporting the Ends of Long Thermowells

When temperature control instruments are provided with very long thermobulbs, they must be installed in suitable thermowells. An example of this is shown in Fig. 91. Thermowells 15 to 24 in. long, if installed and held in place only by the pipe thread connection, could cause the well to vibrate with the flow velocity. Eventually the well could bend and possibly fracture.

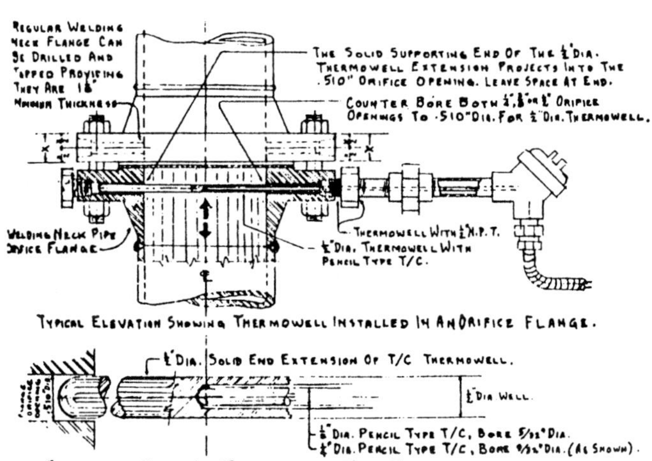

FIG. 90 Orifice flange adapted for installing a T/C well and assembly.

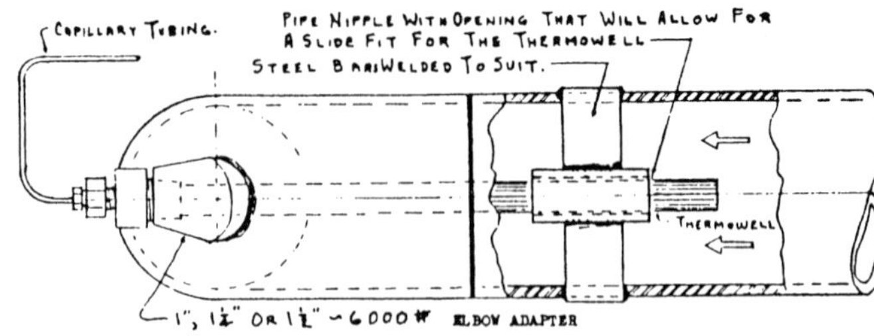

Plan of pipe elbow with internal support for thermowell

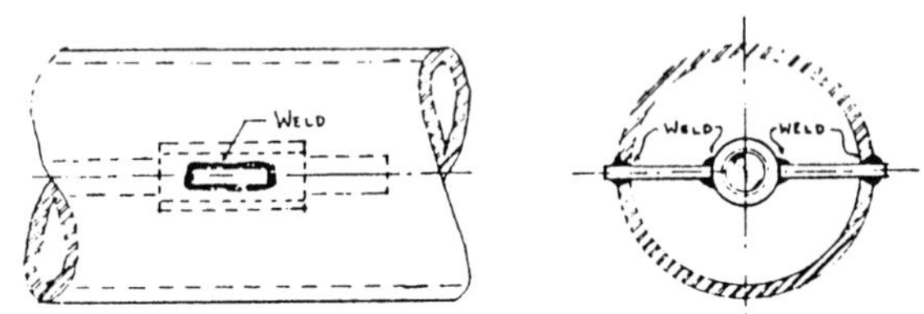

Elevation and section of process pipe showing details for installing internal support for thermowell.

FIG. 91 An internal pipe-supporting bracket for a long thermowell.

Some self-acting types of temperature controllers often require long thermowells. When they are installed in piping, they should be provided with some internal means of supporting and stabilizing the ends. In Fig. 91, such a supporting arrangement is shown.

Protecting Thermowell with Deflector

In some process pipelines the flowing product may carry a mixture of entrained solids. Such an installation is shown in Fig. 92, which also requires a long thermowell. This process line will require a 4-in. flange nozzle made at the outlet of the main pipeline tee, on which the welding neck flange is installed.

A 4-in. blind flange is drilled and tapped for the 1-in. thermowell. To protect the thermowell from the flow velocity, a 2 × 2 × ⅜-in. angle iron flow deflector is welded to the bottom of the flange. This angle iron flow deflector is installed ahead of the well and in this way takes the shock and diverts the flowing product in the pipe, thus protecting the thermowell from bending and from erosion damage. The blind flange supporting the thermowell provides an easy arrangement for inspection of the inside of the process pipeline. If a leak occurs in service, a plug or valve may be installed as shown in Fig. 93.

Backwelding Thermowells

Threaded-type thermowells, when installed on process piping for some services, may have to be backwelded. This is often done on high pressure

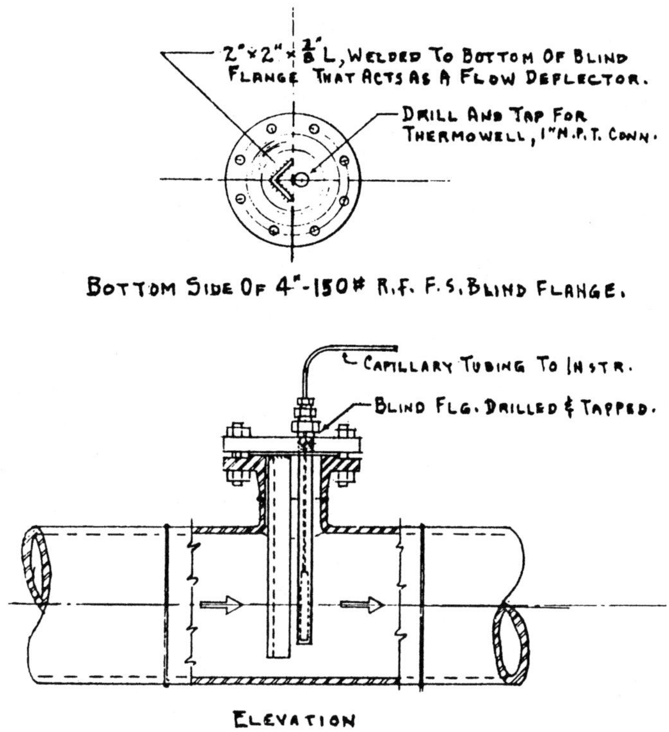

FIG. 92 Thermowell protected with an angle iron deflector.

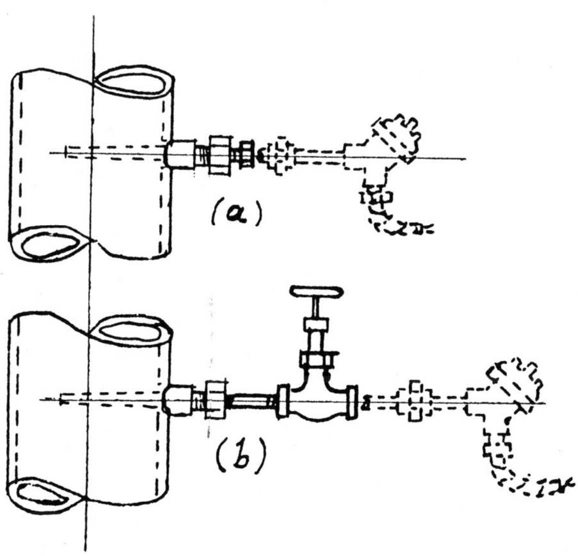

FIG. 93 (a) Pipe plug stops small leaks in a thermowell. (b) Use a shut-off valve and a pipe nipple for a large thermowell.

steam, hydrocarbon, acid, or caustic lines, and on other service pipelines carrying toxic products. Backwelding a thermowell at the threads will prevent its removal for inspection. When backwelding is necessary, it is a good practice to locate and install the thermowell connection and thermowell near pipeline flanges, preferably not more than two or three pipe diameters back of the face of the flange as shown in Fig. 94. This pipe flange becomes a means for inspecting the inside of the pipe and thermowell. In designing the piping, it will be necessary to have the adjoining pipe in the form of a spoolpiece between two pipe flanges to enable this section to be removed.

Pressure Gauge and Thermowell Connections

The process flow sheet will often show a pressure gauge and thermowell connection that must be installed in the same piping. It is considered good practice to install the pressure gauge connection ahead of the thermowell, as shown in Fig. 94. Because the thermowell will create a turbulence in the direction of the flow, this could have an effect on the pressure gauge connection.

On some critical processes requiring alloy steel piping on which a pressure gauge connection must be provided, this too is located near a pipe flange in the line, about two pipe diameters away from the face of the flange. This puts the connection where it can be examined and inspected from the inside of the pipe for possible erosion.

Thermowell Insertion Lengths

Either of two methods can be used to provide for thermowells in the design of process piping. In one the process piping is designed, the necessary 1-in. thermowell connections are provided in the piping or for the enlarged pipe

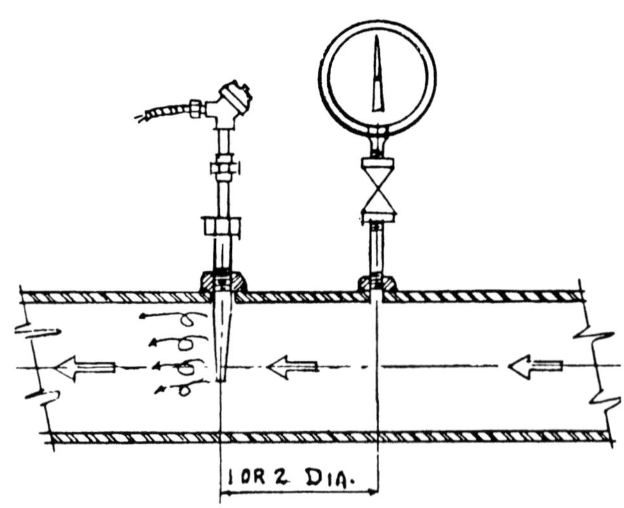

FIG. 94 Locate a thermowell downstream of a pressure gauge connection.

sections wherever they are required, and later the thermowell insertion lengths are selected from the completed drawing to suit each connection, depending upon the piping and configuration.

In the other method, thermowell insertion lengths are decided in advance for all thermowells, including dial thermometers, thermobulbs, resistance bulbs, and thermocouples. The process piping design must then be made to accommodate the assigned thermowell insertion lengths. Thermowells so handled usually have 9, 12, 15-in., or longer insertion lengths. One advantage is that complete immersion will be assured for even the longer thermowells.

The other advantage of this method is that it will also reduce the number of types of thermowells that have to be ordered. The types that must be carried in stock for future maintenance work are kept to a minimum. When the piping designer knows in advance the thermowell insertion lengths preferred, he will be in a better position to provide for their installation in the process piping. Time will be saved and he will be able to complete his drawing sooner.

In Figs. 95–108 the many piping details shown will suggest to the piping designer the various schemes that can be used. The importance of these details is that each has been provided with a scale in inches shown on the centerline run of the thermowell. This scale will enable the instrumentation engineer to select a thermowell. With it he can pinpoint the active effective flow area, preferably in the turbulence zone, where the thermal element of the instrument will be located.

When the thermowell is installed perpendicular or at a 45° angle to the pipe wall, the thermowell tip should be at or near the centerline of the pipe. Thermowells and their connections should point counter to the flow in the process line whether installed on pipe elbows or on a 45° angle. The well perpendicular to the pipe will naturally be in the turbulent flow section of the pipe. Such an arrangement will not always be possible because of the piping arrangement and direction of the process flow.

Because the accessories attached to the thermowell may protrude, they must be so installed on the piping that they will be parallel with walkways and platforms. If not, a workman carrying tools on his shoulder could accidently knock off the head of a dial thermometer. An operator in an emergency rush could be injured by a T/C head sticking out in the aisle. Capillary tubing should not dangle from the well but should be attached and supported with 1 × 1 × ⅛-in. angle iron or placed inside conduit.

In Tables 13–16, lists are given of the various types of thermowells that apply to the details in Figs. 95–108. These are recommended for use for dial thermometers, temperature transmitter thermobulbs, and thermocouples. The proper insertion lengths for special thermowells can be determined from the scale shown on each of the figures. In this way, other thermowell lengths can also be verified for their insertion and immersion.

Piping Details

The long radius welding pipe bend is shown in Fig. 96. The point of maximum flow turbulence is at the end of the elbow where the weld line occurs. Every

TABLE 13 Thermowells with 1-in. Pipe Thread, Giving Immersion and Ordering Lengths

Fig. 95: 1-in., 6000 lb elbow adapter installed on $2\frac{1}{2}$ to 12″ welding pipe elbow for thermowell
Fig. 96: 1-in., 6000 lb elbow adapter installed at welding line of 2 to 10″ pipe elbow for thermowell
Fig. 97: 1-in., 6000 lb elbow adapter installed on 2″ reducing elbow swaged section for 1 and $1\frac{1}{2}$″ process lines
Fig. 98: 1-in., 6000 lb elbow adapter installed on $2\frac{1}{2}$″ elbow swaged up section in $\frac{3}{4}$ to 2″ process lines

Figure Number of Pipe Detail Followed by Reference to the Specific Nominal Pipe Size (in.)	Indicating Dial Thermometer With $\frac{1}{4}$″ Diameter Bimetallic Stem: Dial Thermometer with $\frac{1}{2}$″ Pipe Thread. Stem Ordering Lengths Given Below (in.)	Indicating Dial Thermometer With $\frac{1}{4}$″ Diameter Bimetallic Stem: Immersion Length of Thermowell. No Insulation. See Scale (in.)	Indicating Dial Thermometer With $\frac{1}{4}$″ Diameter Bimetallic Stem: Dial Thermometer With $\frac{1}{2}$″ Pipe Thread. Stem Ordering Lengths Given Below (in.)	Indicating Dial Thermometer With $\frac{1}{4}$″ Diameter Bimetallic Stem: Immersion Length of Thermowell. With 2″ Insulation Extension. See Scale (in.)	Thermowells for Thermobulbs. Immersion Length Starts with Dimension Given Below. Check with the Scale Shown on the Figure. Add the Length Required for the Thermobulb. See Scale for Determining Ordering Length (in.)	Thermocouple Well. Insertion and Ordering Length (in.)
Fig. 95—$2\frac{1}{2}$	6	$4\frac{1}{2}$	9	$5\frac{1}{2}$	5 plus bulb	6
Fig. 95—3	6	$4\frac{1}{2}$	9	$5\frac{1}{2}$	6 plus bulb	6
Fig. 95—4	9	$7\frac{1}{2}$	12	$8\frac{1}{2}$	8 plus bulb	8
Fig. 95—6	9	$7\frac{1}{2}$	15	$11\frac{1}{2}$	10 plus bulb	10
Fig. 95—8	12	$10\frac{1}{2}$	15	$11\frac{1}{2}$	13 plus bulb	12
Fig. 95—10	12	$10\frac{1}{2}$	15	$11\frac{1}{2}$	15 plus bulb	12
Fig. 95—12	12	$10\frac{1}{2}$	15	$11\frac{1}{2}$	18 plus bulb	12
Fig. 96—2	6	$4\frac{1}{2}$	9	$5\frac{1}{2}$	$2\frac{1}{2}$ plus bulb	$3\frac{1}{2}$
Fig. 96—$2\frac{1}{2}$	6	$4\frac{1}{2}$	9	$5\frac{1}{2}$	$4\frac{1}{2}$ plus bulb	6
Fig. 96—3	6	$4\frac{1}{2}$	9	$5\frac{1}{2}$	$4\frac{1}{2}$ plus bulb	6
Fig. 96—4	9	$7\frac{1}{2}$	12	$8\frac{1}{2}$	$4\frac{1}{2}$ plus bulb	8
Fig. 96—6	12	$10\frac{1}{2}$	15	$11\frac{1}{2}$	$7\frac{1}{2}$ plus bulb	10
Fig. 96—8	12	$10\frac{1}{2}$	15	$11\frac{1}{2}$	$7\frac{1}{2}$ plus bulb	12
Fig. 96—10	12	$10\frac{1}{2}$	18	$14\frac{1}{2}$	$10\frac{1}{2}$ plus bulb	12
Fig. 97	6	$4\frac{1}{2}$	—	—	—	$4\frac{1}{2}$
Fig. 97	6	$4\frac{1}{2}$	9	$5\frac{1}{2}$	4 plus bulb	6
Fig. 98	6	$4\frac{1}{2}$	9	$5\frac{1}{2}$	4 plus bulb	6
Fig. 98	6	$4\frac{1}{2}$	9	$5\frac{1}{2}$	4 plus bulb	6

TABLE 14 Thermowells with 1-in. Pipe Thread, Giving Immersion and Ordering Lengths

Fig. 99: 1-in., 6000 lb elbow adapter installed on 2, $2\frac{1}{2}$, 3, and 4″—45° pipe offsets
Fig. 100: 1-in., 6000 lb thread adapter installed perpendicular on 2 to 12″ process pipe lines

Figure Number of Pipe Detail Followed by Reference to the Specific Nominal Pipe Size (in.)	Indicating Dial Thermometer With $\frac{1}{4}$″ Diameter Bimetallic Stem				Thermowells for Thermobulbs. Immersion Length Starts with Dimension Given Below. Check with the Scale Shown on the Figure. Add the Length Required for the Thermobulb. See Scale for Determining Ordering Length (in.)	Thermocouple Well. Insertion and Ordering Length (in.)
	Dial Thermometer with $\frac{1}{2}$″ Pipe Thread. Stem Ordering Lengths Given Below (in.)	Immersion Length of Thermowell. No Insulation. See Scale (in.)	Dial Thermometer With $\frac{1}{2}$″ Pipe Thread. Stem Ordering Lengths Given Below (in.)	Immersion Length of Thermowell. With 2″ Insulation Extension. See Scale (in.)		
Fig. 99a—2	6	$4\frac{1}{2}$	9	$5\frac{1}{2}$	—	$4\frac{1}{2}$
Fig. 99b—$2\frac{1}{2}$	6	$4\frac{1}{2}$	9	$5\frac{1}{2}$	4 plus bulb	$4\frac{1}{2}$
Fig. 99c—3	9	$7\frac{1}{2}$	9	$5\frac{1}{2}$	4 plus bulb	6
Fig. 99d—4	9	$7\frac{1}{2}$	12	$8\frac{1}{2}$	4 plus bulb	6
Fig. 99e—2	6	$4\frac{1}{2}$	9	$5\frac{1}{2}$	—	6
Fig. 99f—$2\frac{1}{2}$	6	$4\frac{1}{2}$	9	$5\frac{1}{2}$	5 plus bulb	6
Fig. 99g—3	9	$7\frac{1}{2}$	9	$5\frac{1}{2}$	5 plus bulb	8
Fig. 99h—4	12	$10\frac{1}{2}$	12	$8\frac{1}{2}$	6 plus bulb	8
Fig. 100—2	4	$1\frac{1}{2}$[a]	6	$1\frac{1}{2}$[a]	—	—
Fig. 100—$2\frac{1}{2}$	4	$1\frac{1}{2}$[a]	6	$1\frac{1}{2}$[a]	—	—
Fig. 100—3	4	$1\frac{1}{2}$[a]	6	$1\frac{1}{2}$[a]	—	$3\frac{1}{2}$[b]
Fig. 100—4	6	$4\frac{1}{2}$	6	$1\frac{1}{2}$[a]	—	$3\frac{1}{2}$[b]
Fig. 100—6	6	$4\frac{1}{2}$	9	$5\frac{1}{2}$	2 to 6 for bulb	$4\frac{1}{2}$
Fig. 100—8	6	$4\frac{1}{2}$	9	$5\frac{1}{2}$	2 to 8 for bulb	6
Fig. 100—10	6	$4\frac{1}{2}$	9	$5\frac{1}{2}$	2 to 10 for bulb	6
Fig. 100—12	9	$7\frac{1}{2}$	9	$5\frac{1}{2}$	2 to 12 for bulb	8

[a]The $2\frac{1}{2}$″ sensitive element of the bimetallic thermometer is immersed only 1 and $1\frac{1}{2}$″ in the flowing stream within the pipeline. Use when indicating temperature is of secondary importance.

[b]Select another piping arrangement so a longer thermowell with more immersion can be installed.

TABLE 15 Thermowells with 1-in. Pipe Thread, Giving Immersion and Ordering Lengths

Fig. 101: 1-in., 6000 lb elbow adapter installed 45° on wall of 2½ to 12″ pipe
Fig. 102: 1-in., 6000 lb elbow adapter installed on 3″ swaged up pipe section for smaller process lines
Fig. 103: 1-in., 6000 lb elbow adapter installed on 4″ swaged up pipe section for smaller process lines
Fig. 104: 1-in., 6000 lb thread adapter installed in back of 2½, 3, and 4″ welding pipe tee
Fig. 105: 1-in., 6000 lb thread adapter or 1″, 6000 lb half coupling installed on side outlet of 1½ to 4″ pipe tees

Figure Number of Pipe Detail Followed by Reference to the Specific Nominal Pipe Size (in.)	Indicating Dial Thermometer with ¼″ Diameter Bimetallic Stem				Thermowells for Thermobulbs. Immersion Length Starts with Dimension Given Below.[a] Check with the Scale Shown on the Figure. Add the Length Required for the Thermobulb. See Scale for Determining Ordering Length (in.)	Thermocouple Well. Insertion and Ordering Length (in.)
	Dial Thermometer with ½″ Pipe Thread. Stem Ordering Lengths Given Below (in.)	Immersion Length of Thermowell. No Insulation. See Scale (in.)	Dial Thermometer with ½″ Pipe Thread. Stem Ordering Lengths Given Below (in.)	Immersion Length of Thermowell. With 2″ Insulation Extension. See Scale (in.)		
Fig. 101—2½	4	1	6	5½	—	—
Fig. 101—3	6	4	9	5½	2 to 4½ for bulb	4½
Fig. 101—4	6	4½	9	5½	2 to 6 for bulb	4½
Fig. 101—6	9	7½	9	8½	2 to 7½ for bulb	6
Fig. 101—8	9	7½	12	8½	2 to 10½ for bulb	8
Fig. 101—10	12	10½	12	11½	2 to 16 for bulb	10
Fig. 101—12	12	10½	15	5½	2 to 16 for bulb	10
Fig. 102a, b—3	6	4½	9	1½	2 to 4½ for bulb	4½
Fig. 102c—3	4	1½	6	5½	—	4½
Fig. 103a, b, c—4	6	4½	9	8½	2 to 6 for bulb	4½
Fig. 103—2½	9	7½	12	8½	7 plus bulb	8
Fig. 104—3	12	10½	15	11½	8 plus bulb	8
Fig. 104—4	12	10½	15	11½	9 plus bulb	10
Fig. 105—1½	9	7½	12	8½	7 plus bulb	8
Fig. 105—2	9	7½	12	8½	7 plus bulb	8
Fig. 105—2½	9	7½	12	8½	8 plus bulb	12
Fig. 105—3	12	10½	15	11½	9 plus bulb	12
Fig. 105—4	15	13½	18	14½	11 plus bulb	12

[a]The 2¼″ sensitive element of the bimetallic thermometer is immersed only 1½″ in the flowing stream within the pipeline. Use when indicating the temperature is of secondary importance.

TABLE 16 Thermowells with 1-in. Pipe Thread, Giving Immersion and Ordering Lengths

Fig. 106: 1″ thermowell installed in threaded pipe fittings and pipe flange fittings
Fig. 107: 1″ thermowell installed in piping arrangement made up with threaded fittings and pipe lengths to suit requirements

Figure Number of Pipe Detail Followed by Reference to the Specific Nominal Pipe Size (in.)	Indicating Dial Thermometer With ¼″ Diameter Bimetallic Stem				Thermowells for Thermobulbs. Immersion Length Starts with Dimension Given Below. Check with the Scale Shown on the Figure. Add the Length Required for the Thermobulb. See Scale for Determining Ordering Length (in.)	Thermocouple Well. Insertion and Ordering Length (in.)
	Dial Thermometer with ½″ Pipe Thread. Stem Ordering Lengths Given Below (in.)	Immersion Length of Thermowell. No Insulation. See Scale (in.)	Dial Thermometer with ½″ Pipe Thread. Stem Ordering Lengths Given Below (in.)	Immersion Length of Thermowell. With 2″ Insulation Extension. See Scale (in.)		
Fig. 106a—1½	6	4½	9	5½	3 plus bulb	4½
Fig. 106b—2	6	4½	9	5½	4 plus bulb	6
Fig. 106b—2½	6	4½	9	5½	4 plus bulb	6
Fig. 106c—1½	9	7½	12	8½	5 plus bulb	6
Fig. 106c—2	9	7½	12	8½	6 plus bulb	8
Fig. 106c—2½	12	10½	12	8½	7 plus bulb	8
Fig. 106c—3	12	10½	15	11½	8 plus bulb	10
Fig. 107	As noted above					

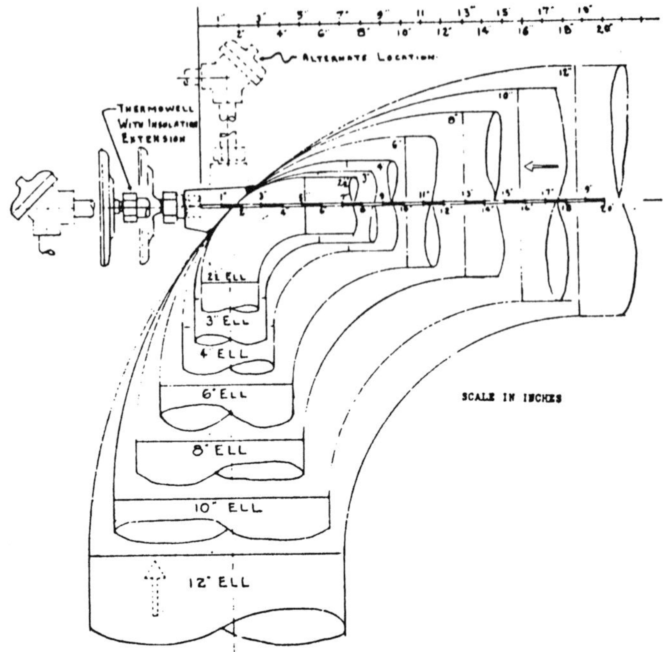

FIG. 95 Thermowell connections made with 1-in., 6000 lb elbow adapter installed on $2\frac{1}{2}$ to 12-in. pipe weld elbows.

effort should be made to use the quarter pipe bend or welding pipe elbow for locating thermowells and their connections. In Fig. 97 the welding pipe elbow has a thermowell 1-in. elbow adapter connection near the welding line. The point of flow turbulence is at the centerline. This arrangement is used when accessibility and observation of the dial thermometer make it desirable. In Fig. 98, two reducing-type quarter pipe bends are used to make up a 2-in. swaged-up pipe section in 1½-in. or smaller process line for the installation of thermowells.

In Fig. 99 the 2½-in. welding pipe elbows are used for swaging-up a section of the piping. This applies to ¾ to 2-in. process line thermowells. By installing a pipe spool piece between the pipe elbows and welding 1-in. elbow adapters on both elbows, two temperature instruments can be handled for the same service.

Figure 100 shows how vertical process pipelines can be provided with short offsets by welding two 45° pipe elbows and providing the 1-in. elbow adapter for the well connection. This arrangement is suitable for a dial thermometer or T/C well and assembly. When a thermowell with a long insertion thermobulb is required, a pipe spool piece can be welded between the 45° elbows. In this way the longest type thermowell can be installed with a minimum pressure drop with a good streamline flow in the process piping.

Thermowells can be installed perpendicular to the process pipeline (Fig. 101), but this is practical only on pipelines 4 in. and larger. When process

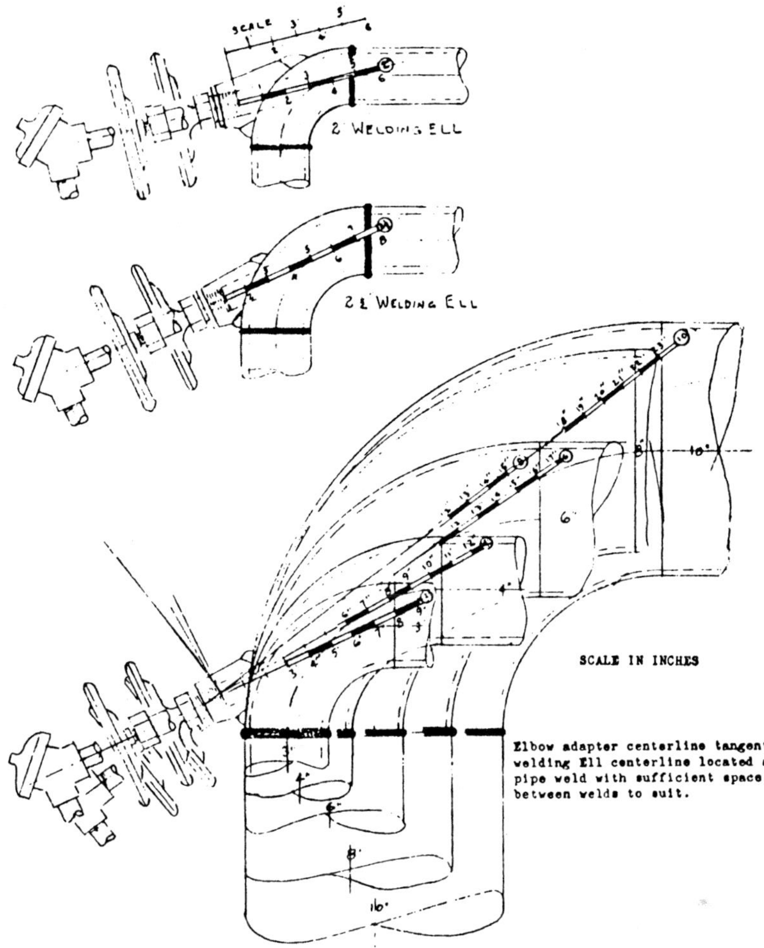

FIG. 96 Thermowell connections made with 1-in., 6000 lb elbow adapter installed on 2 to 10-in. welding elbow.

lines are smaller, it is necessary to swage-up the thermowell section to 4 in. The 1-in. thread adapter connection is welded on the piping. See Fig. 108 for other arrangements.

Installing thermowells on a 45° angle by welding on a 1-in. elbow adapter as a lateral is a good arrangement when it is desirable to place a thermometer where it can be easily seen (Fig. 102). Sometimes this arrangement may provide a means of installing a thermowell that was intended to be perpendicular to the line but was too long by a fraction of an inch. The thermowell can thus be installed on a 45° angle and made to suit the installation.

When small-size process lines have to be swaged-up to 3 in. (Fig. 103), the arrangements shown and their adaptation in several positions can be used.

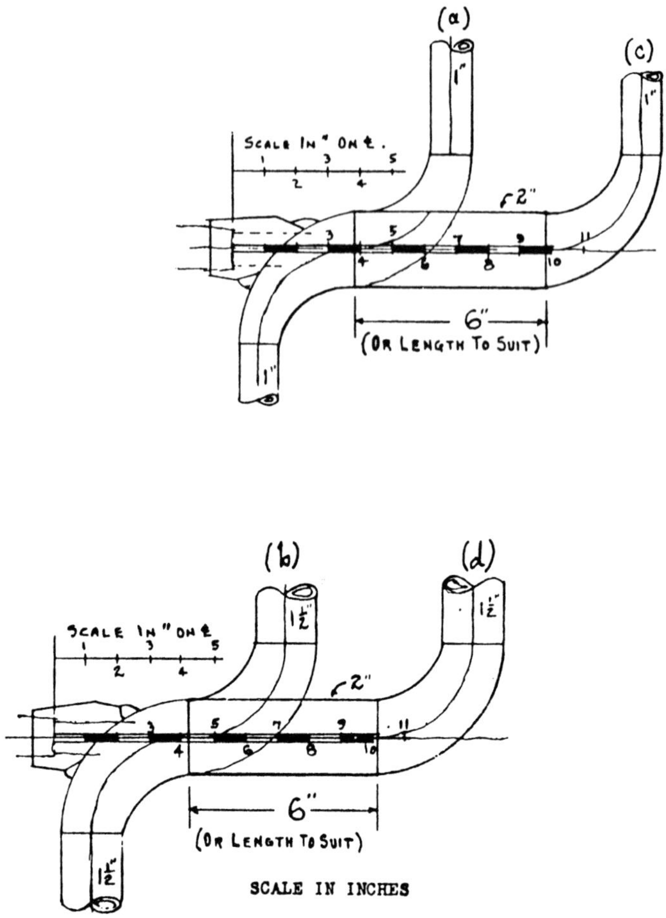

FIG. 97 Elbow adapter installed on a 2-in. reducing elbow for 1 and $1\frac{1}{2}$-in. process piping.

The 4-in. swaged-up pipe section with reducers in Fig. 104 can be connected into small-sized process pipelines where short thermowells can be installed for dial thermometers and T/C assemblies.

The back of a process pipe tee can occasionally be used for the installation of a thermowell (Fig. 105). This takes care of the branch of the process piping into which a thermowell has to be installed. In this arrangement the thermowell must project into the pipe far enough for the thermal element to be in the flow through the line.

Figure 106 shows an alternate arrangement for the pipe elbow. The pipe tee is arranged to allow a possible future branch connection. It is important that the thermowell insertion length project beyond the dead space and be in the flowing product of the line.

When screwed pipe and fittings are specified for use in process piping, the threaded tee and screwed fittings can be arranged with a reducing pipe

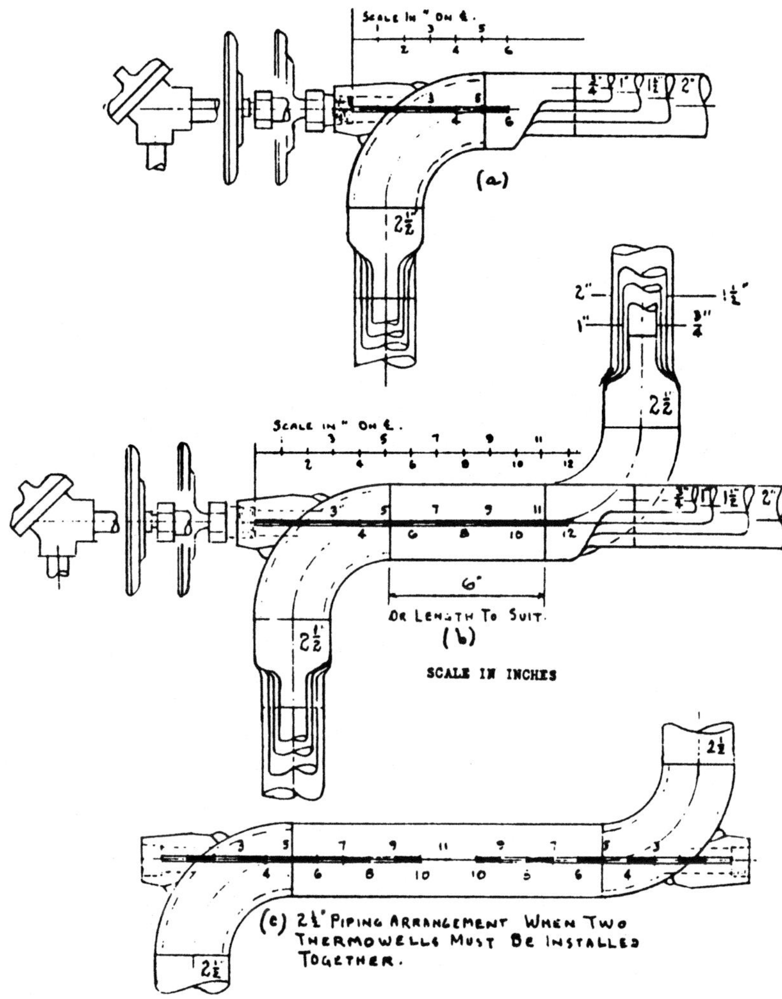

FIG. 98 Thermowell connection for ¾ to 2-in. piping swaged up to 2½-in. with 1-in., 6000 lb elbow adapter installed in elbow.

bushing for the thermowell connection (Fig. 107). These enlarged sections in the screwed piping should be big enough to assure unrestricted flow inside the line where the thermowell is installed.

When a straight run of piping with screwed fittings requires a thermowell, the run of the piping can be made with an offset as shown in Fig. 108 for the sake of the thermowell.

One way to improve response speed is to extend the insertion and immersion length of the thermowell and make sure the thermal element is at the inside tip of the well stem. Figure 108 shows six ways thermowells can be installed in piping. The best possible thermowell connection is the 1-in.

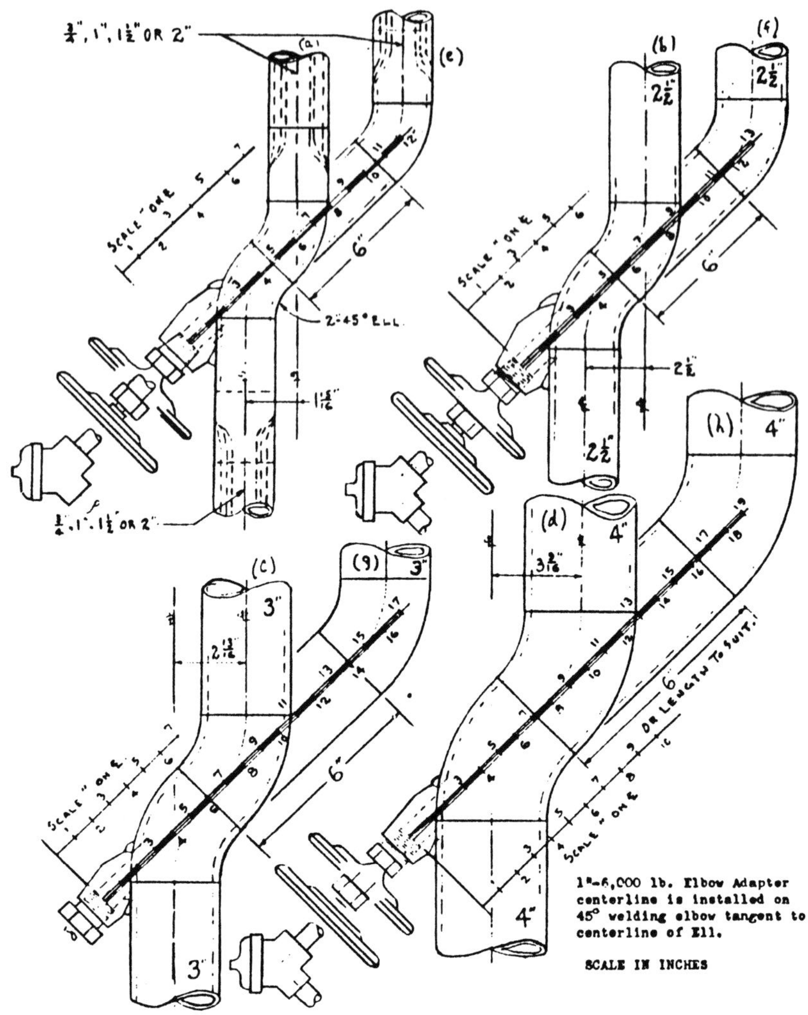

FIG. 99 Thermowell connection installed on 2, 2½, 3, and 4-in. piping, 45° offsets using 1-in., 6000 lb elbow adapters.

elbow adapter installed in the heel of the pipe elbow. It is shown in Fig. 108 in 3-in. (a), 4-in. (a), and 6-in. (a) sizes. The advantage of this arrangement is that it can provide for all thermowell lengths that may be required for most temperature instruments. A long-enough thermowell helps minimize heat conduction and radiation losses to the outside, because the thermal element is completely submerged and away from its mounting connection, so the overall speed of response is improved. In Fig. 108, the 6-in. (a) detail will naturally have a better speed of response than in the smaller 3-in. process pipeline.

The thermowell arrangements shown in Fig. 108 present other methods that can be used in making the 1-in. connections in process piping. To

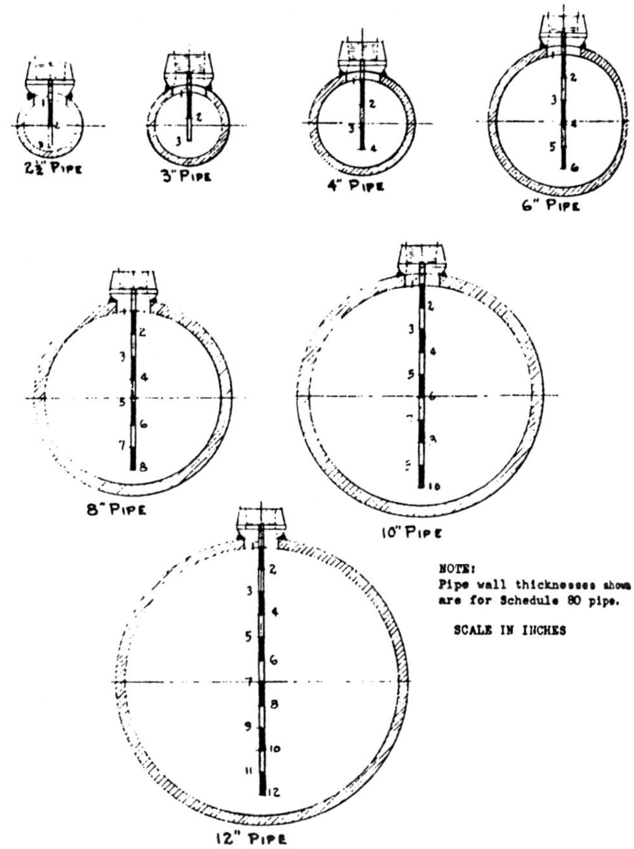

FIG. 100 Thermowell connections perpendicular to pipe wall.

evaluate each of the thermowell details shown, these are summarized and appraised in Table 17. This table presents an analysis showing the active, effective thermal lengths inside the pipe and how they compare in speed of response, and to what instruments they are best applied. It also shows that it is not practical to use certain well arrangements for temperature instruments.

Expansion Joints

Three proven means for absorbing pipe expansion and contraction are (1) expansion bends, (2) corrugated or bellows-type expansion joints, and (3) slip-type expansion joints.

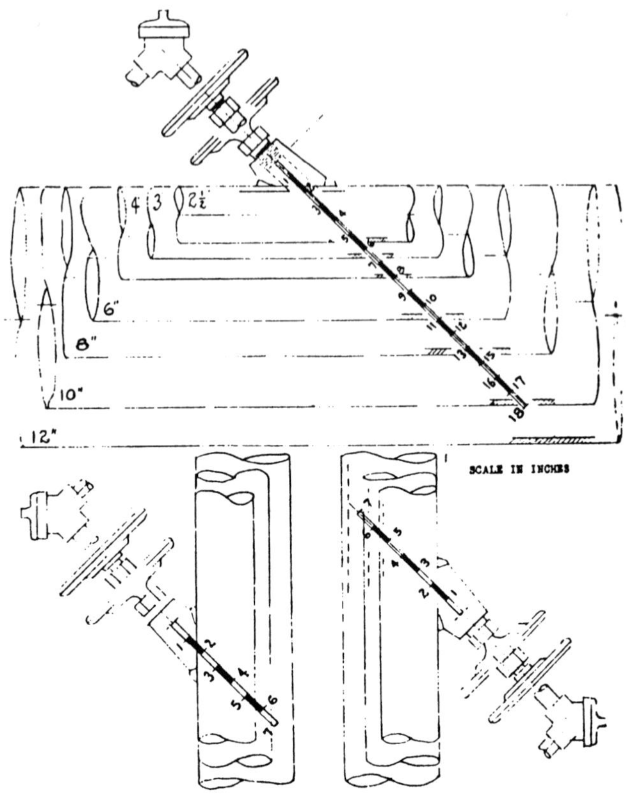

FIG. 101 Thermowell connection installed 45° to pipe wall on 2½ to 12-in. pipe using 1-in., 6000 lb elbow adapter.

Expansion Bends

The common types of expansion bends are either fabricated or built-up. Fabricated bends may be of simple-U, double-offset-U (Fig. 109), circle, or single-offset-U shapes. Creased bends, though once used to some extent, have become less popular in recent years. Built-up bends are made in a number of configurations from welding elbows.

All expansion bends, whether fabricated or not, require more installation space than either corrugated or slip-type expansion joints. This is often an important consideration in crowded refinery or petrochemical plant areas but can usually be overlooked on tank farms and other locations where a large amount of space is available.

The first cost of a fabricated expansion bend may exceed the first cost of a bellows or slip-type expansion joint. A built-up expansion bend usually costs less than a fabricated bend or either type of expansion joint, unless a great deal of labor is required to build the bend.

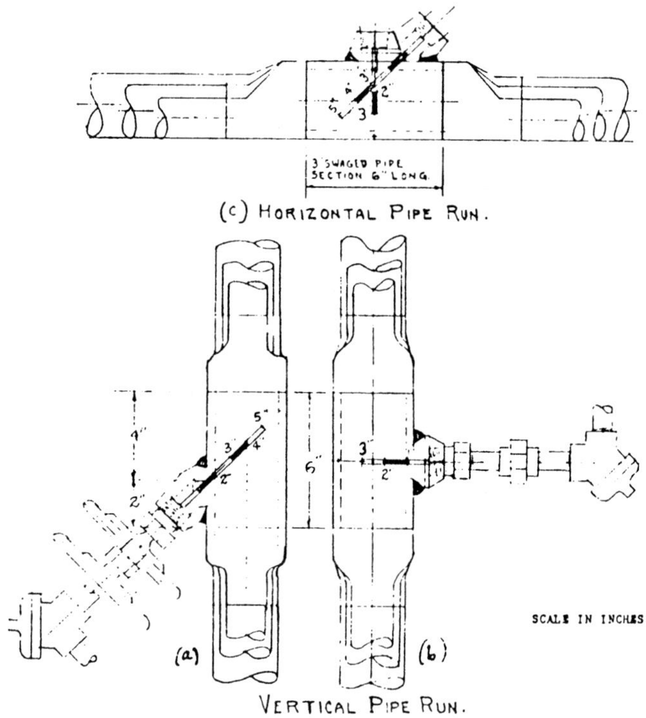

FIG. 102 Thermowell connection installed on 3-in. swaged up pipe section using 1-in., 6000 lb thread adapter for $1\frac{1}{2}$, 2, and $2\frac{1}{2}$-in. process piping.

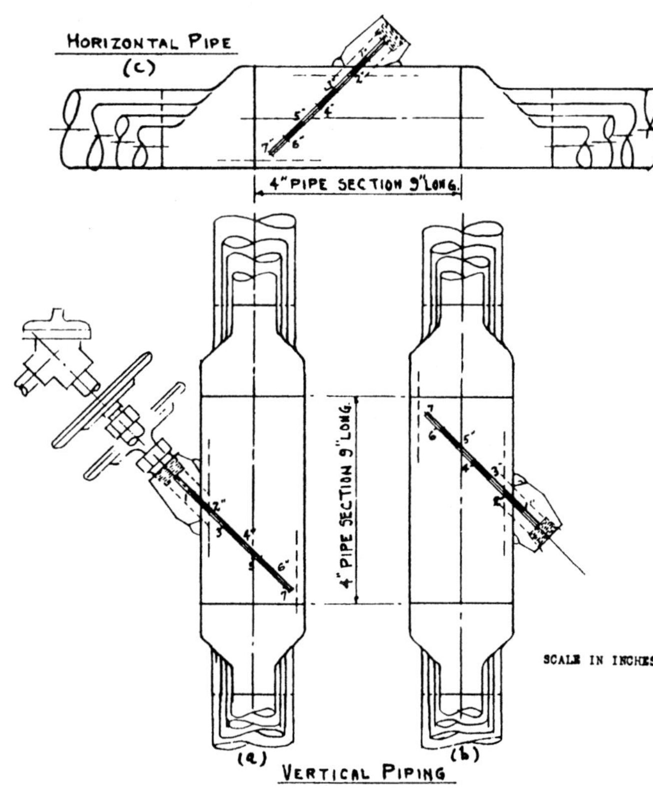

FIG. 103 Thermowell connection installed on a 4-in. swaged up pipe section using a 1-in., 6000 lb elbow adapter for $1\frac{1}{2}$ to 3-in. piping.

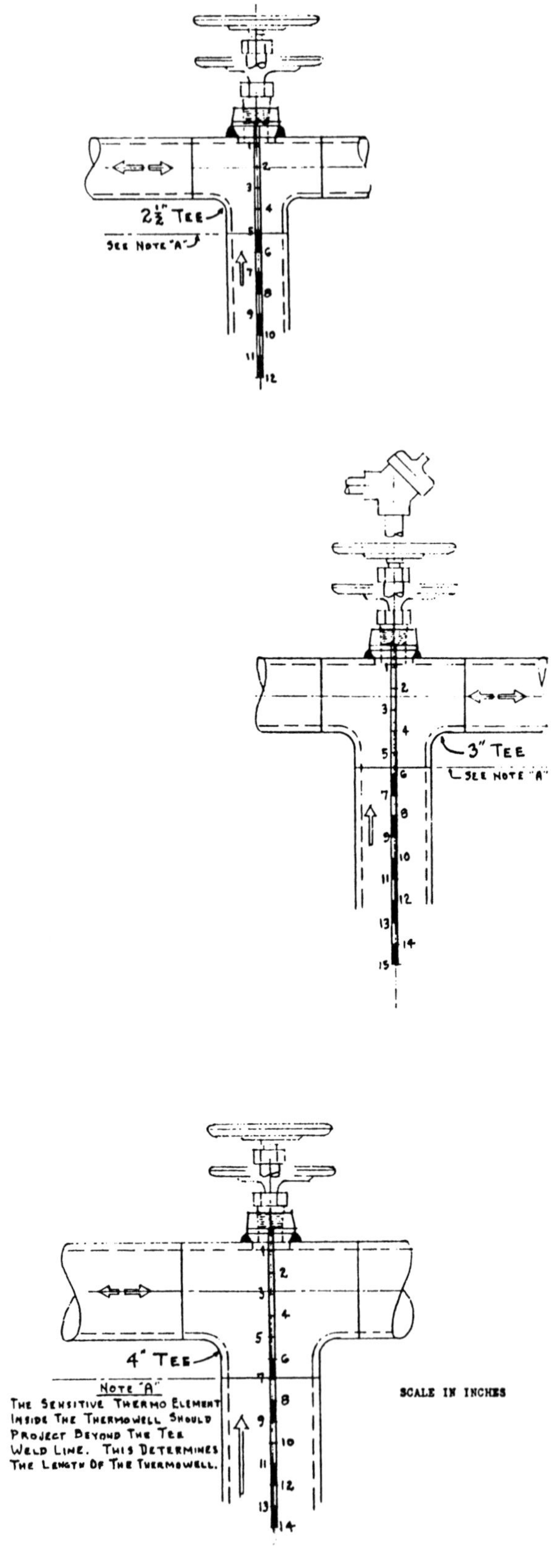

FIG. 104 Thermowell connection installed in back of 2½, 3, and 4-in. pipe tees with a 1-in., 6000 lb thread adapter.

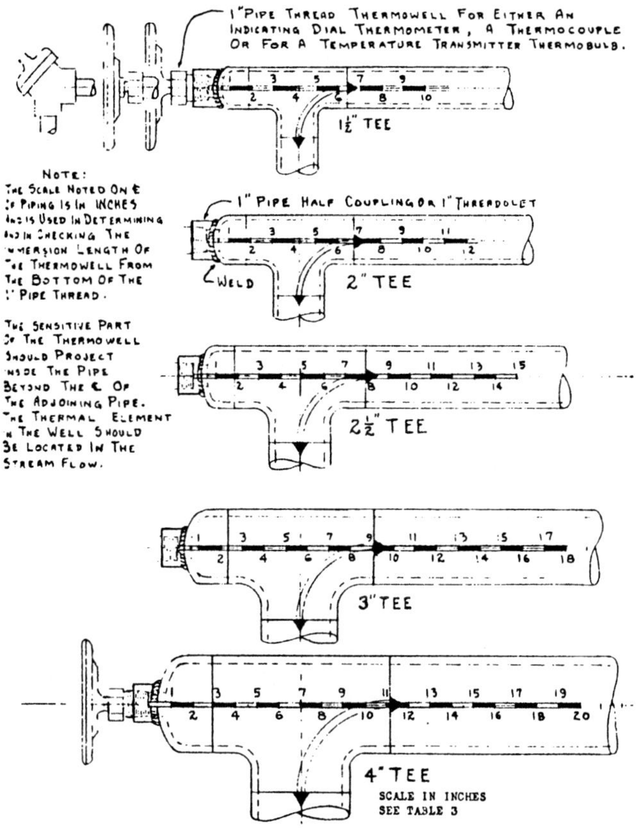

FIG. 105 Thermowells installed at outlet of 1½ to 4-in. welding pipe tees with either a 1-in. thread adapter or a 1-in. half coupling, 6000 lb rating.

Corrugated Joints

These joints consist of one or more corrugations in a metal suitable for the temperature and pressure in the pipe (Fig. 110). Typical materials used for corrugated joints include copper, stainless steel, or Inconel in the bellows, and cast-iron or cast-steel end flanges, or steel welding nipples.

The corrugated joint, when well designed and constructed, has advantages if the installation is inaccessible and maintenance of the joint is difficult or impossible. It is also useful when the piping is subject to lateral and angular misalignment as well as axial motion. However, it is important to select and install the joint properly so that the corrugations will not be overstressed.

In recent years, corrugated expansion joints have been used increasingly as hinge, universal, and gimbal joints. These are specialized applications of the bellows principle which have definite advantages for certain types of installations. One of the most useful is the hinge joint (Fig. 111), which,

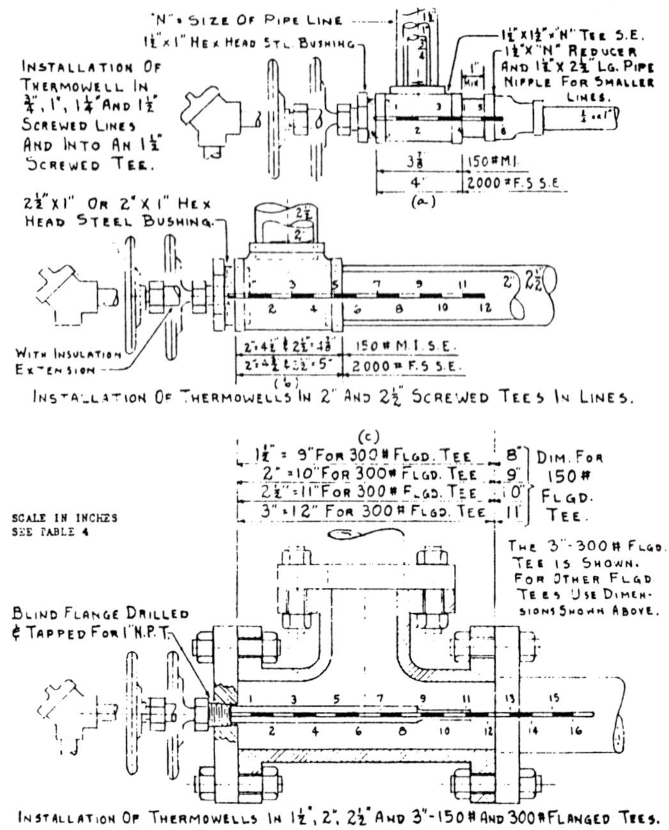

FIG. 106 Installation of thermowells in $1\frac{1}{2}$, 2, $2\frac{1}{2}$, and 3-in. 150 and 300 lb flanged tees.

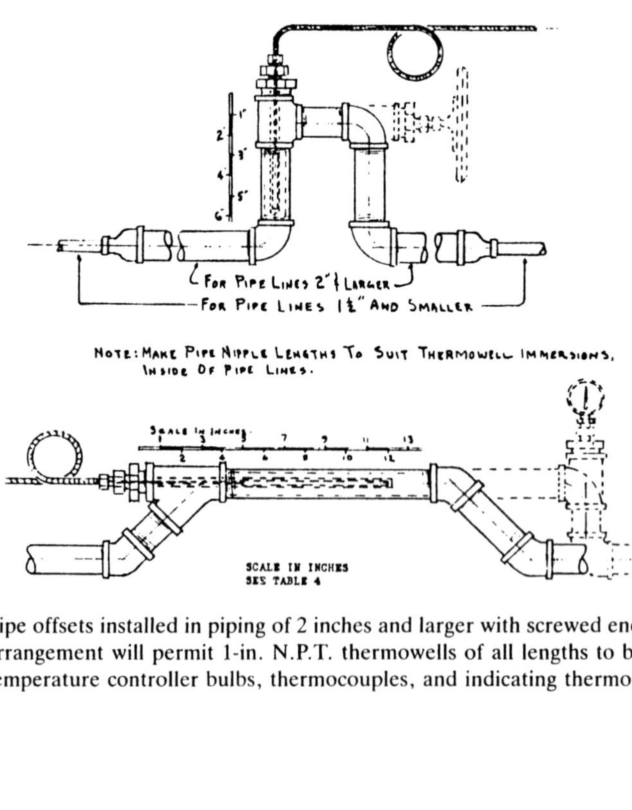

FIG. 107 Pipe offsets installed in piping of 2 inches and larger with screwed end fittings. This arrangement will permit 1-in. N.P.T. thermowells of all lengths to be installed for temperature controller bulbs, thermocouples, and indicating thermometers.

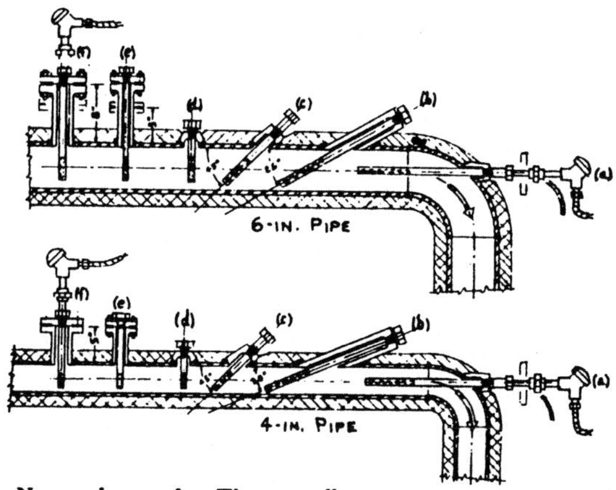

Nomenclature for Thermowell Connections

(a) 1-inch Elbow Adapter, 6,000 lb. rating.
(b) 1-inch specially designed nozzle connection, welded to suit.
(c) 1-inch Elbow Adapter as a lateral connection.
(d) 1-inch Thread Adapter, 6,000 lb. rating.

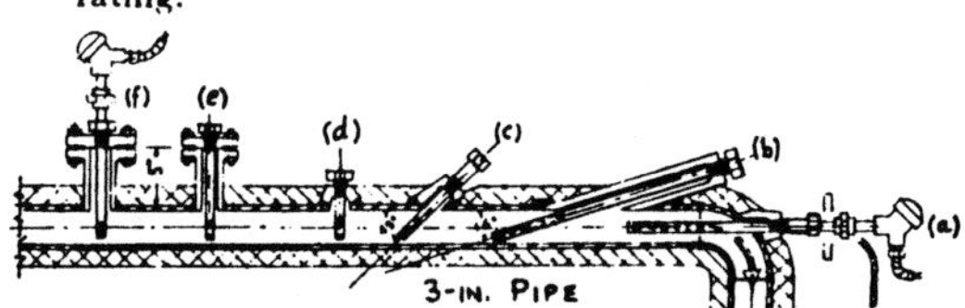

(e) 1-inch Flange Nozzle or 1-inch socket weld flange nozzle made with a 1-inch Sch. 80 pipe nipple. Flange drilled and tapped for Thermowell. Use detail (f) in preference to this.
(f) 1½-inch Flange Nozzle. Blind flange is drilled and tapped for Thermowell. Use this detail rather than (e).

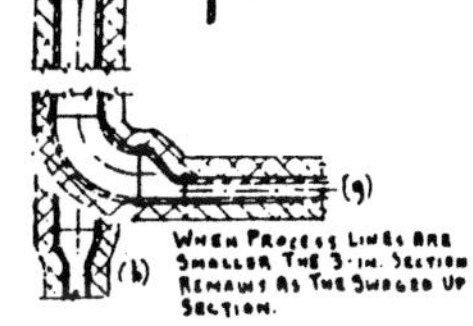

Effective Length of Thermal Element

	DETAIL		
Line Size	**a, b**	**c**	**d, e, f**
3-Inch Pipe	6"	2"	1"
4-Inch Pipe	6"	3"	2"
6-Inch Pipe	6"	4"	3"

FIG. 108 Composite arrangement showing six typical ways thermowells can be installed in 3, 4, and 6-in. process piping (see Table 5).

when circumstances permit, can be used to absorb expansion without the need for anchors.

Slip-Type Joints

The slip-type joint (Fig. 112) is particularly useful in steam and hot-water transmission lines where expansions and contractions are frequent and of

TABLE 17 Evaluating Thermowell Installation Arrangements for Use with Temperature Instruments

Nominal Pipe Size (in.)	Best Temperature Speed of Response is Rated as No. 1 and Follows as Shown	1	2	3	4	6	5
	Identifying the 2-in. thermowell connection shown in Fig. 108	a	b	c	d	f	e
6	Thermobulb	6 in. or longer	3 to 6 in.	3 in.	No	No	No
	Dial thermometer	Yes	Yes	Yes	Yes	Yes	Yes
	Thermocouple on center line of pipe	Yes	Yes	Yes	Yes	Yes	Yes
4	Thermobulb	6 in. or longer	3 to 6 in.	No	No	No	No
	Dial thermometer	Yes	Yes	Yes	Yes	Yes	Yes
	Thermocouple on center line of pipe	Yes	Yes	Yes	Yes	Yes	Yes
3	Thermobulb	6 in. or longer	3 to 4 in.	No	No	No	No
	Dial thermometer	Yes	Yes	Yes	No	No	No
	Thermocouple on center line of pipe	Yes	Yes	Yes	Yes	Yes	Yes

large magnitude. Slip joints contain no highly stressed flexing element subject to failure after a finite number of cycles.

In one design, the gun-packed joint (Fig. 113), packing can be inserted into the stuffing box at any time with full temperature and pressure on the line. This eliminates shutdown and minimizes maintenance time.

A slip joint fitted with a sliding member at only one end is known as a single-end joint. When sliding members are used at both ends, the joint can

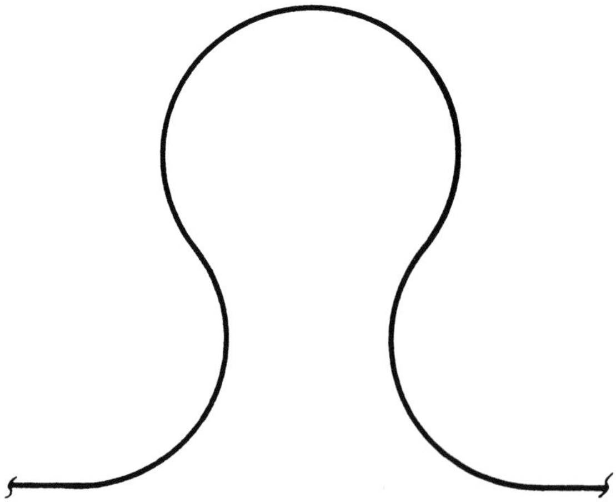

FIG. 109 Double-offset U-type expansion joint.

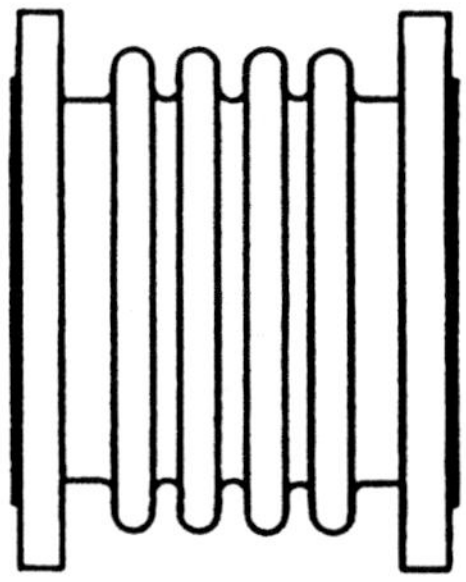

FIG. 110 Simplified sketch showing a typical corrugated-type expansion joint suitable for refinery piping systems.

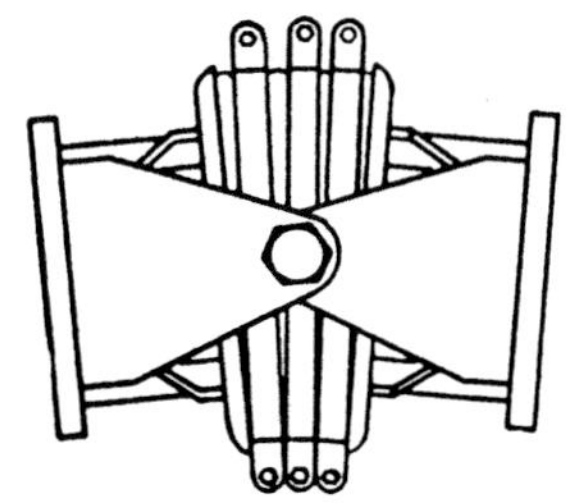

FIG. 111 Simplified sketch showing a hinge-type corrugated expansion joint suitable for refinery piping systems.

absorb twice the expansion of a single-end joint and is known as a double-end joint.

Where repacking under pressure is not a consideration, conventional gland-packed joints (Fig. 112) may be used. These resemble gun-packed joints but have conventional asbestos or rubber and duck ring packing. Cast-iron glands are generally used with wrought steel body and sleeve. The sleeve is polished and chromium-plated.

Joint and Bend Selection

When choosing any type of expansion joint or bend, it is essential that a number of important factors be considered. These include the required traverse, presetting, pipe expansion, anchors, alignment, and supports.

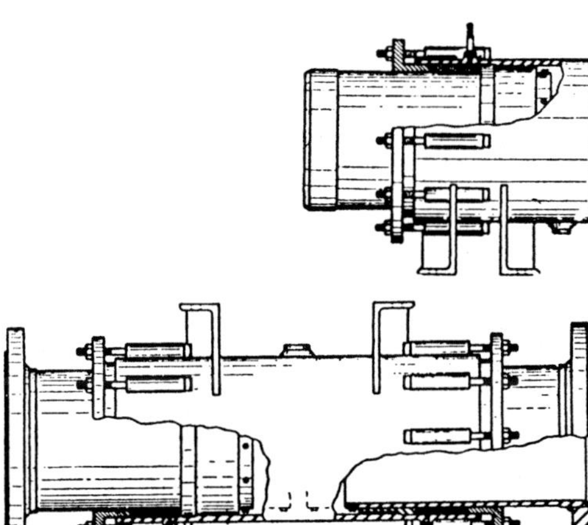

FIG. 112 Single-end and double-end gland-packed slip-type expansion joints use conventional asbestos rubber and duck ring packing.

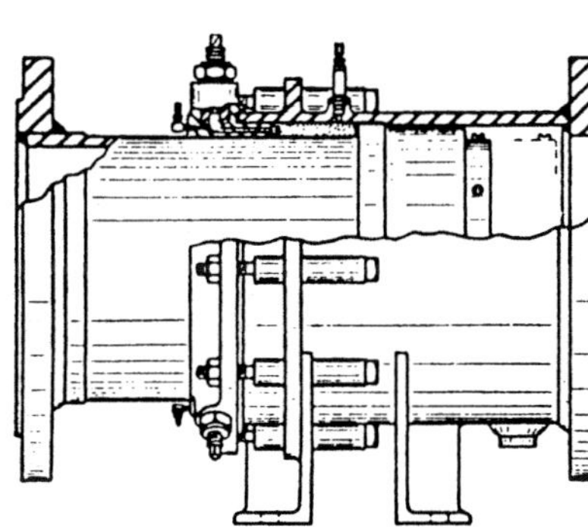

FIG. 113 Single-end gun-packed slip-type expansion joint. This is also built with double ends.

Traverse

The traverse is the distance, expressed in inches, that the joint or bend must contract in order to absorb the expansion of the pipe served by the particular joint or bend. The required traverse is numerically equal to the distance the pipe expands over its full temperature range. The lowest temperature that will be encountered is as important as the maximum temperature. If the expansion tables available start at 70°F (as some of them do) and the line is to be installed outdoors where winter temperatures can drop as low as 20°F below zero, traverse must be provided to cover the additional 90°F range below the reference temperature of 70°F.

The traverse of the joint or the bend is the maximum safe movement specified by the manufacturer. The numerical value of this traverse should always exceed the computed expansion of the pipe (i.e., the required traverse). Thus, with a 20-in. expansion, a joint with a total traverse of 24 in. is usually chosen.

Presetting

Where traverse in both directions is expected, it is essential when actually placing the joint to preset it for the temperature of the pipe at the time of installation.

Pipe Expansion

Figure 114 shows the expansion of steel pipe for any temperature up to 700°F. For convenience, one scale also provides the saturation temperatures for various steam pressures. Given the temperature range and pipe length involved, the total traverse and amount of presetting can be readily determined. It should be noted that when the minimum temperature is below 0°F, the temperature range used in determining expansion must be increased accordingly. Values taken from the curve include a safety factor of 10% over the actual pipe expansion to provide for discrepancies in installation and unforeseen temperature extremes.

Anchors are important in any piping system, but there are some special considerations necessary when expansion joints or bends are used. In general, anchors are installed to stabilize the piping at certain vital points such as valves or other equipment, junctions of two or more pipes, and terminal points.

With expansion joints, anchors also serve to divide the system into sections, so that each joint absorbs only the expansion in its own section. This may seem elementary, but it is frequently overlooked. There are numerous instances where two joints are installed in the same section without intermediate anchors with the result that one joint becomes overloaded and subject to damage while the other one is underloaded. The converse of this is that two anchors should never be placed in a straight run of pipe without an expansion joint or other device to absorb the expansion between them.

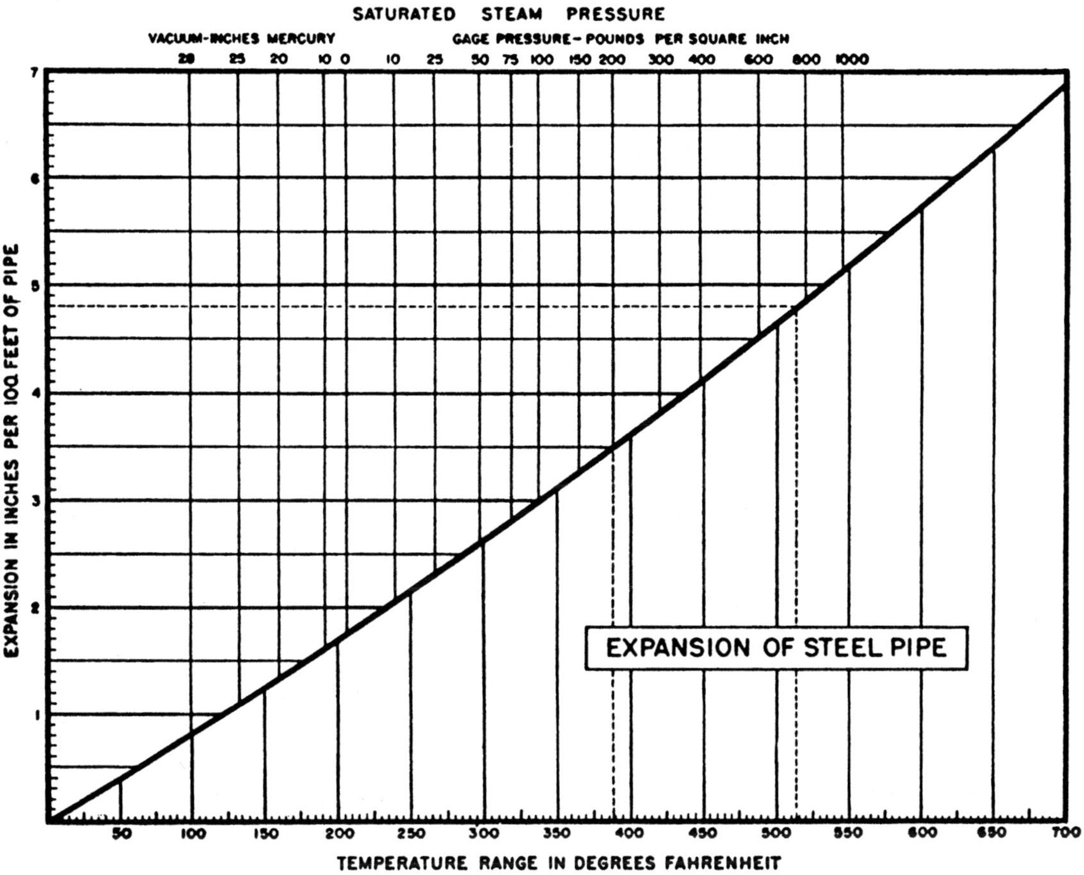

FIG. 114 Chart for quickly determining linear expansion of steel pipe for various steam pressures and temperatures.

Anchor design is particularly important with high pressures and large pipe sizes because of the high thrusts that develop at end anchors. End anchors are so-called because they occur at terminal points and at changes in direction of the pipe. At these points the pressure acting on the inside area of the pipe would tend, if unrestrained by anchors, to pull an expansion joint apart. End anchors must absorb this pressure reaction as well as the forces required to activate the expansion joint and to overcome friction in supports and guides. Intermediate anchors, on the other hand, are subject to only the latter forces.

The curves in Fig. 115 illustrate the comparative magnitude of end and intermediate anchor loads for slip-type expansion joints, and they also provide reasonably accurate anchor load figures for various pipe sizes and pressures. Values for corrugated joints are comparable, except that end anchor loads are somewhat higher in the larger pipe sizes because of the larger area in the bellows. Intermediate anchor loads are somewhat lower because bellows require less force to actuate them than do slip joints.

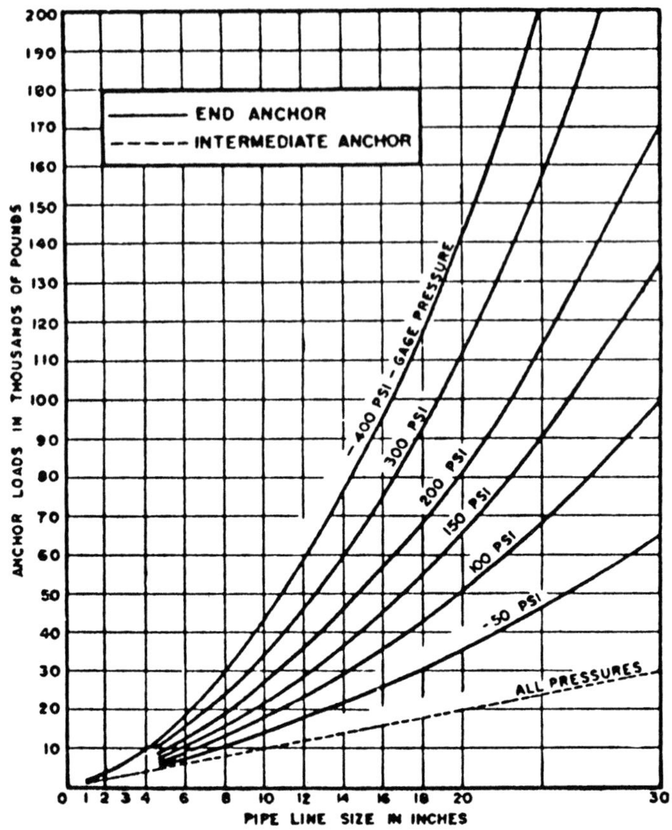

FIG. 115 Anchor loads in intermediate and end anchors used with slip and other types of expansion joints.

Pipe Guides

Except for some specialized applications of the corrugated type, expansion joints and bends require good pipe alignment to perform satisfactorily. Guides serve two important purposes. When installed near an expansion joint, they prevent cocking of the sleeve in a slip-type joint or distortion and possible buckling of a corrugated-type joint or an expansion bend.

Also, guides are usually necessary in long spans to prevent the pipe from buckling as it expands. The tendency to buckle increases as the pipe becomes smaller in diameter, longer, or both. Hence, the need for guides is greatest with long spans of relatively small pipe. Table 18 gives recommended spacing of guides from expansion joints and also between guides for various pipe sizes.

Pipe Supports

Table 18 also provides generally recommended spacing of supports for steel pipe filled with water. The importance of adequate support is well recog-

TABLE 18 Recommended Spacing for Pipe Alignment Guides and Supports, in Feet

Nominal Pipe Size (in.)	$1\frac{1}{2}$	2	$2\frac{1}{2}$	3	$3\frac{1}{2}$	4	5	6	8	10	12	14	16	18	20	24
Distance between guide and expansion joint:																
For internally guided joints	5	5	6	6	7	7	8	8	9	9	10	10	11	11	12	12
For internally-externally guided joints	8	10	11	12	13	14	15	16	18	20	21	22	23	24	25	26
Distance between alignment guides	10	13	15	19	22	25	30	35	45	60	70	80	90	100	105	110
Maximum distance between pipe supports	8	10	11	12	13	14	15	16	18	20	21	22	23	24	25	26

nized. It should be pointed out, however, that supports must be designed to withstand the thrust resulting from motion of the pipe over the supports as well as the vertical load due to gravity. Failure to provide for this thrust has resulted in failure of the supports in some installations.

System Design

Figure 116 shows a typical refinery piping system using slip joints and consisting of 8-in. steel pipe with a total length of 770 ft. It carries steam at 125 lb/in.^{2}gauge, with 25°F superheat. The procedure for finding traverses, anchor loads, and guide support locations is as follows.

Traverses

Saturation temperature for 125 lb/in.^{2}gauge steam is 355°F (Fig. 114). Add 25°F for a total temperature range (assuming minimum temperature to be 0°F) of 380°F. Expansion for this range is 3.4 in./100 ft, from Fig. 114.

Span AB = 140 ft; expansion = (140)(3.4)/100 = 4.8 in. A single-end joint with an 8-in. traverse is suitable at point A. Span BC is 90 ft long and has an expansion of 3.1 in. when figured as above. Hence, the joint C is a single-end with a 4-in. traverse. The span CE is 430 ft long with 14.7 in. of expansion. A double-end joint with an 8-in. traverse at each end is suitable at D, the approximate midpoint. Span EF is 100 ft long with 3.7 in. of expansion for which a single-end, 4-in. traverse joint will be adequate.

Anchor Loads

Intermediate anchors at A, D, and F are subject to possible maximum loads of 8000 lb in both directions. These values are obtained from Fig. 115 for

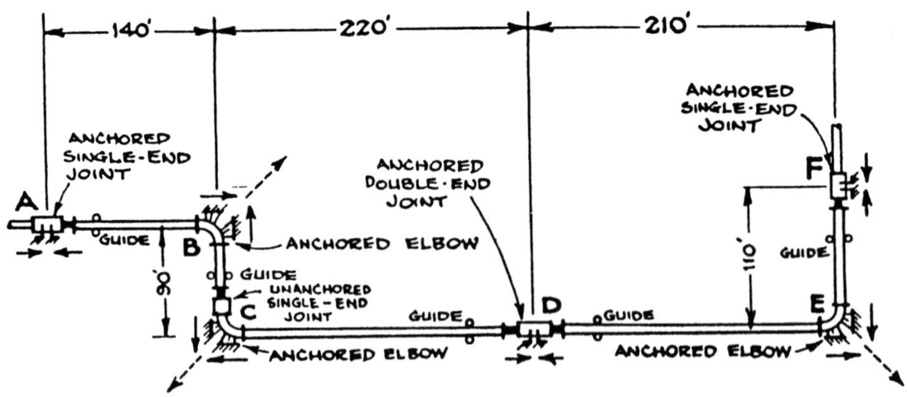

FIG. 116 Typical refinery piping system using slip joints to absorb the expansion and contraction.

8-in. pipe. The contraction loads equal the expansion loads but are in the opposite direction.

End anchors at B, C, and E have possible maximum axial loads of 15,000 lb in each leg (solid arrows, Fig. 116). The resultant anchor load is the vector sum of the thrusts in the two legs, or 21,000 lb acting along an axis 45° from the pipe axes (dotted arrows, Fig. 116). Maximum contraction loads in the reverse direction consist of friction only and amount to 8000 lb along the solid arrows with an 11,300-lb resultant.

Note: Anchor loads with corrugated joints will be somewhat higher or lower depending on pipe size, pressure, etc. Consult manufacturer's instructions.

Supports and Guides

From Table 18, assuming the joints are the internally-externally guides type, install alignment guides not more than 18 ft from the slip end of each joint and at intervals of 45 ft, more or less, along each span. Also, from Table 18, the pipe should be supported at intervals of about 18 ft and wherever concentrated loading exists by valves and fittings.

Ball Joints

Engineering handbooks provide tables and factors for calculating amounts of expansion (e) in a given length of piping between any two temperatures: T_1 (cold) to T_2 (hot) (see Fig. 117).

The offset (O) needed to accommodate (e) depends on amount of allowable angular movement (Θ) in the ball joints used. O is measured as the distance between ball centerpoints of the joints.

Except for smaller size joints, the nominal Θ for all ball joints is 15°. To avoid exceeding this angle, normal practice is to increase O more than the actual minimum as a safety factor.

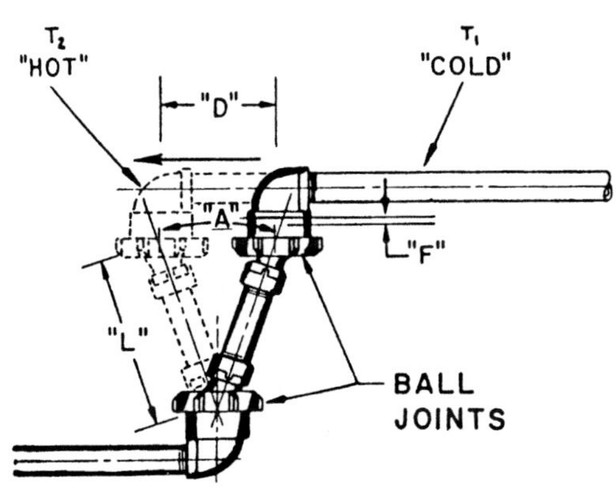

FIG. 117 Linear pipe expansion.

Values for O are easy to calculate graphically using a scale, compass, and protractor to set up the angles. For example, with a 15° angle, it will be found that each 12 in. of offset O will accommodate 3 in. of expansion (e) and that the foreshortening effect (l_f) in the arc is slightly less than 1/8 in. Increasing O to 2 ft as a safety factor has the effect of reducing Θ to 7½° and further reduces l_f to less than 1/16 in.

Hinged expansion joints consist of one or more thin corrugated elements welded to short lengths of pipe with brackets containing pins (Fig. 118). The hinges limit the corrugated bellows to bending or to purely angular movement. The loss of complete flexibility using hinges is offset by the gain in rigidity of the joint protecting the bellows against axial and lateral forces and torsion. The entire end-load caused by pressure, weight, etc. and the transverse loading caused by wind, etc. must be carried by the hinge pins unless relieved by counterweights or springs.

The application of hinged expansion joints is especially useful when nozzle design loads are limited by the equipment manufacturer. Here is a list of conditions warranting consideration of hinged expansion joints.

1. When the natural flexibility of the piping components cannot be used economically (supports, guides, and directional anchors should be figured in the cost comparison)
2. When space considerations make rerouting unworkable and uneconomical
3. When pressure drop limits the routing
4. When temperature, process, line size, service life, or hazard prohibits the use of flexible couplings
5. When the allowable forces and moments on sensitive equipment are excessive and cannot be isolated from the equipment by the use of restraints such as directional anchors and guides

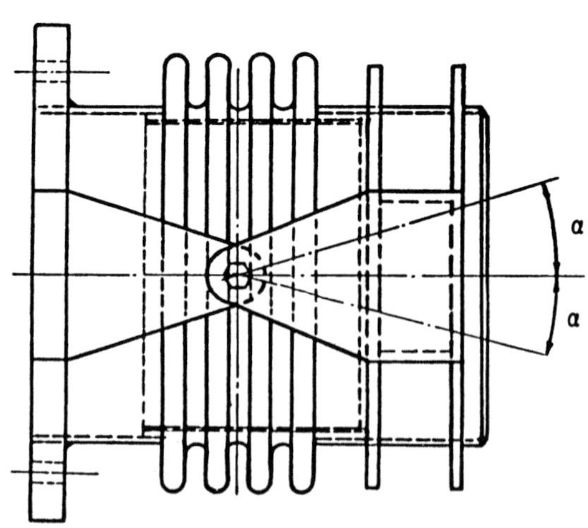

FIG. 118 Typical hinged expansion joint design.

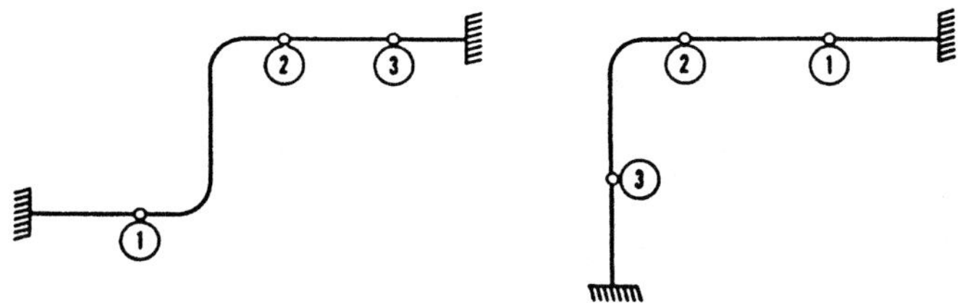

FIG. 119 An offset is required in hinged expansion joints.

Why joints are furnished with hinges:

1. To avoid end thrust and the need for costly anchor designs and additional guides
2. To avoid additional support requirements, springs, etc.
3. To provide protection against abuse of joints in installation

Selection Procedure

Hinged expansion joints will not have the required effect when applied in a straight line. Therefore, an initial off-set in the layout is required (see Fig. 119). After thermal expansion has taken place, the configurations of Fig. 119 will alter as shown in Fig. 120. Joint 3 in configuration A–B takes care of the inevitable deflection h. However, if B-3 is a long flexible length, joint 3 may be omitted.

To select the number of corrugations required, the induced angular rotation must be determined. The moment required to rotate the joint must be calculated if end-moments and forces (transmitted to sensitive equipment nozzles) are required.

Angular Rotation

Generally, the angular rotation is obtained as follows. Temporarily disregard the joints at B and F (Fig. 121), and disconnect the system at E (Fig. 122).

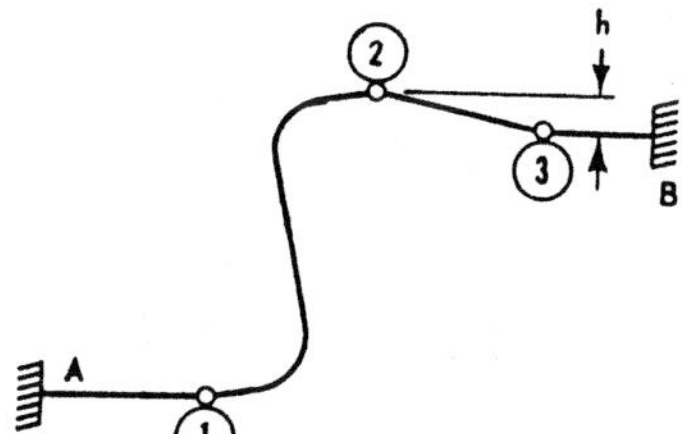

FIG. 120 After expansion, the configuration of Fig. 119 will look like this.

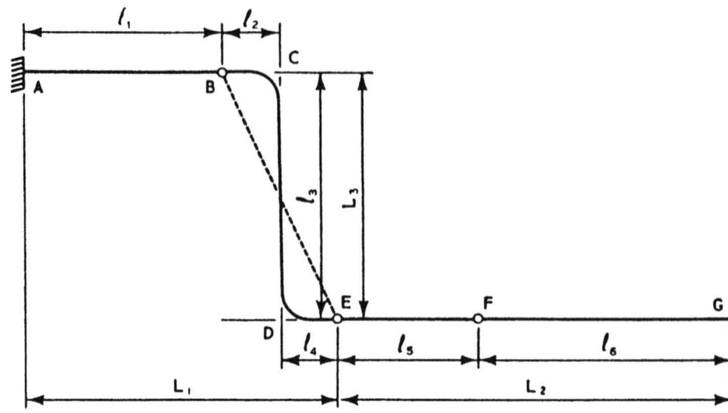

FIG. 121 Piping layout to find angular rotation.

After expansion, the new positions of E_1 and E_2 become $E_1'$and E_2'. Δ_1, Δ_2, and Δ_3 represent the respective expansions of L_1, L_2, and L_3 (Fig. 122).

Now let us reinstall the joints at B and F (Fig. 123). The points E_1' and E_2' will not be able to travel along the arcs of circles with radii BE_1' and FE_2', respectively.

If the ratio arc/radius is small (e.g., 1/20), the arc may be considered to be a straight line. Applying this to E_1' and E_2', the arcs will be represented by lines through E_1' and E_2', perpendicular to BE_1' and FE_2', respectively (Fig. 124).

Applying the above to the initial point E, we obtain the following. If the computed thermal expansions Δ_1, Δ_2, and Δ_3 are drawn to scale, the lengths of arcs $E_1'E_3$ and $E_2'E_3$ may be readily scaled from Fig. 125; E_3 being the ultimate position of E. The arcs represent the angular rotation at B and F. The matching angles may be determined as follows:

$$\theta_B = \frac{\text{arc (in.)}}{\text{radius (in.)}} = \frac{E_1'E_3}{BE} \qquad \text{(in radians)}$$

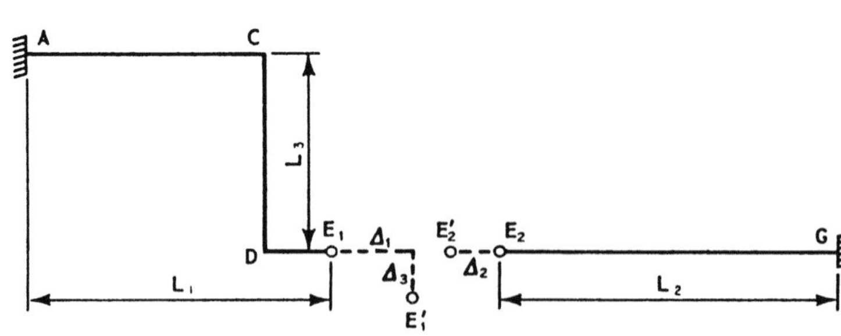

FIG. 122 Temporarily disregard Joints B ad F and disconnect system at E.

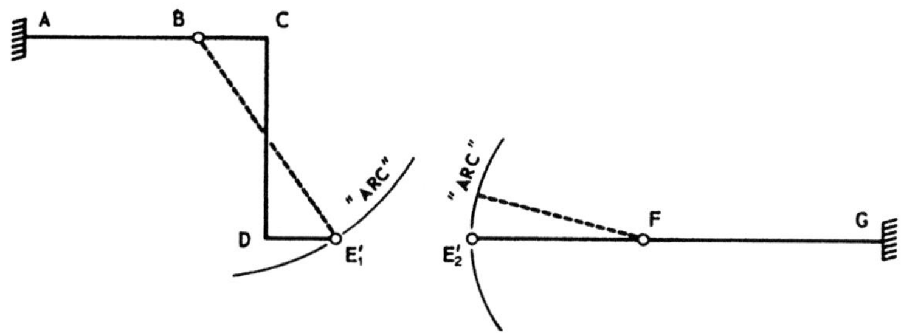

FIG. 123 Reinstall Joints B and F and strike arc through E.

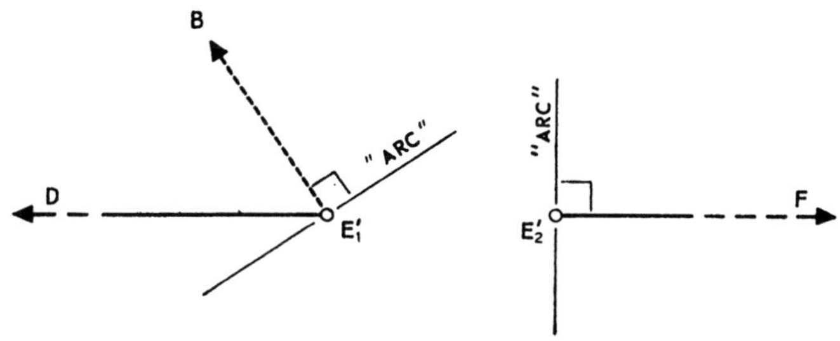

FIG. 124 If the arc/radius ratio is small (1/20), consider the arc a straight line.

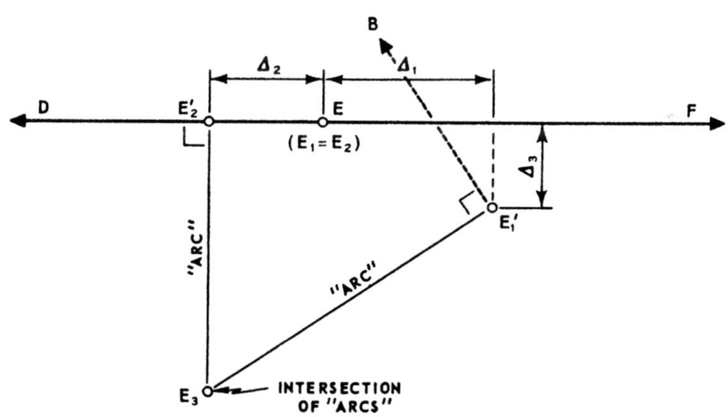

FIG. 125 Scale arc lengths from this drawing.

or

$$\theta_B = \frac{E_1'E_3(57.3)}{BE} \qquad \text{(in degrees)}$$

θ_F may be determined similarly.

The angular rotation at point E is

$$\theta_E = \theta_B + \theta_F$$

This relationship can easily be proved geometrically. Assuming that the maximum allowable Θ for one corrugation is known from manufacturer's data, the number of corrugations required per joint can now be determined.

$$n_c = \theta_{tot}/\theta_{max} \qquad \text{(take to next whole number)}$$

The moment M required to rotate the joint can also be computed provided M_{max} is known:

$$M = \frac{\theta_{tot}}{\theta_{max}}\left(\frac{1}{n}\right)(M_{max})$$

where θ_{tot} = total angular rotation to be accommodated by the hinged expansion joint
θ_{max} = maximum allowable rotation for one corrugation
n_c = selected number of corrugations
M_{max} = moment required to rotate a corrugation over θ_{max}

For applications occurring frequently, the above may be represented by formula arrived at geometrically.

Example 1. Z-Bend (Fig. 126)

$$\theta_B = \frac{e(L_1 + L_2) \pm \Delta_x}{l_3} \qquad \text{(in radians)}$$

$$\theta_F = \frac{eL_3 \pm \Delta_y}{l_5} + \frac{[e(L_1 + L_2) \pm \Delta_x](l_2 + l_4)}{l_3 l_5} \qquad \text{(in radians)}$$

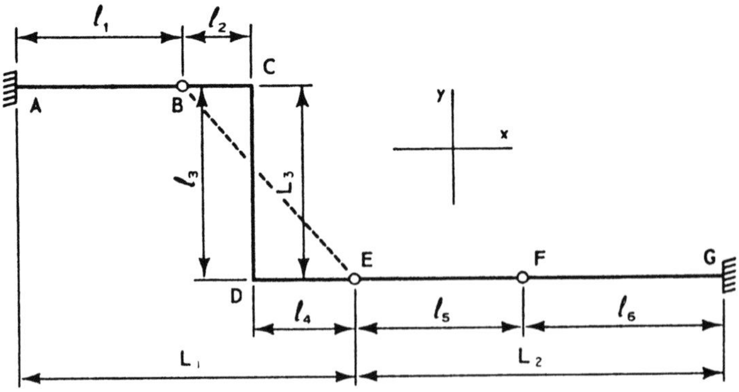

FIG. 126 A Z-bend piping configuration.

$$\theta_E = \theta_B + \theta_F \qquad \text{(in radians)}$$

where e = linear thermal expansion (inches per inch). l_1 through l_6 and L_1, L_2, and L_3 are shown in Fig. 125 and are in inches

$\pm\Delta_x$ = total extraneous movement in x-direction at anchors in inches

$\pm\Delta_y$ = total extraneous movement in y-direction at anchors in inches

Example 2. L-Bend (Fig. 127)

$$\theta_B = \frac{eL_2 \pm \Delta_x}{l_2} \qquad \text{(in radians)}$$

$$\theta_E = \frac{eL_1 \pm \Delta_y}{l_4} + \frac{(eL_2 \pm \Delta_x)l_3}{l_2 l_4} \qquad \text{(in radians)}$$

$$\theta_D = \theta_B + \theta_E \qquad \text{(in radians)}$$

If $L_1 + L_2$ is of considerable length in Example 1, the thermal growth will be large, A 50% cold spring is therefore advisable. The formulas then alter to

$$\theta_B = \frac{e(L_1 + L_2) \pm \Delta_x}{l_3}$$

$$\theta_F = \frac{eL_3 \pm \Delta_y}{l_5} + \frac{[e(L_1 + L_2) \pm \Delta_x](l_2 + l_4)}{2l_3 l_5}$$

If cold spring (50%) is also applied to L_3, then

$$\theta_B = \frac{e(L_1 + L_2) + \Delta_x}{l_3}$$

$$\theta_F = \frac{eL_3 \pm \Delta_y}{2l_5} + \frac{[e(L_1 + L_2) \pm \Delta_x](l_2 + l_4)}{2l_3 l_5}$$

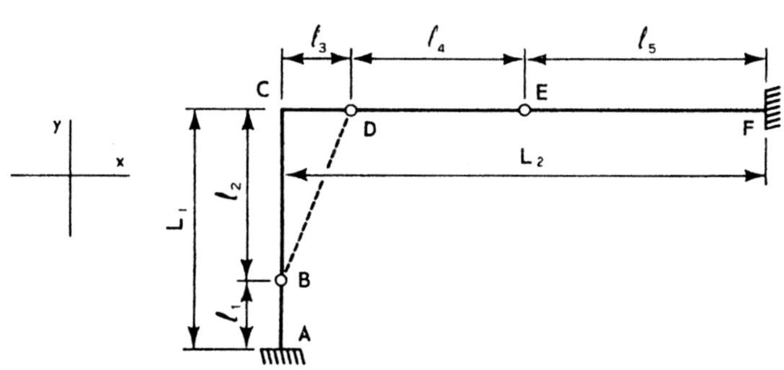

FIG. 127 An L-bend piping configuration.

A 50% cold spring on L_1 and L_2 in Example 2 gives

$$\theta_B = \frac{eL_2 \pm \Delta_x}{2l_2}$$

$$\theta_E = \frac{eL_1 \pm \Delta_y}{2l_4} = \frac{(eL_2 \pm \Delta_x)l_3}{2l_2l_4}$$

To minimize the angular rotations and, therefore, the number of corrugations per joint, it is apparent from the formulas of Example 1 that l_2 and l_4 are to be minimum and l_3 and l_5 are to be maximum.

For Example 2, l_3 is minimum and l_2 and l_4 are maximum.

Observing the preceding, the most advantageous location of the joints in the system may be determined.

Multiplane piping systems requiring hinged expansion joints may be dealt with similarly. Obviously the problem will be more involved and in many applications the use of gimbal-type joints is necessary.

In view of the fact that in general the flexibility of the pipe itself is negligible compared to that of the hinged expansion joint, the pipe is assumed to be infinitely rigid in the method described above. Any bending in the piping system should have a reducing effect on the determined angular rotations of the joints. As a result, the selected expansion joints will always be on the safe side.

Example 3. The following sample problem illustrates the application of the method. It also shows to what extent end-forces and moments are reduced when hinged expansion joints are incorporated.

A flexibility calculation of the piping configuration shown in Fig. 128 yields the following forces and moment on the compressor nozzle:

$$F_H = 2740 \text{ lb}$$

$$F_V = 1250 \text{ lb}$$

$$M_n = 13{,}100 \text{ ft-lb}$$

These are shown on Fig. 129.

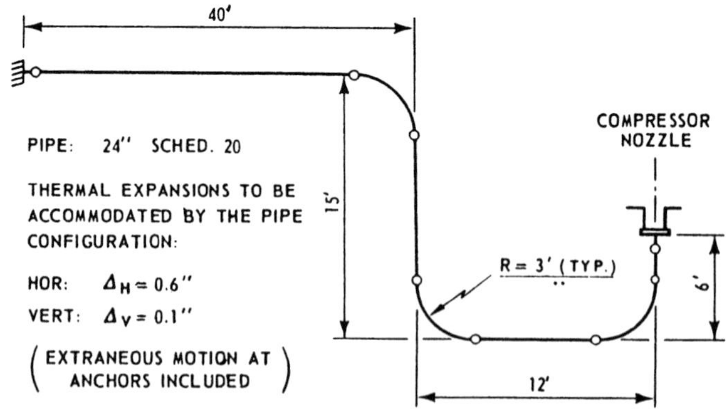

FIG. 128 Example of piping layout to a compressor nozzle.

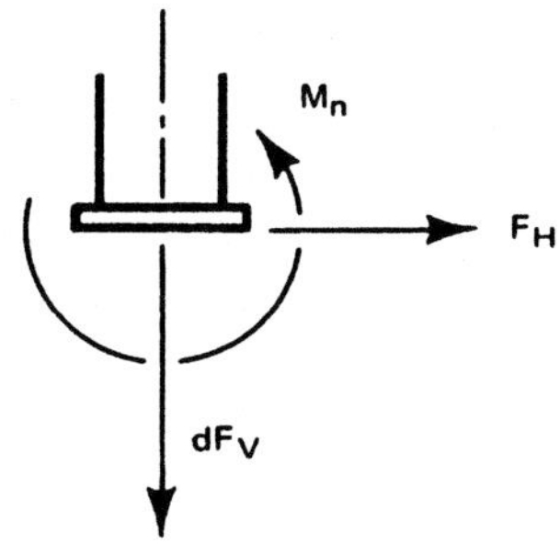

FIG. 129 End-forces and moment about a compressor nozzle.

When hinged expansion joints are inserted, located as shown on Fig. 130 and the theory applied, we obtain the following from Fig. 131. A_2 is the ultimate position of joint 2, the angular rotations of 1 and 3 being represented respectively by A–A_2 and A'–A_2. A–A_2 = 0.65 in.; A'–A_2 = 0.112 in. (scaled). Respective radii are: 7 ft = 84 in., 8.94 ft = 107.2 in.

$$\theta_1 = (0.65/84)(57.3) = 0.44°$$

$$\theta_2 = (0.112/107.2)(57.3) = 0.06°$$

$$\theta_3 = \theta_1 + \theta_2 = 0.44° + 0.06° = 0.5°$$

From manufacturer's data, θ_{max} = 1.5°. Moment to rotate one corrugation over 1°, M = 5350 lb-ft.

Assuming one corrugation per joint:

$$M_1 = 0.44(5350) = 2354 \text{ ft-lb}$$

$$M_2 = 0.5(5350) = 2675 \text{ ft-lb}$$

$$M_3 = 0.06(5350) = 321 \text{ ft-lb}$$

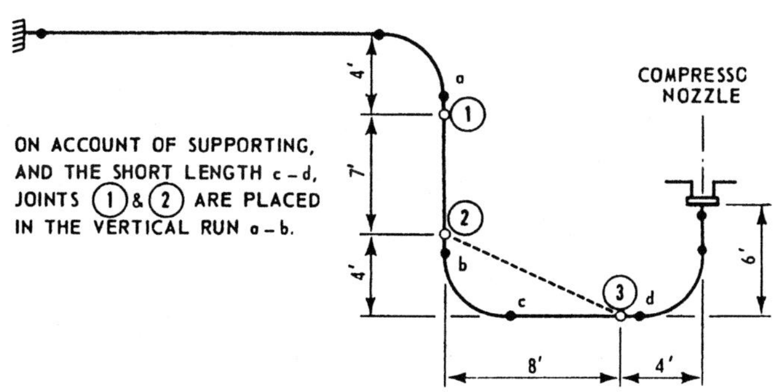

FIG. 130 Install hinged expansion joints at positions 1, 2, and 3.

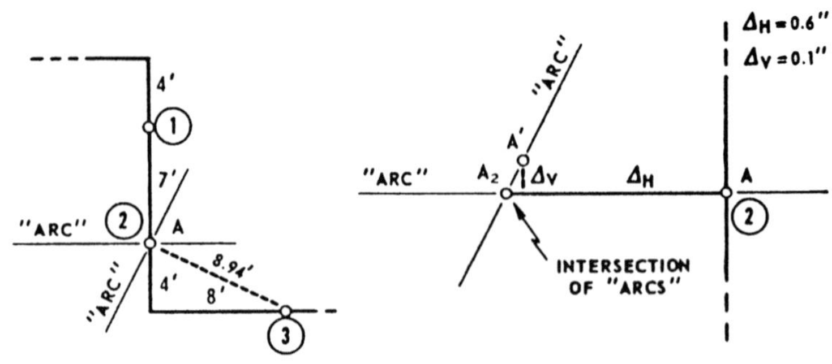

FIG. 131 The angular rotation of the joints after expansion can be measured.

In order to obtain equilibrium, the following equations must be satisfied (see Fig. 132):

$$11F_H + M_{n-3} = (8F_V) + M_{n-1}$$

$$7F_H = M_{n-1} + M_{n-2}$$

$$= 2354 + 2675$$

$$F_H = 5029/7 = 718.4 \text{ lb}$$

$$F_V = [(11)718.4 + 321 - 2354]/8$$

$$= 733.7 \text{ lb}$$

The forces and moment transmitted to the compressor nozzle (Fig. 133) are

$$F_{V(\text{nozzles})} = 733.7 \text{ lb}, F_{H(\text{nozzles})} = 718.4 \text{ lb}$$

$$M_{n(\text{nozzles})} = 6F_H + 4F_V + M_{n-3}$$

$$= 4310.4 + 2934.8 + 321$$

$$= 7566.2 \text{ ft-lb}$$

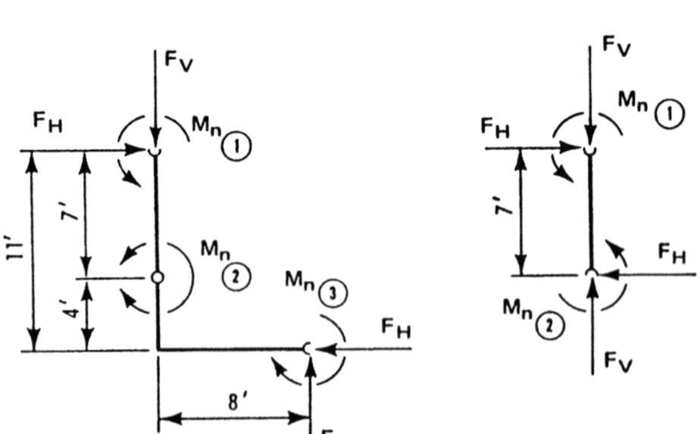

FIG. 132 Use these dimensions to find system equilibrium.

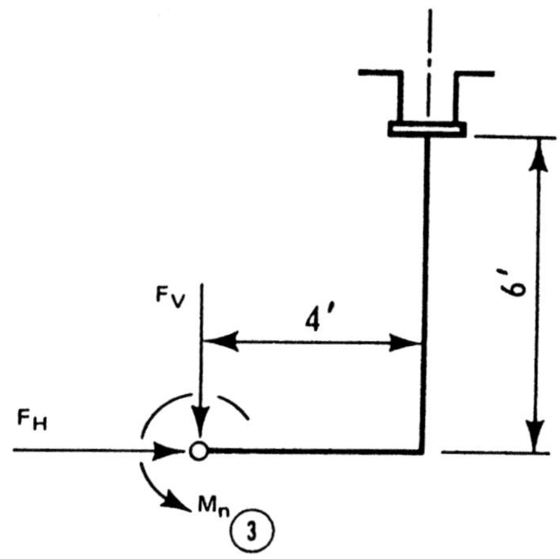

FIG. 133 Use these dimensions to find the forces transmitted to the compressor nozzle.

When assuming two corrugations at joints 1 and 2 and one corrugation at joint 3:

$$M_{n-1} = \tfrac{1}{2}(2354) = 1177 \text{ ft-lb}$$
$$M_{n-2} = \tfrac{1}{2}(2675) = 1337.5 \text{ ft-lb}$$
$$M_{n-3} = 321 \text{ ft-lb}$$

Substitute these values into the above equations:

$$7F_H = 1177 + 1337.5$$
$$F_H = 359.2 \text{ lb}$$
$$8F_V = 11(359.2) + 321 - 1177$$
$$F_V = 386.9 \text{ lb}$$

Transmitted to the compressor nozzle:

$$F_{V(\text{nozzle})} = 386.9 \text{ lb}$$
$$F_{H(\text{nozzle})} = 359.2 \text{ lb}$$
$$\begin{aligned} M_{n(\text{nozzle})} &= 6F_H + 4F_V + M_{n-3} \\ &= 2155.2 + 1547.6 + 321 \\ &= 4023.8 \text{ ft-lb} \end{aligned}$$

It appears from the sample problem above that by introducing a set of hinged expansion joints, a considerable reduction of end-forces and moment is achieved.

If weight loads have not been taken into account, as in the example, careful attention should be paid to the design and location of supports in order to minimize the influence of dead weight on the equipment nozzle.

The following illustrates the necessity to restrain joints by means of hinges or otherwise. Assume that the internal design pressure for the piping system of the previous example is 15 lb/in.^{2}gauge. The axial force set up by this internal pressure in a section of the pipe containing a joint is

$$PA_B = (15)(550) = 8250 \text{ lb}$$

where P = internal design pressure (lb/in.^{2}guage)
A_B = effective cross-sectional area of the bellows obtained from manufacturer's data (in.^{2})

The convolutions are not able to accommodate this force. Therefore, the pipe must be restrained on both sides of the joint.

If no hinges are applied, the joint must be protected by a set of adequate anchors and guides. In view of the magnitude of end-thrusts, anchors will have to be massive and often costly.

Spring Pipe Hangers

Spring pipe hangers or constant-support pipe hanger design can be simplified by using a scale plan piping drawing. This is usually available, or easily made, and simplifies the consideration given to the location and weight of various components of a piping system. The results are well within the limits correctible by adjustment at the time the supports are installed.

Sample Problem. Determine the load required to be supported by a constant-support located at point D of the piping system (Fig. 134).

Assume:

1. The unsupported pipe between A–C and D–F is safe and not overstressed.
2. The piping system between A and F is adequately flexible provided the total vertical movement is shared by A–C and D–F.
3. The pipe weighs 100 lb/ft.

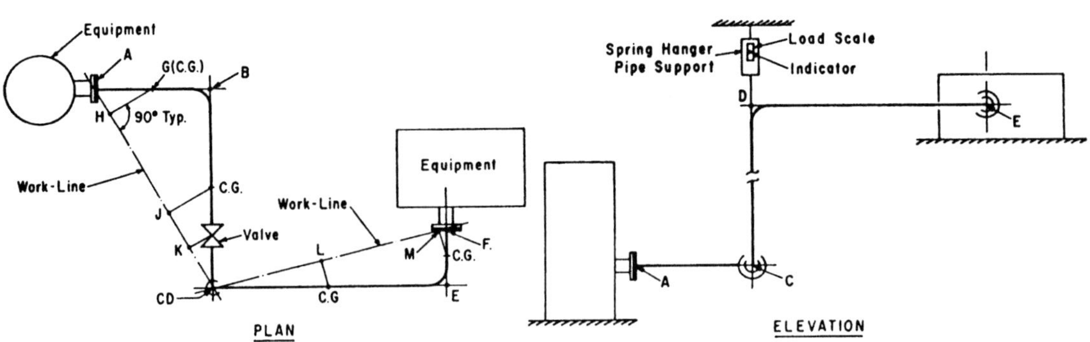

FIG. 134 Use the scale plans of piping (left) and elevation (right) to simplify hanger design.

Procedure:

1. Determine the weights of all components of the piping system between A and F. Disregard the weights of elbows, instead consider the system as made up of straight sections of pipe with mitered elbows. Consider B–C to be one continuous pipe even though it is interrupted by a valve. Include the weight of flanges and bolts connecting to a valve as part of the valve. Add the weights of the flanges at A and F to points A and F and the weight of pipe C–D to fall at D.
2. Layout the piping system to scale and draw worklines in plan between the points of support, A–CD and CD–F.
3. Transpose the weights of the components from their center of gravity to the work-lines, at right angles to the work-lines, such as the weight of pipe A–B, whose center of gravity is at G to fall at H, the weight of the valve to fall at K, and the weight of B–C to fall at J.
4. Calculate the loads at A, D, and F considering that the transposed weights at H, J, L, and M are loads falling on simple beams whose ends are at A and C, and D and F.
5. Determine total weight of the system from the figure and known information:

A–B = H =	400
B–C = J =	700
Valve = K =	500
C–D =	1500
D–E = L =	800
E–F = M =	200
Flange at A =	100
Flange at V =	100
Total	4300 lb

6. Determine the following dimensions by scaling the plan in Fig. 134.

C–K = 1.5
C–J = 3.0
C–H = 7.0
C–A = 8.0
D–L = 3.85
D–M = 8.0
D–F = 8.25

Calculation:

1. Take moments about C to find the load at A:

$$F_A = [1.5(500) + 3(700) + 7(400)]/8 + 100$$
$$= [750 + 2100 + 2800]/8 + 100 = 806 \text{ lb}$$

2. Take moments about D to find the load at F:

$$F_F = [3.85(800) + 8(200)]/8.25 + 100$$

$$= [3080 + 1600]/8.25 + 100 = 667 \text{ lb}$$

3. Find load at D: D = total system load minus the loads at A and F.

$$F_D = 4300 - 806 - 667 = 2827 \text{ lb (answer)}$$

Select a support from a manufacturer's catalog. One spring support suitable for this application has a nominal rating of 2900 lb and a load-range of 2030 to 4060 lb over its working deflection range, which tends to justify the short-cut method described herein.

Note. Where it is desired to load the spring-supports to compensate for the liquid load, proceed as follows:

1. Calculate the load imposed on each support by the liquid alone, using the method given in the example above.

2. Adjust each constant support so that it lifts the additional weight designated for it. The amount of this weight is usually arbitrarily set at 50% of the calculated liquid-load. This will make the prestress in the pipe about equal to the stress during operation and is about half of that which would result if the spring-support were adjusted for the full liquid-load.

Piping Tierods

Piping tierods are used to restrain the anchor force produced by the particular group of pipe joints which tend to pull apart when subjected to internal pressure. Figure 135 shows a typical tierod design.

The absence of tierods where they should be used is one of the most frequent hazards in existing piping systems. This section will show the conditions which require tierods to make a piping system safe and provide the tools to simplify the design of tierods and other anchor systems.

Where tierods support a dead load in a long vertical run-of-pipe, the

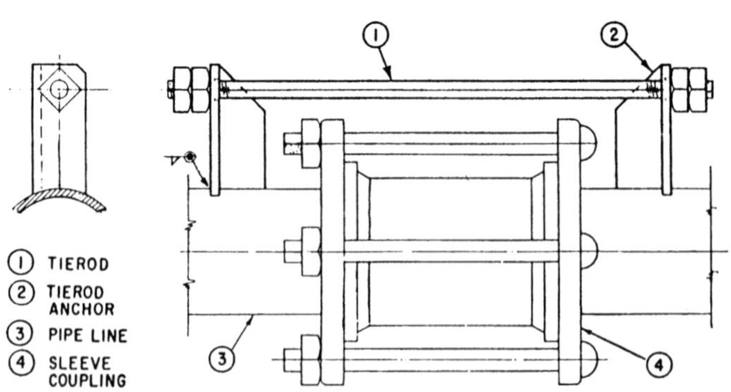

FIG. 135 Typical tierod design detail.

TABLE 19 Effective Area of Pipe Joints

Nominal Pipe Size	Slip Joints	Convoluted Expansion Joints[a]
2	4	13
3	9	20
4	15	29
6	34	51
8	58	97
10	90	135
12	128	180
14	154	231
16	200	289
18	254	353
20	314	424
24	452	583
30	702	871

[a]Areas representative of the product of several manufacturers.

effect of the dead load on the design conditions of the tierods should be checked.

The longitudinal force in a pipe joint, caused by internal pressure, is

$$F = A_J P$$

where A_J = effective area of the pipe joint
P = pressure in the pipe

The effective area of a sleeve coupling or a bell and spigot-type joint is considered to be the area of a circle having a diameter equal to the outside diameter of the pipe. The effective area of convoluted expansion joints is determined by pressure tests, which are made by the manufacturer and published in his catalogs. The effective areas given in Table 19 are representative of the products of several manufacturers.

Design Procedure

1. Determine the longitudinal force in the pipe joint from the above equation, using the effective area from Table 19 and the specified test pressure, or 1½ times the working pressure of the pipeline, whichever is greater. Add to this force any other longitudinal force in the joint to obtain the total longitudinal design force.
2. With this design force, enter Table 20 and select the number and size of tierods most suited to the operating conditions. The joints which are required to have a hinge action should have two tierods.
3. Enter Table 21 with the size of the tierod selected, which determines

TABLE 20 Tierod Selection Data[a,b]

Longitudinal Force, P (lb)	Number of Tierods 1	2	3	4	6
500	1/4				
1,000	1/2	3/8	5/16	1/4	
1,500	9/16	7/16	3/8	5/16	
2,000	5/8	1/2	3/8	3/8	
2,500	3/4	1/2	7/16	3/8	
3,000		9/16	1/2	7/16	
3,500		5/8	1/2	7/16	
4,000		3/4	9/16	1/2	
4,500		3/4	9/16	1/2	
5,000		3/4	5/8	1/2	
6,000		3/4	5/8	9/16	
7,000		7/8	3/4	5/8	
8,000		7/8	3/4	5/8	
9,000		1	3/4	3/4	
10,000		1	7/8	3/4	
12,000		1 1/8	7/8	3/4	
16,000		1 1/4	1	7/8	
20,000		1 3/8	1 1/8	1	
24,000		1 1/2	1 1/4	1 1/8	7/8
30,000		1 3/4	1 3/8	1 1/4	1
40,000		2	1 1/2	1 3/8	1 1/8
50,000			1 3/4	1 1/2	1 1/4

[a]The recommended choice of tierods is within the heavy lines.
[b]The size of tierods is based on a tensile strength of bolts given in Table 22.

the size of the tierod anchor required. The required section modulus of the anchor, given in the second column, has been determined using an allowable fiber stress of 10,000 lb/in.^{2} and a distance of 3½ in. from the rod to the pipe, which is adequate for all joints in common use. The structural member size and shape given in the third column is merely a suggestion; other structural members may be used provided their section modulus is adequate.

Table 22 is provided to give data for the design of tierods to suit conditions other than those given in Table 21.

Design Notes

Steel for tierods should be selected to have an ultimate strength of 40,000 lb/in.^{2} or more.

1. Tierods should have national-course threads and at least two nuts on each end.

TABLE 21 Tierod Anchor Selection Table

Rod Size (in.)	Required Section Modulus (in.^{3})[a]	Structural Member Size and Shape[b]
$\frac{1}{4}$	0.09	$2 \times 1\frac{1}{2} \times \frac{1}{8}$ L
$\frac{5}{16}$	0.15	$2 \times 1\frac{1}{2} \times \frac{3}{16}$ L
$\frac{3}{8}$	0.23	$2 \times 1\frac{1}{2} \times \frac{1}{4}$ L
$\frac{7}{16}$	0.32	$2\frac{1}{2} \times 1\frac{1}{2} \times \frac{1}{4}$ L
$\frac{1}{2}$	0.44	$2\frac{1}{2} \times 2 \times \frac{5}{16}$ L
$\frac{9}{16}$	0.56	$3 \times 2 \times \frac{1}{4}$ L
$\frac{5}{8}$	0.70	$3 \times 2 \times \frac{5}{16}$ L
$\frac{3}{4}$	1.06	$3 \times 2 \times 3 \times \frac{1}{4}$ U
$\frac{7}{8}$	1.47	$4 \times 2 \times 4 \times \frac{1}{4}$ U
1	1.93	$4 \times 2 \times 4 \times \frac{1}{4}$ U
$1\frac{1}{8}$	2.42	$4 \times 2 \times 4 \times \frac{5}{16}$ U
$1\frac{1}{4}$	3.10	
$1\frac{3}{8}$	3.70	
$1\frac{1}{2}$	4.50	
$1\frac{3}{4}$	6.10	
2	8.00	

[a]Based on an allowable fiber stress of 10,000 lb/in.^{2}.
[b]Other shapes may be used if their section modulus is adequate.

2. Sufficient weld-metal should be used to develop the full strength of the tierod anchor where it connects to the pipe.
3. Where an expansion joint is used in a vacuum line, or where any other compressive force is required to be restrained, a pipe sleeve should be

TABLE 22 Tierod Design Data

Bolt Diameter (in.)	No. of Threads per Inch	Areas: Full Bolt (in.^{2})	Areas: Bottom of Threads (in.^{2})	Tensile Strength (pounds at 10,000 lb/in.^{2})
$\frac{1}{4}$	20	0.049	0.027	270
$\frac{5}{16}$	18	0.077	0.045	450
$\frac{3}{8}$	16	0.110	0.068	680
$\frac{7}{16}$	14	0.150	0.093	930
$\frac{1}{2}$	13	0.196	0.126	1,260
$\frac{9}{16}$	12	0.248	0.162	1,620
$\frac{5}{8}$	11	0.307	0.202	2,020
$\frac{3}{4}$	10	0.442	0.302	3,020
$\frac{7}{8}$	9	0.691	0.419	4,190
1	8	0.785	0.551	5,510
$1\frac{1}{8}$	7	0.994	0.693	6,930
$1\frac{1}{4}$	7	1.227	0.890	8,890
$1\frac{3}{8}$	6	1.485	1.054	10,540
$1\frac{1}{2}$	6	1.767	1.294	12,940
$1\frac{3}{4}$	5	2.405	1.745	17,450
2	$4\frac{1}{2}$	3.142	2.300	23,000

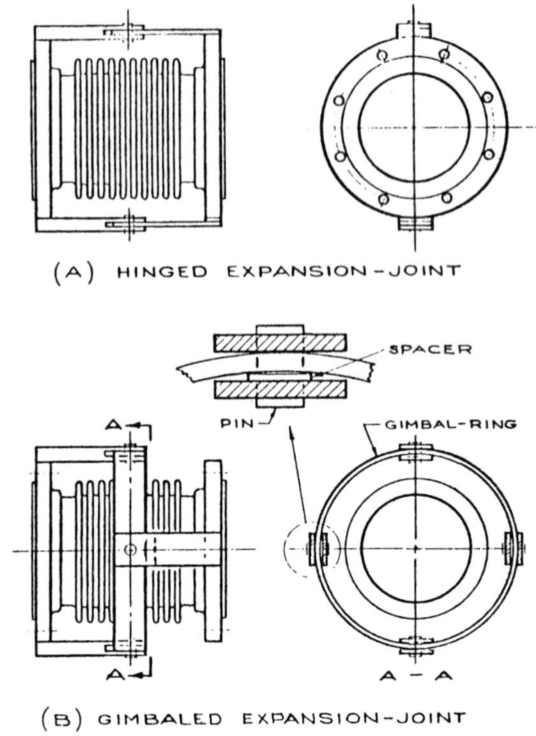

FIG. 136 Gimbaled and hinged expansion joint detail.

placed over each tierod to span the distance between the anchors, with the expansion joint in the free condition, as shown in Fig. 136.

4. The design details of tierods around an expansion joint shall be such that the required function of the expansion joint is not jeopardized. In certain cases it is necessary to specify a particular gap between the tierod anchor and the tierod stop. See Fig. 135 for typical tierod design detail.

Design of Other Anchor Systems

Figure 136 shows the design features of gimbaled and hinged anchor systems around expansion joints. These anchors are particularly suitable where a hinge action is required, while at the same time the joint is subjected to forces tending to produce lateral displacement in the joint. While the gimbaled joint has four pin connections, each pin is subjected to one-half the total force, just as in the hinged joint. The required cross-sectional area of one pin is

$$A_g = 2F/3S_s$$

where F = total longitudinal design force in the expansion joint
S_s = allowable shear stress

Where the force is appreciable, the pin connection should be designed to place the pin in double shear, as shown in Fig. 136.

The maximum stress in the gimbal ring occurs at the pin connections, where the bending moment is

$$M = FR_{\text{o.d.}}/8$$

where F = total longitudinal design force in the expansion joint
$R_{\text{o.d.}}$ = outside diameter of the gimbal ring

This equation takes into consideration the fact that the ring is not an ideal beam, and it is intended to give a conservative design.

The tension bar at the pin connection should be designed to be stronger than the pin.

The holes for the pins should be drilled and reamed to a Class 3 (medium) fit.

Piping Materials

In refinery or petrochemical plant piping, carbon steel, the low-alloy steels containing up to 9% chromium with 0.5 or 1.0% molybdenum, the straight chromium ferritic stainless steels (400 series), and the chromium-nickel austenitic stainless steels (300 series) are about the only ones which have been used successfully.

Nevertheless, materials selection for process pipe is not as simple as this might suggest. The temptation to spend a little more for pipe to avoid possible trouble is great. Do not select a more expensive material than is actually required unless you know that it has been used successfully before or you are fully aware of all the additional problems that could arise.

Carbon Steel

Carbon steel is the most common pipe material, but do not look down on it simply because it is relatively inexpensive. There are other reasons why the process industries use it so widely. One of its big advantages is how close it comes to being foolproof—most of the time. Do not forget either that *carbon steel* is a generic term, not a specific one, even if it is of specification grade.

Brittle Failure

Below some limiting temperature, steel pipe is notch sensitive and can crack at lower-than-yield-point stress with little or no deformation or absorption of energy. Do not use carbon steel pipe at temperatures even as low as 0°F unless the service is completely nonhazardous or you have specific evidence that a particular lot of pipe is resistant to brittle failure at the service temperature.

ERW Pipe

If run-of-the-mill pipe inconsistencies disturb you, and they should, inquire about ERW (electrical resistance welded) carbon steel pipe. This has more consistent properties than the usual steel pipe. You should be able to buy ERW pipe for essentially the same price as pipe meeting your old specifications.

It is made from fine-grain, fully aluminum-killed (0.02% minimum residual), basic-oxygen steel containing 0.08 to 0.15% carbon, 0.27 to 0.63% manganese, 0.05% maximum sulfur, and 0.035% maximum phosphorus. In the usual normalized condition, it has a minimum tensile strength of 40,000 lb/in.^{2}, a minimum yield strength of 30,000 lb/in.^{2}, and a minimum elongation of 40% in 2 in.

Low-temperature impact strengths have bene good, although not guaranteed, down to −50°F. The combination of low carbon content and high manganese-to-carbon ratio increases resistance to cracking and brittle fracture but still provides adequate control of mechanical properties so maximum values need not be specified.

Corrosion

In petroleum refineries, process streams containing hydrogen also frequently contain hydrogen sulfide. This causes sulfidic corrosion. You know from experience that increasing the chromium content of a steel increases its resistance to corrosion by high-sulfur crudes. However, do not jump to the conclusion that chromium alloying always improves resistance to sulfidic corrosion. It does so if the operation is dirty, as it usually is in crude streams, or if the corrodants are elemental sulfur or sulfur compounds that do not decompose to release hydrogen sulfide. This increased resistance to sulfur corrosion depends on formation of a protective scale. With such scales, the corrosion rate is parabolic—it decreases with exposure time.

If the operation is clean, as it usually is when hydrogen is present, the iron sulfide corrosion product is not protective. Under these conditions, the corrosion rate is linear and does not decrease with time, and chromium additions are not beneficial. Carbon steel and 9% chromium steel corrode at substantially the same rate, and a 5% chromium steel may corrode even faster than the other two. So don't waste your money by picking a chromium content higher than you need to resist hydrogen attack.

Welding Problems with Cr-Mo Pipe

Do not use low-chromium steel pipe unless you are willing to pay for more careful welding and postweld heat treatments. At a given hardness, low-alloy chromium-molybdenum steels have somewhat more ductility than carbon steels. However, because they air harden so much, they usually require postweld heat treatments to toughen the weld metal and heat-affected zone. This heat treatment complicates field welding.

Do not always be subservient to convention, because intelligent materials selection should decrease some of these welding complications. For example, when carbon steel pipe is welded to a higher alloy, such as 5-Cr, ½-Mo steel, some engineers require weld metal corresponding to the higher alloy and a postweld heat treatment. Should not carbon steel weld metal be equally satisfactory? A postweld heat treatment might then be unnecessary. Such joints clearly have to be in a process zone where carbon steel is acceptable.

Often the very engineers who insist on postweld heat-treatment with carbon steel or low-alloy weld metals would be willing to omit the postweld heat treatment if an austenitic stainless steel electrode, particularly type 309 (25-Cr, 12-Ni), were used. Ironically, their reasoning that the hardened, heat-affected zone of the 5-Cr steel would have ductile, austenitic stainless steel weld metal on one side of it and ductile, 5-Cr steel parent metal on the other, is equally valid in the rejected case where the weld metal is carbon steel.

Higher Alloys

If low or intermediate-chromium steel pipe will not resist corrosion adequately in refinery streams, it will be necessary to use aluminum-coatings or high-chromium ferritic (Type 400) or austenitic (Type 300) stainless steels. Do not try to justify anything more expensive. It is almost impossible. When trying to choose between these possibilities, look at the overall picture. Do not even consider corrosion resistance. The difference is slight. As a matter of fact, the difference in cost may also be less than expected.

Stainless steels have been most successful at operating temperatures lower or higher than most of those in petroleum refining. Do not expect plain 11 to 13% chromium ferritic stainless steel pipe (Type 410) to be much of an improvement over 9-Cr, 1-Mo steel. It has only borderline corrosion resistance. Type 410 welds also air harden and require postweld heat treatment. Do not put this alloy, or any of the lower content alloys, into service if the hardness is above Rockwell C22. It is likely to hydrogen stress-crack. The aluminum-containing grade (Type 405) is completely ferritic up to the melting point. It does not harden on welding, but it develops extremely coarse grains adjacent to the fusion line, which lower ductility.

Do not use ferritic stainless steels containing more than 16% chromium in the 750 to 1000°F temperature zone. They invariably embrittle because of precipitation of a chromium-rich constituent. Even the 11 to 13% chromium alloy sometimes seems to embrittle in this way, for reasons not clear. Unfortunately, this is a common temperature zone for many refinery and chemical plant processes. These ferritic stainless steels should not be used above 1000°F, either. They will embrittle for a somewhat different reason: the formation of an iron-chromium intermetallic compound called *sigma phase*.

Austenitic stainless steels resist oxidation by either oxygen-bearing or sulfur-bearing process streams. Austenitic stainless steel pipe is not foolproof simply because it has corrosion resistance. Many plant operators are too optimistic, feeling that tight temperature controls are unnecessary with austenitic stainless steels. If temperatures climb too high, even these steels will oxidize and sulfidize significantly.

Relatively short-time tests indicate that austenitic stainless steels are ductile at all temperatures. This conclusion should not be considered valid for steels exposed to thousands of hours at process temperatures. Long exposures between, roughly, 800 and 1600°F precipitate chromium carbide and sigma phase in many stainless steels, causing a significant loss in atmospheric-temperature ductility. Ductility at higher temperatures is less affected.

Stress-Corrosion Cracking

The greatest problems with austenitic stainless steel piping usually arise when the unit is off-stream rather than when it is operating. Such problems must be anticipated. The use of stainless steels requires that the necessary steps be taken to avoid them. Chlorides and caustics can cause any austenitic stainless steel pipe to crack transgranularly under some conditions. Plain chromium stainless steels do not crack in chloride solutions, but they usually pit badly enough to be only moderately satisfactory.

Strictly speaking, chloride stress-corrosion cracking will not occur unless there is contact with an aqueous solution of suitable chloride concentration, a favorable temperature, and strain or residual stress. These requirements may, however, be met rather unpredictably.

For example, the small amounts of chlorides in most external pipe insulations can be leached out by exposure to weather and become concentrated at the pipe wall. Temperature may be difficult to measure, let alone control, especially during startup or shutdown when gradients exist. Residual stresses usually are present in a relatively low yield strength material like annealed stainless steel pipe. A pipe bumped in shipment or sprung or cold bent in fitting can have all the stress needed.

In fact, circumferential weld shrinkage alone, particularly in heavy-wall pipe, may create complex bending stresses at the joint. Although postweld heat treatments should relieve many of these stresses, the subsequent cooling can reintroduce harmful stresses if there is much restraint. The fact that the much higher thermal expansion and contraction of the austenitic stainless steels may introduce unexpected restraint stresses, as well as being troublesome in piping layouts, should not be overlooked.

When the normal carbon (0.08% maximum) grades of austenitic stainless steel pipe are used in the temperature range of 800 to 1500°F, chromium carbides precipitate in the grain boundaries. This sensitizes the material and makes it susceptible to intergranular corrosion in many acid media. As long as the unit stays on-stream, there is no real deterioration from this precipitate except some loss in ductility. In fact, if the material stays at these temperatures for a long enough time (the necessary time decreases as the temperature increases to approximately 1650°F), it will heal itself and lose its sensitivity to intergranular corrosion. Nevertheless, this long-time exposure must not be relied upon. Something, if only a check to be sure everything is working properly, generally will shut down initial runs of a new unit prematurely.

Sensitized material is also susceptible to intergranular stress-corrosion cracking in the polythionic acids formed by the reaction of iron sulfide, air, and moisture. Any type of stress-corrosion cracking is troublesome, because it seldom is noticed until the unit is being brought back on stream. Then it is invariably attributed to something that happened during the startup. Shutdown costs may be increased still further by futile attempts to weld-repair the leaks. With cracking like this, weld shrinkage will open up another crack as rapidly as one leak is repaired.

The extra low-carbon (L grades), the chemically stabilized (Types 347 or 321), or the controlled-ferrite (centrifugally-cast, usually) varieties of austenitic stainless steels cannot be expected to solve these potential cracking problems. They may help, but success is unpredictable.

For example, in the low-carbon types, some carbide precipitation still can occur toward the low side of the 800 to 1500°F temperature range. This may be enough to sensitize the structure to intergranular corrosion. The higher the carbon content, the greater the danger.

The chemically stabilized types (containing columbium-tantalum or titanium) are excellent if carbide precipitation occurs during welding and if subsequent service is near atmospheric temperature (at least below 700°F). Then only randomly distributed columbium, tantalum, or titanium carbides precipitate during cooling. However, if service is between 800 and 1000°F for long times, the carbon left in solution, even after the stabilized carbides form during postweld cooling, precipitates conventionally as chromium carbide at grain boundaries.

Here again, if there is enough precipitation at any one location, and if these grain-boundary particles are properly spaced, sensitization still can occur. In the controlled-ferrite grades, the ferrite won't crack, although it may embrittle or corrode selectively in some media, but the austenitic matrix is subject to all the disabilities mentioned above.

The higher-nickel stainless steels offer some hope. But these alloys are expensive and they may not have adequate resistance to either stress-corrosion cracking or sulfidic corrosion if conditions are severe.

There is no certain way of preventing transgranular (chloride) stress cracking, and the only real solution to intergranular (polythionic acid) stress cracking is to heat-treat the piping, including welds, in order to precipitate as much carbon in a stabilized, nonsensitizing form as possible before service. Even then, the operators must be cautioned to open equipment as infrequently as possible, to leave it open as short a time as possible, and to keep it either dry and blanketed with inert gas or flooded with an alkaline solution when it is off-stream.

Low temperature metallurgy divides naturally into ranges dictated by material impact properties, as shown in Fig. 137. The first, extending down through −50°F, is widely involved in ammonia and propylene production and requires impact-tested carbon steel.

The second, from −51°F down through −150°F, is required to isolate acetylene and ethane and employs low nickel steel and other low alloy steels such as Cr-Cu-Ni steel.

The third, from −151°F down through −325°F, is necessary to liquefy methane and nitrogen and uses more highly alloyed steels such as 9% nickel, stainless steel, or nonferrous metals and alloys such as copper, aluminum, nickel, etc.

The last, from −326°F down, isolates neon, hydrogen, and helium, and uses austenitic stainless steel, copper, or aluminum.

These temperature limits are chosen to be sure that the metals are operating above their transition temperatures.

Insulation

Table 23 lists the thickness of conventional low temperature piping insulation for all the common materials except the foamed resins. Before beginning the piping design, the types and thickness of insulation must be firmly established so that proper line clearances may be set. Because the clearances required are much larger than conventional process piping, this aspect is particularly important with secondary effects on such things as pipe support and structural standards.

Over and above insulation clearance, an allowance must be provided for thermal moment. Illustrations of minimum clearances are shown in Figs. 138 and 139. It should be noted from the control-valve assembly in Fig. 139 that it is not always possible to have a fitting-to-fitting arrangement although it is otherwise economically desirable. This is sacrificed so that maximum

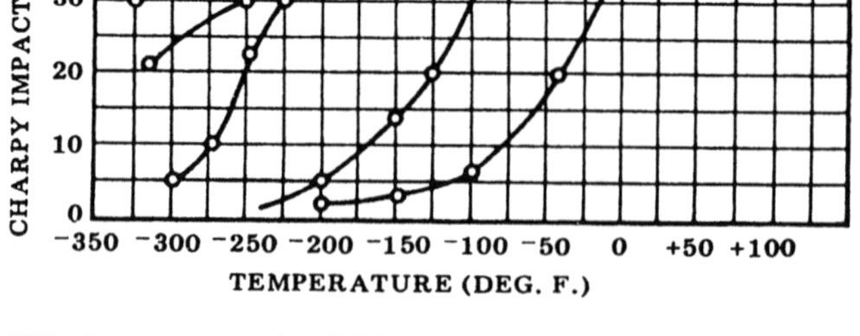

FIG. 137 Impact properties divide materials into low temperature ranges.

TABLE 23 Thickness of Insulation for Piping, Inches

Temperature Range (°F)[a]	Nominal Pipe Size (in.)																	
	½	¾	1	1½	2	2½	3	4	6	8	10	12	14	16	18	20	24	26
+60 to +40	1	1	1	1	1	1	1	1	1	1½	1½	1½	1½	1½	1½	1½	1½	1½
+40 to +30	1	1	1	1½	1½	1½	1½	1½	1½	1½	1½	1½	1½	1½	1½	1½	1½	1½
+30 to +20	1½	1½	1½	1½	1½	1½	1½	1½	2	2	2	2	2	2	2	2	2	2
+20 to +10	1½	1½	1½	1½	2	2	2	2	2	2	2½	2½	2½	2½	2½	2½	2½	2½
+10 to 0	1½	2	2	2	2	2	2	2½	2½	2½	2½	2½	2½	2½	2½	2½	3	3
0 to −10	2	2	2	2	2	2½	2½	2½	2½	3	3	3	3	3	3	3	3	3
−10 to −20	2	2	2	2½	2½	2½	2½	2½	3	3	3	3	3	3½	3½	3½	3½	3½
−20 to −30	2	2	2½	2½	2½	2½	3	3	3	3½	3½	3½	3½	3½	3½	3½	3½	3½
−30 to −40	2½	2½	2½	2½	3	3	3	3	3½	3½	3½	4	4	4	4	4	4	4
−40 to −50	2½	2½	2½	3	3	3	3	3½	3½	4	4	4	4	4	4	4	4½	4½
−50 to −60	2½	2½	3	3	3	3½	3½	3½	4	4	4	4½	4½	4½	4½	4½	4½	4½
−60 to −70	2½	3	3	3	3½	3½	3½	4	4	4½	4½	4½	4½	4½	5	5	5	5
−70 to −80	3	3	3	3½	3½	3½	4	4	4½	4½	4½	5	5	5	5	5	5	5
−80 to −90	3	3	3	3½	3½	4	4	4	4½	4½	5	5	5	5	5	5½	5½	5½
−90 to −100	3	3	3½	3½	4	4	4	4½	4½	5	5	5	5½	5½	5½	5½	5½	5½
−100 to −110	3	3½	3½	4	4	4	4½	4½	5	5	5½	5½	5½	5½	5½	6	6	6
−110 to −120	3½	3½	3½	4	4	4½	4½	4½	5	5½	5½	5½	6	6	6	6	6	6
−120 to −130	3½	3½	3½	4	4½	4½	4½	5	5	5½	6	6	6	6	6	6½	6½	6½
−130 to −140	3½	3½	4	4	4½	4½	5	5	5½	5½	6	6	6	6½	6½	6½	7	7
−140 to −150	3½	4	4	4½	4½	4½	5	5	5½	6	6	6½	6½	6½	6½	7	7	7
−150 to −160	3½	4	4	4½	4½	5	5	5½	5½	6	6½	6½	6½	6½	7	7	7	7
−160 to −170	4	4	4	4½	5	5	5	5½	6	6	6½	6½	7	7	7	7	7½	7½
−170 to −180	4	4	4½	4½	5	5	5½	5½	6	6½	6½	7	7	7	7½	7½	7½	7½
−180 to −190	4	4	4½	5	5	5	5½	5½	6	6½	7	7	7	7½	7½	7½	8	8
−190 to −200	4	4½	4½	5	5	5½	5½	6	6	6½	7	7	7½	7½	7½	7½	8	8
−200 to −210	4	4½	4½	5	5	5½	5½	6	6½	7	7	7½	7½	7½	8	8	8	8
−210 to −220	4½	4½	4½	5	5½	5½	6	6	6½	7	7½	7½	7½	8	8	8	8½	8½
−220 to −230	4½	4½	5	5	5½	5½	6	6	7	7	7½	7½	8	8	8	8½	8½	8½
−230 to −240	4½	4½	5	5½	5½	6	6	6½	7	7½	7½	8	8	8	8½	8½	9	9
−240 to −250	4½	5	5	5½	5½	6	6	6½	7	7½	8	8	8	8½	8½	8½	9	9

[a]Below the first value and including the second.

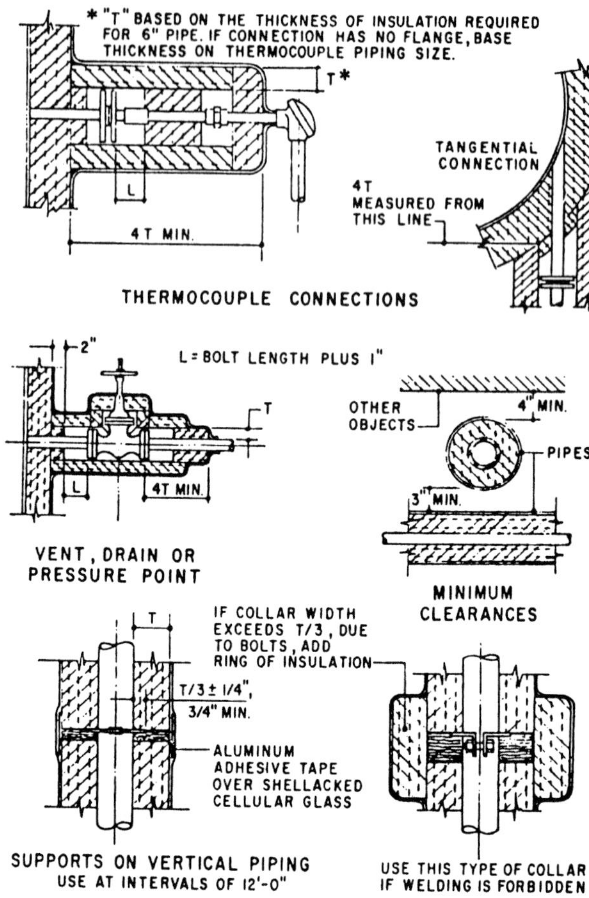

FIG. 138 Insulation and clearance details for low temperature piping.

insulation efficiency can be attained through proper vapor sealing, maximum use of pre-formed sections, reduced installation and insulation material costs, and minimum disruption of insulation when removing bolting and valves.

Low Temperature Piping Materials

Carbon steel piping as applied for warm services may, in accordance with ANSI B31.3, be used at usual allowable stress down to a minimum temperature of −20°F. With additional requirements as to chemistry, melting practice, heat treatment, and impact properties in accordance with ASTM A333 (Grade O), carbon steel is usable down to a minimum temperature of −50°F. This specification limits manufacture to seamless or welded pipe without filler metal addition.

The hazard of brittle fracture is lessened as the average and local stress levels are lowered. Both the ASME Unfired Pressure Vessel Code (Par. UCS-66) and the Code for Pressure Piping (Par. 323.2.2) recognize this by

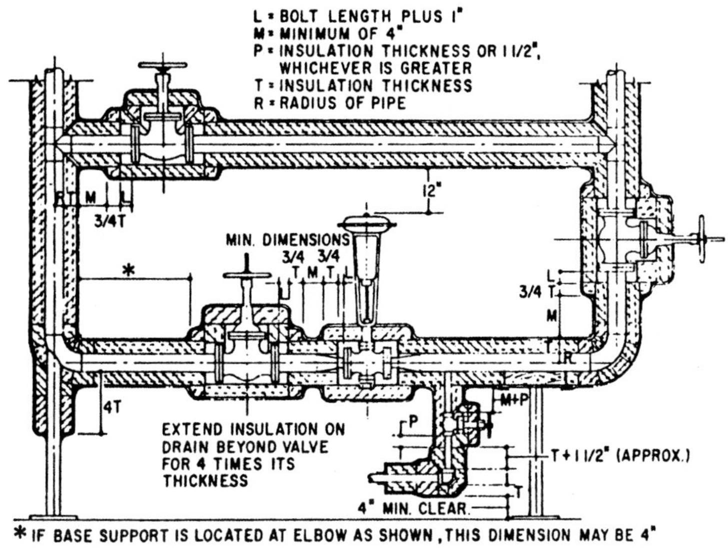

FIG. 139 Insulation details for low temperature control valve piping.

allowing materials to be used below the transition temperature. Where the allowable stress is reduced to 40% of the normal allowable, the ASME Code permits such material to be used without limitation. The code for pressure piping sets an allowable stress of 15% of the maximum allowable without impact test.

Nickel Steel

Steels containing 3½, 5, and 9% nickel also require impact tests to determine their suitability for low temperature operation. The 3½ and 5% nickel steels may be used down to −150°F and require stress relief after welds and forming operations. Nine percent nickel steel requires no stress relief and is usable down to −325°F; however, valves, fittings, and piping are not readily available.

Stainless Steel

Austenitic stainless steels of the AISI 300 series are well suited for low temperature service by their strength and impact resistance properties. They are easy to weld, have a low heat conductivity, and require no stress relief. Partially offsetting these factors is their higher cost. The commonly used analyses of stainless steel are available in thin wall piping schedules 5S and 10S as well as in the normal schedules. Thinner walls are allowable for stainless steel because it is not subject to corrosion. All of its thickness is available for strength.

Nonferrous materials are well suited for low temperature use because of their excellent ductility. They have been widely used in air plants where operating pressures are relatively low.

Copper, in the wrought form, presents a dense, impervious barrier that is good for holding vacuum. The copper-nickel alloys, including the Monels, all have good mechanical properties and low thermal conductivities and are easily soldered. Some of the high nickel alloys have strengths comparable to that of stainless steel.

Aluminum piping is very attractive from a price standpoint and can be welded by means of the inert gas shielded electric arc. Use of aluminum in low temperature service is increasing. Protection against external fire should be considered because of the relatively low melting point. Strength of welds in nonheat-treatable alloys is at least equal to that of the annealed parent metal. This is of special practical significance for the stronger, magnesium alloyed, 5000 series. Strength, ductility, and toughness are retained down to −450°F without heat treatment. The lack of availability of aluminum valves ia a serious handicap to full exploitation of the potential of the piping.

United States vs European Piping Specifications

In selecting a suitable piping specification for refinery or process plant service, the average engineer rarely needs more than a superficial understanding of what the specifications cover. If the chemical composition is of the desired type and the strength characteristics and dimensions are known, the purchaser relies on the specification title and scope to testify that the pipe will be suitable for the intended service. This reliance is justified because national specifications have been prepared by groups of experts who have extensive knowledge of the production and usage of the pipe.

For this average engineer, the comparison Tables 24–32 will give sufficient advice on which outside-United-States specifications are comparable to the more familiar ASTM and API specifications. The specifications tabulated are limited to pipe such as is commonly used in process plants for pressure applications.

Steelmaking Process and Piping Specs

ANSI B31.3 specifically limits steelmaking processes to electric furnace, open hearth, or deoxidized acid Bessemer steel. On the other hand, both API 5L and ASTM A53 permit undeoxidized Bessemer steel. In using these specifications, it is necessary to clearly state any limitations of steelmaking practice that may be considered desirable to ensure adequate quality.

General speaking, the improvement of the old and the continued multiplication of the new steelmaking techniques means that specifying the type of steelmaking furnace or converter is no longer of much value as a means of defining the end product. In the future, specification writing will need to depend more on a definition of the required mechanical properties and chemical composition—particularly nitrogen and residual elements such as

TABLE 24 Seamless Carbon Steel Pipe for Normal Duties—About 50,000 lb/in.^{2} Ultimate Strength

			Chemical Composition (%)						Mechanical Properties (minimum)					
											Flattening Test			
Country	Standard	Steelmaking Practice	C max	Mn	Si	P max	S max	Other Maxima	Yield Stress (lb/in.^{2})	Ultimate Stress (lb/in.^{2})	A[a]	B[b]	Heat Treatment	Maximum Recommended Temperature (°F)
France	GAPAVE 411 A 37C	Open hearth or electric furnace	—	—	—	—	—	—	—	52,500	—	4t	Cold-drawn pipe to be normalized	662
Germany	DIN 1629 St. 35	Electric furnace, open hearth, or oxygen converter	0.18	—	—	0.05	0.05	—	34,000	50,000	—	—	None—generally hot drawn	572 (see DIN 2401)
Italy	Aq 35 UNI 663 C	Not specified	—	—	—	—	—	—	30,000	50,000	2t	—	None	752
Sweden	SIS 1233–05	Killed steel, process unspecified	0.17	−0.5	0.10 to 0.40	0.05	0.05	Cr 0.2 Cu 0.3 N 0.009	30,000	50,000	—	$\frac{1.07t}{0.07 + (t/D)}$	Cold-drawn pipe to be heat treated	752
United Kingdom	BS.3601 HFS 22 CDS 22	Open hearth, electric furnace, or oxygen converter	0.21	0.70 max	—	0.05	0.05	—	30,000	50,000	4t or 2D/3	—	Cold-drawn pipe to be heat treated	850 (see BS.3351)
United States	ASTM A 53 seamless	Type S, Grade A: Open hearth, basic oxygen, acid Bessemer, or electric furnace	—	—	—	0.048 0.11[c]	—	—	30,000	48,000	—	$\frac{1.09t}{0.09 + (t/D)}$	None	1100 (see ASA B31.3)
	API 5L Line pipe seamless	Grade A: Open hearth, electric furnace, or basic oxygen	0.22		—	0.04	0.05	—	30,000	48,000	—	—	None	1100 (see ASA B31.3)

[a]A: Distance between inside surfaces of tube. [b]B: Distance between platens of press. [c]As limited by ASA B31.3:1962.

TABLE 25 Seamless Carbon Steel Pipe Suitable for Higher Temperatures and Pressures—About 60,000 lb/in.^{2} Ultimate Strength

Country	Standard	Steelmaking Process	Deoxidation Practice	Chemical Composition (%) C max	Mn	Si	P max	S max	Other Maxima
France	GAPAVE 421 A 42 C (Si-killed)	Open hearth or electric furnace	Killed steel	—	—	—	—	—	—
Germany	DIN 1629 St 45.4	Open hearth, electric furnace, or oxygen converter	Killed steel	0.22	0.40 min	0.10 to 0.35	0.05	0.05	Cr 0.3
	DIN 17175[c] St 45.8	Not specified	Killed steel	0.22	0.45 min	0.10 to 0.35	0.05	0.05	N 0.010 for basic Bessemer or O_2 converter
Italy	Aq 45 UNI 663 D (Si 0.1% min.)	Not specified	Killed steel	—	—	—	—	—	—
Sweden	SIS 1435–05	Not specified	Killed steel	0.22	0.60 min	0.10 to 0.40	0.05	0.05	Cr 0.2 Cu 0.3 N 0.009
United Kingdom	BS.3602 HFS 27 (Si-killed)	Open hearth, electric furnace, or basic osygen.	Killed steel	0.25	0.30 to 0.70	0.10 to 0.35	0.05	0.05	—
United States	ASTM A106, Grade B	Open hearth, electric furnace, or basic oxygen	Killed steel	0.30	0.29 to 1.06	0.10 min	0.048	0.058	—

[a]A: Distance between inside surfaces of tube.
[b]B: Distance between platens of press.
[c]Similar to DIN 1629 Class 4, but with guaranteed elevated temperature properties.

chromium, nickel, and molybdenum—than on limiting the permissible means of making the steels.

Deoxidation Practice

As far as deoxidation is concerned, carbon steel may be divided into three categories: rimming, killed, and semikilled steel.

Rimming steel is often used for furnace-welded pipe. The low carbon skin of skelp rolled out from a rimmed ingot is particularly good for pressure

Mechanical Properties (minimum)					
Yield Stress (lb/in.^{2})	Ultimate Stress (lb/in.^{2})	Flattening Test A[a]	Flattening Test B[b]	Heat Treatment	Maximum Temperature (°F)
—	60,000	—	5t	Normalize	662
37,000	64,000	—	$\frac{1.07t}{0.07 + (t/D)}$	Not specified	572
37,000	64,000	—	$\frac{1.07t}{0.07 + (t/D)}$	Normalize, anneal, or quench and temper	752
34,000	64,000	4t	—	Not specified	752
37,000	64,000	—	$\frac{1.05t}{0.05 + (t/D)}$	Cold-drawn pipe to be heat treated	887
36,000	60,000	6t or 3D/4	—	Cold-drawn pipe to be heat treated	950 (see BS .3351)
35,000	60,000	—	$\frac{1.07t}{0.07 + (t/D)}$	Cold-drawn pipe to be heat treated	1100 (see ASA B31.3)

welding, and the finished pipe has good corrosion resistance on both inside and outside surfaces. Seamless pipe made from rimming steel is, however, only acceptable for noncorrosive service.

National standards for carbon steel pipe used at higher temperatures and pressures commonly require killed steel. For normal duties the deoxidation practice is not normally specified. Nevertheless, most seamless pipe is either killed or semikilled, except where specially ordered. An exception to this generalization is Germany, where rimmed seamless pipe is a commercial product and where both DIN 1629 St 35 and DIN 17175 St 35.8 would permit such material to be supplied. Although responsible manufacturers would not supply rimmed seamless pipe for refinery or petrochemical service, it is wise to specifically prohibit rimmed steel for seamless pipe when ordering

TABLE 26 Welded Carbon Steel Pipe for Normal Duties—About 60,000 lb/in.^{2} Ultimate Strength

Country	Standard		Plate Specification	Steelmaking Process	Welding Process	Heat Treatment after Welding
Germany	DIN 1626 Blatt 3		DIN 17100 St 34–2	Killed or rimming steel: any process	Double-side fusion weld or any type of pressure weld (including butt weld)	Not required
Sweden	SIS 1233–06		SIS 1233	Killed steel	Electric resistance weld	Required
United Kingdom	BS.3601	BW 22	—	Not specified	Continuous furnace butt weld	Not required
		ERW 22	—	Open hearth, electric furnace, or oxygen converter	Electric resistance weld	Not required
		EFW	—	Open hearth, electric furnace, or oxygen converter	Electric fusion weld (single or double side)	Not required
		SFW	—	Open hearth, electric furnace, or oxygen converter	Spiral seam double-side electric fusion weld	Not required
United States	API 5L[c]	Butt weld	—	Open hearth, electric furnace, or oxygen converter Cl, 1 Acid Bessemer	Continuous furnace butt	Not required
	API 5L[c]	Electric weld or submerged arc weld Grade A	—	Open hearth, electric furnace, or basic oxygen	Electric resistance weld or submerged are weld, double side	Not required
	A 53[c]	Type F Furnace-welded	—	Open hearth, electric furnace, acid-oxygen-steam, or basic oxygen Acid Bessemer	Continuous furnace butt weld	Not required
		Type E Grade A-ERW	—	Open hearth, basic oxygen	Electric resistance weld	Not required
	A 134[c]	Fusion welded plate pipe 16″ and over diameter	A 245 Grade B A283 Grade B A285 Grade B	Open hearth, basic oxygen, or electric furnace	Electric fusion (automatic, single or double side)	Not required
	A 135[c]	ERW 30″ and under Grade A	—	Open hearth, basic oxygen, or electric furnace	Electric resistance welded	Not required
	A 139[c]	Fusion welded Pipe 4″ and over Grade A	—	Open hearth, basic oxygen, or electric furnace	Electric fusion (automatic, single or double side)	Not required

[a]A: Distance between inside surfaces of tube. [b]B: Distance between platens of press. [c]As limited by ASA B31.3:1962.

Chemical Composition (%)						Mechanical Properties (minimum)				Maximum Recommended Temperature (°F)
						Yield Stress (lb/in.^{2})	Ultimate Stress (lb/in.^{2})	Flattening Test		
C max	Mn	Si	P max	S max	Other Maxima			A[a]	B[b]	
0.17	—	—	0.05	0.05	—	30,000	48,000	—	Material: $\frac{1.09t}{0.09 + (t/D)}$ Weld: 2D/3	572
0.17	−0.5	0.10 to 0.40	0.05	0.05	Cr 0.2 Cu 0.3 N 0.009	31,000	50,000	Not required		752
—	—	—	0.06	0.06	—	—	50,000	Material: 5t Weld: D/2	—	200 (BS.3351)
0.20	0.70	—	0.05	0.05	—	30,000	50,000	4t or 2D/2	—	850
—	—	—	0.06	0.06	—	—	58,000	Bend test: Material: 2t rad. Weld: 3t rad.		300
—	—	—	0.06	0.06	—	—	58,000	Bend test: Material: 2t rad. Weld: 3t rad.		—
—	0.30 to 0.60	—	0.045	0.060	—	25,000	45,000	—	Material: 0.6D Weld: 3D/4	400
—	0.30/0.60	—	0.110	0.065	—	30,000	50,000			400
0.21	0.90 max	—	0.04	0.05	—	30,000	48,000	—	Material: D/3 Weld: 2D/3	1100 (ERW only)
—	—	—	0.08	—	—	25,000	45,000	—	Weld: 3D/4 Material: 0.6D	400
—	—	—	0.13	—	—	30,000	50,000			
—	—	—	0.05	—	—	30,000	48,000	—	Material: D/3 Weld: 2D/3	1100
0.25	—	—	0.04	0.05	—	30,000	49,000	—	—	300
0.20	0.80	—	Arid 0.06; Basic 0.04	0.05	—	27,000	50,000	—	—	300
			See specs	FBQ 0.04; flange 0.05		27,000	50,000	—	—	300
—	—	—	0.05	0.06	—	30,000	48,000	—	Material: D/3 Weld: 2D/3	1100
—	0.30 to 1.00	—	0.04	0.05	—	30,000	48,000	—	—	300

TABLE 27 Welded Carbon Steel Pipe Suitable for Higher Temperatures and Pressures—About 60,000 lb/in.² Ultimate Strength

Country	Standard	Plate Specification	Steel Process	Welding Process	Heat Treatment
Germany	DIN 1626 Blatt 4	DIN 17100 St 42–2	Killed, rimming steel, any process	Dbl. side elec fusn weld or elec rest weld 100% nondestruct testing	Not required
Sweden	SIS 1435–06	SIS 1435	Killed steel	Elect. resist. weld	Required
United Kingdom	BS.3602 EFW 28S	—	Open hearth, elect. furn., or O_2 convtr. Silicon killed	Dbl. side elec. fusion weld, spot radiographed	Finished by hot roll
United States	ASTM A.55[c] KC 60 Class 2[d]	A 201 Gr. B	Open hearth, elect. furn., or O_2 convtr. Si. kill	Dbl. side manual or auto. fusion weld, non destruct not reqd.	Stress relief over 3/4″ wall

[a]A: Distance between inside surfaces of tube.
[b]B: Distance between platens of press.
[c]As limited by ASA B31.3: 1962.
[d]Class 1 requires stress relief and 100% radiography of welds.

from European sources. Table 32 includes such a requirement under Footnote 2.

Pipe-Making

In many respects the tube-making methods used in Europe are the same as those used in the United States. There are some differences. Flash-butt welding of line pipe is unknown in Europe. More importantly, electric-resistance-welded pipe, which can be obtained in quite large diameters in the United States, is sometimes available only in small diameter tubes in Europe. If large size ERW pipe is ordered in Europe, fusion-welded pipe may be delivered; fusion-welded pipe may or may not be an acceptable alternative.

Spiral-butt-welded pipe is being increasingly used in Europe. This product must not be confused with spiral-welded pipe to ASTM A211, which may be lap or lock-seam jointed. The European material (as specified, for example, in BS 3601 SFW) is a double-submerged arc-welded pipe which, manufactured with adequate quality control, has been applied successfully to refinery offside duties.

C max	Mn	Si	P max	S max	Other Maxima	Yield Stress (lb/in.^{2})	Ultimate Stress (lb/in.^{2})	Flattening Test A[a]	Flattening Test B[b]	Maximum Temperature (°F)
	Chemical Composition (%)					Mechanical Properties (minimum)				
0.25	—	—	0.06	0.05	—	37,000	60,000	—	Material: $\frac{1.07t}{0.07 + (t/D)}$ Weld: 2D/3	572
0.22	0.6 min	0.10 to 0.40	0.05	0.05	Cr 0.2 Cu 0.3 N 0.009	37,000	64,000	Not required		887
0.25	0.65 to 1.20	0.10 to 0.35	0.05	0.05	Ni 0.30 Cr 0.25 Mo 0.10 Cu 0.20	36,000	62,500	Bend test: 2t radius on weld		950
0.24	0.80 max	0.15 to 0.30	See specs		—	32,000	60,000	Bend test: 2t radius on weld		1100

It is always worthwhile to pay attention to the quality of welded pipe by quoting the correct specification and by adequate shop inspection. API 5L and ASTM A155 require electric fusion welds to be double side welds, as do DIN 1626 Blatt 3 and Blatt 4 and BS 3601 SFW (spiral weld). ASTM A134, DIN 1626 Blatt 1 and Blatt 2, and BS 3601 EFW, on the other hand, permit single side welds. The quality control measures called for in the specifications permitting single side welds do not guarantee freedom from defects, so they are not suitable for more severe duties.

Carbon and Carbon-Manganese Steel Pipe

Tables 24–27 list United States and European specifications for seamless and welded carbon and carbon-manganese steel pipe. Each table is intended to contain material that is equivalent in tensile strength and similar in overall characteristics.

There is a general difference in character between most of the carbon steels of Continental Europe and those of Britain and the United States. Continental specifications indicate a slightly lower level of carbon content

TABLE 28 Carbon and Low Alloy Steel Pipe for Low Temperature Duties

Country	Alloy	Standard	Steel Process	Chemical Composition (%) C max	Mn	Si	Ni	Cr	S and P max
Germany	CS	VDEh 680 TT St 35 N impact tested	Killed, normal, opn hrth elec furn O_2 convtr	0.16	0.40 to 0.60	0.10 to 0.35	—	—	0.05
	5 Ni	VDEh 680 12 Ni 19	Open hearth, elec furn	0.20	0.30 to 0.50	0.15 to 0.35	4.5 to 5.0	—	0.035
United Kingdom	CS	BS:3603 HFS or CDS: 27 LT 50	Open hearth, elec. furn or O_2 convtr steel, fully killed made to fine grade	0.20	0.90 to 1.20	0.10 to 0.20	—	—	0.05
	3½ Ni	BS:3603 HFS or CDS. 503 LT 100		0.15	0.30 to 0.60	0.10 to 0.35	3.25 to 3.75	0.30 max	0.04
United States	CS	ASTM A333 Grade 1	Open hearth, elect. furn or O_2 conv.	0.30	0.40 to 1.06	—	—	—	P: 0.05 S: 0.06
	2¼ Ni	ASTM A333 Grade 7	Open hearth, elect furn	0.19	0.90 max	0.13 to 0.32	2.03 to 2.57	—	P: 0.04 S: 0.05
	3½ Ni	ASTM A333 Grade 3	Open hearth, elect furn	0.19	0.31 to 0.64	0.18 to 0.37	3.18 to 3.82	—	0.05
	9 Ni	ASTM A333 Grade 8	Open hearth, elect furn	0.13	0.90 max	0.13 to 0.32	8.40 to 9.60	—	0.45

[a]According to AD-Merkblatt W. 10. In practice, German carbon steel pipe may be obtained impact tested at −50°C for use down to this temperature.
[b]With the addition of no filler metal in the welding operation.

and higher yield to ultimate strength ratio for quality carbon steels. This tendency has been influenced by an increasing use of yield strength as a basis for design in Continental countries and a desire to achieve optimum weldability.

In general, the Continental tendency is to achieve the required tensile properties through slightly higher manganese and lower carbon content than in the corresponding American and British materials. Improved yield to ultimate ratio is frequently obtained by aluminum treatment. Whereas British and American carbon steels may be semikilled, silicon killed, or silicon/aluminum killed, there is a tendency in Continental European practice (particularly in Germany) to produce either fully killed aluminum-treated steel or rimming quality.

Mechanical Properties (minimum)				Minimum Operating Temperature (°F)	Seamless or Weld	Heat Treatment
		Impact				
Yield Stress psi	Ultimate Stress psi	Temperature (°F)	Value			
32,700	50,000	68	12 kg/cm² DVM min average	−31.5[a]	Seam	Normalized
64,000	85,000	−274	5 kg/cm² DVM min average	−256 unweld −184 Wwlded	Seam	Quenched and tempered
35,800	60,000	−58	20 ft lb min, average 15 ft. lb min, Charpy V	−58	Seam	As-roll or normal
37,000	65,000	−148	15 ft lb min, average 10 ft. lb minimum	−148	Seam	Normalized and tempered
30,000	55,000	−50	15 ft lb min, average 10 ft. lb minimum	−50	Seam or weld[b]	Normalized or normalized and tempered
35,000	65,000	−190	15 ft lb min, average 10 ft. lb, minimum	−100	Seam or weld[b]	Normalized or normalized and tempered
35,000	65,000	−150	15 ft lb min, average 10 ft. lb minimum	−150	Seam or weld[b]	Normalized or normalized and tempered
75,000	100,000	−320	25 ft lb min, average 20 ft. lb minimum	−320	Seam or weld[b]	Quenched and tempered or double normalized and tempered

Such differences in deoxidation practice (which are applicable to both plate and pipe) may have an effect on the properties of carbon steel. Aluminum treatment, for example, combined with the appropriate rolling technique, can reduce the grain size and thus improve the notch ductility as well as increase the yield strength for any given ultimate strength. It has also been suggested that by fixing free nitrogen as aluminum nitride, this deoxidation practice may reduce the creep strength of carbon steel.

The elevated temperature properties of steels produced in different countries may not be identical and, in particular, the ASTM values for creep properties may not be applicable to some European steels.

The need to specify deoxidation and steel-making techniques more closely for carbon steel plate material is now recognized to some degree in standard specifications.

TABLE 29 Carbon-Molybdenum and Chromium-Molybdenum Alloy Steel Pipe for Elevated

			Chemical			
Alloy	Country	Standard	C max	Mn	Si	Ni
C 0.3-Mo	France	AFNOR 15 D 3 GAPAVE 222	0.20	0.50/0.80	0.15/0.35	—
	Germany	DIN 17175 15 Mo 3	0.20	0.50/0.80	0.15/0.35	—
	Sweden	SIS 2912–05	0.20	0.40/0.80	0.15/0.35	—
C ½-Mo	United States	ASTM A335 Grade P1	0.20	0.30/0.80	0.10/0.50	—
½-Cr ½-Mo	France	AFNOR 15 CD 2–05 GAPAVE 222	0.18	0.50/0.90	0.50 max.	—
	United States	ASTM A335 Grade P2	0.20	0.30/0.61	0.10/0.30	—
1-Cr ½-Mo	Germany	DIN 17175 13 CrMo 44	0.18	0.40/0.70	0.15/0.35	—
	Sweden	SIS 2216–05	0.15	0.40/0.90	0.15/0.35	—
	United Kingdom	BS.3604 HF 620 or CD 620	0.15	0.40/0.70	0.10/0.35	—
	United States	ASTM A335 Grade P12	0.15	0.30/0.61	0.50 max	—
1¼-Cr ½-Mo	France	AFNOR 10 CD 5–05 GAPAVE 222	0.15	0.30/0.60	0.50/1.00	—
	United Kingdom	BS.3604 HF 621 or CD 621	0.15	0.30/0.60	0.50/1.00	—
	United States	ASTM A335 Grade P11	0.15	0.30/0.60	0.50/1.00	—
2¼-Cr 1-Mo	France	AFNOR 10 CD 9–10 GAPAVE 222	0.15	0.30/0.70	0.50 max	—
	Germany	DIN 17175 10 CrMo9 10	0.15	0.40/0.60	0.15/0.50	—
	Sweden	SIS 2218–05	0.15	0.30/0.60	0.15/0.50	—
	United Kingdom	BS.3604 HF 622 or CD 622	0.15	0.40/0.70	0.50 max	—
	United States	ASTM A335 Grade P22	0.15	0.30/0.60	0.50 max	—
5-Cr ½-Mo	United Kingdom	BS.3604 HF 625 or CD 625	0.15	0.40/0.70	0.50 max	—
	United States	ASTM A335 Grade P5	0.15	0.30/0.60	0.50 max	—
9-Cr 1-Mo	France	AFNOR Z 10 CD 9 GAPAVE 222	0.15	0.30/0.60	0.25/1.00	—
	Sweden	SIS 2203–05	0.12	0.30/0.60	0.50/0.80	—
	United States	ASTM A335 Grade P9	0.15	0.30/0.60	0.25/1.00	—

Temperature Duties

Composition (%)			Mechanical Properties (min.)			
Cr	Mo	S and P	Yield Stress (lb/in.^{2})	Ultimate Stress (lb/in.^{2})	Heat Treatment	Maximum Recommended Temperature (°F)
0.30 max	0.25/0.35	0.04	38,000	62,500	Annealed or normalized and tempered	977
—	0.25/0.35	0.04	41,000	64,000	Normalized	842
—	0.25/0.50	0.04	41,000	64,000	Normalized and tempered	968
—	0.44/0.65	0.045	30,000	55,000	Annealed or normalized and tempered	1100
0.40/0.65	0.45/0.60	0.03	39,750	64,000	Normalized and tempered	1022
0.50/0.81	0.44/0.65	0.045	30,000	55,000	Annealed or normalized and tempered	1100
0.70/1.00	0.40/0.50	0.04	42,500	64,000	Normalized and tempered	968
0.70/1.10	0.40/0.70	0.04	42,500	64,000	Normalized and tempered	1022
0.70/1.10	0.45/0.65	0.05	33,600	60,500	Normalized	1200
0.80/1.25	0.44/0.65	0.045	30,000	60,000	Annealed or normalized and tempered	1200
1.0C/1.50	0.45/0.65	0.03	30,000	64,000	Annealed	1067
			50,000	71,000	Normalized and tempered	
1.00/1.50	0.45/0.65	0.04	33,600	60,500	Normalized	1200
1.00/1.50	0.44/0.65	0.03	30,000	60,000	Annealed or normalized and tempered	1200
2.0/2.5	0.90/1.10	0.03	30,000	60,000	Annealed	1112
2.0/2.5	0.90/1.10	0.04	38,300	64,000	Normalized and tempered	1022
2.0/2.5	0.90/1.10	0.04	38,300	64,000	Normalized and tempered	1076
2.0/2.5	0.90/1.20	0.04	33,600	60,500	Annealed	1200
			42,500	78,500	Normalized and tempered	
1.90/2.6	0.87/1.13	0.03	30,000	60,000	Annealed or normalized and tempered	1200
4.00/6.00	0.45/0.65	0.04	30,000	60,000	Not specified	1200
4.00/6.00	0.45/0.65	0.03	30,000	60,000	Annealed or normalized and tempered	1200
8.0/10.0	0.90/1.10	0.03	30,000	60,000	Annealed	1157
8.0/10.0	0.80/1.20	0.03	48,000	78,000	Normalized and tempered	1202
8.0/10.0	0.90/1.10	0.03	30,000	60,000	Annealed or normalized and tempered	1200

TABLE 30 Designation of Corrosion-Resistant Austenitic Chromium-Nickel Alloy Steel Used for Piping

Nominal Composition	AISI Type	Italy	Germany Designation	Germany Werkstoff Number	Sweden	United Kingdom
0.08% max C 19-Cr 10-Ni	304	× 8 CN 19 10	× 5 CrNi 18 9	4301	2332	801
0.03% max C 19-Cr 10-Ni	304L	× 3 CN 19 11	—	—	2352	801L
Titanium-stabilized 18-Cr 11-Ni	321	× 8 CNT 18 10	× 10 CrNiTi 18 9	4541	2337	822 Ti
Columbium-stablized 18-Cr 11-Ni	347	× 8 CNNb 18 11	× 10 CrNiNb 18 9	4550	2338	822 Nb
18-Cr 12-Ni 2½-Mo	316	× 8 CND 17 12	× 5 CrNiMo 18 10	4401	2343	845
			× 5 CrNiMo 18 12	4436		
Extra low carbon, 18-Cr 12-Ni 2½-Mo	316L	—	× 2 CrNiMo 18 10	4404 or 4435	—	845L
18-Cr 12-Ni 3½-Mo	317	—	—	—	—	846

TABLE 31 Designations of Heat-Resistant Austenitic Chromium-Nickel Alloy Steels Used for Piping

Nominal Composition	AISI Type	Italy	Germany: Designation	Germany: Werkstoff Number	Sweden	United Kingdom
0.04/0.10-C 19-Cr 10-Ni	304H	—	—	—	—	811
0.04/0.10-C Titanium-stabilized 18/8	321H	—	× 12 CrNiSi 18 9	4878	—	832 Ti
0.04/0.10-C Columbium-stabilized 18/8	347H	—	—	—	—	832 Nb
0.04/0.10-C 18-Cr 12-Ni 2½-Mo	316H	—	—	—	—	855
23-Cr 12-Ni	309	× 20 CN 24 12	—	—	—	—
25-Cr 20-Ni	310	× 25 CN 25 20	× 12 CrNi 25 21	4845	2361	805

TABLE 32 British, French, German, Italian, and Swedish Equivalents to ASTM Specifications

Material	U.S. Specification		British	French
Carbon steel line pipe	API 5L	Seamless:	BS.3601	GAPAVE 411
		Grade A	HFS 22 or CDS 22	A 37 C
		Grade B	HFS 27 or CDS 27	A 42 C
		Electric resistance welded:	BS.3601	—
		Grade A	ERW 22	—
		Grade B	ERW 27	—
		Electric fusion welded:	BS.3601 (Double welded)	—
		Grade A	EFW 22	—
		Grade B	EFW 27	—
		Furnace butt	BS.3601	—
		welded	BW 22	—
Carbon steel pipe	ASTM A53	Seamless:	BS.3601	GAPAVE 411
		Grade A	HFS 22 or CDS 22	A 37 C
		Grade B	HFS 27 or CDS 27	A 42 C
		Electric resistance welded:	BS.3601	—
		Grade A	ERW 22	—
		Grade B	ERW 27	—
		Furnace butt	BS.3601	—
		welded	BW 22	—
Carbon steel boiler tube, seamless		ASTM A83	BS.3059/1 or 2	GAPAVE 211 A 37 C
Silicon-killed carbon steel pipe for high temperature service		ASTM A106:	BS.3602	GAPAVE 421
		Grade A	HFS 23	A 37 C
		Grade B	HFS 27	A 42 C
		Grade C	HFS 35	A 48 C
Electric fusion welded steel pipe		ASTM A134	BS.3601 EFW	—
Electric resistance welded steel pipe		ASTM A135:	BS.3601	—
		Grade A	ERW 22	—
		Grade B	ERW 27	—
Electric fusion welded steel pipe		ASTM A139:	BS.3601	—
		Grade A	EFW 22	—
		Grade B	EFW 27	—

German	Italian	Swedish	Footnote
DIN 1629:			
St 35	Aq 35 UNI 663C	SIS 1233–05	1, 2
St 45	Aq 45 UNI 663C	SIS 1434–05	3, 4
DIN 1625:			
Blatt 3 St 34–2 } Electric resistance welded	—	SIS 1233–06	4, 5
Blatt 4 St 37–2 }	—	SIS 1434–06	
DIN 1626:	—	—	
Blatt 3 St 34–2 } Fusion welded	—	—	2, 4, 5
Blatt 4 St 37–2 }	—	—	
DIN 1626	—	—	4
Blatt 3 St 34–2 Furnace butt welded	—	—	
DIN 1629			1, 2, 3, 4
St 35	Aq 35 UNI 663C	SIS 1233–05	
St 45	Aq 45 UNI 663C	SIS 1434–05	
DIN 1626 Blatt 3	—	—	4
St 34–2 } Electric resistance	—	—	
st 37–2 } welded	—	—	
DIN 1626 Blatt 3	—	—	
St 34–2 Furnace butt welded	—	—	
DIN 1629	—	—	4
St 35	—	—	
DIN 17175			
St 35.8	Aq 35 UNI 663C	SIS 1234–05	2, 3
St 45.8	Aq 45 UNI 663C	SiS 1435–05	4, 6
—	—	—	
DIN 1626 Blatt 2 Electric fusion welded	—	—	2, 4
DIN 1626 Blatt 3	—		4
St 34–2 } Electric resistance	—	SIS 1233–06 or	
St 37–2 } welded	—	SIS 1434–06	
DIN 1626 Blatt 2	—	—	2, 4
St 37	—	—	
St 42	—	—	

(*continued*)

TABLE 32 *(continued)*

Material	U.S. Specification	British	French
Electric fusion welded pipe for high temperature service	ASTM A155:	—	—
	Class 2		
	C 45	—	—
	C 50	—	—
	C 55	BS.3602 EFW 28	—
	KC 55	—	—
	KC 60	BS.3602 EFW 28S	—
	KC 65	—	—
	KC 70	—	—
Austenitic stainless steel pipe	ASTM A312:	BS.3605:	—
	TP 304	Grade 801	—
	TP 304H	Grade 811	—
	TP 304L	Grade 801L	—
	TP 310	Grade 805	—
	TP 316	Grade 845	—
	TP 316H	Grade 855	—
	TP 316L	Grade 845L	—
	TP 317	Grade 846	—
	TP 321	Grade 822 Ti	—
	TP 321H	Grade 832 Ti	—
	TP 347	Grade 822 Nb	—
	TP 347H	Grade 832 Nb	—
Pipe for low temperature service	ASTM A333:	BS.3603	—
	Grade 1	27 LT 50	—
	Grade 3	503 LT 100	—
Seamless ferritic alloy pipe for elevated temperature service	ASTM A335 —	BS.3604 —	GAPAVE 222
	—	—	AFNOR 15 D 3
	P1	—	—
	P2	—	AFNOR 15 CD 2–05
	P12	HF 620 or CD 620	—
	P11	HF 621 or CD 621	AFNOR 10 CD 5–05
	P22	HF 622 or CD 622	AFNOR 10 CD 9–10
	P5	HF 625 or CD 625	—
	P9	—	AFNOR Z 10 CD 9
Aluminum alloy pipe	ASTM B241 3003 H112	—	—
	5154 H112	BS.1471 NT5 or BS.1474 NT5	—
	6061 T6	BS.1471 HT 20 WP or BS.1474 HV 20 Wp	—

1. For pipe fabricated to ASA B31.3, steel should be specified to be open hearth, electric furnace, or basic oxygen. Alternatively, Thomas steel is acceptable if fully killed or if it meets the following composition requirements: S 0.05% max, P 0.05% max, N 0.009% max.
2. Rimming steel is not acceptable for seamless or fusion welded pipe.
3. Analysis and test certificates are required.
4. Above 650°F use mechanical properties quoted in the appropriate national standard as a basis for design in critical applications.

German		Italian	Swedish	Footnote
DIN 1626 Blatt 3 mit Abnahme-zeugnis C		—	—	
St 34–2		—	—	
St 37–2		—	—	
St 42–2		—	—	
St 42–2 Si-killed		—	—	2, 4
St 42–2 Si-killed		—	—	7, 8
St 52–3		—	—	
St 52–3		—	—	
WSN	Designation	—	—	
4301	× 5 CrNi 18 9	× 8 CN 19 10	SIS 2333–02	
—	—	—	—	
4306	× 2 CrNi 18 9	× 3 CN 19 11	SIS 2352–02	
4841	× 15 CrNiSi 25 20	× 25 CN 25 20	SIS 2361–02	
4401/4436	× 5 CrNiMo 18 10	× 8 CND 17 12	SIS 2343–02	
—	—	—	—	
4404	× 2 CrNiMo 18 10	—	SIS 2353–02	
—	—	—	—	
4541	× 10 CrNiTi 18 9	× 8 CNT 18 10	SIS 2337–02	
—	—	—	—	
4550	× 10 CrNiNb 18 9	× 8 CNNb 18 11	SIS 2338–02	
—	—	—	—	
WSN	Designation			
0437	SEW 680 TT St 41	—	—	9
5637	SEW 10 Ni 14	—	—	
—		—	—	
DIN 17175 15 Mo 3		—	SIS 2912–05	
WSN 5423 16 Mo 5		—	—	
—		—	—	
DIN 17175 13 CrMo 44		—	SIS 2216–05	
—		—	—	4, 7
DIN 17175 10 CrMo 9 10		—	SIS 2218–05	
—		—	—	
—		—	SIS 2203–05	
DIN 1746 Al Mn F10		—	—	
DIN 1746 Al Mg 3 FIS		—	—	
DIN 1746 Al Mg Si 1 F32		—	—	

5. For British and Swedish standard welded pipe supplied as equivalent to API 5L Grade B specify: "Welded seams to be nondestructively tested in accordance with paras. 11.5 and 11.6 of API 5L." DIN 1626 Blatt 4 already requires an equivalent degree of testing.
6. Specify "Silicon-killed" for GAPAVE 421, DIN 17175, and UNI 663.
7. Above 1000°F use mechanical properties accepted by the national code-writing body as a basis for design in critical applications.
8. DIN 1626 Blatt 4 may be used as equivalent to ASTM A155 Class 1.
9. SEW = Stabl-Eisen Werkstoffblatt.

The German DIN 1626 specified for welded pipe lists the designations U (rimming), R (killed or semikilled), and RR (specially killed). For quality grades the purchaser may specify the deoxidation practice according to these designations.

The recent ASTM standards for carbon steel plate ASTM A515 and A516 do not attempt to define deoxidation practice, but rely on control of chemical composition and inherent (austenite) grain size to achieve the required properties. ASTM A515 is intended for elevated temperature use and specifies coarse grained steel with silicon between 0.15 and 0.30%. A516, for atmospheric and lower temperature service, requires fine grain and controls both managanese and silicon contents. ASTM A524 is a pipe specification covering fine-grained steel similar to plate specification A516.

The specification for killed steel pipe for elevated temperature service, A106, does not contain any requirement for coarse grain and would not prohibit the supplying of an aluminum-killed fine grain steel. It is possible that in countries where aluminum treatment is the rule, steel ordered to ASTM specifications may also be aluminum-treated with potentially lower elevated temperature properties. A conservative practice for critical piping operating at elevated temperature is to use the nationally accepted design stresses of the country of purchase whenever these are lower than the ANSI B31.3 values (Footnote 4 of Table 32). The list of equivalents for carbon steel holds good up to 650°F, but above this temperature special consideration must be given to material according to the country in which the piping is bought.

The country in which the piping is erected will also influence material selection. In France, Germany, Holland, and Italy, steam piping comes under the jurisdiction of the local Code authorities, who are not so liberal as ANSI in the upper temperature limit permitted for carbon steel. Where reduced design stresses for carbon steel are in force, it becomes economic to change to alloy at a lower temperature than would be the case with ANSI B31.3.

The argument set out above applies with greater force, if anything, to plate—and therefore to large diameter welded pipe.

Ferritic Steel for Low Temperature Duties

British and German standards offer carbon and nickel steel for use at subzero temperature. The British steels are impact tested carbon steel (down to −58°F), and 3½% double-extra-strong would, in most cases, be outside the limits of the DIN standard. In practice, pipe may be obtained to schedule sizes without difficulty in Europe, so that it is not usually necessary to apply local standards. Tolerances are generally similar to or within ASTM requirements for the standards under consideration.

Table of Equivalents

In Table 32, European grades have been selected to give equal or greater strength than ASTM, and to show equivalent corrosion resistance. The British steels are impact-tested carbon steel (to −58°F) and 3½% nickel steel, both virtually identical with similar ASTM grades. German materials are impact-tested carbon steel and 5% nickel alloy, and are standardized in Vereine Deutsche Eisenhuttenleute Stahl-Eisen Werkstoffblatt 680. In practice, 3½% nickel steel is equally available in Germany—or equally unavailable. Nickel alloy is difficult to procure anywhere in Western Europe. For short delivery it is often necessary to substitute austenitic chromium nickel steel.

ASTM A333 covers six grades of impact tested pipe. Two grades are carbon-manganese steel tested at −50°F, the other grades are 2¼, 3½, and 9% nickel steels and a Cr-Cu-Al steel. Although the nickel steels are available in the United States, as in Europe the procurement of all the needed piping components in one grade of steel is frequently so difficult that the use of a stainless steel may be preferred.

Low Alloy Steel for High Temperature Service

The chromium molybdenum steels currently listed by European specifications are almost identical with the corresponding ASTM composition. The only difference of any consequence is that German and Swedish standards call for normalized and tempered pipe and list higher yield and ultimate strength figures than the ASTM specifications, which permit pipe to be annealed or normalized and tempered. The creep rupture properties reported for German steels are lower than those given in ASTM publications at temperatures over about 1000°F. Thus, in selecting European equivalents for ASTM chrom-moly grades operating at temperatures over 1000°F, the considerations already outlined for carbon steel pipe operating at elevated temperatures will apply.

Carbon-molybdenum steel also presents problems. BS.3604 does not include the alloy at all. French, German, and Swedish standards do have this type of steel, but the minimum molybdenum steel is used as a hydrogen-resistant material. This difference may be significant. French, German, and Swedish carbon-molybdenum steels are not equivalent to the American for hydrogen service.

For strength, the low-molybdenum German steel 15 Mo 3 is, like the chromium-molybdenum grades, reported to be lower than Grade Pl over 1000°F. On the other hand, the minimum specified yield strength is within the scatter band for the ASTM material.

Corrosion and Heat Resisting Steels

Generally speaking, the common grades of austenitic chromium nickel steel are readily available in Europe. European specifications (BS.3605 in particular) list a number of compositions that are identical with ASTM stainless steels, including the H grades, which are 0.03 to 0.10% carbon steels intended for high temperature duties. Extra low carbon steel is also in better supply in the United Kingdom than it once was. It is also available on the Continent.

Dimensions and Tolerances

Standard dimensions for pipe are given in BS.3601-5 for seamless and welded carbon and alloy steel pipe, in DIN 2448 (seamless pipe), and in DIN 2458 (welded pipe). All these standard sizes have been written to fall in line with ISO Recommendation R64, which gives outside diameters only. Up to an outside diameter of 5½-in., ISO R 64 follows ISO R7. Above this dimension it follows ASA B2.1:1945 and API 5L. The British standards for wall thicknesses include some of the commonly used schedule sizes, as do the two German standards. The German standard wall thickness is less than the API standard weight (schedule 40). To specify schedule 40 pipe to DIN 2448, it is necessary to select one of the special wall thicknesses given in that standard. In this way a fairly good match for API standard weight pipe may be obtained.

Much of this article was originally written by John J. McKetta for *Equipment Design Handbook,* Gulf Publishing Co., Houston, Texas 77252, 1981, and is used with the permission of the editor of *Hydrocarbon Processing.*

Bibliography

Arnold, M. L., "Vibration in Compressor Plant Piping," *Pet. Refiner,* p. 59 (February 1945).

Thiel, B. C., "Plant Design of Compressor Piping," *Pet. Refiner,* p. 259 (July 1945).

Yeakel, J., "Analysis of Stresses in U-Bend Pipe Frame," *Pet. Refiner,* p. 147 (March 1947).

Yeakel, J., "Analysis of Stresses in Unsymmetrical Pipe Frame," *Pet. Refiner,* p. 541 (June 1947).

Kiester, D. F., "Design of Steam Condensate Lines," *Pet. Refiner,* p. 616 (November 1948).

Hicks, T. G. "Transmission Piping for Refinery Steam System, Part 1," *Pet. Refiner,* p. 135 (April 1949); "Part 2," p. 103 (June 1949); "Part 3," p. 143 (July 1949).

Buthod, P., "Pressure Drop in Gas and Vapor Lines," *Pet. Refiner,* p. 157 (April 1950).

Solosewick, F. E., "Expansion Joints and Their Application," *Pet. Refiner,* p. 146 (May 1950).

Blick, R. G., "Expansion Joints—Their Maximum Spacing," *Pet. Refiner,* p. 135 (July 1950).

Solosewick, F. E., "Forces in Pipe Bends from Expansion," *Pet. Refiner,* p. 103 (October 1950).
DeForest, E. M., "Fluid Flow Pipe Friction Loss Determination," *Pet. Refiner,* p. 97 (March 1951).
Manley, H. B., "Guide to Economic Line Sizing for Plant Piping," *Pet. Refiner,* p. 151 (July 1951).
Wilbur, W. E., "Thermal Stresses in Piping Systems, Part 1," *Pet. Refiner,* p. 143 (March 1953); "Part 2," p. 163 (April 1953); "Part 3," p. 174 (May 1953).
Rase, H. F., "Take Another Look at Economic Pipe Sizing," *Pet. Refiner,* p. 141 (August 1953).
Blick, R. G., "Pipe Flexibility Calculations," *Pet. Refiner,* p. 123 (February 1954).
Peiser, A. M., and Katz, S., "Pipe Stress Calculations," *Pet. Refiner,* p. 153 (February 1954).
Kern, R., "How to Design Tower Piping," *Pet. Refiner,* p. 136 (March 1958).
Lammers, G. C., and Otis, W. G., "Use These Charts for Fast Pipe Sizing," *Pet. Refiner,* p. 127 (October 1959).
Moloney, J. S., "Piping Stress Analysis Simplified," *Pet. Refiner,* p. 133 (October 1959).
Kern, R., "How to Design Heat Exchanger Piping," *Pet. Refiner,* p. 137 (February 1960).
Kern, R., "How to Design Yard Piping," *Pet. Refiner,* p. 139 (December 1960).
Lewis, J. D., "Find Pipe Pressure Drop by Nomograph," *Hydrocarbon Process. Pet. Refiner,* p. 169 (March 1961).
Ellison, G. L., "Find Best Expansion Loop Quickly," *Hydrocarbon Process. Pet. Refiner,* p. 151 (January 1962).
Hilker, R. W., "Which to Use: Ball Joints or Expansion Loops," *Hydrocarbon Process. Pet. Refiner,* p. 185 (February 1962).
Haque, M. S., and Starczewski, J., "Graphical Shortcuts to Pipe Stress Analysis," *Hydrocarbon Process. Pet. Refiner,* p. 135 (June 1962).
Bower, J. N., and Peterson, H. R., "Guide to Steam Tracing Design," *Hydrocarbon Process. Pet. Refiner,* p. 149 (March 1963).
Masek, J. A., "Thermowell Design for Process Piping, Part 1," *Hydrocarbon Process. Pet. Refiner,* p. 119 (February 1964); "Part 2," p. 119 (March 1964); "Part 3," p. 165 (April 1964).
Doyle, W. E., "Piping Tierod Design Made Simple," *Hydrocarbon Process. Pet. Refiner,* p. 118 (August 1965).
Judson, R. W., "What Information is Essential for Good Piping Design," *Hydrocarbon Process.,* p. 114 (October 1966).
Kern, R., "Plant Layout and Piping Design for Minimum Cost Systems," *Hydrocarbon Process.,* p. 119 (October 1966).
Lancaster, J. F., and Hoyt, W. B., "U.S. vs. British and European Piping Specifications," *Hydrocarbon Process.,* p. 127 (October 1966).
Eland, K. G., "New Guide to Steam Tracing Design," p. 218 (November 1966).
Kent, G. R., "New Approach to Pipe Reactions," *Hydrocarbon Process.,* p. 127 (February 1967).
Haque, M. S., and Starczewski, J., "Piping Design Method Beats Computers," *Hydrocarbon Process.,* p. 195 (March 1967).
Shull, W. W., and Bogel, G. N., "A Simplified Computer Program for Pipe Expansion Loop Design," *Hydrocarbon Process.,* p. 183 (September 1967).

Kern, R., "Thermosyphon Reboiler Piping Simplified," *Hydrocarbon Process.*, p. 118 (December 1968).
Smith, B., "Charts Used for Easier Pipe Sizing," *Hydrocarbon Process.*, p. 173 (May 1969).
Kern, R., "How to Size Process Piping for Two-Phase Flow," *Hydrocarbon Process.*, p. 105 (October 1969).

JOHN J. McKETTA

Economic Diameter

The optimum economic pipe-diameter equation in Perry's *Chemical Engineers' Handbook* (Eq. 5-90, p. 5-32, 5th ed.) includes a variable *F:*

$$\frac{D^{4.84+n}}{1 + 0.794L_e'D} = \frac{0.000189YKq^{2.84}\rho^{0.84}\mu'^{0.16}\left[(1 + M)(1 - \phi) + \dfrac{ZM}{a' + b'}\right]}{nXE(1 + F)[Z + (a + b)(1 - \phi)]} \tag{1}$$

This variable is defined as the ratio of the cost of fittings plus the installation cost of pipe and fittings to the cost of pipe only.

Drawbacks of the *Handbook* Approach

A table in the *Handbook* (Table 5-16) lists typical values for the variables, including *F*, in the equation. However, using the *F* values in this table presents a problem, because *F* cannot be accurately selected until the pipe diameter is known. Therefore, one must set up an array of *F* values for each pipe diameter guessed. Next, one must draw a straight line between the points by some method of interpolation; simply feeding in *F* values in quantum jumps will not work.

Another problem arises because Table 5-16 only gives *F* values for pipe diameters up to 18 in. Although most plant piping is less than 18 in. diameter, a considerable amount of it ranges from 18 to 24 in.

Of greater significance, however, is the lag in the Table 5-16 cost basis. The factors in the table are based on estimated June 1968 costs. In the past 11 years, the *Chemical Engineering* Index for Construction Labor has climbed 70 points, and 162 points for material ("Pipe, Valves, and Fittings").

This article proposes a method for determining *F* values that avoids the three foregoing problems. In it, we derive an equation for expressing *F* as a function of (1) diameter, (2) the *Chemical Engineering* Pipe, Valves and Fittings Index, and (3) the *Chemical Engineering* Construction Labor Index. This approach eliminates the three aforementioned problems: the first two by supplying an equation that is continuous for all diameter values, and the last by making it quick and simple to update *F* values to reflect changes in material and labor costs. In this approach, simplicity takes precedence over pinpoint accuracy.

Development of the Equation

The *Handbook* development of *F* values is based on one gate valve, one check valve, one tee, and four elbows per 100 ft of painted overhead pipe. Although much simplified, it does avoid problems generated by more-realistic models. For example, in reality, the smaller the pipe diameter, the more fittings there are. Whereas 2-in. pipe may have 10 to 17 elbows per 100 ft, 12-in. pipe may have only 3 to 7.

However, introducing typical quantity ranges for each type of fitting for each diameter size would only compound programming problems. Therefore, we will retain the *Handbook* equation for *F*:

$$F = (C_{VF} + C_I)/C_P \tag{2}$$

Let us first develop an expression for the cost of pipe, C_P, as a function of diameter. Fitting the data in Table 1 via regression analysis to an equation of the form $y = mx + b$, we get

$$C_P = 1.818D_i - 2.4116 \tag{3}$$

Equation (3) has a correlation coefficient of 0.999. Here, D_i is the diameter in inches.

Table 2 lists the data necessary to derive an expression for C_{VF}. This is the most difficult of the three equations to keep simple, but we can derive

$$\log C_{VF} = 1.64 \log D_i + 0.153 \tag{4}$$

For Eq. (4), the correlation coefficient is 0.985.

Table 3 gives the data for developing an expression for C_i, the cost of installing valves and fittings. From this, we can derive an equation for I_H, the installation man-hours:

$$I_H = 0.128D_i + 0.17 \tag{5}$$

The correlation coefficient is 0.999.

TABLE 1 Data for Developing Cost-of-Pipe Equation

Diameter (in.)	Cost ($/ft)[a]
2	1.90
3	3.20
4	4.50
6	7.60
8	11.50
10	16.00
12	19.90
14	23.20
16	26.70
18	30.90
20	34.30
24	40.50

[a]Based on carbon steel standard-weight pipe.

TABLE 2 Data for Deriving Equation for Cost of Pipe Plus Fittings

Diameter (in.)	Cost ($)[a] Four Tees	One Elbow	One Gate Valve	One Globe Valve	C_{VF}/100 ft Pipe Length	C_{VF}/ft Pipe Length
2	42.00	3.30	275.00	352.00	672.30	6.72
3	64.00	5.50	396.00	456.00	921.50	9.21
4	89.20	9.60	462.00	662.00	1,223.00	12.23
6	191.00	25.90	748.00	1,003.00	1,968.00	19.68
8	308.00	43.90	1,172.00	1,800.00	3,324.00	33.24
10	504.00	91.10	1,834.00	2,568.00	4,997.00	49.97
12	820.00	118.50	2,444.00	3,422.00	6,804.00	68.04
14	1,200.00	194.00	4,030.00	5,642.00	11,066.00	110.66
16	1,480.00	271.00	5,500.00	7,700.00	14,951.00	149.51
18	2,360.00	362.00	6,525.00	9,135.00	18,382.00	183.82
20	3,520.00	484.00	8,400.00	11,760.00	24,164.00	241.64
24	4,680.00	707.00	11,300.00	15,820.00	32,507.00	325.07

[a]Based on carbon steel standard weight pipe.

Now, we can state:

$$C_I = C_L(0.128D_i + 0.17) \tag{6}$$

Here, C_L = cost of labor, $/h.

Finally, we will add cost escalation indexes to these expressions so they will not have to be recalculated frequently. The indexes used are *Chemical Engineering*'s Construction Labor and Pipe, Valves, and Fittings. We will not change the base year. Therefore, updating the labor and material components of the *F* equation, Eq. (2), only involves putting in the appropriate cost index.

TABLE 3 Data for Developing Expression for Installation Man-hours

Diameter (in.)	Man-hours (h) Per Fitting	For All Fittings	Per Valve	Per 2 Valves	Per Weld	For All Welds[a]	Total	Per Foot of Installation
2	1.4	12.6	3.0	6	1.8	27.0	45.6	0.456
3	1.8	16.2	5.0	10	2.0	30.0	56.2	0.562
4	2.5	22.5	8.0	16	2.3	34.5	73.0	0.73
6	3.0	27.0	9.0	18	3.0	45.0	90.0	0.90
8	4.0	36.0	9.5	19	4.0	60.0	115.0	1.15
10	5.0	45.0	13.0	26	5.0	75.0	146.0	1.46
12	6.0	54.0	14.0	28	6.0	90.0	172.0	1.72
14	7.0	63.0	15.0	30	6.5	97.5	190.5	1.90
16	8.0	72.0	18.0	36	7.5	112.5	220.5	2.20
18	9.0	81.0	19.0	38	8.5	127.5	246.5	2.46
20	10.0	90.0	24.0	48	9.5	142.5	280.5	2.80
24	12.0	108.0	26.0	52	11.0	165.0	325.0	3.25

[a]Based on 15 welds.

For C_P and C_{VF}, we can simply incorporate the term $I_M/279.8$. Here, I_M is the material cost index, and 279.8 is the value of the Pipe, Valves, and Fittings Index as of December 1978:

$$C_P = (1.818D_i - 2.4116)(I_M/279.8) \tag{7}$$

Because:

$$\begin{aligned} C_{VF} &= \text{antilog}\ (1.64 \log D_i + 0.153) \\ &= 10^{(1.64\log D_i + 0.153)} \end{aligned} \tag{8}$$

We can now write

$$C_{VF} = (I_M/279.8)10^{(1.64\log D_i + 0.153)} \tag{9}$$

For C_I, we have an option. Equation (5), our derived expression for I_H, installation man-hours, is not a function of cost, as are the other derived equations. Equation (5) becomes Eq. (6) when we add the term C_L, the cost of labor. We can update C_I by either of two ways—by putting in the current cost for C_L in Eq. (6), or via Eq. (10):

$$\begin{aligned} C_I &= (I_L/191.2)(15)(0.128D_i + 0.17) \\ &= (I_L/191.2)(1.92D_i + 2.55) \end{aligned} \tag{10}$$

In Eq. (10), C_L is set at \$15/h, and the ratio $I_L/191.2$ is added for future use. In this equation, I_L is the labor cost index, and 191.2 is its value for December 1978. (For the remainder of the article, we will use the second equation, although, as noted, either would serve.)

To be compatible with the *Handbook* equation, Eq. (1), diameter must be in feet, not inches. Therefore:

$$C_P = (21.816D - 2.412)(I_M/279.8) \tag{11}$$

$$C_{VF} = (I_M/279.8)10^{(1.639\log D + 1.923)} \tag{12}$$

$$C_I = (I_L/191.2)(23.04D + 2.55) \tag{13}$$

Now, in Eqs. (11), (12), and (13), D = diameter, ft.

Our final F factor equation then looks like

$$F = \frac{(I_M/279.8)10^{1.639\log D + 1.923)} + (I_L/191.2)(23.04D + 2.55)}{(21.816D - 2.412)(I_M/279.8)} \tag{14}$$

Multiplying Eq. (14) through by 191.2, and further reducing the denominator:

$$F = \frac{(0.68I_M)10^{(1.64\log D + 1.92)} + I_L(23D + 2.55)}{I_M(14.9D - 1.64)} \tag{15}$$

Comparing the *F* Values

The results of Eq. (15), assuming $I_M = 279.8$ and $I_L = 191.2$, are plotted in Fig. 1, on which are also plotted the F values listed in the *Handbook* table.

Note that both curves have the same shape and differ only in that the Eq. (15) F values plot lower. This difference can be explained by examining the performance of the cost indexes since 1968.

Let us assume that the 1968 values in Table 5-16 were obtained by solving Eq. (2) for each diameter. In 1968, the Pipe, Valves, and Fittings Index was 117.4. In 1968, the Construction Labor Index was 120.9. Therefore, F for 1979, in terms of 1968 dollars, would be

$$F' = \frac{\left(\frac{279.8}{117.4}\right)C_{VF} + \left(\frac{191.2}{120.9}\right)C_I}{\left(\frac{279.8}{117.4}\right)C_P}$$

$$= (C_{VF} + 0.66C_I)/C_P \tag{16}$$

In other words, the numerator becomes smaller while the denominator stays the same. Therefore, it is reasonable to expect the new F values to plot lower.

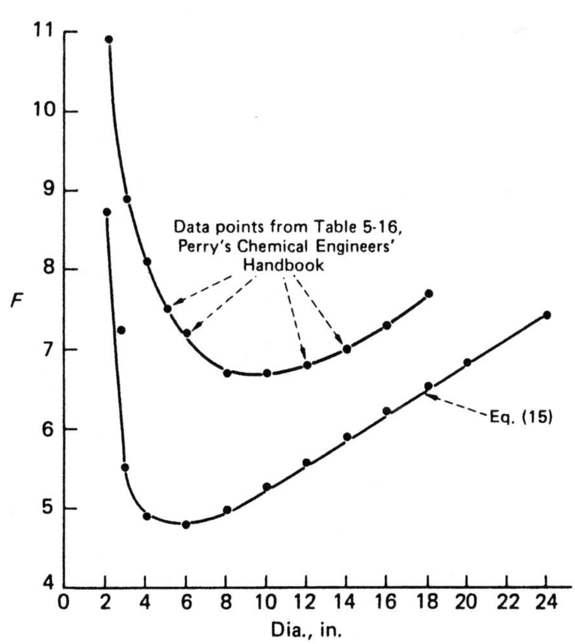

FIG. 1 *Handbook F* values compared to those calculated via Eq. (15).

Other Terms in Eq. (1)

If one uses Table 5-16 values for the other variables in Eq. (1), some of these may also need to be updated.

Table 5-16 values can be used for E, L_e', μ', and (unless operations differ) Y. Values for Z are generated internally; in the absence of a better figure, 0.1 is acceptable. Corporate taxes run about 50%, so $\phi = 0.55$ is adequate. A value for K can easily be ascertained. We already have an expression for X—Eq. (11).

Both a and a' are depreciation terms. Maintenance costs, represented by the term b, are generally time-averaged values. Therefore, setting $a + b = 0.2$, and $a' + b' = 0.4$, should be satisfactory.

For most pipe, a plot of the log of the pipe diameter vs the log of the purchase price per foot is essentially a straight line having a slope of η. A plot of Table 1 data would show η to be 1.31.

The only variables left are P and M, and the second is a function of the first. Rather than deal with P separately, let us look at what happens to M as P changes—note that only M appears in Eq. (1).

By definition, $M = (a' + b')EP/(17.9KY)$. If we set $a' + b' = 0.4$, $E = 0.5$, $K = 0.04$, and $Y = 365$, then $M = 0.00076P$. If P remains at 150 (as listed in Table 5-16), $M = 0.114$. If P rises to, for instance, 350, $M = 0.26$. In other words, a large increment in P does not greatly change M.

LANCE W. CHONTOS

Equivalent Length Estimation

You can easily and quickly determine the equivalent line lengths of loops containing fittings, valves, and pipes of various lengths. All that is needed is an isometric drawing of the piping system and this program for the TI-59 calculator.

The program calculates equivalent pipe lengths for eight types of valves, eight types of fittings, and entrance and exit losses. It then sums the equivalent pipe lengths and prints the total. Next, it adds pipe lengths in feet and inches (tedious ft/in. conversions are not necessary), and prints the total equivalent feet in decimal form. With this information, the total pressure drop through a circuit can be readily determined.

Each valve, fitting, and loss is identified, and the number of each is listed in the printout. The program prompts the user for part of the input.

Although written for the TI-59 calculator and the PC-100C printer, the program could be used without a printer, because all totals are stored in data registers.

Program Development

The pressure drop through valves and fittings is related to velocity changes in the flowing fluid:

$$h = KV^2/2g$$

Here, h = pressure loss in head of fluid, ft; K = experimental coefficient (number of velocity heads); V = average velocity in pipe, ft/s; and g = 32.17 ft/s^2.

Pressure drops calculated via the foregoing equation (which is for turbulent flow) usually give accurate pressure losses for valves and fittings. However, velocity and K data are required to solve the equation.

The equivalent-length method, though less accurate, is very convenient and has gained wide acceptance by piping designers for most work. In it, the fitting or valve is taken to be equivalent to so many feet of pipe. By adding this calculated length to the line length, the pressure drop for an entire loop can be found at once.

Data from a table of representative equivalent lengths to pipe diameters for various valves and fittings [1] were rearranged and expanded by Kern [2]. Factors for this program were taken from the Kern article. When a single factor was not applicable, an equation was developed to represent the data.

Caution is urged in using the part of the program dealing with piping reduction/enlargement. In most cases the correlations presented are only valid for $d_1/d_2 < 1.5$. (Negative values may be obtained for very small lines.) For values beyond 1.5, refer to Kern [2].

Using the Program

Table 1 gives the program operating instructions and the user-defined keys. Table 2 provides the program itself, which occupies both sides of two magnetic cards. The partitioning is 719.29. Label addresses and the data that must be stored are also listed in Table 2.

After the pipe internal diameter has been keyed in and **R/S** pressed, the subroutine for each type of valve or fitting to be converted is called up. Enter the number of valves or fittings and press **R/S** to calculate the equiv-

TABLE 1 User Instructions and Key Definitions for Calculating Equivalent Line Lengths

Step	Procedure	Enter	Press	Display
1.	Partition calculator at 719.29	3	2nd OP 17	719.29
2.	Read in both sides of two magnetic cards	CLR		1,2,3,4
3.	Press A to begin computation		A	*Print*
4.	When calculator stops, key in pipe I.D. and press R/S		R/S	0
5.	Press subroutine label corresponding to type of valve, fitting or loss		Lbl, SBR*	*Print*
6.	Key in number of valves or fittings and press R/S		R/S	0
7.	Repeat Steps 5 and 6 until equivalent lengths of all valves and fittings have been calculated		Lbl, SBR*	*Print*
8.	Press C′ for sum of equivalent length of pipe		C′	0
9.	Press E′ to activate sum of pipe-length program		E′	*Print*
10.	Enter pipe length as feet; for example 8 ft, 10½ in. as 8.105			*Length*
11.	Press R/S			*in/ft*
12.	After calculation has stopped, repeat Steps 10 and 11 until all pipe lengths have been entered			0
13.	Press C′ for sum of pipe lengths		C′	0

*To activate user labels A through E (or A′ through E′), press only the appropriate key. To call other labels, press SBR, then the label; (for example, Label X^2 is called by SBR X^2.

User-defined keys

Key	Definition	Key	Definition
A	Starts program	CLR	Ball check valves
B	Gate valves	x⇆t	Butterfly valves
C	Long-radius 90-deg. elbows	X^2	Three-way straight-through valves
D	Straight-through tees	$\sqrt{X}$	Three-way flow-through branch valves
E	Reduction/enlargement	1/X	Short-radius 90-deg. elbows
A′	Entrance loss	STO	Short-radius 45-deg. elbows
B′	Exit loss	RCL	90-deg. miter bends
C′	Sum of equivalent feet	SUM	45-deg. miter bends
INV	Globe valves	Y^x	Flow-through branch tees
ln x	Plug valves	E′	Activates sum-of-pipe-length program
CE	Swing check valves		

alent feet of pipe. After all the valves, fittings, and entrance and exit losses have been converted, the total equivalent length is recalled and printed by pressing **C′**.

To sum pipe lengths, inch dimensions need not be converted into actual equivalents in foot-decimal form. For example, a pipe 12 ft 2¼ in. long is entered as 12.0225, and one 2 ft 11 in. long as 2.11.

To begin the pipe-length summing section, first press **E′**. After each pipe length is keyed in as indicated, press **R/S** to convert and store the number.

TABLE 2 Program Calculates Equivalent Line Lengths for Piping Loops Containing Valves, Fittings, and Piping of Various Lengths

Step	Code	Key
000	76	LBL
001	99	PRT
002	65	×
003	43	RCL
004	07	07
005	95	=
006	44	SUM
007	00	00
008	42	STO
009	19	19
010	43	RCL
011	22	22
012	69	OP
013	04	04
014	43	RCL
015	19	19
016	69	OP
017	06	06
018	00	0
019	69	OP
020	04	04
021	91	R/S
Pipe I.D.		
022	76	LBL
023	11	A
024	69	OP
025	00	00
026	43	RCL
027	01	01
028	69	OP
029	01	01
030	43	RCL
031	02	02
032	69	OP
033	02	02
034	43	RCL
035	03	03
036	69	OP
037	03	03
038	69	OP
039	05	05
040	43	RCL
041	04	04
042	69	OP
043	01	01
044	43	RCL
045	05	05
046	69	OP
047	02	02
048	43	RCL
049	06	06
050	69	OP
051	03	03
052	69	OP
053	05	05
054	69	OP
055	00	00
056	03	3
057	03	3
058	02	2
059	04	4
060	03	3
061	03	3
062	01	1
063	07	7
064	00	0
065	00	0
066	69	OP
067	01	01
068	02	2
069	04	4
070	01	1
071	06	6
072	00	0
073	00	0
074	07	7
075	01	1
076	00	0
077	00	0
078	69	OP
079	02	02
080	69	OP
081	05	05
082	02	2
083	04	4
084	03	3
085	01	1
086	69	OP
087	04	04
088	91	R/S
089	42	STO
090	07	07
091	69	OP
092	06	06
093	00	0
094	42	STO
095	00	00
096	69	OP
097	04	04
098	91	R/S
Gate valves		
099	76	LBL
100	12	B
101	02	2
102	02	2
103	01	1
104	03	3
105	03	3
106	07	7
107	01	1
108	07	7
109	00	0
110	00	0
111	69	OP
112	01	01
113	71	SBR
114	52	EE
115	91	R/S
116	99	PRT
117	65	×
118	01	1
119	93	.
120	01	1
121	00	0
122	61	GTO
123	99	PRT
Valves		
124	76	LBL
125	52	EE
126	43	RCL
127	20	20
128	69	OP
129	02	02
130	43	RCL
131	21	21
132	69	OP
133	03	03
134	69	OP
135	05	05
136	92	RTN
Globe valves		
137	76	LBL
138	22	INV
139	02	2
140	02	2
141	02	2
142	07	7
143	01	1
144	04	4
145	01	1
146	07	7
147	00	0
148	00	0
149	69	OP
150	01	01
151	71	SBR
152	52	EE
153	91	R/S
154	99	PRT
155	65	×
156	02	2
157	08	8
158	93	.
159	03	3
160	03	3
161	61	GTO
162	99	PRT
Plug valves		
163	76	LBL
164	23	LNX
165	03	3
166	03	3
167	02	2
168	07	7
169	04	4
170	01	1
171	02	2
172	02	2
173	00	0
174	00	0
175	69	OP
176	01	01
177	71	SBR
178	52	EE
179	91	R/S
180	99	PRT
181	65	×
182	01	1
183	93	.
184	05	5
185	61	GTO
186	99	PRT
Swing check valves		
187	76	LBL
188	24	CE
189	03	3
190	06	6
191	04	4
192	03	3
193	01	1
194	05	5
195	02	2
196	06	6
197	00	0
198	00	0
199	69	OP
200	01	01
201	71	SBR
202	52	EE
203	91	R/S
204	99	PRT
205	65	×
206	01	1
207	00	0
208	93	.
209	09	9
210	04	4
211	61	GTO
212	99	PRT
Ball check valves		
213	76	LBL
214	25	CLR
215	01	1
216	04	4
217	02	2
218	07	7
219	01	1
220	05	5
221	02	2
222	06	6
223	00	0
224	00	0
225	69	OP
226	01	01
227	71	SBR
228	52	EE
229	91	R/S
230	99	PRT
231	65	×
232	01	1
233	02	2
234	93	.
235	04	4
236	06	6
237	61	GTO
238	99	PRT
Butterfly valves		
239	76	LBL
240	32	X:T
241	01	1
242	04	4
243	04	4
244	01	1
245	03	3
246	07	7
247	02	2
248	01	1
249	00	0
250	00	0
251	69	OP
252	01	01
253	71	SBR
254	52	EE
255	91	R/S
256	99	PRT
257	65	×
258	03	3
259	93	.
260	06	6
261	00	0
262	61	GTO
263	99	PRT
3-way straight-through valves		
264	76	LBL
265	33	X^2
266	00	0
267	04	4
268	04	4
269	03	3
270	03	3
271	06	6
272	03	3
273	07	7
274	00	0
275	00	0
276	69	OP
277	01	01
278	71	SBR
279	52	EE
280	91	R/S
281	99	PRT
282	65	×
283	03	3
284	93	.
285	06	6
286	07	7
287	61	GTO
288	99	PRT
3-way flow-through branch valves		
289	76	LBL
290	34	√X
291	00	0
292	04	4
293	04	4
294	03	3
295	01	1
296	04	4
297	02	2
298	01	1
299	00	0
300	00	0
301	69	OP
302	01	01
303	71	SBR
304	52	EE
305	91	R/S
306	99	PRT
307	65	×
308	01	1
309	01	1
310	93	.
311	06	6
312	07	7
313	61	GTO
314	99	PRT
Long-radius 90-deg. elbows		
315	76	LBL
316	13	C
317	02	2
318	07	7
319	03	3
320	05	5
321	01	1
322	02	2
323	00	0
324	01	1
325	00	0
326	00	0
327	69	OP
328	01	01
329	71	SBR
330	65	×
331	91	R/S
332	99	PRT
333	65	×
334	01	1
335	93	.
336	06	6
337	07	7
338	61	GTO
339	99	PRT
Print elbows		
340	76	LBL
341	65	×
342	43	RCL
343	23	23
344	69	OP
345	02	02
346	43	RCL
347	24	24
348	69	OP
349	03	03
350	69	OP

(continued)

TABLE 2 (*continued*)

Step	Code	Key
351	05	05
352	92	RTN
Short-radius 90-deg. elbows		
353	76	LBL
354	35	1/X
355	03	3
356	06	6
357	03	3
358	05	5
359	01	1
360	02	2
361	00	0
362	01	1
363	00	0
364	00	0
365	69	OP
366	01	01
367	71	SBR
368	65	×
369	91	R/S
370	99	PRT
371	65	×
372	02	2
373	93	.
374	05	5
375	61	GTO
376	99	PRT
Short-radius 45-deg. elbows		
377	76	LBL
378	42	STO
379	03	3
380	06	6
381	03	3
382	05	5
383	00	0
384	05	5
385	00	0
386	06	6
387	00	0
388	00	0
389	69	OP
390	01	01
391	71	SBR
392	65	×
393	91	R/S
394	99	PRT
395	65	×
396	01	1
397	93	.
398	03	3
399	03	3
400	61	GTO
401	99	PRT
90-deg. miter bends		
402	76	LBL
403	43	RCL
404	01	1
405	02	2

Step	Code	Key
406	00	0
407	01	1
408	00	0
409	00	0
410	03	3
411	00	0
412	02	2
413	04	4
414	69	OP
415	01	01
416	43	RCL
417	25	25
418	69	OP
419	02	02
420	43	RCL
421	26	26
422	69	OP
423	03	03
424	69	OP
425	05	05
426	91	R/S
427	99	PRT
428	65	×
429	04	4
430	93	.
431	08	8
432	03	3
433	61	GTO
434	99	PRT
45-deg. miter bends		
435	76	LBL
436	44	SUM
437	00	0
438	05	5
439	00	0
440	06	6
441	00	0
442	00	0
443	03	3
444	00	0
445	02	2
446	04	4
447	69	OP
448	01	01
449	43	RCL
450	25	25
451	69	OP
452	02	02
453	43	RCL
454	26	26
455	69	OP
456	03	03
457	69	OP
458	05	05
459	91	R/S
460	99	PRT
461	65	×
462	01	1
463	93	.
464	02	2
465	05	5
466	61	GTO

Step	Code	Key
467	99	PRT
Straight-through tees		
468	76	LBL
469	14	D
470	69	OP
471	00	00
472	03	3
473	06	6
474	03	3
475	07	7
476	00	0
477	00	0
478	03	3
479	07	7
480	01	1
481	07	7
482	69	OP
483	01	01
484	43	RCL
485	27	27
486	69	OP
487	02	02
488	43	RCL
489	18	18
490	69	OP
491	03	03
492	69	OP
493	05	05
494	91	R/S
495	99	PRT
496	65	×
497	01	1
498	93	.
499	06	6
500	07	7
501	61	GTO
502	99	PRT
Flow-through branch tees		
503	76	LBL
504	45	YX
505	01	1
506	04	4
507	02	2
508	01	1
509	00	0
510	00	0
511	03	3
512	07	7
513	01	1
514	07	7
515	69	OP
516	01	01
517	43	RCL
518	27	27
519	69	OP
520	02	02
521	43	RCL
522	18	18
523	69	OP
524	03	03

Step	Code	Key
525	69	OP
526	05	05
527	91	R/S
528	99	PRT
529	65	×
530	05	5
531	61	GTO
532	99	PRT
533	91	R/S
Reduction/enlargement		
534	76	LBL
535	15	E
536	69	OP
537	00	00
538	43	RCL
539	12	12
540	69	OP
541	01	01
542	43	RCL
543	13	13
544	69	OP
545	02	02
546	69	OP
547	05	05
548	02	2
549	01	1
550	03	3
551	07	7
552	69	OP
553	04	04
554	91	R/S
555	99	PRT
556	42	STO
557	19	19
558	43	RCL
559	07	07
560	23	LNX
561	65	×
562	43	RCL
563	14	14
564	85	+
565	43	RCL
566	15	15
567	95	=
568	65	×
569	43	RCL
570	19	19
571	95	=
572	44	SUM
573	00	00
574	69	OP
575	06	06
576	91	R/S
Entrance losses		
577	76	LBL
578	16	A'
579	69	OP
580	00	00
581	43	RCL
582	08	08
583	69	OP

Step	Code	Key
584	01	01
585	43	RCL
586	09	09
587	69	OP
588	02	02
589	69	OP
590	05	05
591	02	2
592	01	1
593	03	3
594	07	7
595	69	OP
596	04	04
597	43	RCL
598	07	07
599	45	YX
600	43	RCL
601	16	16
602	65	×
603	43	RCL
604	17	17
605	95	=
606	44	SUM
607	00	00
608	69	OP
609	06	06
610	91	R/S
Exit losses		
611	76	LBL
612	17	B'
613	69	OP
614	00	00
615	43	RCL
616	10	10
617	69	OP
618	01	01
619	69	OP
620	05	05
621	02	2
622	01	1
623	03	3
624	07	7
625	69	OP
626	04	04
627	43	RCL
628	07	07
629	45	YX
630	43	RCL
631	16	16
632	65	×
633	43	RCL
634	17	17
635	65	×
636	02	2
637	95	=
638	44	SUM
639	00	00
640	69	OP
641	06	06
642	91	R/S

Step	Code	Key
Sum of equivalent line lengths		
643	76	LBL
644	18	C'
645	69	OP
646	00	00
647	03	3
648	06	6
649	04	4
650	01	1
651	03	3
652	00	0
653	02	2
654	00	0
655	01	1
656	07	7
657	69	OP
658	01	01
659	03	3
660	04	4
661	02	2
662	07	7
663	00	0
664	00	0
665	02	2
666	01	1
667	03	3
668	07	7
669	69	OP
670	02	02
671	69	OP
672	05	05
673	43	RCL
674	00	00
675	99	PRT
676	00	0
677	42	STO
678	00	00
679	91	R/S
Line length		
680	76	LBL
681	10	E'
682	00	0
683	42	STO
684	00	00
685	69	OP
686	00	00
687	43	RCL
688	28	28
689	69	OP
690	01	01
691	43	RCL
692	29	29
693	69	OP
694	02	02
695	69	OP
696	05	05
697	91	R/S
698	99	PRT
699	42	STO
700	19	19

(continued next page)

TABLE 2 (*continued*)

Line length (cont'd)

Step	Code	Key
701	59	INT
702	44	SUM
703	00	00
704	43	RCL
705	19	19
706	22	INV
707	59	INT
708	65	×
709	43	RCL
710	11	11
711	95	=
712	44	SUM
713	00	00
714	61	GTO
715	06	06
716	97	97
717	91	R/S

Label addresses

Label	Key
GATE VALVES	B
GLBE VALVES	INV
PLUG VALVES	LNX
SWCK VALVES	CE
BLCK VALVES	CLR
BUTF VALVES	X⇄T
3WST VALVES	X^2
3WBF VALVES	√X
SR90 ELBOWS	1/X
LR90 ELBOWS	C
SR45 ELBOWS	STO
90 MITER BEND	RCL
45 MITER BEND	SUM
ST TEES	D
BF TEES	Y^X
RED/ENLR	E
ENTRANCE	A'

Label addresses (cont'd)

Label	Key
EXIT	B'
SUM-EQL FT	C'
LINE LNGTH	E'

Data registers*

Contents	Register
Sum register	00
1734412442.	01
1327173137.	02
27243117.	03
2717312237.	04
2300333532.	05
2235133000.	06
Pipe I.D.	07
1731073513.	08
3115170000.	09
1744243700.	10
8.333333333	11
3517166317.	12
3127350071.	13
2.99761	14
-1.06205	15
1.23787	16
1.83343	17
0.	18
Pipe length	19
4213274217.	20
3600710000.	21
2137.	22
1727143243.	23
3600710000.	24
3717350014.	25
1731160071.	26
1736007100.	27
2724311700.	28
2731223723.	29

*Numbers must be entered

After all the lengths have been entered, press **C′** to obtain the total line length. This feature is also handy for the addition of any linear measurements in feet and inches, such as vessel, tank, tower, and plot-plan dimensions.

If a printer is not available, all calculations can still be performed and the totals recalled from storage register **00**. All sums for equivalent lengths of valves, fittings, and entrance and exit losses, as well as of line lengths, are stored in this register.

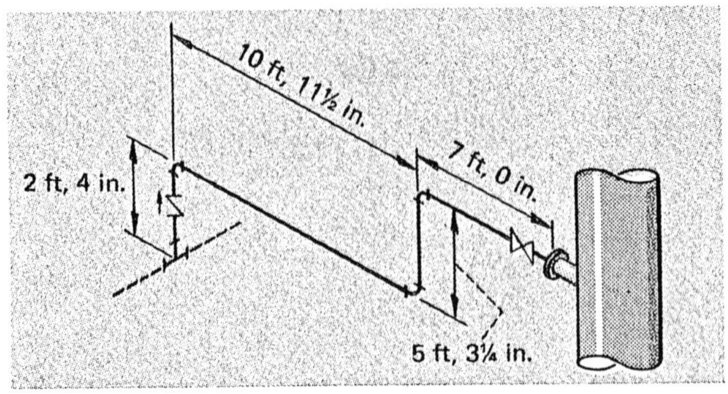

FIG. 1 Piping isometric for the sample calculation.

Do not press **C′** to obtain a total without the printer, because register **00** is cleared by the calculator after the sum has been printed. This register is also automatically cleared when either **E′** or **A** is pressed.

Sample Problem. A piping isometric (see Fig. 1) shows a 4-in. Schedule 40 line containing three 90° long-radius elbows, one flowthrough branch-tee, one gate valve (fully open), one swing check-valve, and an exit loss. The piping section lengths are 2 ft 4 in., 10 ft 11½ in., 7 ft 0 in., and 5 ft 3¼ in. What is the total equivalent length?

The calculator printouts are given in Table 3. The total equivalent length of the valves, fittings, and exit loss is 109 ft. The sum of the line lengths is 26 ft. This yields a total equivalent line length of 135 ft.

TABLE 3 Sample Problem Yields Total Equivalent Line Length of 135 ft [109 ft (fittings) + 26 ft (pipe)]

EQUIVALENT LINE		LINE LNGTH
LENGTH PROGRAM		2.04
PIPE ID ?		10.115
4.026	IN	7.
LR90 ELBOWS ?		5.0325
3.		SUM-EQL FT
20.17026	FT	25.5625
BF TEES ?		
1.		
20.13	FT	
GATE VALVES ?		
1.		
4.4206	FT	
SWCK VALVES ?		
1.		
44.04444	FT	
EXIT		
20.56124573	FT	
SUM-EQL FT		
109.3345457		

References

1. *Flow of Fluids,* Technical Paper 410, Crane Co., New York, 1957.
2. R. Kern, "How to Compute Pipe Size," *Chem. Eng.*, pp. 115–120 (January 6, 1975).

W. WAYNE BLACKWELL

Fittings, Number and Types

The number and types of pipe fittings can be estimated by this method long before the piping isometrics are done. Pipe size and a general idea of the system's complexity are all that is needed.

Typical uses of this method include: (1) making allowance for the cost of valves and fittings in the piping section of a cost estimate, (2) estimating the contribution of fittings to line loss in order to establish preliminary pump heads, and (3) setting up an initial bulk order for fittings for a large project. The method gives only average countings of fittings, so the number of fittings estimated may need to be revised when the design is finalized.

The data base used to develop the method came from 46 projects executed by Monsanto Co.'s Corporate Engineering Dept. The base represents a mixture of small to very large plant projects, and includes piping systems of carbon, galvanized, and stainless steel. The data are summarized in Table 1.

Fitting Frequency

Figure 1 is a plot of total number of fittings per 100 ft of pipe vs nominal pipe size. Curves for three systems of differing complexity are shown. As might be expected, the frequency of fittings decreases as size increases. A given length of 24-in. pipe cannot contain as many fittings as does a normal 2-in. run.

Note that the curves are approximately linear on a log-log plot, with a slope of about $-½$. Above a pipe size of about 12 in., the number of fittings

TABLE 1 Data Base for Correlating Frequencies of Fittings and Valves

Nominal Pipe Diameter (in.)	Total Length of Pipe (ft)	Total Number of Fittings by Type: Elbows 90°	Elbows 45°	Tees	Reducers	Flanges	Unions and Couplings	Caps and Plugs	Grand Total of Fittings	Grand Total of Valves
½	33,990	5,884	364	2,325	847	1,818	4,240	756	20,474	11,589
¾	33,123	5,260	303	2,293	1,473	2,973	2,270	760	17,602	7,551
1	124,513	13,008	791	6,730	4,582	12,552	4,554	833	47,604	10,363
1½	121,212	7,465	609	5,731	3,928	7,299	2,040	441	29,553	3,313
2	142,891	6,197	622	4,120	2,625	11,727	369	640	26,669	4,199
3	125,550	5,018	516	2,546	2,123	10,427	144	554	21,472	2,441
4	84,705	3,301	397	1,714	1,679	6,608	33	367	14,132	1,346
6	77,717	3,219	388	1,501	1,155	4,578	12	264	11,129	898
8	67,667	2,304	283	1,515	769	3,592	3	180	8,649	466
10	39,225	1,155	190	781	393	1,613		80	4,212	301
12	16,445	553	65	580	163	762		51	2,174	162
14	3,997	203	23	117	79	342		17	781	72
16	10,292	335	54	138	121	506	2	38	1,196	90
18	3,530	191	17	116	82	362		23	791	41
20	5,698	466	49	273	186	804		47	1,825	34
24	5,983	214	19	112	79	357		19	800	40
30	3,121	134	14	80	55	255		14	552	13
36	1,608	25	2	16	14	66		3	126	12

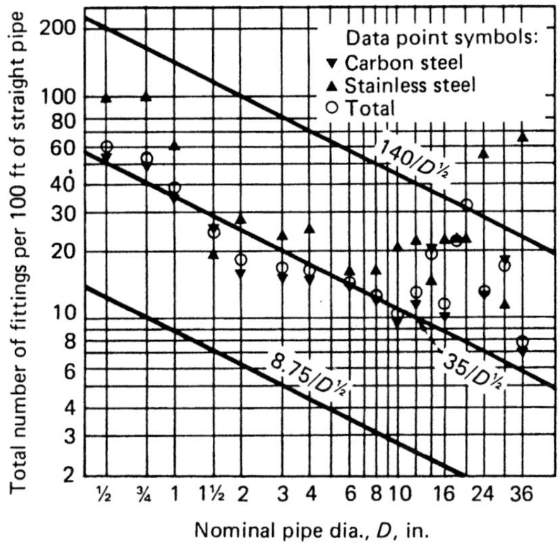

FIG. 1 Fittings for different diameter piping.

becomes greater than what might be expected from extrapolation. This happens because of the smaller number of normal runs in the data base. Also, more of the large pipe is in complex headers. Pipe runs of "normal" complexity fall on, or near, the center curve.

The spread of data away from the center curve makes clear the need to account for piping system complexity.

Complexity Factor

The easiest way to account for variations in the complexity of piping systems is by means of a simple multiplying factor. This factor, F_c, is defined as 1 for normal inside-battery-limits piping. The F_c range needed to correlate the piping runs in the data base extends from ¼ to 4.

Typical values for F_c are: 4—very complex manifolds; 2—manifold-type piping; 1—normal piping; ½—long, straight piping run; and ¼—utility supply lines, outside battery limits. To help quantify the concept of piping complexity, representative systems of 100 ft of 4-in. nominal-size piping for complexity factors of ¼, 1, and 4 are illustrated in Fig. 2.

Correlation Equations

The center curve of Fig. 1 gives the total number of fittings per 100 ft of straight pipe for $F_c = 1$. The correlation equation for this curve is

$$F_{100} = 35F_c/D^{1/2} \tag{1}$$

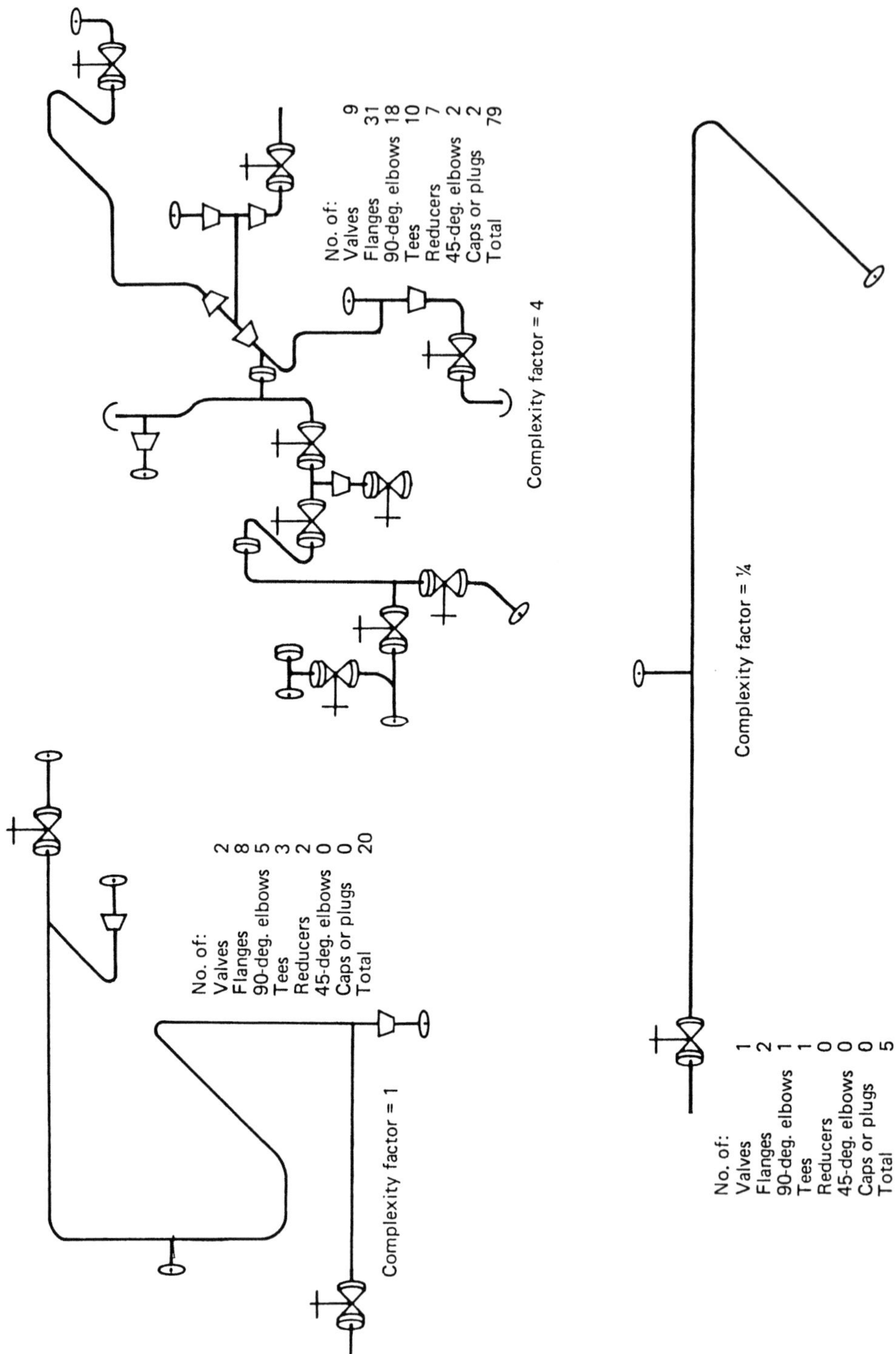

FIG. 2 Complexity factors illustrated for 100 ft of 4-in.-nominal-diameter piping.

Equation (1) says that the total number of fittings (excluding valves) per 100 ft of pipe, F_{100}, is 35 times the complexity factor, F_c, divided by the square root of the nominal pipe diameter, D.

The bottom curve yields the number of fittings for $F_c = 1/4$, which makes Eq. (1) into $F_{100} = 8.75/D^{1/2}$. The top curve is for $F_c = 4$.

Fitting Breakdown

Figure 3 represents plots of the numbers of the different types of fittings (not, as in Fig. 1, the total number of all fittings) vs nominal pipe size—data from the table.

"Flanges" includes couplings and unions as well as flanges, because all are used to join pipe. However, because it takes two flanges to replace one coupling or union, "total flanges" represents the number of flanges plus twice the number of couplings and unions. With this adjustment it can be seen that there is no significant trend with size for type of fitting. Although there is a fair amount of scatter in Fig. 3, an adequate correlation for type of fitting breakdown in piping systems is

Type of fitting	*Percent of total*
"Flanges"	45
90° elbows	25
Tees	15
Reducers	10
45° elbows	2½
Plugs and caps	2½

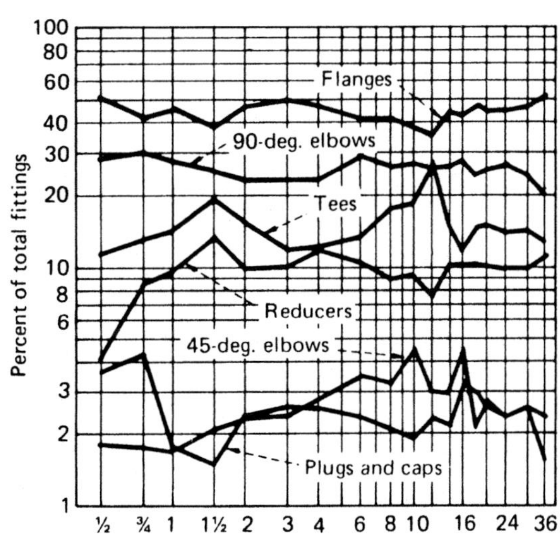

FIG. 3 Percentage breakdown of fitting types.

Valve Frequency

Figure 4 plots total number of valves per 100 ft of piping vs nominal pipe size. Note that the slope is much steeper than for fittings. This may be due to the fact that the cost of valves rises more sharply with increasing size.

The best correlation for normal systems is

$$V_{100} = 9F_c/D \tag{2}$$

Equation (2) says that the total number of valves per 100 ft of straight piping is nine times the complexity factor, F_c, divided by the nominal pipe diameter, D.

An approximate percentage breakdown of valve types, based on the table data, yields: block valves (gate, plug, butterfly, etc.)—90%; check valves (both swing and lift)—8%; and globe valves—2%.

Example Illustrates Computations. Estimate the fittings and valves in a 6-in.-nominal-diameter system that is 300 ft long. Assume normal inside-battery-limit piping—i.e., $F_c = 1$.

From Eq. (1): $F_{100} = 14.3$.

For 300 ft of piping, $F_{300} = 43$.

Using the previously listed percentage breakdown of fitting types, find the number of each type of fitting: "flanges" = 43 × 0.45 = 19; 90°

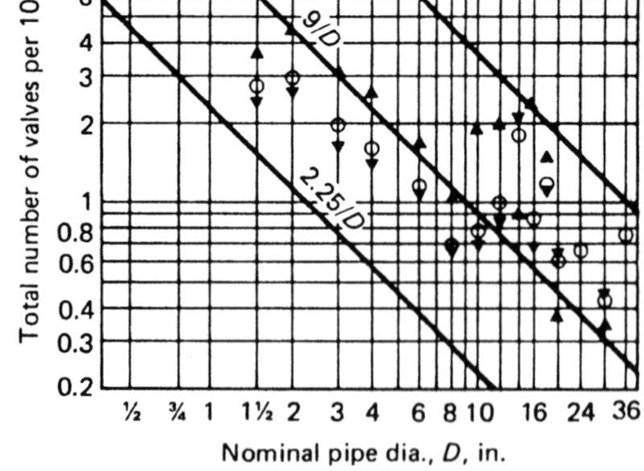

FIG. 4 Valves for different diameter piping.

elbows = 11; tees = 7; reducers = 4; 45° elbows = 1; and plugs and caps = 1. These numbers have been rounded.

From Eq. (2): $V_{100} = 1.5$.

For 300 ft of piping, $V_{300} = 5$.

From the previously given percentage breakdown of valve types, find the number of each type of valve: block valve = 5 × 0.90 = 4; check valve = 1 (rounding up 0.4); and globe valve = 0.

Therefore, this 300 ft of 6-in.-diameter straight piping could be expected to contain 43 fittings and 5 valves, with the foregoing breakdown of types of fittings and valves.

WILLIAM B. HOOPER

Fittings, Pressure Drop

Forcing a fluid through a pipe fitting consumes energy, which is provided by a drop in pressure across the fitting. This pressure drop—or head loss—is caused by friction between the fluid and the fitting wall and by creation of turbulence in the body of the fluid.

The loss due to wall friction is best handled by treating the fitting as a piece of straight pipe of the same physical length as the fitting. All common prediction methods, including this two-K method, do this. But each method predicts the remaining "excess" head loss a different way.

Equivalent Length

The equivalent-length method adds some hypothetical length of pipe to the actual length of the fitting, yielding an "equivalent length" of pipe (L_e) that has the same total loss as the fitting. The unfortunate drawback to this simple approach is that the equivalent length for a given fitting is not constant but depends on Reynolds number and roughness as well as size and geometry. Therefore, use of the equivalent-length method requires consideration of all these factors.

The *excess* head loss in a fitting is due mostly to turbulence caused by abrupt changes in the direction and speed of flow. Thus it is best to predict this loss by using a velocity-head approach.

Velocity Head

The amount of kinetic energy contained in a stream is the velocity head. An equivalent statement is that the velocity head is the amount of potential energy (head) necessary to accelerate a fluid to its flowing velocity.

For example: Pressure gauges on both sides of a gradual, friction-free pipe entrance would show that the pressure in the flowing fluid is lower than the pressure in the feed tank by one velocity head. (This is why an eductor works.) The potential (pressure) energy of the fluid in the tank is not lost; it has been converted to kinetic energy. The number of velocity heads (H_d) in a flowing stream is calculated directly from the velocity of the stream (v):

$$H_d = v^2/2g$$

With this background, consider a square elbow. The entering fluid experiences a pipelike frictional head loss as it moves down the inlet leg. At the turn, the flow stops abruptly and starts in a new direction. Since the inlet velocity vector has no component in the outlet direction, all of the inlet kinetic energy is lost. Thus, this part of the loss in a square elbow is close to one velocity head. The remaining losses are the frictional losses in the turn and the outlet leg.

The total head loss in the elbow is the sum of the frictional and directional losses. The excess head loss (ΔH) is less than the total by the amount of frictional loss that would be experienced by straight pipe of the same physical length. (Of course, the *actual* frictional loss in the fitting will be different than the loss in a pipe.) The excess loss in a fitting is normally expressed by a dimensionless "K factor":

$$\Delta H = KH_d$$

The Two-*K* Method

K is a dimensionless factor defined as the *excess* head loss in a pipe fitting, expressed in velocity heads. In general, it does not depend on the roughness of the fitting (or the attached pipe) or the size of the system, but it is a function of Reynolds number and of the exact geometry of the fitting. The two-K method takes these dependencies into account in the following equation:

$$K = K_1/N_{Re} + K_\infty(1 + 1/ID)$$

where K_1 = K for the fitting at $N_{Re} = 1$

K_∞ = K for a large fitting at $N_{Re} = \infty$

ID = internal diameter of attached pipe, in.

How N_{Re} and Fitting Size Affect *K*

Why two *K*s, when the literature usually reports a single *K* value? Most published *K* values apply to fully-developed turbulent flow. This is convenient because *K* is independent of N_{Re} when N_{Re} is sufficiently high. However, *K* starts to rise as N_{Re} decreases toward 1000, and it becomes inversely proportional to N_{Re} when N_{Re} is below 100.

Figure 1 is a plot of *K* vs N_{Re} for short-radius elbows of 0.6274 in. *ID* [2]. Note that the two-*K* expression, with 800 for K_1 and 0.40 for K_∞, fits the points accurately in all flow regimes. In this case, K_1 has no effect on the predicted *K* at N_{Re} above 10,000; K_∞ is negligible below an N_{Re} of 50.

Theoretically, *K* should be the same for all fittings that are geometrically similar. In fact, smaller fittings are more sensitive to surface roughness and have more abrupt changes in cross-section. Thus *K* is greater for smaller fittings of a given type.

The $1/ID$ correction in the two-*K* expression accounts for the size differences: *K* is higher for small sizes but nearly constant for large sizes. Figure 2 is a plot of *K* vs pipe size data for long-radius ($R/D = 1.5$) elbows [1, 3, 4]. The solid line shows how the two-*K* correlation fits these points; the other lines are correlations that will be discussed later.

Recommended Values

Table 1 lists values of K_1 and K_∞ derived from plots of *K* vs N_{Re} and size (similar to Figs. 1 and 2). The reader is encouraged to keep this and use it, because it is the heart of the two-*K* method.

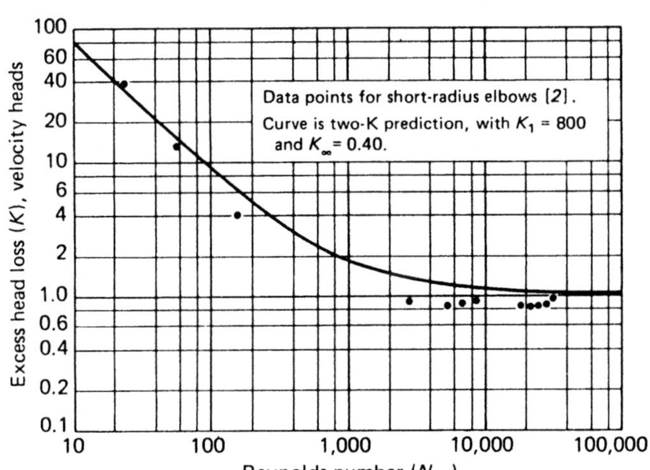

FIG. 1 The two-*K* method fits head-loss data for laminar, transitional, and turbulent flow.

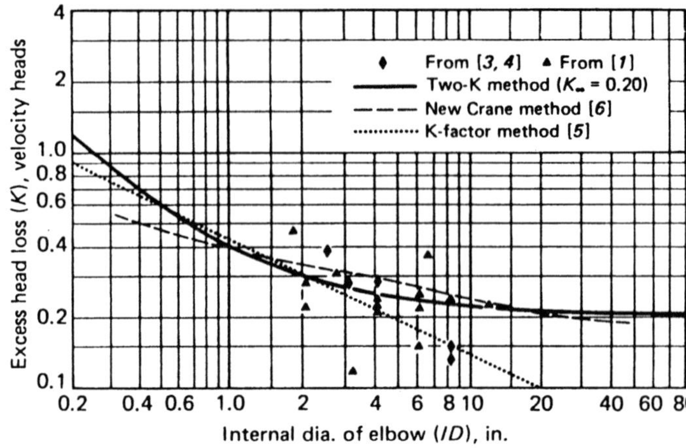

FIG. 2 Size of elbow affects K.

A fuller treatment of all expansions and contractions is given by "Calculate Head Loss Caused by Change in Pipe Size," W. B. Hooper, *Chem. Eng.*, pp. 89–92 (November 7, 1988).

Two-*K* vs Equivalent Length

Why use the two-K method when the equivalent-length method is more familiar and easier to use? This classic method, in which each type of fitting has one "equivalent length," is reliable for 1–6 in. carbon-steel piping in normal runs (see the dashed line in Fig. 2). In large, complex alloy systems, the method could predict head losses 1.5–3 times too high. That means oversized pumps and a large waste of energy and capital. In laminar flow, on the other hand, it could predict head losses a whole order of magnitude too low.

The equivalent-length concept also contains a booby trap for the unwary. Every equivalent length has a specific friction factor (f) associated with it, because the equivalent lengths were originally developed from K factors by the formula $L_e = KD/f$. This is why the latest version of the eqivalent-length method (the 1976 edition of Crane Technical Paper 410 [6]) properly requires the use of two friction factors. The first is the actual friction factor for flow in the straight pipe (f), and the second is a "standard" friction factor for the particular fitting (f_T). Thus the two-K method is as easy to use and as accurate as the updated equivalent-length method.

What about the widely used K-factor graphs published by the Hydraulic Institute? (See Ref. 5 for a good presentation of these graphs.) The graphs are good for 1–8 in. pipe in fully turbulent flow (see dotted line in Fig. 2), but extrapolation to larger sizes can cause errors. For example, the K-factor

TABLE 1 Constants for Two-K Method[a]

		Fitting Type		K_1	K_∞
Elbows	90°	Standard (R/D = 1), screwed		800	0.40
		Standard (R/D = 1), flanged/welded		800	0.25
		Long-radius (R/D = 1.5), all types		800	0.20
		Mitered elbows	1 Weld (90° angle)	1000	1.15
		(R/D = 1.5)	2 Weld (45° angles)	800	0.35
			3 Weld (30° angles)	800	0.30
			4 Weld (22.5° angles)	800	0.27
			5 Weld (18° angles)	800	0.25
	45°	Standard (R/D = 1), all types		500	0.20
		Long-radius (R/D = 1.5), all types		500	0.15
		Mitered, 1 weld, 45° angle		500	0.25
		Mitered, 2 weld, 22.5° angles		500	0.15
	180°	Standard (R/D = 1), screwed		1000	0.60
		Standard (R/D = 1), flanged/welded		1000	0.35
		Long radius (R/D = 1.5), all types		1000	0.30
Tees	Used as elbow	Standard, screwed		500	0.70
		Long-radius, screwed		800	0.40
		Standard, flanged or welded		800	0.80
		Stub-in-type branch		1000	1.00
	Run-through tee	Screwed		200	0.10
		Flanged or welded		150	0.05
		Stub-in-type branch		100	0.00
Valves	Gate, ball, plug	Full line size, β = 1.0		300	0.10
		Reduced trim, β = 0.9		500	0.15
		Reduced trim, β = 0.8		1000	0.25
	Globe, standard			1500	4.00
	Globe, angle or Y-type			1000	2.00
	Diaphragm, dam type			1000	2.00
	Butterfly			800	0.25
	Check	Lift		2000	10.00
		Swing		1500	1.50
		Tilting-disk		1000	0.50

[a]Note: Use R/D = 1.5 values for R/D = 5 pipe bends, 45 to 180°. Use appropriate tee values for flow through crosses.

line in Fig. 2 shows a K of 0.075 for a 36-in. elbow, but the actual K is about 0.200. Of course, these charts greatly underestimate laminar head losses and should not be used for N_{Re} below 10,000.

Example. Consider a 16-in. Schedule 10S stainless-steel system as shown in Fig. 3. The system contains 100 actual ft of pipe. 6 long-radius (normal for most systems) elbows; 2 side-outlet tees; 2 gate valves, and an exit into a tank. The fluid has a viscosity of 1 cP, a specific gravity of 1, and is flowing at 10 ft/s. What is the head loss through this system?

Pipe and fittings: 100 ft of 16-in. Sch.10S stainless-steel pipe
6 long-radius (R/D = 1.5) elbows
2 side-outlet tees
2 gate valves

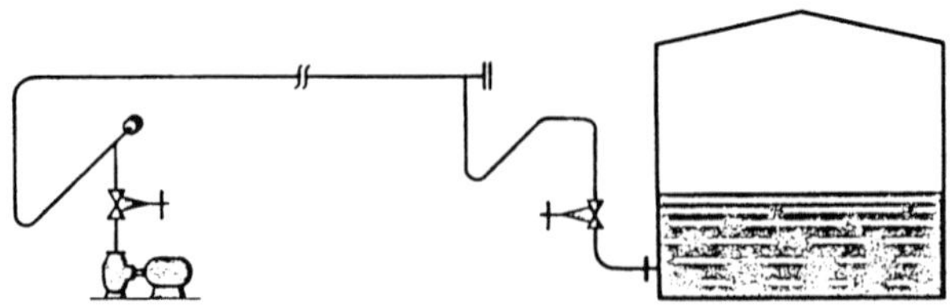

First calculate and convert the given data to get the needed information:

$$\rho = 1 \times 62.43 = 62.43 \text{ lb/ft}^3$$
$$\mu = 1 \times 6.72 \times 10^{-4} = 6.72 \times 10^{-4} \text{ lb/ft-s}$$
$$ID = 15.624 \text{ in. for Schedule 10S pipe}$$
$$D = 15.624/12 = 1.302 \text{ ft}$$
$$N_{Re} = (10)(1.302)(62.43)/(6.72 \times 10^{-4}) = 1{,}210{,}000$$
$$H_d = v^2/2g = 10^2/64.34 = 1.554 \text{ ft of fluid}$$

Given ϵ = 0.00005 ft for stainless pipe, we can find f from the Colebrook equation: f = 0.0122. Thus, fL/D = (0.0122)(100)/(1.302) = 0.937 (this is the K value for the pipe itself).

TABLE 2 Two-K Method

Form: $\Delta H = KH_d$; $K = K_1/N_{Re} + K_\infty(1 + 1/ID)$

Find K for fittings:

Fittings	n	K_1	nK_1	K_∞	nK_∞
90° elbows	6	800	4800	0.20	1.20
Tees (side outlet)	2	800	1600	0.80	1.60
Gate valves	2	500	1000	0.15	0.30
Totals			7400		3.10

$$K = 7400/1{,}210{,}000 + 3.10(1 + 1/15.624) = 3.305$$

Find K for exit and straight pipe:

$$K = 1.0 \text{ for normal exit; } K = fL/D = 0.937 \text{ for pipe}$$

Find head loss:

$$\begin{aligned}\Delta H &= KH_d \\ &= (3.305 + 1.0 + 0.937)(1.554) \\ &= 8.15 \text{ ft}\end{aligned}$$

TABLE 3 Old Equivalent-Length Method [1]

Form: $\Delta H = (fL_e/D)H_d$

Find equivalent lengths:

Fittings	n	L_e	nL_e
90° elbows	6	42	252
Tees (side outlet)	2	89	178
Gate valves	2	9	18
Exit	1	89	89
Straight pipe			100
Total L_e			637 ft

Find head loss:

$$\begin{aligned}\Delta H &= (fL_e/D)H_d\\ &= (0.0122 \times (637/1.302))(1.554)\\ &= 9.28 \text{ ft}\end{aligned}$$

Tables 2–5 show how to calculate the total head loss by the two-K method and three other methods. The results:

1. Two-K method: $\Delta H = 8.15$ ft.
2. Old equivalent-length method: $\Delta H = 9.28$ ft (14% high).
3. K-factor method: $\Delta H = 6.52$ ft (20% low).
4. Revised Crane method: $\Delta H = 8.18$ ft.

TABLE 4 K-Factor Method [5]

Form: $\Delta H = ((fL/D) + K)H_d$

Find K for fittings and exit:

Fittings	n	K	nK
90° elbows	6	0.22	1.32
Tees (side outlet)	2	0.44	0.88
Gate valves	2	0.03	0.06
Exit	1	1.0	1.00
Total			3.26

Find K for straight pipe:

$$K = fL/D = 0.937 \text{ (given)}$$

Find head loss:

$$\begin{aligned}\Delta H &= KH_d\\ &= (3.26 + 0.937)(1.554)\\ &= 6.52 \text{ ft}\end{aligned}$$

Table 5 New Crane Method [6]

Form: $\Delta H = ((fL/D) + K)H_d$

f_T for this system is 0.013 (p. A-26)

Find K for fittings and exit:

Fittings		K	n	nK
90° elbows	$K = 20f_T$	0.260	6	1.560
Tees (side outlet)	$K = 60f_T$	0.780	2	1.560
Gate valves	$K = 8f_T$	0.104	2	0.208
Exit		1.00	1	1.00
Total				4.328

Find K for straight pipe:

$$K = fL/D = 0.937 \text{ (given)}$$

Find head loss:

$$\begin{aligned}\Delta H &= ((fL/D) + K)H_d \\ &= (0.937 + 4.328)(1.554) \\ &= 8.18 \text{ ft}\end{aligned}$$

Note that flow was fully turbulent in this example. For laminar flow, the equivalent-length and K-factor methods would have been off considerably more.

Symbols

D	inside pipe diameter (ft)
f	Moody friction factor ($f = 64/N_{Re}$ for laminar flow)
f_T	"standard" friction factor for head loss in fitting
g	acceleration due to gravity, 32.17 ft/s^2
H_d	velocity head (ft of fluid)
ΔH	head loss (ft of fluid)
ID	inside pipe diameter (in.)
K	excess head loss for a fitting, velocity heads
K_1	K for fitting at $N_{Re} = 1$, velocity heads
K_∞	K for very large fitting at $N_{Re} = \infty$, velocity heads
L	length of pipe, including physical length of fittings (ft)
L_e	equivalent length of a fitting ($L_e = KD/f$) (ft)
N_{Re}	Reynolds number for flow ($N_{Re} = \rho Dv/\mu$)
n	number of fittings of a given type
ΔP	pressure drop ($\Delta P = \rho\Delta H/144$) (lb/in.^{2})

R/D	bend radius of an elbow divided by inside diameter of pipe
v	fluid velocity (ft/s)
β	ratio of orifice diameter to pipe inside diameter
ϵ	roughness of pipe wall (ft)
μ	viscosity of fluid (lb/ft-s)
ρ	density of fluid (lb/ft^3)

References

1. J. R. Freeman, *Experiments upon the Flow of Water in Pipe and Pipe Fittings,* American Society of Mechanical Engineers, New York, 1941.
2. C. P. Kittridge and D. S. Rowley, "Resistance Coefficients for Laminar and Turbulent Flow through 1/2 Inch Valves and Fittings," *Trans. ASME, 79,* 1759 (November 1957).
3. R. J. S. Pigott, "Pressure Losses in Tubing, Pipe and Fittings," *Trans. ASME, 72,* 679 (July 1950).
4. R. J. S. Pigott, "Losses in Pipe and Fittings," *Trans. ASME, 79,* 1767 (November 1957).
5. L. L. Simpson, "Sizing Piping for Process Plants," *Chem. Eng.*, p. 192 (June 17, 1968).
6. Crane Co., *Flow of Fluid through Valves* (Crane Technical Paper 410), 15th printing, Chicago, 1976.

WILLIAM B. HOOPER

Flashing Steam Condensate

Flashing steam-condensate lines can be quickly sized, and flow velocities derived, by using this method. Estimated pressure drops are conservative.

Condensate-return lines in refineries and chemical plants present a classic application for techniques that calculate pressure drop with two-phase flow. More than 20 correlations for such calculations are presented in the literature. Many of these have a theoretical basis, and most have constants derived from curve-fitting to experimental data. Several authors, including Paige [8], Kordyban [6], and Bankhoff [2], deal specifically with steam/water two-phase flow.

Two-phase-flow pressure-drop methods proposed by Baker [1], Lockhart and Martinelli [7], Chenoweth and Martin [4], and Dukler et al. [5] are often referenced. Dukler's method has a semitheoretical basis and is accepted as the most accurate of available correlations by virtue of its close agreement to experimental data.

However, this and most of the other methods are rather difficult to apply, due both to the amount of physical data and to the lengthy computation required.

Figure 1 provides a rapid estimate of the pressure drop of flashing condensate, along with a determination of fluid velocity. This figure is based on the simplifying assumption of a single homogeneous phase of fine liquid droplets dispersed in the flashed vapor. Pressure drop has been calculated accordingly by Darcy's equation for single-phase flow:

$$\Delta P = 0.000336(fw^2)/(d^5\rho) \tag{1}$$

Steam-table data have been used to calculate the isenthalpic flash of liquid condensate from a saturation pressure to a lower end-pressure, and the average density of the resulting liquid–vapor mixture has been used as the assumed homogeneous fluid density, ρ.

Flows within the regime of the figure are characterized as either in complete turbulence or in the transition zone near complete turbulence. The friction factor, f, for Eq. (1) has been calculated by the relation

$$f = 0.25[-\log(0.000486/d)]^{-2.0}$$

This relation is valid for complete turbulence flows in commercial steel and wrought-iron pipe.

Pressure drops for steam-condensate lines can be determined by assuming that the vapor–liquid mix throughout the lines is represented by the mix for conditions at the end-pressure. This assumption conforms to conditions typical of most actual condensate systems, since condensate lines are sized for low-pressure drop, with most flashing occurring across the steam trap or control valve at the entrance.

If the condensate line is to be sized for a considerable pressure drop, so that continuous flashing occurs throughout its length, end-conditions will be quite different from those immediately downstream of the trap. In such cases an iterative calculation should be performed, involving a series of pressure-drop determinations across given incremental lengths.

This iteration is begun at the downstream end-pressure and worked back to the trap, taking into account the slightly higher pressure, and thus the changing liquid–vapor mix, in each successive upstream incremental pipe length. The calculation is complete when the total equivalent length for the incremental lengths equals the equivalent length between the trap and the end-pressure point. This operation can be performed by using the figure.

Example. A reboiler is condensing 1000 lb/h of steam at 600 lb/in.^{2}gauge and returning the condensate to a nearby condensate return header nominally at 200 lb/in.^{2}gauge. What size of condensate line will give a pressure drop of 1.0 (lb/in.^{2})/100 ft or less?

Step 1: Enter the figure at 600 lb/in.^{2}gauge below the insert near the right-hand side, and read down to the 200-lb/in.^{2}gauge end-pressure.

Step 2: Proceed left horizontally across the chart to the intersection, with:

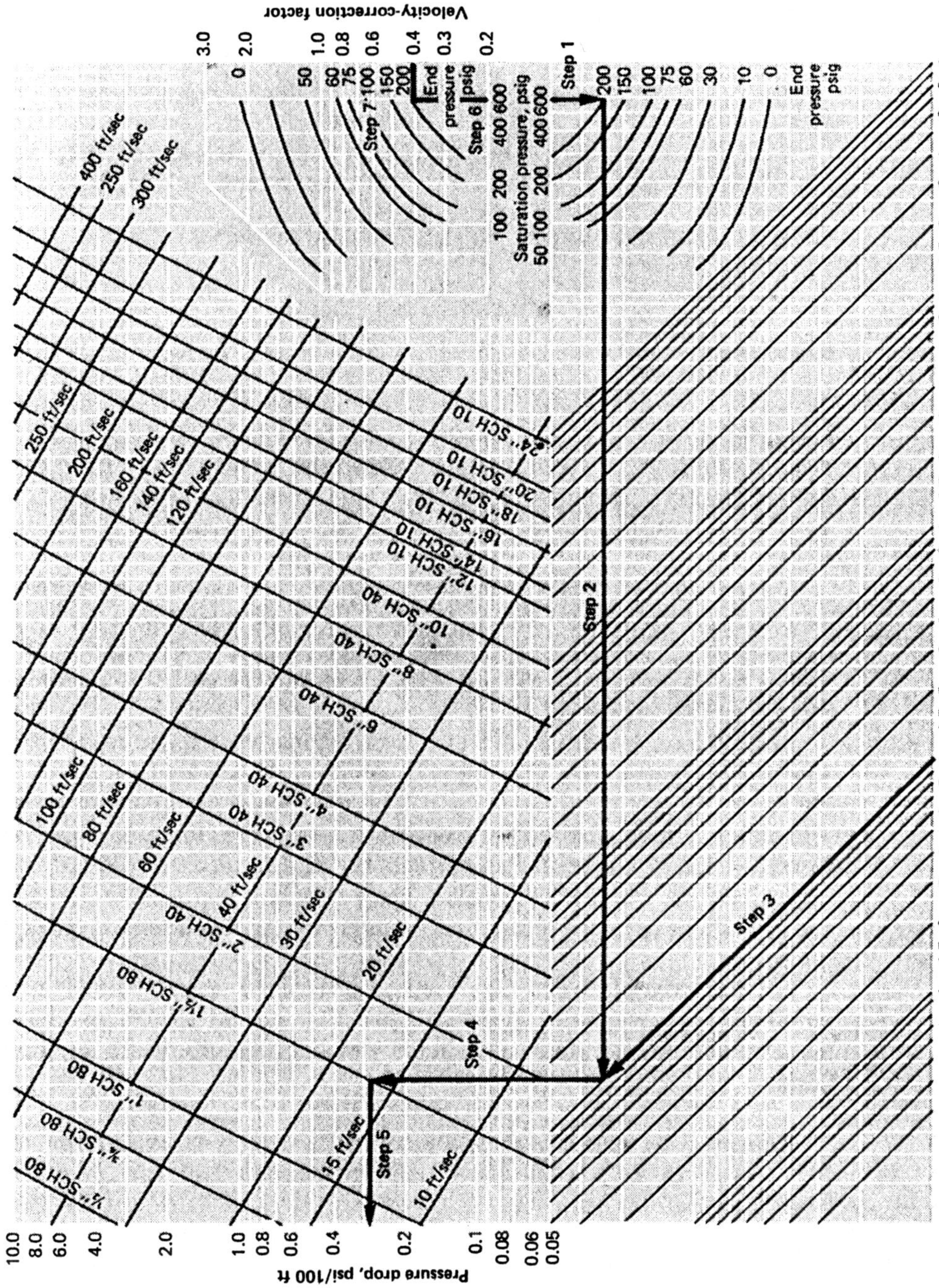

FIG. 1 Flashing steam condensate linesizing chart.

Step 3: The 1000-lb/h flow rate projected diagonally up from the bottom scale.

Step 4: Reading vertically up from this intersection, it can be seen that a 1-in. line will produce more than the allowed pressure drop, so a 1½ in. size is chosen.

Step 5: Read left horizontally to a pressure drop of 0.28 (lb/in.^{2})/100 ft on the left-hand scale.

Step 6: Note the velocity given by this line as 16.5 ft/s, then proceed to the insert on the right, and read upward from 600 lb/in.^{2}gauge to find the velocity correction factor as 0.41.

Step 7: Multiply 0.41 by 16.5 to get a corrected velocity of 6.8 ft/s.

Accuracy

This rapid method provides pressure drops comparable to those computed by more-sophisticated techniques for two-phase flow. Thus, for the conditions of the above example, the Dukler no-slip method estimates 0.22 (lb/in.^{2})/100 ft and the Dukler constant-slip method estimates 0.25 (lb/in.^{2})/100 ft.

Comparisons for this and 12 other examples, representing all regimes of the chart, are shown in Table 1. Pressure drops predicted with the Dukler methods are consistently smaller than those estimated by the chart, with the Dukler no-slip method averaging 23% lower and the constant-slip method averaging 32% lower.

Results from the chart have also been compared to those calculated by a method suggested by Paige [8] and based on the work of Benjamin and Miller [3]. Paige's method assumed a homogeneous liquid–vapor mixture with no liquid holdup, and thus is similar in approach to the present method. However, Paige suggests calculation of the liquid–vapor mix based on an isentropic flash whereas the present chart is based on an isenthalpic flash, and this is believed to be more representative of steam-condensate collecting systems.

For the example, the Paige method gives 0.26 (lb/in.^{2})/100 ft at the terminal pressure—and 0.25 (lb/in.^{2})/100 ft at a point 1000 ft upstream of the terminal pressure, due to the slightly higher pressure, which suppresses flashing.

Symbols

d	pipe inside diameter (in.)
f	friction factor (dimensionless)
ΔP	pressure drop [(lb/in.^{2})/100 ft]

TABLE 1 Predicted Pressure Drops Compared with Dukler

Flow Rate (lb/h)	Pipe Size	Saturation Pressure (lb/in.^{2}gauge)	End Pressure (lb/in.^{2}gauge)	Pressure Drop [(lb/in.^{2})/100 ft]: Presented Method	Pressure Drop: Dukler: No Slip	Pressure Drop: Dukler: Constant Slip	Horizontal Flow Pattern (Baker Chart)
1,000	1½″ SCH 80	600	200	0.28	0.22	0.25	Slug
100,000	16″ SCH 10	30	0	0.08	0.068	0.034	Annular
100,000	18″ SCH 10	600	0	0.19	0.17	0.15	Annular
1,000	1″ SCH 80	400	75	7.4	4.6	4.5	Annular
1,000,000	12″ SCH 10	200	100	3.7	2.3	3.1	Dispersed
100	¾″ SCH 80	400	0	2.5	2.2	1.8	Annular
1,000,000	24″ SCH 10	400	200	0.10	0.075	0.093	Annular
10,000	6″ SCH 40	100	0	0.22	0.19	0.11	Annular
1,000,000	16″ SCH 10	600	75	4.7	3.0	4.2	Dispersed
10,000	3″ SCH 40	150	0	9.4	6.1	4.8	Annular
10,000	4″ SCH 40	150	0	2.3	1.6	1.2	Annular
10,000	6″ SCH 40	150	0	0.27	0:23	0.15	Annular
10,000	8″ SCH 40	150	0	0.064	0.061	0.039	Wave

V velocity (ft/s)
w flow rate (lb/h)
ρ density (lb/ft^3)

References

1. O. Baker, *Oil Gas J.*, July 26, 1954.
2. S. G. Bankoff, *Trans. ASME, C82,* 265 (1960).
3. M. W. Benjamin and J. G. Miller, *Trans. ASME, 64,* 657 (1942).
4. J. M. Chenoweth and M. W. Martin, *Pet. Refiner, 34,* 151 (1955).
5. A. E. Dukler, M. Wicks, and R. G. Cleveland, *AIChE J., 10,* 44, (1964).
6. E. S. Kordyban, *Trans. ASME, D83,* 613 (1961).
7. R. W. Lockhart and R. C. Martinelli, *Chem. Eng. Prog., 45,* 39 (1949).
8. P. M. Paige, *Chem. Eng.*, p. 159 (August 14, 1967).

RICHARD P. RUSKIN

Gravity Flow

Gas entrained in liquid flowing by means of gravity from a vessel can reduce the outlet pipe's capacity and cause flow to surge cyclically. These problems can be avoided by carefully designing for either full-liquid or two-phase flow.

Entrainment curtails liquid gravity flow from vessels by raising the pressure drop (above that for single-phase flow) through the outlet piping and by reducing the static head available for overcoming the pressure drop. A similar problem can arise when a liquid is near its boiling point or contains dissolved gas, especially if the absolute pressure at any point in the piping falls below atmospheric pressure, as occurs in a syphon.

Consider the case of liquid flowing from the bottom of an absorption column through a pipe that has been sized for full-liquid flow (Fig. 1).

When the liquid level in the column is low enough, the liquid entrains gas (Fig. 1a). The resulting increase in pressure drop and reduction of head restrict the flow rate, and the liquid level rises (Fig. 1b). Eventually, the level rises high enough to stop entrainment (Fig. 1c). However, gas still in the outlet pipe causes the level to continue to rise until the gas is all swept out (Fig. 1d). Now, the outlet pipe is running full flow (as was assumed in the design), but the static head, becoming higher than was assumed, creates excessive flow, which causes the level to fall until entrainment occurs again and the cycle is repeated (Fig. 1e).

Such oscillations can be severe, depending on system geometry. In one

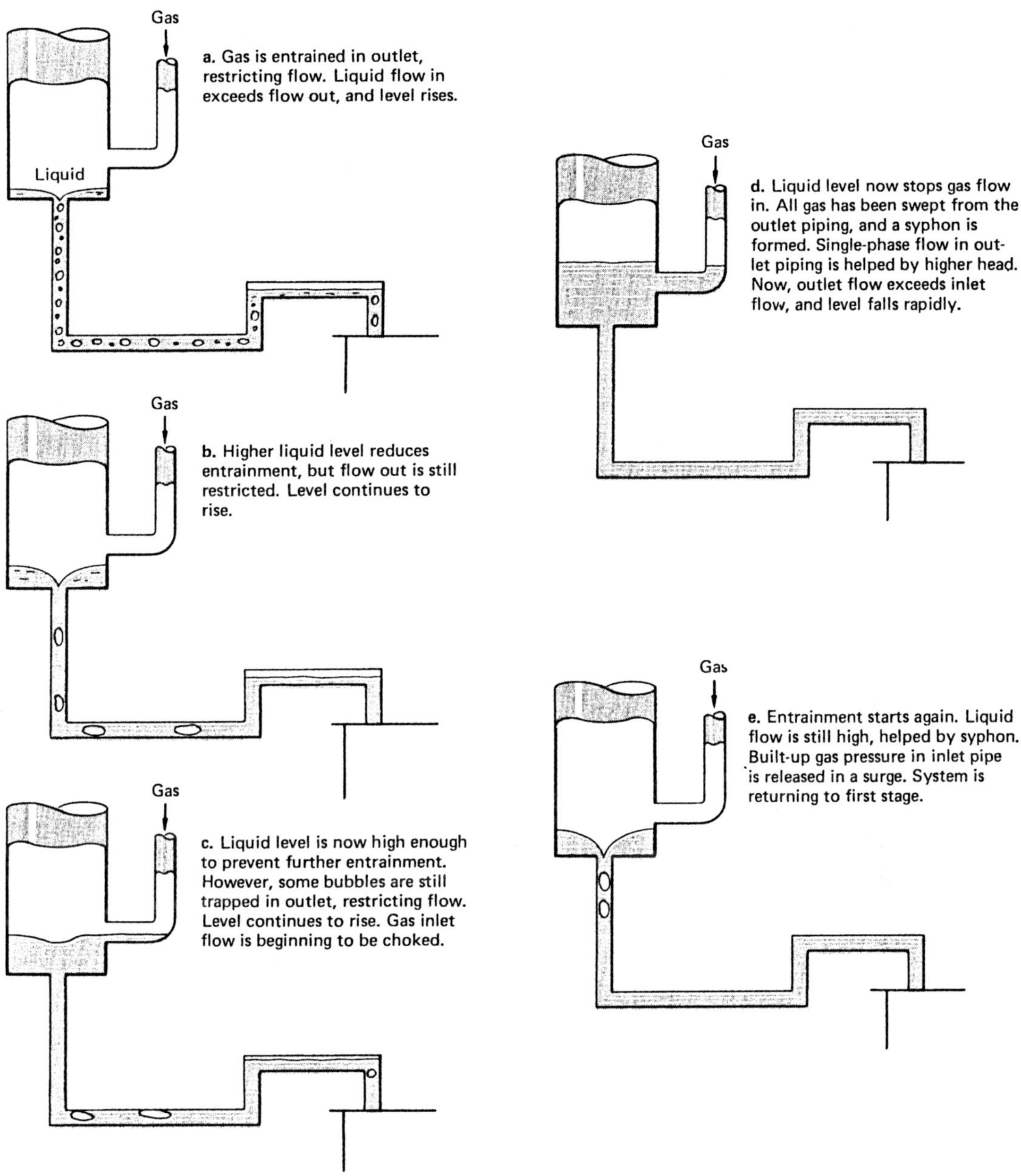

FIG. 1 Entrainment causes surging by increasing pressure drop in piping and lowering head in vessel.

case the peak flow from a tank exceeded the capacity of the vacuum breaker sufficiently to collapse it.

General Expression for Liquid Flow Rate

In this article, liquid flow rates are generally expressed in terms of a dimensionless superficial volumetric flux, J_L^*, which is defined by

$$J_L^* = 4Q_L/\pi d^2(gd)^{1/2} \tag{1}$$

Here, Q_L is the volumetric flow rate, d is the pipe i.d., and g is the gravitational acceleration. Equation (1) is similar to the Froude number. It is used in preference to the Froude number because the latter's definition varies, depending on circumstances. All equations in this article are in consistent units.

Designing for Gravity Flow

Three approaches to the design of gravity drainage systems are possible:

1. For full flow, with the outlet piping size based on single-phase criteria.
2. For self-venting, with the liquid velocity in the outlet pipe kept low enough to allow gas to flow countercurrently to the liquid.
3. For gas entrainment, but with the system designed to accommodate it.

In general, the first approach can be expected to result in the smallest pipe diameter and should be given preference. However, in many instances it is not possible to ensure full pipe flow—in which case the alternatives may have to be adopted.

Designing for Flooded Flow

To avoid gas entrainment in the full-pipe-flow design, the liquid level in the vessel must always be high enough to keep the pipe inlet flooded. To achieve this, some form of control will be necessary, such as via a control valve (Fig. 2a) or a vertical loop in the piping (Fig. 2b). If the latter is used, a syphon break will be necessary (shown in Fig. 2b), and the piping downstream of the syphon break cannot be assumed to run flooded because gas is likely to be entrained at the syphon break. Of course, either arrangement will increase

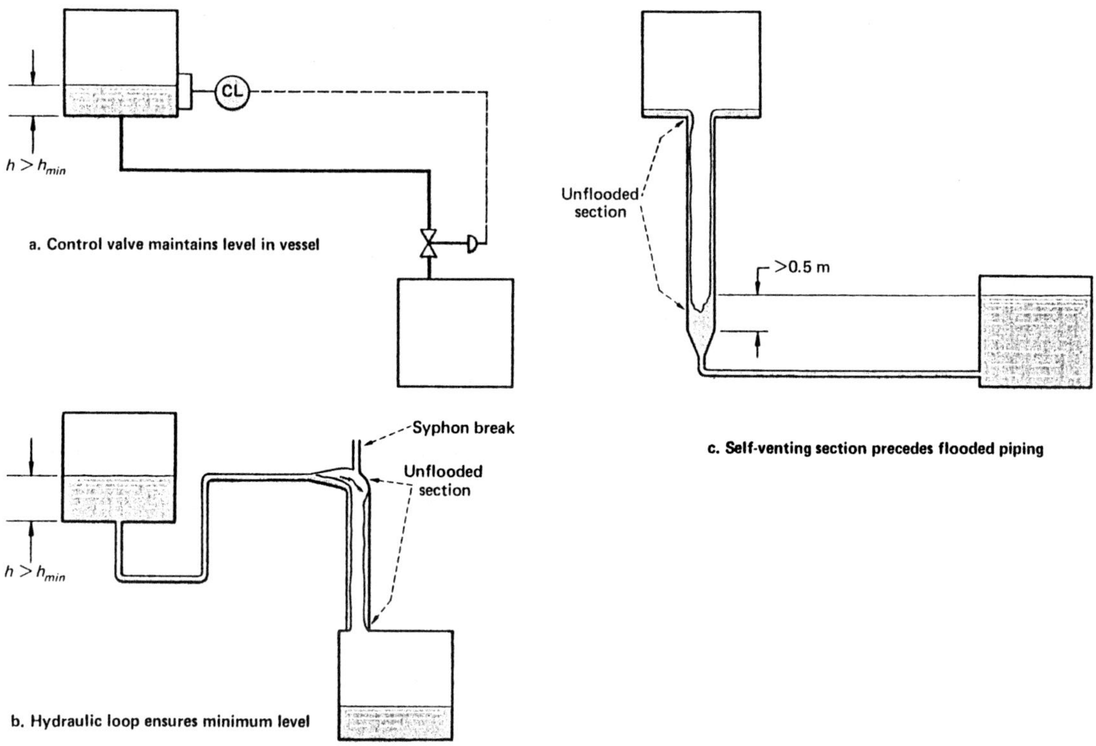

FIG. 2 Designs ensure flooded flow in outlet piping.

the system pressure drop and reduce somewhat the benefits of the flooded-flow design.

Single-phase criteria can be applied to designing sections of outlet piping in which flow can be expected to be flooded. If piping that is certain to be flooded is preceded by a self-venting section, the self-venting section's minimum length should be 0.5 m, to provide for gas disengagement, before the piping is reduced for single-phase flow (Fig. 2c).

The criteria for flooded outlets are Eq. (2) for outlets from the base of vessels and Eq. (3) for outlets from the side of vessels:

$$J_L^* < 1.6(h/d)^2 : h > 0.892[(Q_L)^2/gd]^{0.25} \tag{2}$$

Here, h is the liquid depth in the vessel away from the region of the outlet.

$$J_L^* < (2h/d)^{1/2} : h > 0.811(Q_L)^2/gd^4 \tag{3}$$

Here, h is the liquid height above the top of the outlet away from the region of the outlet.

Designing Unflooded (self-venting) Piping

Side-outlet piping—Coming off from the side of a vessel, piping should be sized such that

$$J_L^* < 0.3 : d > (4Q_L/(0.3\pi g^{1/2})^{0.4} \tag{4}$$

This ensures that the line will run less than half full at its entrance. The level in the vessel away from the outlet will be less than 0.8*d* above the base of the line. The capacity of such an overflow line can be found from Curve 1 in Fig. 3.

Near-horizontal piping—If such a pipeline will run only partially full, it must be inclined to provide the static head to overcome friction losses. A minimum slope of 1:40 is recommended.

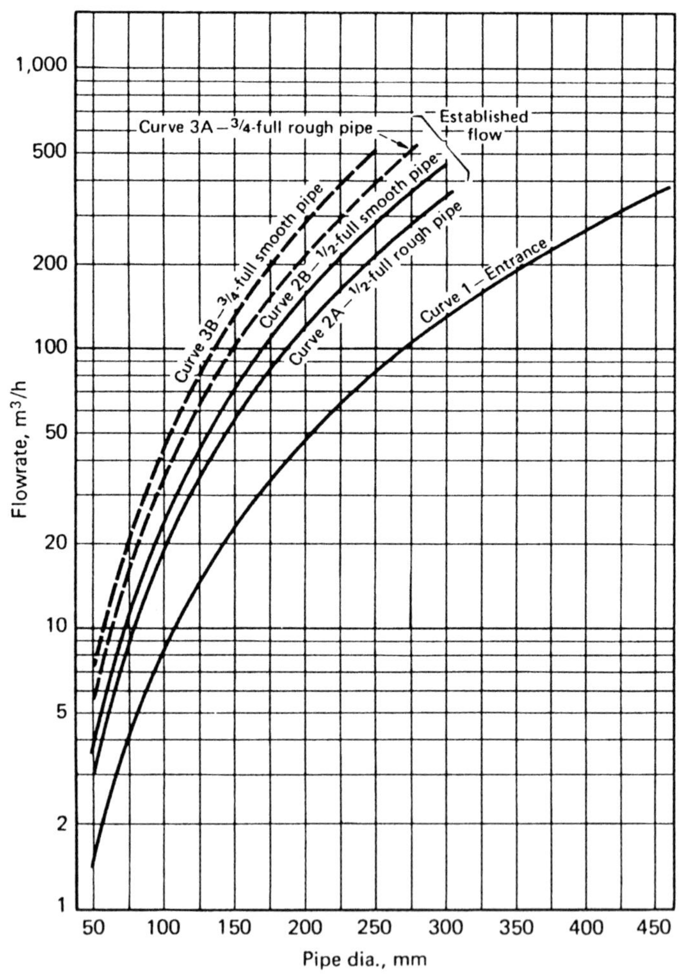

FIG. 3 Capacities for established flows in unflooded pipelines.

To avoid having the liquid carrying gas forward, adequate free area must be left in the pipe to allow gas to pass backward. For pipes up to 200 mm diameter, liquid depths should not be more than half the pipe diameter. For larger pipes, depths up to three-fourths of the diameter may be possible.

When flow in a partially filled pipe is uniform (i.e., constant depth), the energy lost through friction is balanced by the potential-energy change due to the inclination and the pipe. In such a case the mean velocity, $\overline{V}_L$, is related to the inclination and the depth of flowing liquid by Eq. (5) [1]:

$$\overline{V}_L = (32gmi)^{1/2} \log \{[\epsilon/14.8m] + [0.22\nu/m(g_n mi)^{1/2}]\} \tag{5}$$

Here, m is the hydraulic mean depth (flow area/wetted perimeter), i is the inclination of the pipe from the horizontal, ϵ is the pipe roughness, and ν is the kinematic viscosity.

Figure 3 gives the volumetric capacity for established flow in half-full and three-quarters-full rough and smooth pipes. The curves were calculated via Eq. (5) for pipes of slope 1:40 and a fluid having a kinematic viscosity of 10^{-6} m^2/s (e.g., water at 20°C). The absolute roughness used for the rough pipes was 0.25 mm (moderately rusty mild steel). The results are not very sensitive to liquid viscosity. The capacity of a rough pipe is increased by about 1% for a totally inviscid liquid, and it is only reduced by about 10% for a liquid having a kinematic viscosity of 10^{-5} m^2/s. Thus, the Fig. 3 water curves can be safely used for most liquids.

The initial velocity in an outlet line designed to run half full is less than the equilibrium velocity in a pipe having a slope of 1:40. As the liquid accelerates down the pipe, the liquid depth diminishes with distance to that of the depth corresponding to the established flow at a given flow rate. To maintain a constant relative depth, a tapered pipe would be necessary. As this is impractical, reducing the pipe diameter stepwise is recommended. Tapered reducers should be installed to avoid sudden disturbances in the flow.

For long lengths of pipes, the following design approach is suggested:

1. Size the outlet line on the side of a vessel for $J_L^* = 0.3$ (Curve 1 of Fig. 3). If the resulting pipe size is not standard, choose the standard size higher than the calculated size. Continue the size so chosen for at least 10 pipe diameters.
2. Determine the pipe diameter corresponding to half-full established flow for the required flow rate (using Curve 2A or 2B of Fig. 3). Again, select the nearest standard pipe size higher than the calculated size.
3. Reduce the pipe diameter from the outlet size to the established-flow size, using an eccentric reducer that will not change the slope of the bottom of the pipe. Preferably, the reducer's minimum length should be twice that of the upstream pipe diameter.

If the foregoing procedure is followed for pipes of 1:40 slope, the liquid depth after the reducer will not exceed 75% of the pipe diameter.

For long, large-diameter (>200 mm) inclined pipes, it may be worth considering a second reduction down to the size corresponding to an established-flow relative depth of 75%. This reduction can be made after 50 pipe diameters (see Curve 3A or 3B of Fig. 3).

For short pipe runs, the additional cost of tapered reducers—especially if of a gentle angle, as recommended (which may not be standard), or of lined pipe—may exceed the savings in going to smaller-diameter piping. In such cases the entire length of the pipe should be of the large size.

Self-Venting Flow in Vertical Pipes

Liquid flowing vertically down does so as an annular film. In such cases, low superficial velocities are necessary to avoid gas being sucked down with the liquid. Simpson's suggestion of basing pipe outlet diameters on a limiting Froude number of 0.3 is recommended [2]:

$$J_L^* < 0.3 \tag{6}$$

Because Eq. (6) is the same as Eq. (4), pipe diameters can be determined from Curve 1 in Fig. 3.

This approach should be adopted when gas entrainment is to be avoided, as when a vertical pipe extends into a vessel to below the liquid surface, or when the downstream piping must be designed for flooded flow. Smaller pipe than that dictated by Eq. (6) can be expected to cause surging.

Self-Venting Flow in Complex Systems

Little information is available on unflooded flow in systems that include bends, especially for flow changes from vertical to nearly horizontal, and vice versa. Limited evidence suggests that even if the pipe diameter is chosen for self-venting flow (as in a prior section on designing unflooded piping), entrainment and surging may still occur due to the effects of the bends. The design recommendations now given are, therefore, offered only tentatively.

Bends in the horizontal (or nearly horizontal) plane will not necessarily cause problems if the 1:40 slope is continued with the bend and the bend is gentle (preferably, the radius equaling five diameters).

In the vertical plane, the number of bends should be limited as much as possible. Gently sloping piping is preferable to vertical runs. The radius of bends should be at least five diameters.

Bends from, or to, vertical sections should be sized as for vertical piping. Inclined piping following a vertical section can be sized for half-full established flow via the criteria for near-horizontal piping in the previous discussion on designing unflooded (self-venting) piping. Changes in diameter should be made by means of asymmetric tapered reducers whose lengths are equal to twice the larger diameter, and which are installed so that the bottom of the reducer has a slope equal to that of the piping at either end.

If Entrainment Is Acceptable

There are many occasions when it is not necessary to prevent entrainment. Sometimes, moderate surging will not present a problem. In such cases, piping can be sized for smaller diameters at considerable savings.

Sometimes, surging caused by gas entrainment can be reduced by providing a means for the gas to escape at a point downstream in the outlet pipe, such as via some type of gas–liquid separator. If this is practical, the piping can be of smaller diameter. However, because it is not possible to predict the extent of entrainment—and, hence, calculate the pressure drop with certainty—any such approach should be adopted cautiously.

References

1. P. Ackers, *Tables for the Hydraulic Design of Storm Drains, Sewers and Pipelines,* Her Majesty's Stationery Office, 1969.
2. L. L. Simpson, "Sizing Piping for Process Plants," *Chem. Eng.,* June 17, 1968.

P. D. HILLS

Liquid Carbon Dioxide

There is a high interest in tertiary oil recovery techniques. One tertiary recovery technique that has received wide attention recently is carbon dioxide flooding of suitable reservoirs. New schemes have been proposed in Canada, the Middle East, Russia, and the United States.

Carbon dioxide flood proposals typically involve collection, purification, and liquefaction of carbon dioxide at a remote site, and long distance transportation by high pressure buried pipeline followed by distribution and injection facilities at the producing formations.

The SACROC (Scurry Area Canyon Reef Operators Committee) Unit in West Texas was completed in 1972. The main pipeline section of the system consists of 290 km (180 mi) of pipe with a diameter of 406.4 mm (16 in.) and a maximum design pressure of 14 MPa (2035 lb/in.^{2}gauge). Its successful operation demonstrates that it is possible to design, build, and operate carbon dioxide systems safely.

Pipeline Leaks

Line breaks do not normally occur in properly designed and operated pipelines, but third party damage is always a possibility. In addition, major land movements can occur which threaten pipeline integrity, and minor flaws can escape the most stringent quality assurance programs.

A major burst would release carbon dioxide at high rates for short periods of time into the atmosphere at ground level. Because the velocity of carbon dioxide escaping from a burst pipe is very much slower than for a gas pipeline, there will be less physical damage to men and the surrounding environment. Danger from fire is nonexistent, and on this account a carbon dioxide pipeline burst is less hazardous.

Carbon dioxide is about 50% heavier than air and tends to stay close to the ground under stable atmospheric conditions, whereas natural gas tends to rise. In this regard it is similar in behavior to propane. Relatively large percentages of carbon dioxide can be present in the atmosphere without threatening life. It is a principal product of combustion and is vented directly to the atmosphere from homes, cars, factories, power stations, and nearly all forms of life. Plants exposed to sunlight assimilate carbon dioxide from the atmosphere by photosynthesis.

Physiological reports indicate that 2% carbon dioxide in air is subjectively acceptable and compatible with ordinary physical and mental activities [1]. Suppression of shivering has been observed in men inhaling 6% carbon dioxide in air at a temperature of 5°C (41°F) and in refrigerated rabbits breathing 15 to 20% carbon dioxide.

In 1878 Friedlander and Herter, studying the toxicity of carbon dioxide, noted that rabbits could breathe 20% carbon dioxide for 1 h with no toxic effects. Concentrations as high as these have been used routinely in physiological experiments with live animals [1]. Carbon dioxide in concentrations higher than 30% in air has been used as an anaesthetic for minor surgery in humans.

It is clear that high concentrations for short periods of time are unlikely to be hazardous. In the absence of governmental regulations, a concentration of 10 mol% carbon dioxide in the atmosphere was arbitrarily selected as the upper limit for short-term exposure in design work for a proposed carbon dioxide pipeline in Alberta, Canada.

A detailed analysis of carbon dioxide concentrations around leaks, pipeline blowdowns, and bursts was unable to demonstrate significant risk of exposure to concentrations of more than 10% provided normal precautions for high vapor pressure pipelines were followed.

Impurities

Impurities influence the vapor pressure of carbon dioxide and this in turn affects the design in two ways.

First, in order to prevent cavitation, the minimum pump suction pressure needs to be set higher than the fluid's vapor pressure. A high pump suction pressure requires a correspondingly higher maximum operating pressure so that optimum flow rate can be attained.

Second, the vapor pressure sets the decompressed pressure at a pipe break, and this determines whether ductile fractures can propagate or not. It is therefore helpful to specify maximum impurities early in the design work and to be able to predict phase and thermodynamic behavior with sufficient precision to design the pipeline to handle the impurities.

Carbon dioxide is normally collected at near-atmospheric pressure from gas conditioning plants, petroleum refineries, fertilizer plants, and from flue gases. The particular source of carbon dioxide influences the kind and amount of impurities. For example, if the source is natural gas containing high percentages of carbon dioxide, the pipeline-quality carbon dioxide can be expected to contain a residual amount of natural gas. The SACROC pipe line, for example, contains 10 mol% natural gas. A proposed system in Alberta, Canada, would collect carbon dioxide containing hydrogen from several tar sands processing plants.

The thermodynamic behavior of mixtures containing hydrogen has been traditionally difficult to predict. Only a small amount of information concerning hydrogen–carbon dioxide systems appears in the literature. Prausnitz and Gunn studied hydrogen–carbon dioxide systems at pressures higher than those envisaged for proposed carbon dioxide pipelines [2]. They obtained good theoretical results by using pseudocritical constants of −229.7°C (−381.5°F) and 20.0 kmol/m^3 (1.25 lb-mol/ft^3) with an acentric factor of zero rather than true criticals to characterize the behavior of hydrogen.

Three mixtures were tested to find out whether this approach could be used successfully to predict the effect of hydrogen on the phase behavior of carbon dioxide at pipeline pressures. The mixtures contained close to 97.50 mol% carbon dioxide and 2.50 mol% hydrogen.

Each mixture was tested by Core Laboratories in a windowed PVT cell to find the bubble point pressure at various temperatures. The bubble point line was then predicted using the 11 parameter BWRS equation of state. Starling and Han's generalized correlation with a binary interaction coefficient of zero was used to generate BWRS constants [3].

Results of the experimental phase boundary measurements and the theoretical predictions are compared in Fig. 1. Good agreement between experiment and the BWRS equation was obtained. Other equations of state

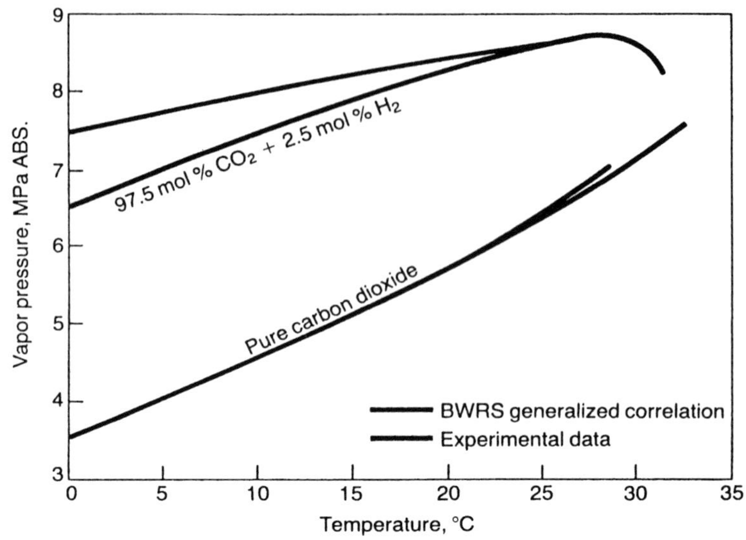

FIG. 1 Comparison of experimental phase boundary with predictions of BWRS equation of state of pure carbon dioxide and a mixture containing 2.5 mol% hydrogen.

which are based on reliable experimental data and which can predict liquid densities accurately should give good results also if the pseudocritical approach tested here were used. As a general rule, it is advisable to conduct bubble point experiments during final design to validate the equation of state being used for the particular impurities under consideration.

The vapor pressure of the mixture with 2.5 mol% hydrogen at 15°C (59°F) was some 2.8 MPa (400 lb/in.^{2}) greater than pure carbon dioxide. Since impurities such as hydrogen have such a dramatic effect on vapor pressure, it is important to establish impurity specifications early in the design work before specifying operating pressures and ordering pipe.

Propagating Fractures

A propagating ductile fracture is driven by fluid pressure which acts on the unrestrained walls of fractured pipe. This concentrates large stresses in the hoop direction at the fracture tip. If these stresses exceed the ultimate strength of the pipe steel in tension, they cause the pipe to tear longitudinally and form a propagating fracture. Since the fluid pressure at the fracture tip provides the driving force for this particular failure mode, it is helpful to be able to predict the decompressed pressure at the fracture before designing the pipe to resist it.

The problem of propagating fractures in gas pipelines is recognized by the CSA (Canadian Standards Association) Standard Z184 which specifies that supplementary design measures, such as increased notch toughness, reduced operating stress levels, or mechanical fracture arrest devices, shall be used to provide positive control of ductile fracture lengths if pipe steel stress levels exceed certain values based on the decompression behavior of an ideal gas.

Although it could be argued that a high-pressure carbon dioxide pipeline contains liquid rather than gas and therefore is not subject to the requirements of the gas code, a ductile fracture could still propagate along a high vapor pressure liquids pipeline if it decompressed as slowly as, or more slowly than, a gas pipeline. Sound engineering practice indicates that propagating fractures be investigated in the design of high-pressure carbon dioxide pipelines.

Figure 2 shows a pressure–enthalpy diagram for pure carbon dioxide based on the BWRS equation. If a pure carbon dioxide pipeline operating in northern Alberta at 13.1 MPa (1900 lb/in.^{2}) and 5°C (41°F) were to burst and form a propagating fracture, the fluid inside the pipe would decompress along the isentropic path indicated in Fig. 2 until heat flow from the surroundings and frictional effects grew to be large enough to have an effect. The relationship between density and pressure along this path is shown in Fig. 3. The velocity of sound is defined thermodynamically as

$$c_s = (\partial P/\partial \rho)_s^{1/2} \tag{1}$$

In other words, the velocity of sound is the square root of the slope of the pressure–density curve at constant entropy. The discontinuities in slope

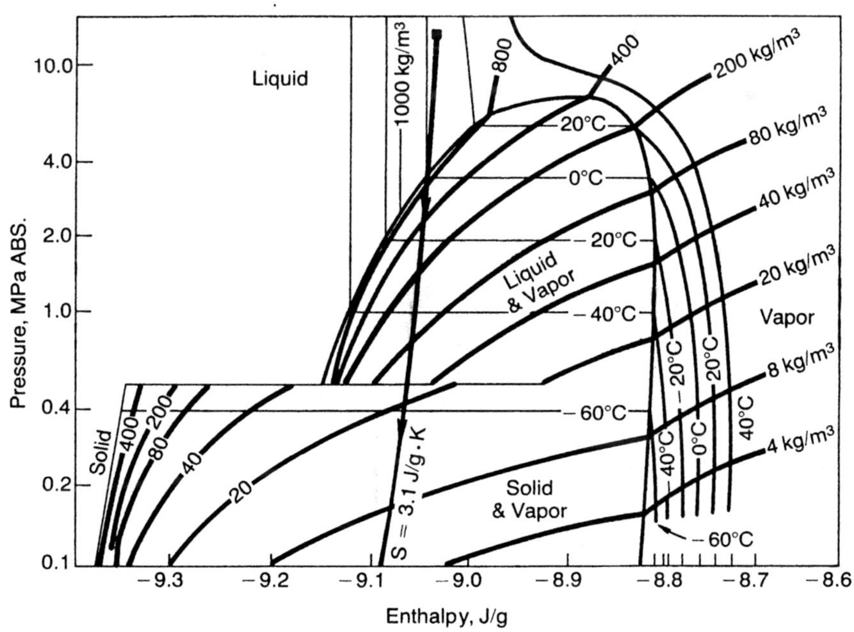

FIG. 2 Pressure–enthalpy diagram for pure carbon dioxide showing decompression path from 13.1 MPa (1900 lb/in.^{2}) and 5°C (41°F).

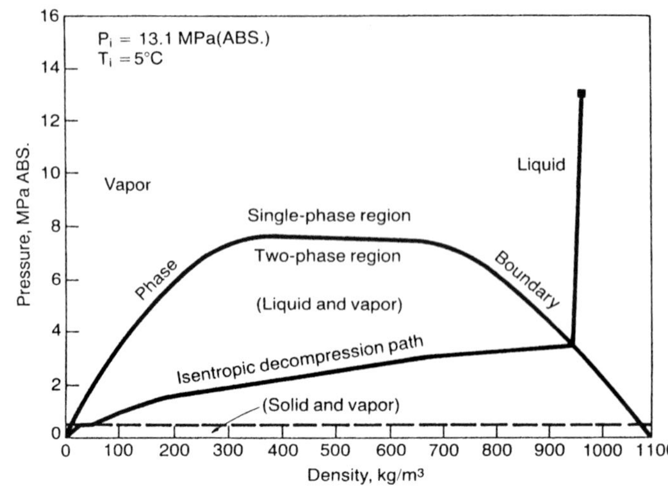

FIG. 3 Pressure–density behavior of pure carbon dioxide during rapid decompression from 13.1 MPa (1900 lb/in.²) and 5°C (41°F).

(Fig. 3) at the phase boundaries between liquid, liquid and vapor, and solid and vapor signify sharp changes in the velocity of sound at the phase boundaries. This is evident in Fig. 4 which shows the square root of the slope of the curve given in Fig. 3. In the single-phase liquid region the sonic velocity is quite high at 730 m/s, but the moment the decompression enters the two-phase liquid–vapor region, the sonic velocity falls to about 50 m/s.

Since fluid escapes from the pipe at sonic velocity, this sudden change in its value at the phase boundary restricts the escape of fluid and prolongs the depressurization of the pipe. In a region with constant sonic velocity it

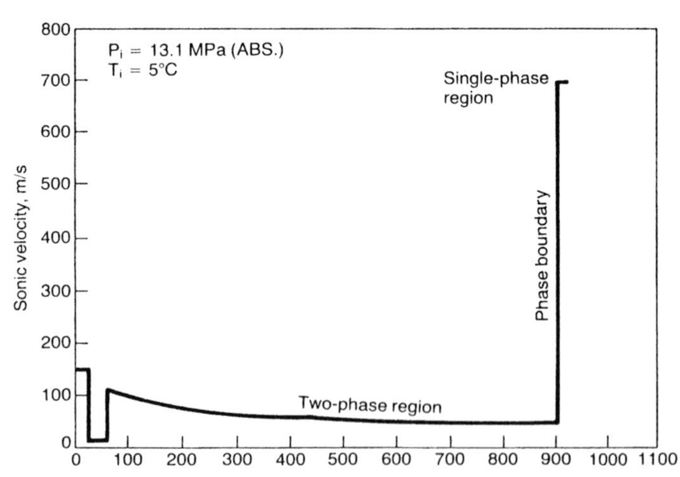

FIG. 4 Sonic velocity of pure carbon dioxide along decompression path.

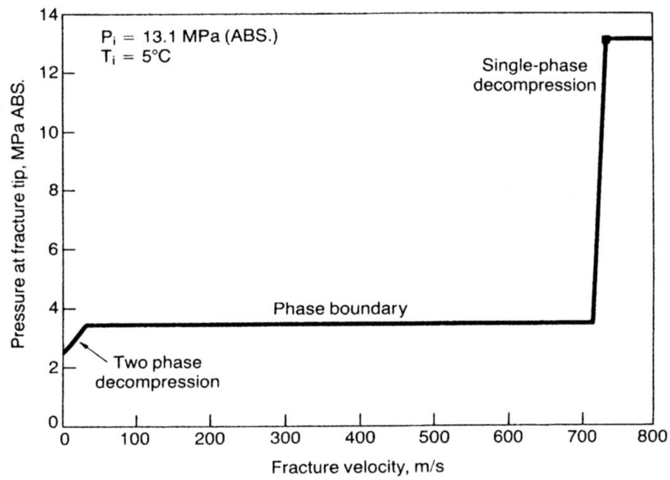

FIG. 5 Decompressed pressure of pure carbon dioxide for differing fracture speeds.

has been demonstrated that fluid pressure, fluid velocity, fracture velocity, fluid density, and sonic velocity are related to each other by the following equations [4]:

$$P = P_i - c_s^2(\rho_i - \rho) \tag{2}$$

$$v = v_i + c_s \log (\rho/\rho_i) \tag{3}$$

$$c_f = c_s + v \tag{4}$$

Figure 5 was constructed by using Eqs. (2), (3), and (4), and it shows fluid pressure at the fracture tip over a range of fracture velocities. Table 1 shows parameters obtained from Fig. 4 that were used to characterize the properties of carbon dioxide in the calculations. Fluid pressure and fluid velocity are continuous across the phase boundary. The final values of fluid pressure and fluid velocity at the phase boundary in the single-phase region are therefore used as initial values at the phase boundary for decompression calculations in the two-phase region.

It is clear from Fig. 5 that a long sustained pressure plateau will be observed during depressurization following a line break in carbon dioxide

TABLE 1 Decompression Parameters for Carbon Dioxide

Parameter	Units	1-Phase Region	2-Phase Region
c_s	m/s	730	50
ρ_i	kg/m^3	918	900
P_i	MPa	13.1	3.5[a]
v_i	m/s	0	−14.5[a]

[a]From Eqs. (2), (3), and (4).

pipelines. Propagating ductile fractures normally have velocities in the range 100–250 m/s (about 350–800 ft/s) so that it may be concluded from Fig. 5 that the pressure at the tip of a propagating ductile fracture will be equal to the vapor pressure of carbon dioxide after isentropic decompression from pipeline operation conditions.

If carbon dioxide pipelines are to be designed to control propagating ductile fractures, they need to be able to withstand stresses induced at the fracture tip by the vapor pressure. This pressure is considerably higher for liquefied carbon dioxide pipelines than for natural gas pipelines and very much higher than for conventional liquid pipelines.

Pipe Selection

The arrest of propagating ductile fractures in pipelines has been researched extensively by the Battelle Columbus Laboratories over the last 15 years by using fluids such as high-pressure water, LNG, and gas with pipe diameters ranging from 190.5 mm (7.5 in.) to 1219.2 mm (48 in.) and yield strengths from around 207 MPa (30 ksi) to 758 MPa (110 ksi). A hypothesis has been developed which links fracture arrest to pipe wall thickness, steel strength, notch toughness, and decompressed pressure at the tip of the propagating fracture [5, 6].

Fractures were observed to arrest in about 90% of test sections, which satisfied the following arrest criterion. Even though some 10% of the fractures did not arrest, the criterion is judged to be suitable for design work because steel strength and toughness vary along each joint of pipe and are generally higher than the specified minimum values. The fracture arrest criterion states that ductile fractures will not propagate if the pipeline is designed so that:

$$3.33\sigma_d/\sigma_f \not> \frac{2}{\pi}\cos^{-1}\exp\left(-\pi E_N/24\right) \tag{5}$$

where

$$\sigma_d = P_d D/2t \tag{6}$$

and

$$E_N = EC_v/A\sigma_f^2(Dt/2)^{1/2} \tag{7}$$

Since the inverse cosine term on the right-hand side of Eq. (5) must lie between $-\pi/2$ and $+\pi/2$, it follows after multiplying by $2/\pi$, as indicated, that the right-hand side of Eq. (5) cannot exceed unity. Therefore, the flow stress of the pipeline steel needs to be at least 3.33 times the decompressed hoop stress. It should be noted at this point that it is possible to satisfy this requirement by either increasing the wall thickness of the pipe or the strength of the steel. Flow stress is equal to the yield stress plus 68.95 MPa (10 ksi).

If the flow stress is equal to 3.33 times the decompressed hoop stress so that the left-hand side of Eq. (5) is unity, the inverse cosine term will be $\pi/2$ and the exponential term will therefore be zero. This means that its exponent $(-\pi E_N/24)$ would be a very large negative number which implies that steel toughness would be very great. Since Charpy notch toughness is limited in practice, the flow stress has to be some amount greater than 3.33 times the decompressed hoop stress. To find out by how much and to find out whether it is better to increase steel strength or pipe wall thickness, it is necessary to investigate Eqs. (5), (6), and (7) in greater detail.

Figure 6 is a graphical representation of Eqs. (5), (6), and (7), assuming that pipe steel toughness is 27 J (20 ft-lb). It was developed for a pipeline carrying pure carbon dioxide in northern Alberta where maximum annual ground temperatures at pipeline depth do not exceed 15°C (59°F). Decompression from pump discharge conditions encounters the bubble point line at 5.51 MPa (800 lb/in.^{2}).

Figure 6 shows minimum pipe wall thickness as a function of specified minimum yield strength for various pipe diameters. Superimposed on Fig. 6 is the maximum design operating pressure based on a manufacturing underthickness of 12.5% combined with a steel stress factor of 82% allowed by CSA Standard Z183 for high vapor pressure pipelines in Class 1 locations. Pressure testing could result in maximum allowable operating pressures slightly higher than the design values.

Figure 6 shows that increases in steel strength do not result in proportional decreases in wall thickness in order to satisfy Battelle's fracture arrest hypothesis. To obtain a feel for the comparative economics of higher strength steel versus thicker wall pipe, it is necessary to introduce the cost of pipe for differing wall thicknesses, diameters, and steel strengths.

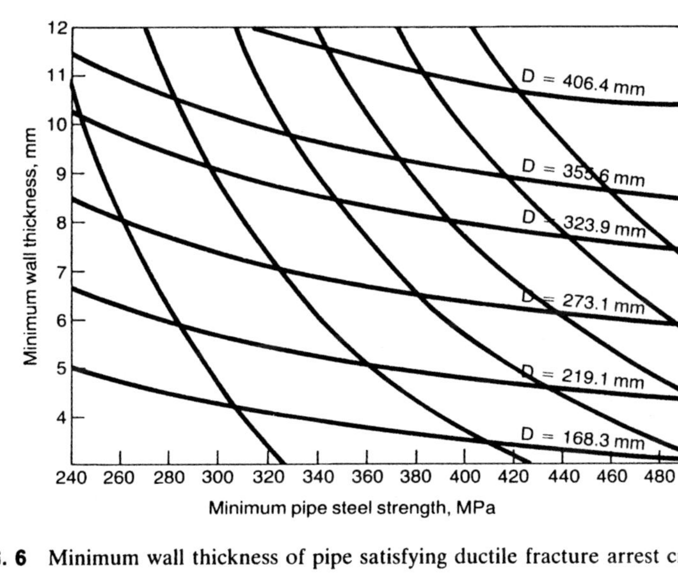

FIG. 6 Minimum wall thickness of pipe satisfying ductile fracture arrest criterion.

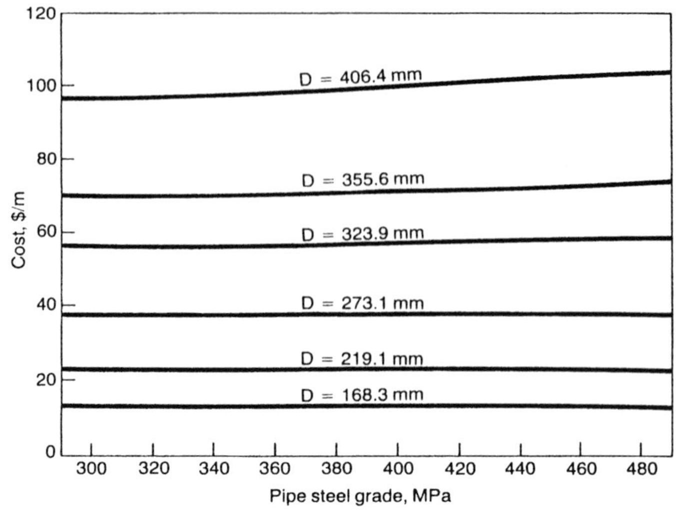

FIG. 7 Costs of pipe satisfying ductile fracture arrest criterion for decompressed pressure of 5.51 MPa (800 lb/in.^{2}).

Figure 7 shows the cost per meter for pipe satisfying Battelle's fracture arrest hypothesis plotted against minimum pipe steel strength. The calculation of cost was based on one manufacturer's recent price list. It was assumed in the calculation that sufficient length of pipe would be ordered to make it feasible to specify pipe with nonstandard wall thicknesses. An allowance for freight and incremental welding costs for pipe of the same diameter but different wall thickness was included.

Higher strength pipe satisfying the ductile fracture arrest criterion permits higher operating pressures. Class 900 flanges are rated at 15.3 MPa (2220 lb/in.^{2}gauge) while the next heavier Class 1500 flanges are rated at 25.5 MPa (3705 lb/in.^{2}gauge). The costs of Class 1500 flanges, fittings, and pump casings when pressures exceed 15.3 MPa (2220 lb/in.^{2}gauge) have not been reflected in this cost analysis. In many cases the cost penalty could tend to limit the maximum operating pressure to 15.3 MPa (2220 lb/in.^{2}gauge), which in turn could make higher strength steels unnecessary for the larger diameter pipe shown in Fig. 6.

A study of Fig. 7 shows that cost of pipe satisfying Battelle's hypothesis does not decrease with increasing steel strength. This eliminates the economic incentive to use higher strength pipe for carbon dioxide pipelines if they are designed to control propagating fractures.

Conclusions

Carbon dioxide's high density with respect to air at atmospheric pressure increases the danger of asphyxiation because it will not dissipate upward into the atmosphere like natural gas.

However, very large concentrations of carbon dioxide can be tolerated without danger. Risks can be further reduced by taking precautions such as installing carbon dioxide detectors, controlling blowdown rates, constructing valves above-grade, and using closely spaced remotely operated sectionalizing valves.

Impurities such as hydrogen have a pronounced effect on the vapor pressure of liquefied carbon dioxide. Knowledge of its vapor pressure is important in design work for controlling propagating fractures and for setting minimum pump suction pressures and hence station sizes and spacings. It is therefore important to finalize the specification of allowable impurities early in the design work.

Very high vapor pressure of CO_2 prevents rapid depressurization in the event the pipeline breaks. High sustained pressures increase the chance of forming propagating ductile fractures. To avoid risks associated with this failure mode, it is best to select lower strength steels and thicker wall pipe than might otherwise be selected.

This material, which appeared in *Pipe Line Industry,* pp. 125 ff., November 1981, has been updated with the permission of the editor.

Symbols

A	area beneath Charpy notch (m^2)
c_f	propagating fracture velocity (m/s)
c_s	fluid sonic velocity (m/s)
C_v	Charpy notch toughness (J)
D	pipe outside diameter (m)
E	Young's modulus of elasticity (Pa)
E_N	normalized toughness parameter (—)
P	fluid pressure (Pa)
P_d	decompressed pressure (Pa)
P_i	fluid initial pressure (Pa)
S	fluid entropy (J/kg-K)
t	pipe wall thickness (m)
T_i	fluid initial temperature (°C)
v	fluid velocity (m/s)
v_i	fluid initial velocity (m/s)
σ_d	decompressed pipe hoop stress (Pa)
σ_f	pipe steel flow stress (Pa)
ρ	fluid density (kg/m^3)
ρ_i	fluid initial density (kg/m^3)

References

1. G. Nahas and K. E. Schaefer (eds.), *Carbon Dioxide and Metabolic Regulation* (25th International Congress of Physiology, Monte Carlo, July 1971), Springer-Verlag, New York, 1974.

2. J. M. Prausnitz and R. D. Gunn, "Volumetric Properties of Nonpolar Gaseous Mixtures," *AIChE J.*, *4*(4), 430–435 (December 1958).
3. K. E. Starling and M. S. Han, "Thermo Data Refined for LPG, Part 14: Mixtures," *Hydrocarbon Process.*, pp. 129–132 (May 1972).
4. G. G. King, "Decompression of Gas Pipelines during Longitudinal Ductile Fractures," *ASME J. Energy Resour. Technol.*, *101*(1), 66–73 (March 1979).
5. W. A. Maxey, R. J. Podlasek, R. J. Eiber, and A. R. Duffy, *Observations on Shear Fracture Propagation Behaviour,* Presented at International Symposium on Crack Propagation, Newcastle, England, January 1974.
6. W. A. Maxey, *Experimental Investigation of Ductile Fractures in Piping,* Presented at 12th World Gas Conference, Nice, 1973.

GRAEME G. KING

Maximum Pressure for Steel Piping

Here is a method to estimate the maximum allowable pressure for ferrous piping according to ASME and ANSI codes without referring to the actual codes' stress tables. In addition, specific pipe dimensions (o.d. and thickness) need not be known, as the schedule number and pipe material are all that are necessary.

Table 1 gives the maximum allowable working pressure, in lb/in.^{2}gauge, based on an allowable stress of 15,000 lb/in.^{2}. To correct for the specific material and temperature, multiply the value obtained from Table 1 by the appropriate F factor from Fig. 1. Table 2 lists the various materials by composition and ASME code specification.

TABLE 1 Maximum Allowable Pressure[a]

Nominal Pipe Size (in.)	Schedule 40	Schedule 80	Schedule 160
¼	4830	6833	—
½	3750	5235	6928
1	2857	3947	5769
1½	2112	3000	4329
2	1782	2575	4225
2½	1948	2702	3749
3	1693	2394	3601
4	1435	2074	3370
5	1258	1857	3191
6	1145	1796	3076
8	1006	1587	2970

[a]Based on allowable stress of 15,000 lb/in.^{2}.

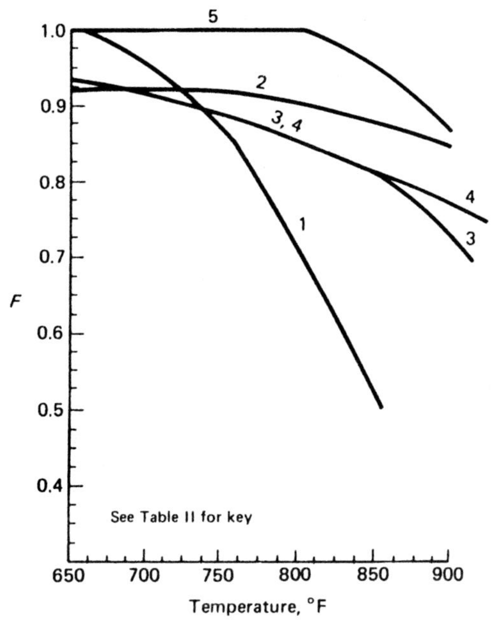

FIG. 1

TABLE 2 ASME Specifications for Various Steels[a]

Composition	Code Designation	Curve on Fig. 1
Carbon-manganese	SA 53B	1
Carbon-manganese-silicon	SA 106B	1
Carbon-½Mo (½% molybdenum)	SA 335 P1	2
5Cr-½Mo (5% chromium, ½% molybdenum)	SA 335 P5	3
9Cr-1Mo	SA 335 P9	4
1¼Cr-½Mo	SA 335 P11	5
2¼Cr-1Mo	SA 335 P22	5

[a]Source: *ASME Boiler and Pressure Vessel Code,* Sec. 1, 1980, p. 184, Table PG23.1.

Example. Estimate the maximum allowable working pressure for a 2-in., Schedule 40 carbon-manganese (SA 53B) pipe at 750°F.

Solution: From Table 1, a value of 1782 lb/in.^{2} is obtained. From the figure, for SA 53B (Curve 1) at 750°F, F is 0.86. Hence, the maximum allowable pressure is $0.86 \times 1782 = 1532$ lb/in.^{2}gauge.

V. GANAPATHY

Offshore Considerations

Transportation of natural gas from an offshore facility with minimal processing facilities requires choosing between dense-phase and two-phase flow design alternatives.

Results of a comparative analysis considering the principal elements of design, construction, and economics indicate that distance of transportation is the critical factor determining a choice between dense-phase and two-phase.

The analysis also examined metallurgical criteria, marine design, and installation considerations which affect design.

Safety, Cost Constraints

Choosing among alternative modes of gas transport from offshore installations is significantly affected by the extent of processing required offshore.

Of the available alternatives, transport in the vapor phase as a "pipeline-quality" gas poses fewer problems for pipelining. However, processing requirements are made more complex to the extent that the gas conditioning on the platform must be sufficient to avoid partial condensation as the gas stream cools and expands in the pipeline.

Since safety and cost constraints encourage adoption of a minimum process philosophy offshore, alternative pipeline systems which require minimal processing on the platform must be considered. These pipeline systems include dense-phase flow and two-phase flow.

A standard cash flow analysis computer program was adopted for the comparative study. The model develops capital and operating costs, itemizing financial charges year by year instead of with a flat capital-recovery factor.

The computer runs include pipeline throughput and the resulting unit cost of service.

In the evaluation of different modes of transport, many factors have an influence outside of the immediate transport system. The total production, transport, and marketing system must be considered.

The comparison of alternative pipeline systems here assumes that this overall analysis has defined the basic onshore and offshore processing requirements and the specifications of oil and gas streams.

The comparison further assumes that certain onshore and offshore separation processing and treating systems would be common to both the dense-phase and the two-phase alternatives. And the comparison assumes that the gas composition would be identical for both pipeline options.

These assumptions may not always be valid, but they are sufficiently so for this evaluation. The basis for the comparison is a 300-MMSCFD, 250-mile subsea pipeline system in about 400 ft of water where sea bottom temperatures are very cold and sea conditions might approximate those experienced in the North Sea.

On the preceding basis, pipeline systems are broadly defined for transport of fluid in the two alternative conditions.

Dense-Phase System

Dense-phase or supercritical conditions can be described by a typical phase envelope for natural gas and gas mixtures. Essentially, the cricondenbar represents the highest pressure at which separate liquid and gas phases can exist.

The cricondentherm is the highest temperature at which liquid can exist regardless of pressure. Dense-phase conditions which are neither liquid nor gas exist in a region confined by the critical point and the cricondenbar.

Phase Envelope

A phase envelope was developed for the given gas composition. Several compositions were considered for testing their effect on the shape of the phase envelope.

The gas composition is shown in Table 1. Phase diagrams for two compositions (variations of the C_6+ fraction) are plotted in Fig. 1.

These calculations indicated that a minimum pipeline pressure of 1800 lb/in.^{2} would allow an operating margin and assure dense-phase flow under pipeline operating conditions. According to the phase diagram, these conditions would be achieved by a hydrocarbon dew point at about 100°F and 800 lb/in.^{2}.

A base case diameter of 24 in. is used with a nominal 0.625-in. W.T. and steel grade of API 5L X60. The maximum allowable operating pressure is calculated to be 2250 lb/in.^{2} using a hoop stress level of 72% of specified

TABLE 1 Gas Composition

N_2	0.3	Distribution taken for C_6:	
CO_2	1.1	nC_6	0.20
C_1	80.2	C_7	0.16
C_2	8.9	C_8	0.03
C_3	6.0	C_9	0.01
iC_4	0.7	Pseudocritical temperature, °R	398.0
nC_4	1.7	Pseudocritical pressure, lb/in.^{2}abs	667.0
iC_5	0.3	Molecular weight	21.0
nC_5	0.4	Specific gravity (at standard conditions)	0.7256
C_6+	0.4		
	100%		

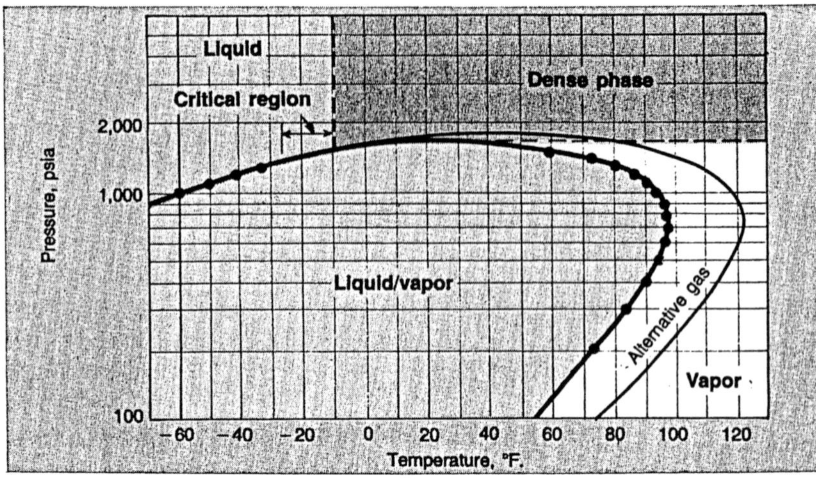

FIG. 1 Phase envelope.

minimum yield strength (SMYS). A maximum diameter-to-wall-thickness ratio of 40:1 was maintained.

The implications of pipe material and diameter-to-wall-thickness ratio are discussed later with metallurgy and marine construction requirements.

Flow Cases

The dense-phase flow cases considered are summarized in Table 2.

Steady-state hydraulic calculations were performed using a gas-pipeline computer program with the ability to define completely a pipeline system at every point of interest.

The program allows the user to change any data variable at any point in the system. An option allows the downstream, or any lowest system pressure, to be specified to obtain pipeline pressure profile and initial pressure.

The basis for pressure computations is the rational flow equation with correction for elevation difference. The pipeline is divided into short sections for linearizing of physical properties and conditions, and calculations are done sequentially from section to section.

The program is designed to handle CO_2 and ethylene as well as natural gas mixtures. The compressibility factor is calculated with the critical pressure and temperature and gas composition. Viscosity is also computed.

Thermal gradient is input by subroutine, with the adiabatic expansion coefficient, for the Joule-Thomson effect, fluid properties, and the overall pipe-to-environment heat-transfer coefficient.

Results of the computer simulation are given in Table 3, and those for the 24-in. pipeline for different throughputs are plotted in Fig. 2.

TABLE 2 Cases Considered

Outside Diameter, (in.)	Throughput (MMSCFD)	Onshore Pressure (lb/in.^{2})
Dense-phase flow		
24[a]	300	1800
24	200	1800
24	120	1800
24	90	1800
24	60	1800
20[a]	300	1800
16[a]	300	1800
Two-phase flow		
24[a]	300	1100
24	300	500
24	200	1000
20[a]	300	1000
20[a]	300	500
16[a]	300	1000
16	200	1000
16	300	500

[a]Included in cash flow analysis, Table 5.

Results of flow calculations for other pipe sizes are also tabulated in Table 3 and plotted in Fig. 3 for a throughput of 300 MMSCFD. The smallest pipeline, 16 in., requires an initial pressure of 2700 lb/in.^{2}, which coincides with the maximum allowable operating pressure of the pipe selected.

Initial pressures for the 20 and 24-in. cases are shown to be 2450 and 2100 lb/in.^{2}, respectively. The pipeline end pressure for all cases and sizes was set at 1800 lb/in.^{2}.

The lowest temperature occurs in the 20-in.-diameter pipeline reaching 24.7°F relative to a minimum seabed temperature of 29°F. The larger-diameter pipelines have temperatures closer to seabed temperature.

TABLE 3 Dense-Phase Flow Results

Pipeline Size (in.)	Throughput (MMSCFD)	Inlet Pressure (lb/in.^{2})	Outlet Pressure (lb/in.^{2})	ΔP (lb/in.^{2})	Inlet Temperature (°F)	Outlet Temperature (°F)
24	300	2089	1800	289	50	27.19
	200	1947	1800	147	50	28.94[a]
	120	1873	1800	73	50	28.94[a]
	90	1854	1800	54	50	28.94[a]
	60	1841	1800	41	50	28.94[a]
20	300	2449	1800	649	50	24.7
16	300	2700	1800	900	50	28.94[a]

[a]Isothermal analysis only.

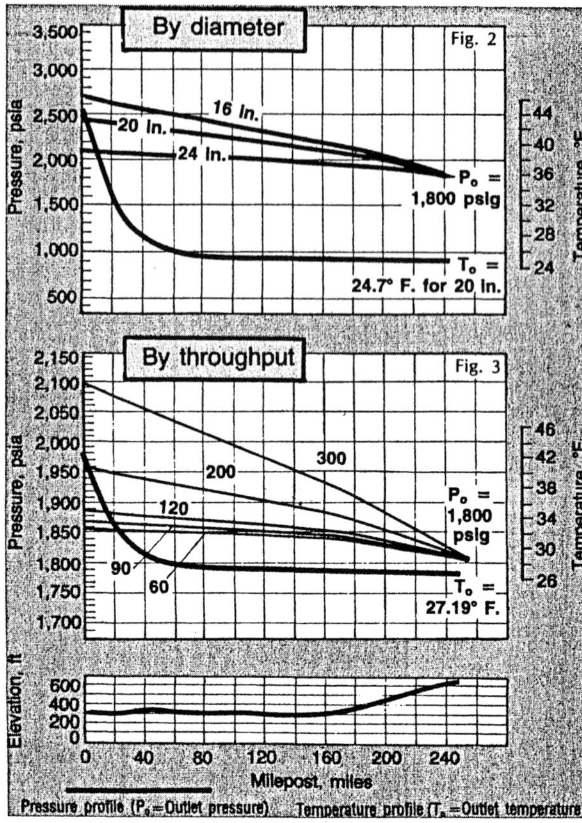

FIGS. 2 AND 3 Dense-phase gradients.

Because of the nature of the comparison, the selection of overall thermal conductivity values was given only cursory attention. Varying thermal conductivity values, input to the calculations to test sensitivity, yielded results which varied only as much as 1.0°F.

The difference in dehydration requirements for dense and two-phase alternatives on the platform is not considered, although at the higher dense-phase pressures and gas gravity involved, significant dewpoint depression would be required to achieve 2–3 lb/MMSCFD, considering the minimum pipeline temperature of 24.7°F for the 20-in. case.

Two-Phase System

It is assumed that the gas composition for the two-phase flow option would be identical to that used in the dense-phase analysis. Hydraulic analysis is based on the adoption of two pressure levels, 500 and 1000 lb/in.^{2}, for arrival of the gas onshore.

Base Case

The 24-in. pipeline with a throughput of 300 MMSCFD is taken as the base case, and diameter and steel grade are the same as those used in the dense-phase analysis. Wall thicknesses reflect the lower operating pressures possible in the two-phase cases.

The two-phase flow cases considered are summarized in Table 2.

Steady-state hydraulic calculations were performed with a two-phase computer program. The pipeline is divided into short sections for linearizing of physical properties and conditions, and calculations are done sequentially from section to section.

Physical properties of the gas are generated for the operating range of temperature and pressure conditions by using the Peng-Robinson equation of state. Predictions of flow regime are according to Mandhane, and viscosity is also computed for gas composition.

Calculations

The program calculates pipeline pressure drop and estimates liquid hold-up and gas/liquid throughput to the onshore receipt facilities. The pipeline thermal gradient used in the dense-phase alternative was used for the two-phase option.

The results of the simulation are given in Table 4 which summarizes pipeline flow conditions of flow regimes for the different cases. Pressure profiles for the different diameters and throughputs are plotted in Fig. 4 for the 1000-lb/in.^{2} onshore case and Fig. 5 for the 500-lb/in.^{2} onshore case.

Upstream and downstream pressures for the principal cases are plotted in Fig. 6 which shows the difference in onshore arrival pressure is reflected in a much smaller difference of pressure on the platform. This effect is emphasized for reduced pipeline diameters where longer upstream sections of the pipeline exhibit single-phase flow.

Estimation of onshore slug catcher requirements used in the comparison was based on the flow conditions modeled by the two-phase program and the liquid holdup difference between initial and final conditions resulting from changes in operation.

Metallurgy, Marine Design

Optimum line pipe metallurgy for transportation in the vapor-phase, dense-phase, or two-phase condition requires consideration of strength, toughness, weldability, and resistance to environmental degradation. An understanding of the steel-making practice, fabrication techniques, nondestructive examination, and corrosion control leads to the development of appropriate specifications.

TABLE 4 Two-Phase Flow Conditions

Diameter (in.)	Throughput (MMSCFD)	Inlet Pressure, (lb/in.^{2}abs)	Outlet Pressure, (lb/in.^{2}abs)	Superficial Velocities (ft/s) Outlet		Liquid Hold-up (bbl)	Outlet Liquids (bbl/d)	Flow Regime[a]
				Gas	Liquid			
24	300	1,665	1,100	10.0	0.36	81,398	15,341	SP/STR/EB
	200	1,334	1,000	7.5	0.23	79,067	9,878	SP/STR/EB
	300	1,350	500	29.2	0.22	48,942	9,343	SP/STR/EB
20	300	2,108	1,000	16.2	0.50	38,851	14,846	SP/STR/EB
	300	1,972	500	42.2	0.32	40,227	9,328	SP/STR/EB
16	300	3,259	1,000	25.4	0.79	7,343	14,851	SP/STR/EB/SL
	200	2,367	1,000	16.9	0.52	18,183	9,898	SP/EB
	300	3,168	500	65.7	0.50	8,121	9,351	SP/STR/EB/SL

[a]SP = single phase, STR = stratified, EB = elongated bubble, SL = slug.

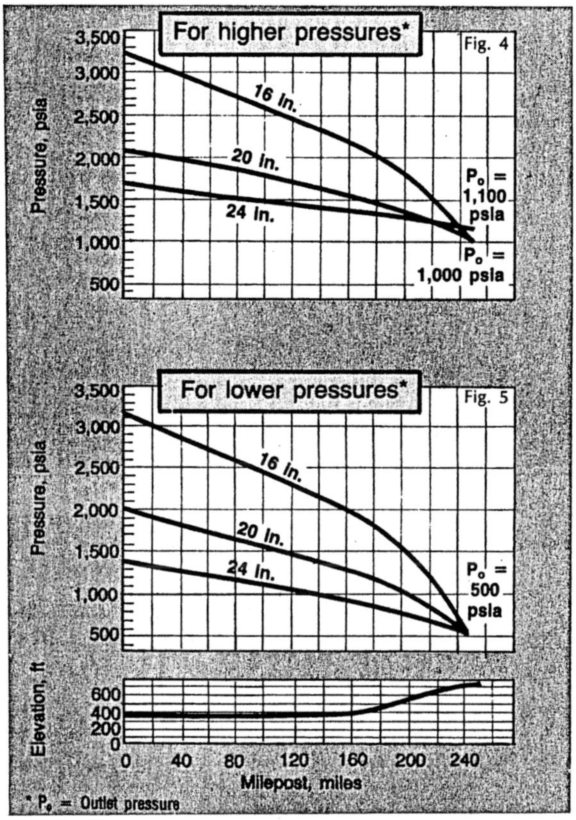

FIGS. 4 AND 5 Two-phase gradients.

Selection Criteria

Strength and toughness of line pipe steel are influenced by a number of factors. In general, however, the approach to selection and specification of material is the same for both the dense-phase and two-phase flow options.

For the dense-phase alternative with the higher pressures required, there are advantages in the use of high-grade line pipe. Grade X70 offers potentially significant cost savings over X60. Reductions in wall thickness may significantly affect economics not only from the steel-tonnage aspect but also in field welding time. And, metallurgically speaking, the use of X70 material poses no particular problems.

Additionally, no significant changes in weldability are expected with the use of X70 grade steel compared with the X60 grade. In this evaluation, however, no benefit could be derived from the use of higher grade steel in order not to exceed a maximum diameter-to-wall-thickness ratio of 40 needed to meet the marine laying conditions.

Pipeline failures due to leaks and fractures were considered. Any leaks that will occur will most likely result from such environmental effects as

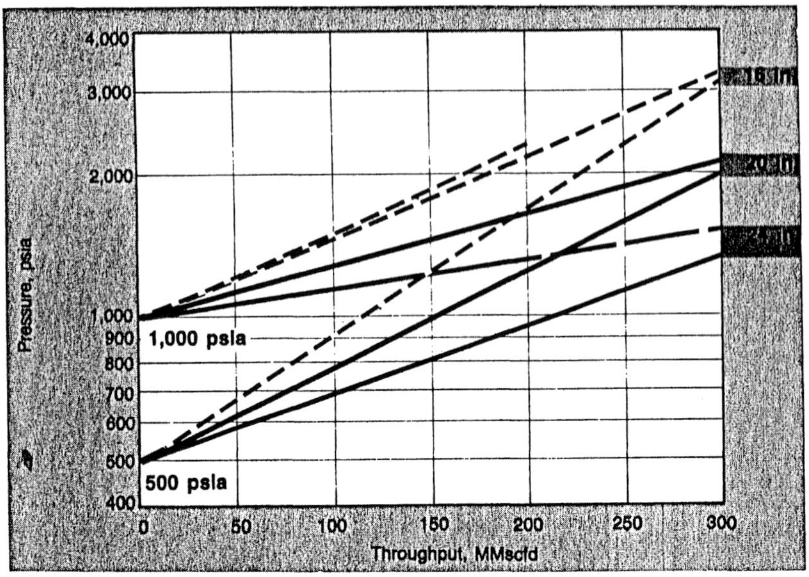

FIG. 6 Two-phase flow pressures: upstream, downstream.

corrosion. External corrosion protection would utilize conventional pipeline corrosion coating, weight coating, and cathodic protection.

Where sour gas is present, steel composition must be specified to control hydrogen-induced cracking and sulfide stress corrosion cracking. Large-diameter, high-pressure pipelines can develop long running fractures, and the level of toughness must be specified to avoid such failure.

The actual level of toughness would be determined by detailed study of pipe metallurgy and the underwater decompression behavior of the different fluid alternatives. Some safety margin over the level of toughness is required, especially to provide crack-arrest capability.

Although these aspects require rigorous examination at the design stage, differences influencing this comparison of the dense-phase and two-phase cases had no significance.

Soft Sand

For the purpose of the study, it is assumed that the pipeline would be routed through sections of soft sand and sections of bedrock exposed to the seabed.

In sand sections, where wave action is severe enough to affect the seabed, it is preferred that the pipeline be partly buried. In this condition the pipeline would be exposed to wave and current forces and at the same time be stabilized against lateral movement.

By adjustments to the specific gravity of the pipe relative to the specific gravities of seawater and saturated seabed sand, equilibrium on the seabed can be attained with half the pipe buried and half the pipe exposed.

Pipeline corrosion coating and concrete weight coating would be required to produce an overall 1.6 specific gravity relative to seawater. This specific gravity is preferred for large pipelines and provides good stability even where liquefaction of surrounding sand is possible.

The bare rock sections of the route do not provide a good foundation for submarine lines. The route should attempt to bypass these exposures. A route through rubble material would be preferable.

If bare rock area cannot be avoided, however, then special anchoring procedures must be considered. Shore landings should be made in a protected inlet wherever possible.

Considering the relatively shallow water assumed for the purpose of the comparison, it is probable that buckle arrestors would not be required. It is unlikely that buckles would propagate in such heavy wall pipe at these shallow water depths.

In general, marine design and construction methods would not be affected by the dense-phase or two-phase requirements except in terms of the pipe wall thickness required for internal pressure.

The dense-phase alternative requires higher pressure and heavier wall thickness than the two-phase alternative. Pipeline design would thus be oriented toward the dense-phase option with the understanding that two-phase flow could be accommodated with reduced wall thickness with the exception of those cases where the diameter-to-wall-thickness ratio exceeds the recommended maximum.

Alternatives

The pipeline throughputs and terminal pressures are identical; only the pressure profiles differ. The objective is to determine which type of operation, dense phase or two phase, is superior.

Gas Pressure

All cases are based on an initial gas pressure of 1350 lb/in.^{2}gauge on the platform and a final pressure of 1000 lb/in.^{2}gauge onshore. Marine construction requirements limit the maximum pipe-diameter-to-wall-thickness ratio to 40. This requirement is no problem in most cases, but it does cause the wall of the two-phase, 24-in. pipeline (Case 4) to be oversized. The 0.625-in. W.T. could be 0.500 in. if based solely on the operating pressure.

All of the pipe has a 60,000-lb/in.^{2} yield strength and is weight coated with 150 lb/ft^3 concrete to provide a negative buoyancy of at least 20 lb/ft of pipe. All gas compressors are turbine-driven centrifugals in order to reduce the space and weight requirements for the offshore platform.

The three dense-phase cases deliver gas onshore at 1800 lb/in.^{2}gauge. This pressure is reduced to the final 1000 lb/in.^{2}gauge in the onshore process facilities.

One two-phase case is included with a 500-lb/in.^{2}gauge delivery pressure that is recompressed to the standard 1000 lb/in.^{2}gauge delivery pressure by an onshore compressor. The other two-phase cases deliver gas onshore at 1000 lb/in.^{2}gauge.

All two-phase cases include slug catchers to handle liquids ahead of the onshore plant.

Cost Comparisons

Capital costs, based on 1985 prices, are shown in the cash flow summary, Table 5. They include current estimates for a 250-mile concrete-coated submarine pipeline, plus compressors and other equipment. Operating costs are 1½% of capital cost for the pipelines and 8% of capital cost for the platform and onshore facilities.

Compressor fuel is priced at $2/MCF. An allowance is made for energy recovery as a result of the high onshore arrival pressure in the dense-phase cases.

Cash flow projections are calculated by computer which facilitate generation of data for a sensitivity analysis. Details for the 24-in., dense-flow case are shown in Table 6 to illustrate the method of calculation.

The basis of any petroleum-oriented economic analysis must be the production profile for the life of the field. A typical profile was assumed for a 1500 BCF gas reserve. Production begins in the second year after construction start-up, and full production (300 MMCFD) occurs after 4 years.

TABLE 5 Cash Flow Comparisons

	Dense Phase			Two Phase			
Case	1	2	3	4	5	6	7
Pipe diameter, in.	24	20	16	24	20	20	16
Capital costs, $1,000:							
Pipeline	261,494	203,351	158,485	261,494	204,480	208,064	164,600
Compressors	15,600	21,600	25,350	4,800	15,900	33,000	31,500
Other	3,440	3,400	3,360	1,440	1,200	1,200	960
Subtotal direct costs	280,534	228,351	187,195	267,734	221,580	242,264	197,060
Indirect costs	112,214	91,340	74,878	107,094	88,632	96,906	78,824
Total cost	392,748	319,691	262,073	374,828	310,212	339,170	275,894
Equity	98,187	79,923	65,518	93,707	77,553	84,792	68,971
Cash flow, $1,000/year:							
Unit cost of service, /MCF[a]	1.3350	1.1011	0.9150	1.2632	1.0662	1.1928	0.9784
Production, MMSCFD	300	300	300	300	300	300	300
Total cost of service, $	146,183	120,570	100,193	138,320	116,749	130,612	107,135
Operating expense, $	4,511	5,039	5,292	3,357	4,790	7,861	7,183
Capital expense, $	66,281	53,950	44,228	63,256	52,351	57,239	46,558
Cash flow, $	75,391	61,581	50,673	71,707	59,608	65,512	53,394

[a]Generates 25% DCF rate of return (cash flows are for Year 5).

TABLE 6 Case 1 Cash Flow[a]

Year	Rate (MMCFD)	Gas Production Revenue (million $/yr)	Operating Expenses ($1,000/yr)	Interest in Annuity ($1,000/yr)	Debt Repaid ($1,000/yr)	Petroleum Income Taxes ($1,000/yr)	Net Cash Flow ($1,000/yr)
1	0.0	0.00	0	0	0	0	(68,731)
2	50.0	24.36	3,110	45,898	20,382	0	(75,955)
3	150.0	73.09	3,632	43,045	23,235	0	3,179
4	250.0	121.82	4,162	39,792	26,488	0	51,377
5	300.0	146.18	4,511	36,084	30,196	0	75,391
6	300.0	146.18	4,680	31,856	34,424	3,425	71,797
7	300.0	146.18	4,858	27,037	39,243	39,684	35,360
8	300.0	146.18	5,044	21,543	44,737	44,703	30,154
9	300.0	146.18	5,240	15,280	51,000	49,557	25,106
10	300.0	146.18	5,446	8,140	58,140	54,396	20,061
11	300.0	146.18	5,662	0	0	38,245	102,276
12	300.0	146.18	5,888	0	0	64,535	75,759
13	250.0	121.82	5,938	0	0	53,305	62,576
14	150.0	73.09	5,812	0	0	30,949	36,331
15	50.0	24.36	5,698	0	0	8,586	10,080
	1,204.5 (BCF)	$1,608.01	$69,682	$268,675	$327,846	$387,387	$454,758

[a]Dense-phase flow in 24-in. pipeline.

Production then begins to decline after the twelfth year, terminating completely after the fifteenth year.

Analyses of the cases are based on calculation of the discounted cash flow (DCF) internal rates of return that result from various transportation rates. The end result is a comparison of transportation costs in $/MCF for each pipeline design that yields the same rate of return, 25%, for each case. Indirect costs and operating expenses are functions of the direct capital costs. A 75/25 debt/equity ratio was used with 14% interest on a 10-year loan. A 15% factor is used for discounting the cash flow. Debt financing includes interest during construction.

The uniform annuity payments include interest on debt with the balance used to retire debt. Income taxes, at 46% rate, include tax loss carried forward and double declining accelerated depreciation.

The program can include the impact of inflation factors on operating costs and revenue but only the former was used in this study.

Higher-Pressure Lines

Examination of results summarized in Table 5 shows the advantage of higher-pressure pipelines. The 16-in., dense-phase line operating at 2700 $lb/in.^2$, the highest pressure considered, requires the least investment and cost of service.

Additionally, the 16-in., two-phase pipeline is undersized, and better performance would be expected for the 20-in. pipeline. It should be noted that the pipelines are not optimized in the dense-phase and two-phase categories; instead, all cases are compared in this economic analysis.

Costs of service are plotted against pipeline diameter in Fig. 7. Costs of the two-phase pipeline design with onshore compression are higher than either the dense-phase or two-phase lines of the same diameter.

The frequently used plot of unit transportation cost vs throughput is shown in Fig. 8. This plot is a truncated version of the normal family of overlapping parabolic curves. Apparently these curves do not converge, crossover, or reverse direction due to the influence of having a single compressor station located on the platform. The usual version of this graph is based on land pipelines with intermediate compressor stations added when needed to increase throughput. When compression becomes excessive, the unit costs reverse direction and start increasing.

Variables

The evaluation confirms the economic advantage of high-pressure pipelines for large-diameter pipelines, where distances to shore are great. It can be seen that the economics for other configurations may favor two-phase systems, and it is apparent that each project needs careful consideration due to the large number of variables.

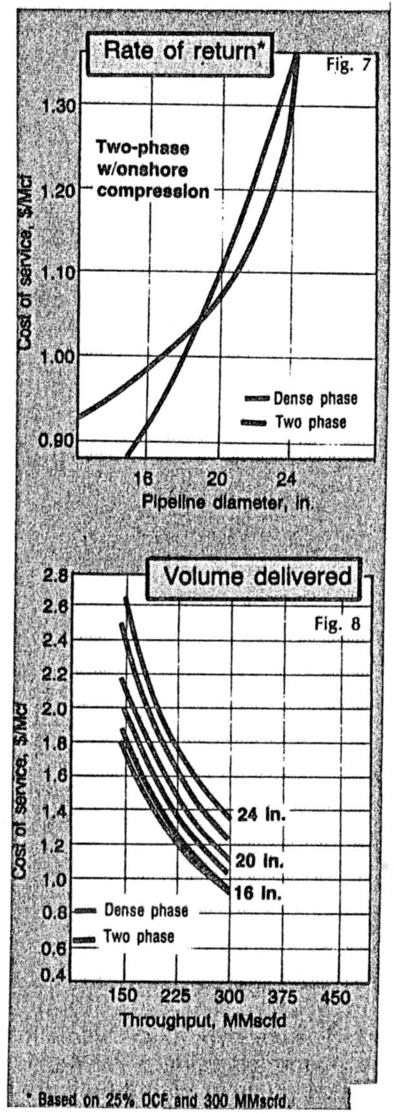

FIGS. 7 AND 8 Economics analysis.

The principal disadvantage of two-phase-flow systems is the absence of sufficiently analytical design techniques for pipeline sizing and prediction of slugging behavior, leading to uneconomic overdesign.

Further, there is the problem of achieving stable flow conditions over a range of flow conditions as production changes with field development and subsequent depletion. Pipelines operating in the dense phase can be designed accurately and are much less sensitive to operational variations.

Two-phase-flow systems may be economically feasible for short pipelines and relatively shallow water where advantage can be taken of the lower-pressure system and reduced pipe wall thickness. There is further advantage to two-phase pipelines where the high residual energy inherent in dense-phase systems cannot be economically recovered.

Also, improved techniques should become available as a result of the ongoing work under the sponsorship of the American Gas Association and the establishment of the two-phase data bank, as well as the contribution of the other researchers.

RONALD E. INGHAM
ARNOLD J. CARRICO

Photogrammetry and Computer-Aided Piping Design

For the most part, photogrammetric technology has been applied to the task of producing topographic maps. This application can be traced as far back as the Civil War. After limited applications in aerospace, photogrammetry's potential later came to be fully exploited in shipbuilding.

Application in Piping Design

Other applications lie in mapping out complex distributive systems, such as those depicted by design models of machinery spaces and of chemical process plants. For example, piping arrangements in process plants may be created in a three-dimensional model, sometimes separated into sections for access to central areas. Pipes may be modeled to scale or their centerlines represented by color-coded wires, with diameters indicated by sliding disks.

Among the advantages of such a model are that it eliminates hundreds of general arrangement drawings, it facilitates quick understanding of the design, and it reduces interferences during construction.

Upon the completion of a design model, pipe shapes and dimensions must still be generated on paper for pipe fabricators and plant erectors. This is normally done by sketching isometrics as the model is assembled. General arrangement drawings are then developed from these sketches by referring to the model.

However, difficulties in measuring the model can force drafters to estimate many dimensions. Manual measurements are inherently inaccurate because access is impeded by the congestion of model details, pipe-bend intersection points are virtual, and locations of bend-intersection points are subject to interpretation, particularly in the case of bends smaller than 90°.

Now, with three-dimensional digitizing hardware, it is possible to bypass manual measurement and mathematically deduce accurate measurements from photographs. Such measurements are reliable because they depend on photographic access (anything that can be seen can be measured, regardless of congestion), and pipe centerlines and bend-intersection points can be precisely calculated from digitized points on pipe surfaces.

Some designers bridge the gap between model and computer by digitizing from orthographic prints obtained via orthography or the laser scanning of model sections. Such prints are digitized *two*-dimensionally, with great care taken to coordinate the manual collection of data for the third dimension at the computer input terminal. However, such a method is as prone to errors as the manual take-off of data.

Via photogrammetry, however, *three* dimensions can be digitized and directly entered into a computer (Fig. 1). This greatly reduces the possibility of error. With photogrammetry, it is possible to use design modeling for creative work while taking advantage of the computer to more speedily and accurately produce design details and material lists.

(There are two methods of photogrammetry that can be applied to measurement: the stereo method and the convergent method. These complement one another, because each is better suited to a different kind of measurement task.)

Equipment and Procedural Suggestions

For dimensioning from models, the following equipment capabilities and procedural techniques are recommended:

1. The camera should be focusable over a range of distances, and should have liberal depth-of-field.
2. Photographs should be taken rapidly so as not to interfere with the use of the model by others.
3. Procedures for digitizing from photographs should be simple so that an expert photogrammetrist is not needed.
4. The digitizing instrument should not be significantly limited by focal length, allowable distance between camera stations, or lack of parallelism between optical axes of adjacent photographs.
5. The output should consist of coordinates of pipe-bend intersection points and centerline locations of pipes, valves, and fittings.
6. The accuracy of coordinate data produced should be compatible with fabrication and assembly requirements.

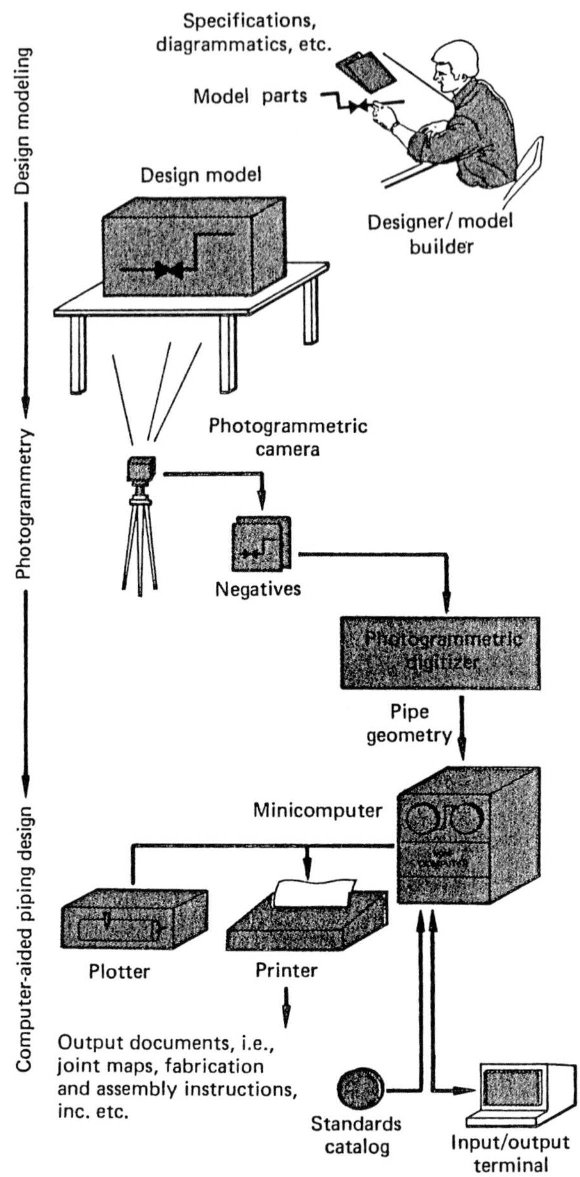

FIG. 1 Dimensions from photographs are digitized and entered directly into a computer.

7. The format for the data obtained should be compatible with input requirements of existing computer-aided pipe detailing and fabrication programs.
8. The equipment should be usable for other measurement, such as dimensioning large structures (tanks, towers, etc.).

The foregoing was the basis for determining that, in the case of a model, the best solution would be to use a computer-controlled stereoplotter to encode and record three-dimension digitized data from a stereomodel. The stereoplotter would create a three-dimensional optical model (a stereomodel) from negatives of a pair of photographs of the same scene (from slightly different vantage points and with their optical axes nominally parallel). Pipe centerlines, bend intersection points, and valve and fitting centerline locations would be accurately calculated from selected points on model surfaces. This indirect approach permits molded-plastic construction of the model's piping and components.

A Dimensioning Application

The application of photogrammetry to dimensioning a complex piping and machinery arrangement was tested by the National Shipbuilding Research Program with part of a model of a floating nuclear power plant. Also tested were the computer programs prepared for reducing the digital data of pipe surfaces, valves, and fittings into needed coordinates, such as for pipe-bend intersection points and centerline locations of pipe fittings and valves.

The design model had been built in sections. These provided easy visual and photographic access. A reference grid was scribed on the base of the sections selected for the test and, where practical, on vertical surfaces. These grids provided absolute reference for the photogrammetric work. Targets were fixed on selected grid intersections, so that these known locations could be easily seen on photographs.

Photographs were taken with a Wild P31 Universal Terrestrial camera because of its ready availability and closeup capability. This camera has a virtually distortion-free lens and can accept sensitized glass plates as well as film.

The procedures followed were simple, and they can be repeated anywhere without a specially prepared room. Because precise measurements were not required, they were made with an ordinary carpenter's tape. "Bounce" lighting was achieved without special precautions, and "eyeball" aiming of the camera was adequate. Each model section was photographed (one pair of stereo photos) from four cardinal directions and in an oblique manner so as to minimize photographic obscurations as much as possible.

The graphics and symbol instructions were prepared in advance by a person having nominal experience in arranging details. This permitted the stereodigitizer operator to collect the needed data rapidly. Thus, the operator was relieved of the need to make many decisions, such as what to digitize, what identification to attach to digitized data, and what detail had to be digitized. The instruction contributed to maximum productivity from the stereodigitizer operator.

Detail Procedures for Pipe Segments

Experience disclosed that only six points need be digitized for each segment, in approximate locations as shown in Fig. 2(a). This was typically done in each of the four stereo views of a model section.

The computer program that best fitted a cylinder to all the points digitized on a pipe-segment's surface was based on rigid assumptions as to the location of the points. It followed the digitizing sequences shown in Fig. 2: (1) the diameter of the pipe segment was estimated from Points 1 and 2, and (2) the location and orientation of the segment was estimated from Points 1 and 4.

Estimates thus made were then refined in the cylinder fitting process. Even when there were obscurations, the approximations for diameter, location, and orientation of a segment were usually obtained by carefully selecting and sequencing point locations (see Fig. 2b).

Digitized points were located reasonably close to bends, because the points of most interest are, in fact, the intersections of the centerlines of the cylinders that best fit to adjacent pipe segments—i.e., the bend-intersection points.

Accuracies of the calculated centerlines were better when cylinders were fitted to points that were widely separated lengthwise. Theoretically, accuracies would be further enhanced if additional points were digitized. However, the remaining choices for locations (toward midlength of a section) were undesirable because unintentional curvature, common in molded pipe segments that are relatively long and small in diameter, invalidates the concept of cylinder fitting.

Detail Procedures for Valves and Fittings

Unlike pipe segments, valves and fittings were usually fixed sufficiently by digitizing from only one stereomodel rather than all four. Decisions as to

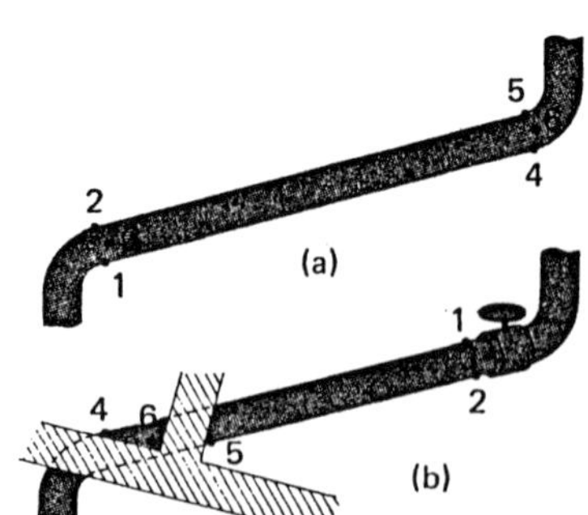

FIG. 2 Digitizing sequence estimates diameter, location, and orientation of pipe segment.

which valves and fittings are to be digitized, and in which stereomodels, are best made during the preparation stage. Because surfaces are not fitted to valves and fittings during data processing, digitizing one or two points is generally sufficient to fix each valve or fitting location and orientation—i.e., there is no need to digitize data from all sides.

The pertinent data-processing program simply constructed a line in space through the digitized points perpendicular to the previously computed location and orientation of the pipe-segment centerline to which the fitting or valve belonged. For a valve or fitting having a single digitized point, the location at which the perpendicular strikes the centerline is the centerline location of the valve or fitting. If two points were digitized for a valve or fitting, the program computed their average centerline location.

Experience indicated that it was best to let the stereodigitizer operator decide whether a valve or fitting should be a one- or two-point valve or fitting. Having this freedom of choice, the operator digitized a symmetrical valve simply by using one point on its stem. Asymmetric valves and fittings were defined as two-point (e.g., one point on each flange of a nonstandard check valve).

Data Processing

Many data-processing functions have already been explained because they influence all tasks, from model preparation to stereodigitizing. Nevertheless, further explanation is needed for a full understanding of the logical progression of calculations.

For each model section, the data-processing steps were:

1. By means of a three-dimensional coordinate transformation program, all digitized information from stereomodel No. 2 was put into the coordinate system of stereomodel No. 1. This was necessary because each view (stereomodel) of a given model section was digitized in its own arbitrary coordinate system. All data from all stereomodels must eventually be in the same coordinate system. Data were transferred from one stereomodel to the coordinate system of another by the best fitting of two sets of data at the tie-in points common to both sets.
2. With the same three-dimensional transfer program, all data resulting from Step 1 were transferred into a coordinate system of the overall structure. Seven transfer constants were determined by best fitting the coordinates (from Step 1) of digitized grid-intersection points to the corresponding known, or design, plant coordinates. The transfer constants were then applied to all data from Step 1 in order to produce plant coordinates for every digitized point.
3. The data collected for a given pipe segment were processed through a cylinder-fitting program to determine the location and orientation of the centerline of the cylinder that best fitted all points (i.e., the cal-

culation found the radius, centerline location, and orientation of the perfect cylindrical surface that minimized the perpendicular departures of all points from the cylindrical surface). The computed centerline locations and orientations of all such pipe segments were stored in a separate file for use in the following two data-processing steps.

4. Where there were two centerlines from adjacent segments (e.g., on each side of a bend), the centerlines were numerically extended in space to determine their point of intersection. More precisely, each "intersection" was actually the closest approach because it is improbable that two such lines would exactly intersect. The calculated intersection is the so-called bend-intersection point (see Fig. 3).
5. To determine the centerline locations of valves and fittings, data for a given valve or fitting, after processing through Step 2, were matched with centerline data contained in the file as a result of Step 4. This was done simply by finding the proper pipe-run-and-segment number in the centerline data file. Computation of the centerline location of the valve or fitting then proceeded as has been described in the section "Detail Procedures for Valves and Fittings."

Models, Photogrammetry, Computer Design

The computed bend-intersection points and centerline locations of valves and fittings (described in the foregoing Steps 4 and 5) are the primary products required of any system for dimensioning from models. They fix pipe-run geometries in formats that can be assimilated by any computer-aided piping design program. Moreover, because they are digital they are consistent with (1) modern methods for preparing numerical-control instructions for fabricating pipe pieces and related equipment (e.g., pipe supports), (2) automatic generation of material lists, and (3) current development to digitize assembly-work instruction.

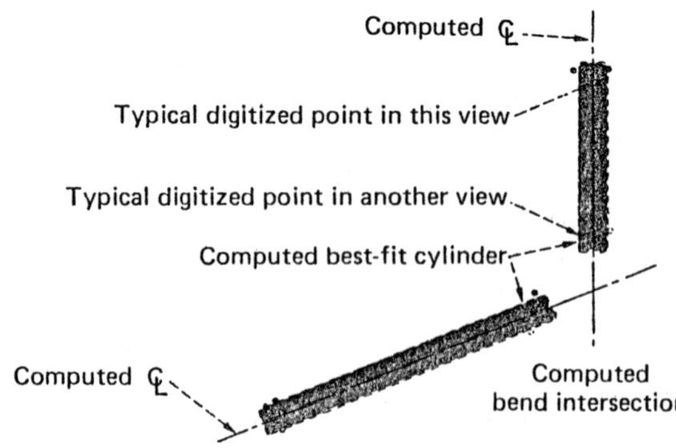

FIG. 3 Calculated intersection point identifies bend-intersection point.

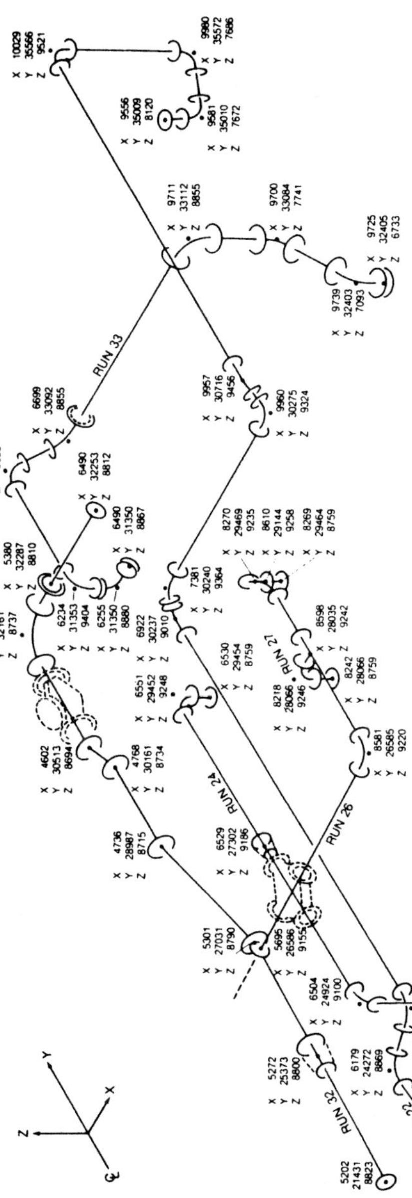

FIG. 4 In this photogrammetrically produced numerical drawing of a piping system, *xyz* coordinates are in millimeters at full scale.

Because computer programs work with numerical data (e.g., dimensions of structures and piping), there is less need for the accuracy of geometrical drawings. On the other hand, only major points on numerical drawings need be accurate. The marriage of photogrammetry and computer-aided piping design can economically produce such numerical drawings. Because the end products of photogrammetry are only accurately determined major points, even geometric drawings produced by this marriage would be accurate. Figure 4 shows a photogrammetrically produced numerical drawing.

Plant Repair, Alteration, and Expansion

Up to now, the focus has been on model engineering, but all the foregoing is also applicable to full-scale structures. To the camera's "eye" and the computer's "mind," scale does not matter. Therefore, photogrammetry holds forth the promise of an economical way for surveying existing plants for repair, alteration, or expansion.

Especially where up-to-date, as-built drawings of plant structures do not exist, photogrammetric surveys could determine the configuration of a structure and could provide, in conjunction with automatic drawing devices, accurate as-built drawings as well as work instructions or, preferably, digitized data for the numerical-control manufacturing of equipment.

JOHN F. KENEFICK
RICHARD D. CHIRILLO

Polytropic Compression

Estimating outlet temperature, flow rates, and so on for a polytropic compressor can be tedious and time-consuming. Here is a program written for the Texas Instruments TI-59/PC-100C calculator that easily finds such quantities in English, metric, or SI units for a multicomponent mixture of gases.

The computations employ the "N" method of The Elliott Co. [1], modified to use the specific-heat ratio evaluated at the average of the suction and discharge temperatures, as recommended by Urick and Odom [2].

Calculation Method

The entering of data will be covered in Table 2. However, after data are entered, the following are calculated: average molecular weight, pseudo-critical temperature and pressure, and average molar heat capacity. Next, the program is run, and total weight flow is found by multiplying average molecular weight by total molar flow. Constants are calculated for average molar heat capacity as a linear function of temperature. Using the punched-in inlet conditions, the program calculates the inlet volumetric flow:

$$Q_1 = Z_1(R/MW)(T_1/P_1)(W) \tag{1}$$

where

$$Z_1 = 1 + \rho_r(C_1 + C_2/T_r + C_3/T_r^3) + \rho_r^2(C_4 + C_5/T_r + C_6/T_r^3) \tag{2}$$

and

$$\rho_r = (Z_c/Z)(P/P_c)(T_c/T) \tag{3}$$

The solution of these equations is iterative and is done by employing a subroutine that uses direct substitution until the difference between successive iterations is <0.001. At this point, the user must supply the polytropic efficiency, η_p.

If data relating polytropic efficiency and inlet volumetric flow are unavailable, the following can be used: $\eta_p = 0.76$ for 500–8000 ft^3/min (850–13,600 m^3/h); $\eta_p = 0.77$ for 8,000–50,000 ft^3/min (13,600–85,000 m^3/h); and $\eta_p = 0.78$ for flows over 50,000 ft^3/min (over 85,000 m^3/h).

Next, the outlet pressure must be supplied. Then the program completes the calculation by first finding temperature:

$$T_2 = T_1(P_2/P_1)^{\gamma_{avg}} \tag{4}$$

where

$$\gamma_{avg} = (k_{avg} - 1)/k_{avg}\eta_p \tag{5}$$

and

$$k_{avg} = c_{p,avg}/(c_{p,avg} - R) \tag{6}$$

The molar heat capacity is the average of that at T_1 and T_2. The solution of the above equations is iterative and is found by using a subroutine that employs direct substitution until the difference between two successive iterations for finding temperature is less than one degree.

TABLE 1 TI-59 Program Performs Multistage, Multicomponent Calculations for Polytropic Compression

Step	Code	Key
Compressibility-factor subroutine		
000	76	LBL
001	22	INV
002	43	RCL
003	33	33
004	32	X:T
005	43	RCL
006	31	31
007	65	×
008	43	RCL
009	21	21
010	55	÷
011	43	RCL
012	20	20
013	55	÷
014	43	RCL
015	32	32
016	95	=
017	42	STO
018	22	22
019	33	X^2
020	65	×
021	53	(
022	43	RCL
023	39	39
024	55	÷
025	43	RCL
026	20	20
027	45	Y^x
028	03	3
029	85	+
030	43	RCL
031	38	38
032	55	÷
033	43	RCL
034	20	20
035	85	+
036	43	RCL
037	37	37
038	54	)
039	85	+
040	43	RCL
041	22	22
042	65	×
043	53	(
044	43	RCL
045	36	36
046	55	÷
047	43	RCL
048	20	20
049	45	Y^x
050	03	3
051	85	+
052	43	RCL
053	35	35
054	55	÷
055	43	RCL
056	20	20
057	85	+
058	43	RCL
059	34	34
060	54	)
061	85	+
062	01	1
063	95	=
064	48	EXC
065	32	32
066	75	−
067	43	RCL
068	32	32
069	95	=
070	50	|X|
071	77	GE
072	22	INV
073	43	RCL
074	32	32
075	92	RTN
Outlet-temperature subroutine		
076	76	LBL
077	23	LNX
078	43	RCL
079	05	05
080	75	−
081	43	RCL
082	27	27
083	95	=
084	65	×
085	43	RCL
086	18	18
087	85	+
088	43	RCL
089	19	19
090	95	=
091	55	÷
092	53	(
093	24	CE
094	75	−
095	43	RCL
096	23	23
097	54	)
098	95	=
099	55	÷
100	53	(
101	24	CE
102	75	−
103	01	1
104	54	)
105	95	=
106	35	1/X
107	42	STO
108	06	06
109	53	(
110	43	RCL
111	14	14
112	55	÷
113	43	RCL
114	10	10
115	54	)
116	45	Y^x
117	53	(
118	43	RCL
119	06	06
120	55	÷
121	43	RCL
122	13	13
123	95	=
124	65	×
125	43	RCL
126	09	09
127	95	=
128	42	STO
129	15	15
130	85	+
131	43	RCL
132	09	09
133	95	=
134	55	÷
135	02	2
136	95	=
137	48	EXC
138	05	05
139	75	−
140	43	RCL
141	05	05
142	95	=
143	50	|X|
144	77	GE
145	23	LNX
146	92	RTN
Initializes program		
147	76	LBL
148	10	E'
149	32	X:T
150	02	2
151	02	2
152	36	PGM
153	01	01
154	71	SBR
155	00	00
156	12	12
157	42	STO
158	00	00
159	04	4
160	69	OP
161	17	17
162	32	X:T
163	42	STO
164	07	07
165	03	3
166	00	0
167	03	3
168	02	2
169	02	2
170	07	7
171	03	3
172	06	6
173	69	OP
174	04	04
175	43	RCL
176	07	07
177	69	OP
178	06	06
179	91	R/S
Component-data entry		
180	76	LBL
181	11	A
182	98	ADV
183	32	X:T
184	69	OP
185	20	20
186	43	RCL
187	00	00
188	99	PRT
189	03	3
190	00	0
191	03	3
192	02	2
193	02	2
194	07	7
195	03	3
196	06	6
197	69	OP
198	04	04
199	32	X:T
200	69	OP
201	06	06
202	55	÷
203	43	RCL
204	07	07
205	95	=
206	42	STO
207	01	01
208	25	CLR
209	91	R/S
210	76	LBL
211	12	B
212	32	X:T
213	03	3
214	00	0
215	04	4
216	03	3
217	69	OP
218	04	04
219	32	X:T
220	69	OP
221	06	06
222	65	×
223	43	RCL
224	01	01
225	95	=
226	44	SUM
227	02	02
228	25	CLR
229	91	R/S
230	76	LBL
231	13	C
232	32	X:T
233	03	3
234	07	7
235	01	1
236	05	5
237	69	OP
238	04	04
239	32	X:T
240	69	OP
241	06	06
242	65	×
243	43	RCL
244	01	01
245	95	=
246	44	SUM
247	03	03
248	25	CLR
249	91	R/S
250	76	LBL
251	14	D
252	32	X:T
253	03	3
254	03	3
255	01	1
256	05	5
257	69	OP
258	04	04
259	32	X:T
260	69	OP
261	06	06
262	65	×
263	43	RCL
264	01	01
265	95	=
266	44	SUM
267	04	04
268	25	CLR
269	91	R/S
270	78	LBL
271	15	E
272	32	X:T
273	01	1
274	05	5
275	03	3
276	03	3
277	00	0
278	02	2
279	69	OP
280	04	04
281	32	X:T
282	69	OP
283	06	06
284	65	×
285	43	RCL
286	01	01
287	95	=
288	44	SUM
289	05	05
290	25	CLR
291	91	R/S
292	32	X:T
293	01	1
294	05	5
295	03	3
296	03	3
297	00	0
298	03	3
299	69	OP
300	04	04
301	32	X:T
302	69	OP
303	06	06
304	65	×
305	43	RCL
306	01	01
307	95	=
308	44	SUM
309	06	06
310	25	CLR
311	91	R/S
312	76	LBL
313	16	A'
Calculates weight flow		
314	43	RCL
315	07	07
316	65	×
317	43	RCL
318	02	02
319	55	÷
320	43	RCL
321	24	24
322	95	=
323	42	STO
324	06	08
Calculates slope and intercept for molar-heat-capacity equation		
325	53	(
326	43	RCL
327	05	05
328	75	−
329	43	RCL
330	06	06
331	54	)
332	55	÷
333	53	(
334	43	RCL
335	29	29
336	75	−
337	43	RCL
338	30	30
339	95	=
340	42	STO
341	18	18

Step	Code	Key
342	65	×
343	43	RCL
344	29	29
345	94	+/-
346	85	+
347	43	RCL
348	05	05
349	95	=
350	42	STO
351	19	19
352	76	LBL
353	17	B'
Requests inlet temperature		
354	98	ADV
355	69	OP
356	00	00
357	01	1
358	07	7
359	03	3
360	01	1
361	03	3
362	07	7
363	01	1
364	07	7
365	03	3
366	05	5
367	69	OP
368	01	01
369	03	3
370	07	7
371	00	0
372	02	2
373	69	OP
374	02	02
375	69	OP
376	05	05
377	25	CLR
378	91	R/S
Calculates reduced temperature		
379	99	PRT
380	85	+
381	43	RCL
382	27	27
383	95	=
384	42	STO
385	09	09
386	42	STO
387	05	05
388	55	÷
389	43	RCL
390	03	03
391	95	=
392	42	STO
393	20	20
Requests inlet absolute pressure		
394	03	3
395	03	3
396	00	0
397	02	2
398	69	OP
399	02	02
400	00	0
401	69	OP
402	04	04
403	69	OP
404	05	05
405	25	CLR
406	91	R/S
Calculates inlet volumetric flow		
407	99	PRT
408	42	STO
409	10	10
410	55	÷
411	43	RCL
412	04	04
413	95	=
414	42	STO
415	21	21
416	71	SBR
417	22	INV
418	42	STO
419	11	11
420	65	×
421	43	RCL
422	26	26
423	55	÷
424	43	RCL
425	02	02
426	65	×
427	43	RCL
428	09	09
429	55	÷
430	43	RCL
431	25	25
432	55	÷
433	43	RCL
434	10	10
435	65	×
436	43	RCL
437	08	08
438	95	=
439	42	STO
440	12	12
441	03	3
442	04	4
443	00	0
444	02	2
445	69	OP
446	04	04
447	43	RCL
448	12	12
449	69	OP
450	06	06
Requests polytropic efficiency		
451	03	3
452	01	1
453	03	3
454	03	3
455	69	OP
456	02	02
457	00	0
458	69	OP
459	04	04
460	69	OP
461	05	05
462	25	CLR
463	91	R/S
464	99	PRT
465	42	STO
466	13	13
467	76	LBL
468	18	C'
Requests outlet absolute pressure		
469	03	3
470	03	3
471	00	0
472	03	3
473	69	OP
474	02	02
475	00	0
476	69	OP
477	04	04
478	69	OP
479	05	05
480	25	CLR
481	91	R/S
482	99	PRT
Calculates outlet temperature		
483	42	STO
484	14	14
485	55	÷
486	43	RCL
487	04	04
488	95	=
489	42	STO
490	21	21
491	01	1
492	32	X⇌T
493	71	SBR
494	23	LNX
495	03	3
496	07	7
497	00	0
498	03	3
499	69	OP
500	04	04
501	43	RCL
502	15	15
503	75	-
504	43	RCL
505	27	27
506	95	=
507	69	OP
508	06	06
Calculates outlet volumetric-flow		
509	85	+
510	43	RCL
511	27	27
512	95	=
513	55	÷
514	43	RCL
515	03	03
516	95	=
517	42	STO
518	20	20
519	71	SBR
520	22	INV
521	42	STO
522	16	16
523	55	÷
524	43	RCL
525	11	11
526	65	×
527	43	RCL
528	15	15
529	55	÷
530	43	RCL
531	09	09
532	65	×
533	43	RCL
534	10	10
535	55	÷
536	43	RCL
537	14	14
538	65	×
539	43	RCL
540	12	12
541	95	=
542	42	STO
543	17	17
544	03	3
545	04	4
546	00	0
547	03	3
548	69	OP
549	04	04
550	43	RCL
551	17	17
552	69	OP
553	06	06
Calculates polytropic head		
554	53	(
555	43	RCL
556	11	11
557	85	+
558	43	RCL
559	16	16
560	54	)
561	55	÷
562	02	2
563	95	=
564	65	×
565	43	RCL
566	26	26
567	55	÷
568	43	RCL
569	02	02
570	65	×
571	43	RCL
572	09	09
573	55	÷
574	53	(
575	43	RCL
576	06	06
577	55	÷
578	43	RCL
579	13	13
580	54	)
581	42	STO
582	01	01
583	65	×
584	53	(
585	53	(
586	43	RCL
587	14	14
588	55	÷
589	43	RCL
590	10	10
591	54	)
592	45	Y^X
593	43	RCL
594	01	01
595	75	-
596	01	1
597	95	=
598	32	X⇌T
599	02	2
600	03	3
601	01	1
602	07	7
603	01	1
604	03	3
605	01	1
606	06	6
607	69	OP
608	04	04
609	32	X⇌T
610	69	OP
611	06	06
Calculates power		
612	65	×
613	43	RCL
614	08	08
615	55	÷
616	43	RCL
617	13	13
618	55	÷
619	43	RCL
620	28	28
621	95	=
622	32	X⇌T
623	03	3
624	03	3
625	04	4
626	03	3
627	03	3
628	05	5
629	69	OP
630	04	04
631	32	X⇌T
632	69	OP
633	06	06
634	25	CLR
635	98	ADV
636	98	ADV
637	98	ADV
638	91	R/S
639	00	0

Using the calculated outlet temperature, the program finds the outlet volumetric flow rate:

$$Q_2 = Q_1(Z_2/Z_1)(T_2/T_1)(P_1/P_2) \tag{7}$$

As before, the program uses a subroutine, this time to find Z_2. Next, the polytropic head is calculated:

$$HEAD = (Z_1 + Z_2)(RT_1/2MW\gamma_{avg})((P_2/P_1)^{\gamma_{avg}-1}) \tag{8}$$

Power input is found from

$$PWR = \text{weight flow} \times \text{head}/C_7\eta_p \tag{9}$$

The program is listed in Table 1. Table 2 presents user's instructions. Two magnetic cards and the Master Library chip are required. Storage registers appear in Table 3.

TABLE 2 User's Instructions for Running the Program

Step	Procedure[a]	Enter	Press	Display
1	Clear calculator		CLR	0
2	Prepare calculator to read		INV 2nd Fix	0
3	Read Card 1, Bank 1	1	INV 2nd Write	1.
4	Read Card 1, Bank 2	2	INV 2nd Write	2.
5	Read Card 2, Bank 3[b]	3	INV 2nd Write	3.
6	Read Card 2, Bank 4[b]	4	INV 2nd Write	4.
7	Initialize program by entering total molar flow	M	2nd E	MOLS
8	Enter molar flow of first component	M	A	MOLS
9	Enter molecular weight	MW	B	MW
10	Enter critical temperature	T_c	C	TC
11	Enter critical pressure	P_c	D	PC
12	Enter first molar heat capacity	$C_{p,1}$	E	CP1
13	Enter second molar heat capacity	$C_{p,2}$	R/S	CP2
14	Repeat Steps 8 through 13 for each component			
15	Run program		2nd A	ENTER T1
16	Enter inlet temperature	t_1	R/S	ENTER P1
17	Enter inlet pressure	P_1	R/S	Q1 ENTER NP
18	Enter polytropic efficiency	η_p	R/S	ENTER P2
19	Enter outlet pressure	P_2	R/S	T2 Q2 HEAD PWR
20	Change inlet temperature		2nd B	ENTER T1
21	Change outlet pressure		2nd C	ENTER P2

[a]Master Library module is required.
[b]Use card for desired system of units.

TABLE 3 Contents of Storage Registers

Register	Contents		
00	Counter		
01	Used for calculations		
02	MW		
03	T_c		
04	P_c		
05	Used for calculations		
06	Used for calculations		
07	M		
08	W		
09	T_1		
10	P_1		
11	Z_1		
12	Q_1		
13	η_p		
14	P_2		
15	T_2		
16	Z_2		
17	Q_2		
18	*Slope of equation for c_p		
19	*Intercept of equation for c_p		
20	T_r		
21	P_r		
22	ρ_r		
	English	Metric	SI
23	1.987	1.987	8.314
24	60	1	1
25	144	98,065	1,000
26	1,545	8,314	8,314
27	459.67	273.15	273.15
28	33,000	3,600,000	3,600,000
29	50	0	0
30	300	100	100
	Constants—Same for All Three Systems		
31		0.27	
32		0.90	
33		0.001	
34		0.31506 (C_1)	
35		−1.0467 (C_2)	
36		−0.5783 (C_3)	
37		0.5353 (C_4)	
38		−0.6123 (C_5)	
39		0.6895 (C_6)	

*Note: c_p = slope × t_{avg} + intercept.

To enter the program: Allow the machine to partition according to the number of registers used. Do this by pressing **4 2nd Op 17** before keying in the program. Store appropriate constants in **Reg. 23–39.** Three versions of the program can be created for the different systems of units by creating three versions of Card 2, each with different constants in **Reg. 23–30.**

To run the program: See Table 2. Entering the molar flow rate and pressing **2nd E** (Step 7) initializes the program—it partitions the calculator to 639.39, clears memory **Reg. 1–22,** and stores molar flow in **Reg. 07.** Enter the molar heat capacities (Step 12 for $c_{p,1}$ and Step 13 for $c_{p,2}$). If only one value of the heat capacity is available, enter it twice. Values should be at 50°F or 0°C for the lower temperature, and 300°F or 100°C for the higher. These temperatures correspond to those used in The Elliott Co. data. If heat capacities are at other temperatures, either find them at the above temperatures or change the values in **Reg. 29** and **30.**

Once data are entered, the program is run by pressing **2nd A.** Then, all data requested by the program will be entered by pressing **R/S.** First, the program calculates the weight flow by multiplying total molar flow (stored in **Reg. 07**) by the average molecular weight (**Reg. 02**) and dividing by units of time (**Reg. 24**). The results are stored in **Reg. 08.** Next, the two calculated average molar heat-capacities and two temperatures (**Reg. 20** and **30**) are used to find the slope (**Reg. 18**) and y-intercept (**Reg. 19**) for molar specific heat as a function of average temperature (see note in Table 3).

The program requests the inlet temperature (in °F or °C; stored in **Reg. 09**) and converts it to absolute temperature (**Reg. 05**). The reduced temperature is calculated (**Reg. 20**), and the inlet pressure is requested. The entered value of absolute pressure (**Reg. 10**) and calculated reduced pressure (**Reg. 21**) are stored and **SBR INV** is employed to calculate the inlet compressibility factor (**Reg. 11**).

Using this value, the inlet volumetric flow (**Reg. 12**) is found and printed, and the polytropic efficiency is requested. This value is stored in **Reg. 13** and the outlet pressure is requested. This absolute pressure is entered (**Reg. 14**) and the reduced pressure is calculated (**Reg. 21**). **SBR LNX** calculates and prints the outlet temperature (**Reg. 15**). This value, along with the user-supplied outlet pressure, is used to find the outlet compressibility factor (calculated in **SBR INV** and stored in **Reg. 16**). The outlet volumetric flow (**Reg. 17**), polytropic head, head, and required power are printed.

After the program is run, the user can account for interstage drawoffs and interstage coolers (see the example).

Example. Determine the total horsepower required and the duty of the interstage cooler for the compressor shown in Fig. 1. Process gas temperature is not to exceed 250°F. Space does not permit presenting a table of the inlet gases and their properties. There are seven gases. These, their flow rates, and their properties appear in Table 4; the program prints the data. Labels for the gases were added. MOLS indicates lb-mol/h; other values are in the corresponding English units.

Initialize the program by entering 2700 **2nd E.** Then enter data for each component, and press **2nd A.** The program then prompts the user to supply data for the

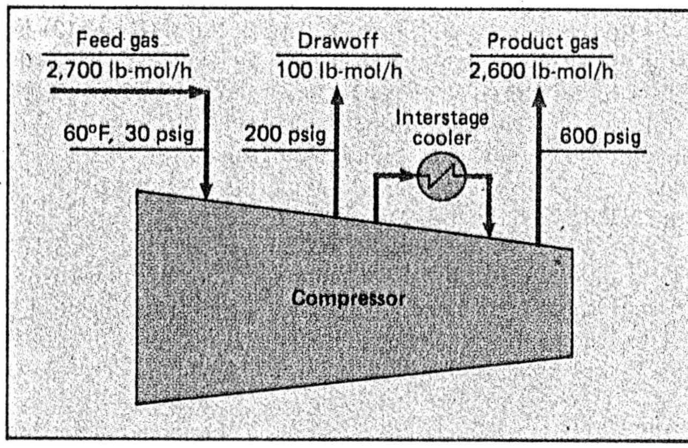

FIG. 1 Example of seven-component mixture for compressor with drawoff and interstage cooler.

first section of the compressor from the inlet (60°F, 44.7 lb/in.^{2}abs) to the drawoff 214.7 lb/in.^{2}abs).

The first option is to account for the drawoff. To do this, recall the total weight flow (**RCL 08**) and subtract the equivalent weight flow of 100 mol/h (− 100 × **RCL 02** ÷ **RCL 24** =) and store the results (**STO 08**). When making such calculations, switch the printer to the **TRACE** model so that a printed record is obtained. Take the printer out of **TRACE** when the program is continued.

Next, an interstage pressure must be picked that would correspond to an interstage temperature of less than 250°F. First, the user returns the program to its requesting the inlet temperature (**2nd B**) and supplies the requested information (203°F, which is the T_1 of the previous run, and 214.7 lb/in.^{2}abs), including the first guess for interstage pressure (350 lb/in.^{2}gauge = 364.7 lb/in.^{2}abs). The program then calculates an interstage temperature of 253°F, which is above the desired 250°F. So, **2nd C** is pressed and a second guess is made for outlet pressure, here 15 lb/in.^{2} lower = 349.7 lb/in.^{2}abs. This procedure is repeated until the calculated interstage temperature is below 250°F.

The next option is to account for an interstage cooler. Once satisfactory interstage conditions have been found, the user assumes a pressure drop of 5 lb/in.^{2} and an outlet temperature of 150°F for the interstage cooler; then, by pressing **2nd B,** the program requests the inlet temperature and other data. The dew point of the gas at the interstage-cooler outlet must be checked to ensure that the stream is all vapor; this requires a separate calculation. The required compressor power is the sum of the individual powers = 2350 + 773 = 3766 hp, plus any seal losses. This procedure could be applied to any number of interstage coolers.

Next, the duty of the interstage cooler is calculated by finding the average molar heat capacity. This is done by using the constants stored in **Reg. 18** and **19** at the average temperature (199.5°F). The heat capacity is then multiplied by the actual molar flow (**RCL 08** ÷ **RCL 02** × **RCL 24** =) and the temperature difference.

TABLE 4 Input Data and Calculations for Example of Seven-Component Mixture—See Fig. 1

```
2700.   MOLS
 Methane
    1.
   18.  MOLS
16.043    MW
  343.    TC
667.8     PC
 8.38    CP1
10.25    CP2
 Ethylene
    2.
   76.  MOLS
28.054    MW
508.3     TC
729.8     PC
10.02    CP1
13.41    CP2
 Ethane
    3.
  321.  MOLS
30.07     MW
549.8     TC
707.8     PC
12,13    CP1
16.33    CP2
```

```
 Propylene
     4.
  1643.   MOLS
42.081      MW
 656.6      TC
  669.      PC
 14.75     CP1
 19.91     CP2
 Propane
     5.
  456.    MOLS
44.097      MW
 665.7      TC
 616.3      PC
 16.82     CP1
 23.57     CP2
 Isobutane
     6.
  176.    MOLS
58.124      MW
 734.7      TC
 529.1      PC
  22.1     CP1
 31.11     CP2
```

```
 n-Butane
            7.
           10.   MOLS
       58.124      MW
        765.3      TC
        550.7      PC
        22.83     CP1
        31.09     CP2
ENTER      T1
          60.
ENTER      P1
         44.7
 5335.577644      Q1
ENTER      NP
         0.76
ENTER      P2
        214.7
 202.8529609      T2
 1319.477461      Q2
 31526.98184    HEAD
  2349.26563     PWR
```

```
           0.    RCL
                   8
 1868.862117
 1868.862117      -
        100.      x
        100.     RCL
                   2
 41.53026926
 41.53026926      ÷
 4153.026926     RCL
                  24
         60.
         60.      =
 1799.645001
 1799.645001     STO
                   8
 1799.645001
ENTER      T1
         203.
ENTER      P1
        214.7
 1270.859809      Q1
ENTER      NP
         0.76
(Continued next col.)
```

```
ENTER      P2
      ,  364.7
 252.9593609      T2
 765.0224153      Q2
 11703.56657    HEAD
 839.8032326     PWR
ENTER      T1
         203.
ENTER      P1
        214.7
 1270.732636      Q1
ENTER      NP
         0.76
ENTER      P2
        349.7
 248.9813977      T2
  797.356017      Q2
 10771.03802    HEAD
 772.8885459     PWR
```

```
ENTER      T1
         150.
ENTER      P1
        344.7
 596.6916962      Q1
ENTER      NP
         0.76
ENTER      P2
        614.7
 203.6373117      T2
 292.1031666      Q2
 8951.561148    HEAD
 642.3298355     PWR

           0.      (
         249.      +
         150.      )
         399.
         399.      ÷
           2.      =
        199.5
        199.5      x
        199.5    RCL
                  18
(Continued next col.)
```

```
 .0220203556
 .0220203556      +
 4.393060933    RCL
                  19
 14.02052296
 14.02052296      =
  18.4135839
  18.4135839      x
  18.4135839    RCL
                   8
 1799.645001
 1799.645001      ÷
 33137.91421    RCL
                   2
 41.53026926
 41.53026926      x
 797.9219688    RCL
                  24
         60.
         60.      x
 47875.31813      (
        249.      -
        150.      =
 4739656.495
```

Symbols

		English	Metric	SI
C_1–C_6	constants for Eq. (2), stored in **Reg. 34–39**			
C_7	constant for Eq. (9)*			
c_p	molar heat capacity at constant pressure	Btu/(lb·mol)(°R)	kcal/(kg·mol)(K)	kJ/(kg·mol)(K)
HEAD	polytropic head	$ft \cdot lb_f/lb_m$	$N \cdot m/kg_m$	$N \cdot m/kg_m$
k	specific-heat ratio			
M	molar flow	lb·mol/h	kg·mol/h	kg·mol/h
MW	molecular weight			
P	absolute pressure	$lb/in.^2abs$	kg/cm^2(abs.)†	kPa
PWR	power	hp	kW	kW
Q	volumetric flow	ft^3/min	m^3/h	m^3/h
R	gas constant	$ft \cdot lb_f/(lb \cdot mol)(°R)$	J/(kg)(K)	J/(kg)(K)
t	temperature	°F	°C	°C
T	absolute temperature	°R	K	K
W	weight flow	lb/min	kg/h	kg/h
Z	compressibility factor			
Greek Letters				
γ	function of k, see Eq. (5)			
η_p	polytropic efficiency			
ρ	gas density	lb/ft^3	kg/m^3	kg/m^3
Subscripts				
avg	average condition			
c	critical state			
r	reduced state			
1	suction condition			
2	discharge condition			

*Converts power into proper units (hp or kW); stored in **Reg. 28.**
†If bar is preferred unit, the constant in **Reg. 25** should be 100,000.

References

1. *Quick Selection Methods for Elliott Multistage Compressors,* Bulletin P-255, The Elliott Co., Jeannette, Pennsylvania, 1975.
2. J. Urick and F. Odom, "Calculator Analyzes Compressor Performance," *Oil Gas J., 78*(2), 60 (January 14, 1980).

JOSEPH W. STANECKI

Piping Design, Sizing Economics

Piping may be sized by one of three ways. When a pump or compressor is involved, an economically optimum pipe diameter is derived by balancing system operating cost against installed cost. When a prime mover is not involved, the piping is sized to accommodate the desired flow with whatever pressure drop is available. Lastly, a velocity constraint—to minimize erosion, maintain particle suspension, or stabilize multiphase flow—dictates pipe diameter.

The latter two approaches are sufficiently constrained that size can be evaluated directly. Economical sizing, however, depends on many factors. Piping system length, complexity, and energy requirements must be simplified to reduce the sizing problem to quantifiable terms. Representative values of these may then be integrated into a model that yields a usable result.

Piping System Length and Complexity

Piping length and complexity dominate piping system economics, and are the least constant design factors. This problem of variability is lessened by assuming that piping cost and pressure loss in one section of piping is proportional to those for the entire circuit. Therefore, optimizing the diameter for a unit length of piping does so for the total system. The foregoing assumption depends on two others: (1) that the average installed cost of piping is not affected by length (beyond some minimum point), and (2) that the pressure drop per length of piping is the same as the average pressure drop over the whole system.

The latter assumption is valid for all liquid flow and is within acceptable limits for compressible fluids as long as the overall pressure drop does not exceed 10% of the initial absolute pressure [1, 2].

For a given length of piping, the installed cost and energy loss must be adjusted to account for the number of valves and fittings (i.e., system complexity). The contribution of valves and fittings to cost and pressure loss must be evaluated separately, however, because the cost of pressure loss changes differently for each as pipe diameter varies. Valve and fitting frequency may be expressed as a function of pipe diameter [3].

Depending on the general location of the piping, the frequencies of valves and fittings, respectively, per 100 ft of pipe may be predicted from

$$V_{100} = 9F_c/D \tag{1}$$

$$F_{100} = 35F_c/D^{1/2} \tag{2}$$

Here, F_c, the piping complexity factor, is equal to 4 for very complex manifolds, 2 for manifolds, 1 for piping within unit boundaries, ½ or long straight pipe, and ¼ for utility supply lines outside unit boundaries. Fittings are typically distributed thus: flanges, 45%; 90° elbows, 25%; tees, 15%; reducers, 10%; 45° elbows, 2½%; and caps, 2½%.

Installed-Cost Equation

In the development of the installed-cost equation, adjustments to cost data are made assuming the piping to be inside unit boundaries (i.e., $F_c = 1$). Hydraulic equivalents are determined for flow through the run of tees and for concentric reducers swaged to one size smaller. The valves are taken to be fully opened gate valves. The values for equivalent piping length are from Crane Co.'s *Flow of Fluids* [4].

Based on the foregoing, the installed cost of piping can be expressed as a function of diameter. The equation is hyperbolic in character, with cost approaching a linear relationship with diameter for standard-wall piping [5]:

$$C_i = H[4.25(D^2 + 51.15)^{1/2} - 20.42]X \tag{3}$$

Equation (3) is based on field fabrication of 300 lb/in.^{2} carbon steel piping (A-53) and is referenced to the base material cost for 1 ft of 2-in. pipe, X. The H factor is an adjustment that covers the costs of design, procurement, and construction administration.

Energy Loss from Pressure Drop

Energy requirements for a piping system result from process pressure increases as well as from piping friction losses. Only piping loss is considered in optimizing pipe diameter, because process needs must be met regardless of the pipe size selected. Similarly, losses for process control are assumed to be relatively constant for a particular service.

To relate energy costs to pipe diameter for a particular flow rate, the friction factor from the Moody diagram must first be put into equation form [1]:

$$f = 0.14/(N_{\mathrm{Re}})^{0.16} \tag{4}$$

Here, $N_{\mathrm{Re}} = 50.6Q\rho/D\mu$.

Equation (4) applies to turbulent flow in commercial steel pipe, giving somewhat conservative results with larger pipe sizes. Inserting Eq. (4) into the Darcy-Weisbach equation, $h = f(12L/D)(V^2/2g)$ [6], yields

$$\Delta p = \rho h/144 = \mu^{0.16}Q^{1.84}\rho^{0.84}/61{,}918D^{4.84} \tag{5}$$

The overall piping loss must be adjusted to account for fitting pressure-drops [4]:

$$\Delta p_t = \Delta p[1 + F_c(0.245D^{1/2} + 0.098)] \tag{6}$$

Of course, pressure loss must be converted into energy loss: equivalent energy, in kWh, $= Q\Delta p_t/2298.72E$, which can be converted into annual power cost via

$$C_e = Q\Delta p_t YK/2298.72E \tag{7}$$

Economic Factors

The capital expenditure is partially offset by a 10% investment tax credit a year later, plus depreciation (sum-of-years-digits) credits during the first 5 years. The rate of return is fixed at 25% to cover the cost of capital and provide an investment return. A 15-yr life is assumed. All future cash flows are discounted accordingly to establish a present value. Pipe cost is, therefore, deflated by a capital adjustment factor of 67.4% of the initial investment. The present value of power costs is 208.4% of annual expenditures. Figure 1 shows this cash flow model.

Piping and power costs can be combined into a total expected cost:

$$C_t = 0.674C_i + 2.084C_e \tag{8}$$

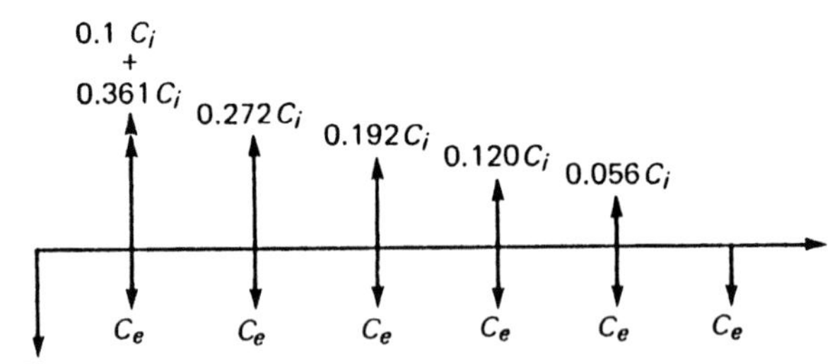

FIG. 1 Cash flow model for a 15-yr-life piping system.

Substituting the terms of Eqs. (3) and (7) into Eq. (8) yields a generalized equation for piping system cost:

$$C_t = 0.674H[4.25(D^2 + 51.15)^{1/2} - 20.42]X + [2.084Q^{2.84}\mu^{0.16}\rho^{0.84}YK(1.098 + 0.245D^{1/2}) \div (1.423 \times 10^8)D^{4.84}E] \quad (9)$$

Equation (9) is depicted graphically in Fig. 2 for a 375-gal/d water system.

An Equation for Compressible Fluids

Although developed with liquid-flow terms, Eq. (9) can also be applied with compressible fluids if the pressure drop does not exceed 10% of the absolute inlet pressure. Fluid properties should be the average of the inlet and outlet conditions:

$$C_t = 0.674H[4.25(D^2 + 51.15)^{1/2} - 20.42]X + [2.084M^{2.84}G^{0.84}\mu^{0.16}Z_a{}^2T^2YK(1.098 + 0.245D^{1/2}] \div \rho^2(1.434 \times 10^{10})D^{4.84}EZ_s \quad (10)$$

Comparing the cost of alternative line sizes and establishing a relative minimum cost involves trial-and-error analyses, which can be cumbersome when done for each line to be installed. Liquid systems can be considerably simplified by assuming fluid properties similar to those of water. A single pump efficiency is not meaningful, however, because pump efficiency im-

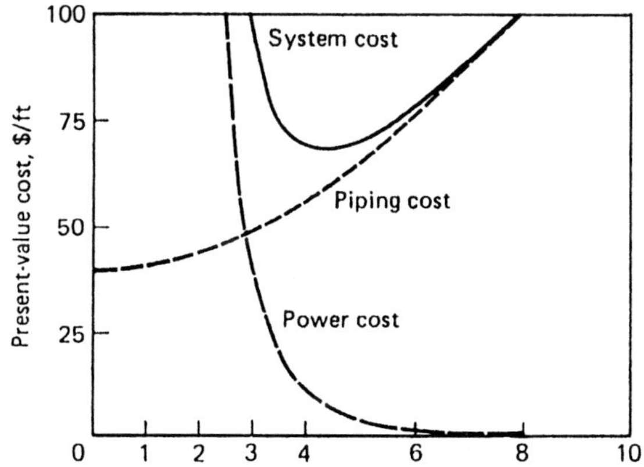

FIG. 2 Minimum cost for a 375-gal/min piping system.

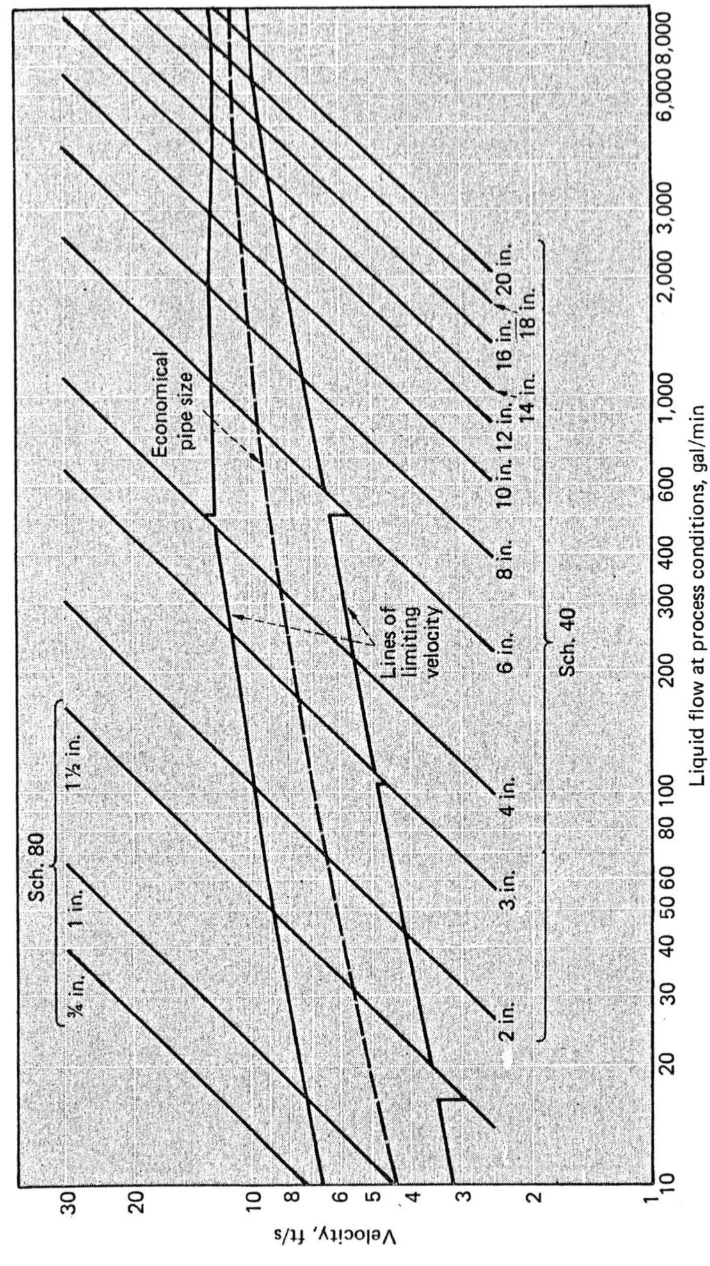

FIG. 3 Preliminary sizing of pipe diameter for pumped liquid within unit boundary.

proves markedly with increasing capacity [7]. If specific speed is greater than 2000, maximum pump efficiency can be expressed as a function of capacity by means of

$$E = 0.98 - e^{-(0.517 + \ln Q)/3.93} \tag{11}$$

Assuming that 75% of E will accommodate motor efficiency losses and applications less efficient than the maximum, the general liquid equation becomes

$$C_t = X[5.729(D^2 + 51.15)^{1/2} - 27.526] + Q^{2.84}(4.480 + D^{1/2}) \div D^{4.84}18{,}643[0.98 - e^{-(0.517+\ln Q)/3.93}] \tag{12}$$

Equation (12) has been solved for standard pipe sizes and is shown graphically in Fig. 3. The economic pipe diameter was determined by differentiating Eq. (12) and setting it equal to zero. The pipe cost used was \$2.98/ft for 2-in. Schedule 40 carbon steel pipe (A-53). Energy was valued at \$0.0425/kWh.

Figure 3 applies only to A-53 piping. Economically optimum velocities are higher for alloy and lined piping (i.e., diameters will be smaller). In using Fig. 3, select the pipe diameter from between lines of limiting velocity. The effects of water-hammer surging should be considered in the final selection of lines 10 in. diameter and larger.

Figure 3 is valid for 1982 costs. Because these factors can change rapidly, it is worthwhile to examine their relative effects on the determination of optimum pipe diameter. It can be shown that $D^{5.34} \approx$ energy cost/piping cost—i.e., the optimum pipe diameter is fairly insensitive; doubling of the cost of energy relative to material increases the optimum diameter by only 10%.

Symbols

C_e	energy cost to overcome piping friction loss (\$/yr)
C_i	installed cost of piping (\$/ft)
C_t	present value cost of piping system (\$/equivalent ft)
D	pipe inside diameter (in.)
E	efficiency of pump or compressor (fraction)
f_c	piping complexity factor
F_{100}	number of fittings per 100 ft of pipe
f	friction factor for fluid flow
G	gas specific gravity
g	gravitational constant [32.174 (ft)(lb$_f$)/(lb$_m$)(s^2)]

H	cost factor covering piping design, procurement, and administrative overhead (typically 2.0)
h	head loss from friction per foot of pipe [(ft)(lb_f)/lb_m]
K	power cost (\$/kWh)
L	piping equivalent length (ft)
M	gas flow (1000 std ft^3/d)
N_{Re}	Reynolds number
p	gas pressure (lb/in.^{2}abs)
Δp	pressure loss from friction [(lb/in.^{2})/ft of pipe]
Δp_t	pressure loss from friction, including fitting losses [(lb/in.^{2})/ft of pipe]
Q	flow rate at operating conditions (gal/min)
T	gas temperature (°R)
V	fluid velocity (ft/s)
V_{100}	number of valves per 100 ft of pipe
X	material cost for 2-in.-diameter pipe (\$/ft)
Y	number of hours of operation per year
Z_a	gas compressibility factor at average flow conditions
Z_s	gas compressibility factor at standard flow conditions
μ	fluid viscosity (cP)
ρ	fluid density (lb_m/ft^3)

References

1. C. B. Nolte, *Optimum Pipe Size Selection,* Trans. Tech. Publications, Clausthal, West Germany, 1978.
2. S. P. Marshall and J. L. Brandt, "1974 Installed Cost of Corrosion-Resistant Piping," *Chem. Eng.,* pp. 94–106 (October 28, 1974).
3. W. B. Hooper, "Predict Fittings for Piping Systems," *Chem. Eng.,* pp. 127–129 (May 17, 1982).
4. *Flow of Fluids,* Technical Paper 410, Crane Co., New York, 1969.
5. *Process Plant Construction Estimating Standards,* Richardson Engineering Services, San Marcos, California, 1982.
6. V. L. Streeter, *Fluid Mechanics,* McGraw-Hill, New York, 1971, p. 277.
7. I. Karassik et al., *Fluid Handbook,* McGraw-Hill, New York, 1976.

D. W. CULLER

Steam Lines, Optimum Diameter

Piping accounts for as much as 30% of the total plant costs, and selection of sizes must be made with care. Though general recommendations are available for line velocities for various fluids, engineers would do well to make some preliminary analysis before arriving at a particular size.

Too high a line velocity results in a high pressure drop, which means large operating costs. Computed over the life of the plant, this could be a substantial expenditure.

On the other hand, if a low line velocity is used, a large pipe has to be chosen. This means higher material cost, higher insulation cost, larger fabrication cost, higher erection costs, and higher pipe hanger and support costs.

An optimum line velocity may be determined for each case. The following example illustrates the basics behind choice of an economical steam piping system for a process boiler plant.

The Problem

Assume 6,000,000 lb/h of steam is required at 800 lb/in.^{2}abs and 900°F in a process plant. Design pressure at the exit of the boiler superheater is 1000 lb/in.^{2}abs, thus a maximum drop of 200 lb/in.^{2} may be tolerated in the piping between the boiler and process steam header.

SA 335 P11 pipes of 8⅝, 10¾, 12¾, 14, 16, and 18 in. o.d. are available. It is estimated that the developed length of pipe required will be around 100 ft. Also, two 90° bends and a swing check valve have to be considered while computing the pressure drop.

Determine the optimum pipe size, if the preliminary cost of piping, including insulation, hangers, fabrication, and erection, is as given in Table 1. The cost of power is 35 mills/kWh. The plant works for 6000 h/yr and has a life of 20 years. Interest rate is 9% and escalation rate is 6%.

Solution Procedure

The design pressure of the piping is taken as 1000 lb/in.^{2}gauge, and the effect of steam line pressure drop on design pressure is considered marginal.

TABLE 1 Study of Optimum Pipe Size

d (in.)	t (in.)[a]	f	V (ft/s)	L_e (ft)	ΔP (lb/in.^{2})	N (kW)	$C_a \times F$ ($)	Fixed Cost ($)[b]	C_c ($)
8.625	0.406	0.0141	465	212	115	100	317,100	17,000	334,100
10.75	0.50	0.0137	299	248	43.4	38	120,498	23,000	143,498
12.75	0.562	0.0133	210	272	19.1	17	53,273	29,000	82,273
14	0.593	0.013	173	288	12.2	11	33,930	35,000	68,930
16	0.656	0.0127	131	312	6.53	6	18,138	54,000	72,130
18	0.75	0.0124	104	328	3.74	3.3	10,400	65,000	76,400

[a]Thickness used is the nearest available thickness to that required by ASME code.
[b]Includes cost of pipe, insulation, valves, pipe hangers, fabrication, and erection.

Also, the impact on boiler costs for small differences in design pressure arising out of line pressure variations will be neglected in this preliminary analysis.

Step 1. ASME code formula is used to determine the minimum pipe thickness:

$$t = Pd/(2SE + 0.8P) \tag{1}$$

S, the allowable stress from ASME Boiler and Pressure Vessel Code, Sect. I, for SA 335 P11 material at 900°F is 13,100 lb/in.^{2}. Using $P = 1000$ and $S = 13{,}100$, $E = 1$. Then t was computed for each pipe and the nearest standard thickness available from the market was chosen.

Step 2. Pressure drop for each alternative has to be determined. The following equation was used [1]:

$$\Delta P = 3.36 \times 10^{-6}\, fw^2 L_e v / d_i^5 \tag{2}$$

Friction factor f depends on the Reynolds number which was computed by using a chart [2]; f was then looked up in Moody's chart [3].

The total equivalent length has to consider the effect of bends and the swing check value, the resistances for which were taken from standard tables [3].

Step 3. The pressure drop is then computed. The additional feed pump power associated with the pressure drop for each case is computed using the following equation, assuming 70% efficiency for the feed pump drive:

$$N = \frac{w}{3600}\,\frac{0.746}{550}\,\frac{144}{\eta}\,\frac{\Delta P}{\rho_w} = 0.88\Delta P \tag{3}$$

The annual cost of operation is then

$$C_a = N \times 6000 \times 0.35 = 210N \tag{4}$$

Step 4. The life cycle cost of operation is obtained by multiplying C_a by F, where

$$F = \left(\frac{1+e}{1+i}\right)\left[\frac{1 - \left(\dfrac{1+e}{1+i}\right)^{Ny}}{1 - \left(\dfrac{1+e}{1+i}\right)}\right] \tag{5}$$

For $i = 9\%$, $e = 6\%$, and $N = 20$, $F = 15.1$. Thus

$$C_a \times F = 3171N \tag{6}$$

The capitalized cost of operation is then $C_a \times F$.

The total capital cost C_c of the piping system is given by adding the fixed cost with the capitalized operating cost:

$$C_c = (C_a \times F + \text{fixed cost}) \quad (7)$$

The results are summarized in Table 1.

It is seen that the 14-in. o.d. pipe gives the lowest life cycle cost and hence is recommended. Optimum line velocity is 173 ft/s.

In practice, a deeper analysis is warranted. For high pressure boilers, pressure drop in steam lines is critical and alters the boiler design pressure.

For pressures in the neighborhood of 2400 lb/in.^{2}, the boiler circulation system is likely to be affected as even a small change affects the process of circulation in the furnace walls. Routing of lines and flexibility analysis should be performed before arriving at the exact length of piping.

Fabrication costs associated with local site conditions and experience would be helpful in arriving at the investment costs on a particular size of pipe.

Nevertheless, this article is aimed at giving an idea of the preliminary analysis involved in arriving at an optimum size of piping. The logic may be applied for large capacity plants, too, where there may be two or even four lines of smaller size instead of one line, as in this case.

Symbols

e	escalation rate
E	efficiency of joint
f	friction factor
F	capitalization factor
i	interest rate
d	outer diameter of pipe (in.)
d_i	inner diameter of pipe (in.)
N	additional power required by boiler feed pump (kW)
P	design pressure of pipe (lb/in.^{2}gauge)
C_a	annual cost of operation ($)
C_c	total capitalized cost ($)
S	allowable stress per code (lb/in.^{2})
L_e	total equivalent of pipe (ft)
t	thickness of pipe (in.)
v	specific volume of steam (ft^3/lb)
ΔP	steam line pressure drop (lb/in.^{2})
ρ_w	density of water at feed pump (lb/ft^3)
V	line velocity (ft/s)
η	efficiency of pump drive system

Ny	life of the plant (years)
w	steam flow (pph or lb/h)

References

1. V. Ganapathy, "Chart Gives Pressure Loss in Steam Lines," *Oil Gas J.*, p. 95 (January 9, 1978).
2. V. Ganapathy, "Reynolds Number Given for Superheated Steam," *Oil Gas J.*, p. 65 (November 6, 1978).
3. R. Kern, *Practical Piping Design, Chemical Engineering* Reprint, 1975.

V. GANAPATHY

Temperature (High) Considerations

High-temperature operation introduces unique stress patterns in piping. This article describes the principal stresses involved and shows how to make the most efficient use of materials.

The most common criteria currently used in the United States for the design of hot piping in petrochemical plants are found in ANSI B31.3.* This code prescribes minimum requirements for the materials, design, fabrication, assembly, erection, inspection, and testing of piping systems subject to pressure or vacuum over a range of temperatures up to 1500°F.

Allowable Stress

Improperly designed piping may fail as a result of excessive stress levels generated during service. The "basic allowable stress" for a particular material at a particular temperature is based on several criteria, any of which

**ANSI B31.3, American National Standards Institute Code for Chemical Plant and Petroleum Refinery Piping,* published by the American Society of Mechanical Engineers, 1976.

may govern. In accordance with the *Code,* this basic allowable stress shall not exceed the lowest of the following:

One-third of the material's minimum tensile strength at room temperature.
One-third of the material's minimum tensile strength at design temperature.
Two-thirds of the material's minimum yield strength at room temperature.
Two-thirds of the material's minimum yield strength at the design temperature.
The average stress for a creep rate of 0.01% per 1000 h.
67% of the average stress for rupture at the end of 100,000 h.
80% of the minimum stress for rupture at the end of 100,000 h.

The various failure theories considered in selecting the above bases for allowable stress are beyond the scope of this discussion.

An exception to the fourth criterion above is that for austenitic stainless steels (generally the 300 series) and for some of the nickel alloys (such as Incoloy 800), when used at temperatures below 1100°F, the limit may be as high as 90% of the minimum yield strength at the design temperature. However, this high allowable stress is not recommended for flanged joints or other applications where slight deformation could cause failure.

Principal Stresses

For most piping design problems, where the pipe wall is thin (less than one-tenth of the radius) in comparison with the other dimensions, we can assume that stresses in the pipe wall exist in only two directions, longitudinal and circumferential. This assumption is not valid for thick-wall pipe where radial stresses may become appreciable.

Longitudinal and circumferential "principal stresses" can be divided into two broad categories: primary stresses and secondary stresses.

Primary stresses are those that would not be relieved by local yielding of the material if the piping were stressed beyond the yield point. For example, internal pressure in a pipe causes longitudinal and circumferential tensile stresses to be generated. If the pressure is high enough, the yield strength of the material will be exceeded and the pipe will be deformed by bulging, which causes the pipe wall to become thinner, as well as larger in diameter. This will cause a further increase in stress, due to the internal pressure. Since stresses are not relieved by yielding of the material, the stresses caused by the pressure are said to be primary ones. Other causes of primary stresses are weight (of piping, contents, and insulation) and, of course, external loads.

Secondary stresses are relieved by local yielding of the material in the highly stressed areas. For example, thermal stresses (primarily those caused by expansion of the metal in the pipe wall when the pipe is operated at a temperature higher than that at which it was installed) are relieved when the pipe material yields.

The *B31.3 Code* has different allowable design limits for primary stresses and secondary stresses. It is obvious that the limit for primary stresses will be less than that for secondary stresses, since if primary stresses exceed the yield point, failure by rupture is inevitable in almost all cases.

Due to the cylindrical shape of pipe, the circumferential tensile stresses due to the internal pressure will be twice as great as the longitudinal tensile stresses. The minimum wall thickness for internal pressure will, therefore, be determined by the allowable *circumferential stress.* It can be shown that the circumferential stress in a thin-wall cylinder under internal pressure is

$$\text{Stress} = (\text{pressure} \times \text{radius})/\text{wall thickness}$$

which is basically the equation used to determine the minimum wall-thickness requirement in accordance with the *Code.* The basic allowable stress values tabulated in the *Code* may be used directly in the procedure for determining the required wall thickness.

In some piping systems, failures may occur as a result of excessive *longitudinal stress* levels generated by the combination of stresses from internal pressure, weight, and other sustained loadings (primary stresses). Since the magnitude of combined stresses will vary according to the design of a particular system, the *Code* cannot present a simple equation to ensure a safe design, as it can in the case of wall thickness for internal pressure. Instead, the *Code* specifies that the sum of longitudinal stresses due to pressure, weight, and other sustained loadings shall not exceed the basic allowable stress for the maximum metal temperature expected during the cycle under analysis.

Secondary stresses, on the other hand, have a higher allowable level because they are, in effect, self-limiting; that is, they will not cause immediate failure of the system as excessive primary stresses will.

Deformation and Springing in Cyclic Loading

Under conditions of cyclic loading, there will be a maximum stress and a minimum stress (assume tensile stresses to be positive and compressive stresses to be negative) generated during each cycle. If these stresses stay within the elastic range for the material, no permanent deformation will occur. If, however, the combination of primary and secondary stresses exceeds the elastic limit, some local yielding (deformation) will occur in the highly stressed areas, and the piping system will acquire a "permanent set." If the system undergoes this plastic deformation on every cycle, failure will eventually occur. If a relatively few (say, less than 7000) cycles are expected during the life of the system (as is the case for most plant piping), the allowable stress range will be greater than the basic allowable stress, and is calculated according to the equation given in Section 302.3.5(d) of the *Code:*

$$S_A = 1.25S_c + 0.25S_h$$

where S_A = allowable-displacement-stress range; S_c = basic allowable stress, lb/in.^{2}, at minimum (cold) metal temperature expected during the displacement cycle under analysis; and S_h = basic allowable stress, lb/in.^{2}, at the maximum (hot) metal temperature expected.

This means that, in some cases, piping may be designed to undergo plastic deformation in service when the yielding is the result of self-limiting secondary stresses, such as those resulting from thermal expansion. (See Section 319.2.3 of the *Code.*) It is possible, of course, for failure to occur as a result of excessive yielding, and a properly designed system will not permit gross deformation to such an extent that failure or misalignment of piping guides and supports results.

One way that stresses resulting from thermal expansion can be offset is by *cold springing*. Cold spring is the intentional deformation of piping during assembly to produce a desired initial displacement, and stresses opposite to those expected in service.

Cold spring may be produced by "cutting short" or "cutting long" rather than by permanently deforming the system. This is the intentional fabrication of the piping to dimensions slightly different from the installed dimensions. The difference between the installed and fabricated dimensions is generally about half of the thermal expansion, but the arrangement of the particular system, the effects of weight, supports, and allowable nozzle loadings or anchor reactions will determine the actual amounts of cold spring required.

A variation on cold springing is *self-springing,* in which the piping is designed to undergo local yielding and consequent permanent deformation in the first cycle of operation. This leaves a residual offset stress that in subsequent cycles allows the piping to operate within the elastic limit.

Although self-springing may eliminate additional yielding on subsequent cycles,the high stress (approaching the yield strength) may cause appreciable creep in the high-temperature range.

Expansion-Loop Example

The following example, although quite specialized, illustrates several important principles of design. The example concerns a steam/hydrocarbon reformer outlet header, which is shown in Fig. 1. The reformer catalyst tubes run vertically through a high-temperature furnace. A circular offtake header rests on top of the cylindrical furnace and is connected to each of the catalyst tubes by a small-bore expansion-loop tube, otherwise known as a pigtail or omega because of its shape.

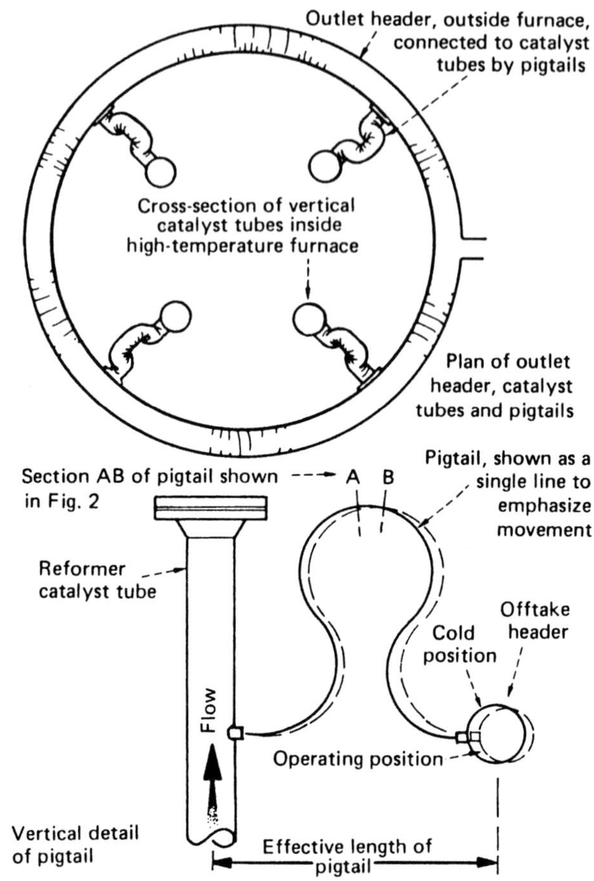

FIG. 1 Catalyst tubes, outlet header, and pigtails.

In service, the catalyst tube remains essentially stationary, while the circular outlet header expands radially. The amount of radial displacement depends on the material, temperature, and circle radius. The pigtail also expands when heated, but since the straight-line distance between its ends is less than the radius of the header, the thermal growth of the pigtail is less than the displacement of the header. The result is that the pigtail is subjected to mechanical extension as well as thermal expansion at operating temperature.

The pigtails are cold-formed from straight lengths of seamless pipe (Fig. 2a). In the forming operation, the material is stressed beyond its yield point, and permanent deformation (bending) occurs. When the forming-die forces are released, the pigtail has some "springback," but is left with residual stresses in the deformed (bent) areas—tensile on the outside of bends and compressive on the inside (Fig. 2b). The minimum bend-radius for small pipe sizes is in the range of two to three pipe diameters. Tighter bends will result in excessive deformation.

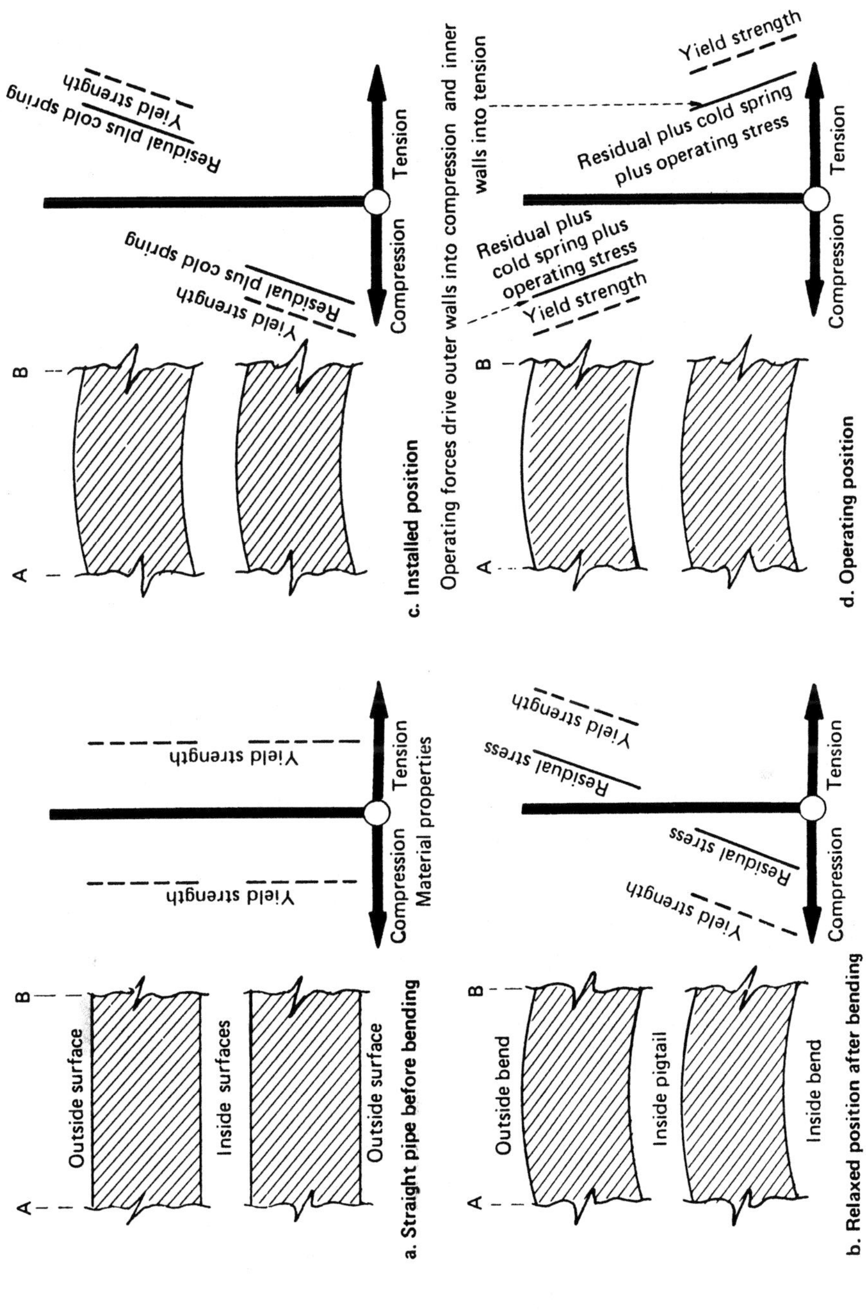

FIG. 2 Cold-formed pigtail in various positions.

High-nickel alloys, such as Incoloy 800, are subject to "strain-hardening," and as a result, the yield strength of the material increases as deformation takes place. Since the outer surface of the pipe undergoes more yielding than does the inside surface, the hardness will be greater on the outside surface, and the yield point will also be highest on the outside. The strain-hardening also decreases the ductility of the material; therefore, the outer surface is also the least ductile.

Under operating conditions, the pigtails are mechanically extended, with the result that new compressive stresses occur on the outside of bends and tensile stresses occur on the inside. When added to the existing stresses in these locations, which are of the opposite direction, the new resultant operating stresses are less than they would have been had the material been in the annealed (unstressed) condition after forming.

The pigtails are compressed for installation, so that the tensile and compressive stresses in the outer and inner sides of the bends are greater than in the "relaxed" position (Fig. 2c). This allows a greater mechanical extension in the operating position without exceeding the yield strength, which would result in further deformation (Fig. 2d).

The design of the pigtail and header system is such that the residual stresses from the forming operation combine favorably with those arising from operating conditions, and the result is an efficient use of materials at minimum cost.

Thermal Gradients

Thermal stresses are the result of differential expansion in a material. A high thermal gradient will, in general, cause more differential expansion and consequent thermal stresses than will a lower thermal gradient. Also, the differential expansion (and, therefore, the thermal stresses) in an irregularly shaped area will be greater than that in an area of uniform cross-section, and so high thermal gradients are particularly to be avoided in areas of irregular cross-section. Other factors in stress design, such as sharp corners, which have a stress-concentrating effect, are even more critical in high-temperature systems than they are in lower-temperature applications, because they tend to aggravate the thermal gradient problem.

The thermal conductivity of stainless steels is less than that of carbon and low-alloy steels; therefore, high thermal gradients will occur more readily in stainless steels. The thermal expansion of stainless steel, however, is much greater (on the order of 40%) than that of carbon steel, so it is obvious that merely using a high-temperature material without regard to its different physical properties is not the solution to high-temperature design.

Allowable stress values at temperatures greater than 1500°F are not tabulated in the *Code,* but some materials are suitable for use in higher-temperature systems, and appropriate allowable stress levels at these higher temperatures can be determined in accordance with *Code* Paragraph 323.2.1.

Two higher-temperature materials are the austenitic stainless steels and nickel-chrome alloys.

At these high temperatures the allowable stress decreases rapidly with increases in temperature. For example, for Incoloy 800H, a nickel-iron-chrome alloy, an increase of 100°F in the design temperature results in a 40% reduction in the allowable stress.

Other Considerations

In the critical design of high-temperature piping, proper specification and application of *insulation* is very important and is based on more than simply conservation of heat. After any maintenance work requiring insulation removal, the insulation should be carefully reinstalled, making sure that all large cracks are packed with loose fill. The thicknesses and types of insulation should not be changed arbitrarily, as they are a part of the overall design, and altering them without considering the rest of the system may decrease the reliability or life of the system. If the insulation is allowed to deteriorate or is not replaced when removed for maintenance, high thermal stresses can develop in the areas of discontinuity. These resultant secondary stresses may cause excessive distortion or possible failure.

Location of *welds* in high-temperature piping designs is also important, since residual stresses in and adjacent to the welds will combine with the operational stresses and may either improve or impair the design.

Digital-computer techniques have made feasible more detailed analysis of piping design and helped to ensure the most economical design consistent with the service requirements, design codes, and good engineering judgment.

Proper *support and guiding* of any piping system is important, especially if it is for hot service, where thermal expansion is significant. If the system is designed to undergo local yielding or plastic flow, the proper guiding and support is not only important—it is absolutely essential. Improper or inadequate guides and supports will permit unintended deformation that may distribute forces and moments on anchor points in such a way that excessive stresses and consequent failure will result.

Many simplifying assumptions have been made in the foregoing discussion, and the graphs are intended to show directional trends only, not precise engineering data. For this reason, persons involved in designing systems of this type should refer to published design properties for appropriate criteria.

JOHN D. DAWSON

Temperature (Low) Considerations

In recent years, design and construction of pipelines in permafrost regions, buried LNG pipelines, hot-oil pipelines, and large-diameter pipelines have focused attention on the need to make accurate flowing-temperature predictions.

This article outlines an equation which has been used successfully to predict pipeline temperatures in a wide variety of pipelines in tropical, temperate, and Arctic environments.

Comparison with temperature data from operating pipelines demonstrates that temperature losses along a pipe section can be predicted during design work with an accuracy around 10%.

Temperature Profile Equation

The following temperature profile equation has been derived from fundamental thermodynamics [1] and is well suited for hand calculations. In the derivation, average values were selected for fluid properties, pressure gradient ($\Delta P/L$), elevation gradient ($\Delta y/L$), and the heat-transfer constant of proportionality (s).

$$T_2 = (T_1 - T_a)e^{-aL} + T_a \quad (1)$$

where

$$T_a = T_g - (J\Delta P + \Delta y/jC_P)/aL \quad (1a)$$

$$a = s/mC_P \quad (1b)$$

$$m = 112{,}800QGP_b/T_b \text{ for gas} \quad (1c)$$

$$m = 14{,}600QG \text{ for oil} \quad (1d)$$

A program for solving these equations using the Texas Instruments TI-58/59 programmable calculator is listed in Fig. 1 together with sample input and output data for a 10-in. heavy-oil pipeline.

The oil in this example starts out at ground temperature and has a Joule–Thomson coefficient of $-5°F/10^3$ lb/in.^{2} which causes it to heat several Fahrenheit degrees as it flows along the pipe.

The program begins with a data-entry routine which prompts the user by displaying the storage-register number for each input in turn. Alterna-

The program flashes the fluid temperature profile at 8 equal intervals along the pipeline, and then continuously displays the end-of-line temper-

```
         Sample Input Data                           Sample Output Data

T1   50      STO 01  (°F)                  Flash    1   50.6   T1/8   (°F)
D    10.75   STO 02  (in)                  Flash    2   51.1   T1/4   (°F)
L    24.25   STO 03  (miles)               Flash    3   51.5   T3/8   (°F)
ΔP   944     STO 04  (psi)                 Flash    4   51.9   T1/2   (°F)
Δy   -7      STO 05  (ft)                  Flash    5   52.3   T5/8   (°F)
c    48      STO 06  (in)                  Flash    6   52.6   T3/4   (°F)
K    0.8     STO 07  (btu/ft-°F-hr)        Flash    7   52.9   T7/8   (°F)
Tg   50      STO 08  (°F)                  Display      53.1   T2     (°F)
cp   .415    STO 09  (btu/lb°-F)           RCL     15   .584   m      (MMlb/hr)
J    -5      STO 10  (°F/10³psi)           RCL     16   53.4   h      (in)
Q    43.2    STO 11  (Mbbl/d)              RCL     17   1.68   s      (btu/ft-°F-hr)
G    .926    STO 12  (water=1)             RCL     18   6.95   a      (/10⁶ ft)
Pb   0       STO 13  (psia)                RCL     19   .890   aL     (-)
Tb   0       STO 14  (°R)                  RCL     20   55.3   Ta     (°F)

Input        034  2      071  ×      106  2         141  0        176  +/-
000  1       035  1      072  RCL    107  ×         142  0        177  INV
001  4       036  6      073  12     108  π         143  0        178  lnx
002  x→t     037  ×      074  =      109  ×         144  ×        179  STO
003  0       038  RCL    075  STO    110  RCL       145  RCL      180  22
004  STO     039  11     076  15     111  07        146  04       181  RCL
005  00      040  ×                  112  =         147  +/-      182  01
006  Op      041  RCL    Find s      113  STO       148  -        183  STO
007  20      042  12     077  Lbl    114  17        149  RCL      184  21
008  RCL     043  ×      078  D                     150  05       185  Fix
009  00      044  RCL    079  RCL    Find a,Ta      151  ÷        186  1
010  R/S     045  13     080  02     115  ÷         152  RCL      187  Lbl
011  STO     046  ÷      081  ÷      116  RCL       153  09       188  E
012  Ind     047  RCL    082  2      117  15        154  ÷        189  RCL
013  00      048  14     083  +      118  ÷         155  7        190  21
014  RCL     049  =      084  RCL    119  RCL       156  7        191  -
015  00      050  STO    085  06     120  09        157  8        192  RCL
016  INV     051  15     086  =      121  =         158  =        193  20
017  x=t     052  D      087  STO    122  STO       159  ÷        194  =
018  006     053  Lbl    088  16     123  18        160  RCL      195  ×
             054  B      089  ×      124  ×         161  19       196  RCL
Mass Rate    055  RCL    090  2      125  RCL       162  +        197  22
019  Lbl     056  11     091  ÷      126  03        163  RCL      198  +
020  A       057  STO    092  RCL    127  ×         164  08       199  RCL
021  0       058  15     093  02     128  ·         165  =        200  20
022  x→t     059  D      094  =      129  0         166  STO      201  =
023  RCL     060  Lbl    095  +      130  0         167  20       202  STO
024  12      061  C      096  (      131  5                       203  21
025  x=t     062  ·      097  x²     132  2         Profile       204  Pause
026  B       063  0      098  -      133  8         168  8        205  Pause
027  RCL     064  1      099  1      134  =         169  STO      206  Dsz
028  13      065  4      100  )      135  STO       170  00       207  0
029  x=t     066  5      101  √x     136  19        171  1/x      208  E
030  C       067  9      102  =      137  RCL       172  ×        209  INV
031  ·       068  ×      103  lnx    138  10        173  RCL      210  Fix
032  1       069  RCL    104  1/x    139  ÷         174  19       211  R/S
033  1       070  11     105  ×      140  1         175  =        212  A
```

FIG. 1 Oil and gas temperature profile program for TI-58/59.

tively, data can be manually entered into storage registers 01 to 14 and modified as desired and the program run by keying A.

The flow rate (STO 11) is expressed in Mb/d for oil and MMSCFD for gas (M = 1000). If the pressure base (STO 13) is zero, the program assumes that the fluid in the pipeline is oil. If the pressure base (STO 13) is nonzero, the program assumes that the fluid is gas.

The program uses a supercompressibility of 0.997 at base conditions for gas measurements. If gravity (STO 12) is zero, the program assumes that flow rate (STO 11) has been expressed in MMlb/h. In this case the program passes over calculations of mass flow rate.

ature. Intermediate results of interest are stored in registers 15 to 20 and can be manually inspected if desired at the completion of the program.

The program can be connected interactively with a pressure-loss program so that the relationship between temperature and pressure loss can be accounted for automatically if that is required.

Joule–Thomson Effect

Early experiments by Thomson (Kelvin) measured the rate of change of temperature with pressure in adiabatic pipeline flow. The phenomenon has been called the Joule–Thomson effect and is the net result of cooling due to isentropic expansion plus heating due to frictional dissipation of pressure energy within the fluid.

For gases, cooling due to isentropic expansion usually exceeds heating due to friction, and there is a net temperature loss along the pipe. For liquids, on the other hand, frictional heating usually exceeds expansion cooling, and there is an increase in temperature as it flows.

Empirical data in the form of charts, tables, or equations of state are normally used to evaluate the Joule–Thomson coefficient for each fluid at the pressure and temperature of interest. Figure 2 shows how the Joule–Thomson coefficient (J) can vary over a range of pressures and temperatures for a light hydrocarbon gas.

The temperature of flowing fluids in uninsulated pipelines which are not horizontal is also affected by the fluid's heat capacity. Figure 3 shows the specific heat at constant pressure for the same light hydrocarbon gas. Calculations were made using Starling's BWRS data [2].

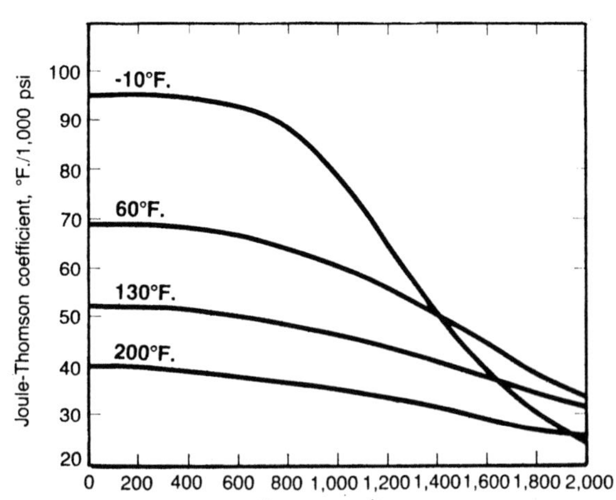

FIG. 2 Joule–Thomson coefficient for a light hydrocarbon gas.

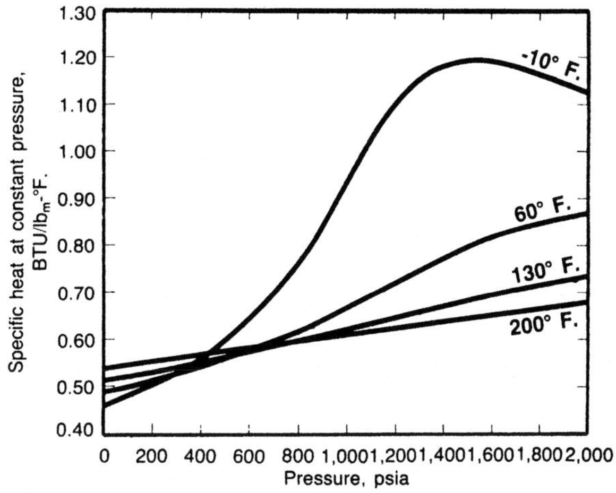

FIG. 3 Specific heats at constant pressure for a light hydrocarbon gas.

It is important when making calculations of flowing temperature to base all gas-property evaluations on the same technique to assure thermodynamic consistency. For example, if temperature loss along the pipeline is calculated using Joule–Thomson data from one source, and the temperature rise through the compressors is evaluated using temperature rise data from a second source which is inconsistent with the first, then significant error can develop in evaluating the temperature cascade from station to station along the pipeline.

Ground Heat Flow

It has been shown [3] that the accuracy of the following equations for predicting heat flow from buried pipe is acceptable for engineering design of buried pipelines:

$$q = s(T_p - T_g) \tag{2}$$

where

$$s = 2\pi K/\cosh^{-1}(2h/D) \tag{2a}$$

Not all hand-held calculators have the inverse hyperbolic cosine function built in. In such cases it can be evaluated by using

$$\cosh^{-1}(x) = \ln [x + (x^2 - 1)^{1/2}] \tag{3}$$

or

$$\cosh^{-1}(x) \approx \ln (2x) \tag{3a}$$

Other heat-transfer equations have been developed [1] for cases where there are two or more pipes in close proximity in the same trench or where pipes are insulated or heat-traced, etc. Experience has shown that, if good predictions of the heat-transfer factor can be made, good predictions of flowing temperatures are then possible.

Diameter Effects

Figure 4 shows typical temperature profiles for 12, 24, 36, and 48-in. pipelines carrying a light hydrocarbon gas. The asymptotic temperature (T_a) is approached exponentially at large distances along the pipe. It is equal to the

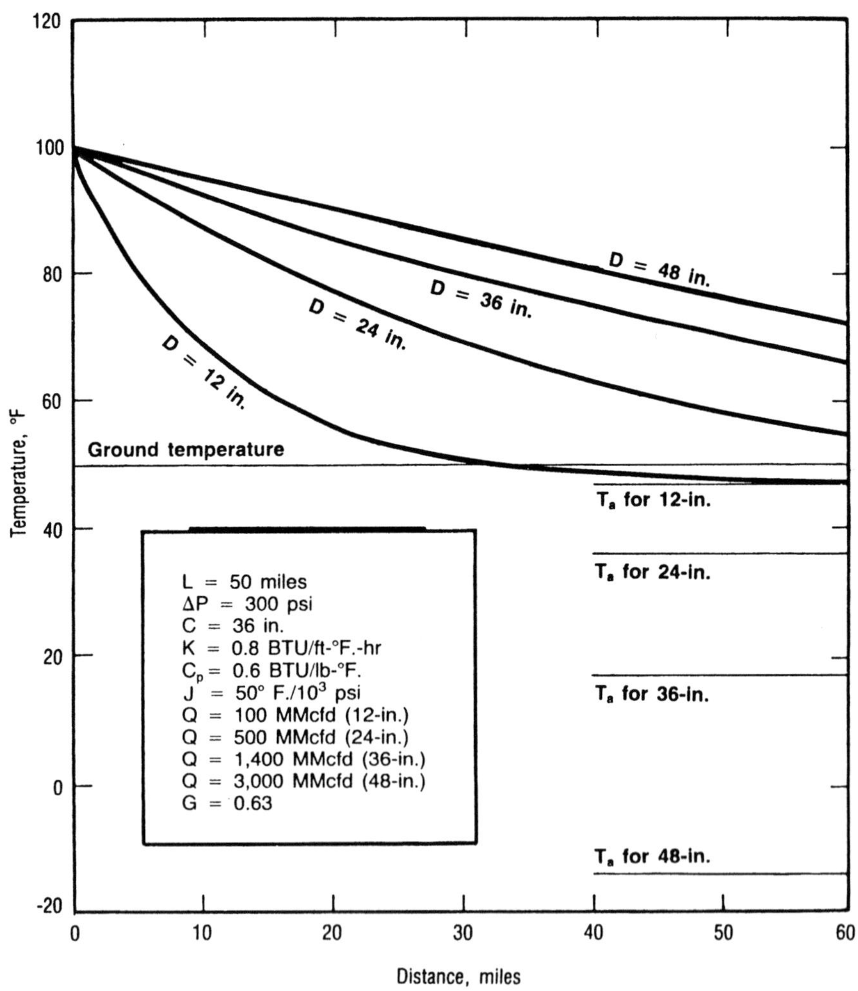

FIG. 4 Typical temperature profiles for 12, 24, 36, and 48 in. gas pipelines.

ground temperature minus a correction term which takes account of elevation gain and Joule–Thomson cooling. For small-diameter pipelines, the correction term is small.

Figure 4 shows that pipes smaller than about 16-in. diameter have asymptotic temperatures very nearly equal to ground temperature. Therefore, for gas pipelines with diameters less than about 16 in., good results can be obtained by assuming that the asymptotic temperature is equal to the ground temperature.

Although oil is normally an order of magnitude denser than gas, it flows more slowly and has a small Joule–Thomson coefficient. A similar conclusion can, therefore, be drawn for oil pipelines.

The heat-flow ratio (a) is the heat flow from the pipe per degree Fahrenheit divided by the heat capacity of the flowing stream. For small-diameter pipes, where heat flow from the pipe is large in comparison to the heat capacity of the flowing stream, the heat-flow ratio (a) is large, and the pipe approaches its asymptotic temperature in a short length. For large-diameter pipes, the converse is true.

This phenomenon can be seen in Fig. 4. The 12-in. pipe has a heat-flow ratio of $16.2/10^6$ ft, while the 48-in. pipe, which carries 30 times more gas, has a heat-flow ratio of only $0.89/10^6$ ft. The 12-in. line is very nearly at ground temperature after 20 miles while the 48-in. pipe is still more than 20°F above ground temperature after 50 miles.

Since 48-in. pipelines with high flowing volumes do not cool off adequately between compressor stations, there is a possibility, unless aftercoolers are built, that flowing temperatures will cascade from station to station and become unacceptably high.

More than half the temperature loss in the 48-in. pipe is due to Joule–Thomson cooling rather than due to heat loss to the ground. The gas temperature in a 48-in. pipeline is, therefore, relatively independent of ground temperature.

It is for these reasons that flowing temperatures in large-diameter pipelines are difficult to estimate and need to be calculated in some detail during design.

Accuracy Studies

Equations (1) and (2) have been used with good results in both hand and computer calculations for predicting temperatures in a number of different pipelines. Table 1 shows four cases that are typical of 30 and 36 in. gas pipelines. The asymptotic temperature is 10 to 20°F below ground temperature in the four cases. Predictions were within 2°F of measured values in these cases.

Not all predictions of downstream temperature agree with field measurements as well as these. Figure 5 compares predicted and observed temperature losses along 30 and 36 in. gas pipelines.

TABLE 1 Four Temperature-Profile Studies

			Pipeline 1		Pipeline 2	
No.	Parameter	Units	Jaunary	July	March	September
01	Upstream temperature	°F	118	119	92	110
02	Outside diameter	in.	30	30	36	36
03	Section length	miles	39.5	39.5	63.4	63.4
04	Pressure loss	lb/in.^{2}	258	211	224	192
05	Elevation gain	ft	0	0	0	0
06	Depth of cover	in.	35	35	36	36
07	Ground conductivity	Btu/ft·°F·h	0.8	0.8	0.8	0.8
08	Ground temperature	°F	26	54	30	56
09	Fluid specific heat	Btu/lb$_m$·°F	0.60	0.59	0.62	0.61
10	Joule–Thomson coefficient	°F/10^3 lb/in.^{2}	49	47	52	48
11	Volumetric flow	MMSCFD	766	703	982	927
12	Specific gravity	air = 1	0.593	0.593	0.593	0.593
13	Pressure base	lb/in.^{2}abs	14.7	14.7	14.7	14.7
14	Temperature base	°R	520	520	520	520
15	Mass flow rate	MMlb/h	1.44	1.32	1.85	1.74
16	Depth to pipe center	in.	50	50	54	54
17	Heat transfer factor	Btu/ft·°F·h	2.68	2.68	2.85	2.85
18	Heat flow ratio	/10^6 ft	3.10	3.44	2.49	2.68
19	Temperature exponent	—	0.647	0.717	0.834	0.898
20	Asymptotic temperature	°F	6	40	16	46
21	Average temperature	°F	89	96	68	88
22	Predicted temperature	°F	65	79	49	72
23	Measured temperature	°F	65	79	51	71

Sections varied in length from 37 to 175 miles and flows varied from 600 to 1000 MMCFD. Gas properties such as the Joule–Thomson coefficient and specific heat were evaluated using an equation of state at the average flowing temperature and pressure for each case. A ground conductivity of 0.8 Btu/ft·°F·h was used in all cases, and data were obtained in all months of the year.

About 70% of the temperature loss predictions were within 10% of the measured values. When it is realized that ground conductivity is difficult to predict with an accuracy better than 10 or 20% and that ground temperature along the right-of-way can only be estimated within about 5°F, it becomes clear that temperature loss along a pipeline cannot be predicted in advance with greater precision than about 10%.

For small-diameter pipes, temperature predictions are affected mainly by the ground temperature, but for large-diameter pipes the ground conductivity, elevation gain, and Joule–Thomson effect are more important.

After the pipeline has been built, it is possible to calibrate Eqs. (1) and (2) by adjusting ground conductivity and ground temperature so that predictions agree with observed downstream temperatures throughout the year. When this has been done, it is possible to predict temperatures at different flow rates and in other pipelines along the same general right-of-way with much greater precision.

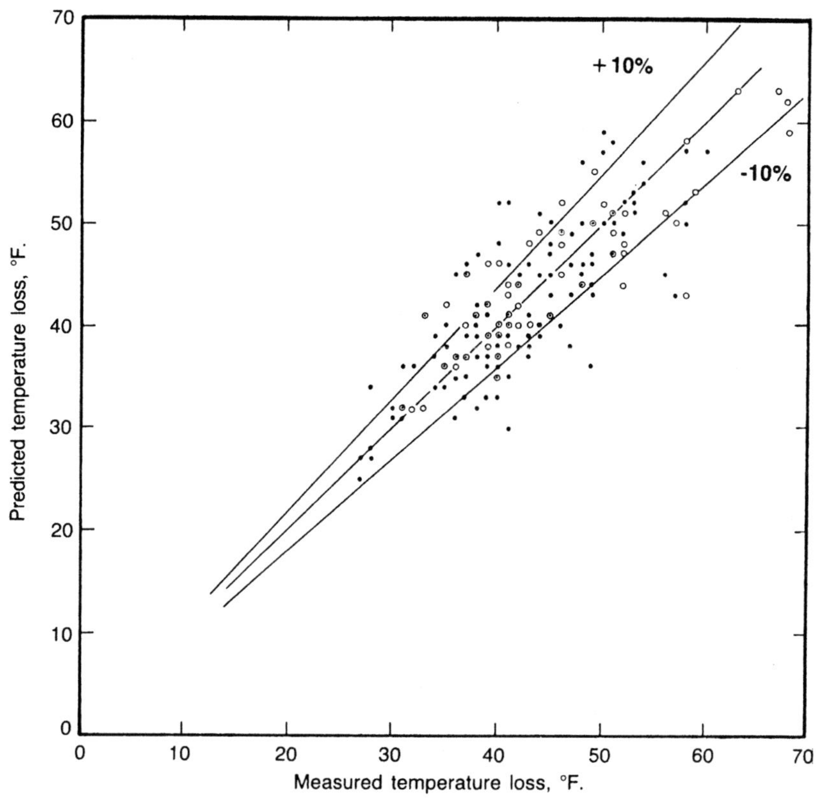

FIG. 5 Comparison between measured and predicted temperature loss in 30 and 36 in. gas pipelines.

Conclusion

The pipeline temperature-profile equation presented in this article is based on fundamental thermodynamic principles and has been used with success for predicting temperatures in a number of different pipelines. The major factor which defeats precise predictions during design, before the pipeline goes into operation, is the accuracy of geothermal parameters such as ground conductivity and ground temperature.

The accuracy of the temperature-profile equation exceeds the accuracy with which input parameters are known. It is, therefore, sufficiently precise for engineering purposes.

This material, which appeared in *Oil & Gas Journal,* pp. 65 ff., March 16, 1981, copyright © 1981 by Pennwell Publishing Co., Tulsa, Oklahoma 74121, has been updated by special permission.

Symbols

a	heat flow ratio per unit length/ft
C	depth to top of pipe (in.)
C_P	fluid specific heat (Btu/lb$_m$·°F)
D	pipe outside diameter (in.)
G	gas (air = 1) or oil (water = 1) gravity
h	depth of burial to pipe center line (in.)
J	fluid Joule–Thomson coefficient [°F/(lb/in.^{2})]
j	Joule's constant (778 ft·lb$_f$/Btu)
K	soil thermal conductivity (Btu/ft·°F·h)
L	length of pipe section (ft)
m	fluid mass flow rate (lb$_m$/h)
P_b	pressure base (lb/in.^{2}abs)
ΔP	fluid pressure loss (lb/in.^{2})
q	heat flow from pipe (Btu/h·ft)
Q	fluid flow rate (MMCFD or Mbbl/d)
s	heat transfer factor (Btu/ft·°F·h)
T_1	upstream temperature (°F)
T_2	downstream temperature (°F)
T_a	asymptotic temperature (°F)
T_b	temperature base (°R)
T_g	ground temperature (°F)
T_p	pipe temperature (°F)
Δy	change in elevation (ft)

References

1. G. G. King, *Geothermal Design of Buried Pipelines,* Presented at ASME Energy Conference and Exhibition, New Orleans, February 1980.
2. K. E. Starling, *Fluid Thermodynamic Properties for Light Hydrocarbon Systems,* Gulf, Houston, Texas, 1973.
3. G. G. King, "Cooling Arctic Pipelines Can Increase Flow, Avoid Thaw," *Oil Gas J.,* August 15, 1977.

GRAEME G. KING

Tracing Costs

For the engineer faced with recommending tracing, here are data and a method for estimating installed and operating costs of steam and common electric systems.

Until the early 1970s, pipes in process plants were traced almost exclusively with steam. Since then, higher energy and labor costs, and improvements in electric heat tracers, have led to increased application of electric heat tracing.*

Steam tracing's chief advantage is its high reliability. Leaks and failed traps may waste a lot of energy, but these problems do not prevent the tracing from maintaining a line at or above the desired temperature. Steam tracing can heat a line quickly, particularly when thermal contact between the pipe and tracer is enhanced by heat-transfer cement.

Its major disadvantages are high costs and the inability to closely control pipe temperature. Installation is labor-intensive and often represents 70–80% of total system cost. Operating cost is high because of the difficulty of controlling temperature despite advances in temperature-sensitive, self-actuated valves. Energy lost from failed steam traps, and the maintenance required to replace such traps, add to operating cost.

Types of Electrical Tracing

The three most common types of electrical tracing are mineral-insulated cable, zone heaters, and self-regulating heaters. The first has been available for over 50 years, the others for about two decades.

Mineral-insulated cable consists of an outer sheath (of copper, stainless steel, Inconel, or some other metal), insulation of magnesium oxide (hence the name), and one or more heating conductors (Fig. 1a).

Its major advantage is its exceptionally wide temperature range (up to 1500°F with special alloys). If a line is to be maintained above 500°F, there are few practical alternatives, this being beyond the range of existing zone and self-regulating heaters. Such temperatures could be maintained with steam tracing but would require high-pressure steam (above 600 lb/in.^{2} for 500°F, which is not practical). Another advantage is that it resists impact well.

Mineral-insulated cable's chief disadvantage is that it cannot be cut to length or spliced in the field. This makes design, installation, and maintenance more difficult with it than with other types of electric tracers. Also, because it is a series heater, an entire circuit is lost if any section fails.

The zone heater consists of two insulated bus wires wrapped with a thin (38–40 American Wire Gage) Nichrome heating wire and covered with polymer insulation. The heating wire is connected to alternate bus wires at nodes every 12 to 48 in. The distance between nodes constitutes a heating zone. A metallic braid or outer jacket, or both, is usually optional (Fig. 1b).

The main advantage of this tracer is that it can be cut to length in the field. However, one must be aware of node location when cutting, because

*Only the more common types of electric tracing—mineral-insulated cable, zone heaters, and self-regulating heaters—are discussed. Less frequently used types of tracing, such as skin-effect current tracing, circulating-liquid systems, and impedance heating, are not considered.

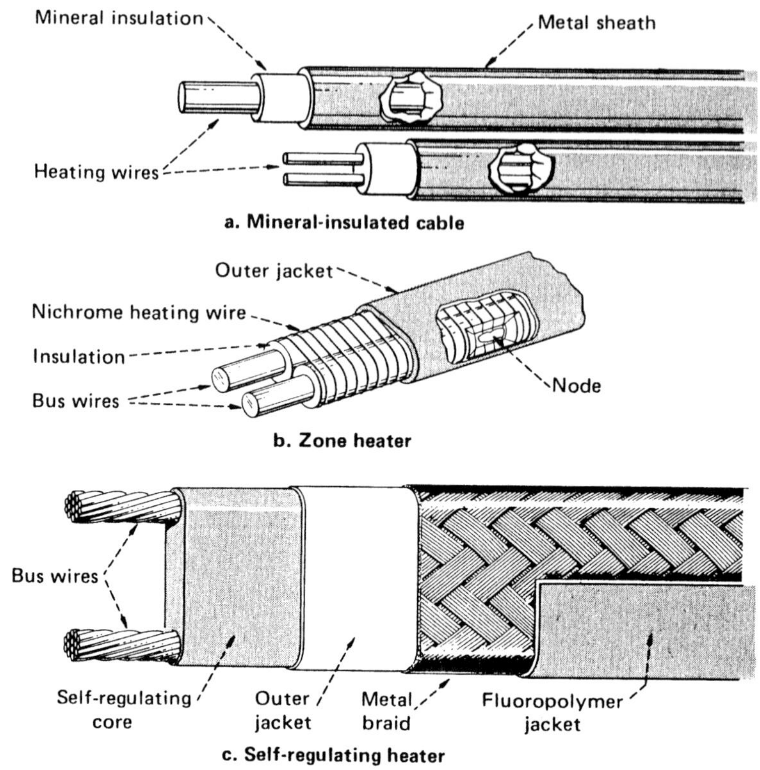

FIG. 1 Typical construction of three common types of electrical tracing for process piping.

there will be no heating from the cut to the nearest node. Another advantage is that it operates on standard voltages.

Because it is jacketed with polymer, the zone heater is more flexible than the mineral-insulated cable. However, it usually is not recommended for service above 400°F. Also, it is more likely to fail from impact or abuse, because of its thin heating element, than other types of electric tracers.

The self-regulating heater consists of two current-carrying bus wires connected by a conductive polymer over which a polymeric insulator is extruded (Fig. 1c). A metallic braid or fluoropolymeric outer jacket, or both, is optional.

This heater operates on standard voltages and may be cut to length in the field without concern about nodes (because it has none). However, its main feature is its self-regulation. The resistance of the conductive polymer core between the bus wires increases as the temperature of the core rises. This causes the power output to decrease with rising pipe temperature, and vice versa.

Power output changes independently at each point on the heater. If the temperature of a section of pipe drops off, only the tracing there will respond by decreasing its resistance and increasing its heat output. This helps prevent low-temperature failures.

Still another important feature of this heater is that it cannot destroy itself with its own heat output, as can a constant-wattage heater. When the latter is trapped inside the insulation or crossed over itself, it will continue to generate a constant amount of heat and may reach a temperature that destroys it. In such a situation, the self-regulating heater will automatically cut back its heat output.

The self-regulating heater's major disadvantage is its limited temperature range: its maximum heat-maintenance temperature being 300°F and its maximum intermittent exposure temperature being 420°F.

Evaluating System Requirements

The first step in designing a heat-tracing system is to calculate the heat loss from the insulated pipe. This can be done via

$$Q = \frac{(T_p - T_a)}{(1/\pi D_2 h_o) + (\ln (D_2/D_1)/2\pi K)} \tag{1}$$

Because it is necessary to design for maximum heat loss, the ambient temperature, T_a, in Eq. (1) should be the minimum expected, and the heat-transfer coefficient, h_o, should be chosen for high wind. The heat-transfer coefficient is usually calculated from an empirical relationship between the Nusselt number and the Reynolds number [2]. Figure 2, which was calculated from this relationship, simplifies finding h_o.

The tracing selected must, of course, be able to replace the maximum heat loss. For example, the heat loss from Line 5 (Table 1) is, via Eq. (1), 30 Btu/h/ft, or 8.8 W/ft. A constant-wattage heater must provide this output to supply the required heat. A self-regulating heater capable of at least 8.8

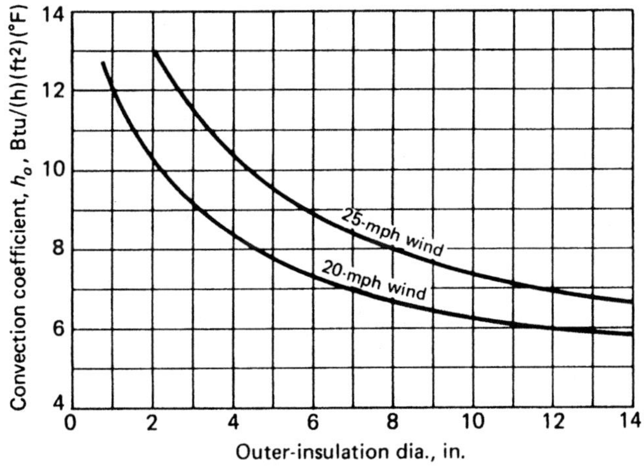

FIG. 2 Forced-convection coefficient for flow normal to cylindrical surface.

TABLE 1 Example List of Lines to Be Traced, with Data

Line No.	Pipe Diameter (in.)	Insulation Thickness (in.)	Pipe Length (ft)	Set Temperature (°F)	Maximum Exposure Temperature (°F)
1	4	2	120	50	100
2	4	2	100	50	100
3	3	1.5	50	140	150
4	4	2	170	80	90
5	3	1.5	75	120	200
6	3	1.5	125	120	200
7	2	2	100	500	700

W/ft at 120°F would be adequate, because the power output of this type of heater varies as a function of pipe temperature.

The power output from a steam tracer is calculated from Eq. (2) [1]:

$$Q = C_t(T_s - T_p) \tag{2}$$

Values for C_t are furnished by Kohli [1].

For ½-in. copper tubing without heat-transfer cement, C_t = 0.393 Btu/°F/ft. Using 50 lb/in.^{2} (298°F) steam would result in a heat output of 70 Btu/ft/h on a 120°F pipe, via Eq. (2). This could be decreased to 53 Btu/ft/h with ⅜-in. tracing, C_t = 0.295 Btu/(h)(°F)(ft), and reduced still further through the use of lower-pressure steam; however, both of these changes would decrease the allowable circuit length and increase installed costs.

One steam tracer can provide much more heat than necessary. This is an advantage in that one tracer can handle a wide range of pipe sizes and set temperatures.

Estimating Installed Costs*

The total installed cost of steam tracing is usually 50–150% greater than that for equivalent electric tracing, principally because of the labor required to install the steam-supply, tracing, and condensate-return lines.

Use of the following technique will produce a preliminary-type estimate (±25%) of the costs of a steam or electric tracing system.

The lines to be traced, along with the data necessary to calculate the heat loss from each, should be listed (as in Table 1).

Estimates of the average circuit length of both the steam and electric systems are necessary for analyzing installed costs. This length is a function of the piping layout and the allowable maximum circuit length of the tracer. For steam tracing, this maximum is usually 150 ft for ½-in. copper tubing and 60 ft for the ⅜-in. size. For electric tracers, the maximum typically varies from 200 to 800 ft, depending on the type of tracer, power output required, and operating voltage.

*These are 1982 costs and should be updated by using current cost indices.

The average circuit length of long, straight piping will approach the maximum allowable circuit length of the tracer selected. However, with increasing piping complexity, other considerations (such as the length and position of each line) will limit the actual tracer circuit-lengths.

The best way to estimate the average circuit length is to choose a number of lines from the line list and actually design the tracer circuits for these lines. The lines chosen should approximate the entire system in length and complexity. (See the article "Steam, Tracing," in this *Encyclopedia* for the design of steam-tracing systems.)

Steam and Condensate Lines

Steam-tracing installed costs strongly depend on the lengths of the steam-supply and condensate-return systems. Accurate estimates of these lengths are the keys to a reliable installed-cost estimate. For estimating purposes, the piping connecting the mains and the tracing should be handled in two parts:

1. *The piping required to bring the steam and condensate mains to within 20–40 ft of the lines to be traced.* The lengths of these steam-distribution and condensate-collection lines can be estimated by sketching them on the plan view. If the process area consists of several levels (there are three in the example), the piping that runs vertically, and that on all the levels, must be considered. Items a and g in Table 2 represent the costs of these lines. The insulation of the lines is also a relevant cost (Items b and h in Table 2). The required diameters of the steam-distribution lines can be estimated from Table 3. Condensate-collection lines are generally somewhat smaller than steam-distribution lines.

2. *The steam-supply lines, which take the steam from the distribution lines to the tracing, and the condensate-return lines, which convey the condensate from the steam traps to the collection lines.* The average length of these lines is estimated by designing them to and from a number of tracing circuits. (The tracing circuits designed previously to estimate the average circuit length may be used for this purpose.) The costs of these lines for the example are noted in Items c and i in Table 2. Their insulation (Items d and j) must also be considered.

After the piping lengths between the mains and the tracing have been estimated, the total average cost of tracing per foot of traced pipe can be determined from standard data in Table 2. Of course, installed costs will vary with labor rate and productivity, supervisory efficiency, material cost, and piping complexity.

Installed Cost of Electric Tracing

Estimating the installed cost of an electric system is easier than estimating that of a steam system, because fewer parts are involved. The cost of the

TABLE 2 Unit Costs and Method for Estimating the Installed Cost of Steam Tracing[a]

Description and Cost	Tracing Served (ft)	Cost/ft of Tracing ($)
a. Steam-distribution lines:		
300 ft, bottom level		
600 ft, 2nd level		
600 ft, 3rd level		
1,500 ft × $25/ft = $37,500	6,000 (entire system)	6.25
b. Insulation of steam-distribution lines:		
1,500 ft × $10/ft = $15,000	6,000 (entire system)	2.50
c. Steam-supply lines:		
20 ft × $16/ft = $320	One circuit, 125 ft	2.56
d. Insulation of steam-supply lines:		
20 ft × $8/ft = $160	One circuit, 125 ft	1.28
e. Tracers (1/2-in. copper tubing with all fittings):		
$10/ft	1 ft	10.00
f. Steam-trap assemblies, including 1 trap, 2 valves and 6 ft of piping:		
$300 per assembly	One circuit, 125 ft	2.40
g. Condensate-collection system:		
1,500 ft × $20/ft = $30,000	6,000 (entire system)	5.00
h. Insulation of condensate-collection system:		
1,500 ft × $10/ft = $15,000	6,000 (entire system)	2.50
i. Condensate-return lines:		
20 ft × $16/ft = $320	One circuit, 125 ft	2.56
j. Insulation of condensate-return line:		
20 ft × $8/ft = $160	One circuit, 125 ft	1.28
Total cost/ft of tracing		$36.33

[a]The example system consists of 6,000 ft of process piping in an area of 100 ft by 200 ft and of three levels. Steam is available on the lowest level. The average circuit length of steam tracing is 125 ft. The maximum allowable circuit length is 150 ft.

tracer itself constitutes only a small part of the total cost of an electrical system (Table 4). The average distance from the electric-distribution panel to the tracing circuit (i.e., the average power-feeder cable length) is critical to installed cost, particularly in hazardous areas, where the wiring must be run in conduit at a cost as high as $10/ft; elsewhere, the power cable can

TABLE 3 Sizing Steam-Distribution Lines

No. of ½-in. Tracers	Line Size (in.)
1–2	$\frac{3}{4}$
3–5	1
6–15	1½
16–30	2
31–60	2½

TABLE 4 Installed Cost of Example Electric Tracing System[a]

Description and Cost	Tracing Served (ft)	Cost/ft of Tracing ($)
a. Power-supply panel: $8,000	6,000 (entire system)	1.33
b. Panel, 48 circuits: $5,000	6,000 (entire system)	0.83
c. Power supply to circuits: 300 ft × $10/ft = $3,000	One circuit, 250 ft	12.00
d. Thermostat (line-sensing): $250	One circuit, 250 ft	1.00
e. Electric tracer: $4.50/ft	1 ft	4.50
Total cost/ft		$19.66

[a]Example system consists of 6,000 ft of piping in an area 100 ft by 200 ft, three levels. Power is available 50 ft from area. Average circuit length is 250 ft, and maximum allowable length is 400 ft. Area is Class 1, Div. 2 (hazardous).

be run in trays at considerably less cost. This distance also is best estimated from a plan view.

If power and space are already available at a panel of an existing motor-control center, the cost of supplying power to the panel will be negligible. If, however, power has to be brought from a distance and transformed down to the proper voltage, supplying power can be the most expensive part of an electric system.

Installation time for electric tracing can vary from 4 h/100 ft to more than 10 h/100 ft, depending on piping complexity and type of tracer. Self-regulating heaters can be installed somewhat faster than zone heaters, because the installer need not be concerned with nodes when making connections, or with the tracer destroying itself if it is either caught inside the insulation or crossed over itself. Mineral-insulated cable takes the longest to install because it is more difficult to bend, cannot be crossed over itself, and each prefabricated length must be fitted precisely on each circuit (usually by spiraling the last 25%, because a safety factor is included in the length of each circuit).

The components of an electrical system are easily estimated. How the parts can be combined to obtain an average cost per foot of tracing is shown in Table 4.

Estimating Operating Costs

Operating costs—energy and maintenance—can be large enough to influence the selection of a system. The energy-cost tradeoff lies between a less expensive but less efficient form of energy (steam) and a more expensive but more efficient form (electricity).

Total operating cost is most affected by the desired traced-line temperatures, energy cost, and the percentage of time that fluid flows in the traced piping. When fluid is not flowing, the energy consumption of a steam tracer can be calculated by the simultaneous solution of Eqs. (1) and (2). However, this is more easily determined graphically, as in Fig. 3.

At an ambient temperature of 50°F, the equilibrium temperature is 210°F, and as Fig. 3 indicates, the tracer heat loss is 35 Btu/ft/h. If there is no flow 50% of the time, the heat loss will be 17.5 Btu/ft/h. When fluid is flowing at 130°F, the tracer heat loss can be calculated directly via Eq. (2): Q_t = [0.393 Btu/(ft)(h)(°F)] (298 − 130) = 66 Btu/h/ft. As there is flow 50% of the time, the energy consumption will be 33 Btu/ft/h.

When there is no flow in the traced line, the steam tracing transfers more energy than is necessary and keeps the line at a higher than desired temperature. When there is flow, the lower temperature of the line draws significantly more energy from the tracer. The energy lost from the steam-distribution and steam-supply lines can be calculated by means of Eq. (1), using the steam temperature as the pipe temperature.

The energy use of operating traps can vary from 700 to 2000 Btu/h, depending on the steam pressure and type of trap. Traps for steam tracing usually operate at the low end of this scale; for the example, therefore: 1000 Btu/h for one trap per 125 ft of tracing = 8 Btu/h/ft.

A failed trap can waste 25,000 Btu/h. The percentage of time that a trap is likely to be passing steam is a critical variable. In a good maintenance program, traps will be checked at least every 3 months and replaced as necessary. At a typical failure rate of 3%/mo, failures will average about 4½%, with the energy loss = 0.045(25,000 Btu/h/125 ft) = 9.0 Btu/ft/h.

If steam leaks are reported and promptly fixed, the waste from them can be kept to a minimum. Annual maintenance cost for the example steam tracing system is determined as follows:

1. Check traps four times per year (10 min each, a trap at every 125 ft of tracing): (40/60)($25/h/125 ft) = $0.13/ft/yr.

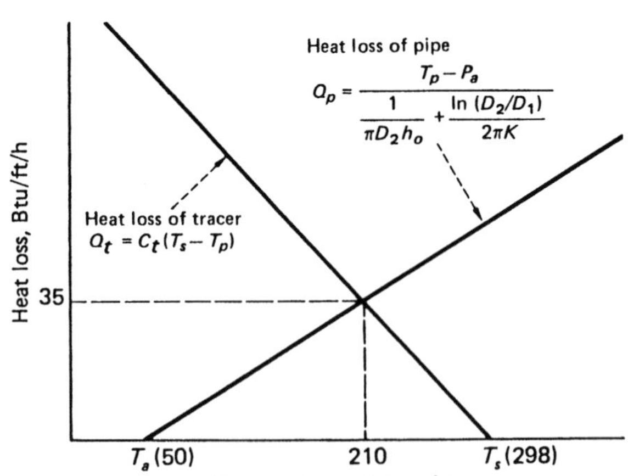

FIG. 3 Steam-tracer loss depends on piping flow time.

2. Replace trap (in-line replaceable traps are assumed) every third year: $40/(125 × 3) = $0.11/ft/yr.
3. Repair leaks and blow strainers (one leak every 2 years): $40/(125 × 2) = $0.16/ft/yr.
4. Total maintenance cost per foot of tracing (adding Items 1, 2, and 3) = $0.40 yr.

For the example steam tracing system, the calculated energy transfer, in Btu/h per foot of traced pipe, is as follows: traced line, no flow—17.5; traced line, with flow—33.0; steam-distribution lines—8.8; and steam-supply lines—5.0; steam traps, operating—8.0; and steam traps, failed—9.0; for a total of 81.3 Btu/ft/h. The annual cost of 50 lb/in.^{2} steam (latent heat of vaporization = 912 Btu/lb) per foot of traced pipe is calculated as follows: [(81.3 Btu/ft/h)/(912 Btu/lb)] ($5/1000 lb steam) (8800 h/yr) = $3.92/ft/yr, which is about 11% of the total installed cost.

Therefore, total operating cost = $4.32/ft/yr.

Note the dependency between maintenance and energy costs. If, for example, trap inspections were cut to one per year, the maintenance cost would be reduced by $0.10/ft of tracing but the energy cost would increase by $1.30/ft. A trap inspection program is usually cost-effective.

Operating Cost of Electric Tracing

The energy used by an electric tracing system can be calculated directly from Eq. (1). The pipe temperature will be the set temperature because the thermostat will cycle the heater to maintain this temperature.* The thermostat should be set slightly below the typical flowing-fluid temperature so that the heater will be off when fluid is flowing.

Via Eq. (1), the energy use for the example is 15.2 Btu/ft/h, or 4.44 W/ft. Because fluid is flowing half the time, the actual use is 2.22 W/ft. Therefore, the energy operating cost per foot of tracing is: (2.22 W/ft)($0.04/1,000 W/h)(8800 h/yr) = $0.78/yr/ft.

Maintenance cost per foot of tracing (checking thermostat and meggering circuit once a year) is: 0.5 h($25/h/200-ft circuit) = $0.06/yr/ft.

Therefore, total operating cost = $0.84/ft/yr. The $0.84/ft/yr for electric tracing vs the $4.32/ft/yr for steam tracing is typical for the example pipe size and maintenance temperature.

To compare total operating costs, an analysis must be made for each temperature range. The operating-cost advantage of electric tracing will be

*Line-sensing thermostats are not always the preferred method of temperature control, particularly for freeze protection. Often, an ambient-sensing thermostat is used to turn off an entire panel when the ambient temperature exceeds 40°F. This can offer a considerable installed-cost saving over using a line-sensing thermostat for each circuit. Self-regulating heaters, which decrease their power drawing as piping becomes warmer, can keep extra energy consumption to a minimum.

less at higher set temperatures. Pipes less than 6 in. diameter are usually kept heated at much less cost with electric tracing. For holding pipes 12 in. and up above 200°F, steam tracing may be cheaper to operate, depending on energy costs and the percentage of time that fluid is flowing.

Sometimes the availability of excess low-pressure steam is given as justification for using steam tracing. This is not always valid. Consider Tables 2 and 4: the steam system costs $16.67/ft more to install than the electric system. Assuming no cost for steam, and maintenance cost reduced by 50% (because of little concern about wasting steam), the operating cost advantage of the steam system becomes $0.64/ft/yr (i.e., $0.84 − $0.20). This represents a pretax return of only 3.8% on the extra money spent to install the steam system.

Added in Proof. Additional work done since 1982 [3, 4] has confirmed that electric tracing is considerably less expensive to install and to operate than steam tracing. Installed cost ratios (steam tracing costs/electric tracing cost) as high as 5.6 have been reported, although the most definitive study showed a cost ratio of 1.4 for a freeze protection system and 2.7 for a process temperature system.

A computer program is now available [3] to compare the costs of steam and electric tracing systems using the labor costs, energy costs, material costs, and system geometry of the user. The accuracy of this program is confirmed by the detailed engineering study done in Ref. 4 by a major tracing user.

Symbols

C_t heat-transfer coefficient from steam tracer to traced pipe [Btu/(h)(ft)(°F)]
D_1 inside diameter of insulation (ft)
D_2 outside diameter of insulation (ft)
h_o heat-transfer coefficient from outside of insulation to air [Btu/(h)(ft²)(°F)]
K insulation thermal conductivity [Btu-in./(h)(ft²)(°F)]
Q_p heat flow from pipe (Btu/ft/h)
Q_t heat flow from tracer (Btu/ft/h)
T_a temperature of air (°F)
T_p temperature of pipe (°F)
T_s temperature of steam (°F)

References

1. I. P. Kohle, "Steam Tracing of Pipelines," *Chem. Eng.*, p. 156 (March 26, 1979).
2. E. R. G. Eckert, *Heat and Mass Transfer,* McGraw-Hill, New York, 1963.

3. C. Sandberg, "Heat Tracing: Steam or Electric," *Hydrocarbon Process.*, March 1989.
4. C. J. Erickson, *A Study of Steam versus Electrical Pipeline Heating Costs on a Typical Petro-Chemical Plant Project* (Paper PCIC-90-02), Presented at the IEEE Conference, Houston, Texas, September 1990.

JOSEPH T. LONSDALE
JERRY E. MUNDY

Vacuum Considerations

Piping systems in petrochemical plants and refineries can be subjected to rapid external depressurization. Sound engineering practice prescribes that stability of pressure vessels and piping be verified at full vacuum condition regardless of what the actual depressurization value may be.

ASME/ANSI codes require iterative calculations as well as the use of graphs. With systems requiring many checks, this method is time-consuming, expensive, tedious, and subject to mistakes. The method presented here is much faster without losing calculation accuracy, and also takes into proper account the geometrical and physical characteristics of the piping.

Traditional ANSI/ASME Calculation Procedure [1, 2]

First, a trial thickness is assumed. The allowable external pressure is then determined for a pipe with the assumed thickness and compared with the atmospheric pressure (assumed at 15 lb/in.^{2}). The assumed thickness is then modified accordingly until the allowable pressure is reasonably greater than 15. Except when the D_o/t_o ratio is lower than 10 (in this case a different analytical procedure applies), this trial-and-error calculation involves:

Reading on a log-log diagram the A factor, which is a function of the L/D_o and D_o/t ratios, thus taking into account the geometrical characteristics of the system

Reading on a log-log diagram factor B, a function of factor A and of design temperature, T, which takes into account the mechanical characteristics of the selected material. Materials have been divided into homogeneous groups, for each of which a diagram is made available.

Calculation of allowable external pressure then is made using the formula [1]

$$Pa = 4B/(3D_o/t) \tag{1}$$

New Method

For piping design, the following hypotheses are assumed:

Nominal pipe diameter within ½ to 24-in. NPS range
L/D_o ratio equal to or greater than 50. This hypothesis is justified because it is sufficiently realistic, reasonably conservative, and the piping layouts are not yet available at this stage and, therefore, the L value cannot be estimated.

The following consequences derive from the above hypothesis:

The A values always fall within the linear part of the B diagram
The B value can be analytically evaluated from the A value, i.e., $B = AE/2$
Groups of materials headed in separate B diagrams but with the same modulus can also be grouped
The D_o/t ratio is always greater than 10

On the basis of the above hypothesis and findings, Eq. (1) can be rewritten as

$$Pa = 2AE/(3D_o/t) \tag{2}$$

where Pa, taken equal to 15, is a function of the D_o/t ratio only.

Using a pocket computer, the above formula has been solved for t by a trial-and-error routine for groups of materials most frequently used.

The t values thus evaluated have been plotted versus pipe nominal diameters and design temperatures. These graphs have been inserted in the upper left part of a four-quadrant nomograph (Figs. 1 to 3). In the other three quadrants, corrosion allowance, minus fabrication tolerances, and schedule lines have been plotted in such a way that determination of the required minimum commercial pipe thickness is made quick and easy by means of a graphical method. An example is given in Fig. 1: An ASTM A 335 GRP12 18-in. pipe at 900°F design temperature with a corrosion allowance of 3⁄16-in. and a minus fabrication tolerance of 12.5% requires a Schedule 30 thickness at full vacuum condition.

In designing the presented diagrams, an additional result was obtained. For a given design temperature and material group, maximum allowable internal pressure for a pipe with thickness equal to design thickness at full vacuum condition varies only slightly with its nominal diameter variation. As a consequence, the check at full vacuum condition is no longer needed at design pressures equal to or higher than some typical values only depending on pipe material.

In Table 1 the minimum values of these pressures are shown for the considered diameter range. These values have been calculated considering

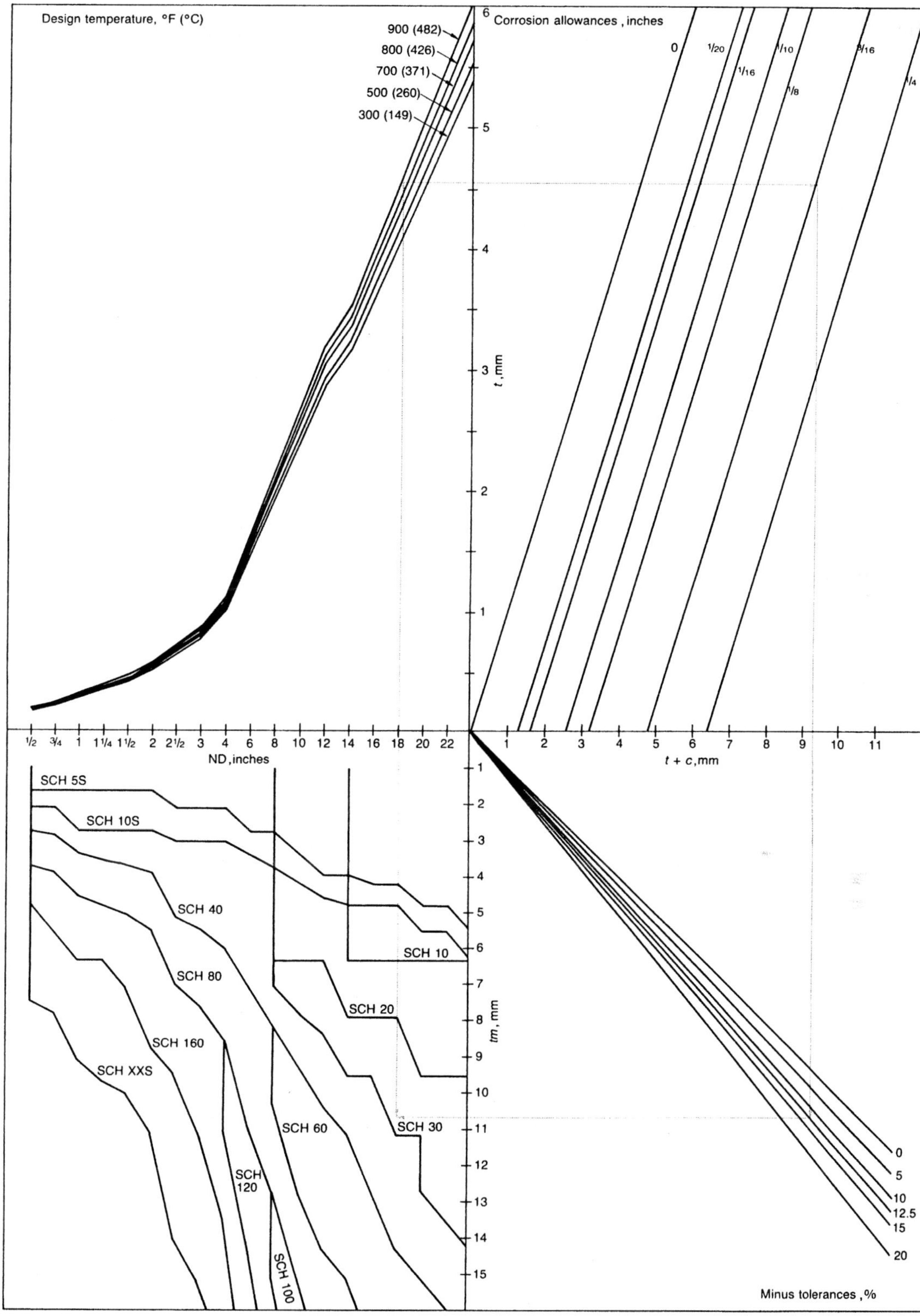

FIG. 1 Nomograph for carbon steel, low alloy steel, and ferritic stanless steel AISI types 405 and 410.

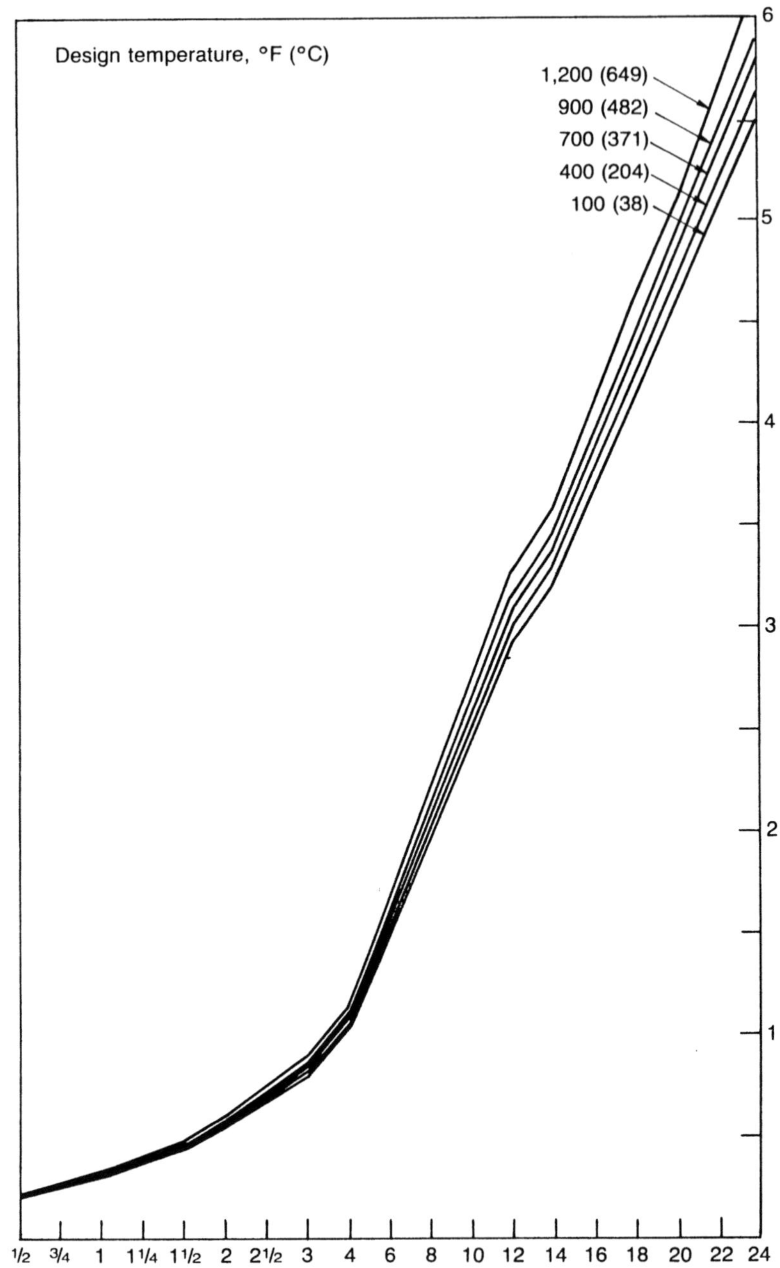

FIG. 2 Design temperature substitution for austenitic stainless steel AISI.

allowable stress at ambient temperature and refer to the strongest seamless material of the group. The importance of the table is that, for a pipe (to be designed/checked at full vacuum condition) with an internal design pressure higher than those presented, the design thickness also satisfies full vacuum requirements at any temperature.

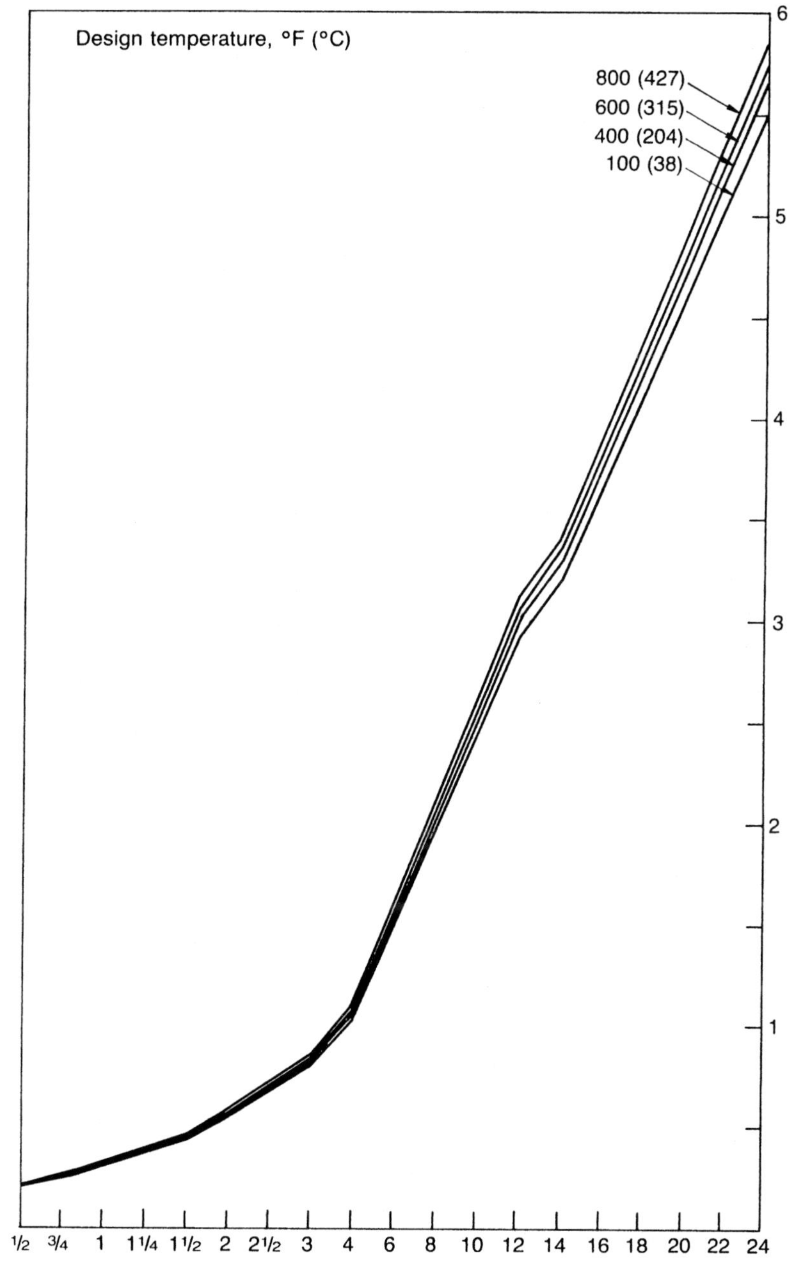

FIG. 3 Design temperature substitution for low carbon austenitic stainless steel AISI.

Conclusion

Using these diagrams allows very good approximation, at least as good as the traditional one. Its advantage is that no iterations are needed and, above all, all necessary data for completing the calculations are found within the

TABLE 1 Minimum Internal Design Pressure Resulting in a Thickness Which Also Withstands Vacuum Conditions

Material Groups	P_{im} (lb/in.^{2})
Carbon steel	500
Alloy steel	500
Ferritic stainless steel	500
Austenitic stainless steel	410
Low carbon austenitic stainless steel	330

presented diagrams. Moreover, more than 100 pipes under vacuum condition can be quickly calculated in no more than 45 min.

Methods and diagrams as published here have been widely checked by Technipetrol specialists since 1979, and are in normal use with all the advantages here emphasized.

Symbols

A	Factor A from ASME Sect. VIII, Div. 1, Appendix 5 (—)
B	Factor B from ASME Sect. VIII, Div. 1, Appendix 5 (lb/in.^{2})
c	corrosion allowance (in.)
D_o	external diameter (mm)
E	modulus of elasticity (lb/in.^{2})
L	unreinforced pipe length (mm)
ND	nominal diameter (in.)
P_{im}	minimum internal design pressure (lb/in.^{2})
t	design thickness (mm)
t_o	minimum installed thickness (mm)
T	design temperature (°C or °F)

References

1. ASME, Sect. VIII, Div. 1, Para. UG.28.
2. ANSI B 31.3, Para. 304.1.3.

PAOLO CONTE
MAURIZIO FELICI

Piping and Instrumentation Diagrams

Piping and instrumentation diagrams, commonly called P&IDs, are used to guide detailed design and construction of process plants. The engineering activities of all the design disciplines—piping, electrical, instrumentation, etc.—must conform to the P&IDs. P&IDs are also a principal form of communication between those who are designing the plant and those who are going to own and operate it. To ensure success and satisfaction, it is important that P&IDs be complete and clear. This article introduces P&IDs and shows how good ones are developed.

Figures 1–3 are examples that show how a P&ID for a portion of a plant passes through several stages before the design is fixed. We will refer to these figures later on.

The process is a batch reaction in which a solution is heated to form an insoluble product. It works as shown in Fig. 1: Feed is pumped to the agitated, jacketed reactor (R-101) and heated with steam under temperature control. All key instruments are tied into a process-control computer. Volatile by-products distill overhead and are partially condensed (E-101) with cooling water. Noncondensables go to a scrubber; by-products flow to a storage tank. The slurry accumulates to a predetermined level, measured by an indicator, then feed and steam are stopped and the slurry is pumped out. As the design progresses, more controls will be added to this scheme.

What a P&ID Contains

A P&ID is a pictorial representation of a plant which shows all the equipment, including installed spares, plus the associated piping, valving, insulation, and instrumentation. It is normally an elevation (side) view; tankfarms are sometimes shown in plan (top) view. There are separate P&IDs for different systems, and the plant may be broken up into sections since a single drawing cannot typically describe an entire plant. Designs are likely to change as the project moves ahead, so the P&IDs go through several revisions, or issues, by the time the plant is actually built.

There are standard symbols for most items in a P&ID; this allows for compact presentation without sacrificing readability. Relative sizes of equipment may be reflected by the sizes of the symbols. The layout of a P&ID conforms roughly to the actual physical layout. Tower and vessel connections are depicted near their actual locations; pipes that run onto other sheets are often shown at the same elevation, for continuity.

All piping should be marked with line designations: symbols showing service, pipe size, and pipe specification; identification number; and some indication of insulation or tracing. The line-designation table, prepared later, will have all the design data and will list the source and destination of each line. An accompanying legend (usually on a cover sheet) should explain the line designations as well as symbols for valves and other elements in the pipeline.

In laying out P&IDs, it is important to allow room for adding details. Consider, too, the complexity of the installation and the number of lines, vessels, and instruments. For example: One reactor, with its instruments and auxiliary equipment, is typically enough for one drawing.

Types of P&IDs

There are three basic types of piping and instrumentation diagrams—systems, distribution, and auxiliary systems:

Systems P&IDs show the production, utility, and pollution-control processes; this category includes:

Process P&IDs. These drawings are for the actual manufacturing process: Reaction, purification, materials handling, separation, and other unit operations from the process flow diagram are detailed here. Process P&IDs show all the process equipment, piping, and controls; the main process flow is left to right.

Utility-generation P&IDs. These drawings show utility systems such as boilers, cooling towers, heat-transfer-fluid heaters, refrigeration/brine coolers, and air compressors, but they do not typically show how these utilities are distributed through the plant.

Environmental P&IDs. These P&IDs for pollution-control processes such as scrubbing, incineration, and wastewater treatment follow the same format as process P&IDs. Vent and discharge-collection systems should follow the format of distribution P&IDs, the next category.

Distribution P&IDs show how utilities, chemicals, and other nonprocess streams are distributed through the plant. These drawings include all headers and subheaders, laid out to approximate their actual arrangement. Vents and relief systems are also in this category:

Utility-distribution P&IDs. These show how steam, cooling water, and other utilities are distributed. Root valves at branches should be included, but controls and check valves on equipment lines will be shown on the process P&IDs.

Safety-system P&IDs. These define the depressuring and safety-relief systems, and include the piping networks that run from relieving devices through blowdown drums or gas holders to vents, stacks, and other destinations.

Chemicals-distribution P&IDs. Distribution is shown as for utilities.

Auxiliary-system P&IDs show compressor lubrication and cooling systems, hydraulic systems, pump seals, and other auxiliaries related to major equipment.

All these types are developed through several stages, or issues, during the conception, design, and construction processes. P&ID development begins with a preliminary issue that defines the plant concept, and ends with final-construction or as-built issues.

Approval Issue

Before generating P&IDs for client approval, engineers sometimes develop preliminary ones to use for conceptual estimates or preliminary design work, or to speed things up in fast-track jobs.

Then the approval issue. These P&IDs are the ones the client checks, and revises if necessary, before giving approval to proceed with engineering. An approval P&ID is often designated Revision 0. Figure 1 shows an example of an approval P&ID for a reactor system. (This is not to scale: A real P&ID would be on a large sheet, so its symbols and type would be relatively smaller, and it would have more notes.) Such a drawing usually depicts:

Equipment, including drivers and piped-in spares. Equipment that is already in place, or planned for the future, may be drawn with dotted lines to contrast with new equipment, drawn with solid lines.

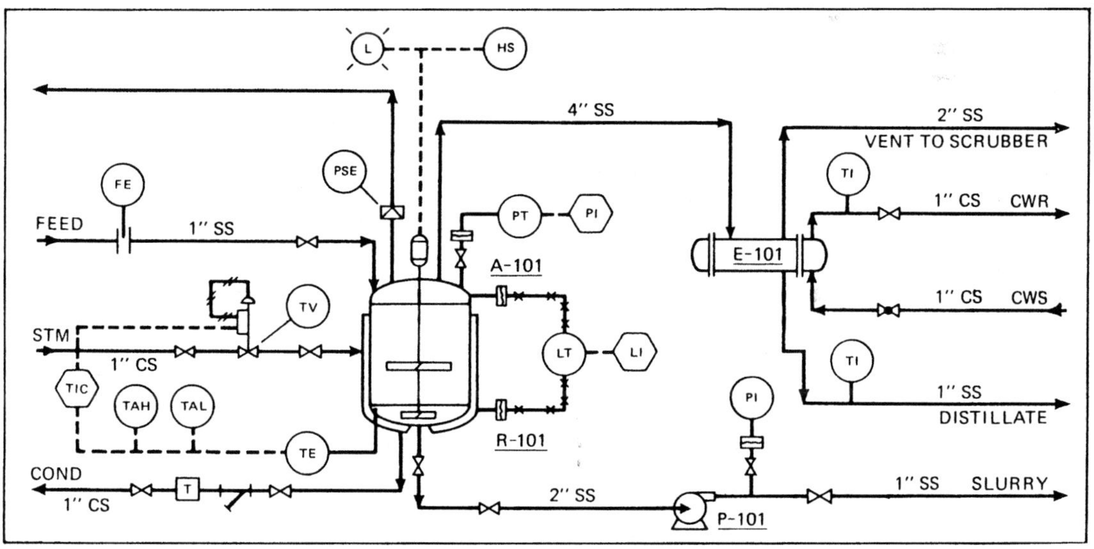

FIG. 1 Approval-issue P&ID for reactor and associated equipment.

Equipment numbers, names, materials of construction, sizes, and types, as available.
Process lines with line sizes and piping specifications or materials of construction.
Instrumentation, including functional designations (but not tag numbers) and showing panel mounting or local mounting.
Valving, including bypasses, drains, vents, sampling connections, and instrument block valves, plus designations for out-of-specification valves.
Utility connections to equipment and controls.
Safety-relieving devices.
Insulation and tracing.
Special notations such as sloped lines, critical elevations, or barometric legs.

The following information is *not* normally shown on an approval P&ID but is added later: instrument numbers, numbers for piping specialties, line numbers, motor horsepowers, details of packaged equipment, tie-in information, control-valve and nonline-size valve information, and safety-system line and valve sizes.

Going back to Fig. 1, suppose that the client reviews this reactor P&ID and approves it subject to several changes: automatic rather than manual feed-flow rate control, reaction at pressure above atmospheric, and liquid-level setpoint control. These changes would be made on the engineering issue, the next step.

Engineering Issue

This issue incorporates the client's changes and so is often called Revision 1. It is the formal document used to start design engineering and procurement. Specifically: piping or model design, material takeoffs, procurement of valves and piping specialties, design and procurement of instruments, and control panels.

Engineering-issue P&IDs normally include everything shown on the approval issue plus instrument numbers; line numbers; full descriptions of equipment, including motor horsepowers; piping-specialty numbers; tie-ins; sizes of control valves and nonline-size valves; and sizes of safety-system lines and valves. Details of packaged equipment and of special-equipment drawings not yet submitted by vendors are not normally included.

Figure 2 shows an engineering-issue P&ID for the example reactor system. Compared with Fig. 1, this has more detail (line and instrument numbers) plus the required revisions: a flow controller for the feed (FIC-101), a backpressure controller on the condenser vent (PCV-103), and a high-level alarm (LAH-101).

Engineering-issue P&IDs are usually accompanied by several documents; line-designation tables, emission and discharge summary, and utility sum-

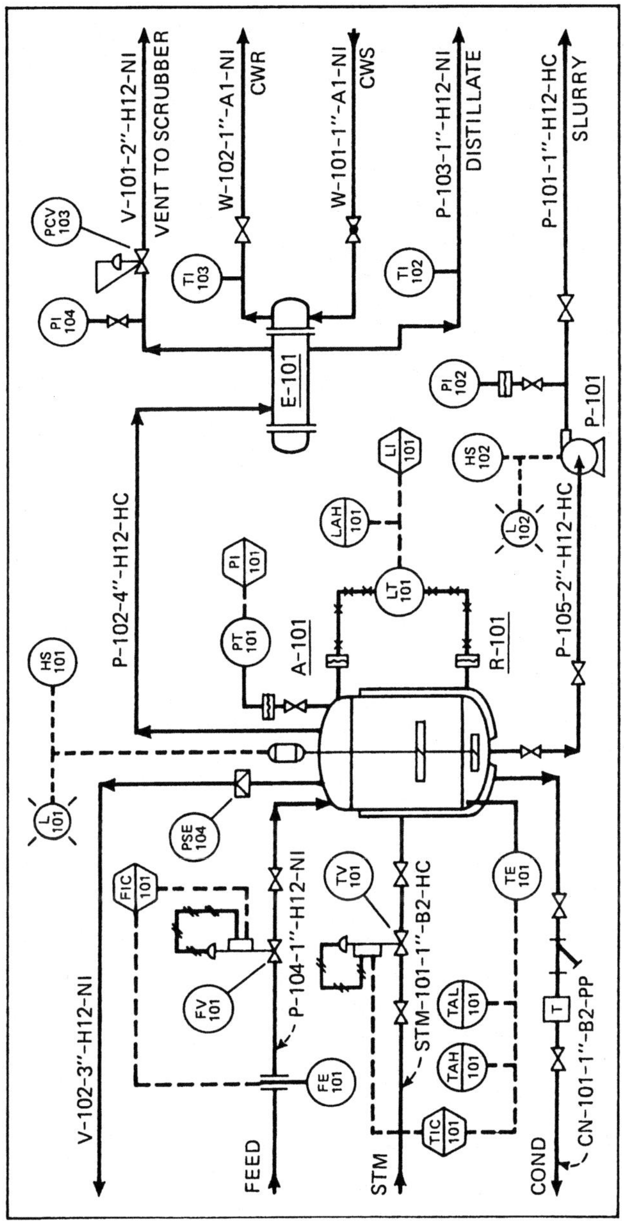

FIG. 2 Engineering-issue P&ID includes line and instrument numbers.

mary. Other documents may be issued soon after: specialty-item summary, tie-in index, instrument data sheets, and preliminary interlock descriptions.

There may be subsequent engineering issues, used internally to inform the various departments of current changes. The next step is the construction issue.

Construction Issue

The first construction issue is used for final checking of designs and drawings. Frequently, the client staff will review this issue line by line.

These completed P&IDs include all the information on the engineering issues plus details of all packaged equipment, details for vendor drawings, and up-to-date input from detailed design (e.g., extra valves or subheaders). The first construction issue is typically accompanied by several documents including updated line-designation tables; updated (if required) specialty items summary, instrument data sheets, and tie-in index; and preliminary steam-trap summary. There may also be a list of revisions made since the previous issue; alternatively, the revisions may be indicated on the drawings (e.g., by circling them on the back of a tracing).

Figure 3 shows a construction-issue P&ID for the example reactor. Since the previous issue, the design was revised, based on vendor test data:

To handle the product slurry, the pump-discharge line was upscaled to 1½ in., and the gate valves replaced with ball valves. A blanked-off cleanout port was added at the pump suction. To achieve better agitation, the reactor was revised to have one impeller, plus baffles. A high-current alarm was added to the agitator to warn of excessive viscosity. As a safety feature, a second temperature element (TI-104) was installed in the reactor bottom. A drain and vent were added to the jacket.

Subsequent construction issues keep the design up to date until the project is completed. The final construction issue will incorporate changes that surfaced during the final model-vs-P&ID check, or, in jobs not involving models, those changes that surface during checking of piping drawings.

At the request of the client, an as-built issue may also be prepared. This reflects the plant as built, including field changes, and is useful in operation and in planning future revamps.

Table 1 is a working checklist that shows the contents of the three issues.

P&ID Presentation

To guarantee understanding, P&IDs should be presented in clear format, using standard symbols and descriptions. Different engineering firms and departments may have somewhat different styles, but the basics are fairly universal.

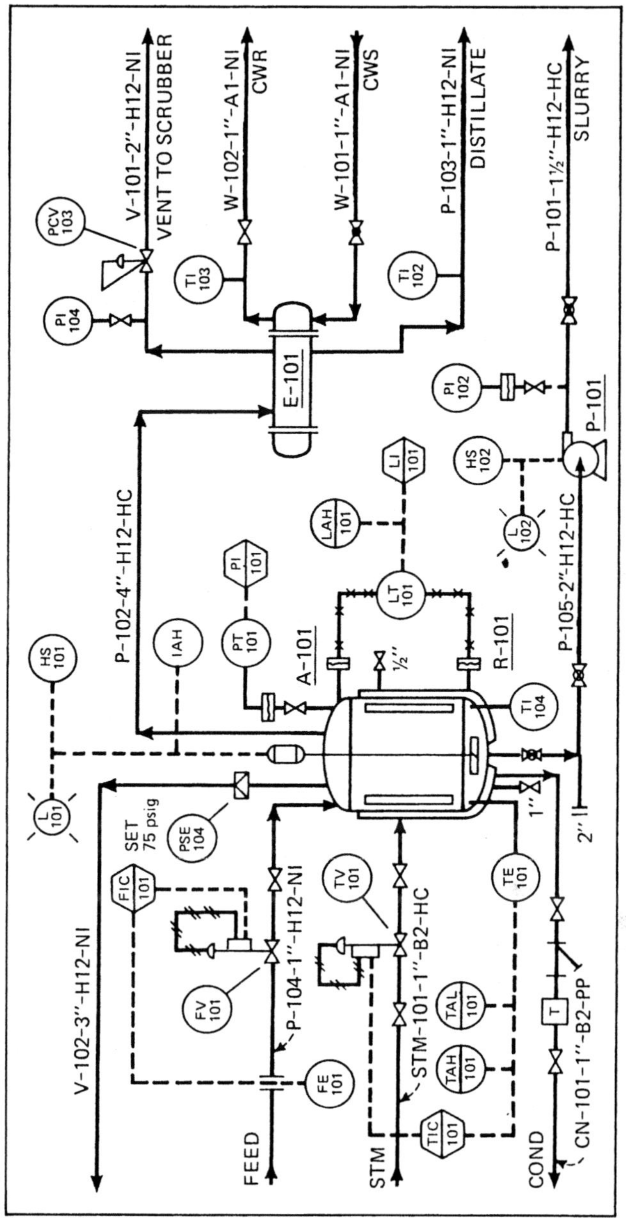

FIG. 3 Construction-issue P&ID shows near-final design and includes all previous revisions.

TABLE 1 P&ID Checklist Shows Contents of Approval, Engineering, and Construction Issues

Contents	Approval Issue	Engineering Issues	Construction Issues
Equipment:	☐	☐	☐
Spares	☐	☐	☐
Drivers	☐	☐	☐
Numbers	☐	☐	☐
Names	☐	☐	☐
Materials of construction	☐	☐	☐
Sizes	☐	☐	☐
Types	☐	☐	☐
Motor hp		☐	☐
Packaged-equipment details			☐
Vendor-drawing details			☐
Process lines:	☐	☐	☐
Piping specifications	☐	☐	☐
Materials of construction	☐	☐	☐
Sizes	☐	☐	☐
Numbers		☐	☐
Tie-in designations		☐	☐
Detailed design			☐
Instrumentation:	☐	☐	☐
Functional designation	☐	☐	☐
Panel vs local	☐	☐	☐
Numbers		☐	☐
Detailed design			☐
Valves:	☐	☐	☐
Bypasses	☐	☐	☐
Equipment drains, vents	☐	☐	☐
Instrument block valves	☐	☐	☐
Control valve sizes		☐	☐
Nonline-size valve sizes		☐	☐
Detailed design			☐
Safety systems:	☐	☐	☐
Relief valves	☐	☐	☐
Line sizes		☐	☐
Valve sizes		☐	☐
Detailed design			☐
Insulation/tracing:	☐	☐	☐
Detailed design			☐
Piping specialties:	☐	☐	☐
Numbers		☐	☐
Utilities:	☐	☐	☐
Connections	☐	☐	☐
Controls	☐	☐	☐
Detailed design			☐
Vendors' input			☐
Special notations:	☐	☐	☐
Sloped lines	☐	☐	☐
Critical elevations	☐	☐	☐
Barometric legs	☐	☐	☐
Detailed design			☐

(*continued*)

TABLE 1 *(continued)*

Contents	Approval Issue	Engineering Issues	Construction Issues
Related documents:		☐	☐
Line designation tables		☐	☐
Emissions/discharge summary		☐	
Utility summary		☐	
Specialty-items summary		☐	☐
Tie-in index		☐	☐
Instrument data sheets		☐	☐
Interlock description		☐	☐
Steam-trap summary			☐

Cover sheet. This defines all the symbols, legends, and nomenclature used on the P&IDs, with enough detail to enable quick interpretation. A company will usually use a standard sheet unless a job demands special items or a client offers specific instructions.

Equipment designations. Each piece of equipment on a P&ID must be identified by number, name, and description according to the company (or client-specified) standard. Whenever possible, the number should be shown only within the equipment outline; this prevents confusion. All equipment numbers, names, and descriptions should be tabulated at the bottom of the P&ID.

Packaged equipment. Some equipment may be specified and purchased as a package of several connected items, with their associated piping and instrumentation. For example: A brine chiller will include a compressor, condenser, and evaporator, and possibly pumps and tanks. Initial issues of the P&ID, when details are not known, should show for such packages an area enclosed in dash-dot lines. When details become available, they are entered in this area. If an item in the package (e.g., an instrument or pump) is to be located elsewhere, it should be designated as "furnished with the package" by marking it with an "F." Packaged-equipment numbers should follow the company standard. Components should have subnumbers—e.g., XE1-101 for the condenser in the brine chiller package numbered X-101.

Holds. When information that would usually be on an issue is missing, the missing part should be circled and marked "hold." This is better than delaying a drawing.

Flowlines. Each flowline on a P&ID should be identified by a numerical code. This will typically include a service classification, line number, nominal size, specification code, and insulation tracing code. For example: The code P-5004-3"-A01-ST might describe a hot process line in Area 5 having a 3-in. diameter, made of carbon steel, with steam tracing and insulation. Each line should show the direction of flow; two-way or gravity flow should be clearly noted. The direction of flow should also be shown for fittings, such as three-way valves, whose hookup depends on it. Main process streams are typically drawn in heavy lines on the engineering issue and after. Piping used only at times—e.g., for startup or shutdown—should be identified as such.

If possible, flowlines should be at least ⅜ in. apart. Where lines cross, one must be broken. Break an instrument lead before a flowline, a utility line before a process line, and any other line before a main process line. If the lines are the same type, break the vertical one.

Instrumentation. Instrument symbols and identification should follow Instrument Society of America (ISA) standards; instrument bubbles are usually ½-in. diameter. All sensing elements (e.g., for pressure or temperature) must be correctly located relative to valves and equipment. Instrument block valves should be shown on piping and instrumentation diagrams, but should not duplicate those on instrument-installation diagrams.

Clarifying notes. Design requirements not explained by symbols or codes should be put into notes on the P&IDs. Notes might cover: tangent elevation above grade of vertical vessels, bottom elevation above grade of horizontal vessels, locations of control or shutoff valves, top elevation of vents and relief stacks, slope of gravity-flow lines, elevation of vacuum equipment requiring barometric legs, relative elevation of equipment in gravity-flow systems, and piping symmetry to ensure equal flow.

In addition, notes at valves can tell whether they need to be in horizontal or vertical runs, or whether there is some minimum safe distance from equipment or instruments. Notes at piping tie-ins can show that a connection must be at the top—e.g., for a pressure-relief line.

MONTE L. SCHWARTZ
JOSEPH KOSLOV

Thermoplastic Hose

Thermoplastic hose is made in two nonhydraulic configurations: Small-i.d. hose is used for air, water, and chemical spraying—such hose is rated for high-pressure service and is specially reinforced to provide safety and durability. Large-i.d. hose is used for materials handling, bulk-commodity transfer, and water and chemical suction and discharge—such hose is easy to handle and provides high flow rates.

Advantages of Thermoplastic Hose

The major ones are:

Weather resistance—Resistance to weathering, ozone, and sunlight is excellent.

Abrasion/friction resistance—The polyvinyl chloride (PVC) cover commonly used on thermoplastic hose has low drag resistance and slides easily over rough surfaces. As to internal friction, the thermoplastic tube is smooth, reducing friction and increasing the flow rate.

Lightweight—Thermoplastic hose weighs 15–50% less than the equivalent rubber hose. The lighter weight, of course, makes handling easier, which reduces worker fatigue and increases productivity. For example, a 4-in.-i.d., 20-ft-long rubber hose weighs 60–65 lb, while a thermoplastic hose providing equal performance for a similar application weighs only 40–45 lb.

Color-coded covers—Hoses can be color-coded to indicate air, water, and various chemicals. Also, the covers resist chemicals and oils, and they are durable.

Thin-wall construction—Typical high-pressure thermoplastic hoses have a smaller o.d., allowing for easier positioning in tight areas. Also, the thin walls permit full flow when the hose is coiled on a reel or bent into a tight radius. This helps prevent collapse and kinking.

Nonwire reinforcement—Either a fabric braid or a plastic helix is used as a reinforcement. Both of these are noncorrosive and durable.

Chemical resistance—Most general-purpose thermoplastic hoses will handle many mild chemicals in concentrations to 10%. In general, resistance is excellent (see Table 1). (For a discussion of rubber hose that includes chemical-resistance tables, see "Beat Corrosion with Rubber Hose," p. 105 of the September 8, 1980 issue of *Chemical Engineering*.)

Thermoplastic vs Rubber Hose

Thermoplastic hose thus offers advantages over standard rubber hose in certain situations:

1. Using flexible lines on portable pneumatic tools. The cover can be perforated to allow air penetration, thereby reducing the chance that cover blisters will occur.
2. Multiple-type applications where hoses must have distinctive colors for coding and identification. This is especially helpful when two hoses of the same i.d. are used in the same area.

TABLE 1 Chemical Resistance of Thermoplastic Hose

This is a guide for selecting thermoplastic hose for specific applications. Space limits the size of the table, and data for many chemicals have had to be omitted. Whenever possible, test samples of candidate thermoplastics under actual process conditions.

Key to symbols used in the table:

For hose stock:
- A = Plasticized polyvinyl chloride
- B = Polyamide resins (nylon base)
- C = Polyester elastomer
- D = Ethylene vinyl acetate
- E = Polyurethane

For chemical resistance:
- 1 = Excellent resistance. The fluid is expected to have minor or no effect.
- 2 = Good resistance. This polymer should give reasonably satisfactory service. Under prolonged continuous exposure, it may show minor to moderate deterioration and/or solution discoloration. Environmental changes (e.g., temperature, concentration) may promote increased degradation.
- X = Not recommended. Unsatisfactory; do not use.
- — = Insufficient or no data available.

The normal upper temperature limit for corrosives being transferred is 100°F. For higher temperatures, consult the supplier or manufacturer.

NOTE: The materials listed here have been plasticized and compounded for use in flexible industrial hoses. The chemical-resistance data were compiled for these special materials.

Chemical	Hose Polymer Types				
	A	B	C	D	E
A					
Acetaldehyde	X	2	1	2	1
Acetamide	—	—	1	—	—
Acetone	X	1	1	X	X
Acetophenone	—	—	2	X	—
Acrylonitrile	1	—	1	—	—
Air, ambient	1	1	1	1	1
Air, 180°F	X	2	1	—	—
Alcohol, amyl	1	—	1	—	—
Alcohol, butyl	1	—	—	2	—
Alcohol, furfural	—	1	1	—	—
Alcohol, ethyl	1	—	2	2	—
Alcohol, isopropyl	2	—	2	2	—
Alcohol, methyl (6%)	1	—	2	—	—
Alcohol, methyl (100%)	2	—	2	—	—
Alkazene	—	—	—	—	—
Benzyl alcohol	1	1	2	—	2
Borax (sodium borate)	1	1	1	2	1
Bordeaux mixture	1	1	1	—	1
Boric acid	1	X	1	2	1
Boric copper sulfate	1	1	1	—	1
Brake fluid (petroleum)	2	1	1	—	—
Brake fluid (synthetic)	2	2	1	—	—
Brine (salt)	1	1	1	—	1
Butyl acetate	—	1	2	X	—
Butyl alcohol	1	1	1	1	1
C					
Calcium arsenate	1	2	1	2	1
Calcium bisulfate	2	—	—	—	—
Calcium bisulfide	2	1	1	—	1
Calcium carbonate	1	1	1	1	1

Chemical					
Aluminum chloride	1	—	1	2	—
Aluminum fluoride	1	—	—	1	—
Aluminum hydroxide	1	—	1	—	—
Aluminum nitrate	1	—	—	—	—
Aluminum sulfate	1	1	—	1	—
Ammonia, aqueous	1	1	1	1	—
Ammonium acetate	1	—	1	1	—
Ammonium bicarbonate	1	—	1	—	—
Ammonium carbonate	1	2	1	1	2
Ammonium chloride	1	X	1	1	1
Ammonium metaphosphate	2	—	—	—	—
Ammonium nitrite	—	—	—	—	—
Ammonium persulfate	—	—	—	—	—
Ammonium phosphate	1	1	1	1	1
Ammonium sulfate	1	1	1	1	1
Ammonium sulfide	1	—	—	—	—
Ammonium thiocyanate	—	—	—	—	—
Amyl acetate	X	X	2	2	2
Amyl borate	—	—	—	—	—
Amyl chloride	2	—	2	X	—
Amyl chloronaphthalene	—	—	—	—	—
Amyl naphthalene	—	—	—	—	—
Amyl phenol	—	—	—	—	—
Anethole	—	—	—	—	—
Aniline	—	2	X	X	X
Aniline oils	2	2	1	—	1
Antimony chloride (50%)	1	X	X	—	—
Aromatic hydrocarbons	—	1	2	—	2
B					
Banvel, concentrated	—	2	—	—	—
Barium carbonate	1	—	1	—	—
Barium chloride	1	X	1	—	1
Barium sulfate	1	1	2	—	—
Barium sulfide	1	2	2	—	—
Basic copper arsenate	1	1	1	—	1
Benzaldehyde	X	2	2	X	1
Benzene	X	1	2	X	2
Calcium chlorate	1	—	—	—	—
Calcium chloride	1	X	1	—	1
Calcium hypochlorite (5%)	1	2	2	1	—
Calcium hypochlorite (15%)	2	X	—	1	—
Calcium nitrate	1	1	—	—	—
Calcium silicate	1	—	—	—	—
Calcium sulfide	2	—	—	—	—
Carbon disulfide	X	1	1	—	2
Carbon dioxide (dry)	1	1	1	1	1
Carbon dioxide (wet)	1	1	1	1	—
Carbon tetrachloride	X	1	2	X	X
Chlordane	2	—	—	—	—
Chlorine water (25%)	2	2	2	—	2
Chlorine trifluoride	X	—	—	—	—
Chlorobenzene	X	—	X	—	—
Chlorobromomethane	X	—	X	—	—
Chloroform	X	—	—	—	—
Chrome alum	1	—	—	2	—
Chromium salts	1	—	1	—	1
Coal gas	1	—	1	—	1
Coal tar	X	—	—	—	—
Copper chloride	1	X	1	2	1
Copper cyanide	1	—	2	—	—
Copper nitrate	1	—	2	2	—
Copper sulfate	1	1	1	2	1
Crude petroleum oil	1	2	1	—	1
Cyclohexane	X	2	1	—	1
Cyclohexanone	X	—	2	X	1
Cymene	X	—	—	—	—
D					
Diacetone alcohol	X	1	2	—	—
Diammonium phosphate	1	1	2	—	—
Dibutyl phthalate	X	2	1	X	2
Dichlorobenzene	X	—	X	X	X
Dichloroethylene	—	2	2	X	—
Diethanolamine (20%)	2	2	2	—	2
Diethylamine	1	—	2	—	—

(*continued*)

TABLE 1 *(continued)*

Chemical	Hose Polymer Types				
	A	B	C	D	E
Diethyl ether	2	—	2	—	2
Diethyl glycol	1	1	1	1	1
Dioctyl phosphate	X	—	2	—	2
E					
Enamels	2	1	1	—	—
Ethanolamine	2	—	1	—	—
Ethers	2	X	2	—	2
Ethyl acetate	X	1	2	1	2
Ethyl alcohol	2	1	1	1	2
Ethyl chloride	X	—	X	—	X
Ethyl ether	X	—	2	—	—
Ethyl mercaptan	X	—	—	—	—
Ethylene chloride	X	1	X	—	X
Ethylene chlorohydrin	X	—	X	—	X
Ethylene glycol	1	1	1	1	1
F					
Ferric chloride	1	—	1	1	—
Ferric sulfate	1	—	1	1	—
Ferrous chloride	1	—	1	1	—
Ferrous nitrate	2	—	2	—	—
Ferrous sulfate	1	—	1	1	—
Formaldehyde (37%)	1	1	2	1	2
G					
Gasoline, regular, unleaded	X	—	1	X	
Gasoline, regular, leaded	X	—	1	X	X
Gasoline, premium	X	—	2	X	X
Gasohol	X	1	2	X	X
Glucose	1	1	1	1	1
Glycerin (glycerol)	1	1	1	1	1
Grease	1	1	1	—	1
M					
Machine oil	2	1	1	X	—
Magnesium carbonate	1	1	1	—	1
Magnesium chloride	1	1	1	1	1
Magnesium nitrate	1	—	1	—	—
Magnesium sulfate	1	1	1	1	1
Manganese salts	1	1	1	—	—
Manganese sulfate	1	—	—	—	—
Melamine varnish	—	—	—	—	—
Mercuric chloride	2	1	1	1	2
Mercuric cyanide	—	—	—	—	—
Mercurous nitrate	—	—	—	—	—
Mercury	1	1	1	1	2
Mercury salts	1	—	—	2	—
Mesityl oxide	X	—	—	—	—
Metallic soaps	1	—	—	—	—
Methyl acetate	X	1	2	—	—
Methyl acrylate	—	—	—	—	—
Methyl alcohol	1	1	1	2	2
Methyl amine (25% aqueous solution)	—	—	2	—	—
Methyl amine (60%)	—	—	2	—	—
Methyl amine (99%)	—	—	—	—	—
Methyl amyl carbinol	—	—	2	—	—
Methyl bromide	X	—	X	—	X
Methyl butyl ketone	X	1	2	—	—
Methyl cellosolve	—	1	1	—	—
Methyl chloride	X	1	—	—	—
Methyl ethyl ketone	X	2	X	—	X
Methyl formate	—	—	—	—	—
Methyl isobutyl ketone	X	2	X	—	X
Methyl isopropyl ketone	X	2	X	—	X
Methyl methacrylate	—	—	—	—	—
Methyl salicylate	—	—	—	—	—
Methyl sulfate	—	1	1	—	1

Chemical					
H					
Heptane	2	2	1	X	1
Hexane	2	—	1	—	1
Hydrogen fluoride	—	—	—	—	—
Hydrogen peroxide (10°)	1	2	—	2	1
I					
Iodine	X	—	X	—	—
Iodine in alcohol	1	1	1	2	—
Isobutyl alcohol	2	2	2	—	—
Isooctane	X	1	1	—	2
Isopropyl acetate	X	2	2	—	2
Isopropyl alcohol	2	1	1	2	2
Isopropyl ether	X	—	1	—	—
Isocyanate (toluene diisocyanate)	—	—	2	—	—
K					
Kerosene	2	—	2	—	—
Ketones	—	—	2	—	—
L					
Lacquers	X	1	2	X	—
Lacquer solvents	X	1	2	X	2
Lead acetate	1	2	2	1	—
Lead arsenate	1	1	1	—	1
Lead sulfate	2	1	1	—	1
Lead, tetraethyl	—	—	1	—	—
Lead, tetramethyl	—	—	2	—	1
Lime	1	1	1	—	1
Lime bleach	2	2	2	1	—
Lime, sulfur	2	2	1	—	—
Linseed cake	—	—	1	—	1
Lindseed oil (boiled)	1	2	1	X	1
Liquid soap	2	—	2	—	—
Lubricating oils	2	1	1	—	2
Lubricating oils (diester)	X	1	2	—	X
Methylene chloride	X	X	X	X	—
Methylene dichloride	X	X	X	X	—
Monochlorobenzene	X	X	X	X	X
Motor oils	2	1	1	—	2
N					
Naphtha (low aromatic content)	X	2	2	—	—
Nickel acetate	—	—	2	—	—
Nickel chloride	1	—	2	1	—
Nickel nitrate	2	2	2	—	—
Nickel salts	2	—	2	1	—
Nickel sulfate	1	2	1	1	—
Nicotine	1	2	1	—	1
Niter cake (sodium bisulfate)	1	1	1	—	1
Nitrobenzene	X	X	X	X	X
Nitroethane	—	—	—	—	—
Nitrogen	1	1	1	1	1
Nitrogen oxide, to 50%	1	1	1	—	—
Nitromethane	—	—	2	X	—
Nitropropane	—	—	2	X	—
O					
Octyl alcohol	2	—	2	—	—
Oil (SAE)	2	1	1	—	1
Oleic acid	2	1	1	X	1
Oxygen	X	—	—	1	—
Ozone	1	1	2	—	1
P					
Paint	—	1	—	—	—
Paint solvents (oil base)	—	2	2	—	—
Paints (oil base)	—	1	—	—	—
Palm oil	—	1	1	—	—
Palmitic acid	—	1	1	X	—
Paraffin (petroleum)	—	1	1	—	—
Paraformaldehyde	1	1	2	1	2
Perchloroethylene	X	—	X	—	—

(continued)

TABLE 1 *(continued)*

Chemical	Hose Polymer Types				
	A	B	C	D	E
Phenolates	2	2	2	—	2
Phorone	—	—	—	—	—
Phosphate esters (to 150°F)	—	1	1	—	—
Phosphate esters (above 150°F)	—	2	2	—	—
Picric acid (water solution)	1	2	2	—	—
Pinene	—	—	2	—	—
Piperazine hydrochloride solution (34%)	—	—	—	—	—
Polyester resin	—	2	2	—	—
Polyurethane (to 125%)	—	—	—	—	—
Potassium acetate	1	—	—	—	—
Potassium bicarbonate	1	1	1	—	—
Potassium bisulfite	1	—	1	—	—
Potassium bromate	1	—	1	—	—
Potassium bromide	1	1	1	—	2
Potassium carbonate	1	1	1	1	2
Potassium chlorate	1	2	1	1	2
Potassium chromate	1	2	—	—	—
Potassium chloride	1	1	1	1	2
Potassium cuprocyanide	1	2	1	—	—
Potassium dichromate	1	2	2	2	—
Potassium ferrocyanide	1	—	2	2	—
Potassium fluoride	1	—	2	—	—
Potassium nitrate	1	1	1	—	1
Potassium permanganate	X	X	X	—	—
Potassium permanganate (5%)	X	X	X	—	—
Potassium persulfate	1	—	—	—	—
Potassium phosphate	—	—	—	—	—
Potassium sulfate	1	1	1	—	1
Potassium sulfide	1	—	1	—	—
Potassium sulfite	2	—	—	—	—
Potassium thiosulfate	1	—	—	—	—
Propyl acetate	—	—	—	—	—
Propyl alcohol	2	1	1	—	2
Propylene glycol	—	2	2	—	—

Chemical	Hose Polymer Types				
	A	B	C	D	E
Sodium hypochlorite (5%)	1	1	2	1	X
Sodium hypochlorite (20%)	1	2	—	—	X
Sodium hyposulfate	1	—	—	—	—
Sodium metaphosphate	—	—	—	—	—
Sodium nitrate	1	1	1	1	1
Sodium perborate	—	2	1	1	1
Sodium peroxide	1	X	—	—	—
Sodium phosphates	1	1	1	1	1
Sodium silicate	1	1	1	1	1
Sodium sulfate	1	1	1	1	1
Sodium sulfide	1	1	1	1	1
Sodium sulfite	1	1	1	2	1
Sodium thiosulfate	1	1	1	1	1
Sodium tripolyphosphate	—	—	—	—	—
Stannic chloride	2	X	2	—	—
Stannous chloride	1	2	1	—	1
Stearic acid	1	—	1	X	—
Stearin	—	2	1	—	1
Stoddard solvent	2	2	X	—	X
Straight synthetic oils (phosphate ester, phosphate ester base)	2	1	1	—	2
Sulfur chloride	2	2	—	—	—
Sulfur dioxide (dry)	1	X	X	1	—
Sulfur dioxide (liquid)	X	—	—	—	—
Sulfur dioxide (moist)	X	—	—	—	—
Sulfur hexafluoride (gas)	2	1	1	—	—
Sulfur trioxide (dry)	1	—	1	—	—
T					
Tallow	2	1	1	—	—
Tannic acid (10%)	1	1	2	X	—
Tar (bituminous)	—	—	2	—	—
Tar oil	2	1	1	—	2

Pyrene (carbon tetrachloride)	X	—	X	—	—
Pyrethrum	1	—	—	—	1
Pyridine (50%)	X	—	X	2	1
R					
Red oil (commercial oleic acid)	—	1	1	—	—
S					
Salicylic acid	—	—	X	—	—
Salt water (sea water)	1	1	1	1	1
Sewage	2	2	2	—	—
Shellac	—	1	1	—	—
Silicone grease	2	1	1	—	—
Silicone oils	2	1	1	—	—
Silver cyanide	1	—	—	—	—
Silver nitrate	2	1	1	2	—
Soap solutions	1	1	1	2	1
Soda ash (sodium carbonate)	1	1	1	1	1
Soda water	1	1	1	1	1
Sodium acetate	1	1	1	—	2
Sodium benzoate	1	—	—	—	—
Sodium bicarbonate	1	1	1	1	1
Sodium bisulfate (niter cake)	1	1	1	1	1
Sodium bisulfite	1	1	1	1	1
Sodium borate	1	1	1	1	1
Sodium carbonate	1	1	1	1	1
Sodium chlorate	1	1	1	2	1
Sodium chloride	1	1	1	1	1
Sodium cyanide	1	1	1	1	1
Sodium dichromate	1	—	1	—	—
Sodium ferricyanide	1	—	2	—	—
Sodium ferrocyanide	1	—	2	—	—
Sodium fluoride (70%)	1	—	2	—	—
Sodium hydrosulfide	1	—	—	—	—
Sodium hydrosulfite	2	—	—	—	—

Tartaric acid	1	1	1	—	—
Tellus oils	2	2	1	—	—
Tenol oils	2	1	1	—	2
Terpineol	2	2	1	—	—
Terresstic	2	1	1	—	—
Tetraethyllead (TEL)	—	2	2	—	—
Tetrahydrofuran	2	1	2	X	—
Tetralin	—	2	2	X	—
Thiopen	—	—	2	X	—
Toluene (toluol)	—	2	2	—	—
Tributoxyethyl phosphate	—	2	2	—	—
Tributyl phosphate	—	—	2	—	X
Tricresyl phosphate (Skydrol)	—	2	2	—	—
Triethanolamine (TEA)	2	—	X	—	—
Tripolyphosphate (STPP)	—	—	—	—	—
U					
Urea solution	2	2	1	—	2
V					
Varnish	—	1	2	—	—
Vinyl chloride (monomer)	X	—	1	—	—
W					
Water	1	1	1	1	1
X					
Xylene	X	2	2	X	—
Z					
Zinc acetate	1	X	2	—	—
Zinc chloride solutions	2	1	2	—	—
Zinc hydrate	1	—	2	—	2
Zinc oxide	1	—	—	—	—
Zinc sulfate solutions	2	2	2	—	—

3. Where the hose is outdoors and is constantly exposed to sunlight and the weather. For outdoor exposure, thermoplastic hose is more resistant than most rubber hose to cracking of the cover.
4. When oil resistance is required for both tube and cover. Because tube and cover of general-purpose hose are made of similar PVC material, they meet oil-resistance requirements.
5. When long hoses must be moved around and minimum weight is needed. Thermoplastic hose is offered in longer continuous lengths, and fewer couplings are thereby needed. Such hose can be up to 40% lighter tha the rubber type of hose.
6. When high electrical resistance is required. The tube and cover are, in general, less conductive than rubber hose. The user should check with the hose manufacturer for values of electrical conductance or resistance.
7. Handling a variety of solvents, such as those used in paints and lacquers. Thermoplastic hose offers good resistance to many petroleum-based fluids. This reduces the number of types of hose needed in the plant.

Limitations

There are some limitations in using thermoplastic hose. Depending on the type of hose and thermoplastic used, the hose may be stiffer than an equivalent rubber hose. For some types of thermoplastics, the high and low temperatures may be limited; some materials become too soft at high temperatures or stiffen at low ones. Yet, there are thermoplastics that offer broader temperature ranges than do some rubbers. Also, special kinds of thermoplastics can be used to transfer aromatics or hot, concentrated acids. Again, check with the manufacturer for these special hoses.

If uncertain about a particular use, consult the hose supplier or manufacturer.

Maintenance Tips

Once the correct hose is selected, it must be properly cared for. Follow these recommendations:

Always use hose and couplings as recommended by the manufacturer.
Do not exceed stated working pressures.
Keep the hose out of work or traffic areas where it could be struck or run over.
Avoid excessive end-pull.
Use hose supports whenever possible to avoid strain on the couplings.
Store new hose so as to protect it from the weather. Keep it in the original shipping carton, preferably on pallets.

Flush and coil hose in between extended storage.

As soon as possible after each use, wipe off any caustics or other corrosive chemicals that contact the hose cover.

Inspect the hose periodically. Look at the ends—make sure that the couplings are secure. Check the hose cover for signs of excessive abrasion or cuts that could result in failure. If this happens, replace the hose.

Couplings

Couplings (not covered in Table 1) used with thermoplastic hose are typically those also used with rubber hose. Carbon steel, stainless steel, brass, bronze, and aluminum are the most common ones. Hose makers publish tables of corrosion resistance for these metals and alloys. Also, standard corrosion tables are available. Reference to such literature will help the engineer make the proper selection.

CHARLES H. ARTUS

Around Control Valves

Control valves often have to be removed from pipelines for maintenance, so the piping around them has to be flexible enough to permit continued operations under such conditions.

Usually, the control valve is one line-size smaller than the main line itself. If it is the same as the line size, or more than two line sizes smaller, the system flow characteristics should be reexamined, and the control valve resized if necessary.

For the best setup from the operating and maintenance points of view, block valves should be installed, one on each side of the controller, together with a throttling-type globe valve for use as a bypass. During normal operation, the bypass valve remains closed, with flow taking place through the controller. Figure 1(a)–1(l) shows various arrangements of piping around

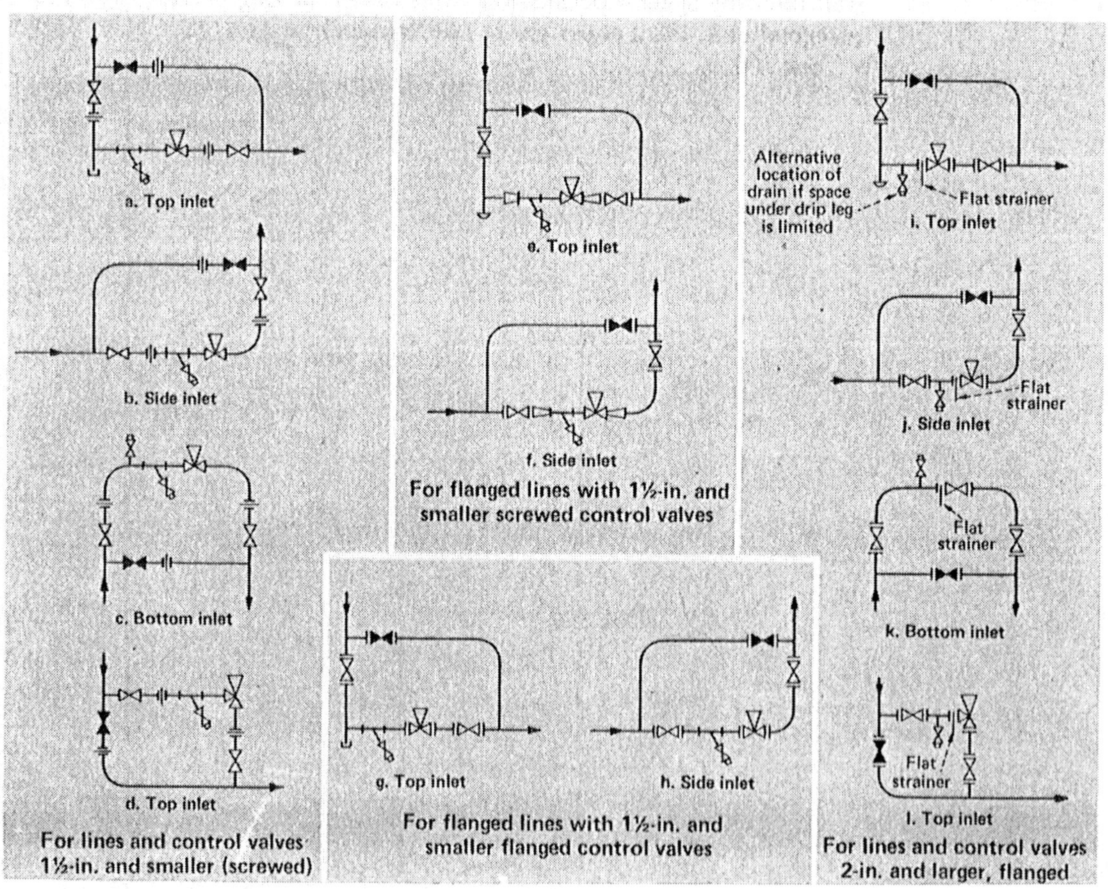

FIG. 1 Control-valve bypass.

control valves, from those using small screwed-pipe connections to flanged valves and pipe, and combinations of both.

The greatest source of trouble with control valve operation is from foreign matter in the pipeline. Therefore, strainers should be installed ahead of all control valves, with drip legs ahead of the strainers. Y-type strainers of reputable manufacture with 20-mesh screens should be used with all types of control valves 1½ in. and smaller, as shown in Fig. 1(a)–1(h). Flat-type strainers with 14-mesh screening, as shown in Fig. 2, should be used with control valves 2 in. and larger, as shown in Fig. 1(i)–1(l).

Sizing Block and Bypass Valves

The block valves should be full line size, except when the control valve is two or more sizes smaller than that, in which case the block valves should be one size smaller than line size. The block valve material and rating should be for the service specified.

Bypass valves and bypass lines at control valves should be main-line size for lines 2 in. and smaller. For lines 3 in. and larger, the bypass line and bypass valves should be one normal pipe size smaller than main-line size (except for gravity flow, where they should be line size).

Bypass valves should be globe valves in sizes up to and including 4 in.; above this size, they should be gates or, in special cases, gear-operated cocks where low pressure-drops in large lines are encountered. Figure 3 provides

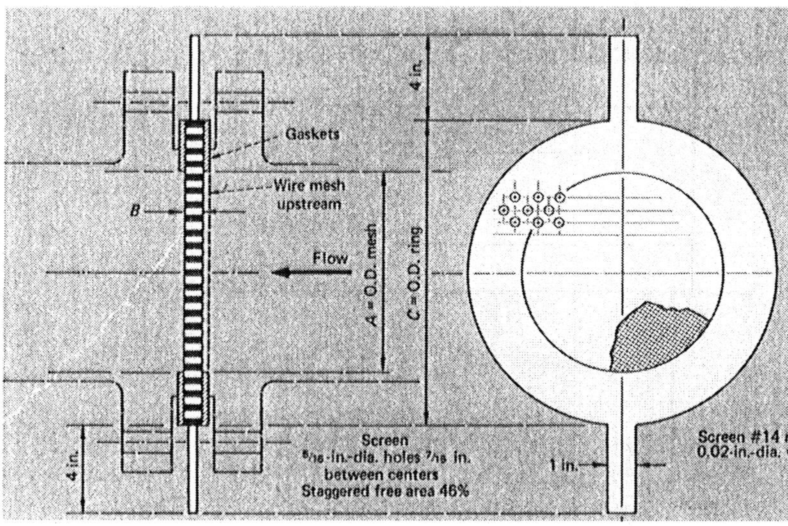

150 and 300 psi ANSI std. raised face

Pipe size, in.	A, in.	B, in.	C, in. 150 psi	C, in. 300 psi
2	1¾	#10 USG	3¾	4
2½	2¼	#10 USG	4½	4¾
3	2¾	#10 USG	5	5½
4	3¾	#10 USG	6½	6¾
6	5¾	3/16	8¾	9½
8	7¾	3/16	10⅝	11¾
10	9¾	¼	13	13⅞
12	11¾	¼	15¾	16¼
14	13	¼	17⅜	18¾
16	15	¼	19⅞	20⅞
18	17	¼	21¼	23⅛
20	19	¼	23½	25⅜

FIG. 2 Flat-type strainers for control valves.

a quick indication of block and bypass valve sizing for control valves from 1 in. through 12 in.

Swaging at Control Valves

With screwed control valves, a swage nipple should be screwed into the control valve body, with the unions at each side of the control valve (used for its removal) located at the large end of the swage nipple. No heavier or lighter swages than Schedule 80 should be used. Where valves are flanged and less than line size, the swaging should be immediately adjacent to the control valve.

Piping around Control Valves

The line piping adjacent to control valves should be arranged to provide flexibility for removing them. For example, the block valves on either side

Main line size, in.	Control valve size, *A*, in.																	
	1		1½		2		3		4		6		8		10		12	
	B	*C*	*B*	*C*	*B*	*C*	*B*	*C*	*B*	*C*	*B*	*C*	*B*	*C*	*B*	*C*	*B*	*C*
1	1	1																
1½	1½	1½	1½	1½														
2	1½	2	2	2	2	2												
3			2	2	3	2	3	2										
4					3	3	4	3	4	3								
6							4	4	6	4	6	4						
8									6	6	8	6	8	6				
10											8	8	10	8	10	8		
12													10	10	12	10	12	10

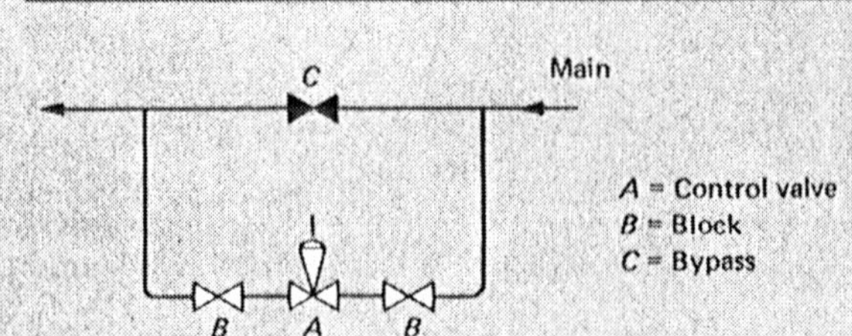

FIG. 3 Piping around control valves.

of a control valve should neither be at the same elevation as the control valve nor both in the same plane with it.

One block valve may be in the same horizontal plane as the control valve, and the other block valve in the vertical, with the bypass below the control valve. Or, the two block valves may be in the vertical plane, with the control valve horizontal, and with the bypass below the control valve.

Several arrangements are permissible, depending on the direction of flow and on the elevation of the control valve with respect to the flow line. However, the piping around control valves should be arranged so that the lines need not be sprung to remove the control valve. Special attention should be given to the location of the control valve so that it is not pocketed in steam, air, or gas flow and will not catch such foreign materials as oil scale, welding beads, and the like.

One-inch valved connections should be provided in the bottom of the line downstream of the first block valve and before the downstream block valve to serve as telltale indicators, and to permit checking the tightness of the block valves should it be required that the control valve be bypassed for removal.

The piping around control valves should be self-supporting or should be permanently supported. This is required in order that when the control valve or block valve is removed, the lines should remain in place without the need of building scaffolding or providing other temporary supports.

Control-Valve Location

All control valves should be installed where they are accessible from equipment platforms and walkways, or at grade, so that they can be easily maintained. Sufficient clearance should be provided above and below control valves so that the bottom flange and plug (or the topwork and plug) may be removed, for repair or replacement, with the valve in place.

When vaporization occurs in control valves, they should be placed as close as possible to the destination of the steam in order to reduce pressure drop, surging, and excess lengths of large pipeline downstream of the control valve.

JOHN D. CONSTANCE

4

Pipeline Design

Dynamic Programming

Introduction

The objective of optimal gas pipeline design is to determine the optimal number of compressor stations and their locations; to select the model, number, and configuration of compressor units in each station; and to select the optimal pipe diameter and maximum allowable operating pressure (MAOP). For capacity expansions, only pipe loopings, new compressor stations, and/or possible additional compressors need to optimized. Nonlinear objective function and nonlinear constraints make pipeline design optimization complicated. A number of different methods have been proposed and tried with limited success. The method proposed is a modification to Bellman's dynamic programming. Each stage of the optimization procedure is based on a potential location for a compressor station. However, the distance covered is variable and can be fixed only if the compressor station location considered establishes an optimal trajectory.

General Notes and Definitions

McClure [1] has observed that "A single mile of large diameter loop line can cost over a million dollars." Thus, a reduction in cost of only a few percent due to a system design optimization effort is generally more than sufficient to compensate for the cost of the effort.

Generally, a gas pipeline can be described topologically as a tree network with many roots. Each root may be a supply source. The terminals are gas consumers. The pressures at the roots and at the consumers are the main system boundary conditions. The gas loads which are functions of time are described by load profiles. The system has to be capable of supplying a specified maximum amount of gas to each consumer. However, the operating cost is a function of the composite load profile for consumers located downstream.

Generally, a number of compressor stations are to be located along the pipeline. Each station contains one or more centrifugal or reciprocating compressor units and auxiliary equipment to generate electricity, cool discharge gas, control the station, etc. It is possible to configure two or more compressors at a station in parallel or in series. Use of larger compression ratios is prohibited because gas temperature increases with larger ratios which eventually exceed the pipe coating thermal limitations. In order to reduce the temperature, air or water cooling systems can be used or a lower compression ratio can be specified.

Pipes of different diameters, wall thicknesses, and materials can be used depending upon the flow rate, anticipated pressure, and temperature distributions within the system, and the proximity of populated areas. However, diameter changes must be compatible with pigging plans. It is not uncommon to have two or more pipes in parallel. In most cases, parallel pipes are used only when the capacity of the pipeline was or is to be increased after installation by looping to reduce hydraulic resistance.

Terminology

Let us call the portion of the pipeline between two nodes a "section." Each section may be divided into many auxiliary stages. It is possible that one section has no compressor stations while another has many. From a topological point of view, each system under consideration has many pipe sections connected at nodes but no circuits are permitted. In graph theory, such systems are called directed trees [3].

First, tree networks with only one root, the gas supply node, will be analyzed. Of course, real gas pipelines usually do have many sources and many consumers, so they must be considered as multiroot directed trees. For convenience, some terminology from data structures is used [4]. The pipeline system in Fig. 1 has 13 sections (arcs) and 14 nodes, where node 14 is the *root node*. The nodes from 1 to 7 are called *terminal nodes*. All sections linked by the same node are divided into parents and children. Node 8 has the *parent* section 8–10 and the *children* sections 5–8, 6–8, 7–8. Similarly, sections 8–10 and 9–10 are children for node 10, and section 10–11 is their parent. Terminal nodes have no children and the root node has no parents. The *degree of the node* is the number of children at the node. The maximum node degree in a tree is called the *degree of the tree*. All of the sections connected to a given node from the childrens side are called *descendants*, and the part of the tree made by the descendants is called a *subtree*. The ancestors for node 10 make the following subtree: 8–5, 8–6, 8–7, 9–3, 9–4, 10–8, 10–9. A number of descendants connected in series is called a *path*. There are 5 paths from node 10: 3–9–10, 4–9–10, 5–8–10, 6–8–10, 7–8–10. The maximum number of nodes on all paths is called the *node level*. For the subtree starting from node 10, the node level is 3. The maximum node level for the whole tree is called the *tree depth*. For the system in Fig. 1, the tree depth is 6.

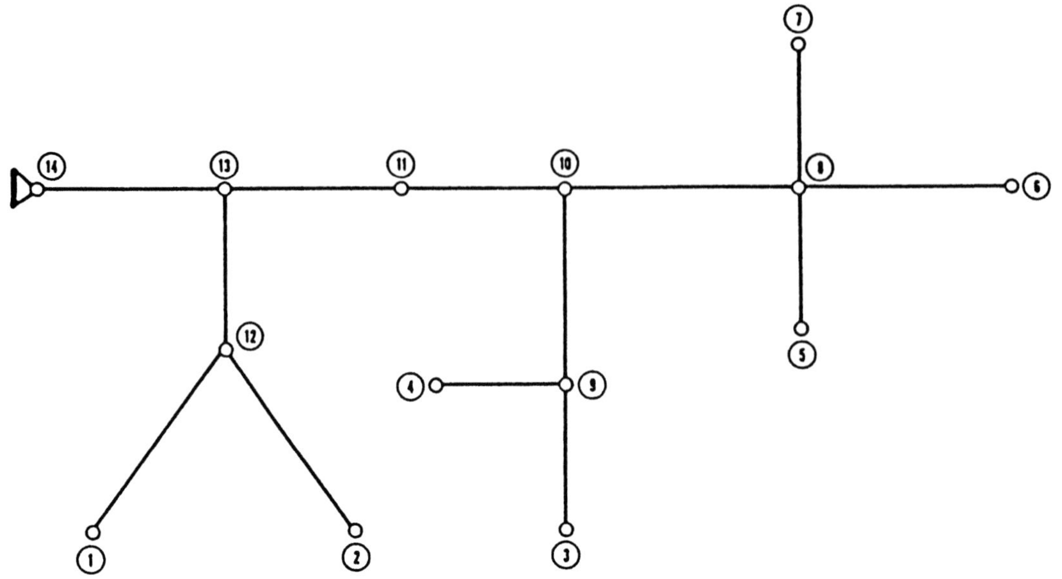

FIG. 1. Tree network.

Optimization Problems

The following classes of optimization problems are under consideration:

1. Design problem. A new pipeline has to be designed for transporting a specified maximum gas volume as a peak load. The objective is to determine the values for the interdependent variables which minimize the net present worth of the life-cycle cost based on specified yearly load profiles. These variables are:

   Number and location of the compressor stations, and the number of compressors, and their size and configuration at each compressor station
   Suction and discharge pressures for each station at peak load conditions
   Material, diameter, and thickness for all pipes

2. Expansion problem. This problem considers an existing pipeline where the volume of gas to be transported is to be increased. This means design and construction of new compressor stations, addition of units at existing compressor stations, and/or addition of new pipes parallel to existing pipes (loopings). The net present worth of the life-cycle costs, which

includes system operation under the new conditions and the initial cost of all additional labor, equipment, and materials, is to be minimized.

The above-mentioned problems include the optimization dilemma: What combination of pipes and compressors will minimize the system life-cycle cost. A larger numbers of compressors will increase the cost of compressor stations but will reduce the cost of pipes. However, the problem becomes more complicated when the configuration of compressors in a compressor station, location of compressor stations, terrain conditions, maximum allowable gas temperature, and other constraints must be taken into consideration.

In general, the optimum number, sizes, and locations of all pipeline facilities have to be considered over a given planning horizon, not at a single time since the gas sales are likely to increase with time.

Flow Considerations

The main requirement for design is that the system must be capable of supplying the specified peak demand for gas. However, the yearly fuel gas consumption is a function of the load profiles for the consumers collectively.

The fuel to run compressors and auxiliary equipment is normally taken from the pipeline. For this reason, the net volume of gas delivered by the pipeline to consumers is less than the total fuel supplied to the pipeline. This means that flows for many sections are unknown as long as pressures and temperatures are unknown. More importantly, the different load profiles must be considered. For a large gas pipeline, the load profile for each consumer can be different, which means that the maximum loads at different nodes in the system may be distributed in time. The gas "packing" and "unpacking" within the pipeline affects daily peaks.

It can be shown that if load profiles are ignored, the flow going in is not equal to the flow going out. In order to satisfy Kirchhoff's first law, the time factor must always be considered. Even if the maximum loads are taken as independent variables, the pressure at the upstream node must be considered as it should be for these loads.

Consider the simple 3-section system illustrated in Fig. 2. The peak month for consumer 4 is May, the peak load for this month is $Q3 = 300$ MMSCF/d, and the pressure at the upstream node 3 is $P3 = 900$ lb/in.^{2}abs. The peak load for node 2 is in July when $Q1 = 220$ MMSCF/d. The sum of the loads creates a peak in June at node 1. Certainly in June the pressure at node 3 will not be equal to $P1 = 900$ lb/in.^{2}abs since in June the flow through section 3–4 is only 270 MMSCF/d. Therefore, a new pressure for July must be calculated at node 3, and this pressure has to be used for section 1–3. Also, if all loads reach their maxima at different times, the computer program must be capable of

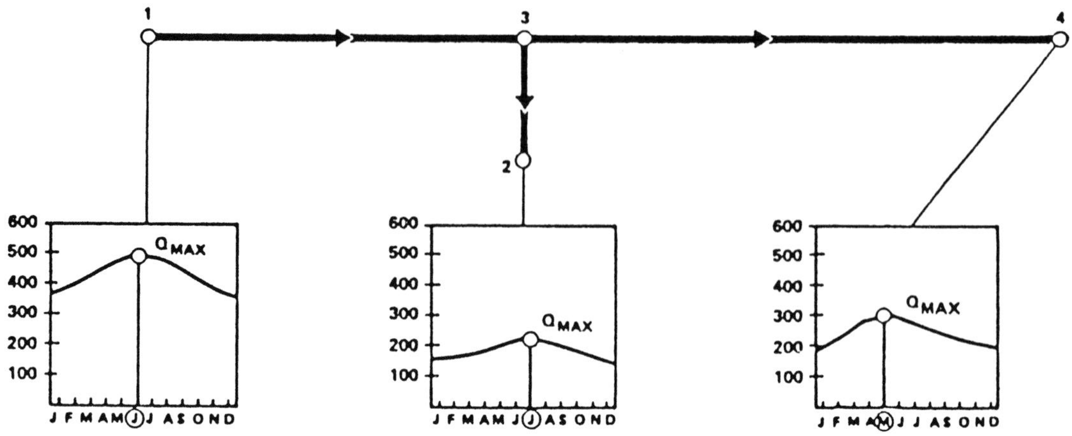

FIG. 2. Flow profiles.

analyzing this situation. Thus the monthly load variations are a general problem which must also be considered in pipeline design optimization.

Objective Function

The following objective function and constraints were developed so that both design and expansion problems can be solved by the same computer program. The objective function to be optimized is the net present worth of the life-cycle system cost:

$$E = E_p(\text{PWEF}) + E_s \tag{1}$$

where E = net present worth of ownership and operating costs
E_p = first year operating cost
E_s = system net present worth
PWEF = present worth factor allowing for escalation of operating costs

$$\text{PWEF} = \frac{[(1 + \text{esc})/(1 + \text{dis})]^a - 1}{1 - [(1 + \text{dis})/(1 + \text{esc})]} \tag{2}$$

where esc = annual escalation rate for operating costs, decimal
dis = annual discount rate, decimal
a = amortization period, years

The *operating costs* include fixed cost based on station investment (depreciation, property tax, insurance, and allowable return on investment) and variable costs (fuel, labor, maintenance, lubrication oil, and operating overhead):

$$E_p = (E\text{depr}_{ij} + E_p\text{tax}_{ij} + E\text{insr}_{ij} + E\text{retr}_{ij})E_s + (1 + E\text{ovhd}) \sum_{i=1}^{n} \sum_{j=1}^{m} \left[\int_{t_b}^{t_e} E\text{fuel}(Q(t),P(t))_{ij} dt + E\text{labr}_{ij} + E\text{main}_{ij} + E\text{lub}_{ij}) \right] \quad (3)$$

where i = index for sections which vary from 1 to n
j = index for compressor stations within the ith section
m = number of compressor stations/platforms in the ith section
n = number of sections in a pipeline network
t_b, t_e = beginning and ending time that corresponds with load profiles

Fixed cost on station/platform Efc is the sum of

Edepr = depreciation cost as a percent of investment, decimal
$E_p$tax = property tax as a percent of investment, decimal
Einsr = insurance as a percent of investment, decimal
Eretr = allowable return on investment, decimal

It must to be pointed out that the integral of fuel with respect to time describes the real fuel consumption for a gas pipeline; however, the pipe diameters and compressor ratios have to allow the delivery of maximum demand for gas. The presented operation cost can describe an expansion problem when an additional amount of gas has to be supplied after some period of time.

Variable cost on station/platform investment:

$E\text{fuel}_{ij}$ = fuel cost, calculated as the cost of 1000 standard cubic feet of natural gas multiplied by the amount of yearly fuel consumed by station j located in section i. The time period t depends on the sum of the loads for a period (day, month, quarter), which is based on the load profiles, $Q_{ij}(t)$, for section i. The difference between loads for different stations located in the same section depends on the fuel consumption for upstream compressor stations. The pressure loss in pipes, $P(t)$, also becomes a function of time, US $

$E\text{labr}_{ij}$ = yearly labor cost, calculated as a function of total station hp and number of units, US $

$E\text{main}_{ij}$ = maintenance cost, calculated as a function of total station hp, type, and number of units, US $

$E\text{lub}_{ij}$ = lubrication oil cost, calculated as a function of total station hp, type, and number of units, US $

$E\text{ovhd}$ = station operating overhead, recommended value is 40% of the sum of the previous variable costs, decimal

The *present worth of the initial cost* includes those for the pipes and compressors, their freight and installation, buildings and foundations, land, etc., and can be described as

$$E_s = \sum_{i=1}^{n} \sum_{j=1}^{m} [(E\text{pip1}_{ij} + E\text{pfe1}_{ij} + E\text{pip2}_{ij} + E\text{pfe2}_{ij} + E\text{pli}_{ij}) + (E\text{mee}_{ij} + E\text{meei}_{ij} + E\text{efou}_{ij} + E\text{buil}_{ij} + E\text{sli}_{ij} + E\text{mgp}_{ij} + E\text{misc}_{ij})] \qquad (4)$$

Piping initial cost: In order to make the problem statement general, two pipes are always considered for each section. For design problems, one of the pipes is equal to zero (imaginary pipe). For the expansion problem, one pipe exists and a second pipe is to be optimized. If more than two parallel pipes are already present in a section, they may be represented as one pipe with the equivalent friction and heat transfer.

$E\text{pip1}_{ij}$ = pipe cost and laying cost for pipe 1, depends on pipe diameter, thickness, and material (pipe 1 of section i, and stage j between compressors), US $

$E\text{pfe1}_{ij}$ = pipe freight cost, depends on pipe weight and freight distance (pipe 1), US $

$E\text{pip2}_{ij}$ = pipe cost and layout cost for pipe 2, US $

$E\text{pfe2}_{ij}$ = pipe freight cost for pipe 2, US $

$E\text{pli}_{ij}$ = pipe land and improvements cost (roads, fences, gradings, etc.) fix data for each pipe terrain, US $

Compressor station or sea platform initial cost:

$E\text{mee}_{ij}$ = mechanical and electrical equipment cost, calculated as a function of the number of units, unit type, and size. For a platform this is the module cost, US $

$E\text{meei}_{ij}$ = mechanical and electrical equipment installation cost, calculated as a function of the number of units, unit type, and size, US $

$E\text{efou}_{ij}$ = equipment foundation cost can be calculated as a function of the number of units, unit type, and size, and the terrain conditions, US $

$E\text{buil}_{ij}$ = building cost, calculated as a function of the total station hp and of the terrain conditions, US $

$E\text{sli}_{ij}$ = station land and improvement cost should be calculated individually for each terrain, US $

$E\text{mgp}_{ij}$ = main gas piping cost, calculated as a function of the total station hp, US $

$E\text{misc}_{ij}$ = miscellaneous; this includes equipment freight, general station overhead, engineering, and contingency. Freight should be separated from the rest and calculated as a function of terrain conditions and hp

$$E\text{misc} = (1 + E\text{sfrt} + E\text{sovh})(E\text{mee}_{ij} + E\text{meei}_{ij} + E\text{efou}_{ij} + E\text{buil}_{ij} + E\text{sli}_{ij} + E\text{mgp}_{ij}) \quad (5)$$

Esfrt = station equipment freight as a percent of total station cost, decimal

Esovh = station overhead as a percent of total station cost, decimal

Substitution of these variables and coefficients for fixed and overhead costs into Eq. (1) gives

$$E = [1 + (E\text{fc})(\text{PWEF})]\sum_{i=1}^{n}\sum_{j=1}^{m}[(E\text{pip1}_{ij} + E\text{pfe1}_{ij} + E\text{pip2}_{ij} + E\text{pfe2}_{ij} + E\text{pli}_{ij} + E\text{mee}_{ij}) + (1 + E\text{sfrt} + E\text{sovh})(E\text{mee}_{ij} + E\text{meei}_{ij} + E\text{efou}_{ij} + E\text{buil}_{ij} + E\text{sli}_{ij} + E\text{mgp}_{ij})] + (1 + E\text{ovhd})(\text{PWEF})\sum_{i=1}^{n}\sum_{j=1}^{m}\left[\int_{t_b}^{t_e}(E\text{fuel}(Q(t),P(t))_{ij}dt + E\text{labr}_{ij} + E\text{main}_{ij} + E\text{lub}_{ij}\right] \quad (6)$$

It is convenient to separate the objective function into two equations (described below for one stage):

a) Piping cost (initial and operating fixed costs as a function of investment):

$$E\text{pipe} = [1 + (E\text{fc})(\text{PWEF})](1 + E\text{sfrt} + E\text{sovh})(E\text{pip1} + E\text{pfe1} + E\text{pip2} + E\text{pfe2} + E\text{pli}) \quad (7)$$

b) Station/platform cost:

$$E\text{stat} = [1 + (E\text{fc})(\text{PWEF})](1 + E\text{sfrt} + E\text{sovh})(E\text{mee} + E\text{meei} + E\text{efou} + E\text{buil} + E\text{sli} + E\text{mgp}) + (1 + E\text{ovhd})(\text{PWEF})\left[\int_{t_b}^{t_e}(E\text{fuel}(Q(t),P(t))dt + E\text{labr} + E\text{main} + E\text{lub}\right] \quad (8)$$

Constraints

A number of constraints must be satisfied:

1. Kirchhoff's first law. For each system node, the flow coming in is equal to the flow going out. The use of this obvious rule is questionable if the time load considerations are not taken into account.

2. Pressure balancing. For each junction, the pressures of intersecting pipes must be equal. This means that the sums of pressure differences for any subtree connected to the same node must be equal.
3. Prohibited terrains. There are many places along the pipeline where the location of compressor stations or platforms will not be possible. For example, in a lake. Some special conditions, such as sensitive environments, can be covered by the use of penalty coefficients although incremental costs are likely to be sufficiently discouraging.
4. Standard pipes. All available pipe diameters, thicknesses, and materials are standard and must be selected from a table. This table must also include maximum allowable pressure and total pipe laying costs for each pipeline terrain type.
5. Pressure at terminals. There is a minimum allowable delivery pressure for each terminal.
6. There are minimum and maximum pipe diameters.
7. Standard turbines. All turbines must be selected only from a standard rating table.
8. Standard compressors. All compressor models, sizes, and maximum compression ratios are presented in table form.
9. Number of compressors. The total number of compressors in a station can be restricted, as well as the number of compressors connected in parallel and/or in series.
10. Pressure in pipes. The maximum pressure in pipes is limited by the pipe strength expressed as its maximum allowable operating pressure (MAOP).
11. Pressure ratio at compressor stations. The maximum pressure ratios for a station can be constrained. Also, the maximum discharge pressure is usually limited by the MAOP. The minimum suction pressure is limited by other considerations such as those for turbine start-up.
12. Maximum velocity. The velocity of the gas cannot exceed a maximum. However, the pressure drop is likely to be excessive long before any velocity maximum is reached.
13. Maximum gas temperature. This constraint usually depends on the type of pipe coating. The same coating is used for all pipe sections. However, some special conditions can exist, such as for the Alaskan pipeline where the maximum gas temperature should not destroy the nearby permafrost. This condition requires the use of refrigeration and/or insulation.
14. Existing pipes. This is a very important limitation since it must be used if a part of the system already exists. Also, this limitation allows the simulation of the design pipe for expansion problems, as well as operability problems. Known pipes are those pipes with a known inner diameter, roughness, and material. Sometime, special requirements such as specific pipe strength or thickness can be applicable. As was mentioned above, each section has two pipes. Therefore, this constraint has to be satisfied for each pipe independently.
15. Existing compressors. Compressors may be preselected or exist, and their parameters are known as well as the configuration in which they are connected.

16. Turbine rating. Minimum and maximum power from turbines can be specified.

Existing Methods and Optimization Procedures

Gas pipeline optimal design is a mathematical programming problem with nonlinear objective functions and mixed nonlinear/integer constraints. A number of techniques have been developed to solve this complicated problem.

Trial-and-error is the most commonly used procedure for pipeline optimization. Lee states "...no pipeline project anywhere in the world has been analyzed more thoroughly than the Canadian section of the Alaska pipeline" [5]. The cost of the Canadian portion of this line was about $8 billion and the total cost was $26 billion. Optimization was divided into two steps: a) Pipe optimization where the cost of four combinations (two pipe diameters and two pressures) was analyzed and the lowest cost was selected, and b) where the compressor was optimized for a given pipe diameter by analyzing a different number of compressor stations (from 8 to 15) with the same distance between the stations. Most major pipelines are optimized in the same way, through the trial-and-error method.

Many things in this method are questionable: The two-step calculation as optimization should be made simultaneously for pipes and compressor stations; the equal distance between stations in spite of different terrain conditions; the limited number of pressures; and the unexplained pressure limits. However, trial-and-error methods have been used for many years because there is no adequate and efficient optimization program in existence.

One of the best surveys of optimization in gas pipeline engineering was published by Huang and Seireg [6]. Mainly, studies have been limited to the optimization of compressor ratios for given compressor stations, pipe lengths, and pipe diameters. Wong and Larson were the first who used dynamic programming to find the solution for this problem [7]. Many programs use some variation of the trial-and-error method or some coordinate descent method. Such a procedure was developed by Cheeseman [8] and was called the step-by-step cost-evaluating method. That author understands that his solution can find a local optimum, therefore he repeats the optimization starting from different points and comparing costs. Cheeseman actually called one of his programs "an experience routine because it uses simple rules of thumb adopted by experienced engineers and cost estimators."

More advanced technique based on the steepest descent method was developed by Flanigan in 1972 [15]. However, the number and the location of compressor stations were prespecified.

An effort to define the problem in the same way as the present approach was undertaken by Bickel, Himmelblau, and Edgar [9]. However, the reduced gradient (by Abadie) and the branch-and-bounds methods used by them were unable to solve the entire problem. These reduced gradient methods have convergence problems and difficulties in introducing a fixed

TABLE 1 Literature Survey

Optimum Number of Compression Stations	Location of Compression Stations	Select Diameter from Tables	Optimum Pressure: Suction Discharge	Any Const-raints	Problems: Design, Expansion, Both	Net-work
No	No	Yes	Yes	Yes	Both	Yes
No	No	Yes	No	Yes	Both	No
Yes	Yes	No	Yes	No	Expansion	Yes
No	No	No	Yes	No	Design	Yes
No	No	Yes	Yes	No	Design	No
No	No	Yes	Yes	Yes	Both	Yes
No	No	No	Yes	No	Design	Yes
No	No	Yes	Yes	Yes	Both	Yes
No	No	Yes	Yes	Yes	Design	Yes
Yes	Yes	No	Yes	No	Design	Yes
No	No	No	Yes	No	Design	Yes
Yes	Yes	No	Yes	No	Both	No
Yes	Yes	No	Yes	No	Design	No
Yes		Yes	No	No	Both	Yes
Yes	Yes		Yes		Design	Yes
Yes	No	Yes	Yes		Expansion	Yes
Yes	Yes	No		No	Both	Yes

cost. Edgar, Dunn, and Murray state that "In examining the problem posed by Bickel et al., it is evident that the length constraints they used were unrealistic."

The program HCOMP, developed by SSI [10], is capable of "selecting" the number of compressor stations and their locations. However, the optimization is limited because of a major simplification, which is that the discharge pressure at each of the compressor stations is assumed by the program to be the downstream pipeline MAOP. Suction pressure is then calculated by dividing the discharge pressure by the specified compression ratio. A compression ratio of 2.1 is used in the example problem.

The use of iterative optimization by linearizing the gas-flow equations for a constrained problem was introduced by Pratt and Wilson [11]. Linearization is not a good way to solve this problem since many constraints are nonlinear. This method should have a problem of converging and searching to a local optimal instead of the desired global optimum.

Therefore, in existing methods based on Bellman's dynamic programming, the number of pressures (branch solution) is organized only for nodes [2] or for pipelines between existing stations [7]. None of these methods allows selection of the number and location of compressor stations.

Any Cost Function or Table	Conver-gency	Global Optimi-zation	Multi-period Optimi-zation	Methods	Year	Ref.
Yes	Good	No	No	Dynamic programming	1971	12
Yes		No	No	Coordinate descent	1971	13
No		No	No	Steepest descent	1971	14
No		No	No	Steepest descent	1972	15
Yes	Good	No	No	Dynamic programming	1972	7
Yes	Good	No	No	Dynamic programming	1972	2
No		No	No	Gradient + random diameters	1972	16
Yes	Good	No	No	Dynamic programming	1972	17
Yes	Good	No	No	Dynamic programming + sorting	1973	18
No		No	No	Reduced gradient + branch-and-bounds	1976	9
No		No	No	Linear programming	1979	19
No		No	No	Reduced gradient	1979	20
No		No	No	Reduced gradient	1980	21
Yes	Good	No	No	HCOMP by SSI	1980	10
No		No		Minimax Legrange + quasi-Newton	1982	22
No			Yes	Heuristic technique	1982	23
No	Good	No	No	Linear programming	1982	24

This short survey points out the many degrees of freedom present in pipeline design and analysis. A large number of independent variables are involved in the optimization. Some of these variables are: pipe diameters, location of compression stations, compressor models, configuration and number, compression ratios, gas temperatures, etc.

The results of a literature survey are presented in Table 1. The information in some publications mentioned below does not clearly address all questions in Table 1. Therefore, no claim to a precise comparison is made. Also, because of this, some questions were left without answers.

Dynamic Programming with Variable Stages

1. General

Dynamic programming was invented in the 1950s by Richard Bellman [25]. This powerful method determines the optimum for multistage decision processes. It is capable of finding a global minimum for a wide variety of problems. It has no convergence problems and generally yields an integer solution capable of coping with many constraints.

Also, the computation process is based on a number of recurrent sequences which are very convenient for digital computer solution. It conducts the process of selecting an optimal strategy by accumulating the best relatively-optimal decisions for each stage. Thus it avoids consideration of the vast number of nonefficient alternatives.

Dynamic programming is a method which is based on the calculus-of-variations which generates an optimum function rather than an optimum point. A number of functions called "trajectories" are created as continuations of a multistage dynamic process where local optimization is performed for each stage. The trajectory with the minimum overall value of the objective function is the optimum solution for the entire pipeline.

The dynamic programming parameters that must be defined are: the physical system, time, control, phase variable, and variation. For our problem the physical system under control is a gas pipeline. Each stage is set by the length of pipes beginning from the terminals to the root node (or vice versa). The control decisions are the selection of compressors and pipes. The phase is the variable that determines system conditions.

In general, optimal design of gas pipelines is an excellent application for the dynamic programming method since the objective function can be calculated as a sum of section and station costs, and also the pressure losses can be calculated as sums of pressure losses for sections connected in series. An excellent theoretical substantiation of the dynamic programming method was made by its inventor, Richard Bellman [25]. However, in spite of many attempts made by different specialists using dynamic programming for gas pipeline optimum design, the best results previously obtained are for the selection of optimals in diameters and pressures for a given number and location of all compressor stations (see references by Rothfarb et al. [12], by Wong and Larson [7], by Graham [14], by Martch and McCall [17], and by Zadeh [18]).

There are two possible explanations for such limited application. The first is that developers applied the dynamic programming method in the traditional way by creation of an optimum control strategy in regular section lengths. For a pipeline this means selection of equipment and pipe diameters (optimal trajectories) along fixed pipe lengths. These lengths could be the distance between the existing compressor stations or the entire pipe sections where stages are in system nodes (see Ref. 26).

The second corresponds to the "Achilles heel" of dynamic programming. This is the dimensionality curse which would occur if telescoping of pipe diameters were considered. Today, the diameter of the pipe between compressor stations tends to be constant. Otherwise "pigging" becomes difficult or impossible. However, the change in pressure due to large changes in elevation or the proximity to populated areas may require selection of different pipe thicknesses between compressor stations. For the dynamic programming method, this means that the selected condition for a trajectory at the current stage may depend to some extent on the conditions in a previous stage. This would introduce nonadditivity.

However, dynamic programming in general is not efficient for solving nonadditive problems. The proposed enhancement of dynamic programming by creating variable stages is considered as an advance to the method which allows solution of more complicated problems.

2. Optimization Method

The modifications proposed to the dynamic programming method are that each new stage is considered as a potential location for a new compressor station, and the distance between the previously specified node and a new stage is variable. This distance can be fixed only if the location of a new compressor station establishes an optimal trajectory.

The following is a short explanation of the proposed method. Gas is transmitted from one or many gas sources to several terminals through an oriented graph in a form of tree. There are a number of fixed nodes j, including pipe junctions, gas sources, terminals, and existing compressor stations. Also, let us assign a number of k-nodes at each potential location for new compressor stations and call them variable nodes. All fixed and variable nodes are stages of the dynamic programming process.

A vector $\mathbf{U}_{jk}$ of feasible decision alternatives is assigned for each arc between the fixed node j and the variable node k. This vector contains the code for the design alternatives being considered for each pipe. Another vector $\mathbf{W}_k$ represents the design alternatives available at the new compressor station which might be built at node k.

Both vectors are discrete functions which can represent design details such as pipe diameter and thickness, compressor models, numbers of compressors, or configuration. The objective function for each arc is

$$\phi_{jk}(U)_{jk} + \psi_k(W)_k \tag{9}$$

In addition to the vectors $\mathbf{U}_{jk}$ and $\mathbf{W}_k$, a phase coordinate P_k which represents possible pressures of n at k is established.

The objective is to select such components for vectors $\mathbf{U}_{jk}$ and $\mathbf{W}_k$ where

$$\sum_{(p_j,k)} \phi_{jk}(U_{jk},p_j,p_k) + \sum_{(p_k)} \psi_k(W_k,P_k) = \min \tag{10}$$

The main constraints are: Minimum pressure at terminals M:

$$p_k(U_{jk}, W_k) \geq p_i^{(\min)}, \qquad i = 1, 2, \ldots, n \quad k \in \mathrm{M} \tag{11}$$

Although temperature constraints generally are relevant only at compressor stations, a general gas temperature constraint for pipes is defined:

$$T_k(U_{jk}, W_k) \leq T^{(\max)} \tag{12}$$

Maximum gas velocity in pipes:

$$v_k(U_{jk}, W_k) \leq v^{(\max)} \tag{13}$$

Pipe pressure rating limitation:

$$p_k(U_{jk}, W_k) \leq \text{MAOP} \tag{14}$$

Any two different combinations (a) and (b) for vectors $\mathbf{U}_{jk}$ and $\mathbf{W}_k$ must be compared if at the node k they yield the same downstream pressure level.

$$p_k(U_{jk}^{(a)}, W_k^{(a)}) = p_k(U_{jk}^{(b)}, W_k^{(b)}) \tag{15}$$

If for that condition

$$\Big(\sum_{(p_j,k)} \phi_{jk}(U_{jk}^{(a)}) + \sum_{(p_k)} \psi_k(W_k^{(a)})\Big) > \Big(\sum_{(p_j,k)} \phi_{jk}(U_{jk}^{(b)}) + \sum_{(p_k)} \psi_k(W_k^{(b)})\Big)$$

or

$$\Big(\sum_{(p_j,k)} \phi_{jk}(U_{jk}^{(a)}\Big) > \Big(\sum_{(p_j,k)} \phi_{jk}(U_{jk}^{(b)}) + \sum_{(p_k)} \psi_k(W_k^{(b)})\Big) \tag{16}$$

then the a-combination is less favorable than b and should be eliminated from further consideration. After checking all possible combinations valid to dynamic programming and performing elimination with the above rules, the remaining one is the optimum.

If the optimal solution at node k has a second component which is a compressor station, the k-node becomes fixed and all components for vectors $\mathbf{U}_{jk}$ and $\mathbf{W}_k$ are stored. Otherwise the node k is ignored.

In other words, the node k is considered as fixed only if at the same pressure level a compressor station with a small diameter pipe becomes cheaper then a large diameter pipe with no compressor station.

3. Algorithm

Let us assume that the range of pressures possible at each j-node is divided into a number of discrete values $P_j^{(s)}$, $s = 1, 2, \ldots, \zeta$. Then a number of feasible combinations of $\mathbf{U}$ and $\mathbf{W}$ are considered for each $P_j^{(s)}$ and an optimal solution selected for each of the $\overline{P_j^{(s)}}$, s = 1, 2, . . ., ζ.

If a solution was found for a section $(0,j)$ between nodes 0 and j, the search for an optimal solution between the node 0 and the node $J + 1$ can be performed by adding the optimal solution for the section $(j,J + 1)$ to the previous solution. Thus, an optimal solution can be found for each section. The algorithm for this procedure follows (Fig. 3).

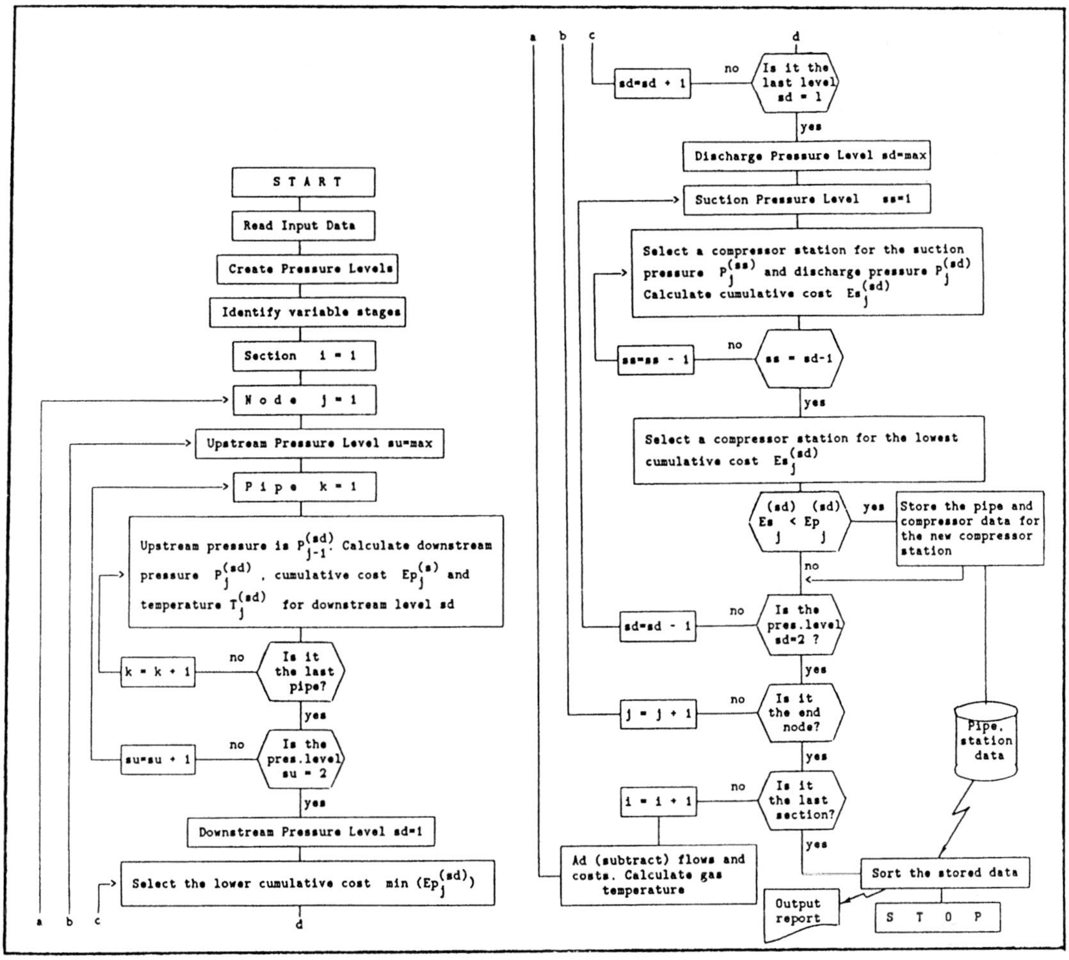

FIG. 3. Flow chart.

A. Create Pressure Levels at the Upstream Node and Initialize Variable Stages

B. Create Candidate Pipe Trajectories

1. Pressure at the starting node $P_0^{(s)}$ is assumed as the maximum available pressure from its gas source.
2. For a given upstream pressure $P_0^{(su)}$, temperature $T_0^{(su)}$, and flow $Q_0^{(su)}$, select a pipe diameter and thickness and calculate the downstream pressure $P_1^{(sd)}$ at node 1.

Violation of any constraint such as velocity maximum, minimum pressure, or maximum pressure eliminates the trajectories. Also, other trajectories will be eliminated by prespecified pipe data.

3. Calculate the cumulative cost E_{p_1} as a sum of the previous E_0 and the present worth of the initial cost for the selected pipe (see Eq. 7).
4. Perform similar calculations for all possible pipe diameters and minimal required thicknesses. For most pipeline situations, variation in nominal pipe diameter is not permitted since that would greatly complicate pigging. In that case, far fewer alternatives need to be considered.
5. Repeat the same process for the rest of the upstream pressure levels $P_0^{(su)}$, su $= 2, 3, \ldots, \zeta$
6. Split downstream pressures at node 1 into pressure levels. For each pressure level, compare costs, and store the solution which has the least cumulative cost $\min(E_{p_1}^{(s)})$. For capacity expansions compare also the new pipe cost with the cost of a larger amount of gas transmitted through the old pipe, without a new loop.

C. Select the Optimal Compressor Stations

1. For node 1 select a compressor station for the highest pressure level $P_1^{(\zeta)}$ as the discharge pressure, and the lowest pressure level $P_1^{(1)}$ as the suction pressure. Calculate the initial and operating costs and also calculate the cumulative cost $E_{s_1}^{(s)}$ by adding the cost of the previous solution for the pressure level used as a suction pressure to the station cost (see Eq. 8).
2. Repeat the same process for different pressure levels as the suction pressures $P_1^{(ss)}$, where ss $= 2, 3, \ldots, \zeta - 1$ are the pressure levels.
3. Select the lowest cumulative cost $\min(E_{s_1}^{(s)})$ for the compressor station which has the discharge pressure $P_1^{(\zeta)}$.
4. Compare the cumulative cost E_{s_1} with the previously stored cumulative pipe cost for the same pressure level $E_{p_1}^{(\zeta)}$ (if it exists). If $E_{s_1}^{(\zeta)} < E_{p_1}^{(\zeta)}$, then identify the pressure $P_1^{(\zeta)}$ at node 1 as the fixed node for a new compressor station and store the data corresponding to the new compressor station. Otherwise erase the data and ignore this pressure. A compressor station must be inserted any time no solution from a pipe exists.
5. Repeat the same process for discharge pressure at the pressure levels $P_1^{(\zeta-1)}$, $P_1^{(\zeta-2)}, \ldots, P_1^{(2)}$.

D. Create Optimal Trajectories

1. Switch to the process for node 2 by using only the pressure from the initial and the developed fixed nodes as an upstream pressure for selecting optimal pipes.

2. Continue the same process for all assigned nodes in the section i including the downstream junction node. The optimal trajectory for one section has been developed.
3. Go to the next child section. If no child section is left, jump to the parent section, calculate the joint amount of gas, cumulative cost as a cost from the joint children sections, and the common gas temperature for each pressure level. For a multiroot system, selection of a next section is more complicated.
4. Continue this process to the last system node by developing a number of optimal trajectories for each pressure level at each node of the system. Then select the least cost solution.

E. Perform the "Back Process" by Reading the Parameters for the Optimal Trajectory and Print Output Data

4. Example

A. Create Optimal Pipe Trajectories

In Fig. 4 the stage j has 6 pressure levels between pressures $P1^{(0)}$ to $P1^{(6)}$. Let us assume that from the previous process a relatively-optimal solution has been developed between stages $j-1$ and j, and pressures and objective functions for the previous trajectories are $(P_j^{(1)}, E_j^{(1)}), \ldots, (P_j^{(6)}, E_j^{(6)})$, where $P_j^{(1)}$ is the pressure for the optimal solution between pressure levels $P1^{(0)}$ and $P1^{(1)}$, and $E_j^{(1)}$ is the cumulative value of the objective function which corresponds to the pressure $P_j^{(1)}$. The same is for $P_j^{(2)}$, $P_j^{(4)}$, $P_j^{(5)}$, and $P_j^{(6)}$. There is no solution for the third pressure level. As was mentioned earlier, two pipes are always considered at each stage, but one of them can be zero or it can be presized.

Starting from level 6 of stage j where the optimal pressure is $P_j^{(6)}$ and the total cost is $E_j^{(6)}$, the dynamic programming process must create all possible trajectories from $P_j^{(6)}$ to stage $j+1$ by selecting different pipe materials and diameters. The selection has to be eliminated if any constraint such as velocity maximum, minimum pressure, or maximum pressure was violated as it was at level 3.

Also, if some pipe data were prespecified, they cannot be changed, and trajectories have to be made for each level only for prespecified attributes. This means that the more constraints specified, the less computer time is required to find the optimum. However, the solution moves farther from the theoretical optimum.

For design problems the second pipe is always zero, therefore only one pipe has to be selected. For expansion problems the first pipe exists and only the second pipe has to be selected and pressure loss has to be calculated with both pipes working in parallel. It is very important for expansion problems that the selection of the second pipe has to start from zero-pipe conditions when no second pipe exists and the total flow has to be transmitted only through the existing pipe. This zero-second-pipe trajectory must be included

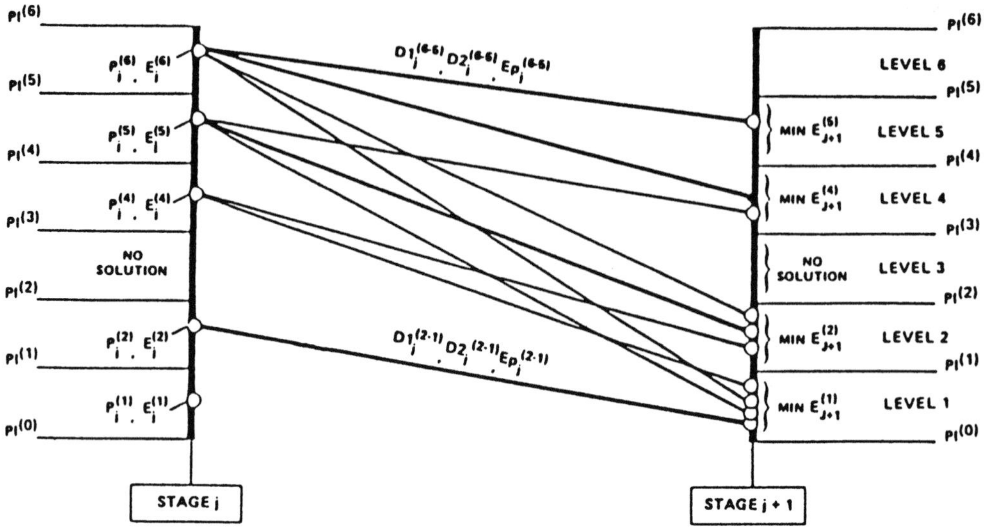

FIG. 4. Pipe trajectories.

into the cost competition since it will identify the solution where no loopings are recommended. The trajectories cannot be created from a particular pressure level if it has no previous solution, such as has happened with the third level of stage j (Fig. 4). The ending pressure, temperature, gas density, and cumulative cost for each pipe trajectory have to be calculated. The maximum number of pipe trajectories is

$$\Omega = \zeta(\zeta - 1)/2 \tag{17}$$

where ζ is the number of pressure levels.

In Fig. 4 this process creates four trajectories at level 1, and three trajectories at level 2, and two trajectories at level 4, and one trajectory at level 5, and no trajectories for levels 3 and 6. In the next step each pipe objective function has to be added to the cost of the corresponding trajectory j and be grouped under each pressure level of the stage $j + 1$:

Level 6: No trajectories
Level 5:

$$E_{j+1}^{(6-5)} = E_{p_j}^{(6-5)} + E_j^{(6)}$$

Level 4:

$$E_{j+1}^{(6-4)} = E_{p_j}^{(6-4)} + E_j^{(6)};\ E_{j+1}^{(5-4)} = E_{p_j}^{(5-4)} + E_j^{(5)}$$

Level 3: No solution

Level 2:

$$E_{j+1}^{(6-2)} = E_{p_i}^{(6-2)} + E_j^{(6)};\ E_{j+1}^{(5-2)} = E_{p_i}^{(5-2)} + E_j^{(5)};\ E_{j+1}^{(4-2)} = E_{p_i}^{(4-2)} + E_j^{(4)}$$

Level 1:

$$E_{j+1}^{(6-1)} = E_{p_i}^{(6-1)} + E_j^{(6)};\ E_{j+1}^{(5-1)} = E_{p_i}^{(5-1)} + E_j^{(5)};$$
$$E_{j+1}^{(4-1)} = E_{p_i}^{(4-1)} + E_j^{(4)};\ E_j^{(2-1)} = E_{p_i}^{(2-1)} + E_j^{(2)}$$

Then the optimal trajectory has to be selected for each pressure level of stage $j + 1$:

$$E1_{j+1}^{(u-w)} = \min(E_{p_i}^{(u-w)} + E_j^{(u)}), \qquad u = 2, 3, \ldots, \zeta \quad w = 1, 2, \ldots, \zeta - 1 \tag{18}$$

where $E_{p_i}^{(u-w)}$ = cost of pipe with upstream pressure at the level u (stage j) and the downstream pressure at the level w (stage $j + 1$)
$E_j^{(u)}$ = cumulative cost at j-stage and u-pressure level

The above-mentioned procedure was explained for a clear understanding of the computation technique. In reality, the selected pipe is stored after checking for constraints only if the value of its objective function is lower then the previously stored value for the same node k and pressure level. As a result of that step, one optimal trajectory for each pressure level has been stored into temporary storage:

Level 1: $D1_j^{(2-1)}, D2_j^{(2-1)}, E_{p_i}^{(2-1)}, E_j^{(1)}$
Level 2: $D1_j^{(5-2)}, D2_j^{(5-2)}, E_{p_j}^{(5-2)}, E_j^{(2)}$
Level 3: No solution
Level 4: $D1_j^{(6-4)}, D2_j^{(6-4)}, E_{p_j}^{(6-4)}, E_j^{(4)}$
Level 5: $D1_j^{(6-5)}, D2_j^{(6-5)}, E_{p_i}^{(6-5)}, E_j^{(5)}$

B. Select The Optimal Compressor Stations

In Fig. 5 the stage $j + 1$ has 6 pressure levels between $P1_{j+1}^{(0)}$ to $P1_{j+1}^{(6)}$. Let us assume that previously a process relatively-optimal solution was developed between the stages (j) and ($j + 1$), and the pressures and objective functions for the previous pipe trajectories are ($P_{j+1}^{(1)}$, $E_{j+1}^{(1)}$), ..., ($P_{j+1}^{(5)}$, $E_{j+1}^{(5)}$). There was no solution for pressure level 3 and for pressure level 6.

Station trajectories have to be developed for each pressure level starting from level 6.

Let us assume that the pressure $P_{j+1}^{(6)}$ at pressure level 6 is the discharge pressure for the compressor station at stage $j + 1$. Then different compressor ratios can be selected for this discharge pressure starting from $P_{j+1}^{(5)}$ to $P_{j+1}^{(1)}$ and for the following compressor ratios:

$$R_{j+1}^{(5-6)} = P_{j+1}^{(6)}/P_{j+1}^{(5)}$$

$$R_{j+1}^{(4-6)} = P_{j+1}^{(6)}/P_{j+1}^{(4)}$$
$$R_{j+1}^{(2-6)} = P_{j+1}^{(6)}/P_{j+1}^{(2)}$$
$$R_{j+1}^{(1-6)} = P_{j+1}^{(6)}/P_{j+1}^{(1)}$$

There is no solution for $R_{j+1}^{(3-6)}$ for level 3.

The costs of the compressor station $Es_{j+1}^{(5-6)}, Es_{j+1}^{(4-6)}, \ldots, Es_{j+1}^{(1-6)}$, as well as the selected equipment identification and station configuration have to be saved temporarily. The station trajectory can be selected only if: a) the solution for the suction pressure level exists, and b) the restriction for maximum compression ratio was not violated.

For each station the actual discharge pressure, discharge gas temperature, station cost, and the cumulative objective function are calculated. If there are a number of compressors with compressor ratios which fall into the same discharge pressure levels, the one with the lowest cost is selected.

The optimum compressor station for each d-pressure level is the one for which

$$E2_{j+1}^{(d)} = \min(Es_{j+1}^{(s-d)} + E_{j+1}^{(s)}), \qquad d = 2, 3, \ldots, \zeta \qquad s = 1, 2, \ldots, \zeta - 1 \tag{19}$$

where $Es_{j+1}^{(s-d)}$ = cost of the compressor station designed for the suction pressure s and for the discharge pressure d

$E_{j+1}^{(s)}$ = cumulative cost for the trajectory at the suction pressure level s

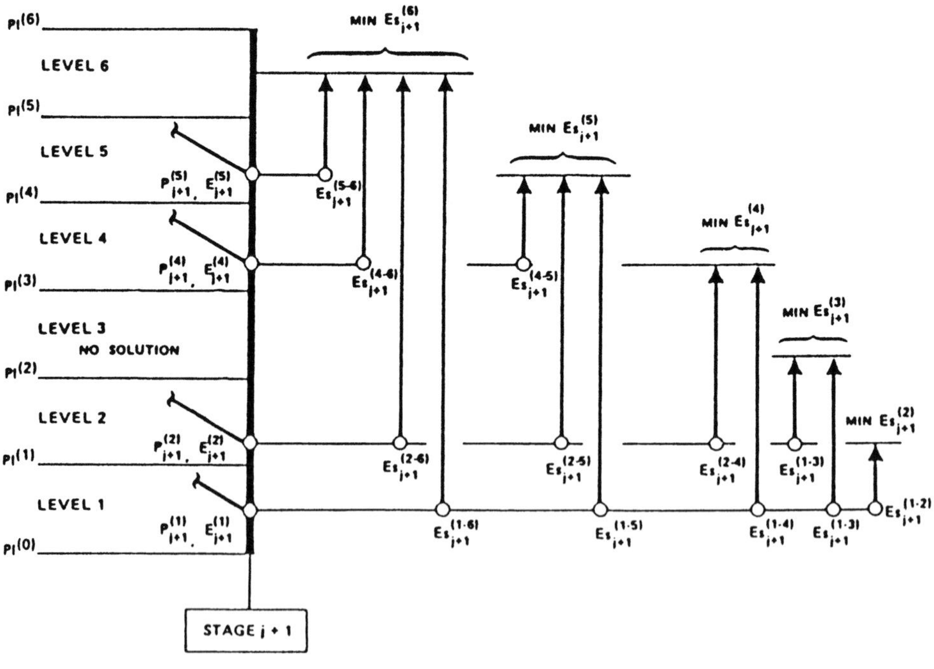

FIG. 5. Compressor station trajectories.

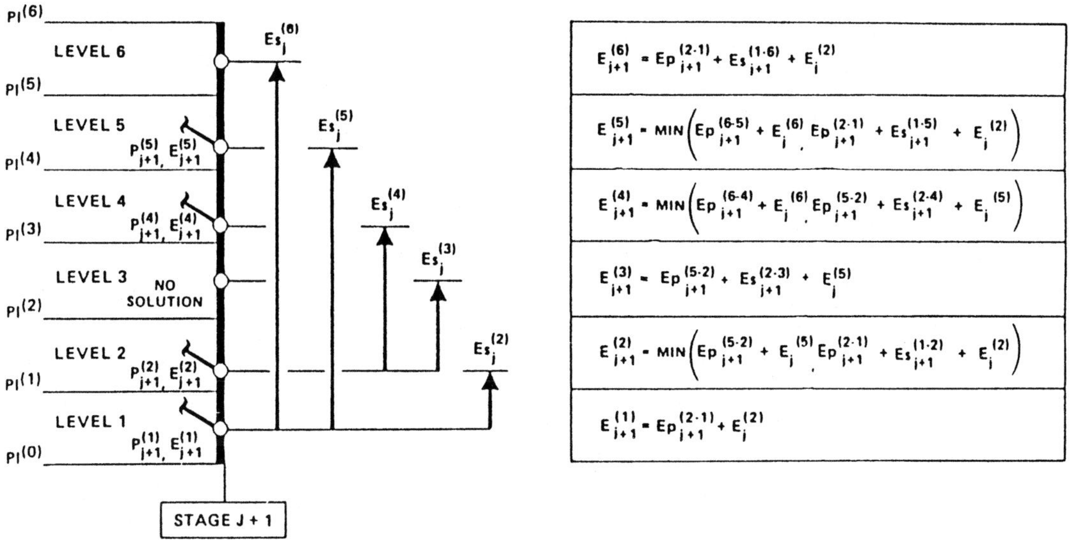

FIG. 6. Optimal cumulative trajectories.

The third level has no trajectory, therefore it must be considered only as a possible discharge pressure level, not suction. The final solution for stage $j + 1$ which includes cumulative trajectories is presented in Fig. 6.

The costs of the optimal trajectories at the downstream pipe level are:

$$E_{j+1}^{(w)} \in (E_{j+1}^{(1)}, E_{j+1}^{(2)}, E_{j+1}^{(4)}, E_{j+1}^{(5)})$$

The costs of the optimal trajectories at the discharge compressor station are

$$E_{j+1}^{(d)} \in (E_{j+1}^{(2)}, E_{j+1}^{(3)}, E_{j+1}^{(4)}, E_{j+1}^{(5)}, E_{j+1}^{(6)})$$

C. Identify a New Compressor Station

The compressor station is considered as a fixed node only when

$$E1_{j+1}^{(x)} > E2_{j+1}^{(x)}$$

where $E1_{j+1}^{(x)}$ = cumulative cost of the trajectory at the downstream pressure level x

$E2_{j+1}^{(x)}$ = cumulative cost of the trajectory at some discharge pressure level x

The results of this comparison are presented in Fig. 6. There was no previous trajectory for level 6. This is considered as $E1_{j+1}^{(6)} = 0$. Therefore the minimal cost accrues at the compressor station with suction pressure $P_{j+1}^{(1)}$.

However, the discharge pressure $P_{j+1}^{(6)}$ is established as the new fixed node. The cumulative objective function E_{j+1}^{6} includes the cost of the previous $E_j^{(2)}$, the cost of pipe (or pipes for the expansion problem) $E_{p_j}^{(2-1)}$, and the cost of the compressor station $E_{j+1}^{(1-6)}$ from pressure level 1 (suction) to pressure level 6 (discharge).

There are two optimal solutions for the pressure level 5: $E1_{j+1}^{(5)}$ for the downstream pipe and $E2_{j+1}^{(5)}$ for the compressor station at discharge pressure level 5. If

$$E_{p_{j+1}}^{(2-1)} + E_{s_{j+1}}^{(1-5)} + E_j^{(2)} < E_{p_{j+1}}^{(6-5)} + E_j^{(6)} \tag{20}$$

then level 5 of the node $j + 1$ is also considered as a fixed node, otherwise it is ignored.

The same comparison must be made for the pressure levels 4 and 2. At level 3 the only compressor station that can be considered is one which defines a fixed node at this level. No compressor station can be established at pressure level 1; however, it can never be a fixed node.

D. Trajectories for Junctions and Terminal Nodes

Each pipe junction is always considered as a fixed node for all pressure levels. Comparison between pipe and compressor station costs have been made in the same way they were made for the variable nodes. However, if the pipe cost is lower than the cost of compression, the pipe data cannot be ignored as was done for the variable stage but must be saved as an initial node for the parent section.

The dynamic process for the terminal node is presented in Fig. 7. Four optimal pipe trajectories between stages j and $j + 1$ were kept in temporary storage. Their costs are $E_{j+1}^{(1)}$, $E_{j+1}^{(2)}$, $E_{j+1}^{(3)}$, and $E_{j+1}^{(4)}$. However, the terminal node has a constraint of $P_{j+1}^{(\min)}$ which is the minimum required pressure in the fourth pressure level. This pressure must be used as the discharge pressure. Three compressor stations are available for this pressure and for the different suction pressures. The following search will select an optimal trajectory:

$$\min(E_{j+1}^{(w)} + E_{s_{j+1}}^{(s-4)}, E_{j+1}^{(4)}), \qquad w = s = 1, 2, 3 \tag{21}$$

5. Important Considerations

Staging. The optimization process starts by dividing sections of a pipeline, including the initial fixed nodes such as terminals, sources, junctions, and new variable nodes, into stages. Each fixed or variable node has to be considered as a potential location for a compressor station. The number of levels influences only the accuracy of optimization and is limited only by the computer run time.

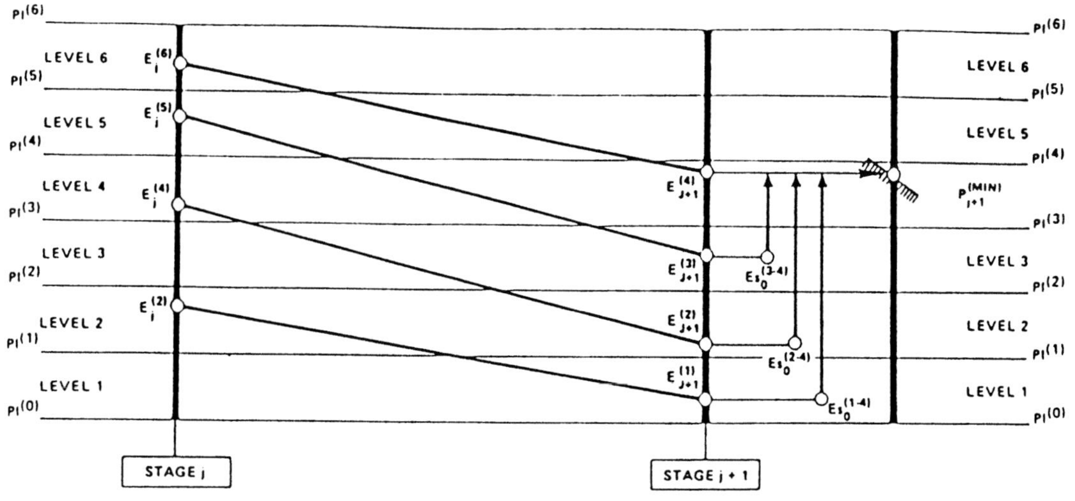

FIG. 7. Trajectories at the terminal node.

Also, a stage must be established for all existing stations. It is convenient to set stages equidistant through a number of miles, but changes in distances can be used.

Starting Node. According to Bellman, the dynamic optimization process has to be started from the end-stage. For tree-network applications this requirement is very important since there is no other way to calculate the costs of the subtree trajectories linked at junction nodes.

However, in order to calculate the heat loss/gain, the calculation should be in the direction of gas flow.

The starting node is always a problem since the amount of gas consumed in compressor stations is also a dependent variable. The factors affecting selection of the start-node also depends on where the flow requirements are. Sometimes the supply flow is known and the gas has to be distributed through the paths as a percent of total flow. It could happen that consumer demands for the amount of gas and the total supply flow have to be calculated. Also, it can happen that supply pressure is an independent variable and has to be optimized as well.

Therefore, requirements for flow rates, for gas temperatures, and for desirable optimization level are the major considerations in selecting a starting node.

It was found that the most convenient way for organizing the dynamic programming algorithm is to start from the upstream supply node. A new procedure which minimizes (but does not eliminate) these disadvantages has been developed. The optimization process always starts from the source nodes where gas temperatures are known. The first node is considered as the lowest degree. A switch to a higher degree of subtree is made only when the trajectories for all lower degree sections have been developed.

A specific procedure was developed for sink sections. The trajectories are started from an upstream node as for source sections. However, prehistory for previous sections is ignored and trajectories are developed from zero cost but for known gas temperatures at each pressure level. After the dynamic process reaches the terminal node, the minimum cost is stored and assigned to the upstream node.

The Difference between Design and Expansion Problems. For the design problem, one pipe is considered as a link between stages. For expansion problems it must be two pipes: one is for the existing pipe and the second pipe (looking) has to be optimized or eliminated. In order to have an opportunity for optimum sizing both problems by using the same algorithm, two pipes are always considered as a link between stages but for the design problems one of these pipes is zero.

It is very important for expansion problems that the selection of the second pipe has to start from zero-pipe when no second pipe exists and expanded flow has to be transmitted only through the existing pipe. This zero-second-pipe trajectory also has to be included into the cost competition since it will identify the sections where no looping is recommended. In other respects the algorithms for design and expansion problems are identical.

Diameter Telescoping. "Pigging" creates a requirement which prevents telescoping of diameters in-between stages. It can be easily proven that the simple treatment of telescoped solutions, elimination of the trajectories where diameters were changed, can cause the loss of solutions. It has even resulted in no solutions for a stage. The new "*Variable Stage*" technique takes care of this requirement.

A simple example shown in Fig. 8 explains this phenomenon. Two trajectories are available from stage 1 to stage 2 pressure level 3: the first, $P_1^{(4)}$–$P_2^{(3)}$, with pipe diameter $D_1^{(4-3)}$ and cost $E_1^{(4-3)}$, and the the second, P_1^5–P_2^3, with pipe diameter $D_1^{(5-3)}$ and cost $E_1^{(5-3)}$. Let us assume that the cost comparison results show that the objective function for the second pipe is less than for the first pipe: $E_1^{(5-3)} < E_1^{(4-3)}$. Following the algorithm, the trajectory $P_1^{(4)}$–$P_2^{(3)}$ is replaced.

For the second stage, the only available trajectory is the one which must have the same diameter as the previous stage, therefore the diameter $D_2^{(3-1)} = D_1^{(5-3)}$ must be selected. But it can happen that this is impossible because of constraint violations (for example, the pressure is too low). As a result, no trajectory from stage 2 level 3 will be developed.

However, such a trajectory does exist. If in the previous stage the more expensive trajectory $P_1^{(4)}$–$P_2^{(3)}$ was selected instead of the cheaper trajectory $P_1^{(5)}$–$P_2^{(3)}$, the trajectory $P_1^{(4)}$–$P_2^{(3)}$–$P_3^{(2)}$ from the stage 2 level 3 will be available. This example shows that the traditional dynamic programming technique is not able to solve the problem.

The "variable stages" technique solves the telescoping problem. As it was shown, the idea of this technique is to create pipe pressure loss trajectories not between a new and a previous stage but between a new stage and previous compressor stations. This means that if comparison between solutions has

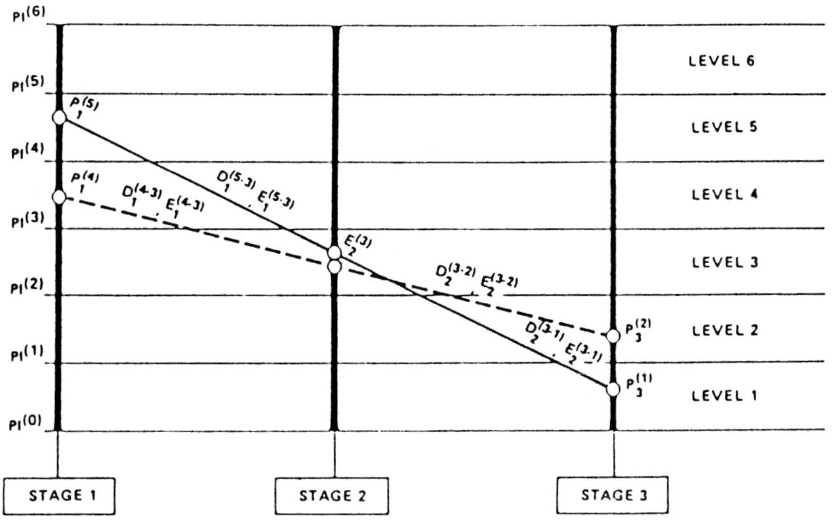

FIG. 8. Possible trajectories for an internal node.

not resulted in selecting a new compressor station, the stage has to be eliminated and the distance from the previous station has to be enlarged to the next stage.

Shadowing. Shadowing is a process where the nonprospective trajectories for a fixed stage are identified and terminated. In Fig. 9a three nodes at stage 2b are fixed for the levels 3, 5, and 6. This means that all trajectories which are "shadowed" by the fixed nodes are not economically reasonable and have to be rejected. This include the trajectories: $P_1^{(6)}-P_{2a}^{(3)}$, $P_1^{(5)}-P_{2a}^{(3)}$, and $P_1^{4}-P_{2a}^{(3)}$. Also, all trajectories which hit the boundary conditions where the downstream pressure is less then $P1^{(0)}$ have to be eliminated as well.

In Fig. 9b the process is continued to the next stage 3. One can see that only the trajectories which were out of the shadows are continued as well as the new trajectories from the fixed nodes at stage 2.

Pressure Levels (states). The pressure that is selected as a variable for dynamic programming has to be divided at each stage into a number of states or pressure levels. The maximum number selected for our problem is 50. The higher this number the more efficient the optimization will be, but it will increase run time. It is possible to run the problem with a limited number of pressure levels, then analyze the boundary conditions and run the problem over for a reduced pressure range in order to get a more accurate solution.

Presized System Elements. If some station equipment or pipe sizes are specified, trajectories must be created from each level only for those equipment sizes or pipe diameters. It is possible to prespecify pipe thicknesses or only a minimum number of compressors connected in parallel. This is very important for station operating considerations.

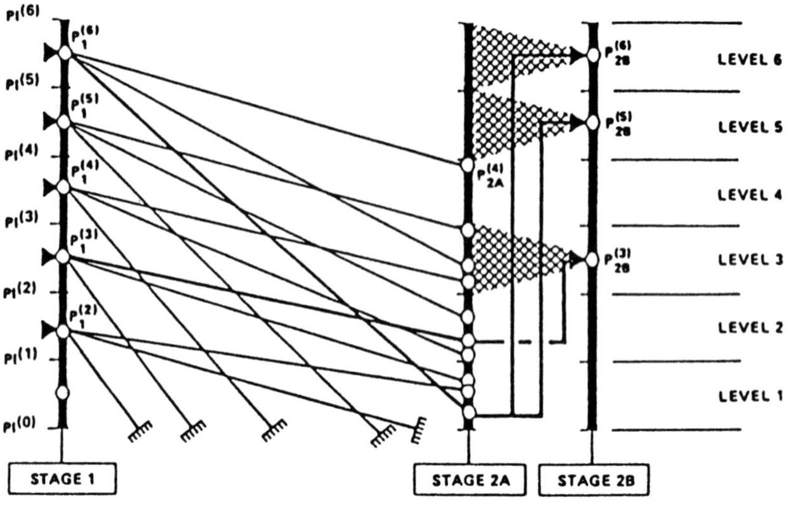

FIG. 9a. Variable stages and shadows.

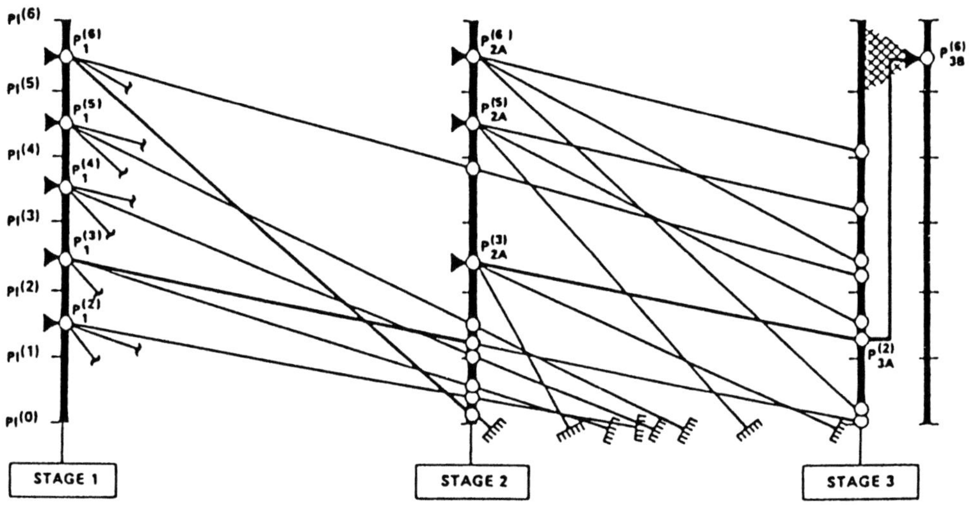

FIG. 9b. Variable stages and shadows.

Trajectories at Junctions. The pressures at each junction have to be the same for all connected sections. For our purposes only tee-junctions are taken into account. The local resistance coefficient in junctions is ignored. After both sets of optimal trajectories are created for descendants, their costs and flows have to be added and the gas temperature in the parent section has to be recalculated at each pressure level.

If one child was connected to the parent at a pressure level where the second child has no trajectory, the trajectors for the first child must also be

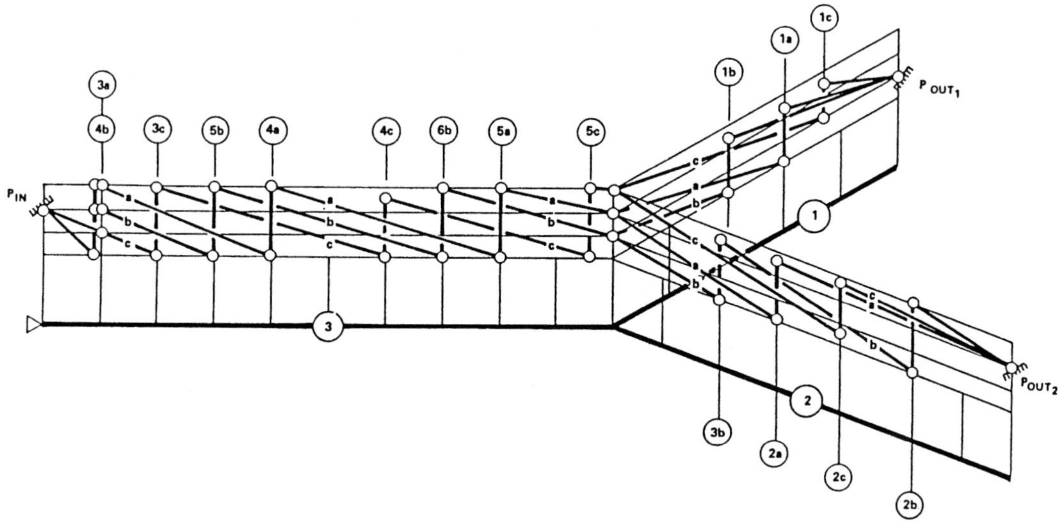

FIG. 10. Trajectories at the junction.

eliminated. A connection of three trajectories *a*, *b*, and *c* at a junction is shown in Fig. 10. The trajectories start from the given pressure *P*out1 at section 1 and are created to the junction. Then section 2 is optimized and three new trajectories are created. Then at the junction the flows from sections 1 and 2 are added, and the combined gas temperatures are calculated for each pressure level. The optimal trajectories for section 3 complete the optimization process.

Advantages of the Proposed Method

The proposed method has many advantages:

Optimization will be performed for all major independent variables, such as the number of compressor stations and their location, pipe diameters, suction and discharge pressures, and compressor types with their number and configuration. The parameters which are not optimized are the discharge temperatures. This is a problem for future consideration. Currently these temperatures must be input.

No necessity for concave objective functions.

The solution is an optimal trajectory which represents a global minimum.

Only the number of stages and pressure levels can influence the accuracy of optimization, therefore the accuracy is a function of run time, not of the quality of a method.

The optimized equipment and pipes can be represented as integers.
The dynamic process is not iterative, therefore it has no convergency problems.
A distribution network or a gathering system can be optimized as well as a long pipeline.
Optimal design or optimal capacity expansion can be performed by using the same program.
Not only optimal solutions but many solutions which are close to optimal can be analyzed.

The program can be effectively used in order to analyze:

The efficiency of pipe thickness telescoping.
The efficiency of pipe diameter telescoping.
The use of two or more pipes in parallel with different diameters in order to get more flexible pressure resistances in pipes. Theoretically, two lower diameter pipes should not be cheaper than one with a larger diameter, but in the real world, where cost is a very complicated function, it can happen.
The use of alternative fuels at peak loads.
From a paper [27] it is known that sometimes the use of lower specific-fuel-consumption reciprocating engines instead of turbine units can increase the efficiency and save gas. This fact should be checked.
The algorithms described are capable of simultaneously finding not only the optimal solution but also all solutions close to the optimal. For example, optimum solutions can be trajectories for different pressures at the supply node or at the consumer nodes.

Conclusion

The method given in this article is a base for development of a computer program. Such a program could be used to substantially reduce costs by generating optimal solutions under different load profiles, and by analyzing the effects of different operating conditions.

The development of a computer program for the IBM AT based on dynamic programming with variable stages is underway. This program will be able to find optimum solutions for onshore or offshore gas pipelines. The program is planned to be tested on an existing system, such as the Alaskan pipeline, to analyze the efficiency of the optimization technique.

In the future the optimization procedure should be extended by considering the gas temperature as an independent variable in addition to the pressure. This will allow optimization of cooling towers and refrigerators. A future enhancement for the prepared method is the addition of multiperiod optimization.

This material was presented at the Pipeline Simulation Interest Group (PSIG) Annual Meeting, October 30–31, 1986, and is used with the permission of PSIG.

References

1. D. C. McClure, *Using Linear Programming to Optimize Pipeline Analysis*, PSIG Annual Meeting, St. Louis, October 1982.
2. G. E. Graham, "Optimizing Gas-System Design," *Oil Gas J.*, November 13, 1972.
3. R. Ore, *Theory of Graphs*, American Mathematical Society, Providence, Rhode Island, 1962.
4. E. Horowitz and S. Sahni, *Fundamentals of Data Structures*, Computer Science Press, New York, 1976.
5. J. B. Lee, "Optimizing the Design for Canadian Section of Alaskan Gas Pipeline," *Oil Gas J.*, May 18, 1981.
6. Z. Huang and A. Seireg, "Optimization in Oil and Gas Pipeline Engineering," *Trans. ASME*, *107*, 264 (June 1985).
7. P. J. Wong and R. E. Larson, "Optimization of Natural-Gas Pipeline Systems via Dynamic Programming," *IEEE Trans. Autom. Control*, October 1968.
8. A. P. Cheeseman, "How to Optimize Pipeline Design by Computer," *Oil Gas J.* December 1971.
9. T. C. Bickel, D. M. Himmelblau, and T. F. Edgar, *The Optimal Design of a Long Gas Transmission Line with Compressors in Series*, Presented at the 81st AIChE National Meeting, Kansas City, Missouri, April 12, 1976
10. Scientific Software-Intercomp, *Compositional Multiphase Pipeline Model (HCOMP)*.
11. K. F. Pratt and J. G. Wilson, "Optimization of the Operation of Gas Transmission System," *IEEE Trans.*, *6*(5) (October 1984).
12. B. Rothfarb, H. Frank, D. M. Rosenbaum, K. Steiglitz, and D. J. Kleitman, "Optimal Design of Offshore Natural Gas Pipeline Systems," *J. Oper. Res. Soc.*, *18*, 992–1020 (1970).
13. A. P. Cheeseman, "How to Optimize Pipeline Design by Computer Appears Possible," *Oil Gas J.*, December 20, 1971.
14. G. E. Graham, D. A. Maxwell, and A. Vollone, "How to Optimize Gas Pipe Line Networks," *AGA Trans.*, *Pipe Line Ind.*, June 1971.
15. O. Flanigan, "Constrained Derivatives in Natural Gas Pipeline System Optimization," *J. Pet. Technol.*, pp. 549–556 (May 1972).
16. G. G. Boyne, "The Optimum Design of Fluid Distribution Networks with Particular Reference to Low Pressure Gas Distribution Networks," *Int. J. Numer. Methods Eng.*, *5*, 253–270 (1972).
17. H. B. Martch and N. J. McCall, *Optimization of the Design and Operation of Natural Gas Pipeline Systems*, Presented at the SPE 47th Annual Meeting, San Antonio, October 8–11, 1972, Paper SPE4006.
18. N. Zadeh, "Construction of Efficient Tree Networks: The Pipeline Problem," *Networks*, *3*, 1–31 (1973).
19. S. Bhaskaran and J. M. Salzborn, "Optimal Design of Gas Pipeline Networks," *J. Oper. Res. Soc.*, *30*(12), 1047–1060 (1979).

20. A. D. Waren and L. S. Lasdon, "The Status of Nonlinear Programming Software," *Oper. Res.*, *27*, 1979.
21. E. Sandgren and K. M. Ragsdell, "The Utility of Nonlinear Programming Algorithms: A Comparative Study—Part I and Part II," *ASME Trans.*, *102*, 1980.
22. F. I. Soliman and B. A. Murtagh, "The Solution of Large-Scale Gas Pipeline Design Problems," *Eng. Optim.*, *6*(2), December 1982.
23. F. O. Olorunniwo and P. A. Jensen, "Optimal Capacity Expansion Policy for Natural Gas Transmission Network—A Decomposition Approach," *Eng. Optim.*, *6*, 13–30 (1982).
24. D. C. McClure and T. Miller, "Linear Programming Offers Way to Optimize Pipeline Analysis," *Oil Gas J.*, July 18, 1983.
25. R. E. Bellman, *Dynamic Programming*, Princeton University Press, New York, 1957.
26. R. J. Tsal and M. S. Adler, *Optimization of Pressure, Temperature, and Flow Rates in Heating and Cooling Tree-Networks*, Presented at World Congress on Heating, Ventilation and Air Conditioning, Copenhagen, Denmark, 1985.
27. R. C. Hesje, "AGTL Program Boosts Energy Efficiency in Gas Transmission," *Oil Gas, J.*, July 23, 1979.

ROBERT J. TSAL
EDWARD GORDON
KENNETH O. SIMPSON
ROBERT R. OLSON

Manifolds

Manifolds disperse or collect flows, and can be characterized as dividing, combining, reverse, or parallel (Fig. 1). Whatever the case, the engineer's objective is to make the flows approximately the same through all the branches without wasteful pressure drops or excessive capital cost.

If the header diameter and the pressure drop through each lateral are made large, the flows are likely to be uniform. Indeed, there are design "rules of thumb" that are more or less based upon this approach, and manifolds so designed are often satisfactory. However, if operating costs could be significant, the engineer should design the manifold more precisely.

The mathematical definition of the flow distribution in a manifold results in a nonlinear second-order differential equation that constitutes a boundary-value problem, which must be solved via an iterative numerical procedure that is very cumbersome. This article presents a more-direct procedure for designing a manifold. The following is assumed: the manifold headers and

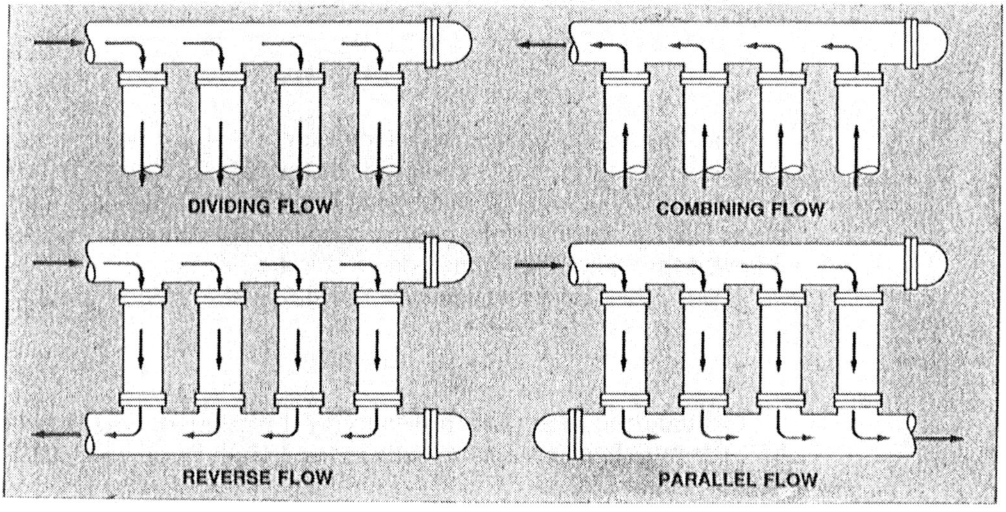

FIG. 1. Four types of pipe manifolds.

laterals are cylindrical pipe of constant cross-section; the fluids are incompressible; flow is turbulent.

Two factors primarily control flow distribution in a manifold: the pressure drop in the header (the primary pipe) due to frictional losses, and the pressure changes in the header due to pressure recovery. A drop in flow velocity in the header results in pressure recovery; in a dividing manifold, not all the momentum is converted into static pressure, because some of it gets carried into the laterals; the momentum correction factor, θ, takes this effect into account.

Combining-Flow Manifold

The fractional variation in the flow through the laterals of a combining-flow manifold can be calculated via Eq. (1) [4]:

$$F_v = \frac{N^2 A_3^2}{HA_2^2}\left[\left(\frac{8}{11}\right)\left(\frac{0.0791}{\mathrm{Re}_{o_2}{}^{1/4}}\right)\left(\frac{L}{D_2}\right) + \frac{\theta_2}{2}\right] \tag{1}$$

Here,

$$F_v = [Q_3(O) - Q_3(L)]/(Q_o/N)$$

Equation (1) is based on the Blasius formula [2]:

$$f = 0.0791/\mathrm{Re}^{1/4} \tag{2}$$

Here,

$$\text{Re} = 4\rho Q_2/\pi\mu D_2$$

The Blasius formula is based on smooth pipes, but the header of a manifold, of course, has openings for the laterals. Multiplying the value of f from Eq. (2) by 1.2 provides a reasonable approximation of the friction factor in the header of a manifold, according to Pigford et al. [3]. This approach is followed throughout the remainder of this article.

The pressure drop through each lateral is assumed to be given by

$$\Delta P = H\rho v^2 \tag{3}$$

Calculate the lateral flow resistance, H, by means of Eq. (3). Note that v is here the superficial average velocity in the laterals based on D_3 (the diameter of the laterals).

The pressure drop through the laterals results from losses due to friction, flow through an orifice, and turning losses at lateral entrances and exits. This pressure drop is assumed to vary directly with the square of the superficial velocity through the lateral. Therefore, H can be calculated directly by means of Eq. (4) if the pressure drop through the lateral, ΔP, is known for a lateral flow of Q_3:

$$H = (g_c \Delta P \pi^2 D_3^{\,4})/(16\rho Q_3^{\,2}) \tag{4}$$

The overall momentum-correction factor, θ, in Eqs. (1) and (7) takes into account the pressure recovery. For dividing-flow manifolds, θ_1 ranges from 0.7 to 1.2, and θ_2 typically varies from 1.5 to 2.2. The shape of the lateral opening can significantly affect the value of θ. For laterals at right angles to the header, θ_1 averages 0.8, and 2 for θ_2, according to published values [3]. Clearly, uncertainly in θ_1, and θ_2 can significantly affect manifold design. Rely on the foregoing average values for θ_1 and θ_2 only if experimental values are not available.

A comparison showed Eq. (1) to agree closely with the numerical solution of the complete model, except when the number of laterals (N) was less than 30 [1]. A study of the two limiting cases (i.e., $\theta = 0$, and $\text{Re}_{o_2} \rightarrow \infty$) led to the incorporation of the following correction factors in Eq. (1):

$$F_v = F_2\left[1 - \left(\frac{1.33 + 3F_2}{N}\right)\right] + R_2\left[1 - \left(\frac{1.05 + 0.9R_2}{N}\right)\right] \tag{5}$$

$$F_2 = \frac{N^2 A_3^{\,2}}{HA_2}\left[\frac{0.0949}{\text{Re}_{o_2}^{\;1/4}}\left(\frac{8}{11}\right)\left(\frac{L}{D_2}\right)\right] \tag{6}$$

$$R_2 = (N^2 A_3^{\,2}/HA_2^{\,2})(\theta_2/2) \tag{7}$$

Note that as $N \rightarrow \infty$, Eq. (5) approaches Eq. (1). Also note that the friction factor for the friction term F_2 is 1.2 times that from the Blasius formula. Equation (5) was compared with values calculated from the numerical solution of a boundary-value problem [1] for the following range of parameters:

$$5 \le N \le 25$$

$$2 \times 10^{-4} \le F_2 \le 1.6$$

$$4 \times 10^{-4} \le R_2 \le 1.0$$

For the 300 cases that resulted in $0.01 \le F_v \le 0.10$, the average deviation was less than 1.0%.

Dividing-Flow Manifold

In the case of a dividing-flow manifold, the pressure recovery and friction work against each other—that is, friction in the header causes the pressure to drop along the header while the pressure-recovery effect increases the pressure down the header. This results in the following version of Eq. (1) for a dividing-flow manifold:

$$F_v = \frac{N^2 A_3^2}{HA_1^2}\left[-\left(\frac{8}{11}\right)\frac{0.0949}{\mathrm{Re}_{o_1}^{1/4}}\left(\frac{L}{D_1}\right) + \frac{\theta_1}{2}\right] \tag{8}$$

Note that the major difference between Eq. (1) and Eq. (8) is the minus sign in front of the friction term of the latter. Applying the same correction terms to Eq. (8) as for Eq. (1) yields

$$F_v = -F_1\left[1 - \left(\frac{1.33 + 3F_1}{N}\right)\right] + R_1\left[1 - \left(\frac{1.05 + 0.9R_1}{N}\right)\right] \tag{9}$$

$$F_1 = \frac{N^2 A_3^2}{HA_1^2}\left[\left(\frac{0.0949}{\mathrm{Re}_{o_1}^{1/4}}\right)\left(\frac{8}{11}\right)\left(\frac{L}{D_1}\right)\right] \tag{10}$$

$$R_1 = (N^2 A_3^2/HA_1^2)(\theta_1/2) \tag{11}$$

When F_1 is nearly equal to R_1, the error resulting from the approximation of these terms rises exponentially (Fig. 2). Here, ϵ is defined as

$$\epsilon = 1 - \frac{|F_v|}{|S_1| + |T_1|} \tag{12}$$

When ϵ approaches 0.0, the friction term becomes larger than the pressure-recovery term, or vice versa. As ϵ becomes larger, the difference between the two terms becomes smaller. From Fig. 2, as ϵ becomes larger, the error between F_v and F_{NS} becomes larger, due to error amplification. Expressions for F_v for dividing-, combining-, reverse-, and parallel-flow manifolds, and

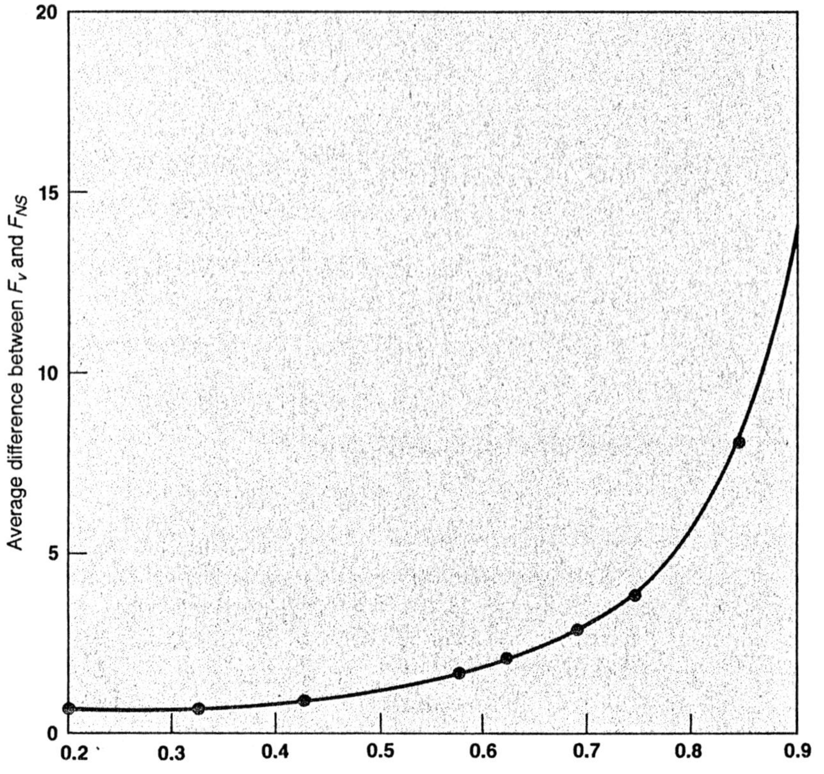

FIG. 2. Error amplification for F_v for a dividing-flow manifold.

average percent deviations from F_{NS} for a range of ϵ, are summarized in Table 1.

For a parallel- or reverse-flow manifold, ϵ is defined as

$$\epsilon = 1 - \frac{|F_v|}{|S_1| + |S_2| + |T_1| + |T_2|} \tag{13}$$

TABLE 1 Design Parameters for F_v for Each Type of Manifold[a]

Flow	F_v	Valid Range of ϵ	Average Relative Error (%)
Combining	$S_2 + T_2$	All $\epsilon = 0$	0.8
Dividing	$T_1 - S_1$	$0 < \epsilon < 0.7$	1.9
Reverse	$-S_2 - T_2 - S_1 + T_1$	$0 < \epsilon < 0.7$	1.2
Parallel	$T_1 - S_1 + T_2 + S_2$	$0 < \epsilon < 0.7$	1.3

[a] $S_i = F_i\{1 - [(1.33 + 3F_i)/N]\}$
$T_i = R_i\{1 - [(1.05+0.9R_i)/N]\}$
$F_i = (N^2A_3^2/HA_i^2)(0.0949/\mathrm{Re}_{o_i}^{1/4})(8/11)(L/D_i)$
$R_i = (N^2A_3^2/HA_i^2)/(\theta_i/2)$

Uncertainty in Design

Although the design equations (for F_v listed in Table 1) provide a good approximation of F_{NS}, considerable uncertainty is associated with certain parameters involved in calculating F_v. This arises primarily from the overall momentum coefficient, θ, the lateral flow resistance, H, and the friction factor, f. Therefore, when F_v is used to predict the behavior of, or to design, a manifold, the uncertainty associated with θ, H, and f must be factored into the analysis. For dividing-, reverse-, and parallel-flow manifolds, the uncertainty in F_v increases as ϵ gets larger.

Assume that H, f_1, and θ_1 are known to within ±95% confidence by

$$H = H \pm d_H$$

$$\theta = \theta_1 \pm d_\theta$$

$$f_1 = f_1 \pm d_f$$

For a dividing-flow manifold, there are two possible worst cases:

$$\theta_1 = \theta_1 + d_\theta$$

$$f_1 = f_1 - d_f$$

$$H = H - d_H$$

or:

$$\theta_1 = \theta_1 - d_\theta$$

$$f_1 = f_1 + d_f$$

$$H = H - d_H$$

Evaluate F_v for both cases and use the larger value for manifold design.

For designing a reverse- or parallel-flow manifold, use the maximum F_v based on a range of uncertainties for θ_1, H, and f_1. Results are more accurate if based on experimentally measured θ_1, H, and f_1.

When designing a manifold by means of F_v, first use the form of F_v that is similar to Eq. (8) and then verify or modify the results via the complete form for F_v in Table 1. This worst-case approach will eliminate the problem of error amplification at large values of ϵ.

Design Examples

The following cases demonstrate the utility of the manifold design parameter, F_v. Except for the first case, it is assumed that the uncertainty associated with θ, H, and f will be factored into the analysis, as in the worst-case analysis described. The following cases are based on the following design conditions for a dividing-flow manifold through which water will flow: $N = 10$; $H = 1.0$; $D_1 = 3$ in.; $D_3 = 0.5$ in.; $Q_o = 75$ gal/min; $L = 20$ ft; and $\theta_1 = 1.0$.

Example 1. *Determine the performance of an existing manifold.* Assume ±20% uncertainty in θ_1, H, and f_1, and perform a worst-case analysis. Under these conditions, the inlet Reynolds number, Re_{o_1}, equals 7.91×10^4. First determine whether friction or pressure recovery has the largest effect.

Friction term: $(0.0949/\mathrm{Re}_{o_1}^{1/4})(8/11)(L/D_1) = 0.329$. Pressure-recovery term: $\theta_1/2 = 0.5$.

The pressure-recovery effect being larger, the pressure should increase along the header. Therefore, we need consider only the first worst case:

$$\theta_1 = \theta_1 + d_\theta = 1.2$$

$$f_1 = f_1 - d_f = 0.0759/(\mathrm{Re}_{o_1})^{1/4}$$

$$H = H - d_H = 0.8$$

Therefore: $F_1 = 0.0254$ and $R_1 = 0.0579$; then $S_1 = 0.0218$ and $T_1 = 0.0515$. Finally, $F_v = 0.0297$. From the definition of F_v, the largest lateral flow rate is expected to be 7.61 gal/min in the last lateral, and the smallest lateral flow rate will be 7.39 gal/min in the first lateral.

Example 2. *Determine lateral flow resistance*, H, *necessary to obtain a specified flow uniformity*. (For this example, and the following one, parameter uncertainty is ignored because its determination has been demonstrated in Example 1.)

Assume that we want $F_v = 0.02$. Note that H appears in both F_1 and R_1, and in the correction terms for S_1 and T_1; therefore, as a first estimate, neglect the correction terms. Solving for H using only F_1 and R_1 yields

$$H = \frac{1}{F_v}\left(\frac{N^2 D_3^4}{D_1^4}\right)\left[\frac{\theta_1}{2} - \left(\frac{0.0949}{\mathrm{Re}_{o_1}^{1/4}}\right)\left(\frac{8}{11}\right)\left(\frac{L}{D_1}\right)\right]$$

The numerical solution of the problem yields

$$H = 0.659$$

With this estimate of H, calculate the correction terms for S_1 and T_1. The correction terms tend to reduce the value of H (they reduced it approximately 12% in Example 1). Therefore, reduce the calculated H value of 0.659 by 12% to 0.580. For this value of H, $F_1 = 0.438$ and $R_1 = 0.665$. Solving for H by using the correction terms yields $H = 0.598$, which is quite close to the reduced 0.580 value. Now design the lateral to meet this value. Remember that H is based on the superficial velocity through the lateral (i.e., Q_3 and D_3) even if the pressure drop through the lateral results from an orifice in the line.

Example 3. *Determine the maximum number of laterals for a specific uniformity in flow*. This case is similar to the last example in that N is embedded in the correction terms. For this example, determine the maximum number of lateral paths that would maintain $F < 0.1$. First calculate N_{max} without the correction terms. Solving the following equation yields $N = 27.5$:

$$N = \left\{ F_v H \left(\frac{D_1}{D_3}\right)^4 \bigg/ \left[\frac{\theta_2}{2} - \left(\frac{0.0949}{\text{Re}_{o_1}{}^{1/4}}\right)\left(\frac{8}{11}\right)\left(\frac{L}{D_1}\right)\right]\right\}^{1/2}$$

Estimate the correction term to be about 96%. Then, the first guess for N would be 26.4; i.e., $N_{max} = 26$. This value of N_{max} would be used to evaluate the correction terms and to again solve for N_{max} to verify the initial estimate.

Example 4. *Choose between a reverse- and a parallel-flow manifold.* An analysis of the values of F_v for a reverse- and parallel-flow manifold will show that the first is preferred if pressure recovery predominates in controlling pressure changes in the header—that is, if F_1 and F_2 are smaller than R_1 and R_2, F_v would be the sum of R_1 and R_2 for a parallel-flow manifold, and the difference between the two for a reverse-flow manifold. Therefore, if frictional effects dominate, a parallel-flow manifold is preferred [4].

Let us examine the conditions under which friction or pressure recovery is controlling, assuming $\theta_1 = 1.0$ and $\theta_2 = 2.0$. Further, let us neglect the correction terms, because both are approximately equal.

A reverse-flow manifold would be more suitable than a parallel-flow one if $|-F_1 + R_1 - F_2 - R_2| < |R_1 - F_1 + R_2 + F_2|$. But, assuming that $D_1 = D_2$, $2R_1 = R_2$, and $F_1 = F_2$, this inequality reduces to $R_1 > F_1$. Substituting into the equations in Table 1 yields

$$\frac{L}{D} < \frac{1}{2}\left(\frac{\text{Re}_{o_1}{}^{1/4}}{0.0949}\right)\left(\frac{11}{8}\right)$$

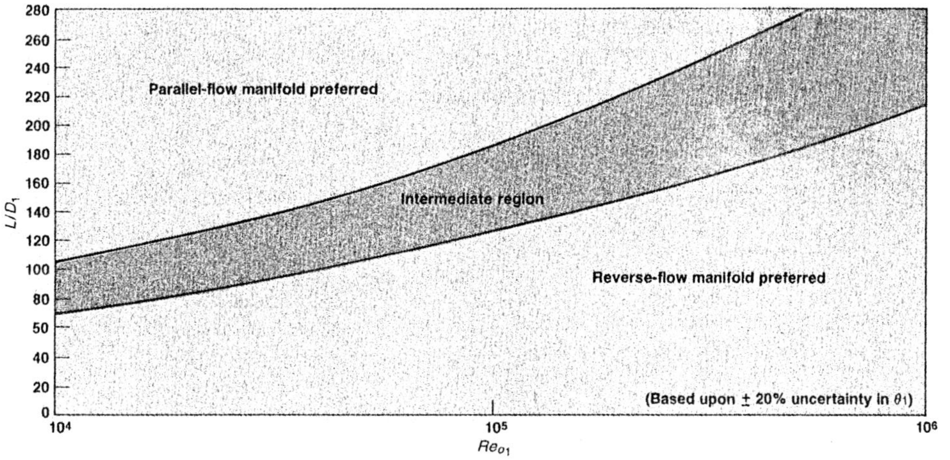

FIG. 3. Criteria for choosing between a parallel-flow or a reverse-flow manifold.

For Re_{o_1} = 10,000, $L/D < 73$. For Re_{o_1} = 300,000, $L/D < 170$. These are the conditions for which a reverse-flow manifold is preferred over a parallel-flow one.

Figure 3 presents selection criteria for reverse- and parallel-flow manifolds based upon the L/D ratio and Re_{o_1}.

Symbols

A_2	header cross-sectional area
A_3	cross-sectional area of a lateral
D	diameter
d	differential
F_v	fractional variation in flow through laterals
f	friction factor
H	lateral flow resistance, dimensionless
L	length of header
N	number of lateral paths
ΔP	pressure difference
Q_0	total flow to or from the manifold
Q_3	total lateral flow rate
$Q_3(L)$	lateral flow at end of manifold
$Q_3(O)$	lateral flow at entrance of manifold
Re_{o_2}	Reynolds number at outlet of header (i.e., maximum velocity condition)
v	average flow velocity in lateral

Greek Letters

θ_2	overall momentum correction factor [1]
μ	viscosity
π	pi (3.14159)
ρ	density

Subscripts

NS	numerical solution
o	inlet condition
1	refers to dividing-flow manifold
2	refers to combining-flow header
3	refers to lateral

References

1. R. A. Bajura and E. H. Jones, *ASME J. Fluid Eng.*, *98*, 654 (1976).
2. R. B. Bird, W. E. Stewart, and D. N. Lightfoot, *Transport Phenomena*, Wiley, New York, 1960, p. 187.
3. R. L. Pigford, M. Ashraf, and Y. D. Miron, *Ind. Eng. Chem., Fundam.*, *22*, 463 (1983).
4. J. B. Riggs, *Ind. Eng. Chem., Res.*, *26*(1), 129 (1987).

JAMES B. RIGGS

Non-Newtonian Flow

Engineers occasionally need to determine pipe size—to establish the diameter that will be the economically optimal or that will accommodate the necessary flow with the available pressure drop. Guidance for sizing pipe for Newtonian liquids can be found in *The Cameron Hydraulic Data Book* [2]. There is no comparable source for non-Newtonian fluids.

To size a pipe diameter for non-Newtonian laminar flow, the engineer needs only a plot of shear stress vs $8V/D$—i.e., a non-Newtonian fluid "flow" diagram—and to know that the fluid is time-independent. (For the definition of this and other pertinent terms, see the Appendix and the Symbols Section.) In some cases, however, knowledge about the type of fluid is also desirable.

Shear stress at the wall of a pipe or tube, T_w, is equal to $D\Delta P/4L$, in Pascals. This relationship holds for all fluids, Newtonian or non-Newtonian.

The velocity-diameter term, $8V/D$, is related to the shear rate at the wall of a pipe or tube, which is $(dv/dr)_w$, in 1/s; here, v is the local velocity, in m/s, at r (the radial or linear distance in meters). For Newtonian fluids, $(dv/dr)_w = 8V/D$; with V the mean linear velocity in m/s and D the pipe or tube diameter in meters. For non-Newtonian fluids, $(dv/dr)_w = [(3n' + 1)/4n'](8V/D)$; here, n' is a non-Newtonian rheological constant numerically equal to a slope of the curve of a plot of $T_w = D\Delta P/4L$ as ordinate vs $8V/D$ as abscissa.

A plot of shear stress, T, vs shear rate, dv/dr, is a shear diagram, or rheogram. Data from an extrusion rheometer yield a plot of $T_w = D\Delta P/4L$ vs $8V/D$. Such a plot is a shear diagram for a Newtonian fluid but not for a non-Newtonian one, because $8V/D$ does not represent rate of shear.

Dividing $D\Delta P/4L$ by $8V/D$ yields Poiseuille's viscosity equation:

$$\mu = (D\Delta P/4L)/(8V/D) \tag{1}$$

Rearranging:

$$\Delta P = 32\mu LV/D^2 \tag{2}$$

For a Newtonian fluid, μ represents the actual viscosity of the fluid, which is constant for all values of $8V/D$ in the laminar flow region. In the case of a non-Newtonian fluid, μ represents μ_a, the apparent viscosity, which is the ratio of two variables, because μ varies with each value of $8V/D$. Because the term $8V/D$, when applied to non-Newtonian fluids, is a flow function, a plot of $D\Delta P/4L$ vs $8V/D$ yields a flow (i.e., not a shear) diagram.

A pipe diameter for a particular pressure drop, or a pressure drop for a given pipe diameter, at any flow rate, can be determined by means of a $D\Delta P/4L$ vs $8V/D$ flow diagram. Another way of presenting results from extrusion rheometer tests is via a plot of apparent viscosity vs $8V/D$, with $8V/D$ calculated from flow rate and pipe size, viscosity read from the plot, and pressure drop then calculated via Eq. (2).

Characterizing Fluid Type

A material can usually be placed into one of the general rheological classifications from an inspection of its shear diagram. As has been noted, a T-vs-dv/dr diagram must be obtained directly via a rotational viscometer or by

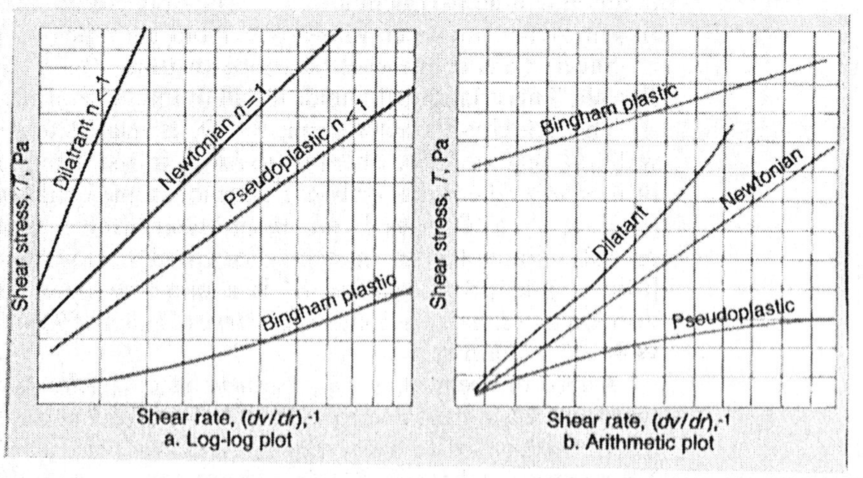

FIG. 1

TABLE 1 Shear-Diagram Curve Characterizes Fluids by Type

	Logarithmic Plot		Arithmetic Plot	
Fluid Type	Line Type	Line Slope	Line Type	o-o?
Newtonian	Straight	1	Straight	Yes
Pseudoplastic	Straight	<1	Curved	Yes
Dilatant	Straight	>1	Curved	Yes
Bingham plastic	Curved		Almost straight	No

converting $8V/D$-vs-$D\Delta P/4L$ data obtained with an extrusion rheometer into the T_w-vs-$(dv/dr)_w$ form, which is the true shear diagram. A fluid can be characterized as to type by comparing its shear diagram with either of the aforementioned plots on logarithmic or arithmetic grids (as in Fig. 1; see the logarithmic and arithmetic relationships in Table 1).

For Newtonian fluids (which include all gases, and most liquids of low molecular weight), the slope, n, equals 1. Values of n between zero and unity characterize pseudoplastic fluids for which the apparent viscosity decreases with increasing shear rate. Conversely, values of n greater than 1 correspond to dilatant fluids for which the lines on a log plot and for which the apparent viscosity increases with the shear rate. The slope, n, can be thought of as an index to the degree of non-Newtonian behavior, in that the farther that n is from unity (above or below), the more pronounced is the non-Newtonian characteristics of the fluid.

Pipe sizing by means of the power-law equation (discussed later) can be applied to pseudoplastic and dilatant fluids for which the lines on a log plot are straight over a large shear range.

Scaling up Rheometer Data

Sizing pipes for time-independent non-Newtonian fluids by scale-up requires a flow or shear diagram, which must be determined by laboratory measurements, preferably via an extrusion rheometer. This instrument consists of a reservoir having a small-bore tube fixed to the bottom. The reservoir is filled with the test fluid, which is forced through the tube by gas pressure or a weight acting on a free-floating piston. Flow rate is determined by weighing the extrudate collected over a specific time period. Fluids of viscosities up to 100 kPa·s can be characterized with the rheometer.

The driving force on a non-Newtonian fluid in a pipe or tube is $(\pi/4)D^2\Delta P$, and the force tending to retard fluid flow is πDLT_w. Here, T_w is the viscous resistance per unit area. If no slippage at the wall is assumed and the gravitational effect is neglected, the two forces will be equal in steady-state laminar flow. Therefore

$$(\pi/4)D^2\Delta P = \pi DLT_w \tag{3}$$

or the shear stress at the wall of a circular pipe:

$$T_w = D\Delta P/4L \tag{4}$$

The shear rate of a Newtonian fluid in laminar flow at the wall of a pipe is $8V/D$, and that of a non-Newtonian fluid is some function of $8V/D$, or

$$D\Delta P/4L = \phi(8V/D) \tag{5}$$

A plot of different values of $D\Delta P/4L$ vs $8V/D$ for a *time-independent* non-Newtonian fluid will fall on the same curve regardless of the pipe's length and diameter, or the flow rate. The curves will be different for each tube length and diameter if the fluid is *time-dependent*. The scale-up method now presented is not suitable for time-dependent fluids.

The extrusion-rheometer measurements required to obtain the data for constructing the plots are the fluid density, the pressure on the fluid, and sample weight. Velocity need not be calculated directly because

$$8V/D = 32Q/\pi D^3 = 4Q/\pi R^3 \tag{6}$$

Scale-Up Example

Calculate the pressure drop for a flow of 6.309×10^{-3} m^3/s through 3, 4, and 6 in. (0.0762, 0.1015, and 0.1524 m) Schedule 40 pipes. The fluid is a viscous plastic consisting of 60% by weight of pigment dispersed in an oil-type plasticizer. The data in Table 2(a) were obtained with an extrusion rheometer at ambient temperature.

1. Construct a plot of $8V/D$ vs $D\Delta P/4L$. Calculate $D\Delta P/4L$ and $8V/D$ (the latter via Eq. 6; for the calculated values, see Table 2b).

For a sample calculation, take $\Delta P = 66{,}192$ Pa (Table 2a):

$$D\Delta P/4L = \Delta P/384 = 66{,}192/384 = 172.375 \text{ Pa}$$

$$\begin{aligned} Q &= 5 \times 10^{-3} \text{ kg}/(1393.74 \text{ kg/m}^3)(913.4 \text{ s}) \\ &= (3.5875 \times 10^{-3} \text{ m}^3)(913.4 \text{ s}) \\ &= 3.9276 \times 10^{-9} \text{ m}^3/\text{s} \end{aligned}$$

$$32Q/\pi D^{-3} = (2.546 \times 10^9)(3.9276 \times 10^{-9}) = 9.9998 \text{ s}^{-1}$$

TABLE 2 Steps in Sizing Pipe via Scale-Up Method[a]

a. Data from Extrusion Rheometer

Gas Pressure on Reservoir, ΔP (Pa)	Time to Collect 5×10^{-3} kg of Fluid(s)	
	Tube A	Tube B
66,192	913.4	111.4
132,384	302.5	39.1
323,686	91.2	11.6
827,400	30.5	3.9

b. Calculated values of $D\Delta P/4L$ and $8V/D$

	Tube A	Tube B
D, m	0.0015875	0.003175
L, m	0.1524	0.3048
$D\Delta P/4L$, Pa	$\Delta P/384$	$\Delta P/384$
$32Q/\pi D^3$, s^{-1}	$(2.546 \times 10^9)Q$	$(3.18248 \times 10^8)Q$

[a]Measured fluid density, ρ_f = 1393.74 kg/m^3. Tube A: 1/16-in. diameter by 6-in. long (0.0015877-m diameter by 0.1524-m long). Tube B: 1/8-in. diameter × 12-in. long (0.003175-m diameter × 0.3048-m long).

(For the smoothed results at this and other gas-reservoir pressures, see Table 3.)

On log-log paper, plot $8V/D$ vs $D\Delta P/4L$. Note that the data obtained by means of the two different sized tubes fall on the same curve (Fig. 2). This indicates that the fluid is time-independent. Because the curve is not a straight line in either a logarithmic or arithmetic plot, the fluid does not resemble a pseudoplastic or Bingham plastic fluid.

2. Convert viscosity to apparent viscosity, and plot $8V/D$ vs μ_a. Convert $D\Delta P/4L$ to μ_a by means of Eq. (1). For a sample calculation, use the first set of

TABLE 3 Smoothed Flow Factors for Example Shear Rates

	Time to Collect 5×10^{-3} kg (s)		$D\Delta P/4L$,	$Q \times 10^9$ (m^3s)		$32Q/\pi D$ (s^{-1})		Smoothed
ΔP (Pa)	Tube A	Tube B	$\Delta P/384$	Tube A	Tube B	Tube A	Tube B	$32Q/\pi D^3$
66,192	913.4	114.4	172.375	3.9	31.3593	9.9998	9.98	10
132,384	302.5	39.1	344.75	11.8595	91.17519	30.19	29.02	30
323,686	91.2	11.6	842.932	39.3366	309.2672	100.151	98.42	100
827,400	30.5	3.9	2154.688	117.623	919.8718	299.4682	292.75	300

variables in Table 3. From a smoothed plot (Fig. 2), $D\Delta P/4L = 172.375$ Pa, and $8V/D = 10\ s^{-1}$. Therefore

$$\mu_a = 172.375/10 = 17.2375 \text{ Pa·s}$$

Apparent viscosities (μ_a) at other shear stresses ($D\Delta P/4L$) are tabulated in Table 4(a).

With values from Table 4(a), plot $8V/D$ vs μ_a on log-log paper. Calculate the flow function $8V/D = 32Q/\pi D^3$: $Q = 6.309 \times 10^{-3}$ (m^3/s). The apparent viscosities determined from the graph are listed in Table 4(b).

3. Calculate pressure drops via Eq. (2), using apparent viscosities. For a sample calculation, take the 3-in. (0.0762 m) pipe:

$$\Delta P = 32\mu_a LV/D^2$$

$$V = Q/A = 4Q/\pi D^2 = 4(6.309 \times 10^{-3})/(\pi)(0.0779)^2$$
$$= 1.32372 \text{ m/s}$$

$$\Delta P = 32 \times 7.9 \times 1 \times 1.32372/(0.0779)^2$$
$$= 55{,}144 \text{ Pa/m}$$

Pressure drops for the three pipe diameters are listed in Table 4(c).

4. Construct a pipe flow chart. Calculate the flow rate, Q, in m/s using the smoothed data for $32Q/\pi D^3$ from Step 2 for the three pipe diameters (see

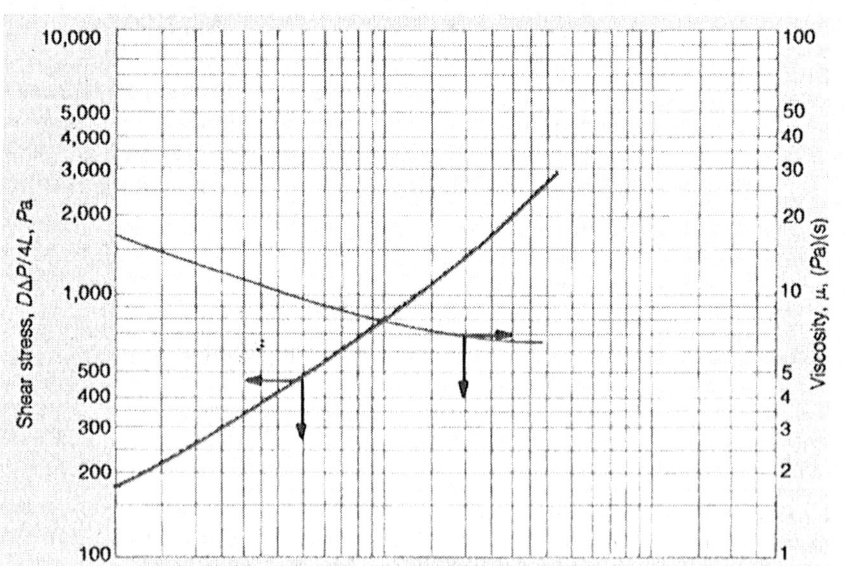

FIG. 2

TABLE 4 Steps in the Calculation of Pipe Pressure Drop

a. Calculated Apparent Viscosities

$D\Delta P/4L$ (Pa)	$8V/D$ (s^{-1})	μ_a (Pa·s)
172.375	10	17.2375
344.75	30	11.492
842.932	100	8.429
2154.688	300	7.1823

b. Apparent Viscosity from Graph

Nominal Pipe Size		Actual Pipe i.d.			
in.	m	ft	m	$32Q/\pi D^3$ (s^{-1})	μ_a (Pa·s) from Graph
3	0.0762	0.2557	0.0779	135.7	7.9
4	0.1015	0.3355	0.10226	60.1	9.4
6	0.1524	0.5054	0.154	17.6	13.8

c. Calculated Pressure Drops

Nominal Pipe Size		Actual Pipe i.d.			
in.	m	(m)	V (m/s)	μ^a (Pa·s)	ΔP (Pa/m)
3	0.0762	0.0779	1.32372	7.9	55,144.0
4	0.1015	0.10226	0.7682	9.4	22,097.4
6	0.1524	0.154	0.3387	13.8	6,306.7

Table 3). Also calculate ΔP in Pascals for $L = 1$ m from the data for $D\Delta P/4L$ determined in Step 1 (see Table 3).

Taking as the example the 6-in. -diameter pipe (nominal size = 0.1524 m; actual i.d. = 0.54 m) for $32Q/\pi D^3 = 10\ s^{-1}$, and $D\Delta P/4L = 172.375$ Pa.

$$\Delta P = (D\Delta P/4L)(4L/D)$$
$$= (172.375)(4)(1)/0.154 = 4477.273 \text{ Pa/m}$$

$$Q = (32Q/\Delta D^3)(\pi D^3/32)$$
$$= [10\pi(0.154)^3]/32 = 3.5856 \times 10^{-3} \text{ m/s}$$

For the 3-in.-diameter pipe (nominal size = 0.0762 m; actual i.d. = 0.0779 m):

$$\Delta P = (D\Delta P/4L)(4L/D)$$
$$= (D\Delta P/4L)[(4)(1)/(0.0779)]$$
$$= (D\Delta P/4L)(51.347882)$$

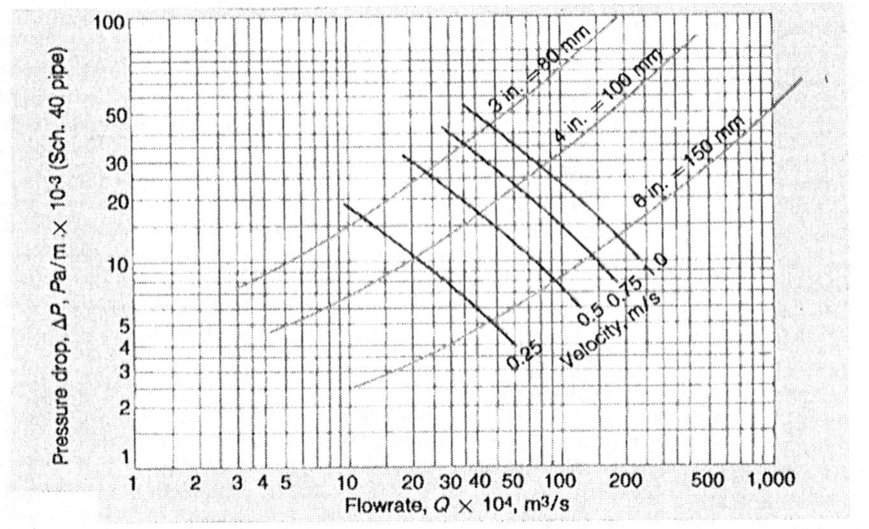

FIG. 3

$$Q = (32Q/\pi D^3)(\pi D^3/32)$$
$$= (32Q/\pi D^3)(\pi/32)(0.0779)^3$$
$$= (32Q/\pi D^3)(4.64 \times 10^{-5})$$

Via the foregoing procedure, also calculate ΔP and Q for the 4-in.-diameter pipe (nominal size = 0.1015 m; actual i.d. = 0.10226 m).

On log-log paper, plot flow rate, Q, in m^3/s vs pressure drop, ΔP, in Pa/m of pipe (Fig. 3). Velocity lines for the different pipe diameters can be calculated and added to further define the system, as shown in Fig. 3.

The scale-up method that has been illustrated can be applied to any type of time-independent non-Newtonian fluid in the laminar-flow region. Only that the fluid is time-independent is all that need be known about its rheological properties, in addition to having experimental data or a plot of $D\Delta P/4L$ vs $8V/D$.

The foregoing procedure and examples pertain solely to scale-up for laminar flow. Experimental data taken in the laminar region cannot be used for scale-up in the turbulent flow region. Stable laminar flow for Newtonian fluids ends at a Reynolds number of about 2100, or a friction factor of 0.0076 (related by $f = 16/N_{Re}$ in the laminar region).

A generalized Reynolds number for non-Newtonian fluids can be expressed as $N_{Re} = DV\rho/\mu_a$, with μ_a being the apparent viscosity of the fluid as calculated via Poiseuille's equation, Eq. (1). For an approximation of μ_a, assume that the onset of turbulence for non-Newtonian fluids occurs at N_{Re} = 2100, the same as with Newtonian fluids.

Power-Law Sizing

For many non-Newtonian fluids, a log-log plot of shear stress vs shear rate or the flow function $8V/D$ is a straight line for shear rates from 10 to 1000 s^{-1}. The relationship between shear stress and rate of shear for these fluids can be represented by a two-constant power function of the form

$$T = K(-dv/dr)^n \tag{6}$$

where dv/dr = rate of shear in circular duct or rotational viscometer, s^{-1}; K = fluid consistency index; and n = non-Newtonian rheological constant. Fluids of this type are called power-law fluids, for which Eq. (7) applies:

$$T_w = D\Delta P/4L = K'(8V/D)^{n'} \tag{7}$$

where n' = slope of line on a logarithmic plot of $D\Delta P/4L$ vs $8V/D$; and K' is a consistency index, which equals the T_w intercept at $8V/D = 1$.

Power-Law Example

For a polymer solution, a log-log plot of T_w vs $8V/D$ yields a straight line having the slope $n = 0.586$ (Fig. 4). Via Eq. (7), calculate the pressure drop for a flow of 3.2803×10^{-3} m^3/s through a 6-in. Schedule 40 pipe (nominal diameter = 0.1524 m; actual i.d. = 0.154 m):

$$Q = 3.2803 \times 10^{-3} \text{ m}^3/\text{s}$$

$$8V/D = 32Q/\pi D^3 = (32)(3.2803) \times 10^{-3}/\pi(0.154)^3$$
$$= 9.1404 \text{ s}^{-1}$$

From the curve for $8V/D = 1$:

$$T_w = D\Delta P/4L = K' = 143.65 \text{ Pa}$$

$$\Delta P = (4L/D)K'(8V/D)^{n'}$$
$$= (4)(1)(143.65/0.154)(9.1404)^{0.586}$$
$$= 13{,}644.51 \text{ Pa/m}$$

From Fig. 3, via the scale-up method, $D\Delta P/4L$ = 526.7 Pa; therefore:

$$\Delta P = 526.7 \times 4 \times 1/0.154$$
$$= 13{,}680.56 \text{ Pa}$$

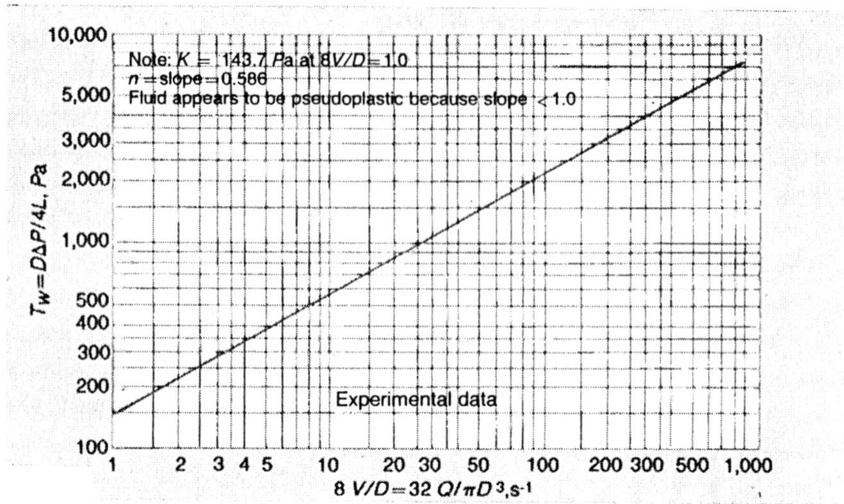

FIG. 4

The foregoing illustrates the greater convenience of the power-law equation over the graphical scale-up method. However, the latter is more widely applicable.

Metzner-Reed Method

A general correlation developed by Metzner and Reed [1] is independent of the rheological properties of fluids that are time-independent. It is based on a modified form of the conventional friction factor and a Reynolds number plot for Newtonian fluids.

The friction factor comes from the Fanning equation:

$$f = D(\Delta P/4LV)/(\rho V^2/2) = D\Delta P/2\rho V^2 L \tag{8}$$

The modified Reynolds number is defined as

$$N_{\text{Rem}} = D^{n'} V^{2-n'} \rho / K' 8^{n'-1} \tag{9}$$

where K' = fluid consistency index, and n' = non-Newtonian rheological constant.

The constants K' and n' are determined via a logarithmic plot of $D\Delta P/4L$ vs $8V/D$. This diagram is based on experimental data taken with an extrusion rheometer. Here, n' is the slope and K' is the value of $D\Delta P/4L$ at an $8V/D$ of unity. For Newtonian fluids, $n' = 1$ and N_{Rem} is reduced to the conventional $N_{\text{Re}} = DV^2\rho/\mu$, with $\mu = K'$. (When $n' = 1$, $D\Delta P/4L = K'8V/D$, with $K' = \mu_a$.)

The recommended design procedure involves:

1. Determine n' and K' from a flow diagram. If the plotted data do not fit a straight line, point values of K' and n' can be used as determined from tangents at the desired points.
2. Calculate the modified Reynolds number, and read off the friction factor from the usual chart of friction factor vs Reynolds number.
3. Calculate the pressure drop.

For laminar flow, $f = 16/N_{\mathrm{Re}}$ is only valid for $N_{\mathrm{Re}} < 2100$ when parameters K' and n' are evaluated at the desired value of $8V/D$. No assumptions were made about the constancy of these parameters over the range of shear rates encountered in the pipe. A generalized form of Poiseuille's equation that is valid for all time-independent fluids can be derived by combining the foregoing equations:

$$\Delta P = 32K'8^{n'-1}LV^{n'}/D^{n'+1} \tag{10}$$

An Example Application

Calculate the pressure drops for the previous example by the Metzner and Reed method. The following are known: $n' = 0.586$; $K' = 143.65$ Pa; $Q = 3.28 \times 10^{-3}$ m^3/s; $D = 0.154$ m; and $\rho = 727.31$ kg/m.

1. Calculate the modified Reynolds number, Eq. (9):

$$V = Q/A = (3.28 \times 10^{-3})/(\pi/4)(0.154)^2$$
$$= 0.1761 \text{ m/s}$$

$$N_{\mathrm{Rem}} = (0.154)^{0.586}(0.1761)^{(2-0.586)}(727.01)/(143.65)(8^{0.586-1})$$
$$= 0.3433$$

2. Calculate the friction factor:

$$f = 16/N_{\mathrm{Rem}} = 16/0.3433$$
$$= 46.61$$

3. Calculate the pressure drop via the Fanning equation—Eq. (8):

$$\Delta P = (2)(46.61)(727.31)(0.1761)^2(1/0.154)$$
$$= 13{,}636.54 \text{ Pa/m}$$

4. Calculate the pressure drop via the generalized Poiseuille equation—Eq. (10):

$$\Delta P = (32)(143.65)(8^{-0.414})(1)(0.176^{0.586})/0.154^{1.586}$$
$$= 13{,}646.13 \text{ Pa/m}$$

Isothermal Turbulent Flow

The turbulent flow of non-Newtonian fluids (as with Newtonian ones) is characterized by the presence of random eddies and whorls of fluid that cause the instantaneous values of velocity and pressure at any point in the system to fluctuate wildly. Because of these fluctuations, flow problems cannot be easily solved via equations of continuity and motions. Non-Newtonian turbulent flow has not received much attention, and a universal correlation has so far not been proposed.

Most non-Newtonian fluids are viscous, their equivalent Reynolds numbers seldom exceeding 100,000. Blasius' friction factor equation for Newtonian flow between Reynolds numbers 4000 to 100,000 is expressed as

$$f = 0.079/(N_{\text{Re}})^{0.25} \tag{11}$$

Equation (11) provides good approximations if the data can be correlated on a Reynolds number chart, and this equation can be fitted with a straight line. Substituting the equivalent terms for Eq. (11), as per Blasius' equation for Newtonian fluids, yields

$$D\Delta P/2\rho L V^2 = 0.079(DV\rho/\mu)^{-0.25} \tag{12}$$

Rearranging the terms of Eq. (12) and introducing the constant K:

$$D^{1.25}\Delta P/L = KV^{1.75} \tag{13}$$

Equation (13) indicates that $D\Delta P/4L$ would have to be multiplied by $D^{0.25}$ for the data for turbulent flow to correlate when plotted against velocity. It further indicates that if $D^{1.25}\Delta P/L$ vs V (flow velocity) were plotted logarithmically, a straight line having a slope of 1.75 would result.

Equation (12) suggests that Eq. (13) has the general form

$$D^{1+b}\Delta P/L = KV^{2-b} = KV^{c} \tag{14}$$

The term b is the exponent in

$$f = a/(N_{\text{Re}})^{b} \tag{15}$$

where a is constant in a Blasius-type equation.

It is evident that c is either the slope defined by Eq. (14) or the slopes of individual turbulent branches when $D\Delta P/L$ vs $8V/D$ data are plotted logarithmically. Equation (14) can also be written as

$$D^{3-c}\Delta P/L = LV^c \tag{16}$$

where $c = 2 - b$.

Because slopes b and c are related, Eq. (16) can be defined if either is known; also, the numerical sum of two slopes must be equal to 2. Equations (14) and (16) are applicable to all fluids if viscosity remains constant and the Reynolds number is defined by Eq. (12).

Pipe diameter or pressure drop can be calculated via this method as follows:

1. Draw a shear diagram ($D\Delta P/L$ vs $8V/D$) for several pipe diameters.
2. Determine slope c of the turbulent branches for various pipe sizes.
3. Plot data on a friction-factor chart, using the constant plastic viscosity term in the Reynolds number.
4. If the data correlate, determine slope b from the plot.
5. If slopes b and c add up to 2.0, a correlation can be expected from a plot of $D^{1+b}\Delta P/4L$ vs V when the data should have a slope of c.
6. After the Step 5 correlation has been established, take values for $D^{1+b}\Delta P/4L$ and V from two points on the smooth curve and convert them to, respectively, $D\Delta P/4L$ and $8V/D$ for several pipe sizes.
7. Plot the Step 6 values on $D\Delta P/4L$-vs-$8V/D$ charts, along with the curve for laminar flow.
8. The intersection of the turbulent branches with the laminar curve represents the point where flow changes from laminar to turbulent for a particular pipe size.
9. Calculate pipe diameter or pressure drop, following the procedure for laminar flow.

Appendix: Classifying Newtonian and Non-Newtonian Fluids

Viscosity is the property of a fluid that resists a shearing force. It can be thought of as the friction resulting when one layer of fluid moves relative to another. As Fig. A1 (used by Isaac Newton in first defining viscosity) shows: two parallel planes of fluid of area A separated by a distance dx are moving in the same direction at different velocities, V_1 and V_2.

The velocity distribution will be linear over the distance dx. Experiments have shown that the velocity gradient, dv/dx, is directly proportional to the force unit per area: $F/A = \mu(dv/dx)$; with μ the viscosity, a constant for a given liquid.

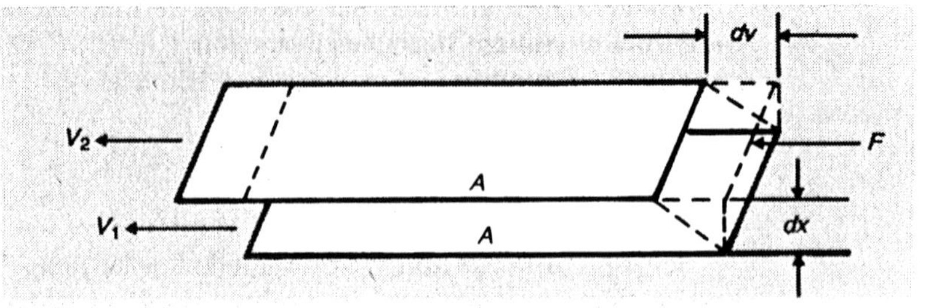

FIG. A1

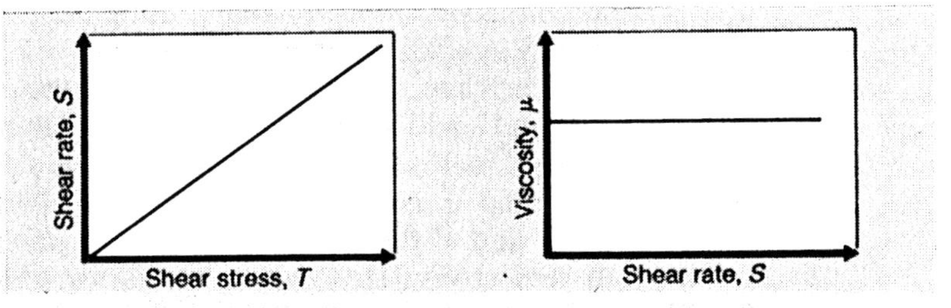

FIG. A2

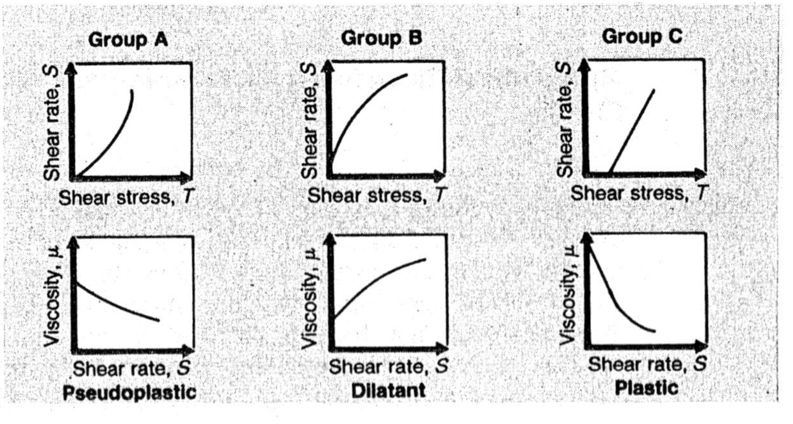

FIG. A3

The velocity gradient, dv/dx, describes the shearing undergone by the intermediate layers as they move relative to one another. Therefore, it can be called the rate of shear, S. Also, the force per unit area, F/A, can be simplified and called the shear force or shear stress, T. With these simplifications, viscosity can be defined as T/S (shear stress/shear rate).

Viscosity—Newtonian Fluids

A Newtonian fluid is one whose viscosity at a given temperature is independent of S, the rate of shear. Figure A2 depicts the linear relationship between shear stress (T) and rate of shear (S) for a Newtonian fluid. Note that the viscosity of a Newtonian fluid at a given temperature is constant regardless of the velocity, previous agitation, or shearing of the fluid.

Figure A2 also shows the relationships between shear stress (T), rate of shear (S), and viscosity (μ) for a Newtonian liquid. The viscosity remains constant, as shown on the right; in absolute units, it is the inverse slope of the line on the left. Water and light oils are good examples of Newtonian liquids.

Viscosity—Non-Newtonian Fluids

A non-Newtonian fluid is one whose viscosity at a given temperature is dependent on the rate of shear. The viscosity may increase or decrease, depending on the type of fluid. A fluid having a viscosity greater than 20 Pa·s is likely to be non-Newtonian. Non-Newtonian fluids can be classed as time-independent, time-dependent, and viscoelastic.

A non-Newtonian fluid is time-independent if the shear stress at any rate of shear is constant with time—that is, the properties of the fluid depend only on the magnitude of the imposed shear stresses and not on the duration of the stresses. Figure A3 characterizes common types of non-Newtonian liquids.

Group A—Viscosity decreases with increasing rate of shear. This is known as a pseudoplastic fluid. This behavior is generally restricted to a certain range of shear rates. At a very low or high shear rate, flow may be Newtonian (e.g., molten polymer, polymeric solutions, slurries of pigments and paper, and greases).

Group B—Viscosity increases with increasing rate of shear (e.g., water slurries of clay and starch, and candy compounds).

Group C—These fluids (curves for a Bingham-plastic material are shown) exhibit a definite yield strength, and a stress below which no flow occurs (the behavior is that of a solid). A certain force must be applied to produce movement. Bingham-plastic fluids differ from Newtonian fluids only in that the linear relationship between rate of shear and shear stress does not pass through the origin. Examples include ketchup, sewage sludge, and water suspensions of rock and grain.

A non-Newtonian fluid is said to be time-dependent if the shear stress changes with the duration of shear—in other words, the viscosity at any time depends on the amount of previous agitation or shearing of the liquid. A liquid whose viscosity decreases with time at a given shear rate is called thixotropic (e.g., asphalts, glues, molasses, mayonnaise, drilling muds, and starches). A liquid whose viscosity increases with time is called rheopectic (e.g., bentonite soils and gypsum suspensions in water).

Viscoelastic fluids exhibit many characteristics of solids. Their resistance to deformation is proportional to the usual viscous effect, plus an elastic effect that is a function of time. When the rate of strain of such a fluid is suddenly increased, there is a relaxation time during which the stress changes from its original to a new steady-state value. Polymeric liquids comprise the largest group of fluids in this class. Equations developed for pseudoplastic fluids can be applied to the steady-state flow of viscoelastic fluids.

There is no straightforward method for determining the pressure drop of time-dependent and viscoelastic fluids. Many theoretical models have been presented by numerous researchers, but none has received wide acceptance. In any case, most of the non-Newtonian fluids in industrial processes are time-independent.

Symbols

A	area, m^2
D	pipe or tube diameter, m
$D\Delta P/4L$	shear stress at wall of pipe or tube, T_w, Pa
F	force, N
F/A	shear stress, T, Pa
f	fanning friction factor, dimensionless
g_c	gravitational constant, 9.80665 m/s^2
K	fluid consistency index
K'	fluid consistency index, T_w (intercept on logarithmic plot at $8V/D = 1$)
L	pipe or tube length, m
N_{Re}	conventional Reynolds number, $N_{Re} = DV\rho/\mu$, dimensionless
N_{Rem}	modified Reynolds number, dimensionless [1]
n	non-Newtonian rheological constant
n'	non-Newtonian rheological constant, n = slope of line on logarithmic plot of $D\Delta P/4L$ vs $8V/D$
ΔP	pressure drop, Pa
Q	flow rate, m^3/s
r	pipe or tube radius, m
S	shear rate, dv/dx, s^{-1}

T	shear stress, Pa
T_w	shear stress at wall of pipe or tube, $D\Delta P/4L$, Pa
V	mean linear velocity, m/s
v	local velocity, m/s
(dv/dx)	local velocity gradient, S, s^{-1}
$(dv/dx)_w$	local velocity gradient or shear rate in a rotational viscometer, s^{-1}
$(dv/dr)_w$	shear rate at wall of pipe or tube, s^{-1}
ϕ	functional relationship
μ	viscosity of fluid, Pa · s
μ_a	apparent viscosity of non-Newtonian fluid, $(D\Delta P/4L)/(8V/D)$, Pa·s
ρ	density of fluid, kg/m^3

References

1. A. B. Metzner and J. C. Reed, "Flow of Non-Newtonian Fluids—Correlation of Laminar, Transition and Turbulent Flow Regions," *AIChE J.*, *1*, 434–440 (December 1955).
2. C. R. Westaway and A. W. Loomis (eds.), *The Cameron Hydraulic Data Book*, Ingersoll-Rand, Woodcliff Lake, New Jersey, 1979.

A. A. SULTAN

Slurry Systems

Design of equipment for handling slurries is based on the type of slurry involved. The four basic kinds are:

Settling slurries
Nonsettling slurries that behave as homogeneous non-Newtonian fluids
Stabilized slurries
Slurries that show thixotropic properties (generally nonsettling)

For settling slurries, design methods are used to determine the minimum transport velocity and to predict the pressure gradient. For slurries that behave as homogeneous non-Newtonian fluids, design methods predict the laminar/turbulent transition and the pressure gradient. Stabilized slurries contain large particles to be conveyed. These particles are supported by a dense or heavy medium which consists of flocculated, much finer particles and which impart non-Newtonian shear-thinning flow behavior to the heavy medium. Shear-thinning media are highly suitable for transporting coarse particles. In the low shear region near the center of the pipe, the apparent viscosity is high and the settling velocity of the suspended coarse particles is either low or zero if the medium has a sufficiently large yield stress. Near the pipe wall, shear rates are high, apparent viscosities are low, and in consequence, pressure gradients in the pipeline are not excessively great.

For slurries with thixotropic properties, methods are also required for predicting the start-up pressures after shutdown and the time required to clear the pipe of gelled material so that steady flow can be resumed.

Two approaches are generally used in an attempt to distinguish between settling and nonsettling slurries, neither of which has proved particularly satisfactory. Criteria have been put forward on the basis of either particle size or settling rate. With the former, particle/liquid density difference is not taken into account and hence different particle size systems can have similar settling rates or vice versa and, in addition, hindered settling effects arising from solids concentration variation are not allowed for. On the other hand, settling rate criteria are usually based on tests executed under quiescent conditions in small laboratory vessels whereas the information is often required for turbulent flow of settling slurries. Unfortunately, neither approach relates settling rates to the length of a pipeline along which the slurry is to be pumped or to the mean flow velocity in the pipe and hence to the residence time of slurry in the pipe length.

Unless a slurry settles very rapidly or not at all, neither settling nor nonsettling slurries are absolute concepts. The propensity for a slurry to settle, which determines which pipeline design procedures to adopt, must always be related to slurry residence time in a pipe. To take two extremes, a settling slurry defined by either of the two approaches mentioned above may be considered a "nonsettling" slurry if the pipeline is sufficiently short and the flow velocity high, while a "nonsettling" slurry may be considered settling if the residence time in a long pipeline is sufficient large.

One way of overcoming these discrepancies is to compare the time required, t_s, for solids to settle (at velocity U_t) a vertical distance of a pipe diameter, D:

$$t_s = D/U_t \tag{1}$$

with the residence time, t_f, of slurry in the pipeline of length L flowing at a mean velocity V:

$$t_f = L/V \tag{2}$$

If $t_f << t_s$, design equations developed for nonsettling slurries should be employed, while if $t_f >> t_s$, these equations developed for settling slurries should be used. When t_f is of the same order as t_s, both sets of design equations should be used and predictions for operating variables compared.

In addition to head loss estimation for straight pipe runs, estimates sometimes need to be made for additional loss of head which arises from flow through various types of pipe fittings. There are also a number of techniques which can, in some circumstances, be employed to reduce frictional head loss arising from friction in straight pipe runs. Both these topics will be discussed later.

Settling Slurries

When handling settling slurries, the main design problem is predicting a design velocity high enough so that there is no possibility of blockage, but not so high that the pressure gradient and wear rates are excessive. Having decided on a design velocity, it is then necessary to have a method of predicting pressure gradient.

Most researchers have tried to find universal methods for predicting design velocity and pressure gradient by correlating experimental results in terms of a few selected parameters—usually pipe diameter, solids concentration, and one or two particle parameters. These particle parameters are sometimes density and particle size, although these two are often combined and expressed in terms of terminal velocity or drag coefficient. All that has resulted from this is that the designer is now faced with a wealth of correlations, but without a method of selecting which one, if any, will yield a satisfactory prediction.

The flow of settling slurries in horizontal pipes can be classified into various flow regimes as shown in Fig. 1. These flow regimes refer to the *in-situ* vertical solids concentration profile (VSCP) and whether all solids are suspended or a proportion are conveyed as a bed sliding along the pipe bottom. At high mean flow velocities it may be possible to convey some coarse slurries so that there exists no discernible VSCP (homogeneous flow), but it is usually uneconomic to operate pipelines carrying settling slurries at high velocity. Heterogeneous flow occurs when there exists a pronounced VSCP but where all particles are suspended in the continuous phase. At lower velocities still, a bed of solid particles may form. This bed may be stationary or may slide. It is dangerous to operate long pipelines in either of these regimes as solids build-up may occur and block the pipeline. The consequence of a pronounced VSCP in heterogeneous flow or, in addition, the presence of particle beds is that the *in-situ* solids concentration is higher than the discharge solids concentration.

Design Velocity

For predicting design velocity for horizontal pipe flow the engineer has the choice of, at the latest count, over 60 correlations. To add to his confusion, there are a number of definitions of design velocity and it is not always made

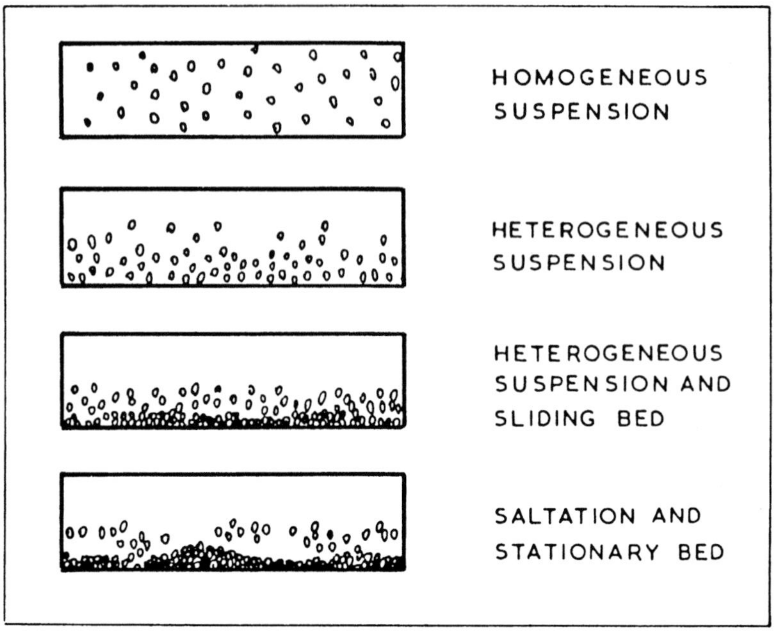

FIG. 1. Flow patterns for settling slurries in horizontal pipe flow. (Taken from Ref. 17.)

clear as to which definition a particular correlation applies. Correlations are available based on the following velocities.

Sliding-bed velocity. This is the velocity at which the shearing forces in the liquid are just sufficient to move particles that lie on the floor of the pipe. This is normally an inefficient method of transporting the solids, but it may well be the mechanism by which solids are carried in high-concentration conveying.

Saltating velocity. At this velocity, particles are repeatedly picked up by the liquid and deposited further along the pipe. This form of transport is not used for long-distance lines carrying fine particles, but for short lines carrying coarse particles, it may be necessary to operate with saltating flow.

Suspending velocity. This is the lowest velocity at which all the particles are picked up and remain in suspension. This velocity is used for designing most pipelines but it is difficult to determine with precision, particularly when the particles have a wide size distribution and when fine particles suspended in the liquid make it opaque.

Deposit velocity. The velocity at which particles start to settle out as the flow is lowered. The particles may settle to a static or a sliding bed. This velocity is not necessarily the same as the suspending velocity.

Velocity corresponding to a minimum in the pressure gradient vs velocity curve. This is often known as the critical velocity. Its determination does not require observations of the flow regime but the minimum point is difficult to locate with precision because the curve is often shallow and not necessarily continuous. Thus, correlations in which the position of the minimum has been derived by differentiation of an analytical expression derived from experimental points must be used with great caution. It is usually assumed that the critical velocity is higher than the suspending velocity so that its use leads to a safe design. Although this is usually true, there can be no guarantee that it is always so.

Velocity for homogeneous flow. In theory, this is the velocity at which the particles become evenly distributed throughout the pipe. In practice, it is defined as the velocity at which the concentration profile across the pipe attains some arbitrary degree of uniformity. Alternative definitions for homogeneous flow are based either on the assumption of the pressure gradient in the pipeline being equal to (1) that for a fluid having the same density as that of the suspension and the viscosity of water (the standard velocity), or (2) that predicted from viscometer measurements on the homogeneous suspension. This is discussed by Cheng and Whittaker [1]. However, velocities corresponding to homogeneous flow will normally lead to a design that is too conservative.

To illustrate how the different correlations for suspending velocity compare with each other, we can examine the variation in the predictions given by 16 assorted correlations demonstrated by Wiedenroth [2] (see Fig. 2). The vastly different velocities predicted need no further comment.

A second comparison was made by Carleton and Cheng [3], who took values calculated from some 50 or so correlations and compared them with measured values for two industrial pipelines. Table 1 compares predictions with measured values for a 25% (by weight) suspension of 50 μm iron oxide in water flowing in a 53-mm pipe. Table 2 shows predictions for the Savage River pipeline in Tasmania [4] which conveys a 60% (by weight) suspension

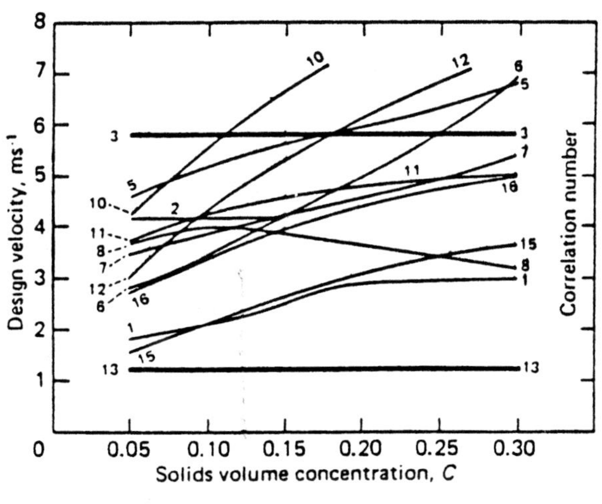

FIG. 2. How 16 correlations predict suspending velocities for a coarse gravel in a 300-mm diameter pipeline.(Taken from Ref.2.)

of iron ore in a 244-mm diameter line. The minimum transport velocity for this line has been found to be 1.2 m/s, and the pipeline is operated at 1.4 to 1.7 m/s. It can be seen from Tables 1 and 2 that the agreement between theory and practice is not good. On average, the correlations underpredict by about 30% for the 53-mm line; however, for the 244-mm line there is an average overprediction of about 120%.

Tables 1 and 2 also show that there is a greater variation, among predictions given by the various correlations, for the large pipe than for the small pipe. The reason is believed to be that, because the correlations are derived from experimental results mostly obtained in small-diameter pipes, the underlying physical phenomenon may not be relevant to flow in large pipes.

As Thomas [5] pointed out in his semitheoretical analysis, suspension is a complex phenomenon and can occur by a number of different mechanisms. He suggested that the size of the particles as compared with the thickness of the laminar sublayer is an important factor—and that different equations apply depending on whether the particle size is larger or smaller than the sublayer thickness. His correlations, therefore, are probably more universally applicable, but even these are based on considerable simplifications. Many workers have relied simply on dimensional analysis, so their correlations amount to no more than simple curve-fitting.

In addition, most correlations include only particle size and density among the particle parameters, quite ignoring the fact that in practice suspension will depend on a number of other factors such as particle-size distribution and shape. It is generally felt that only the coarser particles have an effect on design velocity. In fact, many suspensions contain enough fine particles so that the admixture of these fine particles with the suspending

TABLE 1 Design Velocities for Iron Oxide in 53 mm Pipe

Design Velocity	Number of Correlations	Mean of Predictions (ms^{-1})	Standard Deviation (ms^{-1})	Measured Value (ms^{-1})
Laminar/turbulent transition	2	0.41	—	1.01 (calculated from viscometric data)
Sliding-bed velocity	7	1.47	1.51	—
Suspending velocity	11	0.58	0.50	1.10–1.80 (not sharply defined)
Deposit velocity	11	0.83	0.31	As for suspending velocity
Minimum in pressure-gradient curve (critical velocity)	10	0.66	0.19	0.29 (not sharply defined)
Heterogeneous/homogeneous transition	5	1.74	1.37	1.46 (as defined in Ref. 1)
Standard velocity	4	1.10	—	—

TABLE 2 Design Velocities for Savage River Line—244 mm Pipe

Design Velocity	Number of Correlations	Mean of Predictions (ms^{-1})	Standard Deviation (ms^{-1})
Sliding-bed velocity	6	4.2	5.1
Suspending velocity	9	1.7	2.0
Deposit velocity	11	2.1	1.5
Minimum in pressure-gradient curve (critical velocity)	11	2.4	0.9
Heterogeneous/homogeneous transition	4	3.6	3.5
Standard velocity	4	3.3	1.9

liquid will show significant non-Newtonian properties. This will, of course, affect the settling of the coarser particles in the suspension, but most correlations are based on narrow size ranges of particles.

Carleton and Cheng [3] found that most correlations for the variously defined design velocities could be summarized by the equation

$$V_D = d^\alpha D^\beta C^\gamma (S - 1)^\delta \tag{3}$$

and suggested that because of the wide variation in the values of α, β, γ, and δ from different correlations, Eq. (3) should be used to scale-up V_D data from small-scale testwork in which the exponents α, β, γ, and δ have been determined for a particular slurry type. The terminal settling velocity of some average particle size is the more relevant parameter than d, and this implies that α ranges from a value of unity for the Stokes settling regime ($C_D = 24/\mathrm{Re}_p$) to a value of zero for the Newtonian settling regime ($C_D = 0.44$). The value of β appeared from the survey to be around 0.4 for small pipes, possibly falling to zero for large ratios of D/d, while the values of δ lay between about 0.4 and 0.5. The effect of solids concentration is less clear, but work by Parzonka et al. [6] quantified both the effect of C and, to a certain extent, particle size distribution effect using the Durand and Condolios [15] dimensionless coefficient F_L defined as

$$V_D = F_L \left[2gD \frac{(\rho_{sol} - \rho_L)}{\rho_L} \right]^{0.5} \tag{4}$$

where V_D is the suspending velocity in this case.

Parzonka et al. divided their collated experimental data into five categories according to the type of solid and particle-size range, as follows:

1. Small size sand particles (0.1 mm < d < 0.28 mm)
2. Medium size sand particles (0.4 mm < d < 0.85 mm)
3. Coarse size sand and gravel (1.15 mm < d < 19 mm)
4. Small size high density (2.7 to 5.3 g/cm^3) materials (50 μm < d < 300 μm)
5. Coal particles (1 mm < d < 2.26 mm)

To compare the data, the suspending velocities were reexpressed in terms of the dimensionless coefficient F_L defined in Eq. (4).

The overall picture of the effects of solids concentration on F_L at different mean particle diameters for sand–water systems with varying proportions of fine material present is shown in Fig. 3. This shows that without fines, F_L reaches a maximum and then remains almost constant when C is in the range 0.1 to 0.15, depending on the particle diameter. However, in the presence of fine material (defined as $d < 75$ μm), F_L decreases once it has passed its maximum. Depending on the proportion of fine material present, a greater fall occurs with a greater proportion of fines. As well as the guidelines for sand–water systems, Parzonka et al. also include figures which provide guidelines for estimating F_L for very fine materials (such as iron ore) and coal.

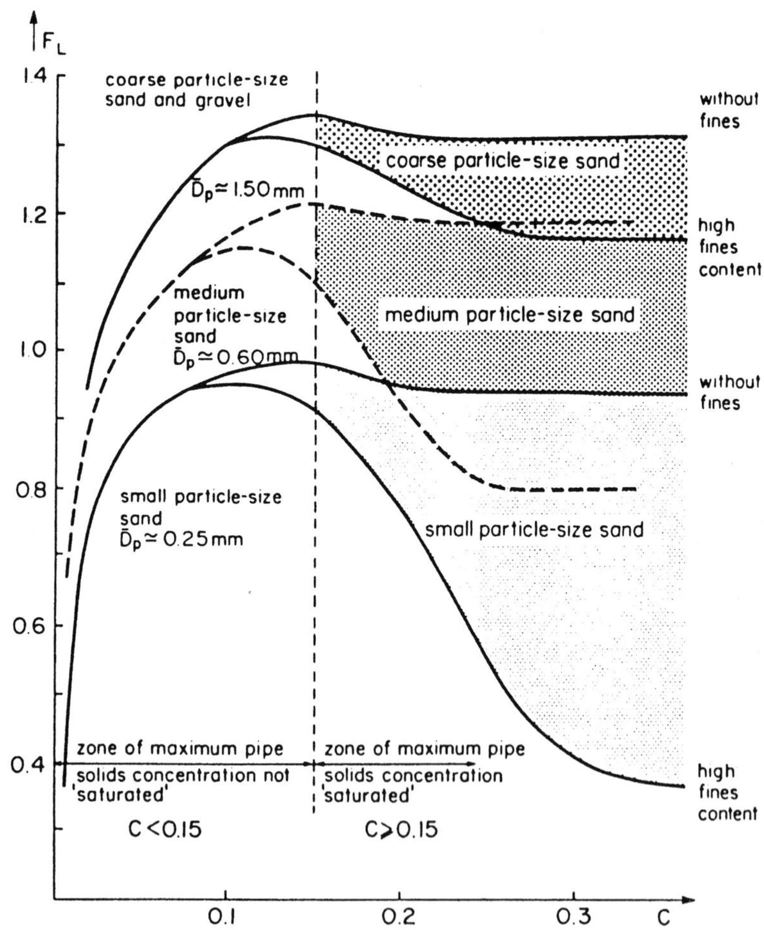

FIG. 3. Overall variation of the Durand parameter F_L with solids concentration and particle size for sand and gravel slurries. (Taken from Ref. 6.)

The difficulty in using Fig. 3 is the determination of the proportion of fine material which affects the rheology of the carrier fluid and the magnitude of the effect on the carrier fluid in terms of changed rheology. This would have to be assessed by viscometric tests on the mixtures used in pilot plant tests. However, it is probably preferable to use the scale-up techniques from small-scale tests instead of attempting to analyze the overall flow behavior in terms of its constituent parts.

Some attempts have also been made to analyze the suspension of particles in a turbulent fluid in order to predict design velocity from suspending velocity. Hanks and Sloan [7] used a model based on the dynamic equilibrium of solid particles at the bottom of the pipe and obtained expressions which were apparently able to correlate data obtained by the Saskatchewan Research Council. Roco and Shook [8] used an approach which attempts to identify the effective flow structure in the pipe and hence critical deposit velocities. This approach aims to predict the point at which a stationary bed begins to form in the pipe and is therefore similar in essence to the Wilson [9] approach which attempts to predict the maximum velocity at which a stationary bed of solids will remain on the pipe bottom. Wilson argues that deposition (the velocity below which a stationary bed of solids appears at the bottom of the pipe) occurs when the forces driving the sliding bed are no longer sufficient to overcome the solid–liquid friction between the bed and the pipe wall. In 1979, Wilson [9] published the major predictions from his model in nomographic form. Thomas [10] has reviewed much of the theory of the sliding bed model and has supplied some valuable experimental data to justify its potential usefulness. In addition, he extended the range of applicability of the sliding bed model to describe deposition of particles smaller than 0.15 mm in diameter in turbulent laminar flow [11] and transitional flow [12].

Oroskar and Turian [13] developed a correlation for the prediction of the "critical velocity" by using an analysis based on balancing the energy required to suspend the solid particles with that derived from dissipation of an appropriate fraction of the turbulent eddies of the flow. They defined the critical velocity as the minimum velocity demarcating flows in which the solids form a bed at the bottom of the pipe from fully suspended flows. Their final correlation is

$$V_c = 1.85[gd(S-1)]^{0.5}C^{0.154}(1-C)^{0.356}\left(\frac{d}{D}\right)^{-0.378}\mathrm{Re}_{\mathrm{OT}}{}^{0.09}X^{0.3} \tag{5}$$

in which

$$\mathrm{Re}_{\mathrm{OT}} = D\rho_L\left(\frac{gd(S-1)}{\rho_L}\right)^{0.5}$$

and X is the fraction of eddies with velocities exceeding the hindered particle settling velocity. A total of 357 data points from their own and other studies was used to estimate the constant and exponents in Eq. (5). In most cases the value of X was found to be close to unity. Oroskar and Turian found an overall rms deviation of 22% between experimental and predicted values from their correlation compared with rms deviations ranging from 50 to 73% using seven other correlations.

However, in general, most correlations tend to be very unreliable—outside the range of experimental conditions on which they are obtained.

There still seems to be no universally "best" correlation that can be used indiscriminately although the work of Oroskar and Turian is a marked improvement. A correlation can be considered adequate only if based on a pipe diameter and on material characteristics closely similar to those involved in the pipeline design itself.

It is true that blockages due to underdesign are rare (or at least unpublicized), but this may be due to the uncertainties in design resulting in predictions that are too conservative. Constructors and operators of large pipelines cannot afford large overdesign factors, so the designer must carry out large-scale trials with the suspension to be transported. Even then, the pipeline on completion has to be operated at a velocity, say, about 20% higher than the minimum value found in the trials. Operation too near the minimum velocity results in instability and possible blockage, particularly when using centrifugal pumps.

However, large-scale testing to obtain design data can be expensive if a number of trials have to be carried out to optimize pipe size, solids concentration, and particle size (the particle size is optimized when the total of grinding plus pipeline costs is minimized). Indeed, the costs of large-scale testing can be justified only when planning very large installations. What is required, therefore, is a routine small-scale or laboratory test that can be carried out on the actual material to be conveyed in the full-scale line. It has been suggested that the best way to develop the test method would be to determine the different flow regions that may exist in a range of pipe diameters and then to develop scale-up procedures. For details of the proposals, see Ref. 3.

Pressure Gradient for Pipeline Flow of Settling Slurries

As in the case of predictions for minimum transport velocity, a number of workers have made experimental measurements of the pressure gradient from which they have derived correlations. The designer is again faced with a wealth of correlations to choose from. Once more, it is clear that they are strictly applicable only to the conditions of the experiments, which are usually on a small scale and use narrow size ranges of material.

A number of correlations and approaches apply to certain flow regimes. For long distance pipelines which transport relatively fine material at design velocities just above the suspending velocity, the flow is heterogeneous and the Durand and Condolios [14, 15] approach is most common. At much higher velocities, where pseudohomogeneous flow is approached, Cheng and Whittaker [1] have applied viscometric data to pipeline design. At lower flow velocities where a sliding bed is formed, or when very large particles are being transported, Wilson et al. [22, 23] have developed an analysis based on the frictional forces between the contact load of particles and the pipe wall. When a very widely sized material is being transported, one promising approach developed by Clift et al. [25] combines the heterogeneous flow and contact load approaches. Each of these approaches will now be discussed briefly.

Durand and Condolios [14, 15] worked on the transportation of gravel in pipes of different sizes and found that their head loss data could be correlated for the *heterogeneous flow regime* by

$$\frac{i_m - i_w}{Ci_w} = K'\left[\frac{V^2\sqrt{C_D}}{gD(S-1)}\right]^{-1.5} \tag{6}$$

Newitt et al. [16] used an expression similar to Eq. (6) but which involved the particle terminal settling velocity, U_t, instead of the particle drag coefficient, C_D.

Charles [17] suggested that a much improved version of Eq. (6) which also gives pressure gradient predictions for the pseudohomogeneous flow regime at higher mean flow velocities is

$$\frac{i_m - i_w}{Ci_w} = K'\left[\frac{V^2\sqrt{C_D}}{gD(S-1)}\right]^{-1.5} + (S-1) \tag{7}$$

with the constant K' having a value of approximately 120.

However, Khan et al. [18] recently argued that there is considerable evidence [19, 20] that whereas the excess head loss i_m i_w is directly proportional to C for transportation with bed formation, this relation is no longer applicable for heterogeneous flow and that an improved representation of data is obtained by incorporating C into the modified Froude number and plotting $(i_m - i_w)/i_w$ versus $V^2\sqrt{C_D}/(CgD(S-1))$. They suggest that the advantage of plotting head loss data in this form is that the modified Froude number marks the transition between sliding bed flow and heterogeneous flow. At values of Froude number below 40, the slope of a log-log plot is −1 and a sliding bed exists, while values above 40 give a slope of around −1.4 (compare Durand and Condolios' value of −1.5) and corresponds to heterogeneous flow. Zandi and Govatos [21] have also suggested that this marks the transition between the two flow patterns.

Khan et al. [18] found that pressure gradient data from a number of studies gave discrepancies between actual values of pressure gradient under given conditions by a factor of up to 4. They argued that one important reason for this is that the parameters used in the correlation are all external properties. In particular, discharge concentration, C, is used whereas the more relevant parameter determining the nature of the flow is the *in-situ* solids concentration, C_x. In addition, it is the velocity of the liquid relative to the particles which determines the hydrodynamic drag on the particles and the energy transfer rate. As a result, Khan et al. measured both C_x and the actual linear velocity of the liquid, V_L. For experiments carried out using 3.5 mm gravel flowing in a 38-mm diameter pipe, V_L and C_x could be correlated by

$$V_L - V = aV^bC_x^c \tag{8}$$

and for their experimental results,

$$V_L - V = 0.08\,V^{-0.64}C_x^{\,0.27} \tag{9}$$

It is possible to show that

$$C = 1 - (1 - C_x)\frac{V_L}{V} \tag{10}$$

and from Eqs. (9) and (10)

$$\frac{(1 - C)}{(1 - C_x)} = (1 + 0.08\,V^{-1.64}C_x^{\,0.27}) \tag{11}$$

In subsequent, currently unpublished work, Khan et al. show that the use of C_x in the modified Froude number instead of C correlates pressure gradient data much better. Thus, if C is known, an estimate for C_x may be obtained by iteration using a relationship of the form of Eq. (11). The constant and exponents in Eq. (11) will, in general, be a function of the properties of the solids (size, shape, density) and liquid (rheology and density) and must be evaluated for different systems in order for the procedure to be more widely applicable. It may then be possible to correlate these adjustable parameters with the physical properties of the slurries in order to use the procedure for slurries not previously investigated.

For flow in which the majority of particles are transported in the form of a *sliding bed*, Wilson [22] analyzed the bed motion by using a coefficient of friction between the bed and the pipe, μ_s, and presented his results graphically as a plot of i_m/i_p against i_w/i_p, where i_p is the hydraulic gradient for a pipe full of solids and is given by

$$i_p = 2\mu_s(S - 1)C_B \tag{12}$$

where C_B is the concentration of solids in the bed. Wilson and Judge [23] expanded this analysis to include flow in which there are suspended particles as well. They suggest that the delivered concentration of solids transported in contact load, C_c, can be found from

$$\frac{C_c}{c} = \left(\frac{V_u}{V}\right)^{\epsilon} = R \tag{13}$$

in which V_u is the pipe flow velocity at full stratification ($R = 1$), given by [24]

$$V_u = 0.6U_t\left(\frac{8}{f_f}\right)^{1/2} \exp\left(\frac{45d}{D}\right) \tag{14}$$

The exponent ϵ in Eq. (13) is stated by Wilson to be slightly less than 2, and Cliff et al. [25] state ϵ is close to 1.7. The constants in Eq. (14) were originally evaluated using data for $0.01 \leqslant d/D \leqslant 0.028$.

Clift et al. [25] employed Eqs. (13) and (14) in their analysis of *combined heterogeneous and sliding bed flows*. The pressure gradient of mixed flow is taken to be

$$i_m = Ri_{ms} + (1 - R)i_{mh} \tag{15}$$

where i_m is the hydraulic gradient (measured as head of slurry per length of pipe) for the mixed flow, and i_{ms} and i_{mh} are the hydraulic gradients which could occur if the solids were fully stratified or fully suspended. The stratification ratio, R, is given empirically by Eq. (13).

From a combination of detailed analysis of the flow situation and empiricism, the two gradients i_{mh} and i_{ms} are expressed as

$$i_{mh} = i_f[1 + A(\rho_s/\rho_L - 1)] \tag{16}$$

and

$$i_{ms} = i_f + B(\rho_s/\rho_L - 1) \tag{17}$$

in which A and B are both material properties of the slurry. A has a value typically between zero and unity while B is of the order of unity.

Defining the friction factors,

$$f_m = \frac{Di_m g}{\frac{1}{2}V^2}, \; f_f = \frac{Di_f g}{\frac{1}{2}V^2}$$

and combining Eqs. (4) to (7),

$$\frac{f_m - f_f}{\dfrac{\rho_s}{\rho_L} - 1} = \frac{2gD}{V^2}B\left(\frac{V_u}{V}\right)^m + Af_f\left[1 - \left(\frac{V_u}{V}\right)^m\right] \tag{18}$$

In this equation, A, B, V_u, and m are material parameters depending on particle characteristics such as particle size and shape. V_u also depends on pipe diameter according to the Wilson Eq. (14). Clift et al. showed how they can be determined from pipeline experiments. Equation (18) is used to determine the effective particle size and for scaling to other pipe diameters.

The effect of non-Newtonian properties has been taken into account by Cheng and Whittaker [1] for both the heterogeneous and pseudohomogeneous flow regimes. These researchers measured the rheological properties of coarse suspensions in a viscometer, from which the pressure gradient for laminar and turbulent flow can be predicted (by a method to be given later). This procedure is applicable only if the suspension does not settle so fast that homogeneity cannot be maintained in the viscometer. A simple experimental test for determining the settling rate is included in the method.

Figure 4 compares experimental values for a 53-mm pipe containing a 5% (by volume) iron-oxide suspension with predictions given by both the Durand [15, 16] correlations (the most widely used of the correlations) and

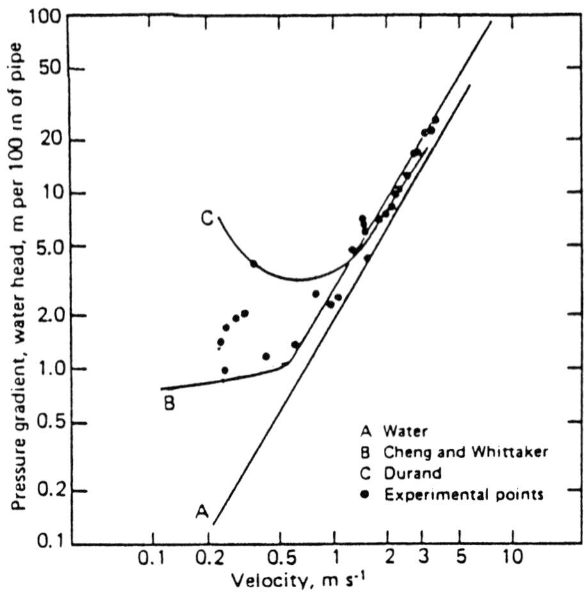

FIG. 4. Pressure gradient for a 50-μm iron oxide in a 35-mm diameter pipe at a 5% by volume concentration of solids. (Taken from Ref. 1.)

with values given by the method of Cheng and Whittaker [1]. The suspension contained particles with a mean size of 50 μm and was significantly non-Newtonian. It can be seen that the experimental results are in good agreement with the Cheng and Whittaker predictions as long as the flow is homogeneous; for this particular suspension of fine particles, this is true at all velocities above the laminar/turbulent transition. For coarser particles the deviation from the predicted curve will occur at a higher velocity. Curves for other materials are given by Cheng and Whittaker. The Durand predictions are seen to overestimate the experimental results for velocities below about 1 m/s.

In summary, the conclusion is that as with the correlations for design velocity, those for pressure gradient are particularly unreliable for large-diameter pipes and the only reliable method is to carry out a large-scale trial. This should be done on a scale as near the full size as is practical and on the actual material to be transported, because the effect of scale-up is uncertain.

The designer of a large installation will also be interested in wear rates which affect the life of the line, but, at present, laboratory methods of predicting wear rates are extremely unreliable and the only practical method is to use either a large-scale test or data from an existing line.

It is clear that, as with design velocity, there is a need for large-scale experiments on which the scale-up rule for pressure gradient and wear rates can be established, so that bench or pilot-scale data can be reliably scaled up in design.

Nonsettling Slurries

Solid–liquid mixtures that settle (when not in flow) can, if the settling rate is low, be treated by the methods to be discussed in this section. Examples of application have already been anticipated in the preceding section. Other slurries are nonsettling because of chemical forces acting between solids and liquid and between the solid particles themselves. These slurries can be treated as single-phase homogeneous fluids; consequently, they can be characterized in a laboratory viscometer with the rheological parameters obtained being used for full-scale design. Alternatively, bench or pilot-scale pipe-flow data can be scaled up directly without reference to rheological characterization. The many methods and correlations that have been developed for design calculations have been reviewed by Cheng and Heywood [26–30]. Here, we shall concentrate on the few that are most generally applicable.

Design from Viscometric Data for Pipeline Flow of Nonsettling Slurries

Laminar Flow

For laminar flow, the pipeline characteristics can be calculated from first principles for any given rheological data or model fitted to the data. The only limitation on the accuracy is the accuracy of the data and the closeness with which the characteristics of the slurry fit the chosen model. The model chosen should not be too simple, and Cheng [30] has suggested the use of the generalized Bingham plastic (also known as the Herschel-Bulkley) as a universal model-fluid. This is a three-parameter model that appears to give a good fit for many industrial slurries:

$$\tau = \tau_y + K\dot{\gamma}^n \tag{19}$$

Two different methods for the estimation of the three parameters are often used: one involves nonlinear least-squares regression on unweighted experimental data while the other involves a nonlinear least-squares regression on weighted data. Heywood and Cheng [28] have shown that while both methods will give best estimates of the three parameters which will then allow prediction of τ to within 2% over the original viscometric shear rate range, extrapolation well outside this shear rate range leads to very different predictions using the two methods. Figure 5 shows the large differences in τ prediction obtained for a 8% digested sewage sludge.

Clearly, extrapolation outside the experimental range should be done only with caution and, if shown to be unsatisfactory, should not be undertaken. As a rule of thumb, it is generally unsatisfactory to extrapolate the flow curve to zero shear rate if the lowest shear rate used in viscometric tests is greater than about 25% of the maximum shear rate, i.e., the nominal wall shear rate, $8V/D$. Extrapolation to larger shear rates beyond $8V/D$ is not required when scaling-up from tube viscometer laminar data to laminar flow in the full-scale

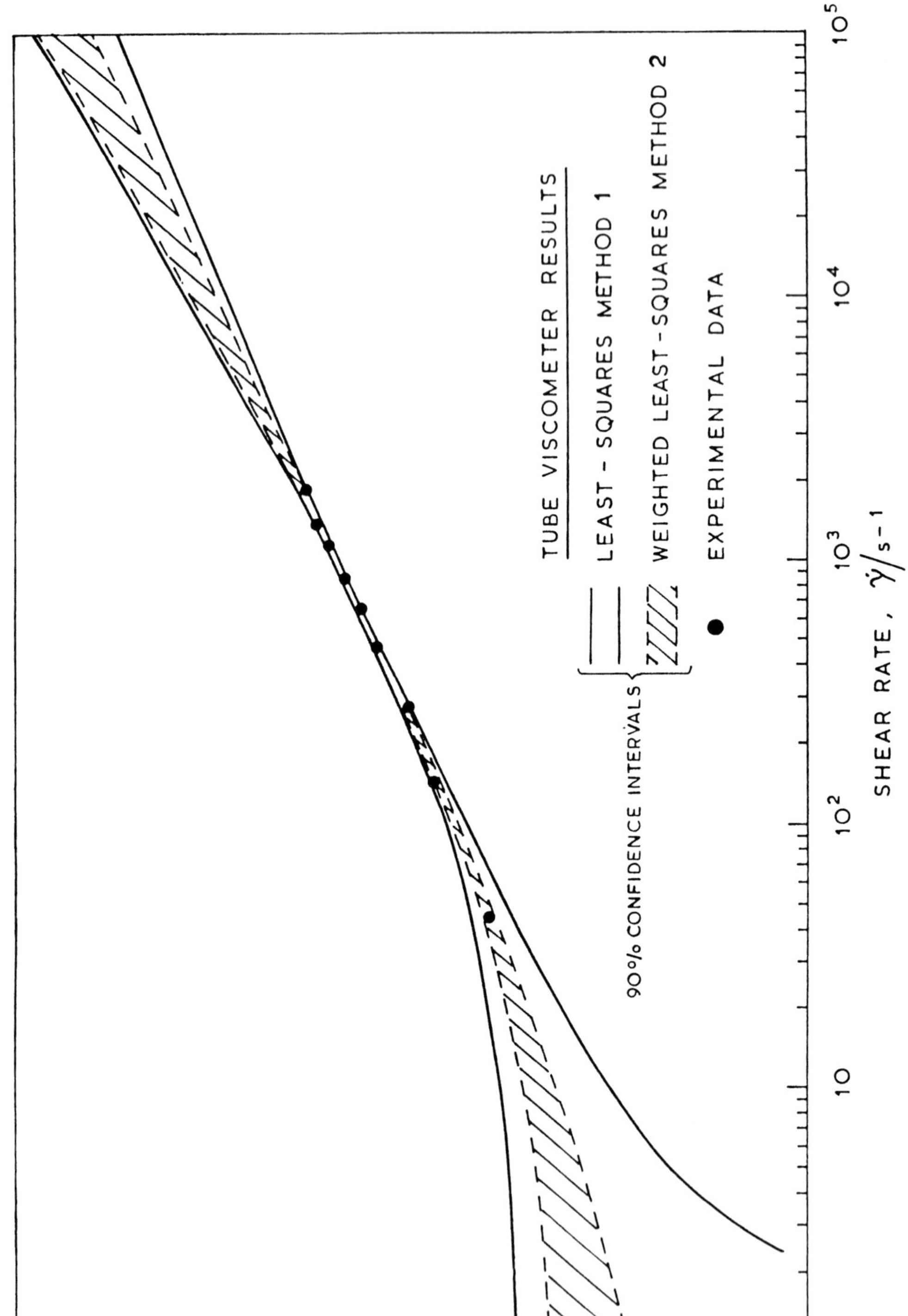

FIG. 5. 90% confidence intervals for flow curve of 8% by weight digested sewage sludge. (Taken from Ref. 28.)

pipeline but can be necessary when scaling-up to turbulent flow. However, experimental ranges employed in viscometric measurement are rarely so wide unless adequate funds are available for such work. Thus it may be worthwhile sometimes to assess the two methods of flow model parameter estimation to determine which method gives the least uncertainty in predictions outside the range used in viscometric tests.

Turbulent Flow

For turbulent flow of Bingham fluids (or fluids with a definite limiting viscosity at high shear rates), Hedstrom [31] suggested that the von Karman equation (see Fig. 6) could be used. However, Fig. 6 shows that the friction coefficient for Bingham fluids is about 12% lower than for Newtonian fluids. For power-law fluids, Dodge and Metzner [32] recommend a generalized form of the von Karman equation in which a generalized Reynolds number is used and in which the coefficients are functions of the power-law index, n. Cheng [30] suggests that this expression can also be used for the generalized Bingham plastic, Eq. (19). This gives a 12% overestimate for $n = 1$ and probably similar overestimates for other values of n, but this is a useful safety margin. Thus, turbulent pressure gradients can be predicted from laminar viscometer data without the expense of pilot-scale pipeline experiments.

The method is not applicable to all fluids and cannot be expected to apply if a fluid is appreciably elastic or thixotropic. However, the method yields

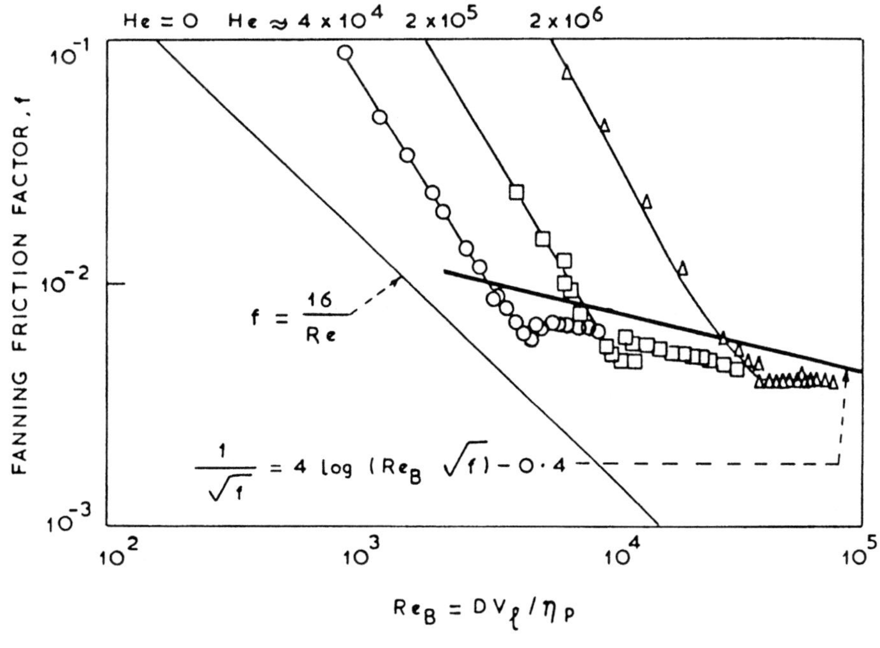

FIG. 6. Plot of friction factor vs Reynolds number for pipeline of a Bingham plastic fluid.

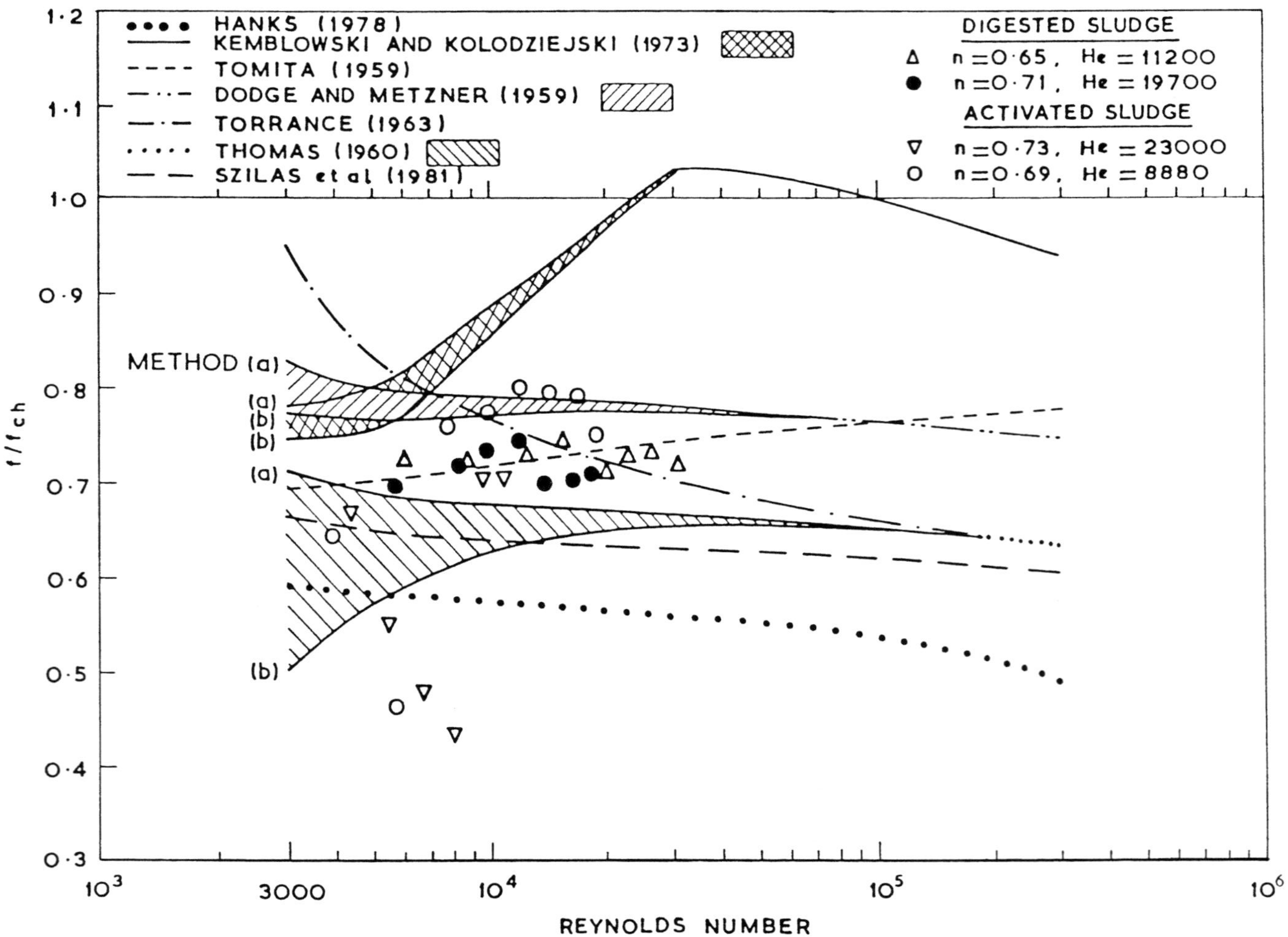

FIG. 7. Comparison of methods for predicting turbulent flow head loss for a generalized Bingham plastic fluid with $n = 0.7$ and He = 104. (Taken from Ref. 28.)

satisfactory predictions for many industrial fluids. Some examples are mentioned by Cheng [33]. For a number of china clay slurries in pipes up to 75 mm diameter, there was an underprediction in pressure of not more than 40%. For a crude oil in a 300-mm pipe, an underestimate of 20% was found. It should be pointed out that the test conditions that produced these comparative data were far from ideal. In many cases these deviations are no greater than those due to the variability of the material (which was apparent even in the viscometric results).

The uncertainty in pressure drop prediction is further increased if other predictive equations are considered. Figure 7 gives a comparison of the Fanning friction factor, predicted by using a number of published equations and normalized by dividing by the friction factor predicted using the Churchill approach [34]. The predictions are compared in the plot with experimental data for the flow of digested and activated sewage sludge in a 100-mm diameter pipeline. Considerable variation in the predictions is evident, and it is necessary to apply engineering judgment of the degree of conservatism to be adopted in head loss estimation.

Methods (a) and (b) in Fig. 7 refer to two alternative approaches which can be used when applying a predictive equation for f developed for a power law flow model to the three-parameter model Eq. (19). With Method (a) the importance of yield stress in turbulent flow is assumed to be negligible, and estimates for K and n in Eq. (19) are used directly in equations for f. With Method (b) the Metzner-Reed approach [35] is applied: yield stress is taken into account indirectly by using localized values K' and n' corresponding to the relevant wall shear stress level for turbulent flow. The hatched areas in Fig. 7 indicate the difference in f-values using the two methods but the same predictive equation for f. This variation diminishes as either the Reynolds number Re′ increases or the Hedstrom number He decreases.

The uncertainty in predicting f is further increased if confidence limits are assigned to results from each predictive equation as shown in Fig. 8. The confidence interval reflects the degree of accuracy of the original viscometric data and the validity of the flow model adopted.

It seems clear that the design engineer must be aware of the many prediction methods available and that they can lead to widely differing f-values. By using as many of these methods as possible, upper and lower bounds for f can be assigned, and then an appropriate f value or f range can be adopted for further detailed design.

It is sometimes also important to be aware of pipe wall roughness. Most predictive equations for f for non-Newtonian slurries assume smooth wall pipe, and while it is well-established that the magnitudes of irregularities on the inside wall of a pipe have no effect on the pressure drop of materials in laminar flow, the situation for turbulent flow is rather different and is well documented for Newtonian fluids. Here an estimate for the relative roughness, e/D, is required as it is an additional variable in equations for the prediction of the friction factor. Heywood and Cheng [28] have also compared the estimates given for f from various equations for Newtonian fluids for two levels of relative roughness, e/D, of 10^{-4} and 10^{-2}. They conclude that the spread in f prediction is wider than for the smooth pipe, being about ±6%

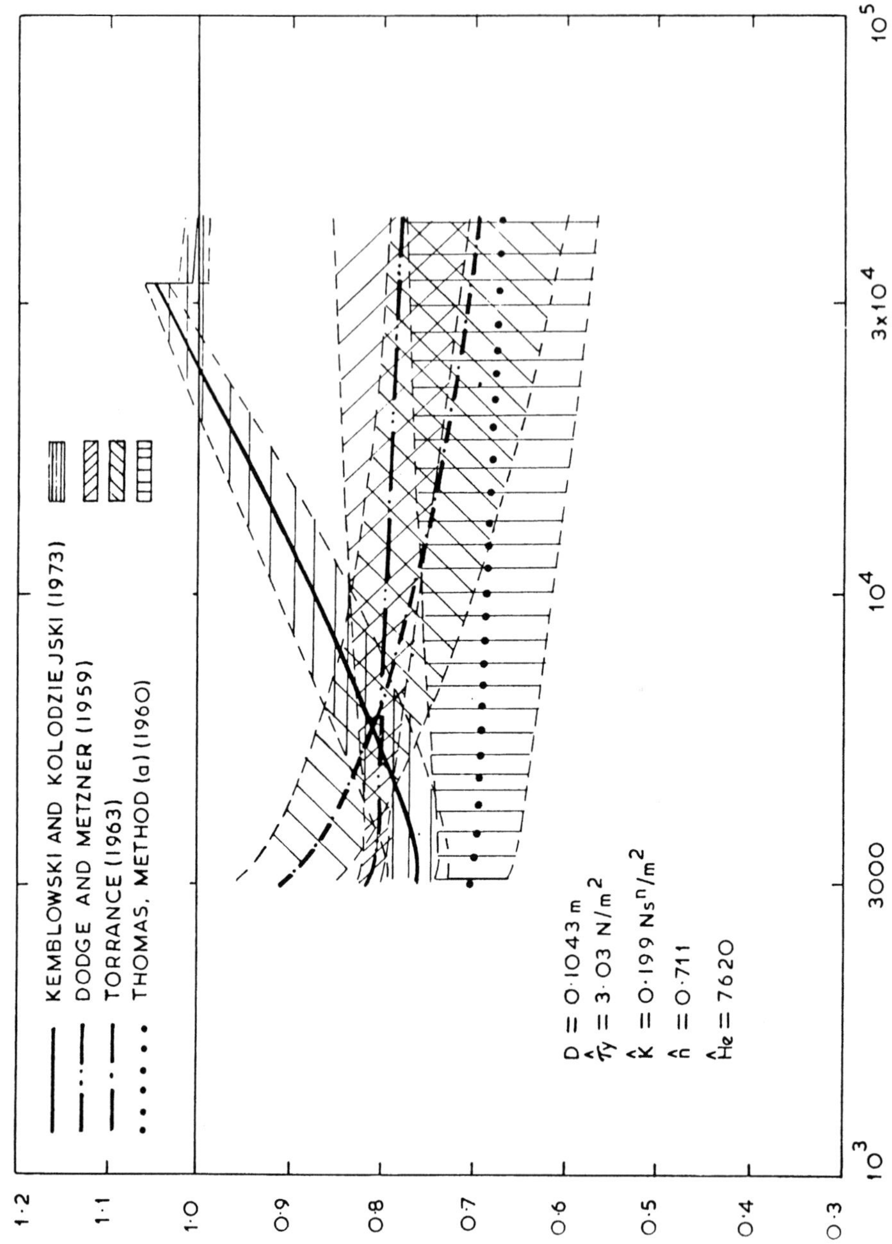

FIG. 8. 90% confidence limits for predicted friction factor using various prediction methods. (Taken from Ref. 28.)

from the mean. In general, the larger the pipe surface irregularities for a given pipe diameter and Reynolds number, the higher is the pressure drop.

Govier and Aziz [36] categorized the roughness effect into three classes: (1) smooth wall pipe (or SW) turbulence; (2) partially rough wall pipe (or PRW) turbulence, where friction factors are larger than for smooth pipe and are dependent on Reynolds number; and (3) fully rough wall pipe (or FRW) turbulence, where friction factors are independent of Reynolds number but are dependent upon relative roughness, e/D.

In their recommended design methods, Govier and Aziz suggest that if the pipe is rough, the pressure gradient calculated by the various methods for smooth pipes should be multiplied by the ratio of the friction factor for rough pipe to the friction factor for smooth pipe, as determined from the Moody chart for Newtonian materials at the appropriate Reynolds number consistent with the fluid model being used; i.e., Re_B for the Bingham plastic model or Re' for either the power law or generalized Bingham models. If the relative pipe roughness is such that FRW turbulence may be expected, Govier and Aziz recommend for all three flow models an equation developed by Torrance [37]:

$$\frac{1}{\sqrt{f}} = \frac{1.77}{n} \log_e \frac{D}{2e} + 6.0 - \frac{2.65}{n} \tag{20}$$

When pipe roughness is expected to be an important factor, the relative roughness, e/D, may be estimated from standard texts [36, 38] or, alternatively, an estimate of e/D may be obtained from the Moody chart [40] by measuring friction factors for the turbulent flow of water through the pipe at known Reynolds numbers.

Laminar/Turbulent Transition

The simplest approach toward dealing with the transitional region between laminar and turbulent flow is to assume that it does not exist, and that the transition, as the flow velocity is progressively increased, occurs completely at a single operating point. In this case only one critical Reynolds number need be predicted and this can be achieved by taking the intersection of the f–Re relationships for laminar and turbulent flow for the fluid model under consideration. However, this approach will not yield accurate results. It is well-known that for Newtonian fluids the lower critical Reynolds number for the initial breakdown of laminar flow, Re_1, is 2100 and the upper value, Re_2, for the commencement of fully turbulent flow is approximately 3000.

However, for non-Newtonian fluids, the individual values of Re_1, and Re_2 are much more difficult to predict. Various attempts have been made to develop methods for Re_1 prediction based on different flow models, but Cheng [30] has offered an approach for predicting both Re_1 and Re_2 valid for both power law slurries and generalized Bingham slurries. There are few data with which to verify Cheng's approach, because industrial pipelines tend to avoid operating in this regime. However, the most economical design often suggests that flow should be in the transitional regime and some pipelines operate just above Re_2. One of the reasons why Re_2 is difficult to define more precisely is that the fluctuations in pressure gradient prevalent in the transitional regime diminish progressively as flow velocity is increased, rather than abruptly when the fully turbulent flow regime is reached.

Design of Nonsettling Slurries by Scale-Up

Laminar Flow

An alternative method for predicting the performance of a large pipeline containing a homogeneous non-Newtonian fluid is by scale-up from small pipeline data. For laminar flow, a plot of shear rate ($8V/D$) against shear stress ($D\Delta P/4L$) is independent of pipeline diameter provided there are no significant wall-slip effects, so small-scale data can be converted directly to data to be used for calculating the full-scale line. Although this method is often used for the treatment of small-scale data, it is not widely used for scale-up to large diameter pipes.

Turbulent Flow (Bowen Method)

For turbulent flow, Bowen [41] suggests a modification of the Blasius equation:

$$D^x \tau = k V^w \tag{21}$$

A plot of shear stress τ against flow rate allows w to be determined. Then, by replotting in the form τ/V^w against D, the values of k and x can be obtained. The transition is given by the intercept of the curves for laminar and turbulent flow, though of course this yields a single point rather than a range. The Bowen method does not depend on whether or not rheological properties of the fluid fit a given model, and the method should also be valid for elastic fluids.

Kenchington [42] has used this method for scaling up from 27, 35, and 53 mm pipes to a 329-mm pipe. His results for a clay slurry are shown in Fig. 9, in which for each line of best fit there are two lines giving 90% confidence limits. After rejection of a few rogue points, it was found that the small-scale data could be correlated to within 7%; but as the data are scaled up, the confidence limits become wider. For the 329-mm pipe, the 90% confidence limits are 63% below and 165% above the best-fit line, so that scale-up on the basis of the best-fit line could lead to serious underdesign—though, for the particular case shown in Fig. 9, the limited amount of experimental data for the 329-mm line appears to fall about 25% below the best-fit line.

The reason for the wide confidence limits is the scatter of experimental results from the small pipes, which is partly due to degradation of the particles

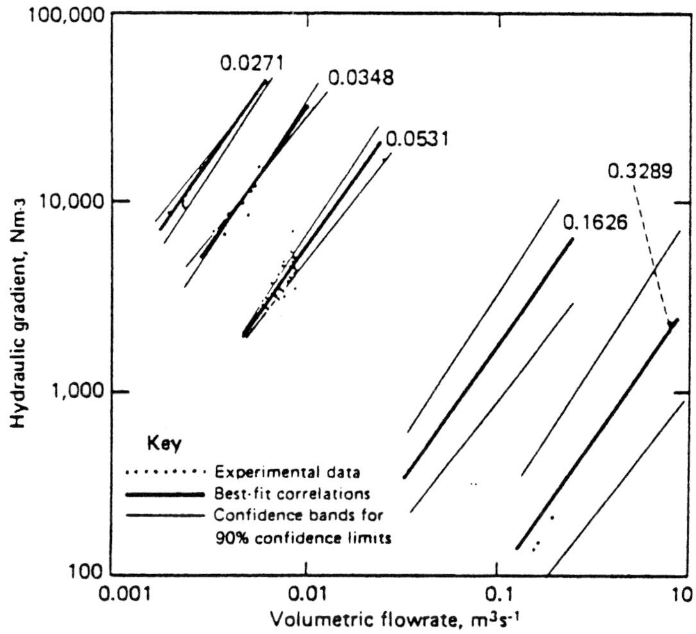

FIG. 9. Comparison of predicted and experimental hydraulic gradients for a range of pipe diameters in meters. (Taken from Ref. 42.)

during the tests. It was also shown that if the amount of test work was reduced by doing tests only on 27 and 53 mm pipes, the confidence limits were widened from −63 to +165% of the best-fit line to −92 to +1300% of the line.

Kenchington's technique is very powerful for high accuracy in pipeline design if it is used in reverse to determine the range of flow rates and precision required in pilot-scale experiments in order that specified accuracy in prediction can be obtained at the design flow rates for the full-scale pipeline.

It may be that, because industrial suspensions do show variable properties, unattainably high precision is necessary at the pilot scale. If this is so, the uncertainties found by Kenchington are inevitable when scaling up by a factor of 6 to 12 on pipe diameter and of the order of 100 on flow rate. So, if high accuracy is essential, large-scale test work is unavoidable. This is yet another aspect of pipeline flow behavior that requires further research to resolve.

The degree of pipe roughness will be reflected in both exponents for D and V in Eq. (21). The problem arises, however, if scale-up from smooth-walled small bore (e.g., uPVC pipe) to rough wall large-scale pipe (e.g., galvanized iron) is undertaken, or vice versa. A corrected pressure gradient could be estimated in these cases, presumably by using the Govier and Aziz [36] approach of multiplying the predicted pressure gradient by a ratio of friction factors obtained from the Moody chart, although such a procedure has not yet been worked out. The friction factor ratio would obviously depend upon the wall roughnesses of the small bore and full-scale pipe and would neglect any wall slippage effects arising from flow in the smooth-walled pipe.

Stabilized Slurries

During the last decade many studies have been centered on the development of stabilized or semistabilized slurries. For the hydraulic transport of settling slurries containing large particles [43–45], relatively high flow velocities are required to prevent pipe blockage. Even so, significant asymmetric solids concentration profiles mean that delivered solids concentrations are often low (typically no more than 15 to 20% by volume) compared with the *in-situ* solids concentration. In addition, pipe wall wear rates are high and specific energy requirements (the energy consumption per unit mass of solids transported per unit distance) are usually much in excess of those for the hydraulic transport of nonsettling slurries. For these reasons very few long distance pipelines (i.e., in excess of 10 km) transport particles larger than around 25 mm.

There are a number of ways in which this specific energy requirement may be reduced. Some have been reviewed recently by Heywood [46] and will be outlined below. Two important methods described in detail here are through the use of "dense phase conveying" or, alternatively, by both increasing the density of the carrier fluid (thereby decreasing the density difference between solids and the carrier fluid) and imparting the non-Newtonian shear-thinning property to the suspending medium through the presence of flocculated fine particles. The term "stabilized slurry" is generally used with reference to the latter slurry type, but both slurry types may be considered stabilized.

Dense Phase Conveying [47, 48]

By increasing the solids concentration in the pipe to such an extent that the flow within the pipe resembles that of a sliding bed occupying practically the entire pipe cross-section, it is possible to achieve a delivered solids concentration between 25 and 50% by volume. The mean slurry flow velocity can be over a wide range because the coarse particles are not being supported by the energy derived from turbulent eddies but are mutually supported by numerous particle–particle contacts throughout the slurry. However, some slurries tend to exhibit dilatant (or shear-thickening) flow behavior at relatively low flow velocities, and this may explain in part the resistance to using such mixtures for pipeline conveying because their susceptibility to pipeline blockage has not been adequately investigated.

Use of a Heavy Medium

Large particles, which would normally settle out rapidly in a low viscosity carrier liquid such as water, can be incorporated into a "heavy medium," which is usually a homogeneous suspension of colloidal particles in a liquid. These colloidal particles may often be flocculated, resulting in the carrier medium possessing the shear-thinning non-Newtonian flow property. Such property can reduce or almost eliminate the settling of large particles as a result of the high carrier medium viscosity at the relevant low shear rates, while, at the higher shear rates close to the pipe wall, the viscosity of the carrier fluid and of the slurry as a whole (including the large particles) is reduced because of the shear-thinning behavior.

The presence of fines is seen generally to decrease the critical velocity for suspending the larger particles as shown in Fig. 3. This is mainly due to the increase in viscosity of the carrier fluid, but this effect is partly offset by a reduction in the magnitude of turbulent eddies supporting the particles when turbulent flow conditions prevail.

However, stabilized slurries will generally be transported in the laminar flow regime for the following reasons. Optimal slurry velocities are determined by minimizing the energy consumption per km per tonne of the coarse material to be conveyed [48] and by attempts to minimize wear of the pipe wall. The high overall solids concentration resulting from both coarse and fine material gives rise to correspondingly higher viscosities than if the coarse solids were transported on their own and this, combined with typical slurry velocities in the range of 1 to 2 m/s, will generally give Reynolds number substantially below 2000.

A useful method for representing the slurry compositions giving rise to either settling or nonsettling slurry for a bimodal particle size distribution has been given by Charles and Charles [49], who used a triangular diagram where the apexes represent 100% water, 100% sand (coarse), and 100% clay (fines) (Fig. 10). Various regions of the diagram denoted either settling or nonsettling states, and they concluded that nonsettling slurries could often be more economically conveyed under laminar flow conditions.

However, any reduction in specific energy requirements arising from the use of bimodal, or multimodal, particle size distributions in order to formulate stabilized slurries must be weighed against possible increases in costs arising from additional facilities for slurry preparation and recovery at the ends of the slurry pipeline. Even so, by appropriate slurry formulation, substantial reductions in head losses can be achieved. Hisamitsu et al. [50] reported that silica sand transported at 20% solids volume concentration in a carrier slurry containing 5% by volume of fine clay resulted in an excess head loss compared to the transport of water alone of only one-third that expected for the sand alone. Brookes et al. [51, 52] carried out successful large-scale trials in the British Petroleum STABFLOW project on coarse coal stabilized in a fine coal slurry medium which was pumped at high concentration through several kilometers of 300 mm diameter pipeline. One-third of the coal particles comprised sub-200 μm fine coal which was found to give stable horizontal flow and a near-uniform vertical solids concentration over a wide mean slurry velocity range. This proportion of fines was held constant for a range of overall solids contents.

Large-scale tests have also been undertaken by Duckworth et al. [53, 54] over a number of years at CSIRO, Melbourne, Australia. Coarse coal with a maximum particle size of 20 mm has been transported [54] through a 152-mm diameter pipe over a range of concentrations (at coarse coal fractions

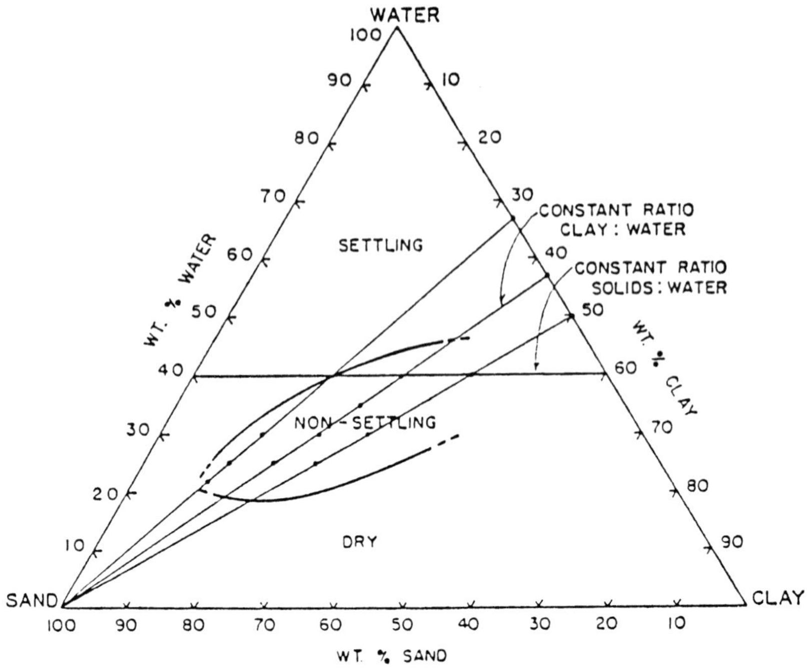

FIG. 10. Use of a triangular diagram to identify operating regimes for the application of heavy media to coarse solids transport. (Taken from Ref. 49.)

ranging from 0 to 0.45 and at total coal concentrations ranging from 53 to 67%) in a heavy medium composed of fine coal in water. The flow properties of the stabilized slurry were found to be described well by using the Bingham plastic model. They were able to show how the yield stress and the plastic viscosity of a coarse coal/fine coal slurry varies with the corresponding values for the fine coal carrier and with the coarse fraction.

An attempt has been made [55] to develop an optimal particle size distribution for a stabilized slurry based on achieving maximum particle packing density. This was achieved by using a statistical procedure with a limited number of experiments on the compaction of various size fractions using the Procter test from soil mechanics. The result was a slurry formulated to allow maximum pipeline capacity and minimum segregation on storage or during pipeline shutdown.

Some stabilized slurries have been observed [56, 57] to exhibit a reduction in head loss in turbulent flow compared with that observed with fines alone forming the heavy medium. Thomas [57] argues that this effect depends on the ratio of the large particle size, d_{coarse}, to the viscous sublayer thickness close to the pipe wall δ and occurs if d_{coarse}/δ is of the order of 5.

Research on stabilized slurries has thus demonstrated that the transport of coarse particles by pipeline over long distances is feasible practically. Most of the work has concentrated on coal for commercial reasons. The transport of coarse coal over long distances, particularly in the United States, Canada, and

Australia, is likely to save considerably on operating costs. However, a commercial long distance pipeline carrying stabilized slurries has yet to be built, and the promise of one in the future is subject to sharp changes in coal prices and to problems with railroad companies and interstate transfer of water in the United States.

For fine coal transport, the energy costs per tonne for grinding and preparation, dewatering, and transporting 100 km are in the ratio 5:3:2, and thus the use of stabilized slurries offers the prospect of significant cost savings over fine coal transport. Duckworth [54] suggests that such technology is particularly attractive in cases where coal is to be exported overseas because slurry dewatering to a transportable moisture content of 10% is likely to be facilitated. In addition, it has been demonstrated [54] that stabilized slurry transport gives rise to relatively low pressure gradients for operating velocities between 1 and 1.5 m/s, low wear rates, low energy consumption of 0.05 kWh/tonne km, and the reliability of restarting a pipeline following shutdown arising from a power failure.

Thixotropic Slurries

For slurries that possess strong interparticle attraction and form a gel when left to stand in a pipeline, the start-up pressures are very much greater than the steady-state pressure. A method for predicting both start-up pressures and the time taken for the gelled material to be expelled from the pipeline and attain steady flow has been developed by Carleton, Cheng, and Whittaker [58]. It is based on a generalized Bingham fluid with thixotropic build-up and breakdown following the Moore model [59]. The method is available as a computer program and can be used to predict the relationship between pressure, position in the pipe, and time, both for conditions of constant applied pressure and for a sawtooth variation in flow rate (the latter represents the flow from a positive displacement pump). The program for constant applied pressure is particularly useful for estimating thixotropic parameters from measurements in a tube viscometer.

The program for constant applied pressure has been used [60] for sewage sludges in pipes of up to 26 mm diameter. The time to empty the pipe, using water as the expelling liquid, was generally −50 to +100% of the predicted value. This order of discrepancy is understandable in view of the gross inhomogeneity of the material involved.

As yet, this method has not been applied to larger diameter pipes. It can readily be anticipated that experimental verification will not be easy, not only because of the simplifying assumptions made about thixotropic behavior and pipeline flow, but also because of the intervention of other phenomena in large pipes. These include temperature or other effects resulting in nonuniform build-up, inclusion of air or liquid vapor pockets (leading to overall compressibility), and surge effects in full-scale installations.

The problems of gelling of both crude and fuel oils and of drilling mud and bentonite clay slurries are of particular concern to oil and civil engineers.

Little published information is available for clay slurries. Some recent analyses of this problem for waxy crude oils have assumed Bingham plastic behavior by using a value for the yield stress measured after the crude oil has stood in a tube viscometer. It was found that yield stresses measured in a 6-mm tube were about 30% higher than those found in the field in a 610-mm line [61]. This was regarded as providing a safety margin of the order of magnitude desired. However, it is also known [62] that the thermal history of the oil has a considerable effect on its rheological properties and that yield values measured in a small pipe do not give realistic predictions when applied to full-scale (203 mm) lines.

Cheng [63] conducted a study into the possibility of using scale-up procedures to predict the pipe flow characteristics of thixotropic fluids. This study was based on the assumption that experimental results obtained on a small, laboratory-scale pipe rig can be scaled up to a full-scale industrial pipeline without specific reference to the detailed thixotropic property of the fluid, that the fluid obeys single-structured thixotropic constitutive equations (see reference), and that quasi-steady-state flow pertains. Three different cases for start-up were discussed depending on whether the flow rate or the pumping pressure is prescribed, or if the pump characteristics (described by a functional relationship between flow rate and pump pressure) are prescribed. Cheng showed that for geometrically similar pipelines (i.e., D/L constant), if the initial out-going fluid and the inlet incoming fluid are similar and the flow rate is prescribed as a function of time only, then the pumping pressure is a function of time only and independent of L. The situation arising if the pumping pressure is prescribed, or if the pump characteristics are prescribed, is more complex but was also analyzed. Cheng also considered how the scale-up rules might be achieved in practice. A rather impractical situation which has to be overcome is that D/L should be constant between the full-scale pipeline and the pilot or laboratory-scale rig. This means that in order to match a 0.6-m diameter by 50 km pipeline, for example, where $D/L = 12 \times 10^{-6}$, a 0.1-m diameter pilot-scale rig would have to be 8 km long, and a 0.01-m diameter laboratory rig would have to be 800 m long. Because of this, and other impracticalities, Cheng concluded his study by suggesting certain experimental work which could be conducted to check the validity of a simpler approach toward developing scale-up rules for predicting the start-up characteristics for pipelines transporting thixotropic fluids. No report of this experimental work has appeared in the open literature.

Head Losses from Pipe Fittings

Estimates of head losses arising from flow through pipe fittings are required before total system head loss estimation can be made. This then allows the selection and sizing of an appropriate pump or pumps. Pipeline fittings may be classified as branching, reducing, expanding, or deflecting types. Tees and crosses are branch fittings. Reducers, bushes, sudden enlargements, and contractions are fittings which change the area of the fluid passage and belong to the reducing or expanding class of pipe fittings. Elbows, bends, return bends, and other such fittings which cause a change in the direction of flow are of the deflecting type. A number of fittings will, of course, possess combinations of the attributes of the various general classes described above, and other fittings, such as unions and couplings, will ordinarily offer negligible resistance to flow.

Information on losses may be classified for two applications: settling and nonsettling slurries. Much information exists for the turbulent flow of Newtonian nonsettling slurries, with head loss often expressed in terms of the number of equivalent pipe diameters of straight pipe giving the same head loss [38, 39]. For the turbulent flow of non-Newtonian slurries, both Cheng [30] and Turian and Hsu [64] have concluded that the losses do not depend significantly on the non-Newtonian character of the slurry.

Normally, head loss is expressed in terms of the loss of number of velocity heads through a loss coefficient, K_L:

$$h_L = K_L(V_f^2/2g) \tag{22}$$

where V_f is the characteristic slurry velocity in the fitting. K_L is approximately constant and independent of pipe Reynolds number for the turbulent flow of either Newtonian or non-Newtonian slurries, but K_L is inversely proportional to Reynolds number for laminar flow in both cases.

Much less information is available for the laminar flow of Newtonian slurries through fittings than for turbulent flow. A summary is provided by Perry [38], and the available information has been reviewed by Edwards et al. [65], who have also provided some much-needed data for the laminar flow of non-Newtonian slurries. Their overall conclusion is that correlations relating K_L to Re for Newtonian slurries may also be used for non-Newtonian, power law slurries modeled by the Ostwald-de Waele flow model:

$$\tau = K\dot{\gamma}^n \tag{23}$$

provided the non-Newtonian Reynolds number Re′ is used in these correlations instead of the Newtonian Re.

It is important to note that the Reynolds number ranges over which these correlations apply are generally much smaller than the usual range for laminar pipe flow. Thus the maximum Reynolds number in the case of globe valves is around 12, while for a 90° elbow the value is 900. This compares with the usual value of 2100 for the breakdown of Newtonian laminar flow in a straight pipe. The reason for this, of course, is that the actual velocity in the fitting will often be much higher than the pipe flow velocity, and hence the breakdown of laminar flow will, in general, occur earlier. It is the actual velocity which is taken as the characteristic velocity in the correlations.

Losses in pipe fittings for the flow of settling slurries or dense phase suspensions and stabilized slurries are much more difficult to estimate. Only Iwanami and Suu [66, 67] appear to provide any data. It is suggested that fittings for use with settling slurries should be designed so as not to encourage solids deposition and possible pipeline blockage. This is particularly important with respect to pipeline bends, changes in angles of incline and decline, and the placement of any essential valves.

Techniques for Reducing Energy Consumption in Slurry Pipelining

Heywood [46] has reviewed the methods available to reduce head loss and hence energy consumption incurred when slurries are pumped through horizontal pipes. Some of the more obvious ways of reducing slurry viscosity levels directly include:

a. Reducing slurry concentration
b. Increasing particle size (if Brownian motion and particle surface effects control viscosity levels)
c. Broadening the particle size distribution at constant total solids concentration
d. Reducing the angularity of particle shape while maintaining the particle size distribution and solids concentration essentially constant. Also, adding high aspect ratio fibers
e. Adding deflocculants (soluble ionic compounds) to disperse flocculated slurries (important when a significant percentage, say 5–10%, of solids are below approximately 5 μm in size)
f. Addition of soaps
g. Addition of high molecular weight polymer
h. Injection of water (or other suspending medium) through the pipe wall periodically along the pipeline length to create a water–film lubricating layer
i. Development of stabilized or semistabilized slurries by supporting coarse particles in a fine slurry medium made up of flocculated particles of a lower aspect ratio or consisting of a fibrous particle network

Most of the above methods can alter a suspension or slurry formulation irreversibly in the sense that a significant cost arising from extra capital and operating requirements may be incurred to return the slurry to its original state (if it is possible at all) once the slurry has been transported through the pipeline.

A further set of techniques tends not to have this disadvantage:

a. Oscillation of the slurry flow rate or pressure gradient
b. Vibration or oscillation of the pipeline in the direction of the pipe axis while maintaining a constant slurry flow rate

TABLE 3 Classification of Techniques to Reduce Head Loss

Slurry Classification	Flow Regime	
	Laminar	Turbulent
Nonsettling	1. Reduce slurry concentration 2. Increase particle size 3. Broaden particle size distribution 4. Reduce particle angularity 5. Add deflocculants 6. Inject air into pipeline 7. Oscillate slurry flow or applied pressure gradient 8. Vibrate pipe axially 9. Inject water periodically along pipeline length	Little scope for energy reduction
Settling	Not applicable	1. Add fibers 2. Add soap 3. Add soluble, long-chain polymer 4. Oscillate slurry flow rate 5. Use helical ribs attached to inner pipe wall 6. Use segmented pipe

c. Injecting air (or other gas, if appropriate) into the pipeline to create a three-phase mixture which would generally be readily separated on discharge from the pipeline
d. Use of spiral ribs on the inner pipe wall to reduce the deposit velocity of settling slurries and hence power requirements
e. Use of segmented pipe (a horizontal flat plate welded to the pipe bottom) to reduce deposit velocity and hence power requirements

The methods listed above are effective under certain conditions only. In determining whether a method may be appropriate for a particular application, it is important to decide whether

1. A settling or essentially "nonsettling" slurry is to be pumped
2. Turbulent or laminar flow conditions will prevail during pipe flow
3. Newtonian or non-Newtonian flow property is important, if the flow is in the laminar regime

The methods can be classified essentially into those applicable to the laminar pipe flow of "nonsettling" slurries and the turbulent pipe flow of settling slurries, as shown in Table 3.

Conclusions

This article has considered the industrial and other practical experience in the application of correlations and methods available in the design of slurry pipelines. Four classes of fluids are discussed in terms of their relevant pipeline characteristics. The first is settling slurry, for which the pipeline characteristics are minimum transport velocity and pressure gradient. The second is nonsettling slurry, for which laminar/turbulent transition and pressure gradient are relevant. The third is stabilized slurry, where nonsettling slurry design equations can be used. The fourth is thixotropic fluid, for which start-up pressure and time required to clear the pipe of gelled material are of particular importance.

With settling and nonsettling slurries a vast variety of correlations and methods have been developed to provide universal equations or procedures for predicting full-scale pipeline characteristics from the minimum design data. These include solids concentration, specific gravity, particle size or terminal velocity or drag coefficient, viscometric data, and pilot-scale pipeline flow data. However, experience shows that whereas the design procedures give reasonably good predictions that correspond well with actual pipeline performance properties of the slurry of interest are similar to those used to develop the procedure, they are as a general rule unreliable when applied to novel materials. For long-distance and large-diameter pipelines they can lead to gross errors in the predictions. The reasons for this inadequacy are because the material properties, particularly particle size distribution, particle shape, and the exact trace chemical and additive composition of the suspensions, are not known and that their effects are not properly understood. Also, the correlations, usually based on small-scale experiments, may be relevant only to a flow regime that does not exist in large diameter pipes. Limited data exist for estimating additional head losses arising from slurry flow through pipe fittings. However, for long distance pipelines these are relatively unimportant compared with losses arising from friction and static head changes. In situations where fittings losses may be a significant proportion of the total head loss, such as short pipe runs in process plant, the overall head loss may give an operating cost that is relatively small compared with other costs.

As a consequence of the above experience, the accurate design of large-scale installations should be based on correlations and design methods that are derived from materials and pipe diameters closely similar to those involved in the design exercise. Because there is no general guideline as to which of the existing correlations is best for each particular circumstance, design data should be obtained by pipeline trials using the actual material to be transported in the pipeline under design. Also, where large pipe diameters are involved, the experimental pipe sizes should be as close as possible to the full size. An assessment should be made of the available methods to minimize head loss to determine if any can be applied with advantage to the design application.

However, pipeline trials are expensive activities, especially when the test rig has to be specially erected. Thus, it is often difficult for the design engineer to justify such activities financially. Many a pipeline has no doubt been overdesigned and many more will be. There is therefore a clear need for further research into the phenomenon of pipeline flow in larger diameter pipes and for the subsequent development of accurate design procedures that require only bench-scale (or at most pilot-scale) experiments to provide the design data.

Symbols

A, B	material properties of a coarse slurry, in Eqs. (16) and (17)
C	total discharge volumetric solids concentration
C_B	concentration of solids in a sliding bed
C_C	discharge volumetric solids concentration in contact load
C_D	particle drag coefficient
C_x	total *in-situ* volumetric solids concentration
d	particle diameter
D	inside pipe diameter
e	absolute pipe wall roughness
f	Fanning friction factor, $= \tau_w/(^1/_2\rho_s V^2)$

f_f	Fanning friction factor for equivalent discharge of clear water at velocity V_u
f_m	Fanning friction factor for slurry flour
f_0	Fanning friction factor for water flow alone in Eq. (14)
F_L	Durand and Condolios factor for suspending velocity in Eq. (4)
g	gravitational acceleration
He	Hedstrom number, $= (D^2\rho_s/\tau_y)(\tau_y/K)^{2/n}$
i_f	hydraulic gradient for carrier fluid flow alone, at same mean velocity as for i_{ms} and i_{mh}
i_m	hydraulic gradient for slurry flow in terms of head of slurry per length of pipe
i_{ms}	hydraulic gradient for fully stratified flow
i_{mh}	hydraulic gradient for fully suspended flow
i_p	hydraulic gradient for a pipe full of solids
i_w	pressure gradient for water flow alone in terms of head of water per length of pipe
k	constant in Bowen Eq. (21)
K	consistency coefficient in generalized Bingham plastic model or power law flow model
K'	constant in Eq. (6)
K_L	head loss coefficient for pipe fitting
L	pipe length
m	exponent in expression for stratification ratio
n	exponent on shear rate in generalized Bingham plastic model or in power law flow model
ΔP	pressure drop
R	proportion of head loss arising from fully stratified flow, or "stratification ratio," i.e., Eq. (15)
Re	Reynolds number for Newtonian pipe flow, $= DV\rho_s/\eta$
Re_1	lower critical Reynolds number for breakdown of laminar flow
Re_2	upper critical Reynolds number for commencement of fully turbulent flow
Re_B	Bingham plastic Reynolds number, $= DV\rho_s/\eta_p$
Re_{OT}	Reynolds number defined by Oroskar and Turian
Re′	non-Newtonian power law Reynolds number, $= \left(\frac{D^n V^{2-n}\rho_s}{8^{n-1}K}\right)\left(\frac{4n}{1+3n}\right)^n$
S	ratio of solids density to carrier liquid density
t_f	residence time of slurry in a pipe defined by Eq. (2)
t_s	time required for solids to settle over a vertical distance D
U_t	terminal settling velocity for largest single particle in the slurry medium
V	mean slurry flow velocity
V_C	Oroskar and Turian critical velocity
V_D	design velocity
V_L	average liquid velocity during slurry flow
V_u	"threshold velocity for turbulent uplift" used in Clift et al. model and defined by Wilson Eq. (14)
w, x	indices in Bowen Eq. (21)
X	fraction of eddies with velocities exceeding the hindered particle settling velocity, in Eq. (5)

$\alpha, \beta, \gamma, \delta$	exponents in Eq. (3)
ϵ	exponent in Eq. (13)
$\dot{\gamma}$	shear rate
η	Newtonian viscosity
η_p	Bingham plastic viscosity
μ_s	coefficient of friction for sliding bed
ρ_L	liquid density
ρ_{sol}	solids density
ρ_s	slurry density
τ	shear stress
τ_w	wall shear stress
τ_y	yield stress

References

1. D. C.-H. Cheng and W. Whittaker, *Proceedings, 2nd International Conference on Hydraulic Transport of Solids in Pipes*, 1972, Paper C3.
2. W. Weidenroth and H. Kirchner, *Proceedings, 2nd International Conference on Hydraulic Transport of Solids in Pipes*, 1972, Paper E1.
3. A. J. Carleton and D. C.-H. Cheng, *Proceedings, 3rd International Conference on Hydraulic Transport of Solids in Pipes*, 1974, Paper E5.
4. W. F. McDermott, *Can. Min. Metall. Bull.*, pp. 1378–1383 (1970).
5. D. G. Thomas, *AIChE J.*, *10*, 303–308 (1964).
6. W. Parzonka, J. M. Kenchington, and M. E. Charles, *Can. J. Chem. Eng.*, *59*, 291–296 (1981).
7. R. W. Hanks and D. G. Sloan, *Proceedings, 6th Slurry Transport Association Conference*, March 1981, pp. 107–120.
8. M. C. Roco and C. A. Shook, *AIChE J.*, *31* 1401–1404 (1985).
9. K. C. Wilson, *Proceedings, 6th International Conference on Hydraulic Conveying of Solids in Pipes*, BHRA Engineering, Cranfield, England, 1979, Paper A1.
10. A. D. Thomas, *Proceedings, 4th Slurry Transportation Association Conference*, March 1979.
11. A. D. Thomas, *Int. J. Multiphase Flow*, *5*, 113–129 (1979).
12. A. D. Thomas, *Proceedings, 6th International Conference on Hydraulic Conveying*, BHRA Fluid Engineering, Cranfield, England, 1979, Paper A2.
13. A. R. Oroskar and R. M. Turian, *AIChE J.*, *26*, 550–558 (1980).
14. R. Durand and E. Condolios, *Deuxiemes Journees de l'Hydraulique*, Soc. Hydrautech de France, Grenoble, 1952.
15. R. Durand and E. Condolios, *Proceedings, Colloquium on the Hydraulic Transport of Coal*, National Coal Board, London, 1952, Paper IV, pp. 39–52.
16. D. M. Newitt, J. F. Richardson, M. Abbott, and R. B. Turtle, *Trans. Inst. Chem. Eng.*, *33*, 93–115 (1955).
17. M. E. Charles, *Proceedings, 1st International Conference on Hydraulic Transport of Solids in Pipes*, BHRA Fluid Engineering, Cranfield, England, 1970, Paper A3.
18. A. R. Khan, R. L. Pirie, and J. F. Richardson, *Chem. Eng. Sci.*, *42*(4), 767–778 (1987).
19. R. P. Chhabra and J. F. Richardson, *Chem. Eng. Res. Des.*, *61*, 313 (1983).

20. R. P. Chhabra and J. F. Richardson, *Chem. Eng. Res. Des.*, *63*, 300 (1985).
21. I. Zandi and G. Govatos, *J. Hydraul. Div., Am. Soc. Civ. Eng.*, *93*, 145 (1967).
22. K. C. Wilson, *Proceedings, International Symposium on Dredging Technology*, BHRA Fluid Engineering, Cranfield, England, 1975, Paper C3.
23. K. C. Wilson and D. G. Judge, *J. Powder Bulk Solids Technol.*, *4*(1) 15–22 (1980).
24. K. C. Wilson and W. E. Watt, *Proceedings, 3rd International Conference on the Hydraulic Transport of Solids in Pipes*, BHRA Fluid Engineering, Cranfield, England, 1974, Paper D1.
25. R. Clift, K. C. Wilson, G. R. Addie, and M. R. Carstens, *Proceedings, 8th International Conference on the Hydraulic Transport of Solids in Pipes*, BHRA Fluid Engineering, Cranfield, England, 1982, Paper B1.
26. D. C.-H. Cheng, *Chem. Eng.* (*London*), p. 525 (1975).
27. N. I. Heywood, *Inst. Chem. Eng. Symp. Ser.*, *60*, 33–52 (1980).
28. N. I. Heywood and D. C.-H. Cheng, *Trans. Inst. Meas. Control.*, *6*(1), 33–45 (1984).
29. D. C.-H. Cheng and N. I. Heywood, *Warren Spring Laboratory Report LR 502* (*MH*), Stevenage, England, 1984.
30. D. C.-H. Cheng, *Proceedings*, *1st International Conference on Hydraulic Transport of Solids in Pipes*, BHRA Fluid Engineering, Cranfield, England, 1970, Paper J5.
31. B. O. A. Hedstrom, *Ind. Eng. Chem.*, *44*, 651–656 (1952).
32. D. W. Dodge and A. B. Metzner, *AIChE J.*, *5*, 189–204 (1959).
33. D. C.-H. Cheng, *Br. Chem. Eng. Process Technol.*, *16*, 955–956 (1971).
34. S. W. Churchill, *Chem. Eng.*, pp. 91–92 (November 7, 1977).
35. A. B. Metzner and J. C. Reed, *AIChE J.*, *1*, 434 (1955).
36. G. W. Govier and K. Aziz, *The Flow of Complex Mixtures in Pipes*, Van Nostrand-Reinhold, New York, 1972.
37. B. M. Torrance, *S. Afr. Mech. Eng.*, *13*, 89 (1963).
38. R. H. Perry and C. H. Chilton (eds.), *Chemical Engineers' Handbook*, 6th ed., Section 5.
39. V. L. Streeter (ed.), *Handbook of Fluid Mechanics*, McGraw-Hill, New York.
40. L. F. Moody, *Trans. ASME*, *66*, 671 (1944).
41. R. L. Bowen, *Chem. Eng.*, pp. 143–150 (July 24, 1961).
42. J. M. Kenchington, *Proceedings, 2nd International Conference on Hydraulic Transport of Solids in Pipes*, 1972, Paper C4.
43. A. J. Carleton, C. A. Shook, M. Streat, and Y. Televantos, *Can. J. Chem. Eng.*, *57*, 255 (1979).
44. A. J. Carleton and R. J. French, *Warren Spring Laboratory Report LR 290* (*MH*), Stevenage, England, 1978.
45. A. J. Carleton, R. J. French, J. G. James, B. A. Broad, and M. Streat, *Proceedings, 5th International Conference on Hydraulic Transport of Solids in Pipes*, BHRA Fluid Engineering, Cranfield, England, 1978, Paper D2.
46. N. I. Heywood, *Proceedings, 10th International Conference on Hydraulic Transport of Solids in Pipes*, BHRA Fluid Engineering, Cranfield, England, 1986, Paper K3.
47. D. E. Elliott and B. J. Gliddon, *Proceedings, First International Conference on Hydraulic Transport of Solids in Pipes*, BHRA Fluid Engineering, Cranfield, England, 1970, Paper G2.
48. M. Streat, *Proceedings, 8th International Conference on Hydraulic Transport of Solids in Pipes*, BHRA Fluid Engineering, Cranfield, England, 1982, Paper B3.
49. M. E. Charles and R. A. Charles, in *Advances in Solid–Liquid Flow in Pipes and Its Applications* (I. Zandi, ed.), Pergamon, 1971, pp. 187–197.

50. N. Hisamitsu, Y. Shoji, and S. Kosugi, *Proceedings, 5th International Conference on Hydraulic Transport of Solids in Pipes*, BHRA Fluid Engineering, Cranfield, England, 1978, Paper D3.
51. D. A. Brookes and C. Dodwell, *Proceedings, 9th International Conference on Hydraulic Transport of Solids in Pipes*, BHRA Fluid Engineering, Cranfield, England, 1984, Paper A1.
52. D. A. Brookes and P. E. Snoek, *Proceedings, 10th International Conference on Hydraulic Transport of Solids in Pipes*, BHRA Fluid Engineering, Cranfield, England, 1986, Paper C3.
53. R. A. Duckworth, L. Pullum, and C. F. Lockyear, *Proceedings, 10th International Conference on Hydraulic Transport of Solids in Pipes*, BHRA Fluid Engineering, Cranfield, England, 1986, Paper C2.
54. R. A. Duckworth, L. Pullum, C. F. Lockyear, and J. Lenard, *Bulk Solids Handling*, *3*(4), 817–824 (1983).
55. C. Asszanyi, Y. Kapolyi, C. Kantas, and T. Meggyes, *Proceedings, 2nd International Conference on Hydraulic Transport of Solids in Pipes*, BHRA Fluid Engineering, Cranfield, England, 1972, Paper D3.
56. J. M. Kenchington, *Proceedings, 4th International Conference on Hydraulic Transport of Solids in Pipes*, BHRA Fluid Engineering, Cranfield, England, 1976, Paper D3.
57. A. D. Thomas, *Proceedings, 5th International Conference on Hydraulic Transport of Solids in Pipes*, BHRA Fluid Engineering, Cranfield, England, 1978, Paper D5.
58. A. J. Carleton, D. C.-H. Cheng, and W. Whittaker, *Determination of the Rheological Properties and Start-Up Pipeline Flow Characteristics of Waxy Crude and Fuel Oils*, Inst. Petroleum Publication Paper No. IP74-009, 1974.
59. F. Moore, *Trans. Br. Ceram. Soc.*, *58*, 407–494 (1959).
60. D. C.-H. Cheng and W. Whittaker, *Proceedings, 2nd International Conference on Hydraulic Transport of Solids in Pipes*, BHRA Fluid Engineering, Cranfield, England, 1972, Paper B4.
61. J. T. Knegtel and E. Zeilinga, *J. Inst. Pet.*, *57*, 165–174 (1971).
62. E. Verschuur et al., *J. Inst. Pet.*, *57*, 139–146 (1971).
63. D. C.-H. Cheng, *Warren Spring Laboratory Report LR 317* (*MH*), Stevenage, England, 1979.
64. R. M. Turian and F. L. Hsu, *Part. Sci. Technol.*, *1*, 365–392 (1983).
65. M. F. Edwards, M. S. M. Jadallah, and R. Smith, *Chem. Eng. Res. Des.*, *63*(1), 43–50 (1985).
66. S. Iwanami and T. Suu, *Proceedings, First International Conference on Hydraulic Transport of Solids in Pipes*, BHRA Fluid Engineering, Cranfield, England, 1970, Paper J1.
67. T. Suu, *Proceedings, 2nd International Conference on Hydraulic Transport of Solids in Pipes*, BHRA Fluid Engineering, Cranfield, England, 1972 Paper X3.

NIGEL I. HEYWOOD
DAVID C.-H. CHENG
A. J. CARLETON

Slurry Systems, Energy Reduction

Introduction

Over the last 20 years or so a number of techniques for reducing head loss in slurry pipe flow have been demonstrated, in many cases at both pilot-plant and full scales. These techniques are capable of reducing frictional pressure drop either by altering the slurry rheological properties directly or by reducing the impact of adverse flow properties indirectly.

Some of the more obvious ways of reducing slurry viscosity levels directly include:

a. Reducing slurry concentration
b. Increasing particle size, if Brownian motion and particle surface effects control viscosity levels
c. Broadening the particle size distribution at constant total solids concentration
d. Reducing the angularity of particle shape while maintaining the particle size distribution and solids concentration essentially constant. Also, adding high aspect ratio fibers
e. Adding deflocculants (soluble ionic compounds) to disperse flocculated slurries (important when a significant percentage, say 5–10%, of solids are below approximately 5 μm in size)
f. Addition of soaps
g. Addition of high molecular weight polymer
h. Injection of water (or other suspending medium) into the flowing slurry periodically along the pipeline length.

Most of the above methods can alter a suspension or slurry formulation irreversibly in the sense that a significant cost arising from extra capital and operating requirements may be incurred to return the slurry to its original state (if it is possible at all) once the slurry has been transported through the pipeline.

A further set of techniques tends not to have this disadvantage:

a. Oscillation of the slurry flow rate or pressure gradient
b. Vibration or oscillation of the pipeline about the pipe axis while maintaining a constant slurry flow rate
c. Injecting air (or other gas, if appropriate) into the pipeline to create a three-phase mixture which would generally be readily separated on discharge from the pipeline

d. Use of spiral ribs on the inner pipe wall to reduce the deposit velocity of settling slurries, and hence power requirements
e. Use of segmented pipe (a horizontal flat plate welded to the pipe bottom) to reduce deposit velocity and hence power requirements

The methods listed above are effective under certain conditions only. In determining whether a method may be appropriate for a particular application, it is important to decide whether:

1. A settling or essentially "nonsettling" slurry is to be pumped
2. Turbulent or laminar flow conditions will prevail during pipe flow
3. Newtonian or non-Newtonian flow property is important, if the flow is in the laminar regime

The methods can be classified essentially into those applicable to the laminar pipe flow of "nonsettling" slurries (see Table 1) and the turbulent pipe flow of settling slurries (see Table 2).

In the following sections, each of the important influences on determining head loss and hence methods of reducing power requirements will be discussed. The review begins by considering how modification of three important particle properties can reduce slurry viscosity: particle size, size distribution, and shape.

Effect of Modification to Particle Properties

Particle Size

Both from experimental observations and from theoretical considerations, the absolute size of particles has no effect on suspension rheology provided that the particles are well-dispersed in suspension and are influenced by hydrodynamic forces only. In practice, however, the effect of variation in particle size can be seen in rheological data for a variety of reasons.

First, when particles are in aqueous suspension, an adsorbed layer of molecules on the particle surfaces causes the suspension to behave rheologically as though the volume of solids in suspension is greater than a value calculated from a knowledge of the weight and density of the particles. If the thickness of this adsorbed layer remains constant, and is independent of particle size, suspension viscosities will tend to increase with decreasing particle size at constant particle volume concentration and shear rate. An adsorbed layer generally occurs in aqueous suspensions in which stabilizers are frequently used, but can also occur in nonaqueous suspensions.

Second, as particle size is reduced to micron sizes or below, Brownian movement forces play a significant part in determining suspension flow properties. Owing to particle collisions arising from Brownian motion, non-

TABLE 1 Summary of Techniques to Reduce Head Loss for Pipeline Flow of "Nonsettling" Slurries in the Laminar Regime

Technique	Requirement for Suspension Flow Property	Essential Mechanism
Increase in particle size	Newtonian or non-Newtonian	Reduction in Brownian motion contribution to viscosity and/or reduction in increase in effective volume fraction of solids arising from reduced importance of adsorbed layer effect
Broader particle size distribution	Newtonian or non-Newtonian	Wider particle size distribution at fixed solids concentration leads to lower suspension viscosity. (Alternatively, higher solids loading is achievable at a fixed suspension viscosity and hence fixed head loss)
Reduce particle angularity	Newtonian or non-Newtonian	Reduction in viscosity arising from reduced surface friction contacts between particles at high solids concentrations
Addition of soluble ionic compounds as dispersants	Usually non-Newtonian, shear-thinning	Changes pH and/or zeta potential on fine particle surfaces, causing particle dispersion and hence reduces suspension viscosity
Air injection into flowing slurry	Must be non-Newtonian, shear-thinning with or without yield stress	Presence of air reduces wetted pipe wall area on which suspension shear stress acts, so reducing drag and head loss. Increase in slurry velocity at fixed slurry flow rate tends to increase head loss marginally owing to shear-thinning behavior
Oscillation of slurry flow or applied pressure gradient	Non-Newtonian, shear-thinning. Viscoelastic property may enhance effect	Probable creation of annulus adjacent to pipe wall which is depleted in particles and hence lower wall shear stress occurs
Pipe vibration	Non-Newtonian, shear-thinning	Possible creation of slip layer near to wall depleted in solids. May also be thixotropic structure breakdown effect occurring
Periodic injection of water (or other suspending medium) along pipeline length	Newtonian or non-Newtonian but probably latter with high yield stress	Water injection gives local reduced solids concentration at pipe wall and hence lower head loss

TABLE 2 Summary of Techniques to Reduce Head Loss for Pipeline Flow of Settling Slurries in the Turbulent Flow Regime

Technique	Essential Mechanism
Fiber addition	Two mechanisms possible: with 10 to 1000 ppm of fibers turbulent drag reduction occurs, probably because of dampening of turbulent eddies. With fiber concentrations up to 5%, fibers form a central plug with water annulus. In both cases, technique is also applicable to "nonsettling" slurries
Addition of soaps	Probable reduction in interparticle friction
Addition of soluble, long-chain polymers	Toms' effect occurs in turbulently-flowing suspending medium. Mechanism still unclear but probably dampening of turbulent eddies
Oscillation of slurry flow rate	Reduction in deposit velocity arising from more effective particle suspension in oscillating flow
Use of helical ribs attached to inner pipe wall	Ribs reduce deposit velocity by continuously picking up particles tending to settle. Lower velocity means lower head loss
Use of segmented pipe	Mechanism unclear but deposit velocity reduced and hence pipeline can be operated at lower slurry flow rates and hence lower head losses are incurred

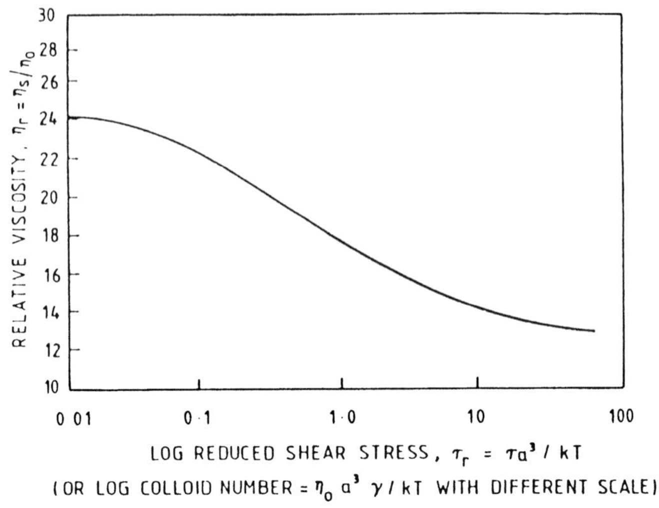

FIG. 1. Effect of particle size on suspension viscosity when Brownian motion effects are important. (After Krieger [6] and Papir and Krieger [3].)

Newtonian behavior is common and can consist of both shear-thinning and shear-thickening characteristics, as Wagstaff and Chaffey [1, 2] have demonstrated. The effect of Brownian motion forces is similar to the effect of the presence of an adsorbed layer, in that suspension viscosity increases with decreasing particle size [3], and it is not clear at present how the two effects may be quantified separately, although the treatment by Woods and Krieger [4] shows promise. Figure 1 indicates how the suspension relative viscosity rises as particle size is reduced at constant total solids concentration.

Third, as particle size is reduced to the order of a μm or below, Coulombic forces will become increasingly important, no matter how successful one is at suppression of the double layer of charge on the particle surfaces using appropriate types and concentration levels of electrolyte. The relative importance of these additional processes, besides hydrodynamic effects, can be assessed as a function of particle size by the method of Chaffey [5], who has defined and listed the relevant dimensionless groups and extended the pioneering analysis of Krieger [6].

It can be concluded, therefore, that the proportion of solids having a particle size below a few μm should be reduced if suspension viscosity is to be reduced. However, there are instances when a high viscosity, shear-thinning suspension is desirable to act as a suspending medium for the transport of coarse particles. The optimum solution is to reduce the proportion of fines while maintaining an essentially "nonsettling" slurry.

Particle Size Distribution

The importance of particle size distribution in determining viscosity levels of essentially "nonsettling" suspensions has generally been overlooked in the past. Yet there now exists ample theoretical evidence and a limited amount of experimental data which show that an economic advantage can be gained by adjusting particle size distribution. This usually means that, where particle surface effects are insignificant, a wide particle size distribution should be strived for. Then for a given solids concentration, viscosities will be reduced and hence head losses for laminar pipe flow will be minimized. The effect will be smaller for turbulent pipe flow and will depend upon the friction factor–Reynolds number relationship for the "nonsettling" suspension. On the other hand, the effect of particle size distribution on head losses for settling slurries is far from clear.

Theoretical predictions of the viscosity reduction effect by Farris [7] suggest that its significance emerges only for slurries with total solids concentrations exceeding some 30 to 40% by volume solids. Figure 2 shows predicted viscosities for a Newtonian slurry consisting of a variable mixture of fine and coarse particles. If a slurry consists only of either fine or coarse particles, slurry viscosities are relatively high, but an appropriate mixture of the two particle fractions will lead to a minimum in viscosity. This minimum is likely to become more pronounced and its position dependent on the proportions of fine and coarse material as the overall solids concentration is raised.

Such concepts and the Farris analysis can be extended to a slurry having many size fractions and also to continuous particle size distributions. The importance of theoretical predictions is that they are useful to determine approximately an optimal particle size distribution which can then be defined more precisely through experimentation.

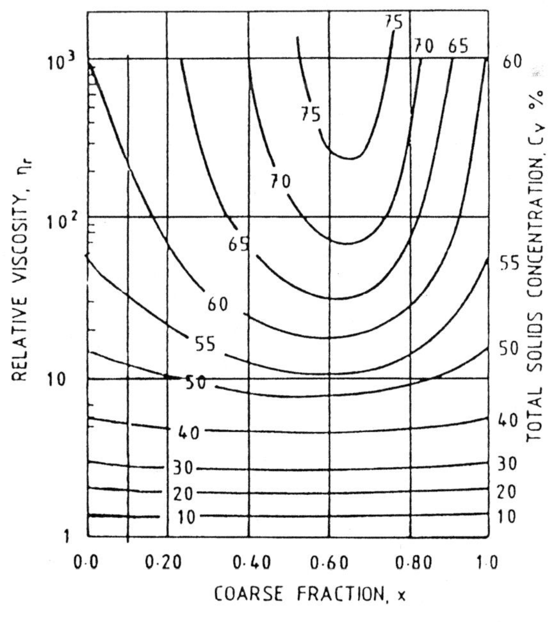

FIG. 2. Predictions of suspension viscosity for a mixture of coarse and fine particle fractions. (From Farris [7].)

Another way of looking at the particle size distribution effect is as follows. When the main concern is to maximize the dry solids rate through the pipe, this can be achieved by maximizing solids concentration while keeping viscosity levels constant by adjusting particle size distribution. This conclusion obviously has a number of implications, including not only a reduction in cost per tonne of dry material moved but also a reduction in water usage, important when water supplies are limited or when restrictions on water transport across state or national boundaries are present, such as those which occur in the United States and parts of Africa.

The basic concepts of the theory were first discussed by Eveson et al. [8] and Mooney [9] in 1951 and Brinkman [10] and Roscoe [11] in 1952. More recently a different theoretical approach has been taken by Lee and Lee [12]. Experimental evidence for the effect has been provided by Chong et al. [13], Skvara and Vancourova [14], Parkinson et al. [15], and Goto and Kuno [16]. Investigations relevant to pipeline flows have been carried out by Ferrini et al. [17] and Klose [18]. However, despite these studies, more experimental effort needs to be aimed at higher solids concentrations than those used up to now so that the limits of the Newtonian viscosity and even granular property can be studied. The importance of the granular property is particularly in evidence for compressible pastes and cakes which can be pumped over relatively short distances using high-pressure positive displacement pumps originally developed for pipelining concrete.

This area is currently under investigation at Warren Spring Laboratory and a cooperative project with companies is underway.

Particle Shape

It is well-established that reduction in particle angularity or anisometry generally leads to lower viscosities at a given overall solids concentration and while maintaining a similar particle size distribution, provided that only physical effects are occurring. Lower viscosities lead to lower head losses in laminar pipe flow. Clarke [19] has provided ample evidence to support this conclusion. In addition, particles with high aspect ratios, typically polymeric or wood pulp fibers, can exhibit two separate effects:

a. A reduction in head losses for turbulent pipe flow when fibers are present in typically low concentrations, i.e., hundreds of ppm (see Refs. 20–22)
b. A reduction in head losses for both laminar and turbulent pipe flow when they are present in sufficiently high concentrations (but often no more than 1 to 5% by volume) to give an unsheared plug of fibrous suspension flowing in the central pipe core (see Refs. 23 and 24)

Drag reductions effected in fiber systems are usually rather lower than those obtained by polymer solutions. The advantage of a fiber-laden system is that the fibers do not generally degrade, unless they are particularly friable or subjected to very high shear stresses in tortuous flow fields, and hence the drag reduction effect does not diminish with time on progressive cycling within a pipeline loop or for flow in a long pipeline. A further advantage is that, because effects within the turbulent core of the pipe dominate the drag reduction process, in Case **a** above, the effect is not diminished on scaling up from small to large pipe diameters. With polymer solutions, the effect is diminished on scale-up. However, in Case **b** it still remains to be established whether a pipe diameter effect exists or not.

The extent of drag reduction increases with either an increase in fiber aspect ratio while holding solids concentration constant, or an increase in solids concentration at constant fiber aspect ratio; in the latter case, up to a maximum effect.

Both the laminar and turbulent flow behavior of more concentrated fiber suspensions have some interesting features. Much effort has been devoted in the paper industry to the examination of the flow of paper pulp suspensions at weight concentrations typically from 0.5 to 3%. Although these concentrations may appear low, the highly fibrous nature of the pulp causes entanglement and the formation of large flocs and makes the material flow essentially as a plug, with a lubricating water annulus adjacent to the pipe wall. With these materials, unlike ordinary single-phase liquids, it is possible for the pressure drop–flow velocity plot to pass through both a maximum and a minimum in the laminar region. This behavior may be attributable in part to changes in the width of the lubricating water annulus as the flow velocity is varied; this is governed by the degree to which water is expressed from the matted plug of paper pulp as the hydrodynamic conditions vary. At higher flow velocities still, the pressure drop can fall below that for water flow at the same velocity, indicative of a drag reduction phenomenon, whose underlying cause is probably different from the similar effect produced by very low concentration fibers.

The consequence of such behavior in the turbulent flow regime is that for many fibrous suspensions, a knowledge of their laminar flow properties either in a tube, pipeline, or rotational viscometer will not aid in the estimation of turbulent head losses, nor in determining whether drag reduction will occur or not. In fact, there currently exists no method which allows this prediction. Thus two slurry systems may possess near identical laminar flow behavior, but the fibrous slurry may exhibit drag reduction in the turbulent flow regime whereas the slurry containing nonfibrous particles normally would not show a drag reduction effect.

Work has been carried out on turbulent velocity profiles and turbulent radial and longitudinal intensities but is rather inconclusive, and more effort is required here to obtain a proper understanding of the mechanism of drag reduction in fiber suspensions. Despite numerous experimental studies, conclusions drawn by various workers remain muddled and confusing. However, this, of course, should not prevent the exploitation of the drag reduction phenomenon in industrial applications. A precise understanding of the drag reduction mechanism is naturally not a prerequisite for the exploitation of the effect, although predictions of the extent of the effect are almost impossible.

Effect of Additives

Use of Soluble Ionic Compounds as Deflocculants

It has been known for some time that the addition of certain types of soluble, ionic compound to flocculated suspensions can result in substantial pressure drop reductions in pipe flow in the laminar flow regime. Usually, however, little effect is noticeable in the turbulent regime. The ionic compounds disperse the particles, thereby breaking up the flocculated particle network which had previously given rise to higher shear stresses in pipe flow. In order to disperse particles, the charge on their surfaces needs to be of the same sign, and the higher the charge the greater the repulsive forces between particles.

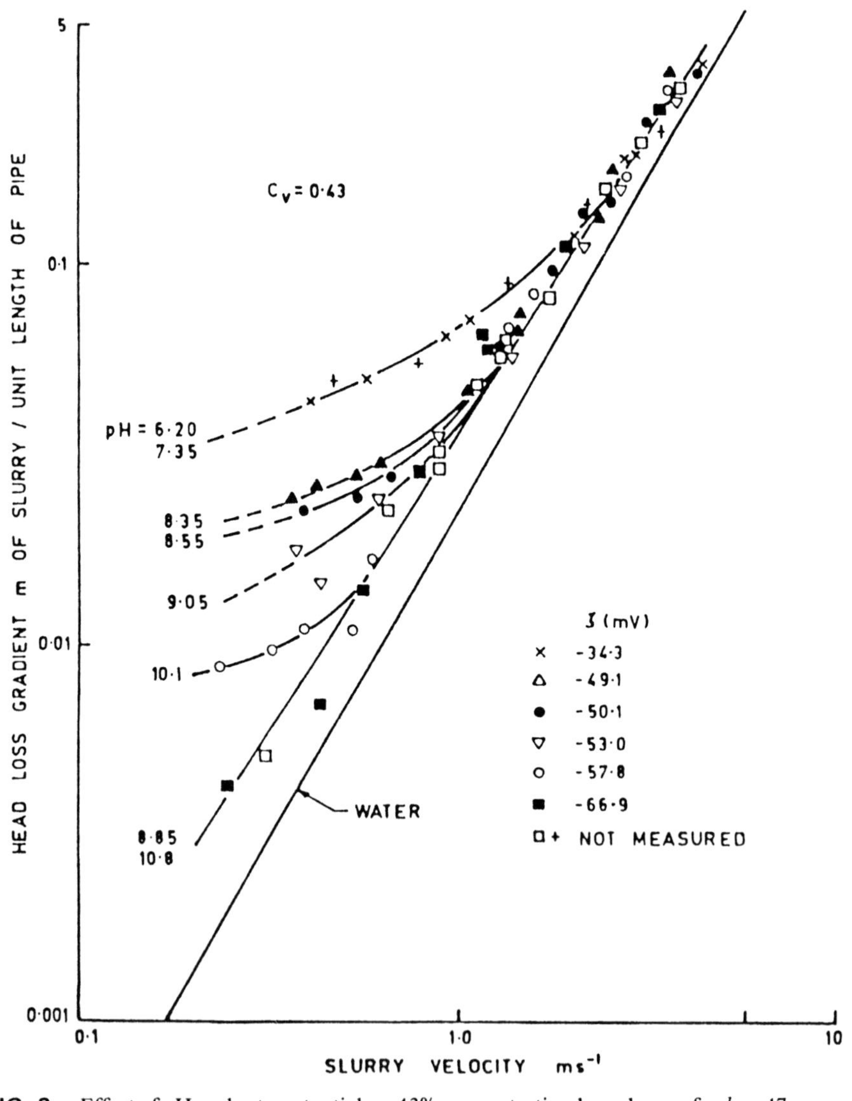

FIG. 3. Effect of pH and zeta potential on 43% concentration by volume of a $d_{50} = 17\ \mu m$ sand slurry in pipe flow. (From Horsley and Reizes [25].)

It is only relatively recently that the reduction in pressure drop has been related to the degree of deflocculation in quantitative terms. Horsley et al. [25–27] carried out a number of studies in which the zeta potential on the particle surfaces, defined as the potential in the electrochemical double layer at the interface between a particle moving in an electric field and the surrounding liquid, was measured and correlated with pressure drop reduction. The zeta potential is a measure of the repulsive forces acting on the particles, and thus is useful in measuring the dispersing effects of chemical additives.

Figure 3 shows some typical head loss–slurry velocity data obtained by Horsley and Reizes [25] for the flow of a 43% by volume sand slurry (37% by weight of particles less than 10 μm) at different pH levels in the range 6.2 to 10.8; the pH was controlled by the addition of nitric acid and sodium hydroxide. The changes in pressure drop in the laminar regime were observed to be greater as the sand concentration was raised. Figure 4 shows a plot of head loss against zeta potential. This illustrates that the higher the zeta potential (i.e., the larger the negative value), the lower the pressure drop.

Sikorski et al. [28] studied the effects of chemical additives on a number of mineral slurries including drilling muds (thinners such as sodium acid pyrophosphate and sodium hexametaphosphate), phosphate rock slurries (caustic soda), limestone cement feed (combination of sodium tripolyphosphate and sodium carbonate), and coal slurry (sodium tripolyphosphate, sodium dioctyl sulfosuccinate, and sodium carbonate), while Shook et al. [29, 30] studied the effects of various alkaline additives on the flow properties of coal–water slurries.

Addition of Soaps

Drag reduction using soap solutions which do not undergo irreversible mechanical degradation has been obtained using liquid and suspension systems.

Zakin et al. [31] used various aqueous mixtures of 1-naphthol and cetyl trimethylammonium bromide (CTAB) with sand slurries (particle size 0.9 and 1 mm). A 0.07% naphthol and 0.15% CTAB combination was found to give the lowest pressure drop reductions of up to 80% for flow through a 20-mm tube. Temperature and aging effects were also studied in the absence of sand particles. For quartz sand, 0.09% soap caused a 60% drag reduction of critical velocities, while doubling the soap concentration to 0.18% gave an additional improvement of less than 10%. It was also found that the critical velocities for the complex soap solution suspensions were about half those for the water–sand suspension, and this was examined through settling rate data in graduated cylinders where settling rates were found to be much lower than for water–sand suspensions.

Bakelite sand suspensions showed a less marked drag reduction at 0.09% soap concentration, while at 0.18% a 40% drag reduction was obtained and at 0.225% soap concentration, 70%.

The significance of the Zakin et al. study is that the soap does not appear to have been degraded mechanically. A critical shear stress was found with

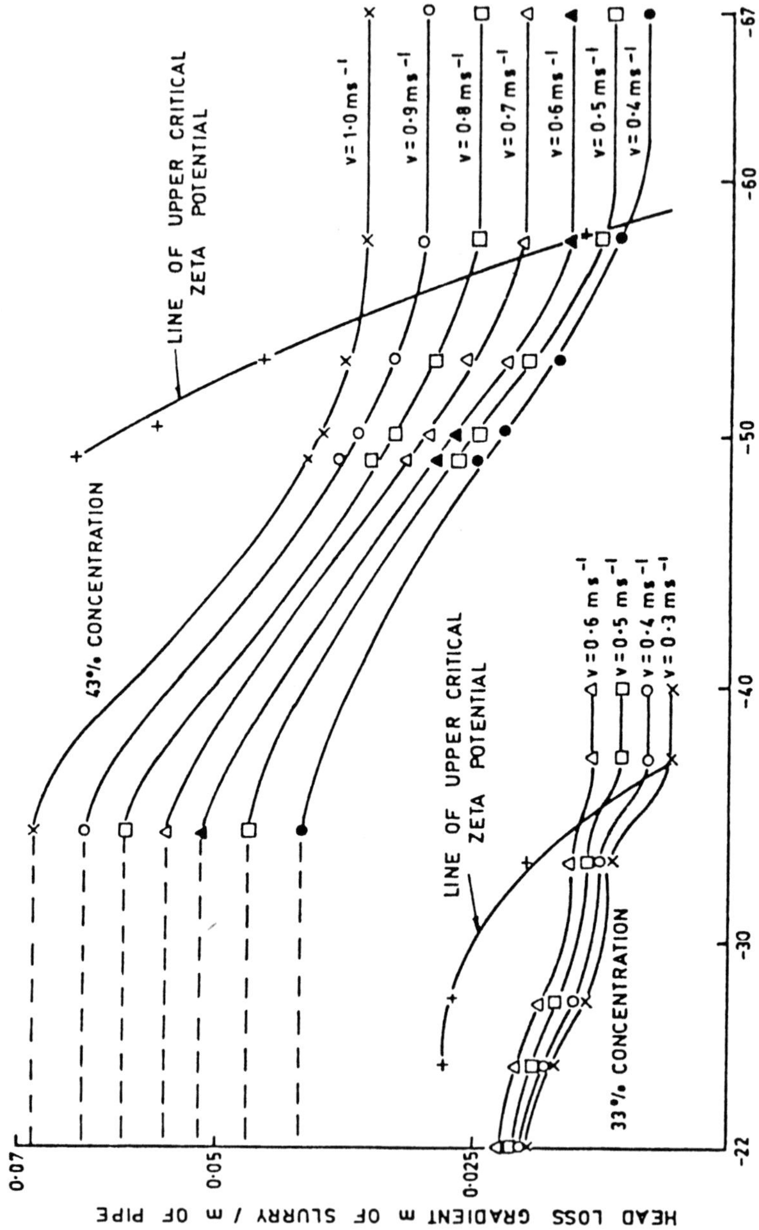

FIG. 4. Effect of zeta potential on head loss gradient for 33 and 43% concentrations by volume of a $d_{50} = 17\ \mu m$ sand slurry. (From Horsley and Reizes [25].)

sand suspensions above which no drag reduction occurred, but on reducing the wall shear stress the same levels of drag reduction were returned to as previously. Thus the potential of complex soaps for drag reduction in suspension flow would appear to be greater than high molecular weight polymer. Further work is required to assess whether the addition of this, or related, complex soap would give an economic as well as technical advantage.

Use of Soluble High Molecular Weight Polymer

Ever since the famous experiments of B. A. Toms in the late 1940s, it has been known that the addition of small amounts (10's to 100's ppm) of soluble, high molecular weight polymer to a liquid flowing in the turbulent regime will cause a reduction in the head loss. Typical polymers used to demonstrate this phenomenon have been polyacrylamides, polyacrylic acid, and polyethylene oxide with molecular weights up to six million. One of the main drawbacks to the successful application of this phenomenon to numerous situations is that the polymer can degrade, often quite rapidly, under mechanical shear with a consequent reduction in average molecular weight of the polymer and hence reduction in effectiveness in reducing head losses. This is particularly acute in recirculating test flow loops which often incorporate many different types of fittings as well as repeated passage of the polymer solution through the pump. Degradation may prove to be much less of a problem in once-through applications, but can also occur through attack from microorganisms present in water.

Despite the lack of industrial applications so far, the amount of research effort into this drag reduction effect for single-phase liquids has been immense during the last 20 years or so. More recently attention has been turned to the potential of polymers to reduce head loss in solid–liquid flows. Radin et al. [32] reviewed some of the work done in slurry flows up to 1975. Zakin et al. [31] used polyethylene oxide and polyacrylamide with sand suspensions flowing in a 20-mm diameter pipe. These polymers gave very high drag reduction, but excessive mechanical degradation occurred and the effect was lost in 2 min. However, they also used guar gum, a high molecular weight natural gum, which is much more stabled to mechanical degradation but which requires up to 100 times greater concentrations than polyethylene oxide for appreciable drag reduction.

Golda [33] recently undertook a systematic study of the effectiveness of six different polyacrylamides (all sold under the Dow Chemical trade name Separan) in reducing the head loss of a 2.91% coal slurry (mean particle size of 2.8 mm flowing in a 40-mm diameter pipe). A fixed polymer concentration of 80 ppm was used, and mechanical stability tests were carried out by monitoring drag reduction initially and after 60 min. Separan AP45 performed best and consequently was further used in a series of tests to explore the effects of both polymer and coal concentration on the level of drag reduction. The degree of drag reduction was less for coal slurry than for water flow alone, but in both cases the drag reduction passed through a maximum as polymer concentration was progressively increased. The polymer concentration giving the maximum drag reduction was higher for coal slurry than for water flow alone. The maximum drag reduction level also decreased progressively with increasing coal concentration.

Other studies include those by Pollert [34], who obtained a maximum drag reduction of 32% using polyacrylamide (Separan AP273) dissolved in fly-ash slurry; Kolar [35], who used mica and kaolin in polyacrylamide solution; and Sifferman and Greenkorn [36], who also used guar gum with sand suspensions flowing in three differently sized pipes.

From the limited studies so far, it would appear that, despite the problem of mechanical degradation, the use of polymers may well facilitate substantial reductions in power consumption.

Air Injection into Flowing "Nonsettling" Slurry

For all Newtonian slurries in either laminar or turbulent flow and for all non-Newtonian slurries in turbulent flow, introduction of gas into a horizontal pipeline carrying slurry at a constant rate will invariably increase the pressure gradient at any point along the pipeline. This is because the average velocity is raised in the pipe, owing to the reduced mean cross-sectional area for liquid flow brought about by the gas presence, and this in turn creates higher shear stresses acting on the pipe wall which are responsible for the pressure drop.

In contrast, pseudoplastic non-Newtonian fluids, where the viscosity decreases with increasing flow (i.e., shear-thinning behavior), can be in laminar flow concurrently with gas in a horizontal pipe, and the presence of gas results in a decrease in average pressure gradient. This was first reported by Ward and Dallavalle [37]. Some typical experimental results for shear-thinning flocculated kaolin suspensions are shown in Fig. 5. Here the parameter ϕ_s^2 (termed the drag ratio in the figure) is plotted against the superficial air velocity, with the superficial slurry velocity in the pipe as a parameter. ϕ_s^2 is defined as the ratio of the pressure gradient along the pipe with gas injection to the pressure gradient for flow of suspension alone at the same suspension flow rate. It can be seen that provided the slurry flow rate is sufficiently low, i.e., the Reynolds number is low and in the laminar region, the presence of gas causes a reduction in the ϕ_s^2 parameter below unity, indicating a reduced pressure gradient over that for slurry flow alone at the same slurry rate. At any sufficiently low fixed slurry rate, this reduction progressively increases with increasing gas rate until a maximum effect is reached. Thereafter the pressure gradient rises with further gas rate increases. The highly non-Newtonian, shear-thinning character of these kaolin suspensions is indicated by the flow curves in Fig. 6 (taken from Ref. 40).

This phenomenon has important implications for pipelines carrying shear-thinning slurries. Three cases arise. In the design of a new pipeline installation, the injection of air downstream from the slurry pump may result either in an overall operating cost saving or facilitate the transport of high viscosity paste-like material which a practical slurry pump could not other-

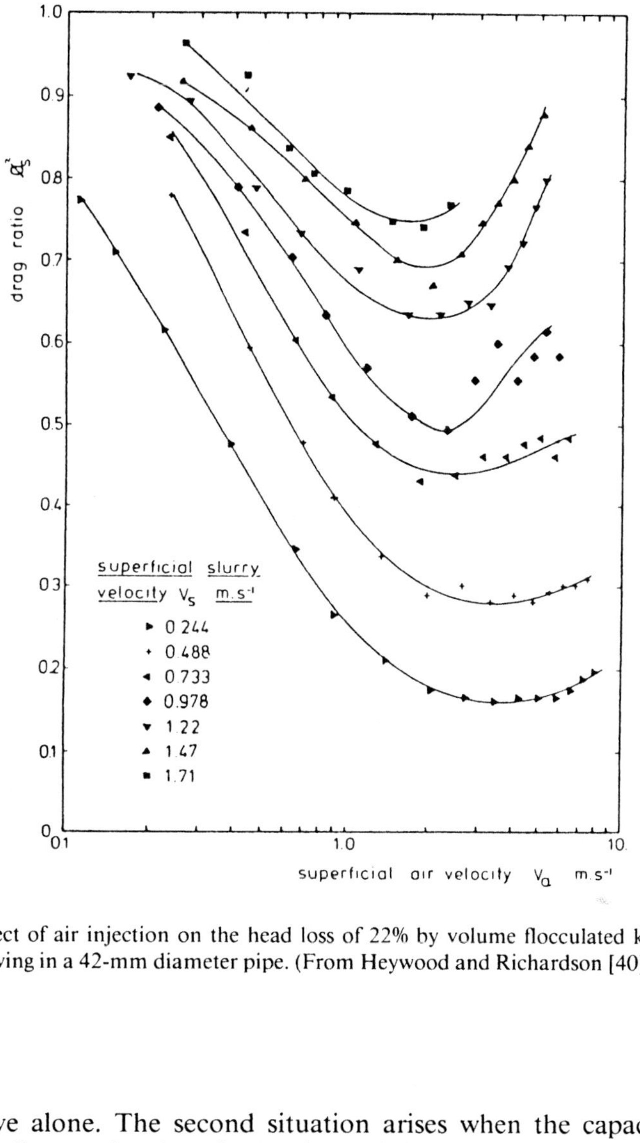

FIG. 5. Effect of air injection on the head loss of 22% by volume flocculated kaolin slurry flowing in a 42-mm diameter pipe. (From Heywood and Richardson [40].)

wise achieve alone. The second situation arises when the capacity of an existing pipeline needs to be raised while retaining the same pump system. Air injection will reduce the frictional head against which the pump acts at the old slurry rate, and the result will be an increase in slurry throughput for the same pump speed. Third, an existing pipeline can be lengthened to transport the same slurry with the same pipe diameter.

In taking advantage of these benefits it is obviously important to be able to predict the maximum extent of drag reduction achievable for a given slurry and the air flow required to obtain maximum reduction in head loss.

It has been shown [40, 43] that the maximum reductions could be collapsed onto a single curve either by plotting minimum values of ϕ_s^2 against the slurry Reynolds number with flow behavior index n in the non-Newto-

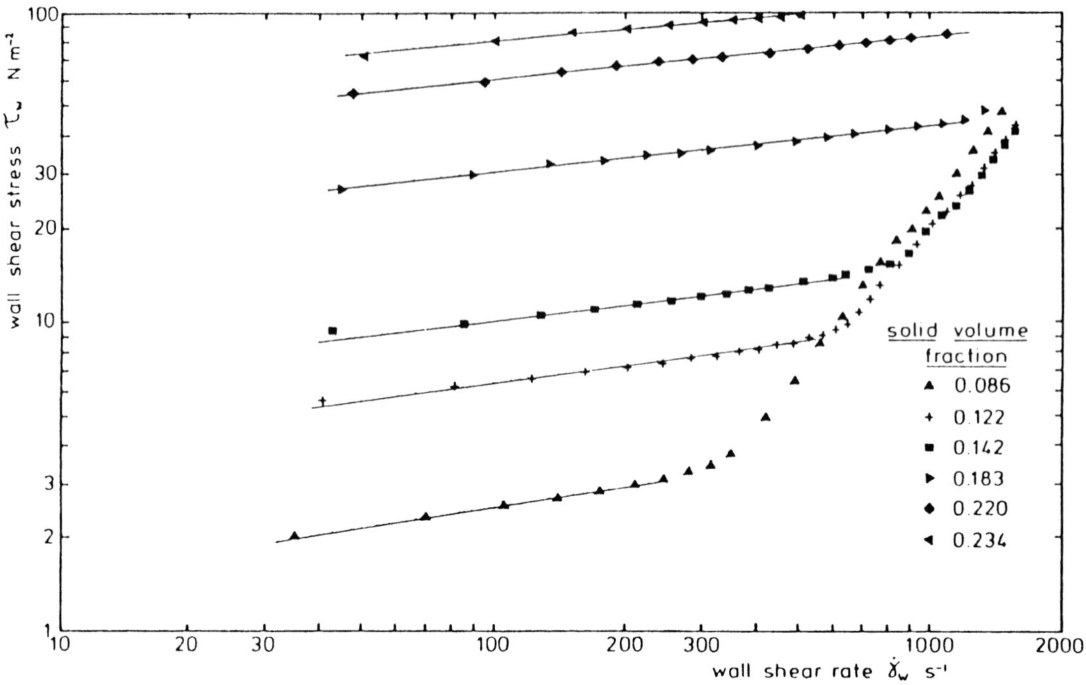

FIG. 6. Flow curves for flocculated kaolin slurries obtained by using a 42-mm diameter pipe. (From Heywood and Richardson [40].)

nian power law mode as a parameter or by plotting ϕ_s^2 against a parameter λ_c defined by

$$\lambda_c = \left[\frac{V_{SL}}{(V_{SL})_c}\right]^{1-n} \tag{1}$$

where V_{SL} is the superficial slurry velocity and $(V_{SL})_c$ is the critical superficial slurry velocity at the breakdown of laminar flow for slurry flow alone.

Further work [42, 46] has involved investigation of these effects in vertical pipe flow, where the introduction of air always introduces a reduction in head loss because of the decrease in the static head component to the total head loss which arises due to a decrease in the density of the mixture.

Although gas injection into a pipeline has yet to be exploited widely and its benefits under some conditions yet to be appreciated, some examples of practical applications already exist. Three are cited here:

a. A transfer system for asbestos slurry was designed at Warren Spring Laboratory [38, 39] for a brake manufacturer. The system, involving the injection of air into a 90-m long pipe carrying the slurry, has been in operation in the client's factory for over 10 years

b. The distance over which waste residue is transferred at a sugar-producing factory has been increased by some 50% by injecting air into the pipeline immediately downstream from the pump, while retaining the same pump
c. Putzmeister Ltd. markets a MIXOKRET pumping system [48] for transfer of concrete on building sites. The system consists of a single-shaft positive-action mixer which also acts as a conveyor. The pressure vessel lid to the mixer which is hinged on one side is closed with a quick-action clamp and air above the concrete is compressed, pushing the mix out of the vessel. Compressed air is added from a second line directly to the delivery line connection on the vessel outlet, giving "air cushions" which transport the mix in an intermittent flow through the hose line. The compressed air to the vessel and to the delivery line connection can be controlled separately. A discharge stand with pot or curved discharge device lets the conveying air escape at the end of the line, allowing uniform discharge of the concrete

Recently an attempt was made [47] to define more precisely the economically beneficial operating conditions when using air injection. A simple method was presented for evaluating the power saving achieved and the volume of air required to obtain maximum power saving. There is no doubt that this is an important technique for head loss reduction when there exists no or little scope for reducing suspension viscosities by particle deflocculation, and hence turning a non-Newtonian, shear-thinning suspension into an often much lower viscosity Newtonian suspension. One of the main advantages of air injection is that the separation of the phases on discharge from the pipeline is relatively straightforward since the air normally flows in the form of discrete slugs separately from the suspension. Hence no alteration of suspension properties using this power-saving technique occurs during pipeline transport.

Disturbances to Slurry Flow or Pipe

Oscillation of Slurry Flow Rate or Pressure Gradient

Oscillation of either the flow of slurry or the applied pressure gradient have been shown to result in a decrease in head loss under certain circumstances. When the fluid flow is oscillated, then typically a sine wave pulsation is superimposed on a steady flow with the frequency lying in the 0–10 Hz range. Edwards and Wilkinson [49] observed both experimentally and from theoretical considerations that this pulsation of the flow affects the profile of the amplitude of velocity across a pipe. The result is an increase in flow rate under a constant pressure gradient for some non-Newtonian fluids, or the energy for pumping could be minimized at constant flow rate by suitably dividing it between that required to provide a steady pressure gradient and that required to produce an oscillating component. This is known as "split-energy pumping" and may have many potential industrial applications.

Most experimental verification of the effect has involved aqueous polymer solutions which in general exhibit viscoelastic behavior. However, the simplified theory of Barnes et al. [50] does not include viscoelastic effects but nevertheless predicts pressure gradient reductions. It appears that the presence of viscoelastic properties enhances the drag reduction effect. It has been hypothesized [49] that for the flow of concentrated slurries the thickness of a particle-depleted layer at the pipe wall may be increased by the oscillating component of the pressure gradient, and the central core of slurry then moves as a plug on a lubricating layer of liquid, leading to a possible reduction in power consumption. The problem in the exploitation of this phenomenon is the dampening of the oscillatory component of the applied pressure gradient at large distances downstream from the point at which oscillations to the flow are imposed.

Round et al. [51–53] pointed out that flowing systems, typically fluids in pipelines, all have some sort of pulsations inherent in them. These are usually regarded as a nuisance which should be eliminated as simply and as economically as possible, and pulsation dampers are often used to achieve this. However, situations in which these pulsations could be exploited mean that the capital costs of both a pulsation damper and a device for imparting pulsations to the flow could be saved.

Benefits can be gained for either "settling" or "nonsettling" suspensions. Round [51] noted that the pulsation of "settling" suspensions in turbulent flow resulted in:

1. A decrease in the terminal settling velocity
2. A decrease in the energy required to suspend particles

This enhancement of suspension occurs particularly for flow in inclined pipes where there exists a component of the flow opposing the sedimentation direction of the particles, but even in horizontal pipes it has been observed by Chan et al. [54] that particle suspension is easier to achieve in a pulsating flow than in a steady flow. Round [51] considers the application of pulsations with the most potential is to settling suspensions and, in fact, particles could be transported in a purely symmetric oscillatory flow without a net steady flow component by suitable adjustment of the pipe internal geometry. This has been done by using baffles in a pipe.

Without pipe internals, purely oscillatory flow in a regular manner does not produce a net flow. However, asymmetric oscillations which do produce a net flow may be considered to consist of steady flow plus oscillatory components. Colamussi and Merli [55] have indicated the advantages of asymmetric pulsing in a solid–liquid system by means of flow interruptions. Such a system has been successfully applied to a dredging operation which transported a mixture of gravel, soil, and water.

Vibration or Oscillation of Pipe

Oscillation of the pipe rather than the slurry obviously has less practical appeal, but a handful of studies has provided information on reduction in head loss. Deysarkar and Turner [56] transported a normally stiff paste of fine iron ore containing 16% by weight water through a mechanically vibrated tube. The paste appeared to liquefy and flowed through the tube under a "very modest" pressure head. If no slip of the paste was assumed to exist at the tube wall, then the calculated viscosity for different tube diameters could be correlated with the peak cyclic acceleration given by Af^2, where A is the peak-to-peak amplitude and f is the angular frequency.

Manero and Mena [57, 58] passed Newtonian, nonelastic and non-Newtonian, elastic aqueous polymer solutions through relatively small diameter tubes which were oscillated using a variable speed motor with an eccentric shaft. No change in flow rate under a constant pressure gradient was observed for the Newtonian liquids, but an increase in flow rate was found for elastic fluids, the percentage increase being greater at small flow rates.

It is difficult to envisage that it would be technically feasible or economic to oscillate a pipeline along its axis over any significant distance, and hence the exploitation of such a phenomenon is probably limited to short pipe lengths. Since in-factory pipelines often have vibrating equipment attached to them, it may well be that the effect is, sometimes unknowingly, being exploited. Frictional energy losses resulting from pipe flow within chemical plant are relatively minor compared with other energy requirements, and hence it appears doubtful and it would ever be worthwhile to install dedicated vibrational equipment, unless it is used specifically to prevent some pipework blocking.

Use of Modified Pipe Cross-Section Geometries for Settling Slurries

Helical Ribs

Pipes containing helical ribs attached to the inner pipe wall surface have been shown in several studies [59–66] to have significant advantages for some slurry pipeline transport requirements. For rapidly-settling slurries, head losses can be lower than for smooth pipes at low velocities. Charles [60] used ribs with pitch-to-diameter ratios of 1.79, 3.3, and 5.2 for a 51-mm diameter pipe (see Fig. 7). At high velocities the ribs increase the head loss. It seems feasible though to operate ribbed pipes at lower velocities with reduced risk of blockage compared with plain pipes at similar velocities, and lower operating velocities will aid in reducing energy consumption.

The pitch-to-diameter ratio is related to the rib angle, θ, measured from the pipe axis by

$$\tan \theta = \pi D/p \tag{2}$$

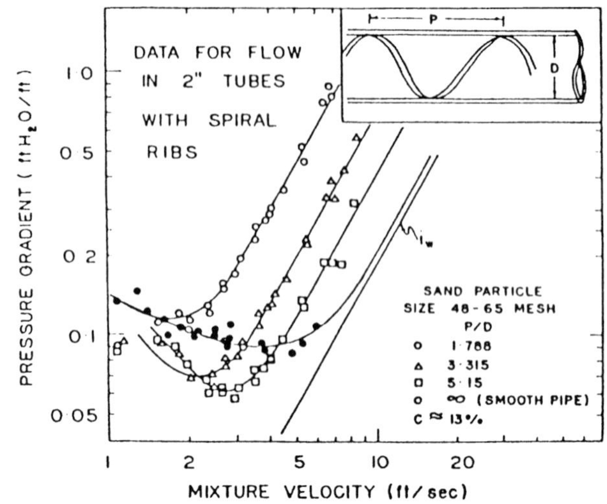

FIG. 7. Reduction in head losses obtained in a 48-mm diameter pipe for a 13% settling sand slurry. (From Charles [60].)

Charles [60] suggested that a p/D ratio of 5 should prove useful in many applications but, in a later study, Singh and Charles [64] found that the optimum p/D ratio seems to be about 8 ($\theta = 31.4°$) with a rib height of 10 to 15% of the pipe diameter.

Both Charles et al. [61] and Schriek et al. [63] looked at how power consumption can be reduced by using helical ribbed pipes. Because the advantage of the ribbed pipe is only realized at velocities below the critical deposit velocity in the smooth pipe, in order to provide the same solids flow rate, either the diameter of the ribbed pipe must be larger than the diameter of the smooth pipe if the delivered concentration is the same in both cases, or the slurry flowing through the ribbed pipe must be more concentrated if the diameters of the pipes are the same in both cases. Using two somewhat arbitrary comparisons, Charles et al. [61] were able to show that the power consumption for a 13% slurry flowing through a 48-mm pipe (with p/D = 5.15) at a velocity of 1 m/s was some 60% of that required for the same slurry concentration passing through a 36-mm diameter smooth pipe at 1.68 m/s. A p/D ratio of 3.32 gave a similar power consumption, while a p/D ratio of 1.80 gave a higher power consumption.

In the second comparison, the slurry concentrations must be different if the pipe diameter is to remain approximately the same, again for a fixed solids throughput. The power consumption for a 5% slurry through a 50-mm pipe at 2.1 m/s is greater than that for a 13% slurry flowing through a 48-mm ribbed pipe at 1 m/s at either p/D = 5.15 or 3.3, but is slightly less for p/D = 1.80.

Schriek et al. [63] obtained head loss and energy consumption data for sand slurry flow in a 150-mm diameter pipe which showed similar trends to the data of Charles et al. using a 48-mm diameter pipe. Energy consumption calculations showed that for $p/D < 4$, ribbed pipes are not very attractive in reducing energy requirements, whereas in the range $5 \leqslant p/D \leqslant 10$, the minimum energy consumption is relatively insensitive to p/D.

It is probable that there are numerous situations where ribbed pipes would be desirable. Capital cost of a ribbed pipe is obviously going to be larger but it is technically relatively straightforward to provide a plastic pipe with ribs. In the Schriek et al. study the 150-mm diameter ribbed pipe was manufactured by extruding over a mandrel which could be rotated at various speeds to alter the p/D ratio. Thus, the ribs were an integral part of the pipe itself. However, extruded plastic pipe can withstand only low pressures, but it may be possible to develop a liner with the ribs for installation into steel pipe.

Since slurry pipelines can be operated at lower velocities when helical ribs are incorporated, pipe wall wear rates can be expected to be lower, but erosion of the ribs could be a potential problem and this must be evaluated as far as possible in any feasibility study owing to the high capital cost investment in a ribbed pipe. Smith et al. [66] measured wear rates for a ribbed pipe ($p/D = 4$) and smooth pipe using a weight loss method. For the same velocity, higher erosion rates were observed with the ribbed pipe, but when comparison is made at the corresponding deposit velocities for the ribbed and smooth pipes, wear rates are similar. Also, Smith et al. point out that in analyzing wear rate data, it is the maximum local penetration rate which determines pipe durability, and so the distribution of wear rates must be considered. Although the flows in ribbed and smooth pipes are most similar at the lowest velocities, one might expect the wear to be more evenly distributed in the ribbed pipe.

Unfortunately, there currently exists no systematic design approach for helically ribbed pipe, and each system must be evaluated both by desk study and probably pilot-scale testwork to investigate such variables as the viscosity of the slurry, the size distribution of the solids, and the density ratio of solid to suspending medium. However, there exists quite a large volume of information for sand slurries, and starting points, such as a p/D ratio lying in the range 5 to 8, can be used. Schriek et al. [63] state that it is arbitrarily assumed that one rib for every 50 mm of pipe diameter will produce adequate agitation, but this seems a reasonable "rule-of-thumb" since their data for three ribs in a 150-mm diameter pipe and one rib in a 50-mm diameter pipe are in "reasonably good agreement."

Despite the lack of information concerning the effects of a number of variables, such as the optimization of rib shape and size, ribbed pipe has special advantages for relatively coarse material and for intermittent operation when unavoidable shutdowns are anticipated. The ribs could aid in the restarting of such lines over uneven terrain and on steep gradients, and there appears to be no reason why ribs should not be installed only in sections of a line where blockage is of concern.

Use of Segmented Pipe

A number of studies have indicated that a circular cross-section to a pipe is not the most favorable for minimizing head losses in slurry pipe flow, and

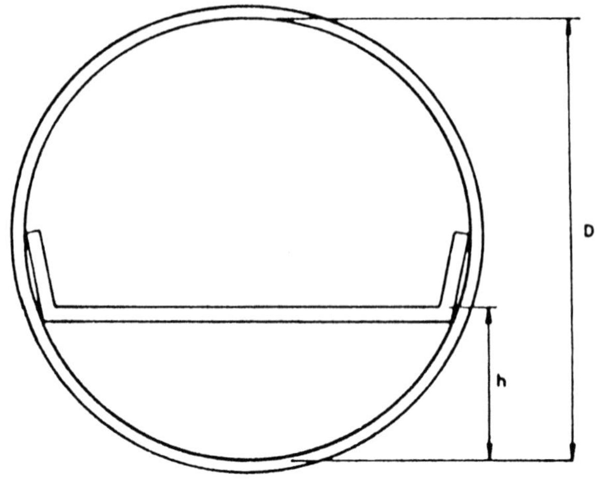

FIG. 8. Design of segmented pipe used by Sauermann [69].

experiments on noncircular cross-section pipes have been undertaken. Although the results of these studies appear attractive from an operating cost saving point-of-view, it is probable that few, if any, industrial installations exploit this saving owing to technical and cost disincentives in manufacturing noncircular geometry pipework.

Wang and Seman [67], by using six pipe geometries, showed without doubt that pipe geometry and particularly the base area of the pipe significantly affect the slurry flow. The investigation revealed that the head loss at the minimum carrying velocity in a wide-base rectangular pipe was approximately 20% less than that in a circular pipe of the same cross-sectional area. All noncircular cross-sections, however, suffer from the disadvantage that points of stress intensification occur in the pipe wall, and thicker walled pipes become necessary. To avoid this, Führböter [68] proposed that a segment plate be placed inside the pipe which would at the same time act as a wear plate. By equalizing the pressure above and below the segment plate by means of, for instance, a few small holes in the plate, the advantage of uniform stress distribution in a circular pipe is retained.

Führböter [68] calculated that transport in a segmented pipe is particularly beneficial when transporting large solids, and he predicted that the best position of a segment plate is at approximately one-third of the internal pipe diameter above the bottom level of the pipe as shown in Fig. 8, that is, at $h/D = 1/3$.

In order to verify this theoretical finding, Sauermann [69, 70] carried out tests in pipes of 100 mm nominal diameter. Segment plates were placed at three different heights approximately at, above, and below the level $h/D = 1/3$. By using both sand and anthracite slurries, Sauermann showed that a segmented pipe could reduce the power consumption, and that in both cases the least power consumption was required when $h/D = 0.313$ (see Fig. 9 for anthracite slurry). For the highest solids concentrations, reductions in power consumption were of the order of one-sixth.

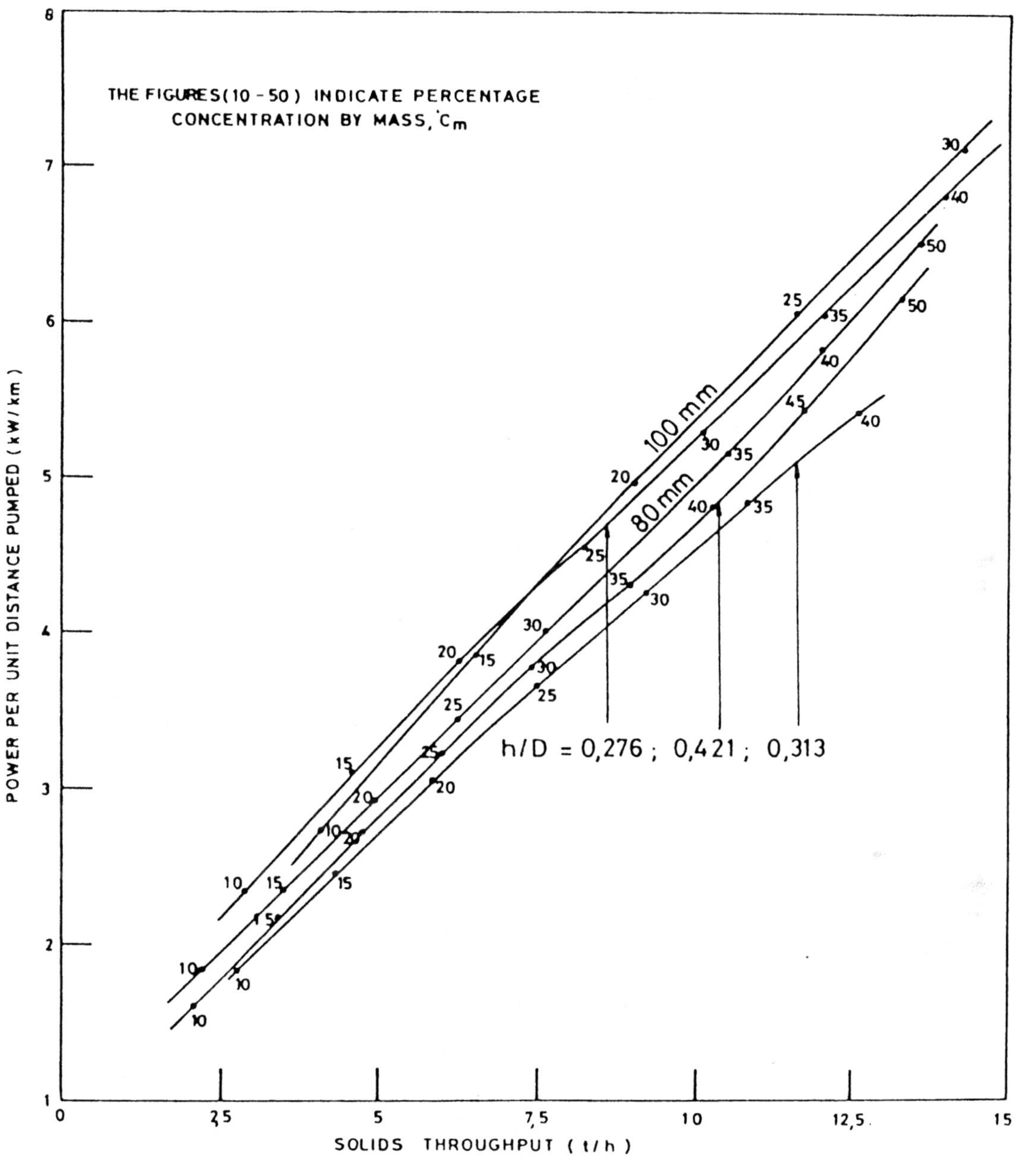

FIG. 9. Power reductions for anthracite slurry using a segmented pipe at three alternative positions in 80 and 100 mm diameter pipe. (From Sauermann [69].)

References

Effect of Changing Particle Size

1. I. Wagstaff and C. E. Chaffey, *J. Colloid Interface Sci.*, *59*, 53–62 (1977).
2. C. E. Chaffey and I. Wagstaff, *J. Colloid Interface Sci.*, *59*, 63–75 (1977):
3. Y. S. Papir and I. M. Krieger, *J. Colloid Interface Sci.*, *34*, 126 (1970).
4. M. E. Woods and I. M. Krieger, *J. Colloid Interface Sci.*, *34*, 91 (1970).
5. C. E. Chaffey, *Colloid Polym. Sci.*, *255*, 691 (1977).
6. I. M. Krieger, *Trans. Soc. Rheol.*, *7*, 101 (1963).

Effect of Changing Particle Size Distribution

7. R. J. Farris, *Trans. Soc. Rheol.*, *12*, 281–301 (1968).
8. G. F. Eveson, S. G. Ward, and R. L. Whitmore, *Discuss. Faraday Soc.*, *11*, 11–14 (1951).
9. M. Mooney, *J. Colloid Sci.*, *6*, 162–170 (1951).
10. H. C. Brinkman, *J. Chem. Phys.*, *20*, 571 (1952).
11. R. Roscoe, *Br. J. Appl. Phys.*, *3*, 267–269 (1952).
12. J. W. Lee and K. J. Lee, *Proceedings of the 3rd Congress of Fluid Mechanics Fundamentals II*, 1983, pp. 32–37.
13. J. S. Chong, E. B. Christiansen, and A. D. Baer, *J. Appl. Polym. Sci.*, *15*, 2007–2021 (1971).
14. F. Skvara and M. Vancourova, *Silikaty*, *17*, 9–18 (1973).
15. C. Parkinson, S. Matsumoto, and P. Sherman, *J. Colloid Interface Sci.*, *33*, 150–160 (1970).
16. H. Goto and H. Kuno, *J. Rheol.*, *28*, 197–205 (1984).
17. F. Ferrini, V. Battarra, E. Donati, and C. Piccini, *Proceedings of Hydrotransport 9*, Organized by BHRA Fluid Engineering, Cranfield, Beds., October 1984, Paper B2.
18. R. B. Klose, *Proceedings of Hydrotransport 9*, Organized by BHRA Fluid Engineering, Cranfield, Beds., October 1984, Paper C3.

Effect of Changing Particle Shape, Including High Aspect Ratio Fibers

19. B. Clarke, *Trans. Inst. Chem. Eng.*, *45*, 251 (1967).
20. R. J. E. Kerekes and W. J. M. Douglas, *Can. J. Chem. Eng.*, *50*, 228–231 (1972).
21. R. C. Vaseleski and A. B. Metzner, *AIChE J.*, *20*, 301–306 (1974).
22. D. D. Kale and A. B. Metzner, *AIChE J.*, *20*, 1218–1219 (1974).
23. A. J. Bobkowicz and W. H. Gauvin, *Can. J. Chem. Eng.*, *43*, 87–91 (1965).
24. A. A. Robertson and S. G. Mason, *Tappi*, *40*, 326 (1957).

Addition of Soluble Ionic Compounds

25. R. R. Horsley and J. A. Reizes, *Proceedings of Hydrotransport 7 Conference*, Organized by BHRA Fluid Engineering, November 1980, Paper D3.
26. R. R. Horsley, *Proceedings of Hydrotransport 8 Conference*, Organized by BHRA Fluid Engineering, August 1982, Paper H1.

27. R. R. Horsley, *J. Pipelines*, *3*, 87–96 (1982).
28. C. F. Sikorski, R. L. Lehman, and J. A. Shepherd, *Proceedings of 7th Slurry Transport Association Conference*, 1982, pp. 163–173.
29. C. A. Shook and J. Nurkowski, *Can. J. Chem. Eng.*, *55*, 510–515 (1977).
30. N. Teckchandani and C. A. Shook, *J. Pipelines*, *2*, 43–47 (1982).

Addition of Soaps

31. J. L. Zakin, M. Poreh, A. Brosh, and M. Warshavsky, *Chem. Eng. Prog., Symp. Ser. No. 11*, *67*, 85–89 (1971).

Addition of Soluble High Molecular Weight Polymer

32. I. Radin, J. L. Zakin, and G. R. Patterson, *AIChE J.*, *21*, 358–371 (1975).
33. J. Golda, *Proceedings of 9th Slurry Transport Association Conference*, 1984, pp. 55–61.
34. J. Pollert, *Proceedings of 2nd Conference on Drag Reduction*, Organized by BHRA Fluid Engineering, Cranfield, Beds., 1977, Paper B3.
35. V. Kolar, *Proceedings of Hydrotransport 1 Conference*, Organized by BHRA Fluid Engineering, Cranfield, Beds., 1970, Paper F2.
36. T. R. Sifferman and R. A. Greenkorn, *Soc. Pet. Eng. J.*, *21*(6), 663–669 (1981). (See also Ref. 31.)

Air Injection into Flowing Slurry

37. H. C. Ward and J. M. Dallavallc, *Chem. Eng. Prog., Symp. Ser.*, *10*, 1–14 (1954).
38. D. C.-H. Cheng, P. E. Jones, E. F. Keen, K. G. Laws, and R. J. French, *Proceedings of Pneumotransport 1 Conference*, Organized by BHRA Fluid Engineering, Cranfield, Beds., September 1971, Paper A1.
39. A. J. Carleton, D. C.-H. Cheng, and R. J. French, *Proceedings of Pneumotransport 2 Conference*, Organized by BHRA Fluid Engineering, Cranfield, Beds., September 1973, Paper F2.
40. N. I. Heywood and J. F. Richardson, *Proceedings of Hydrotransport 5 Conference*, Organized by BHRA Fluid Engineering, Cranfield, Beds., May 1978, Paper C1.
41. S. I. Farooqi, N. I. Heywood, and J. F. Richardson, *Trans. Inst. Chem. Eng.*, *58*, 16–27 (1980).
42. N. I. Heywood and M. E. Charles, *Proceedings of Hydrotransport 7 Conference*, Organized by BHRA Fluid Engineering, November 1980, Paper E1.
43. S. I. Farooqi and J. F. Richardson, *Trans. Inst. Chem. Eng.*, *60*, 323–333 (1982).
44. R. P. Chhabra, S. I. Farooqi, Z. Khatib, and J. F. Richardson, *J. Pipelines*, *2*, 169–185 (1982).
45. R. P. Chhabra, S. I. Farooqi, J. F. Richardson, and A. P. Wardle, *Chem. Eng. Res. Des.*, *61*, 56–61 (1983).
46. Z. Khatib and J. F. Richardson, *Chem. Eng. Res. Des.*, *62*, 139–154 (1984).
47. M. Dziubinski and J. F. Richardson, *J. Pipelines*, *5*, 107–111 (1985).
48. *MIXOKRET*, Putzmeister Ltd. Commercial Brochure, DF 520-2GB.

Oscillation of Slurry Flow Rate or Pressure Gradient

49. M. F. Edwards and W. L. Wilkinson, *Trans. Inst. Chem. Eng.*, *49*, 785 (1971).
50. H. A. Barnes, P. Townsend, and K. Walters, *Nature*, *244*, 585 (1969).
51. G. F. Round, *Proceedings of Hydrotransport 3 Conference*, Organized by BHRA Fluid Engineering, Cranfield, Beds., 1974, Paper B3.
52. G. F. Round, B. Latto, and K.-Y. Lau, *Proceedings of Hydrotransport 4 Conference*, Organized by BHRA Fluid Engineering, Cranfield, Beds., 1976, Paper D1.
53. G. F. Round, A. Hameed, and B. Latto, *Proceedings of 6th Slurry Transport Association Conference*, 1981, pp. 121–130.
54. K. W. Chan, M. H. I. Baird, and G. F. Round, *Proc. R. Soc. London*, *330*, 537–559 (1972).
55. A. Colamussi and V. Merli, *Dock Harbour Auth.*, pp. 471–482 (March 1972).

Vibration or Oscillation of Pipe

56. A. K. Deysarkar and G. A. Turner, *J. Rheol.*, *25*(1), 41–54 (1981).
57. O. Manero and B. Mena, *Rheol. Acta*, *16*, 573–576 (1977).
58. O. Manero, B. Mena, and R. Valenzeula, *Rheol. Acta*, *17*, 693–697 (1978).

Use of Spiral Ribs

59. S. E. Wolfe, *CIM Bull.*, *60*(658), 221–223 (1967).
60. M. E. Charles, *Proceedings of Hydrotransport 1 Conference*, Organized by BHRA Fluid Engineering, Cranfield, Beds., 1970, Paper A3.
61. M. E. Charles et al., *Can. J. Chem. Eng.*, *49*, 737–741 (1971).
62. S. Kuzuhara, *Proceedings of Hydrotransport 3 Conference*, Organized by BHRA Fluid Engineering, Cranfield, Beds., 1974, Paper F2.
63. W. Schriek et al., *Sask. Res. Counc. Rep. E72-4, CIM Bull.*, *67*(750), 84 (1974).
64. V. P. Singh and M. E. Charles, *Can. J. Chem. Eng.*, *54*, 249–254 (1976).
65. S. Kuzuhara and T. Shakouchi, *Proceedings of Hydrotransport 5 Conference*, Organized by BHRA Fluid Engineering, Cranfield, Beds., May 1978, Paper H3.
66. L. G. Smith, C. A. Shook, D. B. Haas, and W. H. W. Husband, *Proceedings of Hydrotransport 5 Conference*, Organized by BHRA Fluid Engineering, Cranfield, Beds., May 1978, Paper B2.

Use of Segmented Pipe

67. R. C. Wang and J. J. Seman, *U. S., Bur. Mines, Rep. Invest. 7725* (1973).
68. A. Führböter, *Mitteilungen des Franzius—Instituts für Grund und Wasserbau der Technischen Hochschule*, Hannover, Heft 19, 1961.
69. H. B. Sauermann, *Proceedings Hydrotransport 5 Conference*, Organized by BHRA Fluid Engineering, Cranfield, Beds., May 1978, Paper A4.
70. H. B. Sauermann, *Proceedings of 7th Slurry Transport Association Conference*, 1982, pp. 57–60.

NIGEL I. HEYWOOD

Two-Phase Flow, Pressure Drop and Holdup Equations

New equations for pressure drop and holdup in a two-phase pipeline not only show good agreement with a broad range of large-diameter data but also can be easily applied.

These equations are based upon a fundamental model for both pressure drop and liquid holdup, and they are compared to large-diameter field data encompassing a wide range in operating parameters.

Flow Parameters

The development of hydrocarbon resources often requires the transportation of vapor and liquid states simultaneously in a pipeline.

Economics may dictate that facilities be minimized and, where practical, a single two-phase pipeline be employed in lieu of separate gas and liquid pipelines. Two-phase flow is also commonly encountered in onshore gathering flowlines upstream of separating facilities.

In design of a two-phase pipeline system, consideration must be given to gas and liquid fluid properties, flow regimes, pressure drop, and liquid holdup. A detailed study of these parameters under various operating conditions is necessary to size the pipeline, size and configure the liquid slugcatcher, develop pigging strategies, and assess the effects of random or flow-induced slugging.

Of these phenomena, only the equilibrium pressure drop and holdup aspects of horizontal or slightly inclined pipelines are addressed here.

Numerous technical papers are currently available in the literature dealing with pipeline two-phase flow [1–9]. Dozens of correlations purporting to characterize two-phase pressure drop and holdup have been proposed over the past three decades.

In most cases, correlation validity is based upon comparison with field and/or laboratory measurements. Unfortunately, due in part to the general scarcity of accessible large-diameter (greater than 6 in.) operating data, most of these comparisons have been made with relatively small-diameter conduits.

The application of the resulting correlations to actual large-diameter pipelines has proven, in most cases, to be unsatisfactory.

Familiar Configuration

The two-phase model used here is the familiar one-dimensional stratified-flow configuration as shown in Fig. 1.

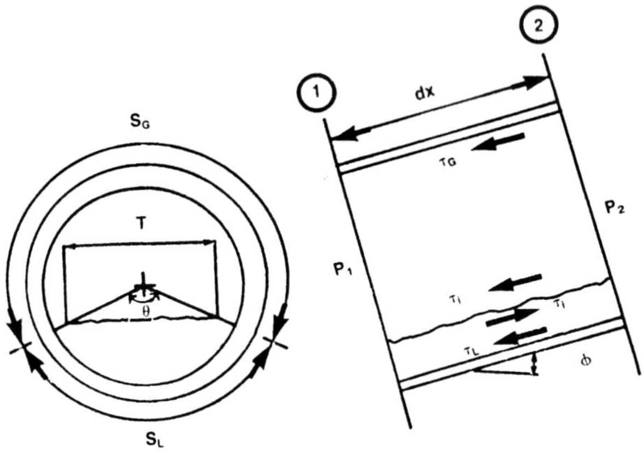

FIG. 1. One-dimensional stratified model.

It is generally accepted that two-phase flow can be best characterized with a flow regime-dependent mechanistic approach. Fortunately, the stratified-flow regime which lends itself most readily to model development is also the regime in which most actual pipelines operate (along with the intermittent regime).

The favorable comparison between the proposed stratified-based approach and a wide range of field data seems to justify its general applicability to horizontal and slightly inclined pipelines.

The proposed model does not apply to vertical/near vertical flow, as in a riser, nor to severe slug flow, as is encountered just upstream of a riser.

A force balance (Fig. 1) across the gas and liquid, respectively, yields Eqs. (1) and (2). (See Tables 1 and 2 for the equations.) The shear stresses, τ_G, τ_L, and τ_i, are defined in Eqs. (3), (4), and (5). The Fanning friction factors, f_{wG} and f_{wL}, are determined with a conventional friction equation such as the Colebrook [10] expression for fully turbulent flow, Eq. (6).

D_H in Eq. (6) is the hydraulic diameter which is defined for the gas and liquid phases, respectively, as given in Eqs. (7) and (8). These hydraulic diameters are also used in the calculation of Reynolds number for each phase.

Determining the friction factor at the interface (f_i) and the interface velocity (V_i) presents more of a challenge. Oliemans suggested setting V_i equal to V_L. For f_i, Oliemans proposed using the Colebrook equation (Eq. 6), but related the interface roughness to the interface wave height [11].

Notwithstanding this and other proposed techniques regarding the interfacial relationship [12, 13], this remains largely an undefined area.

Pressure-Drop, Holdup Equations

Fortunately, when Eqs. (1) and (2) are added together for the total force balance, the interfacial-shear terms cancel out. When the resulting equation is divided through by Adx, Eq. (9) results.

Substituting Eqs. (3) and (4) for τ_G and τ_L, the expressions $\rho = \gamma/G$, $A_G = (1 - H_L)A$ and $A_L = H_L A$, and rearranging result in the pressure-drop relationship given in Eq. (10), it is relatively simple to apply but requires that the holdup fraction be known.

TABLE 1 Equations

$(P_1 - P_2)A_G - \tau_G S_G dx - \tau_1 T dx - \rho_G A_G G \sin\phi dx = 0$	(1)
$(P_1 - P_2)A_L - \tau_L S_L dx + \tau_1 T dx - \rho_L A_l G \sin\phi dx = 0$	(2)
$\tau_G = f_{wG}\rho_G V_G^2/2$	(3)
$\tau_L = f_{wL}\rho_L V_L^2/2$	(4)
$\tau_i = f_i\rho_G(V_G - V_i)^2/2$	(5)
$f = 1/[3.48 - 4\log(2k/D_H + 9.35/(\mathrm{Re}\, f^{0.5}))]^2$	(6)
$D_{HG} = 4A_G/(S_G + T)$	(7)
$D_{HL} = 4A_L/(S_L + T)$	(8)
$(P_1 - P_2)/dx - \tau_G S_G/A - \tau_L S_L/A - (\rho_G A_G/A + \rho_L A_L/A)G \sin\phi = 0$	(9)
$dP/dx = f_{wG}\gamma_G V_G^2 S_G/(2GA) + f_{wL}\gamma_L V_L^2 S_L/(2GA) + [\gamma_G(1 - H_L) + \gamma_L H_L] \sin\phi$	(10)
$\tau_G S_G/A_G - \tau_L S_L/A_L + \tau_i T(1/A_G + 1/A_L) - (\rho_L - \rho_G)G \sin\phi = 0$	(11)
$H_L = (\tau_L S_L - \tau_i T)/[\tau_G S_G + \tau_L S_L - (\rho_L - \rho_G)G \sin\phi\, A_G)]$	(12)
$\mathrm{Fr} = (1 - H_L)/(1 - \lambda)$	(13)
$\mathrm{Fr} = 1/2 m_{NS} V_G^2/(m_{NS} G y_{NS}/\cos\phi)$	(14)
$\mathrm{Fr} = V_G^2 \cos\phi/2 y_{NS} G$	(15)
$y_{NS} = D/2(A_{NS}/A)^{0.5}$	(16a)
$\quad = D/2(\lambda)^{0.5}$	(16b)
$(1 - H_L)/(H_L - \lambda) = V_G^2(\cos\phi)/GD\lambda^{0.5}$	(17)
$H_L = \lambda + (1 - \lambda)/[1 + V_G^2(\cos\phi)/GD\lambda^{0.5}]$	(18)

TABLE 2 Two-phase flow relationships

$\lambda = Q_L/(Q_L + Q_G)$	$S_G = D(\pi - \theta/2)$
$r = 10^6/GOR$	$S_L = D(\theta/2)$
$V_M = V_{SG} + V_{SL}$	$T = D \sin(\theta/2)$
$= V_{SG}/(1 - \lambda)$	$A_L = D^2/4(\theta/2 - \sin(\theta/2)\cos(\theta/2))$
$V_G = V_{SG}/(1 - H_L)$	$A_G = D^2/4(\pi - \theta/2 + \sin(\theta/2)\cos(\theta/2))$
$= (1 - \lambda)V_M/(1 - H_L)$	$H_L = (\theta/2 - \sin(\theta/2)\cos(\theta/2))/\pi$
$V_L = (\lambda/H_L)V_M$	$D_{HG} = (1 - H_L)\pi D^2/(S_G + T)$
$= (\lambda/H_L)(1 - H_L)V_G/(1 - \lambda)$	$D_{HL} = H_L\pi D^2/(S_L + T)$

The following relationships use the imperial system of units and assume base conditions of 60°F and 14.73 lb/in.^{2} abs

$C = (TZ/P)_{\text{avg}}$	°R lb/in.^{2} abs
$\lambda = r/(r + 5045C)$	
$Q_G = (3.279 \times 10^{-7})Q_{\text{std}}C$	ft^3/s
$Q_L = (6.487 \times 10^{-5})Q_{\text{std}}r$	ft^3/s
$V_{SG} = (6.01 \times 10^{-5})Q_{\text{std}}C/D^2$	ft/s
$V_{SL} = (1.19 \times 10^{-8}Q_{\text{std}}r/D^2$	ft/s

Two equations for liquid holdup will be presented, the first based upon the fundamental model previously given and the second based upon a simple Froude-number relationship.

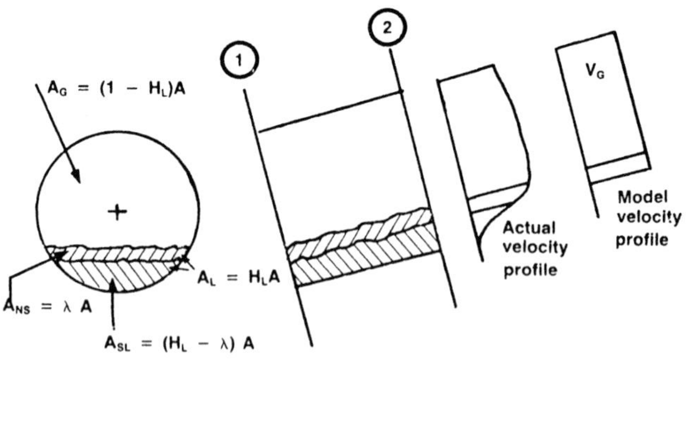

FIG. 2. Liquid slippage model.

The first holdup equation is derived by dividing Eq. (1) by A_G, dividing Eq. (2) by A_L, then subtracting Eq. (1) from Eq. (2), thereby eliminating the pressure-gradient term. The result is given in Eq. (11). Rearranging and substituting $A_G = (1 - H_L)A$ and $A_L = H_L A$ yield the fundamental equation for holdup, Eq. (12). It is implicit because H_L must be known to determine S_L, T, S_G, and A_G as well as the shear-stress terms. However, the real complexity of Eq. (12) lies in determining the interfacial shear stress (τ_i) as mentioned previously.

A fundamentally derived holdup equation such as Eq. (12) is desirable. However, data comparisons using Eq. (12) in conjunction with various proposed interfacial shear equations have proven unsatisfactory.

Until future developmental and/or experimental work more adequately defines the interfacial relationship, Eq. (12) is of little practical use.

Froude-Number Equation

The second holdup equation is much simpler to apply. Before a discussion of the second equation, it is useful to review two-phase slippage concepts.

The term "slip" in two-phase flow terminology refers to the lag in average liquid velocity relative to the gas.

In reality, a velocity profile as shown in Fig. 2 exists wherein the liquid at the gas–liquid interface moves at an accelerated rate approaching the gas velocity. As the distance from the interface increases, the local liquid velocity decreases to a limit approaching zero at the pipe wall.

Most two-phase flow theories and correlations recognize this slip between the gas and liquid phases. Some correlations, however, are based upon a homogeneous or "no-slip" model.

The no-slip concept, when applied to the entire fluid cross section, implies that the gas and liquid are traveling at the same velocity with no lag between phases.

Another concept, which is mathematically convenient in two-phase modeling, employs a "slip/no-slip" approach wherein a portion of the liquid is assumed to travel at the gas velocity, i.e., not slipping, while the remaining liquid is assumed at rest, i.e., complete slippage (Fig. 2).

The second holdup equation to be presented utilizes this "slip/no-slip" concept.

The equation is derived by equating two simple relationships for the Froude number (Fr) of the flowing portion of the liquid in the model.

The Froude number is a dimensionless grouping of variables found by ratioing the kinetic or inertial energy and the gravitational energy of a flowing stream. It is similar to the Reynolds number, which relates inertial to viscous energy.

Just as the Reynolds number is utilized to characterize enclosed fluid systems, the Froude number is typically used for "free-surface" flow characterization.

Energy Ratio

The first relationship for the Froude number follows from the postulation that the ratio of inertial to gravitational energy can be represented by the ratio of gas flow area to slipped-liquid flow area, as follows:

$$\text{Fr} = (\text{gas flow area})/(\text{slipped liquid flow area})$$

This can be expressed as Eq. (13).

No theoretical derivation can be offered for this relationship. It is simply a statement that the equilibrium flow area for the gas which provides the motivational force for liquid motion is a measure of the kinetic/inertial energy present.

Likewise, the slipped liquid area is an indicator of the level of gravitational energy present.

A second relationship for the Froude number is derived by ratioing the inertial energy of the no-slip liquid by its gravitational energy as given in Eq. (14). Rearranging Eq. (14) yields Eq. (15).

The cos ϕ term in Eq. (15) relates the vertical fluid depth to y_{NS}, which is normal to the pipe wall. The no-slip flow depth is defined as given in Eq. (16a) or Eq. (16b).

Now, setting Eq. (13) equal to Eq. (14) and substituting Eq. (16b) for y_{NS} yields Eq. (17), which can be rearranged to give the final holdup equation, Eq. (18).

In addition to being a relatively simple relationship, Eq. (18) satisfies the following boundary conditions illustrated in Fig. 3:

$$\text{As } V_G \longrightarrow \infty,\ H_L \longrightarrow \lambda$$
$$\text{As } \lambda \longrightarrow 0,\ H_L \longrightarrow 0.0$$
$$\text{As } V_G \longrightarrow 0,\ H_L \longrightarrow 1.0$$

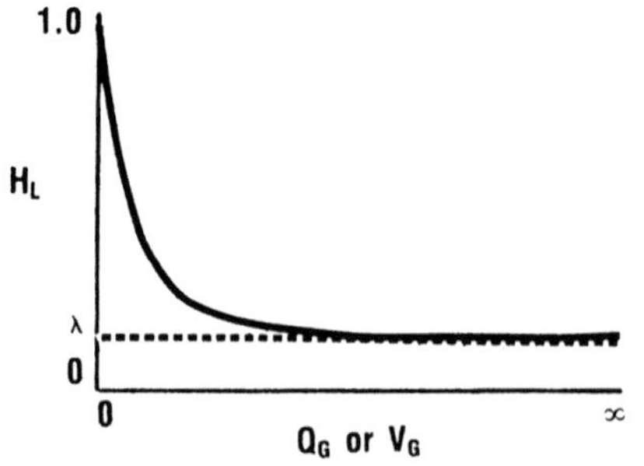

FIG. 3. Boundary conditions (Eq. 18).

Application of Eq. (18) requires the use of the actual gas velocity which, of course, is not known until H_L is determined. An initial assumption is usually made that $V_G = V_{SG}$ and H_L is then calculated.

A new gas velocity corresponding to this H_L is then calculated as $V_{SG}/(1 - H_L)$ and compared with the initial assumption.

If the two velocities are not sufficiently close, then a new assumption for V_G is made by averaging the previous two values. Convergence is usually found in a few iterations.

Large-Diameter Field Data

The only useful gauge of large-diameter two-phase correlation validity is comparison with field data from large-diameter systems encompassing a wide range in fluid and operating parameters.

Figures 4 and 5 present such a comparison between measured and calculated values for the fundamental presure-drop equation, Eq. (10), and the simple Froude number-based holdup equation, Eq. (18), respectively.

The field data represented in these figures are tabulated in Tables 3 and 4.

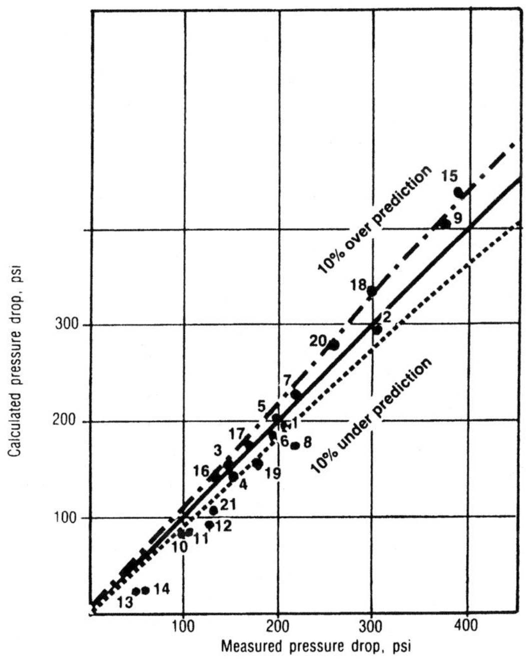

FIG. 4. Fundamental pressure drop (Eq. 10).

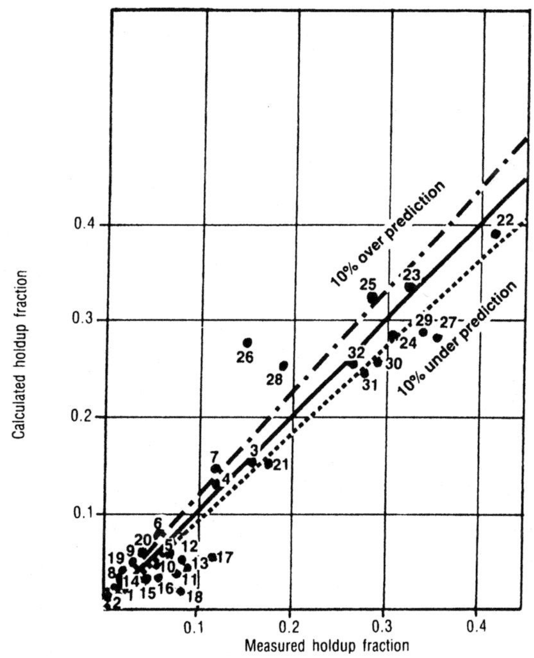

FIG. 5. Froude number holdup equation (Eq. 18).

TABLE 3 Pressure-Drop Field Data[a]

Data point	Nominal diameter (in.)	Average mixture velocity (ft/s)	Average liquid loading (bbl/MMSCF)	Measured pressure drop (lb/in.^{2})	Calculated pressure drop (lb/in.^{2})	Percent error
1	36	19.5	10	207	193	−6.76
2	36	22.8	8	305	296	−3.15
3	30	24.3	15	146	152	+4.11
4	30	25.2	7	151	142	−5.96
5	30	27.5	8	200	199	−0.50
6	30	29.0	5	198	190	−4.04
7	30	28.7	9	217	224	+3.23
8	20	9.2	65	216	149	−31.02
9	20	14.8	65	375	401	+6.93
10	20	13.6	19	97	83	−14.43
11	20	13.0	28	103	83	−19.42
12	20	13.6	35	123	93	−24.39
13	20	8.6	34	52	24	−53.85
14	20	9.9	21	55	23	−58.18
15	12	30.7	1321	390	435	+11.54
16	12	18.1	1284	137	138	+0.73
17	12	20.5	1217	171	172	+0.59
18	12	26.9	1290	302	331	+9.60
19	12	22.9	1292	174	159	−8.62
20	12	29.6	1221	260	278	+6.92
21	12	19.3	1258	131	107	−18.32

Average percent error = −9.76
Standard deviation = 19.10

[a]Large-diameter, two-phase pipeline.

The tables contain operating data either available in the open literature or obtained during the execution of flow test or consulting work. Several operators' pipelines are represented which are operating in the Gulf of Mexico off the Texas and Louisiana coasts, in the Australian Bass Strait, in the Gulf of Thailand, in the Prudhoe Bay area, and in the North Sea.

The data were culled to include only significant liquid loadings, i.e., at least 5 bbl/MMSCF.

The tables also present the calculated percent error determined as (calculated-measured)/measured × 100%.

As given in Table 3, the average error for pressure-drop prediction is −9.76% with a standard deviation of 19.10. For holdup prediction per Table 4, the average error is −1.60% with a standard deviation of 32.79.

It should be noted that these comparisons were made in most cases without precise data, particularly as regards pipeline profile and fluid composition.

TABLE 4 Holdup Field Data[a]

Data point	Nominal diameter (in.)	Average mixture velocity (ft/s)	Average liquid loading (bbl/MMSCF)	Measured holdup fraction	Calculated holdup fraction	Percent error
1	36	19.5	10	0.0165	0.0182	+10.30
2	36	22.8	8	0.0124	0.0129	+4.03
3	24	4.3	10	0.1570	0.1510	−3.82
4	24	4.9	10	0.1250	0.1270	+1.60
5	24	9.8	10	0.0460	0.0450	−2.17
6	20	16.4	65	0.0580	0.0760	+31.03
7	20	8.3	65	0.1200	0.1410	+17.50
8	20	13.2	14	0.0180	0.0290	+61.11
9	20	12.2	20	0.0370	0.0410	+10.81
10	20	12.9	25	0.0430	0.0420	−2.33
11	20	9.0	7	0.0780	0.0380	−51.28
12	20	8.4	10	0.0840	0.0480	−42.86
13	20	9.2	11	0.0850	0.0450	−47.06
14	20	13.6	19	0.0197	0.0235	+19.29
15	20	13.0	28	0.0405	0.0322	−20.49
16	20	13.6	35	0.0474	0.0334	−29.54
17	20	8.6	34	0.1135	0.0504	−55.59
18	20	9.9	21	0.0781	0.0284	−63.64
19	18	14.8	15	0.0200	0.0290	+45.00
20	18	9.3	15	0.0400	0.0550	+37.50
21	16	4.1	24	0.1690	0.1520	−10.06
22	16	3.1	176	0.4140	0.3900	−5.80
23	16	3.9	174	0.3250	0.3310	+1.85
24	16	4.8	170	0.3090	0.2810	−9.06
25	16	5.0	169	0.2840	0.2700	−4.93
26	12	30.7	1320	0.1500	0.2760	+84.00
27	12	18.1	1280	0.3510	0.2810	−19.94
28	12	20.5	1220	0.1900	0.2510	+32.11
29	12	26.9	1290	0.3400	0.2850	−16.18
30	12	22.9	1290	0.2900	0.2550	−12.07
31	12	29.6	1220	0.2700	0.2460	−8.89
32	12	19.3	1260	0.2600	0.2560	−1.54

Average percent error = −1.60
Standard deviation = 32.79

[a]Large-diameter, two-phase pipeline.

In many applications, however, this is exactly the situation the pipeline engineer faces.

Given only the gas and liquid flow rates and gravities, the inlet and sink temperatures, the major profile features, and a pressure constraint, the pipeline engineer must make a reasonable assessment of the pipeline's two-phase characteristics. Obviously, if more time and precise data are available, a more accurate assessment results.

Iterative Procedure

A calculation procedure using the equations presented is outlined here for a given pipeline section. Due to its iterative nature, the procedure would need to be incorporated into a personal computer or other computer program.

Pipeline heat-transfer aspects are not addressed here but should also be a part of the overall calculation method.

The recommended steps for determining holdup and pressure drop with Eq. (18) and Eq. (10) are the following:

1. Calculate V_{SG} and λ
2. Assume $V_{G1} = V_{SG}$
3. Calculate H_L using Eq. (18)
4. Calculate $V_{G2} = V_{SG}/(1 - H_L)$
5. If $|V_{G1} - V_{G2}| <$ tolerance, go to Step 7; otherwise go to Step 6
6. $V_{G1} = (V_{G1} + V_{G2})/2$, go to Step 3
7. Assume θ
8. Calculate $H_{L2} = (\theta/2 - \sin(\theta/2)\cos(\theta/2))/\pi$
9. If $|(H_L - H_{L2})/H_L| <$ tolerance, then go to Step 11; otherwise go to Step 10
10. Assume new θ, go to Step 8
11. Calculate S_G, S_L, T, V_L, V_G, D_{HG}, D_{HL}, Re_G, Re_L, f_{wG}, and f_{wL}
12. Calculate dP/dx using Eq. (10)

The pressure drop and holdup equations here are based upon rational models rather than upon a fit of laboratory or field data.

These equations (Eq. 10 for pressure drop and Eq. 18 for holdup) have compared favorably with a wide range of large-diameter operating data encompassing both liquid-dominated and vapor-dominated systems.

As two-phase boundaries are approached, i.e., dry gas or all liquid, the equations remain valid, yielding the familiar fundamental single-phase equations.

As the interface relationship becomes better defined, it is anticipated that a more fundamental holdup equation, such as Eq. (12), can be utilized.

Symbols

D	inside diameter
S_G	gas-wall perimeter
A	total X-section area
S_L	liquid-wall perimeter
A_G	gas area

T	width of free surface
A_L	liquid area
ϕ	angle of pipe inclination
A_{Ns}	no-slip liquid area
θ	liquid sector angle
A_{SL}	slipped liquid area
τ_G	gas-wall shear stress
H_L	liquid holdup fraction
τ_L	liquid-wall shear stress
λ	volumetric flow rate ratio no-slip holdup fraction $Q_L/(Q_L + Q_G)$
P_1	pressure at upstream end
P_2	pressure at downstream end
τ_i	interfacial shear stress
V_G	gas velocity
G	gravity constant
V_L	liquid velocity
Q_L	liquid volumetric flow rate
V_i	interface velocity
Q_G	gas volumetric flow rate
V_{SG}	superficial gas velocity
γ_G	gas density
V_{SL}	superficial liquid velocity
γ_L	liquid density
V_M	mixture velocity
ρ_G	gas mass density
f_{wG}	gas-wall friction factor
ρ_L	liquid mass density
f_{wL}	liquid-wall friction factor
D_{HG}	gas hydraulic diameter
f_i	interface friction factor
D_{HL}	liquid hydraulic diameter
r	liquid loading, bbl/MMSCF
Fr	no-slip Froude number
GOR	gas-oil ratio, scf/st-tk bbl
y_{NS}	no-slip flow depth
Q_{STD}	gas volumetric flow rate at standard conditions, SCFD

References

1. H. D. Beggs and J. P. Brill, "A Study of Two-Phase Flow in Inclined Pipes," *J. Pet. Technol.*, pp. 607–617 (May 1973).
2. J. M. Mandhane, G. A. Gregory, and K. Aziz, "Critical Evaluation of Friction Pressure Drop Prediction Methods for Gas–Liquid Flow in Horizontal Pipes," *J. Pet. Technol.*, *29*, 1348–1358 (1977).

3. J. M. Mandhane, G. A. Gregory, and K. Aziz, "Critical Evaluation of Holdup Prediction Methods for Gas–Liquid Flow in Horizontal Pipes," *J. Pet. Technol.*, pp. 1017–1025 (August 1975).
4. G. A. Hughmark, "Holdup in Gas–Liquid Flow," *Chem. Eng. Prog.*, *58*(4) (1949).
5. R. S. Cunliffe, "Condensate Flow in Wet Gas Lines Can Be Predicted," *Oil Gas J.*, pp. 100–108 (October 30, 1978).
6. O. Flanigan, "Effect of Uphill Flow on Pressure Drop in Design of Two-Phase Gathering Systems," *Oil Gas J.*, pp. 132–141 (March 10, 1958).
7. Y. Taitel and A. E. Dukler, "A Model for Predicting Flow Regime Transitions in Horizontal and Near Horizontal Gas–Liquid Flow," *AIChE J.*, *22*(1), 47–55 (1976).
8. R. W. Lockhart and R. C. Martinelli, "Proposed Correlation of Data for Isothermal Two-Phase, Two-Component Flow in Pipes," *Chem. Eng. Prog.*, *45*(1), 39–48 (January 1949).
9. B. A. Eaton, "The Prediction of Flow Patterns, Liquid Holdup, and Pressure Losses Occurring during Continuous Two-Phase Flow in Horizontal Pipelines," Ph.D. Thesis, University of Texas, 1966.
10. C. F. Colebrook, "Turbulent Flow in Pipes with Particular Reference to the Transition Region between the Smooth and Rough Pipe Laws," *J. Inst. Civ. Eng.*, *11*, 133–156 (1939).
11. R. V. A. Oliemans, *Modelling of Gas-Condensate Flow in Horizontal and Inclined Pipes*, Presented at the 1987 ASME Pipeline Engineering Symposium–ETCE, Dallas, PD-Vol. 6, pp. 73–81.
12. T. N. Smith and R. W. F. Tait, "Interfacial Shear Stress and Momentum Transfer in Horizontal Gas–Liquid Flow," *Chem. Eng. Sci.*, *21*, 63–75 (1966).
13. E. Kordyban, "Interfacial Shear in Two-Phase Wavy Flow in Closed Horizontal Conduits," *J. Fluids Eng.*, pp. 97–102 (June 1974).

JOHN A. BARNETTE

Waxy Crude Oils

Introduction

Transportation of waxy crude oil poses serious handling problems that historically have been very expensive. Heating, blending with low used cutting stock, or transporting by truck or by rail are very costly means of handling them. The crystallization of wax in crude oils causes severe difficulties in pipelining and storage [1].

This article discusses the problems encountered in pipelining waxy crude oils and the concept of flow improvement by wax crystal modifiers [2].

Waxy Crude Oils

The crude oils pumped in pipelines until the early 1960s generally showed normal characteristics in respect to pumping conditions such as viscosity and low pour point. However, the opening up of remote oil fields in North Africa and India in the 1960s [3] to exploit the low sulfur (but waxy) crudes in these locations, and the need to pump these oils through the much colder pipelines has led oil producers to study the pumpability of waxy crudes at temperatures below the pour point limit.

Characteristics of some of these crudes are given in Table 1.

In contrast, most of the Middle East crude oils have a pour point below 0°C and wax content less than 7% and therefore pose no problems in pipeline transportation even at low temperatures.

Flow Properties of Waxy Crude Oils

The viscosity of a crude oil is perhaps its most important physical property. For most crudes, at sufficiently high temperature, the viscosity at a given temperature is constant and the crude, although chemically very complex, is a simple Newtonian liquid. As the temperature is reduced, however, the flow properties of a crude oil can readily change from the simple Newtonian to very complex flow behavior due to the crystallization of waxes and the colloidal association of asphaltenes [4]. The waxes basically consist of n-alkanes (nC_{17}–nC_{43}) which crystallize to form an interlocking structure of plates, needle, or malformed crystals [5]. These crystals can entrap the oil into a gel-like structure that is capable of forming thick deposits in pipes and increasing pumping pressures to the point where flow ceases. Asphaltenes, on the other hand, are very large heterogeneous molecules with condensed aromatic nuclei [6] that can associate to form colloidal-sized particles that strongly influence the viscosity of the oil medium and affect the crystallization of wax [7, 8].

TABLE 1 Characteristics of Waxy Crude Oils

Crude Oil	API Gravity	Pour Point (°C)	Wax Content (wt.%)
Nigeria	38	12	9
C-65 Libya	36	24	18
Nahorkatia (India)	31.3	30	10.1
Moran (India)	34.9	30	13.5
Bahia (Brazil)	40	35	28
Minas (Sumatra)	36	35	32
Bombay High (India)	38	30	12.5

The flow properties of an oil containing crystallized wax are distinctly non-Newtonian. A yield stress (the minimum pressure required to restart flow, also termed gel strength [9]) can be detected which, under some circumstances, can be many times higher than the normal pumping pressure.

Upon yielding, the flow properties show time-dependency (the measured stress is not a constant at constant shear rate) indicating a degradation of structure with continued shear (termed, "thixotropy" or "line clearing behavior") [10], finally giving equilibrium or time-dependent flow properties (under certain circumstances) which still exhibit a yield stress and pseudoplastic (shear-thinning) behavior. For such a flow curve (the plot of shear stress vs shear rate), the viscosity is a function of the shear rate, so that the use of a single value for viscosity becomes meaningless.

The flow properties of waxy crude oils are also dependent upon the shear and temperature history of the oil.

Methods for Pipeline Transportation of Waxy Crude Oils

One of the following methods for pipelining waxy crude oils may be considered:

1. Select pumps to allow a parallel/series arrangement which could transport at slower rates and higher pressures when required. The piping could be manifold so that parallel arrangement would be accommodated by repositioning of valves to handle higher flow rates.
2. Use of separate low flow high head pumps for restarting.
3. Side traps at frequent intervals to allow short sections to be started separately.
4. Reverse pumping to create back and forth pumping sequence which prohibits static cool down.
5. Use of pour point depressants/flow improvers.
6. Adding hydrocarbon diluent such as a less waxy crude or light distillate.
7. Injection of water to form a layer between pipe wall and crude.
8. Mixing water with crude to form an emulsion.
9. Displacement with water or light hydrocarbon liquid in case of shutdown of pipeline.
10. Separation at higher than normal pressure to allow as much gas and light hydrocarbons as possible to remain in the crude.
11. Conditioning the crude before pipelining to change the wax crystal structure and reduce pour point and viscosity.
12. Further subdivision of pipeline into smaller segments or reducing batch length of waxy crude to increase maximum shear stress available.
13. Combination of above methods.

Oil producers have long been aware of the difficulties of pipelining waxy crude oil and fuel oils. Traditionally the answer has been to avoid the problem by heating the crude and/or the pipelines, thus holding the wax in solution [11], or by frequently emptying and clearing the lines [12].

To exploit the low sulfur waxy crude oils, the pumpability characteristics of waxy crudes at temperatures below the pour point limit have been studied. It has been found possible to improve the flow of waxy crude oils by a number of methods. Pipelining the crude as an oil-in-water (O/W) emulsion reduces the flow properties to nearly the viscosity of the continuous water phase [13]. An O/W emulsion pipeline handling 40,000 BOPD (265 m^3/h) of 70 vol.% crude oil has been operating in Kalimantan (Borneo) in Indonesia since 1962 [14]. Blending with a less waxy crude oil or distillate improves the flow properties by altering the wax solubility relationships. Both of these methods have the disadvantage of reducing the crude oil carrying capacity of the pipeline [11]. Note that separation at the well head to include more condensate in the crude oil (if available) has the same effect as dilution. There is one interesting case of a shear and temperature treatment being used to favorably alter the flow properties of a waxy crude oil in Assam (India) [15]. More recently, pour point depressants/flow improvers have been developed that, in small concentration, affect the crystal growth [16, 17] and as a result improve the flow properties.

Of the various methods developed, the use of pour point depressants/flow improvers is found to be more attractive. The main attraction of this method is its relative cheapness and variability of dosage with respect to the temperature and desired viscosity requirements.

Use of Pour Point Depressants/Flow Improvers

The injection of pour point depressant/flow improver additives [18] appears to hold the greatest promise of achieving the desired overall objectives of:

1. Operational safety, i.e., protection of the line against blockage by the setting of the oil into a strong gel.
2. Operating economy, i.e., maintenance of reasonable flowing viscosity with resulting economical level of power consumption.

Flow improvers should have the capacity to:

1. Reduce the pour point viscosity, and yield stress under dynamic conditions.
2. Restart the pumping after a shutdown with the available shear stress and aid in fast clearance.

Chemical pour point depressants/flow improvers are ashless polymeric additives [19] which when added into the crude oil at the 300–600 ppm level reduce the pour point and viscosity of the crude oil. Polymeric materials widely used as pour point depressants/flow improvers are (a) alkyl acrylate polymers and copolymers, (b) olefin alkyl maleate copolymers, (c) vinyl ester polymers and copolymers, and (d) alkylated polystyrene. Normally the average molecular weight of the commercially available pour point depressants for crude oils is between 2000 to 20,000.

Various flow improvers [19] developed at RRL Jorhat (India) are SWAT-104, SWAT-106, FIRI, and FIRJ-B. These are polymers/copolymers, easily soluble in crude oil around 40–45°C, and noncorrosive. They are used in pipelining crudes of Bombay High and Assam (India).

Mechanism of Flow Improvement

When a waxy crude oil is cooled below its cloud point, the wax crystals form and begin to agglomerate, and with further temperature reduction crystal agglomeration reaches a point at which a gel structure is formed below the pour point due to interlocking of the growing crystals and is dependent on constituents like resins, asphaltenes, asphalts, paraffin and microcrystalline waxes, etc.; their molecular weight, structure, and quantity; and also on the rate of cooling and degree of agitation during cooling. When the additives or flow improvers are added, they alter the wax crystal size and shape in some manner and prevent the tendency to interlock [18]. The flow improvers or pour point depressants act by retarding the growth of the wax crystals in the XY crystallographic plane, thereby producing smaller crystals of higher volume/surface ratio. It appears that the flow improvers cocrystallize with the growing wax crystal, leading to the formation of a fault in the otherwise compact regular wax crystal and resulting in diminished gel strength, so that by coating onto a growing wax crystal, the flow improvers reduce the tendency of wax crystals to interlock.

This material appeared in *Chemical Age of India*, *38*(12), 673–675 (1987), and is reprinted with the permission of the editor.

References

1. P. E. Ford, J. W. Ells, and R. J. Russel, *Oil Gas J.*, *63*(9), 1983 (1985).
2. J. Singh, 8th Short Term Course for IOC Chemical Engineers, I.I.P. Dehradun, August-November 1980.
3. P. E. Ford, J. W. Ells, and R. J. Russel, *Oil Gas J.*, 63(16), 88(1965); *63*(17), 107 (1965); *63*(19), 183 (1965); *63*(20), 134 (1965).
4. K. C. Khilar, B. M. A. Rao, and S. P. Mahajan, *37th Annual Session of I.I.Ch.E.*, New Delhi, December 17–20, 1984, Paper 141.
5. S. W. Ferris and H. C. Cowles, *Ind. Eng. Chem.*, *37*(11), 1054 (1945).
6. J. W. Bunger and N. C. Li (eds.), *Chemistry of Asphaltenes* (American Chemical Society, Advances in Chemistry Series), 1985.

7. J. A. Fernandez-Lozano and Y. M. Rodriquez, *Ind. Eng. Chem., Process Des. Dev.*, *23*, 115 (1984).
8. M. Brod, B. C. Deane, and F. Rossi, *J. Inst. Pet.*, *57*(554), 110 (1971).
9. T. R. Sifferman, *J. Pet. Technol.*, p. 1042 (August 1979).
10. T. C. Davenport and R. S. H. Somper, *J. Inst. Pet.*, *57*(554), 86 (1971).
11. A. Uhde and G. Kopp, *J. Inst. Pet.*, *57*(554), 63 (1971).
12. M. N. Shaw, *APEA J.*, Part 1, 153 (1984).
13. S. S. Marsden and R. Raghavan, *J. Inst. Pet.*, *59*(570), 273 (1973).
14. M. J. Lamb and W. C. Simpson, *Proceedings of the 6th World Petroleum Conference*, Frankfurt, 1963, Section VII, Paper 13, p. 23.
15. R. J. Russel and E. D. Chapman, *J. Inst. Pet.*, *57*(554), 117 (1971).
16. G. A. Holder and J. Winkler, *J. Inst. Pet.*, *51*(499), 228 (1965).
17. R. C. Prince, *J. Inst. Pet.*, *57*(554), 106 (1971).
18. "Flow Improvers for Pipeline Transport of Crude Oil," *Urja*, *19*(3), 199 (1986).
19. "CSIR Shifts Rheology Work from Assam to Deccan," *Urja*, *19*(3), 231 (1986).

RAM PRASAD

Permafrost Considerations

A refrigerated gas pipeline which is insulated and heat traced at frost-susceptible locations can overcome troublesome thermal problems that arise from operating pipelines in permafrost terrain.

Equations have been developed for fast analysis and prediction of heat flow and its effect on soil surrounding buried pipes operated in permafrost terrain. They have been found to agree, within the accuracy of input parameters, with experiments and more-detailed computer models.

Cooling gas pipelines increases flow considerably and avoids thaw of permanently frozen ground in the Arctic [1]. But pipelines in permafrost terrain need to pass through patches of unfrozen ground, particularly along the southern fringes of permafrost where interfaces between frozen and unfrozen ground can occur many times per mile.

Frost heave, which would threaten the safety of the line, could occur wherever a refrigerated pipeline passed through interfaces between frozen and unfrozen, frost-susceptible ground.

It has been proposed [1] that refrigerated pipelines can be insulated and heat traced in unfrozen areas of the permafrost zone to prevent frost heave. This article discusses frost heave and thaw settlement and shows why heat-tracing is needed in addition to insulation.

It has been estimated that insulation and heat tracing would increase overall costs of a 48-in. Arctic pipeline by about 1%. This cost is more than counterbalanced by the increase in throughput attained by refrigerating the gas below 32°F (0°C).

The equations presented are useful for fast and accurate hand calculation of thaw and frost-bulb growth, frost heave rates, insulation thicknesses, and heat-tracing power for buried pipelines.

Thaw and Frost Bulbs

In cases where thaw and frost bulb development cause settlement or heave, soil contains sufficient moisture that its specific heat is small in comparison with its latent heat. In addition, soil is at or near 32°F (0°C), and the thaw or frost front does not lose or gain heat from the surrounding ground.

The rate of advance of the thaw or frost bulb below the pipes can therefore be obtained as follows [2]:

$$t = \frac{Lr^2(\log_e (2r/D)^2 - 1 + (D/2r)^2)}{4K(T_f - T_p)} \tag{1}$$

See the nomenclature listing for the meaning of terms.

Many refinements of this pseudosteady-state procedure are possible to improve its precision. However, ground conductivity and soil moisture have a significant impact on thaw and frost-bulb growth. They are seldom known with sufficient accuracy at every location along the right-of-way to make more complicated calculations worthwhile in practice.

Figure 1 compares Eq. (1) with reported results [3] from a computer program which uses variational algebra to obtain solutions to general geothermal-conduction problems involving change of phase.

The program can accurately handle variations of temperature with time, nonhomogeneous soils, irregular geometries, and a variety of boundary conditions.

In the analysis, the computer program used a sinusoidal air temperature, a ground-surface, heat-transfer coefficient that varied with surface temperature, different values for frozen and thawed soil properties, a realistic initial soil-temperature distribution, and an unfrozen moisture content allowance at temperatures below 32°F (0°C).

The silt used in the analysis had an average temperature of 30.6°F (-0.8°C) at depth. Latent heat was 3456 Btu/ft^3 (130 MJ/m^3), and thawed conductivity was 0.80 Btu/ft·°F·h (1.38 W/m·K).

The pipe was maintained at a temperature of 150°F (66°C) and had a diameter of 4 ft (1.22 m). Equation (1) predicted about 4% more thaw after 5 years and about 7% more thaw after 10 years.

Since ground conductivity and latent heat are not normally known with this precision along the right-of-way, it is reasonable to conclude that Eq. (1) is, in practice, as accurate as the more-detailed computer program.

Permafrost frequently contains quantities of excess ice which cause settlement when it thaws. Resulting loss of support, particularly for hot pipes in thermally induced compression, can cause buckling and failure [4].

As the pipe settles or buckles downwards, thaw bulb growth can be much faster than predicted. In addition, if the thawing soil liquefies and flows along the trench, convective heat transfer can increase long-term thaw rates by several orders of magnitude [5].

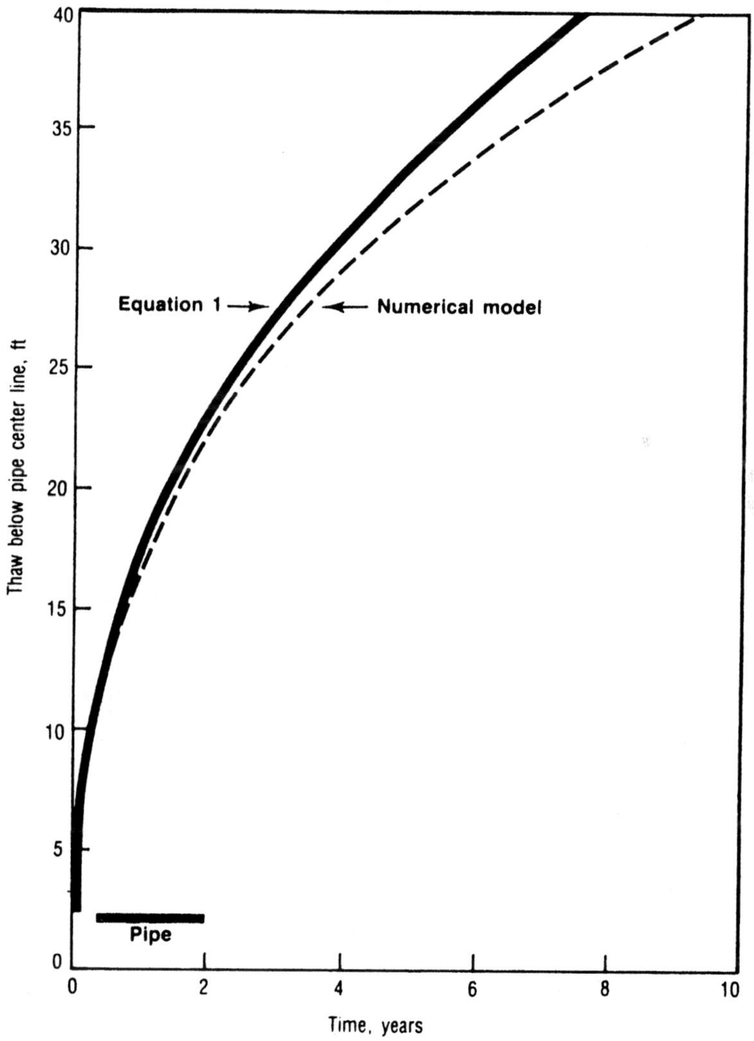

FIG. 1. Comparison of thaw-bulb predictions.

Long-term changes in weather and effect of construction disturbances can also increase thaw rate [6].

Detailed studies using numerical models have shown that frost bulbs around refrigerated pipes in unfrozen soil can also grow faster than is predicted by Eq. (1). In summer, unfrozen soil at the ground surface has a low conductivity which helps to insulate the subsurface from the effects of summer heat.

In winter, when seasonal frost makes contact with the frost bulb, it acts something like a cooling fin. Because frozen ground is a better conductor than unfrozen ground, heat is conducted more quickly than normal from the ground at depth, through the frost bulb to the frozen ground surface.

Equation (1), therefore, does not provide an upper limit for thaw and frost-bulb growth rate. Calculations, because they do not normally include all the relevant parameters, provide order of magnitude estimates only.

Careful judgment, which takes account of long-term changes in weather, construction disturbance, settlement, heave, convection, variations along the right-of-way, etc., is advisable for designs which rely for their success on accurate predictions of thaw and frost-bulb growth rates.

Frost Heave

When a refrigerated pipeline passes through unfrozen areas, it freezes the soil and can interrupt cross drainage which can cause troublesome springs and icings above the pipe. In addition, freezing causes the soil to expand.

The amount of expansion can be aggravated by migration of water to the frost front and the formation of ice lenses around the pipe.

Figure 2 shows a buried refrigerated pipe passing from permanently frozen or frost-stable ground, through a frost-susceptible unfrozen area, and back into permanently frozen or frost-stable ground. On one side of each interface between frost stable and frost-susceptible ground, the pipe is held rigidly.

On the other side of each interface, frost-heave forces lift the pipe, resulting in severe bending and shearing forces at the interfaces. These interfaces can occur many times per mile in permafrost terrain.

Recent experiments with large-diameter refrigerated pipes buried in unfrozen ground [7] show that stresses on the frost front can be as high as 2500 lb/ft^2 (120 kPa) without arresting lensing heave. Stresses as high as these at the frost front, if they are concentrated on the pipe at interfaces between heaving and nonheaving soils, can clearly threaten pipe integrity.

Order of magnitude rate of heave can be estimated by multiplying the fraction of excess ice (I) formed in the frost bulb by the frost-bulb growth beneath the pipe:

$$Y = I(r - D/2) \tag{2}$$

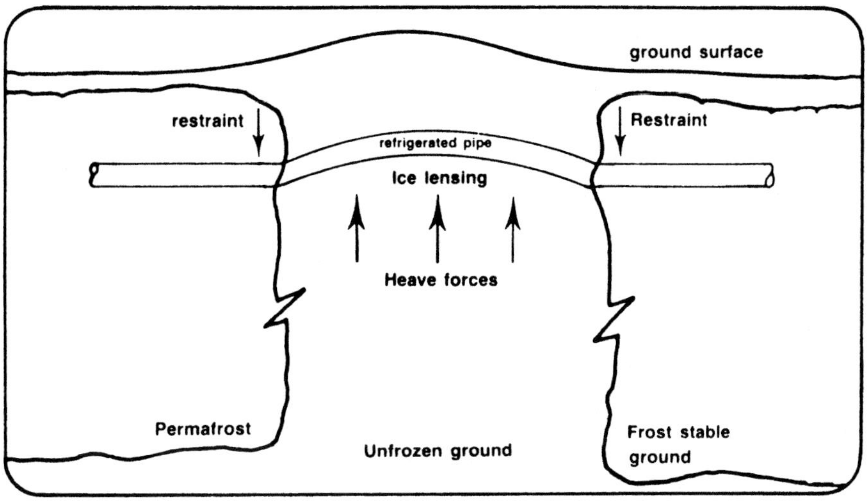

FIG. 2. Differential frost heave schematic.

When predicting frost-bulb depth, the latent heat of ground moisture needs to be adjusted for the water that migrates to the frost front and forms ice. Figure 3 compares predictions of Eqs. (1) and (2) with experimental heave measurements [7].

Calculations were made for an excess ice fraction of 0.2 and a corresponding latent heat of 3,780 Btu/ft^3 (140 MJ/m^3). A value of 1.2 Btu/ft•°F•h (2 W/m•K) was used for frost-bulb conductivity. Pipe diameter in the experiment was 4 ft (1.22 m), and temperature was maintained at 15°F (-9°C).

Not enough is known about ice-lensing mechanisms to accurately predict the excess ice fraction that will form in the frost bulb. Uncertainties are also encountered when obtaining soil properties at all points along the right-of-way.

Long-term weather changes over the life of the project and different soil properties at a different site could result in measured heave rates exceeding those predicted here. Conservative design assumptions, e.g., $I = 1$, are, therefore, advisable for designs which rely on frost-heave predictions.

Insulation

Insulation can prevent frost penetration of unfrozen ground below refrigerated pipes if the ground temperatures are warm enough. A good approximation for the thickness of insulation to keep its outer surface at an acceptable temperature (T_i) that is normally within the accuracy of input parameters for practical cases is:

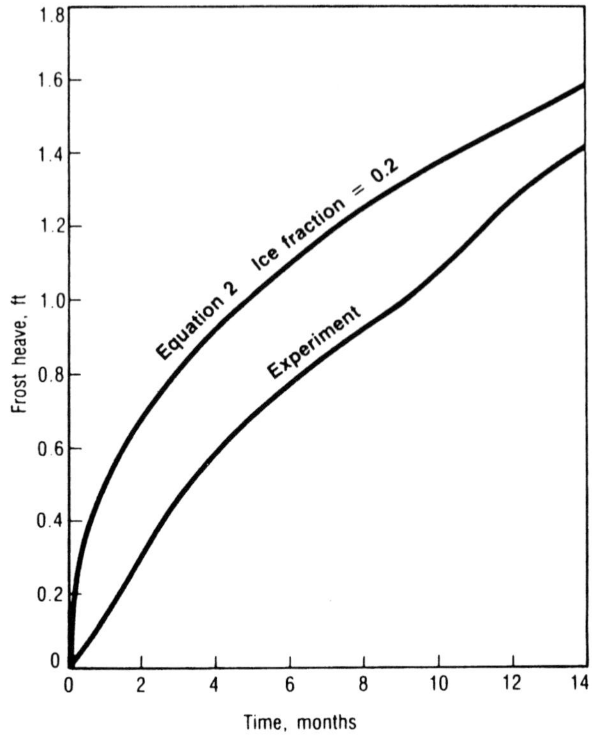

FIG. 3. Experimental and theoretical frost heave.

$$w = \frac{DK_i(T_p - T_i)}{2K(T_i - T_g)} \cosh^{-1}(2h/D) \tag{3}$$

Construction disturbances, annual variations in ground temperature and variations in ground temperature and soil properties, changes in weather and pipe operating temperature over the life of the project, and a factor of safety need to be considered when selecting a safe design temperature for the outer surface of insulation (T_i).

Figure 4 shows how much insulation is required for a 48-in. (1.22 m) diameter pipe operated at 15°F (-9°C) in unfrozen soil for a range of soil temperatures. A safe design temperature for the surface of the insulation of 41°F (5°C) was selected for this example.

It is clear that insulation is not the answer to the problem when average ground temperatures are lower than 41°F (5°C) because infinite thicknesses of insulation would be required. In discontinuous permafrost terrain where there are interfaces between frozen and unfrozen ground, ground temperature is close to 32°F (0°C). Insulation, on its own, will be ineffective in preventing heave or thaw.

Insulation can, however, slow the rate of heave or thaw. Figure 5 shows results of calculations, for a refrigerated 48-in. pipe with 6 in. of insulation, using theory similar to that used for Eqs. (1) and (2).

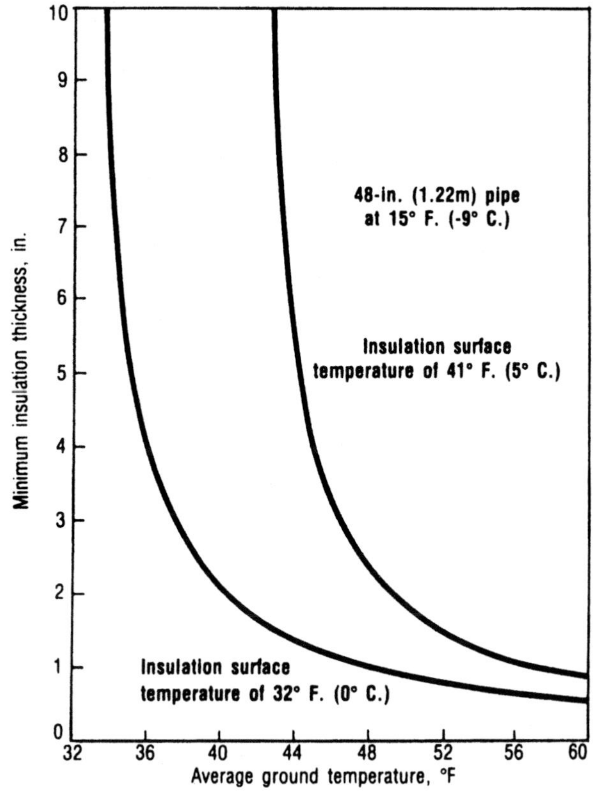

FIG. 4. Insulation thickness required to prevent frost penetration of surrounding soil.

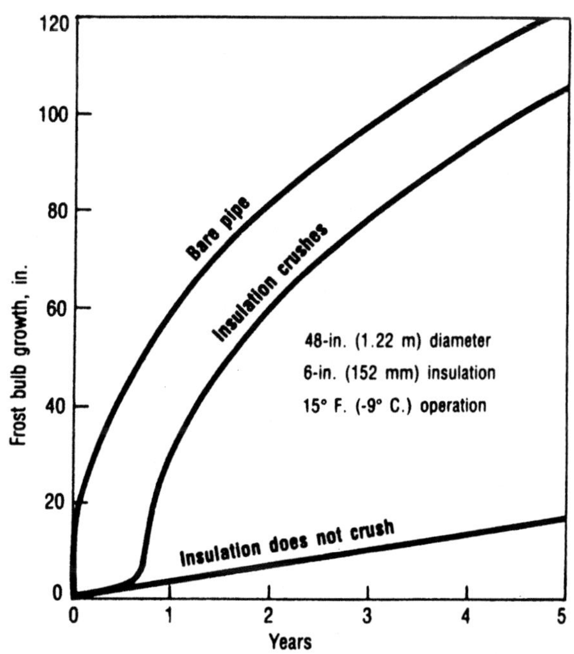

FIG. 5. Frost bulb growth in cold unfrozen ground.

If the insulation does not crush, heave could be about 1.5 ft (0.5 m) after 5 years of operation at 15°F (-10°C). Likelihood is good that insulation would be crushed at interfaces. Crushing would lead to high local heave rates which could overstress the pipe within a relatively short time.

A larger factor of safety is, therefore, required when determining the safe design temperature for the outer surface of insulation for cold pipes in unfrozen ground than for hot pipes in permafrost. Phenomena which reduce insulation efficiency, such as crushing at field bends, water penetration, and migration to cold surfaces, are also more important in insulated refrigerated-pipe designs than they are in insulated hot-pipe designs.

Heat Tracing and Insulation

Insulation slows heave rate but cannot eliminate it if the ground temperature is near 32°F (0°C). Heating the ground, in addition to insulating the pipe, can effectively prevent frost penetration beneath refrigerated pipe and, therefore, can eliminate frost heave and its attendant problems.

Figure 6 shows temperature distribution around an insulated refrigerated pipe with heat tracing. Two heat-tracing cables are used, one on each side of the pipe in the bottom of the trench.

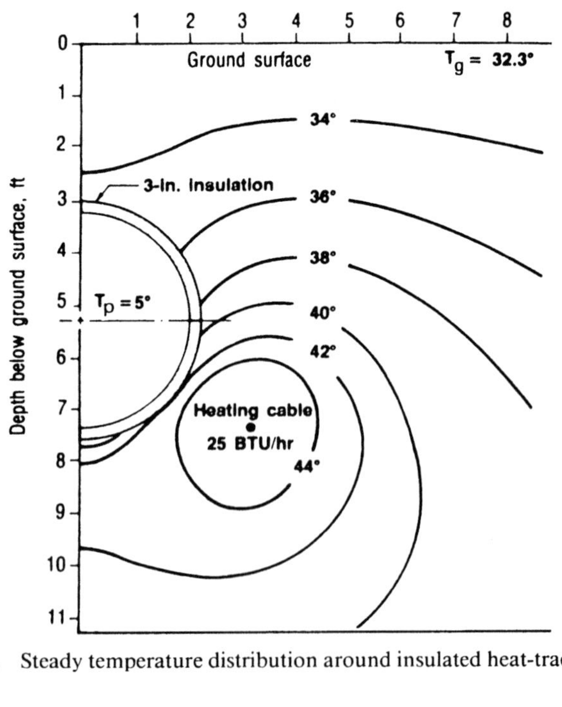

FIG. 6. Steady temperature distribution around insulated heat-traced pipe.

Power output of each heat-tracing cable which will increase the temperature of the outside of the insulation to a safe design level to accommodate variations in pipe temperatures, ground temperatures, soil conductivities, etc. over the life of the project can be found from

$$Q_t = 2\pi K \frac{(T_i - T_g)/C_1 + (T_g - T_p)/C_2}{(U_1 - U_2)/C_1 + (U_2 - U_3)/C_2} \quad (4)$$

where

$$C_1 = \log_e (4h/D_i) \quad (4a)$$

$$C_2 = \log_e (4h/D_e) \quad (4b)$$

$$U_1 = \log_e (b + h)^2 \quad (4c)$$

$$U_2 = \log_e (a^4 + D_i^4/4)^{1/2} \quad (4d)$$

$$U_3 = \log_e (a^2 + D_i^2/4) \quad (4e)$$

$$D_e = 2h/\cosh \left(\frac{K}{K_i} \log_e \left(\frac{2w}{D} + 1\right)\right) \quad (4f)$$

These equations are useful for investigating the sensitivity of heat-tracing power requirements to changes in design parameters and for investigating the economic trade-off between thickness of insulation, heat tracing power, depth of pipe burial, heat cable separation, etc.

Equation (4) was developed for the case where pipe cover is greater than pipe diameter, there is at least 1 in. of insulation on the pipe, cable separation is less than trench depth, and the heat cables are equidistant from the pipe in the bottom of the trench. Other equations for different cases can be developed from first principles [8].

For the case shown in Fig. 6, with two heating cables emitting 25 Btu/h·ft each, it would take 1700 hp to heat 25% of the right-of-way for 30 miles on each side of a compressor station (0.25 × 60 × 5280 × 25 × 2/0.92).

Proposed Arctic gas pipelines have 30,000-hp compressors with 15,000-hp refrigeration units at each station. The electrical generator for the heat-tracing system at this one station would, therefore, consume less than 5% of that station's fuel on an annual basis.

Cost estimates indicate that total direct costs of heat tracing and insulation for both labor and materials would increase overall project costs by about 1%.

When gas pipelines are operated at low temperatures, the increase in gas density increases the throughput of the pipeline or reduces horsepower requirements. The increased throughput more than counterbalances the costs of insulating and heat tracing those parts of the pipeline which pass through unfrozen soil at river crossings and along the southern fringes of the permafrost zone.

Nomenclature

a	half separation of heat cables (ft)
b	depth of trench and heat cable (ft)
c	cover over pipe (ft)
D	pipe outside diameter (ft)
D_e	thermally equivalent diameter (ft)
D_i	outside diameter of insulation (ft)
h	depth to pipe center line (ft)
I	excess ice fraction.
K	soil conductivity (Btu/ft·°F·h)
K_i	insulation conductivity (Btu/ft·°F·h)
L	latent heat of soil (Btu/ft^3)
Q_t	heat tracing power (Btu/h·ft)
r	phase boundary below pipe center (ft)
t	time (h)
T_f	temperature of phase boundary (°F)
T_g	ground temperature (°F)
T_i	insulation surface temperature (°F)
T_p	pipe temperature (°F)
w	thickness of insulation (ft)
Y	heave or settlement of pipe (ft)

This material appeared in *Oil & Gas Journal*, pp. 80–84, December 15, 1980, and is reprinted with the permission of the editor.

References

1. G. G. King, "Cooling Arctic Pipelines Can Increase Flow, Avoid Thaw," *Oil Gas J.*, August 15, 1977.
2. H. S. Carslaw and J. C. Jaeger, *Conduction of Heat in Solids*, 2nd ed., Oxford, 1959.
3. J. A. Wheeler, *Simulation of Heat Transfer from a Warm Pipeline Buried in Permafrost*, Presented at 74th National Meeting, AIChE, New Orleans, Louisiana, March 11–15, 1973.
4. *Oil & Gas Journal* "Thawing May Cause Relaying of Part of Alyeska Line," pp. 24–26 (July 23, 1979).
5. A. H. Lachenbruch, *Some Estimates of the Thermal Effects of a Heated Pipeline in Permafrost*, Geological Survey Circular 632, U.S. Department of the Interior, 1970.
6. L. A. Linell, "Long Term Effects of Vegetative Cover on Permafrost Stability in an Area of Discontinuous Permafrost," in *Second International Conference on Permafrost*, North American Contribution, National Academy of Sciences, July 1973, pp. 688–693.
7. W. A. Slusarchuk et al., "Field Test Results of a Chilled Pipeline Buried in Unfrozen Ground," in *Proceedings of Third International Conference on Permafrost*, Vol. 1, National Research Council, Canada, July 1978, pp. 878–883.
8. G. G. King, *Geothermal Design of Buried Pipelines*, Presented at ASME Energy Technology Conference, New Orleans, Louisiana, February 1980.

GRAEME G. KING

Flow Basics

Pipelines are used to convey a wide range of single phase and multiphase materials, both within chemical and process plant and over thousands of kilometers across international frontiers. If the material is pumpable, then transport by pipe is often the most economic and environmentally acceptable answer. This two-part article reviews the various classes of material transported, their industrial significance, and the fundamental fluid mechanics involved in different applications. Part 1 considers the steady-state flow of homogeneous incompressible fluids, turbulent drag reduction, and start-up flow of thixotropic fluids. Part 2 covers the flow of multiphase materials.

Part 1: Flow of Homogeneous Fluids

Flow in pipes is a varied and fascinating subject; while the simpler flow situations are well understood, there are many facets still to be researched. Pipe flow is found in industry, the environment, and medicine. It is used for conveying materials not just within a factory but over long distances such as transcontinental pipelining of oil, coal, and ores. It is often the most economic and environmentally acceptable method of transport compared with rail and road haulage. It forms part of boilers and tube-and-shell heat exchangers and is used in mass transfer operations such as tubular reactors. In the environment, water, gas, as well as sewage sludges are pipelined. In medicine, there is the flow of blood and other body fluids. The study of flow in pipes is therefore not only intellectually challenging, but also of considerable practical importance.

The simplest pipe is straight and has a circular cross section, but a pipeline can also be curved and of noncircular section. Whereas flow in a straight circular pipe can be telescopic for a viscous fluid, in a curved pipe secondary flow is developed under the same conditions. If the fluid is rheologically complex by possessing viscoelasticity, telescopic flow is not tenable in noncircular sections even in a straight pipe. Most pipes have rigid walls, but rubber hoses are flexible and in the human and animal bodies the blood vessels, for example, have flexible walls. In addition, under the pulsatile pumping of the heart, the pipe cross section varies in dimension along its length and peristaltic flow is thus obtained in the alimentary canal. Peristaltic motion is employed in certain metering pumps, and recent research into pulsatile flow suggests that there are advantages that could be exploited in industry.

The material flowing in a pipe can be gas, liquid, or solid. Gas is compressible and its flow characteristics depend on the Mach number. Liquids, whether pure chemicals or solutions, are incompressible under steady or slowly varying flow, but in surge situations (obtained during start-up of a fire-fighting sprinkler system or the water hammer in a long-distance water pipeline) the compressibility of the liquid has to be taken into account, and also the elasticity of the pipe wall. Solids can be conveyed in pipes if pulverized and supported by air (pneumatic transport, lean or dense phase) or liquid (hydraulic transport). If the suspended solids are of the form of long flexible fibers (such as paperstock and fermentation broth or other biomasses), the flow obeys different laws than for suspensions of hard particles. Alternatively, the solid can be in the form of a capsule of nearly the same diameter as the pipe bore; for example, an ingot of metal or solid sulfur can be supported by a liquid and propelled by pressurizing the liquid. Small hollow capsules are propelled by air in department stores to convey cash; now larger container capsules, supplied with wheels, are being developed to transport goods as well as people. Other mixtures include gas (or vapor)–liquid two-phase flow in boiler tube bundles or in North Sea gas pipelines. Liquid–liquid mixtures can exist in the form of emulsions or be stratified such as the mixtures obtained when water is injected into a crude oil pipeline to reduce pressure drop and in the coextrusion of molten polymers.

Mixtures are in general composed of two or more phases which interact but move separately from each other. To understand the overall flow of a mixture, the behavior of the individual components must be studied. The importance of this approach will become clear in the main part of this article, and it is true even for neutrally buoyant suspensions of solids in liquids. However, if the dispersed phase is very finely divided in a liquid, the mixture may be considered to be homogeneous. Such mixtures are colloidal dispersions of solids or emulsions, but ones which show complex rheological properties. The viscosity may not be constant with respect to shear rate but decrease with it (shear thinning), or a yield stress may exist so that the mixture does not flow until the stress exceeds the yield value (viscoplasticity).

Unsteady pipe flow is found on start-up or shut-down, over the entrance length or exit length, or when the pipe section is changed. Other unsteady flows are obtained when the pumping pressure or flow rate is varied periodically (pulsatile flow) or when the pipe section changes periodically along its length. The effect is governed by fluid inertia in the first instance but it is also modified if the fluid shows a time-dependent property such as thixotropy or viscoelasticity.

It is clearly not possible to describe all those different topics of flow in pipes in this article, however briefly. We have chosen to concentrate on steady-state flows in straight, rigid pipes of circular cross section. Part 1 considers homogeneous, incompressible liquids and mixtures with particular emphasis on the effect of non-Newtonian viscosity properties, drag reduction in high molecular weight polymer solutions and fibrous suspensions, and one example of a transient flow, namely the start-up of a pipeline containing a thixotropic fluid. Part 2 covers multiphase flow of solid–liquid (settling suspensions or hydraulic transport), gas–liquid (so-called two-phase flow), liquid–liquid, and gas–solid (pneumatic transport) mixtures.

There will not be sufficient space in this article to discuss detailed theoretical considerations or to review the experimental results of the topics chosen. Instead, the salient features will be stated with some explanation in words of the underlying physics. However, emphasis will be given to what design data are needed for the prediction and design of large-scale pipeline installations.

Steady-State Flow of Homogeneous Incompressible Fluids

Liquids composed of low molecular weight species, whether pure chemicals or mixtures, are widely found in industry. They are known as Newtonian liquids; their flow property is simply governed by the viscosity, and the pipe flow characteristics are well understood. To design a pipeline the engineer simply determines the viscosity from standard references or by using estimation equations. The pipeline calculations can then be carried out according to well-known design equations or graphs to be found in engineering handbooks. Specialized slide rules are available, and programs for hand-held calculators or microcomputers are becoming common.

A wide range of other industrial fluids are more complicated in composition and rheological properties. Some are composed of high molecular weight species, such as solutions and melts of macromolecules, surfactants, thickeners, and thermoplastics. Others contain finely divided particles, such as colloidal dispersions, slurries, emulsions, and foamed material. If the dispersed species are large in size and/or of high density (or low density), the mixture will behave as multiphased species (discussed in Part 2 of this article). But if they are finely divided and remain suspended by Brownian motion, the dispersions will behave as homogeneous fluids. These fluids do not possess a constant viscosity with respect to shear and are known as non-Newtonian fluids. Because their viscosity is very sensitive to the exact chemical and physical compositions (such as molecular weight or particle size and shape distributions), there are no standard references for fluid properties—the viscosity of each fluid has to be separately determined experimentally. Design equations are then available in the literature for pipeline design. Some details of Newtonian and non-Newtonian pipeline flow characteristics are now given.

Newtonian Liquids

The flow of Newtonian liquids through pipes is governed by the Hagen-Poiseuille equation

$$Q = \frac{\pi \Delta P D^4}{128 \eta L}$$

where Q is the volume flow-rate, ΔP is the pressure drop over a pipe of length L and diameter D, and η is the viscosity of the liquid. This flow, tenable at low velocities, is known as telescopic or laminar flow because cylindrical elements of liquid at radius r move axially as rigid bodies. The velocity distribution is parabolic as illustrated in Fig. 1, in which V is the average velocity.

At high velocities the laminar flow breaks down to become turbulent. Theoretical investigations of the stability show that Poiseuille flow is infinitely stable to a wide variety of disturbances. This is supported by experiments in which the pipe is carefully insulated from extraneous vibrations from the mountings and the pump. Laminar flow can be maintained for Reynolds numbers (Re $= DV\rho/\eta$, where ρ is the liquid density) as large as 15,000. However, under the usual conditions pertaining in a laboratory or industrial environment, the breakdown of laminar flow takes place at a Reynolds number of about 2100.

The liquid velocity is then no longer axial and takes on random characteristics. The study of the nature and properties of the random velocity components is still a subject of active research. Whereas the laminar velocity distribution can be derived from simple theory, the average turbulent velocity $\bar{u}(r)$ has to be determined by experiment. It is now established that the velocity distribution is determined by a laminar boundary layer adjacent to the pipe wall, and a turbulent core, with a buffer layer in between.

Empirical equations have been established for the velocity profile, one of which is known as the Prandtl logarithmic velocity distribution. Figure 1 shows the turbulent velocity profile in comparison with the laminar profile. The relation between pressure drop and flow rate is usually expressed in the form of a dimensionless equation. Two popular versions are the von Karman equation and the Blasius equation. They relate the friction factor

$$C_{\text{f}} = \frac{\frac{1}{4}D\Delta P/L}{\frac{1}{2}\rho V^2}$$

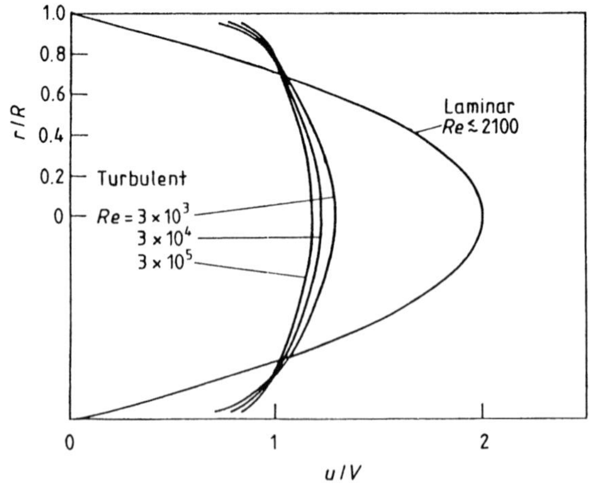

FIG. 1 Laminar and turbulent velocity profiles for Newtonian liquid $V = (\eta/\rho D)\text{Re}$.

to the Reynolds number

$$\mathrm{Re} = \rho VD/\eta$$

A plot of the turbulent flow equation is shown in Fig. 2(a). This also includes the dimensionless form of the Hagen-Poiseuille equation

$$C_{\mathrm{f}} = 16/\mathrm{Re}$$

Whereas laminar flow is not affected by wall roughness, turbulent flow is. The effect of wall roughness is usually depicted as a Moody chart which can be found in standard engineering handbooks.

The breakdown of laminar flow takes place at a Reynolds number of about 2100. There is a distinct transition regime before the flow becomes fully turbulent. Careful experiments show that the flow in transition is com-

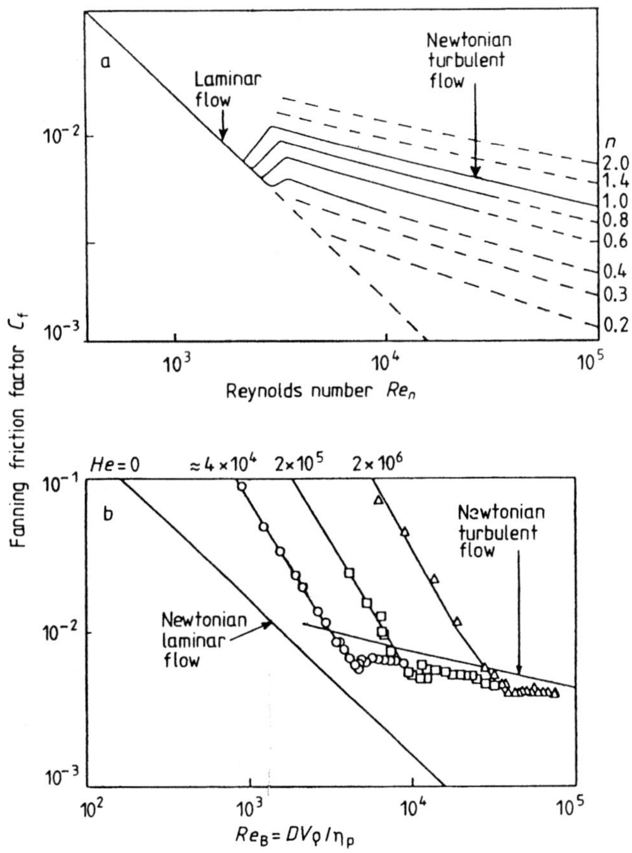

FIG. 2 Friction factor plotted against Reynolds number for (a) power law fluid: (——) experimental regions, (— —) extrapolated regions [from D. W. Dodge and A. B. Metzner, *Am. Inst. Chem. Eng. J.*, 5, 189 (1959)]; (b) Bingham plastic fluid [from D. G. Thomas, *Am. Inst. Chem. Eng. J.*, 6, 631 (1960)].

posed of alternate laminar and turbulent plugs of liquid which follow each other down the pipe length. The pressure measured at a given point along the pipe can be observed to fluctuate regularly as the different plugs pass that station.

Non-Newtonian Fluids

Two classes of variable viscosity behavior can be recognized in non-Newtonian fluids (Fig. 3). The first is shear thinning: the viscosity decreases from η_0 to η_∞ as the shear rate increases. The other, viscoplastic behavior, is characterized by a yield stress τ_y: such a fluid does not flow unless the shear stress acting on it exceeds the yield value. It is usual to model the shear thinning behavior by means of the Ostwald–de Waele power law flow curve, relating shear stress τ to shear rate $\dot{\gamma}$

$$\tau = k\dot{\gamma}^n, \qquad n < 1$$

where k and n are the power law coefficient and index. Viscoplasticity is modeled by the Bingham plastic model

$$\tau = \tau_B + \eta_p\dot{\gamma}$$

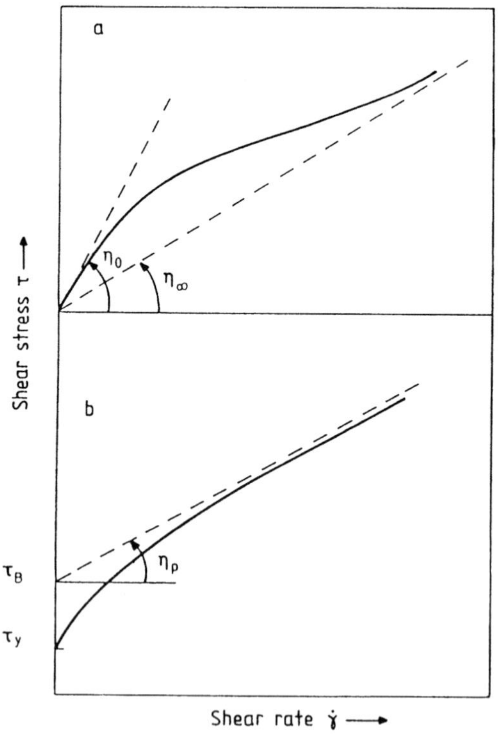

FIG. 3 Two types of variable viscosity behavior in non-Newtonian fluids: (a) shear thinning, and (b) viscoplastic.

where τ_B is the Bingham yield stress and η_p the plastic viscosity. A more general model, the Herschel–Bulkley flow curve, incorporates both features

$$\tau = \tau_y + K\dot{\gamma}^n, \qquad n < 1$$

The pipe flow of a non-Newtonian fluid depends on the flow curve.

The laminar velocity distribution for the power law fluid is shown in Fig. 4(a). It can be seen that as the departure from Newtonian behavior increases (i.e., as n decreases), the velocity profile becomes progressively flattened. (For $n > 1$, the fluid is said to be dilatant or shear thickening and the velocity profile is sharpened. Such fluids are not, however, common, and in any case dilatancy is usually accompanied by sufficient other complex rheological behavior to render a discussion of dilatancy in isolation unrealistic.) Figure 4(b) shows the velocity profile for a viscoplastic fluid. This contains a central plug of unsheared material through which the shear stress is less than yield. As the velocity increases, the plug radius is reduced.

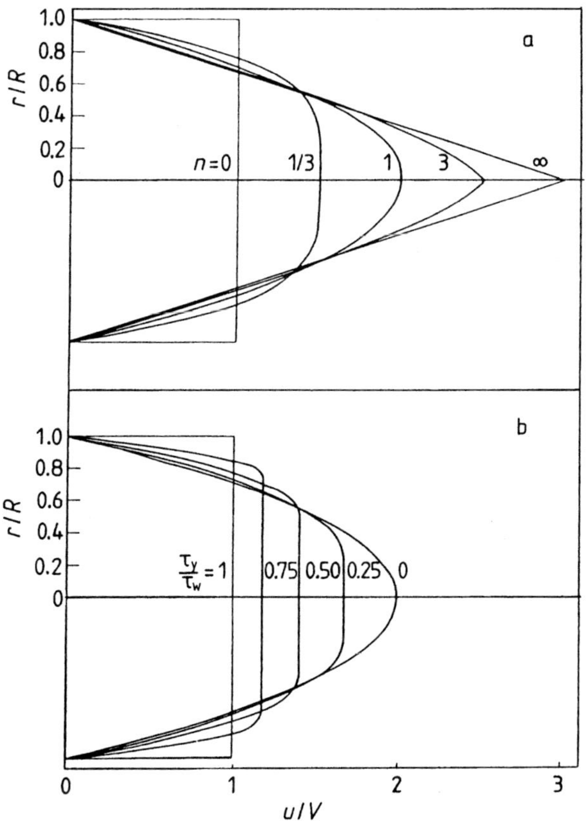

FIG. 4 Laminar velocity profiles for non-Newtonian fluids: (a) power law fluid, and (b) Bingham plastic fluid.

The pressure drop–flow rate relationship can again be expressed in dimensionless form. For the power law fluid, a modified Reynolds number is used

$$\mathrm{Re}_n = \frac{\rho V D}{K[3n + 1)/4n]^n (8V/D)^{n-1}}$$

For laminar flow, a modified Hagen–Poiseuille equation applies

$$C_\mathrm{f} = 16/\mathrm{Re}_n$$

and for turbulent flow, an extended von Karman equation has been proposed

$$C_\mathrm{f}^{-1/2} = \frac{4.0}{n^{3/4}} \lg\left(\mathrm{Re}_n C_f^{1-1/2n}\right) - \frac{0.40}{n^{6/5}}$$

The constant-n loci are shown in Fig. 2(a). The transitional relationship is established semiempirically. For the Bingham plastic fluid the Reynolds number is defined in terms of the plastic viscosity

$$\mathrm{Re}_\mathrm{B} = \frac{\rho V D}{\eta_\mathrm{p}}$$

Then Fig. 2(b) shows that the laminar relationship depends on a third dimensionless group which involves the yield stress, the Hedstrom number

$$\mathrm{He} = \frac{\tau_\mathrm{y}}{\rho V^2} \mathrm{Re}_\mathrm{B}{}^2$$

However, the turbulent conditions fall on a master curve which is very close to the Newtonian relationship.

In general, if the flow curve of a non-Newtonian fluid is known from viscometric measurement, say in the form

$$\tau = g(\dot\gamma) \qquad \text{or} \qquad \dot\gamma = f(\tau)$$

then the laminar velocity profile can be calculated from the equation

$$u(r) = \frac{R}{\tau_\mathrm{w}} \int_\tau^{\tau_\mathrm{w}} f(\tau)\,d\tau$$

and the pressure drop–flow rate relationship from

$$Q = \frac{\pi D^3}{8\tau_{\mathrm{w}^3}} \int_0^{\tau_\mathrm{w}} \tau^2 f(\tau)\,d\tau$$

where

$$\tau_w = \tfrac{1}{4} D\Delta P/L$$

It is not possible, however, to predict turbulent flow characteristics unless the flow curve is approximated by the power law, the Bingham, or the Herschel–Bulkley equation.

Pilot-scale pipeline data are not so restricted. For any fluid, laminar data collapse on to a master plot of τ_w against $8V/D$, and turbulent data conform to the Bowen equation

$$D^x \tau_w = k' V^w$$

where x, k', and w are constants for the fluid. The master curve and the Bowen equation allow small-scale pipeline data to be used to predict large-scale flow without the need for a detailed knowledge of the flow curve.

To sum up: The flow of a homogeneous, incompressible fluid is governed by the non-Newtonian viscosity or flow curve. A knowledge of the pressure drop–flow rate relationship is required to size the pipeline transport system and pump rating. A knowledge of the velocity profile will assist in the understanding of heat and mass transfer operations.

The flow curve can be measured using viscometers or the fluid property can be determined on pilot-scale pipeline rigs. Depending on the type of data available, different equations are used to predict velocity and pressure drop–flow rate relationships.

Turbulent Drag Reduction

Since 1947 when Toms performed his classic experiment using polymethyl methacrylate dissolved in monochlorobenzene, it has been known that very small amounts of high molecular weight polymer (MW typically 5×10^4–6×10^6) dissolved in either aqueous or nonaqueous solvents can give rise to substantial reductions in pressure drop during turbulent pipe flow compared with the pressure drop incurred at the same flow rate for a solvent without dissolved polymer. Reductions of up to 80% are commonplace, and Virk (1975) has shown how the maximum drag reduction asymptote (MDRA) may be estimated for a given system.

The greatest effects have been obtained in numerous, usually small-scale, experimental studies using aqueous solutions of either polyethylene oxide or polyacrylamide (see Fig. 5). Normally the most effective polymer initially can be one of the least effective if sheared in a pipeline for long times because the polymer will degrade, causing its mean molecular weight to reduce progressively with time. As well as this mechanical degradation through molecule bond rupture, some systems are prone to biological degradation, and even tap water will frequently contain sufficient *Aspergillus niger* and other microorganisms to reduce polymer molecular weight.

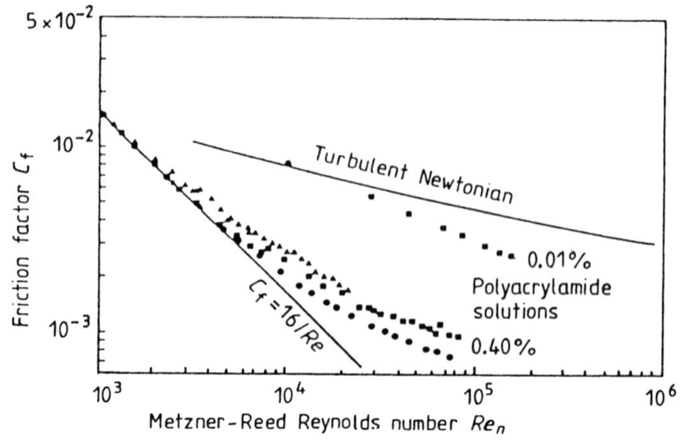

FIG. 5 Turbulent drag reduction for typical viscoelastic fluids–aqueous polacrylamide solutions. Pipe diameter: (●) 0.5 in., (■) 1 in., (▲) 2 in. [from F. A. Seyer and A. B. Metzner, *Can. J. Chem. Eng.*, *B*, *45*, 121 (1967)].

These problems, and others such as the often slow dissolution of high molecular weight polymer in solvents, have meant that the application of the drag reduction phenomenon to industrial needs has yet to be realized. The mechanical degradation, exacerbated by the high shear stress levels to which the polymer solutions are likely to be subjected in the turbulent flows of practical importance, means that the phenomenon is more likely to be exploited in once-through applications, such as flow through a firehose, rather than in flow through long pipelines, although dissolved polymer is used to reduce pressure drop in crude oil pipelines in the North Sea.

Ranges (i.e., the distances over which the water is thrown) of existing fire pumps have been increased by 2½ times by dissolving polymer rapidly in a high intensity mixer before the water is pumped through the hose, and an additional benefit is the creation of a more coherent jet for greater distances after discharge from the hose. This allows greater accuracy in directing water onto flames, which is particularly important when aiming jets through open windows in high-rise buildings. Further practical applications may emerge as new shear degradation resistant polymers are developed. More stable to mechanical degradation but rather less effective than polyethylene oxide and polyacrylamide are polyacrylic acid and hydroxyethylcellulose (HEC).

Although some considerable effort has been directed toward the understanding of the mechanisms responsible for polymer drag reduction, it is still not clear exactly what occurs. Measurement of turbulence intensities throughout the pipe cross section suggests that turbulence is damped by the presence of polymer. In addition, as pipe diameter is increased, the drag reduction effect is reduced, so there is an apparent wall effect involved. Discussion in recent years has focused on extensional viscosity as a relevant variable. Unlike purely viscous, nonelastic liquids where the extensional viscosity is three times the shear viscosity, the extensional viscosities of elastic polymer solutions or melts can be thousands of times greater than the shear viscosities.

Suspensions of solid particles in a liquid or gas have long been known to decrease the shearing stresses developed as the fluid moves past a solid surface under turbulent flow conditions. The most effective particle shape for achieving this drag reduction effect is one having a high aspect ratio (i.e., the ratio of the longest to the shortest particle dimension of at least 10:1), for example, fibers. For particles having much lower aspect ratios than many fiber systems (e.g., less than 5:1), there still remains considerable controversy as to whether drag reduction is possible. The current store of experimental information on this appears contradictory.

Drag reductions effected by fiber systems are usually rather lower than those obtained by polymer solutions, but again typical concentrations are of the order of 10s or 100s of parts per million. The advantage of a fiber-laden system is that the fibers do not generally degrade unless they are particularly friable or subjected to very high shear stresses in tortuous flow fields, and hence the drag reduction effect does not diminish with time on progressive cycling within a pipeline loop or for flow in a long pipeline. A further advantage is that, because effects within the turbulent core of the pipe dominate the drag reduction process, the effect is not diminished on scaling up from small to large pipe diameters.

The extent of the drag reduction increases with either an increase in fiber aspect ratio while holding solids concentration constant, or an increase in solids concentration at constant fiber aspect ratio; in the latter case, up to a maximum effect.

Both the laminar and turbulent flow behavior of more concentrated fiber suspensions have some interesting features. Much effort has been devoted in the paper industry to the examination of the flow of paper pulp suspensions at weight concentrations typically from 0.5 to 3%. Although these concentrations may appear low, the highly fibrous nature of the pulp causes entanglement and the formation of large flocs and makes the material flow essentially as a plug, with a lubricating water annulus adjacent to the pipe wall. With these materials, unlike ordinary single-phase liquids, it is possible for the pressure drop–flow velocity plot to pass through both a maximum and a minimum in the laminar region as shown in Fig. 6. This behavior may be attributable in part to changes in the width of the lubricating water annulus as the flow velocity is varied; this is governed by the degree to which water is expressed from the matted plug of paper pulp as the hydrodynamic conditions vary. At higher flow velocities still, the pressure drop can fall below that for water flow at the same velocity indicative of a drag reduction phenomenon, whose underlying cause is probably different from the similar effect produced by very low concentration fibers.

Start-Up Flow of Thixotropic Fluids

The main problem with thixotropic fluids is the start-up of a pipe filled with gelled material after a prolonged period of shut down. To predict the excess

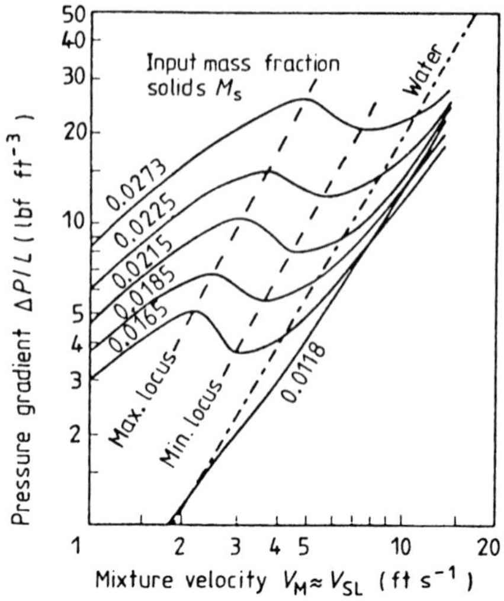

FIG. 6 Pressure gradient data for horizontal pipe flow of unbleached sulfite pulp fibers in a 25.4-mm smooth pipe. Data show drag reduction effect at higher velocities [from T. E. Guenter and N. H. Ceaglske, *Tappi J.*, *34*, 140 (1951)].

pressure needed to initiate flow and the time needed to clear the pipeline of the thixotropic gel, a knowledge of the thixotropic property is required. The simplest thixotropic fluid is one with a viscosity η_0 when fully built up ($\lambda = 1$) and η_∞ when fully broken down ($\lambda = 0$). At intermediate thixotropic structural levels λ, it has a viscosity given by the equation of state

$$\eta = \eta_\infty + (\eta_0 - \eta_\infty)\lambda$$

The time-dependent behavior is governed by the build up and break down of structure, and this is described by the rate equation

$$d\lambda/dt = a(1 - \lambda) - b\lambda\dot{\gamma}$$

where a is the build up rate constant, b is the break down rate constant, and t is the time. More complicated viscosity (or flow curve) and rate equations can be devised. They can be measured by using appropriate viscometric tests. Such constitutive equations are used to predict start-up flow.

The model of pipe flow shown in Fig. 7 is used. The fluid is divided into two sections with the incoming fluid (ICF) displacing the outgoing fluid (OGF). If the OGF is assumed to be uniformly and fully built up initially, its structure will remain uniform along the pipe during the flow; then the fluid state and flow conditions are functions of time and radius and not of axial position. If the ICF is assumed to be fully broken down in the pump as it enters the pipe, then it would build up as it travels down it. In general, fluid state and flow conditions are functions of time and of radial and axial positions. But if the inlet flow rate is kept constant, a steady state will prevail

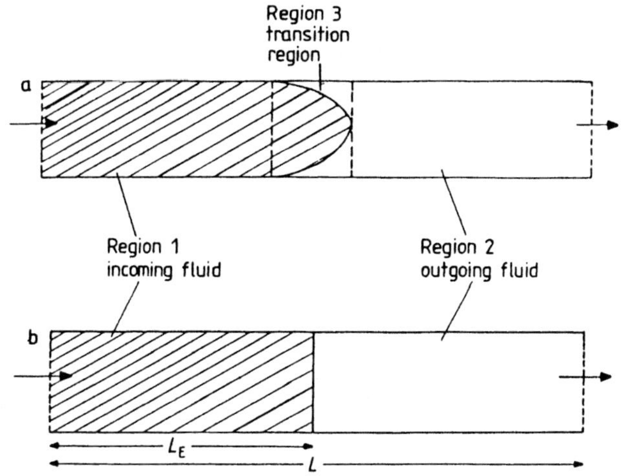

FIG. 7 Distribution of fluids in pipeline during start-up flow: (a) actual distribution, and (b) approximate distribution assumed.

when the dependence is on axial position and not on time. In all cases, by considering the changes in the fluid state and flow condition for all positions in the pipe with time according to the equations of state and rate, and matching the flow rate of the ICF and OGF and adding the pressure drops along the two sections to give the overall figure, the start-up flow can be calculated.

The calculation is a complicated one and has to be carried out on a computer. The results depend on whether the pumping pressure is held constant or the flow rate is held constant. If the former, the flow rate increases with time as the OGF is displaced by ICF of a lower structure; also the pressure gradient and therefore wall shear stress in both the OGF and the ICF increase with time as shown in Fig. 8. If the flow rate is held constant, the overall pressure drop decreases with time (Fig. 9); the OGF pressure gradient decreases with time as it is broken down but the ICF pressure profile maintains its shape while being lowered with time. $\Delta P(0)$ gives the peak pressure needed to initiate the flow after shut down. The calculations give the position of the interface L_E with time, from which the time needed to clear the pipe of gelled material can be determined.

For materials that cannot be conveniently tested with viscometers, it has been suggested that small-scale pipe tests under constant pressure be used to determine the constitutive equations. A computer program written for a sawtooth shaped flow rate profile can then be used to predict start-up flow using a positive displacement pump. The application of this procedure has been illustrated by calculations carried out for a pipeline of the same conditions as the Rotterdam–Rhine crude oil pipeline.

A direct scale-up approach has also been studied. This could allow large-scale pipeline behavior to be predicted from small-scale pipeline testing

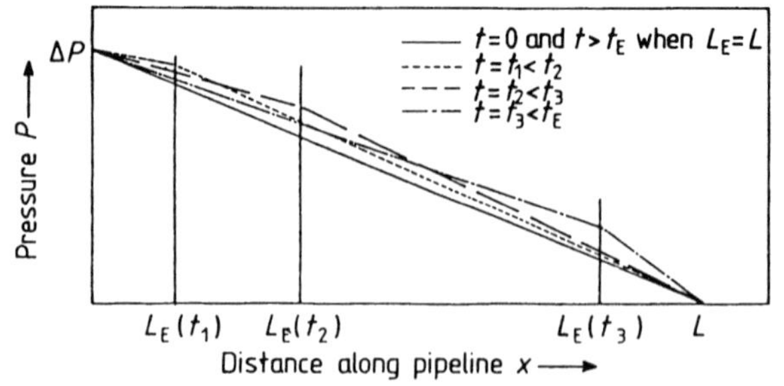

FIG. 8 Pressure distribution in the pipeline under constant pressure drop for thixotropic fluids.

without the need to convert results into constitutive equations as an intermediate step, and without lengthy computer calculations.

Concluding Remarks for Part 1

This brief review of the flow of homogeneous, or nominally homogeneous, incompressible fluids shows that the steady flow of Newtonian liquids in straight pipes of circular cross section is well understood and the viscosity data required for pipeline design can be obtained from the literature. For non-Newtonian fluids, the flow curve has to be determined by experiment for the fluid concerned. For thixotropic fluids, the flow properties needed for design include a knowledge of viscosity as a function of structure and a rate equation governing the structural build up and break down; start-up flow can then be modeled. If the fluid contains long flexible species, such as a macromolecule or suspended fibers, turbulent flow may be suppressed, leading to the drag reduction phenomenon. This illustrates the complex flow behavior that can be found when the fluid is not composed of low molecular weight species. Part 2 of this article examines the flow of multiphase materials.

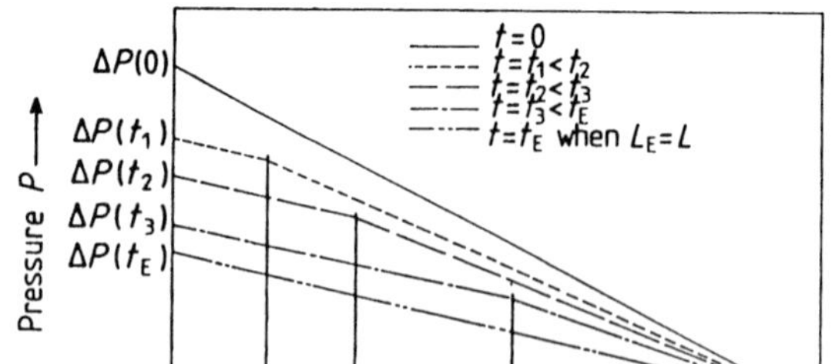

FIG. 9 Pressure distribution in the pipeline under constant flow rate for thixotropic fluid.

Further Reading

Carleton, A. J., and Cheng, D. C.-H., "Pipeline Design for Industrial Slurries," *Chem. Eng.*, *84*, 95 (1977).

Carleton, A. J., Cheng, D. C.-H., and Whittaker, W., *Determination of the Rheological Properties and Start-Up Pipeline Flow Characteristics of Waxy Crude and Fuel Oils*, Institute of Petroleum Publication IP 74-009, London, 1974.

Cheng, D. C.-H., "Pipeline Design for Non-Newtonian Fluids," *Chem. Eng.*, *301*, 525 (1975); *302*, 587 (1975).

Cheng, D. C.-H., *A Study into the Possibility of Scale-Up in Thixotropic Flow*, Warren Spring Laboratory Report LR 317 (MH), Stevenage, Herts, 1979.

Cheng, D. C.-H., and Whittaker, W., *The Start-Up Flow of Thixotropic Fluids in Pipelines*, Warren Spring Laboratory Report LR 155 (MH), Stevenage, Herts, 1971.

Heywood, N. I., "Pipeline Design for Non-Newtonian Fluids," *Inst. Chem. Eng. Symp. Ser.*, *60*, 33–52 (1980).

Heywood, N. I., and Cheng, D. C.-H., "Comparison of Methods for Predicting Head Loss in Turbulent Pipe Flow of Non-Newtonian Fluids," *Trans. Inst. Meas. Control*, *6*, 33–45 (1984).

Hydrotransport 1 to 11, 1970–88 series of conferences on the hydraulic conveying of solids in pipes, BHRA Fluid Engineering, Cranfield, Beds.

Patterson, G. K., Zakin, J. L., and Rodriguez, J. M., "Drag Reduction: Polymer Solutions, Soap Solutions and Solid Particle Suspensions in Pipe Flow," *Ind. Eng. Chem.*, *61*, 22 (1969).

Skelland, A. H. P., *Non-Newtonian Flow and Heat Transfer*, Wiley, New York, 1967.

Virk, P. S., "Drag Reduction Fundamentals," *Am. Inst. Chem. Eng. J.*, *21*, 625 (1975).

Wilkinson, W. L., *Non-Newtonian Fluids*, Pergamon, Oxford, 1960.

Part 2: Multiphase Flow (see also Fluid Flow, Slurry Systems)

Numerous examples of two or more phases flowing simultaneously in a pipeline are encountered in industry. Perhaps the most frequently met are suspensions or slurries (an example of solid–liquid flow) and gas–liquid flows, which are of particular importance at oil-well heads and in electricity generating plants. Two other important combinations are liquid–liquid (e.g., stratified flow or emulsions) and gas–solid (e.g., dense or lean phase pneumatic conveying).

Because of the large differences in the physical properties of the two or more phases involved, such as differences in density and viscosity, one phase will normally flow more quickly in the pipe than the other. Thus, when large particles (e.g., larger than 1 or 2 mm sand particles in water) are transported in a horizontal pipeline, the flow velocity is frequently inadequate to prevent a nonuniform vertical solids concentration profile from being formed. As a result, liquid flows faster than the solids on average and there is a *hold-up* of solid material. This means that the *in-situ* solids concentration in the pipe could be greater than either the fed or discharge solids concentration. Under steady-state conditions, the fed and discharge concentrations are equal. The effects of both gravity and the velocity profile in the pipe contribute to this hold-up phenomenon.

Hold-up of one phase with respect to other phases in any multiphase flow is an important design variable. It is almost always measured in any experimental study, and numerous correlations exist for hold-up prediction. Unfortunately, many of these correlations have been obtained with relatively small-scale equipment, and their use when designing larger-scale equipment is often questionable. Nevertheless, predictions of hold-up are often necessary since hold-up determines the rates of processes, such as heat and mass transfer occurring between the phases, as well as the pressure drop which is incurred when multiphase mixtures are pumped along a pipeline.

During the last 35 years or so, considerable advances have been made in the modeling of multiphase flow in pipes. These models are generally based on an appreciation of how the phases are distributed with respect to each other in the pipe. The various classifications of the phase distributions are normally referred to as *flow regimes*. Thus in hydraulic conveying along a horizontal pipe the solid and liquid may flow as a pseudohomogeneous mixture, or a substantial proportion of the solids may be moving as a sliding bed along the bottom of the pipe with other particles saltating above the sliding bed, or a proportion of the solid particles may be present in the form of a stationary bed. Similar flow regimes may be encountered in gas–solid flows, whereas in gas–liquid or liquid–liquid flows various flow regimes exist including dispersed bubble (or droplet), stratified, wavy, plug (or elongated bubble), slug, annular, and annular mist flows. When the flow is in a vertical pipe, a certain level of time-averaged flow symmetry occurs since gravity acts in the direction of the pipe axis rather than normal to it as in horizontal flows. The result is that the number of readily definable flow regimes is reduced, typically from around seven for horizontal to around four for vertical gas–liquid flows.

A great deal of research effort has been directed toward establishing the transition regions between the different flow regimes. Because of the difficulty in objectively defining flow patterns, either visually or through measurements of fluctuations in local hold-up and pressure in a pipe, these transitions are normally fairly broad and often occur over quite a wide range of flow velocity of each phase. The information is often displayed in the form of flow regime or flow pattern maps whose ordinates involve the relevant variables such as mean flow velocity, average flow velocity of each phase, and the viscosities and densities of the phases. If the flow pattern can be predicted confidently from a knowledge of input conditions, pipe diameter, etc., it is then possible to proceed to correlations for hold-up and pressure drop appropriate for that flow regime.

The introduction of a dispersed phase, such as solid particles into a continuous gas or liquid phase flowing in a horizontal pipe, while keeping the flow rate of the continuous phase constant, almost always increases the pressure drop over a given length of pipe. This is because the presence of the dispersed phase increases the average velocity of the mixture, so raising shear stresses acting on the pipe wall. However, there are a few notable exceptions to this.

If particles with large aspect ratio (typically fibers of aspect ratio greater than 10:1) are present in the turbulently flowing liquid at concentrations of hundreds of parts per million (ppm), the pressure drop is often reduced. No such drag reduction occurs if the liquid is in laminar flow. On the other hand, reduction in the pressure drop of a laminar flowing liquid is possible under appropriate conditions if gas is introduced into a horizontal pipe. In this situation the liquid must possess shear-thinning (or pseudoplastic) non-Newtonian flow characteristics, which are frequently met in industrially important materials.

In vertical pipe flow the situation is somewhat different. Unlike horizontal flow where the total pressure drop is composed mainly of frictional energy dissipation (and a rather smaller accelerational contribution if a gas is present and expands as the pressure along the pipe falls, thereby accelerating the other phases present), the main component of pressure drop in vertical flow arises from the static head contribution. Frictional and accelerational components frequently account for no more than 10 or 15% of the total pressure drop. If a liquid is flowing in a vertical pipe at a constant rate while gas is introduced, there will be an immediate initial reduction in pressure drop since the resulting increase in the frictional component will almost invariably be less than the reduction in static head. On the other hand, the introduction of solid particles, or liquid droplets having a greater density than the continuous liquid phase density, will raise the overall pressure drop. This will also occur for a continuous gas flow. Thus in vertical flow the relative densities of the phases are of prime importance.

Although it is possible to generalize to a certain extent on the fundamentals of multiphase pipe flows, a more detailed discussion requires that reference be made to flows involving particular phase combinations. The following four sections will therefore briefly outline the essential features of four two-phase flows: solid–liquid, gas–liquid, liquid–liquid, and gas–solid.

Solid–Liquid Flow

Liquids, chiefly water, are used as carrier media for a range of different solids both over long distances and within chemical and processing plant. Important materials such as gypsum, cement grout, iron ore, limestone, and clay are often transported by pipeline over long distances in different parts of the world, but it is the resurgence of coal as a major source of energy since the oil crisis of 1973 that has largely been the impetus for substantial effort devoted to feasibility studies for interstate transfer of coal over distances of up to 1600 km (1000 miles) in the United States. A major success story is the Black Mesa coal slurry pipeline which has operated continuously since 1969, carrying coal from mines in North Arizona to the Mohave power

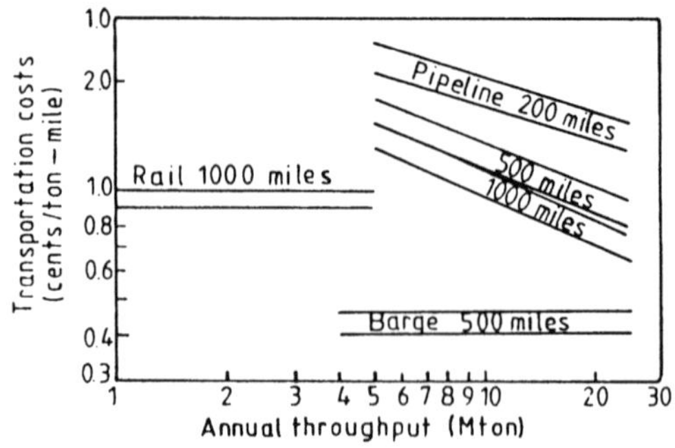

FIG. 10 Typical unit transportation costs for the conveying of coal in pipes. 1976 prices (from E. J. Wasp, J. P. Kenny, and R. L. Gandhi, *Solid–Liquid Flow: Slurry Pipeline Transportation*, Trans. Tech., Clausthal, 1977).

station in southern Nevada over a distance of 440 km (273 miles) and at a rate of some 6 Mt/yr.

As a transportation system, slurry pipelines have to compete economically with railway and barge. A proposed pipeline must convey solids at a sufficient rate and for a long enough time before it is preferable to rail transport. Figure 10 shows how the cost per tonne-mile reduces as the annual throughput is raised. However, the significant railroad lobby in the United States has resulted in rail freight tariffs being reduced temporarily in order to stop a promising slurry pipeline venture which might compete with the railroad company. Apart from the often favorable economics of slurry pipelines, other advantages over rail or barge transport include a low labor component to operating costs and hence a partial shield from inflationary effects, a low impact on the environment (the pipeline is often buried), and no dust nuisance or loss of valuable product such as may occur during open car rail transport.

Most slurry pipelines transport particles which have a top size of a few millimeters. Particles larger than this require uneconomically high flow velocities, with correspondingly high frictional head losses, in order that the particles remain suspended through the turbulence generated in the liquid phase. Depending on the flow velocity, the mean particle size, and the density difference between the particles and the suspending liquid, there exists a range of flow regimes as indicated in Fig. 11. At low flow velocities a stationary bed of particles may form at the bottom of the pipe and some particles may be transported in the liquid flow above the bed by saltating over the top of the bed. The bed height may achieve a certain unstable equilibrium temporarily but will tend to increase with time until the pipe may finally block. This is one of the chief concerns of any slurry pipeline operator, and it is often believed that because of this potential blockage, the characteristics of positive displacement pumps are more suited to slurry pipelines than to centrifugal pump characteristics. Some centrifugal pump manufacturers naturally dispute this!

At higher flow velocities the bed may slide along the pipe bottom, and

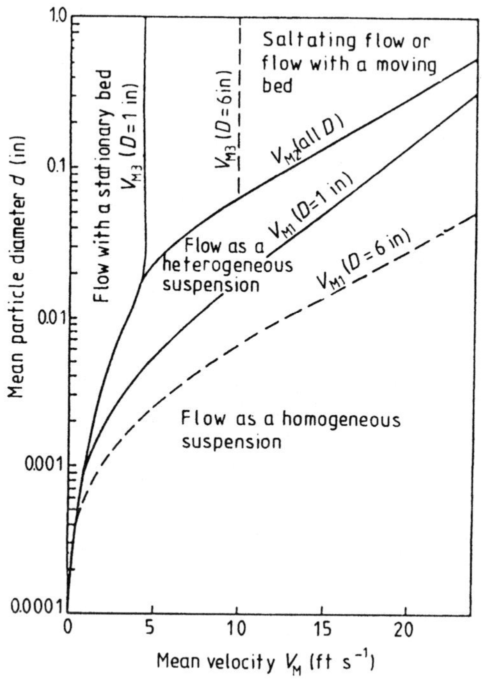

FIG. 11 Variation of flow regimes with particle size for sand and gravel in 1 and 6 in. diameter pipes [from D. M. Newitt et al., *Trans. Inst. Chem. Eng.*, *33*, 93 (1955)].

as the flow velocity is progressively raised, the bed will slide more rapidly and its height reduce until no more particles remain permanently located on the pipe bottom but with some saltating. The flow velocity at which this occurs, V_L, is variously referred to as the critical velocity, limit deposit velocity, or other terms, and it is an important design variable. Many correlations relating this flow velocity to solids volume concentration C, particle size d_p, density difference $\Delta\rho$, and pipe diameter D exist, but there is little, if any, real agreement between them. They are generally of the form

$$V_L = KC^a d_p{}^b \Delta\rho^c D^d$$

The greatest disagreement appears to lie in what effect solids concentration has on the critical velocity. This is not surprising since, at sufficiently high concentrations, changes in solids concentration will impart different Newtonian or non-Newtonian flow properties to the slurry depending on the physical and chemical nature of the particles and their surfaces. These flow properties are crucial in determining the critical velocity.

If pressure drop is plotted against flow velocity for a settling slurry, a minimum is generally observed at, or near to, the critical velocity (Fig. 12). The most economic flow velocity is usually at this point, but slurry pipelines are normally operated at slightly higher flow velocities so that the risks of bed formation and eventual pipe blockage are reduced. At still higher flow

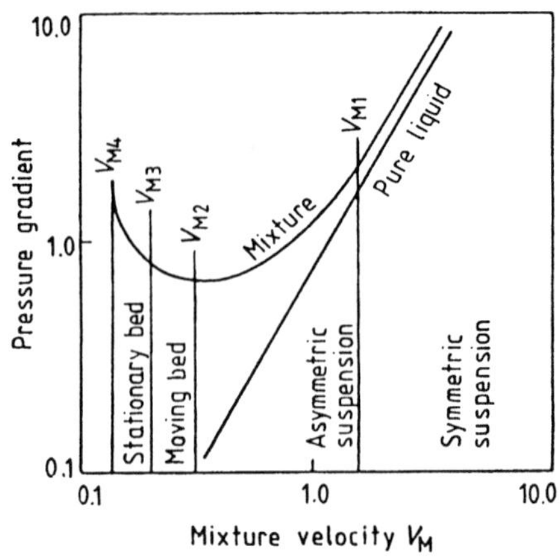

FIG. 12 Schematic representation of flow regimes for settling suspension pipe flow on a pressure gradient–velocity plot (from G. W. Govier and K. Aziz, *The Flow of Complex Mixtures in Pipes*, Van Nostrand Reinhold, Wokingham, 1972).

velocities, pressure drop rises rapidly with velocity, and vertical solids concentration profiles become increasingly less marked. In the limit at very high flow velocities, the slurry can be considered pseudohomogeneous, but this condition is of no practical significance if the slurry settles rapidly under gravity in stagnant conditions.

If the particles are very fine (say typically less than 20–50 μm) and/or if the density difference between particles and suspending liquid is small, slurries are often considered pseudohomogeneous even at relatively low flow velocities. Neither sliding nor stationary beds are formed and hence there is no minimum in the pressure drop–flow velocity plot. In these cases, small-scale tube or rotational viscometric tests may be undertaken in the laboratory on representative slurry samples and the results used to design full-scale pipelines. However, pilot-scale tests are often advisable to check on confidence limits on scaling up and to determine the importance, or otherwise, of any "wall-slip" effects.

The "wall-slip" phenomenon is commonly encountered in many solid-liquid mixtures flowing in pipes. Because of the shear rate distribution across the pipe, from a maximum value at the pipe wall to zero at the pipe center, there frequently exists a radial particle concentration profile which is most pronounced near the pipe wall. The result is a layer of slurry adjacent to the pipe wall depleted of particles, and this gives rise to a reduced pressure gradient along the pipe compared with that occurring if the solids were uniformly distributed. Such an effect can obviously be beneficial, but its importance reduces with increasing pipe diameter and is only really quantifiable for laminar flow and not for turbulently flowing slurries. The general phenomenon of the creation of a radial particle concentration profile is often referred to as the Segre–Silberberg effect. In their original studies, neutrally buoyant spheres in a 4% suspension were found to concentrate at a dimensionless radial position of 0.6.

Considerable effort has been devoted during the last 40 years to the estimation of pressure drop for the flow of settling slurries. Earlier attempts related the increase in head loss for slurry flow compared with water flow alone at the same flow velocity ($i - i_w$) to various physical properties of the slurry (including particle drag coefficients) and to system variables. Many expressions take the form

$$\frac{i - i_w}{Ci_w} = f\left(\frac{gD}{V^2}, \frac{\rho_s}{\rho_w} - 1, \text{particle size/shape parameters}\right)$$

in which C is the volumetric solids concentration, ρ_s is the solids density, and ρ_w is the water density. Such expressions relate to fully suspended flow, but later on more detailed models incorporated other flow regimes in which the frictional wall shear stresses generated by the sliding bed were taken into account.

Recent interest has focused on the feasibility of transporting much larger particle sizes than those currently pipelined, without the need for excessive flow velocities. Investigations are underway to see if large particles (typically 25–50 mm in size, or larger) can be suspended in fine particle slurry, which may often be flocculated. The fine, flocculated particles would impart shear-thinning (i.e., pseudoplastic) flow property to the suspending medium, which would have high viscosities to reduce large particle deposition in the pipe and low viscosities at the higher shear rates at the pipe wall to minimize pipeline head losses.

Gas–Liquid Flow

The literature on gas–liquid flows in pipes is vast. A survey by Hewitt (1978) of the measurement techniques available for the most important design variables referred to 1100 articles, and there are many thousands in addition covering other aspects. Identification of the flow pattern present in a pipe (see Figs. 13 and 14) is vital in the prediction of the most important design variables, such as hold-up, pressure drop, and critical heat flux, from experimental correlations. The first flow pattern map for horizontal flow was developed by Baker in 1954 and has remained the standard, particularly in the petroleum industry, although many attempts have been made to improve on this. The ordinates Baker chose are shown in Fig. 15 and are G/λ and $L\lambda\psi/G$. These parameters are proportional, respectively, to the mass velocity of the gas G and the ratio of the mass velocity of the liquid to that of the gas L/G. The quantities λ and ψ are defined as

$$\lambda = \left(\frac{\rho_G}{1.2}\right)\left(\frac{\rho_L}{1000}\right)^{1/2}$$

$$\psi = \frac{0.073}{\sigma}\left[\left(\frac{\mu_L}{10^{-3}}\right)\left(\frac{1000}{\rho_L}\right)^2\right]^{1/3}$$

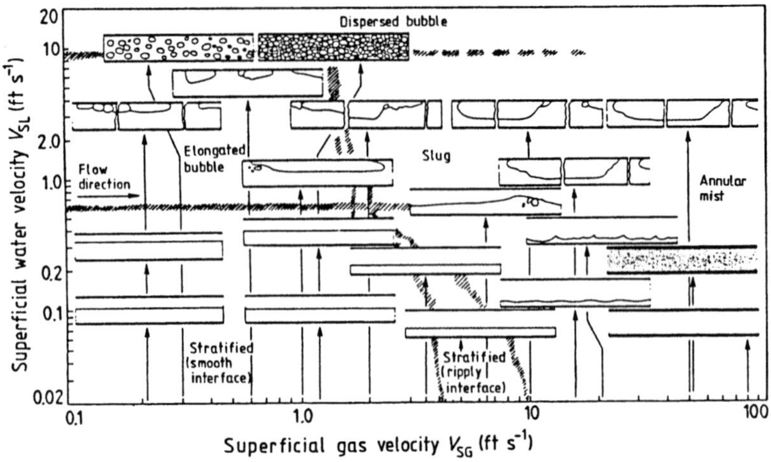

FIG. 13 Flow patterns for air–water mixtures flowing in a 25-mm diameter horizontal pipe based on observations of Govier and Omer and calculations (from G. W. Govier and K. Aziz, *The Flow of Complex Mixtures in Pipes*, Van Nostrand Reinhold, Wokingham, 1972).

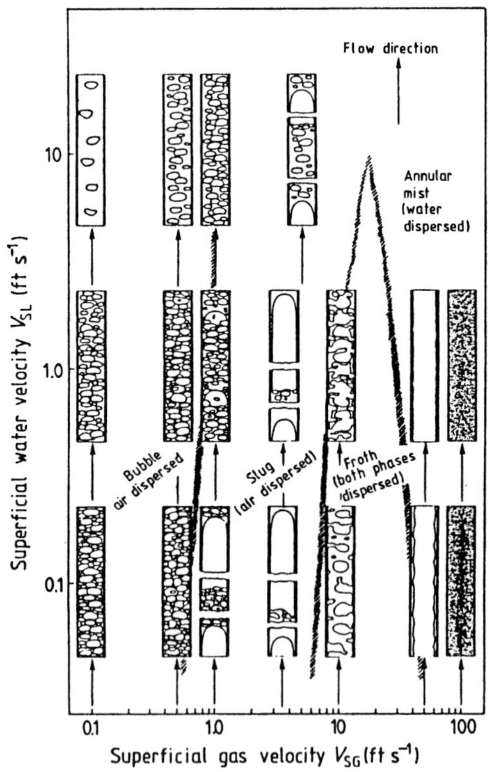

FIG. 14 Flow patterns for air–water mixtures flowing in a 25-mm diameter vertical pipe based on observations of Govier, Radford, and Dunn and calculations (from G. W. Govier and K. Aziz, *The Flow of Complex Mixtures in Pipes*, Van Nostrand Reinhold, Wokingham, 1972).

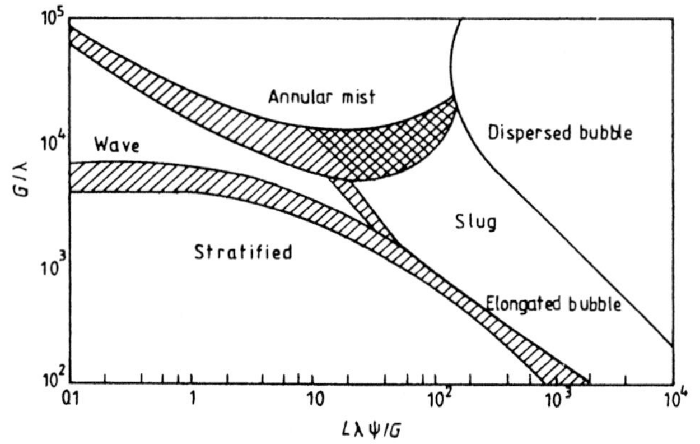

FIG. 15 Baker's flow pattern map for horizontal gas–liquid pipe flow as modified by Scott (from D. S. Scott, in *Advances in Chemical Engineering*, Vol. 4, Academic, New York, 1963, p. 200).

in which ρ_G, ρ_L, μ_L, and σ are gas density, liquid density, liquid viscosity, and surface tension, respectively (all in SI units). The quantities λ and ψ are entirely empirical devices designed to bring the transition lines for systems other than air and water into coincidence with the air–water system.

Lockhart and Martinelli (1949) were the first to present a general pressure drop correlation for gas–liquid flow in horizontal pipes. The method, although empirical, has some theoretical basis and applies surprisingly well over almost all the flow regimes for horizontal flow. The correlation is based upon the concept that the pressure drop for the liquid phase must equal the pressure drop for the gas phase regardless of the flow pattern. They produced empirical plots of ϕ_L (or ϕ_G) versus χ (Fig. 16) where

$$\phi_L{}^2 = \frac{\Delta P_{TP}/L}{\Delta P_L/L}, \qquad \phi_G{}^2 = \frac{\Delta P_{TP}/L}{\Delta P_G/L}$$

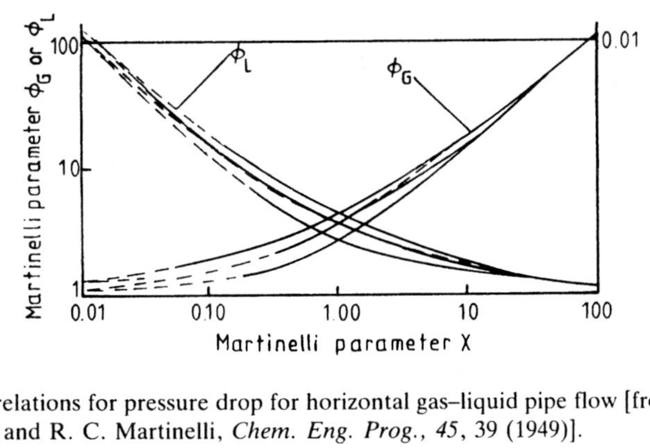

FIG. 16 Correlations for pressure drop for horizontal gas–liquid pipe flow [from R. W. Lockhart and R. C. Martinelli, *Chem. Eng. Prog.*, *45*, 39 (1949)].

and

$$\chi^2 = \frac{\Delta P_L/L}{\Delta P_G/L}$$

Thus

$$\phi_G^2 = \chi^2 \phi_L^2$$

ΔP_{TP} is the two-phase pressure drop, ΔP_L is the single liquid-phase pressure drop at the same liquid flow rate as for two-phase flow, ΔP_G is the single gas-phase pressure drop at the same gas flow rate as for the two-phase flow, and L is the pipe length. Four regimes were identified by liquid- and gas-phase Reynolds numbers according to whether the liquid and gas were nominally in laminar or turbulent flow.

Lockhart and Martinelli also correlated the liquid hold-up α_L with χ for all flow patterns, but later experimental evidence suggests that the single curve they produced is in fact a function of liquid flow rate. Instead, the approach by Zuber and Findlay (1965) involving the drift flux model appears more generally applicable. The basis of the approach is that the average gas velocity V_G may be related to the mixture velocity (the sum of the gas and liquid superficial velocities) by

$$V_G = \frac{V_{SG}}{(1 - \alpha_L)} = C_0 V_M + V_d$$

where V_{SG} is the superficial gas velocity, and hence liquid hold-up

$$\alpha_L = 1 - \frac{V_{SG}}{C_0 V_M + V_d}$$

where C_0 is called the distribution parameter and V_d is the drift velocity; these two are experimental quantities whose values generally depend on the flow regime. Once these are known, estimates for α_L can often be made much more accurately than by using the Lockhart–Martinelli approach.

For all Newtonian liquids in either laminar or turbulent flow and for all non-Newtonian liquids in turbulent flow, introduction of gas into a horizontal pipeline carrying liquid at a constant rate will invariably increase the pressure gradient at any point along the pipeline. This is because the average velocity is raised in the pipe, owing to the reduced mean cross-sectional area for liquid flow brought about by the gas presence, and this in turn creates higher shear stresses acting on the pipe wall which are responsible for the pressure drop.

In contrast, pseudoplastic non-Newtonian fluids, where the viscosity decreases with increasing flow (i.e., shear-thinning), can be in laminar flow concurrently with gas in a horizontal pipe, and the presence of gas results in a decrease in average pressure gradient. Some typical experimental results

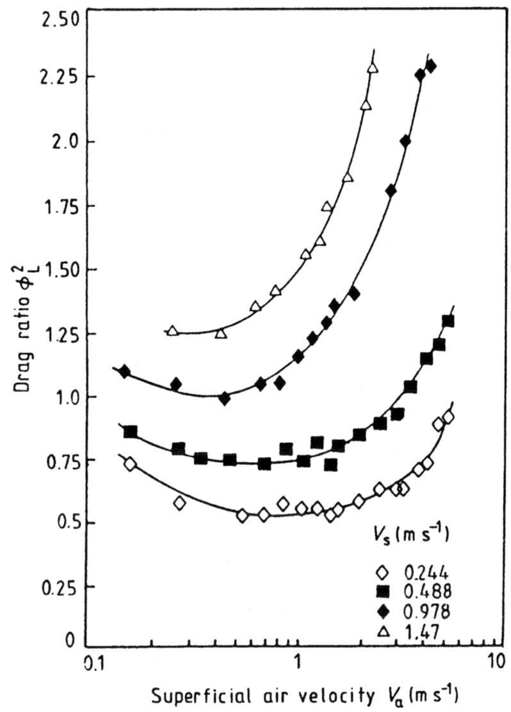

FIG. 17 Pressure drop data for 33.5% by weight anthracite slurry flowing in a 42-mm diameter horizontal pipe with air present. Values of ϕ_L^2 less than unity indicate reduced pressure drop below that for slurry flow alone [from S. I. Farooqi, N. I. Heywood, and J. F. Richardson, *Trans. Inst. Chem. Eng.*, *58*, 16 (1980)].

for shear-thinning anthracite coal slurry are shown in Fig. 17. Here the Lockhart–Martinelli parameter ϕ_L^2 (termed the drag ratio in the figure) is plotted against the superficial air velocity, with the superficial slurry velocity in the pipe as a parameter. It can be seen that provided the slurry flow rate is sufficiently low, i.e., the Reynolds number is low and in the laminar region, the presence of gas causes a reduction in the ϕ_L^2 parameter below unity, indicating a reduced pressure gradient over that for slurry flow alone at the same slurry rate. At any sufficiently low fixed slurry rate this reduction progressively increases with increasing gas rate until a maximum effect is reached. Thereafter the pressure gradient rises with further gas rate increases.

This phenomenon has important implications for pipelines carrying shear-thinning slurries. Two cases arise. In the design of a new pipeline installation, the injection of air downstream from the slurry pump may result either in an overall operating cost saving or facilitate the transport of high viscosity paste-like material which a practical slurry pump could not otherwise achieve. The second situation arises when the capacity of an existing pipeline needs to be raised while retaining the same pump system. Air injection will reduce the frictional head against which the pump acts at the old slurry rate, and the result will be an increase in throughput until the head once again equals that which existed prior to air injection.

Air-lift of either liquids or slurries finds application in a variety of industries. The basic principle is shown in Fig. 18. Air is injected into a vertical, or inclined, riser at a point which is often located below the surface of the fluid. This causes a reduced average density of the two-phase mixture in the riser, and the result is continuous movement of gas and fluid up through the riser. The principle is attractive because there are no mechanical moving parts and hence toxic and corrosive liquids can be readily pumped with minimal risk of spillage. However, relatively low efficiencies are obtained compared with well-designed positive displacement or centrifugal pumps, and the venting of used air and its separation from the fluid present problems.

There has nevertheless been renewed interest in air-lift techniques since around 1970, and a number of trials have been carried out to assess the feasibility of recovering manganese nodules from the ocean floor off the east coast of Florida. These nodules are abundant in a number of locations around the world and are an important source of a range of valuable metallic

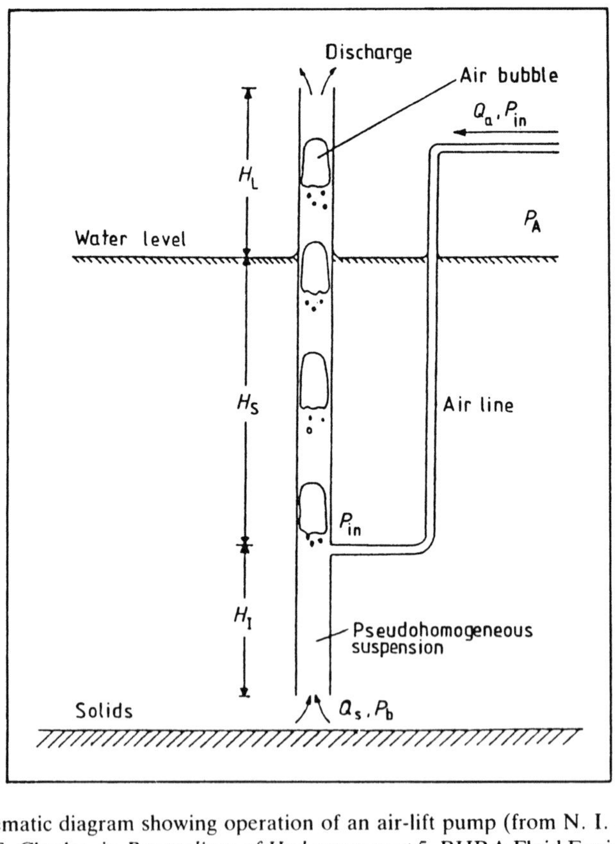

FIG. 18 Schematic diagram showing operation of an air-lift pump (from N. I. Heywood and M. E. Charles, in *Proceedings of Hydrotransport 5*, BHRA Fluid Engineering, Cranfield, Beds, 1978, Paper F5).

elements in addition to manganese. Tests have been carried out using air-lifts to depths of some 1500 m at sea and down disused mine shafts, but the most significant deposits are often up to 6000 m below sea level. Consortia of interested companies have been formed in the last few years to develop the technology required for nodule recovery, but commercial operations have still to be initiated.

Liquid–Liquid Flow

Two of the most common ways in which two immiscible liquids flow concurrently in a pipeline are either as an emulsion or in a separated stratified form. Stable fine-droplet emulsions can normally be treated as pseudohomogeneous and frequently exhibit pronounced non-Newtonian, shear-thinning flow properties. The "mousse," consisting of an oil-in-water or water-in-oil emulsion and pumped from the sea surface when an oil slick is being treated, frequently exhibits very high viscosities in excess of either of the viscosities of the two component liquids. This is because the presence of a high concentration of droplets in the continuous phase can raise the viscosity of the continuous phase by several orders of magnitude and may even impart a yield stress property to the emulsion.

When flowing in a separated form in a pipe, two immiscible liquids commonly flow in the stratified regime with either a smooth or wavy interface or, if the liquids have very similar densities, a central core of one liquid may move within a flowing annulus of the other liquid. The research interest in the oil–water system shown over the last 30 years by many university and government research bodies, particularly in Canada, has been extensive because it relates directly to the pipeline transport of heavy viscous crude oils. While in most situations the presence of a second phase would be considered undesirable, the presence of a low viscosity, immiscible water phase can provide very significant reductions in the pressure gradient and power required for the transport of the crude. Both the technical and economic feasibility of this approach are proven by the fact that several pipelines are in operation throughout the world into which water is deliberately introduced to "lubricate" the flow of crude petroleum.

Typical flow patterns for the horizontal flow of oil–water mixtures are shown in Figs. 19 and 20. Those in Fig. 19 show the patterns produced for equal-density systems; this situation is approximated closely by many heavy oil–water systems. Those in Fig. 20 are typical for when the oil density is less than the water density. It is clear that the relative density has a substantial effect on the flow pattern. In the equal-density case a significant degree of concentricity is apparent while a density difference invariably leads to stratification. For concurrent vertical flow the relative density is much less important (Fig. 21) and the flow patterns are very similar to the equal-density case for horizontal flow.

The pressure gradient reduction factor (PGRF), which is simply the pressure gradient of the oil phase flowing alone at a given rate divided by

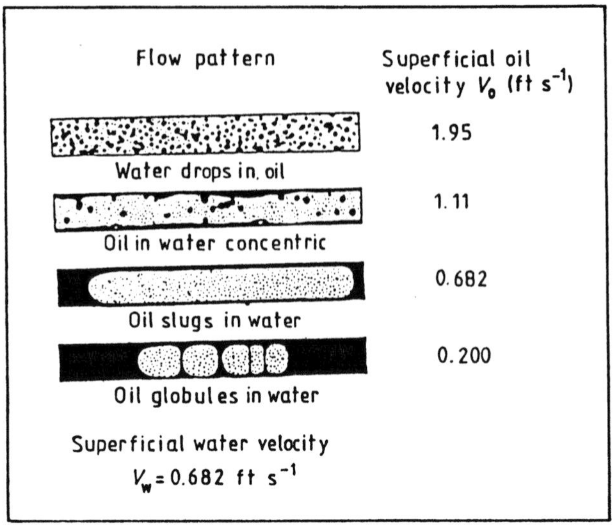

FIG. 19 Flow patterns for horizontal pipe flow of an equal density oil water system [from M. E. Charles, G. W. Govier, and G. W. Hodgson, *Can. J. Chem. Eng.*, *39*, 27 (1961)].

the pressure gradient for the liquid–liquid system with the oil flowing at the same rate, has been used to quantify the reduction in energy consumption through water injection. It is usually not difficult to determine the proportion of water that needs to be combined with the oil in order to minimize the two-phase pressure gradient for either stratified or concentric flow. The

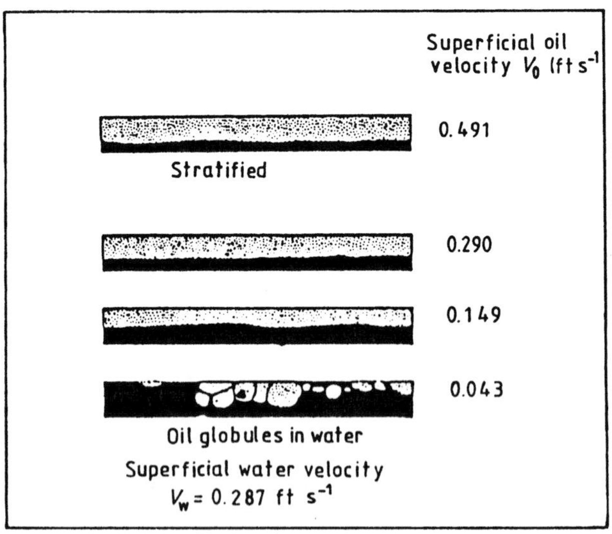

FIG. 20 Flow patterns for horizontal pipe flow of an unequal-density oil–water system [from T. W. F. Russell, G. W. Hodgson, and G. W. Govier, *Can. J. Chem. Eng.*, *37*, 9 (1959)].

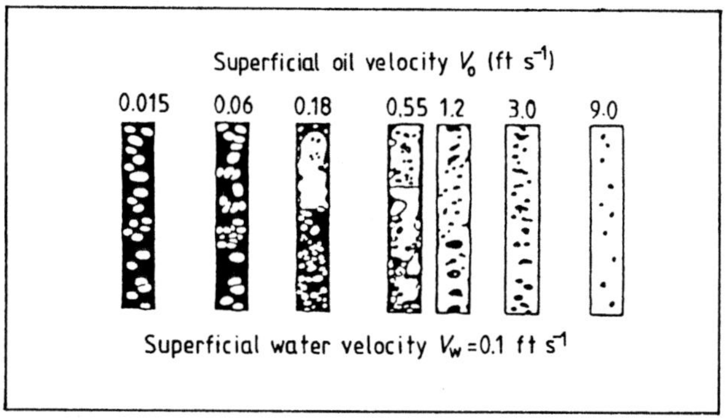

FIG. 21 Flow patterns for vertical pipe flow of an oil–water system [from G. W. Govier, G. A. Sullivan, and R. K. Wood, *Can. J. Chem. Eng.*, *39*, 67 (1961)].

greatest effect is obtained when no oil is in contact with the pipe wall. Even allowing for the adverse effect of turbulence in the less viscous water phase and the concomitant interfacial motions which also dissipate energy, the concentric type of flow pattern still has a significant advantage over the stratified pattern, as seen from the plot of PGRF versus liquid viscosity ratio in Fig. 22. For example, for an oil–water viscosity ratio of 100 the maximum PGRF for stratified flow is just 1.29 whereas for concentric flow it is 50. In vertical flow an approximation to concentric flow is not too difficult to achieve irrespective of relative densities, but in both vertical and horizontal flows it is important that the pipe wall is preferentially wetted by water.

The observations of the horizontal flow patterns for oil–water mixtures made in Alberta, Canada, in the late 1950s suggested that dry solids could

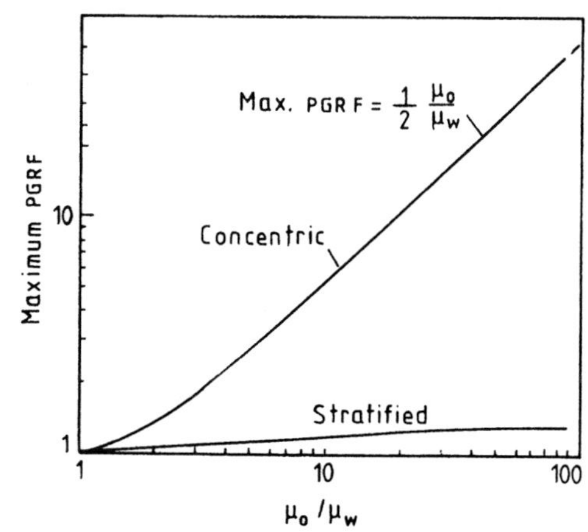

FIG. 22 Maximum pressure gradient reduction factor as a function of the viscosity ratio for horizontal pipe flow of two immiscible liquids [from M. E. Charles, *Can. J. Chem. Eng.*, *55*, 365 (1977)].

be transported within rigid capsules using a carrier liquid. Even in the presence of a density difference between the capsule and the carrier liquid, a lubricating liquid layer would always exist between the capsules and the bottom of the pipe. This heralded the start of academic and industrial research and development of the use of capsules, conveyed either hydraulically or pneumatically. An adequate discussion of these studies falls outside the scope of this article.

Gas–Solid Flow

Gas can be used as a conveying medium in a pipeline as well as liquid. However, owing to considerably lower gas viscosity and density, higher gas velocities are often required to maintain solids suspension. A wide range of materials is conveyed pneumatically, including coal and minerals, grain, foodstuffs, chemicals, plastics, etc. Particle sizes range from submicrometer to 150 mm lumps and greater. The rapid growth in the use of pneumatic conveying systems can probably be attributed to the rapid growth of the plastics industry since 1945, and many thousands of systems are operated in plastics plants, normally over relatively short distances of tens of meters. Larger tonnage and longer pipelines are found in other industries such as the paper industry, where wood chips are conveyed at rates up to 800 t/h over distances which may exceed 200 m.

Many companies in the UK and throughout the world offer design expertise in pneumatic conveying, and there exist a number of proprietary and patented systems. Despite this, much of the available information is distributed throughout the literature and only one or two design guides for pneumatic conveying systems currently exist. A perplexing feature of pneumatic conveying is that the number and range of values of variables describing any particular system—solids density, particle size and size range, shape, moisture content, length and velocity of flow, number of bends in the pipeline and their geometry, direction of the gravitational forces and their relationship to aerodynamic forces, as well as the electrostatic forces a particle might acquire—all combine to give an almost limitless number of possibilities.

Conveying systems can be placed essentially into two main categories depending on the flow regime: suspended and nonsuspended flow. The former covers dilute phase (or lean phase) conveying, where relatively high gas velocities are required to maintain particles in suspension, while the latter covers dense phase systems. Dense phase systems use much less gas; velocities are lower and hence any particle attrition problems are minimized and wear at the pipe wall is reduced. These systems may be operated in a pulse-phase manner in which gas is introduced intermittently into the pipe (instead of continuously for dilute phase conveying) and alternate plugs of gas and solid particles are created. The tendency for cohesive powders to agglomerate is exploited in pulse flow because this aids the retention of the identity of the solid plugs. The material is encouraged to span the pipe diameter and is then transported *en bloc* along the pipeline. Since most of the difficult cohesive materials to be conveyed are relatively impermeable, the gas seal is provided by the material in the plugs which remains in an agglomerated form and can therefore be disengaged easily from the relatively small quantity of air.

Another type of dense phase system occurs when single slugs of powder are transferred from blow tanks. This operation relies on the gas lubricating the solid–pipe interface. With either pulse-phase or single slug dense phase systems the concept of a stable plug of powder retaining its identity along the length of the pipeline is useful for modeling purposes but is rarely, if ever, achieved in practice.

Pneumatic conveying systems may be operated under positive pressures above atmospheric pressure or by applying a vacuum. An obvious example of the latter is the domestic vacuum cleaner. The limits on pipeline length and solids loading are most rapidly reached with vacuum conveying, but vacuum systems offer advantages such as any leakages in the system being inward and hence toxic or explosive powders being conveyed more safely by this means. Vacuum conveying is also advantageous when multiple inlets to the system are required.

Just as with any other two-phase flow, gas–solid flow may be characterized by various flow regimes. Figure 23 shows the various stages of saltation which occur when the gas velocity is progressively reduced, under continuous

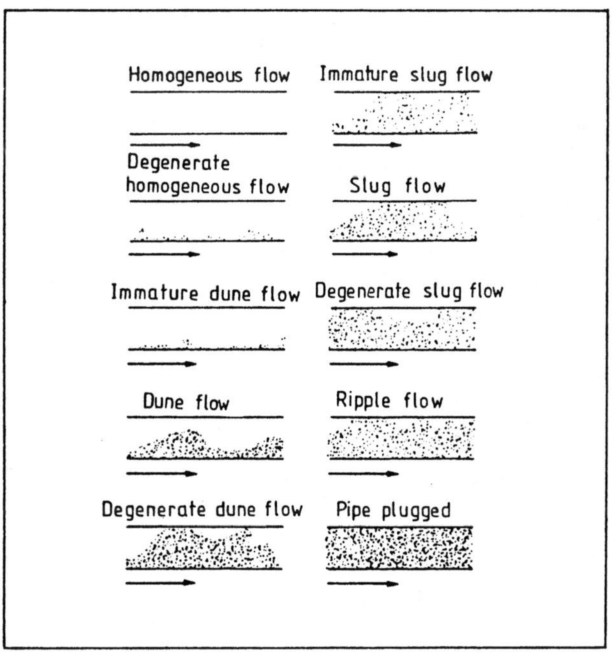

FIG. 23 Flow regimes for horizontal pipe flow of gas–solid mixtures [from C. Y. Wen and H. P. Simons, *Am. Inst. Chem. Eng. J.*, *5*, 263 (1959)].

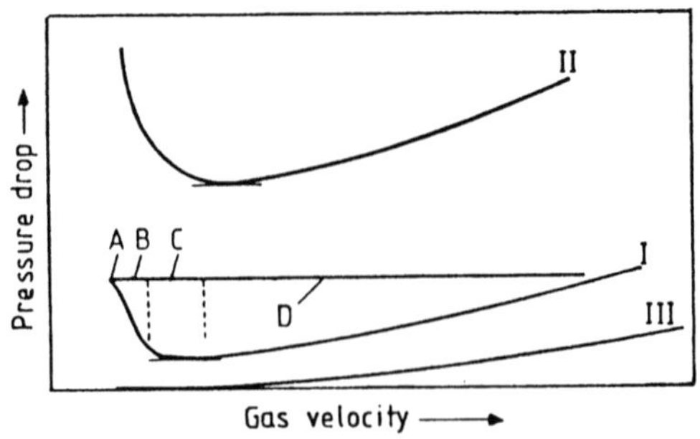

FIG. 24 Schematic representation of flow patterns for gas–solid flow in a horizontal pipe on a pressure drop–velocity plot. I, light solids loading; II, heavy solids loading; III, particle-free gas; A, choking; B, deposition, dune formation, saltation; C, higher concentration in lower part of pipe; D, uniform suspension [from P. R. Owen, *J. Fluid Mech.*, *39*, 407 (1969)].

rather than pulsed gas injection, from a very high value giving essentially homogeneous flow to zero when the pipe becomes plugged. As is the case with the flow of settling slurries in a horizontal pipe, the pressure drop–flow velocity relationship passes through a minimum but there is no distinct flow condition at which this minimum occurs (Fig. 24). Unlike most two-phase systems involving a liquid, where the pressure in the pipe falls approximately linearly along the pipe length, the pressure in a pneumatic conveying pipe carrying high mass ratios falls exponentially. Hence very high pressures would normally have to be generated to transport highly concentrated solids. However, in dense phase systems, which are being increasingly applied in industry, low air velocities—and hence solid particle plug velocities—result in lower energy requirements compared with dilute phase conveying.

Obviously there is a limit to how low an air velocity can be used, and Fig. 25 indicates the limit of conveyability for cement in a specific pipe size.

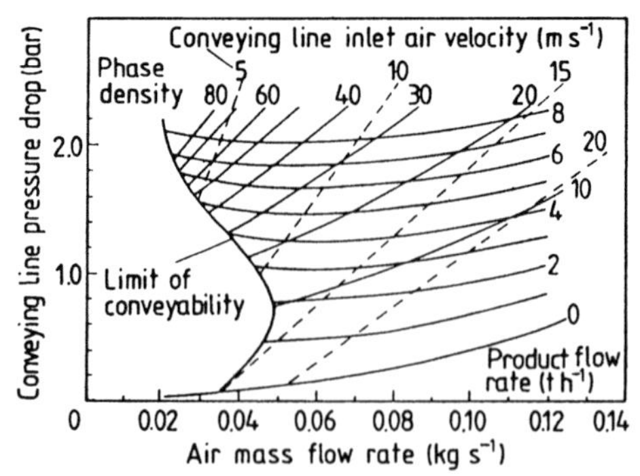

FIG. 25 The influence of product flow rate and phase density on conveying line pressure drop for gas–solid flow (after D. Mills, Thames Polytechnic).

The plot shows the typical relationship between conveying line pressure drop and air mass flow rate with inlet air velocity and phase density as parameters. Phase density is the ratio of the mass flow rate of solids to that of air; it has a value usually below 10 for dilute phase conveying and above 40 for dense phase conveying. For particles which are capable of being conveyed in dense phase, there is no abrupt transition from suspension (dilute phase) to nonsuspension (dense phase) flow, as indicated in Fig. 25. For material which cannot be conveyed in fully developed dense phase flow, pipeline blockage usually occurs when deposited material arising from a reduced air velocity is swept up to fill the full bore of the pipeline, generally at a bend or some other pipeline discontinuity.

Concluding Remarks for Part 2

While the design of pipelines carrying single phase fluids is now well established, much still needs to be done in the modeling, and hence prediction, of design variable data for multiphase flows in pipes. At present, often neither pressure drop nor hold-up can be predicted accurately for full-scale installations from small-scale testwork or models. The situation becomes further complicated and uncertainties increase when multiphase flow is considered in pipe networks or through pipe fittings such as valves, bends, and sudden contractions or expansions in pipe cross section. Inadequate and patchy information exists in these important areas.

A variety of projects aimed at improving the quality of both design data and methods is in progress at a number of university engineering departments, government research institutions, and research associations. The fruits of these efforts will ultimately be manifested in the greater commercial exploitation of multiphase flows and in reduced costs of operating existing installations.

Reprinted by permission from *Physics in Technology, 15*, 244–251, 291–300, 314 (1984).

Further Reading

Bain, A. G., and Bonnington, S. T., *The Hydraulic Transport of Solids by Pipeline*, Pergamon, Oxford, 1970.

Baker, P. J., and Jacobs, B. E. A., *A Guide to Slurry Pipeline Systems*, BHRA Fluid Engineering, Cranfield, 1979.

Brown, N. P., and Heywood, N. I. (eds.), *A Design Handbook for Slurry Transfer Systems*, Elsevier Applied Science, Barking, England, 1990.

Butters, G. (ed.), *Plastics Pneumatic Conveying and Bulk Storage*, Applied Science, Barking, 1981.

Charles, M. E., "Fluid Mechanics and Resource Development," *Can. J. Chem. Eng., 55*, 365 (1977).

Govier, G. W., and Aziz, K., *The Flow of Complex Mixtures in Pipes*, Van Nostrand Reinhold, Wokingham, 1972.

Hetsroni, G. (ed.), *Handbook of Multiphase Systems*, Hemisphere/McGraw-Hill, New York, 1982.

Hewitt, G. F., *Measurement of Two Phase Flow Parameters*, Academic, New York, 1978.

Kraus, M. N., *Pneumatic Conveying of Bulk Solids*, McGraw-Hill, New York, 1981.

Lockhart, R. W., and Martinelli, R. C., "Proposed Correlation for Isothermal Two-Phase, Two-Component Flow in Pipes," *Chem. Eng. Prog.*, *45*, 39 (1949).

Mills, D., *Pneumatic Conveying Design Guide*, Butterworths, London, 1990.

Wasp, E. J., Kenny, J. P., and Gandhi, R. L., *Solid–Liquid Flow: Slurry Pipeline Transportation*, Trans. Tech., Clausthal, 1977.

Zandi, I. (ed.), *Advances in Solid–Liquid Flow in Pipes and Its Application*, Pergamon, Oxford, 1971.

Zuber, N., and Findlay, J. A., *J. Heat Transfer, Trans. ASME Ser. C*, *87*, 453 (1965).

DAVID C.-H. CHENG
NIGEL I. HEYWOOD

Loops or Expansion Joints

A piping system subjected to temperature fluctuations will change in length if free to do so. If not free, it will exert reactive forces and moments on the equipment to which it is attached. When the magnitude of such a reaction would be unacceptable, flexibility must be designed into the piping system.

Before the development of the expansion joint, flexibility was provided by piping configurations that promoted bending. A loop was commonly included in a long run of straight pipe. In recent years, expansion joints have often been installed instead of pipe loops for a variety of reasons.

The Decision Factors

Both the pipe loop and the expansion joint will safely accommodate cyclic thermal movements while retaining pressure integrity. The choice for a particular system may be obvious because of space limitations. In many cases, either will do the job with equal effectiveness and reliability. Too often, the choice is based on personal preference or on the "we did it that way before" principle. The economic aspect, which should be paramount, often is ignored.

The expansion joint most commonly installed in long piping runs is designed so that the pressure is external to the convolutions (Fig. 1). This type of construction makes longer axial movements possible than could normally be accommodated with an internally pressurized expansion joint. Inherent in this design are internal guide rings, a full-thickness cover, self-draining convolutions, and insensitivity to flow direction. No lubrication or packing is required. A drain connection can be installed to remove sediment or condensate, or both.

Pipe loops have proven to be a safe and reliable way of dealing with thermal expansion. Three typical loop configurations are shown in Fig. 2.

An expansion joint may be advantageous, or necessary, for one or more of the following reasons:

1. Space is inadequate for a pipe loop of sufficient flexibility.
2. A minimum pressure drop through the line and the absence of fluid turbulence are essential for process or operating purposes.
3. The fluid is abrasive and flows at high velocity.
4. The available supporting structure is not adequate for the size, shape, or weight of a pipe loop.
5. A pipe loop may be impractical, as in low-pressure, large-diameter piping.
6. The construction schedule does not allow for the additional workhours required to install a loop and its supporting structure.

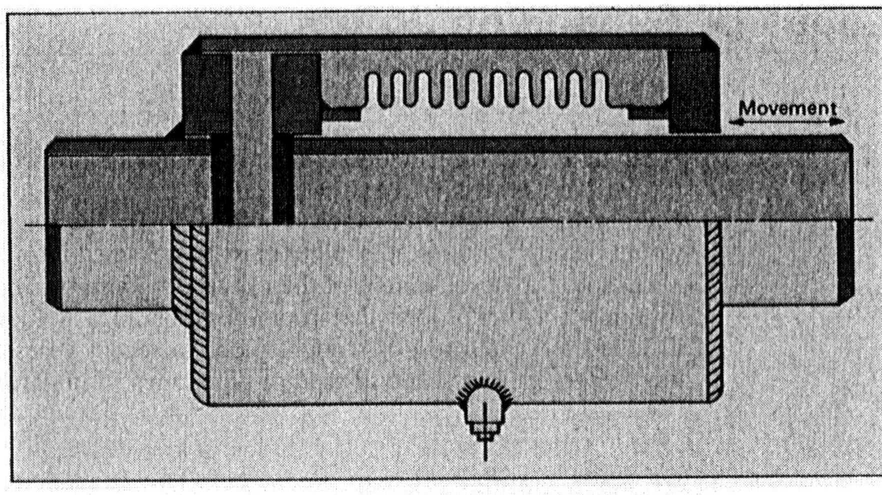

FIG. 1 Externally pressurized single-bellows expansion joint.

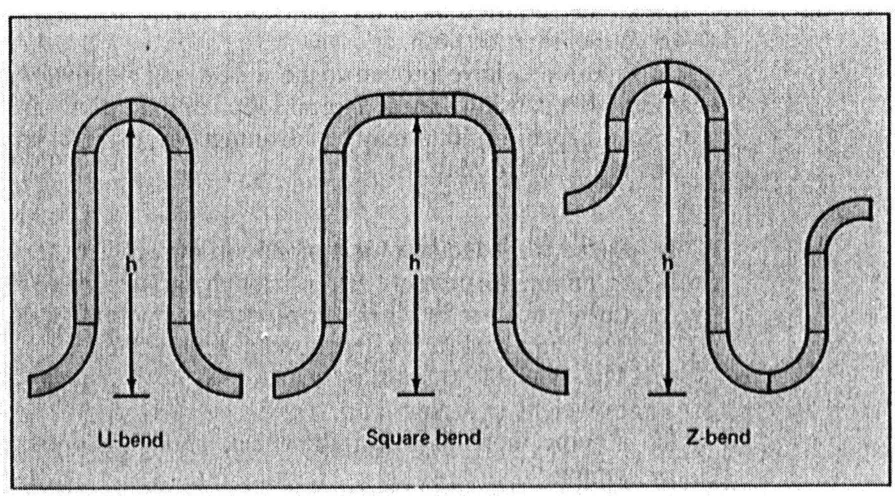

FIG. 2 Three typical pipe-loop configurations.

For any of the following reasons, a pipe loop may be the more appropriate choice:

1. Directional changes are built into the pipe's routing.
2. The pocket in the pipe run that would be created by an expansion joint cannot be tolerated.
3. An expansion joint would be impractical, as in small-diameter, high-pressure piping.
4. Corrosive attack of the bellows element would be a problem.
5. Expansion joints are not permitted by the applicable code.

If none of these reasons applies to a particular piping design, either a pipe loop or an expansion joint could be selected. In such a case, the final decision should be based on economics.

To compare the pipe loop with the expansion joint on an economic basis, one must consider more than just the first cost of the materials. The analysis should also include the cost of labor to cut, bevel, fit, and weld pipe and elbows, as well as the cost of labor to fabricate a supporting structure.

In addition to these first costs, any continuing annual costs associated with loops or expansion joints must be included. For example, the annual costs associated with pumping a liquid, such as hot water, around a loop rather than through an expansion joint can be significant. Each of these cost factors is presented in a simplified manner to aid the engineer in preparing a simple annual cost comparison between pipe loops and expansion joints.

Material and Labor Costs

Because actual costs are always changing, factors are presented for determining the costs of piping material and labor in a simplified analysis. The material and labor cost factors provided in Tables 1 and 2 are based on nominal 8-in.-diameter standard-weight ASTM A53 Grade B seamless pipe having a material cost of \$10/ft [2]—i.e., all the other material-cost and all the labor-cost factors listed in the tables are relative to this single cost. If the current cost of the nominal 8-in.-diameter standard-weight ASTM A53 Grade B seamless pipe were still \$10/ft, all the cost factors in Tables 1 and 2 would be very close to actual costs in dollars.

If, however, this 8-in.-nominal-diameter pipe were now to cost \$15/ft, for example, the current cost in dollars for material and labor could be approximated by adding all the applicable factors for material and labor, and multiplying this sum by 1.50. Thus, knowing the current cost of the nominal 8-in.-diameter standard-weight ASTM A53B seamless pipe, one can quickly estimate the material and labor costs for pipes of other materials and sizes.

TABLE 1 Material Cost Factors for Standard-Weight Carbon Steel Pipe and Fittings[a]

Nominal Size (in.)	Pipe Type A53B	Pipe Type A106B	90° Long-Radius Elbow	90° Short-Radius Elbow
3	2.94	3.23	8.50	11.60
4	4.18	4.60	14.80	19.70
6	7.00	7.70	39.90	46.20
8	10.00	11.00	67.50	88.70
10	14.45	15.90	140.20	152.60
12	17.90	19.78	182.20	194.90
14	20.48	22.53	270.00	311.00
16	25.64	28.20	379.00	424.00
18	29.78	32.76	537.00	612.00
20	32.96	36.26	723.00	718.00
24	39.00	42.90	1065.00	1090.00

[a]These cost factors are presented as a guide for a simplified economic comparison. Greater accuracy can be obtained by using actual cost data supplied by pipe and elbow manufacturers.

The following example demonstrates this approach. The U-bend pipe loop being considered would consist of 20 ft of 12-in. standard-weight A106B pipe and four short-radius elbows. The current cost of 8-in. standard-weight A53B seamless pipe is $18/ft. What would be the approximate cost of this loop?

TABLE 2 Labor Cost Factors for Cutting, Beveling, and Buttwelding Standard-Weight Carbon Steel Pipe

Nominal Size (in.)	Flame Cutting	Beveling	Buttwelding	Radiography
3	2.50	3.00	30.00	35.00
4	3.50	3.50	35.00	39.00
6	5.75	4.50	46.00	48.00
8	8.00	6.00	52.00	54.00
10	11.25	7.75	59.00	61.00
12	12.25	9.75	67.00	69.00
14	17.00	12.00	74.00	75.00
16	19.00	16.00	86.00	84.00
18	24.00	24.00	94.00	93.00
20	29.00	30.00	109.00	103.00
24	44.00	44.00	135.00	127.00

Using Tables 1 and 2, the cost of the pipe loop is estimated as follows.

Material cost—For pipe, \$19.78 × 20 ft = \$395.60; for elbows, \$194.90 × 4 \$779.60. This total is \$1175.20.

Labor cost—For flame-cutting pipe to length, \$12.25 × 2 = \$24.50; for beveling pipe ends, \$9.75 × 4 = \$39.00; for buttwelding pipe to elbows, \$67.00 × 4 = \$268.00; for buttwelding elbow to elbow, \$67.00 × 1 = \$67.00; for radiographing girth welds, \$0. This total comes to \$398.50.

Material and labor costs total \$1573.80. Multiplying this figure by 1.8 (to adjust the Table 1 factor for 8-in. A53B pipe for the present higher cost of \$18/ft) gives a grand total of \$2833.

Annual Cost of Pumping through a Loop

Because the annual cost of pumping a fluid through a pipe loop can be significant, it should be a part of an economic comparison. This cost results from the greater horsepower required to overcome the loop's head loss, which would not be present with the straight-through construction of an expansion joint. The resistance, or head loss, in a pipe loop is assumed to consist of the loss due to curvature and the frictional loss due to length [3]. To account for the curvature and frictional losses, a pipe loop can conveniently be described in terms of equivalent length (L), which is calculated as follows.

For short-radius elbows of U-bend and square-bend types, $L = 2h + 116D$; of the Z-bend type, $L = 2h + 174D$.

For long-radius elbows of the U-bend and square-bend types, $L = 2h + 74D$; of the Z-bend type, $L = 2h + 111D$.

Here, h is the height (or, if horizontal, the width) of the loop, ft, and D is the pipe i.d., ft. These equivalent lengths are based on an L/D ratio of 30 for short-radius elbows and 20 for long-radius elbows [3].

The standard method for calculating head loss in straight pipe is by the Darcy equation:

$$h_L = (0.000483fLW^2)/d^5\rho^2$$

Horsepower requirements are calculated via

$$\text{bhp} = Wh_L/1{,}980{,}000e$$

Combining these equations into an annual pumping-cost formula:

$$C_a = W^3fLuc/(627{,}300e\rho^2d^5)$$

Here, C_a = annual cost of pumping, \$; W = mass flow rate, lb/h; f = friction factor, a function of Reynolds number and the character of the pipe wall (approximate values for fully turbulent liquid flow through smooth pipe are listed in Table 3); L = equivalent length, ft (determined by means of the equations previously provided); u = utilization factor, % (i.e., the percentage of time that the system will be operating during a year); c = average cost of electricity, \$/kWh; e = pump and motor efficiency, % (normally 70–75%); ρ = density of the pumped fluid, lb/ft^3; and d = pipe i.d., in.

Sample Problem—Pipe Loop vs Expansion Joint

A decision on whether to install a pipe loop or an expansion joint can be resolved on the basis of economics. Hot water at 150 lb/in.^{2}gauge and 300°F is to be distributed at a flow rate of 1,459,000 lb/h through nominal 12-in.-diameter standard-weight A106B pipe. The loop would be a short-radius U-bend having a width of 12 ft (i.e., h = 12 ft). Both the pipe run and loop would be horizontal.

An externally pressurized single-bellows expansion joint would cost \$4200. The cost of electricity averages \$0.06/kWh. The system is expected to be in operation an average of 16 h/d. The loop supporting structure would cost \$500. The minimum rate of return, plus taxes and insurance, is 20%. Determine the most economical approach to provide the needed pipe-system flexibility.

As previously calculated, the piping material and labor costs for the loop add up to \$2833. The loop support costs an additional \$500. The equivalent length of the U-bends is 2(12) + 116(1) = 140 ft. Via the pumping-cost equation, the annual cost of pumping hot water through the loop would be: $(1{,}459{,}000)^3(0.013)(140)(0.67)(0.06)/(627{,}300)(0.70)(57.3)^2(12)^5$, or \$633/yr.

TABLE 3 Approximate Friction Factors for Turbulent Fluid Flow through Smooth Pipe

Nominal Pipe Diameter (in.)	Friction Factor, f
3	0.0175
4	0.0162
6	0.0150
8	0.0140
10	0.0135
12	0.0130
14	0.0128
16	0.0123
18	0.0120
20	0.0118
24	0.0150

TABLE 4 Pipe Loop Expansion Joint—An Annual Cost Comparison

Cost Item	Loop	Expansion Joint
Cost of expansion joint, $[a]	—	4200
Material and labor for loop, $	2833	—
Material and labor for loop support structure, $[b]	500	—
Total initial investment	3333	4200
Minimum rate of return, plus taxes and insurance, %	20	20
Annual investment charge	667	840
Annual cost of pumping water through the loop	633	—
Total annual cost	1300	840

[a]Expansion joints are normally custom engineered for each specific installation. Because no two systems are alike, cost factors cannot be tabulated for general application. For an economic determination, actual costs should be obtained from a manufacturer.

[b]The cost of erecting a pipe-loop support structure is usually estimated on the basis of dollars per pound. Because this cost factor can vary considerably, accurate values should be obtained from a contractor for the area where the system is to be installed.

The tabulation in Table 4 shows that, while on the basis of purchase costs the expansion joint is more expensive, it is 35% less expensive based on total annual costs.

References

1. M. W. Kellogg Co., *Design of Piping Systems*, Wiley, New York, 1956, p. 210.
2. R. Weaver, *The Piper's Pocket Handbook*, Gulf Publishing, Houston, 1979.
3. *Flow of Fluids through Valves, Fittings, and Pipe*, Crane Co., New York, 1979.
4 S. Crocker, in *Piping Handbook*, 5th ed. (R. C. King, ed.), McGraw-Hill, New York, 1973.

ROBERT K. BROYLES

Materials—Selection, Costs, and Installation

Part 1: Product Transfer Piping

Published articles provide installed-cost estimates for corrosion-resistant piping of alternative materials in 2-, 4-, and 6-in.-diameter sizes [1–3]. Besides updating these articles, this 5-part article extends the coverage to 12-in. diameter, adds a 20-in.-diameter bleach-plant system, and outlines the fluid suitability guidelines set forth in ANSI/ASME B31.3 [4].

Schedule for this Five-Part Article

This 5-part article shows how pipe size can influence installed cost.

Part 1—Cost estimates are furnished for a simple arrangement represented by a 501-ft-long product-piping system typical of those in which fluids are transferred from one plant to another (Table 1). It consists of one simulated road crossing—8 ft high and 25 ft wide, requiring four 90° ells—that connects straight runs of 180 and 280 ft.

Part 2—Costs for a complex arrangement exemplified by a 406-ft-long process-piping system considered typical of those in chemical process plants.

Part 3—Costs for a 534-ft system modeled after those in modern pulpmill bleach plants.

Part 4—Guidelines for assessing the suitability of the different piping materials for particular fluids and ranges of operating conditions.

Part 5—The information presented in Parts 1–4 is summarized and interpreted.

Material costs are based on what the average plant might reasonably expect to pay for the quantity of pipe and fittings required for each of the three systems if purchased separately in the spring of 1985. The specific cost basis for each type of material will be discussed in Part 2.

Table headings needing definition are:

Material costs: piping—These are based on market cost of pipe in longest lengths available (up to 20 ft) for each material.

Material costs: fittings—These are based on the market costs of fittings, stub ends, and flanges for each material.

Material costs: equipment and miscellaneous—These cover rental of welding machines and cranes, and costs of pipe hangers, filler metal or adhesive, flange bolts, wire brushes, and cleaning gear.

Material costs: painting and color-coding—In the case of painting costs for carbon steel and lined piping, it is assumed that the pipe has been primed and requires only a light sandblasting before two top coats of coal-tar epoxy paint are applied. Stainless steel and nonmetallic piping is only color-coded. The costs of paint and cleanup gear are included. Painting labor involves only time, it being assumed that plant personnel are available for this task.

Shop fabrication—Carbon steel and stainless steel pipe in 2-in. size, and in all sizes for the product-transfer system (except for pipe lined with natural rubber), are field-fabricated and erected. Pipe 4 in. and larger in the process and bleach-plant systems are assumed to be shop-fabricated, to the extent that this is feasible and is normally so done. Estimated costs are based on Midwestern pipe-shop practice, with each size and material considered a separate job.

Field erection—Random 20-ft lengths are assumed for stainless steel pipe, and 10-, 15-, or 20-ft lengths for nonmetallic pipe (i.e., the longest available up to 20 ft). All piping, except for the transfer system, is assumed to be hung from existing racks by means of the recommended number of hangers for each type of material. For both shop and field welding, time is included for quality corrosion-resistant joints. Not included are X-raying, postweld heat treating, and pickling. Workhour estimates are based on standard industry-accepted manuals and on procedures currently followed by pipe contractors. Cost estimates are based on a Midwestern erection rate of \$35/h, which includes contractor's overhead, insurance, and taxes.

Cost ratio—Schedule 5S Type 316L stainless steel was selected as the cost base (i.e., ratio = 1.00) instead of Schedule 40 carbon steel because the latter is now seldom considered if corrosion or product contamination may occur. Schedule 5S rather than 10S is the reference wall thickness because Schedule 5S is rated for greater pressures than 150 lb/in.^{2}, the maximum for Class D fluid service and the pressure at which the nonmetallic piping can be reasonably compared.

References for Part 1

1. S. P. Marshall and J. L. Brandt, "Installed Cost of Corrosion-Resistant Piping," *Chem. Eng.*, p. 68 (August 23, 1971); p. 94 (October 28, 1974).
2. J. Yamartino, "Installed Cost of Corrosion-Resistant Piping—1978," *Chem. Eng.*, p. 138 (November 20, 1978).
3. O. H. Barrett, "Installed Cost of Corrosion-Resistant Piping—1981," *Chem. Eng.*, p. 97 (November 2, 1981).
4. *ANSI/ASME B31.3, Chemical Plant and Petroleum Refinery Piping*, 1984 ed., American Society of Mechanical Engineers.

ARTHUR H. TUTHILL

TABLE 1 Summary of Costs for 2 to 12-in.-Diameter Corrosion-Resistant Piping in a Simple Arrangement Represented by a Product-Transfer System

	Material	Pipe ($)	Fittings ($)	Equipment ($)	Painting ($)	Total Material ($)	Shop ($)	Erection Hours	Installed Cost ($)	Cost Ratio
				2 in. Diameter						
Metallic	Carbon steel, Sch. 40	689	15	1,100	2,600	4,404	—	90	7,554	0.73
	Type 304L, Sch. 5S	1,182	18	1,400	1,600	4,200	—	160	9,800	0.95
	Type 304L, Sch. 10S	1,738	29	1,500	1,600	4,867	—	190	11,517	1.12
	Type 316L, Sch. 5S	1,668	24	1,400	1,600	4,692	—	160	10,292	1.00
	Type 316L, Sch. 10S	2,535	39	1,500	1,600	5,674	—	190	12,324	1.20
	Type 317L, Sch. 5S	2,004	85	1,420	1,600	5,109	—	165	10,884	1.06
	20Cr-25Ni-4.5Mo, Sch. 5S	3,006	120	1,420	1,600	6,146	—	165	11,921	1.16
	20Cr-25Ni-6Mo, Sch. 5S	4,960	121	1,425	1,600	8,106	—	170	14,056	1.37
	Alloy G, Sch. 5S	13,056	480	1,435	1,600	16,571	—	170	22,521	2.19
Nonmetallic	PVC,[a] Sch. 40	251	2	900	1,600	2,753	—	65	5,028	0.49
	PVC,[a] Sch. 80	621	9	900	1,600	3,131	—	65	5,406	0.53
	RTRP,[b] 150 lb/in.^{2}	2,756	96	3,350	1,600	7,802	—	80	10,602	1.03
Lined	SL,[c] 150 lb/in.^{2}	8,124	434	2,100	2,600	13,258	—	270	22,708	2.21
	PPL,[d] 150 lb/in.^{2}	5,416	369	1,200	2,600	9,585	—	210	16,935	1.65
	PVDF,[e] 150 lb/in.^{2}	11,279	434	1,200	2,600	15,513	—	210	22,863	2.22
	PTFE,[f] 150 lb/in.^{2}	15,208	562	1,650	2,600	20,020	—	240	28,420	2.76

4 in. Diameter

Metallic	Carbon steel, Sch. 40	2,019	42	1,510	4,800	8,371	—	170	14,321	0.83
	Type 304L, Sch. 5S	2,600	67	1,910	1,600	6,177	—	290	16,327	0.94
	Type 304L, Sch. 10S	3,482	96	2,110	1,600	7,288	—	340	19,188	1.11
	Type 316L, Sch. 5S	3,547	92	1,910	1,600	7,149	—	290	17,299	1.00
	Type 316L, Sch. 10S	4,845	132	2,110	1,600	8,687	—	340	20,587	1.19
	Type 317L, Sch. 5S	6,012	325	1,940	1,600	9,877	—	300	20,377	1.18
	20 Cr-25Ni-4.5Mo, Sch. 5S	9,018	460	2,120	1,600	13,198	—	300	23,698	1.37
	20Cr-25Ni-6Mo, Sch. 5S	13,787	484	2,130	1,600	18,001	—	300	28,501	1.65
	Alloy G, Sch. 5S	27,139	1,840	2,140	1,600	32,719	—	310	43,569	2.52
Nonmetallic	PVC,[a] Sch. 40	736	38	1,035	1,600	3,410	—	110	7,260	0.42
	PVC,[a] Sch. 80	1,824	42	1,060	1,600	4,526	—	145	9,601	0.55
	RTRP,[b] 150 lb/in.^{2}	3,657	152	3,100	1,600	8,509	—	110	12,359	0.71
Lined	Rubber, 150 lb/in.^{2}	15,527	856	1,510	4,800	22,693	6,700	320	40,593	2.35
	SL,[c] 150 lb/in.^{2}	16,500	781	2,540	4,800	24,622	—	410	38,972	2.25
	PPL,[d] 150 lb/in.^{2}	11,312	689	1,510	4,800	18,311	—	320	29,511	1.71
	PVDF,[e] 150 lb/in.^{2}	26,259	979	1,510	4,800	33,548	—	320	44,748	2.59
	PTFE,[f] 150 lb/in.^{2}	36,791	1,268	2,025	4,800	44,884	—	365	57,659	3.33

(*continued*)

TABLE 1 *(continued)*

	Material	Pipe ($)	Fittings ($)	Equipment ($)	Painting ($)	Total Material ($)	Shop ($)	Erection Hours	Installed Cost ($)	Cost Ratio
				6 in. Diameter						
Metallic	Carbon steel, Sch. 40	6,408	118	1,825	7,100	15,451	—	235	23,676	0.90
	Type 304L, Sch. 5S	4,770	274	2,480	2,100	9,624	—	420	24,324	0.93
	Type 304L, Sch. 10S	5,641	334	2,800	2,100	10,875	—	500	28,375	1.08
	Type 316L, Sch. 5S	6,623	379	2,480	2,100	11,582	—	420	26,282	1.00
	Type 316L, Sch. 10S	7,901	463	2,800	2,100	13,264	—	500	30,764	1.17
	Type 317L, Sch. 125	5,942	393	2,510	2,100	10,945	—	415	25,470	0.97
	Type 317L, Sch. 5S	9,970	659	2,520	2,100	15,249	—	430	30,299	1.15
	20Cr-25Ni-4.5Mo, Sch. 125	10,180	652	2,530	2,100	15,462	—	415	29,987	1.14
	20Cr-25Ni-4.5Mo, Sch. 5S	17,079	1,094	2,790	2,100	23,063	—	430	38,113	1.45
	20Cr-25Ni-6Mo, Sch. 125	18,076	685	2,540	2,100	23,401	—	420	38,101	1.45
	20Cr-25Ni-6Mo, Sch. 5S	30,326	1,148	2,810	2,100	36,384	—	435	51,609	1.96
	Alloy G, Sch. 125	38,011	3,340	2,560	2,100	46,011	—	425	60,886	2.32
	Alloy G, Sch. 5S	64,087	5,604	2,830	2,100	74,621	—	445	90,196	3.43
Nonmetallic	PVC,[a] Sch. 40	1,323	120	1,150	2,100	4,693	—	145	9,768[g]	0.37
	PVC,[a] Sch. 80	3,487	95	1,200	2,100	6,882	—	185	13,357[g]	0.51
	RTRP,[b] 150 lb/in.^{2}	5,160	548	2,700	2,100	10,508	—	115	14,533[g]	0.55
Lined	Rubber, 150 lb/in.^{2}	26,289	1,016	1,900	7,100	36,305	8,100	500	61,905	2.36
	SL,[c] 150 lb/in.^{2}	27,383	1,482	3,150	7,100	39,114	—	650	61,864	2.35
	PPL,[d] 150 lb/in.^{2}	19,370	1,268	1,900	7,100	29,637	—	500	47,138	1.79
	PVDF,[e] 150 lb/in.^{2}	46,704	2,154	1,900	7,100	57,858	—	500	75,358	2.87
	PTFE,[f] 150 lb/in.^{2}	72,647	2,536	3,150	7,100	85,433	—	650	108,183	4.12

8 in. Diameter

Metallic	Carbon steel, Sch. 40	8,402	217	2,450	9,300	20,369	—	300	30,869	0.77
	Type 304L, Sch. 5S	7,254	477	3,800	2,100	13,631	—	685	37,606	0.94
	Type 304L, Sch. 10S	9,008	643	4,050	2,100	15,801	—	760	42,401	1.06
	Type 316L, Sch. 5S	9,534	688	3,800	2,100	16,122	—	685	40,097	1.00
	Type 316L, Sch. 10S	12,169	927	4,050	2,100	19,246	—	760	45,846	1.14
	Type 317L, Sch. 125	8,206	485	3,770	2,100	14,561	—	660	37,661	0.94
	Type 317L, Sch. 5S	13,737	813	3,800	2,100	20,450	—	685	44,425	1.11
	20Cr-25Ni-4.5Mo, Sch. 125	13,813	1,136	3,740	2,100	20,789	—	660	43,889	1.09
	20Cr-25Ni-4.5Mo, Sch. 5S	23,166	1,905	4,300	2,100	31,471	—	695	55,796	1.39
	20Cr-25Ni-6Mo, Sch. 125	24,529	1,193	3,800	2,100	31,622	—	685	55,597	1.39
	20Cr-25Ni-6Mo, Sch. 5S	41,142	2,000	4,330	2,100	49,572	—	705	74,247	1.85
	Alloy G, Sch. 125	51,583	4,444	3,840	2,100	61,967	—	690	86,117	2.15
	Alloy G, Sch. 5S	86,523	7,454	4,350	2,100	100,427	—	726	125,837	3.14
Nonmetallic	PVC,[a] Sch. 40	2,350	273	1,600	2,100	6,323	—	210	13,673	0.34
	PVC,[a] Sch. 80	5,391	190	1,650	2,100	9,331	—	250	18,081	0.45
	RTRP,[b] 150 lb/in.^{2}	8,091	713	3,350	2,100	14,254	—	140	19,154	0.48
Lined	Rubber, 150 lb/in.^{2}	29,315	1,392	2,500	9,300	42,507	12,300	640	77,207	193
	SL,[c] 150 lb/in.^{2}	46,968	2,435	4,200	9,300	62,903	—	790	90,553	2.26
	PPL,[d] 150 lb/in.^{2}	34,261	2,072	2,500	9,300	48,133	—	640	70,533	1.76
	PVDF,[e] 150 lb/in.^{2}	67,730	3,715	2,500	9,300	83,245	—	640	105,645	2.63
	PTFE,[f] 150 lb/in.^{2}	129,812	4,012	4,200	9,300	147,324	—	790	174,974	4.36

(continued)

TABLE 1 *(continued)*

	Material	Pipe ($)	Fittings ($)	Equipment ($)	Painting ($)	Total Material ($)	Shop ($)	Erection Hours	Installed Cost ($)	Cost Ratio
				10 in. Diameter						
Metallic	Carbon steel, Sch. 40	11,904	359	3,100	11,500	26,863	—	365	39,638	0.73
	Type 304L, Sch. 5S	10,386	1,113	5,000	2,100	18,599	—	910	50,449	0.93
	Type 304L, Sch. 10S	12,435	1,366	5,250	2,100	21,151	—	980	55,451	1.02
	Type 316L, Sch. 5S	13,968	1,228	5,000	2,100	22,296	—	910	54,146	1.00
	Type 316L, Sch. 10S	16,563	1,508	5,250	2,100	25,421	—	980	59,721	1.10
	Type 317L, Sch. 125	10,401	760	4,990	2,100	18,251	—	920	50,451	0.93
	Type 317L, Sch. 5S	21,438	1,935	5,100	2,100	30,573	—	940	63,473	1.17
	20Cr-25Ni-4.5Mo, Sch. 125	17,580	2,160	4,990	2,100	26,830	—	925	59,205	1.09
	20Cr-25Ni-4.5Mo, Sch. 5S	36,232	4,452	5,675	2,100	48,459	—	935	81,184	1.50
	20Cr-25Ni-6Mo, Sch. 125	31,217	2,268	5,040	2,100	40,625	—	930	73,175	1.35
	20Cr-25Ni-6Mo, Sch. 5S	64,338	4,676	5,710	2,100	76,824	—	950	100,074	2.03
	Alloy G, Sch. 125	65,651	5,908	5,090	2,100	78,749	—	945	111,824	2.07
	Alloy G, Sch. 5S	135,310	12,176	5,730	2,100	155,316	—	965	189,091	3.49
Nonmetallic	PVC,[a] Sch. 40	6,313	588	2,100	2,100	11,101	—	300	21,601	0.40
	PVC,[a] Sch. 80	8,352	588	2,175	2,100	13,215	—	360	25,815	0.48
	RTRP,[b] 150 lb/in.^{2}	11,899	897	3,750	2,100	18,646	—	145	23,721	0.44
Lined	Rubber, 150 lb/in.^{2}	37,477	2,265	3,400	11,500	54,642	15,600	925	102,617	1.90
	PVDF,[c] 150 lb/in.^{2}	97,512	5,615	3,400	11,500	118,027	—	925	150,402	2.78
	PTFE,[f] 150 lb/in.^{2}	179,156	5,940	5,750	11,500	202,346	—	1,225	245,221	4.53

12 in. Diameter

Metallic	Carbon steel, Sch. 40	14,569	560	3,450	13,700	32,279	—	440	47,679	0.68
	Type 304L, Sch. 5S	13,732	1,560	6,000	2,600	23,892	—	1,170	64,842	0.93
	Type 304L, Sch. 10S	16,092	1,797	6,450	2,600	26,939	—	1,300	72,439	1.04
	Type 316L, Sch. 5S	18,512	1,768	6,000	2,600	28,880	—	1,170	69,830	1.00
	Type 316L, Sch. 10S	20,827	2,037	6,450	2,600	31,914	—	1,300	77,414	1.11
	Type 317L, Sch. 125	12,340	1,100	5,600	2,600	21,640	—	1,090	59,790	0.86
	Type 317L, Sch. 5S	24,680	2,200	6,120	2,600	35,600	—	1,170	76,550	1.10
	20Cr-25Ni-4.5Mo, Sch. 125	20,867	2,720	6,260	2,600	32,447	—	1,100	70,947	1.02
	20Cr-25Ni-4.5Mo, Sch. 5S	41,733	5,440	6,860	2,600	56,633	—	1,200	98,633	1.41
	20Cr-25Ni-6Mo, Sch. 125	37,054	2,856	5,670	2,600	48,180	—	1,100	86,680	1.24
	20Cr-25Ni-6Mo, Sch. 5S	74,107	5,712	6,900	2,600	89,319	—	1,220	132,019	1.89
	Alloy G, Sch. 125	77,915	8,812	5,720	2,600	95,047	—	1,100	133,547	1.91
	Alloy G, Sch. 5S	155,831	17,620	6,920	2,600	182,971	—	1,240	226,371	3.24
Nonmetallic	PVC,[a] Sch. 40	8,317	956	2,325	2,600	14,198	—	410	28,548	0.41
	PVC,[a] Sch. 80	13,737	956	2,425	2,600	19,718	—	500	37,218	0.53
	FRP,[d] 150 lb/in.^{2}	14,354	1,068	3,800	2,600	21,822	—	160	27,422	0.39
Lined	Rubber, 150 lb/in.^{2}	41,959	2,795	3,900	13,700	62,354	20,200	1,150	122,804	1.78
	PTFE,[f] 150 lb/in.^{2}	268,158	9,040	6,300	13,700	297,198	—	1,450	347,948	4.98

(*continued*)

[a]Polyvinyl chloride.
[b]Reinforced resin setting plastic.
[c]Polyvinylidene chloride.
[d]Polypropylene.
[e]Polyvinylidene fluoride.
[f]Polytetrafluoroethylene.
[g]Safeguarded nonmetallic pipe in 6-in. size has been estimated at triple the installed cost; costs in other sizes can be expected to be affected similarly.

Part 2: Process Piping

In this part, installed-cost estimates are based on a 406-ft-long process-piping system similar to those in chemical process plants. It is the same complex system by which such installed costs were presented in earlier articles [1–3] It includes twenty-one 90° ells, four 45° ells, and fifteen tees (see Fig. 1).

For each material, cost estimates are tabulated for the components of the 2 to 12-in.-diameter systems in Table 2. (Table headings were explained in Part 1.) Component-cost totals are expressed as ratios, with Schedule 5S Type 316L stainless steel the reference base (see Part 1).

Descriptions of the Piping Materials

The metallic materials consist of Types 304L, 316L, 317L, 20 Cr-25 Ni-4.5 Mo, 20 Cr-25 Ni-6 Mo, and Alloy G stainless steels, and carbon steel. Polyvinyl chloride (PVC) and reinforced resin-setting plastic (RTRP) are the nonmetallic materials. Natural rubber, polyvinylidene chloride (SL), polypropylene (PPL), polyvinylidene fluoride (PVDF), and polytetrafluoroethylene (PTFE) make up the lining materials.

Types 304L and 316L piping in Schedules 5S and 10S are produced in accordance with ASTM A 312, which requires a final annealing after welding. All the stainless steel material in Schedule 125 (the newer, thinner wall schedule developed for large-diameter papermill piping) is made to ASTM A 778, which does not require final annealing of the low-carbon grades that it covers. PVC pipe is made to ASTM D 1785, RTRP pipe to ASTM D 4163, PPL-lined pipe to ASTM F 492, PVDF-lined pipe to ASTM D 491, and PTFE-lined pipe to ASTM D 423. SL-lined and natural-rubber-lined pipe are fabricated to manufacturer standards.

Cost Basis for Each Material

Again, material costs are based on what the average plant might reasonably expect to pay for the quantity of pipe and fittings required for the system if purchased separately in the spring of 1985.

Costs of carbon-steel, Types 304L and 316L to 12 in. o.d., and PVC pipe are based on distributor prices, with prevailing discounts; costs of 304L and 316L larger than 12 in. o.d., and other alloy pipe, on quoted prices from a major manufacturer for the quantities required for each system; costs of RTRP on list prices for vinyl ester from a major producer; of plastic-lined pipe on list prices from a major producer; and of rubber-lined pipe on quoted prices from a major producer.

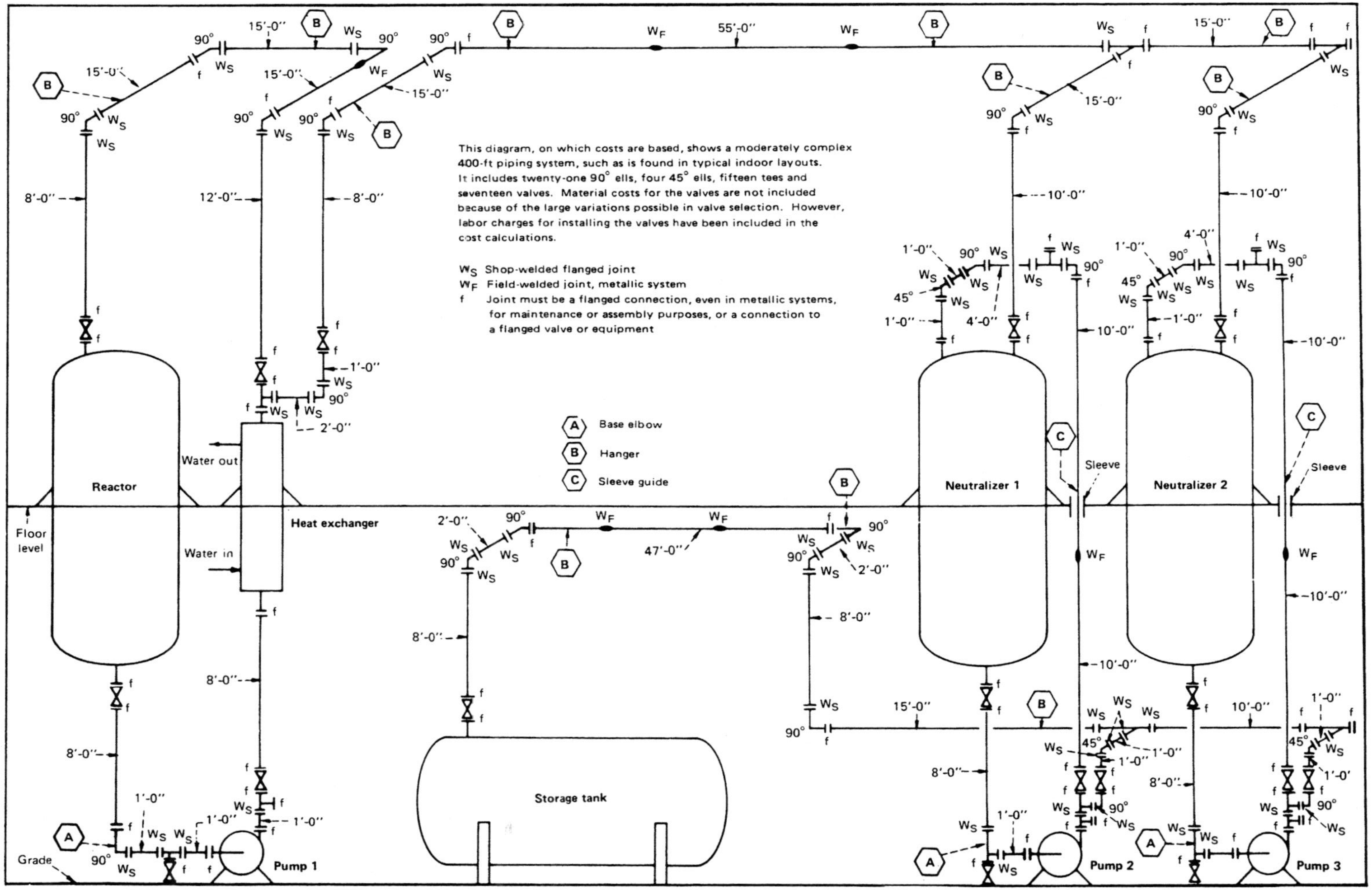

FIG. 1 Representative process system consists of 406 ft of pipe, twenty-one 90° ells, four 45° ells, and fifteen tees.

TABLE 2 Summary of Costs for 2 to 12-in. o.d. Corrosion-Resistant Piping in a Complex Arrangement Represented by a Process-Piping System

	Material	Pipe ($)	Fittings ($)	Equipment ($)	Painting ($)	Total Material ($)	Shop ($)	Field Hours	Installed Cost ($)	Cost Ratio
				2-in. o.d.						
Metallic	Carbon steel, Sch. 40	560	316	1,290	2,100	4,266	—	400	18,267	0.93
	Type 304L, Sch. 5S	958	580	1,490	2,800	5,828	—	380	19,128	0.97
	Type 304L, Sch. 10S	1,409	845	1,590	2,800	6,644	—	400	20,644	1.05
	Type 316L, Sch. 5S	1,352	705	1,490	2,800	6,347	—	380	19,647	1.00
	Type 316L, Sch. 10S	2,054	1,050	1,590	2,800	7,494	—	400	21,494	1.09
	Type 317L, Sch. 5S	1,624	1,779	1,530	2,800	7,733	—	390	21,383	1.09
	20Cr-25Ni-4.5Mo, Sch. 5S	2,306	2,416	2,015	2,800	9,537	—	470	25,987	1.32
	20Cr-25Ni-6Mo, Sch. 5S	4,019	2,528	2,040	2,800	11,387	—	475	28,012	1.43
	Alloy G, Sch. 5S	10,580	9,169	2,070	2,800	24,619	—	495	41,944	2.13
Nonmetallic	PVC,[a] Sch. 40	203	203	440	2,800	3,646	—	185	10,121	0.52[g]
	PVC,[a] Sch. 80	503	299	440	2,800	4,042	—	185	10,517	0.54[g]
	RTRP,[b] 150 lb/in.^{2}	2,233	2,609	530	2,800	8,172	—	130	12,722	0.65[g]
Lined	SL,[c] 150 lb/in.^{2}	8,302	5,188	1,740	2,100	17,330	—	845	46,905	2.39
	PPL,[d] 150 lb/in.^{2}	4,449	4,415	1,440	2,100	12,404	—	735	38,129	1.94
	PVDF,[e] 150 lb/in.^{2}	7,710	5,614	1,440	2,100	16,864	—	735	42,589	2.17
	PTFE,[f] 150 lb/in.^{2}	15,591	8,040	1,590	2,100	27,321	—	790	54,971	2.80

	4-in. o.d.									
Metallic	Carbon steel, Sch. 40	1,636	629	1,160	3,900	7,325	4,400	301	22,260	0.62
	Type 304L, Sch. 5S	2,107	1,455	2,110	2,800	8,472	8,450	510	34,772	0.97
	Type 304L, Sch. 10S	2,822	1,953	2,410	2,800	9,985	8,800	585	39,260	1.09
	Type 316L, Sch. 5S	2,874	1,825	2,110	2,800	9,609	8,450	510	35,909	1.00
	Type 316L, Sch. 10S	3,926	2,481	2,410	2,800	11,617	8,800	585	40,892	1.14
	Type 317L, Sch. 5S	4,872	5,095	2,160	2,800	14,927	8,700	525	42,002	1.17
	20Cr-25Ni-4.5Mo, Sch. 5S	7,308	6,545	2,495	2,800	19,148	9,500	560	48,248	1.34
	20Cr-25Ni-6Mo, Sch. 5S	11,124	6,860	2,510	2,800	23,294	9,585	565	52,654	1.47
	Alloy G, Sch. 5S	21,993	25,216	2,535	2,800	52,544	10,650	585	83,669	2.33
Nonmetallic	PVC,[a] Sch. 40	593	992	780	2,800	5,165	—	350	17,415	0.48[g]
	PVC,[a] Sch. 80	1,478	1,120	830	2,800	6,228	—	420	20,928	0.58[g]
	RTRP,[b] 150 lb/in.^{2}	2,964	3,774	760	2,800	10,298	—	200	17,298	0.48[g]
Lined	Rubber, 150 lb/in.^{2}	11,350	8,606	2,675	3,900	26,531	5,500	1,320	78,231	2.18
	SL,[c] 150 lb/in.^{2}	16,586	9,218	3,430	3,900	33,134	—	1,720	93,334	2.60
	PPL,[d] 150 lb/in.^{2}	8,898	8,009	2,675	3,900	23,482	—	1,320	69,682	1.94
	PVDF,[e] 150 lb/in.^{2}	17,349	10,672	2,675	3,900	34,596	—	1,320	80,796	2.25
	PTFE,[f] 150 lb/in.^{2}	33,818	16,319	3,060	3,900	57,097	—	1,520	110,297	3.07

(*continued*)

TABLE 2 *(continued)*

	Material	Pipe ($)	Fittings ($)	Equipment ($)	Painting ($)	Total Material ($)	Shop ($)	Field Hours	Installed Cost ($)	Cost Ratio
				6-in. o.d.						
Metallic	Carbon steel, Sch. 40	4,990	1,145	1,630	5,700	13,465	6,384	432	34,969	0.51
	Type 304L, Sch. 5S	3,865	4,831	3,730	3,200	15,626	16,420	950	65,296	0.96
	Type 304L, Sch. 10S	4,572	5,797	3,830	3,200	17,399	18,400	935	68,524	1.00
	Type 316L, Sch. 5S	5,367	6,337	3,730	3,200	18,634	16,420	950	68,304	1.00
	Type 316L, Sch. 10S	6,403	7,638	3,830	3,200	21,071	18,400	935	72,196	1.06
	Type 317L, Sch. 125	4,815	6,825	3,465	3,200	18,305	11,125	865	59,705	0.87
	Type 317L, Sch. 5S	8,079	11,123	3,830	3,200	26,232	16,900	975	77,257	1.13
	20Cr-25Ni-4.5Mo, Sch. 125	8,250	10,862	3,980	3,200	26,292	15,900	910	74,042	1.08
	20Cr-25Ni-4.5Mo, Sch. 5S	13,481	17,887	4,705	3,200	39,273	18,400	1,095	95,998	1.41
	20Cr-25Ni-6Mo, Sch. 125	14,648	11,380	4,000	3,200	33,228	16,050	920	81,478	1.19
	20Cr-25Ni-6Mo, Sch. 5S	24,575	22,950	4,740	3,200	55,465	18,565	1,105	112,705	1.65
	Alloy G, Sch. 125	30,803	40,940	4,050	3,200	78,993	18,700	950	130,943	1.92
	Alloy G, Sch. 5S	51,680	68,364	4,795	3,200	128,039	21,325	1,250	193,114	2.83
Nonmetallic	PVC,[a] Sch. 40	1,072	2,992	1,200	3,200	8,464	—	425	23,339[g]	0.34[g]
	PVC,[a] Sch. 80	2,826	3,114	1,250	3,200	10,390	—	475	27,015[g]	0.40[g]
	RTRP,[b] 150 lb/in.^{2}	4,182	8,563	1,000	3,200	16,945	—	245	25,520[g]	0.37[g]
Lined	Rubber, 150 lb/in.^{2}	22,183	10,350	5,940	5,700	44,173	7,980	1,920	119,353	1.75
	SL,[c] 150 lb/in.^{2}	26,813	16,760	6,950	5,700	56,223	—	2,205	133,398	1.95
	PPL,[d] 150 lb/in.^{2}	14,381	14,285	5,940	5,700	40,306	—	1,920	107,506	1.57
	PVDF,[e] 150 lb/in.^{2}	29,845	23,140	5,940	5,700	64,625	—	1,920	131,825	1.93
	PTFE,[f] 150 lb/in.^{2}	64,540	32,315	6,950	5,700	109,505	—	2,205	186,680	2.73

8-in. o.d.

Metallic	Carbon steel, Sch. 40	6,809	1,904	1,920	7,400	18,033	7,888	505	43,596	0.54
	Type 304L, Sch. 5S	5,879	7,857	4,420	3,200	21,356	17,100	1,105	77,131	0.95
	Type 304L, Sch. 10S	7,300	9,821	4,820	3,200	25,141	19,600	1,220	87,441	1.08
	Type 316L, Sch. 5S	7,726	9,838	4,420	3,200	25,184	17,100	1,105	80,959	1.00
	Type 316L, Sch. 10S	9,862	13,022	4,820	3,200	30,904	19,600	1,220	93,204	1.15
	Type 317L, Sch. 125	6,650	8,638	4,390	3,200	22,878	17,250	1,100	78,628	0.97
	Type 317L, Sch. 5S	11,157	13,989	4,540	3,200	32,886	17,600	1,150	90,736	1.12
	20Cr-25Ni-4.5Mo, Sch. 125	11,193	16,237	5,120	3,200	35,750	20,650	1,180	97,700	1.21
	20Cr-25Ni-4.5Mo, Sch. 5S	18,773	26,710	5,245	3,200	53,928	21,800	1,210	118,078	1.46
	20Cr-25Ni-6Mo, Sch. 125	19,878	16,795	5,170	3,200	45,043	20,850	1,190	107,543	1.33
	20Cr-25Ni-6Mo, Sch. 5S	33,341	27,680	5,280	3,200	69,501	22,000	1,220	134,201	1.66
	Alloy G, Sch. 125	41,802	56,757	5,250	3,200	107,009	22,500	1,250	173,259	2.14
	Alloy G, Sch. 5S	70,116	94,699	5,410	3,200	173,425	24,600	1,450	248,775	3.07
Nonmetallic	PVC,[a] Sch. 40	1,904	6,692	2,420	3,200	14,216	—	590	34,866	0.43[g]
	PVC,[a] Sch. 80	4,369	6,692	2,475	3,200	16,736	—	640	39,136	0.48[g]
	RTRP,[b] 150 lb/in.^{2}	6,557	12,372	1,200	3,200	23,329	—	315	34,354	0.42[g]
Lined	Rubber, 150 lb/in.^{2}	25,074	13,017	7,900	7,400	53,391	9,860	2,410	147,601	1.82
	SL,[c] 150 lb/in.^{2}	44,598	26,924	10,300	7,400	89,222	—	3,045	195,797	2.42
	PPL,[d] 150 lb/in.^{2}	23,871	22,912	7,900	7,400	62,083	—	2,410	146,433	1.81
	PVDF,[e] 150 lb/in.^{2}	43,028	35,509	7,900	7,400	93,837	—	2,410	178,187	2.20
	PTFE,[f] 150 lb/in.^{2}	114,106	47,562	10,300	7,400	179,368	—	3,045	285,943	3.53

(continued)

TABLE 2 *(continued)*

	Material	Pipe ($)	Fittings ($)	Equipment ($)	Painting ($)	Total Material ($)	Shop ($)	Field Hours	Installed Cost ($)	Cost Ratio
				10-in. o.d.						
Metallic	Carbon steel, Sch. 40	9,647	3,149	2,930	9,200	24,926	11,154	715	61,105	0.49
	Type 304L, Sch. 5S	8,416	15,981	6,930	3,200	34,527	26,190	1,700	120,217	0.96
	Type 304L, Sch. 10S	10,077	18,959	6,930	3,200	39,166	27,400	1,720	126,766	1.02
	Type 316L, Sch. 5S	11,319	17,699	6,930	3,200	39,148	26,190	1,700	124,838	1.00
	Type 316L, Sch. 10S	13,422	21,348	6,930	3,200	44,900	27,400	1,720	132,500	1.06
	Type 317L, Sch. 125	8,429	12,614	5,930	3,200	30,173	21,500	1,430	101,723	0.81
	Type 317L, Sch. 5S	17,373	24,219	7,110	3,200	51,902	26,902	1,750	140,102	1.12
	20Cr-25Ni-4.5Mo, Sch. 125	14,247	25,931	6,705	3,200	50,083	25,900	1,480	127,783	1.02
	20Cr-25Ni-4.5Mo, Sch. 5S	29,362	49,900	8,280	3,200	90,742	33,500	1,885	190,217	1.52
	20Cr-25Ni-6Mo, Sch. 125	25,298	27,143	6,810	3,200	62,451	26,150	1,495	140,926	1.13
	20Cr-25Ni-6Mo, Sch. 5S	52,139	54,185	8,345	3,200	117,869	33,800	1,900	218,169	1.75
	Alloy G, Sch. 125	53,202	78,491	7,355	3,200	142,248	30,500	1,650	230,498	1.85
	Alloy G, Sch. 5S	109,652	159,975	8,995	3,200	281,822	36,000	1,995	387,647	3.11
Nonmetallic	PVC,[a] Sch. 40	5,116	11,880	3,150	3,200	23,346	—	810	51,696	0.41[g]
	PVC,[a] Sch. 80	6,768	11,880	3,240	3,200	25,088	—	895	56,413	0.45[g]
	RTRP,[b] 150 lb/in.^{2}	9,643	17,955	2,240	3,200	33,038	—	375	46,163	0.37[g]
Lined	Rubber, 150 lb/in.^{2}	32,042	20,612	10,370	9,200	72,224	13,950	3,250	199,924	1.60
	PVDF,[c] 150 lb/in.^{2}	65,155	57,970	10,370	9,200	142,695	—	3,250	256,445	2.05
	PTFE,[f] 150 lb/in.^{2}	157,755	73,905	13,700	9,200	254,560	—	4,100	398,060	3.19

12-in. o.d.

Metallic	Carbon steel, Sch. 40	11,806	5,054	3,145	10,950	30,955	12,200	745	69,230	0.46
	Type 304L, Sch. 5S	11,128	20,840	7,745	4,200	43,913	31,400	1,920	142,513	0.96
	Type 304L, Sch. 10S	13,041	22,755	8,445	4,200	48,441	34,100	2,100	156,041	1.05
	Type 316L, Sch. 5S	15,002	23,498	7,745	4,200	50,445	31,400	1,920	149,045	1.00
	Type 316L, Sch. 10S	16,877	26,760	8,445	4,200	56,282	34,100	2,100	163,882	1.10
	Type 317L, Sch. 125	10,000	16,229	7,170	4,200	37,599	27,900	1,740	126,399	0.85
	Type 317L, Sch. 5S	20,000	30,408	7,945	4,200	62,553	32,350	1,980	164,203	1.10
	20Cr-25Ni-4.5Mo, Sch. 125	16,910	33,857	8,070	4,200	63,037	31,750	1,800	157,787	1.06
	20Cr-25Ni-4.5Mo, Sch. 5S	33,820	65,544	9,295	4,200	112,859	33,900	2,115	220,784	1.48
	20Cr-25Ni-6Mo, Sch. 125	30,028	35,446	8,125	4,200	77,799	32,050	1,815	173,374	1.16
	20Cr-25Ni-6Mo, Sch. 5S	60,056	68,844	9,375	4,200	142,475	39,250	2,135	256,450	1.72
	Alloy G, Sch. 125	63,141	107,225	8,550	4,200	183,116	35,000	1,990	287,766	1.93
	Alloy G, Sch. 5S	126,282	338,683	9,990	4,200	479,155	45,000	2,300	604,655	4.06
Nonmetallic	PVC,[a] Sch. 40	6,740	17,905	3,620	4,200	32,465	—	1,060	69,565	0.47[g]
	PVC,[a] Sch. 80	11,133	17,905	3,740	4,200	36,978	—	1,180	78,278	0.53[g]
	RTRP,[b] 150 lb/in.^{2}	11,632	23,031	2,240	4,200	41,103	—	420	55,803	0.37[g]
Lined	Rubber, 150 lb/in.^{2}	37,665	25,678	12,250	10,950	86,543	15,250	3,915	238,818	1.60
	PTFE,[f] 150 lb/in.^{2}	238,076	112,840	21,300	10,950	383,166	—	4,975	557,291	3.74

[a]Polyvinyl chloride.
[b]Reinforced resin setting plastic.
[c]Polyvinylidene chloride (not available >8-in. o.d.).
[d]Polypropylene (not available >8-in. o.d.).
[e]Polyvinylidene fluoride (not available >10-in. o.d.).
[f]Polytetrafluoroethylene.
[g]Safeguarded nonmetallic pipe in 6-in. size has been estimated at more than triple the installed cost; costs in other sizes can be expected to be affected similarly.

Carbon-steel, standard grades of stainless steel and PVC piping, being true commodities (i.e., standardized, interchangeable, and available from many producers), tend to be discounted substantially from list price according to market conditions. The lighter-schedule, larger-size, and higher-alloyed stainless-steel piping is priced in accordance with each job's requirements. Although widely produced, RTRP pipe is not fully interchangeable because of resin variations among manufacturers. Although standardized, plastic-lined pipe has few producers, list prices prevail, with occasional slight discounting.

Piping lined with SL or PPL is not normally available in larger than 8 in. o.d., nor is PVDF-lined piping in larger than 10 in. o.d. Rubber-lined piping is not available in 2 in. o.d.

References for Part 2

1. S. P. Marshall and J. L. Brandt, "Installed Cost of Corrosion-Resistant Piping," *Chem. Eng.*, p. 68 (August 23, 1971); p. 94 (October 28, 1974).
2. J. Yamartino, "Installed Cost of Corrosion-Resistant Piping—1978," *Chem. Eng.*, p. 138 (November 20, 1978).
3. O. H. Barrett, "Installed Cost of Corrosion-Resistant Piping—1981," *Chem. Eng.*, p. 97 (November 2, 1981).

ARTHUR H. TUTHILL

Part 3: Bleach-Plant Piping

In this part, installed-cost estimates are given for nominal 20-in.-diameter stainless-steel piping, with the focus on the newer, thinner-walled Schedule 125 pipe used in the pulp and paper industry (see Table 3; see Part 1 for explanations of the table column headings).

Bleach-Plant Piping

The estimates are based on a 534-ft system modeled after those in modern pulpmill bleach plants (see Fig. 2). Although pipe diameters in such systems range from 6-to-30 in. o.d., 20-in.-o.d. piping predominates.

TABLE 3 Cost Summary of 20-in.-Diameter Corrosion-Resistant Piping for Modern Pulpmill Bleach Plants

Material	Pipe ($)	Fittings ($)	Equipment ($)	Painting ($)	Total Material ($)	Shop ($)	Erection Hours	Installed Cost ($)	Cost Ratios: 316L, 5S = 1	Cost Ratios: 317L, 125 = 1
Type 304L, Sch. 125	13,674	12,274	12,130	4,600	42,678	60,000	2,790	200,328	0.97	0.93
Type 304L, Sch. 5S	20,185	17,101	13,430	4,600	55,316	62,000	2,890	218,466	1.06	1.01
Type 316L, Sch. 125	17,500	13,753	12,130	4,600	47,983	60,000	2,790	205,633	1.00	0.95
Type 316L, Sch. 5S	26,301	19,437	13,430	4,600	63,768	62,000	2,890	226,918	1.10	1.05
Type 317L, Sch. 125	20,525	16,712	12,330	4,600	54,167	61,700	2,870	216,317	1.05	1.00
Type 317L, Sch. 5S	32,447	24,111	13,730	4,600	74,888	63,800	2,980	242,988	1.18	1.12
20Cr-25Ni-4.5Mo, Sch. 125	32,956	29,050	14,630	4,600	81,236	75,600	3,515	279,861	1.36	1.29
20Cr-25Ni-4.5Mo, Sch. 5S	52,251	47,421	16,230	4,600	120,502	78,100	3,650	326,352	1.59	1.51
20Cr-25Ni-6Mo, Sch. 125	54,623	32,168	14,730	4,600	106,121	76,300	3,550	306,671	1.49	1.42
20Cr-25Ni-6Mo, Sch. 5S	86,912	50,438	16,330	4,600	158,280	78,800	3,680	365,880	1.78	1.69
Alloy G, Sch. 125	123,069	109,875	17,030	4,600	254,574	90,000	4,190	491,224	2.39	2.27
Alloy G, Sch. 5S	195,125	174,659	19,030	4,600	393,414	93,000	4,340	638,314	3.10	2.95
RTRP,[a] 100 lb/in.^{2}	20,194	76,135	7,060	4,600	108,169	45,000	1,500	205,669	1.00	0.95

[a]Reinforced thermosetting resin plastic.

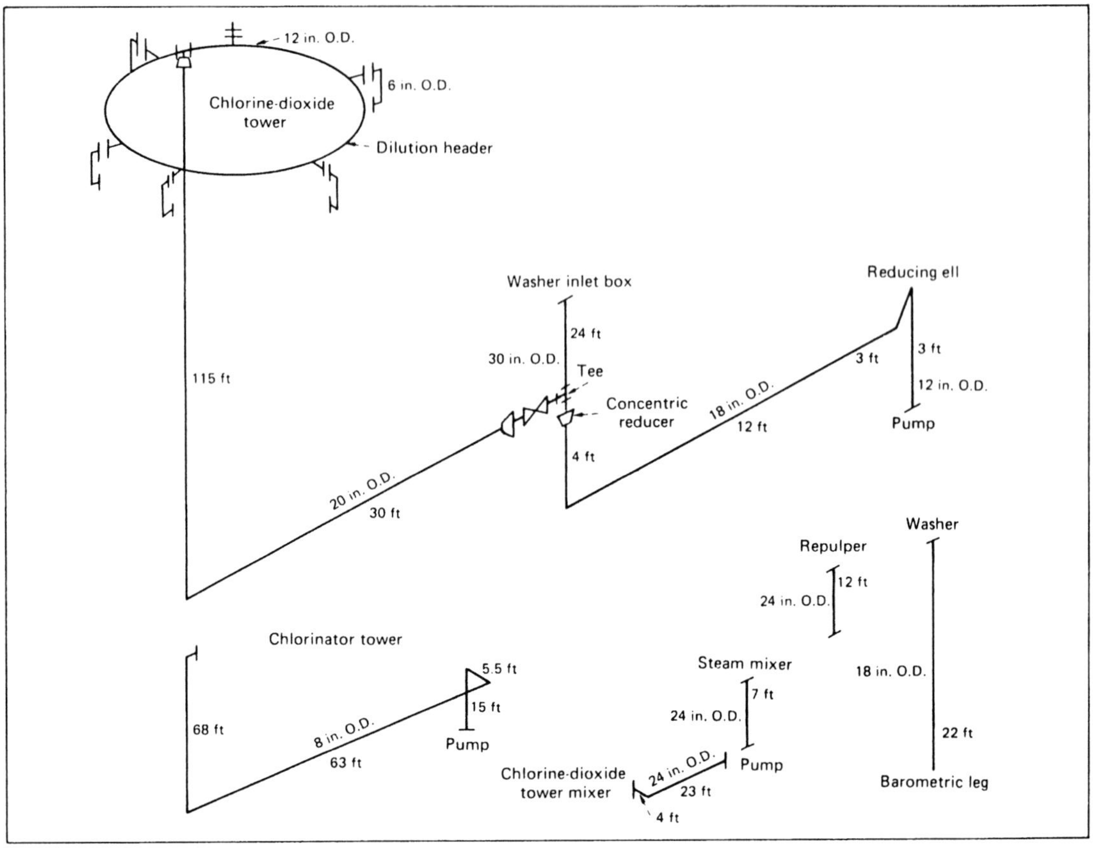

FIG. 2 This 534-ft system, modeled after piping in modern pulpmill bleach plants, consists of a 20-in. o.d. main, 6-to-30-in. o.d. branches, 19 ells, 34 stub ends, 8 reducers, and 1 tee.

Only metallic and reinforced thermosetting resin plastic (RTRP) piping are included in the large diameter systems of bleach plants. Polyvinyl-chloride (PVC) piping is not used. Plastic-lined piping is not available in larger sizes. Although available in larger sizes, rubber-lined piping is rarely used.

Whereas many of the chemical process industries have adopted ANSI/ ASME B31.2 as the basic guideline for piping design [1] the pulp and paper industry has developed its own standard for the large-diameter, thin-wall piping now installed in bleach plants. Wall thicknesses of this Schedule 125 piping are listed in Table 4, in which thicknesses for both Schedule 5S and 10S pipe are included for reference.

In the case of the simple and complex piping systems represented, respectively, by product-transfer piping in Part 1 and process piping in Part 2, installed-cost estimates were based on assumptions that there are no thermal, vibration, waterhammer, or other mechanical stress loadings of significance on the piping, and that nonmetallic piping is coupled by means of bell-and-spigot joints and molded flanges and fittings. These assumptions,

TABLE 4 Pipe Wall-Thickness Specifications of the Technical Association of the Pulp and Paper Industry

Nominal Pipe Size (tube o.d., in., or National Pipe Standard)	Sch. 125 Wall Thickness[a] [2]		Comparable Wall Thickness, in.	
	U.S. Standard Gauge	In.	Sch. 5S	Sch. 10S
½–2	16	0.062	0.065	0.109
2½–4	16	0.062	0.083	0.120
6	16	0.062	0.109	0.134
8	16	0.062	0.109	0.148
10	16	0.062	0.134	0.165
12	14	0.078	0.156	0.180
14	12	0.109	0.156	0.188
16	12	0.109	0.165	0.188
18	11	0.125	0.165	0.188
20	11	0.125	0.188	0.218
24	10	0.140	0.218	0.250
30	8	0.172	0.250	0.312
36	5	0.218		

[a]The wall thicknesses specified for Schedule 125 (125-lb/in.^{2} working pressure) are based on an ultimate tensile strength of 70,000 lb/in.^{2} and a safety factor between working pressure and theoretical bursting pressure of 6.5 to 1. This 6.5-to-1 was arrived at by increasing a 4-to-1 safety factor, considering joint efficiency to be 70%, and the allowable minimum material-thickness to be 87½% of the nominal pipe size in a worst-case condition.

although somewhat biased in favor of RTRP, are customary in installed-cost analyses.

In bleach plants, press-fitted bell-and-spigot joints and molded flanges tend to pull apart under normal operating conditions; therefore, hand layup of flanges and fittings with glass-to-resin ratios ranging between 70/30 to 60/40, butt-wrapped joints, and strict quality control are assumed for bleach-plant RTRP piping. In the absence of vibration, hand layup of fittings and flanges allows RTRP piping to approach equivalency in mechanical properties to metallic piping under other operating conditions.

Piping made to ASTM A 778 of 317L and higher molybdenum content are assumed to be welded with a filler metal of a higher molybdenum content, such as Alloy 625. Filler metal of matching composition is assumed for 304L and 316L.

The 4.5 and 6% molybdenum alloys are available at a 30–40% premium over Type 317L. Antivibration cost for RTRP piping is specific to location; therefore, this cost is not included in Table 3. With it included, the full install cost of RTRP piping seems likely to fall in the same range as the 4.5 and 6% molybdenum-alloy piping.

References for Part 3

1. *ANSI/ASME B31.3, Chemical Plant and Petroleum Refinery Piping*, 1984 ed., American Society of Mechanical Engineers.
2. *Recommended Specifications for Stainless Steel Piping, Fittings and Accessories for the Pulp and Paper Industry*, 3rd ed., Technical Association of the Pulp and Paper Industry, Atlanta, Georgia, 1985.

ARTHUR H. TUTHILL

Part 4: Guidelines for Selection

Fluid Service Suitability

Not all piping materials are equally suitable for handling certain fluids or for service at all operating conditions. The applicability of various piping materials, as governed by ANSI/ASME B31.3, are set forth in the first section of Table 5; their suitability for certain process conditions is indicated in the second section of Table 5.

ANSI/ASME B31.3—Class D fluids are nonflammable, nontoxic, less than 150 lb/in.^{2}, and between −20 and 366°F.

Class M fluid service is defined by B31.3 as one "in which a single exposure to a very small quantity of a toxic fluid, caused by leakage, can produce serious irreversible harm to persons on breathing or bodily contact, even when prompt restorative measures are taken." OSHA's *General Industry Safety and Health Regulations, Part 1910, Sub-part G* provides guidance on handling toxic and carcinogenic substances.

The responsibility for classifying fluids and for the safe design of piping systems in accordance with B31.3 lies with the plant owner. Laws in six U.S. states and seven Canadian provinces mandate that piping be designed according to B31.3.

Piping of polyvinyl chloride (PVC) or reinforced thermosetting resin plastic (RTRP) must be safeguarded when it contains a Class M fluid. B31.3 permits safeguarding to be accomplished in a variety of ways, including by isolation, controlled access, protective barricade, spillage collection and recovery, and design (such as providing insulation, shock protection, armor covering, and double containment). Lined steel piping is considered to be safeguarded.

Piping of thermoplastic materials, such as PVC, is not permitted for aboveground flammable-fluid service. RTRP piping must be safeguarded for handling flammable fluids.

TABLE 5 Service Factors That Influence the Selection of Corrosion-Resistant Piping Materials[a]

Piping Material	ANSI/ASME B31.3					Process Conditions	
	Class D Fluids			Class M Fluids	Flammable Fluids	Vacuum	Steam, 15 lb/in.^{2}-gauge (250°F)
	To 150 lb/in.^{2}	To −20°F	To 366°F				
Stainless steel	OK	OK	OK	OK	OK	See Table 6	OK
Nickel-base alloys	OK	OK	OK	OK	OK	See Table 6	OK
Polytetrafluoroethylene lining	OK	OK	OK	OK	OK	R	OK
Polyvinylidene fluoride lining	OK	OK	OK to 300°F[1]	OK	OK	R	OK
Polypropylene lining	OK	OK	OK to 250°F[1]	OK	OK	R	C
Reinforced thermosetting resin plastic with resins of:							
Bisphenol A and fumaric acid	OK to 12 in. o.d.	C	OK to 285°F[1]	R-SG	R-SG	C	OK
Chlorinated polyesters	OK to 12 in. o.d.	C	OK to 250°F[1]	R-SG	R-SG	C	C
Hydrogenated Bisphenol A and Bisphenol A	OK to 12 in. o.d.	C	OK to 220°F[1]	R-SG	R-SG	C	R
Bisphenol A and isophthalic acid	OK to 12 in. o.d.	C	OK to 180°F[1]	R-SG	R-SG	C	R
Isophthalic acid	OK to 12 in. o.d.	C	OK to 150°F[1]	R-SG	R-SG	C	R
Polyvinylidene chloride lining	OK	OK	OK to 200°F[1]	OK	OK	R	R
Rubber lining	OK	C	OK to 180°F	OK	OK	R	R
Polyvinyl chloride, Type I	OK to 8 in. o.d. [1]	C	OK to 150°F[1]	R-SG	R-AG	OK	R
Type II	OK to 2½ in. o.d. [1]	C	OK to 140°F[1]	R-SG	R-AG	OK	R

[a]OK: Generally satisfactory, with size and temperature limitations as noted. R: Restricted; not generally satisfactory. R-SG: Restricted; unless safeguarded, prohibited by B31.3. R-AG: Restricted; prohibited by B31.3 for aboveground service. C: Consult manufacturer for temperature or vacuum limitations.

Process conditions—Process equipment is frequently scoured with low-temperature steam. If piping is to transport low-pressure steam, it must be able to handle a fluid temperature of 250°F (15 lb/in.^{2}gauge steam), regardless of the temperature of the fluid normally transported.

Lined piping is generally not used for less than atmospheric service. PVC and RTRP piping can withstand negative pressure. Manufacturer recommendations represent the best guidelines. Guidelines for metallic piping in vacuum service are given in Table 6.

Inadvertent as well as steady-state vacuum should be considered when selecting piping. An inadvertent vacuum could result from, for example, a valve at the bottom of a vertical run of pipe being opened (to drain the line) before a valve at the top is opened.

Pressure Rating

Allowable stresses for metallic piping are listed in Appendix A of B31.3. For example, the allowable stress in tension for electric fusion-welded Type 304 (ASTM 358) pipe is given as 17,000 lb/in.^{2} from −425 to 300°F, and as 14,000 lb/in.^{2} at 600°F. For seamless (ASTM A 386) pipe, the comparable allowable stresses are 20,000 and 16,400 lb/in.^{2}, respectively.

From the allowable stress, the pressure rating can be calculated for any wall thickness. Flanges are rated separately. The flange rating, which is usually lower than that for the pipe, generally governs the pressure rating of a metallic piping system.

The pressure rating of nonmetallic piping is established differently. Times-to-failure of thermoplastic and RTRP piping are first determined in water at 73.4°F, in accordance with ASTM D 1598. Then, pressure ratings are developed from either 100,000-h (11.4-yr) stress-rupture strengths or 150-million-cycle pressure-fatigue strengths and appropriate service (design) factors.

Stress-rupture and fatigue strengths are specific to resin formulation, manufacturing procedure, pipe-to-fitting joining method, and the test fluid and its temperature. The service factor is set by the pipe manufacturer on the basis of experience.

This means of determining pipe pressure rating makes the user dependent on the manufacturer for pressure ratings and tend to force the user to return to a manufacturer for future alterations and maintenance if original design safety factors are to be maintained.

The pressure rating of lined pipe is governed by that of the containing pipe, which is normally Schedule 40 carbon-steel with 150-lb/in.^{2} flanges.

Temperature Rating

The lower temperature limit for Class D fluids is −20°F. Piping of PVC and RTRP, in most formulations, is not fully satisfactory at this temperature.

TABLE 6 Wall Thickness and Stiffening-Ring Spacing for Full Vacuum Service[a]

Nominal Thickness	Type 304L and 316L Stainless Steel at 100°F Pipe o.d. (in.)												
Gauge	8	8⅝	10	10¾	12	12¾	14	16	18	20	24	30	36
16	62S	58S	46S	44S	38S	34S	31S	24S	NR	NE	Not for vacuum service		
14	NR	NR	76S	70S	62S	57S	51S	40S	36S	29S			
12	NR	NR	NR	NR	NR	135S	105S	93S	81S	69S	53S	39A	30A
11	NR	NR	NR	NR	NR	NR	NR	121S	110S	92S	76A	54A	43A
10	NR	NR	NR	NR	NR	NR	NR	NR	137S	128S	102A	70A	55A
8	NR	NR	NR	NR	NR	NR	NR	NR	NR	210A	153A	116A	86A
3/16	NR	NR	NR	NR	NR	NR	NR	NR	NR	NR	185A	140A	111A
1/4	NR	NR	NR	NR	NR	NR	NR	NR	NR	NR	NR	292B	222B
3/8	NR	NR	NR	NR	NR	NR	NR	NR	NR	NR	NR	NR	NR

[a]NR: Stiffening rings are not required for sizes smaller than 8 in. o.d. if wall is 16 gauge or heavier, regardless of length. S: Use standard 10-gauge flat-face ring. A: Use angle-face ring. B: Use ½-in. × 2½-in. flat bar.

RTRP piping must be shielded from cold temperatures. This is also true in some instances of natural-rubber lining, especially if the pipe is installed outdoors and aboveground.

Polyvinylidene fluoride (PVDF) and polytetrafluorethylene (PTFE) linings serve acceptably at temperatures below −20°F. However, in services below −20°F, a material such as stainless steel must be substituted for the carbon steel that is normally the containing shell for these linings.

Only metallic and PTFE-lined piping are suitable for service at the 366°F maximum temperature limit for Class D fluids. Approximate temperature limits for other materials are reported by Schweitzer [1].

RTRP piping is suitable for maximum temperatures of from 140 to 285°F, depending on the resin formulation. The maximum temperature limit for RTRP and other nonmetallic piping is determined in water. The manufacturer should be consulted for limits in other fluids.

Reference for Part 4

1. P. A. Schweitzer, *Handbook of Corrosion Resistant Piping*, Krieger, Malabar, Florida, 1985.

ARTHUR H. TUTHILL

Part 5: Summary and Conclusions

Factors to Consider in Evaluating Cost Estimates

Certain assumptions relative to the cost estimates and comparisons made in this series of articles, and in the earlier articles, should be examined.

Job Size

In the earlier articles, installed-cost estimates were based on piping systems approximately 500 ft long. To ensure comparability, the estimates in the present series of articles are based on systems of similar length, even though doing this tends to exaggerate the cost of metallic piping.

For a relatively small system, quantity extras boost the cost of metallic piping significantly over that for nonmetallic or composite piping. Indeed, quantity extras represent an appreciable percentage of the total material cost of the higher-alloyed steels.

Had the estimates been based on 10,000-ft systems, rather than 500-ft ones, these quantity extras would not have been applicable, and the cost comparisons would have shifted in favor of stainless-steel pipe. Moreover, labor costs for installation, setup, and cleanup would have been lower for 10,000 ft systems. Other economies of scale could also be realized—e.g., from the use of bends in place of fittings. These savings would be larger for stainless-steel piping than for nonmetallic or lined piping.

Pipe Size

Valid comparisons among all the materials can only be made for pipe diameters of 6 and 8 in. Two of the lining materials (polyvinylidene chloride and polypropylene) normally are not applied to pipes larger than 8 in. Only rubber-lined steel piping is available in diameters larger than 12 in.

In sizes less than 6 in., the plastic lining significantly reduces the flow area of the pipe. Thus, Schedule 10 4-in. lined pipe provides 15% less flow area than the corresponding metallic or nonmetallic pipe; in 2-in pipe, the reduction is 80%. Because so much of the cross-section is lost due to the lining, comparisons of smaller piping should be based on flow area rather than on nominal pipe size.

Contractor Experience

Piping contractors have had more experience with Schedule 5S and 10S, Type 304L and 316L pipe than with the other corrosion-resistant piping. Their experience with reinforced thermosetting resin (RTRP) piping has been mainly with the press-fit (socket-joined) type. Therefore, costs for RTRP piping in standard cost manuals and in the estimates of individual contractors are based primarily on this type.

Contractors have had much less experience with lined piping, the hand layup and safeguarding of RTRP pipe, and the higher-alloyed metallic piping. Although estimates for installations involving piping of the more common materials can be reasonably accurate, caution must be exercised in evaluating estimates and quotations for the less frequently specified piping materials.

For example, if, for the 6-in.-o.d. simple (product-transfer) and complex (process) arrangements, the field-erection workhours of the earlier articles are substituted for those in this series, the estimates for Schedule 10 304L and 316L and lined piping will be 15–20% lower. For the simple system, the estimate for RTRP pipe will be about 20% higher. For the complex system, the estimate for lined piping will be 40% lower.

The foregoing illustrates how substantial the differences in installed-cost estimates can be. The magnitude of difference usually depends on the extent of experience with a material that has been incorporated into cost estimating tables. Although large, a difference of 40% in proposals from a number of contractors bidding on the same job is not uncommon.

Functional Equivalency

Comparisons of installed costs are valid only if the piping systems are reasonably equivalent functionally. By this criterion, some of the installed-cost comparisons in the earlier articles are misleading because 150-lb/in.^{2} nonmetallic and lined pipe were compared with Schedule 10S stainless-steel pipe. In this series, Schedule 5S has been selected as the reference cost base because pipe of this schedule is rated for service in excess of 150 lb/in.^{2}.

Care must be observed to ensure functional equivalency between metallic and nonmetallic piping. Table 7 presents some of the factors that significantly influence the installed cost of RTRP pipe.

Antivibration measures for RTRP piping begin with a complete vibration analysis of the system to determine whether and where mass must be added. Such additions can be made by sliding a short section of larger diameter pipe over the primary piping and filling the space between the two with resin.

Because antivibration and many safeguarding options are site specific, their costs cannot easily be included in general installed-cost estimates, even though their influence can be major (as Table 7 indicates).

Insurance premiums represent an annually recurring cost. Because insurers tend to prefer integral-metallic to nonmetallic or composite piping, surcharges have been made on insurance premiums for RTRP piping systems, especially when vibration was considered a factor. However, insurance costs, also being site specific, cannot be readily included in general installed-cost estimates.

TABLE 7 Cost Comparison between RTRP and Type 316L Piping

Type of Installation	Approximate Ratio of Installed Cost (316L Sch. 5S = 1.0)
Press fits acceptable	0.5
Press fits not acceptable	1.0
Press fits acceptable with safeguarding	1.5
Press fits not acceptable and antivibration measures required	>1
Press fits not acceptable and both safeguarding and antivibration measures required	>2

Summary and Conclusions

Labor cost, the major part of installed cost, is virtually the same for 304L, 316L, and 317L stainless-steel piping, and about 10% higher for the newer 4.5 and 6% molybdenum alloy pipe. As pipefitters become more experienced in working with the higher-alloy piping, the labor cost of installing it should approach that for 316L piping.

The current premium (in relation to the cost of 316L pipe) for 4.5% molybdenum alloy pipe ranges from 6 to 35%, from 20 to 50% for 6% molybdenum, and from 85 to 120% for Alloy G.

Of the lined piping, polypropylene-lined (PPL) is least expensive, followed by polyvinylidene-chloride-lined (SL), natural rubber, polyvinylidene-fluoride-lined (PVDF), and polytetrafluoroethylene-lined (PTFE), the most expensive.

In the sizes for which lined piping can be considered, PPL-lined piping appears to be reasonably competitive with 6%-molybdenum piping. Pipe lined with natural rubber, SL, or PVDF seems to be competitive with Alloy G piping.

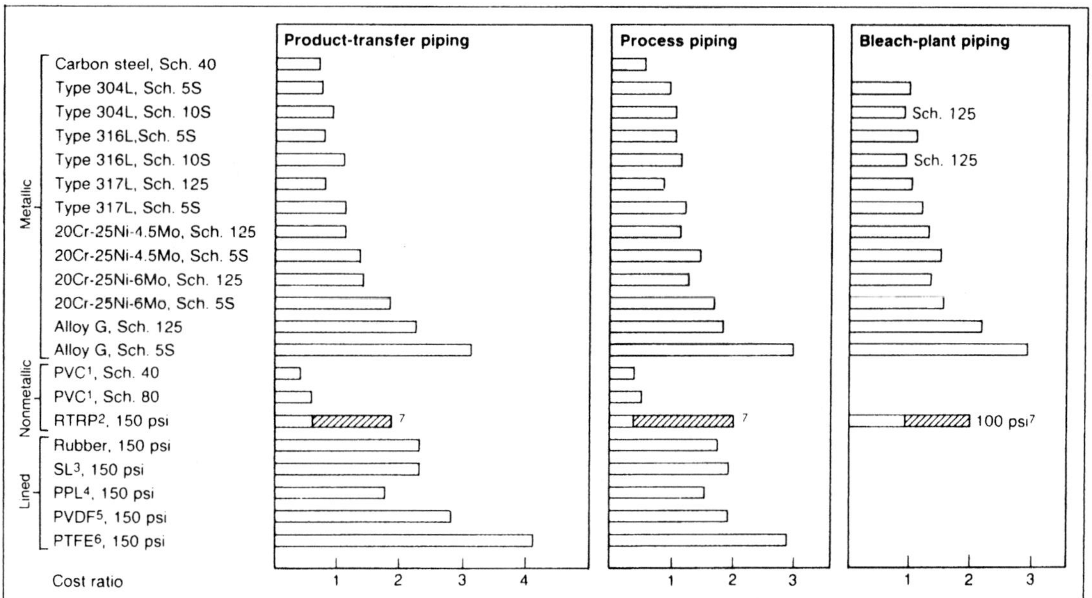

[1]Polyvinyl chloride
[2]Reinforced thermosetting resin plastic
[3]Polyvinylidene chloride
[4]Polypropylene
[5]Polyvinylidene fluoride
[6]Polytetrafluoroethylene
[7]Installed cost of RTRP piping varies substantially, depending on whether press fits are allowed, safeguarding and anti-vibration measures are required, and surcharges on insurance premiums are applicable

FIG. 3 Relative installed costs for 6-in. o.d. simple transfer, 6-in. o.d. complex process, and 20-in. o.d. bleach-plant piping systems.

The cost of labor for installing lined piping appears to be unusually sensitive to the experience of the pipefitters. Specially trained crews can significantly reduce the labor cost of installation.

Of the nonmetallic piping, that of polyvinyl chloride (PVC) Schedule 40 is least expensive. Schedule 80 PVC piping follows, with RTRP piping the most expensive. Because nonmetallic piping is press fitted and lighter in weight, the cost of labor for installing it is less than that for metallic and lined piping. The installed cost of RTRP piping can vary from 50 to 200% of the corresponding cost for 316L piping, depending on the cost necessary to bring RTRP pipe up to functional equivalency with 316L pipe.

A reasonable summary of relative installed costs, suitable for quick estimating, is presented in Fig. 3 for the three piping systems.

ARTHUR H. TUTHILL

5

Buried Pipelines

Corrosion Control

In designing a chemical-process-industries plant, one of the first considerations should be corrosion prevention for underground metal structures. Such plants typically have extensive underground pipelines, tanks, electrical grounds, and other setups which may corrode or result in corrosion of other constructions.

Corrosion problems must be assessed and preventive measures taken while the plant is in the design stage. To help the engineer do this, we shall detail the principal causes of underground corrosion and the methods to eliminate it.

Types of Underground Corrosion

Soil Corrosion

In general, corrosion is caused by a direct current that leaves a metal and enters an electrolyte. For underground pipelines, soil is the electrolyte. At the place where the current leaves the pipeline or other structure to enter the soil, corrosion occurs. If the soil is the sole cause of corrosion, then the corrosion rate depends upon the electrical resistance and driving potential of the electrical circuit.

The resistance varies with the moisture content in the soil (see Fig. 1, from Ref. 1). No corrosion will occur in completely dry soil; however, nearly all soils have enough moisture to sustain corrosion. Some are so dry, however, that the rate is very slow.

The main factor that causes corrosion on pipelines is differential aeration of the soil [2]. If a pipeline is located in soil that is high in oxygen at one location and low at another, an oxygen-concentration cell will be established. A direct current results, with corrosion occurring at the low-oxygen area where the current enters the soil. The current flows back onto the pipe at a high-oxygen area, providing cathodic protection there. The anode and cathode may be inches or hundreds of feet apart, but in either case, serious corrosion may result.

The corrosion mechanism is different in oxygen-rich soils than it is in oxygen-poor ones. When oxygen is plentiful, the initial rate is high, but it is slowed by the corrosion products that adhere tightly to the pipe. These products have a higher resistivity than does the surrounding soil, thus reducing the corrosion rate [3]. Figure 2 shows examples of an oxygen-concentration cell.

It should not be assumed that the corrosion products form a coating that is even remotely comparable to an insulating pipeline-coating. A hard scale is formed, but it does not adhere tightly to the pipeline in all places. In other

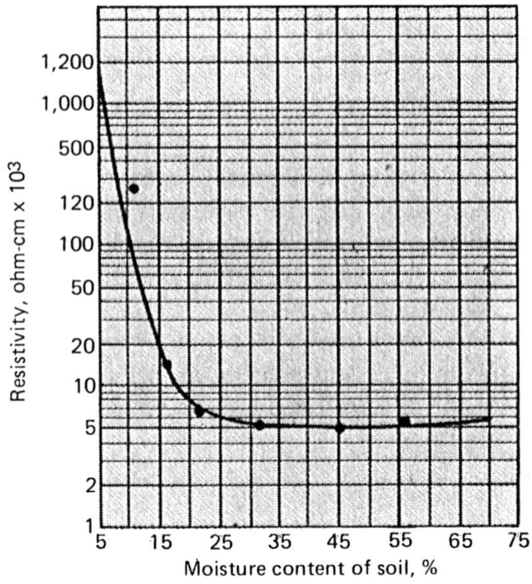

FIG. 1. How well a soil conducts current is a function of its moisture content.

places it is thin, and in still other places there is no scale at all. Oxygen-concentration cells will exist in this well-aerated area, and corrosion will continue. Even so, the scale formed by the corrosion products increases the electrical resistance and often reduces the overall corrosion rate significantly. Sometimes it quickly drops to a low value.

While the corrosion rate in a well-aerated area decreases in the first few months, this is not so in a poorly aerated location. The corrosion products produced in heavy moist soil differ from those formed in well-aerated earth. The corrosion products produced in oxygen-deficient soil do not adhere tightly to the pipe and do not form an insulating barrier that reduces current flow.

Thus, corrosion continues unabated. The effect of this continuing corrosion is cumulative, and the total amount over a few years is much greater than in well-aerated soil.

In these heavy, moist soils, pitting corrosion can worsen the problem. Pits grow rapidly, and a pipe can be penetrated quickly. Often, in such cases, even though more than 95% of the pipe surface shows little corrosion, a section will have to be replaced because of a few deep pits.

Bacterial Corrosion

Most of this type of corrosion is attributed to *Desulfovibrio desulfuricans*, an anaerobic sulfate-reducing bacterium. This organism is often found in heavy,

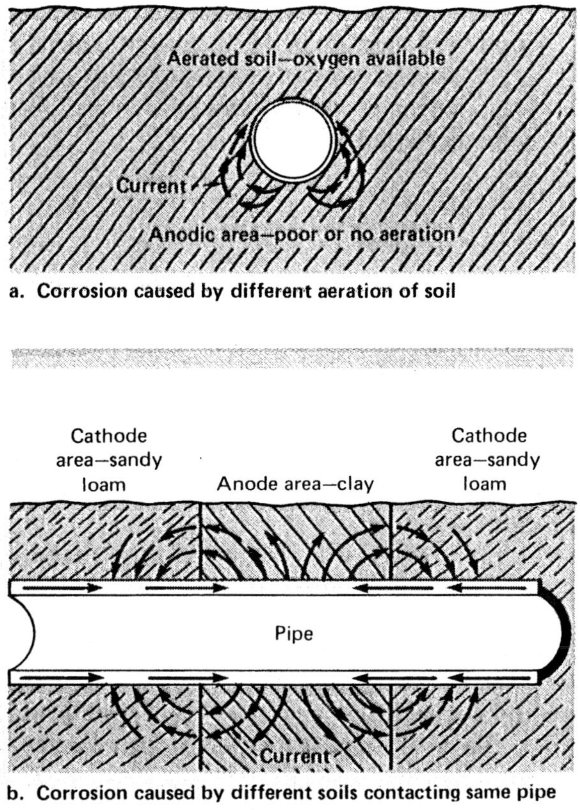

FIG. 2. Here are just two situations that can cause below-ground oxygen-concentration cells.

moist soils, such as the clays, and thrives where there is a lot of organic matter [4].

D. desulfuricans may cause severe corrosion on underground pipelines. Where there are high numbers of sulfate-reducing bacteria in the soil around the pipeline over extended periods, a typical pattern of corrosion is often seen: Deep, craterlike pits develop and the pipe wall may be penetrated in a short time. However, other factors also create this pattern, so it is not always an indication of sulfate-reducing bacteria at work.

Galvanic Corrosion

This type of corrosion occurs when two dissimilar metals are connected together in an electrolyte, such as soil. For instance, copper water pipes are often run to steel pipes or tanks—a powerful corrosion cell is set up and serious corrosion can result in a short time.

Any two metals in the galvanic series will cause a current to flow when connected together in an electrolyte. The less-noble metal will corrode.

The use of insulators eliminates galvanic couples. Also, a sacrificial anode (e.g., one made out of magnesium or zinc) can be connected to a steel pipe to cathodically protect the steel.

Chemical Corrosion

When a plant is expanded or a new one is built and land is scarce, marginal land may be used. This includes dumps, landfills, and abandoned sewage-treatment plants. Chemical dumps are often adjacent to some of the older chemical-processing plants—just the locations where new plants are likely to be built.

Van Eck [5] warns that all of these marginal areas may contain corrosives, and their presence must be determined by chemical tests. Soil samples should be collected at enough locations and depths to provide a good assessment of conditions. Often, however, only a small part of an area will contain corrosives. Most plants will not be built where the soil is contaminated, but some soil tests and chemical analyses should be made anyway.

In most cases, if there is enough foreign chemical in the soil to damage pipelines, there will be some indication of it from a cursory examination. Appearance, odor, and pH will usually flag a warning. A conventional chemical analysis may also indicate the presence of corrosives. If a plant that is manufacturing or using chemicals is nearby, then soil tests should be made for the chemicals found at that plant.

Chemicals that leak into the soil or are applied to it (e.g., fertilizers) can be corrosive. If the concentration is high enough, fertilizers can corrode steel. In most cases where chemicals have damaged substructures, care has simply been inadequate.

Most chemical damage is in the form of pitting corrosion—a costly problem. Stress-corrosion cracking (SCC) may also occur, and it could result in catastrophic failure of a pipeline. Even a small amount of material can cause SCC.

Sulfides, carbonates, nitrates, chlorides, and caustics have all been reported to cause SCC in steel [6]. Although these groups are normally found in soils, they are typically a problem only when they occur in large concentrations, i.e., when they come from manmade chemicals.

The use of coatings, SCC-resistant steels, and cathodic protection have reduced the number of accidents and problems caused by this form of corrosion. Nonetheless, contaminated soil is an undesirable environment for steel structures. If such soil cannot be removed, the pipeline should have one or two layers of pipeline coating applied over the standard coating already on it. Cathodic protection should be used, too.

Tests to Determine Soil Corrosivity

The most common method of doing this is by measuring electrical resistivity. Many factors affect the resistivity of a soil (e.g., moisture, temperature), so this measurement only provides an approximation. Further, the resistivity of a soil may change from time to time.

For example, rainfall may differ from one season to the next. In Southern California, tests made in desert areas during summer often yield values of 50,000–100,000 ohm-cm. During the rainy winter, readings may drop to less than 5000 ohm-cm.

An excellent booklet on earth-resistance testing is published by the James G. Biddle Co. [7]. As described in the booklet, most resistivity tests use the Wenner four-pin method. It has the advantage that only surface measurements need be made; no excavation or hole boring is necessary.

Figure 3 shows a typical setup for making this test. Current is put into the soil through pins C_1 and C_2; potential is measured at pins P_1 and P_2. By spacing the pins at different distances, average resistivity can be found at various depths. For instance, with a 10-ft spacing, resistivity at a 10-ft depth will be found. If the soil is uniform from the surface down to 10 ft, then the measurement will be correct for all depths down to 10 ft.

In most cases, readings are made at several depths, and it is possible to see from these the changes in resistivity with depth. The resistivity at the bottom of the hole where the pipe will be placed can then be estimated. Usually, such a method suffices to determine resistivity.

Resistivity is not the only important measurement; the soil pH has a significant effect on the corrosion rate. Recognizing this, Stratfull [8] developed a test that incorporated pH and soil resistivity.

The reasons for making soil tests have changed. Federal law now requires that interstate pipelines carrying flammables be coated and cathodically protected. Many state and local governments have statutes for pipelines and other structures not covered by federal laws.

So today, tests are done primarily to determine the resistivity of the soil at locations where anodes will be installed. Such tests are particularly useful to obtain data for deep-anode emplacements. Approximate resistivities are

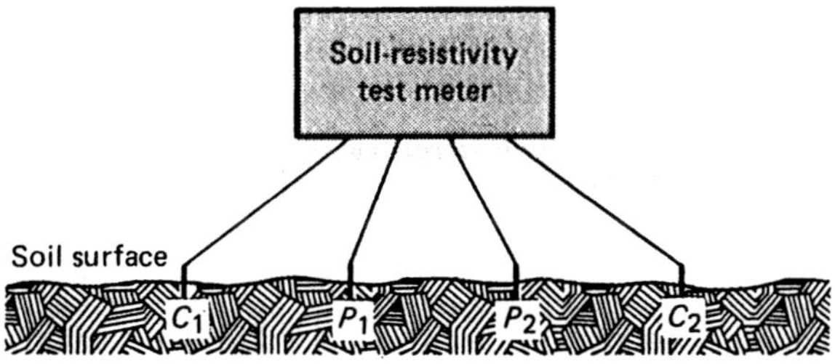

FIG. 3. Wenner four-pin method measures resistivity of soil without need for excavation.

found and are used to sort out suitable locations from unsuitable ones. Afterwards, a hole for the anode is drilled and an electrode is lowered into it. Then, test currents can be applied. From these tests, strata of low and high resistivity can be found. Once this soil profile is known, an anode can be designed.

Methods of Protection

Coatings and cathodic protection are integral to controlling underground corrosion. This was not always so. Some companies simply put a bare pipe in the ground and hoped that it would not corrode. Others coated pipes but did not use cathodic protection. Many plants located near the seacoast installed bare pipes and flooded them with cathodic protection. This was because power was cheap and coastal soils have low resistivity, thus not wasting much power.

However, none of these methods is practical for new construction. Coatings and cathodic protection are needed. Sometimes, law requires this. Also, if there is concern about leakage of noxious liquids, a firm not using both systems might be found negligent if leakage were to occur. If cathodic protection is properly designed and installed, it will be nearly 100% effective.

Coatings

Selecting coatings for plant piping differs from doing it for a cross-country pipeline. Plant structures and belowground plant piping are usually ordered factory-coated while pipelines are coated "over the ditch."

Such differences preclude some types of coatings from being used for plant facilities. Also, some coatings are difficult to use, and some are difficult to obtain in certain locations.

For example, the bitumens are good coatings, but they are messy to apply, and small-diameter pipe coated with them is often unavailable. Typical factory-applied coatings follow:

1. *Heat-cured plastics*. Liquids that are applied to the pipe and heat-cured. Mainly, these are epoxies and phenolics.
2. *Extruded plastics*. Polyethylene is commonly used for extruded coatings. An adhesive, usually butyl rubber, is applied to the pipe and then the polyethylene is extruded over it. The pressure due to extrusion forces the adhesive into good contact with the plastic and the pipe. This process is limited to such shapes as pipes and bars.
3. *Heat-cured powder resins*. Included are epoxies which are applied to a hot pipe (or other metal object) which quickly cures the resin. Such curing yields a hard, dense plastic coating that has good metal-adherence.

The coating is relatively thin, about 12–25 mils. The manufacturer supplies compatible materials for repairs; its directions for field joint-covering *must* be followed.

In selecting coatings, talk to local coatings applicators. They know the types most in demand and can point out their features. They can also provide information on cost and availability [9].

Various organizations have prepared standards and specifications for coatings used in underground service. Such properties should be reviewed and the applicable ones included in a purchase order.

Most coatings withstand temperatures to about 160°F; above that, special coatings must be used. These are often more expensive and require more care during application.

Some structures may be coated at the plant site. These include tanks that will be installed underground and tank bottoms that will rest on the ground or on concrete pads. Such coatings must be self-curing to provide ease of application. Coal-tar epoxies are a favorite here. Many other materials also perform well.

Cathodic Protection

Cathodic protection is not simply a matter of pouring current onto a pipeline. To make it effective, coatings, test leads, and so on are needed. Insulators are used, when needed, to electrically isolate substructures that are to be cathodically protected. A cathodic-protection system should be included at the earliest stage of plant design.

At that stage, many decisions have to be made. For example, it must be decided whether to protect all underground metal structures as a unit or to insulate those that do not need cathodic protection. For an existing facility that has not had cathodic protection before, insulating may not be feasible, and it is better to protect the whole plant as one unit. New plants usually employ insulators and protect only those structures that need it. When an area is isolated and protected, it is referred to as the basic system. Such isolation eliminates most cathodic interference problems and saves power costs. But more care is needed and technicians must be better skilled.

The Basic Cathodic System

The basic system comprises all underground, coated, metal structures, including pipelines and tanks. It does not include reinforcing steel in slabs of foundations, copper ground-rods or mats, or copper water-pipes.

When structures are to be isolated, insulators will have to be installed, and continual diligence will be required to make sure that they are not breached. Tests should be made routinely to ensure that insulators are not shorted and that no piping or wiring has been connected to them. Also, the plant staff involved in maintenance, new construction, or plant modifications should preserve this electrical isolation during such operations.

Care must be taken that electrical controls not be so installed that they provide a path around an insulator. Similarly, be careful of small-diameter pipes. They are plentiful in many plants, and any of them may accidentally be connected around an insulator, shorting it. When this happens, it sometimes takes considerable testing to determine the cause of the short.

Whole-Plant Protection

In some plants, maintaining electrical isolation will be too difficult, and all substructures will therefore be cathodically protected, whether they need it not. While this system is simpler since no insulators are needed, the current required is greater, and this is expensive. This is somewhat compensated for by eliminating the costly process of searching for shorts.

When the plant is protected as one unit, the copper electrical grounds are tied to the underground steel structures. A galvanic couple is set up between copper and steel. This couple could cause corrosion if the cathodic-protection current that reached such areas were not strong enough to overcome it. To guard against this, a magnesium anode should be connected to the steel near each ground rod. The anode will protect the steel and improve the plant's electrical grounding system.

The other drawback to whole-plant protection is cathodic interference, which will be covered later on. Large protection currents may be needed in some plants. Some of this current may cause corrosion on pipelines and cables nearby.

A procedure will be given for finding the proper amount of current needed. It applies whether the whole plant or just the basic system is to be protected. The values shown are for a typical basic system of a very large petrochemical plant, assuming that there would be some damage to the coatings during construction.

Amount of Current Required

After plant construction is completed, tests are needed to design the cathodic-protection system. First, the amount of current needed must be determined. To do this, test current is applied to the substructures to be protected. Pipe-to-soil potentials are measured at strategic plant locations with the current on and off.

Figure 4 shows how a test current is supplied by a temporary cathodic-protection station. The ideal amount of current would create a minimum pipe-to-soil potential—anywhere in the plant—of 1.0 V, and a change in the off-and-on potential of about 0.5 V.

If there is insufficient current to do this, then the resistance of the temporary anode should be lowered by watering it or by adding more rods to it. If it is impossible to bring the pipe-to-soil potential to 1.0 V, then it can be

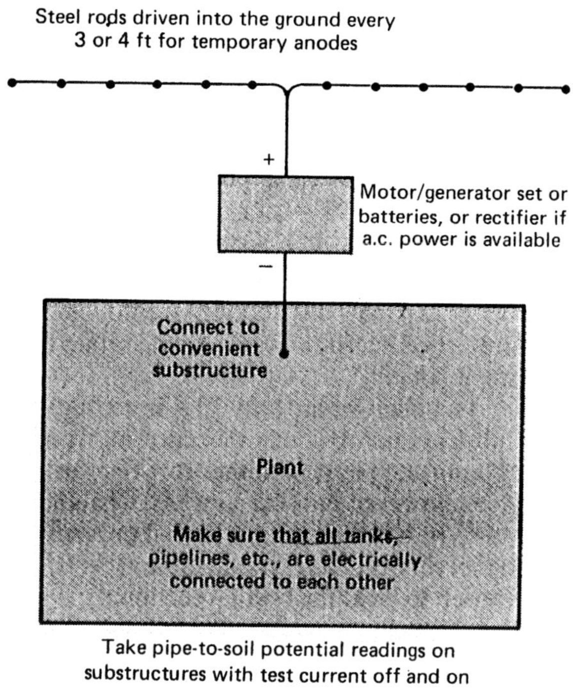

FIG. 4. A temporary cathodic-protection station is used to apply a test current.

extrapolated. If the off-potential is 0.6 V, and when 10 A is applied the potential is 0.8 V, then 20 A should produced about 1.0 V.

For this plant, 20 A is assumed to offer excellent protection. However, most new plants would not initially require as much current unless the coatings were substantially damaged or improperly applied. But in this hypothetical plant, there is a fence around the property line. If nothing were done, the bottom of the steel fence posts would corrode due to cathodic interference. To prevent this, bond wires are connected from the fence to the rectifiers.

Impressed Current vs Anodes

Now, the current source must be selected. Should galvanic anodes or impressed current be used? Galvanic anodes are usually made of zinc and, more often, magnesium, due to the latter's higher driving potential. In most cases when the impressed current is above 3–4 A, an impressed-current system is usually the less expensive of the two.

Further, such a system provides greater flexibility: The current can be adjusted easily, while with galvanic anodes more of them must be added to increase the current. Also, the impressed-current system makes testing easier.

The current can be turned on and off with a switch, whereas anodes must be disconnected and reconnected.

Impressed current can be supplied in several ways. A rectifier can change ac into dc, solar systems or fuel cells can produce dc directly, and gas engines can drive dc generators. However, when commercial electric power is available, it is almost always chosen. It is reasonably efficient, troublefree, and cheap.

More than one rectifier can be used. With large plants, particularly old ones, a large amount of current might be required, possibly 100 A or more. Better current distribution would be obtained if four or five cathodic-protection stations were installed rather than having the current come from a single source.

In a plant where only 20 A is needed, the decision is not so clear. With the substructures coated, the current would probably be evenly distributed around the plant. Another rectifier costs money to install and maintain. It does, however, provide a means of balancing potentials should the need arise. Also, with two rectifiers, partial protection will be afforded if one of them is out of service. Weighing all these aspects, two cathodic-protection stations are chosen to yield the best protection.

Anode Design

For simplicity and to show both types of installation, with two cathodic-protection stations, assume that one anode will be a shallow horizontal type and the other a deep vertical design. Each will have an initial output of 10 A.

Figure 5 shows the location of the plant and some features that affect anode design. Adjoining plants are too close to put anodes on the sides of the plant. Pipelines, cables, etc. in the street rule out the use of shallow anodes in the front of the plant. At the rear, borings taken prior to construction showed a layer of rock about 17 ft below the surface. This dictated that any anode installed here would have to be shallow.

Soil tests will be made in the front and rear of the plant by using the Wenner four-pin method. At the rear, soil samples will be taken from the surface down to a depth of 17 ft. Resistivities of these samples will be measured in the laboratory.

The type of anode must be selected. Graphite or high-silicon cast-iron rods are the most common. Both are usually surrounded with coke breeze (undersized coke screenings about 5/8 in. and smaller) for 4–5 in. If installed in a round hole, the diameter would be 11–12 in. Steel pipes were once a favorite for anodes but are now rarely used. Today's anodes are more expensive and are specifically suited for use in certain soils [10].

For the two cathodic-protection stations, graphite rods with coke breeze are chosen. For the station at the rear of the plant, holes, 10 ft apart, will be drilled in a straight line parallel to the rear lot line. The holes will be 14 ft deep, with the anode occupying 12 ft, allowing 2 ft of soil cover. Each hole will be of

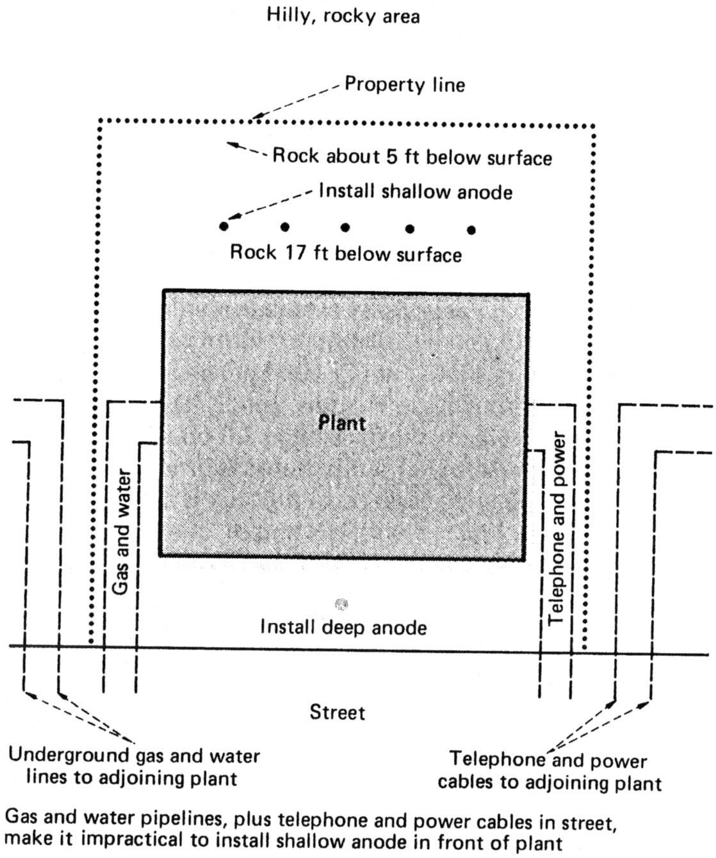

FIG. 5. A rough, typical plant layout shows factors affecting cathodic-protection system design.

12-in. diameter and will contain a single graphite rod, 3 × 60 in., surrounded by coke breeze.

If the soil resistivity is assumed to be equal to 4800 ohm-cm, then the anode's resistivity is 2.2 ohms, as calculated by a formula given by Dwight [11]. With a 10-A output, the minimum required voltage is found to be 22 V.

The deep anode at the front of the plant will be a single unit, 70 to 150 ft deep. The soil resistivity from 70 to 150 ft deep averages 5600 ohm-cm. Graphite rods are used, this anode will also have an output of 10 A.

Seven graphite rods will be installed in a 12-in.-diameter hole and be surrounded by coke breeze. The anode will be 80 ft long, with its top 70 ft below the surface. A vent pipe will be installed from the anode to the surface to carry away any gases that might form in the anode and raise its resistance.

This pipe also serves as a means of watering the anode if the soil around it becomes too dry. The resistance is calculated to be 2.0 ohms; the required voltage is 20 V.

Installation and Start-up

Some engineers delay ordering rectifiers until the anodes have been installed. Then additional tests are made to see exactly how much voltage is needed to supply the required current. It it not always possible, particularly with deep anodes, to accurately calculate anode resistances. Measuring the resistances after anode installation is more accurate. But even using this method, rectifiers should not be sized precisely to match these resistances.

Although the shallow anode has a calculated resistance of 2.2 ohms, measurement shows it to be 2.0 ohms. A driving voltage of at least 20 V is needed. But rectifiers are always ordered with extra capacity for current and voltage; how much extra capacity is a matter of judgment. Here, the rectifier will be ordered with an output of 40 V and 15 A. This will provide extra voltage in case the anode dries out, and extra current if the coating is damaged or additions are made to the plant.

For the deep anode, the measured resistance is 2.3 ohms. The rectifier's output will also be 40 V and 15 A.

Once installed, the rectifiers are turned on and set at 10 A. Pipe-to-soil potentials are measured throughout the plant. Although the ideal protection potential is 1.0–1.2 V, some potentials may fall out of this range. Outputs of the rectifiers will be adjusted to bring all of the low spots to at least 1.0 V and the high spots down to as close to the 1.0–1.2 V range as possible. This may result in an imbalance of currents from the two rectifiers, but this is not a matter of concern since the pipe-to-soil potential is the governing factor.

Even after the currents have been adjusted, some of the potentials may fall outside of the desired range. If they are low, and if it is a matter of one or two spots with potentials of no lower than 0.9 V, a few magnesium anodes can be installed at these spots.

Potentials that are slightly higher than the desired ranges at a few spots are tolerated by most corrosion engineers. However, it is advisable to get the coating manufacturer's recommendations in regard to this. Some coatings have withstood high potentials for long periods, but a few corrosion engineers think that some disbonding may occur if potentials are as high as 1.0 V. If necessary, zinc anodes can be installed at the "high" spots to lower the potential.

Corrosion can result if the potential is too high. The coating will disbond, and then moisture will penetrate under it. The cathodic-protection current will not travel far under a disbonded coating and will not reach all spots. Corrosion occurs and the cathodic-protection system does nothing to stop it.

When oxygen is absent, the corrosion process under disbonded coatings may be complex. Anaerobic bacteria may play a role in this [12]. Disbonding of a coating is just about the worst thing that can happen to it. Most new pipeline coatings can withstand potentials to about 1.6 V without much disbonding. But to be safe, potentials less than this should be the norm.

Notices and Permits

The local cathodic-protection committee should be notified that a system will be installed. So should managers of nearby plants. In some cases they may suggest changes. Notification may avoid trouble later if a cathodic interference problem develops at an adjacent plant. The same people should be sent a notice once the system is in operation. This final notice should state the operating current and voltage of each rectifier.

There are nearly 50 cathodic-protection committees in the United States and Canada. These groups were devised to that companies owning pipelines or other underground facilities would be informed when cathodic-protection stations were being installed.

Such groups do not maintain local offices, and their officers change every year or two. NACE—the National Association of Corrosion Engineers (Houston)—has tried to maintain a mailing list of these committees, but unsuccessfully since they have no fixed addresses. Nonmember companies may send representatives to their meetings. For instance, a plant engineer with a specific interference problem can attend a session and get expert advice.

These groups often use formal notification forms when a cathodic-protection station is being installed. Also, they have developed guidelines for handling cases of cathodic interference. NACE has published a booklet that contains procedures developed by various cathodic-protection committees [13].

Many states and local governments are concerned about pollution of underground aquifers. Some require permits for any hole drilled deeper than 50 ft. California is one, and has written specific guidelines for drilling and installing deep anodes. When adoped by cities and counties in that state, these guidelines become law [14].

For our example, the deep anode will require a permit. In addition, it is usually required that the anode hole be sealed for the top 50 ft or more with concrete or similar material. This keeps out surface water. Sometimes, much deeper sealing is called for to prevent interchange of water between aquifers.

Some local governments require building permits for cathodic-protection stations using commercial power.

Routine Tests and Recordkeeping

After the cathodic-protection system is in operation, pipe-to-soil measurements should be made throughout the entire plant. Minor rectifier adjustments might be needed to keep the potential within the desired range.

After about a month of operation, the pipe-to-soil potentials should be measured again. If they are satisfactory, a routine test schedule should be set up. Rectifier outputs should be read at least once a month. For the first year, pipe-to-soil potentials should be measured every 3 months.

If potentials are stable during the first year, the time between measurements can be lengthened, but the risk of corrosion must be balanced against the money saved. Many companies adopt a modified schedule whereby tests

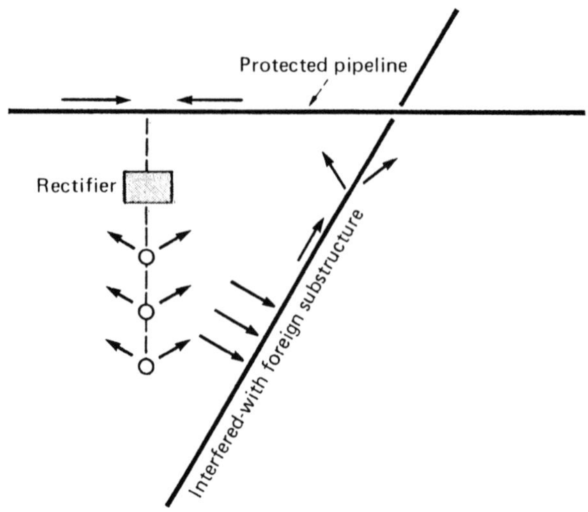

FIG. 6. Current interference is by far the worst kind of cathodic interference.

are made only at key points every 3 months. A complete set of potentials is measured annually. Routine test records should be kept so that results can be compared.

Cathodic Interference

There are two types of cathodic interference, one from current and the other from voltage. Current interference is by far the worse (see Fig. 6). Figure 6 shows a cathodic-protection station with a rectifier and a ground bed. Normally, such a setup will cause several amperes to flow through the soil to the protected pipeline.

Some current collects on a nearby foreign pipeline and travels on it for some distance. As the current nears the cathodically protected pipeline, it discharges into the soil, causing corrosion. If the foreign pipeline is coated, as most are, the interference current leaves the pipeline at holidays. This can cause rapid, severe corrosion.

Figure 7 illustrates voltage interference, in which an excessive voltage may be created between the foreign pipeline and the surrounding soil. If the foreign pipeline is coated, disbonding may occur. The excessive voltage and high current density at the holidays make the soil quite alkaline.

This alkalinity destroys the coating's bond. If the voltage is high enough, hydrogen gas will be formed at the pipe wall and will penetrate the coating, lifting it. Considerable damage can be done to a coating.

Equally bad are the effects on uncoated amphoteric-metal structures, e.g., lead. Such metals are subject to corrosion in highly acidic or alkaline

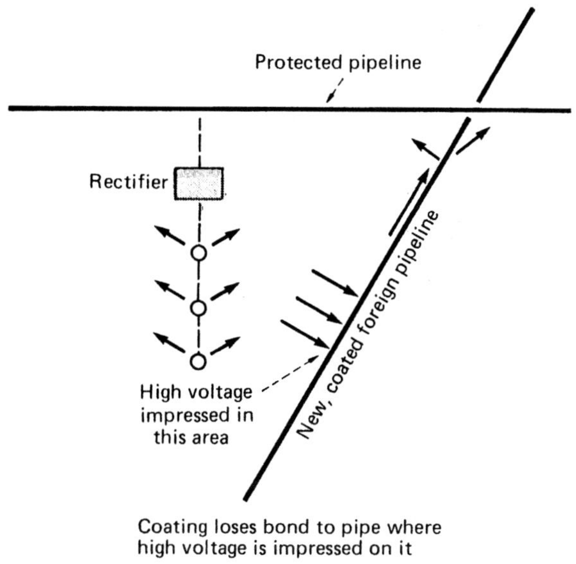

FIG. 7. Voltage interference can result in coating disbondment and corrosion.

environments. Alkaline products may be formed if lead cable-sheaths are made excessively negative, especially in soils containing appreciable amounts of sodium or potassium salts [15].

Another amphoteric metal is aluminum. It is not used much underground, although some pipelines are made of it. Amphoteric metals may suffer much more corrosion than other metals when affected by cathodic interference. Extra precautions must be taken.

And remember that interference can be caused on a plant's substructures by the operation of cathodic-protection stations of other companies.

Avoiding Interference

Whether or not a cathodic-protection station will cause interference depends primarily on the current output of the anode and the proximity of the anode to other substructures. Galvanic anodes produce small currents, usually less than 200 mA. Rectifiers may have outputs of over 100 A.

However, if enough galvanic anodes are installed and spread out over a long pipeline, they could produce a substantial current. Usually, this is not so, because rectifiers are more economical above a certain threshold total current value.

Locating the anode as far away as possible from any substructure that might pick up current will reduce interference. A cardinal error in corrosion engineering is to place the anode so that a foreign substructure is between it and the protected structure.

The second-worst error is to place the anode close to a foreign substructure that offers a good path for the current to follow in its effort to get to the protected pipeline.

Designing to Avoid Interference

Substructure maps for the streets around the plants should be examined. These are kept by most large cities and some county offices. At this time, the engineer should contact nearby companies to determine whether their plants might be affected by cathodic interference, and to arrange for cooperative efforts to prevent any damage. Finally, the local cathodic-protection committee should be notified.

In avoiding cathodic interference, prime responsibility rests with the company installing the cathodic-protection system. Once the system is in, the responsibility shifts to any company that installs a new pipeline or other new substructure that might suffer interference from the system.

The owner of the cathodic-protection system should be notified. Joint tests would then be arranged between the two companies, and a decision made as to how to solve the interference problem.

Legal Aspects

Some consideration should be given to the legal aspects of cathodic interference [16]. The corrosion engineer should keep records of tests and designs of cathodic-protection systems, especially those pertaining to interference.

It is difficult to assign blame in cases of interference, because of the natural tendency of underground pipelines to corrode. It is hard to determine how much corrosion would occur were there no interference. But, in many cases, losses amount to hundreds of thousands of dollars, and lack of documentation may lose a case for a company.

Effects of Alternating Current

Alternating current may cause corrosion on underground steel structures in large plants such as oil refineries and chemical-processing facilities. Electrical grounds are usually connected through some metallic paths to underground tanks, pipelines, etc.

As a result, these steel structures often carry several amperes of ac. While ac is not normally considered to cause significant corrosion, this may not be true with underground steel. Fuchs et al. [17], in laboratory tests, found a relatively high corrosion rate with high current densities. Other investigators have come up with mixed results.

However, even in plants where large amounts of ac flow on underground structures, there is no need for concern if these are cathodically protected.

Bruckner [18] states that cathodic protection will prevent ac corrosion, but cautions that magnesium anodes may not be satisfactory. This is because a reversal in potential of the anodes may sometimes occur. He suggests that impressed-current systems be used.

Other corrosion engineers have recommended that the pipe-to-metal potentials of cathodically protected steel structures be maintained from 0.1–0.2 V higher than would be needed if there were no ac on the structures. This seems to be a reasonable precaution.

Insulators

In most plants, steel substructures will inadvertently be connected by metallic paths to electrical grounds, reinforced steel in building slabs, and various kinds of pipelines and electrical cables coming into the plant. If cathodic protection is installed without using insulators, then the reinforcing steel, copper ground-rods, etc. will soak up large quantities of protective current.

Shielding and Test Connections

In large plants, substructures are often crowded together, and shielding may occur so that a sufficient cathodic-protection current does not reach all parts of a protected structure [19]. In older plants, holidays may develop in coatings, and shielding may begin to cause corrosion. High-resistivity soil tends to make shielding worse [20].

A good way to provide needed current at shielded locations is to use magnesium anodes. During plant design, locations where shielding might occur can be surmised by examining the plans. If a large pipe is in front of a small one, the small pipe might not get enough current to prevent corrosion. One or more magnesium anodes connected to the small pipe and placed between the two pipes would offer protection.

A related precaution: One should install a permanent copper sulfate electrode at each shielded location to provide for future measurements of pipe-to-soil potentials. All wires should be brought to a surface test-box. A few other permanent copper sulfate electrodes should be installed at strategic locations throughout the plant. These locations should be remote spots where the level of protection cannot be estimated with reasonable accuracy.

A final note: Cathodic protection is extremely effective in controlling corrosion, but it must be designed and maintained properly.

References

1. National Bureau of Standards, U.S. Department of Commerce, *Underground Corrosion*, Circular 579, 1957, p. 155.
2. E. Schaschl and G. Marsh, "Some New Views on Soil Corrosion," *Mater. Prot.*, pp. 8–17 (November 1963).
3. M. J. Schiff, *What Is Corrosive Soil?*, Paper Presented at the National Association of Corrosion Engineers Western States Corrosion Seminar, Houston, 1969.
4. H. Uhlig, *The Corrosion Handbook*, Wiley, New York, 1955, pp. 474–476.
5. W. A. Van Eck, "What is Soil?," *Pipe Line News*, pp. 17–22 (March 1964).
6. F. L. LaQue and H. R. Copson, *Corrosion Resistance of Metals and Alloys*, Reinhold, New York, 1963, pp. 315–319.
7. *Getting Down to Earth*, James G. Biddle Co., Plymouth Meeting, Pennsylvania.
8. R. F. Stratfull, *A New Test for Estimating Soil Corrosivity*, Paper Presented at the National Association of Corrosion Engineers Western Region Conference, 1960.
9. R. N. Sloan, "How to Select Mill-Applied Coatings," *Pipeline Gas J.*, pp. 40–48 (February 1982).
10. R. W. Stephens, *Deep Ground Beds—Material Selection and Economics*, Paper Presented at National Association of Corrosion Engineers Corrosion/83, Anaheim, California, April 18–22, 1983.
11. H. B. Dwight, "Calculation of Resistances to Ground," *Electr. Eng.*, December 1936.
12. A. W. Peabody, *Control of Pipeline Corrosion*, NACE, 1967, p. 175.
13. Bulletin IV of the Report of the Correlating Committee on Cathodic Protection, NACE.
14. Department of Water Resources, State of California, *Cathodic Protection Well Standards*, Bulletin No. 74-1, 1973, with revisions.
15. Ref. 4, p. 603.
16. H. M. Hatley, "Pipeline Corrosion: The Legal Aspects," *Anti-Corrosion*, pp. 6–7 (November 1971).
17. W. Fuchs et al., "Corrosion of Iron by Alternating Current with Relation of Current Density and Frequency," *Gas- Wasserfach*, January 1952.
18. W. H. Bruckner, *The Effects of 60-Cycle Alternating Current on the Corrosion of Steel and Other Metals Buried in Soils*, University of Illinois Bulletin, *62*(32), (1964).
19. M. D. Orton, *Fundamentals of Cathodic Protection*, Paper Presented at the National Association of Corrosion Engineers Western States Corrosion Seminar, 1978.
20. B. Martin, "Cathodic Protection Shielding of Pipelines," *Mater. Perform.*, pp. 17–21 (February 1982).

JOSEPH S. DORSEY

Design

Large diameter pipelines with considerable amounts of compression horsepower per mile are needed to economically transport natural gas from the Arctic to the user. These large amounts of horsepower increase gas temperature and require cooling facilities.

At the high operating pressures proposed for Arctic pipelines, natural-gas density is more sensitive to changes in temperature than it is at lower pressures. Lowering gas temperature, therefore, increases gas density more dramatically than in lower-pressure lines and causes greater increases in flow.

Pipelines which will carry Arctic gas across southern Canada and the United States can therefore can benefit economically from cooling. Although flow increases are modest for these air-cooled pipelines, the cost of air-cooling facilities is low enough to permit a reduction in the unit cost of transportation.

Arctic pipelines benefit more significantly due to increases from refrigeration facilities which are required to maintain the integrity of frozen ice-rich right-of-ways. Percentage increases in flow exceed percentage increases in costs of extra refrigeration facilities and special designs to eliminate frost heaving.

Cooling Methods

In 1970 it was noted that gas pipelines from the Arctic would be operated at temperatures below 32°F to avoid thawing frozen ground surrounding the pipe. Except for a thin surface layer in the brief summer, ground in Arctic regions is frozen to great depths.

In many locations the frozen soil consists of fine silts and clays with excess ice content which makes ground unstable when it thaws. Unless gas is cooled below 32°F, the resulting thermal degradation will cause soil erosion and ground settlement which can jeopardize pipeline security and cause environmental difficulties.

Closed-cycle propane refrigeration units have been proposed to chill gas below 32°F and maintain integrity of the pipe and the right-of-way.

In 1971, studies of transporting gas from the Arctic across warmer parts of southern Canada and the United States showed that cooling is desirable for high flow pipelines even if there is no permanently frozen ground to be maintained. Large blocks of horsepower needed to achieve optimum flows through large diameter pipelines cause high compressor discharge temperatures.

Air-cooled heat exchangers have been proposed to control gas temperatures and to prevent temperatures cascading from compressor station to compressor station.

Between the frozen areas of the North and the unfrozen regions in southern Canada is a broad transitional region about 600 miles wide which contains both frozen and unfrozen ground beneath the surface layer which freezes and thaws seasonally (Fig. 1).

Thousands of individual interfaces between frozen and unfrozen ground along pipeline routes from the Arctic create frost heave difficulties where frozen pipe must traverse unfrozen, frost-susceptible ground. Thaw settlement also creates difficulties where a warm pipe must traverse frozen, high-ice content ground.

Since it is not practical to provide gas heaters and gas refrigerators at every individual interface, special designs must be used.

Frost heave can be eliminated by insulating the pipe to reduce heat flow and by heat tracing the unfrozen ground to prevent freezing. This method is

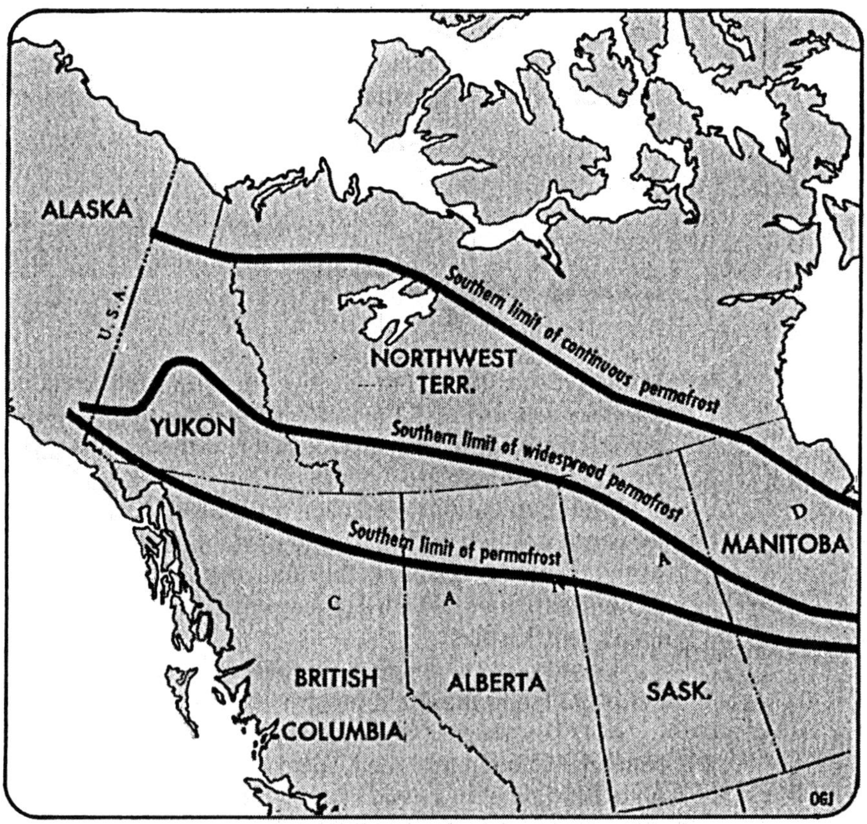

FIG. 1. Permafrost zones in Western Canada.

especially important in the northern half of the transitional region where most of the right-of-way is frozen.

Depth and ice content of frozen ground along the southern fringes of permafrost are much less than in the north, so difficulties of thawing permanently frozen ground are less.

It is generally feasible to thaw frozen ground along the southern fringes of permafrost without environmental or pipeline safety hazards. In those few areas where thaw settlement is excessive, pipe can be supported on belowground piles to prevent settling and unsafe stresses. Surface disturbances can be corrected along the southern fringes by increased right-of-way maintenance.

Special designs such as heat tracing and insulation do not contribute heavily to overall costs of Arctic projects. Increased gas deliveries due to refrigerating gas below 32°F more than counterbalance cost of special designs.

Energy Balance

Figure 2 shows a long distance pipeline connecting a producing area to a consumer area. The pipeline is made up of a large number of pipeline sections, and each section is composed of pipe, a pump or compressor, and a gas cooler.

An envelope has been drawn around one of the pipeline sections in Fig. 2 to isolate it from its surroundings so that energy flow into and out of it can be seen.

Sources of energy that flow across this isolating envelope are:

Total energy of gas flow entering the section, E_1.
Heat flow from the ground to the pipe, q.
Heat equivalent of the compressor horsepower, H_p.
Heat flow across the surface of the gas cooler, H_c.
Total energy of gas flow leaving the section, E_2.

If compressor-station discharge pressures and temperatures are the same and the pipeline is horizontal, energy of flow entering the section equals energy of flow leaving the section ($E_1 = E_2$). The remaining three sources of energy crossing the boundary of the section shown in Fig. 2 can be summed to derive cooler duty as

$$H_c = q + H_p, \text{ if } E_1 = E_2$$

Since no reference was made to fluid properties during derivation of this cooler duty equation, it can be applied simply and effectively to gas, oil, or slurry pipelines.

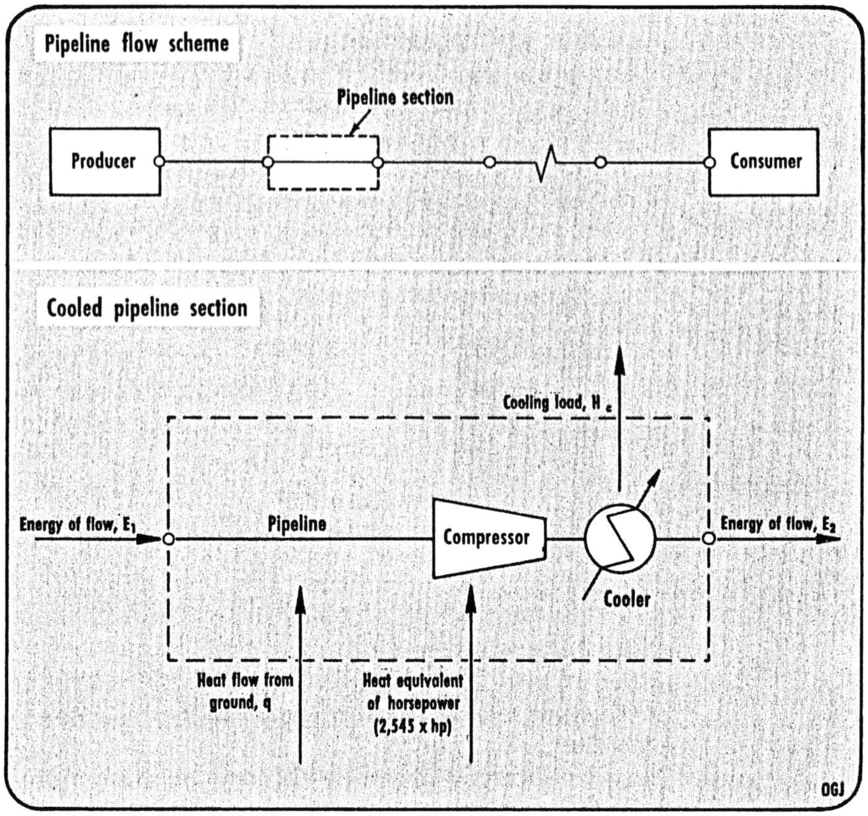

FIG. 2. Pipeline energy balance.

For example, if coolers maintain the average pipeline temperature at ambient environmental temperature, heat flow between the pipe and its surroundings is zero ($q = 0$) and coolers handle the heat equivalent of compressor or pump horsepower ($H_c = H_p$).

If the compressor or pump stations use 30,000 hp, then the cooling load is 77×10^6 Btu/h ($2545 \times 30{,}000$).

Detail design work should consider what happens when a compressor is out of service, the effect of significant elevation differences, and operating condition changes. In these cases, energy of flow entering the section is not exactly equal to energy of flow leaving the section. Change in energy ($E_1 - E_2$) for an infinitesimal segment is obtained in consistent units by

$$\begin{aligned} dE &= dU + v\,dv + dy + d(PV) \\ &= q + H_p - H_c \end{aligned}$$

Flow predictions shown later are based on this general formula and consider differences in elevations, pressures, temperatures, and flow rates.

Ground Heat Flow

Heat flow between a buried pipe and the ground can be derived from equations of potential flow between a source and a sink. Figure 3 shows a heat flow diagram around a buried uninsulated pipe at 0°F in soil which was initially at 30°F.

The ground heat flow equation for the flow pattern shown in Fig. 3 is

$$q = 2\pi K_g(T_g - T_f)L/\cosh^{-1}(2h/D)$$

This equation has been tested against actual operating data from two pipeline companies. A total of 184 sets of data from 23 different pipeline sections covering more than 1200 miles of right-of-way and representing all months of the year were used.

In the analysis, ground conductivity K_g was assumed to be 0.8 Btu/ft · °F · h and ground temperature T_g was evaluated in each month at half the depth to pipe center line. The amount of heat lost by the gas stream between compressor stations was calculated from operating data by using

$$q = m(dH + g/jg_c dy)$$

In this equation, no external work, coolers, or significant changes in kinetic energy occur. Figure 4 compares the ground heat flow equation and

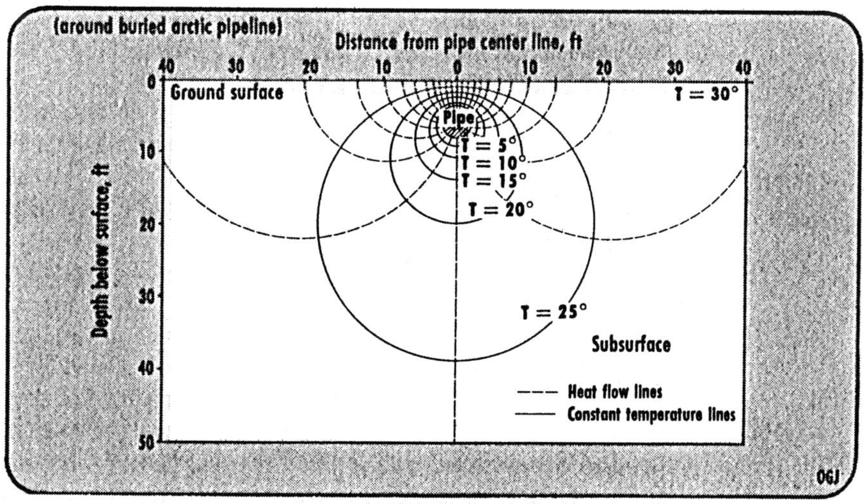

FIG. 3. Heat flow and temperature.

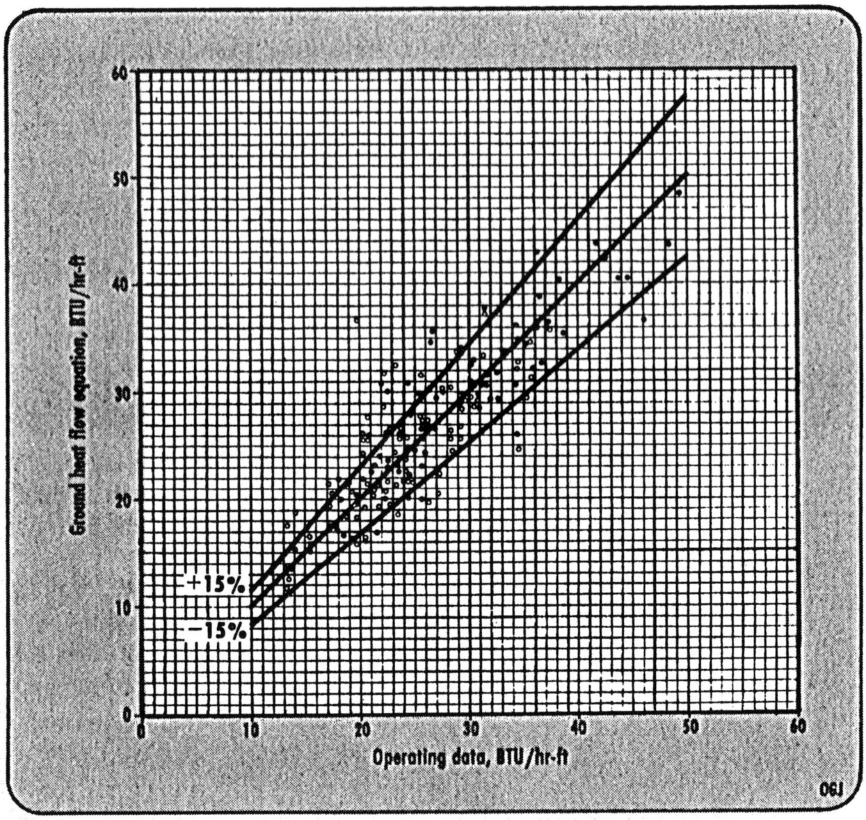

FIG. 4. Accuracy of ground heat flow predictions.

actual operating data. It shows that most ground heat flow predictions of the equation are within 15% of the heat that was actually lost by the gas stream between stations.

A standard deviation of this size is easily accounted for by to variations that exist in estimates of average ground conductivity, average ground temperature, and average burial depth. Since accuracy of this equation depends largely on variations in input parameters, it and the steady heat conduction theory used in its development are sufficiently accurate for pipeline design.

Temperature in a long pipeline will cascade from station to station until it reaches equilibrium. Heat flow between buried pipe and the ground is proportional to the difference between average gas and ground temperatures. From the first equation, if a long pipeline is uncooled ($H_c = 0$), heat flow from pipe to the ground equals the heat equivalent of the horsepower ($q = -H_p$).

For a long transmission line, temperature difference between equilibrium flowing temperature and ground temperature is, therefore, proportional to compression horsepower:

$$T_f - T_g = H_p \cosh^{-1} (2h/D)/2\pi K_g L$$

For most buried gas lines, this can be approximated by $T_f - T_g = 0.35H_p/L$, where $H_p = 2545$ (hp). For example, if the pipeline uses 10,000 hp every 45 miles and $K_g = 0.8$ Btu/ft · °F · h, $h = 54$ in., and $D = 36$ in., then $T_f - T_g = (0.35)(107) = 37$°F.

This means the equilibrium gas temperature in this long uncooled pipeline is 37°F above ground temperature. If more horsepower is installed per mile, the equilibrium flowing temperature will likewise increase.

Large Pipeline Cooling

Large-diameter, high-pressure, long-distance pipelines need cooling due to the relationship between pipe size and optimum horsepower. Increases in diameter and wall thickness of the pipe increase the amount of capital investment and therefore increase capital spent on horsepower to maintain the economic balance between pipe and horsepower.

For a pipeline that is operating at its optimum flow, owning and operating costs that are directly proportional to horsepower (including fuel) are approximately equal to one-third of the total pipeline owning and operating costs.

Air-cooling and refrigeration plants have many cost components which are directly proportional to horsepower. If costs that are directly proportional to horsepower are to be one-third of all owning and operating costs, air-coolers and refrigeration plants reduce optimum horsepower. Coolers and refrigerators replace compression horsepower and improve pipeline fuel efficiency.

Fuel is a major annual cost item that is proportional to horsepower. Therefore, optimum horsepower for any pipeline is sensitive to future fuel costs as compared to other costs. If fuel costs continue to escalate faster than labor and material costs, optimum horsepower will decrease in the future.

Also, since it is difficult to predict future fuel costs, it is difficult to predict optimum pipeline horsepower with accuracy.

For preliminary designs, it can be assumed that most of the costs that are independent of horsepower are roughly proportional to the tons of mainline pipe (pipe steel, pipe transportation, welding, valves, fittings, etc.). Therefore, it can be concluded that optimum horsepower for an uncooled pipeline is roughly proportional to the weight of pipe between stations.

If optimum station size for an uncooled 30-in. pipeline with 0.241 in. wall thickness and a 45-mi spacing is about 10,000 hp, then optimum station size for an uncooled 48-in. pipeline with 0.72-in. wall thickness and 45-mi spacing is about 47,800 hp.

Using the above equation, the equilibrium flowing temperature of the 30-in. pipeline will average 40°F above ground temperature and the equilibrium flowing temperature of the 48-in. line will average 180°F above ground temperature.

Therefore, cooling is necessary to economically carry Arctic natural gas through southern Canada even where there is no environmental reason for it.

Cooling vs Delivery

Natural gas is considered supercompressible at normal pipeline operating conditions. If it is cooled at constant pressure, it reduces volume more quickly than an ideal gas (which reduces in proportion to its absolute temperature). Compressibility factor, Z, is the ratio of real gas volume to ideal gas volume.

Figure 5 shows the effect of temperature on compressibility factor for natural gas. At 200°F and 1500 lb/in.^{2}abs, natural gas occupies 8% less volume than ideal gas ($Z = 0.92$). At 0°F, it occupies almost 40% less volume than the ideal gas ($Z = 0.62$). At 500 lb/in.^{2}abs, the supercompressible effect is less noticeable and reduces the advantage of cooling for lower pressure pipelines.

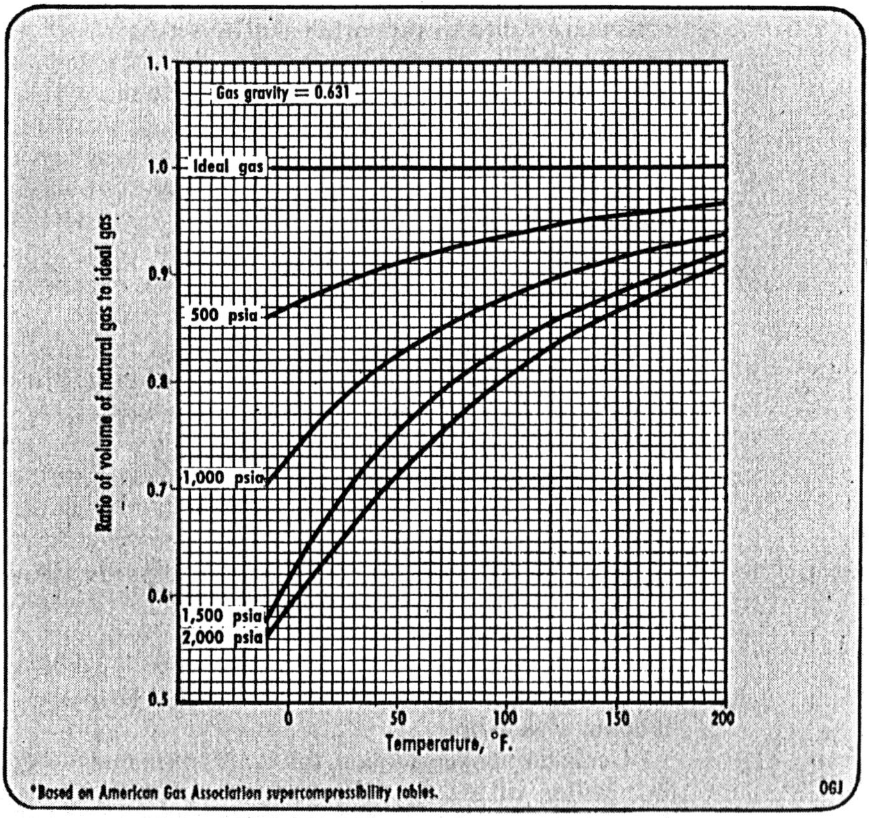

FIG. 5. Compressibility of natural gas. (Based on American Gas Association supercompressibility tables)

The difference in volume between ideal gas and natural gases becomes more pronounced at lower temperatures. Although this tends to encourage operating at cryogenic temperatures, more exotic metallurgy, reductions in refrigeration efficiency, and pipe insulation to control ground heat flow outweigh flow advantages of increased density at cryogenic conditions.

The relationship between amount of flow through a pipeline and horsepower is

$$dH_p = (m^3 2f/Eg_c A^2 \rho^2 d)dx$$

If horsepower, pipe diameter, friction factor, etc. are constant for a pipeline, mass flow through the pipeline is proportional to the density of the fluid as shown:

$$m = a\rho^{2/3} = a(PM_w/ZRT)^{2/3}$$

By integrating with respect to temperature, an expression for rate of change of flow with temperature can be derived which contains the $1/T$ term characteristic of an ideal gas but has an additional term for rate of change of the compressibility factor of a real gas:

$$(1/m)(dm/dT) = (-2/3)(1/T + 1/Z\ dZ/dT)$$

Figure 6 shows this equation graphically for natural gas with a specific gravity of 0.631 and for an ideal gas. Pressure does not affect the rate of increase in flow with decreases in temperature for an ideal gas, but for a real gas it is important.

For example, a 10°F temperature depression at 50°F will increase flow in a pipeline with an average pressure of 500 lb/in.^{2}abs by 1.75%. The same 10°F temperature depression at 50°F in a pipeline with a pressure of 1500 lb/in.^{2}abs will increase flow by 3.2%, or almost twice as much.

Installing gas coolers at compressor stations is advantageous since gas occupies less volume at lower temperatures and therefore causes less pressure loss due to frictional resistance to flow through the pipe. At the same time, since colder gas occupies less volume, it requires less horsepower to compress it by the same pressure ratio.

Lower pressure losses in the pipe and lower horsepower requirements at stations combine to permit more gas flow in the pipeline. This increase helps offset the cost of extra cooling facilities and other special designs.

It is difficult, however, to determine if gas cooling can economically amplify gas deliveries of existing transmission systems, because each pipeline is affected by cost parameters which can vary significantly from region to region.

For example, difficult mountainous or urban terrain lead to higher pipe-related costs which in turn lead to large optimum blocks of horsepower and the desirability of cooling. In addition, easy availability of water for use as a heat sink can reduce the cost of cooling and make it more desirable.

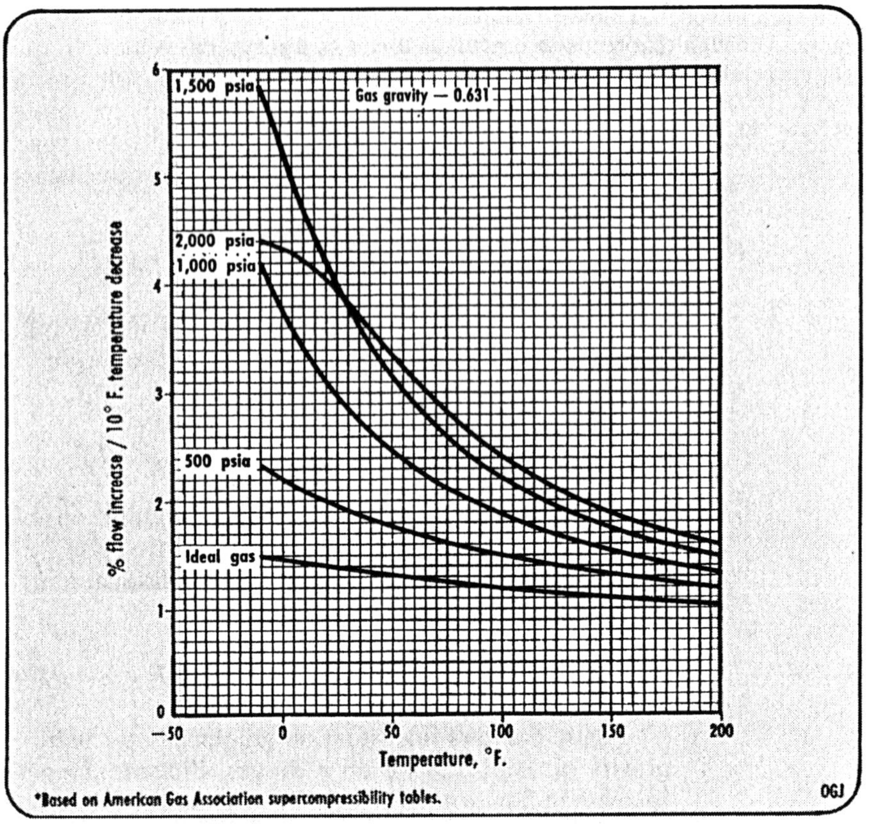

FIG. 6. Effect of 10°F depressions. (Based on American Gas Association supercompressibility tables.)

Case Studies of Cooling

Two large pipelines have been studied in detail to show the effect of cooling. The first case is a 36-in. pipeline designed for a maximum allowable operating pressure of 900 lb/in.^{2}gauge with 10,300-hp stations every 45 mi.

The second case is a 48-in. pipeline designed for a maximum operating pressure of 1680 lb/in.^{2}gauge with 30,000-hp stations every 45 mi.

The pipeline route used in the analysis begins at the Mackenzie Delta and Beaufort Sea area on the northern coast of Canada and follows the Mackenzie Valley south to the 60th parallel. It then runs south and east across Alberta to a point near Empress on the Saskatchewan border near the southern boundary of Canada. Total distance is about 1600 miles from the Mackenzie Delta.

Each pipe diameter has been analyzed in three separate configurations: (1) a refrigerated pipeline north of the 60th parallel where extensive areas of frozen ground exist, (2) an air-cooled pipeline south of the 60th parallel which is predominately frozen, and (3) an uncooled pipeline along the same route as the air-cooled section south of the 60th parallel.

No attempt has been made to match flows between the refrigerated pipeline north of the 60th parallel and the cooled and uncooled pipelines in Alberta.

Table 1 summarizes the six configurations. Although each configuration has the same number of stations, they are not all in the same locations. Station locations have been optimized in each configuration to fully use the available horsepower and to maximize deliveries. This was done so that none of the configurations would have a relative advantage in the comparison.

A 0.63 gravity gas with 1% CO_2 and 0.65% nitrogen was used in the comparative analysis. Gas thermodynamic properties were derived from supercompressibility tables published by the American Gas Association. Air and ground temperatures used in the analysis are the average for six summer months at each point along the route.

The flow model used to simulate the pipelines was developed in the early 1970s and solves continuity, force, and energy equations throughout all components of the system. It uses actual turbine performance curves and heat rates, operating limitations imposed by the refrigeration facilities and hydrocarbon dewpoint of the gas, and it accommodates detailed environmental data such as ground temperature and conductivity for each identifiable terrain unit along the route.

During design of the air-coolers and propane refrigeration plants for these example pipelines, relative costs of the cooling facilities were kept in approxi-

TABLE 1 Summary of 36 and 48-in. Pipelines

	36-in. Pipelines	48-in. Pipelines
Wall thickness, in.	0.29	0.72
Steel SMYS, lb/in.^{2}	70,000	70,000
Operating pressure, lb/in.^{2}gauge	900	1,680
Average station spacing, mi	45	45
Horsepower for gas compression	10,300	30,000
Horsepower for refrigeration	6,400	15,800
Compressor efficiency, %	80	80
Gas gravity	0.631	0.631
Operating season	Summer	Summer
Pipe internal roughness, in.	0.0003	0.0003
Average ground temperature (Alberta), °F	48	48
Average ground temperature (N.W.T.), °F	32	32
Average air temperature (Alberta), °F	52	52
Average air temperature (N.W.T.), °F	46	46
Depth of cover (Alberta), ft	3	3
Depth of cover (N.W.T.), ft	4	4

[a]Canadian code CSA Z-184 allows operation at 80% of the hydrostatic test pressure up to a maximum of 80% of SMYS.

mately the same ratio to the total owning and operating costs for both pipelines.

Heat flow from the ground to the pipeline is almost the same whether it is 36-in. or 48-in. pipe. Since the 36-in. pipeline requires only about 30% as much horsepower as the 48-in. pipeline, ground heat flow makes a larger relative contribution to the refrigeration load.

The 36-in. pipeline, therefore, was refrigerated to only 25°F to keep costs of the refrigeration plant in the same proportion to total costs as for the 48-in. pipeline. Reducing station discharge temperature of the 36-in. pipeline to 12°F does not increase flow significantly because at these flow levels, heat flow from the ground is enough to prevent suction temperatures from falling significantly.

Table 2 shows how the simulated pipelines performed. Pipeline inlet temperature for both uncooled and air-cooled pipelines is 65°F. Gas temperatures cascade from station to station in the uncooled designs and, in the case of the 36-in. pipeline, reach an equilibrium value of about 100°F. In the case of an uncooled 48-in. line, temperatures reach about 160°F.

The 36-in. pipeline takes only four station sections (180 mi) to reach its equilibrium temperature, while the 48-in. line takes 10 sections (450 mi) to reach its equilibrium. Average flowing temperature of the whole 48-in. pipeline is considerably less than the equilibrium temperature.

Turbines which drive gas compressors develop more horsepower in the north where temperature and elevations above sea level are generally lower than in southern Canada. The correction factor used in Table 2 to adjust for this difference and to make all designs comparable was the cube root of the ratios of horsepower as seen in the previous equations.

Both air-cooled and refrigerated pipelines have a 5-lb/in.^{2} pressure loss through the gas coolers which reduces flow by about 1% in the 36-in. pipeline and about 0.5% in the higher-pressure 48-in. pipeline. Table 3 shows the increases in flows and costs due to air cooling and refrigeration.

The air-cooled 36-in. pipeline delivers 4% more gas than an uncooled 36-in. pipeline. The air-cooled 48-in. pipeline delivers 13% more gas than an uncooled 48-in. pipeline.

This difference in the effect of cooling is due largely to the difference between flowing temperature of the uncooled and cooled designs of each pipeline and to a lesser extent to the difference in operating pressures.

The average temperature (of both pipe and compressor) of the 36-in. pipeline was reduced from 86 to 60°F and the average temperature of the 48-in. pipeline was reduced from 118 to 62°F. When reduction in flow due to the gas cooler pressure loss is accounted for, increases in flow due to decreases in temperature compare favorably with Fig. 6.

Costs of the air-cooling facilities increase average owning and operating costs of both pipelines by about 6%. The incremental cost of transporting 4% more gas through the 36-in. pipeline is therefore about 50% more than the average uncooled cost of transportation.

Incremental cost of transporting 13% more gas through the 48-in. pipeline is less than half the average uncooled cost of transportation.

Average cost of service of the 36-in. pipeline increases by about 2% and that of a 48-in. pipeline decreases about 6%.

TABLE 2 Performance of 36 and 48-in. Pipelines

Parameter	36-in. Pipelines			48-in. Pipelines		
	Uncooled	Air-Cooled	Refrigerated	Uncooled	Air-Cooled	Refrigerated
Location	Alberta	Alberta	N.W.T.	Alberta	Alberta	N.W.T.
Pipeline receipt temperature, °F	65	65	25	65	65	12
Pipeline delivery temperature, °F	100	69	25	159	71	12
Average compressor suction, °F	71	50	19	106	52	0
Average compressor discharge, °F	100	79	51	131	76	24
Average station discharge, °F	100	64	25	131	66	12
Average gas temperature, °F	86	61	29	118	62	8
Gas compressor horsepower	9,555	9,573	10,591	27,960	28,088	30,878
Refrigeration horsepower	—	—	2,218	—	—	10,040
Compression ratio	1.21	1.21	1.24	1.17	1.18	1.22
UA factor, MMBtu/h · °F	—	1.74	1.64	—	5.44	5.80
Cooling duty, MMBtu/h	—	24.4	36.6	—	60.8	100.9
Inlet flow, MMcfd	997.0	1,036.7	1,156.7	3,464.6	3,903.9	4,856.9
Fuel, MMcfd	34.6	35.4	60.0	95.2	98.2	156.7
Outlet flow, MMcfd	962.4	1,001.3	1,096.7	3,369.4	3,805.7	4,700.2
Adjusted values:						
Gas compression horsepower	9,555	9,555	9,555	27,960	27,960	27,960
Inlet flow, MMcfd	997.0	1,036.0	1,117.7	3,464.6	3,898.0	4,698.8
Fuel, MMcfd	34.6	35.4	58.0	95.2	98.1	151.6
Outlet flow, MMcfd:	962.4	1,000.6	1,059.7	3,369.4	3,799.9	4,547.2
Percent of uncooled flow	—	104	110	—	113	135

TABLE 3 Flow and Costs of Air-Cooled and Refrigerated Pipelines vs Uncooled Pipelines

Pipeline	Increase in Flow (%)	Increase in Cost[a] (%)	Incremental Cost of Service (%)	Increase in Cost of Service (%)
Air-cooled 36-in.	4	6	+50	+2
Refrigerated 36-in.	10	17	+70	+6
Air-cooled 48-in.	13	6	−54	−6
Refrigerated 48-in.	35	17	−52	−13

[a]Includes fuel and operating costs. Increases in costs could vary significantly for different pipelines depending on local factors.

The propane refrigerated 36-in. pipeline delivers about 10% more gas than the uncooled 36-in. pipeline. The propane refrigerated 48-in. line delivers about 35% more than the uncooled 48-in. pipeline.

This difference is due to differences in flowing temperature between the uncooled and refrigerated designs of each pipeline and differences in operating pressures. Average temperature of the 36-in. pipeline was reduced to 29°F and the 48-in. pipeline was reduced to 8°F.

Costs of propane-refrigeration facilities including fuel at each station increase total owning and operating costs for each pipeline by about 17%. If the need for special designs due to the frozen terrain is ignored, the incremental cost of transporting 10% more gas through the refrigerated 36-in. pipeline will be about 70% more than the normal uncooled cost of transportation. Incremental cost of transporting 35% more gas through the refrigerated 48-in. pipeline will be about half of the uncooled cost of service.

Therefore, the average cost of service increases about 6% for the 36-in. pipeline and decreases by about 13% for the 48-in. pipeline.

Special designs such as insulation, heat tracing, and pipe supports to help overcome intermittent frost heave and thaw settlement can increase capital cost of arctic pipelines by about 2.5%.

Therefore, cost increases due to refrigeration plant, cooling facilities, and other special designs to solve the unique temperature problems of arctic pipelines are far outweighed by increases in flow gained by their use.

Symbols

A internal cross-sectional area of pipe, ft^2
C_H unit cost of horsepower related costs, \$/hp · h
C_1 cost of components not related to horsepower, \$/ft · h
C_T unit cost of gas transportation, $\$/lb_m$

d	pipe inside diameter, ft
D	pipe outside diameter, ft
E	pump or compressor efficiency
E_1	total energy of flow entering section, Btu/h
E_2	total energy of flow leaving section, Btu/h
f	Fanning friction factor
g	acceleration due to gravity, ft/s^2
g_c	conversion constant, 32.174 $lb_m \cdot ft/lb_f \cdot s^2$
h	depth of burial to pipe centerline, ft
H	enthalpy, Btu/lb_m
H_c	cooling or chilling duty, Btu/h
H_p	horsepower, Btu/h, ft · lb_f/s
j	Joule's equivalent, 778 ft · lb_f/Btu
K_g	thermal conductivity of ground, Btu/ft · °F · h
L	length of pipe, ft
m	mass flow rate of fluid in pipe, lb_m/h, lb_m/s
M_w	molecular weight
ρ	density of fluid, lb_m/ft^3
P	pressure of fluid in pipe, lb_f/ft^2
q	heat flow from the ground to the pipe, Btu/h
R	universal gas constant, 1545 lb_f · ft/lb · mol · °R
T	temperature of gas, °R
T_f	average flowing temperature of fluid or gas in pipe, °R
T_g	temperature of ground, °R
U	internal energy of gas, Btu/lb_m
v	velocity of gas, ft/s
V	specific volume of gas, ft^3/lb_m
x	distance along pipe, ft
y	elevation of pipe above zero datum, ft
Z	gas compressibility factor

Bibliography

Brown, R. J. E., *Permafrost in Canada*, Geological Survey of Canada and the Division of Building Research, 1965.

Carslaw and Jaeger, *Conduction of Heat in Solids*, 2nd ed., Oxford University Press, 1959, p. 451.

Ferians, O. J., *Permafrost Map of Alaska*, U.S. Geological Survey, 1965.

Hydrodynamics, 6th ed., Cambridge University Press, 1932, Article 64.

Katz, D. L., and King, G. G., *Dense Phase Transmission of Natural Gas*, Presented at the C.G.A. National Technical Conference, October 1973; *Energy Process./Canada*, pp. 36–47 (December 1973).

Manual for Determination of Supercompressibility Factors for Natural Gas, PAR Research Project NX-19, American Gas Association, 1962.

GRAEME G. KING

Heat Loss

The design and construction of buried LNG pipelines and large diameter hot-oil pipelines in permafrost regions have emphasized the need to make accurate flowing temperature predictions.

When the temperature of the buried pipe in which fluid is flowing is higher than the ground temperature—a permanent condition in the Arctic environment—a constant loss of heat occurs.

It has been shown that the accuracy of the following equations for predicting heat flow from buried pipe is acceptable for engineering design of buried pipelines [1, 2]:

$$q = s(T_p - T_g) \tag{1}$$

where

$$s = 2\pi K/\cosh^{-1}(2h/D) \tag{2}$$

In these equations:

q = the heat flow from pipe, Btu/(h)(longitudinal foot)
T_p = pipe temperature, °F
T_g = ground temperature, °F
K = soil thermal conductivity, Btu/ft · °F · h
h = depth of burial to pipe centerline, in.
D = pipe outside diameter, in.

The function, $\cosh^{-1}$, is the inverse hyperbolic cosine and is rather awkward to handle. It may be replaced by another mathematical expression:

$$\cosh^{-1}(x) = \ln\left[x + (x^2 - 1)^{1/2}\right]$$

where ln is a natural logarithm (base $e = 2.718+$).

Hence, when T_p, T_g, K, h, and D are known, the heat flow from a pipe (heat loss) may be calculated. However, even with the aid of a pocket calculator, the computation is a long and tedious one.

A nomograph (Fig. 1) has been designed to solve this problem quickly, using four sequential movements of a straight edge.

Use of the nomograph involves the following steps:

1. Connect the known values of h and D on the appropriate scales with a ruler. Extend this line up to the intersection with Reference Line 1. This point is labeled I.
2. Transfer point I from Reference Line 1 to Reference Line 2, using the oblique tie-lines as a guide. Mark the transferred point II.

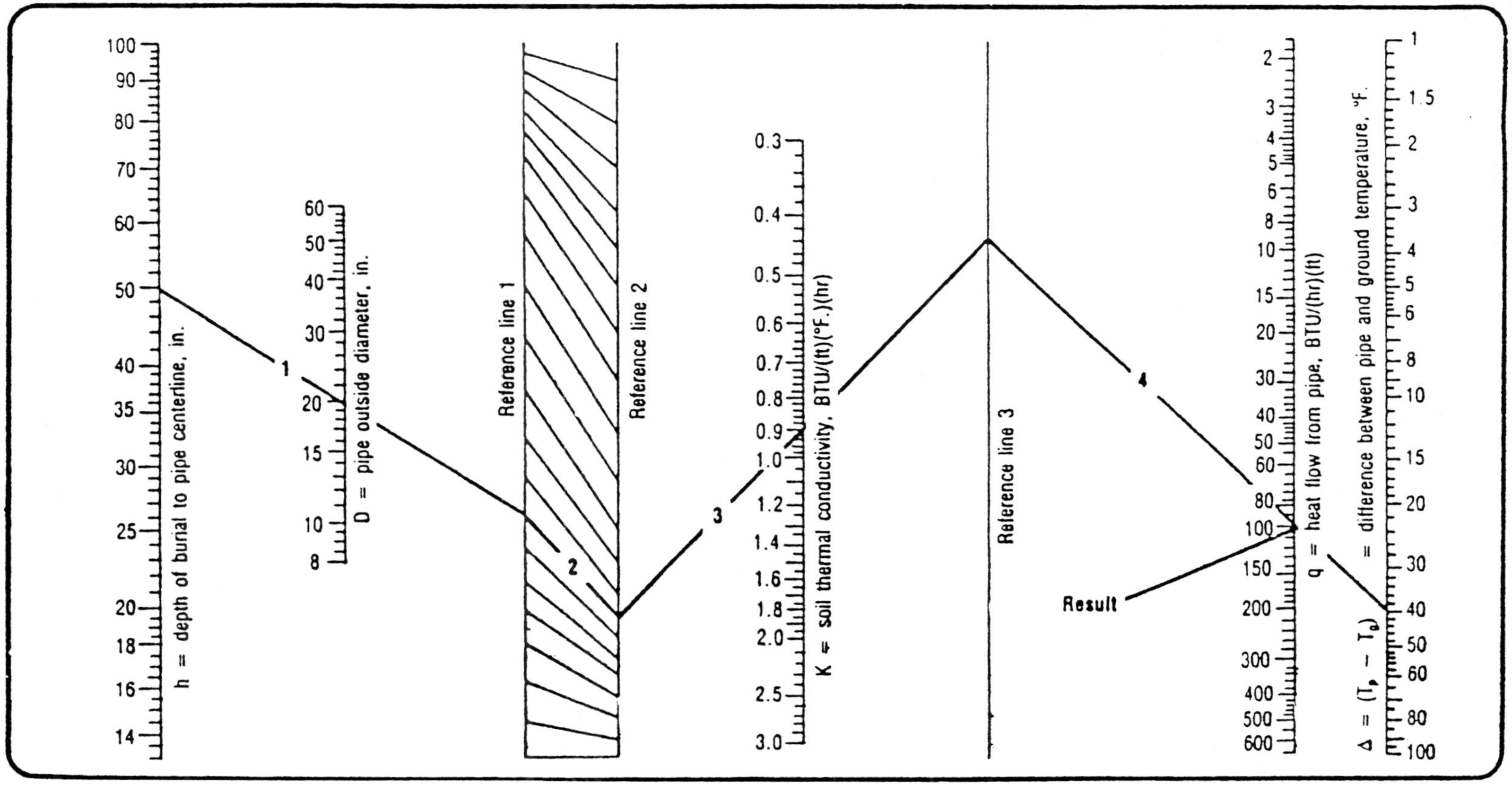

FIG. 1. Heat loss from buried pipeline.

3. Connect point II with a ruler to the known value of K on the appropriate scale; extend this line up to the intersection point with Reference Line 3. This point is labeled III.
4. Connect point III with a ruler to the known value of Δ ($\Delta = T_p - T_g$) on the appropriate scale. Read the result, q, on the intersection point of the ruler with the q scale.

If the K value is not known exactly, it may be assumed, in most cases, as approximately equal to 0.8 Btu/ft · °F · h [1].

The value of Δ has to be calculated prior to using the nomograph.

Example. Use of the nomograph can be described by the following example.

Assume a buried pipe having an outside diameter, D, of 20 in. and a depth of burial to the pipe centerline of 50 in.

The temperature of the pipe, T_p, is 90°F and the ground temperature, T_g, is 50°F.

The soil conductivity, K, equals 0.9 Btu/ft · °F · h.

Calculate the heat loss from this pipe in Btu/(h)(longitudinal ft).

First, calculate Δ:

$$\begin{aligned}\Delta &= T_p - T_g \\ &= 90 - 50 = 40°\text{F}\end{aligned}$$

Then following the procedure described previously, the heat loss from the buried pipeline, q, is 99 Btu/h · ft.

This example problem is shown on the nomograph.

References

1. G. G. King, "Equation Predicts Buried Pipeline Temperatures," *Oil Gas J.*, pp. 65–72 (March 16, 1984).
2. G. G. King, "Cooling Arctic Pipelines Can Increase Flow, Avoid Thaw," *Oil Gas J.*, August 15, 1977.

ADAM ZANKER

6
Pipeline Support Design

Pipeline Support Design

Part 1: Design Criteria

Basic knowledge of structural analysis and design is enough to solve the most complicated type of pipe support structure.

Civil-structural engineers know how difficult it is to come up with a safe, economical design to support the piping loads, especially when final piping drawings are not available. Civil-structural engineers are the last to get all the information but the first group required to produce construction drawings to meet the schedules. Civil engineers must be able to reach into their bag of tricks and dream up a structure to meet all these needs in a hurry.

It is the object of this article to provide the civil-structural engineer with approximate methods that can be used to design almost any type of pipe support structure. Also included are some suggested methods for developing design criteria as an emergency aid for those suffering from a lack of information on design loads.

The entire subject on pipe supports includes: (1) Design Criteria, (2) Design Procedures, and (3) Construction Details for specific types. Only the Design Criteria will be presented in Part 1 under the following headings:

(A) Classification of types, (B) Selection of a type, and (C) Design loads. Simplified design procedures and construction details for single, double, and multiple column pipe supports are presented in Part 2.

Classification of Types

Even though an infinite number of types can be evolved by various combinations of such variables as the number of columns, number of tiers, type of construction, and so on, the majority of common types can be classified as follows:

Class 1 Pipe Supports. Single Column Pipe Supports
Class 2 Pipe Supports. Double Column Pipe Supports
Class 3 Pipe Supports. Multiple Columns And Other Types

Various arrangements included under each type are shown in Fig. 1. Each of these types may be subclassified as free standing or strutted (Fig.

2). Because of their wide popularity, only wide flange, rolled steel sections are implied unless specifically mentioned otherwise.

Selection of Types

Piping drawings more or less dictate the spacing, elevations, and width of the pipe bents.

The Operation and Maintenance Departments often require certain minimum spaces which may restrict or eliminate the use of bracings. Operating companies often specify a certain type of construction which they decide is the most economical system.

Soils and foundation engineers present their unbiased opinion about the general soil conditions.

Now, it is up to the civil-structural engineer to put the pieces together and come with a particular type and design that is not only strong, safe, and economical, but also meets the local and national codes. Regardless of how

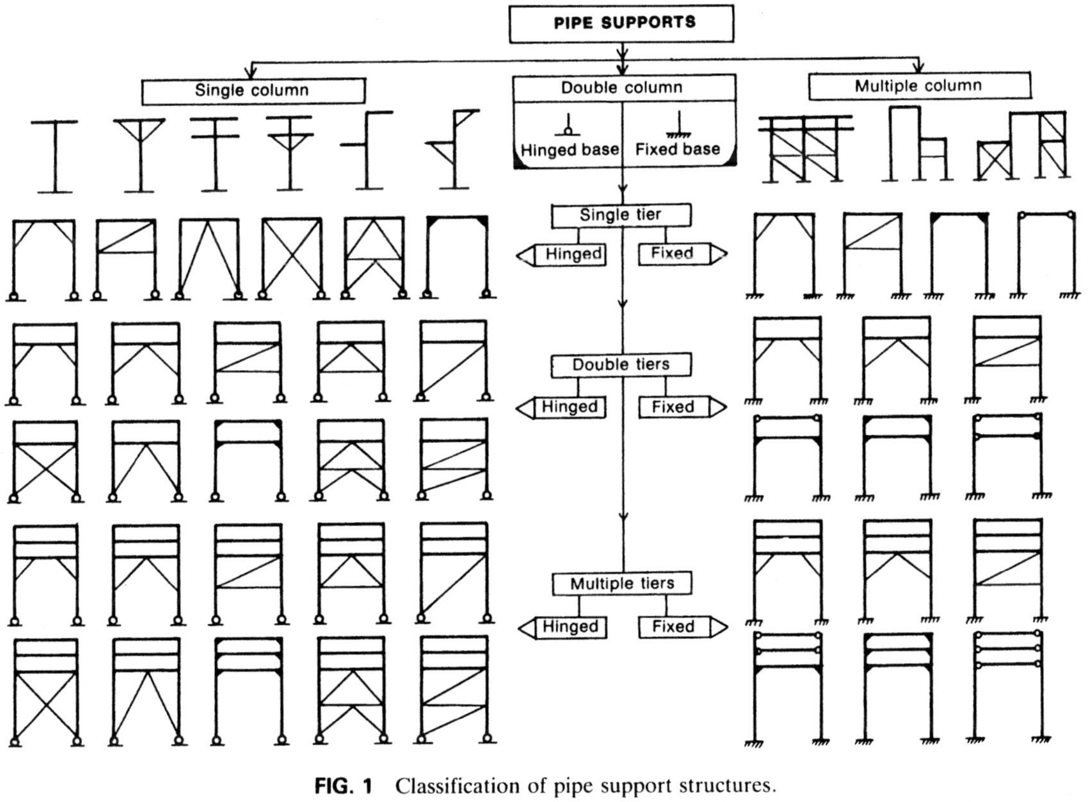

FIG. 1 Classification of pipe support structures.

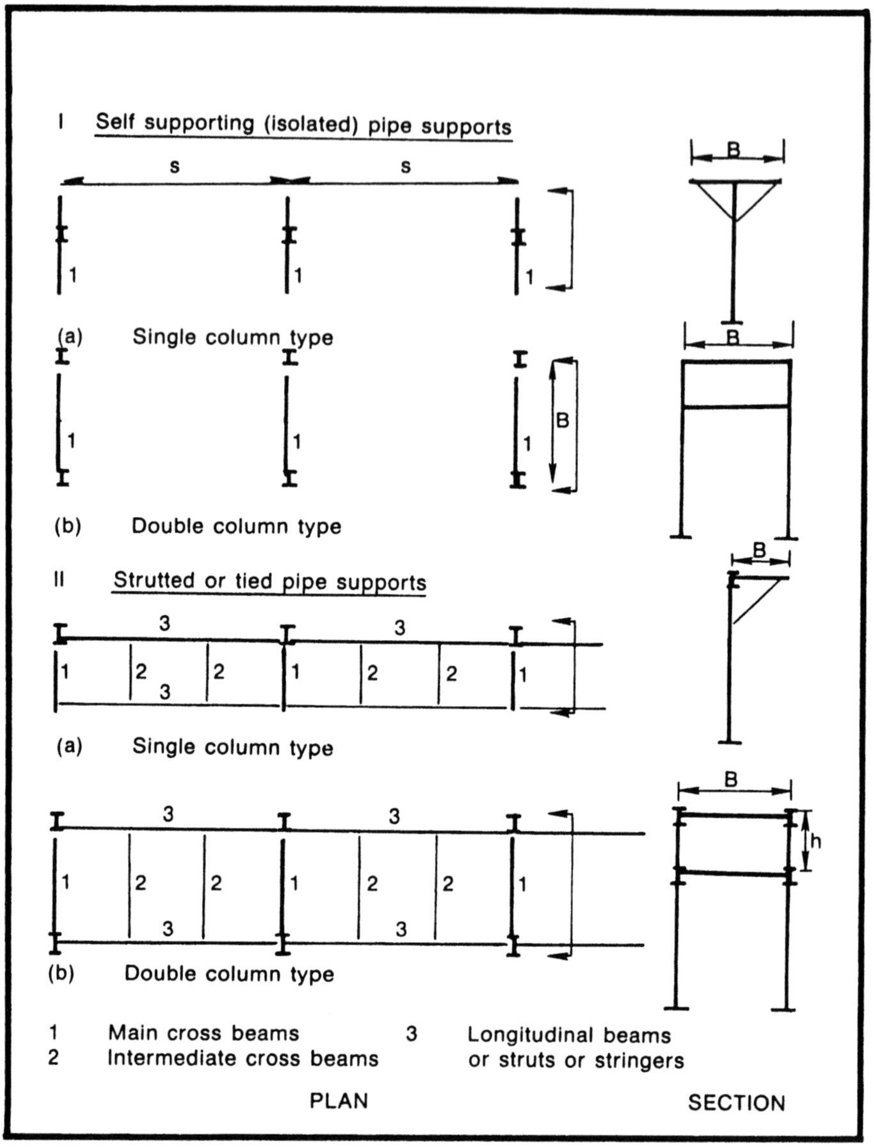

FIG. 2 Pipe support terminology.

restrictive this may sound, there are a few important factors that must always be considered before selecting any particular type.

Braced Structure or Not

From an analysis point of view, the braced structure appears more easy to analyze than a rigid frame because most braced structures are determinate, but rigid frames can be analyzed with almost equal ease.

A braced structure is not desirable when clear access ways are needed. The connections are simple to design and fabricate for a braced structure. Rigid frames offer more open areas but the connections call for special care and must be properly designed. Also, columns and foundations may be subjected to additional moments when a rigid frame structure is used.

Bolted or Fully Welded Structure

When there are no other limitations, it is always desirable to have a fully shop-welded structure with field-bolted connections. However, this cannot always be accomplished. If the size is too big, a fully shop-welded structure may pose shipping and handling problems. If the structure is to be galvanized, the size of the galvanizing tank may affect the cost. Also keep in mind that field welding is an expensive item and should be avoided if possible.

Hinged Base or Fixed Base

A hinged base is desirable when the moments on the foundations are to be kept to a minimum. However, a hinged base has the effect of increasing the column moments in a rigid or knee braced frame. A two bolt base plate is usually considered a hinged base.

A fixed base plate with four bolts should be designed for the column moments induced at the base plate level; also, the foundation should be capable of resisting these additional moments. Just the nature of the soil encountered may suggest the most desirable type of base.

Longitudinal Struts and Bracing

It is good practice to use adequately braced bays at intervals of 200 ft (maximum), whether large forces are anticipated in the longitudinal direction or not. Longitudinal struts (also called struts or stringers) are necessary when the longitudinal forces at intermediate pipe supports are to be transferred to braced bays farther away. Longitudinal members are also needed when piping requirements call for supports at close intervals for small size pipes or glass fiber reinforced plastic lines or similar reasons. Struts will then act as beams and are referred to as longitudinal beams.

Special Considerations

It is sometimes necessary to select a particular type to suit special loading conditions or other requirements. Provision for future expansion is a common requirement wherein the pipe supports need to be designed for additional future tiers.

There are situations where the piping is very sensitive to settlement and pile foundations are mandatory. This might place a restriction on the choice of a pipe support type.

Single column pipe supports are vulnerable for large transverse loads and may need to be replaced by double column pipe supports even though the vertical loads are small. Severe corrosive atmospheres may place restrictions on the choice of a pipe support type and even on member sizes.

Column Orientation

Proper orientation of the column sections is not always obvious. The geometry of the structure, the relative magnitudes of the transverse and longitudinal forces, the limitations on bracings, and the type of end connections are some of the most important factors which must be considered in the selection of proper column orientation.

Design Loads

Dead loads and wind loads are the most common loads to be considered. However, temperature loads, earthquake loads, and other loads may also have to be considered where required.

Dead Loads

The common practice is to assume a suitable value for the Pipe Loading Intensity, p (lb/ft^2). However, this value depends on the pipe schedule, the specific gravity of the contents, G, and the type of insulation, if any. In general, the total weight per unit length of pipe, w_t, is given by

$$w_t = w_e + w_i + w_w G \tag{1}$$

The weight of empty pipe only, w_e, the weight of water only, w_w, and the weight of pipe with water, w_f, are given for three different weights of pipes in Table 1. The weights of insulation, w_i, can be computed using the coefficients given in Table 2.

TABLE 1 Weights and Properties of Pipes[a]

D (o.d.) (in.)	Standard Weight (ST) t (in.)	w_e (empty)	w_w (water)	w_f (full)	Heavy Weight (XS) t (in.)	w_e (empty)	w_w (water)	w_f (full)	Extra Heavy Weight (XX, 160) t (in.)[b]	w_e (empty)	w_w (water)	w_f (full)
1 (1.32)	0.133	1.68	0.37	2.05	0.179	2.17	0.31	2.48	0.358	3.66	0.12	3.78
2 (2.375)	0.145	2.72	0.88	3.60	0.218	5.02	1.28	6.30	0.400	6.41	0.41	6.82
3 (3.5)	0.216	7.6	3.2	10.8	0.300	10.3	2.9	13.2	0.600	18.6	1.8	20.4
4 (4.5)	0.237	10.8	5.5	16.3	0.337	15.0	5.0	20.0	0.674	27.5	3.4	30.9
5 (5.6)	0.258	14.6	8.7	23.3	0.375	20.8	7.9	28.7	0.750	38.6	5.6	44.2
6 (6.625)	0.280	19.0	12.5	31.5	0.432	28.6	11.3	39.9	0.864	53.2	8.2	61.4
8 (8.625)	0.322	28.7	21.7	50.4	0.500	43.4	19.8	63.2	0.906	74.7	15.8	90.5
10 (10.75)	0.365	40.5	34.1	74.6	0.500	54.7	32.3	87.0	1.125	115.7	24.6	140.3
12 (12.75)	0.375	49.6	49.0	98.6	0.500	65.4	47.0	112.4	1.312	160.3	34.9	195.2
14	0.375	54.6	59.7	114.3	0.500	72.1	57.5	129.6	1.406	189.1	42.6	231.7
16	0.375	62.6	79.1	141.7	0.500	82.8	76.5	159.3	1.593	245.1	55.8	300.9
18	0.375	70.6	101.2	171.8	0.500	93.5	98.3	191.8	1.781	308.5	70.9	379.4
20	0.375	78.6	126.0	204.6	0.500	104.1	122.8	226.9	1.968	379.1	87.8	466.9
22	0.375	86.6	153.6	240.2	0.500	114.8	150.0	264.8	*1.000*	224.3	136.0	360.3
24	0.375	94.6	183.8	278.4	0.500	125.5	179.9	305.4	2.343	542.0	126.9	668.9
26	0.375	102.6	216.8	319.4	0.500	136.2	212.5	348.7	*1.000*	267.0	195.9	462.9
28	0.375	110.7	252.5	363.2	0.500	146.9	247.9	394.8	*1.000*	288.4	229.9	518.3
30	0.375	118.7	219.0	409.7	0.500	157.6	286.0	443.6	*1.000*	309.8	266.6	576.4
32	0.375	126.7	332.1	458.8	0.500	168.2	326.8	495.0	*1.000*	331.4	306.1	637.2
34	0.375	134.7	376.0	510.7	0.500	178.9	370.3	549.2	*1.000*	352.5	348.2	700.7
36	0.375	142.7	422.6	565.3	0.500	189.6	416.6	606.2	*1.000*	373.9	393.1	767.0
42	0.375	166.7	578.7	745.4	0.500	221.6	571.7	793.3	*1.000*	437.9	544.1	982.0

[a] t = wall thickness, in.
D = nominal diameter, in.
o.d. = outside diameter, in.
i.d. = inside diameter, in.
A_m = area of metal = $(\pi/4)(\text{o.d.}^2 - \text{i.d.}^2)$
A_f = flow area = $(\pi/4)(\text{i.d.})^2$, in.^{2}

γ_m = unit weight of metal, lb/in.^{3} (0.283 steel)
γ_w = unit weight of water, lb/in.^{3}
w_e = empty weight of pipe = $12A_m\gamma_m$, lb/ft
w_w = weight of water = $12A_f\gamma_w$, lb/ft
w_f = weight of pipe full of water = $w_e + w_w$

S = section modulus = $\dfrac{\pi}{32}\dfrac{(\text{o.d.}^4 - \text{i.d.}^4)}{\text{o.d.}}$

I = moment of inertia = $\dfrac{\pi}{64}(\text{o.d.}^4 - \text{i.d.}^4)$

r = radius of gyration = $\dfrac{\sqrt{\text{o.d.}^2 + \text{i.d.}^2}}{4}$

[b] Values in *italics* indicate maximum stock size.

TABLE 2 Weight of Insulation on Pipes[a]

a. Insulation Volumes, V (ft³/ft)

o.d.	T: 1″	1-1/2″	2″	2-1/2″	3″	3-1/2″	4″	4-1/2″	5″	5-1/2″	6″
3 (3.5)	0.10	0.16	0.24	0.33	0.43	0.54	0.65	0.79	—	—	—
4 (4.5)	0.12	0.20	0.28	0.38	0.49	0.61	0.74	0.88	1.04	—	—
5 (5.56)	0.14	0.23	0.33	0.44	0.56	0.69	0.83	0.99	1.15	—	—
6 (6.63)	0.17	0.27	0.38	0.50	0.63	0.77	0.93	1.09	1.27	—	—
8 (8.63)	—	0.33	0.46	0.61	0.76	0.93	1.11	1.29	1.49	1.70	—
10 (10.75)	—	0.40	0.56	0.72	0.90	1.09	1.29	1.50	1.72	1.95	—
12 (12.75)	—	0.47	0.64	0.83	1.03	1.24	1.46	1.69	1.94	2.19	2.45
14	—	0.51	0.70	0.90	1.11	1.34	1.57	1.82	2.07	2.34	2.62
16	—	0.57	0.79	1.01	1.24	1.49	1.75	2.01	2.29	2.58	2.88
18	—	0.64	0.87	1.12	1.37	1.64	1.92	2.21	2.51	2.82	3.14
20	—	0.70	0.96	1.23	1.51	1.79	2.09	2.41	2.73	3.06	3.40
24	—	0.83	1.13	1.45	1.77	2.10	2.44	2.80	3.16	3.54	3.93

b. Unit Weight, γ_1, of Insulation Materials

	Insulating Material	γ_i (lb/ft³)		Insulating Material	γ_i (lb/ft³)
1	Amosite asbestos	16	7	Kaylo	12.5
2	Calcium silicate	11	8	Mineral wool	8.5
3	Carey temperature	10	9	Perlite	13
4	Fiberglass	7	10	Polyurethane	2.2
5	Foam-glass	9	11	Styrofoam	1.8
6	High temperature	24	12	Super-X	25

c. Weight of Insulation, w_i (lb/ft)

$$w_i = V_i\gamma_i$$

Example: 10-in. pipe with 3-in. Kaylo insulation, $w_i = (0.9)12.5 = 11.25$ lb/ft

[a] $V_i = \frac{\pi}{4}\left(\frac{(\text{o.d.} + 2T)^2 - \text{o.d.}^2}{12^2}\right)$

The values of the Pipe Loading Intensity, p, for some commonly encountered cases can be obtained as follows.

Case 1: Rack Full of Same Size Pipe (Fig. 3a)

$$p = 12w_t/x \tag{2}$$

where x = uniform spacing of the pipes, inches.

Example 1. Calculate p assuming 8-in. standard weight pipe full of water and spaced 16 in. on centers. (Two inch high temperature insulation.)

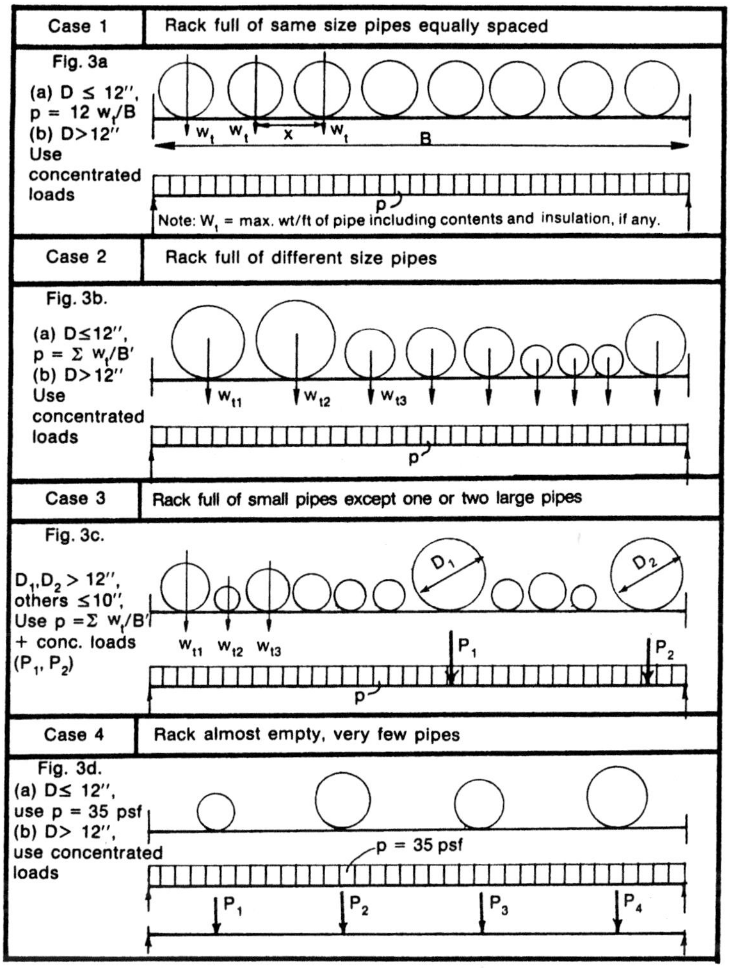

FIG. 3 Actual and assumed pipe loadings.

Solution: From Table 1, $w_e = 28.7$ lb/ft, $w_w = 21.7$ lb/ft. From Table 2, $w_i = 0.46(25) = 11.5$ lb/ft, $w_t = w_e + w_w + w_i = 28.7 + 21.7 + 11.5 = 61.9$ lb/ft.

From Eq. (2),

$$p = \frac{12w_t}{x} = \frac{12(61.9)}{16} = 46.9 \text{ lb/ft}^2$$

Case 2: Rack Full of Different Size Pipes (Fig. 3b)

$$p = (n_1w_1 + n_2w_2 + \cdots)/B = (\Sigma w_t)/B \qquad (3)$$

Example 2. Calculate p for the rack loaded as shown in Fig. 4.

Solution: The weights of pipe, contents, and insulation for each pipe are obtained from Tables 1 and 2 and results are shown on Fig. 4. There are three pipes of 3-in. diameter each, two pipes of 4-in. diameter each, and one each of 8-in. and 10-in. diameter. Using Eq. (3), $n_1 = 3$, $w_1 = 10.8$ lb/ft, $n_2 = 2$, $w_2 = 16.3$ lb/ft, $n_3 = 1$, $w_3 = 68.7$ lb/ft, $n_4 = 1$, $w_4 = 101.9$ lb/ft.

$$p = [3(10.8) + 2(16.3) + 1(68.7) + 1(101.9)]/5 = 47.1 \text{ lb/ft}^2$$

Case 3: Rack Full of Same Size Pipe, Except One or Two Large Pipes (Fig. 3c)

Step 1: Calculate p using Eq. (2) for small pipes.
Step 2: Calculate the concentrated loads due to large pipes.
Step 3: Superimpose the two loadings.

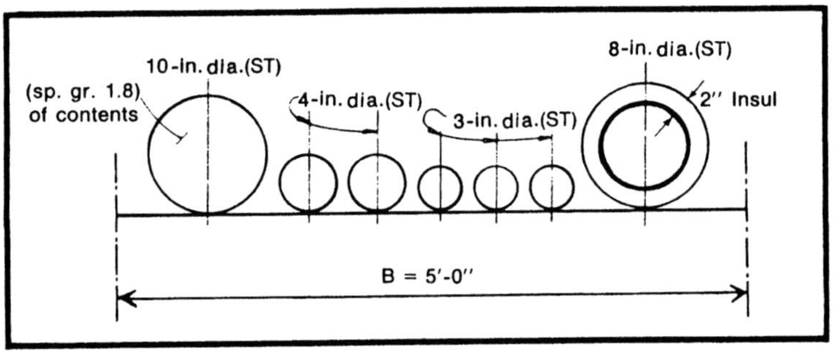

FIG. 4 Design pressure for uniformly spaced pipes.

A simplified graph is show in Fig. 5 to obtain the design pressure, q, for uninsulated pipes full of water and spaced 3-in. clear. The values for q and D on Fig. 5 are presented in both metric and English units.

Case 4: Rack with Only a Few Pipes (Fig. 3d)

Assume $p = 35$ lb/ft^2 for pipes up to 12-in. diameter. For pipes 12-in. or greater in diameter, use actual loads as concentrated loads.

Wind Loads

The wind load computations have been very well standardized for a variety of structures in many building codes and standards. However, pipe support structures do not fall under any specific case and must be treated as special

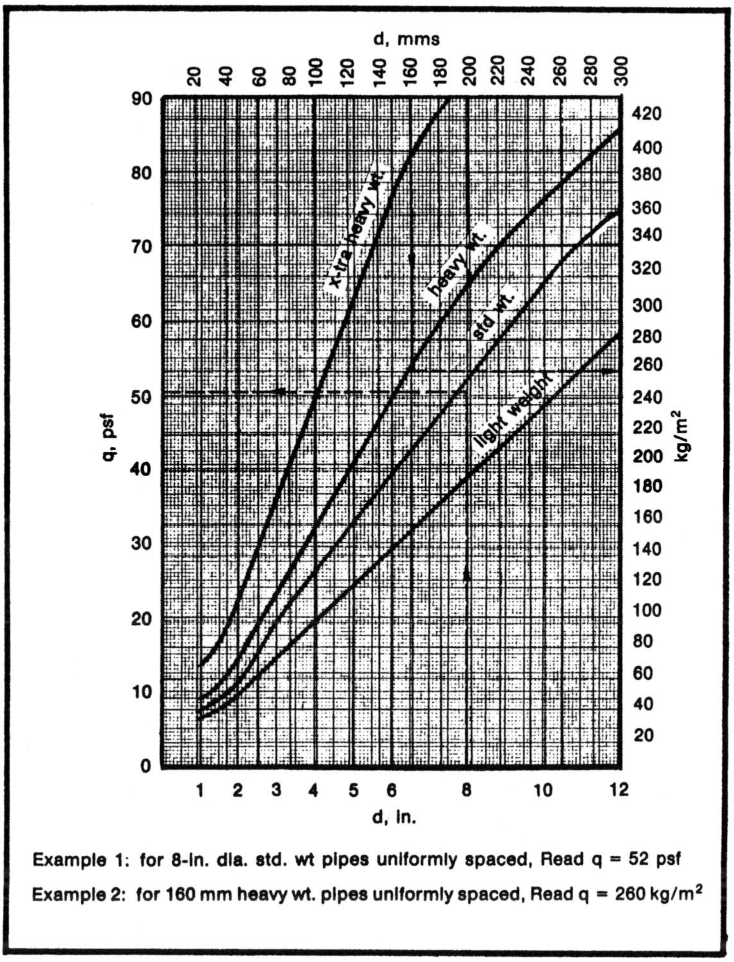

FIG. 5 Example problem 2.

structures. The methods used currently vary widely in practice. A new simplified method is presented in this discussion as appropriate for pipe support structures and is primarily based on the methods presented in the ANSI A58.1 standards [2]. Wind loads are calculated using an equivalent flat area based on an assumed wind strip height, y, as shown in Fig. 6. The general equation for y is given by

$$y = B \sin \alpha \cos \alpha \, C_{f1} C_{f2} \tag{4}$$

The value for the coefficient C_{f1} is taken from the ANSI standards assuming a wind incidence of about 20° and aspect ratio for the pipe supports ($\lambda = s/B$), to vary between 1 and 2. Also, for circular shapes, the value of the shape coefficient C_{f2} is taken as 0.6.

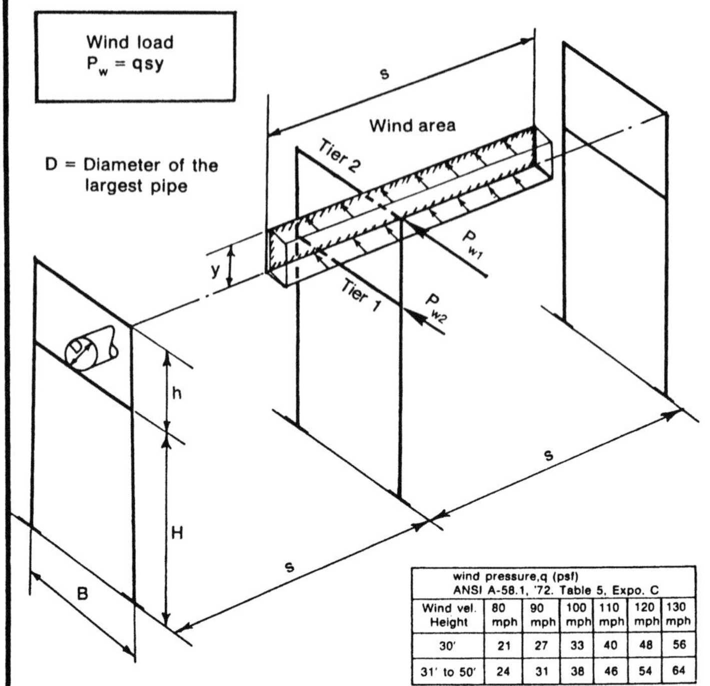

wind pressure, q (psf) ANSI A-58.1, '72. Table 5, Expo. C						
Wind vel. Height	80 mph	90 mph	100 mph	110 mph	120 mph	130 mph
30'	21	27	33	40	48	56
31' to 50'	24	31	38	46	54	64

Values of windstrip height, y (ft)				
Single column P.S.	Double column pipe supports			
	h<4'	h≥4'		
y = 1.5' or y = 0.6 D	B ≤ 25' y = 1.5' or y = 0.6 D	B ≤ 10' y = 1.5 or y = 0.6 D	10'< B ≤15' y = 2.0 or y = 0.6 D	15'< B ≤ 25' y = 2.5 or y = 0.6 D
Note: For strutted pipe supports consider y = 0.6D + 1'				

FIG. 6 Wind loads on pipe support structures.

The new method accounts for the effects of width of rack and height between tiers, and can be summarized as follows (also see Fig. 6):

Step 1: Select a value for the wind pressure, q, for a flat surface based on geographic location and height zone from applicable codes or standards.
Step 2: Select a value for the wind strip height, y, from Fig. 6 for the given width of rack (B), and height between tiers (h).
Step 3: Calculate wind load to be applied at each tier by using the formula

$$P_w = qsy \tag{5}$$

The equation is completely general and valid for metric or English units when appropriate units for q, s, and y are used. Wind on the column section can be neglected but wind on longitudinal struts must not be neglected. Shielding effects felt between pipes and at the level of lower tiers caused by tiers above are not included in this discussion.

Example 3. Calculate the wind load on a double column pipe support using the following data: Wind pressure $q = 32$ lb/ft^2; spacing of pipe supports, $s = 20$ ft; width of rack, $B = 15$ ft; height between tiers, $h = 5$ ft.

Solution: From Fig. 6 for $B = 15$ ft and $h = 5$ ft, read the value of $y = 2.0$ ft. Calculate the wind load using Eq. (5):

$$P_w = qsy = 32(20)2.0 = 1280 \text{ lb}$$

Temperature Loads

These are the loads induced by the frictional resistance between the pipe support surface and the expanding (or contracting) pipe caused by temperature changes. It is very likely that these loads may be insignificant when only a few light pipes are subject to the temperature changes while all the rest of the pipes remain at ambient temperature. On the other hand, when a number of heavy pipes are subject to severe temperature movements, the frictional forces could become very significant. When pipe supports with multi-tiers are subjected to arbitrary temperature movements, the calculation of actual friction forces can become extremely complicated. However, for practical purposes the following approximate method can be used to make a conservative estimate of the forces involved.

Step 1: Calculate the maximum temperature movement:

$$\Delta = \alpha_s(\Delta T)$$

where α = coefficient of linear expansion
s = spacing between pipe supports
ΔT = maximum temperature difference

Important: Use proper units. Also use this approximation only if actual temperature movements are not available from pipe stress analysis.

Step 2: Calculate the force P_Δ required to deflect the column by an amount Δ.

$$P_\Delta = (3EI\Delta)/h^3 \qquad (6)$$

Use appropriate values for I and h.

Step 3: Assume a value for the coefficient of friction μ and calculate the friction force

$$P_f = \mu W \qquad (7)$$

where W = maximum weight of the pipe at that temperature.

Step 4: The temperature force, P_t, is given by the smaller of the values for P_Δ and P_f.

$$P_t = P_\Delta \text{ or } P_f \tag{8}$$

whichever is smaller.

In Table 3 are given some suggested values for the value of the coefficient of friction, μ.

Example 3. Calculate the temperature loads on single column single tier pipe support given the following data.

Temperature movement, $\Delta = 2.75$ in., $E = 30(10^6)$ lb/in.^{2}, $\mu = 0.4$,
Total vertical load, $W = 3600$ lb; height, $h = 10$ ft; $I = 7.44$ in^4.

Solution:

$$P_\Delta = \frac{3EI\Delta}{h^3} = \frac{3(30)(10^6)(7.44)2.75}{[(10)(12)]^3} = 1070 \text{ lb}$$

$$P_f = W = 0.4(3600) = 1440 \text{ lb}$$

$P_\Delta < P_f$. Therefore, temperature force $P_t = P_\Delta = 1070$ lb.

Earthquake Loads

Recommendations of the Lateral Force Requirements, Seismology Committee, Structural Engineers Association of California, will be used only as guidelines for calculating earthquake loads. However, since pipe support structures are very much different from the class of building structures, suitable modifications are suggested. A general procedure will now be presented specifically for pipe supports, but must be used with extreme caution to be sure that specific local code requirements are not violated.

TABLE 3 Values for Coefficient of Friction

Contacting Surface	μ (sliding)
Steel on steel (dry)	0.3
Steel on steel (lubricated)	0.2
Steel on Lubrite	0.1
Steel on concrete (lubricated)	0.4
Metal on wood	0.5
Teflon on Teflon	0.06
Metal on asbestos	0.4

Step 1: Calculate the total lateral force V at each tier.

$$V = ZIKCSW$$

where Z is based on intensity expected, I is based on type of occupancy, K is based on type of framing, C is based on flexibility of structure, and S is based on relative natural period of vibration.

K = Numerical coefficient depending on the type of structure. The following values are recommended for pipe supports: $K = 1.33$ for pipe supports braced longitudinally and transversely to form a box system. $K = 0.8$ for pipe supports braced longitudinally and transversely, and with moment resisting connections. $K = 2.5$ for isolated pipe supports with no bracings. W = Maximum total dead load at the level under consideration.

Step 2: Apply the calculated lateral forces as external concentrated loads at each level in the transverse direction.

Step 3: To obtain the forces in the longitudinal direction, calculate the beam reactions R at each level and use the value of R in place of W.

Step 4: If there are overhangs, treat them separately as cantilever members and calculate lateral force F_p as

$$F_p = ZIC_pSW_p \tag{9}$$

where $C_p = 1.0$ and W_p = maximum dead load on cantilever portion only.

Other Loads

While pipe supports are meant to support the pipes only, several extraneous loads are imposed for a variety of reasons: (1) Guy wires supporting a stack are tied to the pipe support columns, (2) Platforms and walkways are added for convenience of the operating crew, (3) Heavy equipment is supported by pipe support columns, and (4) Erection loads caused by dropping and dragging pipe sections may not always be insignificant—may impose not only loads but also moments on pipe support members.

The only logical way to account for these loads is to check the pipe supports on an individual basis at specific locations.

K-Factors for Pipe Support Columns

Suggested values of K-factors for pipe support columns are presented in Fig. 7 for commonly encountered situations. Distinction is made between single and multiple column pipe supports and also whether the loading is in the transverse (normal to piping run) or longitudinal (along the piping run) direction. For multiple tiers, appropriate modifications should be made, sometimes treating the portion between two consecutive tiers as a separate column.

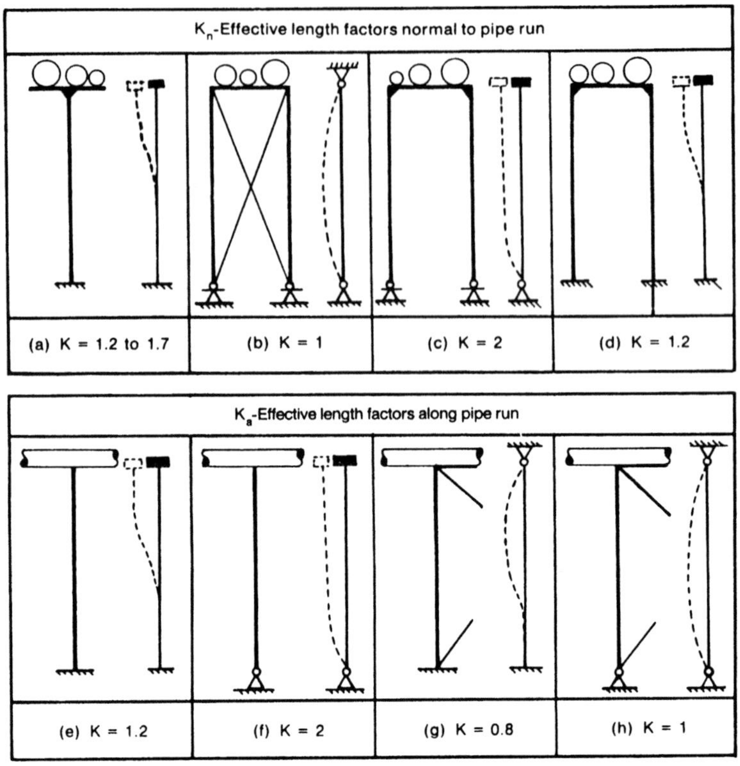

FIG. 7 *K*-factors for pipe support columns.

The basic design criteria described here should be used as a first step toward achieving a final design and in developing custom design standards to meet specific local needs. The actual detailed design should take into consideration such factors as: (1) provisions for future tiers, (2) effect of modifications, (3) severity of the corrosive conditions, (4) need for skilled labor and new construction methods, (5) proper selection of loading combinations, and (6) departure from idealized design conditions.

The detailed design procedures for various types of pipe supports are presented in Part 2 along with tabulated results and numerical examples.

Symbols

C, C_p	numerical coefficient in earthquake analysis
C_1	coefficient for calculating pipe insulation weight

C_f	net wind pressure coefficient
F_p	lateral force in earthquake analysis, lb
G	specific gravity of pipe contents
h	height between consecutive tiers or levels, ft
K	numerical coefficient in earthquake analysis
K_a	effective length factor of column along pipe run
K_n	effective length factor of column normal to pipe run
L_a	maximum unbraced length of column along pipe run, ft
L_n	maximum unbraced length of column normal to pipe run, ft
$n_1, n_2, \ldots$	number of pipes of unit weights $w_1, w_2, \ldots$
P_e	load due to earthquake longitudinal or transverse, lb
P_f	load due to friction between pipe and supporting surface, lb
P_t	load due to effects of temperature movements, lb
P_Δ	force required to deflect by Δ the free end of column, lb
q	design wind load per specification, lb/ft^2
s	spacing between consecutive pipe supports, ft
w_c	weight of pipe contents only, lb/ft $= w_w G$
w_e	weight of empty pipe only, lb/ft
w_f	weight of pipe full of water, lb/ft
w_i	weight of pipe insulation only, lb/ft
w_t	weight of pipe, insulation, and contents, lb/ft
w_w	weight of water in pipe, lb/ft
W, W_1, W_p	maximum vertical loads in earthquake analysis, lb
x	uniform spacing of pipe, in.
y	wind strip height or height of equivalent flat area, ft
α	angle of wind incidence to the horizontal, degrees
Δ	maximum deflection due to temperature movement, in.
μ	coefficient of friction between contacting surfaces
λ	aspect ratio of pipe supports defined as the ratio s/B

References

1. C. V. Char, "Check Pipe Support Orientation," *Hydrocarbon Process., 54*(9), 207 (1975).
2. *Building Code Requirements for Minimum Design Loads in Buildings and Other Structures,* American National Standards Institute, Publication ANSI A58.1–1972.
3. *Piping Handbook,* 5th ed., McGraw-Hill, New York.
4. *Standard Handbook for Mechanical Engineers,* 7th ed., McGraw-Hill, New York.

C. V. CHAR

Part 2: Step-by-Step Design Procedures

The first part of this article [2] included the classification of pipe supports, selection of types, and methods of calculating the design dead loads, temperature loads, wind loads, and earthquake loads. In this part, step-by-step design procedures will be presented for designing Class 1 (Single column), Class 2 (Double column) and Class 3 (Multiple column) pipe support structures.

Class 1 (Single Column) Pipe Supports

The following designation will be used to identify this group of pipe supports:

S1- . . . . Single column, 1 tier (or level), type no. . . . .

S2- . . . . Single column, 2 tier (or level), type no. . . . .

Some commonly encountered types are shown in Fig. 8. For simplicity, other variations are not shown. In designing any type of Class 1 pipe support structures, the following points should be kept in mind:

Isolated pipe supports act as free standing columns and must always have a fixed base (4-bolt base plate) for stability.

Even small transverse loads (caused by wind, temperature, or earthquake) could induce large moments at the base of the column and must be resisted as a concentrated moment by the pedestal and the foundation.

It is usually not economical to design for two or more future tiers, especially when large heights and spacings are involved.

The possibility of eccentric vertical loading should always be considered, that is, one side is empty while the other side is fully loaded.

The K-factor for column design could be as high as 2.1 when the free end of the column is clearly unrestrained against rotation. For all other situations, a value of 1.2 to 1.7 is recommended.

Unless larger longitudinal loads are encountered, it is preferable to orient the columns to resist transverse loads about their strong axes.

Piping anchors should be avoided on this class of supports.

Class 2 (Double Column) Pipe Supports

Perhaps the majority of all pipe support structures fall into one or the other type of this class of pipe supports. The following designation will be used to identify these most common structures:

D1- . . . . Double column, 1 tier (or level), type no. . . . .

D2- . . . . Double column, 2 tier (or level), type no. . . . .

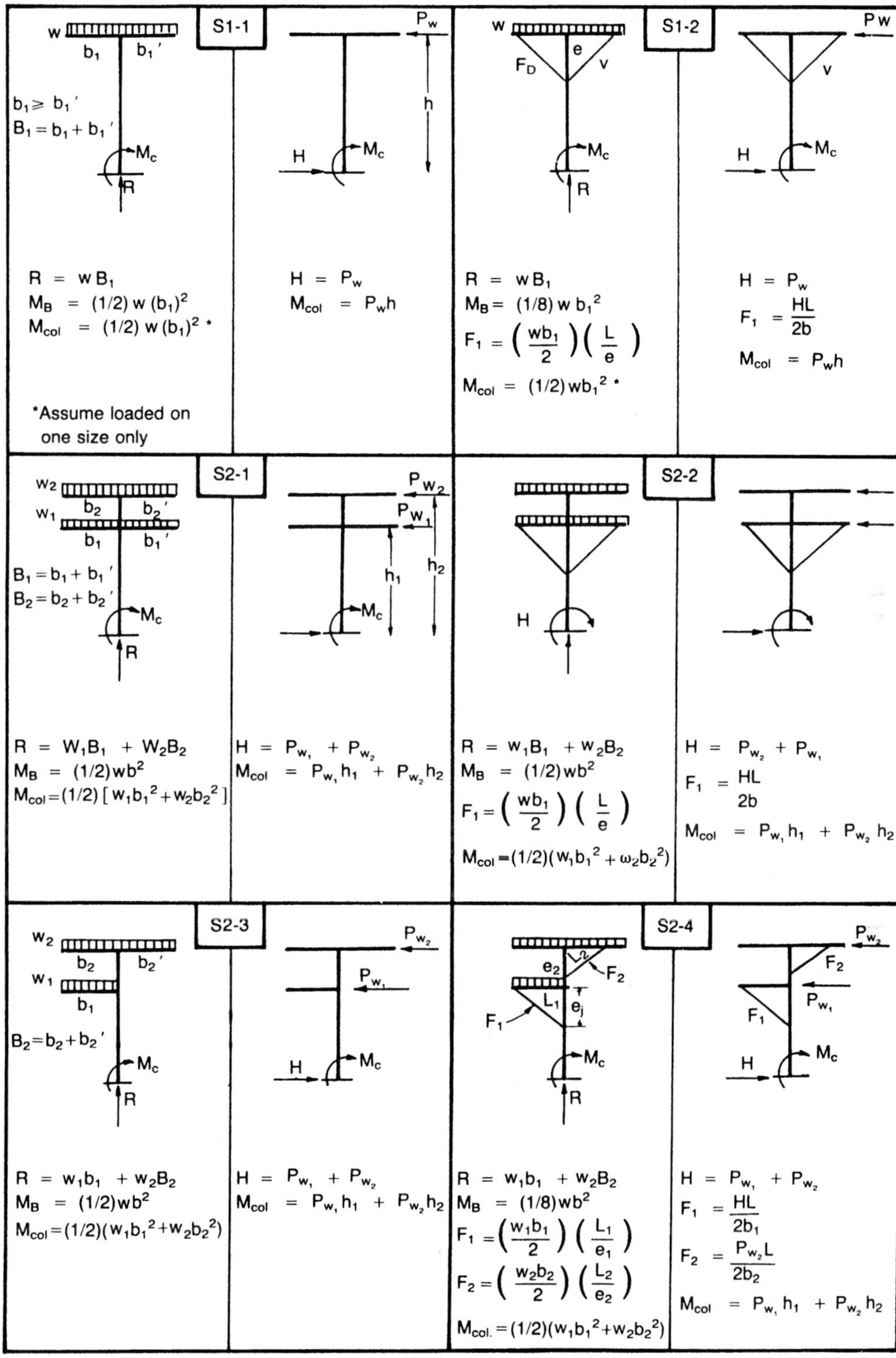

FIG. 8 Load tables for single column pipe supports.

Some of the most commonly encountered types are shown in Figs. 9, 10, and 11. To conserve space and time, a number of related variations have been omitted. In general, the column base may be fixed or hinged. Some significant features of this class of supports are the following:

By a suitable choice of bracing members, the orientation of the columns can be changed.

A completely open bay is most desirable to provide easy access.

A fully braced bay is desirable to effectively transmit the forces from pipe anchors and similar loads to the foundations.

Pipe supports can be designed to withstand the loadings from a large number of future tiers.

When field modifications are made, the resulting structure should be checked to see that it is not unstable.

External loads should preferably not be applied at intermediate points of bracing members or columns.

Since a fixed base induces concentrated moments on the foundations, it is desirable to determine if the soil can withstand these moments before deciding to use a fixed base.

Class 3 (Multiple Column) Pipe Supports

The following designation will be used to identify this group of pipe supports:

M1- . . . . Multiple column, 1 tier (or level), type no. . . . .

M2- . . . . Multiple column, 2 tier (or level), type no. . . . .

Some typical structures of this class are shown in Fig. 12. There is no specific way by which these types are evolved. In general, the use of more than two columns is usually required either because of improper planning for space requirements or because of the need to accommodate a large number of big diameter pipes.

The types shown in Figs. 12(e) and 12(f) are the result of field modifications (or last minute revisions), whereas the types shown in Figs. 12(a) and 12(b) were planned to accommodate the pipes.

There is no general method that can be used to design these structures. However, the following features are of special interest for this class of pipe supports:

The interior columns always carry a greater portion of the loads than the exterior columns.

The all-rigid type structure can be solved by approximate methods using hand computations or by more refined computer methods.

Whenever an adjacent tier is at a different elevation, as in Fig. 12(f), it will induce additional moments in the columns due to transverse loads.

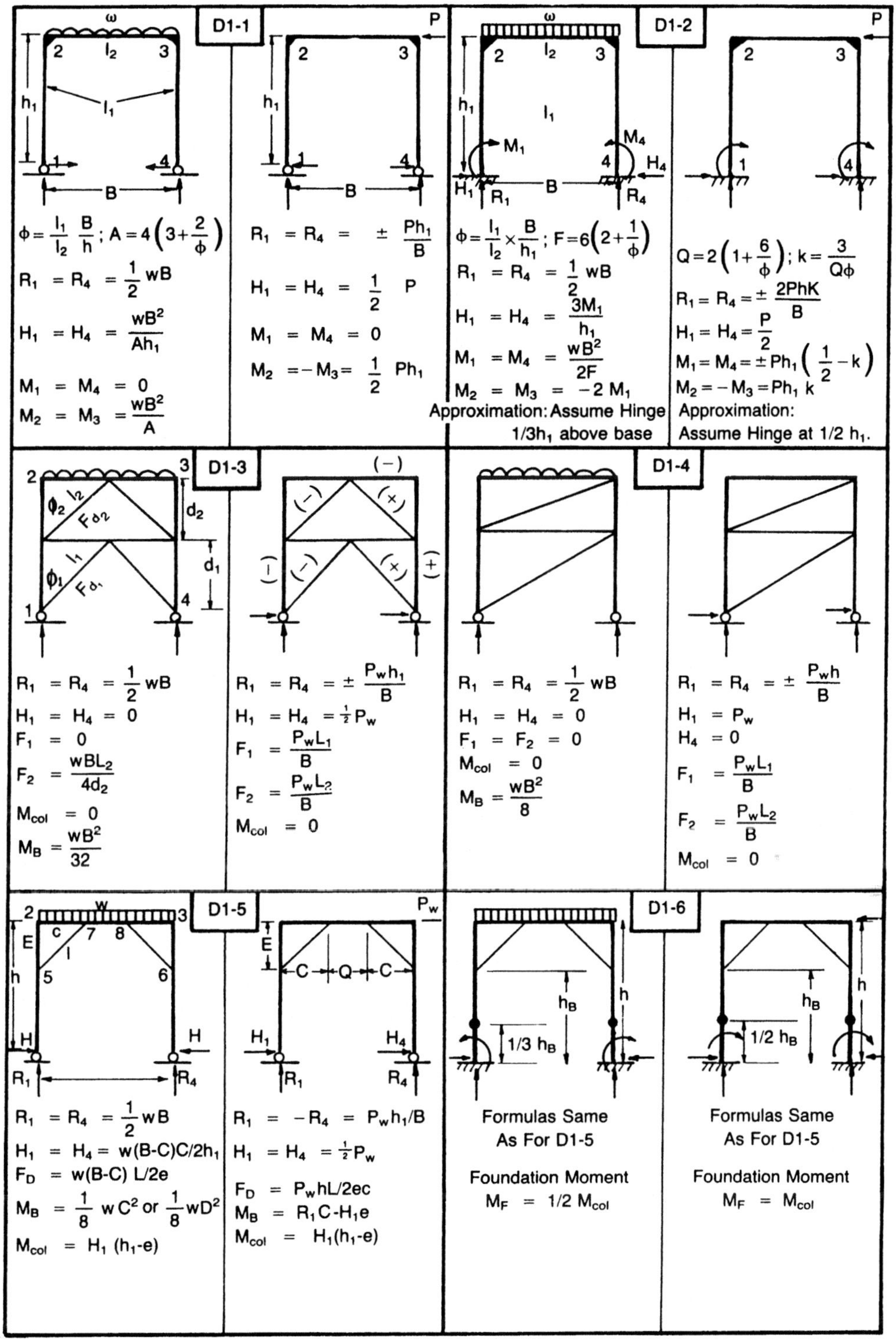

FIG. 9 Load tables for double column, single tier pipe supports.

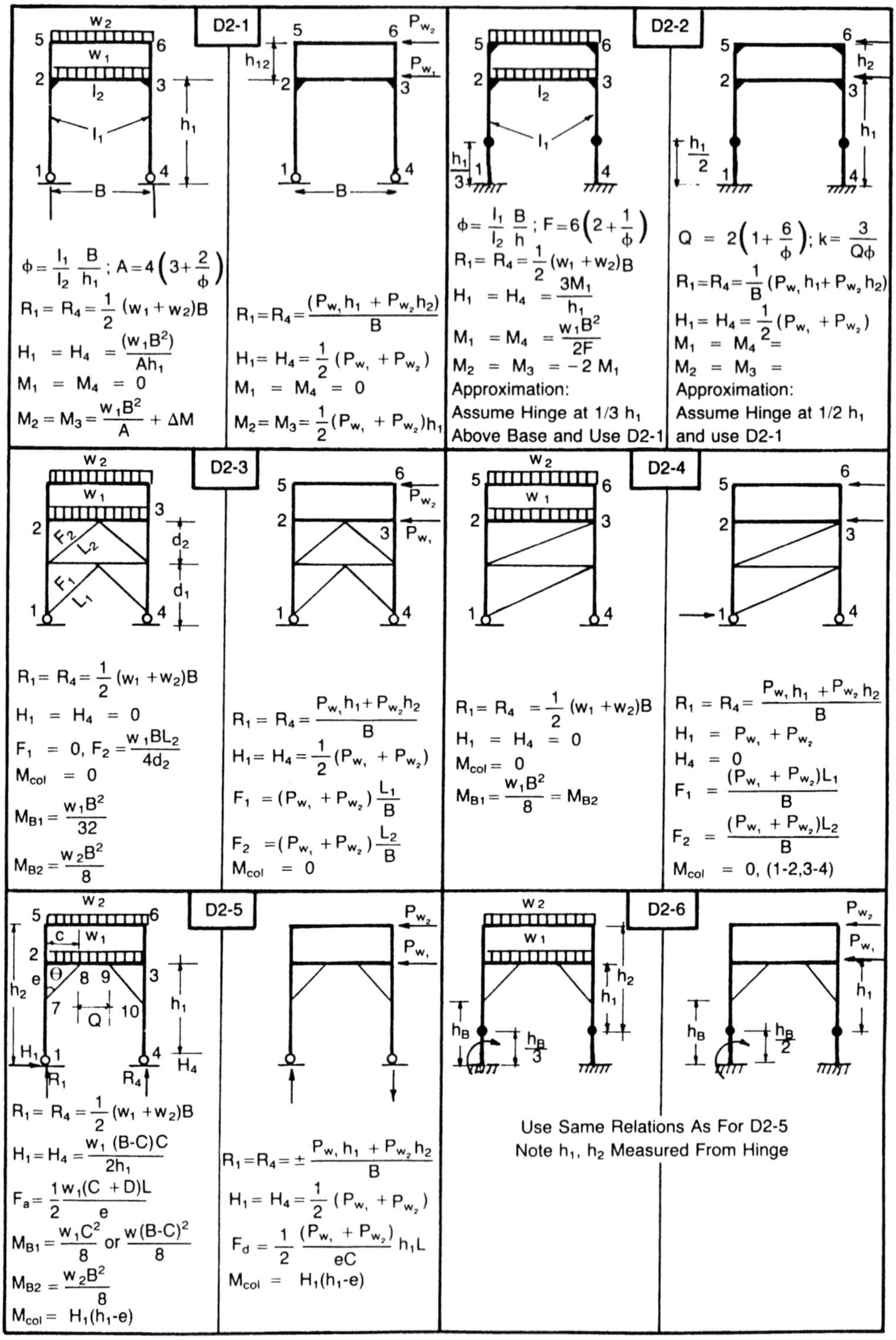

D2-1

$\phi = \frac{l_1}{l_2}\frac{B}{h_1}$; $A = 4\left(3 + \frac{2}{\phi}\right)$

$R_1 = R_4 = \frac{1}{2}(w_1 + w_2)B$

$H_1 = H_4 = \frac{(w_1B^2)}{Ah_1}$

$M_1 = M_4 = 0$

$M_2 = M_3 = \frac{w_1B^2}{A} + \Delta M$

$R_1 = R_4 = \frac{(P_{w_1}h_1 + P_{w_2}h_2)}{B}$

$H_1 = H_4 = \frac{1}{2}(P_{w_1} + P_{w_2})$

$M_1 = M_4 = 0$

$M_2 = M_3 = \frac{1}{2}(P_{w_1} + P_{w_2})h_1$

D2-2

$\phi = \frac{l_1}{l_2}\frac{B}{h}$; $F = 6\left(2 + \frac{1}{\phi}\right)$

$R_1 = R_4 = \frac{1}{2}(w_1 + w_2)B$

$H_1 = H_4 = \frac{3M_1}{h_1}$

$M_1 = M_4 = \frac{w_1B^2}{2F}$

$M_2 = M_3 = -2M_1$

Approximation: Assume Hinge at 1/3 h_1 Above Base and Use D2-1

$Q = 2\left(1 + \frac{6}{\phi}\right)$; $k = \frac{3}{Q\phi}$

$R_1 = R_4 = \frac{1}{B}(P_{w_1}h_1 + P_{w_2}h_2)$

$H_1 = H_4 = \frac{1}{2}(P_{w_1} + P_{w_2})$

$M_1 = M_4 =$

$M_2 = M_3 =$

Approximation: Assume Hinge at 1/2 h_1 and use D2-1

D2-3

$R_1 = R_4 = \frac{1}{2}(w_1 + w_2)B$

$H_1 = H_4 = 0$

$F_1 = 0$, $F_2 = \frac{w_1BL_2}{4d_2}$

$M_{col} = 0$

$M_{B1} = \frac{w_1B^2}{32}$

$M_{B2} = \frac{w_2B^2}{8}$

$R_1 = R_4 = \frac{P_{w_1}h_1 + P_{w_2}h_2}{B}$

$H_1 = H_4 = \frac{1}{2}(P_{w_1} + P_{w_2})$

$F_1 = (P_{w_1} + P_{w_2})\frac{L_1}{B}$

$F_2 = (P_{w_1} + P_{w_2})\frac{L_2}{B}$

$M_{col} = 0$

D2-4

$R_1 = R_4 = \frac{1}{2}(w_1 + w_2)B$

$H_1 = H_4 = 0$

$M_{col} = 0$

$M_{B1} = \frac{w_1B^2}{8} = M_{B2}$

$R_1 = R_4 = \frac{P_{w_1}h_1 + P_{w_2}h_2}{B}$

$H_1 = P_{w_1} + P_{w_2}$

$H_4 = 0$

$F_1 = \frac{(P_{w_1} + P_{w_2})L_1}{B}$

$F_2 = \frac{(P_{w_1} + P_{w_2})L_2}{B}$

$M_{col} = 0$, (1-2,3-4)

D2-5

$R_1 = R_4 = \frac{1}{2}(w_1 + w_2)B$

$H_1 = H_4 = \frac{w_1(B-C)C}{2h_1}$

$F_a = \frac{1}{2}\frac{w_1(C + D)L}{e}$

$M_{B1} = \frac{w_1C^2}{8}$ or $\frac{w(B-C)^2}{8}$

$M_{B2} = \frac{w_2B^2}{8}$

$M_{col} = H_1(h_1 - e)$

$R_1 = R_4 = \pm\frac{P_{w_1}h_1 + P_{w_2}h_2}{B}$

$H_1 = H_4 = \frac{1}{2}(P_{w_1} + P_{w_2})$

$F_d = \frac{1}{2}\frac{(P_{w_1} + P_{w_2})}{eC}h_1L$

$M_{col} = H_1(h_1 - e)$

D2-6

Use Same Relations As For D2-5
Note h_1, h_2 Measured From Hinge

FIG. 10 Load tables for double column, two tier pipe supports.

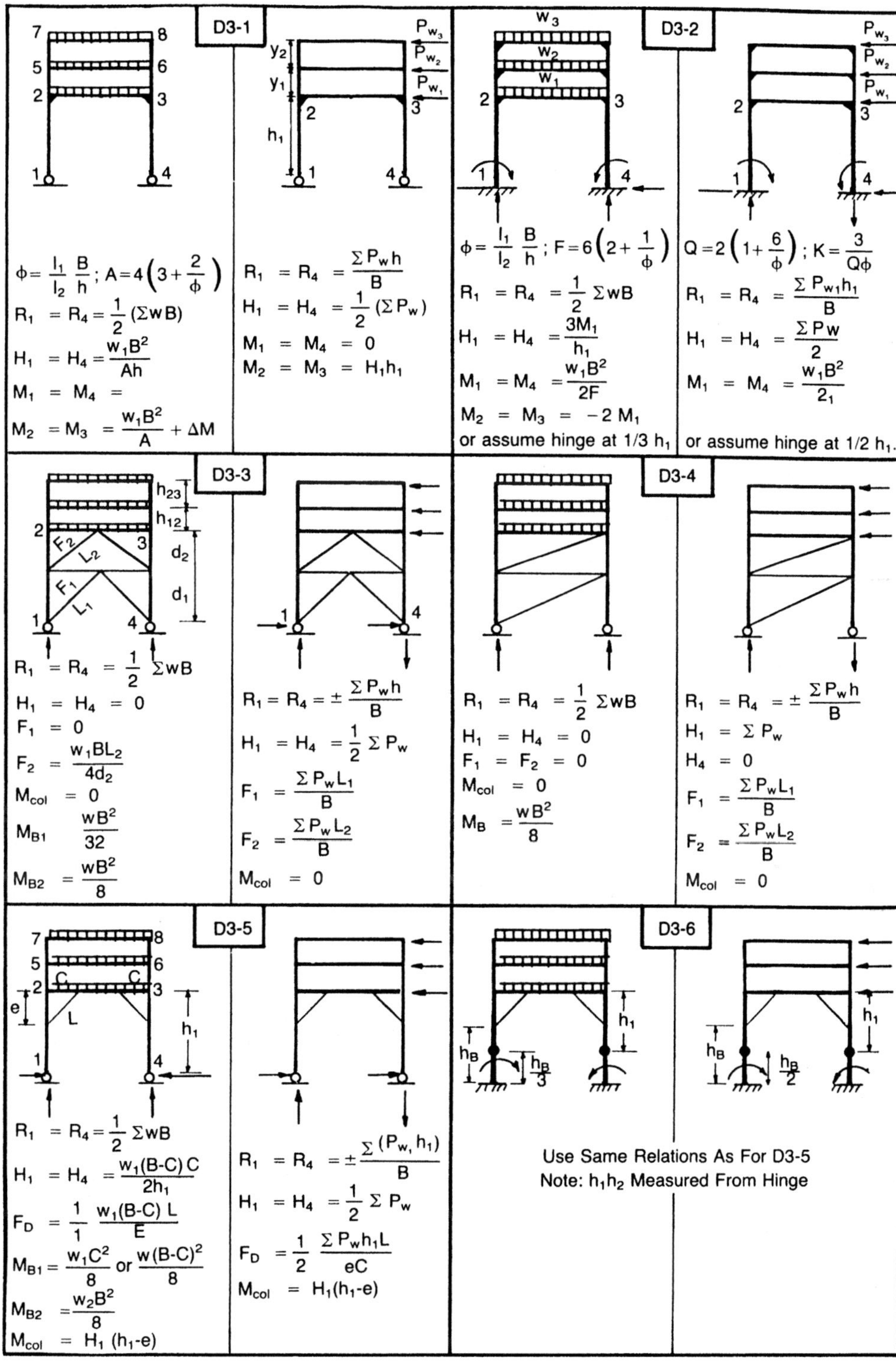

FIG. 11 Load tables for double column, three tier pipe supports.

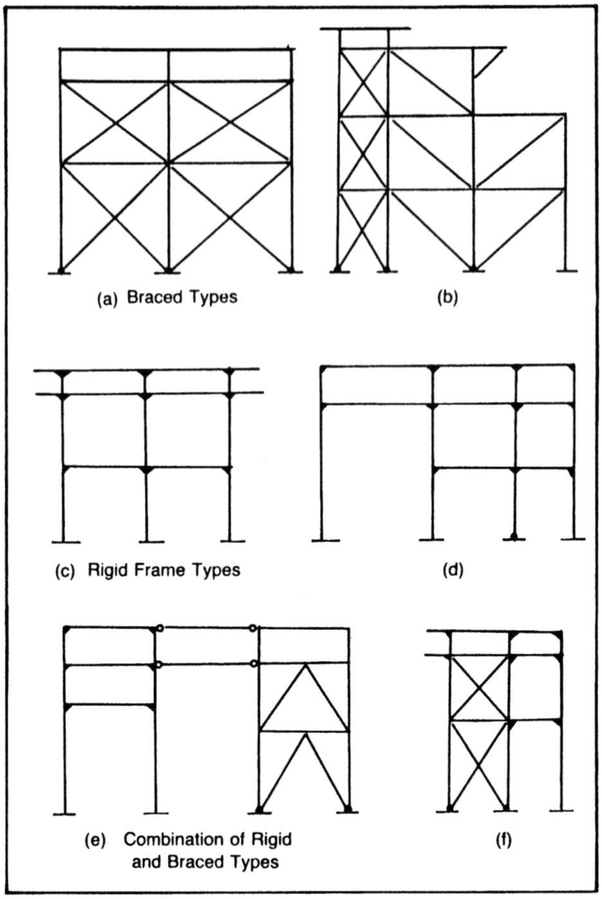

FIG. 12 Multiple column pipe support types.

The overhangs also have the effect of adding moments on the columns as cantilever members.

Design Procedures

Establishing firm design criteria is the first obstacle to be overcome jointly by the design and project teams. The next problem is concerned with definitive design and has to be solved by the engineer alone. Due to diversified and unique plant requirements, the engineer often finds it almost impossible to repeat procedures from his previous designs. The following steps are suggested to help not only the old pros, but also the new engineers who are desperately looking for specific design procedures:

Step 1: Calculate the loads in each component member (cross beam, bracing member and column) for each loading.

Step 2: Calculate the most critical loads using all applicable loading combinations:

Case 1: Max. D.L. only
Case 2: 0.75(Max. D.L. + W.L.)
Case 3: 0.75(Min. D.L. + W.L.)
Case 4: 0.75(Max. D.L. + E.L.)
Case 5: 0.75(Max. D.L. + T.L.)
Case 5: Special loading, if any

Step 3: Design the component members to resist the most critical loads and meet all code requirements and standards.

To obtain the loads in each of the members, the structure must be analyzed for all the different loadings. The most common loadings are the dead loads and the wind loads. To simplify the procedures, the results of analysis are summarized in Figs. 8–11 for the most commonly encountered types of pipe supports. When large number of loadings are to be considered on complicated types of structures, it may be necessary to use detailed analyses.

Cross beams are designed as flextural members. Bending moment caused by vertical dead loads only are usually the most critical. However, other combinations may be critical for rigid frame types. For single column T-type supports and overhangs, the beams should be designed as cantilever beams.

Step 1: Calculate maximum bending moment, $M_{max.}$
Step 2: Calculate allowable moment by using $M_{allow.}$ appropriate unsupported length, L_u
Step 3: Check $M_{max.} < M_{allow.}$

Bracing members are designed to withstand the maximum axial load, F_d. Usually the cross (X) braced members are designed as tension members; all other bracing members are designed as compression members. The slenderness ratio of tension members should be limited to 300. For compression members use the following procedure:

Step 1: Calculate the maximum axial compression, F_d
Step 2: Assume a section and note r and a
Step 3: Check $(KL/r) < 200$
Step 4: Calculate or read from steel tables the allowable axial stress, F_a
Step 5: Check $F_d < P_{allow.} = F_a a$
Note: For tension members check $F_d <$ allowable tensile load, $F_t a$

Columns are designed for axial load only or combined axial load and bending moment. Suggested K-factors for pipe support columns were given in Part 1. In designing multi-tier pipe supports, an approximate procedure would be to treat each section as a separate column. When simple connections are encountered without bracing, the section should be treated as a free standing column. Two methods given here are based on procedures set by AISC specifications.

Method 1. Approximate method using column load tables.

Step 1: Assume a section and note B_x, B_y, r_x, and r_y
Step 2: Calculate $P_{reqd.} = P_{max.} + M_x B_x$ (or $M_y B_y$)
Step 3: Calculate (KL) critical
Step 4: For the assumed section and KL, read $P_{allow.}$ from tables
Step 5: Check $P_{reqd.} < P_{allow.}$
Note: This is a very approximate method and should be used when column moment is about one axis only and its magnitude is small in comparison to the axial load. For a more accurate analysis, use Method 2

Method 2. Using interaction formulas.

Step 1: Assume a section and note section properties
Step 2: Calculate axial stress, $f_a = P/a$
Step 3: Calculate bending stresses, f_{bx} and f_{by}
Step 4: Calculate allowable axial stress, F_a, by using proper value of KL
Step 5: Calculate allowable bending stresses F_{bx} and F_{by}
Step 6: Calculate $F_e' = \dfrac{12\pi^2 E}{23(KL/r)^2}$
Step 7: If $\dfrac{f_a}{F_a} \leq 0.15$, check $\dfrac{f_a}{F_a} + \dfrac{f_{bx}}{F_{bx}} + \dfrac{f_{by}}{F_{by}} \leq 1.0$

Step 8: If $\dfrac{f_a}{F_a} > 0.15$, check $\dfrac{f_a}{0.6F_y} + \dfrac{f_{bx}}{F_{bx}} + \dfrac{f_{by}}{F_{by}} \leq 1.0$

and

$$\frac{f_a}{F_a} + \frac{C_{mx} f_{bx}}{(1 - \frac{f_a}{F_{ex}})F_{bx}} + \frac{C_{my} f_{by}}{(1 - \frac{f_a}{F_{ey}})F_{by}} \leq 1.0$$

Effect of Longitudinal Beams

Recall that the loading on the cross beams is proportional to the spacing between pipe supports. The use of longitudinal and intermediate cross beams reduces the loading in proportion to the spacing between the intermediate cross beams. Referring to Fig. 13, it can be concluded that the vertical loads and bending moments in the cross beams are reduced by 1/2 with the use of one intermediate cross beam and by 2/3 with the use of two intermediate

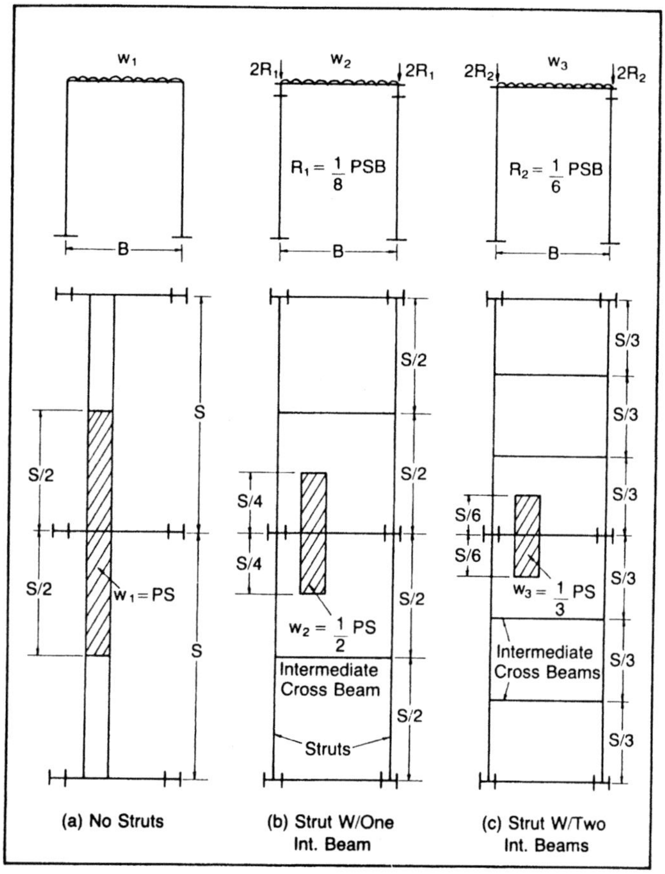

FIG. 13 Effect of struts and intermediate beams on loading.

cross beams. The use of longitudinal beams also helps prevent translational movements of the columns and thus improves the column buckling strength.

The approximate procedures presented here should cover a wide range of types of pipe support structures normally encountered in practice. Care should be exercised in meeting specific local requirements. Where a particular structure is exposed to unusual loadings or unusual environment, the engineer should use his judgment in making appropriate modifications. Haphazard or indiscrete field modifications can be easily detected by using these procedures to determine if there will be a change in the basic structure. The subject of these modifications and suggested construction details constitutes an important part in the design of pipe support structures and should be considered with care.

Example 1. Design a double column pipe support with three tiers, shown in Fig. 14, using the following data: Dead loads; $w_1 = w_2 = w_3 = 2950$ lb/ft., $w_{min.} = 0.5$ $w_{max.}$ for each level. Wind loads; $P_{w1} = 4000$ lb, $P_{w2} = 4500$ lb, $P_{w3} = 5000$ lb.

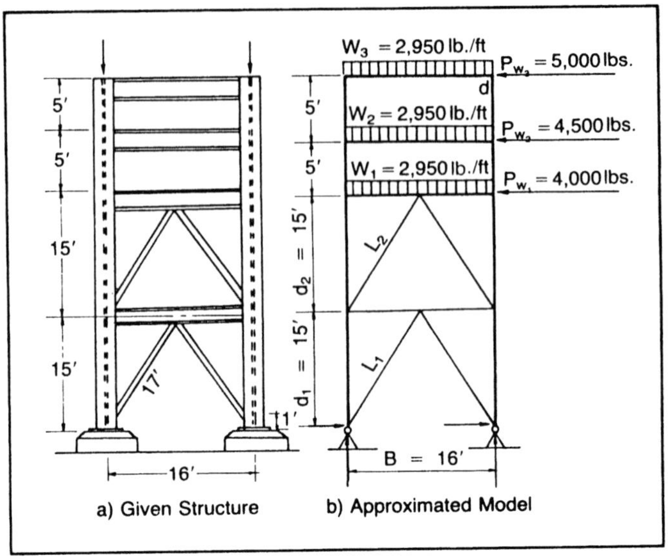

FIG. 14 Example problem of double column, three tier pipe support.

Assume $K = 1$ for column design, $C_m = 0.85$, $F_y = 36$ ksi, $F_b = 22$ ksi. (Class 2 Pipe Support, Type D3-3, shown in Fig. 11.)

Solution. Step 1. Calculate loads in all members for each loading.
Case 1: D.L. Max., from Fig. 11, Type D3-3,

$$R_1 = R_4 = 0.5\Sigma wB = 0.5(3)(2.95)16 = 70.8^k$$

$$H_1 = H_4 = 0,\ F_1 = 0,\ F_2 = \frac{w_1 BL_2}{4d_2} = \frac{2.95(16)17}{4(15)} = 13.37^k$$

W.L. only, from same figure,

$$R_1 = R_4 = \pm\frac{\Sigma P_w h}{B} = \frac{4(30) + 4.5(35) + 5(40)}{16} = \pm 29.84^k$$

$$H_1 = H_4 = 0.5\Sigma P_w = 0.5(4 + 4.5 + 5) = 6.75^k$$

$$F_1 = F_2 = \Sigma(P_w L)/B\ [13.5(17)]/16 = 14.34^k$$

Step 2: Calculate combined loads and note maximum values.
Case 2: 0.75 (Max. D.L. + W.L.)
Cross beams: Max. D.L. only is most critical.
Bracing members:

$$F_1 = 0.75(0 + 14.34) = 10.76^k,\ F_2 = 0.75(13.37 + 14.34) = 20.78^k$$

Columns:

$$P_{max.} = 0.75(70.8 + 29.84) = 75.48^k,\ M_{col.} = 0.75(6.75) = 5.1^{k\text{-}ft}$$

Case 3: 0.75(Min. D.L. + W.L.)
Check uplift on columns.

$$P_{min.} = 0.75[0.5(70.8) - 29.84] = 2.78^k$$

No uplift.

Step 3: Design component members for critical loads and moments.
Cross beams: First tier,

$$M = \left(\frac{1}{32}\right)(w_1B^2) = \left(\frac{1}{32}\right)[2.95(16^2)] = 23.64^{k-ft}$$

From steel handbook, try W6 × 20, for

$$L_u = 16 \text{ ft}, \; M_{allow.} = 24.5^{k-ft}$$

However, a deeper beam is recommended to reduce deflection at center; try M10 × 22.9 or W8 × 24. You might even consider neglecting the support offered at the center by the bracing member, if future modifications are uncertain.

Cross beam; Second tier,

$$M = \left(\frac{1}{8}\right)(wB^2) = \left(\frac{1}{8}\right)[2.95(16^2)] = 94.4^{k-ft}$$

From steel handbook, try W12 × 40 for

$$L_u = 16 \text{ ft}, \; M_{allow.} = 95.0^{k-ft}$$

Bracing members:

$$F_1 = 10.76^k, \text{ try } \quad \text{WT4} \times 8.5 \; (a = 2.5 \text{ in.}^2, r = 1.13)$$

Calculate

$$\frac{KL}{r} = \frac{1(17)12}{1.13} = 180.5; \text{ corresponding } F_a = 6.57 \text{ ksi}$$

Check

$$P_{allow.} = F_a a = 6.57(2.5) = 16.43 > F_1; \; F_2 = 20.78^k; \text{ try WT5} \times 10.5$$

$$(a = 3.1 \text{ in.}^2, r = 1.32 \text{ in.}); \; \frac{KL}{r} = \frac{1(17)12}{1.32} = 154.5$$

Corresponding F_a = 7.56 ksi. Check

$$P_{allow.} = 7.56(3.1) = 23.44^k > F_2$$

Columns:

$$P_{max.} = 75.48^k, \; M_{col.} = M_y = 5.1^{k-ft} = 61.2^{k\text{-}in.}$$

The approximate Method 1 will be used. Assume a trial section, W10 × 39. From steel tables note,

$$B_y = 1.027,\ r_x = 4.27 \text{ in.},\ r_y = 1.98 \text{ in.}$$

Calculate

$$P_{reqd.} = 75.48 + 1.027(61.2) = 138.3^k$$

Calculate (KL) critical.

$$(KL)_{xx} = 1(30) = 30,\ (KL)_{yy} = 1(15) = 15,$$

$$\text{equivalent } (KL)_{xx} = 30\left(\frac{1.98}{4.27}\right) = 13.9;\ (KL)_{yy} = 15 \text{ controls}$$

From column load tables, for W10 × 39, for $KL = 15$,

$$P_{allow.} = 162^k > P_{reqd.}$$

Use W10 × 39 for columns.

Note: The column orientation determines whether applied moment should be taken about xx axis or yy axis. The column moment assumed here is based on the bracing member not extending to column base.

Symbols

A	frame constant = $4(3 + 2/\phi)$
a	cross-sectional area of member, in.^{2}
B	width of pipe rack, ft
B_x, B_y	column bending factors about x or y axis
B_1, B_2, b, b_1, b_2	width of single column supports as noted, ft
C	horizontal projection of knee brace, ft
C_m, C_{mx}, C_{my}	coefficients for use in interaction formulas
D	nominal pipe diameter, in.
d_1, d_2	vertical distance between bracing members as noted, ft
$D.L.$	dead load
$E.L.$	earthquake load
E	modulus of elasticity of pipe support material, lb/in.^{2}
e	vertical projection of knee brace, ft
F	frame constant = $6(2 + 1/\phi)$
F_a	allowable axial stress, lb/in.^{2}
F_b	allowable bending stress, lb/in.^{2}
F_d, F_1, F_2	axial load in bracing member, lb
F_e', F_{ex}', F_{ey}'	$= \dfrac{12\,\pi^2 E}{23(KL/r)^2}$

F_t	allowable tensile stress, lb/in.^{2}
F_y	yield stress of pipe support material, lb/in.^{2}
F_{bx}, F_{by}	allowable bending stress about x (or y) axis, lb/in.^{2}
f_a	actual axial stress, lb/in.^{2}
F_{bx}, F_{by}	actual bending stress about x (or y) axis, lb/in.^{2}
H, H_1, H_4	horizontal reaction at column base (at 1 or 4), lb
h_1, h_2	height above base to tier no. 1, no. 2, ft
h_{12}	height between tiers 1 and 2, ft
h_{23}	height between tiers 2 and 3, ft
I_1	moment of inertia of column section, in.^{4}
I_2	moment of inertia of cross beam, in.^{4}
K	effective length factor in the plane of bending
k	frame constant $= 3/(Q\phi)$
L, L_1, L_2	length of bracing members, ft
L_u	maximum unbraced length of member, ft
M, M_1, M_2, M_3, M_4	bending moments at locations 1, 2, 3, 4, lb-ft
M_B, M_{b1}, M_{b2}	bending moment in cross beam, at tier 1, 2, lb-ft
M_x, M_y	bending moment about x (or y) axis, lb-ft
$M_{allow.}$	allowable bending moment, ft-lb
$M_{col.}$	bending moment in column, ft-lb
$M_{max.}$	maximum bending moment, lb-ft
ΔM	change in bending moment, lb-ft
P	axial load, lb
$P_{allow.}$	allowable axial load, lb
$P_{reqd.}$	required minimum axial load capacity, lb
$P_{max.}$, $P_{min.}$	maximum (or minimum) axial load, lb
P_w, P_{w1}, P_{w2}	wind load on pipe bent, at tier 1, 2, lb
p	pipe loading intensity, lb/ft^2
Q	frame constant $= 2(1 + 6/\phi)$
R, R_1, R_4	vertical column reaction at base, at locations 1, 4, lb
r, r_x, r_y	radius of gyration, about x, y axis, in.
S	section modulus, in.^{3}
$T.L.$	temperature load
$W.L.$	wind load
w, w_1, w_2	uniform vertical load, on tier 1, 2, lb-ft
$w_{max.}$, $w_{min.}$	maximum (or minimum), uniform vertical load, lb-ft
ϕ	frame constant $= (I_1/I_2)(B/h_1)$
θ	vertical angle between column and knee brace, degrees
θ_1, θ_2	vertical angle between column and bracing member 1 (or 2), degrees

References

1. *Manual of Steel Construction,* American Institute of Steel Construction.
2. Part 1 of this article.
3. C. H. Norris and J. B. Wilbur, *Elementary Structural Analysis,* McGraw-Hill, New York, 1960.
4. V. Leontovich, *Frames and Arches,* McGraw-Hill, New York, 1959.

C. V. CHAR

7
Pipeline Shortcut Methods

Collapsing Pressure

Piping and equipment that may experience net external pressure should be designed for such a condition to avoid failure. External pressure needs to be considered in the design of piping, tubing, and vessels for uses such as vacuum distillation columns, autoclaves, and heat exchangers.

For example, if external pressure is not accounted for, loss of tubeside pressure in a shell-and-tube heat exchanger could collapse the tubes. Also, buried and submerged piping must be looked at for the effects of external pressure to prevent failure.

Types of Failure

External pressure can cause two types of failure in pipes and vessels—plastic buckling and elastic collapse. Plastic buckling occurs in thick-walled pipes at stresses greater than the yield stress. Failure takes place in the plastic region of a material's stress-strain curve. This type of problem will not be dealt with here since only thin-walled pipes and vessels (i.e., those with a T/D of 0.10 or less) will be discussed.

For thin-walled pipes, elastic collapse is the predominant mode of failure. In it, collapse occurs in the elastic region of the material's stress-strain curve, below the yield stress. Failure happens here because local stresses exceed the yield stress (although the applied stress is below the yield value) at defects in the structure. Defects include variations in wall thickness due to manufacturing operations or corrosion, and out-of-roundness, also a result of the pipemaking process. In elastic collapse, a pipe will deform into nodes.

A method will be given to determine the strength of the pipe under external pressure, and an example will illustrate the method. Using this method will avoid elastic collapse of piping and equipment.

Elastic Collapse

A pipe's geometric and mechanical properties determine its strength under external pressure. Geometric properties are characterized by T/D and L/D, while mechanical properties are described by E.

T/D and L/D can be represented by n, the number of lobes created during elastic collapse. The minimum number for n is 2. (Here, the pipe's cross-section would somewhat resemble a figure-8.) Since n is a physical quantity, it must be an integer.

Collapse pressure is given by Von Mises' equation [1]:

$$P = \frac{1}{3}\left[n^2 - 1 + \frac{(2n^2 - 1 - \mu)}{\left(n^2\left(\frac{2L}{\pi D}\right)^2 - 1\right)}\right]\frac{2E}{(1 - \mu^2)}(T/D)^3 + \frac{2E}{(n^2 - 1)}\frac{T/D}{\left(n^2\left(\frac{2L}{\pi D}\right)^2 + 1\right)^2} \quad (1)$$

Equation (1) is for a perfectly cylindrical tube that is simply supported, and has no axial loading.

Equation (1) must be solved for n in terms of T/D and L/D. For a given n, there is a corresponding minimum P.

Thus, differentiating Eq (1) with respect to n and setting $dP/dn = 0$ will enable the minimum value of P to be found.

The following substitutions will be made prior to differentiation. Also, we will assume that the quantity n^2 is much greater than 1.

$$a = 1/3\ (T/D)^2 \quad (2)$$

$$b = \pi D/2L \quad (3)$$

$$c = 1 - \mu^2 \quad (4)$$

$$d = 1 + \mu \quad (5)$$

$$m = P(1 - \mu^2)^2/2E(T/D) \quad (6)$$

Making the above substitutions in Eq. (1) yields

$$m = an^2 + \frac{(2n^2 - d)}{\left(\frac{n^2}{b^2} - 1\right)} + \frac{c}{(n^2)\left(\frac{n^2}{b^2} + 1\right)^2} \quad (7)$$

Differentiating Eq. (7) and setting $dm/dn = 0$ gives

$$n^{14}a + n^{12}b^2a + n^{10}(dab^2 - 4ab^4) + n^8(3dab^4 - 8ab^6) + n^6(3dab^6 - 5ab^8 - 2cb^4 - cb^2) + n^4(dab^8 - ab^{10} - 4cb^6 - cb^4) + n^2(cb^6 - 2cb^8) - cb^8 = 0 \quad (8)$$

For Eq. (8), the following conditions apply:

1. $n \geq 2$
2. $L/D > 1$
3. $0.005 \leq T/D \leq 0.10$

Equation (8) then simplifies to

$$n^{14}a + n^6(3dab^6 - 5ab^8 - 2cb^4 - cb^2) = 0 \tag{9}$$

Such a simplification can be made because all terms in Eq. (8) are negligible compared to those terms in it that contain n^{14} or n^6. Solving Eq. (9) for n gives

$$n = \left(\frac{2cb^4 + cb^2}{a} + 5b^8 - 3db^6\right)^{1/8} \tag{10}$$

Replacing a, b, c, and d with their original expressions in Eqs. (2)–(6) gives

$$n = \left(\frac{(1 - \mu^2)}{3(T/D)^2}\left(2\left(\frac{\pi D}{2L}\right)^4 - \left(\frac{\pi D}{2L}\right)^2\right) + 5\left(\frac{\pi D}{2L}\right)^8 - 3(1 + \mu)\left(\frac{\pi D}{2L}\right)^6\right)^{1/8} \tag{11}$$

Graphical Method

Using Eqs. (1) and (11) to calculate the collapse pressure is tedious. These computations can be avoided by using Fig. 1, which presents worked-out solutions to these two equations. The figure allows quick estimates to be made of the collapse pressure and stress on the pipe.

The figure is for cases in which the stress is below the yield stress of the material. The chart can be extended to problems in plastic buckling by replacing E by E_T, the tangential modulus. A method to calculate E_T is given in Ref. 2.

The right side of the figure is used to find the collapse pressure. This part of the chart represents finding solutions to Eqs. (1) and (11). For various values of T/D and L/D. n was calculated from Eq. (11). Then, n was rounded to the next integer and was substituted into Eq. (1) to yield the collapse pressure. P was divided by E to give a nondimensional quantity.

On the left side of the figure, the lines for S/E and P/E were generated by using a modified hoop-stress formula:

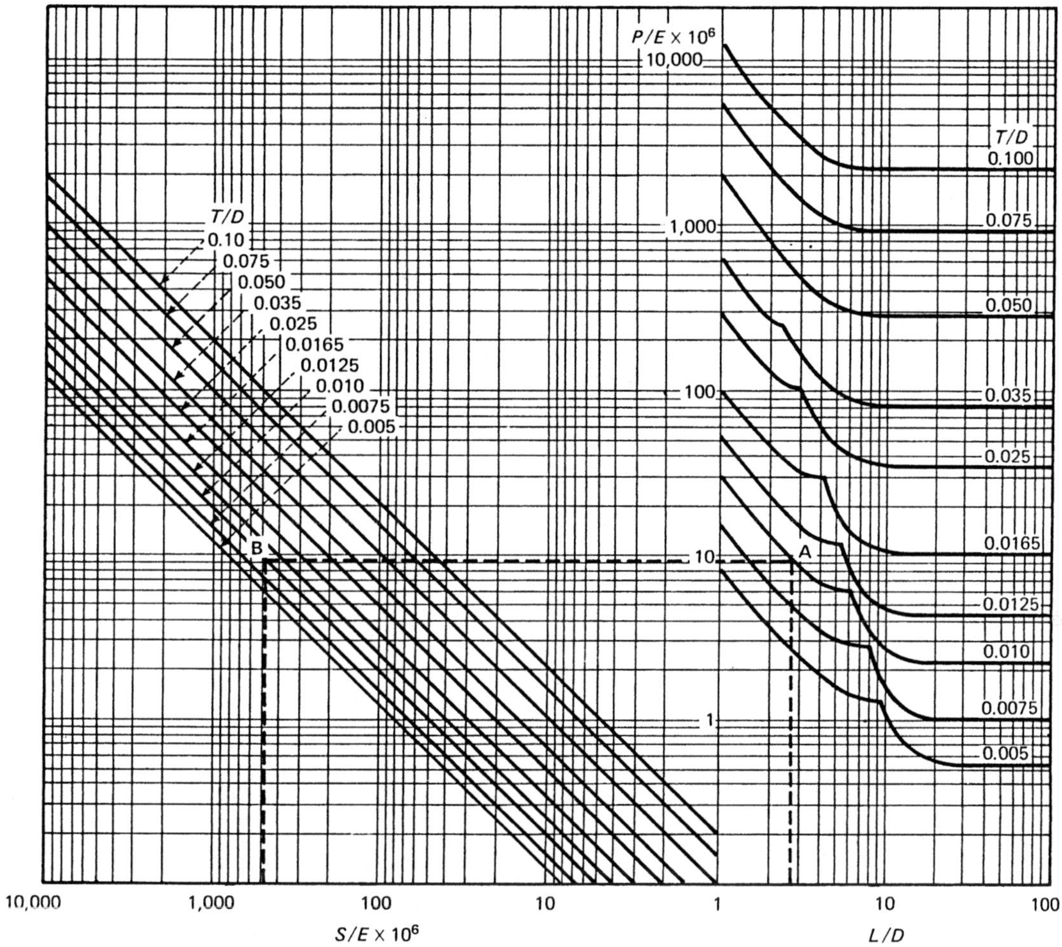

FIG. 1 This chart quickly reveals if a pipe or tube will withstand an external pressure.

$$\frac{S}{E} = \frac{1}{2(T/D)}\frac{P}{E} \tag{12}$$

The left part of the figure can be used to determine the stress in the pipe and whether it is below the allowable level.

The figure is for $\mu = 0.3$. Standard pipe sizes from nominal pipe size (NPS) of ⅛ to 42 in. with a $T/D \leq 0.10$ fall within range of the figure. These sizes correspond to pipe that is Sch. 160 or less. Exactly how to use the chart will be shown in an example.

Stiffening Rings

When a pipe cannot withstand the external pressure, two solutions are possible—either to use a thicker pipe or to add stiffening rings. These devices, while usually not employed for process piping, are sometimes installed in heat exchangers and vacuum equipment. Of course, for large pressure or vacuum vessels, these rings can be a practical and economical alternative to the use of thicker walls.

With piping or tubing, stiffening rings are typically added on the outside of the pipe for obvious reasons. When adding these devices to a pipe is considered, temperature gradients must be looked at. A temperature difference usually exists between the inside and outside of the pipe. Problems could occur if a thick stiffening ring is added. The high temperature of the pipe would result in a relatively large increase in its diameter.

However, if the ring is cool enough, it will not expand sufficiently to accommodate the increase in pipe diameter. The result will be buckling of the pipe around the stiffening ring.

Costs of stiffening rings are a consideration as well. Will it be more expensive to add rings than to go to a thicker pipe? And, if piping must be insulated, it usually is more expensive to wrap it with stiffening rings on it than it is to wrap it if it is plain.

At an L/D greater than about 15, stiffening rings will not have a significant effect on pipe stability. This corresponds to the flat region of the right-hand curves in the figure. From the figure, it is seen that P/E approaches a constant for L/D greater than 15. At such values, use of stiffening rings is not needed.

When stiffening rings are a suitable alternative, they can be sized according to

$$I_s = \frac{1}{28} D^3 L \frac{P}{E} \tag{13}$$

I_s is the minimum required moment of inertia of the ring, measured parallel to the centerline of the pipe. The example will explain how to calculate I_s. If the I for the ring exceeds I_s, the ring will be able to withstand the external pressure. The ring is usually made of the same material as the pipe. In Eq. (13), E accounts for the material of construction of the ring.

Safety Factors

Equation (1) assumes an ideal material (i.e., continuous, isotropic, etc.) that is perfectly cylindrical. If the piping is nonideal, the collapse pressure will be lower than predicted. Two major factors account for nonideality: ovality and wall-thickness variation.

A certain amount of ovality is unavoidable in most pipes or tubes. Thin-walled tubes, for example, become oval during annealing and straightening.

Ovality is the most common and widely studied defect, and is the major contributor to the reduction in collapse pressure [3, 4].

Ovality is more pronounced in thin-walled pipes. Also, the greater the percentage of a pipe's length that is oval, the lower its collapse pressure.

Ovality is expressed as a percentage of the outside diameter. Material specifications for boilers, pressure vessels and piping—such as SA134, SA409, and SA530 [5]—permit ovalities of 1–2%. For a 2% ovality, the collapse pressure will be reduced by 25%.

The effects of wall thickness on collapse pressure are not as great. Typically, a 1% reduction in thickness leads to the same decrease in collapse pressure. Of the specifications mentioned in Ref. 5, SA530 allows the greatest variation in wall thickness: −12.5% of nominal thickness, reducing collapse pressure by about 12%.

However, the wall may become quite thin, unless proper precautions are taken. Substantial drops in thickness may result from corrosion, erosion, and wear. An allowance is needed for these. It depends upon the material of construction, the fluid, corrosiveness, etc. To account for such problems, a safety factor is applied:

$$P = S_F P_d \tag{14}$$

Table 1 lists some common safety factors for external-pressure design.

Any concentrated or point loads on the pipe can contribute to its early collapse under pressure. To avoid this, pipe supports should extend over at least one-third of the pipe circumference.

Example. A 16-in.-diameter carbon steel pipe must withstand an external pressure of 85 lb/in.². The pipe is 0.16 in. thick—it was chosen for ease of calculation. Design data:

$$E = 27.9 \times 10^6 \text{ lb/in.}^2$$

$$S_A = 18 \times 10^3 \text{ lb/in.}^2$$

$$\mu = 0.3$$

$$S_F = 3$$

1. Are stiffening rings required? If so, what length is required between rings? Is the stress below the allowable stress?

$$P = 3 \times 85 = 255 \text{ lb/in.}^2$$

$$P/E \times 10^6 = 255/27.9 = 9.14$$

This establishes the dotted horizontal line in Fig. 1.

$$T/D = 0.16/16 = 0.10$$

This establishes Point A in the figure. Since Point A is *not* on the flap portion of the right-hand curve, stiffening rings are required. Reading downward from A, L/D is found to be 2.75. Therefore, rings are required every

$$2.75 \times 16 \text{ in.} = 44.0 \text{ in.}$$

TABLE 1 Typical Safety Factors Used in Designing for External Pressure [5–8]

Organization	Safety Factor
German Industry Norm	3–4
American Water Works Association	1.5–2.0
British Standards Institution	3

To find whether the stress is below the allowable level, move across the P/E line to Point B to where this line intersects the line for $T/D = 0.010$. Reading down:

$$S/E \times 10^6 = 480$$

Therefore, the stress is

$$480 \times 27.9 = 13.4 \times 10^3 \text{ lb/in.}^2$$

which is below $S_A = 18 \times 10^3$.

2. For the above L/D, *what is the minimum required moment of inertia of the stiffening ring?*

Equation (13) is used for the determination:

$$I_s = \frac{1}{28}\,\frac{(16^3)(44.0)(255)}{27.9 \times 10^6} = 0.058 \text{ in.}^4$$

Therefore, any ring whose moment is equal to or greater than this value can be used. For example, a ring that is ½ in. thick and whose outside radius is 1¼ in. larger than its inside radius has the following moment of inertia, using a standard means of calculation:

$$I = (1/12)(1/2)(1.25)^3 = 0.081 \text{ in.}^4$$

This ring is suitable for use.

3. What pipe thickness can be used so that no stiffening rings are needed? Is the stress below S_A?

A thicker pipe must be chosen. The pipe must fall on the horizontal dotted line in the figure. If the line were extended to the right past Point A, it would intersect with $T/D = 0.0125$, but at a place where the slope of this line is not 0. Thus, a stiffening ring would still be needed. However, if the horizontal dotted line were extended to the right, it would fall below the zero-slope portion of $T/D = 0.0165$. Thus, this size of pipe is certainly suitable for use.

The pipewall thickness is

$$16 \times 0.0165 = 0.264 \text{ in.}$$

From the left part of the figure, for $P/E \times 10^6 = 9.14$ and $T/D = 0.0165$, $S/E \times 10^6 = 260$, and the stress is

$$260 \times 27.9 = 7.25 \times 10^3 \text{ lb/in.}^2$$

which is below the allowable stress.

Symbols

a, b, c, d	as defined in Eqs. (2)–(4)
D	outside pipe diameter, in. (mm)
E	modulus of elasticity, lb/in.^{2} (N/m^2)
E_T	tangential modulus, lb/in.^{2} (N/m^2)
I	moment of inertia, in.^{4} (mm^4)
I_s	moment of inertia of stiffening ring, in.^{4} (mm^4)
L	pipe length between stiffening rings or supports, in. (mm)
m	as defined in Eq. (6)
n	number of lobes
P	collapse pressure, lb/in.^{2} (N/m^2)
P_d	design pressure, lb/in.^{2} (N/m^2)
T	pipewall thickness, in. (mm)
S	stress, lb/in.^{2} (N/m^2)
S_A	allowable stress, lb/in.^{2} (N/m^2)
S_F	safety factor
μ	Poisson's ratio

References

1. D. F. Windenburg and C. Trilling, "Collapse by Instability of Thin Cylindrical Shells under External Pressure," *Trans. ASME, 56*(8), 569 (1934).
2. R. G. Sturm and H. L. Obrien, "Computing Strength of Vessels Subjected to External Pressure," *Trans. ASME,* p. 353 (May 1947).
3. A. Lohmeier and N. C. Small, *Collapse of Ductile Heat Exchanger Tubes with Ovality under External Pressure,* Presented at the 2nd International Conference on Structural Mechanics in Reactor Technology, Berlin, 1973, Paper F6/4.
4. W. O. Livsey and A. A. Juneio, "Collapse of Heat Exchanger Tubes with Ovality and Simulated Defects," *Int. J. Pressure Vessels Piping* (*U.K.*), *4*(1), 47 (1976).
5. American Society of Mechanical Engineers (ASME), New York, *Boiler and Pressure Vessel Code: Section VIII, Pressure Vessels,* 1974, and *Section II, Part A,* 1974.
6. O. Heise and E. P. Esztergar, "Elastoplastic Collapse of Tubes under External Pressure," *ASME J. Eng. Ind.,* p. 735 (November 1970).
7. American Water Works Association (AWWA), Denver, *Steel Pipe Design and Installation,* AWWA M11, Published by New York Chapter.
8. British Standards Institution, London, *British Standard 1515: Specification for Fusion Welded Pressure Vessels,* 1968.

MILETA MIKASINOVIC
PETER A. MARCUCCI

Collapsing Pressure, Nomograph

More and more underwater pipelines are being placed on sea and ocean bottoms to transfer oil and gas to various destinations. As a consequence, these pipelines are subjected to considerable outer hydrostatic pressure during and after installation. These outer hydrostatic pressures may cause collapsing phenomena in the submarine pipes.

It has to be emphasized that in the absence of internal pressure in pipes (which is normally the case during installation, and may sometimes occur during the operating life), a pipe section may become unstable and collapse due to the outer hydrostatic pressure and certain combinations of axial forces and bending moments induced in the pipe.

The critical outer pressure that causes collapse of a nonpressurized tubular pipe may be governed by either elastic or plastic instabilities.

The theoretical critical outer hydrostatic pressure for *elastic* instability may be expressed as

$$P_e = \frac{2E}{(1 - \nu^2)(D/t)(D/t - 1)^2} \tag{1}$$

The theoretical outer hydrostatic pressure for *plastic* instability may be written as

$$P_p = 2\sigma_y(t/D) \tag{2}$$

where E = modulus of elasticity of pipe material, (Young's modulus) lb/in.^{2}
ν = Poisson's ratio, dimensionless.
D = outside pipe diameter, in.
t = pipe wall thickness, in.
σ_y = yield stress, lb/in.^{2}

The real critical hydrostatic pressures may be considerably lower than those calculated by means of the above equations. They may be caused by pipeline imperfections such as wall-thickness variations, the initial out-of-roundness of a pipe, axial tension, and bending.

The elastic instability becomes dominant for very large values of D/t, i.e., in thin-walled pipes [1, 2].

Since most underwater lines are made from various grades of carbon steel, certain simplifications may be introduced into Eq. (1).

If it is assumed that for most carbon steels the values of E and ν are constants ($E \cong 30{,}000{,}000$ lb/in.^{2} and $\nu \cong 0.3$), the simplified Eq. (1) will

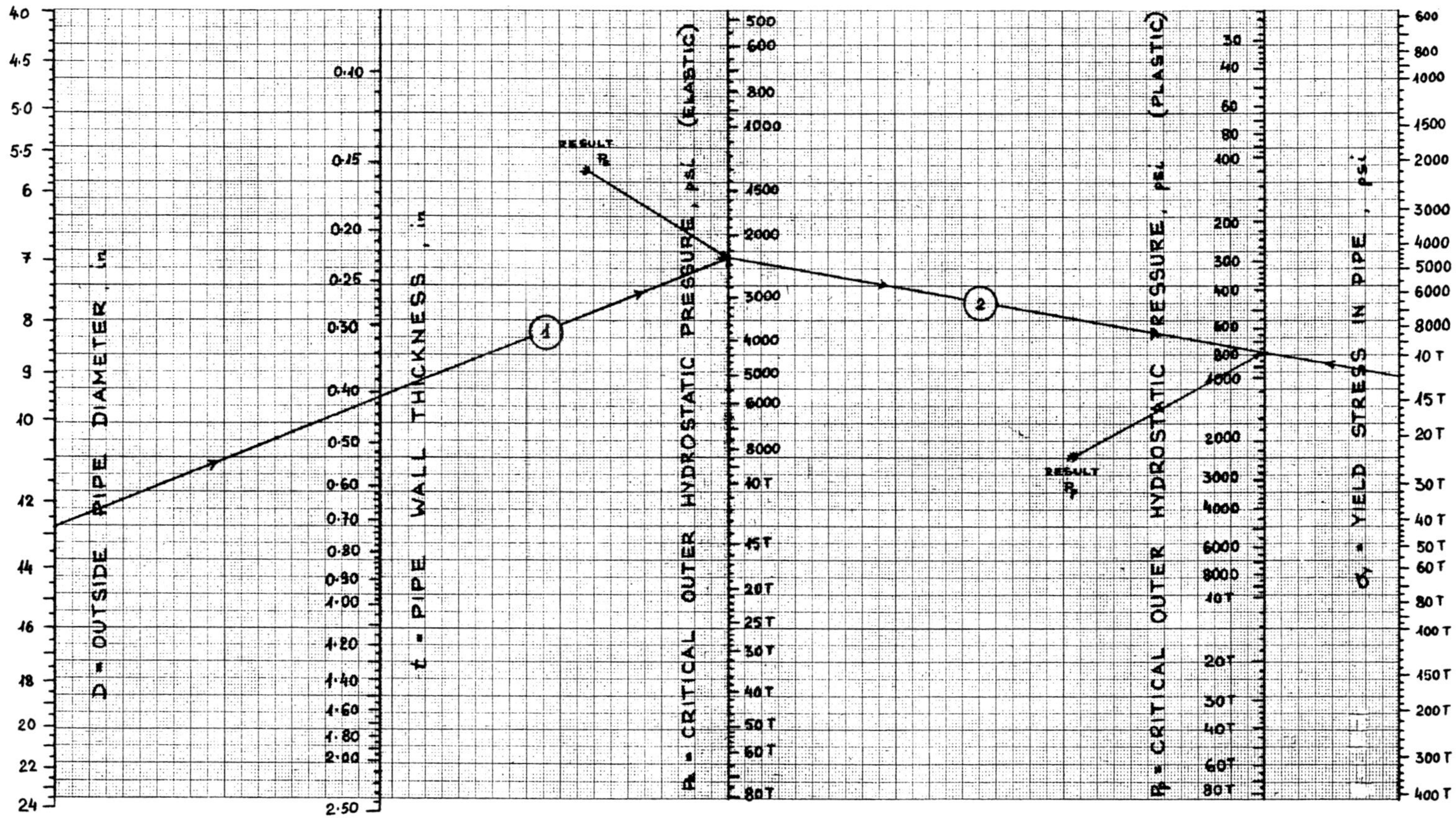

FIG. 1 Nomograph of the collapsing resistance of submarine lines.

be

$$P_e = \frac{65{,}930{,}000}{(D/t)(D/t - 1)^2} \tag{1a}$$

The Nomograph

A nomograph (Fig. 1) has been designed which allows Eqs. (1a) and (2) to be solved rapidly and simultaneously by using only two consecutive movements of a ruler. The use of this nomograph involves two steps:

Step 1. Connect the known values of D and t on the appropriate scales with a ruler. Extend this line up to intersect with the P_e scale. Read the result, P_e, at this intersection.

Step 2. Connect the just found P_e value to the known value of σ_y on the appropriate scale with a ruler. Read the second result, P_p, at the intersection of the ruler with the P_p scale.

Example. The use of this nomograph may be best shown by a typical example: Calculate the critical collapsing pressures P_e and P_p of a pipe with the following dimensions: D = 12.75 in., t = 0.406 in. (standard 12 in. pipe, Schedule 80), and σ_y = 12,000 lb/in.^{2}.

Solution: Following the "procedure" described, we find the following theoretical critical pressures:

$$P_e = 2270 \text{ lb/in.}^2 \text{ and } P_p = 765 \text{ lb/in.}^2$$

Note: This example is drawn on the nomograph.

References

1. S. C. Haagsma and D. Schaap, "Collapsing of Submarine Lines Studied," *Oil Gas J.*, pp. 87–95 (February 2, 1981).
2. S. P. Timoshenko and J. M. Gere, *Theory of Elastic Stability*, 2nd ed., McGraw-Hill–Kogakusha, Tokyo.

ADAM ZANKER

Discharge from Horizontal Pipes

The so-called California pipe flow method for estimating the discharge of water and similar liquids such as water solutions, wastes, etc. requires that a pipe be horizontal and at least 6 pipe diameters in length as shown in Fig. 1. No orifice or nozzle is required for this method, and the only two necessary measurements are the inside diameter of the pipe and the distance from the inside top of the pipe to the surface of the flowing water.

Discharge is calculated by

$$q = 3900(1 - a/d)^{1.88}d^{2.48}$$

where q = the discharge, gal/min
a = distance from inside top of pipe to surface of flowing water, ft
d = inside diameter of pipe, ft

See Fig. 1.

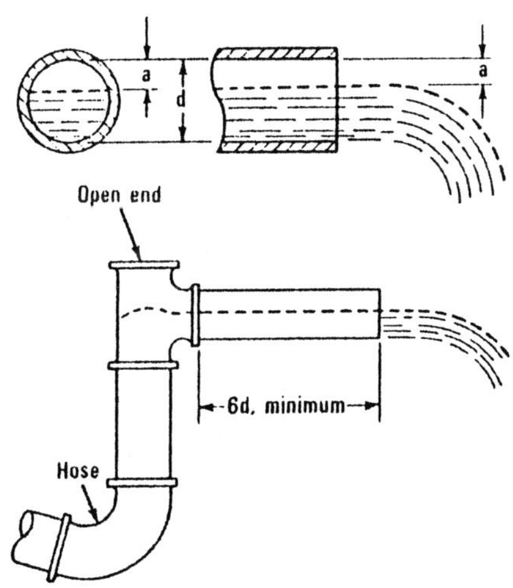

FIG. 1

Although the California method is very simple, it may be troublesome to calculate because of the fractional exponents involved. The two nomographs presented here as Figs. 2 and 3 solve the equation within a matter of seconds.

Example 1. What will be the discharge from a horizontal pipe with an 8 in. i.d., d, and a equal to 2 in?

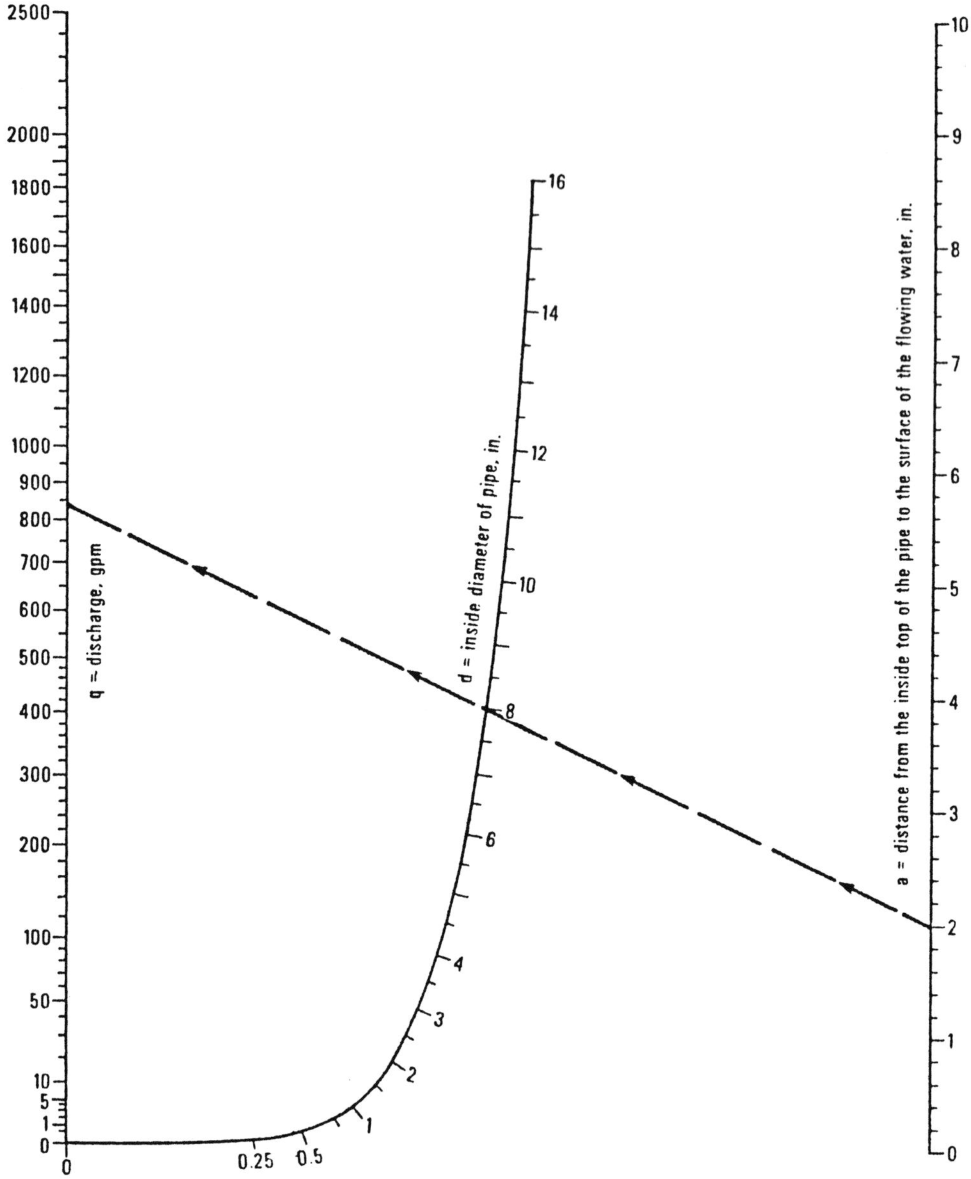

FIG. 2 This nomograph determines the discharge from an open end, horizontal pipe for values of d, a, and q of 0.25–16 in., 0–10 in., and 0–2500 gal/min, respectively. The broken line indicates the solution to Example 1.

Solution: Extend a line from 2 on the a scale of Fig. 2 through 8 on the d scale and extend the line through to the q scale. Read the answer as 840 gal/min at the point of intersection on the q scale.

Example 2. What will be the discharge from a horizontal pipe with a 33 in. i.d., d, and a equal to 11 in?

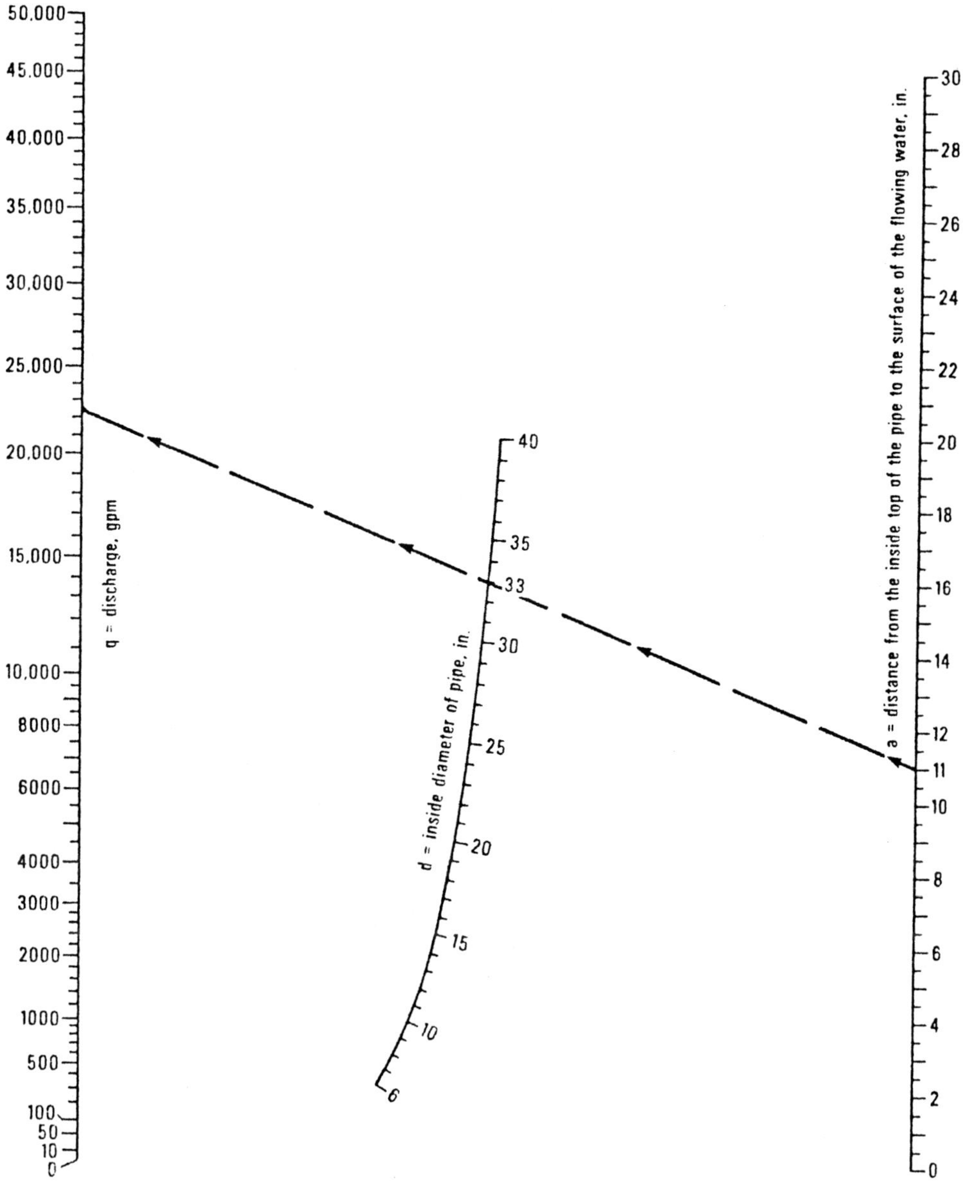

FIG. 3 This nomograph determines the discharge from an open end, horizontal pipe for values of *d*, *a*, and *q* of 6–40 in., 0–30 in., and 0–50,000 gal/min, respectively. The broken line indicates the solution to Example 2.

Solution: Extend a line from 11 on the *a* scale of Fig. 3 through 33 on the *d* scale. Read the answer as 22,400 gal/min at the point of intersection on the *q* scale.

Although the values of *d* and *a* in the equation are expressed in feet, the nomographs' scales are graduated in inches for greater convenience.

This material appeared in *Heating/Piping/Air Conditioning,* p. 69, January 1973, and is reprinted with the permission of the editor.

ADAM ZANKER

Discharge from Vertical Pipes

The discharge from a vertical open-end pipe may be of two types, "weir flow" or "jet flow," as illustrated in Fig. 1.

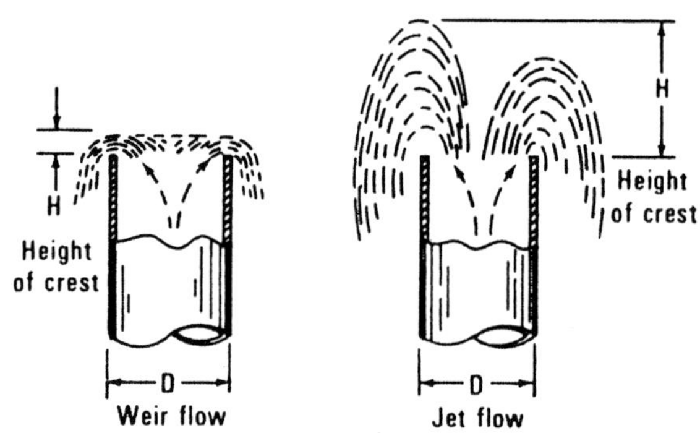

FIG. 1

Discharge rates for water and similar liquids (water solutions, wastes, etc.) may be calculated from the following equations [1]:

For weir flow:

$$Q_w = 8.80D^{1.20}H_w^{1.24} \tag{1}$$

For jet flow:

$$Q_j = 5.84D^{2.025}H_j^{0.53} \tag{2}$$

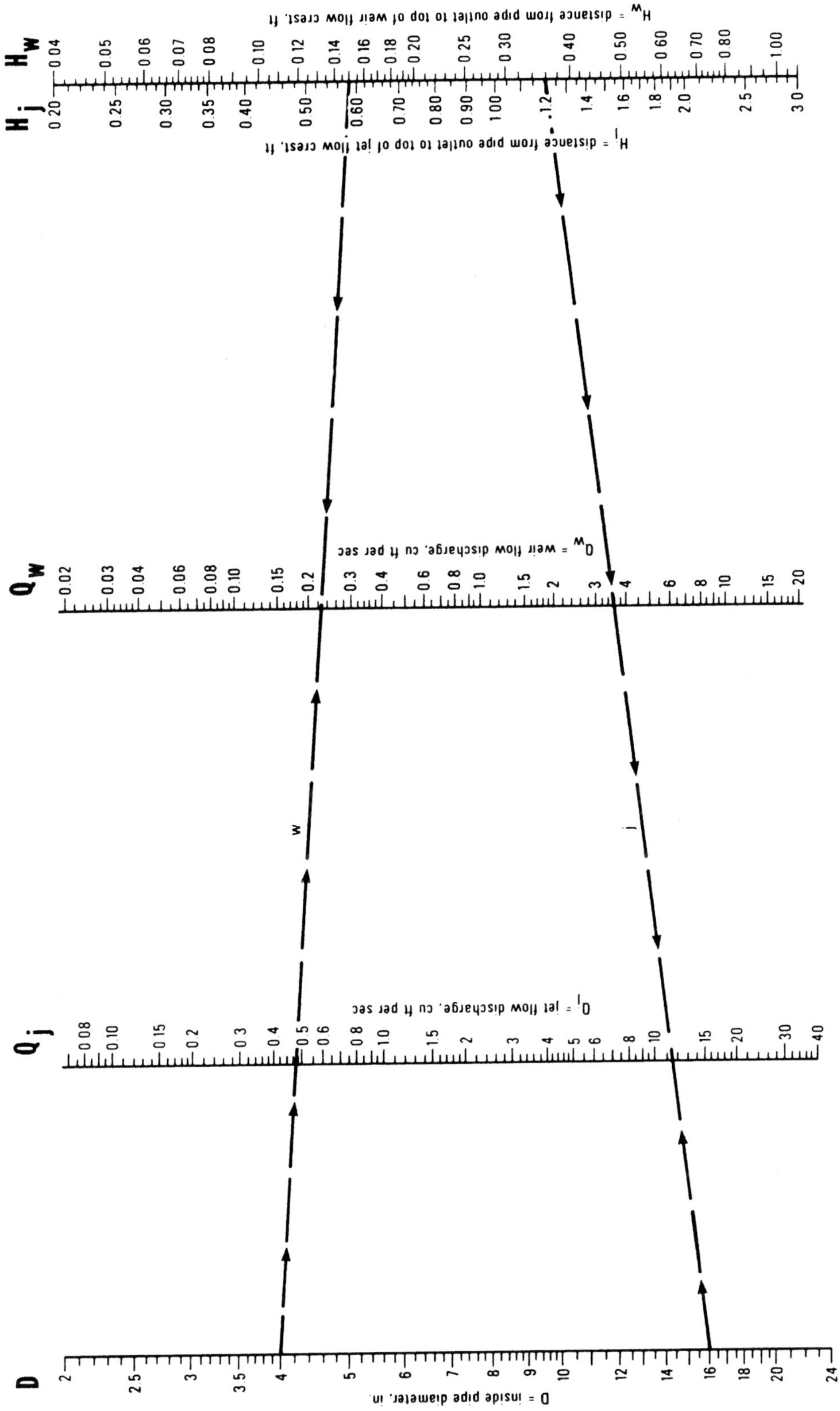

FIG. 2 This nomograph gives the discharge rate from vertical open-end pipes for either weir- or jet-type flow when the inside pipe diameter and the height of the crest are known. The broken lines indicate solutions to the examples in the text.

TABLE 1

Pipe Diameter (in.)	Head at Transition from Weir to Jet Flow (ft)	
	From	To
2	0.04	0.20
4	0.15	0.40
6	0.15	0.50
8	0.30	0.75
12	0.40	1.00
18	0.75	1.00
24	1.00	3.00

where Q = discharge flow rate, ft^3/s
D = inside pipe diameter, ft
H = distance from pipe outlet to top of crest, ft

The subscripts w and j denote weir flow and jet flow, respectively.

For pipes of various diameters, transition from weir flow to jet flow occurs as shown in Table 1 [2].

Because the fractional exponents of D and H values make Eqs. (1) and (2) rather troublesome to solve, a nomograph is presented to permit rapid determinations of discharge rate Q. The nomograph (Fig. 2), which solves both equations, consists of four scales: a single scale for D: separate scales for Q_w and Q_j; and a scale for H, graduated on the left for H_j and on the right for H_w. It should be noted that although D in the equations is expressed in feet, the D scale on the nomograph is graduated in inches for greater convenience.

Weir Flow Example. What is the discharge rate from a vertical pipe with a 4-in. diameter if the distance to the top of the crest is 0.15 ft?

Solution: Extend a line from 4 on the D scale to 0.15 on the H_w scale and read a value of 0.224 ft^3/s at its intersection with the Q_w scale.

Jet Flow Example. What is the discharge rate from a vertical pipe with a 16-in. diameter if the distance to the top of the crest is 1.2 ft?

Solution: Extend a line from 16 on the D scale to 1.2 on the H_j scale and read a value of 11.5 ft^3/s at its intersection with the Q_j scale.

This material appeared in *Heating/Piping/Air Conditioning,* pp. 129–130, January 1973, and is reprinted with the permission of the editor.

References

1. M. Brooke, "Flow File—II," *Chem. Eng.*, February 1957.
2. Manual of Disposal of Refinery Wastes, American Petroleum Institute, New York, 1969, Chap. 4.

ADAM ZANKER

Economic Pipe Sizes

Among the many existing formulas for calculating the most economic pipe size, the formula developed by G. R. Kent [1] appears to be the most accurate. It is based on the average fluid velocity (of a liquid or a gas) and the allowable average pressure drop, and it is designed to allow for turbulent flow, which represents the majority of cases in pipe flow.

The formula is

$$D = 0.627\frac{W^{0.486}}{\rho^{0.342}}\mu^{0.0274}\left(\frac{C}{K}\right)^{0.171}$$

where D = pipe internal diameter, in.
W = fluid flow rate, lb × 1000/h
ρ = fluid density
μ = fluid viscosity, cP
C = the pumping equipment installation cost, £/bhp,
K = the cost of the piping, £/in. diameter/longitudinal ft

The average and typical data on flow conditions may be found in numerous technical sources; among these, the reader is referred to Ref. 2.

The solution of this equation is troublesome and difficult to achieve without spending considerable time. To obviate the necessity for complex calculations, two nomographs have been constructed to solve the equation simply by using four movements of a ruler. Nomograph I solves the flow equation for liquids, while Nomograph II solves for gas, including steam. The use of both nomographs is the same, and they are separated only because they rely upon entirely different ranges of variables.

The Use of the Nomographs

In order to use the graphs, the following variables have to be known: K, C, and W, together with ρ and μ at the average flow conditions. The values

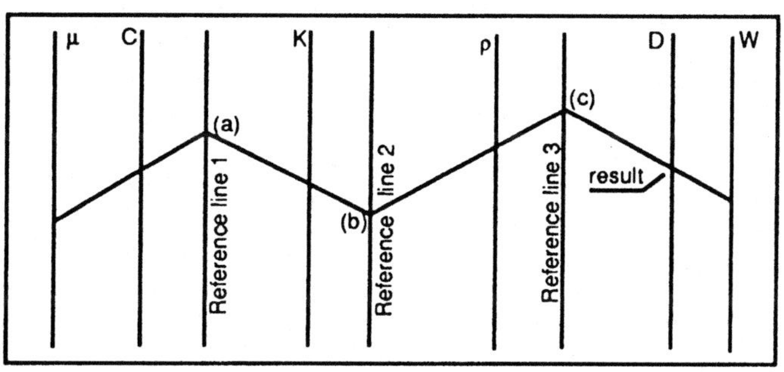

FIG. 1 Illustration of the use of the nomographs.

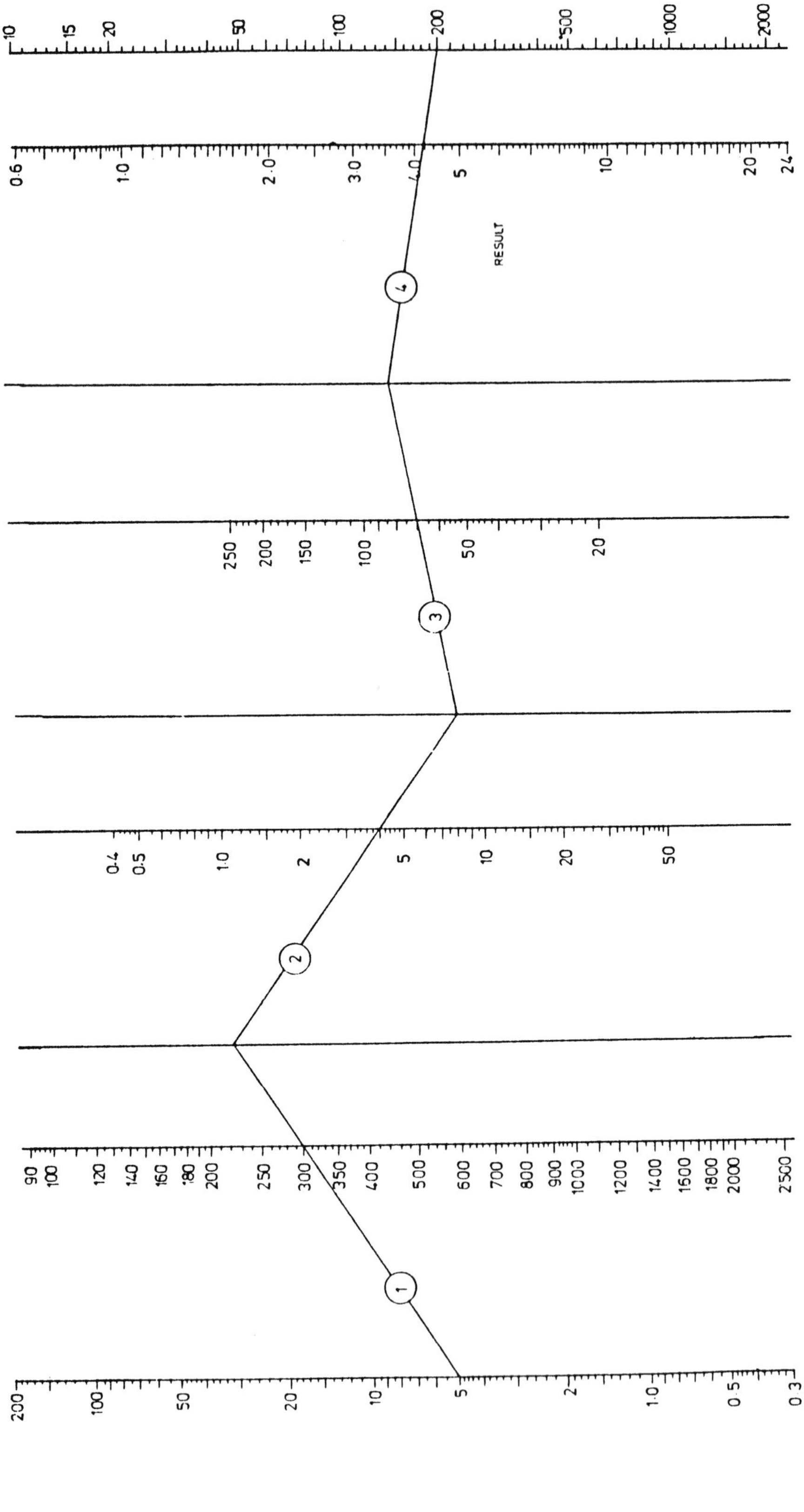

FIG. 2 Nomograph for calculation of the economic pipe size for liquid transmission.

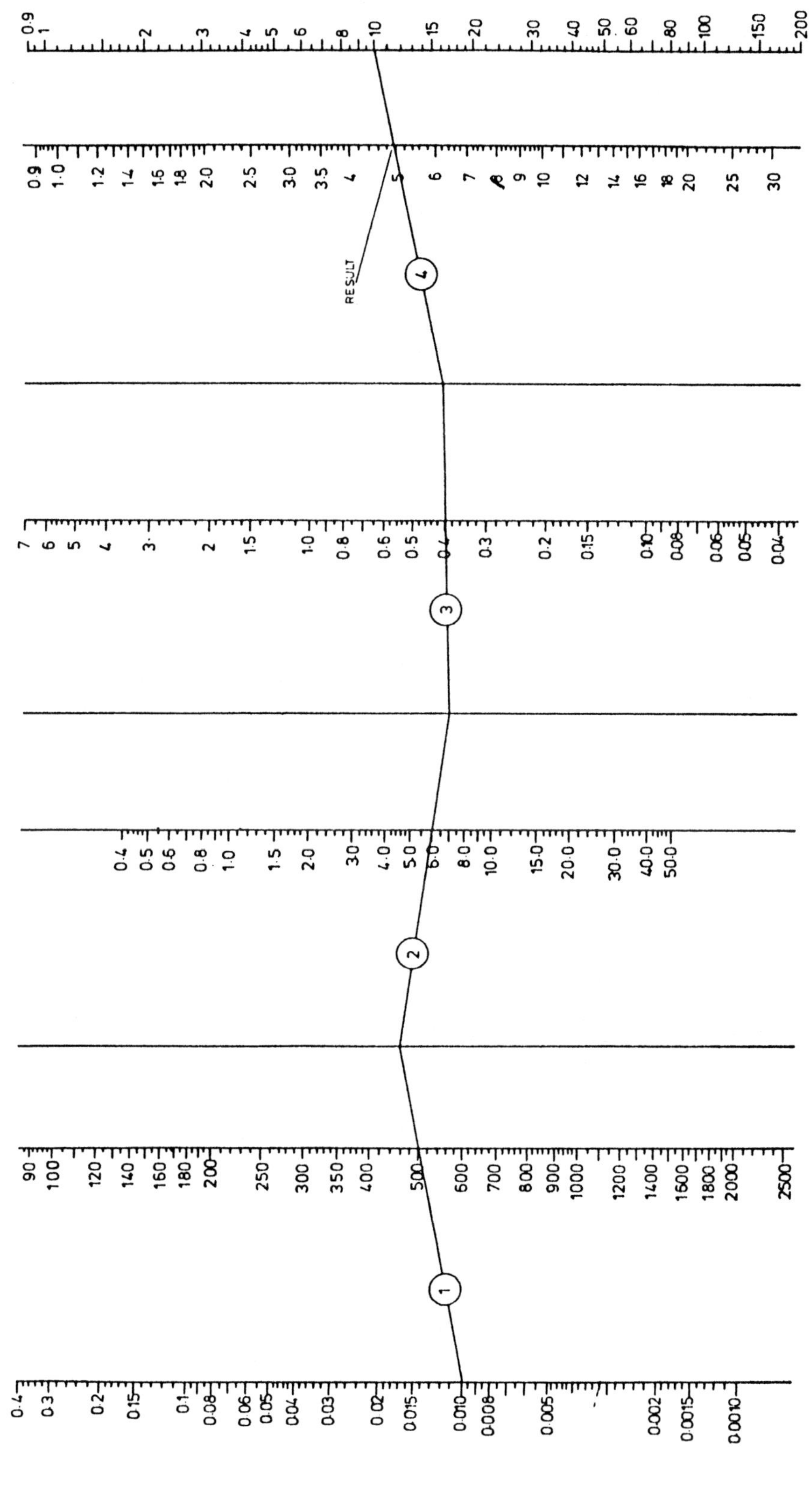

FIG. 3 Nomograph for calculation of the economic pipe size for gas transmission.

of μ and C are first connected with a straight edge on the appropriate graph (see Fig. 1), and this line is produced to an intersection with Reference Line 1 (the Point a). Point a is joined by a straight edge to the value of K on the scale, and this line is produced to intersect Reference Line 2, giving Point b. Point b is connected to the value of ρ and extended to meet Reference Line 3, resulting in Point c. Finally, Point c is connected with the known value of W, and the required value of D is read off the scale.

Worked Examples

Figures 2 and 3 are the nomographs drawn to scale and suitable for accurate use. In order to illustrate the technique involved, two worked examples are given, the lines for which have been drawn on the figures.

1. Liquid Application

It is desired to calculate the economic pipe size for a liquid with density of $\rho = 70$ lb/ft^3 and viscosity of $\mu = 5$ cP. The flow rate is to be $W = 200{,}000$ lb/h.

The cost of installment is known to be $C = 300$ £/bhp, and the cost of piping in this case is given as $K = 4$ £/in./ft.

It can therefore be seen from Fig. 2, following the use illustrated in Fig. 1, that $D = 4.2$ in.

2. Gas Application

The economic pipe size for a gas with density $\rho = 0.4$ lb/ft^3 and viscosity of $\mu = 0.01$ cP is required for flowing conditions with a flow rate of $W = 10{,}000$ lb/h. Cost of installment is given as $C = 500$ £/bhp, and the cost of piping is $K = 6$ £/in./ft.

Use of Nomograph II (Fig. 3) produces the result that $D = 4.93$ in. It should be noted that for both these examples, the nearest standard pipe size should be used, and the most economical situation will follow.

References

1. G. R. Kent, "Preliminary Pipe Sizing," *Chem. Eng.*, pp. 119–120 (September 25, 1978).
2. H. Popper, *Modern Cost-Engineering Techniques,* McGraw-Hill, New York, 1970.

ADAM ZANKER

Expansion of Pipe Materials

To find the thermal expansion of piping materials listed in Table 1, draw a line from the temperature scale at the left in Fig. 1 through the key points and extend it to the scales at the right. Table 1 identifies the key point materials and the proper scale. The result is the thermal expansion of the pipe at the desired temperature in centimeters per 100 meters of pipe.

This material appeared in *Hydrocarbon Processing*, May 1969, copyright © 1969 by Gulf Publishing Co., Houston, Texas 77252, and is reprinted by special permission.

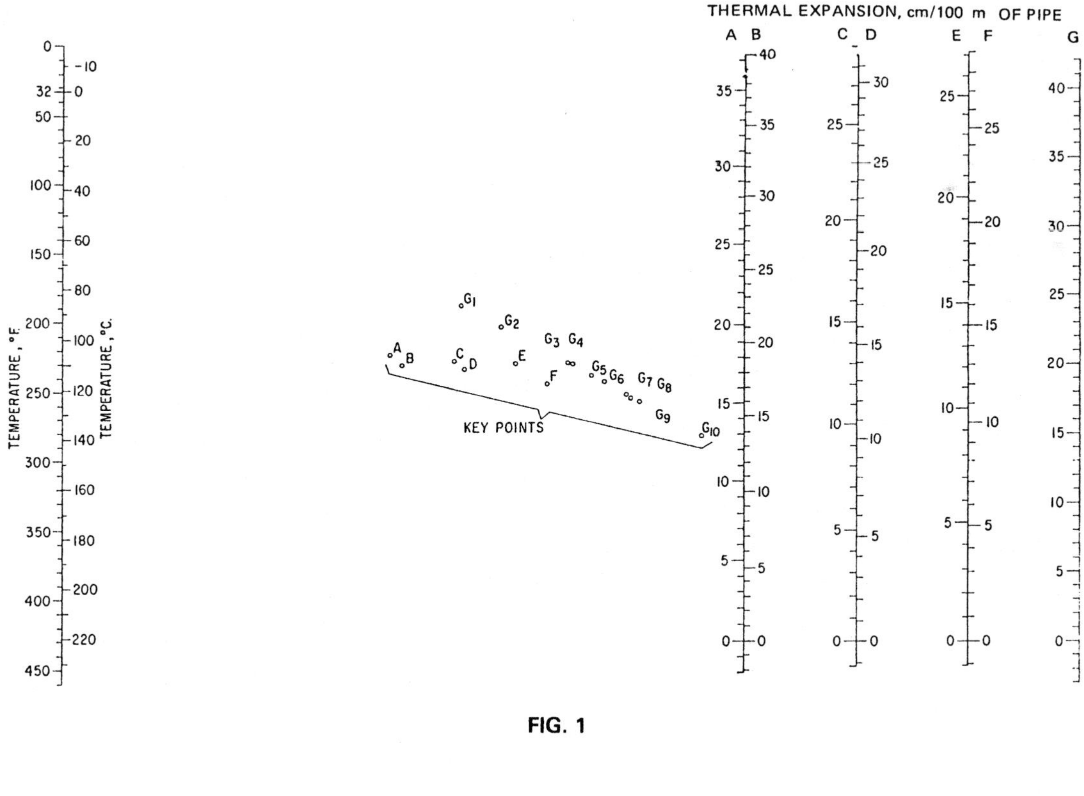

FIG. 1

TABLE 1 Guide to Fig. 1

Key Point	Scale	Material	Refs.
A	A	ASTM 301	1
B	B	ASTM 321	1
C	C	Carbon steel	2
D	D	Wrought iron	2
E	E	Intermediate alloys	3
F	F	Straight chromium stainless steels	3
G_1	G	Lead	4
G_2	G	Aluminum	3
G_3	G	Bronze	3
G_4	G	Brass	3
G_5	G	Copper	4
G_6	G	70Cu–30Ni	3
G_7	G	Monel I (67Ni–30Cu)	3
G_8	G	25Cr–30Ni	3
G_9	G	Monel II (66Ni–29Cu-Al)	3
G_{10}	G	Gray cast iron	3

References

1. *Metals Handbook,* 8th ed., American Society of Metals, pp. 422–423.
2. Holborn and Day, *Smithsonian Physical Tables,* 7th ed., p. 218.
3. J. H. Perry, *Chemical Engineers' Handbook,* 4th ed., p. 24-50.
4. *International Critical Tables.*

ADAM ZANKER

Flow Rates

Water flowing through a smooth tube or pipe is subject to pressure loss in the direction of flow. This loss of head in water piping systems because of friction is a predictable quantity and can be estimated by using the equation

$$\Delta p = \frac{fL_eV^2\rho}{24gd_i} \qquad (1)$$

where Δp = pressure drop per 100 ft of length, lb/in.^{2}
f = friction factor, dimensionless
L_e = equivalent length of pipe or tube, ft
V = velocity of water, ft/s
ρ = density of water, lb/ft^3
g = acceleration of gravity, 32 ft/s^2
d_i = tube or pipe inside diameter, in.

Friction factor f is a function of the Reynolds number:

$$f = 0.133\text{Re}^{-0.174} \tag{2}$$

where Re = Reynolds number, dimensionless. Defining the Reynolds number in Eq. (2) as

$$\text{Re} = \frac{\rho V d_i}{12\mu}$$

where μ = viscosity of water, lb-s/ft^2 and combining with, and simplifying, Eq. (1) gives

$$\frac{\Delta p}{L_e} = \frac{0.000267\rho^{0.826}\mu^{0.174}V^{1.826}}{d_i^{1.174}}$$

Because head losses increase as a function of increasing velocity, it is often desirable to adjust the flow rate to the actual needs of a specific processing or plant operating system. These flow rates are, as a rule, limited by the efficiency and capacity of the pump, or in some instances by the device used for measuring flow.

Even though the maximum possible flow rate that can be sustained in a specific size pipeline is not usually the limiting factor, it is desirable to know. When this flow rate is known, the actual flow rate can be, and generally is, reduced to a value that is significantly less than the maximum.

Data for maximum possible flow rates [1] can be compiled, and a nomograph (Fig. 1) that allows rapid determination of maximum flow rate has been constructed. The results obtained from this nomograph are valid only for flow conditions when the Reynolds number is greater than 10,000. As a practical matter, therefore, the nomograph cannot be used with accuracy for very viscous liquids (25 centistokes or greater under actual flow conditions). Likewise, it cannot be used for pipes smaller than 2 in. in diameter.

The nomograph is constructed in two parts: the left-hand side provides the maximum possible mass-flow rates, and the right-hand side gives the maximum possible volume-flow rates of liquids, the volume being measured at the specific flow conditions (temperature, pressure, and specific gravity of the liquid).

Use of this nomograph requires that the internal diameter of the pipe be expressed in inches and the specific gravity of the liquid be known. The specific gravity of water at 60°F is assumed to be 1.

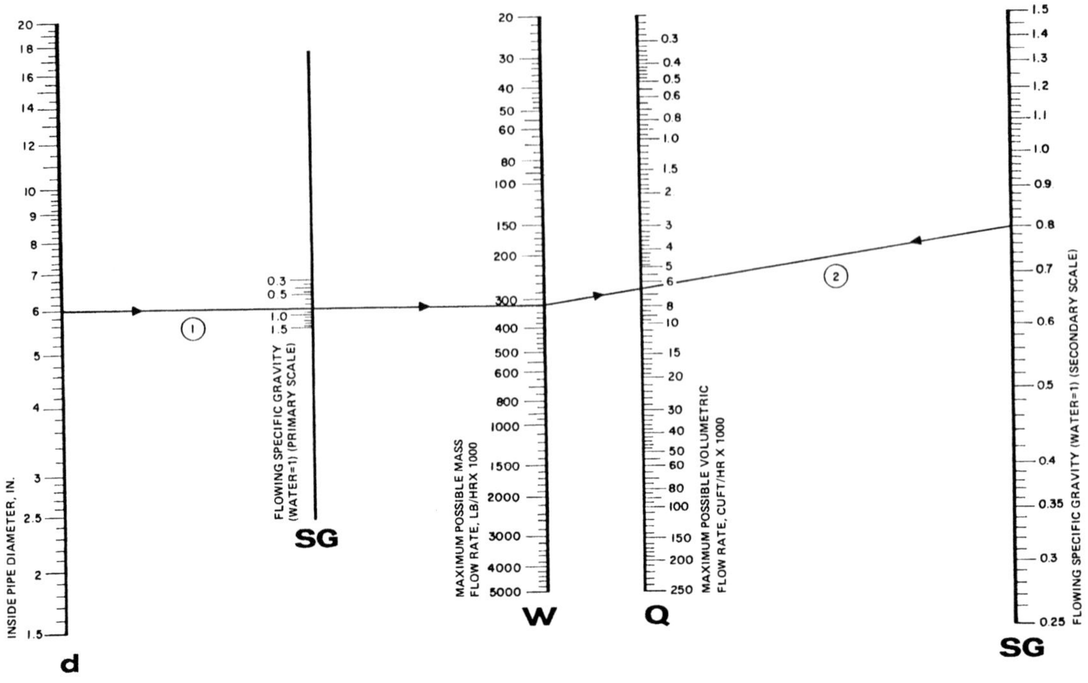

Problem. Find the maximum flow rate possible when a liquid with a specific gravity of 0.8 is flowing through a pipe having an internal diameter of 6 in.

Solution: Use a straightedge to connect 6 on Scale d with 0.8 on Primary Scale SG and extend the line to intersect Scale W, reading the maximum possible flow rate as 320,000 lb/h. Connect 320,000 on Scale W with 0.8 on Secondary Scale SG and, where Scale Q is crossed, read the maximum possible volume flow rate as 6400 ft^3/h.

This material appeared in *Plant Engineering*, November 24, 1982, and is reprinted with the permission of the editor.

Reference

1. R. F. Stearns, *Flow Measurements with Orifice Meters*, Van Nostrand, New York, 1951.

ADAM ZANKER

Friction Losses, Contraction or Expansion

Any change of velocity or direction of flow of fluid in pipes causes a certain friction loss. The main reason for the change of velocity of such fluid is a sudden change in pipe cross-section. There are two separate cases involving a change of velocity: (A) The sudden expansion of cross-section, and (B) The sudden contraction of cross-section.

Case A. Friction Loss from Sudden Expansion of Cross-Section

If the cross-section of the conduit is suddenly enlarged, the fluid stream separates from the wall and issues as a jet into the enlarged section. The jet then expands to fill the entire cross-section of the larger conduit (Fig. 1).

The space between the expanding jet and the conduit wall is filled with fluid in vortex motion, characteristic of boundary-layer separation, and considerable friction is generated within this space.

The friction loss h_{fe} from a sudden expansion of cross-section is proportional to the velocity head of the fluid in the small conduit and can be written

$$h_{fe} = K_e \frac{\overline{V}_a^2}{2g} \tag{1}$$

where h_{fe} = the frictional loss, ft
$\overline{V}_a$ = the average velocity in the smaller (upstream) conduit, ft/s
K_e = expansion-loss coefficient
g = the acceleration of gravity, taken as constant, g = 32.2 ft/s^2

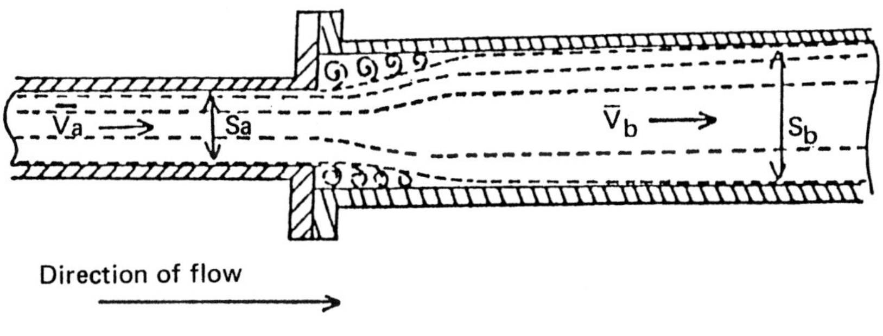

FIG. 1 Diagram illustrating effects of sudden expansion of pipe cross-section.

The expansion-loss coefficient, K_e, is defined by

$$K_e = \left(1 - \frac{S_a}{S_b}\right)^2 \tag{2}$$

where S_a = cross-section area of the upstream duct (in.^{2})
S_b = cross-section area of the downstream duct (in.^{2})

When taking in account that

$$S_a = \frac{d_a{}^2\pi}{4} \tag{3a}$$

and

$$S_b = \frac{d_b{}^2\pi}{4} \tag{3b}$$

where d_a and d_b are the diameters of upstream and downstream ducts, respectively, and combining Eqs. (1), (2), (3a), and (3b), we get the final equation for the calculation of the friction loss from sudden expansion:

$$h_{fe} = \left[1 - \left(\frac{d_a}{d_b}\right)^2\right]^2 \frac{\overline{V}_a{}^2}{2g} \tag{4}$$

Case B. Friction Loss from Sudden Contraction of Cross-Section

When the cross-section of the conduit is suddenly reduced, the fluid stream cannot follow around the sharp corner, and the stream breaks contact with the wall of the conduit (Fig. 2). A jet is formed, which flows into the stagnant

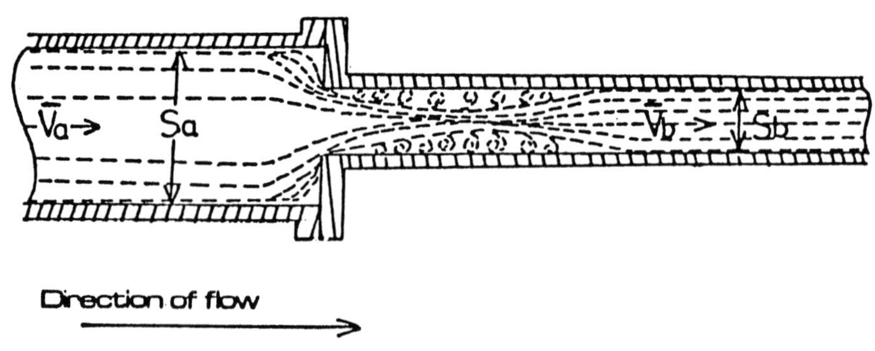

FIG. 2 Diagram illustrating effects of sudden contraction of pipe cross-section.

fluid in the smaller section. The jet first contracts and then expands to fill the smaller cross-section, and downstream from the point of contraction the normal velocity distribution is reestablished.

The friction loss from the sudden contraction is proportional to the velocity head in the smaller (downstream) conduit, and may be calculated by

$$h_{fc} = K_c \frac{\overline{V}_b^2}{2g} \tag{5}$$

where h_{fc} = the frictional loss, ft
K_c = the contraction-loss coefficient
$\overline{V}_b$ = the average velocity in the smaller (downstream) duct, ft/s

The K_c value may be calculated from

$$K_c = 0.4\left(1 - \frac{S_b}{S_a}\right) \tag{6}$$

When taking into account the relationships shown by Eqs. (3a) and (3b), and combining Eqs. (6) and (5), we get the final formula for friction loss at sudden contraction:

$$h_{fc} = 0.4\left(1 - \frac{d_b}{d_a}\right)\frac{\overline{V}_b^2}{2g} \tag{7}$$

It has to be emphasized that the friction loss is *always* proportional to the square of the velocity in the smaller conduit; at expansion it is proportional to the upstream flow, and at contraction to the downstream flow.

All the above equations refer only to turbulent flow. Friction losses due to expansion or contraction of flow are very small—almost negligible—for laminar flow.

These equations only refer to incompressible fluids, i.e., liquids only, and not to gases [1, 2].

The Nomograph

The accompanying nomograph (Fig. 3) enables one to calculate both friction losses—for expansion and for contraction—in a couple of seconds by using one movement of a ruler. The nomograph solves Eq. (4) for expansion friction losses and Eq. (7) for contraction friction losses.

In order to use this nomograph, the following data have to be known:

1. The average linear velocity ($\overline{V}$) in the smaller duct, i.e., the $\overline{V}_a$ value for expansion and $\overline{V}_b$ value for contraction
2. The upstream (d_a) and downstream (d_b) duct diameters

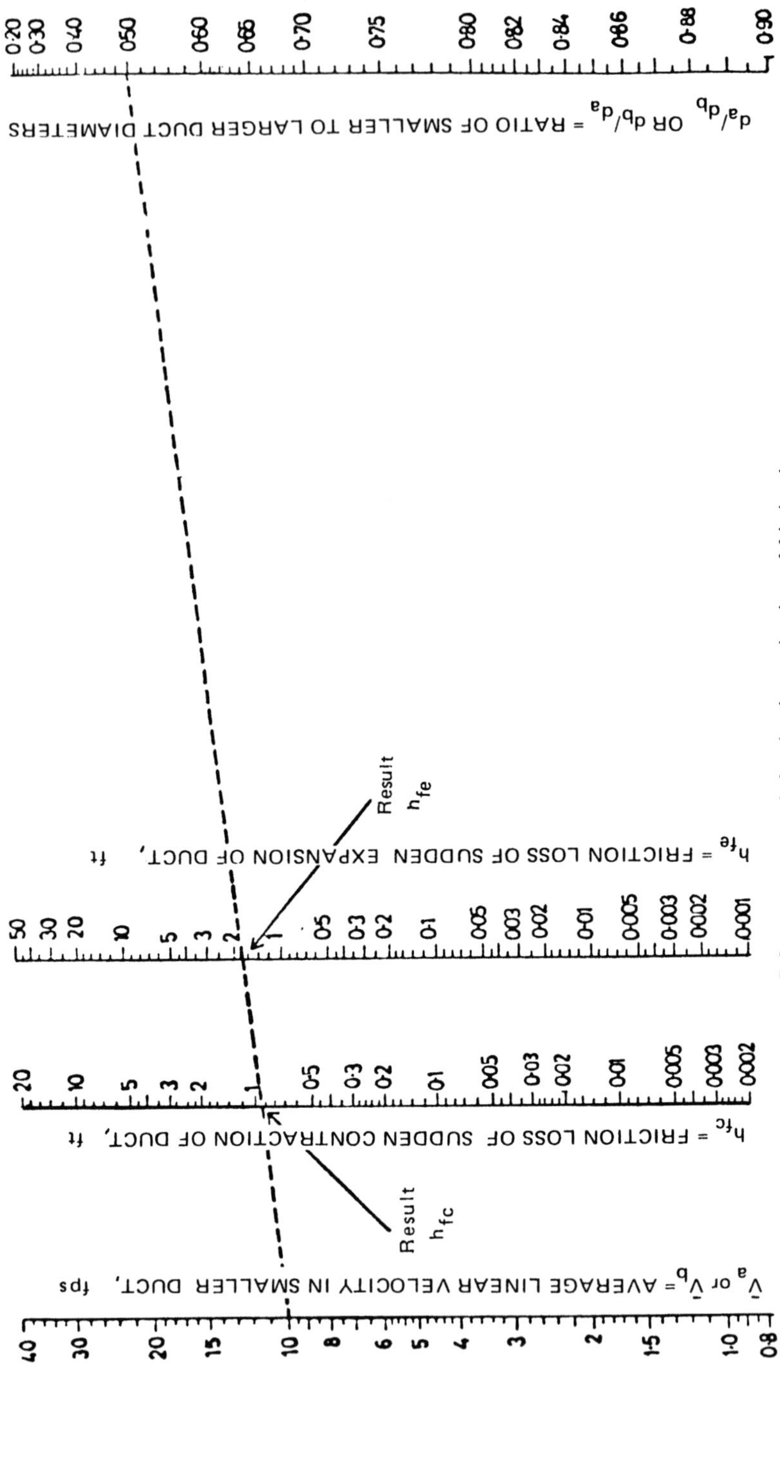

FIG. 3 Nomograph for the determination of friction losses.

The ratio of duct diameters—d_a/d_b for expansion and d_b/d_a for contraction—have to be calculated prior to using the nomograph.

When the density of flowing fluid is known, the pressure drop may be calculated as follows:

$$\Delta p_e = h_{fe}\frac{a}{144} \tag{8a}$$

or

$$\Delta p_c = h_{fc}\frac{a}{144} \tag{8b}$$

where $\Delta p_e, \Delta p_c$ = pressure drop, caused by sudden expansion (contraction), $\text{lb}_f/\text{in.}^2$

a = density of flowing fluid, lb/ft^3

h_{fe}, h_{fc} = friction losses (loss of head), calculated by means of the nomograph, ft

How to Use the Nomograph

The use of the nomograph (Fig. 4) depends on the value wanted. There are two ways to use it.

A. Expansion Friction Loss. Connect the known values of $\overline{V}_a$ and the d_a/d_b ratio (Scales I and IV) with a ruler. Read the final result h_{fe} at the intersection point of the ruler with the h_{fe} (Scale III).

B. Contraction Friction Loss. Connect the known values of $\overline{V}_b$ and the d_b/d_a ratio (Scales I and IV) with a ruler. Read the final result h_{fc} on the intersection point of the ruler with the h_{fc} (Scale II).

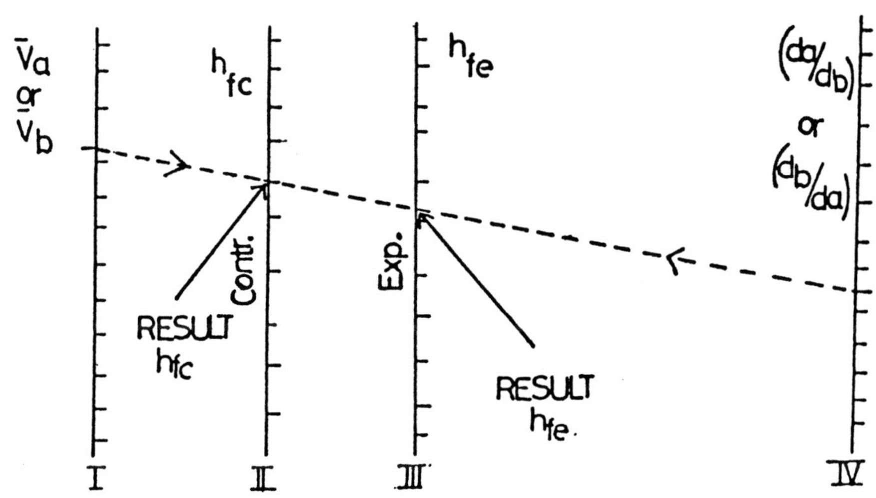

FIG. 4 An example of how the nomograph is used.

Typical Examples

A. For Expansion. Fluid is flowing through the small duct (d_a = 3 in.) and suddenly expands into the large duct (d_b = 6 in.). The average linear velocity $\overline{V}_a$ at the small duct equals $\overline{V}_a$ = 10 ft/s. What will be the friction loss (h_{fe})?

Solution: $\overline{V}_a$ = 10 ft/s, (d_a/d_b) ratio = 3/6 = 0.50. Result: h_{fc} = 1.75 ft.

B. For Contraction. Fluid is flowing through the large duct (d_a = 6 in.) and enters the small duct (d_b = 3 in.). The average linear velocity $\overline{V}_b$ at the small duct equals $\overline{V}_b$ = 10 ft/s. What will be the friction loss (h_{fe})?

Solution: $\overline{V}_b$ = 10 ft/s, (d_b/d_a) ratio = 3/6 = 0.50. Result: h_{fc} = 0.93 ft.

NOTE: Both of these examples are drawn with a broken line on the nomograph.

This material appeared in *Plant Engineering,* pp. 23–24, June 1974, and is reprinted with the permission of the editor.

References

1. W. L. McCabe and J. C. Smith, *Unit Operations of Chemical Engineering,* McGraw-Hill, New York, 1967.
2. T. Baumeister (ed.), *Standard Handbook of Mechanical Engineers,* 7th ed., McGraw-Hill, New York, 1967.

ADAM ZANKER

Friction Losses, Fittings

Because obtaining complete test data on the pressure drop of every available size and type of valve and pipe fitting is impossible, a practical method for extending available information is useful. This technique, known as the equivalent-length method for calculating pressure losses, applies only to single-phase, noncompressible, nonflashing liquids.

In this method of determining total pressure loss in a system, the pressure drop through a valve, fitting, miter elbow, or any pressure-reducing component is equated with the length of straight, round pipe that will have the same pressure loss under the same flow conditions. The values shown in

Table 1 apply only to turbulent flow, the most prevalent condition in general piping systems.

When laminar flow exists, calculations are required. Standard texts should be consulted for empirical relationships covering laminar flow conditions.

Resistance to flow includes the pressure drop from valves, fittings, sudden changes in the pipe cross-section, and miter elbows. Values for pressure loss through system components are expressed as equivalent feet of round, straight pipe in Table 1. Pressure losses also result from flow measurement devices and fluid entry into and exit from a pipe. These losses represent special cases and are not included in the tables.

Pipe characteristics vary considerably, and Table 1 applies only to ordinary commercial steel and iron pipe. For pipe of other materials, Table 2 lists values based on the coefficient of roughness c. If possible, this coefficient should be obtained from the manufacturer.

The effect of pressure drop through valves and fittings is negligible when the ratio of pipeline length to pipe diameter is equal to or greater than 1000 to 1, as is usually the case for long water and oil pipelines with the usual number of fittings and valves.

Example. Crude oil is flowing through a 20-mile long, 12-in. diameter pipeline. Should the effect of valves and fittings in the line be considered in the determination of pressure drop?

Solution:

$$\frac{20 \times 5280}{\frac{12}{12}} = 105{,}600$$

Because the result is greater than 1000, the pressure drop through valves and fittings in this pipeline can be ignored.

Table 3 lists factors for calculating pressure drop when c is some value other than 100.

Example. Pressure drop in a water pipeline with turbulent flow is 25 lb/in.^{2} for ordinary steel pipe. What would the pressure drop be if copper pipe were used?

Solution: Table 2 lists the coefficient of roughness for copper pipe as 130; from Table 3, the multiplier factor for c is found to be 0.6152. Therefore, the pressure drop for copper pipe is $25 \times 0.6152 = 15.38$ lb/in.^{2}.

Example. A piping run consists of 37 ft of 4 in. diameter straight pipe, three short-radius elbows, two wide-open gate valves, and one wide-open globe valve. What total equivalent length of straight pipe is used to calculate head loss?

TABLE I. EQUIVALENT LENGTHS OF VALVES, SUDDEN CROSS-SECTIONAL CHANGES, AND MITER BENDS, FT

Nominal pipe size, in.	Globe valve	Angle valve / Ball check	Return bend / Swing check	Standard tee / Square elbow	Standard elbow / Reduced tee ½	Medium elbow / Reduced tee ¼	Long elbow / Run of standard tee	45 deg elbow	Gate valve	Ball valve	Plug cock straight way Full port
1½	40	21	10	7	4	4	3	2	1	1	2
2	50	27	13	10	5	5	3	2.5	1	2	3
2½	60	33	15	12	6	6	4	3	1.5	2	4
3	80	40	18	15	8	7	5	4	2	2	5
4	115	55	23	20	11	9	6	5	2.5	3	6
6	160	80	36	30	16	14	9	7.5	3.5	4	9
8	225	110	50	40	20	18	14	10	4.5	5	12
10	290	135	60	50	25	22	18	13	7	7	16
12	350	160	70	60	30	25	20	16	8	8	18
14	400	190	85	66	35	30	23	18	9	9	20
16	450	220	100	76	40	35	27	21	10	10	23
18	500	250	110	86	45	40	30	24	12	11	26
20	550	280	125	96	50	45	35	26	13	13	29
22	600	300	155	105	55	50	38	29	15	15	32
24	660	335	190	116	60	56	45	32	18	16	35
30	–	–	–	146	75	–	50	40	21	20	–
36	–	–	–	176	90	–	60	48	25	24	–
42	–	–	–	205	105	–	70	56	30	27	–
48	–	–	–	235	120	–	80	64	35	31	–
54	–	–	–	265	135	–	90	72	40	35	–
60	–	–	–	295	150	–	100	80	45	40	–

TABLE 2 Roughness Values for Pipe of Various Materials

Type of Pipe	Coefficient of Roughness *c*
Cast iron (properly installed):	
Time since installation, years:	
Less than 4	130
4 to 6	120
10 to 12	110
13 to 20	100
21 to 35	80
Asbestos cement	140
Cement lined:	
Hand applied	120
Applied by centrifuge	140
Riveted steel (66 to 144 in. diameter):	
Time since installation, years:	
Less than 1	140
1 to 10	100
More than 10	90
Fiber	140
Bitumastic-lined iron or steel	140
Copper, brass, lead	130
Tin or glass	130
Wood stave	110
Welded and seamless steel	100
Wrought iron	100
Ordinary commercial steel and iron	100
Concrete	100
Vitrified	100
Corrugated steel	60
Tile	110
Brick (sewers)	100
Fire hose:	
Extremely smooth	143
Rubber lined	125 to 140
Mill hose	100 to 120
Unlined linen	85 to 95

Sudden enlargement, Equivalent L's in terms of d: d/D=¼	d/D=½	d/D=¾	Borda entrance	Ordinary entrance	Sudden contraction, Equivalent L's in terms of d: d/D=¼	d/D=½	d/D=¾	Two-miter bend	Three-miter bend	Four-miter bend	Six-miter bend
4.5	3	1	4	2.5	2	1.5	1				
5	3.5	1	5	3	2.5	2	1				
6	4.5	1.5	6	3.5	3	2.5	1.5				
8	5	2	8	4.5	4	3	2				
11	7	2.5	11	6	5	4	2.5				
16	10	3.5	15	9	7.5	5.5	3.5				
20	14	4.5	19	12	10	7.5	4.5				
25	18	7	25	15	13	10	7				
30	20	8	30	17.5	16	12	8	28	21	20	12
35	23	9	35	20	18	13.5	9	32	24	22	18
42	27	10	40	23	21	15	10	38	27	24	20
48	30	12	45	25	24	17	12	42	30	28	22
52	35	13	50	29	26	18.5	13	46	33	32	24
58	38	15	55	31	29	20	15	52	36	34	26
65	42	18	60	35	32	21.5	18	56	39	36	28
75	50	21	–	–	40	26	21	70	51	44	36
100	60	25	–	–	48	28	25	84	60	52	42
110	65	30	–	–	56	40	30	98	69	64	48
140	80	35	–	–	64	45	35	112	81	72	54
150	90	40	–	–	72	50	40	126	90	80	60
160	100	45	–	–	80	60	45	140	99	92	66

Solution: Equivalent lengths are found in Table 1.

Straight pipe	37 ft
3 elbows × 11 ft	33 ft
2 gate valves × 2.5 ft	5 ft
1 glove valve × 115 ft	115 ft
Total	190 ft

NOTE: Calculations are normally made on the basis of wide-open valve position and apply to check valves, foot valves, butterfly valves, and cocks. An orifice meter inserted in the pipeline presents a special case. This restriction causes a permanent loss of pressure across the orifice, measured

TABLE 3 Conversion Factors for Values of Roughness

Coefficient c	Multiplier Factor
60	2.575
70	1.936
80	1.512
90	1.215
100	1.000
110	0.8382
120	0.7135
130	0.6152
140	0.5363
150	0.4683

Interpolate as required

as a percentage of the manometer differential. The loss may be from 50 to 95% of the meter differential, varying inversely with the orifice-to-pipe inside diameter. In a Venturi meter, the permanent loss is much less (10 to 20% of meter differential), because of static pressure regain in the gradual configuration of the Venturi throat and after section.

This material appeared in *Plant Engineering,* pp. 162–164, August 23, 1979, and is reprinted with the permission of the editor.

JOHN D. CONSTANCE

Friction Losses, Incompressible and Compressible Flow

Every design engineer occasionally needs to approximate pipe sizes. For liquids, the information can be found in *The Cameron Hydraulic Data Book* [1] or in Fig. 1. For gases or vapors, there is no such source. The tables available are specific; for example, "Crane Technical Paper No. 410" covers air at 100 lb/in.^{2}gauge but does not supply a procedure for applying the data to other gases [2].

This article provides a procedure for converting flows of process fluids at any temperature and pressure into equivalent air or water flows. Using the equivalent flow and Fig. 1 for liquids, or Fig. 2 for gases, one can graphically determine the pipe size required for a particular flow rate and corresponding pressure drop.

Flow rates vs pressure drops, with pipe diameters and fluid velocities as parameters, were obtained for both Figs. 1 and 2 by solution of the incompressible-fluid flow equation:

$$\Delta P/L = 32fW^2/\pi^2\rho g_c D^5 \tag{1}$$

Here, f, the Fanning friction factor, is defined by the Colebrook equation:

$$1/f^{1/2} = -4 \log [(\epsilon/3.7d) + (1.255/N_{Re}f^{1/2})] \tag{2}$$

Incompressible-flow equations can be used for gases if the system pressure drop is less than 10% of the inlet pressure; if it is more than 10%, the change in density (i.e., volumetric flow rate) becomes significant, and one must resort to compressible-flow or to incremental-incompressible-flow equations.

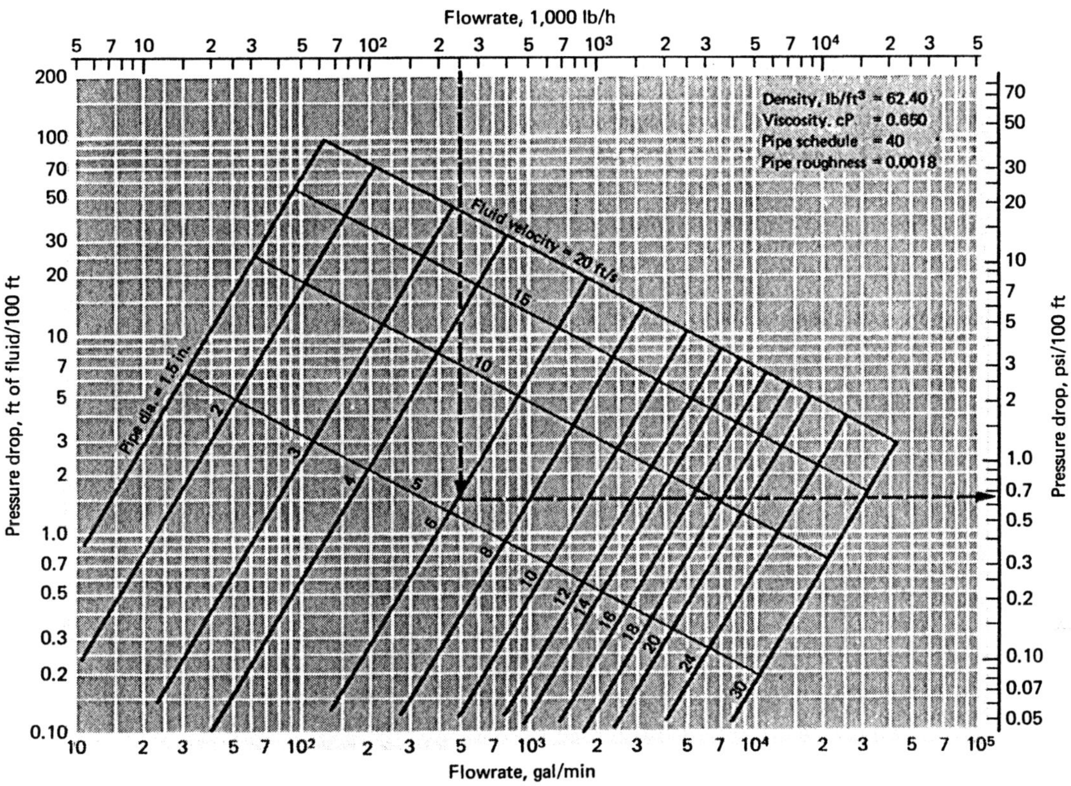

FIG. 1 Pipe diameter, fluid velocity, flow rate, and pressure drop for water at 14.7 lb/in.^{2} abs and 100°F.

Pipe Diameter for Water or Air

To find the required pipe diameter for a given flow rate via Fig. 1 or 2, enter the abscissa at the flow rate and move vertically to the intersection of the appropriate velocity/pipe-diameter line. At this intersection, traverse horizontally to the ordinate and read the corresponding pressure drop.

The appropriate velocity/pipe-diameter line can be determined by applying some simple rules of thumb. For liquids, the fluid velocity should be 5–10 ft/s. For gases, it should be 50–100 ft/s.

To illustrate the application of these rules of thumb, find the pipe diameter for 250,000 lb/h (500 gal/min) of water. Moving vertically from the Fig. 1 abscissa at 250,000 lb/h shows that numerous pipe diameters (3–10 in.) can accommodate this flow. Applying the 5–10 ft/s rule, a 6-in.-diameter pipe, with a corresponding pressure drop of 0.65 lb/in.^{2} per 100 ft of pipe, would be selected.

Diameter for Other Than Water or Air

To use Figs. 1 and 2 for fluids other than water or air, the process fluid must be converted into an equivalent water or air flow. This is done by solving Eq. (1) for pipe diameter, and inserting into it first the process

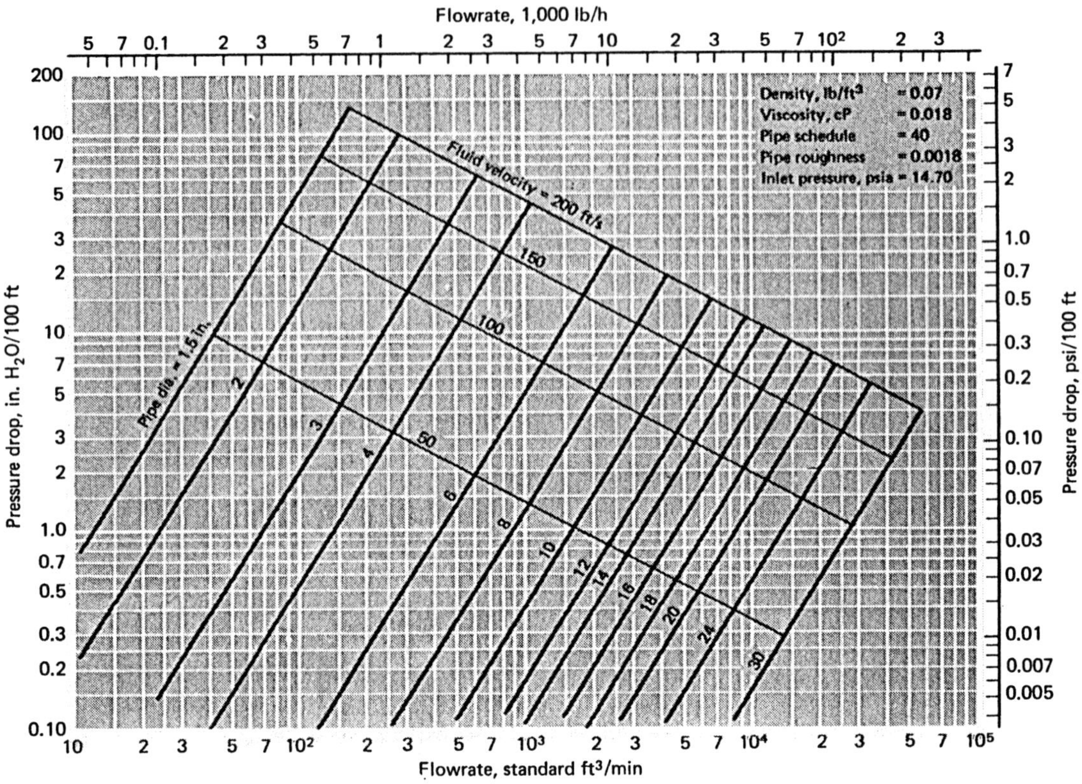

FIG. 2 Pipe diameter, fluid velocity, flow rate, and pressure drop for air at 14.7 lb/in.²abs and 60°F.

parameters and then the water or air parameters. These two forms of Eq. (1) are then equated by holding the pipe diameter constant:

$$\frac{W_e^2}{\rho_s(\Delta P/L)_s)(N_{\text{Re}})_s^{0.16}} = \frac{W_p^2}{\rho_p(\Delta P/L)_p(N_{\text{Re}})_p^{0.16}} \tag{3}$$

Here, $N_{\text{Re}} = 6.31(W/d\mu)$. The term f is approximated by the turbulent-flow friction factor:

$$f = 0.04/(N_{\text{Re}})^{0.16} \tag{4}$$

(Note that the abscissas of Figs. 1 and 2 have dual scales. For a liquid other than water, the lb/h scale of Fig. 1 must be used. For a gas other than air at standard conditions, the lb/h scale of Fig. 2 must be used.)

Some Simplifying Assumptions

To solve Eq. (3) for an equivalent air or water flow, one must know the pressure drop in the two systems. Assume that the pressure drop in both cases is 10% of the inlet pressure (i.e., $\Delta P/L = 0.1P$). With this assumption, the equivalent air or water flow becomes

$$W_e = \left\{\left[\left(\frac{\rho_s}{\rho_p}\right)\left(\frac{P_s}{P_p}\right)\right]^{1/2}\left[\left(\frac{W_e}{W_p}\right)\left(\frac{\mu_p}{\mu_s}\right)\right]^{0.08}\right\}W_p \tag{5}$$

The accuracy of this procedure is limited by one's ability to interpolate the log-log axes of Figs. 1 and 2. Numerous calculations by this procedure have shown that the inclusion of the Reynolds number correction to the 0.08 power is not justified, and the correction can be deleted with little loss in accuracy. Simplifying on this basis, the abscissa scale factor (equivalent flow) for gases becomes

$$W_e = [(\rho_s/\rho_p)(P_s/P_p)]^{1/2}W_p \tag{6}$$

For liquids, the pressure dependence (compressibility) is minimal, so the abscissa scale factor becomes

$$W_e = (\rho_s/\rho_p)^{1/2}W_p \tag{7}$$

A scale factor must also be applied to the ordinate. It is obtained by solving Eq. (3) for $(\Delta P/L)_p$. The ordinate correction factor (pressure drop) is

$$\left(\frac{\Delta P}{L}\right)_e = \left(\frac{P_p}{P_s}\right)\left[\left(\frac{W_e}{W_p}\right)\left(\frac{\mu_p}{\mu_s}\right)\right]^{0.16}\left(\frac{\Delta P}{L}\right)_s \tag{8}$$

Again, because the pressure dependence is minimal, the ordinate correction factor for liquids is

$$\left(\frac{\Delta P}{L}\right)_e = \left[\left(\frac{W_e}{W_p}\right)\left(\frac{\mu_p}{\mu_s}\right)\right]^{0.16}\left(\frac{\Delta P}{L}\right)_s \tag{9}$$

Examples Illustrate Procedure

The versatility of this procedure becomes evident when Eq. (6) or (7) is solved for the flow ratio W_e/W_p, and Eq. (8) or (9) for the pressure-drop ratio. These ratios are scale factors for Figs. 1 and 2. After they have been calculated for a particular process condition, they can be tabulated for future reference.

Example 1. Find the required pipe diameter and the corresponding pressure drop for 10,000 lb/h of a fluorocarbon vapor at 60 lb/in.^{2}gauge and 60°F ($\rho = 1.19$ lb/ft^3, and $\mu = 0.012$ cP). From Eq. (6), the equivalent air flow is

$$W_e/W_p = [(14.7/74.7)(0.07/1.19)]^{1/2} = 0.11$$

$$W_e = 0.11(10{,}000) = 1100 \text{ lb/h}$$

Enter Fig. 2 with an abscissa value of 1100 lb/h and move vertically to the intersection of the 4-in.-diameter-pipe and 50-ft/s-velocity lines. At this intersection, move horizontally to an air pressure drop of 0.11 (lb/in.^{2})/100 ft of pipe. Calculated by means of Eq. (8), the equivalent process pressure drop is

$$\frac{\Delta P_p}{\Delta P_s} = \left(\frac{74.7}{14.7}\right)\left[\left(\frac{1100}{10{,}000}\right)\left(\frac{0.012}{0.018}\right)\right]^{0.16} = 3.36$$

$$\Delta P_p = 3.36(0.11) = 0.37\ (\text{lb/in.}^2)/100\ \text{ft}$$

Solving Eq. (1) gives a ΔP of 0.43 (lb/in.^{2})/100 ft of pipe. Hence, for a 17-fold increase in density and a 4-fold increase in pressure, the error in pressure drop via the equivalence procedure is 15%.

Using the 4-in. pipe as a basis, Fig. 2 shows that the pressure drop will be 20 times higher in a 2-in. pipe, 3 times higher in a 3-in. pipe, and 10 times lower in a 6-in. pipe. Which pipe diameter should be selected? It depends. If the piping run is short and pressure drop is not critical, a smaller diameter would be suitable. (Beware, however, because with short piping, inlet and exit losses are usually large in comparison to friction losses.) If, on the other hand, the pipe run is long and pressure drop is important, a larger diameter would be the more appropriate choice.

Example 2. Suppose that for start-up or emergency blowdown the preceding pipe must handle twice the flow. Enter Fig. 2 at 2200 lb/h, and notice that the pressure drop in 4-in.-diameter pipe increases from 0.1 to 0.4 (lb/in.^{2})/100-ft air equivalent.

(Does this result look familiar? It demonstrates the rule of thumb that in turbulent flow with the same pipe diameter, pressure drop increases with the square of the change in flow rate. So, this result could have been found without resorting to Fig. 2. Figure 2, however, allows one to visually evaluate alternative pipe diameters; it shows "the big picture.")

Returning to the question, what pipe diameter should be chosen for twice the flow rate of Example 1? Again, it depends. For start-ups, pressure drop should be minimized and a larger diameter selected. For emergency blowdowns, one may want a high-pressure-drop system, and so decide upon a smaller diameter.

From the preceding discussion, it should be apparent that the choice of pipe diameter should not be based on steady-state flow.

It should be emphasized that the equivalence procedure is an approximation technique for evaluating alternatives. After the pipe size has been approximated via the procedure, the pressure drop for the final design should be calculated by means of Eq. (1).

If Example 1 is repeated for 100,000 lb/h of 550-lb/in.^{2}gauge steam, the equivalence pressure drop for an 8-in.-diameter pipe is 1.43 (lb/in.^{2})/100 ft. Equation (1) gives 1.53 (lb/in.^{2})/100 ft. Hence, for a 17-fold increase in density and a 38-fold increase in pressure, the discrepancy is 6%.

Example 3. Find the required pipe diameter and resulting pressure drop for a flow rate of 100,000 lb/h of calcium chloride brine (ρ = 77 lb/ft^3 and μ = 6.2 cP). From Eq. (7), the equivalent water flow rate is calculated to be

$$W_e = (62.4/77)^{1/2}100{,}000 = 90{,}000\ \text{lb/h}$$

For this flow rate, Fig. 1 indicates a 3-in.-diameter pipe having an equivalent water ΔP of 2.9 (lb/in.^{2})/100 ft. Correcting the ΔP via Eq. (9) gives

$$\Delta P_e = \left[\left(\frac{90{,}000}{100{,}000}\right)\left(\frac{6.2}{0.65}\right)\right]^{0.16} 2.9 = 4.09 \text{ (lb/in.}^2\text{)/100 ft}^2$$

Solving Eq. (1) yields a ΔP of 3.8 (lb/in.^{2})/100 ft, a discrepancy of 8% for a 10-fold increase in viscosity.

Correcting for Pressure Effect on Velocity

In Example 1, it was shown that pipe sizing should be based on pressure-drop considerations. However, it was earlier indicated that pipe can be sized using rule-of-thumb fluid velocities. The former may be preferable because the thumb rules do not take into account variations in process flows. Requirements for future expansion, upset conditions, plus a host of other reasons may dictate the selection of a pipe size different from that indicated by the velocity rules of thumb.

For the Figs. 1 and 2 velocity lines to be valid for fluids other than water or air at standard conditions, a scale factor must be applied to the constant-velocity lines. For liquids, the effects of pressure and density variations are liquids, the effects of pressure and density variations are negligible, and the Fig. 1 velocity lines can be used without being corrected. In the case of gases, however, increases in pressure compress the gas (and increase density), causing the volumetric flow and velocity to decrease. The actual velocity can be calculated via

$$V = 0.0509(W/\rho d^2) \tag{10}$$

By means of procedures shown earlier, the velocity-equivalence equation can be obtained:

$$V_p = \left[\left(\frac{W_p}{W_e}\right)\left(\frac{\rho_s}{\rho_p}\right)\right] V_s \tag{11}$$

If the air velocity were 50 ft/s in the 4-in.-diameter pipe of Example 1, the actual velocity would be

$$\frac{V_p}{V_s} = \left(\frac{10{,}000}{1100}\right)\left(\frac{0.07}{1.19}\right) = 0.54$$

$$V_p = 0.54(50) = 27 \text{ ft/s}$$

This illustrates the variation in velocity.

Symbols

D, d	inside diameter, feet or inches, respectively
f	Fanning friction factor
g_c	32.17 $(lb_m)(ft)/(s^2)(lb_f)$
L	length of piping, ft
N_{Re}	Reynolds number
P	pressure, lb/in.^{2}abs
ΔP	pressure drop, lb/in.^{2}
V	velocity, ft/s
W, w	mass flow rate, lb/h or lb/s, respectively
ϵ	pipe roughness, in.
ρ	density, lb/ft^3
μ	viscosity, cP

Subscripts

e	equivalent
p	process
s	standard

References

1. C. R. Westaway and A. W. Loomis (eds.), *The Cameron Hydraulic Data Book,* Ingersoll-Rand Co., Woodcliff Lake, New Jersey, 1979.
2. *Flow of Fluids through Valves, Fittings, and Pipe,* Technical Paper 410, Crane Co., New York, New York.

RALPH A. CROZIER, Jr.

Friction Losses, Nonstandard Ducts

There are little data for estimating friction losses when the cross section of the duct is nonstandard (triangular, trapezoidal, flat oval, elliptical, etc.).

The method described here shows how friction losses through these nonstandard ducts can be estimated for both gases and liquids with the help of preestablished standard friction charts that can be found in the literature.

Of course, a definite need for unusual cross-sectional shapes has to be established because they are more costly to fabricate than the standard ones. Triangular ducts or conduits, for instance, require almost twice as much material per foot of length of equivalent capacity. In addition, triangular ducts are more difficult to form because their angles are larger than a standard breaker will handle. Nevertheless, when certain esthetic effects are desired, or when a given structural space is to be used as a duct itself, the nonstandard shapes come into play.

Whenever fluid flow is considered in the more expensive noncircular conduits, it is necessary to determine some characteristic term for calculating the Reynolds number (N_{Re}). For circular conduits this term is the diameter of the conduit in feet, but for other shapes the length term in the Reynolds number equation is equal to four times the hydraulic radius of the conduit.

The hydraulic radius (r) is defined as the ratio of the cross-sectional area of the conduit (running full, as in our case) to the calculated "wetted" perimeter:

$$r = A/P \tag{1}$$

Handbooks provide formulas for A and P, from which hydraulic radii for numerous cross-sections may be calculated. For conduits of triangular and trapezoidal cross-section, the friction factor (f) for laminar flow is obtained from [1]

$$f = 16/N_{Re} \tag{2}$$

The diameter is obtained from

$$N_{Re} = Dv\rho/\mu \tag{3}$$

To calculate the effective velocity (v) for noncircular constant-cross-section conduits, the value of D in Eq. (3) is four times the hydraulic radius. Thus, $D = 4r$, ft; v = velocity, ft/s; μ = viscosity, cP $\times$ 0.000672; and ρ = density of the fluid, lb/ft^3.

In circular conduits the transition from laminar to turbulent flow occurs at $N_{Re} = 2000$ [1]. In conduits of noncircular cross-section, experimental data during turbulent flow indicate that the Blasius equation suitably predicts f from the transition point to $N_{Re} = 100{,}000$;

$$f = 0.079(N_{Re})^{-0.25} \tag{4}$$

Friction factors for noncircular conduits for laminar and turbulent flow have been published [1, 3]. The results may be reduced to Eq. (4) for all conduits of whatever cross-section. Another published graph [2] shows turbulent friction-factor data for air flowing in rectangular and square ducts for N_{Re} values from 20,000 to 500,000. The straight line on which practically all the plotted points fall represents Eq. (4). For all practical purposes, the data are identical and may be used interchangeably.

Thus, it is safe to use the hydraulic radius to calculate the effective diameter, and to apply standard published charts to equal flow conditions to calculate friction loss in circular conduits or ducts.

Figure 1 shows a friction chart for flow of low-pressure (close to atmospheric) air in ducts for temperatures from 30 to 120°F. After determining the equivalent diameter ($4r$), the chart may be used to find the flow of air under like conditions for other-than-circular cross-sections. Circular-conduit charts for other fluids may be used similarly. Figure 2 is an example of a chart for water.

Table 1 has been developed for the equilateral triangle to illustrate the approach for other cross sections. Table 2 lists flat-oval equivalents of circular ducts for equal friction and capacity.

The step-by-step method to determine friction is: (1) find cross-sectional area of conduit, ft^2; (2) find wetted perimeter, ft; (3) calculate hydraulic radius using Eqs. (1) and (2); (4) find equivalent diameter, ft; (5) convert feet to inches, and apply standard flow chart for friction, interpolating for or taking the nearest diameter size.

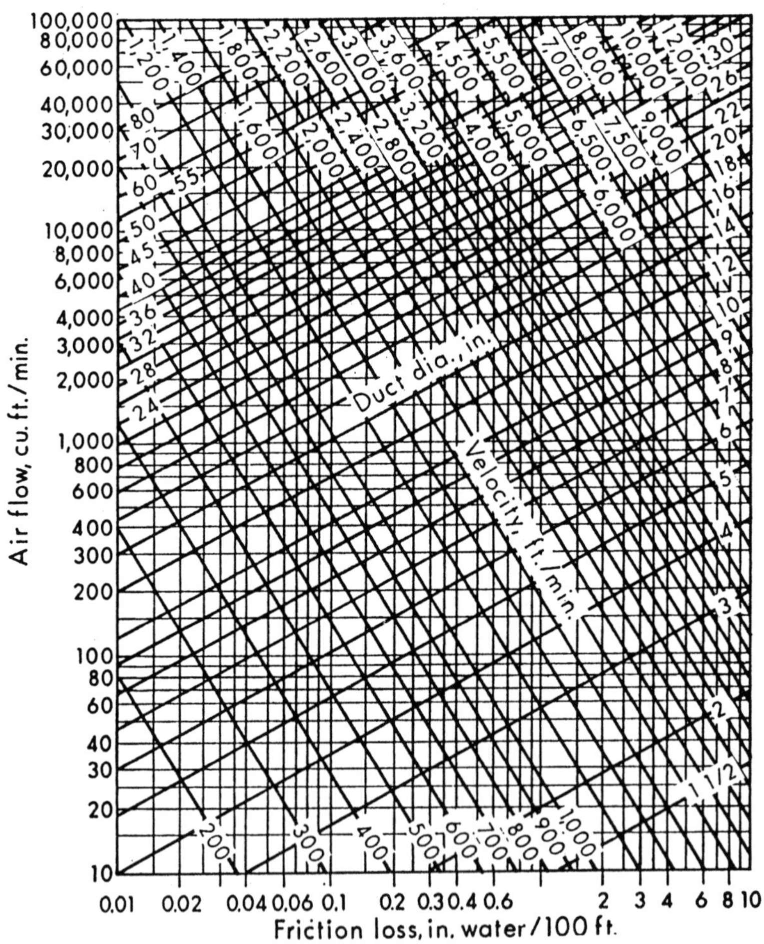

FIG. 1 Friction chart for the flow of low-pressure air in ducts. (Reprinted by permission from *Heating, Ventilating and Air Conditioning Guide*, 1948.)

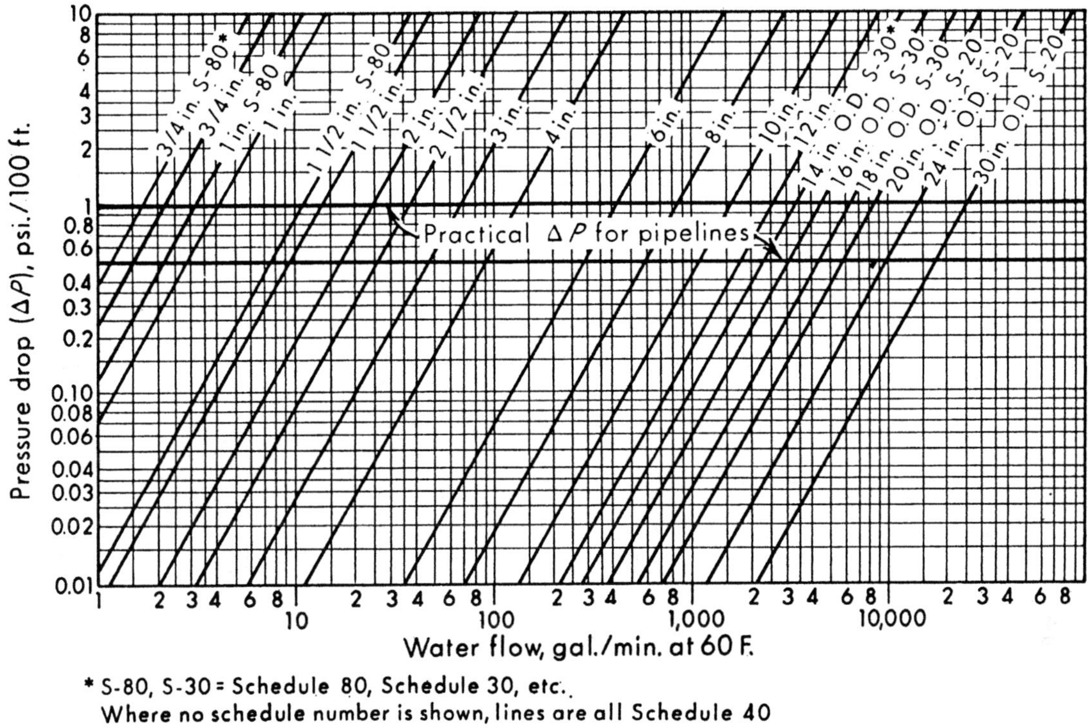

FIG. 2 Friction chart for the flow of water in ducts. (Reprinted with permission from *Heating/Piping/Air Conditioning*, copyright © 1961, Keeney Publishing Co.)

TABLE 1 Values of r for Equilateral-Triangle Ducts

Side of Equilateral Triangle (in.)	Hydraulic Radius, r (in.)	Equivalent Diameter, $4r$ (in.)
12	1.73	6.9
14	2	8
16	2.3	9.2
18	2.6	10.4
20	2.87	11.45
22	3.17	12.68
24	3.47	13.88
26	3.75	15.0
28	4.05	16.20
30	4.42	17.68
32	4.6	18.4
34	4.95	19.80
36	5.21	20.84
38	5.5	22.0
40	5.79	23.2
42	6.08	24.32
44	6.31	25.24
46	6.68	26.72
48	6.90	27.40

TABLE 2 Flat-Oval Equivalents of Circular Ducts, for Equal Friction and Capacity

Circular Duct Diameter (in.)	Flat-Oval Duct Size[a] 30% Flattening, Axes (in.)	20% Flattening, Axes (in.)	100% Flattening, Axes (in.)
10	7 × 14	8 × 12	9 × 11
11	8 × 14	9 × 13	10 × 12
12	9 × 15	10 × 14	11 × 13
14	10 × 18	11 × 17	13 × 15
16	11 × 22	13 × 19	14 × 18
18	13 × 23	14 × 22	16 × 20
20	14 × 26	16 × 24	18 × 22
22	15 × 30	18 × 26	20 × 24
24	17 × 32	19 × 29	22 × 26
26	18 × 35	21 × 31	23 × 29
28	20 × 36	22 × 34	25 × 31
30	21 × 40	24 × 36	27 × 32
32	22 × 44	26 × 37	29 × 35
34	24 × 45	27 × 40	31 × 36
36	25 × 48	29 × 42	32 × 40

[a]Expressed as: minor axis × major axis.

10% Flattening **20% Flattening** **30% Flattening**

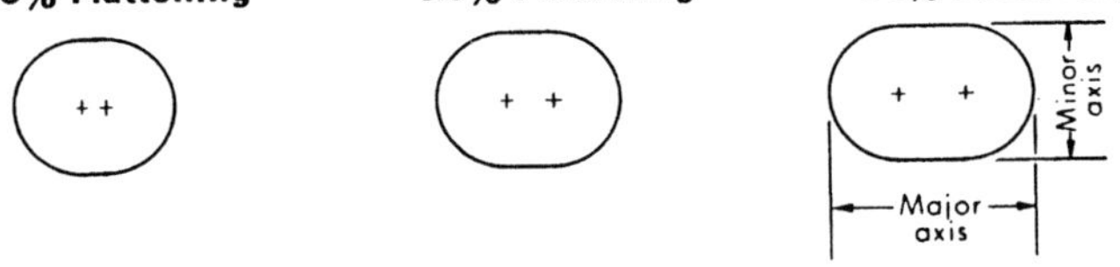

The procedure for the design of a flat-oval duct is the same as for a rectangular one. The system is first designed on the basis of a circular duct, which is converted to its flat-oval equivalent by means of this table.

If flow and friction are known, select the circular conduit size from a chart. Then calculate the hydraulic radius (diameter/4) for the desired section. Table 1 shows *r* values for several sizes of ducts with equilateral-triangle shape. The same procedure may be applied to other cross sections and fluids, so long as flow takes place in full, constant-cross-section conduits or ducts.

Friction losses through fittings must be approximated from available data. Equivalent velocities may be found from equivalent diameters—4*r*—and actual flows.

A graph of Table 1 would show that the equivalent diameter of an equilateral triangle is equal to the diameter of the inscribed circle. This also applies to right-angle triangles. Apparently, the corners of triangular, trapezoidal, or any irregular cross-section are areas of total loss, insofar as fluid-carrying capacity is concerned.

Example 1. Select the size needed for an equilateral triangular duct of constant cross-section to carry 1000 ft^3/min of air at atmospheric conditions. Unit friction here is assumed to be 0.1 in. water for every 100 ft of straight duct.

Solution: From Fig. 1, the circular equivalent is 14 in. Since $r = 14/4 = 3.5$ in., select—from Table 1—the 24-in. triangle.

Example 2. Find the effective velocity through a 36-in. equilateral triangle for a flow of 10,000 ft^3/min standard air.

Solution: From Table 1, $r = 5.21$ in. Since equivalent diameter $= 4 \times 5.21 =$ 20.84 in. (or 21 in.), the effective velocity, as read in Fig. 1, is 4250 ft/min.

Head-loss charts for water or any other fluid may also be used as has been illustrated in these two examples.

References

1. J. Nikuradse, "Untersuchungen über turbulente Stromungen in nicht kreisformigen Rohren," *Ing. Arch.*, *1*, 306 (1930).
2. P. G. Huebscher, "Friction in Round, Square and Rectangular Ducts," *Heat./Piping/Air Cond.*, *19*, 127 (1947).
3. L. Schiller, "Uber den Stromungwiderstand von Rohren verscheidenen Querschnitts und Rauhigkeitsgrades," *Z. Angew. Math. Mech.*, *3*(2) (1923).

JOHN D. CONSTANCE

Friction Losses, Straight Pipes

The loss of head that occurs in water piping systems because of friction is a predictable quantity. Under turbulent flow conditions in filled pipes, this friction head may be calculated from the Darcy-Weisbach formula:

$$h_f = \frac{fLV^2}{2gD}$$

where h_f = friction head, ft
f = friction factor
L = length of pipe, ft
V = linear velocity of water in pipe, ft/s

g = gravitational acceleration, 32.2 ft/s^2
D = inside diameter of pipe, ft

This formula is difficult to use, however, because it incorporates a complex variable. The friction factor, f, is dependent on pipe size, D, and flow rate, V, as well as on the surface condition of the pipe (degree of roughness). Therefore, a set of equations more precise than the Darcy-Weisbach formula has been developed for friction head calculations. These equations eliminate the friction factor, f, of the general formula by dividing the range of pipe roughness normally encountered into four bands, as follows:

Flow in extremely smooth pipes:

$$h_f = \frac{0.30LV^{1.75}}{1000D^{1.25}}$$

Flow in fairly smooth pipes:

$$h_f = \frac{0.38LV^{1.86}}{1000D^{1.25}}$$

Flow in rough pipes:

$$h_f = \frac{0.50LV^{1.95}}{1000D^{1.25}}$$

Flow in extremely rough pipes:

$$h_f = \frac{0.69LV^{2}}{1000D^{1.25}}$$

Used within their defined ranges, these equations yield accurate results. But the calculations can be time consuming because of the fractional exponential powers involved. The nomograph presented here as Fig. 1 is designed to speed and simplify the calculation process. It provides a simultaneous solution of all four equations with only two consecutive alignment motions. Accuracy of the solutions will depend on the size of the scales and the precision to which they are constructed.

The nomograph has four vertical scales and a grid which consists of four vertical lines, corresponding to the degree of pipe roughness, intersecting a family of curved lines representing friction head.

As shown on the nomograph, scale L (length) is graduated for the range from 100 to 1000 ft. If the actual L value lies outside this range, the scale values (and the resulting solution) may be increased or decreased in multiples of 10. For example, if L = 30 or 3000 ft, solve for L = 300 ft, and multiply the result by 0.1 or 10, respectively.

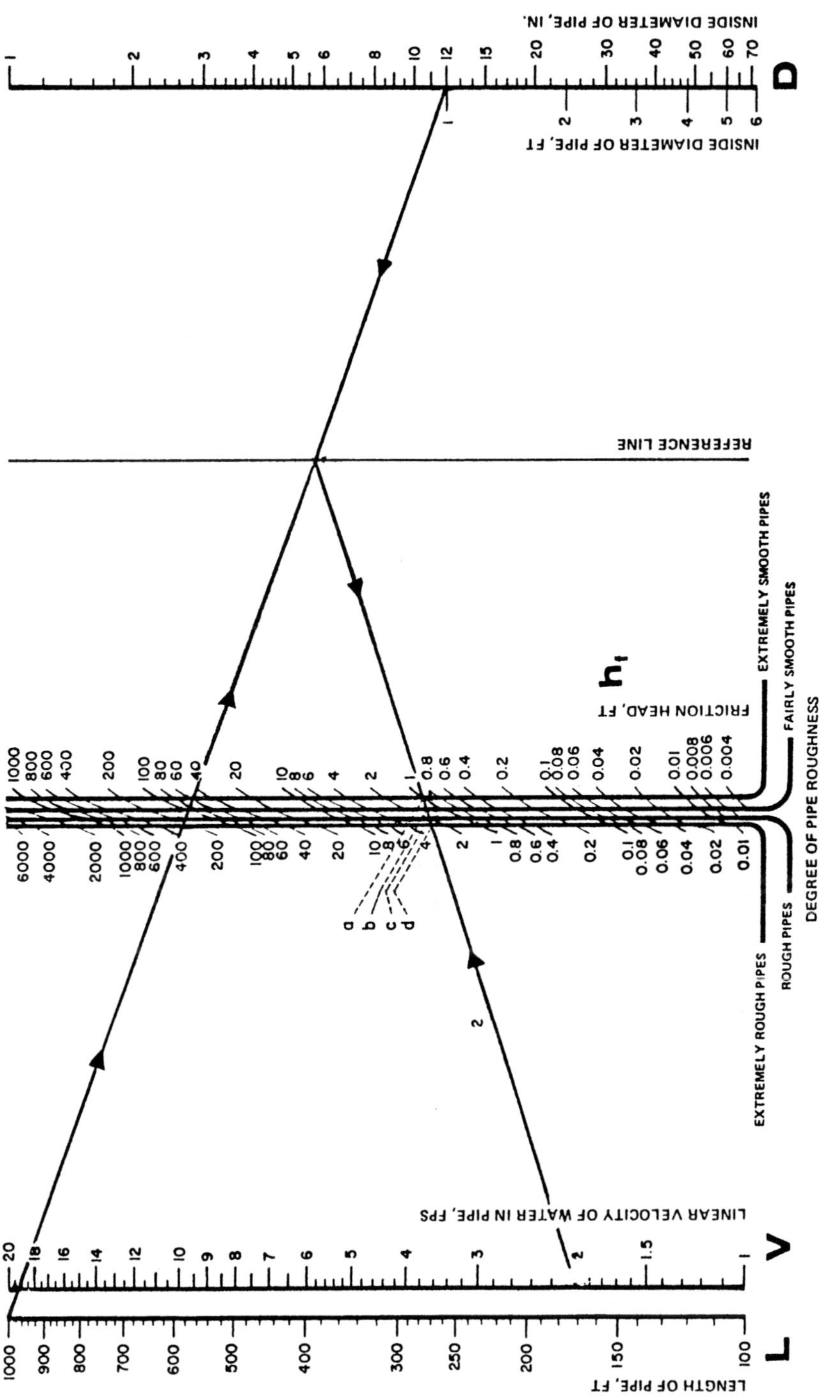

FIG. 1 Nomograph for the determination of friction head losses in water pipes.

Problem. Find the friction-head loss in a 1000-ft-long, 12-in.-diameter pipe with turbulent water flow and linear velocity of 2 ft/s when the pipe surface is (a) extremely smooth, (b) fairly smooth, (c) rough, and (d) extremely rough.

Solution: Connect 1000 ft on Scale *L* with 1 ft on scale *D*, and mark the point of intersection with the reference line. Connect this point with 2 ft/s on Scale V, and read the approximate friction heads from the grid as (a) 1.0 ft, (b) 1.5 ft, (c) 2.0 ft, and (d) 2.75 ft.

This material appeared in *Plant Engineering*, pp. 136–137, June 23, 1983, and is reprinted with the permission of the editor.

ADAM ZANKER

Gas Flow

A simplified method of calculating the gas flow rate through a pipeline is the Panhandle method [1], and it is generally accepted in the gas industry.

The Panhandle formula is

$$Q = 435.87E\left(\frac{T_0}{P_0}\right)^{1.0788}\left(\frac{P_1 - P_2}{G^{0.8539}T_fL}\right)^{0.5394} d^{2.6182} \tag{1}$$

where Q = rate of flow, ft^3/d, at base conditions
E = pipeline flow efficiency, dimensionless decimal fraction (design values of 0.88 to 0.94 are common)
G = specific gravity of gas (air = 1.00)
L = length of pipeline, miles
P_0 = pressure base, lb/in.^{2}abs
P_1 = inlet pressure, lb/in.^{2}abs
P_2 = outlet pressure, lb/in.^{2}abs
T_0 = temperature base, degrees Rankine (°R = °F + 460)
T_f = the average flowing temperature, degrees Rankine
d = the inside diameter of pipeline, in.

If we take T_0 = 60°F = 520°R and P_0 = 14.7 lb/in.^{2}abs (1 atm) as the base conditions, and if we assume that the average efficiency E equals 0.91, and if we define ΔP as the difference between the inlet and outlet pressures ($\Delta P = P_1 - P_2$), then we obtain a simplified form of the Panhandle equation:

$$Q = 18583\Delta P^{0.5394}T_f^{-0.5394}L^{-0.5394}G^{-0.4606}d^{2.6182} \tag{2}$$

This form of the Panhandle equation is much simpler to handle and gives the results with an accuracy of ±3% of those of the "original" formula, which is good enough for the majority of technical calculations.

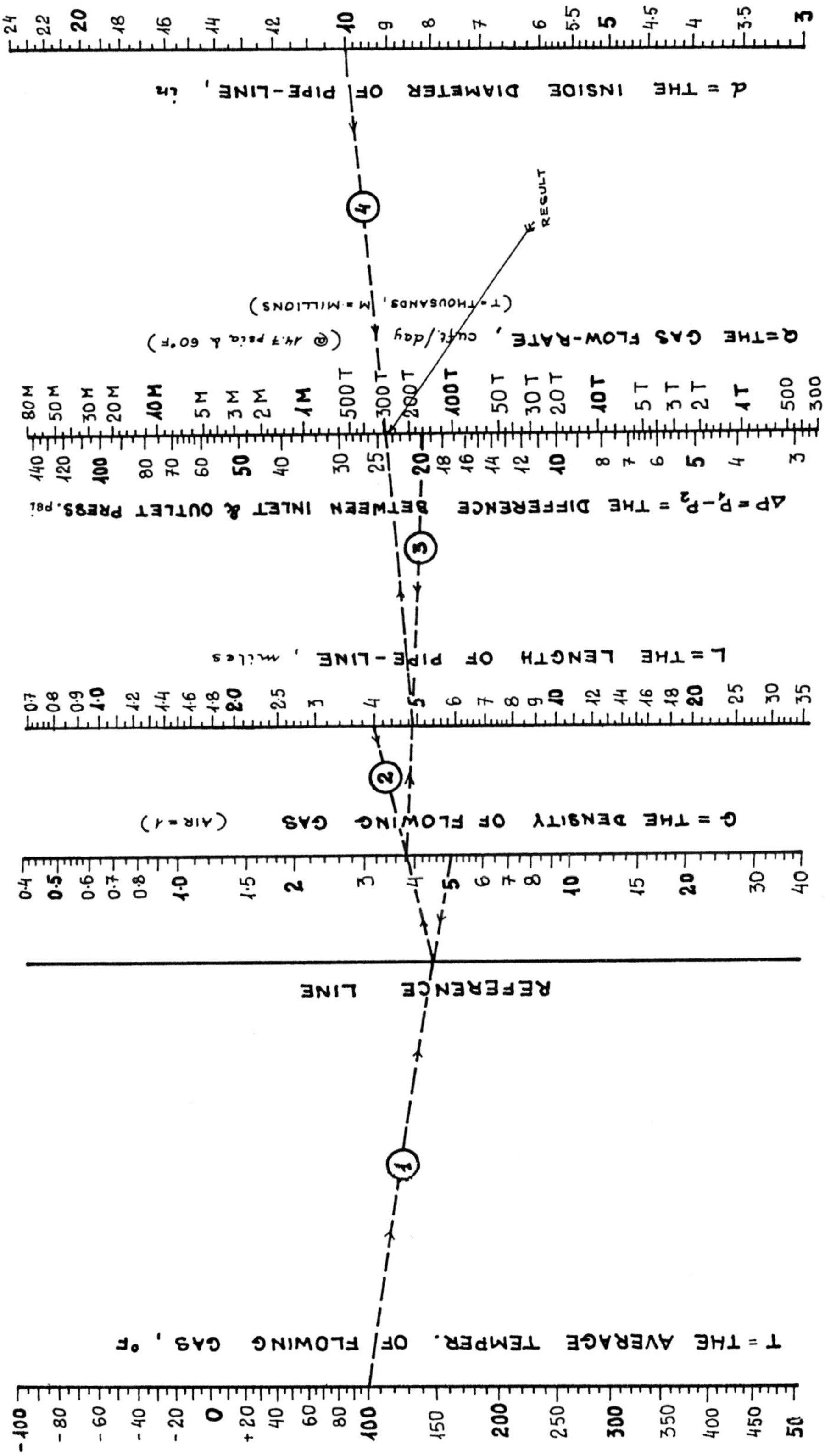

FIG. 1 Nomograph for the calculation of the gas flow rate through pipelines by the Panhandle equation.

The Nomograph

A nomograph (Fig. 1) has been designed which gives a rapid solution to Eq. (2) by using four sequential movements of a ruler.

The use of this nomograph, which is outlined in Fig. 2, is as follows.

1. Connect with a ruler the known values of T_f and G on the appropriate scales. Mark (I) the intersection point of a ruler with the Reference line.
2. Connect point (I) with a ruler to the known value of L on the appropriate scale. Mark (II) the intersection point of a ruler with the G scale.
3. Connect point (II) with a ruler to the known value of ΔP on the appropriate scale. Mark (III) the intersection point of a ruler with the L scale.
4. Connect with a ruler point (III) to the known value of d on the appropriate scale.

Read the final result, Q, on the intersection point of a ruler with the resulting Q scale.

NOTE. The scale for Q is plotted on the opposite side of the ΔP scale, on the same scale support.

Example. The use of this nomograph is best explained with a typical example: A gas is flowing through a pipeline (L = 4 miles) having the inside diameter, d = 10 in.

The entrance pressure is 100 lb/in.^{2}abs and the outlet pressure is 80 lb/in.^{2}abs. The average flowing temperature is 100°F (560°R), and the gas specific weight is 5.

Calculate the flow rate of this gas.

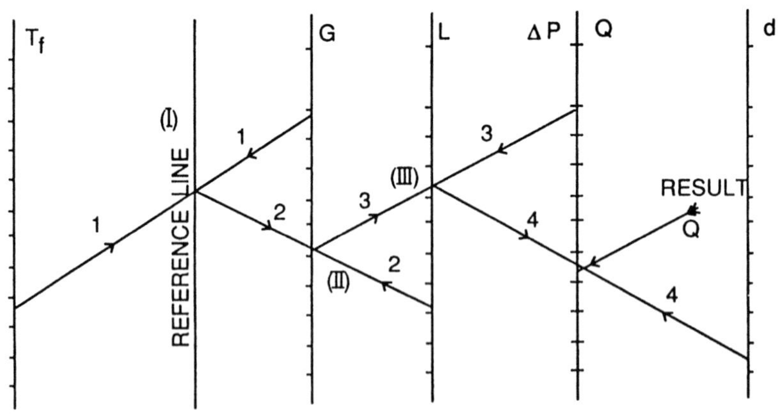

FIG. 2 Key to the use of the nomograph of Fig. 1.

Solution: First, compute the pressure drop:

$$\begin{aligned}\Delta P &= P_1 - P_2 \\ &= 100 - 80 \\ &= 20\ \text{lb/in.}^2\end{aligned}$$

With this value computed, enter the attached nomograph and follow the "procedure" described.

The solution is $Q = 290{,}000\ \text{ft}^3/\text{d}$ (at $P_0 = 14.7\ \text{lb/in.}^2\text{abs}$, $T_0 = 60°\text{F}$).

NOTE. This example is drawn with dashed lines on the nomograph.

Reference

1. T. Baumeister (ed.), *Standard Handbook for Mechanical Engineers,* 7th ed., McGraw-Hill, New York, 1967, pp. 11-180 to 11-185.

ADAM ZANKER

Looped Lines

The following formulas were derived for the design of looped lines.

The looped section can be connected at any two points along the main line (see Fig. 1) but must be connected at both ends of the loop to the main line. The limitations of the formulas are that after a section of the main line has been looped, the pressure drop, temperature, and gravity of the gas are the same as before paralleling.

The first formula is used to determine the fraction of the original line of a given diameter that must be paralleled with the same or different diameter of pipe to increase the volume from Q to Q_1:

$$x = \frac{\left[\left(\dfrac{Q}{Q_1}\right)^2 - 1\right]}{\left(\left\{\dfrac{1}{\left[1 + \left(\dfrac{d_1}{d}\right)^{8/3}\right]^2}\right\} - 1\right)} \tag{1}$$

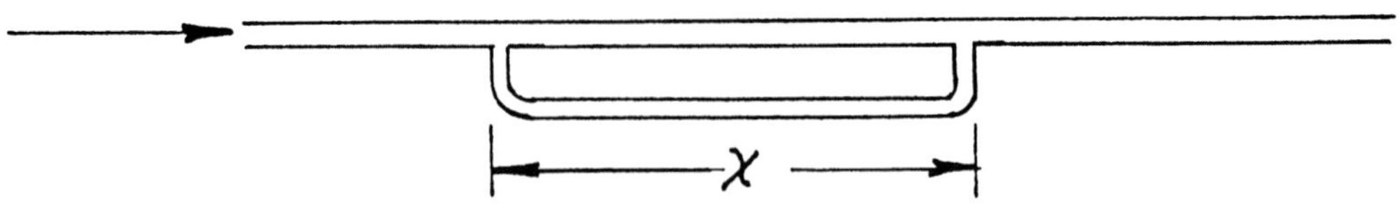

FIG. 1 A looped line.

By merely rearranging the above equation, the volume flowing through a system for any length paralleled with any diameter pipe, d_1:

$$Q_1 = \frac{Q}{\left[x\left(\left\{\frac{1}{\left[1 + \left(\frac{d_1}{d}\right)^{8/3}\right]^2}\right\} - 1\right) + 1\right]^{1/2}} \tag{2}$$

Again rearranging Eq. (1), the diameter of a looping line that will be required to increase the flow of gas a given amount for a given length of line paralleled may be determined as follows:

$$d_1 = d\left(\left\{\frac{x}{\left(\frac{Q}{Q_1}\right)^2 - 1 + x}\right\}^{1/2} - 1\right)^{3/8} \tag{3}$$

Symbols

x	portion of length of line parallel, expressed decimally
d	diameter of original line, in.
d_1	diameter of parallel line, in.
Q	rate of flow through original system at base conditions
Q_1	rate of flow through new system (now paralleled) at same base conditions

JOHN J. McKETTA

Parallel Lines

A piping system with a single line can usually meet small increases in demand by increasing operating pressures. Any sizable demand increase, however, requires looping or paralleling sections of the existing system.

In this discussion, the lines are referred to as *original* line and *parallel* line. A parallel line *may or may not* extend the entire length of the original line, or in the case of a loop, it may be longer than the original line. Both ends of the parallel line, however, are connected to the original line.

Equations (1)–(6) [1] are applicable under the conditions that whenever a section of the original line is paralleled, the temperature and specific gravity of the gas and pressures at the inlet and outlet ends of the original line are the same as they were before paralleling.

In all of the equations the following definitions apply:

x = portion of the original line length that is to be paralleled. This is expressed decimally.
d = internal diameter of original line, in.
d_1 = internal diameter of parallel line, in.
Q = rate of flow through system before paralleling at specified conditions of temperature and pressure
Q_1 = rate of flow through the system after paralleling at specified conditions of temperature and pressure

Equation (1) determines what portion of the entire length of the line must be paralleled with the same or different diameter pipe to increase the volume of flow to the desired amount.

$$x = (Q/Q_1)^2 - 1/\{(1/[1 + (d_1/d)^{8/3}]^2) - 1\} \tag{1}$$

Equation (1) can also be arranged to determine the volume flowing through a system for any value of x (decimal fraction or original line length paralleled) for any diameter pipe.

$$Q_1 = Q/\{x(1/[1 + (d_1/d)^{8/3}]^2 - 1) + 1\}^{1/2} \tag{2}$$

The equation can also be arranged to determine the diameter required for the parallel line to provide a given amount of gas flow for a given portion (x) of the original line paralleled.

$$d/d_1 = \{[x/(Q/Q_1)^2 - 1 + x]^{1/2} - 1\}^{3/8} \tag{3}$$

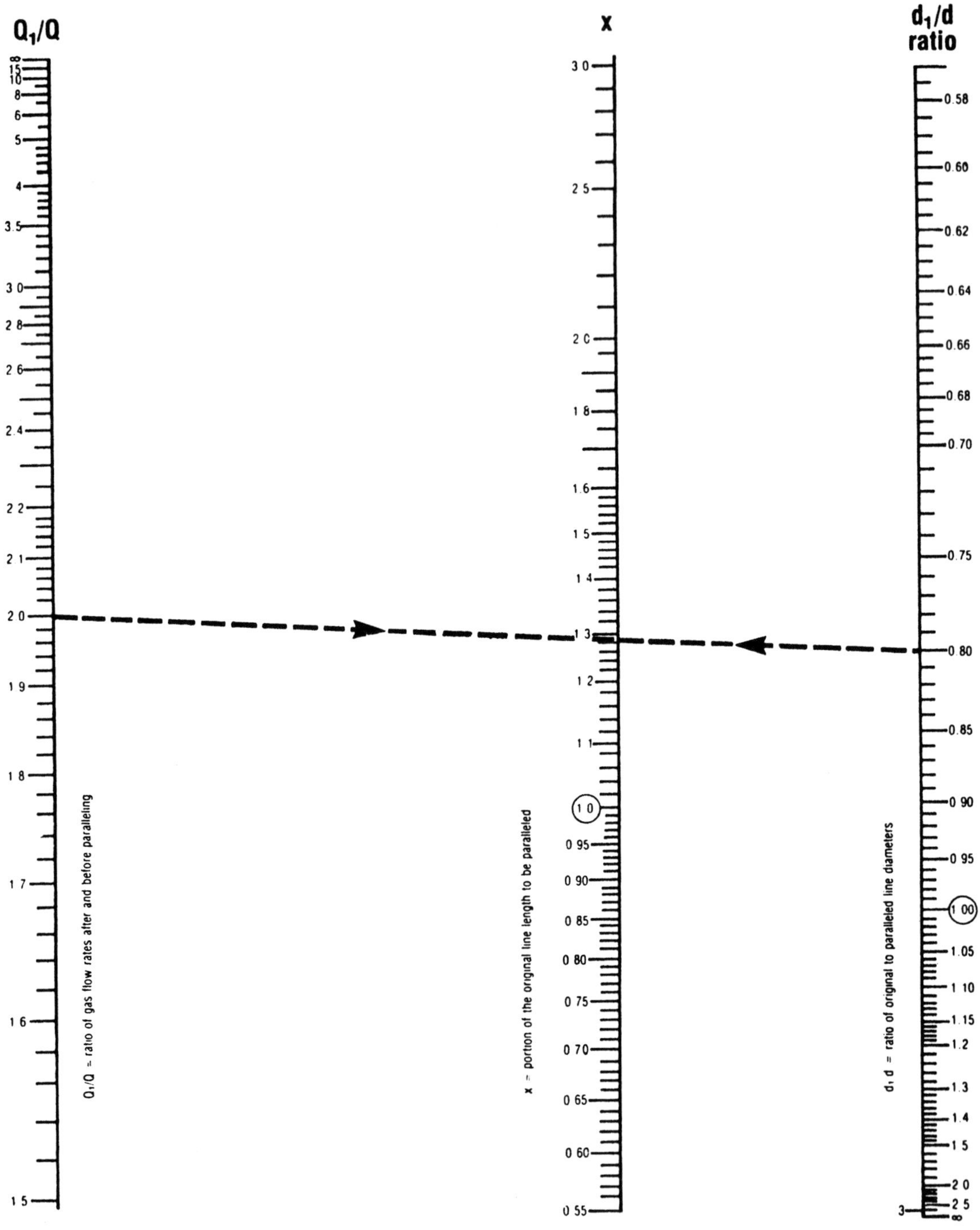

FIG. 1 Nomograph to calculate and design parallel pipeline systems. The broken line indicates the answer to the example problem presented in the text.

When the value of x is 1 (when the entire length of the original line has been paralleled), the following relationship may be used:

$$Q_1/Q = 1 + (d_1/d)^{8/3} \tag{4}$$

When the diameters of the original and parallel lines are the same, the formulas simplify to

$$x = 4[1 - (Q/Q_1)^2]/3 \tag{5}$$

and

$$Q_1/Q = 2(4 - 3x)^{1/2} \tag{6}$$

The nomograph (Fig. 1) solves all six equations. Equations (1) and (3) are solved by aligning the known values on their respective scales, and reading the answer at the point of intersection on the scale of the unknown value. To solve Eq. (2), proceed as usual, and then calculate the Q_1 value according to the following: $Q_1 = Qx(Q_1/Q)$. To solve Eq (4), proceed as usual, but remember that x equals 1. To solve Eqs. (5) and (6), remember that (d_1/d) equals 1.

Example. In a single line pipe system, the gas flow rate must be doubled. Tight clearances do not permit paralleling adjacent to the existant pipe; so a pipe loop must be used for the additional flow. It will be 0.8 of the existing pipe diameter. Which portion of the original line must be paralleled? With total flow doubled and a smaller diameter pipe, the parallel line will be longer than the original one, hence the loop system.

Solution: Extend a line from 2 on the Q_1/Q scale to 0.8 on the d_1/d scale, and read the answer as 1.283 on the x scale.

This material appeared in *Heating/Piping/Air Conditioning,* pp. 89–90, January 1976, and is reprinted with the permission of the editor.

Reference

1. Bell, *Petroleum Transportation Handbook,* McGraw-Hill, New York, 1963, pp. 5.22–5.23.

ADAM ZANKER

Parallel Lines, Computer Method

A new technique has been developed that permits a simple computer program to calculate flow rates in parallel piping.

Conservation of mass for the piping system gives

$$Q_T = Q_1 + Q_2 + \cdots + Q_n \quad (1)$$

where Q_T is the total flow rate entering the system (ft^3/s) and Q_1, Q_2, ..., Q_n represent the flow rates in the individual branches. The pressure difference, h_f, between the two ends (upstream and downstream of the branching junctions) is equal for each branch since all branches are in parallel. Equating the friction terms for each branch (i.e., the head loss) in terms of flow rate yields

$$h_f \frac{(\pi^2)}{16} 2g = \left[\frac{(f_1 l_1)}{D_1^5}\right] Q_1^2 = \left[\frac{(f_2 l_2)}{D_2^5}\right] Q_2^2 = \cdots = \left[\frac{(f_n l_n)}{D_n^5}\right] Q_n^2 \quad (2)$$

where D is pipe diameter (ft), and f_i is the friction factor for each branch. The gravitational constant is g (32.2 ft/s^2).

Procedure

The usual technique to solve the equations is to assume a value for the flow rate in one of the branches, e.g., Q_1, calculate the corresponding Reynolds number, and select a friction factor, f_1, from the Moody diagram or an equivalent formula (e.g., Colebrook), that is an implicit function of f.

Then substitute this friction factor and flow rate into Eq. (2) to calculate the other flow rates and friction factors. The sum of these flow rates must then satisfy Eq. (1). If it is not satisfied, a new value for Q_1 is estimated and the process is continued until reasonable agreement between the assumed flow rate and the calculated flow rate is found.

The difficulty with this procedure is the wide variation in the values of Q_1 that can be assumed (i.e., from zero to Q_T), while the selection of the second guess in the iteration procedure for the flow rate is also very subjective (standard iteration techniques may produce diverging values).

Therefore, producing converging values of flow rate can be time-consuming, especially for piping systems with many branches, and result in inaccurate answers.

New Method

The procedure described here is less variable than that discussed above and leads to more accurate results.

The major difference in the two methods is that the friction factors, f_i, are chosen initially rather than the flow rates, Q_i, and used to relate the various flow rates. Then an explicit rather than an implicit relationship between the friction factor and the flow rate is used in the iteration scheme.

Equation (2) may be rearranged to give

$$Q_1 = \left[\frac{(f_2 l_2)}{D_2^5}\frac{(D_1^5)}{f_1 l_1}\right]^{1/2} Q_2 = \cdots = \left[\frac{(f_n l_n)}{D_n^5}\frac{(D_1^5)}{f_1 l_1}\right]^{1/2} Q_n \tag{3}$$

Substitution of these expressions into Eq. (1) yields

$$Q_1 + \left[\frac{(f_1 l_1)}{D_1^5}\frac{(D_2^5)}{f_2 l_2}\right]^{1/2} Q_1 + \cdots + \left[\frac{(f_1 l_1)}{D_1^5}\frac{(D_n^5)}{l_n f_n}\right]^{1/2} Q_1 = Q_T \tag{4}$$

Rearranging and solving for Q_1 yields

$$Q_1 = Q_T / \left(1 + \left[\frac{(f_1 l_1)}{D_1^5}\frac{(D_2^5)}{f_2 l_2}\right]^{1/2} + \cdots + \left[\frac{(f_1 l_1)}{D_1^5}\frac{(D_n^5)}{f_n l_n}\right]^{1/2}\right) \tag{5}$$

Unknowns

Again the total flow (Q_T), length (ft), diameter (ft), and relative roughness (ϵ) are known for each branch; only the friction factors (f_i) and the branch flow rates (Q_i) are unknown.

For the initial guess of the friction factors, the computer algorithm assumes turbulent flow. After calculating the initial Q_1 from Eq. (5), the Reynolds number is determined. Chen's formula:

$$\frac{1}{\sqrt{f}} = -2.0 \log_{10}\left[\frac{\epsilon}{3.7065D} - \frac{5.0452}{\mathrm{Re}}\log_{10}\left(\frac{1}{2.8257}\left(\frac{\epsilon}{D}\right)^{1.1098} + \frac{5.8506}{\mathrm{Re}^{0.8981}}\right)\right] \tag{6}$$

is then used to calculate f_1 which is compared to the assumed value.

If the comparison is unfavorable, then the friction factor is updated and an iteration process is begun. It should be noted that the range of values for friction factors is much narrower than values for flow rates.

Therefore, the iteration scheme given here will converge much more rapidly. Also, since the square root of the ratio of friction factors appears in the equation, a "smoothing" effect tends to be produced which allows a straightforward iteration producing converging results. The computer program for carrying out the above calculations is listed in Fig. 1. It is written in BASIC.

```
10  DIM S(16),F(16),L(10),D(10),E(10),Q(10),R(10),A(10),B(10),C(10),P(10),F1(10)
,F2(10)
20 DATA 0.000005,0.018,0.0001,0.018,0.0002,0.019,0.0004,0.02,0.0006,0.021,0.0008
,0.022,0.001,0.023,0.0015,0.024,0.002,0.025,0.003,0.027,0.004,0.03,0.006,0.032,0
.008,0.035,0.01,0.037,0.050,0.075
30 FOR J=1 TO 15:READ S(J),F(J):NEXT J
70 PRINT"NUMBER OF BRANCHES";
80 INPUT N
150 FOR I=1 TO N
160 PRINT"INPUT VALUES FOR BRANCH ";I;" IN ENGLISH UNITS"
170 PRINT"LENGTH (FEET)";
180  INPUT L(I)
190 PRINT"DIAMETER (INCHES)";
200 INPUT D(I):D(I)=D(I)/12
210 PRINT"ROUGHNESS FACTOR (EPSILON IN FEET)";
220 INPUT E(I)
230 NEXT I
240 PRINT"OVERALL VOLUMETRIC FLOW RATE (FEET^3/SEC)";
250 INPUT X
280 PRINT"KINEMATIC VISCOSITY (FEET^2/SEC)";
290 INPUT V
300 FOR I=1 TO N
310  P(I)=E(I)/D(I)
320  A(I)=(4)/(3.1415*V*D(I))
330  C(I)=L(I)/(D(I)^5)
340 NEXT I
350 FOR I=1 TO N
360  FOR J=1 TO N
370   IF P(I)<>S(J) THEN 400
380   F1(I)=F(J)
390   GOTO 460
400   IF P(I)<S(J) THEN 420
410  NEXT J
420   H=(S(J)+S(J-1))/2
430   IF P(I)>=H THEN 450
440    J=J-1
450   F1(I)=F(J)
460 NEXT I
470 FOR I=1 TO N:B(I)=0:NEXT I
500 FOR I=1 TO N
510   FOR J=1 TO N
520   B1=SQR((C(I)*F1(I))/(C(J)*F1(J)))
530   B(I)=B(I)+B1
540   NEXT J
550 Q(I)=X/B(I):R(I)=A(I)*Q(I)
580   T1=(P(I)^1.1098)/2.8257
590   T2=5.8506/(R(I)^.8981)
600   T3=(LOG(T1+T2))/.3026
610   T4=P(I)/3.7065
620   T5=(-5.0452*T3)/R(I)
630   T6=(LOG(T4+T5))/2.3026
640 F2(I)=(1/(-2*T6))^2
660  NEXT I
670 Y=Y+1
680 FOR I=1 TO N
690  PRINT"Q(";I;")","F1(";I;")","F2(";I;")"
700  PRINT Q(I),F1(I),F2(I)
710  IF ABS((F1(I)-F2(I))/F1(I))<=.001 THEN 740
720 F1(I)=F2(I):Y=1
740 NEXT I
750 IF Y=1 THEN 470
760  END
```

FIG. 1 Computer program to calculate flow rates in parallel piping.

This material appeared in *Pipe Line Industry*, p. 41, August 1985, and is reprinted with the permission of the editor.

LEIGH A. SMITH
EDWIN P. RUSSO

Pressure Drop Calculations

An engineer usually needs to determine pressure drops for only a few lines at a time, which does not justify programming a computer or calculator and then testing the program for errors.* In such cases, tables come in very handy.

This article reviews several tables already available for liquid and compressible-fluid pressure-drop calculations and points out their limitations. Two new tables for compressible fluids, and instructions for using them, are presented and discussed, and their use is demonstrated through a few simple examples.

Previously Available Tables

Liquids

There are several good tables for determining the pressure drop of liquids in pipelines. Those in Ref. 5 cover the full range of flow conditions—from laminar to turbulent—for all Newtonian liquids, and essentially eliminate the need to perform detailed calculations.

Perhaps the most well-known table for liquids is "Flow of Water Through Sch. 40 Pipe," p. B-14 in Ref. 6. It is easy to use, but not as useful as the tables in Ref. 5 because it is based on water on 60°F; the pressure drop for other liquids, and water at other temperatures, will differ from the values obtained for water.

Compressible Fluids

The tables available [6] for compressible fluids are not as useful as those for liquids. "Flow of Air Through Sch. 40 Pipe," p. B-15, is excellent for pressure drops in air systems at 100 lb/in.^{2}gauge in Sch. 40 pipe. However, most situations do not fall into this narrow category.

Better is "Simplified Flow Formula for Compressible Fluids," pp. 3-22 and 3-23 [6], a table and nomograph combined. It is fairly easy to use, but can be very inaccurate at low flow rates. If the pipe is other than Sch. 40, its internal diameter can be used (instead of schedule number), but this usually means looking up data on yet another table. Although this "Simplifed Flow . . ." table is frequently used, most people unfortunately do not realize its limitations.

"Pressure Drop in Compressible Flow Lines," pp. 3-20 through 3-24 [6], are more accurate nomographs, but also more difficult to use—three lines must be drawn, on the chart itself, to obtain the pressure drop.

*References 1–4 provide good calculator programs for handling a large number of calculations.

New Tables for Compressible Fluids

The tables presented here are free of many of the disadvantages of the earlier tables. They are applicable to any fluid because density has been incorporated into the pressure-drop calculations rather than having been built into the tables. The pipe sizes covered are those most commonly used. Because the tables are for turbulent conditions only, each lists the minimum flow rates for which it can be used.

Table 1, in both English and metric units, is for adiabatic turbulent flow and for pressure drops of up to 20% of the upstream pressure. It can be used for flows significantly less than fully turbulent, but not for laminar or transition flow.

Table 2 (in English units only) is for fully turbulent isothermal flow, and should be used only when the pressures drops are expected to be high enough to ensure that the assumption of turbulence is correct.

To make the tables easier to use, instructions appear on the tables themselves. However, the procedure is basically as follows: (1) determine the vapor density and the mass flow rate, (2) choose a pipe size, (3) check that the flow rate is greater than a given minimum, (4) obtain certain data (described below) from the table, (5) choose a safety factor (discussed below), and (6) calculate a pressure drop by using the equations given at the top of the table. The use of Table 2 requires a trial-and-error calculation based on assuming a downstream pressure.

Table 1: Adiabatic Turbulent Flow

Table 1 is based on the Darcy formula for pressure drop in a pipeline:

$$\text{Pressure drop} = 3.36 \times 10^{-6}(fLW^2/\rho d^5)$$

where f = friction factor for fully turbulent flow, dimensionless; L = line length, ft; W = fluid flow rate, lb/h; ρ = fluid density at average pressure, lb/ft^3; and d = actual internal diameter of pipe, in.

By letting L = 100 ft, this can be simplified to

$$\Delta P = 3.36 \times 10^{-4}(fW^2/\rho d^5) \tag{1}$$

where ΔP = pressure drop per 100 ft, lb/in.^{2}.

In metric units, for L = 100 m, the equivalent of Eq. (1) is

$$\Delta P = 5.9 \times 10^{-3}(fW^2/\rho d^5) \tag{1'}$$

where ΔP = pressure drop per 100 m, bar; W is in kg/h, ρ is in kg/m^3; and d is again in inches.

TABLE 1 Simplified Pressure-Drop Calculations for Adiabatic Turbulent Flow of Compressible Fluids

$$\Delta P = BW^2/\rho \times (\text{safety factor})$$

ΔP = pressure drop per 100 ft or 100 m, lb/in.^{2}abs or bar
B = factor from table below
W = flow rate, lb/h or kg/h
ρ = fluid density (at average pressure), lb/ft^3 or kg/m^3
Safety factor = 1.2 usually (higher in corrosive or plugging service)

Instructions:

1. Calculate the vapor density. For low pressure drops (less than 20% of the upstream pressure), it can be based on the upstream pressure; for high pressure drops (20 to 40% of the upstream pressure), the average pressure should be used (requiring a trial-and-error calculation).
2. Calculate the mass flow rate.
3. Choose a pipe size.
4. Check to make sure the flow rate is greater than W_{min}. If it is significantly less than W_{min}, do not use this procedure.
5. Obtain B from below.
6. Choose a safety factor (usually, 1.2 is used, although for corrosive or plugging conditions, a higher value should be used).
7. Calculate the pressure drop by using the equation above.

Nominal			English Units		Metric Units	
Pipe Size (in.)	Schedule	f	B	W_{min}	B	W_{min}
$\frac{1}{2}$	80	0.026	2.0×10^{-4}	60	3.5×10^{-3}	25
$\frac{3}{4}$	80	0.025	4.0×10^{-5}	90	7.0×10^{-4}	40
1	80	0.023	1.1×10^{-5}	140	1.9×10^{-4}	60
$1\frac{1}{2}$	80	0.021	1.0×10^{-6}	360	1.8×10^{-5}	160
2	80	0.020	2.6×10^{-7}	550	4.6×10^{-6}	250
	40	0.019	1.9×10^{-7}	590	3.3×10^{-6}	270
3	40	0.018	2.4×10^{-8}	1,500	4.2×10^{-7}	660
4	40	0.017	5.8×10^{-9}	2,900	1.0×10^{-7}	1,300
6	40	0.015	6.6×10^{-10}	8,700	1.2×10^{-8}	3,900
8	40	0.014	1.6×10^{-10}	15,000	2.8×10^{-9}	6,900
10	40	0.013	4.9×10^{-11}	20,000	8.6×10^{-10}	9,200
12	3/8-in. wall	0.013	1.9×10^{-11}	29,000	3.3×10^{-10}	13,000
14	30	0.012	1.1×10^{-11}	35,000	1.9×10^{-10}	16,000
16	30	0.012	5.5×10^{-12}	44,000	9.6×10^{-11}	20,000
18	3/8-in. wall	0.012	2.9×10^{-12}	53,000	5.1×10^{-11}	24,000
20	20	0.012	1.7×10^{-12}	64,000	3.0×10^{-11}	29,000
24	20	0.011	6.0×10^{-13}	170,000	1.1×10^{-11}	76,000
30	3/8-in. wall	0.011	1.9×10^{-13}	210,000	3.3×10^{-12}	95,000
36	3/8-in. wall	0.011	7.1×10^{-14}	420,000	1.2×10^{-12}	190,000
42	3/8-in. wall	0.010	3.1×10^{-14}	510,000	5.4×10^{-13}	230,000
48	3/8-in. wall	0.010	1.5×10^{-14}	620,000	2.6×10^{-13}	280,000
54	3/8-in. wall	0.010	8.2×10^{-15}	760,000	1.4×10^{-13}	350,000
60	3/8-in. wall	0.009	4.8×10^{-15}	870,000	8.4×10^{-14}	400,000
66	3/8-in. wall	0.009	2.9×10^{-15}	990,000	5.1×10^{-14}	450,000

The fluid's flow rate and density are independent of pipe diameter. However, both f and d^5 are a function of line size, and can be grouped together with the constant term:

$$B = 3.36 \times 10^{-4}\, f_{110}/d^5 \text{ for English units}$$

$$B = 5.9 \times 10^{-3}\, f_{110}/d^5 \text{ for metric units}$$

TABLE 2 Calculations for Large Isothermal Pressure Drops of Compressible Fluids in Fully Turbulent Flow

$$P_2 = [P_1^2 - X' - Y \ln (P_1/P_2)]^{1/2}$$

P_2 = downstream pressure, lb/in.²abs
P_1 = upstream pressure lb/in.²abs
$X' = (B_1 W^2 P_1 L_e/\rho)$(safety factor)
$Y = B_2 W^2 P_1/\rho$
B_1 and B_2 = factors from table below
W = flow rate, lb/h
L_e = equivalent line length, ft
ρ = fluid density at P_1, lb/ft³
Safety factor = 1.2 usually (higher in corrosive or plugging service)

Instructions:

1. Calculate the vapor density based on the upstream pressure.
2. Calculate the mass flow rate.
3. Choose a pipe size.
4. Check to make sure the flow rate is greater than W_{min}.
5. Obtain B_1 and B_2 from below.
6. Choose a safety factor (usually 1.2, but for corrosive or plugging conditions, a higher value should be used).
7. Calculate P_2 by trial and error: Assume a value for P_2, substitute this into the right-hand side of the equation above, and calculate a value for P_2. If the assumed and calculated values do not agree, assume a new P_2 and repeat.

Nominal Pipe Size, in.	Schedule	f	B_1	B_2	W_{min}
$\frac{1}{2}$	80	0.026	3.6×10^{-6}	6.3×10^{-6}	90
$\frac{3}{4}$	80	0.025	7.3×10^{-7}	1.9×10^{-6}	700
1	80	0.023	$2.0 \times 10-^{7}$	6.7×10^{-7}	1,400
$1\frac{1}{2}$	80	0.021	1.9×10^{-8}	1.1×10^{-7}	3,600
2	80	0.020	4.8×10^{-9}	4.0×10^{-8}	6,900
	40	0.019	3.4×10^{-9}	3.1×10^{-8}	7,400
3	40	0.018	4.3×10^{-10}	6.4×10^{-9}	12,000
4	40	0.017	1.1×10^{-10}	2.1×10^{-9}	33,000
6	40	0.015	1.2×10^{-11}	4.2×10^{-10}	86,000
8	40	0.014	2.9×10^{-12}	1.4×10^{-10}	150,000
10	40	0.013	9.0×10^{-13}	5.6×10^{-11}	220,000
12	3/8-in. wall	0.013	3.5×10^{-13}	2.7×10^{-11}	350,000
14	30	0.012	2.1×10^{-13}	1.8×10^{-11}	420,000
16	30	0.012	1.0×10^{-13}	1.0×10^{-11}	550,000
18	3/8-in. wall	0.012	5.3×10^{-14}	6.3×10^{-12}	690,000
20	20	0.012	3.0×10^{-14}	4.1×10^{-12}	930,000
24	20	0.011	1.1×10^{-14}	1.9×10^{-12}	1,400,000
30	3/8-in. wall	0.011	3.4×10^{-15}	7.7×10^{-13}	2,100,000

where $f_{110} = 1.1f$. This friction factor of 110% of the fully turbulent friction factor was used to extend the range of applicability of this approach down to 10% of fully turbulent flow. (Laminar flow typically occurs at about 0.1% of full turbulence, so there is no danger of being in laminar flow at 10%.)

Equations (1) and (1′) can thus be rewritten as

$$\Delta P = BW^2/\rho \tag{2}$$

One value of this table is the listing of W_{min}, which is the lowest flow rate for which the table should be used. At very low flow rates, this calculation procedure could predict pressure drops 50% too low. W_{min} corresponds to the Reynolds number associated with a friction factor 10% higher than that for fully turbulent flow.

W_{min} was calculated by using a vapor viscosity of 0.015 cP, which would be a normal viscosity for steam, heavy hydrocarbons, or light hydrocarbons at elevated temperatures. For light hydrocarbons, the minimum flow rate for which it would be safe to use the table is a little lower than W_{min}.

Table 1 can be used for a pressure drop of up to 20% of the upstream pressure. It is extremely inaccurate if the pressure drop exceeds 40% of the upstream pressure (because acceleration of the gas significantly affects the calculation).

When the pressure drop is between 20 and 40% of the upstream pressure, Table 1 can be used if the vapor density is based on the average pressure. This requires a trial-and-error calculation: Assume a pressure drop, calculate an average density, then calculate the pressure drop based on the average density, and repeat the procedure until the difference between the assumed and calculated pressure drops is acceptable.

Table 2 has been developed for cases where the pressure drop is expected to be too high to accurately use Table 1. However, since Table 1 is easier to use, it should be tried first if the pressure drop is likely to be less than 20% of the upstream pressure. If the calculated pressure drop is greater than 20% of the upstream pressure, the results should be checked by using the Table 2 procedure (or an average density used with the Table 1 procedure, as described in the previous paragraph).

Table 2: Isothermal, Fully Turbulent Flow

Table 2 is based on the equation for isothermal flow:

$$P_1^2 - P_2^2 = (W^2 P_1/144\rho_1 g A^2)[fL_e/D + \ln(P_1/P_2)] \tag{3}$$

where P_1 and P_2 are the upstream and downstream pressures, respectively, lb/in.^{2}abs; ρ_1 = fluid density at P_1, lb/ft^3; g = acceleration due to gravity = 32.2 ft/s^2; A = inside cross-sectional area of pipe, ft^2; L_e = equivalent line length, ft; and D = internal pipe diameter, ft.

This can be rearranged to

$$P_2 = \left[P_1^2 - \left(\frac{1.67 \times 10^{-11} f}{A^2 D}\right)\left(\frac{W^2 P_1 L_e}{\rho_1} \times \text{safety factor}\right) - \left(\frac{1.67 \times 10^{-11}}{A^2}\right)\left(\frac{W^2 P_1}{\rho_1}\right)\left(\ln \frac{P_1}{P_2}\right)\right]^{1/2} \tag{4}$$

By again grouping the constants and the line-size-dependent terms (f, A, and D) and defining $B_1 = 1.67 \times 10^{-11} f/A^2D$ and $B_2 = 1.67 \times 10^{-11}/A^2$, Eq. (4) can be simplified further to

$$P_2 = [P_1^2 - X - Y \ln (P_1/P_2)]^{1/2} \tag{5}$$

where $X = B_1W^2P_1L_e/\rho_1$ and $Y = B_2W^2P_1/\rho_1$.

In cases of high pressure drop where Table 2 is to be used, the flow is almost always fully turbulent, but W_{min} has been included just to be sure. The table can be used for flows below W_{min}, but only with great caution. Under no circumstances should Table 2 be used at flows less than the minimum flow rates listed in Table 1 for the same size pipe.

Safety Factors

B, B_1, and B_2 were developed for new, clean pipe, so a safety factor must be applied to allow for the accuracy of the friction factor and to correct for fouling or an increase in pipe roughness with age. Equation (2) thus becomes

$$\Delta P = BW^2/\rho \times \text{safety factor} \tag{6}$$

In Eq. (5), the safety factor is incorporated only into the term represented by X, rather than applied to the equation for P_2, because its purpose is to account for inaccuracies in the friction factor, which is a component of B_1 in X. Thus,

$$P_2 = [P_1^2 - X' - Y \ln (P_1/P_2)]^{1/2} \tag{7}$$

where $X' = X(\text{safety factor}) = (B_1W^2P_1L_e/\rho_1)(\text{safety factor})$.

Generally, a safety factor of 1.2 can be used in clean, noncorrosive service. This accounts for the ±10% accuracy of the friction factor, with a small allowance for increased roughness. For fouling or corrosive systems, a higher safety factor may be required—up to 1.86 for untreated water in steel pipe.

The choice of the safety factor, when fouling and/or corrosion-caused roughness is of concern, is really a matter of economics. Using a large safety factor will result in a larger calculated pressure drop, and oversizing of the line. One situation where large safety factors are commonly used is in the design of underground lines, because these are expensive to dig up and replace, and it is generally cheaper to oversize the line initially than replace it prematurely.

In some cases a safety factor less than 1 may be required if a conservatively high estimate of the flow rate is desired. This could occur, for example, in the sizing of a safety valve for flow from a high-pressure vessel to one at a lower pressure. In this case a safety factor of about 0.9 should be used to allow for the fact that the actual flow rate might be greater than what the equations had predicted.

Pipe Schedules

Tables 1 and 2 are based on the pipe schedules commonly encountered by the author, but they can be corrected for different standard schedules. While for large lines any correction for wall thickness would be negligible, the effect of pipe schedule on small lines is significant. This is evidenced by the B values for 2-in. Sch. 40 and Sch. 80 pipe—the pressure drop in Sch. 80 pipe will be 37% higher than that in Sch. 40.

In general, B is proportional to d^5. Thus, to scale to a different pipe schedule, or even to metric sizes, simply multiply the value of B by the ratio of the pipe (internal) diameters to the fifth power. The friction factor is not significantly affected by wall thickness, so no additional correction is required.

The tables can also be corrected for alloy pipe, which is usually thin walled and so will have lower pressure drops than the tables predict. Alloy pipe is also expensive, so it is important for the pressure drops to be accurate.

Example 1. Calculate the pressure drop (per 100 ft) for a fluid having a density of 2.2 lb/ft^3, flowing at a rate of 30,000 lb/h in an 8-in. Sch. 40 pipe. Use the typical safety factor of 1.2 for clean noncorrosive service.

Solution: Since the pressure drop is expected to be low, Table 1 will be tried first. From that table, $B = 1.6 \times 10^{-10}$, and the flow rate is above the minimum flow rate of $W_{min} = 15{,}000$ lb/h. Using Eq. (6), $\Delta P = (1.6 \times 10^{-10})(30{,}000)^2(1.2)/2.2 = 0.0785$ lb/in.^{2}.

Example 2. A fluid with a density of 0.5 lb/ft^3 flows in a 200-ft-equivalent length of 4-in. Sch. 40 pipe at a flow rate of 40,000 lb/h. The upstream pressure is 115 lb/in.^{2}. Using a safety factor of 1.2, calculate the pressure drop.

Solution: In this case the pressure drop is expected to be relatively large, so Table 2 will be used. The flow rate is above $W_{min} = 33{,}000$ lb/h. From the table, $B_1 = 1.1 \times 10^{-10}$ and $B_2 = 2.1 \times 10^{-9}$. $X' = (1.1 \times 10^{-10})(40{,}000)^2(115)(200)(1.2)/0.5 = 9715$, and $Y = (2.1 \times 10^{-9})(40{,}000)^2(115)/0.5 = 773$. Assume first a downstream pressure of $P_2 = 65$ lb/in.^{2}abs, and then use Eq. (7) to calculate P_2: $P_2 = 115^2 - 9715 - 773 \ln (115/65)]^{1/2} = 55$ lb/in.^{2}abs, which is less than the 65 lb/in.^{2}abs assumed. Then try $P_2 = 55$ lb/in.^{3}abs: $P_2 = 54.2$ lb/in.^{2}abs, which is close. Finally, try $P_2 = 54$ lb/in.^{2}abs: $P_2 = 54.1$. Thus, the downstream pressure is about 54 lb/in.^{2}abs, for a pressure drop of 61 lb/in.^{2}abs.

Example 3. Calculate a new pressure drop for the situation described in the previous example, but let $L_e = 300$ ft.

Solution: $X' = (1.1 \times 10^{-10})(40{,}000)^2(115)(300)(1.2) \div 0.5 = 14{,}570$. Note that this is greater than $P_1^2 = 115^2 = 13{,}225$, which means that the term inside the brackets in Eq. (7) would be negative for all assumed values of P_2. This indicates that the pressure drop in the line is excessive, and a larger line should be used.

References

1. L. L. Simpson, "Versatile Calculator Program Eases Piping Design," *Chem. Eng.*, pp. 105–109 (January 29, 1979).
2. P. Kandell, "Program Sizes Pipe and Flare Manifolds for Compressible Flow," *Chem. Eng.*, pp. 89–93 (June 29, 1981).
3. W. W. Blackwell, "Calculating Two-Phase Pressure Drop," *Chem. Eng.*, pp. 121–125 (September 7, 1981).
4. C. R. Brunner, "Program Predicts Pressure Drop for Steam Flow," *Chem. Eng.*, pp. 97–99 (February 22, 1982).
5. *Cameron Hydraulic Data,* Ingersoll-Rand Corp., Woodcliff Lake, New Jersey, 1977.
6. *Flow of Fluids through Valves, Fittings and Pipe,* Technical Paper 410, Crane Co., New York, 1969.

MICHAEL L. BRADFORD

Relative Capacities of Pipes

From a strictly geometric point of view, the cross-sectional area of a pipe opening is proportional to the square of its diameter. This means that two different pipes with internal diameters of, say, 1 and 4 in. have cross-sectional areas of $1^2(\pi/4)$ and $4^2(\pi/4)$, respectively. The ratio of their cross sections will thus equal 1:16. However, their carrying capacities *will not* equal the ratio 1:16.

Flow in smaller pipes is generally less efficient than in larger ones made of similar material due to the variation of the friction factor, f, with the Reynolds number and the pipe diameter. Therefore, the carrying capacity of a larger pipe is much greater than that of a number of smaller pipes combined, having a total cross-sectional area equal to one larger pipe.

Several equations have been proposed to determine the number of smaller pipes needed to equal the carrying capacity of the larger pipe. The two given here have been widely used and have proved to be accurate.

The substance to be carried affects the number of smaller pipes required. Equation (1) determines the number of smaller pipes needed to equal the

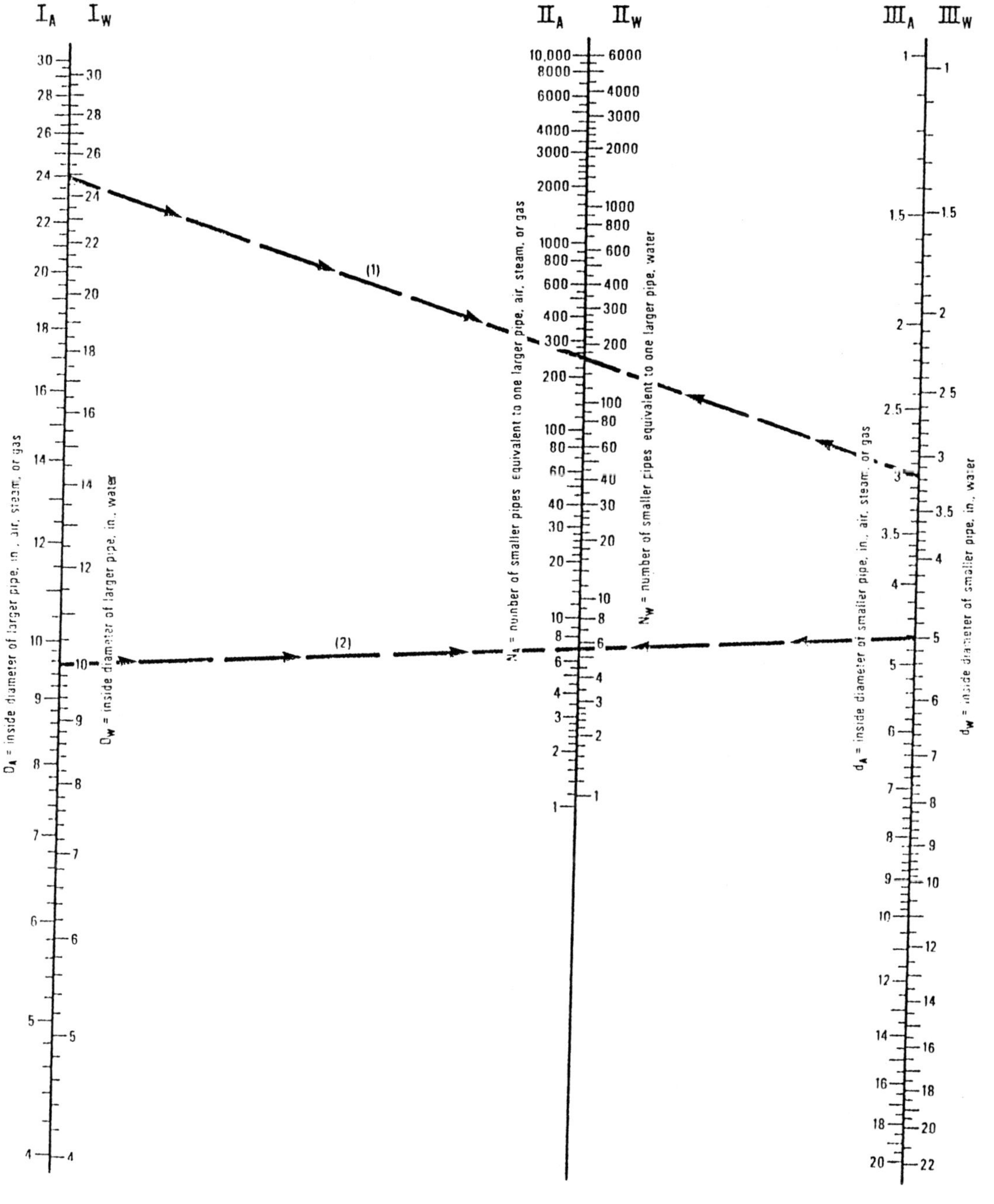

FIG. 1 This nomograph determines the relative capacities of pipes. The broken lines indicate solutions to the example problems in the text; the numbers on the broken lines indicate which problem and equation are solved.

carrying capacity of one larger pipe when air, steam, or gas is to be transported:

$$N = (D^3\sqrt{d + 3.6})/(d^3\sqrt{D + 3.6}) \qquad (1)$$

where N = the number of smaller pipes having a carrying capacity equal to one larger pipe
D = inside diameter of larger pipe, in.
d = inside diameter of smaller pipe, in.

Equation (2) determines the number of smaller pipes needed when water is to be carried:

$$N = (D/d)^{2.5} \qquad (2)$$

The nomograph (Fig. 1) permits rapid solutions to both equations. Equation (1) can be solved by using Scales I_A, II_A, and III_A, and Eq. (2) is calculated by using Scales I_W, II_W, and III_W.

Example 1. How many smaller pipes with an inside diameter of 3 in. (d) will be needed to equal the carrying capacity of a larger pipe having an inside diameter of 24 in (D) to carry air, steam, or gas?

Solution: Extend a line from 3 on the III_A scale to 24 on the I_A scale. Read the number of smaller pipes needed as 250 at the point of intersection on the II_A scale.

Example 2. How many smaller pipes with an inside diameter of 5 in. (d) will be needed to carry an equivalent quantity of water as one larger pipe having an inside diameter of 10 in. (D)?

Solution: Extend a line from 5 on the III_W scale to 10 on the I_W scale. Read the number of smaller pipes needed as 5.7 or 6 at the point of intersection on the II_W scale.

This material appeared in *Heating/Piping/Air Conditioning,* pp. 71–72, March 1974, and is reprinted with the permission of the editor.

Bibliography

Crocker-King, *Piping Handbook,* 5th ed., McGraw-Hill, New York, 1967, Section 3, pp. 133–134 and 186–187.

Hicks, T. G., *Standard Handbook of Engineering Calculations,* McGraw-Hill, New York, 1972, Section 3, pp. 377–378.

ADAM ZANKER

Sizing for Steam Traps

Undersized condensate return-lines create one of the most common problems encountered with process steam traps. Hot condensate passing through a trap orifice loses pressure, which lowers the enthalpy of the condensate. This enthalpy change causes some of the condensate to flash into steam. The volume of the resulting two-phase mixture is usually many times that of the upstream condensate.

The downstream piping must be adequately sized to effectively handle this volume. An undersized condensate return-line results in a high flash-steam velocity, which may cause waterhammer (due to wave formation), hydrodynamic noise, premature erosion, and high backpressure. The latter condition reduces the available working differential pressure and, hence, the condensate removal capacity of the steam trap. In fact, with some traps, excessive backpressure causes partial or full failure.

Due to the much greater volume of flash steam compared with unflashed condensate, sizing of the return line is based solely on the flash steam. It is assumed that all flashing occurs across the steam trap and that the resulting vapor–liquid mixture can be evaluated at the end-pressure conditions. To ensure that the condensate line does not have an appreciable pressure-drop, a low flash-steam velocity is assumed (50 ft/s) [1].

Nomograph

The nomograph (Fig. 1) quickly sizes the recommended condensate return-line.

The nomograph employs an enthalpy balance at the upstream and end-pressure conditions to calculate the weight percentage of flash steam that is formed and the flash-steam flow rate.

$$x_{fs} = \frac{(h_{l_1} - h_{l_2})}{\Delta h_{v_2}} \times 100 \tag{1}$$

$$W_v = W_l \frac{x_{fs}}{100} \tag{2}$$

The flash-steam volumetric flow rate is then determined:

$$Q_v = W_v v_{v_2} \tag{3}$$

Based on the assumed velocity, the required cross-sectional area is calculated as

$$A_{\text{req}} = \frac{Q_v}{3600 \times 50} \tag{4}$$

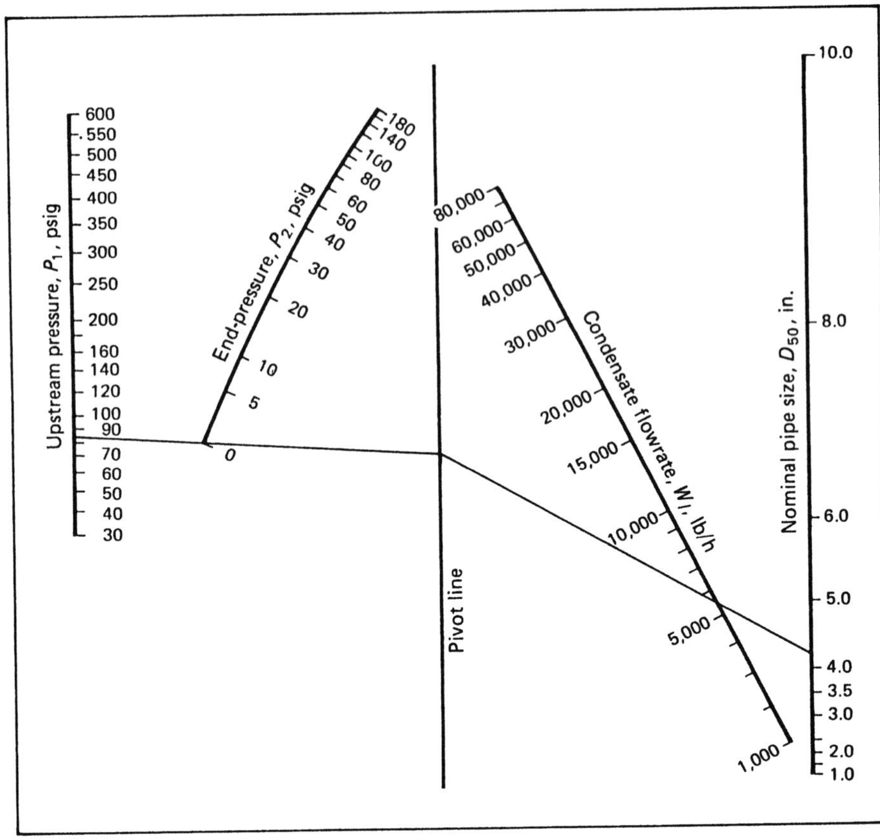

FIG. 1 This nomograph determines the recommended pipe size downstream of steam traps.

To simplify the nomograph, the flow area is converted to a nominal pipe diameter. For cases of low pressure drop or high subcooling, it may be necessary to size the condensate line based on the liquid velocity. Generally, a velocity of 3 ft/s is acceptable.

For flash-steam velocities other than 50 ft/s, the nominal pipe size may be approximated by

$$D_u = \frac{7.07 D_{50}}{\sqrt{u}} \tag{5}$$

Additional Advantages to the Nomograph

1. It yields a single result; decision-making is not required, as with other methods.
2. With minimal training, field maintenance personnel can use it.

3. Not only is it applicable to new construction but, more importantly, it can be used to check existing line sizes when trap performance is questionable.
4. It builds on the concept in Ref. 2 to provide a complete calculation procedure.

Example. An evaporator is condensing 5500 lb/h of steam at 150 lb/in.^{2}gauge supply pressure. During normal operation, a control valve maintains a pressure of 85 lb/in.^{2}gauge upstream of the steam trap. The condensate is returned to a vented tank. What line size is recommended downstream of the trap?

Solution: On the chart, connect upstream pressure of 85 lb/in.^{2}gauge through vented tank pressure (0 lb/in.^{2}gauge) to the pivot line. From pivot line, connect through condensate flow rate of 5500 lb/h to find 4 in. as the closest size.

Symbols

A_{req} required cross-sectional area, ft^2
D_u nominal pipe size, based on velocity u, in.
D_{50} nominal pipe size, based on 50 ft/s, in.
h_{l_1} condensate enthalpy at upstream pressure, P_1, Btu/lb
h_{l_2} condensate enthalpy at end-pressure, P_2, Btu/lb
Δh_{v2} latent heat of vaporization at P_2, Btu/lb
Q_v flash-steam volumetric flow rate, ft^3/h
u new flash-steam velocity, ft/s
v_{v2} flash-steam specific volume at P_2, ft^3/lb
W_l condensate formed at P_1, lb/h
W_v flash steam formed at P_2, lb/h
x_{fs} flash steam, wt.%

References

1. *Condensate Manual 2,* 3rd ed., Gestra AG, Bremen, West Germany, 1986, pp. 95–96.
2. V. K. Pathak, "How Much Condensate Will Flash?" *Chem. Eng.,* p. 89 (May 14, 1984).

MICHAEL V. CALOGERO
ARTHUR W. BROOKS

Steel Pipe Properties

It is frequently necessary to find several basic properties of steel pipes. The following formulas, chart, and nomograms may be used for that purpose.

D = outside pipe diameter, in.
d = inside pipe diameter, in.
y = distance from neutral axis to extreme fiber, $D/2$, in.
D_n = nominal pipe size that is a function of the outside pipe diameter, $f(D)$, in. (1)
A = cross-sectional area of metal part of pipe, $(\pi/4)(D^2 - d^2)$, in.^{2} (2)
a = cross-sectional area of hollow part of pipe, $(\pi/4)d^2$, in.^{2} (3)
G = weight of a foot of plain-end pipe where the density of the pipe material is taken as constant (490 lb/ft^3), $3.4A$, lb (4)
I = moment of inertia, $(\pi/64)(D^4 - d^4)$, in.^{4} (5)
z = section modulus, I/y or $\pi(D^4 - d^4)/32D$, in.^{3} (6)
K = radius of gyration, $\sqrt{I/A}$ or $(\sqrt{D^2 + d^2})/4$, in. (7)

The four nomograms provide a fast soluton of Eqs. (1) through (7). They solve these equations by making use of Table 1.

The only data necessary to use these four nomograms is the outside pipe diameter D and the inside pipe diameter d. All other data on the nomograms are functions of these two values. The use of Nomograms 1 and 2 is identical, as is the use of Nomograms 3 and 4.

TABLE 1

Nomogram Number	Pipe Properties on Nomogram	Nominal Pipe Size Range
1	1) Nominal pipe size 2) Cross-sectional area of metal part of pipe 3) Cross-sectional area of hollow part of pipe 4) Weight of one foot of pipe 5) The radius of gyration	$1/8 < D_n < 4$
2	Same 5 properties as for Nomograph 1	$4 < D_n < 24$
3	1) Nominal pipe size 2) Moment of inertia 3) Section modulus	$3/4 < D_n < 4$
4	Same 3 properties as for Nomograph 3	$4 < D_n < 24$

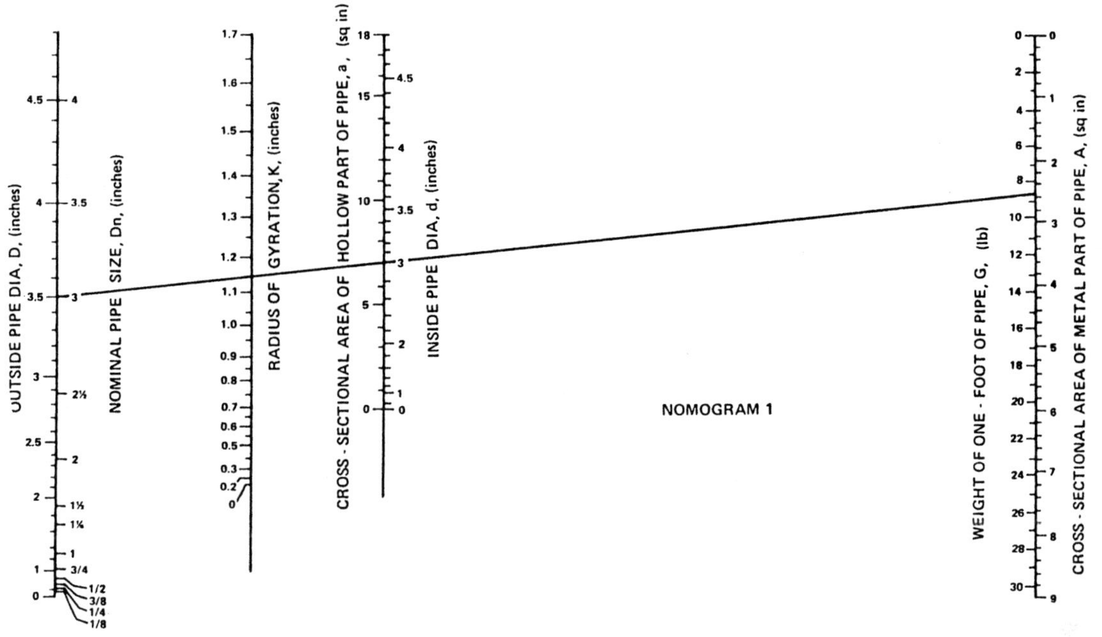

Nomograms 1 and 2

To use the nomograms, connect the known value of *D* on the outside pipe diameter scale to the known values of *d* on the inside pipe diameter scale, continuing this line to the right until it intersects the cross-sectional area of metal part of pipe scale. Now, where this line has intersected the various

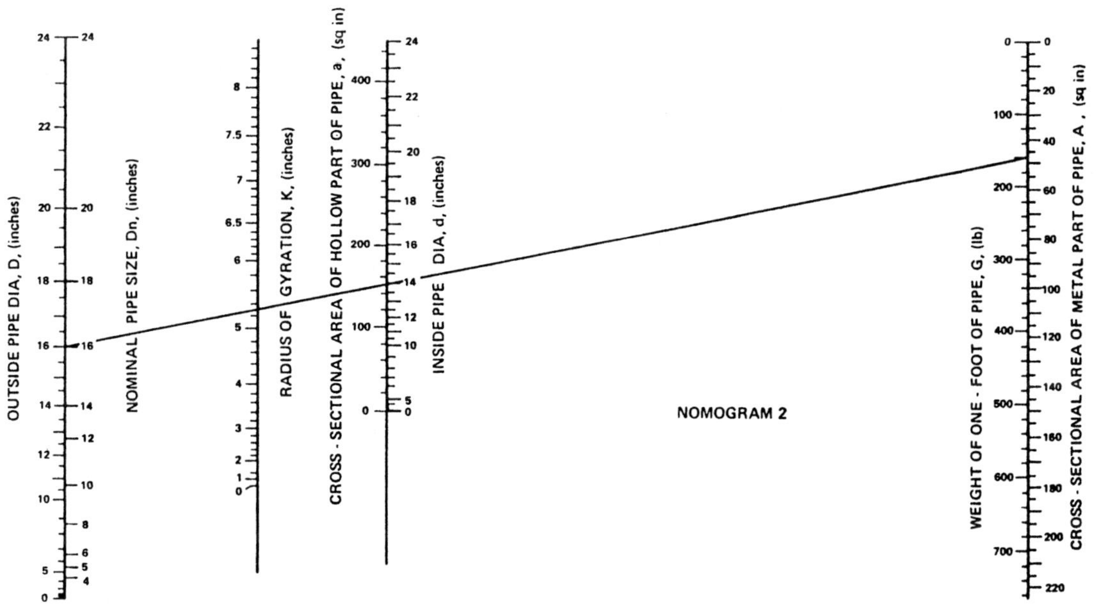

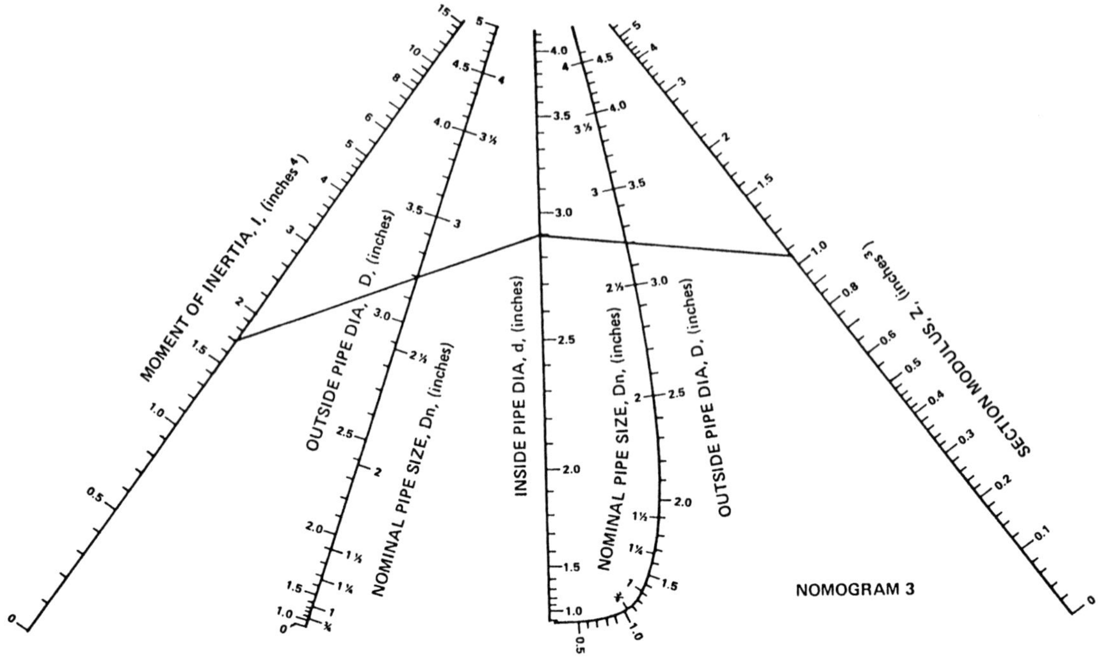

NOMOGRAM 3

vertical scales, the desired basic properties of the steel pipe can be directly read at the intersection points.

Nomograms 3 and 4

To use the nomograms, connect the known value of *d* on the inside pipe diameter scale to the known value of *D* on the outside pipe diameter scale (curved) that is in the right half of the nomogram, continuing this line to the right until it intersects the section modulus scale. The value of the section modulus and nominal pipe size can be directly read at the intersection points.

To use the nomograms, connect the known value of *d* on the inside diameter scale to the known value of *D* on the outside pipe diameter scale (straight) that is in the left half of the nomogram. Continue this line to the left until it intersects the moment inertia scale. The value of the moment of inertia and nominal pipe size can be directly read at the intersection points.

Nomogram 1 Given an outside pipe diameter of 3.5 in. and inside pipe diameter of 3.0 in., determine the following basic properties of the pipe: *a, K, G, A,* and D_n.

Solution: From the above explanation on the use of Nomogram 1, the following values are directly read from the nomogram: a = 7.05 in.^{2}, K = 1.15 in., G = 8.65 lb, A = 2.55 in.^{2}, and D_n = 3 in.

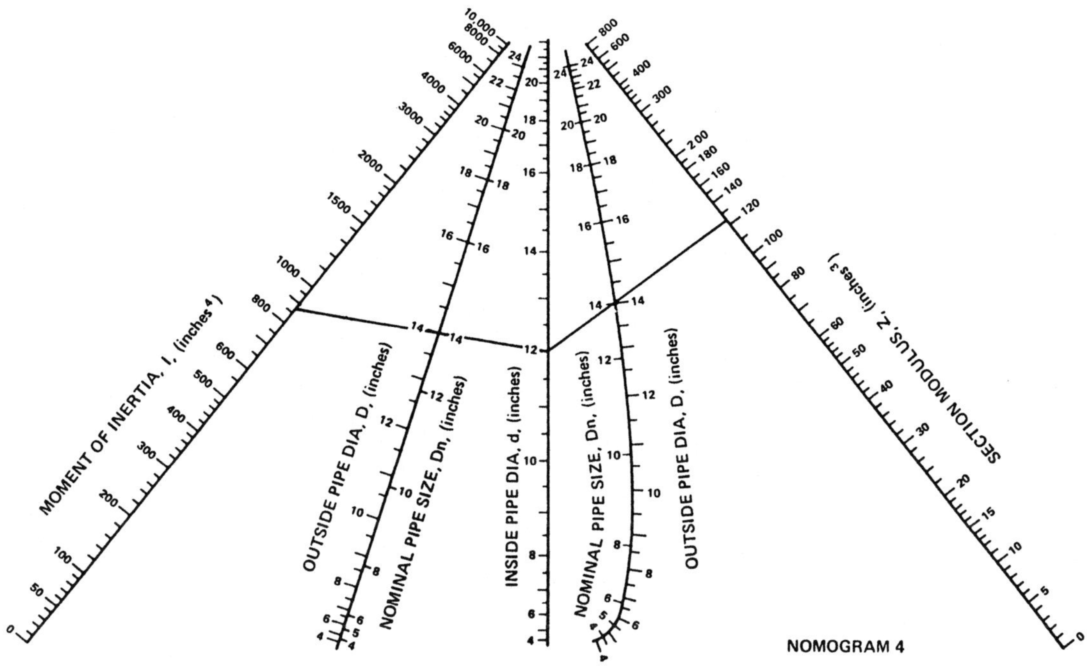

Nomogram 2 Given an outside pipe diameter of 16.0 in. and inside pipe diameter of 14 in., determine the following basic properties of the pipe: *a, K, G, A,* and D_n.

Solution: From the above explanation on the use of Nomogram 2, the following values are directly read from the nomogram: a = 155 in.^{2}, K = 5.3 in., G = 160 lb, A = 47.0 in.^{2}, and D_n = 16 in.

Nomogram 3 Given an outside pipe diameter of 3.2 in. and inside pipe diameter of 2.9 in., determine the following basic properties of the pipe: *Z* and *I*.

Solution: From the above explanation of the use of Nomogram 3, the following values are directly read from the nomogram: Z = 1.06 in.^{3} and I = 1.70 in.^{4}.

Nomogram 4 Given an outside pipe diameter of 14 in. and inside pipe diameter of 12 in., determine the following basic properties of the pipe: *Z, I,* and D_n.

Solution: From the above explanation on the use of Nomogram 4, the following values are directly read from the nomogram: Z = 124 in.^{3}, I = 870 in.^{4}, and D_n = 14 in.

ADAM ZANKER

Two-Phase Flow

When a mixture of a gas and a liquid flows along a horizontal pipe, it is possible to have up to seven different flow patterns. According to Baker [1] these flow patterns are [2]:

1. *Dispersed.* When nearly all the liquid is entrained as spray by the gas.
2. *Annular.* The liquid forms a film around the inside wall of the pipe, and the gas flows at a high velocity as a central core.
3. *Bubble.* Bubbles of gas move along at about the same velocity as the liquid.
4. *Stratified.* The liquid flows along the bottom of the pipe and the gas flows above over a smooth gas–liquid interface.
5. *Wave.* Is similar to stratified except the interface is disturbed by waves moving in the direction of flow.
6. *Slug.* Waves are picked up periodically in the form of frothy slugs that move at a much greater velocity than the average liquid velocity.
7. *Plug.* Alternate plugs of liquid and gas move along the pipe.

Baker [1] has correlated the seven flow patterns by means of two numbers, the "Baker parameters," B_x and B_y. Thus the B_x, B_y plane shows seven regions, one for each flow pattern. Although the borders of the regions are drawn as lines, they are really broad transition zones.

Determination of Two-Phase Flow by Computer

In a recent paper Soliman [3] provided a BASIC program to compute two-phase pressure drop based on the excellent work of Kern [2]. To determine the pattern of two-phase flow, Soliman reduced the Baker graph to four regions instead of the seven found experimentally. This makes the computer programming a great deal easier. However, we think that this is an over-simplification.

We chose to approximate the boundaries of the Baker graph by eight straight lines, as is shown in Fig. 1. The equations of those lines are:

$$\text{C1: } \log B_y = 3.698 - 0.163 \log B_x$$

$$\text{C2: } \log B_y = 4.261 - 0.642 \log B_x$$

$$\text{C3: } \log B_y = 4.959 - 0.410 \log B_x$$

$$\text{C4: } \log B_y = 4.477$$

$$\text{C5: } \log B_y = 4.019 - 0.241 \log B_x$$

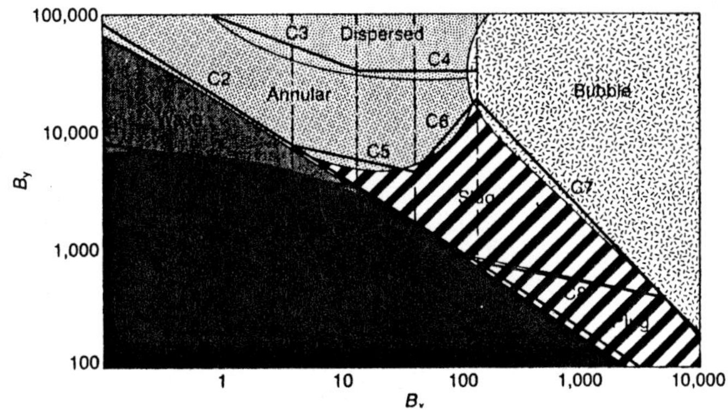

FIG. 1 Program's approximation of the Baker graph.

$$\text{C6: } \log B_y = 1.935 + 1.057 \log B_x$$

$$\text{C7: } \log B_y = 6.527 - 1.072 \log B_x$$

$$\text{C8: } \log B_y = 3.301 - 0.197 \log B_x$$

To simplify the computer programming we divided the range of B_x into the following five zones:

1. $0.1 < B_x > 4.013$
2. $4.013 < B_x > 15.00$
3. $15.000 < B_x > 40.322$
4. $40.322 < B_x > 143.51$
5. $143.51 < B_x > 10{,}000$

The Baker parameters have the following expressions:

$$B_x = K_1(W_L/W_G)(\sqrt{\rho_L \rho_G}/\rho_L^{2/3})(\mu_L^{1/3}/\sigma_L)$$

$$B_y = K_2W_G/(\pi D^2/4) \sqrt{\rho_L \rho_G}$$

where W = flow rate
ρ = density
μ = viscosity
σ = liquid surface tension
D = internal diameter of pipe
L = liquid
G = gas

K_1 and K_2 are constants that involve the conversion factors of units. Two sets are included in our program: $K_1 = 2.1039$ and $K_2 = 25{,}524.62$ when the units of the data are metric:

W = kg/s

TABLE 1

```
10   REM PROGRAMA "BAKER" EN CINTA 01-F
11   REM   IPAKO   ING.DE PROCESO   -1984-
12   REM
20   REM DEFINES THE PRINTER
30   PRINTER IS 7,1,WIDTH(76)
40   OPTION BASE 1
50   REM
60   REM DATA INPUT
70   REM
80   INPUT "ENGLISH UNITS? (Y/N)",S$
90   INPUT "LIQUID FLUW RATE?",G1
100  INPUT "GAS FLOW RATE?",G2
110  INPUT "LIQUID DENSITY?",R1
120  INPUT "GAS DENSITY?",R2
130  INPUT "LIQUID VISCOSITY?",V1
140  INPUT "LIQUID SURFACE TENSION?",T
150  INPUT "PIPE DIAMETER?",D
160  REM
170  REM DATA PRINTING
180  REM
190  U$=" METRIC UNITS"
200  IF S$="Y" THEN U$=" ENGLISH UNITS"
210  PRINT U$,LIN(2)
220  IF S$="Y" THEN 290
230  PRINT "Liquid flow rate(Kg/s):........   ";G1
240  PRINT "Gas flow rate(Kg/s):...........   ";G2
250  PRINT "Liquid density(Kg/m3):.........   ";R1
260  PRINT "Gas density(Kg/m3):............   ";R2
270  PRINT "Pipe diameter(m):.............  ";D
280  GOTO 340
290  PRINT "Liquid flow rate(lb/h):........   ";G1
300  PRINT "Gas flow rate(lb/h):...........   ";G2
310  PRINT "Liquid density(lb/ft3).........   ";R1
320  PRINT "Gas density(lb/ft3):...........   ";R2
330  PRINT "Pipe diameter(inch):...........   ";D
340  PRINT "Liquid viscosity(cp):..........   ";V1
350  PRINT "Liquid surface tension(dynes/cm)  ";T,LIN(2)
360  PRINT "RESULTS"
370  PRINT LIN(1)
371  REM
380  REM  DEFINES SYSTEM OF UNITS
381  REM
390  IF S$="Y" THEN 450
400  T=T/1000
410  V1=V1/1000
420  K1=2.1039
430  K2=25524.617
440  GOTO 480
450   K1=531
460  K2=2.16
470  D=D*.08333
480  A=PI*D*D/4
490  X=K1*(G1/G2)*((R1*R2)^.5/R1^.667)*(V1^.333/T)
500  Y=K2*G2/(A*(R1*R2)^.5)
510 IF X)10000 THEN 1010
520 IF X(.1 THEN 1010
530 IF Y)100000 THEN 1010
540 IF Y(100 THEN 1010
550  L3=LGT(X)
560  L4=LGT(Y)
570  C1=3.698-.163*L3
580  C2=4.261-.642*L3
590  C3=4.959-.410*L3
600  C4=4.477
610  C5=4.019-.241*L3
620  C6=1.935+1.057*L3
630  C7=6.527-1.072*L3
640  C8=3.301-.197*L3
650  IF X)15 THEN 730
660  IF L4(C1 THEN 870
670  IF L4(C2 THEN 890
680  IF L4)C3 THEN 910
690  IF X)4.013 THEN 710
700  GOTO 950
710   IF L4(C5 THEN 930
720  GOTO 950
730  IF L4(C2 THEN 870
740  IF X)143.51 THEN 810
750  IF L4)C4 THEN 910
760  IF X)40.322 THEN 790
770  IF L4)C5 THEN 950
780  GOTO 930
790  IF L4)C6 THEN 950
800  GOTO 930
810 IF L4)C7 THEN 970
820  IF L4)C8 THEN 930
830  GOTO 990
840  REM
850  REM RESULTS
860  REM
870  PRINT "ESTRATIFIED"
880  GOTO 1020
890  PRINT "WAVE"
900  GOTO 1020
910  PRINT "DISPERSE"
920  GOTO 1020
930 PRINT "SLUG"
940  GOTO 1020
950  PRINT "ANNULAR"
960  GOTO 1020
970 PRINT "BUBBLE"
980  GOTO 1020
990  PRINT "PLUG"
1000 GOTO 1020
1010 PRINT "DATA OUTSIDE BAKER GRAPH"
1020 PRINT "X:.............................   ";X
1030 PRINT "Y:.............................   ";Y
1031 PRINT LIN(4)
1040 END
```

μ = cP
D = m
ρ = kg/m^3
σ = dyn/cm

K_1 = 513 and K_2 = 2.16 when the units of the data are English:

W = lb/h
μ = cP
D = ft
ρ = lb/ft^3
σ = dyn/cm

The Computer Program

Table 1 is a listing of the computer program written in BASIC. It has two options for the input of the data: metric or English units. Table 2 is an example of the program output illustrating one zone in the Baker graph. We think that the errors in using our approximations are of the same magnitude as the inherent errors of the original Baker graph.

TABLE 2

```
METRIC UNITS

Liquid flow rate(Kg/s):........           7.488
Gas flow rate(Kg/s):...........           1.176
Liquid density(Kg/m3):.........           499.888
Gas density(Kg/m3):............           29.637
Pipe diameter(m):..............          .102
Liquid viscosity(cp):..........           .11
Liquid surface tension(dynes/cm)          5.07

                        RESULTS
BUBBLE
X:.............................           244.886469269
Y:.............................           30180.2515474

ENGLISH UNITS

Liquid flow rate(lb/h):........           59428.57
Gas flow rate(lb/h):...........           9333.333
Liquid density(lb/ft3).........           31.204
Gas density(lb/ft3):...........           1.85
Pipe diameter(inch):...........           4.01967
Liquid viscosity(cp):..........           .11
Liquid surface tension(dynes/cm)          5.07

                        RESULTS
BUBBLE
X:.............................           244.839562172
Y:.............................           30111.0615956
```

Although the program was originally written for a HP-9835 and IBM PC, we rewrote it in a version simple enough to run in most personal computers. It is easily converted to a subroutine by changing the GOTO 1020's to RETURN's.

This material appeared in *Hydrocarbon Processing*, pp. 46–47, December 1986, copyright © 1986 by Gulf Publishing Co., Houston, Texas 77252, and is reprinted by special permission.

References

1. O. Baker, *Oil Gas J.*, July 26, 1954.
2. R. Kern, *Hydrocarbon Process.*, *48*(10), 105–116 (1969).
3. R. Soliman, *Hydrocarbon Process.*, *63*(4), 155–157 (1984).

C. E. YAMASHIRO
L. G. SALA ESPIELL
I. H. FARINA

Weight of Piping

The weight of plain end pipe may be easily calculated with

$$W = 0.0218(Dx - x^2)L\rho$$

where W = the weight of plain ended pipe, lb
D = the specified (real) outside pipe diameter, in.
x = the pipe wall thickness, in.
L = the length of pipe, ft
ρ = the density of pipe material, lb/ft^3

The nomograph (Fig. 1) solves this equation. On the opposite side of scale D (specified outside pipe diameter) are plotted values of standard pipe

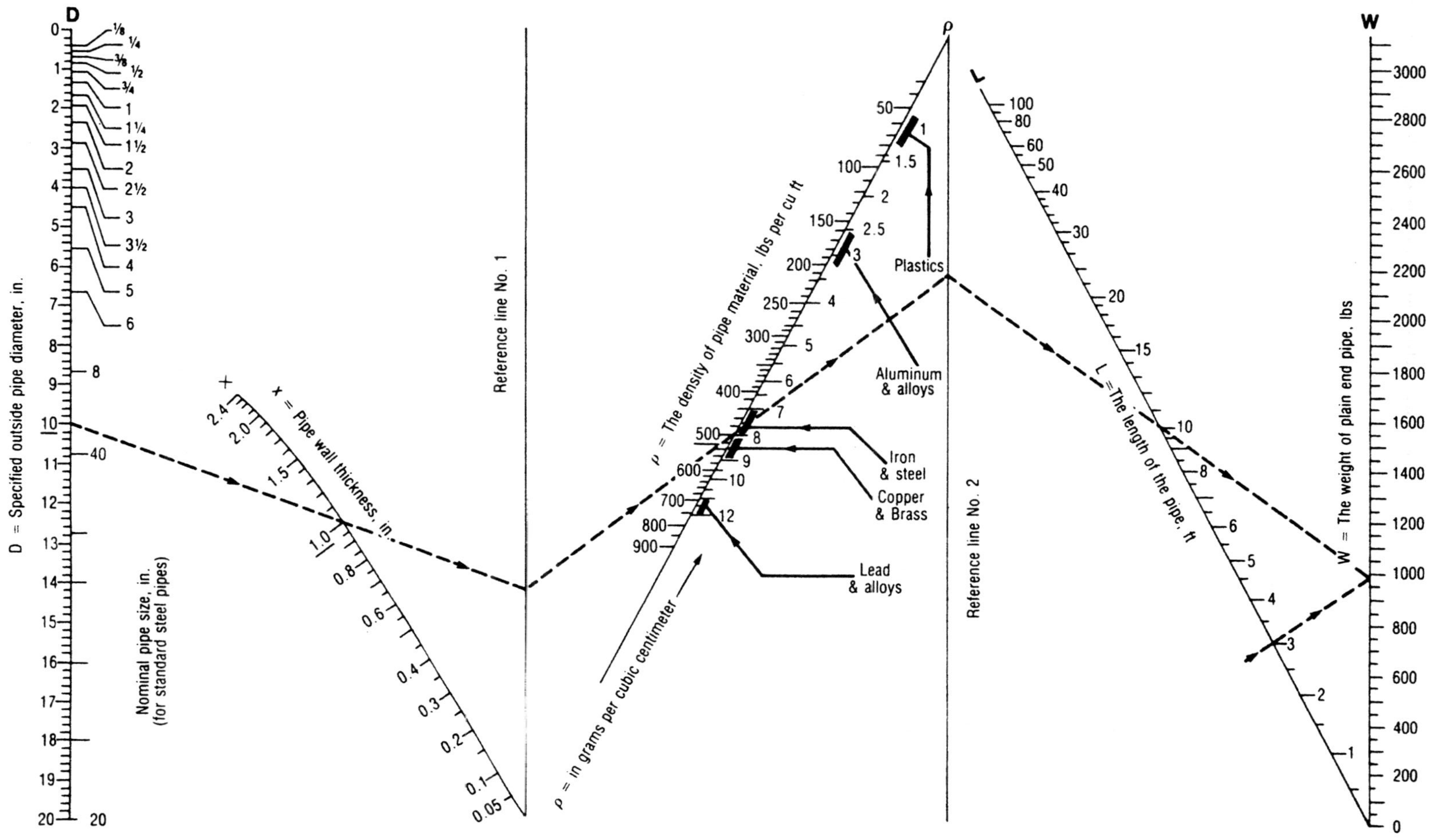

FIG. 1 This nomograph determines the weight of pipe. The broken lines indicate the answer to the example problem presented in the text; the numbers indicate the order of steps in the solution.

sizes; therefore, it is not necessary to know the real outside pipe diameter, just the standard size.

The scale of pipe material density, ρ, is graduated on both sides: one in lb/ft^3 and the other in g/cm^3. Either can be used. This scale is also marked with density ranges for some common pipe materials.

Example. What is the weight of a pipe where D equals 10 in., x equals 1 in., L equals 10 ft, and ρ equals 500 lb/ft^3?

Solution: Extend a line from 10 on the D scale through 1 on the x scale to Reference line No. 1. From that point, extend a line through 500 on the ρ scale to Reference line No. 2. From that point, extend a line through 10 on the L scale to the W scale. Read the answer as 982 lb at the point of intersection on the W scale.

This material appeared in *Heating/Piping/Air Conditioning,* pp. 75–76, October 1976, and is reprinted with the permission of the editor.

ADAM ZANKER

8

Pipeline Operation and Maintenance

Friction Reduction

Part A: The Basics of Drag Reduction

Drag reducing agents (DRA's) can substantially reduce friction loss in most pipelines when flowing most hydrocarbon liquids.

This article is in four parts which will review DRA's, their fundamentals and applications, role in active and passive drag, and incidence as specific products and hardware.

When a DRA is dissolved in a solvent, it produces a solution which in laminar flow has the same pressure drop, and therefore the same Newtonian viscosity, as the solvent. But when the solution is in turbulent flow, a DRA produces a pressure drop smaller than that which would occur with untreated solvent moving at the same flow rate. The turbulent flow friction of the solution is less than that of the solvent.

DRA's are highly viscoelastic and thixotropic. They may have a viscosity at low shear rates in excess of 3,000,000 cP while having a viscosity 100–1000 times less at higher rates. The viscosity is somewhat affected by temperature and increases with decreasing temperature. DRA–solvent solutions are viscoelastic, time-independent, shear degradeable, non-Newtonian fluids. A DRA–solvent solution handles as an ordinary hydrocarbon except in turbulent flow when the reduced friction becomes evident.

DRA's in present use in oil and products pipelines are themselves hydrocarbons and thus should have no effect on refining processes or refined products. Millions of barrels of crude treated with DRA's on the Alyeska system have been refined without incident. Drag reduction installations need not involve large diameter pipelines. The 200,000 b/d increase in the Alyeska system resulting from the injection of some thousands of gallons per day of DRA in this 48-in. line is dramatic.

However, the offshore operator who obtains a 20% increase in capacity in his 8-in. flow line by injecting a few dozen gallons of DRA per day is happy: Without the DRA, he would have to shut in production or loop the flow line. Either alternative would be orders of magnitude more expensive than the injection of a DRA.

And, as production decays over time, the amount of DRA injected will be reduced with the eventual result that production will match the untreated flow line capability and operations can proceed as before but with no shut in wells in past history and no sterilized investment in a loop to be carried on the books of account in future years.

Early Applications

The drag reduction effect as applied to solutions was accidentally discovered by Toms in the UK in 1947 [1]. It had been known for years that concentrated water–coal slurries and wood pulp–water slurries can produce a friction factor lower than that of water alone. But Dr. Toms's 1948 publication was the first indication of drag reduction in otherwise Newtonian solutions.

The first use of DRA's in the oil industry was to reduce pressure loss during pumping of fluids downhole to fracture tight formations. The quantities of DRA used are proportionally large but the savings are substantial. Data published on one DRA–kerosene solution with 600 weight ppm of DRA in kerosene showed an 80% drag reduction over untreated kerosene.

The ongoing use of DRA's to debottleneck the Alyeska 48-in. crude line has been described in several published articles. There are many other uses on commercial pipelines which have not yet been described in published material—one considerably larger than the Alyeska example—and the list of operators using DRA is growing. Use of DRA's in multistation pipelines is a special application because the drag reduction effect does not survive shear stresses raised by centrifugal pumps and appears to be badly degraded going through most positive displacement pumps.

The general rule is that DRA's are always injected on the discharge side of a pump station. If drag reduction is required all along a pipeline with N pump stations, the DRA must be injected at N points.

The effectiveness of a DRA depends to a certain extent on the viscosity of the untreated oil and decreases as viscosity increases. It varies slightly with diameter and decreases as diameter increases. And the effect varies greatly with flow velocity which, for a given oil and line diameter, means it varies greatly with Reynolds number. Drag reduction increases as Reynolds number increases.

However, dilute solutions are believed to disintegrate at shear rates found in commercial pipelines.

There are some as yet little understood practical aspects of DRA application. For instance, there seems to be some relationship between the physical characteristics of the oil and of the DRA. A certain DRA will produce good results in a given crude yet produce poor results in a seemingly similar oil.

Untreated oils which are non-Newtonian may produce non-Newtonian DRA–oil solutions. And a particular DRA may increase, rather than decrease, flowing friction. Sometimes an additional, specially compounded additive is required to obtain the desired drag reduction. And there appears to be something in certain DRA–solvent combinations which has to do with the solubility of the DRA in the oil. Test loop studies verify that for these combinations, drag reduction appears to increase asymptotically as the distance from the point of injection increases. Specially designed DRA injection nozzles help in these cases.

DRA Activity

The precise mechanism describing how a DRA works to reduce friction is not established. It is believed these agents work by directly reducing turbulence or by absorbing and later returning to the flowing stream energy which otherwise would have been wasted in producing the crossflows which comprise turbulence. The absorb-and-return theory has a certain intuitive attractiveness because one of the characteristics of a viscoelastic fluid is its ability to do this on a physically large scale. Whether the effect exists at the molecular scale is unknown.

Pipeliners generally work with the rational flow formula. The friction factor therein, which is a scaling factor applied to velocity head to yield friction loss, is a macro-factor in that it lumps all of the friction-producing effects acting in a flowing system in a single numerical value. The friction factor concept does not provide an insight into how a DRA might work. It is necessary to look at the velocity distribution pattern in a pipeline in turbulent flow to understand the bases which support the concept of the f-factor and understand better how drag reduction might take place.

The concept of turbulent flow in pipelines provides that the point flow may be directed in any direction while maintaining net pipe flow in the direction of decreasing pressure. All flow not in the direction of net pipe flow absorbs energy; the more turbulent the flow becomes, the more energy absorbed in these crossflows.

In flow in hydraulically rough pipes, the f-factor does not vary with increasing flow, but the unit friction loss increases as the square of the velocity of net pipe flow. The model of turbulent pipe flow we will use is based on the Universal Law of the Wall which applies to both smooth and rough pipe flow. It presumes two general modes of flow in a pipe: the wall layer and the turbulent core.

The wall layer is presumed to contain all of the flow where the point velocity varies with distance of the point from the pipe wall. The turbulent core carries all of the flow where variations in point velocity are random and independent of this distance.

The wall layer itself is presumed to be comprised of three sublayers: the laminar sublayer, the buffer zone, and the turbulent sublayer.

The laminar sublayer has as its outer boundary the wall of the pipe, and flow in the entire sublayer is laminar. This definition requires that a film of liquid which is not moving lies immediately at the wall.

The increase in point velocity as the point moves away from the wall is a linear function of the distance from the wall and directed parallel to the wall in the direction of pipe flow. There are no crossflows in this sublayer. The contribution of the laminar sublayer to total friction arises in viscous shear generated in the sublayer.

Pipe roughness acts to inhibit laminar flow; therefore, the definition is only statistically true for rough or partially rough pipe.

In the turbulent sublayer, which has as its inner boundary the turbulent core, it is presumed that the velocity varies logarithmically with distance from the wall but that crossflows in the sublayer may be statistically large and nearly as great as in the core. The contribution of the turbulent sublayer to total friction lies in the turbulent shear generated in this sublayer.

In the buffer zone, which lies between the laminar and turbulent sublayers, variation of point velocity with point position is not estabished. Ideally, it should be the same as the laminar sublayer at the outer boundary and the same as the turbulent sublayer at the inner boundary.

Any number of mathematical functions can be written to approximate these two requirements. Both viscosity and density contribute to shear losses in the buffer zone.

The contribution of the turbulent core to total friction loss lies entirely in turbulent shear; DRA's are presumed not to act on this component of total friction. Therefore the action of a DRA must take place in the laminar sublayer, the buffer zone, or the turbulent sublayer.

That the action of a DRA increases with increasing flowing velocity would indicate that DRA action could not be primarily in the laminar sublayer. And it sometimes would appear that this is so: Liquids treated with a DRA flowing under conditions in which the pipe is hydraulically rough and viscosity presumed to make no contribution to total friction can yield excellent drag reduction. This kind of result would lead to the presumption that a DRA works in the buffer zone.

On the other hand, large doses of a DRA have been observed to keep a liquid in laminar flow at Reynolds numbers much higher than those expected of the same liquid without DRA, which leads to the conclusion that a DRA acts in the laminar sublayer.

This conclusion suggests that a DRA may inhibit the action of crossflows, whether generated in the laminar sublayer or the buffer zone, which move inward into the turbulent sublayer and core and absorb energy which otherwise would go toward producing pipe flow. It appears this is the case.

Reduction in the energy absorbed by crossflows originating at the pipe wall can be accomplished by reducing their number, size, or velocity. Or, the energy in the crossflows may be absorbed by a DRA and later returned to the flowing stream for reuse which, in effect, means reducing their effect if not their size, number, or velocity. And it is here that theory breaks down; no one really knows which effect predominates—if either does. What is known is:

That DRA's act to reduce turbulence.

That to do so, they seem to act in the laminar sublayer or the buffer zone by effectively enlarging them or their effect.

That the net effect is to reduce total friction.

That the *f*-factor is less than would be calculated from the pipeliner's usual sources: Experimental data taken on liquids flowing without a DRA do not translate directly to liquids which have been dosed with a DRA. The rational flow formula still applies but the usual f-factor relationship does not.

Therefore, the concepts of drag reduction and the drag ratio have entered the pipeliner's world.

Reduction Fundamentals

Use of DRA's requires introducing a few equations (Table 1). In the following discussion, we will use the terms treated and untreated to refer to flowing liquids which have or have not been dosed with a DRA.

By common consent, drag reduction (DR) is measured in terms expressed in Eq. (1). (Equations 1–7 are presented in Table 1.) By definition, since

TABLE 1 Equations (1)–(7)

$$\mathrm{DR} = (S_{f\,\mathrm{untreated}} - S_{f\,\mathrm{treated}})/S_{f\,\mathrm{untreated}} \tag{1}$$

where S_f = unit friction loss taken as a slope

$$D/R = (1 - \mathrm{DR}) = (S_{f\,\mathrm{treated}})/(S_{f\,\mathrm{untreated}}) \tag{2}$$

$$f_{\mathrm{treated}} = f_{\mathrm{untreated}} \times D/R \tag{3}$$

$$S \times (1 + A_0 \times ([2^{3/2} \times \log_{10} S + 1.454 \times G_0 \times T_p \times S - 0.8809])^2 = 1 \tag{4}$$

where $S = D/R$

$A_0 = (f_{\mathrm{untreated}}/8)^{1/2}$

G_0 = wall shear rate$_{\mathrm{untreated}}$

T_p = characteristic time of the treated DRA–solvent solution

$$T_p = a \times P^b \times G_p^c \times A_0^d \tag{5}$$

where P = DRA concentration

G_p = wall shear rate$_{\mathrm{treated}}$

a, b, c, d = constants empirically determined for each DRA–solvent solution

$$\mathrm{ppm} = A + B \times (D/R) + C \times (D/R)^2 + D \times (D/R)^3 + E \times (D/R)^4 \tag{6}$$

$$1/\mathrm{DR} = A \times (1/\mathrm{ppm}) + B \tag{7}$$

$S_{f\,\text{untreated}}$ is larger than $S_{f\,\text{treated}}$, DR is always less than unity. DR is often expressed in percent, or $DR_{pct} = 100 \times DR$. Another commonly used factor (Eq. 2), and more useful to pipeliners, is the drag ratio (D/R) which is (1 − DR). A drag reduction of 0.30, or 30%, for example, yields a drag ratio of 0.70, or 70%.

Knowing D/R, one knows the expected f-factor (Eq. 3). This f-factor can be used in any of the rational flow formulas.

We need now to be able to calculate drag reduction itself. The generally accepted form of the drag reduction correlation is based on an extension of the model of turbulent viscoelastic flow of Savins and Seyer as expressed by Burger in Eq. (4) [2].

The characteristic time T_p is specific to a specific DRA–solvent solution. A general form of the characteristic time equation is expressed in Eq. (5).

Up-to-date values for the empirical constants are generally not available in the literature for two reasons. The science of drag reduction is one of the fastest moving disciplines of industrial chemistry, and what is published today is out of date tomorrow. Also, there is a natural reluctance of a producer to make available to his competitors the technical characteristics of his product.

The data used here are based on one formulation of FLO Pipeline Booster produced by ChemLink Inc., a subsidiary of Atlantic Richfield Co., with permission of the producer.

Solving Equations

The Savins and Seyer correlation, taking into account the time constant equation, has been programmed in several languages for several different computers.

Program CBLFLO (Fig. 1) is in HP BASIC for the HP 9845 T. Aside from the use of the REPEAT-UNTIL and SELECT CASE statements, both of which can be replaced with IF-THEN-ELSE statements, the program should run with appropriate I/O statements on almost any BASIC interpreter.

It is heavily commented, uses descriptive names for variables compatible with this text, and should be readable by anyone familiar with most versions of BASIC.

Figure 2 is a plot of parts per million (ppm) of DRA against drag ratio from $D/R = 0.95$ to $D/R = 0.50$. Using Eq. (2), we can derive the ppm-vs-drag reduction curve shown in Fig. 3 for a DR of 0.05 to 0.50. Both curves are of the form expressed in Eq. (6).

Any of the curve-fitting programs will develop the constants given a minimum of five values for ppm–D/R pairs developed using CBLFLO.

```
1000  ! ************************  Program CBLFLO  *****************************
1010  ! CBLA Longview                                                 08 May 84
1020                                        !
1030                                        !
1040  ! Program calculates gpd of hydrocarbon copolymer drag reducer required
1050  ! to produce a given drag reduction in a liquid hydrocarbon pipeline.
1060                                        !
1070  ! The a,b,c,d factors in the characteristic time equation must be
1080  ! entered into the program as evaluated for the drag reducer and pipeline
1090  ! fluid under study.  Otherwise erroneous answers may result.  Enter
1100  ! these factors in the four lines following label Abcd_parameters.
1110                                        !
1120  ! Combined coefficients are CBLA standards so that program finds the same
1130  ! answers for Fo, Sfo and Sfp as other CBLA programs. Conversions from
1140  ! input data units to fps units use base conversion factors directly
1150  ! instead of compound conversion factors to facilitate rewriting the
1160  ! program into other systems of units.
1170                                        !
1180  ! The fundamental algorithm is the Burger interpretation of the
1190  ! Savins and Seyer model of viscoelastic flow written in the form f(x)=1.
1200  !
1210  ! CBLFLO and FLOCBL are mirror programs; output from the first program
1220  ! inputted to the second program yields the input to the first program.
1230                                        !
1240                                        !
1250  !              **********  INPUT ROUTINE  **********
1260                                        !
1270                                        !
1280  INPUT "INPUT DENSITY, VIS CST, DR PERCENT, OD IN, WT IN, ROUGH IN,
      FLOW BPD      ",Den_15,Vis_cst,Dr_pct,Od_in,Wt_in,Rough_in,Q_bpd
1290                                        !
1300                                        !
1310  !          **********  SET UNITS TO FPS SYSTEM  **********
1320                                        !
1330                                        !
1340 Start:                                 ! entry point for re-run using a
1350                                        ! new value for a variable entered
1360                                        ! from the keyboard after the
1370                                        ! program has been run once with
1380                                        ! INPUTTED data.
1390                                        !
1400  Id_in=Od_in-2*Wt_in                   ! pipe id inches
1410  Id_ft=Id_in/12                        ! pipe id feet
1420  A_sqft=PI*(Id_ft/2)^2                 ! pipe area sq ft
1430  Ub_fps=42*231/1728*Q_bpd/(24*3600)/A_sqft   ! bulk velocity ft/sec
1440  Vis_sqftps=Vis_cst*1.076391E-5        ! kinematic viscosity sq ft/sec
1450  Nr=Ub_fps*Id_ft/Vis_sqftps            ! Reynolds number
1460                                        !
1470                                        !
1480 ! **********  USE COLEBROOK-WHITE CORRELATION FOR F-FACTOR  **********
1490                                        !
1500                                        !
1510  C2=Rough_in/(3.7*Id_in)               ! 1st factor in C-W correlation
1520  C3=2.51/Nr                            ! 2nd factor in C-W correlation
1530  T1=10                                 ! initialize transmission factor
1540  REPEAT                                ! start loop for C-W
1550    T2=-2*LGT(C2+C3*T1)                 ! C-W correlation
1560    T1=(T1+T2)/2                        ! refine estimate for T
1570  UNTIL T1-T2<.00001                    ! loop exit condition
1580  Fo=1/(T2*T2)                          ! f = 1/(transmission factor)^2
1590                                        !
1600                                        !
1610 ! **********   COMPUTE VARIABLES FOR BERGER CORRELATION  **********
1620                                        !
1630                                        !
1640  Gammao=Fo*Ub_fps^2/8/Vis_sqftps       ! wall shear rate w/o drag reducer
1650                                        ! = wall shear stress/viscosity
1660                                        ! = (f*Ub^2/8)/vis
1670                                        !
1680  Alphao=SQR(Fo/8)                      ! Alphao= friction vel/bulk vel
1690                                        !        = U_star/Ub
1700                                        !        = sqr(f/8)
1710                                        !
1720  Sigma=1-Dr_pct/100                    ! Sigma = drag ratio as a decimal
1730                                        !
1740                                        !
1750 !        **********  SOLVE BERGER CORRELATION  **********
1760                                        !
1770                                        !
1780 ! Burger's correlation is:
1790                                        !
1800 ! Sigma*[1+Alphao*(2^1.5*LGT(Sigma)+1.454*Gammap*Thetap-0.8809)]^2=1
1810 ! However, Gammap = Sigma*Gammao, by definition, so in terms of Gammao
1820 ! Sigma*[1+(Alphao*(2^1.5*LGT(Sigma)+1.454*Gammao*Thetap*Sigma-0.8809))]^2=1
1830                                        !
1840 ! It can be solved for Thetap, the characteristic time of the viscoelastic
1850 ! solution comprised of the crude oil solvent and the drag reducer polymer.
1860                                        !
1870  Numerator=.8809+(Sigma^(-.5)-1)/Alphao-2^1.5*LGT(Sigma)
1880  Denominator=1.454*Gammao*Sigma
1890  Thetap=Numerator/Denominator          ! characteristic time
1900                                        !
1910                                        !
```

FIG. 1 CBLFLO computes DRA for desired ratio.

```
1920 !  **********  SOLVE CHARACTERISTIC TIME EQUATION FOR PPM  **********
1930                                               !
1940                                               !
1950 ! Given the characteristic time (Thetap) the dosage rate Phi can be found
1960 ! by solving Thetap = A*Phi^B*(Gammao*Sigma)^C*Alpha^D for Phi.
1970                                               !
1980 ! A, B, C and D are set by the drag reducer manufacturer based on tests
1990 ! or other input and take into account the kind of pipeline oil and the
2000 ! characteristics of the drag reducer chemical.
2010                                               !
2020 Abcd_parameters:                              ! entry for inserting a,b,c,d
2030  A=1                                          !
2040  B=1                                          !
2050  C=1                                          !
2060  D=1                                          !
2070                                               !
2080  Phi=(Thetap/(A*(Sigma*Gammao)^C*Alphao^D))^(1/B)  ! ppm of drag reducer
2090                                               !
2100                                               !
2110 !          **********  COMPUTE GPD OF DRAG REDUCER  **********
2120                                               !
2130                                               !
2140  Gpd=Phi*Q_bpd*42*231/1728*62.371718*Den_15/6.59/1E6   ! y = f(x)
2150                                               ! density of drag reducer at
2160                                               ! 15 deg C taken as 6.59 lbs/gal
2170                                               !
2180                                               !
2190 ! ***** COMPUTE UNTREATED AND TREATED UNIT FRICTION LOSSES  *****
2200                                               !
2210                                               !
2220  Sfo=2.647271E-2*Fo*Q_bpd^2/Id_in^5           ! Sf mlc/km or flc/kft(untreated)
2230                                               ! 2.647271E-2 is CBLA combined
2240                                               ! coefficient for Sf 1/1000 base
2250                                               !
2260  Sfp=(1-Dr_pct/100)*Sfo                       ! Sf mlc/km or flc/kft(treated)
2270                                               !
2280                                               !
2290  !            **********  OUTPUT ROUTINE  **********
2300                                               !
2310                                               !
2320  PRINTER IS 0                                 ! set output to printer
2330                                               !
2340  IMAGE "Q   BPD  DR PCT  OD IN  WT IN  VIS CST f w/o FLO     MLC/KM
PPM    GPD  "
2350  PRINT USING 2340
2360  IMAGE "                                                   w/o DRA  w/DRA"
2370  PRINT USING 2360
2380  PRINT
2390  IMAGE DDDDDDD,XX,DDD.DD,XX,DD.DD,XX,Z.DDD,XXX,DD.DD,XXX,XZ.DDDD,XXX,DD.DDD
,X,DD.DDD,XX,DDD.D,XX,DDDDD
2400  PRINT USING 2390;Q_bpd,Dr_pct,Od_in,Wt_in,Vis_cst,Fo,Sfo,Sfp,Phi,Gpd
2410                                               !
2420  PRINTER IS 16                                ! reset output to screen
2430                                               !
2440                                               !
2450  PRINT PAGE                                   ! clear screen
2460                                               !
2480  !            **********  EXIT ROUTINE  **********
2500                                               !
2510  PRINT "YOU MAY CHANGE ANY INPUT PARAMETER AND RE-RUN WITH THE NEW VALUE"
2520  PRINT
2530  PRINT "YOU MAY NOW QUIT THE PROGRAM OR ENTER A NEW VALUE FOR A VARIABLE"
2540  PRINT
2550  PRINT "FOR READY REFERENCE, THE VARIABLES AVAILABLE FOR CHANGE ARE THE FOL
LOWING"
2560  PRINT
2570  PRINT
2580  PRINT "        (1)  DENSITY                     Den_15"
2590  PRINT "        (2)  VIS IN CST                  Vis_cst"
2600  PRINT "        (3)  DRAG REDUCTION PERCENT      Dr_pct"
2610  PRINT "        (4)  OD INCHES                   Od_in"
2620  PRINT "        (5)  WT INCHES                   Wt_in"
2630  PRINT "        (6)  ROUGHNESS INCHES            Rough_in"
2640  PRINT "        (7)  FLOW RATE IN BPD            Q_bpd"
2650  PRINT
2660  PRINT
2670  PRINT "     TO QUIT THE PROGRAM ENTER 0,0 AND HIT CONT"
2680  PRINT
2690  PRINT "     TO CHANGE THE VALUE OF A VARIABLE ENTER ITS NUMBER FROM THE AB
OVE TABLE"
2700  PRINT "     AND ITS NEW VALUE, SEPARATED BY A COMMA, AND HIT CONT."
2710  PRINT
2720  PRINT
2730  INPUT "MAKE YOUR CHOICE",Choice,Value
2740  IF Choice=0 THEN
2750    END
2760  ELSE
2770    SELECT Choice
2780    CASE 1
2790      Den_15=Value
2800    CASE 2
2810      Vis_cst=Value
2820    CASE 3
2830      Dr_pct=Value
2840    CASE 4
2850      Od_in=Value
2860    CASE 5
2870      Wt_in=Value
2880    CASE 6
2890      Rough_in=Value
2900    CASE 7
2910      Q_bpd=Value
2920    END SELECT
2930  END IF
2940  GOTO Start
```

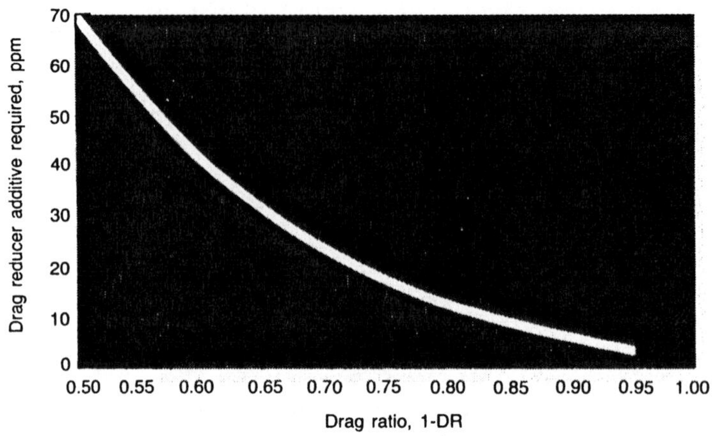

FIG. 2 CBLFLO program listing.

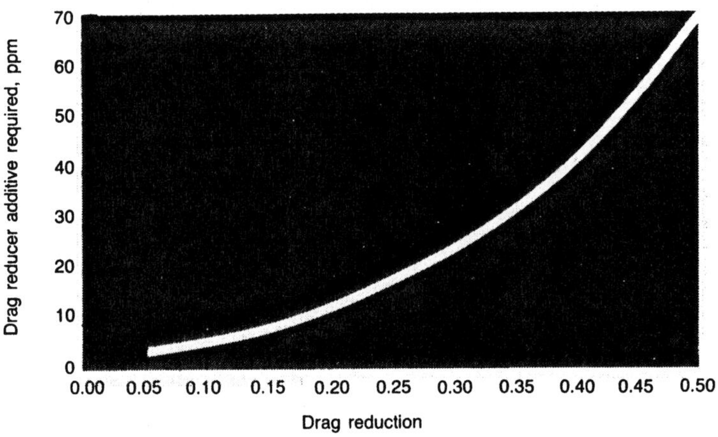

FIG. 3 CBLFLO computes DRA for desired reduction.

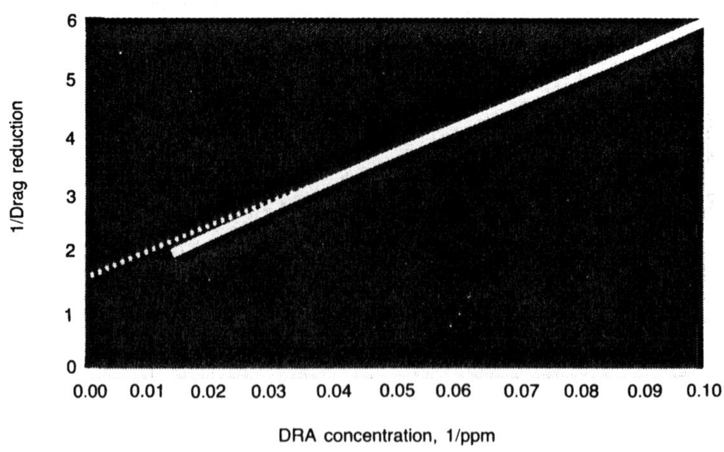

FIG. 4 Effect of increased DRA.

The value of this representation is that it programs well in handheld calculators and runs in seconds. The Savins and Seyer correlation incorporating the characteristic time equation takes many minutes to run on such machines. Another representation commonly seen in published literature which programs well in handheld calculators is found in Eq. (7). Here A and B are determined by regression of the Savins and Seyer correlation output.

This kind of presentation is shown in Fig. 4. The solid line represents the output of the Savins and Seyer correlation; the straight-line Eq. (7) is the dashed line. The y-intercept of the straight line is commonly thought of as the (strictly) theoretical maximum drag reduction obtainable with the given DRA and solvent. In Fig. 4 the intercept is about 1.55, and maximum DR is, therefore, about 0.65. The straight line yields acceptable values up to a DR of about 0.40.

These curves apply only to a pipeline of given parameters flowing crude of given characteristics dosed with a DRA of given characteristic time coefficients for a given flow rate. As yet, with the best present knowledge, the absolute accuracy of the Savins and Seyer correlation, and therefore of the curves, is not what might be desired.

The relative accuracy, however, given two or three experimentally obtained points for a given DRA–solvent solution, is quite good. Therefore, we will proceed with the use of the data in hand and consider that the accuracy indicated is obtainable while realizing it may not be.

For the present, it is the shape of these curves which is important.

First it should be noted that the independent variable is D/R or DR; the inut to the Savins and Seyer correlation is D/R and the output is parts-per-million of DRA. The inverse problem, given parts-per-million to find D/R or DR, requires an iterative solution to the Savins and Seyer correlation.

Second, note the effect is not linear, and the effect of the DRA decreases as concentration increases. With very large doses of DRA, the effective viscosity of the solution becomes so great that drag reduction disappears and is replaced with a drag increasing effect. The Savins and Seyer correlation will not predict this effect, and the y-intercept value of Fig. 4 is purely fictitious.

Last, note that the amount of DRA required to produce a reasonable drag reduction is physically quite small: A DR of 30% requires about 24 weight ppm of DRA.

Thus, if a pipeline is flowing under the conditions used to derive the curves and carries 1,000,000 gal/d of oil which has the same density as the DRA, injecting 24 gal/d of DRA into the line will reduce the friction loss of the same 1,000,000 gal/d to 70% of what it was before.

Caution

The accuracy implied in calculations herein should not be taken as representative of practice. Pipeline hydraulics, and especially the application of

DRA's, are still inexact sciences, neither accurate nor especially precise. We justify the number of significant figures used in our calculations by noting that if calculations in such detail can be produced in two or more ways—which we tried to do—the calculations, themselves, are probably accurate.

References

1. B. A. Toms, "Some Observations on the Flow of Linear Polymers Solutions through Straight Pipe at Large Reynolds Numbers," in *Proceedings of the International Congress on Rheology,* Vol. 2, Scheveningen, The Netherlands, 1948, pp. 135–141.
2. E. D. Burger, W. R. Munk, and H. A. Wahl, *Flow Increase in the Trans Alaska Pipeline Using a Polymeric Drag Reducing Additive,* SPE 9419, 1980.

C. B. LESTER

Part B: DRA vs Looping or Boosting

There often seem only two kinds of oil pipelines: those too small or those too large.

If a pipeline is too large, an operator can worry over capital funds not well utilized. But the more fearful worry—inability to handle business offered—remains with the operator of a pipeline which is too small to take advantage of the market.

Until recently, after discharge pressures had been pushed up to the absolute maximum, the operator had to face reality: To obtain any further increase in throughput capability, his system had to be looped or boosted.

Now there is a third possibility: The operator can inject a drag-reducing agent (DRA).

This part will show how this liquid loop is used as a tool to benefit operators who need only a little more throughput.

Fundamental Relations

A too-small pipeline is one which has no more physical room to produce an increased unit friction loss in the existing pipe. That is, the discharge pressures of the existing pump stations cannot be increased nor the suction pressures decreased. If an operator needs to double the capacity of his pipeline, his obvious solution is to loop it completely.

If he needs an increase of 10–30%, he has the choice of partial loops, booster pump stations, or injecting a DRA. We will look at problems in this general range.

It is usual to consider booster stations and loops in terms of graphical solutions. However, since a graphical solution is not really meaningful for the DRA alternative, we will consider loops and boosters in terms of analytical solutions which can be handled by any engineering calculator.

The most commonly used, though perhaps not the best, friction factor correlation is that of Colebrook and White (C-W). This was almost never used in analytical form until programmable calculators became available but is now in wide use. Its results can also be precisely duplicated. This brings up a problem which seems more difficult with time and involves the little-understood difference between precision and accuracy.

That the C-W correlation can be reproduced by different persons using different calculators means that it is precise but not necessarily accurate.

For purposes here, we will assume it is accurate, while recognizing that it may not be. But this will cause us little trouble because we will, in the end, be comparing flow rates of an untreated fluid with flow rates of the treated fluid in the same pipeline.

In this kind of comparison, absolute accuracy becomes unimportant as long as the relative accuracy—which relates to precision—is acceptable. The C-W correlation has a severe disadvantage when used in analytical solutions because it does require iterative techniques. To overcome this deficiency for purposes here, we will use a regression of the correlation over the range we need to use.

The C-W correlation as plotted by Moody (ASME, 1944) is on log-log coordinates.

Substantial ranges of the correlation can be approximated by a straight line, the slope of which is the derivative of the correlation.

To illustrate a specific situation, we may use the example of a 20 in. o.d., 0.250 in. W.T. pipeline with 0.0007 in. equivalent roughness flowing 160,000 b/d of 5 cSt oil having a density of 0.8388.

The Reynolds number (Re) for this situation is 151,000; the roughness ratio is $0.0007/19.5 = 3.6 \times 10^{-5}$, and $S_f = 4.031$ feet of liquid column (flc)/1000 ft.

We may derive a straight-line approximation for any range of the C-W correlation, but it is best to pick a point near the lower third of the range of use and a Reynolds number which yields a derivative which is a round number of two significant figures. The slope of the f-vs-Re C-W curve for the described conditions is -0.1900 at Re = 147,000, so we select this point. With some manipulation we find:

$$f = 0.161715 \times \mathrm{Re}^{-0.1900} \tag{8}$$

where f = D'Arcy friction factor
Re = Reynolds number

This formula yields f less than 1% from C-W with $e/D = 3.6 \times 10^{-5}$ for a range of Re from 73,500 (half the 147,000 chosen point) to 441,000 (three times the 147,000 point).

The corresponding formula for S_f, the unit friction loss in terms of flc/1000 ft, or meters of liquid column (mlc)/kilometer (the two are numerically the same and represent hydraulic gradeline slope on a 1:1000 basis) is the following:

$$S_f = 4.281034 \times 10^{-3} \times \mathrm{Re}^{-0.1900}(Q^2/D^5) \tag{9}$$

where Q = U.S. oil, b/d
D = i.d., in.

Given S_f, one can find Q directly by

$$Q = 74{,}070 S_f^{0.552486} \tag{10}$$

These formulas are interesting but we are, for the most part, interested in the relations between the S_{fa} and Q_a, the unit friction loss and flow rate under existing conditions a, and S_{fb} and Q_b, the unit friction loss and flow rate under some other conditions b.

Thus we can find the following:

$$S_{fb} = S_{fa}(Q_b/Q_a)^{1.81} \tag{11}$$

and

$$Q_b = Q_a(S_{fb}/S_{fa})^{0.552486} \tag{12}$$

Here we have S_f and Q at described conditions b in terms of S_f and Q at known conditions a.

Note that while the explicit equation for S_f is written in terms flc/1000 ft or mlc/km, the equations involving ratios can use any consistent units, i.e., psig/mile, ksc/km, etc.

Similarly, although the explicit equation for Q is in U.S. barrels per day, the formulas involving ratios can be in terms of gallons per minute, cubic meters per hour, metric tons/annum, and so forth, with equal validity.

We need to know how to calculate loop lengths for a desired increase in flow compared to the single-line flow and to calculate the increase in flow obtained if a specified length of loop is constructed.

These relationships can be derived from the previous two equations if it is assumed that the loop pipe is the same outside diameter, wall thickness, and roughness as the pipe in the original line—which we do assume.

We find:

$$N = (1 - 0.714809 \times M)^{-0.552486} \qquad (1 > N > 2) \tag{13}$$

$$M = (N^{1.81} - 1)/(0.714809 \times N^{1.81}) \qquad (0 > M > 1) \tag{14}$$

where N = flow factor, the ratio of the flow rate without a loop to the flow rate with a loop $M \times L$ long.
M = loop factor, the fractional part of the line length L which is looped.

Figure 5 shows the relationship between N and M.

We need a formula for horsepower in simple terms. Substituting in the basic U.S. horsepower formula hp = PLAN/33,000 = (pressure, psf) × (volume, cu ft)/33,000 yields:

$$\text{hp} = 1.701389 \times 10^{-5} \times (\text{psi}) \times (\text{b/d}) \tag{15}$$

Taking 100's of psi (Cpsi) and 1000's of b/d (Mb/d) and assuming a pump efficiency of 0.85 yields

$$\text{hp} = 2 \times \text{Cpsi} \times \text{Mb/d} \tag{16}$$

which is simple, completely accurate at an efficiency of 0.8506, and accurate in practice for large pipelines using pumps of modern design.

For the 0.8388 density oil selected for our example, 1 psi = 2.75 flc and

$$\text{hp} = 0.7273 \times \text{Cflc} \times \text{Mb/d} \tag{17}$$

where Cflc = 100's flc.

If the line is looped, or if a DRA is used, the horsepower required is a

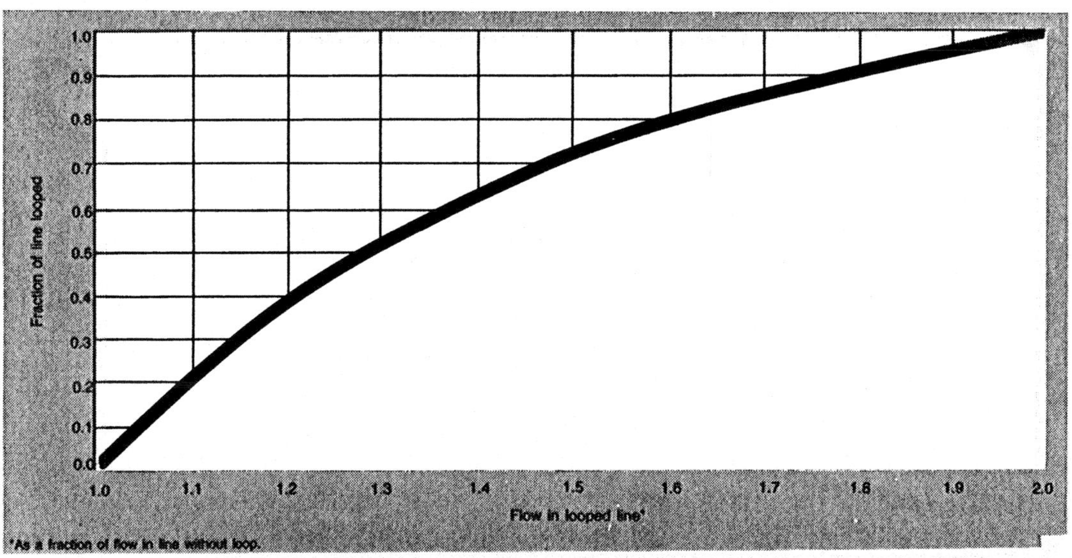

FIG. 5 Flow affected by loop length.

direct function of the flow increase. A 10% increase in flow over Q_a requires 110% of the horsepower required to pump Q_a because the discharge pressure is assumed to have been held the same at Q_b as at Q_a.

Thus the ratio of hp required at Q_b to that required at Q_a is (Q_b/Q_a) and all horsepower is installed at the existing station.

If a booster station is used, the ratio of system horsepower (the sum of horsepower at the existing station and the booster station) at Q_b to that required at Q_a is $(Q_b/Q_a)^{2.81}$. Of this, horsepower in the ratio (Q_b/Q_a), as above, is installed at the existing station and $(Q_b/Q_a)^{2.81} - (Q_b/Q_a)$ is installed at the booster station.

The ratio of the increase in system horsepower at Q_b to that required at Q_a is $[(Q_b/Q_a)^{2.81} - 1]$.

Figure 6 is a plot of the ratio of system horsepower with the booster to system horsepower without the booster-vs-flow with the booster as a fraction of flow without the booster. We can handle drag reduction solutions in a similar manner.

Recall the following definitions of drag reduction for the same flow rate and of drag ratio:

$$\mathrm{DR} = (S_{f\,\mathrm{untreated}} - S_{f\,\mathrm{treated}})/(S_{f\,\mathrm{untreated}}) \tag{1}$$

$$D/R = (1 - \mathrm{DR}) = (S_{f\,\mathrm{treated}})/(S_{f\,\mathrm{untreated}}) \tag{2}$$

Since one can derive ratio formulas involving S_{fa} and S_{fb} without drag reduction, formulas with drag reduction follow directly since D/R operate on S_f directly:

$$Q_{\mathrm{treated}} = (1 - \mathrm{DR})^{-0.552486} \times Q_{\mathrm{untreated}} \tag{18}$$

$$= (D/R)^{-0.552486} \times Q_{\mathrm{untreated}} \tag{18a}$$

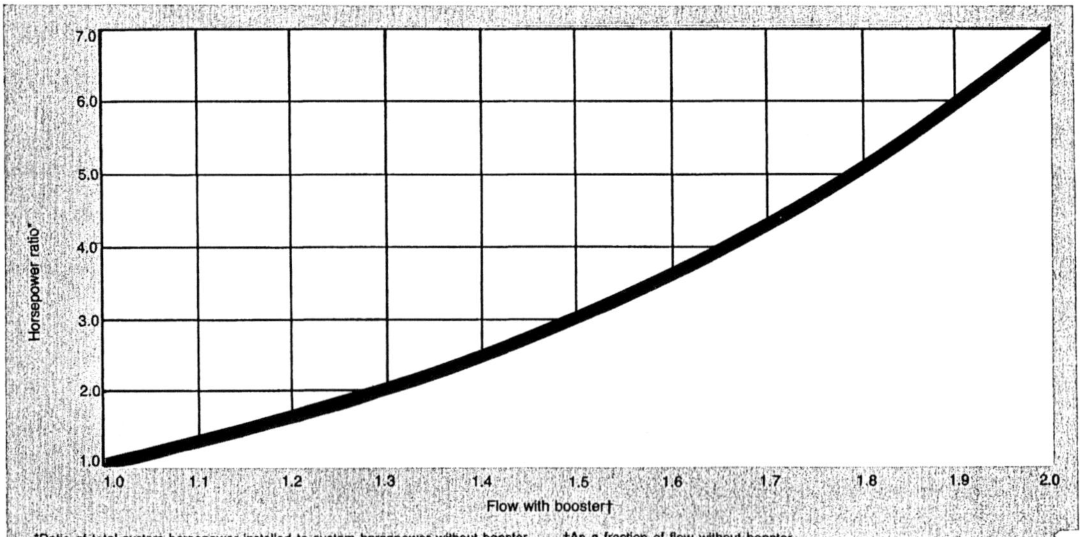

FIG. 6 Flow affected by changes in system horsepower.

$$Q_{\text{untreated}} = (1 - \text{DR})^{0.552486} \times Q_{\text{treated}} \tag{19}$$

$$= (D/R)^{0.552486} \times Q_{\text{treated}} \tag{19b}$$

Note the relationships here are the same as the fundamental relations expressed by Eq. (12).

Having in hand the drag reduction required and the characteristic time coefficients for the DRA–crude oil solution, we can use the computer program of the previous part to compute amount of DRA required directly.

A Specific Application

The example we selected was a section of a 20-in. o.d. × 0.250-in. W.T. × 0.0007 in. roughness pipe flowing 160,000 b/d of 0.8388 density 5 cSt oil yielding a slope of 4.031 flc/1000 ft.

To make the example more specific, we will assume the section is flat (i.e., no static heads are involved start-to-finish) and that it is constructed of API 5 LX 52 pipe having a maximum allowable operating pressure in accordance with ANSI B 31.4 of 936 psig = 2574 flc of oil.

If the friction loss in the piping of one station

$$H_{fs} = 2.15 \times 10^{-3} \times (\text{Mbd})^2 = 55 \text{ flc} \tag{20}$$

and the (reasonable) suction requirements of the pumps

$$\text{NPSH} = 40 \text{ flc} \tag{21}$$

then the total head available to overcome friction is

$$H_f = (2574 - 55 - 40)\text{flc} = 2479 \text{ flc} \tag{22}$$

and the indicated station spacing is

$$L = H_f/4.031 = 615{,}000 \text{ ft} \tag{23}$$

or more than 116 miles.

The mainline pumps will have to put up total dynamic head (tdh):

$$\text{tdh} = 2479 + 55 \text{ flc} = 2534 \text{ flc} \tag{24}$$

We assume the operator of this pipeline wants to boost his throughput capability by 10, 20, or 30%. The degree of expansion depends on its cost and the income to be derived. The kind of expansion depends on the length of time the additional load is expected to be available.

We assume he will install pumping horsepower in the existing station to maintain the 936 psig P_d = 2479 flc H_d under all pumping conditions.

We can first look at loops. With reference to Eq. (14), we find the following: For Q = 176,000 b/d, M = 1.10 and N = 0.2216, L_L = 136,300 ft; for Q = 192,000 b/d, M = 1.20 and N = 0.3932, L_L = 241,800 ft; and, for Q = 208,000 b/d, M = 1.30 and N = 0.5259, L_L = 325,300 ft, where L_L is the length of loop required per 615,000-ft section.

The incremental horsepower required is proportional to the increase in flow and is 295, 590, and 885 hp, respectively.

Next we can look at a booster. Referring to Eq. (11), we find the following: for Q = 176,000 b/d, $S_{fb} = 4.031 \times 1.1^{1.81} = 4.790$, H_f = 2946 flc; for Q = 192,000 b/d, $S_{fb} = 4.031 \times 1.2^{1.81} = 5.607$, H_f = 3448 flc; and for Q = 208,000 b/d, $S_{fb} = 4.031 \times 1.3^{1.81} = 6.481$, H_f = 3986 flc.

We assumed the operator would maintain pressure at the existing station and the difference between the total required head and head produced by the existing station will be picked up in the booster. For the three conditions, the differences are the following: $(2946 - 2534) + 2.15 \times 10^{-3} \times (176)^2 = 412 + 67 = 479$ flc; $(3448 - 2534) + 2.15 \times 10^{-3} \times (192)^2 = 914 + 79 = 993$ flc; and $(3986 - 2534) + 2.15 \times 10^{-3} \times (208)^2 = 1452 + 93 = 1545$ flc.

The important thing, horsepower added at the booster, is as follows: For Q = 176,000 b/d, hp_{added} = 613 hp; for Q = 192,000 b/d, hp_{added} = 1387 hp; and for Q = 208,000 b/d, hp_{added} = 2337 hp.

Horsepower figures were made using Eq. (17).

To evaluate the DRA alternate, it is necessary to compute the dosages (in parts-per-million) of the DRA to achieve 176,000, 192,000, and 208,000 b/d and convert these figures to gallons-per-day (b/d) requirements.

Using the $S_{untreated}$ values listed in the booster calculations and, given the coefficients a, b, c, and d of the characteristic time equation, we can find for the particular formulation of ChemLink's FLO DRA previously used as an example: for Q = 176,000 b/d, D/R = 0.842, ppm = 9; for Q = 192,000 b/d, D/R = 0.719, ppm = 21; and for Q = 208,000 b/d, D/R = 0.622, ppm = 35.

The parts-per-million values are to the nearest ppm. We recognize that this accuracy may not be obtainable without having in hand a few experimentally taken correlation points. Since parts-per-million are calculated in weight units, it is necessary to derive a relationship for the flowing rate of the oil in pounds.

This is, for b/d of 0.8388 sp gr oil,

$$\begin{aligned} \text{lb} &= 1\ \text{b/d} \times 42\ \text{gal} \times 231\ \text{in.}^3/1{,}728\ \text{in.}^3/\text{ft.}^3 \ldots \\ &\times 62.42795\ \text{lb/ft}^3\ H_2O \\ &\times 0.8388\ \text{sp gr oil} = 294 \times \text{b/d} \end{aligned} \tag{25}$$

The density of most DRA's is about 0.790 or 6.59 ppg. Thus the gallons-per-day requirements can be calculated by this equation:

$$\begin{aligned} \text{gpd} &= 294 \times \text{b/d} \times \text{ppm}/1{,}000{,}000/6.59 \\ &= 4.46 \times 10^{-5} \times \text{b/d} \times \text{ppm} \end{aligned} \tag{26}$$

Using this relationship, we find the dosage rates for the three conditions to be: for $Q = 176{,}000$, gpd $= 4.46 \times 10^{-5} \times 176{,}000 \times 9 = 71$ gpd; for $Q = 192{,}000$, gpd $= 4.46 \times 10^{-5} \times 192{,}000 \times 21 = 180$ gpd; and for $Q = 208{,}000$, gpd $= 4.46 \times 10^{-5} \times 208{,}000 \times 35 = 325$ gpd.

Comparisons

Table 2 summarizes the results derived in the preceding discussion.

Note that the same added horsepower is required at the existing station for each kind of solution because of our assumption that the pressure at the existing station would remain constant in each case.

For comparison purposes, therefore, we can forget the increased horsepower required at the existing station.

The comparison, then, comes down to the cost of *X* feet of loop, *Y* booster horsepower, or *Z* gpd of DRA.

Table 3 gives the same information as Table 2, except in terms of incremental throughput of 10% = 16,000 b/d each.

TABLE 2 Capacity Increases of 10, 20, and 30%[a]

	10%	20%	30%
Alternate A—partial loops:			
Loop required, ft	136,300	241,800	325,300
Additional hp required:			
Existing stations	295	590	885
Booster stations	0	0	0
Drag reducer required, gpd	0	0	0
Alternate B—boosters:			
Loop required, ft	0	0	0
Additional hp required:			
Existing stations	295	590	885
Booster stations	613	1,387	2,337
Drag reducer required, gpd	0	0	0
Alternate C—drag reducers:			
Loop required, ft	0	0	0
Additional hp required:			
Existing stations	295	590	885
Booster stations	0	0	0
Drag reducer required, gpd	71	180	325

Comparison of cost/ft, or of booster hp, or of gpd of DRA:

Increase, %	Flow rate, b/d	Loop, ft	Booster, hp	DRA, gpd
10	176,000	136,300	613	71
20	192,000	241,800	1,387	180
30	208,000	325,300	2,337	325

[a]160,000 b/d, 116.5-mile; 20-in. pipeline.

TABLE 3 Incremental Capacity Increases of 30%[a]

	1st 10%	2nd 10%	3rd 10%
Alternate A—partial loops:			
Loop required, ft	136,000	105,000	83,500
Additional hp required:			
Existing stations	295	295	295
Booster stations	0	0	0
Drag reducer required, gpd	0	0	0
Alternate B—boosters:			
Loop required, ft	0	0	0
Additional hp required:			
Existing stations	295	295	295
Booster stations	613	774	950
Drag reducer required, gpd	0	0	0
Alternate C—drag reducers:			
Loop required, ft	0	0	0
Additional hp required:			
Existing stations	295	295	295
Booster stations	0	0	0
Drag reducer required, gpd	71	109	145

Comparison of cost/ft, or of booster hp, or of gpd of DRA:

Increase, %	Flow rate, b/d	Loop, ft	Booster, hp	DRA, gpd
10	16,000	136,300	613	71
10	16,000	105,500	774	109
10	16,000	83,500	950	145

[a]160,000 b/d, 116.5-mile; 20-in. pipeline.

Instinctively, one realizes that a DRA is really a liquid loop, and the partial loop-vs-DRA cases are directly comparable.

To make a decision for a long-term increase, one has to determine if it is better to build X feet of loop, which will have large annual fixed costs and nearly zero operating costs, or to inject Z gpd of DRA and gain the same capacity with nearly zero investment and, therefore, fixed costs while incurring the ongoing cost of DRA consumed.

Instinct fails when making long-term comparisons of boosters with either loops or DRA because boosters have not only a substantial investment, and therefore annual fixed costs, but they can have a high operating cost as well.

A kind of "reductio ad absurdum" comparison exists when comparing to DRA, in that if $365 \times Z$ gpd of DRA costs less than the annual power cost of Y booster hp, the DRA is obviously the more economic solution.

From experience, one knows that the booster station-vs-partial loop comparison depends on the cost and time of recovery of invested capital and on operating costs. The same factors enter into long-term comparisons of the two hardware solutions with the DRA solution, and the relationships are quite complex and sometimes obscure.

As an example of this complexity, note that the efficiency of a partial loop increases as its length increases, so that it takes less loop to go from 20 to 30% increase than from 10 to 20%.

On the other hand, it takes less horsepower to increase from 10 to 20% with boosters than it does to go from 20 to 30%. And, as we have seen, DRA consumption is a higher order function of the drag ratio (D/R) and therefore has the same kind but a different order of increase with increasing capacity as the booster solution.

In every case, now that DRA's are available, DRA's provide the only quickly available, short-term solution to a capacity increase problem. One cannot construct a substantial partial loop, or build a booster station, overnight; there is a certain minimum time required for construction. And, usually one cannot economically construct a loop or a booster station for a 1 or 2-year increase because the cost of capital recovery over a short-term would be prohibitive. DRA's, however, offer an almost instantaneous solution to a capacity increase problem.

Once the line is filled with treated oil, the full capacity increase is in effect. This kind of increase is available in a matter of hours, not months or years. A capacity increase obtained with a DRA can be an on-and-off kind of increase with no investments to salvage at a loss if the period of time the increase is required turns out to be overestimated. If a greater capacity is required this month, use it; if it isn't needed next month, don't use it. It won't go stale in the carton.

We have purposefully not included an economic evaluation of the loop-vs-booster-vs-DRA situation. Every operator has his own ideas as to his direct and indirect costs of construction, the rate at which he wants to recover his invested capital, and his operating costs. He can ascertain the cost of a DRA. He also knows the worth of a capacity increase instantly available as compared to an increase obtained months or years from now. With this information in hand, an economic evaluation comes very quickly.

C. B. LESTER

Part C: How Active, Passive Drag Affect DRA Injections

This part examines applications more deeply and eliminates some of the simplifying assumptions purposely made in Part B to explain fundamental functions of DRA's in pipelines.

Line Drag

Drag reduction was defined in Part A in the following equation:

$$\text{DR} = (S_{f\,\text{untreated}} - S_{f\,\text{treated}})/(S_{f\,\text{untreated}}) \tag{1}$$

Here S_f means unit friction loss in consistent units (feet of liquid column [flc] per 1000 ft, meters of liquid column [mlc] per kilometer, psig/mile, etc.). The terms treated and untreated refer to pipeline liquids dosed or not dosed with a DRA.

A companion, sometimes more useful factor is the drag ratio:

$$D/R = (1 - \mathrm{DR}) = (S_{f\,\mathrm{treated}}/S_{f\,\mathrm{untreated}}) \tag{2}$$

DR and D/R thus computed involve only the parameters of the pipeline, the characteristics of the DRA–solvent solution, and the flow rate. Therefore, they can be related to line drag.

Line drag arises in the pipeline itself and varies with flow as a complex function of Reynolds number and pipe roughness. It also is the only drag acting in a pipeline system which can be reduced with a DRA and is therefore the active drag.

Most workers in the field of drag reduction are concerned only with active drag, thus with S_f only. This definition of drag is, for their purposes, complete and sufficient.

For a pipeliner, however, this definition is incomplete because there are other effects working in a pipeline which, while not a drag *per se,* act as one. These as much as any increased active drag must be overcome if a desired increase in flow is to be achieved. There are two of these passive drags: station and static.

Station Drag

One of the passive drags is the turndown in the H-Q characteristic of centrifugal pumps with increasing flow, and the accompanying increase in station piping loss which increases with flow. Both reduce H_d, the pump station discharge head, as flow increases. We call this kind of loss station drag because it arises within the pump station. It varies as the square of the flow rate.

Figure 7 shows the situation for a section of a pipeline with no static head acting in the system, i.e., static heads do not act start-to-finish in the pipeline section.

We again take the example of a 20-in. × 0.250-in. × 0.0007-in. roughness pipeline flowing 160,000 b/d of 0.8388 density 5 cSt oil with S_f on a 1/1000 basis of 4.031 flc/1000 ft.

Under ANSI B 31.4, the pipeline is allowed to pump at 936 psig = 2574 flc, all available to overcome friction except $2.15 \times 10^{-3} \times (1000\ \mathrm{b/d})^2$ = 55 flc lot in friction in the station piping and an assumed 40 flc suction requirement at the next downstream station.

Thus at 160,000 b/d, 2479 flc are available to overcome friction over the 615,000-ft length of the section. This value is $H_{d\,\max}$, the maximum effective discharge head of the pump station. In Fig. 7 this value is the intersection of the station head curve with the 160,000 b/d flow ordinate.

The value of $H_{d\,\max}$ at the 160,000-b/d initial flow rate is called H_0 and

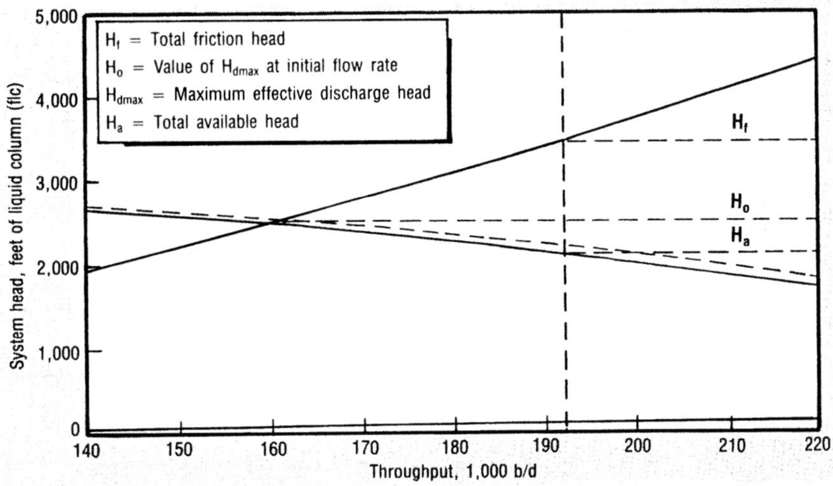

FIG. 7 Example for a 20-in. pipeline system: Initial section (615,000 ft; no static head).

is so marked on the H_d curve. H_0 is the point of reference for computing active and passive drag.

To increase flow from 160,000 to 192,000 b/d, the unit friction loss, S_f, rises from 4.031 to 5.607 flc/1000 ft and total friction head H_f rises to $615 \times 5.607 = 3448$ flc.

This is an increase of $615 \times (5.607 - 4.031) = 969$ flc. Overcoming this increase requires a DR of $(H_f - H_0)/H_f = (3448 - 2479)/3448 = 0.281$ and yields a D/R of 0.609.

Note, however, that there is less effective head available at 192,000 b/d than at 160,000 b/d. Pump turndown has reduced the head available from the pumps, and the increased loss in the station piping has further reduced the head available from the station.

The dashed line pump curve in Fig. 7 is the total dynamic head (tdh) curve for the pumps; the solid line below it is the station discharge head (H_d), which is the tdh curve with station losses, shown as a solid line at the bottom of the graph, subtracted. The head available (h_a) at 192,000 b/d = 2102 flc.

The reduction in the total available head, H_a, from a pump station acts as a loss in the same way in the total flow picture as the increase in total required friction head H_f. It is a loss in the system which must be made up by reducing S_f because, other than changing out pumps or repiping the station manifold, there is no way to increase H_d. We must consider station drag as well as line drag to arrive at total drag and be able to compute the drag reduction required to achieve a given flow rate.

A graphical solution for total drag in multi-station pipelines without active static heads is handy, in a way, in that the problem is immediately seen. With $S_f = H_f/L$ and $S_a = H_a/L$, the required total drag ratio D/R can be written as

$$D/R = (H_a/L)(H_f/L) = H_a/H_f \tag{27}$$

The corresponding required total drag reduction DR becomes

$$\mathrm{DR} = (1 - D/R) = [1 - (H_a/H_f)] \tag{28}$$

The main problem with graphical solutions is that they don't program and are difficult to repeat.

A graph yields a different answer every time it is read, and as the problems become more complex the accuracy of the solutions suffers. Also, the pipeliner's usual way of plotting system curves does not serve well in studying drag-reduction problems. We will therefore use an analytical solution which can be applied to any kind of problem.

For the pump station design we are using, the loss in station piping has been given previously in the following form:

$$H_{fs} = kQ^2 \tag{29}$$

which we took numerically as

$$H_{fs} = 2.15 \times 10^{-3} \times (1000\ \mathrm{b/d})^2 \tag{30}$$

This kind of simple velocity-squared relationship is satisfactory for any kind of problem. It may not be completely accurate, but it will suffice.

The general form of the characteristic curve for a centrifugal pump of the kind used on pipelines can be written in the following form:

$$\mathrm{tdh} = A + BQ + CQ^2 \tag{31}$$

This representation is usually quite accurate for flow rates between $0.75 \times Q_{\mathrm{bep}}$ and $1.25 \times Q_{\mathrm{bep}}$, where Q_{bep} is the flow rate at the best efficiency point on the curve. A regression of the pump curve shown in Fig. 7 is as follows:

$$\mathrm{tdh} = 2666 + 7.67 \times (1000\ \mathrm{b/d}) - 53.10 \times 10^{-3} \times (1000\ \mathrm{b/d})^2 \tag{32}$$

Inasmuch as Eq. (30) subtracts directly from Eq. (32) to yield H_d, we may write

$$H_d = 2666 + 7.67 \times (1000\ \mathrm{b/d}) - 55.25 \times 10^{-3} \times (1000\ \mathrm{b/d})^2 \tag{33}$$

which gives us H_d in terms of the flow rate in 1000's b/d.

At 160,000 b/d, $H_f = H_0 = H_d = 2479$ flc; at 192,000 b/d, $H_f = 3448$ flc, and $H_a = H_d = 2102$.

Station drag and line drag cannot be added algebraically to compute total drag reduction. Note the following, with Fig. 7 as reference. For line drag:

$$\begin{aligned} \mathrm{DR} &= (H_f - H_{d\,\max})/H_f \\ &= (3448 - 2479)/3448 \\ &= 0.281 \end{aligned} \tag{34}$$

and for pump drag:

$$\begin{aligned} \text{DR} &= (H_{d\,\max} - H_a)/H_0 \\ &= (2479 - 2102/2479) \\ &= 0.152 \end{aligned} \tag{35}$$

These yield a total of line drag + pump drag = 0.433, compared to the correct 0.390 calculated previously from the DR = H_a/H_f relationship.

However, without proof:

$$\begin{aligned} D/R_{\text{line}} &= H_0/H_f \\ &= 2479/3448 = 0.719 \end{aligned} \tag{36}$$

$$\begin{aligned} D/R_{\text{station}} &= H_a/H_0 \\ &= 2102/2479 = 0.848 \end{aligned} \tag{37}$$

and

$$\begin{aligned} D/R_{\text{total}} &= D/R_{\text{line}} \times D/R_{\text{station}} \\ &= 0.719 \times 0.848 \\ &= 0.610 \end{aligned} \tag{38}$$

This is the correct value for D/R.

Thus we may take the following as rules:

$$D/R_{\text{total}} = D/R_{\text{line}} \times D/R_{\text{station}} \tag{39}$$

$$\text{DR}_{\text{total}} = \text{DR}_{\text{line}} + \text{DR}_{\text{station}} - \text{DR}_{\text{line}} \times \text{DR}_{\text{station}} \tag{40}$$

Here Eq. (40) follows from substitution of Eq. (2) in Eq. (39).

Having in hand the a, b, c, and d coefficients of the formulation of FLO produced by ChemLink Inc., a subsidiary of Atlantic Richfield Co., we can use the CBLFLO program to estimate the amount of this particular DRA required to bring the pipeline up to 192,000 b/d while taking into account the 377 flc reduction in H_d resulting from this 32,000 b/d increase; thus the following: Q = 192,000 b/d, DR = 0.390, ppm = 38, and gal/d = 326.

When we assumed pump station head stayed constant as flow increased, we obtained Q = 192,000 b/d, DR = 0.281, ppm = 21, and gal/d = 180.

To make up the 377 flc lost in pump drag, the pipeline operator can decide whether to spend the extra 146 gal/d of FLO or to install a pump burning 527 hp in the existing station.

Static Drag

Previous examples have progressed from a simplistic approach to drag reduction on a single pipeline section with constant station pressures without static heads acting end-to-end in the section through an example which takes into account the reality that pump heads and station losses change with changing flow.

Now we can consider the more general problem of a multistation pipeline with operative static heads.

Static drag is a passive drag. It may be a positive drag which acts as any other drag to reduce flow, or it may be a negative drag which increases flow.

It differs from line drag and station drag in that it does not vary with flow. It does, however, directly affect the application of DRA.

For simplicity we will look at a two-station system; using more than two stations could cloud fundamentals and would not improve the example.

Section 1 is the same 160,000-b/d, 615,000-ft, 20-in. crude line without active static heads used in the previous example. Section 2 is an identical section except it is a 367,000-ft uphill section and looks at a 1000-ft positive static head.

Figure 8 shows the system head curves for Section 2. It is important for pipeliners to understand that this set of system hed curves is different from their usual kind of system curves: Instead of the positive H_s being added to H_f to obtain a curve of effective $H_t = H_s + H_f$, the H_s is subtracted from the H_d curve to yield an effective H_d curve.

This kind of representation correctly shows the relationship between energy input and energy use in the system from the DRA standpoint but at the expense of incorrectly showing the absolute value of H_d and H_a.

The absolute value of H_d at 160,000 b/d is 2479 flc as in Section 1, but 1000 flc of this is used to overcome 1000 ft of static head. The effective value of H_d in Section 2 at 160,000 b/d, therefore, is 2479 flc − 1000 ft = 1479 flc and is called H_0.

Illustrated here is the most important point in handling active and passive

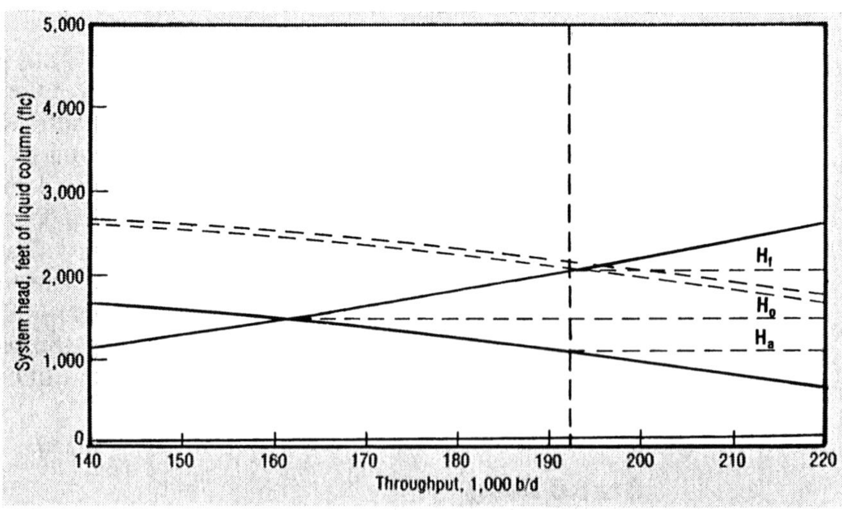

FIG. 8 Example for a 20-in. pipeline system: Uphill second section (367,000 ft; 1000-ft positive static head).

drags in a pipeline problem: Passive drags must be removed from the problem before calculations relating to active drag are carried out.

We did this intuitively for Section 1 when we took into account station drag by using the tdh and H_{fs} equations to arrive at H_d, and this is the pipeliner's usual way. We must now agree that, after station drags are taken into account, static drags must also be taken into account in the H_d representation by a further subtraction (or addition, if a downhill section) of H_s from H_d to yield the effective H_d curve.

In Fig. 8, H_s is not shown as it would lie on the H = 1000 flc abscissa. H_d is shown as a solid line which is the algebraic sum of the pump tdh curve, the H_{fs} curve, and the constant H_s. The curve of H_f is as shown in Fig. 7 for the shorter length of Section 2.

From this representation we can take H_f = 2057 flc, H_0 = 1479, and H_a = 1102 flc. The following comparison results. Section 1: $D/R = H_a/H_f$ = 2102/3448 = 0.610, and DR = 0.390; Section 2: $D/R = H_a/H_f$ = 1102/2057 = 0.536, and DR = 0.464.

Section 2 requires a greater DR to achieve 192,000 b/d than Section 1. The same pump drag must be recovered, but an additional 1000 ft of static drag must be overcome and the active line drag on which DRA's can operate to effect drag reduction is smaller by the 366,000/615,000 ratio of the section line lengths.

The dosage of DRA works out as follows. For Section 1: Q = 192,000 b/d, DR = 0.390, ppm = 38, gal/d = 326; for Section 2: Q = 192,000 b/d, DR = 0.464, ppm = 57, gal/d = 488.

To complete this analysis, we can assume a third 20-in. pipeline section following Section 2 which is a downhill section with a negative H_s of 1000 ft. With the same pumps, this section would be 863,000 ft long, and the conditions working are those shown in Fig. 9. Here H_f = 4839 flc and H_0 and H_a = 3479 and 3102 flc, respectively.

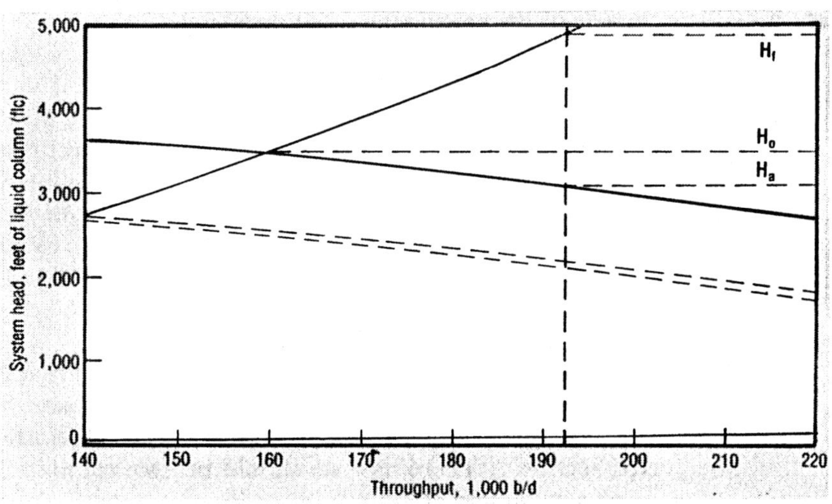

FIG. 9 Example for a 20-in. pipeline system: Downhill third section (863,000 ft; 1000-ft negative static head).

TABLE 4 Obtaining 32,000-b/d Increase in a 20-in. Line[a]

Section 1	Q = 192,000 b/d	DR = 0.390	ppm = 38	gal/d = 326
Section 2	Q = 192,000 b/d	DR = 0.464	ppm = 57	gal/d = 488
Section 3	Q = 192,000 b/d	DR = 0.359	ppm = 33	gal/d = 284
Total				gal/d = 1,098
Average				gal/d = 366

[a]Three-station, 347.5-mi.

The D/R required for Section 3 is $H_a/H_f = 3102/4839 = 0.641$ and DR = $(1 - D/R) = 0.359$. To obtain this drag reduction requires 33 ppm of FLO = 283 gal/d.

We can now obtain the total cost (Table 4) of obtaining a 32,000-b/d increase in flow in the three-station, 1,835,000-ft (347.5-mi) 20-in. line.

Optimizing DRA Dosage

The average DRA dosage rate per station is 366 gal/d, but the rates vary from 76.7% of the average to 133.3% of the average. We might suspect that there is a better way to use DRA than what we have calculated thus far. And this is sometimes true.

Remembering that the curve of ppm-vs-D/R or DR is a higher order function of DR or D/R, we know that 100 gal/d injected in Section 2 does not produce the same incremental increase in flow rate Q as 100 gal/d in Section 1. It produces the same change in S_f in each section, but it does not produce the same change in Q. Thus the effect of passive drags is evinced.

If we could adjust the dosage so that each section receives the same dosage of DRA, we could produce a lower total dosage rate. The mathematics of the calculation are controlling.

We might suspect that we can take advantage of the flexibility of pump stations using centrifugal pumps to carry forward unused head from one section to another; i.e., to dose one section more heavily than required by the hydraulics of that section alone with the object of carrying forward the unused head to and through the succeeding downline station.

Heads thus carried forward would decrease the dosage rate of DRA in the downline section. We can do this subject to two limitations: Pump suction heads cannot drop below minimum required NPSH; and station discharge heads cannot exceed maximum working head.

The following example illustrates.

If we take the three sections together, we can find the results illustrated in Table 5. The D/R for the total line is the same as that of Section 1, something we might have suspected, since H_s for Section 2 balances out H_s for Section 3. Therefore, we should be able to inject 326 gal/d of FLO in each section and attain our objective.

But we can't. If we inject 326 gal/d at Station 1 to obtain a DR of 0.390 in Section 1, we will arrive at Station 2 with a suction head of 40 flc. But a

TABLE 5 Example DRA Injection by Section[a]

	Section 1	Section 2	Section 3	Total
H_f = flc 192,000 b/d	3,448	2,057	4,839	10,344
H_a = flc 160,000 b/d	2,102	1,102	3,102	6,306
$D/R = H_a/H_f$	0.610	0.536	0.641	0.610
DR = 1 − D/R	0.390	0.464	0.359	0.390

[a]flc = feet of liquid column.

DR of 0.390 will not be enough to overcome all of the drags working in Section 2, and we will arrive at the suction of Station 3 with something less than 40 flc which is technically unacceptable. We cannot always combine sections to reduce total DRA dosage.

We can illustrate the method by reversing the order of the sections (Table 6). These calculations, and a recommended form for making such calculations, are shown in Table 7.

Now if we start out with a DR of 0.390 in Section 1, we will come into the suction of Station 2 with a higher suction head than required which will increase the effective H_d of Station 2. The effect will go forward ending up with the required 40 flc at the end of Section 3.

Thus we obtain the advantage of carrying drag reduction forward and our total consumption of DRA will be 3 × 326 gal/d = 978 gal/d, as compared to 1098 gal/d when each section is treated as a separate pipeline. The savings, 120 gal/d, is well worthwhile because while the economic effect of DRA's is sometimes dramatic, they are not inexpensive.

The rule here is that when a pipeline is perfectly hydraulically balanced, i.e., when maximum allowable discharge head and minimum allowable suction head are reached simultaneously in all stations at the same flow rate (a theoretical situation usually but a good basis for making rules), drag reduction to achieve a higher flow rate can be carried forward only as long as a limiting station discharge or suction head is not reached.

For unbalanced lines, sometimes even reaching a limit in discharge or suction does not prevent unbalancing the drag-reduced system with a consequent savings in total DRA.

A long line can sometimes be divided into groups of sections and drag reduction balanced among the sections in the group. Thus while drag reduction cannot be carried forward through all the stations, it can be carried

TABLE 6 Reversed Order of DRA Injection in Example[a]

	Section 1	Section 2	Section 3	Total
H_f = flc 192,000 b/d	4,839	2,057	3,448	10,344
H_a = flc 160,000 b/d	3,102	1,102	2,102	6,306
$D/R = H_a/H_f$	0.641	0.536	0.610	0.610
DR = 1 − D/R	0.359	0.464	0.390	0.390

[a]flc = feet of liquid column.

TABLE 7 Calculation Sheet for Balancing DRA Requirements[a]

Description	Unit	Start/end	Section 1	Section 2	Section 3
Flow rate	b/d		192,000	192,000	192,000
Flowing density	g/mL		0.8388	0.8388	0.8388
Flowing viscosity	cSt		5.0	5.0	5.0
Pipe o.d.	in.		20.0	20.0	20.0
Pipe W.T.	in.		0.250	0.250	0.250
Pipe roughness	in.		0.0007	0.0007	0.0007
Elevation	ft amsl[b]	start	1,000	0	1,000
Station suction	flc[c]	start	40	192	40
Pump suction	flc	start	1,040	192	1,040
Pump thd[d]	flc	start	2,181	2,181	2,181
Pump discharge	ft amsl	start	3,221	2,373	3,221
Station piping loss[e]	flc	start	79	79	79
Station discharge	ft amsl	start	3,142	2,294	3,142
S_f required	flc/1,000 ft		5.607	5.607	5.607
Drag reduction	decimal		0.3904	0.3904	0.3904
Drag ratio	decimal		0.6096	0.6096	0.6096
S_f produced	flc/1,000 ft		3.418	3.418	3.418
Length of section	ft		863,061	366,906	614,984
H_f	flc		2,950	1,254	2,102
Station suction	ft amsl	end	192	1,040	1,040
Elevation	ft amsl	end	0	1,000	1,000
Station suction	flc	end	192	40	40

[a]Three-station system.
[b]Above mean sea level.
[c]Feet of liquid column.
[d]Total dynamic head.
[e]Station pipe loss lumped on discharge side of pumps for simplicity. For accuracy, loss should be divided between suction and discharge side to ensure pump suction does not drop below required NPSH.

forward through a group of stations with the consequent economic advantage of reducing total DRA required to obtain a desired objective.

This material appeared in *Oil & Gas Journal,* pp. 107–110, March 4, 1985, copyright © 1985 by Pennwell Publishing Co., Tulsa, Oklahoma 74121, and is reprinted by special permission.

Part D: Handling DRA's

The first three parts treated the theory and practice of drag reduction using drag reducing agents (DRA's). Those parts considered the fundamentals of the drag reduction phenomenon, the concept of DRA's as liquid loops, and some practical examples of the application of DRA's to multistation pipeline systems.

This concluding part is concerned with the practical aspects of drag reduction, i.e., with the product and the hardware, and with the kind of results that should be expected.

Commercial DRA's

DRA's are solutions of polymers in a light hydrocarbon.

The compositions of DRA's in field use today are proprietary information. Some technical papers describe polymers that can be used to effect drag reduction, but none describes the polymer in the highly efficient DRA's commercially available today.

The solvent is usually described as a kerosene-like hydrocarbon. It has to have a flash point high enough to ensure the DRA can be safely stored and transported. But other than a certain purity requirement, there seems to be nothing special about it.

DRA's may contain the polymer in almost any concentration up to 11 or 12%. Recent technical papers describe tests run with a DRA containing only 0.16% polymer, but this was a premixed DRA purposefully made dilute to ensure quick solution in the pipeline liquid in a test loop.

DRA's commercially available are in concentrations from 2 or 3% up to a maximum of 11 or 12%. The lower concentrations assist in obtaining a more rapid solution in the pipeline liquid.

The upper limit to polymer concentration is set by the need to obtain an acceptable time of solution and by the needs of industrial usage to have a DRA which can be handled with reasonably priced and readily maintained equipment.

With all other factors aside, a high concentration is more to be desired than a low concentration simply because there is less DRA to transport, store, and inject.

DRA's do not seem to age, particularly, but there may be some aging effect over long periods. There is very little written about this characteristic of DRA's. Thus it would seem DRA's could be stored for considerable periods without losing their one desirable characteristic of being able to reduce drag in a pipeline.

Physical Characteristics

DRA's are usually described as a thick, viscous liquid with the appearance of old honey. This is true, but they are more than that.

DRA's are very difficult fluids to handle. They are highly viscoelastic; if one dips a finger in a bucket of a DRA, he can draw a strand from the bucket to shame a spider.

If a flange connection is parted to change hoses or supply tanks, the solution draws out like honey. Unlike honey, however, if the DRA still connecting the parted flanges is quickly cut with scissors, the severed parts of the DRA will immediately retract into the openings from which they came. Scissors are also handy to have around a DRA installation.

The three physical characteristics of interest are flash point, density, and viscosity.

DRA's commercially available hold the flash point over 140°F to meet the requirements of regulatory bodies. Most DRA's have a flash well over 140°F.

The density of DRA's varies with their source and composition. A density in the range 0.75–0.85 is usual, yielding a weight in the order of 6.3–7.1 ppg.

DRA's are highly viscoelastic but also highly thixotropic, and measured viscosity depends on shear rate. Data for one DRA are given below; others should be similar. Data given are for 25°C. Shear rate: 0.125 s^{-1}; 2.5 s^{-1}; 25 s^{-1}; and viscosity, cP: 1,500,000; 90,000; 10,000.

The viscosity of DRA's is also quite sensitive to temperature. The same DRA at a shear rate of 0.125 s^{-1} had a viscosity at 0°C of 1,800,000 cP and at −10°C of 2,600,000 cP.

The thixotropic nature of DRA's actually helps in handling them. If the viscosity remained at the breakout viscosity which, for these liquids, could be several million centipoise, they would be almost impossible to pump.

The thixotropy, however, acts to reduce the effort to maintain a DRA flowing after it has started to flow. And, except for temperature effects, after a DRA flow has been established, there is nothing particularly difficult about keeping it flowing. An exception exists for the pumping equipment itself; more about that later.

Flow Characteristics

Burger (see References, Part A) gave the following formula for calculation of the flowing characteristics of a DRA of the kind in use in 1980:

$$P/L = (K/3D) \times ((3n + 1)/4n)^n \times (96V/D)^n$$

where P = pressure loss, psig
L = length, ft
D = i.d., in.
V = velocity, ft/s
K = power law constant, $lb/s^n/ft^2$
n = power law constant, dimensionless

For the DRA Burger was using, $K = 4.8$ and $n = 0.24$. These constants must be determined for each DRA by laboratory measurements.

As a practical matter, DRA's may be handled at high pressures, i.e., on the discharge side of injection pumps, without major problems. The pressure may rise but the positive displacement pump keeps pumping.

The problem is on the suction side of the injection pumps, where the DRA's must be handled at low pressures.

Transport and Storage

DRA's are commonly transported in rail cars, tank trailers, or tank containers.

Any container transporting DRA's must be capable of being preloaded to some reasonable pressure with an inert gas to force the DRA to move. Otherwise left to its own resources and internal energies, the DRA would take weeks to leave an open container.

The large containers are usually pressured to 30–70 psig using nitrogen.

Smaller containers sometimes allow pressures in excess of 100 psig. And since these containers are usually used in applications in which the dosage rate of DRA in gallons per hour is quite small and the flowing viscosity correspondingly large, the additional pressure is quite well worthwhile.

There are several kinds of generally standard rail cars which will handle DRA's. Any butane or propane carrier, if unbaffled, will serve, and any number of specially built cars are available.

So far as we know, the only DRA user with specially constructed tank trailers is Alyeska Service Co. which uses them to transport DRA to the injection points on its 48-in. line. Other volume users are using commercially available tank containers on container chassis or on flat bed trailers.

There is a large supply of IMCO Type 1 containers available for sale or lease. These are rated at 20 m^3 = 5000 U.S. gal, nominal contents, or 24 m^3, rated 6000 U.S. gal, nominal contents.

The containers may or may not be insulated, equipped with heating coils or elements, or have instrumentation relating to these elements. All have a pressurizing inlet and a manhole on top.

They may or may not have a bottom discharge connection; containers equipped for top-only discharge have, to our knowledge, not been used as DRA transport or storage vessels. Most containers have a 3-in. outlet connection.

IMCO Type 1 containers are usually rated at 3.5 $kg/cm^2 \approx 50$ psig and tested to 4.5 $kg/cm^2 \approx 64$ psig. Some containers are rated at higher pressures, but designers of injection systems shouldn't count on these containers being available.

DRA Transfer Systems

Most DRA's may be handled in mild steel piping. The usual pipeliner's caution of using stainless fitted valves is recommended, however. When you want to close a valve in a DRA system, you want it to close.

Hoses are no problem if the recommendations of the DRA producer are kept in mind. As to hydrocarbon compatibility, any hose which will handle kerosene will handle a DRA. If there are other precautions as to the hose or to the metallurgy of the hose end fittings, the DRA producer should advise you.

Obtain clearance fom the producer before using aluminum, bronze, or brass end fittings.

Pressure ratings for hoses should be carefully selected. Suction hoses, running from the supply tank to the injection pumps, usually may be 75 psig working pressure hoses without vacuum protection. It is improbable that the safety control system will allow the injection pumps to operate if a vacuum exists on pump suction. If this problem does present itself, more expensive pressure-vacuum hoses should be used.

Discharge hoses should be rated to at least 125% of the setting of the discharge relief system in the injection unit, even if the discharge side of the pump is protected with rupture disks. Safety systems have been known to fail, or, more commonly, to be blocked out by an operator who doesn't want to be bothered, and an injector system pumping a DRA out on the station grounds is not a nice thing. A 125% rating won't guarantee to stop this kind of incident, but it will help.

Injector Nozzles

There are pros and cons to the use of specially designed injector nozzles.

About the only con is that they add cost to the injector system. But for some operators, this is sufficient to avoid use of nozzles and inject directly into a tapped port on the pipeline.

The major pro to the use of an injector nozzle as compared to a flush port in the pipe wall is that the DRA is injected in the turbulent core of the pipe and, in large nozzles, several spaced holes can be provided which sprinkle the DRA into the pipeline like the output of a spaghetti machine.

This should assist in more rapid solution of the DRA in the transported solution of the DRA in the transported oil so that drag reduction is at full effect in the shortest possible distance from the point of injection.

If nozzles are used, they should be designed to be inserted and removed while the pipeline is under pressure. This requirement calls for some kind of jack screw, or hydraulic ram, arrangement to overcome the force generated by line pressure acting on the unblanced area of the nozzle.

Injector Pumps

We have purposefully left the most difficult part of handling a DRA until the last. This is the injector pump itself.

There are at least 10 or 20 different injector pumps which have been tried with varying degrees of success. Almost all pumps installed as of June

1984 in commercial applications are some variation on a rotary positive displacement pump.

Process gear pumps designed for the polymer chemistry industry are popular. Since these are positive displacement pumps, they must be driven by some kind of variable speed drive. The characteristics of the DRA preclude using a variable bypass system to control volume output by the pump.

Inasmuch as the pumps must handle a very high viscous liquid, they turn quite slowly; speeds of a few hundred revolutions per minute are usual.

This means that the pumps must be very closely fitted because a pump turning 100 rpm with 10 psig suction and 1200 psig discharge has a very large differential pressure trying to backflow the pump or bypass its internals.

Pumps fitted to a few ten-thousandths of an inch are common. Pumps which are not so closely fitted use loaded wear plates to provide the close tolerances required. Small reciprocating plunger pumps have been tried but pump valving remains a problem.

Closely fitted pumps of any kind are very tender devices, in that they are very sensitive to running dry and may cavitate in the suction nozzle at relatively large positive pressures if the DRA is very viscous and the flow rate is high.

Any kind of solid particle which enters the suction nozzle will cause trouble. Soft particles may or may not be crushed by the pump, but enough of these will build up in the root of the gears and eventually cause the pump to seize.

And hard particles—a piece of welding slag left over in a tank container, a piece of gasket material cut off in an overtightened, misaligned flange connection, or a large grain of hard quartz sand—will cause the pump to seize and tear up the pump, the coupling, or the motor drive, depending on which is the weak link in the chain.

A studious approach to keying coupling halves on the pump and driver shafts is a worthwhile effort to ensure that the key will shear before any of the expensive equipment is badly injured.

Caveat

Each of the preceding parts of this article has been governed by a caution (stated in Part A) that too much should not be read into the use of a large number of significant figures in the calculations. This use allowed solution of each problem in two ways and thus ensured that, if the answers were duplicated, the calculations were correct.

This did not mean the answers were absolutely accurate, however. If relationships between two flows, two unit head losses, or two drag ratios, for instance, were relatively precise, the calculations were useful for any practical purpose we might have.

A further caution is in order. The Savins and Seyer correlation and the characteristic time equation are in the same category but in a lower order of accuracy with the Colebrook-White correlation for f-factor. The absolute accuracy may be poor but the relative precision is acceptable.

The a, b, c, and d coefficients of the characteristic time equation must be obtained from the producer of the DRA being studied or used. The producer, in turn, must develop these coefficients from all available data including histories of performance of the DRA with similar oils and, if available, loop tests or, preferably, field tests on the pipeline to be treated, using the same DRA and the same crude oil or product.

Given the characteristic time equation coefficients, the Savins and Seyer correlation will yield acceptable results. A pipeline operator who keeps good records of the application of a DRA in his system can develop sufficient information to allow the producer of the DRA to refine further the coefficients until, after a time, the operator should be able to predict with commercial accuracy the dosage required to achieve a given drag reduction in a given pipeline flowing a given crude or product.

The key words in the above are tests and records. Tests facilitate the initial application of a DRA, and records allow the application to be refined.

C. B. LESTER

Cleaning

Pipe cleaning is important in any plant because internal buildup restricts flow and increases the power required to force fluid through the line. Product purity can suffer and internal corrosion can result. Lines may get so clogged that they impede production in plants, cause hazards in cooling jackets of high-temperature equipment, and restrict fire-fighting capability.

In water lines, all types of pipe materials experience buildup. The only difference is that some are easier to clean than others, and some pipe materials resist corrosion better than others. Internal linings of coatings or cement will increase the period of time between cleanings but will not eliminate buildup. These linings preserve the pipe, giving it longer life, and they definitely make it easier to clean.

The practice of pipe cleaning received relatively little attention before World War II. Generally, it was easier simply to replace the pipe. With today's high cost of labor, equipment, and materials, owners are looking at the most economical ways of improving flow and reducing corrosion.

Although our discussion here is centered on water lines, the basic principles will apply for other services. Some techniques and tools that are unique to particular services, such as paraffin removal or product separation in petroleum product lines, will be omitted.

Pigging

Pigs—cleaning devices that are forced through the pipe—are made of a variety of materials, ranging from rubber to Styrofoam, and in a multitude of configurations. In all cases, however, these devices employ air or liquid pressure against the back end of the pig to move it through the pipe. (In addition to their use in cleaning, pigs may be employed to separate different fluids moving in the same line, but this article will concentrate on their cleaning function.)

Types of Pigs

Today there is such a variety of pigs on the market that it would be impossible to include all of them. Experienced pipe-cleaning people will even make their own, on-site, to meet the needs of a particular line or its deposits. However, pigs generally fall into one of the following three categories:

Cup or disc type, or a combination of the two
Foam type
Spherical type

Each type has its advantages. There is no such thing as a universally applicable pig.

Cup or Disc Type

This type of pig can be made in a multitude of configurations to suit individual needs (Fig. 1). In general, it consists of at least two cups, one of which acts as the lead (guiding) head and the other as the tail (or pressure) end. A variety of devices can be installed between the cups to perform brushing, scraping, cutting, dewatering, or debris-removal functions. One's imagination and experience must determine which combination is best suited to the particular job needs.

Cups and discs can be manufactured in a variety of materials, of which the following are the most popular:

Polyurethane—Cast from liquid polyurethane, this is a high-strength, tough elastomer with higher abrasion resistance than other materials. It is not as flexible as neoprene or Buna-N.
Neoprene—Most economical and more flexible.
Polyester-urethane—Lighter in weight (not available from most manufacturers).
Buna-N—More difficult to obtain. Is used mainly on LPG and hydrocarbon products.

These pigs are generally designed for one pipe size only. They accommodate to pipe out-of-roundness, but do not usually have the capability of traversing different-size pipes, although several manufacturers have some models that attempt this with varying degrees of success. Small dents or buckles in the pipe will not stop the pig, but large ones may. These pigs can be used in a variety of chemical, hydrocarbon, and internal coating applications. (This article will not cover the application of internal coatings.)

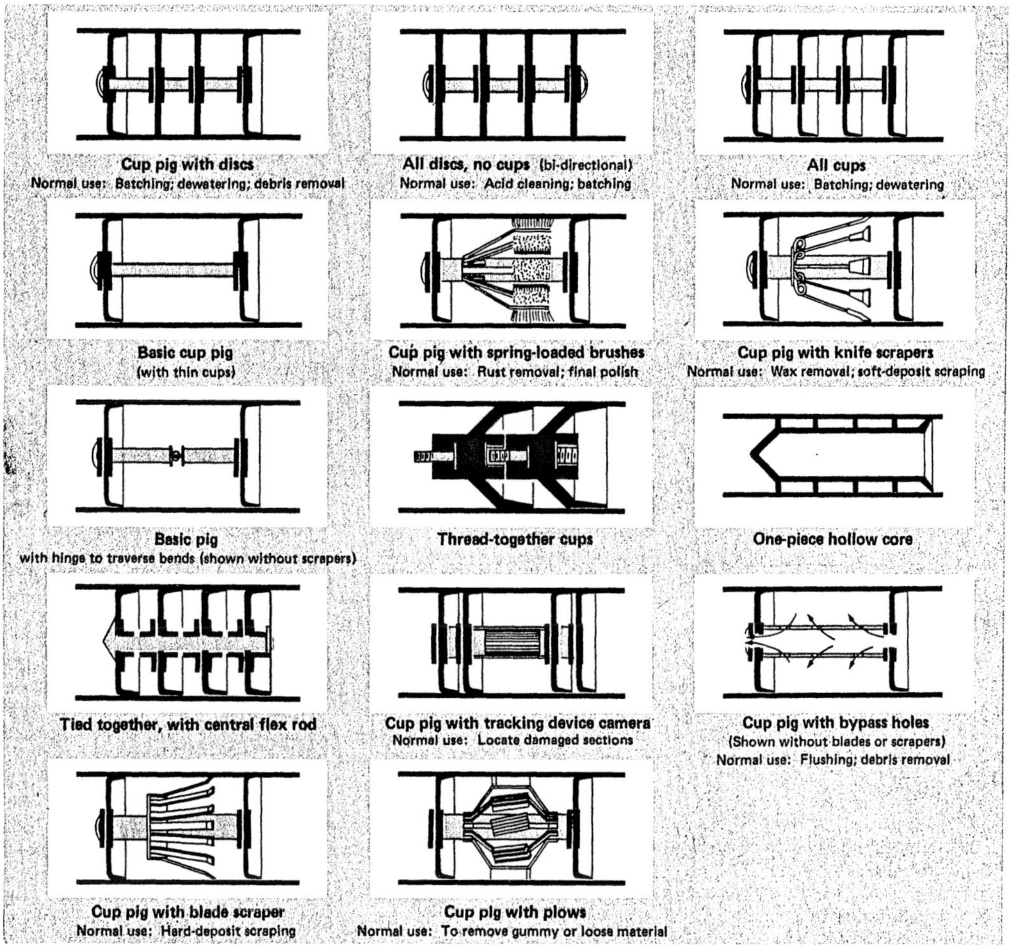

FIG. 1. Cup-and-disc pigs can be had in various combinations and with numerous accessories.

Foam Type

This type of pig can also be made up into many configurations to suit particular needs (see Fig. 2). The basic pig consists of an inner core of open-cell foam in a variety of densities. It can be used in the pipe as plain foam, or it can be provided with a variety of rubber or plastic coverings for wear resistance and turning effects.

Additional options of various patterns of brushes, or silicon carbide granules, plus some protruding cutters, are available. The lengths will also vary according to needs—short for short bends, long for long traversing tees, valves, etc.

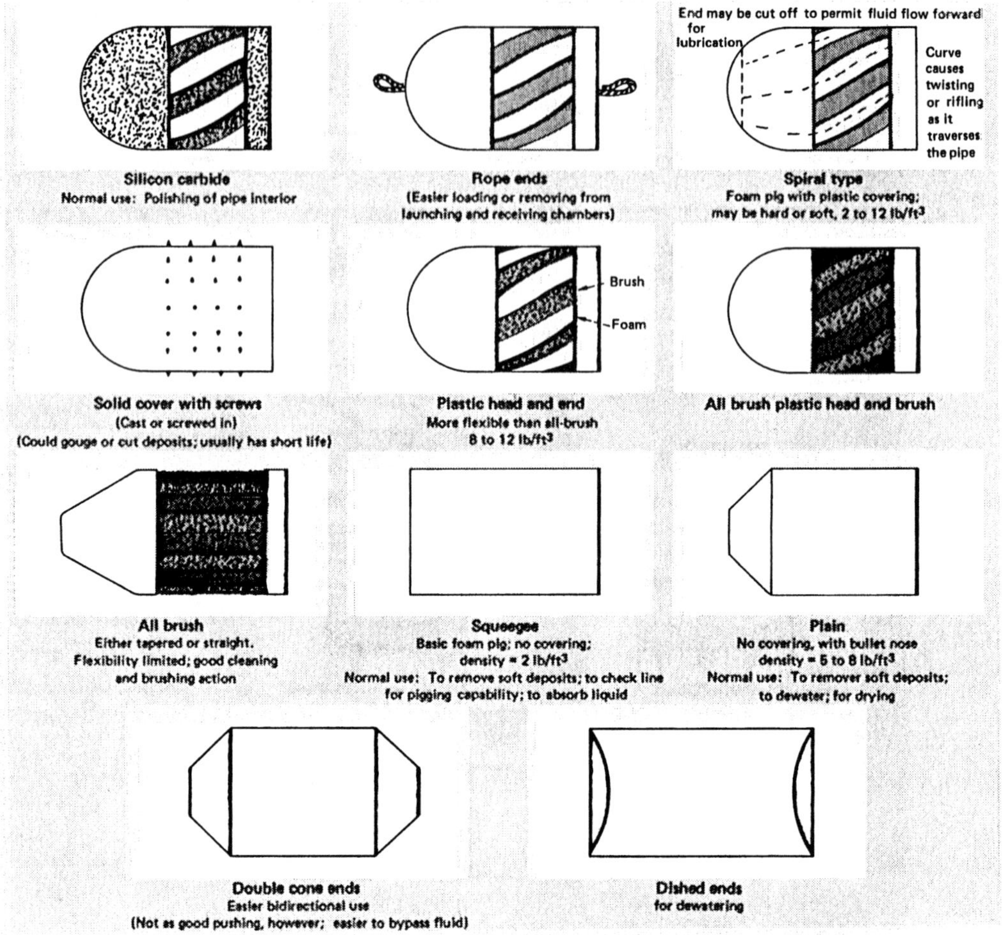

FIG. 2. Foam pigs are available in many configurations to suit specific requirements.

These pigs have the greatest capability of passing through restrictions, and traversing bends and sharp turns (e.g., into tees), and can be used in several different sequential line sizes. The extremely soft ones can be squeezed to half their original size.

Pigs with coverings, however, are restricted in their degree of flexibility, and caution must be exercised in their use; otherwise the pig will either stick in the line, bypass fluid, or tear.

Foam pigs are generally not suitable for strong alkaline or acidic fluids. Most solvents will destroy them. They can be employed for degreasing and for most petroleum products, but they are best suited for use in water-line services.

Spherical Type

The spherical pig is simply a round rubber, or plastic, ball (Fig. 3). It can be solid or hollow. The hollow type can be filled with liquid and expanded (by pumping) to the desired diameter. These balls are primarily used for separation of sequential batches of differing liquids, but are excellent for cleaning. They are ideal for lines where tight bends occur, especially if numerous bends occur within a short distance. They can be expanded to match any pipe wall schedule in their size range and can be used for internally coated pipe. These pigs clean more by crushing the deposits than by scraping.

Pigging Pressures

To move any pig through a pipe requires that a seal or contact be made between the pig and the pipe surface. Pressure upstream of the pig will force it through the line, owing to the differential pressure on its two ends. The pressure required to move a pig is a function of the following:

Friction between the pig and the pipe
Uphill/downhill movement
Lubricating quality of the pipe's fluid
Backpressure on the pig

Each of the pig designs requires a different pressure for movement, and pressure variations are required within each type depending on the pig's

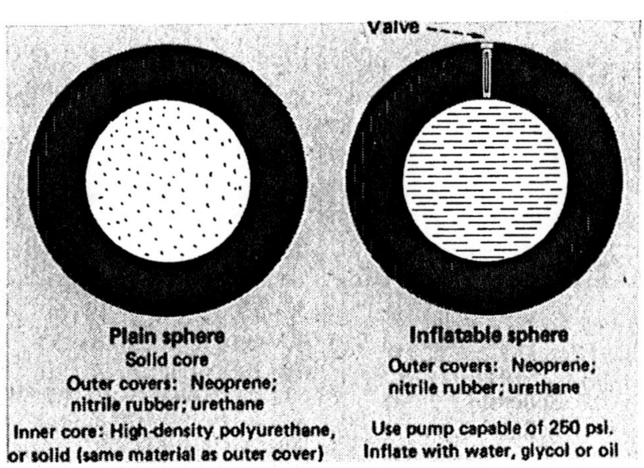

FIG. 3. Solid and inflatable spherical pigs.

hardness. Each type has a different sealing or surface contact that must be considered in the initial selection stages.

The larger the pipe size, the less the differential between seal area pressure and pushing pressure, therefore the easier the movement.

Example. 6-in. pipe with 1-in.-wide disc: push area = 28.5 in.^{2}, and seal area = 18.8 in.^{2}, or 1.5:1 ratio. 12-in. pipe with 1-in.-wide disc: push area = 113.1 in.^{2}, and seal area = 37.7 in.^{2}, or 3 times the seal area.

For batch separation or new-line cleaning, the pressure of the system is normally more than adequate for pig movement. However, for cleaning dirty lines, the pressure must be watched in the early stages to ensure that it does not exceed that allowable for the pipe. This is particularly critical in municipal systems where water mains are weak and the taps are brass or brittle cast iron, or where joints of leads, packing, etc. could blow due to excess pressure. The pressure limitations could dictate the type of pig as well as the sequence of cleaning operations.

The actual pressure required for cleaning cannot be determined exactly, due to the many unknown variables. Some trial and error is required in the initial stage of every job. For reference, however, the figures shown in Table 1 are typical ranges of initial pressures we have encountered on short-length water lines when using various pigs and sizes. Gas or petroleum lines are simpler, but present other problems. There pressures are naturally reduced as deposits are removed and the interior becomes smoother.

In most cases it takes up to two or three times the moving pressure to start the pig moving, either from the launching chamber or after one stalls in the line.

TABLE 1 Typical Pigging Pressures,[a] lb/in.^{2}

Type of Pig	2 in.	6 in.	10 in.	18 in.
Cup/disc:				
Four-cup	50–200	35–140	15–100	10–60
Foam, 5 lb:				
Plain	30–50	15–50	15–35	10–30
Criss-cross	40–100	20–50	15–60	10–50
Brush	40–100	20–80	15–60	10–50
Sphere:				
Ball	30–50	15–60	15–40	10–40

[a]Applications are for pipe with multiple bends, at least 25% restriction. As each pass is made, the pressure reduces. Initial breaking of a high point or spike may require a momentary fluid-pressure spike.

Pig Velocities

Velocity is important, so that the pig hitting the deposits will shatter, shear, and crush them. It is best to run any of the pig types between 2 to 6 mi/h (3 to 6 ft/s) for most cleaning applications. Although higher rates have been employed, they should be used with caution since a line and pig could be damaged by excess velocity. Slower speeds will either stall the pig, especially when air is the propelling fluid, or will not provide effective cleaning of the line.

When cleaning pipes, it is impossible to keep the pig moving at a constant speed, due to the various direction changes, restrictions, and the amount of fluid bypassed for upfront flushing. Speed variation is even more pronounced when air (or some other gas) is employed as the propellant.

The velocity can be controlled somewhat by applying backpressure at the discharge end, but this requires increased pressure on the inlet side.

Propelling Fluids

Water and air are the most common propelling fluids for cleaning operations, with water being the most popular due to availability and better flushing action. Water will also provide a more uniform movement of the pig through the pipe since it does not compress or expand as does air.

Air, however, is more convenient and creates less of a handling problem in larger-sized pipes, but a pig will stall if the volume of air is not sufficient to keep it moving at a steady rate. The pig will rest until sufficient pressure is built up to break the initial hold. Then, once broken loose, the pig will rifle down the line until the upstream pressure drops. The start/stop sequence will continue throughout the pig's travels through the line.

It is also possible to use a combination of air and water for cleaning. This has proven a valuable tool to us, but a severe word of caution: Waterhammer is created in the line and unless careful technique is used by experienced personnel, line damage will occur.

Pig Launching/Receiving Chambers

Pig launchers and receivers can vary in design depending on several factors:

Type of pig
Amount of deposits removed per run
Nature of material being removed
Accessibility to the line

In all cases the basic reason for using such chambers is to provide a starting point from which to launch the pig and a receiving point to catch it (Fig. 4). The larger end of the launching chamber is needed to ease the loading of the pig into the pipe (all pigs are slightly larger than the pipes they are entering). To insert a pig larger than the pipe i.d. would be rather difficult without the launching chamber. The only pigs that can be inserted relatively easily without an enlarged section are soft foam ones, with densities of under 5 lb/ft^3.

The length of the launch chamber is determined by the length of the longest possible pig to be used. The small end will attach to the receiving pipe with a flange or coupling, or it can be welded on. The entrance is provided with a cap, flange, or quick-opening door. Miscellaneous connections for pushing fluid, pressure gauges, etc. can be added as desired.

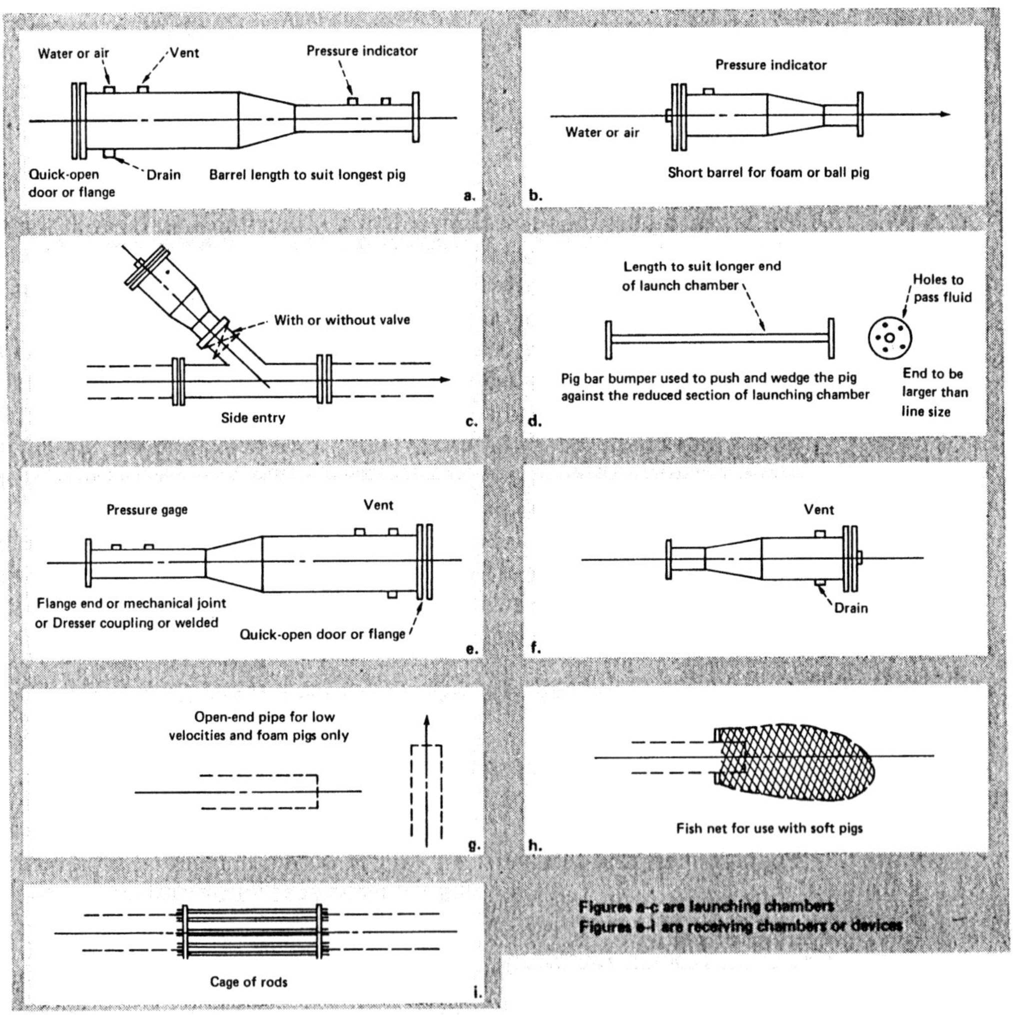

FIG. 4. Launching and receiving chambers for pigs.

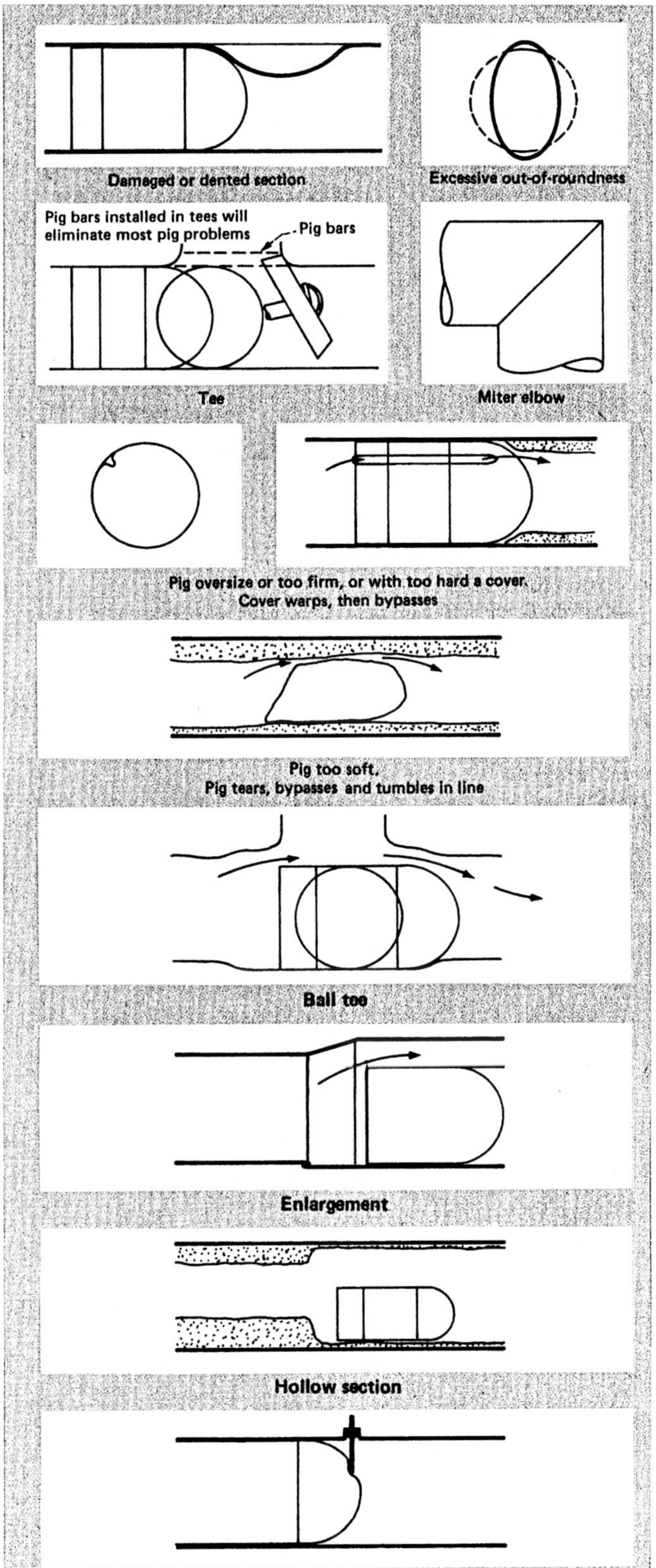

FIG. 5. Pipeline obstacles: Pigs may hang up or fluid may bypass.

The receiving end can be identical to the launching end, especially if dual-direction passage is desired, or may be simplified as shown in Fig. 4. Special designs for mounting on fire hydrants or adapting to a variety of equipment or locations are possible. The basic thought is always the same—get the pig going.

Several points will expedite the launch of the pigs (there are commercial indicators that will signal successful launch and arrival). A pig bar, which is merely a pipe or rod long enough to push a pig firmly against the reducer, will help. The end of the bar should be larger than the pipeline size, otherwise the fluid velocity can carry it into the pipe. The end should also have slots to permit fluid flow.

Launchers and receivers for oil/gas fields are more complex and will not be reviewed here.

Pigging Obstacles

There are limitations to the use of pigs (Fig. 5) that should be carefully looked into before a job begins:

Valves—Unless all valves are full open, a pig can get stuck. Butterfly and globe valves are forbidden. Check the length of valves to make sure the body is not longer than the pig. A pig could stop and remain if the valve is too long.

Tees—Ball-type tees may cause a short pig to drop and bypass. Stiff pigs of all designs can catch on the lips of tees if the pig bars are not installed.

Elbows—The design of the pig must be matched with the radius of bends in the pipe, otherwise disaster will occur. Generally, no pig is able to traverse a one-cut miter.

PETER KIPIN

Coatings

Internal pipe coatings, no matter how applied, definitely provide benefit by reducing power costs, increasing flow, or eliminating corrosion. Rehabilitating a pipeline with in-place internal coating versus replacement returns the line to active service in a minimum of time at a fraction of replacement cost, with less public disturbance or environmental headaches, and minimal capital expenditure.

During research and studies of flow improvement due to pipe cleaning, additional studies were made to determine the improvements to pipeline flow and corrosion protection with the use of coatings.

The conclusion of studies by such organizations as the American Gas Association and extensive field testing by Transcontinental Pipeline Corp. was that both flow improvement and corrosion protection were evident.

Internal pipe-coating applications have become standard practice for many companies. Many pipe manufacturers offer internally coated pipe.

The pipe materials using internal coatings include iron, steel, asbestos, cement, concrete, and clay.

Internally coated pipe done by the in-place method can rehabilitate existing, old, or abandoned pipes. Application of coatings employs either the joint-by-joint (shop) or in-place (*in-situ*) method.

There is no limit to the pipe services which can benefit from coatings. Typical services include crude oil, natural gas, and petroleum-product pipelines. Rehabilitation of existing pipe is commanding greater attention due to the high cost of replacement. Internal, in-place coating is being looked to for the bulk solution.

The problem has brought inquiries coming from Japan, Russia, the Middle East, South America, and Africa.

Slip lining and cement-mortar lining have provided relief for short lengths (500 ft and less), but are prohibitive in long lines. They are also costly and time consuming even on short lengths.

Long-term storage (in excess of 10 years) of pipe has been accomplished by use of coatings. This has been especially important to gas companies who maintain a stock of pipe for emergency use. The pipe, usually of heavy wall thickness, must be available on short notice.

Coating Materials

Many types of coatings have been investigated for use in most pipeline services. The American Gas Association and the National Association of Corrosion Engineers have much documentation on the various coatings, their corrosion resistance, and application.

The types of coatings potentially available for internal coating can be drawn from a wide spectrum of formulas. For specialty applications, such as required in the chemical industry, these can be drawn upon for shop application of short lengths.

However, the coatings which can best meet the needs of the pipeline industry are the polyamide-cured epoxies and amine adduct types, with the polyamide-cured epoxies having the edge.

As a general comparison, the polyamide-cured epoxies are superior in flexibility and water resistance, while the amine type is superior in solvent and chemical resistance. The polyamide-cured epoxies are better in solids at application viscosities, have much less tendency to "crawl" on a metal surface (i.e., better surface wetting), longer pot life, and much less toxicity hazards, particularly skin irritations.

These factors can sometimes make the difference between the success or failure of a coating job. The amine type is more susceptible to producing a porous film which can be patched in shop coating but is critical to in-place coating.

For in-place coating applications, polyamide-cured epoxies are preferred due to their characteristics which are more conductive to in-place coating methods. Where chemical resistance requires the amine-cured type, the in-place application should be repeated after the first coat to ensure a pin-hole-free surface.

In actual practice and observation, rather than scientific documentation, long-term adhesion of polyamide-cured epoxies is superior to that of the amine type. Also, long-term impact resistance and flexibility of polyamides

TABLE 1 Coating Comparison

	Coating	
Property	Amine	Polyamide
Hardness	5H to 6H pencil hardness, faster initial cure	4H to 5H pencil hardness, slower initial cure
Tolerance to inadequately prepared surface	Good	Very good
Brittleness	Very brittle in a few months	Same resiliency remains after a few months
Sag resistance 5–7 mil wet	Good	Good
Adhesion	Good	Good
Application characteristics	Very good	Excellent
Flexibility		Best
Abrasion resistance	Equal	Equal
Water resistance		Best

over amines are observed. This could be most important on above-ground applications and lines subjected to frequent pigging.

A comparison of the two types is presented in Table 1.

Pipe Lengths

Shop-coated pipe is most common in 40-ft lengths. However, some shops can go as long as 80 ft.

In-place (*in-situ*) coated lengths are virtually unlimited in length, but the most economical lengths of pipe to coat in place are those over 3000 ft long because it takes the same amount of time and equipment to set up, prepare, haul, and dismantle for a 30,000-ft run as it does for a 1000-ft run. The only difference is the material cost and some labor.

Short lengths, under 500 ft, if accessible, can best be done manually with a mechanical spin sprayer by using mechanical or chemical cleaning. The mechanical means involves using sand blasting with a special spinning sand-blasting head. Chemical cleaning is usually done with muriatic acid followed by inhibited water flushes and dry-air drying.

In-place coated lines are not affected by pipe bends or elbows. Such items as these can be tolerated. Valves, such as full-bored ball valves, do not interfere with the process. Screwed fittings or socket-welded fittings have also been successfully coated.

Surface Preparation

For steel pipe, the surface must be free of all foreign material. The surface must be cleaned to a white condition for best results.

For concrete, cement, and similar materials, all loose materials must be scraped and blown free, then conditioned with an acid to ensure tight material to bond to.

It has been determined that the coating material will penetrate into the cement surface, thus hardening and conditioning the material after coating.

Shop-coated pipe is normally shot blasted or sand blasted. In-place pipe is best and more economically cleaned to a white surface by acid treatment. It is especially important on existing lines that the products and deposits within the line be determined.

Should the surface contain oily or greasy deposits, these must be removed first by a degreasing solution such as a caustic solution or any one of a number of commercial chemicals available for degreasing. In some cases, regular gasoline has been used as the initial grease and oil-cutting agent. In other cases, special scrapers will accelerate the cleaning job. Foam-type pigs are not efficient in degreasing operations.

Coating Application

Shop-applied coating is applied by use of an airless spray. An airless spray head with multiple nozzles, depending upon pipe size, is inserted into the pipe. Either the pipe is rotated or the spray head is rotated during application.

The nozzle starts at the far end and is withdrawn at a predetermined rate to give an even coat. Such variables as type of coating, viscosity, temperature, tip size, boom speed, rotating speed, and paint pump pressure play an important part in a quality coating job.

In-place coating is applied by placing into the pipe two sets of plugs with coating in between them. All air is vented.

The "train" is moved through the pipe by air pressure with a backpressure on the downstream side to ensure a tight fit is maintained between the two plugs.

Differential pressure, speed, internal temperature, and viscosity play important parts in a quality job. For relatively smooth internal surfaces, in-place coating can be applied in one operation, but good practice dictates that the "train" should run in one direction, the coating plugs be reversed, then run in the opposite direction to ensure that all "dead" spots such as misaligned fittings, welds, etc. are coated.

For badly pitted steel, concrete, or cement pipe, the coating should be applied in two or more coats depending on the pit depths. This is extremely important since solvent entrapment can occur in a thick coating.

Important Considerations

Several items should be considered for in-place coating application. For example, weld icicles in excess of 1/4-in. can cut the coating tools. On new lines these can be discovered or knocked off by running a sizing plate. Under no circumstances should pipe be coated if the ground temperature around the pipe is less than 45°F. The coating will not cure, even with forced hot air.

For temperatures in excess of 95°F, the pipe surface should also be avoided since the pot life could cure the batch before it reaches the other end. This is especially true for above-ground lines exposed to the sun. This can be overcome by delaying the coating phase to evening hours.

Humidity has an effect only during the drying stage prior to coating. This varies extensively by locale and time of year, even on a day-to-day basis. Should drying fall on a day which is humid or raining, dried air or nitrogen should be used for best results.

Extreme out-or-round pipe, due either to the pipe manufacturer or to the pipebending machine, can cause coating to bypass or seize the entire train. On steel lines this is easy to check with a sizing plate which is run through the line after rough cleaning.

All in-place valves, except full-port ball valves, should be replaced with temporary spool pieces. Gate valves, although full port, will have their seats filled with coating. Globe valves and check valves of the piston type will not allow passage of the coating plugs. Swing checks will have their seats filled with coating.

Tees of any size should be avoided since there is no way to really prevent some coating runs. If tees must be kept in place, pig bars at the branch circumference must be installed. Tees of the enlarged ball type should also be avoided to reduce runs.

One-cut miters are absolutely to be avoided. Others can be used provided they maintain a 1½ times diameter radius.

On some gas lines, branches were made by inserting the takeoff into the pipe, then welding. These can penetrate anywhere from a fraction to all the way to the bottom. These are usually found during the cleaning step and must be removed.

PETER KIPIN

Wear

By the very nature of the transport process, pipelines used for pneumatic conveying systems are prone to wear when abrasive products have to be conveyed. In a dilute phase, products are conveyed in suspension in the air, and a high conveying air velocity must be maintained in order to keep the product moving and avoid pipeline blockage.

The main problem relates to the wear of bends in the pipeline and any other surfaces where particles are likely to impact as a result of a change in flow direction. Bends provide pneumatic conveying system pipelines with their flexibility in routing, but if the product is abrasive and the velocity is high, rapid wear can occur.

Influence of Particle Hardness

The main property of a bulk solid, in determining whether erosive wear is likely to be a problem or not, is particle hardness. A graph showing the relationship between particle hardness and specific erosion is presented in Fig. 1. This was derived from test work carried out with a range of bulk solids pneumatically conveyed through a 2-in. bore pipeline. The wear relates to that of 90° mild steel bends in the pipeline [1].

Erosive wear is expressed in terms of the mass of metal eroded from a bend per tonne of product pneumatically conveyed through the bend. It will be seen that there is a plateau with respect to particle hardness, and that for particles harder than silica, or quartz, there is essentially no further increase in erosive wear.

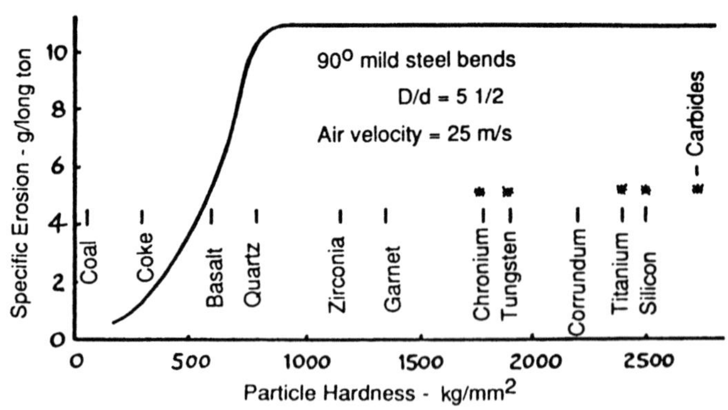

FIG. 1 The influence of particle hardness and the erosion of steel bends.

The hardness values of typical products, both potential conveyed products and bend surface materials, have been superimposed for reference. It will be noticed from this that coal is a very soft material and is unlikely to be a problem with respect to erosion. In reality, of course, both pulverized (pf) and lump coal are erosive products. This, however, is due to the presence of noncombustible minerals, such as quartz and alumina in the coal, and not to the coal itself.

With large tonnage flows of coal, even small percentages of these highly abrasive minerals will cause severe wear. A similar situation applies to pf ash and other products containing small percentages of similar contaminants, such as barites and wood chips.

Influence of Velocity

Of all the properties relating to conveying parameters, conveying air velocity is the most important. If the product being conveyed is abrasive, erosive wear must be expected, but the order of magnitude of the problem is essentially dictated by velocity.

The influence of conveying air velocity on erosive wear is illustrated in Fig 2. This is also drawn from actual data obtained from a pneumatic conveying system pipeline. It relates to 90° mild steel bends, and erosive wear is expressed in specific terms again. The conveyed product was sand.

These bends had a wall thickness of about 4 mm, and failure occurred after approximately 100 g of metal had been eroded. At a velocity of 30 m/s, therefore, the bends would fail after only 5 tonnes of sand had been conveyed. A particular problem with pneumatic conveying systems is that air is compressible and so the air velocity gradually increases from the pick-

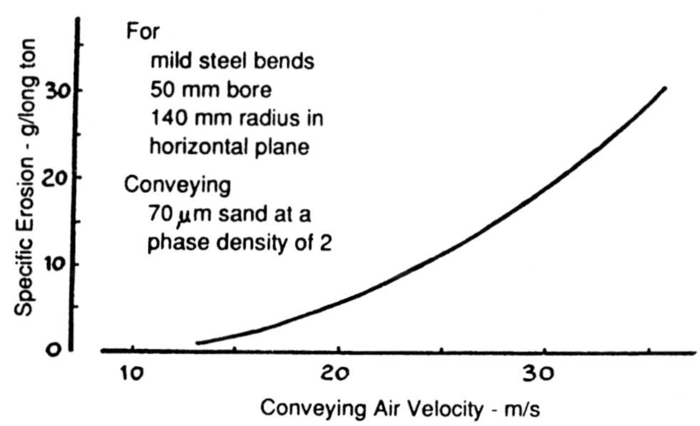

FIG. 2 The influence of velocity on erosion.

up point to the receiving hopper, along the length of the pipeline. This helps to explain why bends at the end of a pipeline generally fail first.

Influence of Bend Geometry

Bends are available in a wide range of geometries in terms of bend curvature, from long radius bends to tight elbows and mitered bends. Because bends are so vulnerable to wear, there have been many developments and innovations for reducing the problem. Much work has been carried out on conventional radiused bends to determine whether there is any optimum geometry.

Brauer and Kriegel investigated the influence of bend geometry, and the results of some of their work are presented in Fig. 3. The curve of Brauer and Kriegel [2] relates to the erosion of steel bends in hydraulic conveying, and that of Kriegel [3] to the erosion of Plexiglas bends in pneumatic conveying. The hydraulic transport work shows a pronounced maximum value of erosion at a bend diameter, D, to pipe bore, d, ratio of about 5.5:1. The pneumatic conveying results are very similar, although the slope of the curves either side of the maximum is not so great.

The author has also investigated the influence of bend geometry [4]. A wider range of D/d values was investigated, and the results for 90° mild steel bends are shown in Fig. 4. The bends were eroded by sand, conveyed at a phase density of 2, and with an air velocity of 25 m/s. The erosive wear is in terms of the mass of metal eroded from the bends, in grams, per tonne of sand conveyed.

The results can, to a certain extent, be predicted from the data presented in Fig. 5 on the influence of impact angle. To explain this a little further, bends with D/d ratios of 2, 5½, and 10 have been drawn to scale in Fig. 6.

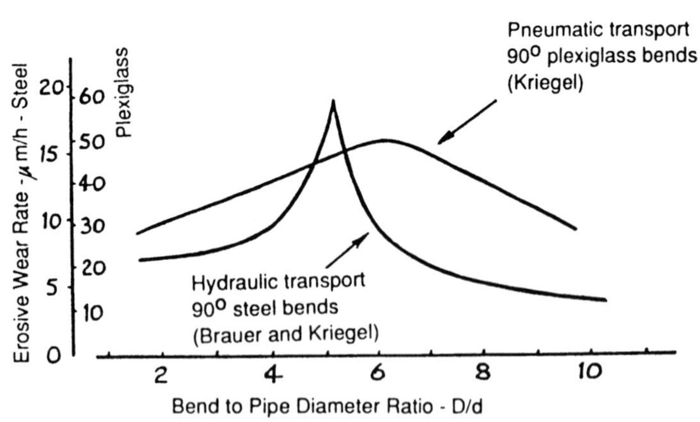

FIG. 3 The influence of bend geometry on erosive wear.

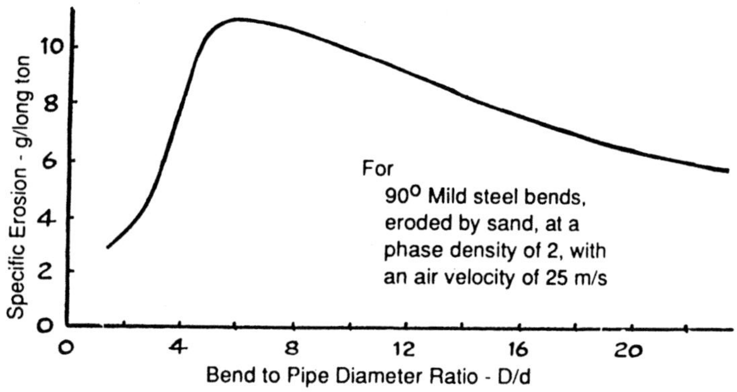

FIG. 4 The influence of bend geometry on erosive wear.

With the sharp bend having a D/d ratio of 2, it can be seen that the majority of the particles will impact against the bend wall at a fairly steep angle. The data presented in Fig. 4 are for mild steel bends, which are ductile. At a high impact angle the erosion is not too severe, and so it can be expected that the bend will not wear too rapidly.

The bend with a D/d ratio of 5½ corresponds to the worst case from the data in Fig. 4. It can be seen that the majority of the particles impact against the bend wall at an impact angle of about 20°. For a ductile material this will result in maximum erosion, and so the bend can be expected to fail quickly.

Effect of Impact Angle

Erosive wear is significantly influenced by the angle of attack of the impacting particles against surface materials, and by the surface material itself. This is illustrated in Fig. 5 which shows the variation of erosion with impact angle for two different surface materials. These materials show vast differences

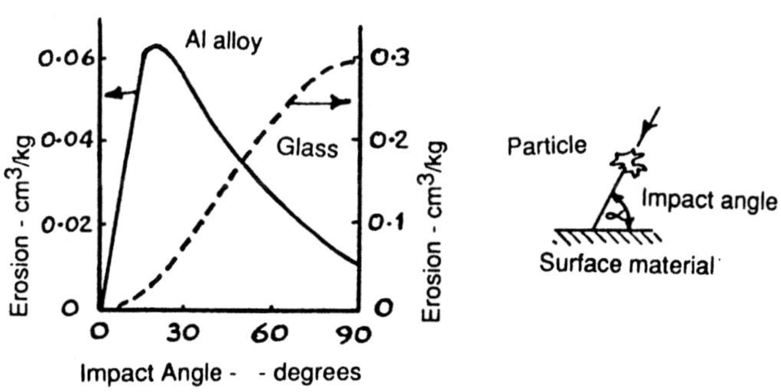

FIG. 5 Variation of erosion with impact angle for various surface materials.

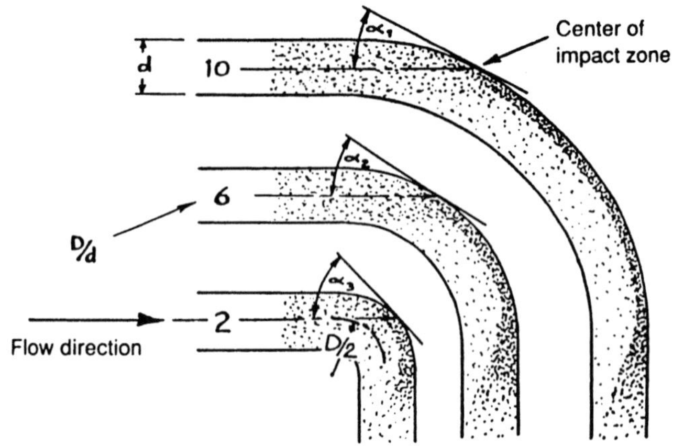

FIG. 6 Influence of bend geometry on particle impact angle, α, for suspension flow. $\alpha_1 < \alpha_2 < \alpha_3$.

in both erosion rate and impact angle. The generally accepted reference angle for impact is also defined on Fig. 5.

The aluminum alloy is typical of ductile materials; it suffers maximum erosion at an impact angle of about 20° and offers good erosion resistance to normal impact. The glass is typical of brittle materials and suffers severe erosion under normal impact.

Influence of Surface Material

A knowledge of the tendency of a bulk solid to cause erosive wear is important at the design stage of a bulk solids handling plant. A knowledge of the potential wear resistance of surface materials is equally important. With respect to erosive wear it is clear that hard and brittle materials should not be used in situations where normal impact of particles is likely to occur. Under conditions of normal impact, however, ductile materials could be used. In glancing, low impact angle situations, brittle materials are ideal and ductile materials should be avoided. A knowledge of flow conditions is also important here.

The particle impact against the wall for the bend with a D/d ratio of 10 is at a much shallower angle. If the impact angle is relatively small, the erosion will not be too severe, and so for this case also it can be expected that the bend will not wear too rapidly.

In pneumatic conveying, therefore, either sharp elbows or long radius bends are used in order to avoid this rapid wear situation. The preference, however, is to use long radius bends. The pressure drop in conveying a product round a tight bend is greater than that for a long radius, or easy bend, and as the minimizing of line pressure drop is also a major design feature in pneumatic conveying plants, the general recommendation is for the installation of long radius bends in conveying lines [5].

If a brittle material is to be used for a bend wall surface, such as basalt, Ni-hard cast iron, or a ceramic material, it is essential that a reasonably long radius bend be used, for short radius bends will result in high impact angles and rapid wear, as shown in Fig. 5.

Wear Patterns

Mason and Smith [6] carried out tests on 25 and 50 mm^2 section 90° bends with a flow of alumina particles from vertical to horizontal. The Perspex bends were constructed with substantial backing pieces in order that the change in flow pattern and wear over a period of time could be visually observed. The results from one of their tests are given in Fig. 7.

On reaching a bend, the particles tend to travel straight on until they impact against the bend wall. After impact they tend to be swept round the outside surface of the bend. They are then gradually entrained in the air in the following straight length of pipe, as shown in Fig. 6.

In Fig. 7 the flow pattern is shown after substantial wear has occurred. This shows quite clearly the gradual wearing process of a bend and the effect of impact angle on the material in the process. Erosion first occurred at a bend angle of 21°, which became the primary wear point, as expected. After a certain depth of wear pocket had been established, however, the particles were deflected sufficiently to promote wear on the inside surface of the bend and then to promote a secondary wear point at a bend angle of 76°.

A small tertiary wear point was subsequently created at an angle of 87°. If such a well-reinforced bend were to be used in industry, in preference to replacing worn bends, the deflection from the latter wear points would probably cause erosion of the straight pipe section downstream of the bend.

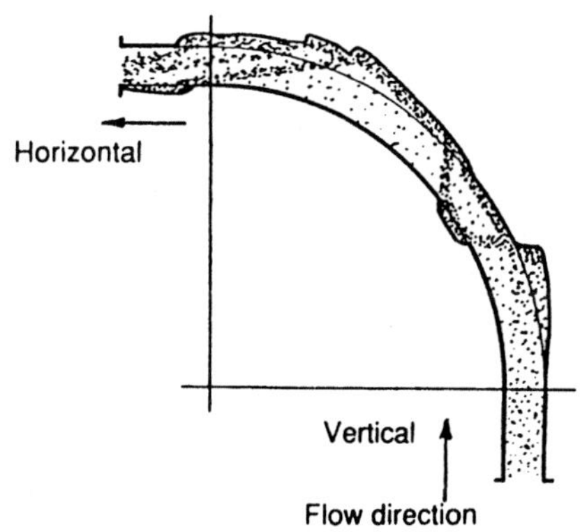

FIG. 7 Wear and flow pattern for an eroded bend.

Effect of Impact Angle

The curve in Fig. 5 of erosion against impact angle again provides a means by which an interpretation of the type of wear produced can be obtained. The outer wall of the bend initially presents a surface at a low impact angle to the particles issuing from the preceding vertical straight pipe run, and as Perspex is a ductile material, rapid erosion takes place.

Gradually the impact angle at this primary wear point changes to almost 90°. From Fig. 5 it can be seen that ductile materials suffer relatively little erosion under normal impact, and this explains why little further erosion takes place at this point.

The conveyed product can be seen quite clearly to be deflected out of this primary wear pocket. Because of this abrupt change in direction, however, it is no longer swept around the bend as before, but impacts on the inside surface of the bend. It is then deflected to the outer wall again, and because the low impact angle is maintained here, the erosion at this point is far greater than that at the primary wear point.

Mason and Smith [6] also mention that a conventional bend design used to avoid plant shut down due to bend wear is to reinforce the outside of the bend with a mild steel channel backing filled with a suitable concrete. They included a radiograph of such a 100-mm bore pipe, and this shows a primary wear pocket developing in precisely the same manner as for the Perspex bend tests. It is believed that the bend ultimately failed through erosion of the inner surface due to product deflection from the primary wear point.

Straight Pipeline

Wear of straight pipeline is rarely a problem, although with large particles it can be severe along the bottom of the pipe. Streaming of particles, as shown in Fig. 7, however, is a common cause of straight pipeline failure. From wear pockets in bends the following section of straight pipeline can be eroded, as mentioned above. Misaligned flange joints, and welded joints with weld metal protruding inside the pipeline, as illustrated in Fig. 8, can often lead to straight pipeline failure, particularly in small bore pipelines.

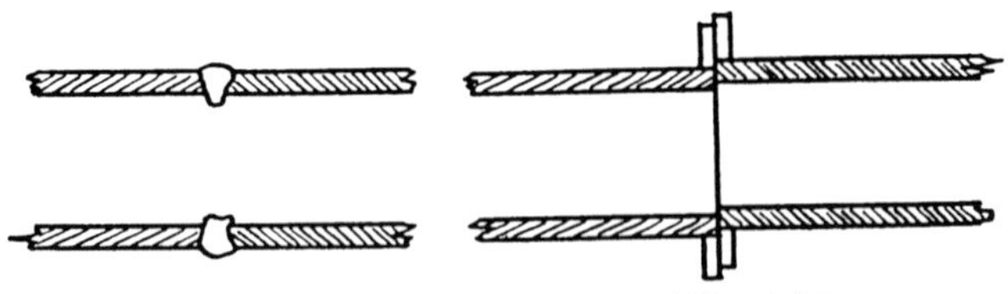

FIG. 8 Examples of erosion promoting sites at poorly jointed pipe sections.

Hot, dust-laden gases from boilers and reactors are often passed through heat exchangers for generating steam. The tubes through which the gases flow often wear, and they are usually very expensive to repair. The wear is usually only at the start of the tube. This is because the dusty gases on entry to the tube are in a very turbulent state and numerous particle impacts occur. After a short distance the flow is effectively straightened out and little further pipeline wear occurs. An effective solution to the problem is to provide a sacrificial extension to the pipe prior to the tube plate and the heat exchange section for flow straightening purposes.

References

1. K. N. Tong, D. Mills, and J. S. Mason, "The Influence of Particle Hardness on the Erosion of Pipe Bends in Pneumatic Conveying Systems," in *Proceedings of the 6th Powder and Bulk Solids Conference,* Chicago, May 1981.
2. H. Brauer and E. Kriegel, "Verschleiss an Rohrleitungen bei Hydraulischer Forderung von Festoffen," *Stahl Eisen, 84,* 1313–1322 (October 1964).
3. E. Kriegel, "Druckverlust und Verschleiss in Rohrkrummern bei Pneumatischen Transport," *Verfahrenstechnik, 4,* 333–339 (August 1970).
4. K. N. Tong, D. Mills, and J. S. Mason, "The Influence of Bend Radius on the Erosion of Pipe Bends in Pneumatic Conveying Systems," in *Proceedings of the 5th Powder and Bulk Solids Conference,* Chicago, May 1980.
5. D. Mills and J. S. Mason, "The Influence of Bend Geometry on Pressure Drop in Pneumatic Conveying System Pipelines," in *Proceedings of the 10th Powder and Bulk Solids Conference,* Chicago, May 1985.
6. J. S. Mason and B. V. Smith, "The Erosion of Bends by Pneumatically Conveyed Suspensions of Abrasive Particles," *Powder Technol., 6,* 323–335 (1973).

DAVID MILLS

Repairing, In-Service

New, improved techniques have been developed for in-service pipeline welding. In the process of developing these procedures, we determined that welding on pressurized pipe should be done only if the wall thickness exceeds 5 mm (at 66 bar operating pressure). In addition, the heat input during the first manual weld pass is restricted by 2.5-mm electrodes. These two requirements minimize the risk of burn-through.

Gas Flow and Weld Quality

The gas flow causes the weld pool and its adjacent zones to cool rapidly. This is detrimental to the weld quality due to the formation of brittle structures and the entrapment of hydrogen. In combination with weld stresses, these create risk of weld cracking. Normal welding calls for preheating to reduce the cooling rate. However, due to the extreme cooling action of the gas flow, preheating is not feasible. Equipment is bulky and requires a lot of power. A different route was therefore followed.

In the past many split tees have been welded on in-service pipe lines. A series of 20 was tested for bursting strength and weld quality. Although the latter was poor by present standards, the split tee never failed before the adjacent pipe sections.

We focused on how to offset adverse effects of the rapid cooling rate and minimize residual stresses. Two projects established welding procedures which, despite the extreme cooling rates, provide good weld quality and establish weld preparations that give good residual stress level and fatigue strength.

Welding Procedure

Four parameters influence cracking susceptibility: the formation of brittle structures in the heat-affected zones of the weld, hydrogen entrapment, residual stresses and weld imperfections. During in-service welding, hardness in the weld zone will remain high (400 HV as compared to the normal acceptance level of 320 HV). This is acceptable only if the other three parameters can be kept at an improved level.

Hardness

During welding the base material adjacent to the weld metal is heated to above its transition temperature. On subsequent cooling, brittle (hard) structures can occur in these so-called heat-affected zones (HAZ). Such brittle zones are prone to hydrogen-induced cracking and are also undesirable because of the reduced toughness. To a large extent, hardness depends on the cooling rate. Faster cooling means harder structures. Since the cooling rate cannot be influenced, a special welding technique is used. This "temper bead" method is based on the fact that a weld pass, if positioned properly, gives a previous pass a heat treatment such that the HAZ hardness of that pass will be reduced. Hardness levels can be kept below 400 HV.

Hydrogen Entrapment

The hydrogen source during manual arc welding is the covering of the welding rod which contains crystallization water. In the weld arc this is split into hydrogen and oxygen. Hydrogen readily diffuses into the weld metal. During cooling, the hydrogen solubility reduces and, if the cooling rate is too high, hydrogen is trapped in the weld metal and adjacent zones. This causes high internal stresses. The cellulose-coated electrodes normally used in pipeline welding give very high hydrogen contents. But proper preheating and postheating reduces the cooling rate and hydrogen is allowed to diffuse out of the metal. Since such treatments are not feasible on in-service pipelines, the only alternative is to use a welding process that gives reduced hydrogen intake.

Semiautomatic welding processes such as TIG, MIG, and MAG are not commonly employed in pipeline welding. And experienced personnel are not available. Automatic or semiautomatic processes are more sensitive to weather conditions and the equipment is mostly very bulky. This creates handling problems in the field. The manual arc process using low-hydrogen electrodes is therefore used. These electrodes should give a hydrogen content of less than 2 mL/100 g. This requirement may seem extreme, but it is necessary to ensure adequate failure stress, especially when welding on older lines which often have a high carbon content.

Figure 1 shows the failure stress as a function of both diffusible hydrogen of the electrode covering and the carbon equivalent, and thus also the steel type and quality. Electrodes were prepacked in airtight vacuum containers in small batches (up to 10 electrodes) to ensure brief exposure to air.

Residual Stress

A number of factors influence residual stresses in both weld metal and HAZ. These are weld preparation, welding procedure, and weld metal/base metal matching. Apart from design criteria based on drilling forces acting during tapping procedures, weld preparations have been designed such that the lowest possible after-weld stresses could be recorded.

Several weld preparation layouts have been tested by both static and cyclic tests, using strain gauges to give information on stress levels. Designs in Fig. 2 gave the best results for fatigue strength and residual stress level, without reduction in weldability. Welding procedures, including buttering and temper bead welding, have been developed that further reduce cracking susceptibility (Fig. 3).

By using an extremely low-yield weld metal, deformations caused by the welding process are absorbed by the weld metal. This means that residual stress levels in the adjacent (brittle) HAZ's will be below its yield strength and will therefore be less harmful.

Weld Imperfections

Nondestructive testing techniques are often used as a final check on weld quality. Due to the geometry of the critical circumferential weld, neither radiography nor ultrasonic test methods are "feasible." Only magnetic particle inspection can be employed, but this reveals surface defects only. The absence of weld imperfections thus depends mainly on the welder's skill. So we decided that in-service welding should be confined to a *selected team* of welders.

Quality Assurance

Good, safe welds come from these factors:

No welding on pipes with wall thickness less than 5 mm
Weld preparation and weld sequence as specified

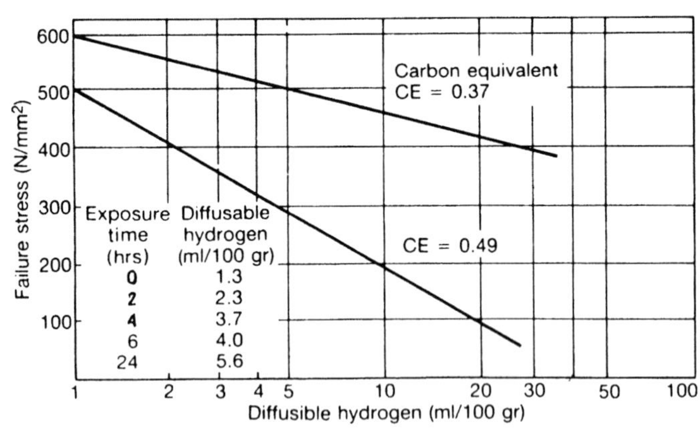

FIG. 1 The effect of diffusible hydrogen in welding rod coating (from humid-air exposure) on failure stress of weld deposit. (Diffusible hydrogen absorption of ϕ4 mm electrode after exposure to air—22°C, 90% RH.)

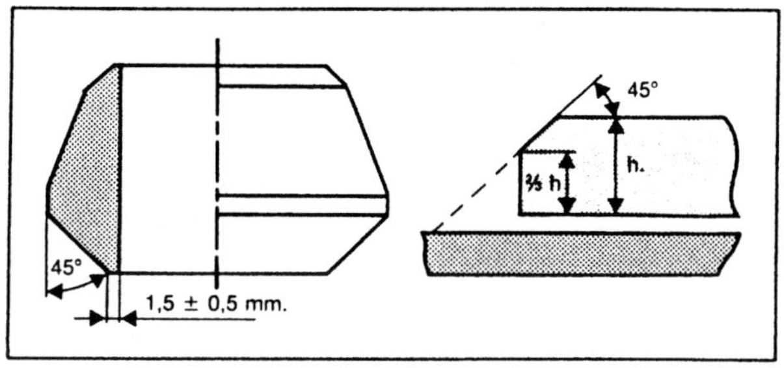

FIG. 2 Mandatory weld preparation for in-service welding. Left: Branch connections up to $\phi 2$ in. Right: Branch connections (split tees) and repair sleeves larger than $\phi 2$ in.

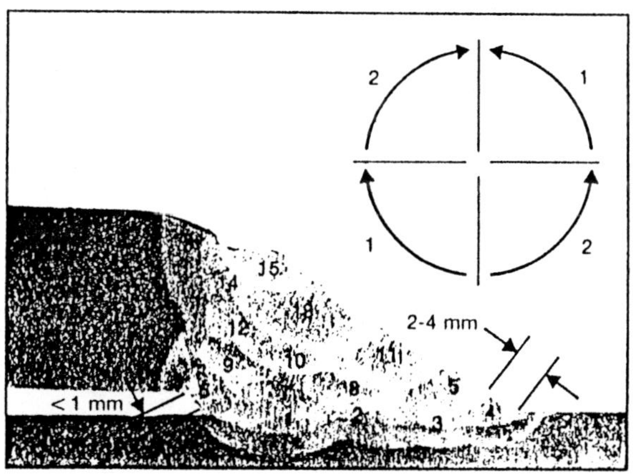

FIG. 3 Weld sequence for split tee, repair sleeve fillet welds.

Electrodes must be of the low-hydrogen low-yield type
Welders must be qualified for in-service welding
Weld passes directly on the pressurized pipe must be made with 2.5-mm ϕ electrodes to restrict heat input
Magnetic particle inspection of circumferential welds both immediately after 24 h and after completion of welding

Sleeve Repair Method

To lower risk, Gasunie has a repair procedure for pipe sections with axial notches and/or dents (the main types of damage) which makes it possible to repair a damaged pipe section in-service without welding into the pipe itself. This method, ESR (epoxy sleeve repair), uses an epoxy-filled sleeve around the defect zone to reduce the hoop stresses so low that neither defect fatigue growth nor plastic collapse can occur. The cost is comparable to that for welded sleeves but it provides a better safety level.

Principle

The basic idea is simple: equalize the pressure on both sides of the defective wall and the pressure-induced stress will be zero. Unfortunately, it was not as simple as that. But an effective method has been developed for the repair of pipeline sections damaged by corrosion or containing sharp axial notches (as caused by plows, caterpillars, etc.). Here are the steps:

Depending on the results of the damage assessment, the pressure in the damaged pipeline is reduced to a safe level
A steel sleeve is positioned and sealed
The gap between sleeve and pipe is filled with an epoxy compound which is forced to cure at a pressure equal to the (reduced) line pressure
After curing, the line pressure is restored to its original level and the pipe is fully operational again

The ESR method is shown in Fig. 4. Two repair sleeve halves are positioned around the pipe and joined by two longitudinal welds or flanges. The sleeves have three grooved rings on either side. These grooves contain gaskets pressed onto the pipe by injection. After pressurizing, these rings provide a pressure-resistant seal. The sleeve contains inlet and outlet ports for filling, pressurizing, etc.

Effectiveness

The method's effectiveness is influenced by repair pressure level, shrinkage and compressibility of the epoxy compound, and pipe wall/sleeve wall thick-

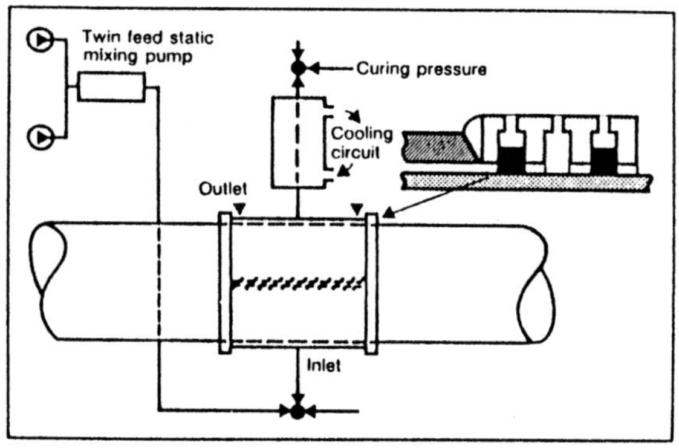

FIG. 4 In-service application of ESR method.

ness ratio (Fig. 5). The following conditions obtain optimum results:

The repair pressure should preferably be equal to the line pressure
Shrinkage of the epoxy during cure should be zero
Compressibility of the epoxy should be zero

It is possible (and safe) to restore the line pressure to normal operating level before the injected epoxy compound is cured. If, at the same time, the epoxy is maintained at the prevailing line pressure, effectiveness of the method is enhanced.

Both shrinkage and compressibility depend on the filling agent specified. Epoxy compounds have low shrinkage and low compressibility, especially when the bulk factor is high. Too high a bulk factor adversely affects workability. So a compromise has to be struck between workability and shrinkage/

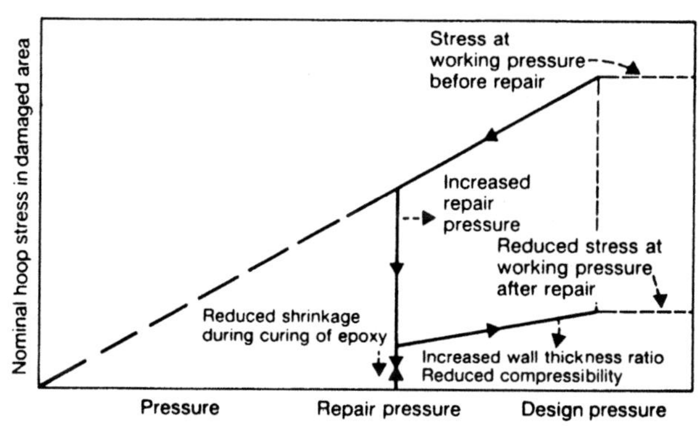

FIG. 5 Parameters affecting stress level after repair.

compressibility. The epoxy compound has been used in similar applications (filling of thin slits). It was thoroughly investigated for long-term behavior.

The sleeve wall thickness is specified as twice the pipe wall thickness, ensuring a sufficiently high wall thickness ratio. A further increase in size is not feasible. Its effect on the resulting stress level is small, handling is difficult on large diameter pipes, and welding time is prolonged.

Test Results

Test models had artificial damage (dents up to 4% of the diameter and notches up to 50% of the wall thickness). They would be unacceptable and would not be repaired under normal operating circumstances, but they do provide extreme test conditions. The pipes were instrumented with strain gauges, pressure transducers, and thermocouples to monitor the defect zone, but during the repair and after (see Fig. 6).

At operating pressure, the hoop stresses in the damaged pipe section could be reduced to 10% of the design level, effectively excluding the possibility of fatigue crack growth starting from notches in the defect zone. The steel encirclement sleeve prevents bulging of the defect zone, providing the repaired section with a bursting strength higher than that of the adjacent pipes. The ESR method thus gives a repaired pipe section with a safety level at least equal to that of the original pipe.

A drawback of the ESR method is that it is effective only for axial and other defects whose behavior is dictated by internal pressure, although these comprise most damage categories. Since no stress-bearing material is applied between sleeve and pipe, the sleeve will not support axial stresses induced by external forces (e.g., due to subsidence). Circumferential defects cannot therefore be repaired by this method. To repair such defects, sleeves have to be welded to the pipe.

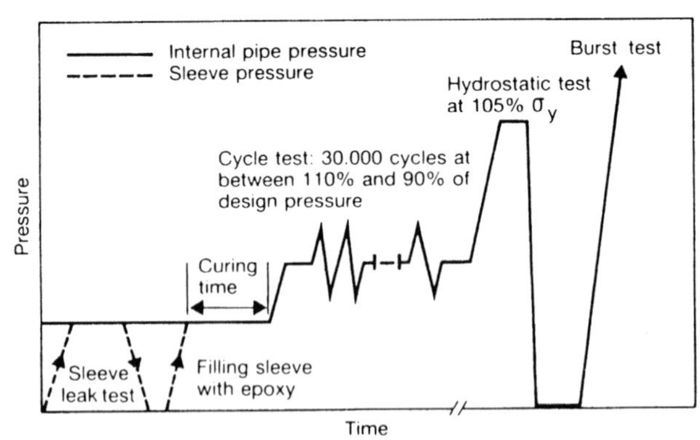

FIG. 6 Repair and test cycle plot for testing effectiveness of repair by ESR method.

Quality Assurance

ESR ensures safe, continuous operation of the pipeline after repair. Establish beyond doubt that the repair meets the required quality level. Two factors could affect this: the curing of the epoxy compound and entrapment of air in the epoxy. In the former case, premature release of pressure could lead to pressure loss and thus jeopardize the repair itself. Curing is therefore monitored by ultrasonic equipment. Only after full cure is recorded can the pressurizing equipment be removed. Air pockets or shrinkage cavities in the defect region are detrimental. They increase compressibility. To ensure proper filling, the defect zone is X-rayed. Only if this is found acceptable can the line pressure be restored to operational levels.

Damage Assessment

Two mechanisms can be distinguished: pressure-controlled failure and strain-controlled failure. The first category includes axial defects, dents, and corrosion damage exceeding specified dimensions which may possibly give rise to bursting. The second category includes circumferential defects and most weld defects in girth welds; these will lead to leakage only. Due to its pressure-controlled nature, the safety of the former type of defect can be guaranteed by pressure reduction. Repair is therefore possible. Assessment of the latter type of defect can only result in a go/no-go situation. Either the defect is acceptable or it is not.

Determination of defect size comes first. Correct measurement of defect dimensions is a first requirement. Where this is not possible, defect dimensions should be conservatively estimated. This is especially true for the determination of defect heights of weld defects. To ensure reliable analysis, additional safety margins should be included. Only if defect size can be determined with certainty may these additional safety margins be omitted.

Pressure-Controlled Failures

Defects such as axial notches and corrosion attack are pressure-controlled. For the development of acceptance criteria, use was made of burst strength data from many sources, based on numerous full-scale tests. For tough pipeline steels, this bursting strength is dependent on defect dimensions and flow stress, which is a material parameter. For each design stress level and material grade (flow stress), a graph of bursting strength can be plotted as a function of both defect depth/wall thickness ratio and defect length/wall thickness ratio (Fig. 7).

For very short defects, there is no risk of bursting and only leakage is possible. The leak line is also plotted. The figure is thus divided into four quarters: if a defect lies in the lower quarters, then neither bursting nor leakage is probable. This does not imply that the pipe is fit. This is deter-

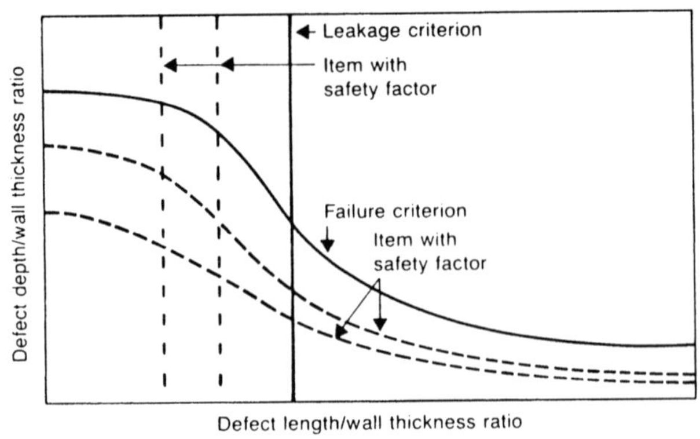

FIG. 7 Damage assessment depicted is for notches based on full-scale test results.

mined by safety criteria. To obtain a workable assessment diagram, safety factors should be included on both defect length and defect depth. This results in a series of "bursting curves" and "leakage lines" that incorporate safety factors (relative to predicted burst or leak levels) of between 1.5 and 8. These safety factors may seem high, but they allow for the fact that the real defect depth is often twice the measured depth due to the formation of microcracks and/or embrittled zones at the defect tip. The safety factors are set such that hydrostatic tests up to 105% yield stress can be carried out (Figs. 7 and 8).

Each letter code represents a damage category. Once a defect has been letter-coded, this indicates both the severity of the damage and the necessary actions (or restrictions). The system of defect classification is based on safety criteria for both normal operation and repair work, availability of repair methods, and known inaccuracies in defect dimensioning.

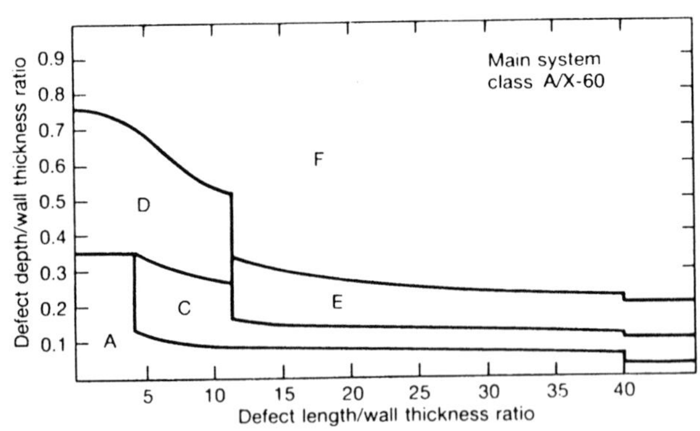

FIG. 8 Example of damage assessment diagram for axial notches. Damage category and required actions: A = acceptable; C = temporarily acceptable, no hydrostatic test allowed, repair by ESR; D = unacceptable, 10% pressure reduction, repair by ESR; E = unacceptable, 40% pressure reduction, repair by ESR; F = unacceptable, immediate replacement necessary.

For dented pipes and combinations of dents with notches, the above classification cannot be used. Smooth dents can be considered "harmless" up to a dent depth of 12% of the pipe diameter. Nevertheless, they must be replaced since pigs might not pass through such a severely dented pipe. Sharp-contoured dents are considered unsafe and call for replacement. Dents combined with notches are dangerous. Acceptance levels on such defects are very strict. The acceptance levels are based on the stress state in the notch tip region where bending stresses caused by the denting make the defect more serious (Fig. 9).

Strain-Controlled Failures

The axial stress component caused by internal pressure is equal to about 30% of the hoop stress. Near bends where free deformation is possible, these could amount to 50% of the hoop stress. Stresses induced by temperature, subsidence, etc. further add to the axial stress level. The total axial stress level is often difficult to quantify and should be estimated conservatively. Basis for acceptance of such strain-controlled defects (such as circumferential notches) is that the surrounding material must yield before the defect itself causes failure.

Since pipeline materials are ductile, failure of such strain-controlled defects can be predicted by plastic collapse criteria. Adding a safety factor, the assessment diagram (Fig. 10) includes defect length and depth and pipe dimensions. Since pressure reductions would make little or no contribution to the reduction of stresses in the defect zone, the assessment merely represents a yes/no situation. The defect is either acceptable or it is not.

Sometimes weld defects are unacceptable under current construction standards. Since these are based mainly on good workmanship criteria, this nonacceptability does not mean that the pipe section in question is not fit.

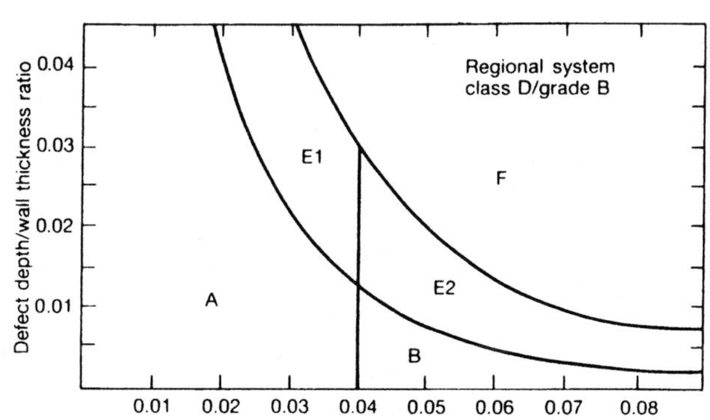

FIG. 9 Example of damage assessment diagram for dent/notch combinations. Damage category and required actions: A = acceptable; B = temporarily acceptable, no pigging, replacement necessary; E = unacceptable, 40% pressure reduction, repair by ESR (E1) or replacement (E2); F = unacceptable, immediate replacement necessary.

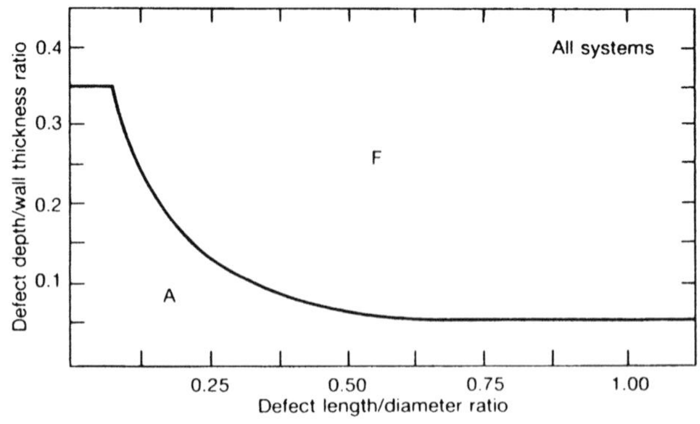

FIG. 10 Damage assessment diagram for circumferential defects. Damage category and required actions: A = acceptable; F = unacceptable, immediate replacement necessary.

To set acceptance criteria for such defects, concepts of fracture mechanics and plastic collapse criteria are employed. On the basis of failure mechanism (strain or pressure controlled) and material properties, defect, and pipe dimensions, a number of assessment diagrams have been developed (Fig. 11). Diagrams are based on the minimum specified yield strength and a conservative lower limit of fracture mechanics tests (CTOD) on removed pipe sections. This ensured a safe estimation of failure levels.

Quality Assurance

Since damage can pose a direct safety hazard, quality assurance is synonymous with safety assurance. The two uncertain factors in the assessment are the defect dimensions and the mechanical properties of the defect zone

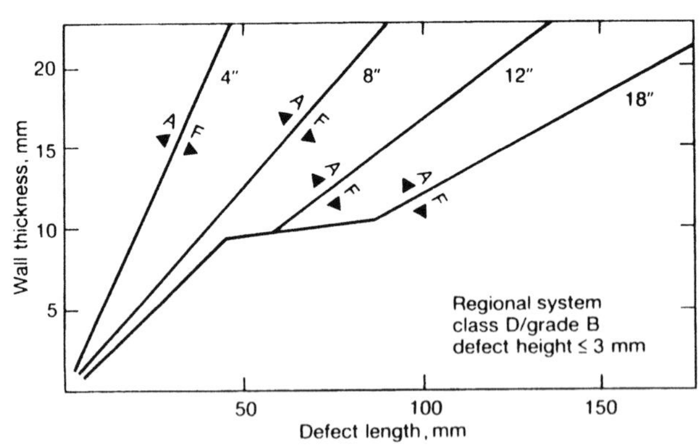

FIG. 11 Example of assessment diagram for surface-breaking weld defects. Damage category and required actions: A = acceptable; F = unacceptable, immediate replacement necessary.

material. The former is due to the fact that present NDT techniques don't always provide exact information on defect dimensions.

In thin-walled pipe in particular, this is very significant because of the small margin of error allowed. Outcome is greatly influenced by the skill and experience of the NDT operator. Although the specification calls for an NDT operator of the highest standard, it is still necessary to include safety margins in the assessment procedure which guarantee that defect dimensions will always be conservative. The second uncertain factor, mechanical properties of the defect zone material, is caused by the fact that establishing the real mechanical properties is impossible since this would require destructive testing. For defects in the pipe material, mechanical properties are known from the materials certificates. These values can be employed as a safe estimation of mechanical properties. For girth weld material properties, an estimation of minimum material properties was made using Gasunie's databank, in which results of all previous materials tests are collected.

Further quality assurance is accomplished in much the same manner as in any good quality assurance program. It includes a damage assessment directive for most types of defects, instruction on the use of NDT equipment, written lines of responsibility and competence, etc.

PAUL RIETJENS

Service Problems

Research before 1979 adequately described the causes of degradation in gas transmission pipelines that have led to failures [1–4]. Although more current research has shown no new failure causes, many variations have been identified.

They will be covered in this article updating failure causes since 1979.

Between 1979 and 1986, the new types of service problems that have been experienced are due to:

Hydrogen effects causing failure of the steel
Long ductile fracture propagation in a 16-in.-diameter line
Bacterial corrosion

Causes of failure have been divided into two categories: those associated with both preservice test failures and with service failures. Table 1 summarizes the types of defects that have caused pipeline incidents.

Failure Investigations

Since 1979, Battelle Columbus Division (BCD) has conducted more than 60 failure investigations for the American Gas Association (AGA). The resulting data do not constitute a statistically valid sample because only the failures BCD was asked to investigate are included in the data base. It is believed, however, that all previously identified causes are included.

In the years since failure data were last reported, pipes with increasing yield strengths have been installed in gas pipelines. In the United States, the highest yield strength used is X70.

Reported failures suggest that the higher strength steels (X65 and X70) may be more susceptible to potential damage from hydrogen introduction into the steel from the cathodic protection systems. The hydrogen can also be induced from either internal or external corrosion.

Mechanical damage has led to failures from a hydrogen-stress-cracking mechanism in X65 steel. Hydrogen stepwise-cracking has occurred at midwall in two steels: X60 and X65.

The lower strength steel exhibited greater hardness because of alloy segregation at midthickness, but both cracked in the plane of the pipe and then a crack propagated through the wall thickness.

Also, hydrogen-stress cracking occurred in a hard spot in Grade B pipe which is the lowest yield-strength line pipe that has failed from hydrogen-stress cracking in a hard spot.

Stress corrosion has continued to cause failures, although the number has been reduced because companies have sought to inspect, test, and control the cathodic potential applied to the pipeline.

One service failure occurred in 16-in.-diameter X52 pipe in which a ductile fracture propagated 1864 ft, having been initiated from a region of

TABLE 1 Causes of Pipeline Incidents

Defects Which Cause Preservice Test Failures	Defects Which Cause Service Failures
1. Defects in the pipe body:	1. Defects in the pipe body:
Mechanical damage	Mechanical damage
Shipping fatigue cracks	Material defects
Material defects	Environmental effects:
2. Longitudinal weld defects:	Corrosion
Submerged arc:	Hydrogen-stress cracks
Weld area cracks	Stress-corrosion cracks
Incomplete fusion	Sulfide-stress cracks
Porosity	Stepwise cracks
Slag inclusions	2. Longitudinal weld defects:
Off-seam	Submerged arc:
Incomplete penetration	Toe cracks
Electric weld:	Fatigue cracks from cyclic loading, such as in compressor stations
Hook cracks	Electric weld:
Weld-line inclusions	Selective corrosion
Cold weld	Hydrogen-stress cracks
Contact burns	3. Field weld defects:
Excessive trim	Lack of penetration
High hardness	Corrosion (usually internal due to flow conditions)
3. Field weld defects:	4. Special causes:
Underbead cracks	Secondary loads from soil movement
Weld metal cracks	Earthquake loads
	Wrinkle bends
	Internal combustion
	Sabotage

mechanical damage. This failure is unusual in the extensive length of ductile-crack propagation.

This is believed to have occurred because a rich gas was present in the pipeline at the time of the failure, which prevented the decompression a lean gas (1000 Btu/SCF of gas) would have allowed.

Four general categories of failures will be described: those due to hydrogen effects, stress-corrosion cracking (SSC), mechanical damage, and anaerobic bacterial corrosion.

Hydrogen-Stress Cracking

A leak was detected in a region of mechanical damage on a 36-in.-diameter (0.322-in. W.T., X70) double-submerged-arc welded pipe that had been in service for approximately 6 years.

The line normally operates at a pressure of 830 lb/in.^{2} [hoop stress of

66% specified minimum yield stress (SMYS)]. However, at the time the leak was detected, the pressure was 666 lb/in.^{2}, or 53% SMYS. The pipe had been coated with primer, polyethylene tape, and nonadhesive kraft paper outer wrap.

The line had been subjected to cathodic protection since installation. The pipe-to-soil potential measured at the leak site prior to removal of the section of pipe containing the leak was −1.3 V ($Cu/CuSO_4$). A holiday in the coating was apparent at the location of the leak when the leak was detected.

The mechanical damage on the pipe surface is believed to have occurred during construction of the pipeline, based on the absence of construction activity and the remoteness of the area.

The leak was approximately 2½-in. long on the outside surface and was essentially parallel to the pipe axis. The laboratory investigation of the failure consisted of a fractographic analysis of the crack surface in the scanning electron microscope (SEM), metallographic examination of sections through the crack, and microhardness measurements.

The nominal composition of the pipe steel was 0.09% C, 1.45% Mn, 0.01% P, 0.016% S, 0.027% Si, 0.05% Cb, and 0.23% Mo.

The surface of the crack was corroded, but no black corrosion products characteristic of SCC were present. The surface of the leak after being cleaned is shown in Fig. 1.

The markings on the leak surface indicated that the crack initiated at or near the outside surface of the pipe and grew through the wall toward the inside pipe surface. On the inside surface a small shear lip was evident, indicating that the final fracture was ductile.

Fracture Mode

The predominant fracture mode appeared to be intergranular based on examination of the cleaned fracture surface in the SEM (Fig. 2). Energy-dispersive X-ray analysis of inclusion particles present on the fracture surface revealed that cerium and lanthanum were present in the inclusions, indicating that the pipe steel had been rare-earth treated for inclusion shape control.

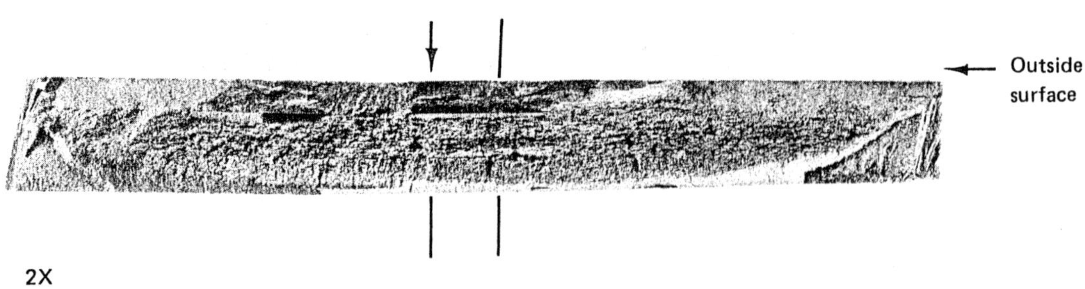

FIG. 1 Hydrogen stress crack surface after cleaning.

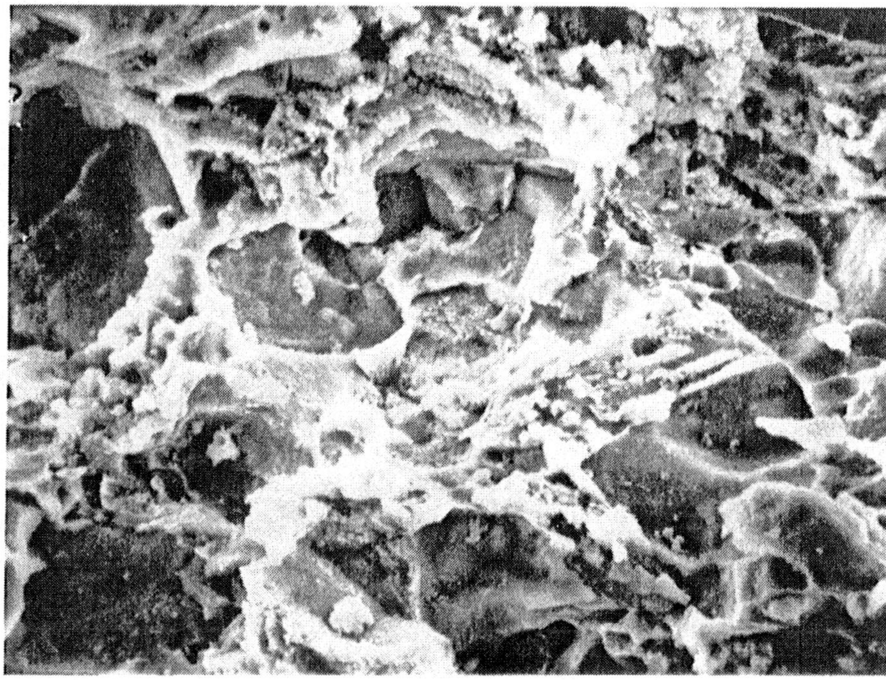

2,000X After cleaning

FIG. 2 Fracture surface near origin.

Three metallographic sections were prepared through the leak, one at the origin and two adjacent to the origin.

Examination of the section through the origin revealed that the steel immediately adjacent to the outside surface had been cold-worked as a result of the mechanical damage, as shown in Fig. 3. The hardness in that cold-worked region, which extended inward for 7 mils from the outside surface, ranged from 300 to 340 Knoop (500 g load), or 28.5 to 34 Rockwell C by conversion.

Below that cold-worked region and at locations away from the mechanically damaged area, the hardness of the steel through the wall thickness ranged from 225 to 250 Knoop (95 to 100 Rockwell B by conversion). Examination of the metallographic sections confirmed that the fracture mode was predominantly intergranular, as shown in Fig. 4.

No secondary cracks extending inward from the outside surface were observed, and the primary crack did not exhibit any branching as it grew through the wall thickness. However, as shown in Fig. 5, several internal, predominantly transgranular cracks were present adjacent to the primary leak crack.

The microstructure of the steel consisted of a matrix of fine grain ferrite with significant amounts of a mixture of bainite and/or acicular ferrite and martensite uniformly dispersed in the matrix. In addition, some small patches

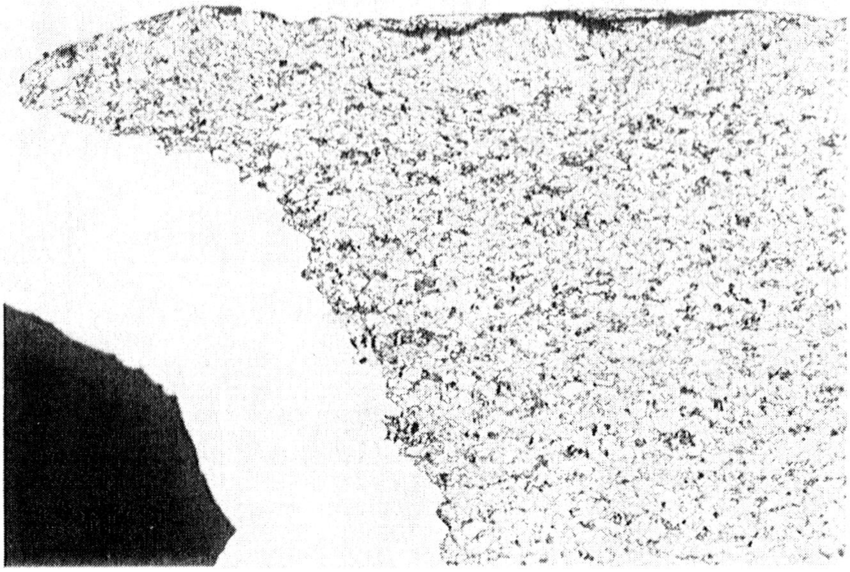

FIG. 3 Crack profile at origin. Mechanical damage at outside surface.

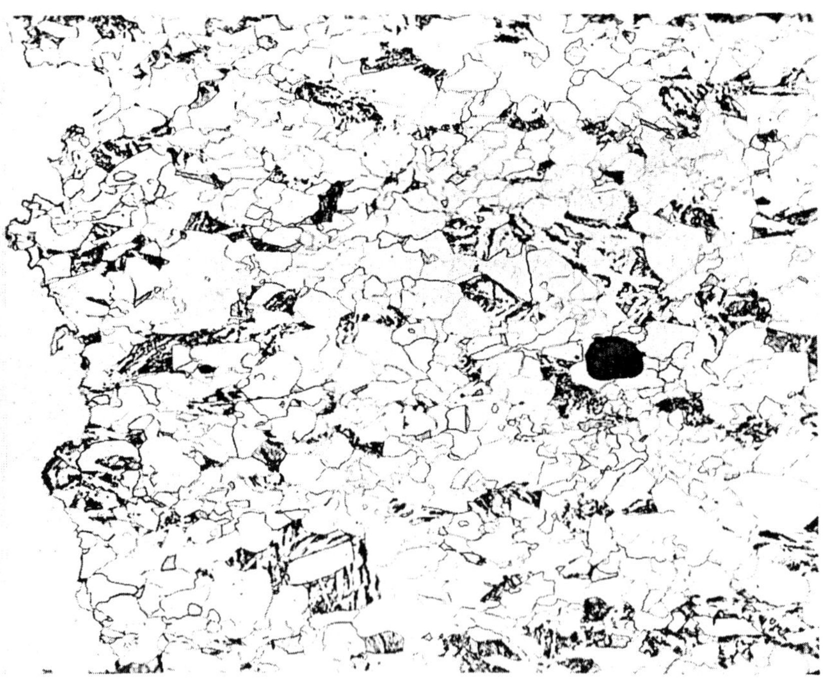

FIG. 4 Intergranular primary crack. Location well below cold-worked region.

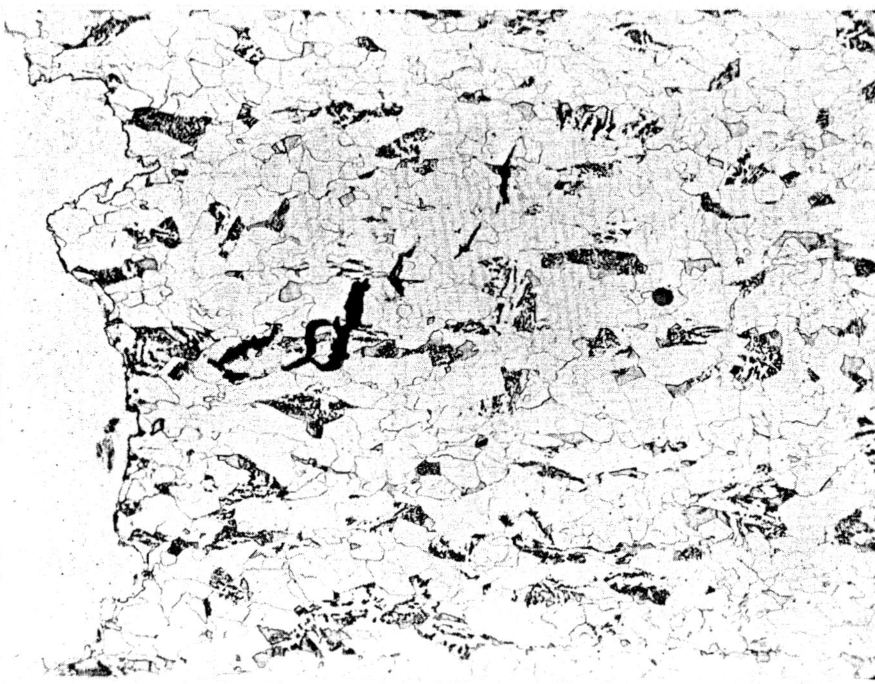

500X Picral and nital etch

FIG. 5 Internal cracks near primary crack. Location farther from outside surface than in Fig. 4.

of free martensite were present. Typical microstructure of the steel is shown in Fig. 6.

To determine the stress state, residual stress measurements were made on the outside surface of the pipe which indicated that relatively high residual tensile stresses were present in the mechanically damaged region.

The magnitudes of the maximum residual tensile stresses were reported to be 33.7 ksi in the hoop direction and 59.8 ksi in the longitudinal direction.

Adding the residual and pressure stresses results in transverse and longitudinal stresses of 70.6 and 69.8 ksi, which are close to the yield strength in the two directions.

Failure Cause

The probable cause of the failure is a crack that developed from a hydrogen-stress cracking mechanism in the highly stressed, cold-worked layer on the outside surface of the pipe. The atomic hydrogen necessary for the hydrogen-stress cracking is believed to have resulted from the cathodic protection applied to the pipe.

The voltage level at which hydrogen is generated is −1.1 V relative to $Cu/CuSO_4$. In this instance the indicated cathodic protection level was −1.3

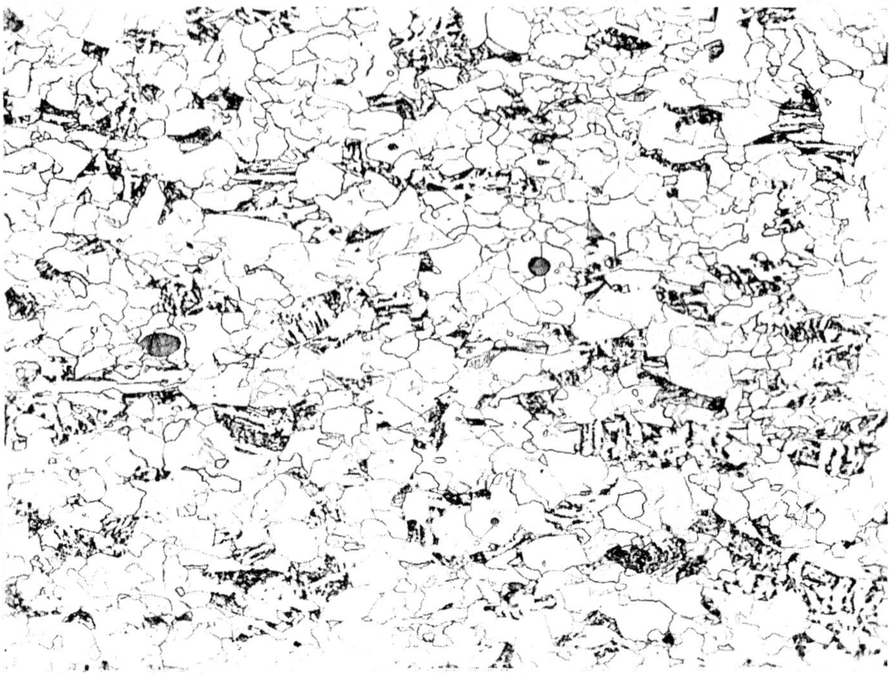

500X Picral and nital etch

FIG. 6 Typical microstructure.

V, more than enough to result in cathodic charging of the pipe surface with hydrogen.

The unusual feature of this failure is that prior studies on higher strength, controlled-rolled steels had not indicated that they would be susceptible to hydrogen-stress cracking in service [*5*].

The results of this investigation, however, suggest that the combination of mechanical damage, the resultant high residual stresses in the damaged region, and atomic hydrogen absorbed by the steel may result in hydrogen-stress cracking failures in these types of line-pipe steels. Recent research into this failure indicates that the strain aging of the cold-worked steel in the mechanically damaged area may be a contributing factor to this steel's sensitivity to hydrogen-stress cracking.

Effects of Hydrogen

Research updating causes of pipeline failures since 1979 has identified a shift in pipe service problems since the last such report.

This part, based on examinations of reported failures, focuses on several because of the effects of hydrogen.

We will later discuss three other major areas of service problems: stress-corrosion cracking, mechanical damage, and anaerobic bacterial corrosion.

Stepwise Cracking

A leak was detected in compressor-station yard 42-in. piping (0.681-in. W.T., X65). The leak was located on the bottom of the line pipe. The nominal hoop stress in the pipe was 40% of the SMYS.

The steel involved was ingot cast steel. The composition was 0.08% C, 1.90% Mn, 0.007% S, 0.07% Cb, 0.05% V, and 0.22% Mo. Examination of the leak surface indicated that a midwall crack in the plane of the pipe wall existed.

As shown in Fig. 7, the through-wall cracks initiated at midwall at the location of the midwall crack. As shown in Fig. 8, cracks then grew to the inside and outside, finally penetrating the outside surface and fracturing on the inside to create the leak path.

As shown in Fig. 9, the transverse crack was primarily transgranular. Examination of the midwall crack indicated its extent to be approximately 100 in.^{2}.

Examination of the corrosion products from the fracture surface did not reveal the presence of any sour gas environment. Because this was a gas transmission pipeline, no presence of sour gas would be anticipated, other than perhaps on rare occasions when the treatment plant was in an upset condition. The fracture surface is predominantly a transgranular fracture surface.

Examination of microstructure (Fig. 10) revealed that it is a normal ferrite-pearlite material and does not appear to contain any martensite, untransformed austenite, or bainite in the microstructure.

This is also reflected in the fact that the hardness of the base metal is a maximum of Rockwell C 24 based on conversion from Knoop microhardness measurements.

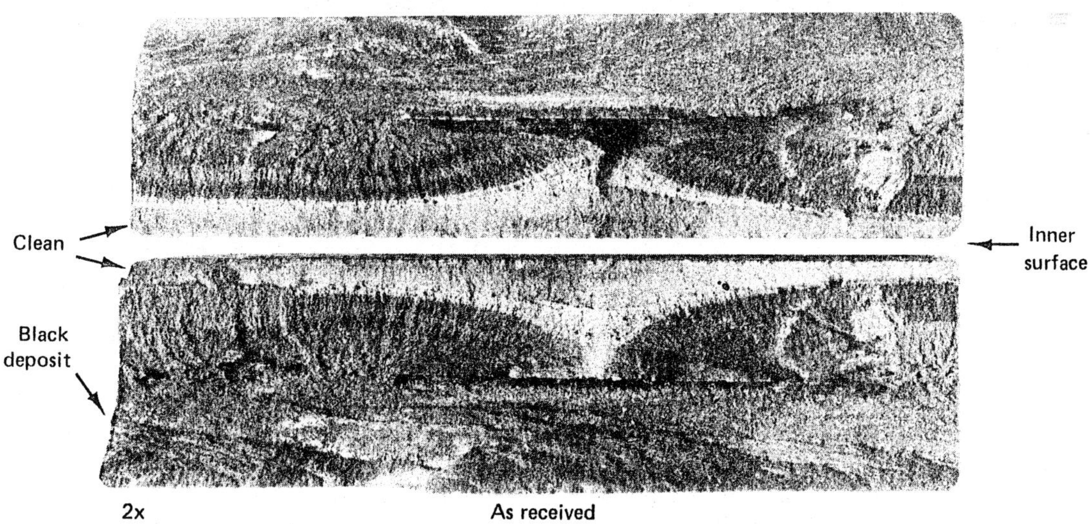

FIG. 7 Fracture surface of leak.

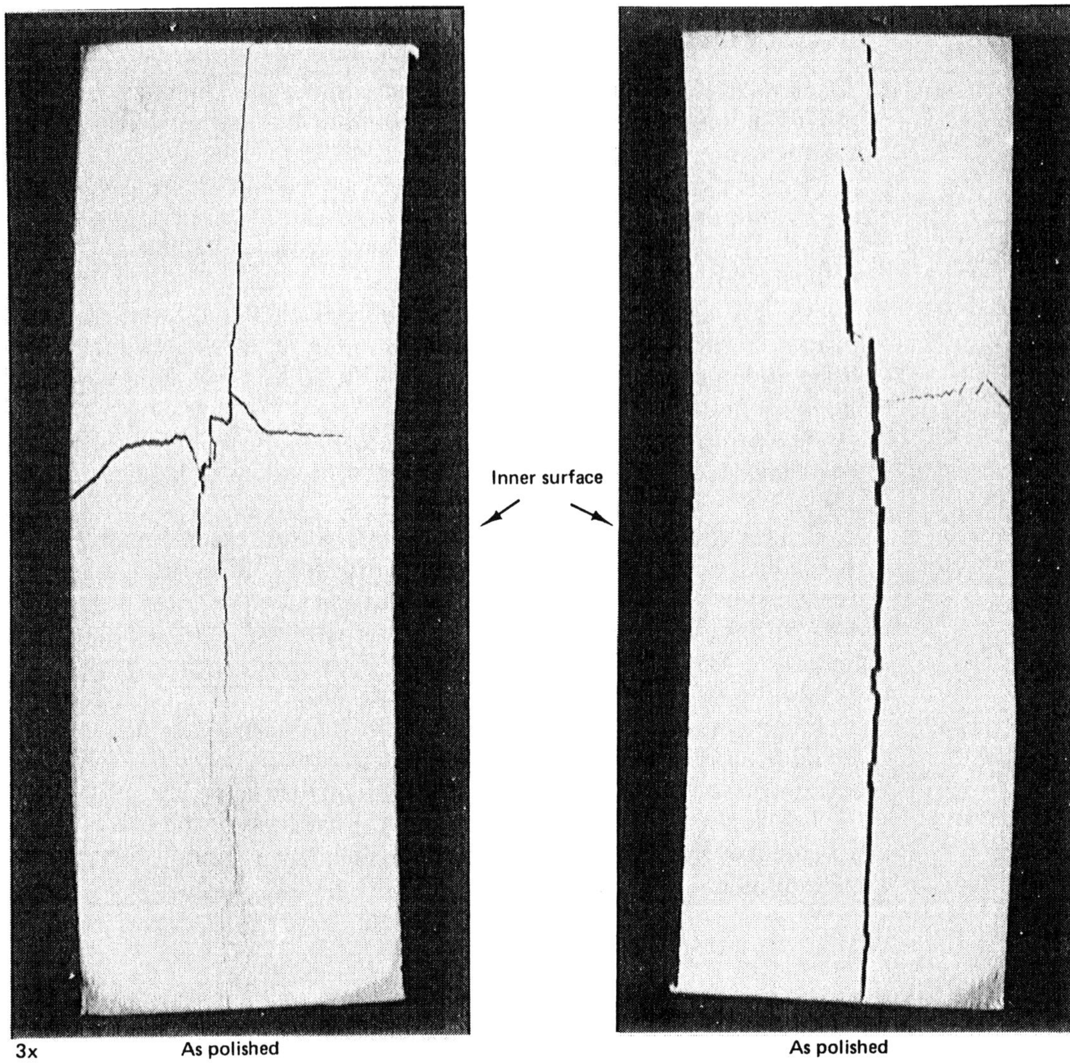

FIG. 8 Suctions through the leakage path.

No Lamination Crack

The midwall-crack surface was examined to determine whether this was an area of inclusions or laminations in the pipe wall. Examination of this midwall-crack surface in the SEM, coupled with X-ray dispersive examination, revealed no inclusions on the plane of the midwall crack. This appears to rule out the probability that the midwall crack was a lamination.

Examination of the variation in composition through the thickness (Fig. 11) indicated that at the plane of the midwall crack there was segregation of the microalloying elements. From the surface to the midwall, the following composition changes were noted: carbon, 0.07 to 0.09; manganese, 1.85 to 1.97; and columbium, 0.065 to 0.075.

Junction of midwall crack; transverse through-wall crack

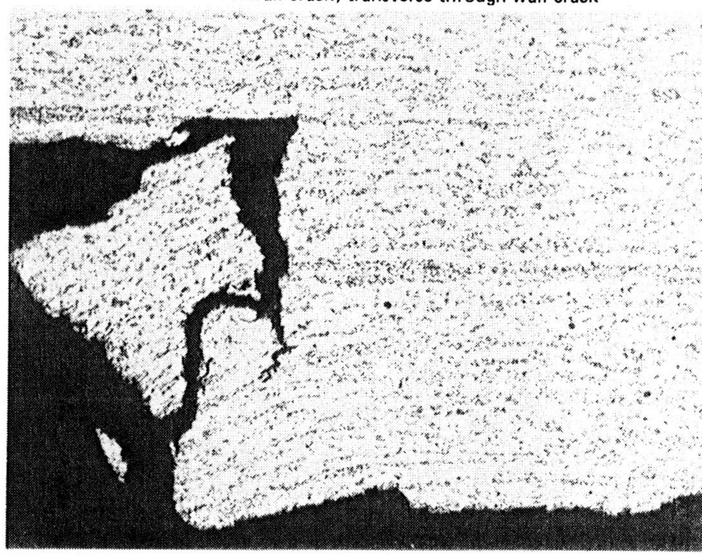

Fig 9a

100x Picral etch

Transverse crack at origin

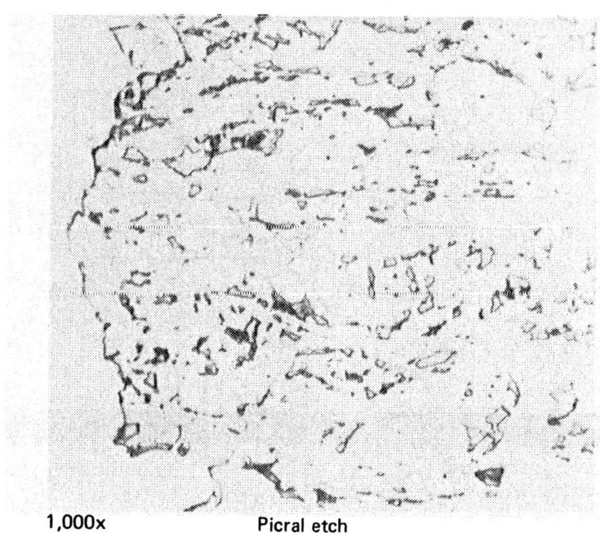

Fig 9b

1,000x Picral etch

The probable cause of the crack was identified as stepwise cracking resulting from hydrogen in the plane of the pipe caused by alloy segregation at the midwall. The indications are that the hydrogen source is from the −1.8 V applied through the cathodic protection system.

FIG. 10 Microstructure of steel.

Hydrogen-Stress Cracking

In another service incident, a rupture initiated at 963 lb/in.^{2}gauge within the 36-in. line pipe (0.93-in. W.T., X60). The fracture in this instance propagated approximately 120 ft in a ductile manner.

As in the previous failure, a midwall crack also existed in this failure.

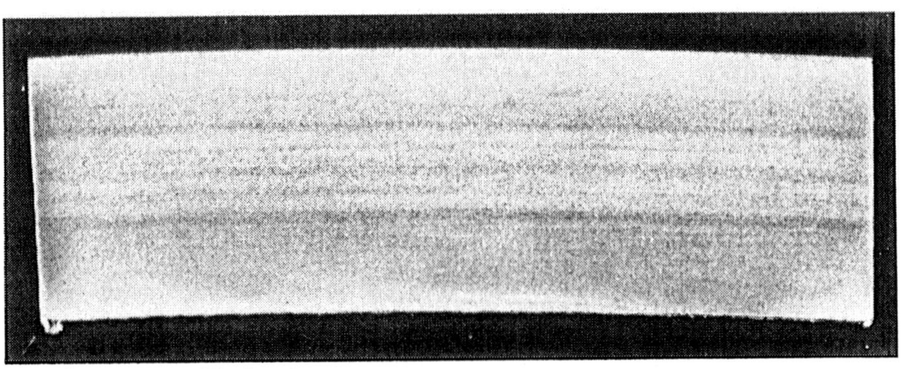

2x Nital etch

FIG. 11 Section through pipe wall showing segregation.

The midwall crack, however, occupied an area of just a few square inches. A transverse crack creating the rupture initiated from the midwall crack and propagated to each surface, creating the critical flaw size, which in turn caused the rupture.

The nominal composition of the pipe steel in this instance was 0.25% C, 1.45% Mn, and 0.07% V.

The fracture at the origin was a cleavage fracture, whereas the remainder of the fracture surface was a ductile fracture surface as the crack propagated away from the origin area. Examination of the microstructure at the origin indicated that the maximum Knoop hardness at midwall immediately adjacent to the plane of the midwall crack reached a maximum hardness of Rockwell C 40, based on conversion from Knoop microhardness impressions.

The area immediately adjacent to the midwall crack was found to contain significant alloy segregation, which contributed to the higher hardness in this region. The microhardness of the base metal was approximately Rockwell C 23 by conversion from Knoop microhardness impressions.

X-ray dispersion spectroscopy revealed the presence of approximately 13% sulfur in the oxide layer on the fracture surface. Presumably, this was a sulfide present on the fracture surface from a contaminant in the gas.

This failure was probably associated with the presence of H_2S in the gas and a layer of inclusion-rich material at mid-thickness that probably cracked due to a hydrogen blister. The additional stress from the blister probably caused a transverse crack to grow from the midwall crack through the wall thickness to the inner and outer surface, eventually causing a rupture.

In the area of the midwall crack, numerous manganese-sulfide inclusions were detected along with the alloy segregation.

ERW Pipe

Hydrogen-stress cracking in a 20-in. Grade B line pipe (0.250-in. W.T.) involved a leak in electric-resistance welded (ERW) pipe that had been in service for approximately 35 years. The line had been coated with coal tar with an external wrap (applied over the ditch) and subjected to cathodic protection since installation.

The leak was located 4.3 miles from the nearest rectifier station, and the pipe-to-soil potential measured near the failure was approximately −1.35 V (vs $Cu/CuSO_4$) during the past 2 years. The leak was located on the bottom of the pipe about 9 in. from the seam weld and 32 in. from the nearest girth weld.

The nominal composition of the pipe was 0.29% C and 0.85% Mn. The yield strength of the steel was 58.7 ksi, more representative of an X52 steel than a Grade B steel. The ultimate tensile strength was 75.7 ksi.

Since a number of hard spots have been investigated in this program in the past, an examination was conducted to determine whether there were any unusual features in this hard spot failure in a Grade B steel in contrast to those previously examined.

Brinnell hardness measurements made on the outside surface of the pipe confirmed that the crack occurred in a hard spot, with the maximum hardness in the hand spot region being 467 Brinnell (BHN), 49.5 Rockwell C by conversion. The hardness of the pipe in the base metal near the edge of the hard spot ranged from 207 to 217 BHN.

Examination of the fracture surface of the crack under low power magnification indicated that the crack was about 2¾ in. long and that it initiated at the outside surface of the pipe (Fig. 12).

Fracture surfaces were relatively flat and essentially perpendicular to the surfaces of the pipe. A small shear lip was present near the inside surface, indicating it was the point of final separation.

Examination of the surfaces of the sections of pipe revealed no secondary cracks and no significant corrosion on either surface. The crack surfaces were corroded, leading to the suspicion that the leak had been present for some time prior to its detection.

Fracture Profile

Figure 13 presents a section cut through the origin region and illustrates the fracture profile through the pipe wall. As shown in Fig. 13, there were no secondary cracks.

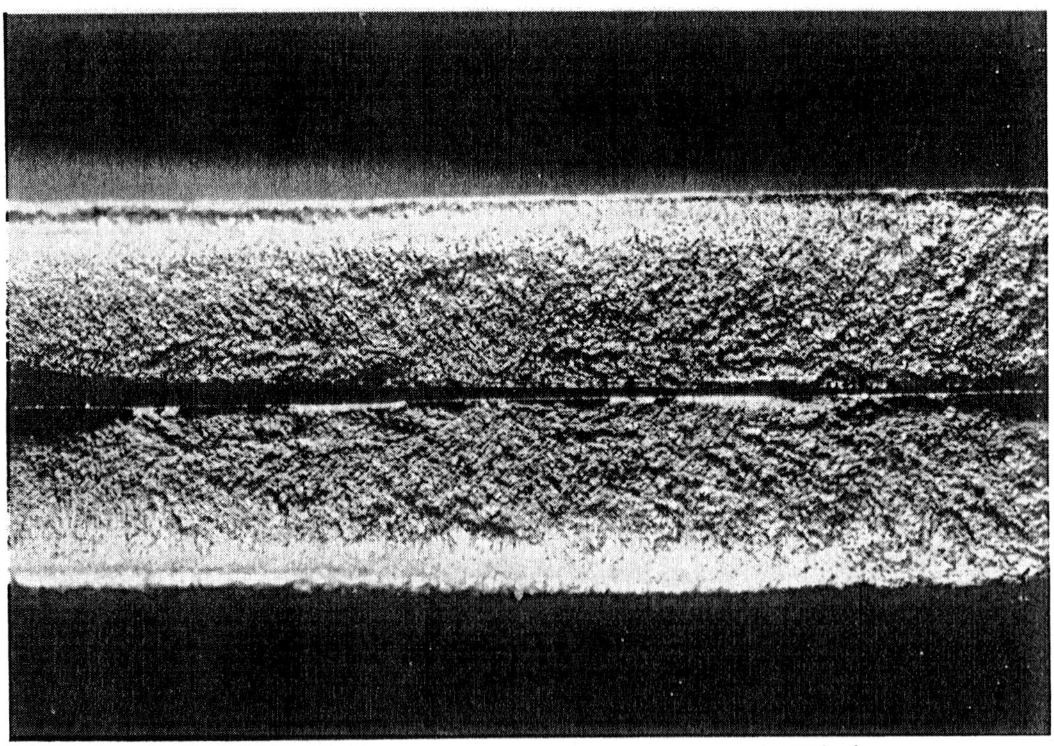

5x As received

FIG. 12 Fracture surfaces: origin region of crack in hardspot. Outside surfaces are back to back at center.

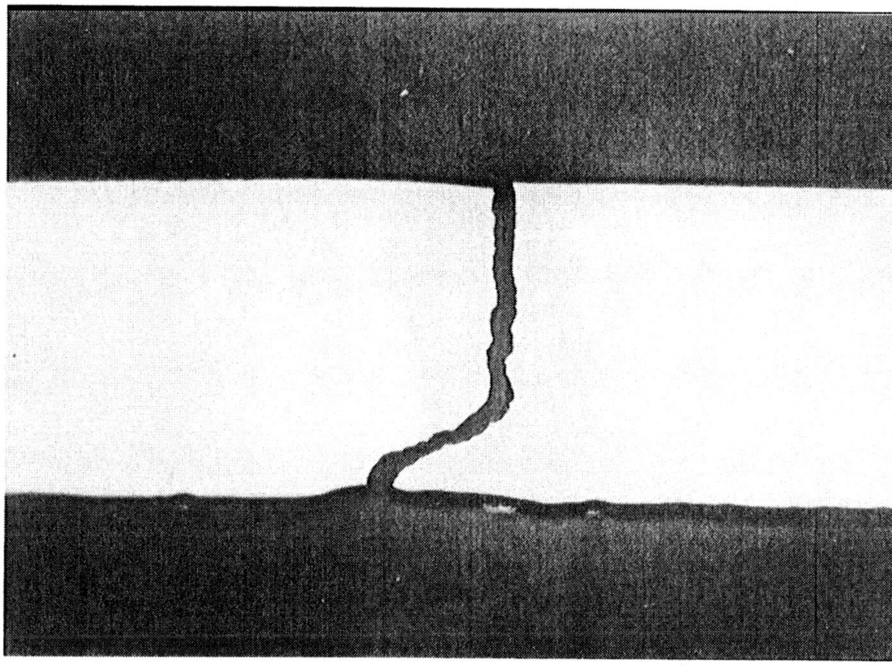

6.5x As polished

FIG. 13 Crack profile at origin. Outside surface at top.

Figure 14 shows the crack profile at the origin at higher magnification and two short, intergranular-appearing secondary cracks that extended inward from the outside surface. These were the only secondary cracks observed in the section examined.

The microstructure of the steel at the origin of the crack (on the outside surface) was essentially 100% martensite (Fig. 15). The results of a Knoop microhardness traverse made adjacent to the surface of the crack indicated hardnesses that ranged from 422 to 550 Knoop (41.5 to 50.5 Rockwell C by conversion).

Thus, the steel was hard all the way through the wall thickness.

This metallurgical examination revealed that the probable cause of the crack was a hydrogen-stress crack in a localized hard spot in the wall of the pipe, the hard spot having been produced by a local quenching of the skelp or plate after hot rolling in the steel mill.

This is the first hard-spot crack that has been found in Grade B pipe, although in reality this pipe is very similar to an X52 steel in which the majority of the hard-spot failures have occurred.

Other Effects

This part addresses failures due to stress-corrosion cracking (SCC), mechanical damage, and anaerobic bacterial corrosion.

100x As polished

FIG. 14 Profile at origin. Secondary cracks at outside surface.

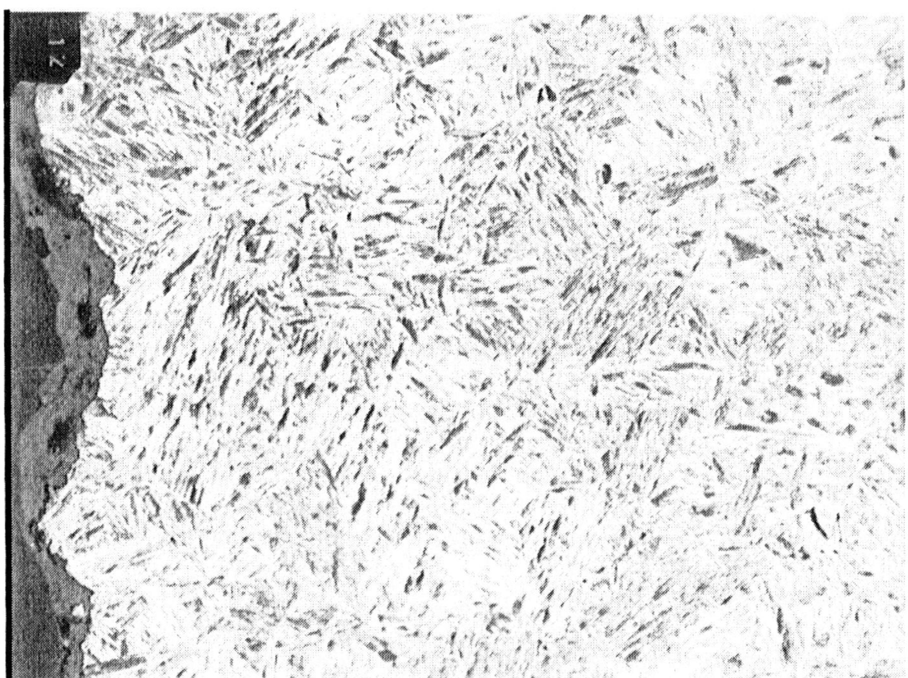

500x Picral etch

FIG. 15 Typical microstructure of crack-origin section.

These causes, along with hydrogen effects discussed earlier, comprise a shift in pipe service problems evident in failures reported since 1979. This shift was identified in a major study by Battelle Columbus Division laboratories.

Stress-Corrosion Cracking

Several failures involving external SCC of line pipe were investigated. One investigation was conducted on a 30-in. (0.344-in. W.T., X52) pipe that experienced a retest failure 40 lb/in.^{2}gauge above the maximum operating pressure of the pipeline.

In this instance, the failure was due to classic stress-corrosion cracks in the pipe in an area that did not appear to be unusual metallurgically. Figure 16 shows the appearance of the stress-corrosion cracks on the external surface of the pipe.

The pipe involved had been coated with coal tar and in service for 24 years prior to the retest failure. No prior stress-corrosion cracks had been observed in this section of the pipeline.

Metallographic examination of the cracks in the pipe indicated that they were intergranular and contained the familiar black oxide on the fracture surface. Figure 17 shows a metallographic section across the main fracture. (Note that part of the fracture surface is missing, apparently as a result of branching of the fracture.)

Upon first examination of the metallographic cross section, it was thought that perhaps a dent might have been present in the area of the cracks. Careful examination, however, revealed no dent had been present.

No unusual features of the chemical composition or of the soil around the pipe were noted.

This retest prevented a service failure, which in all likelihood would have occurred from continued stress-corrosion-crack growth if the retest had not been conducted.

FIG. 16 Secondary cracking at fracture origin: exterior surface.

FIG. 17 Metallographic section across main fracture.

Other stress-corrosion cracks that were investigated have been associated with stress risers in the pipe. Stress-corrosion cracks have been observed in localized hard-spot regions, at the toe of the longitudinal double-submerged arc seam weld, and associated with gouges on the surface of the pipe.

All of these examples are of local areas where the stress has been elevated, contributing to the initiation of SCC in that region. It is somewhat surprising that the local elevation of stress, such as from the reinforcement in a longitudinal weld or a girth weld, has not contributed in a greater way to the formation of stress-corrosion cracks in general on pipelines. However, the more common situation is for the stress-corrosion cracks to be remote from stress concentrations. Probably this is due to the absence of the environment on the pipe surface in these high-stress regions.

When these failures have been experienced at welds, most frequently the coating has been a tape coating, which tends to bridge over these areas. A region of disbondment is thus created where the carbonate–bicarbonate environment can occur under the coating from the action of the cathodic protection system.

Mechanical Damage

One incident occurred in which 16-in. (0.219-in. W.T., 5LX 60) ERW pipe experienced a 100% shear fracture that propagated 1864 ft.

The fracture was initiated from a region of mechanical damage on the pipe surface believed to have been produced during construction of a recently installed adjacent pipeline. The pipe had been thin-film coated, and the damage extended through the thin-film coating into the pipe base metal.

This failure was investigated because of the extensive fracture propagation in a relatively small-diameter pipeline. At the time of the failure, the pressure in the pipe was 1183 lb/in.^{2}gauge or a hoop stress of 56% of the SMYS. Ten samples of pipe were removed for laboratory examination.

The measured Charpy plateau energy using one-half thickness specimens indicated values from 8½ to 11½ ft-lb. The highest energies, of course, occurred in the arrest pipe lengths. The pipe was a 0.13% C, 1.11% Mn steel, with 0.034% V and 0.025% columbium microalloy additions.

A prediction of the fracture arrest toughness was made based on a 100% methane gas and also the gas estimated to be present in the line at the time of the failure. Figure 18 presents the results of the calculations.

It can be observed that the Charpy V-notch one-half thickness energy calculation as being required for fracture arrest with a 100% methane gas is 7½ ft-lb. For the 84% methane gas present in the pipeline at the time of the failure, the half-thickness Charpy plateau energy required for fracture arrest increases by approximately 33% to 10.1 ft-lb. Thus, the extensive propagation of the fracture is predicted by the AGA formulations [4].

It was concluded from the present study that the extensive ductile fracture propagation was predictable with existing formulations.

The unique feature of this extensive propagation is that the arrest toughness requirements were increased approximately 33% by the relatively rich gas contained in the pipeline at the time of the failure. (It should be noted that this pipeline normally does not carry this rich gas but was doing so because of rerouting due to construction.)

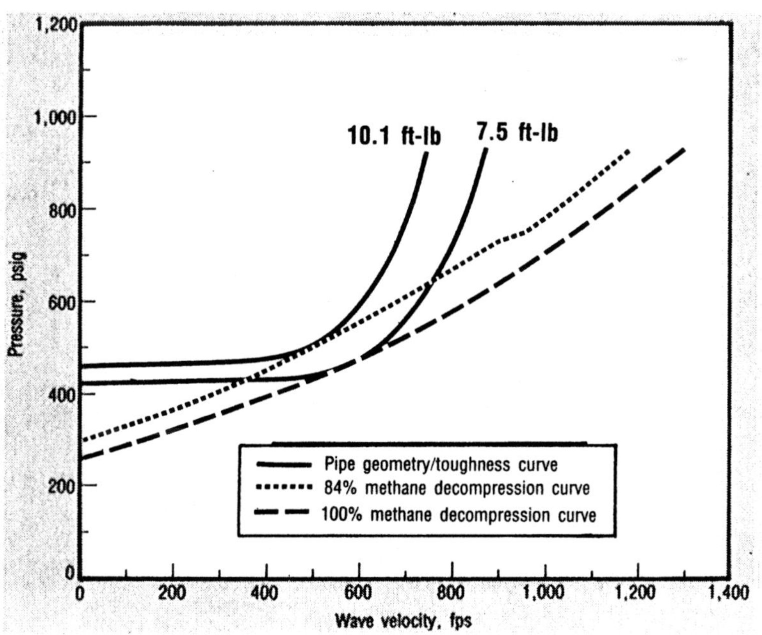

FIG. 18 Two gas systems' pipe toughness values.

This failure reinforces the influence of gas composition on gas decompression and hence toughness required for fracture arrest.

Bacterial Corrosion

A leak was detected in a 20-in. (0.281-in. W.T., X52) double-submerged, arc-welded pipe which was part of a gathering system. The line had been in service for 25 years and was coated with an asphalt enamel coating.

The leak occurred at 5 o'clock on the pipe while the line was operating at a pressure of 520 lb/in.^{2}gauge (a hoop stress of 36% of the SMYS). The leak extended for approximately 2½-in. axially in a corroded area that was approximately 8 in. by 6 in. (Fig. 19).

Energy-dispersive X-ray (EDX) analyses were performed on the outside diameter surface in the corroded area and on the noncorroded area. The results of the EDX analyses are shown in Table 2.

High Chlorine

The results show high amounts of chlorine and relatively high amounts of silicone and sulfur.

3/4x As received

FIG. 19 Corroded OD area. Note longitudinal ridges in corroded area.

TABLE 2 Energy-Dispersive Analyses

Area Analyzed	Element (wt.%)						
	Fe	Mn	Ni	Si	Cl	S	Al
Base material	97.78	1.00	0.60	0.62			
Corrosion Area 1	83.43		1.02	5.24	8.41	1.90	
Corrosion Area 2	92.18	0.44		1.79	4.44	1.10	0.09

The aluminum and silicon are constituents of sand and clay and are found almost universally where ground waters are present. Chlorine is also a constituent of ground water. The sulfur could have come from several sources, including the ground water, soil, or sulfate-reducing anaerobic bacteria. The sodium azide-iodine spot test for sulfides was applied to the corroded outside diameter surface, and positive indications were obtained in the walls of the ridges. This indicated the presence of sulfides. Little or no sulfide-sulfur was found in other areas using this test.

Since chlorine and sulfur were detected in the corrosion product on the pipe by EDX, a soil sample was analyzed from the backfill in the vicinity of the leak. The analysis of the leached sample revealed a pH of 6.5 with a chlorine level of 0.12 wt.% and sulfur of 1.00 wt.%.

Again the sodium azide-iodine test was applied to the soil sample and no indication of sulfide was observed. These results indicate that the sulfur in the soil is probably in the fully oxidized state (SO_4). The moisture in the soil was determined to be 15 wt.% and the resistivity was measured at 10,000 to 12,000 ohms/cm.

These results suggest that the soil is not very corrosive.

A metallographic cross section through the leak is shown in Fig. 20a. The corrosion appears to be localized. There was no evidence of selective attack of either the pearlite or ferrite phases which could be correlated with the ridge-like features of the corrosion attack (Fig. 21).

The most probable cause of the failure appears to be corrosion of the pipe by anaerobic sulfate-reducing bacteria.

Conditions for the growth of bacteria appear to be present in the soil/pipe environment, i.e., the pH of the soil (6.58) is within the pH range for bacteria (pH 5 to 13), the 15% moisture in the soil is conducive to bacterial growth, and the asphalt coatings supply the nutrients which bacteria feed on. The soil in the absence of anaerobic bacteria does not appear to be very corrosive but did contain sulfur and chlorine.

However, the presence of sulfide-sulfur in the corrosion product, but not the soil, suggests that sulfate-reducing bacteria were probably present. When sulfate-reducing bacteria are active, the cathodic protection levels must be made at least 100 mV more negative than when bacteria are absent to control corrosion.

Fig. 20a

10x Cross section

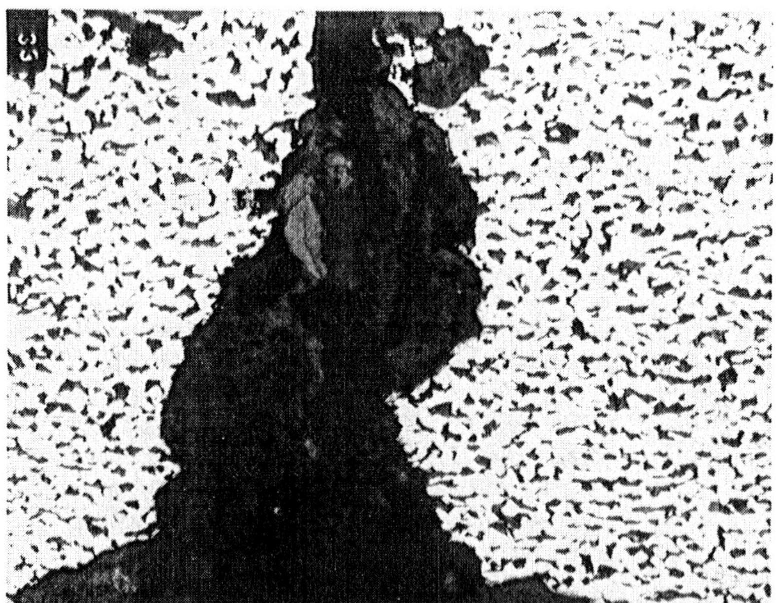

250x

FIG. 20 Through-wall penetration. Etchant: Nital.

Fig. 21a

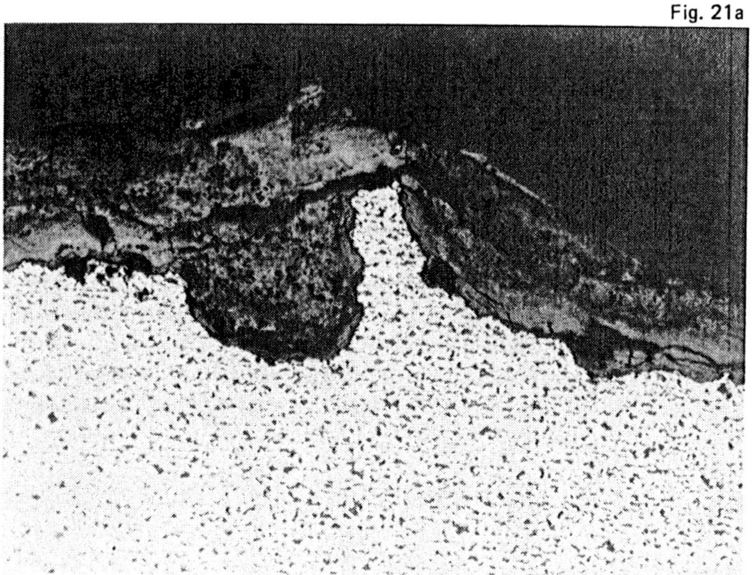

110x General corroded surface

Fig. 21b

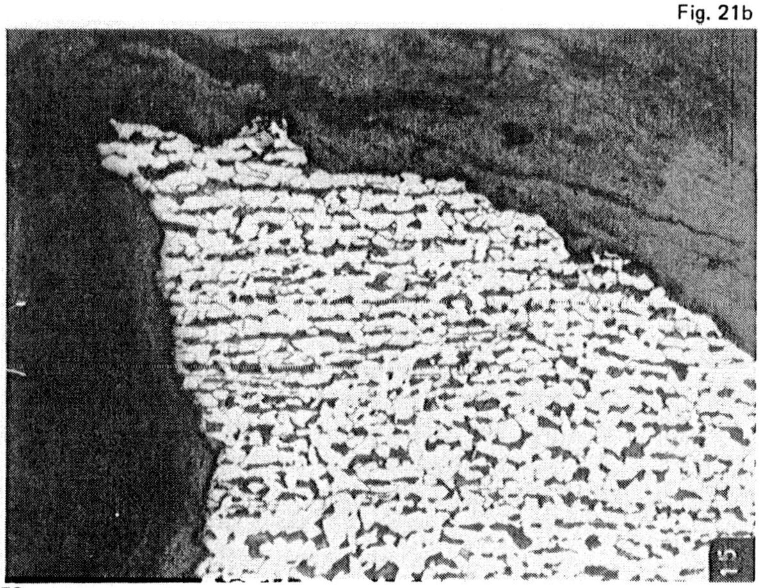

250x

FIG. 21 Corroded area on OD. Etchant: Nital.

References

1. *Third Symposium on Line Pipe Research,* American Gas Association, Catalog No. L3000, November 1965.
2. *Fourth Symposium on Line Pipe Research,* American Gas Association, Catalog No. L30075, November 1969.
3. *Fifth Symposium on Line Pipe Research,* American Gas Association, Catalog No. L30174, November 1974.
4. *Sixth Symposium on Line Pipe Research,* American Gas Association, Catalog No. L30175, October 1979.
5. T. Groeneveld, *Effect of Hydrogen on the Properties of High Strength Controlled-Rolled Line-Pipe Steels,* American Gas Association, Catalog No. L11877, April 1977.

JOHN F. KIEFNER
ROBERT J. EIBER

Prevention of by Bleeding

Exposed water pipe is sometimes left bare and unprotected in plant yards during construction or general plant usage. Since no insulation or steam tracing is used, the only cold-weather protection remaining is to continuously bleed an amount of water from the line.

If the line is dead-ended without bleeding, no amount of insulation will protect it. Insulation will retard heat loss but will not eliminate it altogether. If no flow takes place, it is only a matter of time before the line freezes.

Figure 1 gives an economical bleed flow-rate to protect bare steel pipe under freezing conditions of −20°F and an entering water temperature of 40°F. Bleed rates are read in gallons per minute per 100 ft of pipe in the sizes indicated. For wind velocities over the accepted 15 mi/h, double the bleed rates obtained from Fig. 1. If 2 in. of insulation is provided, reduce the rates to one-half.

Changes in water pressure should also be taken into account because Fig. 1 is based on steady pressure. If pressure fluctuations occur, double the flow rate shown in Fig. 1. Pipe wall thickness has no material effect in Fig. 1.

For temperatures under −20°F, the line should be steam traced for maximum protection. Provide similar protection for lines more than 500 ft long.

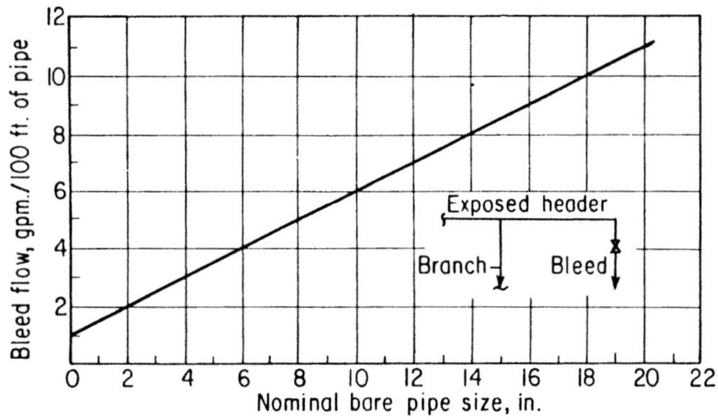

Fig. 1

Example. A dead-ended pipe, exposed to the weather and carrying plant water, is 250 ft long and has a nominal 4 in. diameter. Water is to be bled at the dead end to save insulation and tracing steam. What is the bleed rate required?

From Fig. 1, the bleed rate is 3 (gal/min)/100 ft. Thus the flow from the dead end is 3 × 250/100, or 7.5 gal/min. If insulation is applied, the bleed rate would be about half, or 4 gal/min. With wind velocities in excess of 15 mi/h, expected and no insulation, bleed should be 15 gal/min.

To estimate the bleed rate, use the weight rate method by catching the bleed in a bucket and weighing it.

In the case of a branch line leaving the main exposed header: If the branch is continuously flowing, calculate the bleed from the branch point to the dead end of the header. Check to make sure that the branch flow is sufficient to protect the upstream portion of the header. If the branch is nonflowing, determine the bleed for it. The sum of the branch and header bleeds should meet the need for the upstream header.

JOHN D. CONSTANCE

Leak Detection

The installation in 1987 of a computerized leak-detection system on the 8000 miles of line operated by Williams Pipe Line Co. (WPL) represented the largest such project in the United States.

Williams Pipe Line operates the largest independently owned batched liquid petroleum products pipeline in the United States with more than 8000 miles of lines, 117 pump stations, and 47 delivery terminals.

The system serves a 12-state area with yearly deliveries averaging around 185 million bbl and with a total system line fill of approximately 4.5 million bbl.

Centralizing Movement

Because of an ongoing modernization program, an increasing amount of the Williams system falls under the operational responsibility of the WPL dispatching center in Tulsa, Oklahoma.

More than 90% of all stations are now operated from Tulsa and, with the introduction of programmable controllers, very complex operations can be performed remotely. These include proving meters, lining up manifolds, starting units, opening and closing valves, and other operations which traditionally were handled by location personnel.

As this centralizing shift has taken place, it has become apparent that an on-line system to aid the dispatcher in rapidly finding measurement discontinuities, related either to mainline or to instrument integrity, would be very advantageous. The dispatcher's time could then be used more effectively and, at the same time, provide a better system for monitoring the mainlines.

Beyond aiding the operator, other factors influencing WPL's decision to purchase such a system were the high cost of petroleum products and cleaning up leak sites and the fact that some areas once remotely situated are now densely populated.

As WPL began researching such a mainline monitoring system, it became apparent that this would not be an off-the-shelf item. To incorporate all of the advance features necessary to model the WPL system adequately, the real time modeling and operator presentation would have to be advanced.

The key features required were the ability to receive data from WPL's existing TRW supervisory control and data acquisition (SCADA) system, on-line display generation, color graphics, trending, batch tracking, and, most importantly, an advanced real time modeling system. This modeling system should be capable of accurately simulating all WPL's mainline activities while not placing an excessive additional burden on its operators.

The ability not only to find the existence of a leak but also to estimate the size and location was of prime importance, as was a full-time over- and under-pressure monitoring system.

The system which WPL finally contracted is from Scientific Software Intercomp (SSI), Houston, which is the prime contractor and has turnkey responsibilities. The system was delivered in 1986 with project completion in 1987. This is an on-line system and is incorporated into the WPL SCADA system as the leak detection system (LDS).

Hardware

The LDS computer hardware consists of a Data General MV-10000 minicomputer with a 354 megabyte (Mb) hard disk and 8 Mb of internal memory (Fig. 1). The MV-10000 is rated at 2.5 million instructions executed per second and has a cycle time of 140 ns.

Two Aydin 2000 display generators drive five color monitors with a graphic resolution of 1024 × 1024 pixels and with an alphanumeric overlay channel (ANC) for text. The operator's input functions are performed through five Aydin 5116 keyboards.

The MV-10000 is linked to the existing TRW SCADA computer by a custom-built interface which provides the necessary parallel to serial conversion and handshaking to accomplish the data transfer. The data brought across this interface are in block form and contain several error check points.

Two methods are provided for hard copy: a Data General dot-matrix printer for strictly alphanumeric data and a Seiko RGB color copier for a full-screen printout.

The software consists of five major subsystems: the configuration editor, the real time model, leak detection and location, the man-machine interface, and over/under-pressure protection.

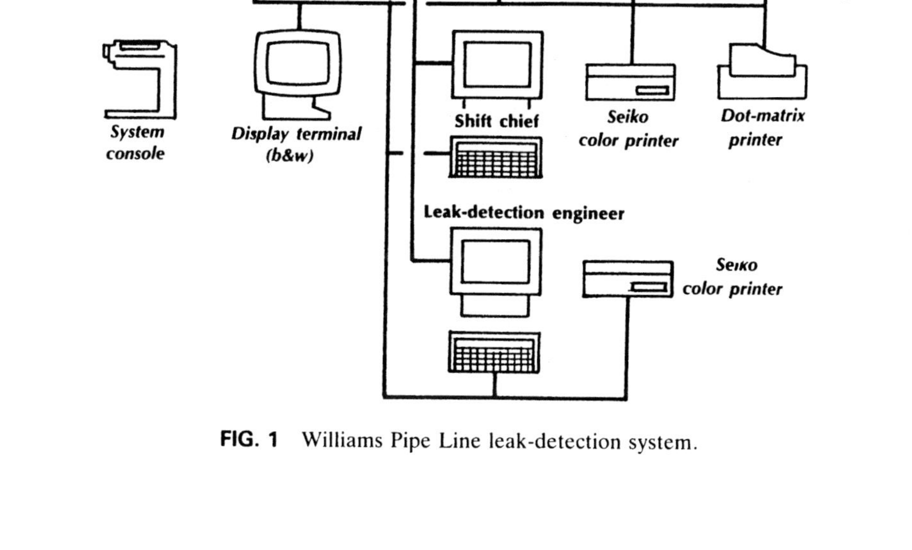

FIG. 1 Williams Pipe Line leak-detection system.

Hardware

The LDS computer hardware consists of a Data General MV-10000 minicomputer with a 354 megabyte (Mb) hard disk and 8 Mb of internal memory (Fig. 1). The MV-10000 is rated at 2.5 million instructions executed per second and has a cycle time of 140 ns.

Two Aydin 2000 display generators drive five color monitors with a graphic resolution of 1024 × 1024 pixels and with an alphanumeric overlay channel (ANC) for text. The operator's input functions are performed through five Aydin 5116 keyboards.

The MV-10000 is linked to the existing TRW SCADA computer by a custom-built interface which provides the necessary parallel to serial conversion and handshaking to accomplish the data transfer. The data brought across this interface are in block form and contain several error check points.

Two methods are provided for hard copy: a Data General dot-matrix printer for strictly alphanumeric data and a Seiko RGB color copier for a full-screen printout.

The software consists of five major subsystems: the configuration editor, the real time model, leak detection and location, the man-machine interface, and over/under-pressure protection.

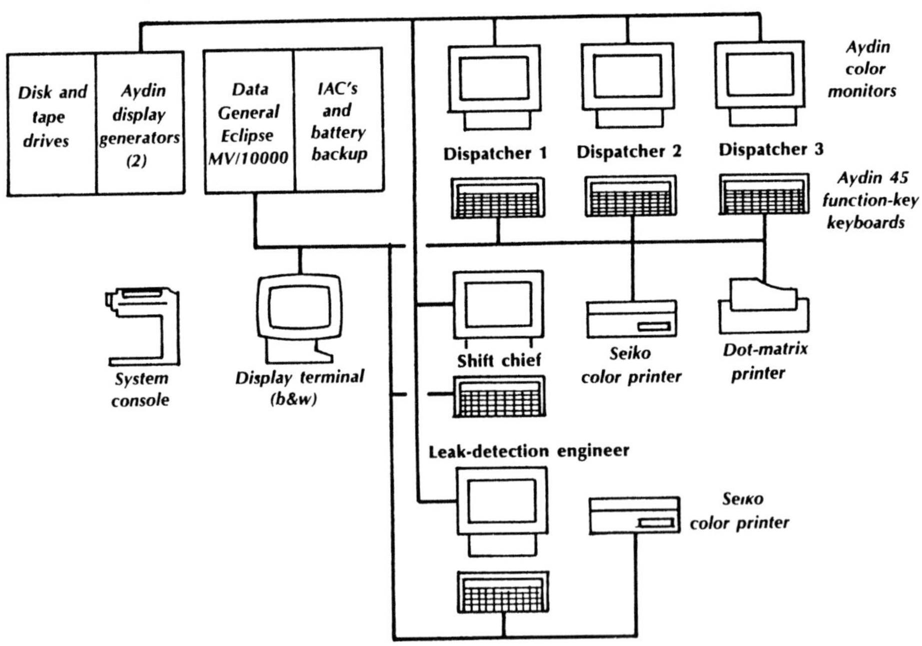

FIG. 1 Williams Pipe Line leak-detection system.

configuration where the model is forced to perform a calculation of state and is on the order of every 5 miles for the Williams system.

This list of physical connectivity is called a "network configuration."

All network configurations must be bounded by meters. Also, at every delivery and receipt point, as well as the suction and discharge side of intermediate pump stations, there must be temperature and pressure transmitters. Generally, instruments must be provided anywhere along the system where the liquid state is expected to change.

For predictive modeling, it is necessary to model the pumping equipment (i.e., entering pump curves).

Accuracy Improvement

For real time modeling, however, accuracy is improved by modeling the pipeline on either side of the equipment as a separate segment (i.e., a pump is simulated in step function form rather than with pump curves)[2].

Once a configuration is built, it can then be used by the real time model to perform leak detection. Configuration data files can be edited at any time, but before changes can be used by the real time model, that particular configuration must be stopped and restarted. This allows the data structure to be updated and also prevents spurious alarms from being generated.

Before the configuration data can be manually compiled, a naming convention must be chosen. This convention will be used as the primary identifier and should be one which is already known or easily learned by the system operators. All reporting, batch tracking, and alarming will refer to this convention.

One of the most difficult aspects of setting up a system as complicated as WPL's is the configuration process.

Some hard decisions must be made when dividing a pipeline system into networks. Since each network is modeled independently, much work by the operator can be saved by judicious choices of locations for the network boundaries.

For instance, assume that two lines cross at a junction point where one line can be switched to the other and that these lines are configured separately. This will force the operator to set up schedules at the junction point and reenter line fills when the operations change.

On the other hand, if both lines were configured together, the model would automatically handle this line switching.

Real Time Model

The function of the real time model (RTM) is to simulate accurately the current conditions within the pipeline as a function of time.

The model receives data from the SCADA system once per SCADA scan and then performs data validation routines and conditioning. These filtered data are then used by the RTM to advance each configuration one time step.

Progressive time steps use the most recent SCADA data to advance each configuration forward in time. The scan rate on the WPL system is approximately 20 to 30 s.

Pressure (including transients), temperature, and flows are calculated on approximately 5-mile intervals (every knot) and for critical points, using the classical equations for continuity, momentum, and energy. These equations take into account both the product-related information (heat capacity, absolute viscosity, bulk modulus, etc.), as well as pipeline-related information (elevations, heat-transfer coefficients, pipe diameter changes, etc.).

When a configuration is initially started or reactivated, it goes through a cold-start period. The length of this period varies according to many different parameters but, for liquid lines, is on the order of 1 h.

During this cold-start period, a validation of the SCADA values is performed and the pipeline profiles are initialized [4]. The RTM starts the cold-start period assuming steady-state flow and ends the cold-start period simulating transient flow. The potential exists for leaks to be masked during this period because the model assumes no leaks during the cold-start period.

Also, during the cold-start period the pressure/flow leak-detection module is inhibited. This allows the RTM to tune to flowing conditions and to generate leak-threshold values by tuning out instrument errors and roughness of the pipe walls.

The pipeline system is divided into many independent configurations or sections. These configurations can be stopped or started individually without affecting the operation of the other sections.

Once the cold-start period is complete, leak detection is enabled and the real time model is in its normal mode of execution. During this mode, the model performs auto tuning. That is, in order to make the model match the actual field conditions, the pipe-wall roughness and ground temperatures are tuned slowly with time.

The threshold values, which were calculated during the cold-start period, are adjusted with use of the ground temperature and roughness values as tuning variables. As transients propagate through the line, the threshold values dynamically change to prevent unit start-ups and other changing conditions from creating false alarms. This tuning process, however, proceeds very slowly so that leaks are not masked as measurement errors.

Leak Detection and Location

The leak-detection routine uses two different schemes to detect leaks: metered volume balance and pressure/flow deviation.

Metered-volume-balance (MVB) leak detection works as the name implies. The metered volumes pumped into the pipeline must equal the metered volumes received out of the pipeline less the change in line pack. The change in line pack or actual net volume of the pipeline is calculated by a breaking of the pipeline into sections in which the head and tail of each section fall on a calculation point (knot). Linear temperature and pressure gradients between these points are assumed, and the inventory of the sections is summed. This value is then compared to the previous line-fill amount. The accuracy of this scheme depends directly upon the accuracy of the meters and the accuracy by which the change in line pack can be determined. Therefore, on large volume lines, the tolerances will be large and the smallest leak which can be detected increased [3].

The metered-volume-balance detection scheme adds to the effectiveness of the system not only by confirming problems identified by the pressure/flow deviation method but also (because the MVB is not inhibited during the cold-start period) by providing full-time leak detection.

The second method for leak detection on the Williams system is the pressure/flow-deviation method. This is a much more sophisticated system than metered volume balance and is more resistant to transients induced by changing pipeline conditions. Under this scheme, each leg in a predefined configuration is treated as an independent segment with the flow being equal across the nodal points (unless there is a meter present). At these nodes, boundary conditions are forced on the model with the actual values received from the SCADA system. Typically, these could include the upstream pressure and downstream flow. With these values given, all other points in this leg can be calculated by the real time model. These calculated point values are then compared to the actual values and discrepancies noted [4]. The discrepancies are monitored and compared to the leak thresholds. A voting algorithm is then used to analyze the information and decide whether a leak should be declared. If a leak is declared, its size is estimated and a separate routine estimates the leak location. These values (leak size and location) are then updated after every SCADA scan.

Operator Link

The man-machine interface (MMI) is the system's link to the operator. As the typical operator will have no technical background, the system must be easy to use, have clear presentation, be responsive to operator request, and, at the same time, be flexible enough to allow for ever-changing display requirements. The Williams system uses a combination of windowing, local and global function keys, and high-resolution color graphics to accomplish these goals.

A window is defined as an arbitrary rectangular area on the screen in which data can be entered and displayed. The size of each window can vary from a full screen to just a few lines. Each of these windows, along with the data contained within its boundaries, is called a format. As many as five formats can be displayed at any one time on a CRT (cathode-ray tube). Each format can overlay other formats, either partially or completely, with different information. The dynamic information on a CRT, which is not occluded by other windows, is updated at refresh time.

Several different methods can be used by the operator to request a format. In all control and operator-entry formats, local function keys have been defined to reduce the number of keystrokes required to perform the necessary edits or operations. These keys can also be defined to bring up the next logical format in a series-type operation. The manual method can also be used to move around in the system by typing "add window" and the name of the desired format in the command field. The command "add window" can also be abbreviated down to the fewest characters necessary for a unique command. Actually, this manual method can be used anywhere in the MMI to execute a command.

Format Builder

Formats are generated and maintained in the system by an on-line package called the format builder. Formats are first built by a description of the static background which consists of any background text or graphic characters which do not change.

Dynamic field descriptors (DFD's) are then placed on this background. The DFD's are packets of information which are required to display the data, such as location in the data base, desired color, high and low limits, etc. DFD's can also be assigned color conventions on dynamic values. For example, a pump symbol on a summary display could change color as the status value assigned to that field in the SCADA data buffer changed. Once this building process is completed, the format can then be requested on the screen.

Trends or plots are built using the display builder. When a variable is displayed against time, it is called a trend; and when displayed against distance, a plot. Up to four variables can be displayed on a trend or plot, with each having its own associated color.

Historical data are saved for 48 h for every nodal point in three different files. The first file contains every data point from every model run for the previous hour. At the end of 1 h the data are merged into 5-min averages and stored in a second file. This file contains the 5-min averages from all nodal data for 8 h. The third file contains 1-h averages for 48 h. All data files are circular so that the oldest data are written over by the latest data when the file is full.

Batch Tracking

Because the Williams system is a batch petroleum pipeline, batch tracking is essential to simulate accurately the pipeline conditions. This information is also very useful to the dispatcher, although it will not replace the manual batch tracking that the dispatchers currently perform.

Batch-tracking displays are included for every line section. These displays actually contain a pressure profile on the top portion of the display and a horizontal batch-tracking bar graph on the bottom portion of the display.

Seven colors have been set up to correspond to the seven different API product ranges. These colors are used on seven different API product ranges. These colors are used on the bar graph to indicate the product type. Also, the head end of each batch is identified with a flag which contains the batch name, volume, and the estimated time of arrival to the next downstream meter. Along the top of this bar graph are tick marks and labels which identify the stations and branch points along the line. By using the pressure profile, the operator can readily see if maximum operating points are being exceeded and if the line is being operated efficiently.

The alarm package supplied with the Williams system has three different alarm displays: alarms by operator, alarms and event by operator, or alarms by configuration. Also, the top three lines of each CRT contain a permanent window, or display, which has the three most current active alarms for that operator. With special sorting keys, any alarm can be routed to any operator or console.

Pressure Protection

The over/under-pressure protection package is a separate system that will continuously monitor all operating configurations for instances of either over-pressure or under-pressure. As the configuration data files contain elevation data, pressures at critical points (high or low points) along the pipeline can be calculated and monitored.

Instrumentation

Second to the software, the instrumentation is the most crucial aspect of a leak-detection project. Without the proper field input in the form of pressure, temperature, and flow, no amount of investment on the master station can provide adequate leak detection.

Preparation of the Williams system for the leak-detection system was extensive. A system-wide study of all instrumentation was conducted, along with the undertaking of breaking down the system into configurations.

Because each mainline has its own operating constraints, each line had to be evaluated separately with a mind toward operations.

At every entry and exit to the pipeline, flow, temperature, and pressure measurement must be provided. Where one meter could pump on several different pipelines at different times, provisions have to be made to isolate this from the model. This isolation is required because the model must have a unique meter associated for each entry and exit. In some cases this is accomplished through software by the setting of an override on a data point. In most cases, however, programmable controllers are used to let one meter act as several.

Emphasis must be placed in the field, not only on installation but also on maintenance and calibration of this equipment. This requires the support from all facets of the company. Monies must be provided to upgrade the existing instrumentation and to keep it within specification. The technical support group staff may need to be increased to manage this increased load.

The engineering department must be educated in the requirements of the leak-detection system and be continuously involved to ensure that future projects provide them. Williams instituted a procedure whereby, before a project entered the final design stages in engineering, a piping and instrumentation drawing was circulated through the systems development group. (Systems development manages the leak detection project.) This provided an open channel of communications between engineering and systems development to ensure that the basic requirements of leak detection are met.

Compensation Tuning

As the quality of the data coming back from the field on a particular line section (configuration) degrades, the model will attempt to compensate by tuning the thresholds. When the preset limits are reached, no further tuning is possible; leak alarms will be generated and leak detection will have to be shut off. Prior to this point, operating policies should be established which will direct dispatching whether to operate a line under these conditions. If the line operations continue and a leak occurs, what are the repercussions? On the other hand, should a perfectly sound line be shut down because one pressure transmitter fails? The best solution, obviously, is to dispatch immediately a technician to either calibrate or replace the failed instrument, but this luxury is not always an option. It is imperative that these questions be addressed and clear-cut decisions laid out before the system is turned over to the operators.

It should be noted that if one of the instruments which has been selected as a forced boundary condition in the pressure/flow deviation scheme fails, leak detection for that line section must be shut off because tuning is not possible under these conditions.

Operation

Every effort has been made to reduce the extra work load which this system will place on the dispatcher (operator). Where possible, commands have been combined to allow one keystroke to perform multiple functions. Many function keys have been set up to reduce cursor manipulation required to operate the system.

Color schemes have been developed to aid the dispatcher's ability rapidly to scan a screen and pick out the necessary information. Color graphics have been used in conjunction with text where possible to take advantage of the system's high-resolution capabilities.

Due to the nature of the system, the data which the model receives from both the operator and the SCADA system must be accurate. Otherwise errors will be introduced and false leak alarms will occur.

The operator will be required to keep the system updated on all product information on the lines in operation. This means that both the line fill and schedule must be current and maintained in real time. This is accomplished with the line fill and schedule displays. Both of these displays have full-screen edit features and assigned local function keys to aid in this frequent process. Both the line fill and the schedule display screens have fields for the operator to enter the volume, batch name, product type, and degrees API. Once this information is entered, the model for that line can be started from the system control display.

References

1. Scientific Software Intercomp, *Functional Design Study—Modeling and Leak Detection System*, Presented to Williams Pipe Line Co., September 30, 1983.
2. T. P. Lindsey and J. C. Vannelli, *Leak Detection on Petroleum Pipelines*, Presented to the 1980 API Measurement School, Norman, Oklahoma.
3. J. C. Vannelli, and T. P. Lindsey, "Real Time Flow Modeling and Applications in Pipeline Measurement and Control," *IEEE Trans. Ind. Appl., 1A-18*(5) (September-October 1982).
4. S. J. Isenhower, *State-of-the-Art Leakage and System Pressure Monitoring during Product Transfer*, Presented to the Fifth Annual ILTA National Operating Conference, July 17–18, 1985.

THOMAS C. LIPPITT

Leak Detection Using SCADA Information

Supervisory control and data acquisition (SCADA) reported pressure and temperature data on typical remote terminal unit (RTU) spacings, along with inlet and delivery flows, can be used to drive a real-time model of a pipeline network. The model can respond to detect, size, and locate leaks. This procedure has been applied successfully in the field to all-liquid systems. Studies show that the approach will work equally well for gas networks, dense-phase ethylene systems, and in liquid-filled portions of networks that are operating in slack-line flow. It is currently being implemented in the field for the latter case.

There is also a related approach for determining the limits of detectability of leaks in a given system before the procedure is actually implemented in the field. This permits an evaluation in advance of weaknesses, if any, in the SCADA data to be used. It enables the determination of realistic expectations of leak detection performance during the design phase of a leak detection project.

The Procedure

To implement the procedure, an accurate computer model of each element of a pipeline network is built. This includes the pipe geometry and a representation of the fluid properties, which would include the bulk modulus, the viscosity, the thermal modulus, and the reference density of each batch of material to be flowing in the pipeline.

The required SCADA data are continual updates of temperatures and pressures at all RTU locations as well as inlet fluid properties, and all inlet and outlet flow rates. Leak detection is accomplished by requiring that the concurrently running network model of the pipeline agrees with the measured pressures at each SCADA scan at the locations corresponding to the measured values. The temperatures and other fluid properties entering the model are caused to match those of the fluids entering the pipeline at the corresponding locations.

Figure 1 is a schematic of a typical pipeline and model system with seven RTU locations. Spacing for RTU locations in commercial pipelines ranges from 25 to 100 miles. The measured pressures (and temperatures, which are not shown) are fed into the model.

After running briefly, a properly constructed model, with temperature profiles and batches (if any) properly located, will be in "lock-step" with the pipeline, i.e., all the modeled pressures, flows, and temperatures agree with their counterparts in the pipeline. The time required for lock-step is usually in the range of 5 min for fluids with high bulk modulus, such as liquids, or 50 min for fluids with low bulk modulus, such as gases.

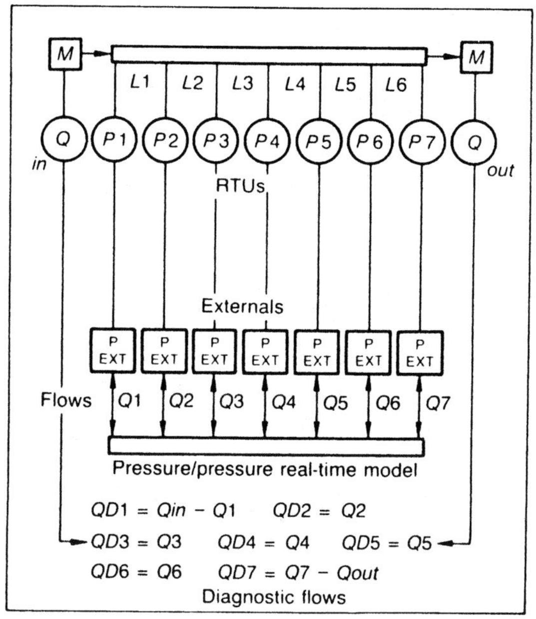

FIG. 1 Typical pipeline with SCADA.

The achievement of lock-step will occur and its maintenance will persist even if the pipeline is undergoing rapidly changing transients, as these will also be induced in the model by the changing pressures. At locations corresponding to pressure measurements (where the model is forced to agree in pressure), the model will necessarily calculate external flows, i.e., flows to or from the outside. These are $Q1, \ldots, Q7$ in Fig. 1. $Q1$ is the model-calculated flow that is positive for flow into the inlet location. $Q2, \ldots, Q7$ are model-calculated flows that are positive for flow out of the corresponding locations.

If the model and the data are accurate, the external flow in the model at the points corresponding to the inlet and outlet locations in the pipeline will continually agree with the measured values at those points. At the other points, the modeled external flows will be zero.

From a lock-step model, a very useful set of quantities, called the diagnostic flows, can be defined. At inlet points, the diagnostic flow is the measured flow less the modeled inlet flow. At outlet points, the diagnostic flow is the modeled outlet flow less the measured outlet flow. At all other points corresponding to RTU locations, the diagnostic flow is the flow leaving the model at the point. With these definitions in a properly constructed lock-step model, the diagnostic flows will always be zero in the absence of a leak condition in the pipeline.

The diagnostic flows, $QD1, \ldots, QD7$, are shown in Fig. 1 as the values $Q2, \ldots, Q6$ for the interior points and as $Qin - Q1$ and $Q7 - Qout$ at the inlet and outlet points, respectively. Qin and $Qout$ are the measured flows. The signs of these flows are chosen so that they are positive if the model detects a leak condition.

Leak Detection

To see how a leak is revealed, consider a leak at a location 40% of the way between the measurement points $P3$ and $P4$. Because it is assumed that the model is accurate, pressures $P2$ and $P3$ in the pipeline imply a flow in the model that continually agrees with the flow in the pipeline in the section of pipe above the leak. This flow in the pipeline includes the flow required to support the leak. Similarly, the two pressures $P4$ and $P5$ imply a flow in the model that agrees with the flow in the section of the pipeline below the leak. This flow, of course, does not include the flow leaving in the leak. From these two observations, we conclude that the sum, $Q3 + Q4$, of the diagnostic flows must be the same as the leak flow.

Leak Location

The relative size of $Q3$ and $Q4$ also serves to locate the leak. For the case of a liquid, it is as follows.

Let x be the distance between the $P3$ measurement point and the leak site. The distance between the leak site and the $P4$ measurement point is $L4 - x$, where the spacing between the $P4$ and $P3$ points is $L4$ miles. Let the flow delivered past the point corresponding to $P4$ be $Q0$, and let the leak flow rate be Q. Then the flow rate delivered into the point corresponding to $P3$ will be $Q0 + Q$.

The pressure difference $P3 - P4$ is made up of two parts. Let P be the pressure at the leak site. Then

$$P3 - P4 = (P3 - P) + (P - P4) \tag{1}$$

Assume Q is small relative to $Q0$. For some constant C, which depends upon pipe roughness, there are two relationships associated with the frictional pressure drop in the pipe above and below the leak site:

$$P3 - P = C * (Q + Q)^2 * x \tag{2}$$

and

$$P - P4 = C * Q0^2(L4 - x) \tag{3}$$

In the model, flow in the section corresponding to that in which the leak occurs equals $Q0 + Q = Q3$. This follows insofar as $Q0 + Q$ is implied by the pressures $P2$ and $P3$ measured in the pipeline, which is carrying that flow, and by the definition of $Q(3)$, which is flow out of the model at the point corresponding to the pressure measurement $P3$. Thus, for the same constant C as used for the pipeline (assuming a faithful model of the pipeline):

$$P3 - P4 = C * (Q0 + Q - Q3)^2 * L4 \tag{4}$$

and as noted earlier:

$$Q3 + Q4 = Q \tag{5}$$

Combining Eqs. (1) through (5) and using Q small relative to $Q0$ gives

$$\frac{x}{L4} = \frac{Q4}{(Q3 + Q4)} \tag{6}$$

The value of $L4$, which appears in Eq. (6), is, of course, the known RTU spacing. The meaning of Eq. (6) is that the ratio of the distance of the leak from the upstream RTU to the RTU spacing is the same as the fraction of the leak flow that appears to come from the downstream end of the modeled interval.

The result is only slightly more complicated if we drop the assumption that Q is small relative to $Q0$. In this case:

$$\frac{x}{L4} = \frac{Q4}{(Q3 + Q4)} * \frac{(2Q0 + Q4)}{(2Q0 + Q3 + Q4)} \tag{7}$$

This analysis shows that the detection and location of the leak depends on the change in frictional pressure drop required to supply the leak.

This leads to the question: Will the procedure work also for a gas pipeline? Consider as follows: If the fluid were a gas, the analysis differs only in that the square of the flow is approximately proportional to the drop in the squares of the absolute pressures. Since the pressure terms are substituted out in the result, Eqs. (6) and (7) apply equally for gases and liquids.

The foregoing statements lead to the conclusion that the procedure described should function for both gases and liquids if flow can be deduced from measured pressure differences with comparable accuracy. In commercial pipelines, the pressure drop at design conditions for both gas and liquid pipelines is comparable, a fact that confirms this conclusion.

Low Flow Case

In the derivation of Eqs. (6) and (7), it was assumed that the underlying flow, $Q0$, or the leak flow, Q, was sufficiently large to produce noticeable frictional pressure changes due to the leak Q. If $Q0$ and Q are both quite small, say only 5% of design flow rate, then the foregoing analysis will not be an accurate representation of the situation. In this case, however, the onset of the leak will produce an effect called the waterhammer or downsurge. The downsurge is related to the pressure changes required to accelerate the fluid which is supplying the leak. The analysis of this situation is somewhat more complex, and will not be described here.

In practice, the detectability threshold for a leak in a liquid line when $Q0$ is small is about four times what it would be in the case in which $Q0$ is near the design flow rates. In a gas line, the detectability of the leak depends strongly on a number of local factors, which must be examined independently for each case.

Autocalibration

The foregoing assumes that the model provides a faithful representation of the fluid flow in the pipeline in each section of pipe between pressure measurements. In practice, there is drift in the transducers that measure temperature, pressure, and flow, and there are changes with time of the line efficiencies in the pipeline. Each of these will produce errors in modeling and thus complicate the rather simple analysis provided here.

While errors cannot be totally eliminated, and thus a more complex analysis must always be undertaken in practice, the size of the errors can be minimized by a process called autocalibration. Autocalibration consists of searching for a consistent set of calibrations for the transducers and the efficiencies of the pipeline sections between RTU locations so that the diagnostic flows are reduced to zero or near zero. It can be shown by a mathematical analysis that such a calibration procedure is always possible for commercial pipelines.

What autocalibration achieves is to reduce the errors in the reported data to something close to the repeatability errors in the measurements. Generally, the repeatability errors are an order of magnitude less than the manufacturer's accuracy ratings of the instruments. So autocalibration yields close to the best possible use of the SCADA information provided. Coupled with the robust procedure described, this translates to providing close to the lowest achievable threshold of detectability using the available SCADA information.

Leak Tester

In the application of these procedures to leak detection in the field, it is helpful to assess in advance the expected performance. This can be achieved by simulating the pipeline and the SCADA system to generate the information that is required for applying the leak detection and calibration procedures described.

In this application, errors can be introduced to simulate errors in the field. These would include errors in the SCADA information, drifts in the pipeline roughness, errors in the thermal environment of the pipe, errors in the fluid properties, such as bulk and thermal moduli, and the fluid viscosities. Also, errors in the pipe description such as length and elevation profiles can be included. The intent in the introduction of errors is to evaluate the sensitivity of the leak detection process to each category of error, as well as to determine the effectiveness of the autocalibration procedure.

In the field applications that follow, the leak tester principle will be illustrated by the introduction of noise in the pressure, which is the most fundamental of the measurements used in the approach.

Field Applications

The leak detection-location technology described has been successfully implemented in the field in at least two applications. The first is a crude oil/products pipeline carrying numerous continuously blended batches that differ widely in density and viscosity. The pipe environment offers many long steep grades, frequently occurring slack-line flow, and significant thermal changes. Implementation of the foregoing leak detection procedure for this line is in progress.

The second field system evaluated is an ethylene line carrying dense-phase ethylene. This line is characterized by very large changes in the properties of the flowing fluid. The isothermal wave speed, which is related to the bulk modulus, varies approximately by a factor of six between the inlet and delivery points. This line is a difficult system for performing leak detection. There have been a number of unsuccessful attempts using other leak detection technologies to achieve a reasonable threshold of detectability for similar lines.

Insofar as significant emphasis is placed on gas lines, a third field example considered is a hypothetical one. It is included primarily to illustrate the types of results that might be expected in a gas system.

In the work presented, the masking effect of the noise introduced into the leak detection process will be limited to noise in the pressure measurements. Other sources of error were evaluated, but interpretation of their effects is more complex and beyond the scope of this article.

Study 1. Trans Mountain Pipeline

Trans Mountain Pipeline Co. carries crude oil and products from Edmonton, Alberta, to Vancouver, British Columbia, with intermediate tankage, supplies, and deliveries. It is 711-mi long and goes over rough mountainous terrain. Elevations range from sea level to over 4000 ft, with many grades in excess of 25%. Some sections regularly operate in slack flow. The pipeline typically carries 30 different batches at a single time, and the batches frequently change composition and properties continuously throughout their length.

A section of the line was chosen as representative for the study. This was a 161-mi section from Kamloops to Sumas. The RTU locations *R*0 (the inlet), *R*4, *R*6, *R*9, and *R*10 (the delivery) are used for simplicity rather than geographical field names. The mileposts of these RTU locations are, respectively, 0, 50, 67, 117, and 161. The RTU numbering system was set up to increase in the direction of flow. The numbers not referenced in the list above were defined to allow for other possible locations. No others of these were used as RTU locations in the studies presented here.

The pipe in the test section is 24 in., with a 0.316-in. weight. It ranges in elevation from 17 ft at *R*10 to 4009 ft about 10 mi from *R*0. The elevation at *R*0 is 2389 ft. It is initialized in steady flow at 150 Mbpd with 33 batch interfaces separating batches of 14 different fluids.

The leak tester was built as illustrated in Fig. 2. The top part of the figure is a schematic of the pipeline. There are three locations indicated as leak sites. At one or more of these sites, fluid may be withdrawn from the pipeline at any desired rate to simulate a leak. This pipeline was simulated using DREM's pipeline simulator (DPS).

At the RTU locations are indicated pressure transmitters *P*0, *P*4, *P*6, *P*9, and *P*10. The noise in the SCADA system pressure is generated by equipment in the DPS called noise relays. The noise relays can generate "white" noise in a data-enterable interval. This means that the input to the relay is perturbed in the relay's output by an amount that is a uniform random variable within the interval. For this study, the intervals were 0.25%

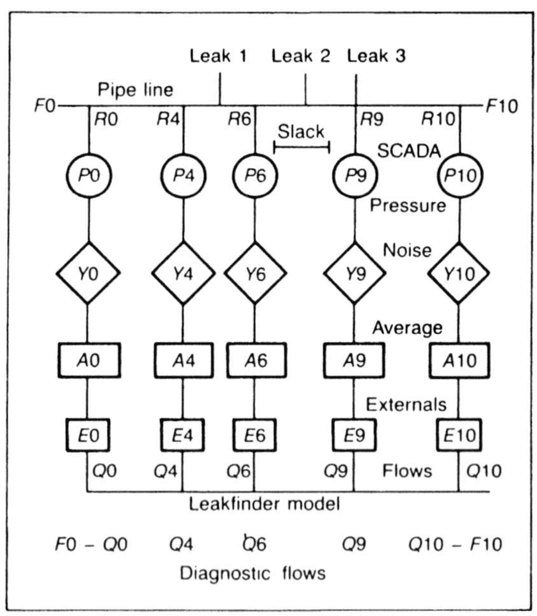

FIG. 2 Schematic for Trans Mountain Pipeline leak test.

of span, where the span for each instrument was zero to P_{max}, the maximum pressure expected at the location during pipeline operation. The noise relays in Fig. 2 are labeled *Y*0, *Y*4, *Y*6, *Y*9, and *Y*10. The noise generation relay will change its perturbation each SCADA scan, which in this case is a period of 6 s.

In operation, the leak detection procedure will function more realistically if the pressure inputs are subjected to a small amount of time-averaging before they are imposed on the model. The time averaging is done by the devices in the schematic indicated as *A*0, *A*4, *A*6, *A*9, and *A*10. Their function is to provide approximately a moving average of the input over a 1-min interval.

The output of the time-averaged signals from *A*0, . . ., *A*10 control the pressure externals for the model of the pipeline whose schematic is shown in the lower part of Fig. 2. These externals are model components of the DPS, which serve to cause flow of fluid to or from the outside to maintain the pressure at the value specified by their input signals. The flow required by the externals at the points in the pipeline model in the lower part of Fig. 2 are defined to be *Q*0, *Q*4, *Q*6, *Q*9, and *Q*10.

To be consistent with the earlier discussion, the sign of *Q*0 will be positive for flow entering the model at the left end. The sign of all other values, *Q*4, *Q*6, *Q*9, and *Q*10, are taken to be positive for flow out of the modeled pipeline. With these definitions, the diagnostic flows will be *QD*0, *QD*4, *QD*6, *QD*9, and *QD*10. *QD*4, *QD*6, and *QD*9 are, of course, equal to *Q*4, *Q*6, *Q*9, respectively. *Q*0 is *F*0 − *Q*0 and *QD*10 is *Q*10 − *F*10, where *F*0 and *F*10 are measured inlet and delivery flows, respectively.

The definition of the system to be modeled for the leak tester function is now complete. All of the elements defined will be modeled by the DPS. This means that for the leak tester the DPS is used to model both the pipeline and the leak detection model, along with the SCADA system and the noise to be introduced into the pressure signals for this study.

The Leak Study

The pipeline was brought to steady state carrying 151 Mbpd of a fluid whose properties were "average" for the batches of material that were to be placed in the line for the study. The leak finder model was turned on and allowed to come into lock step. The temperatures of the fluids in the model were initialized from a steady-state thermal model. In all, 33 batch interfaces between 14 different fluids were required. The batches were placed at the desired locations in both the pipeline and the model by simply redefining the properties of the fluids by location. This abrupt change caused a slight perturbation that quickly settled out.

In field operation, the lock-step model will always be current in simulating temperature profiles and batch locations. This is because it would continually move batches in response to the inlet temperature, composition, and flow rate data. Abrupt batch adjustments would be made in the field only following a period of outage of the model, or to make small adjustments to match data supplied by interface sensors.

The study consists of three cases. Beginning with the state described in the previous paragraph, a leak was introduced successively at each of three locations, one for each case. These locations are shown in Fig. 2 as: (1) between *R*4 and *R*6, 11.2 mi downstream of *R*4; (2) between *R*6 and *R*9, 25.5 mi downstream of *R*6; and (3) at *R*9.

At the 151-Mbpd rate and at the pressure level at which the pipe was operated, a slack section about 9 mi in length exists between *R*6 and *R*9. Leak Site 2 lies below the slack section. All the leaks except at Leak Site 2 were 12 Mbpd. The leak rate at Leak Site 2 was 15 Mbpd.

Case 1 Results

Figure 3 shows the results for Case 1 of the study, i.e., when the leak is introduced at the first leak site. The leak rate is shown as *QL*5, and the diagnostic flows *QD*0, *QD*4, *QD*6, and *QD*9 are plotted on the same scale. *QD*10 was omitted as it is very similar to *QD*0 and *QD*9. It is apparent that the state corresponding to the analysis that led to Eq. (6) is achieved rather quickly. Less than 0.5 min after the onset of the leak rate, *QD*4 and *QD*6

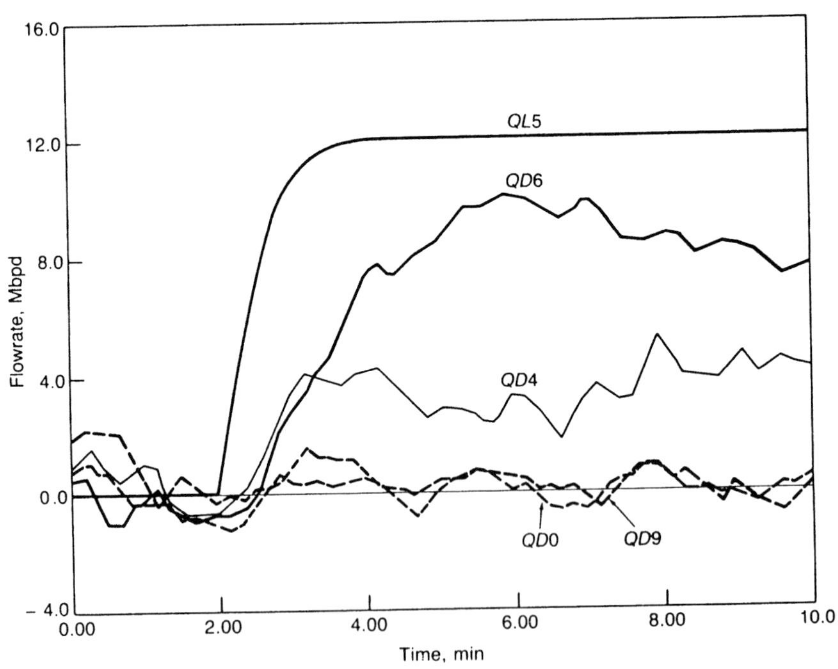

FIG. 3 Trans Mountain Case 1 leak flow at *R*4 + 11.2 mi.

responded by increasing from oscillating about zero. Further, within another 4 min, the sum of $QD4$ and $QD6$ was approximately the leak rate, as deduced in the steady-state analysis which led to Eq. (5).

It is easy to show that the results of Fig. 3 do not depend upon the fact that the state was steady in the pipeline at the time of leak onset. This results because the model and the pipeline are continually maintained in lock-step with respect to the underlying flow, whether it is transient or not. The leak is simply a constant perturbation of the agreement between the model and pipeline, and this difference presents a transient that dies away quickly for the RTU spacing of this example. We now apply Eq. (6) by taking $QD4$ as $Q3$ and $QD6$ as $Q4$. Here $L4$ is the spacing between $R4$ and $R6$, namely 16.8 mi. Then the approach with time to applicability of Eq. (6) can be explored by computing x (which should be 11.2 mi) at several times after the leak onset. These results are shown in Table 1.

For $QD4$ and $QD6$ in Eq. (6) we could have substituted these same flows accumulated in time. This substitution would in effect replace the diagnostic flow rates with the time-averaged diagnostic flow rates and lead to an x-value that becomes increasingly stable in time following the leak onset. We shall use this option in the next section.

As expected, $QD0$ and $QD9$ (and $QD10$, which is not shown) did not respond to the onset of the leak. $QD0$ and $QD9$ were not, however, zero. We note that $QD4$ and $QD6$, which did respond, were also not zero before the leak onset. Each of these would have been zero before leak onset, and $QD0$ and $QD9$ would have been zero thereafter were it not for the errors deliberately introduced in the modeling process by the noise simulated in the SCADA pressure measurements.

The noise introduced into the diagnostic flows by the noise in the pressure measurements could be expected to mask a small leak. The size leak that would be masked due to the pressure noise in this case would be comparable to the noise in the diagnostic flows. In this case this noise is estimated conservatively to be 2 to 3 Mbpd, or about 1.3 to 2% of the underlying flow. Actually, integration with time of the diagnostic flows will reduce the effects of random noise. Practically, in data that are of this quality, and if pressure noise is the solid source of error, leaks on the order of 0.5% would

TABLE 1 Applicability of Eq. (6)

Time (min)	$QD4$ (Mbpd)	$QD6$ (Mbpd)	x (mi)	Error in x (mi)	Error in $QD4 + QD6$ (Mbpd)
6	3.9	9.4	12.0	0.8	1.3
7	2.9	9.5	12.9	1.7	0.4
8	4.3	8.7	11.2	0.0	1.0
9	4.2	7.5	10.8	−0.4	−0.3
10	5.2	7.5	9.9	1.3	0.7

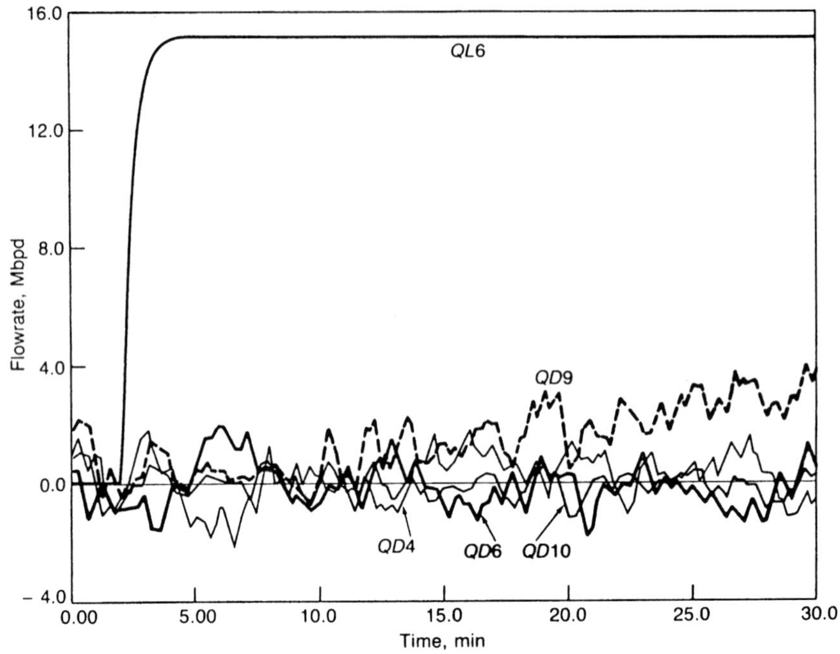

FIG. 4 Trans Mountain Case 2 leak flow at $R6 + 25.5$ mi.

be detectable using time averaging. Unfortunately, pressure errors are not the sole source of leak masking.

Case 2 Results

Figure 4 shows the results from the second leak location 25.5 mi downstream of $R6$. In addition to the leak flow, $QL8$, the diagnostic flows $QD4$, $QD6$, $QD9$, and $QD10$ are plotted.

Between the leak site and $R6$ occurred a slack section about 9 mi in length. One consequence of this slack section is that the pressure at $R6$ could not possibly "feel" the effect of the leak. This follows because the pressure in the slack section depends only upon the vapor pressure of the fluid and not the leak rate. Thus, the only effect that could be expected was at $R9$. And this effect should primarily be from the abnormal rate of fall of the interface between the slack and full sections, which occurred near the leak site.

As seen in Fig. 4, the diagnostic flow, $QD6$ at $R6$, was indeed totally uninfluenced by the presence of the leak. Further, the responses to the leak onset for all diagnostic flows except $QD9$ were also nil. The response at $QD9$ was very much slower in developing than had been the responses of $QD4$ and $QD6$ in Fig. 3. The reason for this, of course, is the much slower time for the pressure changes due to the loss of fluid to be felt at $R9$. This delay is a consequence of the fact that the only pressure effect that could be observed was that due to the dropping of the vapor/liquid interface at the slack section. This interface fell at a rate that was higher because of the leak than it would have been in the absence of the leak.

Again, based upon the effect of pressure errors alone, the 15 Mbpd leak could be adjudged to be detectable in about 20 min, and the "noise" in the pressure measurements would be expected to mask any leak less than about 2 Mbpd. However, many other errors in this slack line situation dominate the effects of the pressure noise, so that practically the detectability of a leak in this section is perhaps restricted to leaks of about 6-fold this size.

Case 3 Results

Figure 5 shows the diagnostic flows for Case 3 in which the leak, $QL9$, is located directly at RTU $R9$. It is apparent that except for a slight delay that was built in by the averaging process, the diagnostic flow $QD9$ tracked the leak flow almost exactly within the limits imposed by the pressure noise. None of the other diagnostic flows responded at all.

The purpose for running this case was primarily to verify that there would be no problems with this approach if the leak coincided with an RTU location. Problems in this case have been observed in other SCADA-based detection systems.

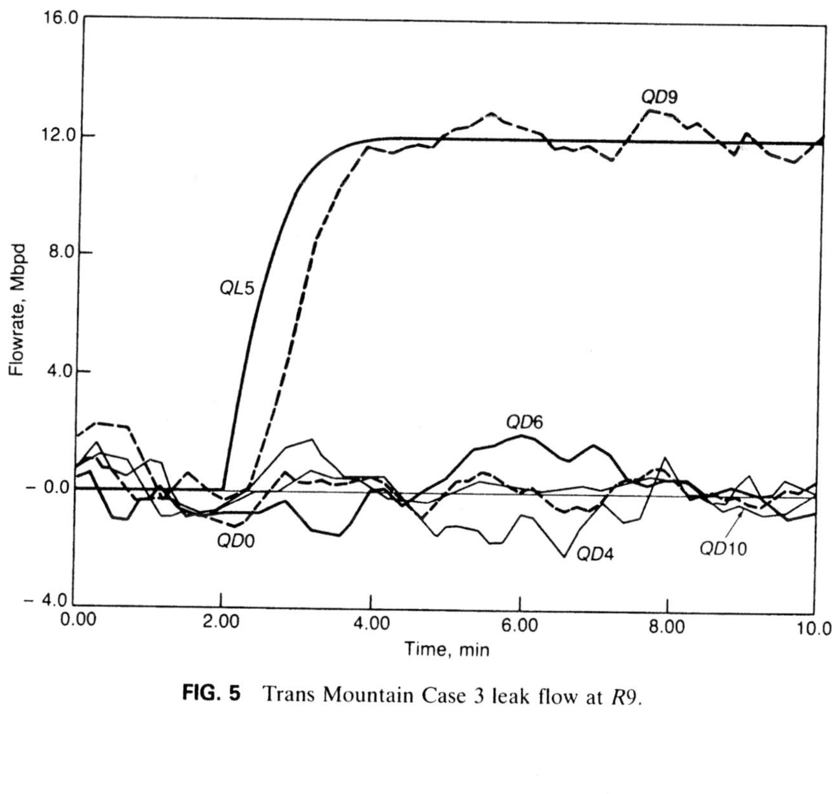

FIG. 5 Trans Mountain Case 3 leak flow at $R9$.

Conclusions from Oil/Products Line Study

The results presented for the foregoing three cases are the basis for the following conclusions considering the effects of pressure noise only:

With 0.25% of pressure level introduced as noise in pressure measurements, leaks in the 2 to 3 Mbpd range can be detected.

Under the foregoing conditions, leaks intermediate between RTU locations in the tight sections can be located within 8% of the RTU spacing within 4 min.

Under the foregoing conditions, leaks between RTUs in which a slack section lies can be detected in about 20 min, but location of a leak close to the vapor–liquid interface requires a different analysis from that of Eq. (6).

Study 2. South Texas Pipeline—Ethylene Study

The second study is of the leak detectability in an ethylene pipeline. The pressure ranges from about 2000 lb/in.^{2}gauge at the inlet to about 900 lb/in.^{2}gauge at the delivery point for the line chosen for study. The temperature seasonally varies from 60 to 80°F.

Under these conditions ethylene is above its critical pressure and is referred to as being in its dense phase. The compressibility of the fluid varies widely from the inlet to the outlet ends. This leads to about a 6-fold difference in wave speed along the line. Near the delivery end of the line, pressure is about 150 lb/in.^{2} above the critical pressure, and in the winter temperature is near the critical point. This location on the Mollier chart reveals that density at constant pressure becomes a rapidly changing function of gas enthalpy. The enthalpy may strongly depend on the thermal interchange of heat between the pipe and its surroundings.

The properties of the fluid under transport conditions lead to a complex modeling situation, and ethylene leak detection systems based upon technologies different from the one described here have been known to fail. For this reason, the leak test simulations described here were considered critical in designing a system which would avoid the known difficulties.

A number of situations were studied. These included the effect of errors in the reported pressure and temperatures, lack of information about and changes in the thermal properties of the pipe surroundings, the pack rate, the leak location relative to the rapidly changing wave speed, and uncertainties in the equation of state.

It is beyond the scope of this article to describe the study of each of these effects in detail. This discussion is limited to the effects of pressure noise with all other factors being held constant. The modeling was done in a fully thermal way which included energy effects as they relate to such items as Joule–Thompson cooling, frictional heating, and the transient in-

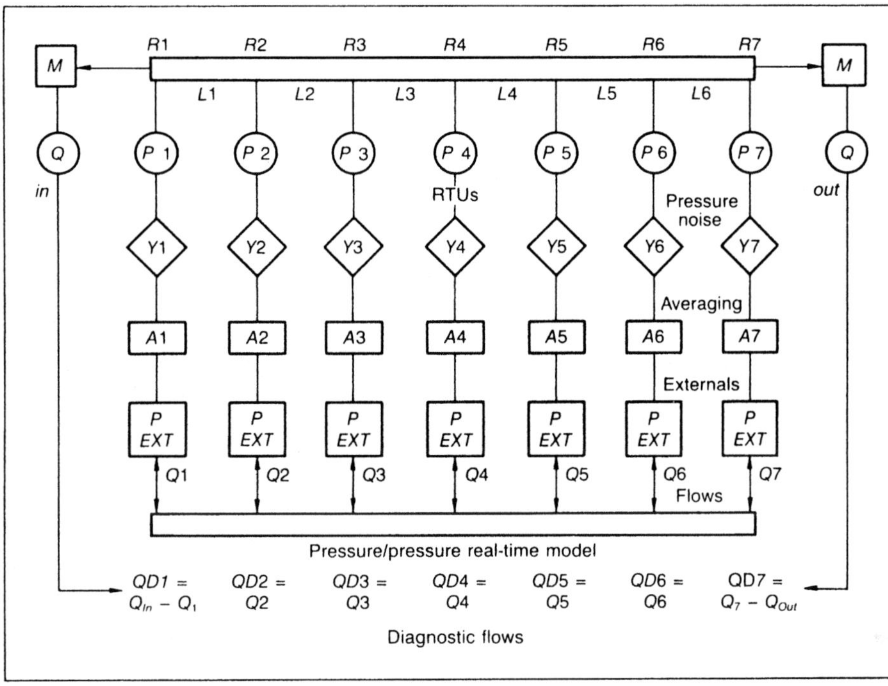

FIG. 6 Typical pipeline with SCADA.

terchange of heat energy with the pipe and surroundings. The latter factors are of considerable importance in the transient flow of ethylene under the test conditions.

The line studied was the South Texas ethylene line from Corpus Christi to Markham, Texas. The line is 8 in. in diameter and 126 mi long. The wall thickness is almost constant at 0.312 in., and the elevation changes are negligible.

There are seven RTU locations. The schematic of the leak tester system is shown in Fig. 6. The modeling is analogous to that in Study 1. The spacings between RTUs are as shown in Table 2. As before, the RTU number increases in the direction of flow.

The leak is at a point 5.2 mi downstream of *R*5. The line was initialized at no flow and 900 lb/in.^{2}gauge, and the pressure was raised to 2000 lb/in.^{2}gauge at the outlet (*R*7). These conditions were maintained until the inlet rate reached 148,000 lb/h and the outlet rate was 115,000 lb/h. Thus

TABLE 2 RTU Locations on South Texas Ethylene Line

RTU location	*R*1	*R*2	*R*3	*R*4	*R*5	*R*6	*R*7
Milepost	0	22.5	47.5	66.4	80.6	98.0	125.7

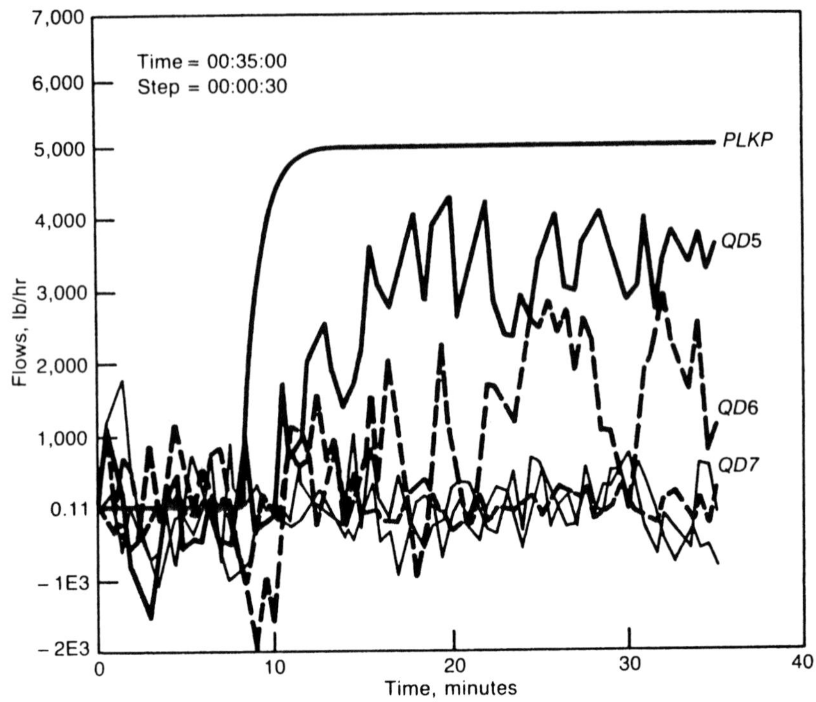

FIG. 7 Ethylene diagnostic flows.

the line was not in a steady state, but it was still packing at a rate equal to one-third the delivered rate. At this point a leak of 5000 lb/h was simulated in the pipe at the leak location. The pressure noise assumed to be introduced by the SCADA system was chosen uniformly from the interval −1 to +1 lb/in.^{2}. The SCADA scan rate was taken to be 30 s. The simulated SCADA pressures were time-averaged over a 2-min. interval and then imposed upon the leak finder model. The diagnostic flows were recorded for plotting.

South Texas Ethylene Line Results

Figure 7 shows as functions of time the diagnostic flows *QD*1, *QD*3, *QD*5, *QD*6, and *QD*7 which resulted. *QD*2 and *QD*4 were similar to *QD*1 and *QD*3, and were omitted to improve the clarity. The time scale was chosen so that the leak was initiated at 8 min. Only two of the diagnostic flows (*QD*5 and *QD*6) respond, and these within about 8 min. The leak flow is plotted as *PLKP*.

A more revealing way of presenting the data is in terms of cumulative flows. This means simply accumulating the values of the diagnostic flows in time. Figure 8 presents the cumulative values of the diagnostic flows *QD*3,

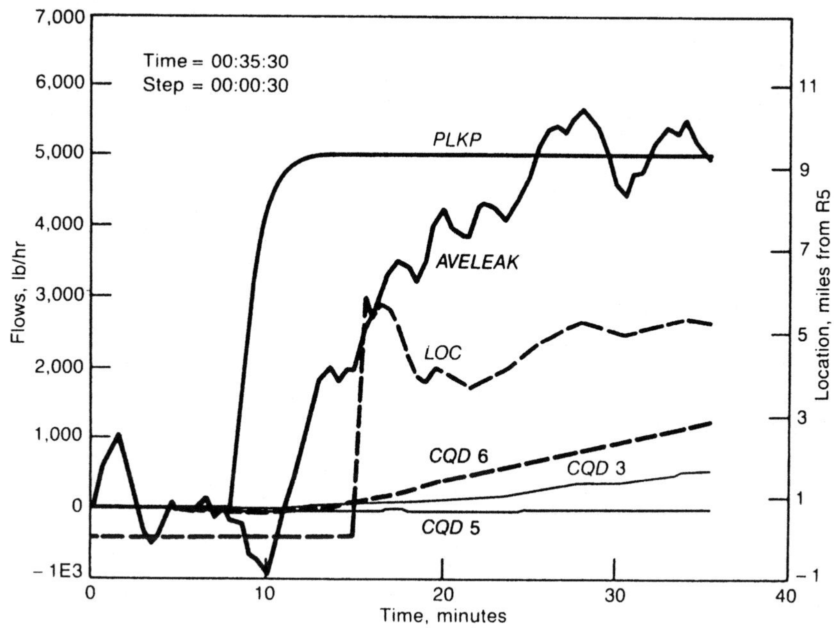

FIG. 8 Ethylene diagnostic output.

*QD*5, and *QD*6 as *CQD*3, *CQD*5, and *CQD*6, respectively. Observing the response of *CQD*5 and *CQD*6 emphasizes the fact that the leak is detectable within 8 min of onset.

Another diagnostically helpful trend plot is the 2.5 min moving time-average of the sum of *QD*5 and *QD*6. This is also plotted in Fig. 8 and labeled *AVELEAK*.

Figure 8 also shows the trend with time of another variable labeled *LOC*. This is defined by Eq. (6), in which *CQD*5 is substituted for *Q*3, and *CQD*6 is substituted for *Q*4. This in effect substitutes what is the equivalent of the ratio of average flows on the right-hand side of Eq. (6) so that *LOC* should be the distance of the leak from *R*5. In practice the computation of the *LOC* variable would occur whenever *AVELEAK* attains a sufficiently positive value. Figure 8 shows that the *LOC* approximation is attained reasonably promptly, as it is correct within 10% of the RTU spacing within 10 min of leak onset.

Conclusions from Ethylene Study

Based upon the assumption that the only source of error in the leak detection procedure is the pressure noise, and that this noise will be introduced from an interval from -1 to 1 lb/in.^{2}, then this study would indicate that for flow rates in the 100,000 to 150,000 lb/h range in this line:

A 5000 lb/h leak can be detected within 8 min of leak onset.
A 5000 lb/h leak can be located within 10% of the RTU spacing within 10 min of leak onset.
A leak of about 2000 lb/h could possibly be masked and its detection delayed by the assumed pressure noise.

As mentioned above, pressure errors are not the only errors which can mask leaks. Among the others are the thermal uncertainties associated with local variations in soil moisture. These factors were found to affect the precision of autocalibration adversely and thus introduce masking errors which are of about the same size as the pressure errors assumed for this study.

Study 3. A Hypothetical Gas Pipeline

The final study concerns the question raised earlier of the applicability of the leak finder procedure for commercial gas transmission lines. One reason for this study is to address the popular misconception that it is impossible to perform gas leak detection using standard SCADA data.

The motivation for the study comes primarily from two sources. One is that a forerunner of TRIM was installed and passed its site acceptance test on a gas line in New Zealand. Second, the analysis leading to Eq. (7) shows that the primary leak response required from the data is the change in frictional pressure drop. This depends upon having a measurable change in pressure drop that results from the leak.

So the purpose here is to choose a typical gas pipeline and RTU spacing, and determine using the leak tester approach what would be expected in terms of detectability. As before, the noise introduced in the measurements by the SCADA system will be confined entirely to that in the pressure measurements.

The Gas Pipeline

The section is 30-in. pipe, 300 mi in length, and the RTU spacing is constant at 50 mi. The pressure measurement noise is introduced in the interval -1 to $+1$ lb/in.^{2}. The SCADA scan period is in 1-min intervals.

The schematic for this study is exactly the same as for the previous study, namely Fig. 6. In this case the period for the time-averaging relays will be 3 min. This choice will give an appropriate smoothing for the pressure data from the 1-min repetition rate from the SCADA system.

The study was carried out by initializing the line at 1000 lb/in.^{2}gauge, maintaining the inlet ($R1$) at 1000 lb/in.^{2}gauge. The gas used is methane.

The leak finder model was turned on and allowed to maintain lock step. A flow of 350 MMSCFD at the delivery point was initiated and maintained (*R*7) for 67 simulated hours, at which time the line was still in an unpacking state. A leak flow was then simulated to occur in the pipeline at milepost 120, or about 40% of the distance between *R*3 and *R*4. The diagnostic flows before and after the leak onset were recorded for analysis.

Gas Line Results

The ethylene study showed that cumulative diagnostic flows are quite effective diagnostic tools to detect leaks, Fig. 8. The leak was initiated at simulated time slightly before 0.5 h. The leak flow is labeled *LEAK*. The cumulative diagnostic flows for only three locations are shown for clarity. These are labeled *CQD*2, *CQD*3, and *CQD*4, corresponding to the RTUs *R*2, *R*3, and *R*4. The leak lies between *R*3 and *R*4.

Figure 9 shows that the cumulative flows *CQD*3 and *CQD*4, which had been oscillating near zero, begin to increase after leak onset. They then continue to rise in contrast with *CQD*2, which continues simply to oscillate near zero. About 30 min after onset, it is obvious that *CQD*3 and *CQD*4 could not respond this way in the absence of a leak.

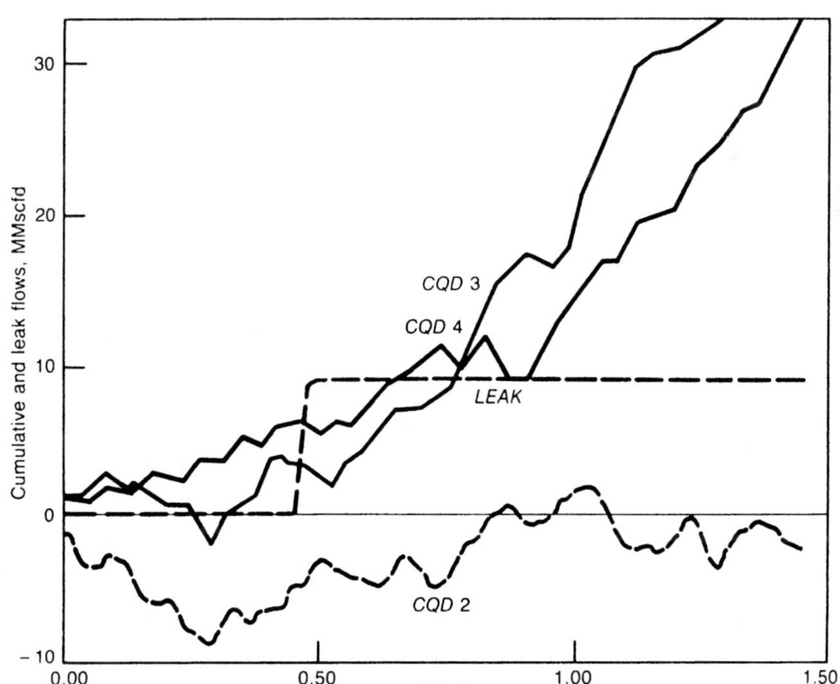

FIG. 9 Gas line, Study 3, leak at MP 120.

The ethylene study also showed that a time-averaged sum of the two diagnostic flows which appear to be involved in a leak is a good diagnostic tool. Figure 10 shows the leak flow labeled *LEAK* and a 2.5-min moving time average of the sum of the flows *QD*3 and *QD*4. This moving average is labeled *AVELEAK*. Prior to leak onset, *AVELEAK* was oscillating about zero. After leak onset, it appears to be oscillating about the leak level.

Figure 10 also shows, as the variable *LOC*, the value computed from Eq. (6) with the cumulative values, *CQD*3 and *CQD*4, substituted for *Q*3 and *Q*4. The variable *LOC* was computed beginning 15 min after leak onset. In the field, its computation would have been keyed on the condition that the variable *AVELEAK* attain a sufficiently positive value.

Within 30 min of leak onset, the diagnostic indicators have confirmed the existence of the leak, and have defined its location within a mile of the correct value, in this case 20 mi from RTU *R*3.

Conclusions from Gas Leak Study

The study of the 30-in. line shows that a 2.5% gas leak intermediate between RTU locations can be confirmed; a 2.5% gas leak intermediate between RTU locations can be located within 2.5% of the RTU spacing within 30 min; and the popularly held view in the United States that leak detection

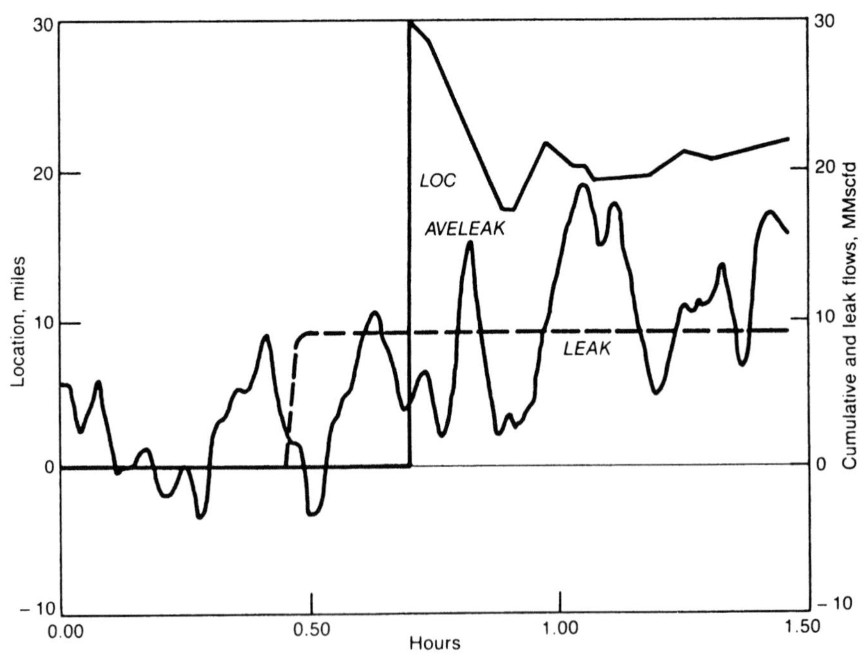

FIG. 10 Gas line, Study 3, leak at MP 120.

cannot effectively be performed in gas pipelines appears, for this example, to be unfounded.

W. R. WADE
H. H. RACHFORD, Jr.

9

Pipeline Failure

Pipeline Failure

Damage from a force outside the pipeline is the primary cause of natural-gas pipeline service failures in the United States. The second and third main causes of line service failures are material failure and corrosion.

These are the major conclusions of an analysis of reports of line service incidents filed with the U.S. Department of Transportation (DOT).

The study, covering incident reports for a recent 14-year period, analyzed reportable incidents from the DOT's Office of Pipeline Safety Regulation (OPSR) 20-day incident report forms to assess the causes of incidents and the safety record of the industry.

The data used for the analysis were collected from 1970 through June 1984 by DOT-OPSR. In June 1984, changes in the report format reduced the amount of data collected. Therefore, 1984 data reported here represent only the first half of that year.

Incident Definition

Reportable incidents are defined by OPSR as those which:

Resulted in a death or injury requiring hospitalization
Required the removal from service of any segment of transmission pipeline
Resulted in gas ignition
Caused an estimated $5000 or more damage to the property of the operator, or others or both
Involved a leak requiring immediate repair
Involved a test failure that occurred during testing either with gas or another test medium
Was significant, in the judgment of the operator, even though it did not meet any of the above criteria.

The data base used for this data analysis was found to contain a number of errors. These data were reviewed to eliminate any possible duplicate entries. Duplicate entries have been a major source of error because telephone reports and written reports of the same incident have been discovered. Therefore, careful review has resulted in the following data set.

Service Incidents

Over the 14.5-year period, 5872 reportable service incidents occurred on natural-gas transmission and gathering lines. The yearly average number of incidents was 404, with the maximum 482 (1979) and the minimum 254 (1976).

In addition, 2013 test failures were reported. Many of these represent failures during retesting to validate the structural integrity of operating lines.

The primary cause of service failures was outside force, which accounted for 53.5% of the failures in the 14.5-year period. The next two causes, material failure and corrosion, are both responsible for about 17% of the failures each (Table 1).

Figure 1 presents the yearly incident numbers by cause. A main conclusion from this figure is that there are wide variations in numbers of service incidents.

The outside-force incidents appear to have peaked in 1973–1974 with a general downward trend evident since that time. The incidents related to

TABLE 1 Service Failures, 1970–1984

Cause	Number	%
Outside force	3144	53.5
Material failure	990	16.9
Corrosion	972	16.6
Other	437	7.4
Construction defect	284	4.8
Construction or material	45	0.8
	5872	100

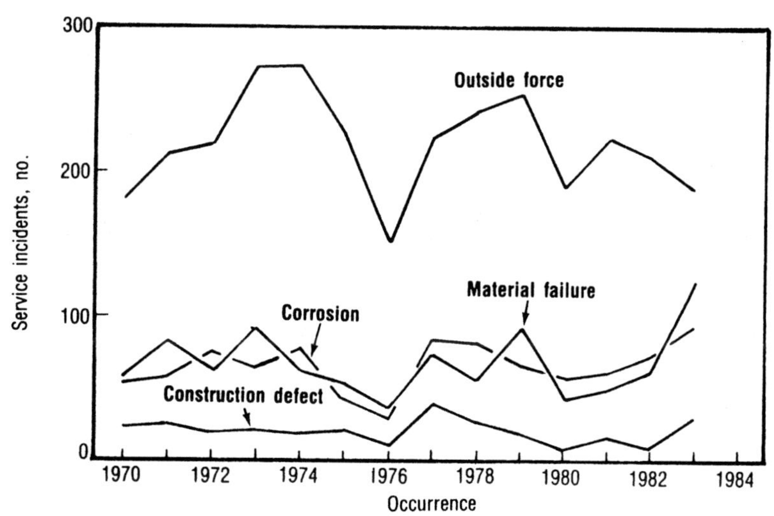

FIG. 1. Yearly incidents by cause.

corrosion and material failure indicate an increasing number over the final 4 years.

This trend will be examined in more depth by normalizing the data per year per mile for the various age categories of pipe. Figure 2 presents data on service incidents by cause and by the year installed. For these data, the older the pipe, the greater the number of incidents per mile-year.

The number of outside-force incidents decreases in those pipes laid after 1940. The pre-1940 lines appear to experience about a constant number of outside-force failures per year at a rate that is about double the rate of the newer lines.

Why the 1950 to 1959 lines exhibit a slightly higher failure rate than the 1940 to 1949 lines is unknown. The 1950 to 1959 period was one of major line installation (approximately 25% of the total), but a similar amount of pipe

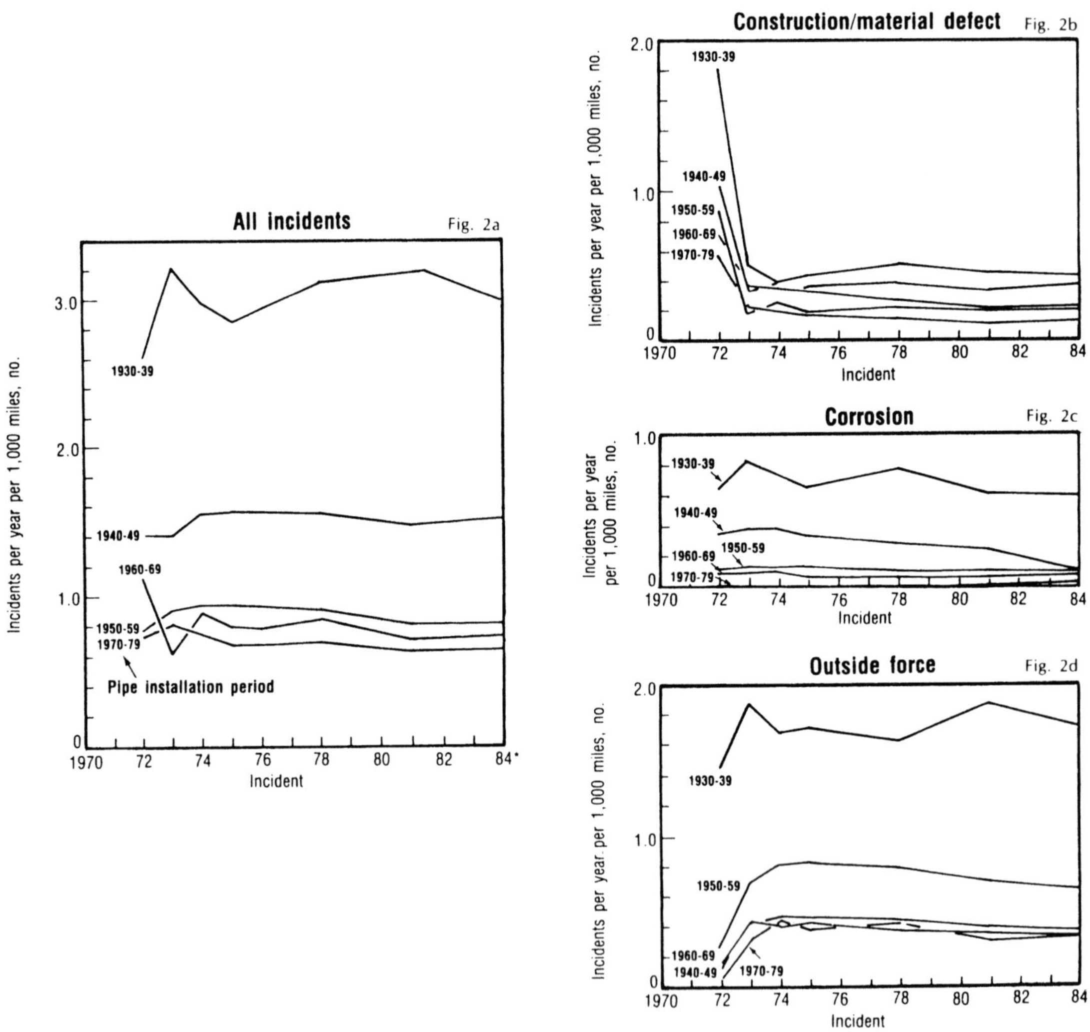

*1984 has only one-half year of data.

FIG. 2. Incidents by installation periods.

was installed in the 1960 to 1969 time period with a much lower number of service incidents.

The trend of the corrosion incident data (Fig. 2c) is about the same as for the outside-force data, which is that lines laid before 1940 exhibit an incident rate about four times that of the post-1940 lines. Also the pre-1940 lines exhibit a constant number of incidents per year, while the others exhibit a decreasing number of incidents with time.

The construction and material-defect incident rates exhibit a sharp drop with the early 1970s and remain relatively constant thereafter.

Where Incidents Occurred

Figure 3 presents data on the types of areas in which service incidents have occurred.

Figure 3 shows that 80% of the incidents in the 1970-through-June-1984 period occurred in rural and undeveloped areas. This result is undoubtedly attributable to the fact that most of the transmission pipeline mileage is in these areas.

The type of area for outside-force incidents is also shown in Fig. 3 which has the same trends as shown for the service incident data.

Figure 4 presents the number of service incidents per mile-year by cause by diameter. These data indicate that the outside-force incidents occur more frequently on a mile-year basis in the smaller-diameter pipes.

This is probably the result of the smaller-diameter pipes with their thinner walls being more susceptible to failure than the large-diameter pipes with heavier wall thicknesses. Also, the smaller pipes are probably older and may not be marked as well, or their locations known as well.

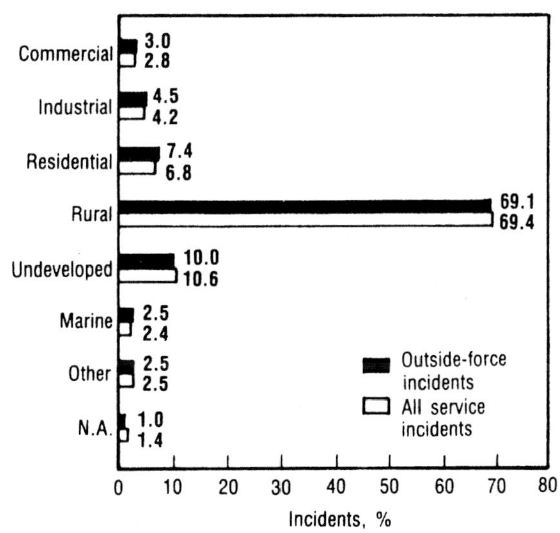

FIG. 3. Distribution of incidents.

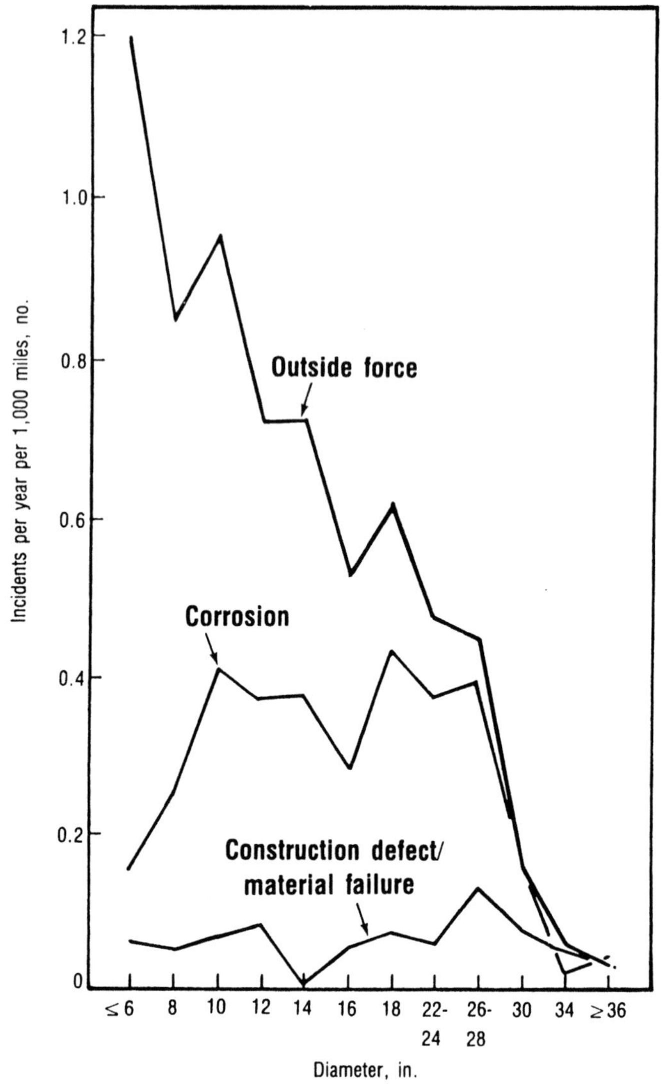

FIG. 4. Line sizes involved.

Corrosion Incidents

The corrosion-caused incidents occurred most frequently in the 10 to 20-in.-diameter pipe range. It is interesting that 40% of the installed pipe is in this size range.

It seems obvious that the larger diameters represent newer pipe that is less susceptible to corrosion because of improved coating techniques. Much of the small-diameter pipe may be in gathering systems which are more likely to experience corrosion problems because many of these extend from wellhead

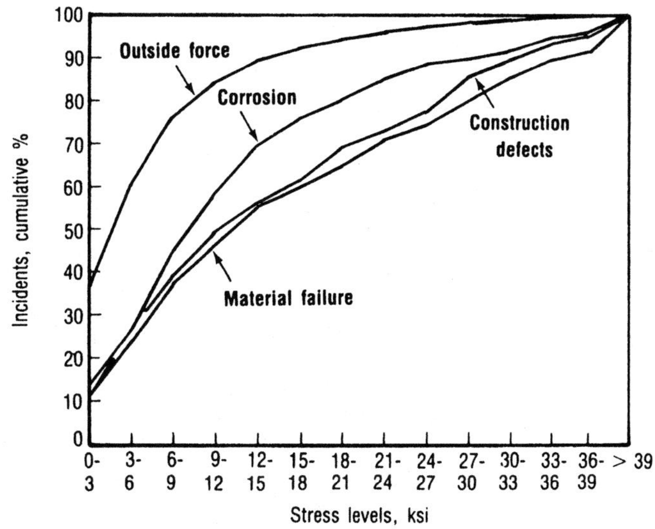

FIG. 5. Stress levels of failures.

to treatment plant which removes any corrosive components that may be present in the gas.

Figure 5 presents the service incidents by stress level. These data indicate that more than 50% of the service incidents occurred at stress levels below 12,000 lb/in.^{2} hoop stress. Since it was previously shown that a high percentage of the incidents occur in small-diameter pipes, it is probable that many of the failures are in Grade B pipes.

Thus, the probable conclusion is that at least 50% of the failures are occurring at stress levels that are less than 35% of the steels' yield strengths. This indicates that high hoop stress is not a primary factor in at least 50% of the incidents.

These conclusions are very evident for the outside-force incidents in which 80% of the incidents occur at hoop stresses below 12,000 lb/in.^{2}.

Table 2 presents an analysis of the corrosion service incidents by cause. It reveals that of 1013 corrosion incidents, 40% involved external corrosion, 27% involved internal corrosion, and 17% involved stress corrosion.

These data indicate (Fig. 2c) that 50% of the corrosion incidents per 1000 miles of pipe occurred in lines placed in service before 1939.

TABLE 2 Causes of Corrosion Incidents

Cause	%
External corrosion	40
Internal corrosion	27
Stress corrosion cracking	17
Other	16
	100

Furthermore, pitting is the primary cause of internal and external corrosion, with about 90% of the incidents occurring from pitting. These data also indicate that 76% of the corrosion incidents occurred on cathodically protected pipelines, which is as expected since most lines are cathodically protected.

Outside Force Incidents

The service incidents resulting from outside forces are identified by cause in Table 3.

TABLE 3 Causes of Outside-Force Incidents

Course	%
Equipment operated by outside party	67.1
Earth movement	13.3
Weather	10.8
Equipment operated by, or for, pipeline operator	7.3
Other	1.5
	100

A large portion of the 3144 outside-force incidents result from equipment operated by an outside party.

This clearly indicates that many incidents result from human error and miscalculations and therefore are potentially preventable.

Data on the depth of burial provide some insight into the outside-force incidents.

Twenty-nine percent of the incidents occurred in aboveground piping.

The remaining 7% of the incidents occurred in buried lines. Thirteen percent of the outside-force incidents occurred in lines buried 6 to 12 in., 17% in lines buried 12 to 24 in., and 41% occurred in pipelines buried from 24 in. to more than 60 in.

Thus, deep burial provides no significant amount of protection increase against outside-force incidents.

ROBERT J. EIBER
DANA J. JONES
GREGORY S. KRAMER

Outside-Force Damage

Damage from a force outside the pipeline has been established as the primary cause of natural gas pipeline service failures in the United States.

Experiments have been conducted that provide means for evaluating the serviceability of line pipe that has suffered outside-force defects.

In the first of three parts detailing that research, characteristics of such line damage are examined as well as factors affecting failure pressure in a pipeline that has suffered mechanical damage. The second part will discuss the effects of temperature on failure pressure in a damaged line. The conclusion will examine dynamic-damage research.

Part A. Characteristics

Damage Incidence

Incidents caused by outside force, or mechanical damage, accounted for 53.5% of all reportable service incidents [1] in gas transmission and gathering lines during the period 1970 through June 1984. These mechanical-damage defects include dents, gouges, and gouges in dents.

The dent-and-gouge defects occur under a variety of conditions, such as during shipping, handling, line construction, line repair, or miscellaneous construction near the line during the service life of the line. Severe mechanical damage incurred during the construction of a pipeline will be "removed" by the hydrotest.

Damage that occurs in post-hydrotest operations, such as third-party contractors hitting the pipeline with mechanical equipment at a later date, clearly cannot be removed by the hydrotest and may cause failure at the time the damage occurs or at some later time depending on the defect's severity and the service conditions.

Damage incurred while a line is in service, such as during pipeline repair or from miscellaneous nearby construction, presents the most common type of service defect and potentially the most hazardous. Therefore, it is desirable to determine the severity of such defects in order to assess the serviceability of the damaged area (and thus avoid costly downtime).

Understanding the basic mechanisms of mechanical-damage failure also will assist in specifying fracture toughness requirements for future line-pipe steels.

Prior studies [2–6] provide in-depth consideration of the separate effect of plain dents and simulated gouges (sharp machine surface flaws with negligible indentation) on gas line-pipe integrity. The failure characteristics of such defects are well understood.

Recent studies on mechanical damage [7–9] have concentrated on the complex behavior of a localized gouge within a dent, the most typical form of damage that occurs in the industry.

However, these experimental studies were hampered in that the defect combinations had to be introduced into the pipe test section at ambient pressure to produce adequate repeatability.

The past research has been included here for completeness. The most recent research differs from past work in that the effects of dynamically produced dent-and-gouge defects in line pipe under pressure are studied. This provides a more realistic representation of the type of outside-force damage that occurs in service.

The specific objective of this research was to develop a means of assessing the severity of mechanical damage defects and the effect of fracture toughness in resisting the failure of these defect types.

Mechanical Damage Research

Nearly all research on this subject has been conducted at two laboratories: Battelle Columbus Division (BCD) and British Gas Corp. (BG) research facilities at Newcastle-upon-Tyne, U.K.

BG conducted most of its research using rings cut from pipe, damaged, then tested on a ring yield-test machine, using the same type of ring tester used by pipe mills to determine pipe yield strength.

BG also conducted tests on pipe that was damaged while not pressurized and later on pressurized and damaged pipe.

Battelle's efforts involved pipe testing both on pipes that were damaged and then pressured to failure and on pipes damaged while pressurized. It is believed that gouge length is a strong influencing parameter, and it is not possible to use gouge length as a variable in the ring test method.

The research is handicapped by the large number of variables that affect the failure pressure of the damaged pipe. Variables examined to date are gouge depth, gouge length (same as primary-dent length), dent depth, pipe size, pipe toughness (as indicated by Charpy upper shelf energy), pipe yield strength, and failure pressure.

Other variables known to affect dent and gouge severity and consequently pipe-failure pressure are coatings and the shape and size of the gouging agent.

Coatings act in different ways. The softer coatings appear to act as lubricants, allowing the tool to glide over the pipe surface, causing more indentation and less gouging. Tougher coatings are more easily gouged but in turn protect the pipe underneath, which is gouged less.

If the tool is sharp and small, it will gouge more (dent less) than if it is wide and blunt. Real damage done in service is random. Damage can be positioned at any orientation relative to the pipe.

Damage also varies by type of equipment being used and the persistence of the equipment operator. It would be impracticable to investigate all these variables independently, and the research done on this project did not examine many variables such as tool variation, construction equipment variation, defect orientation, and coating (generally no coating was tested) types.

Notch-in-Dent Failure

Preliminary experiments showed that a dent alone does not constitute a serious flaw [2]. Pipes containing artificially created dents in the pipe body were tested and observed to fail at or near the burst pressure of nondented pipes.

Therefore, the defect chosen to simulate external damage was a gouge-and-dent combination. It was reproduced for experimentation by machining a V-shaped notch in the pipe wall and then pressing a round bar, laid over the notch, into the pipe to form a dent.

The material directly beneath the root of each notch was severely cold-worked during the denting process, leaving a residual tensile stress at the notch root.

The results of these notch-in-dent experiments are presented in Fig. 1. The failure stress of each notch-and-dent defect is plotted as a percent of the calculated failure stress of the surface flaw without the dent vs the notch length.

The severity of a notch-and-dent defect is greater than that of the notch alone. The longer the notch-and-dent defect, the greater the difference in their severities. All of these notch-and-dent failures were leaks because the failure pressures were below those of the resulting through-wall flaws.

Because the notch-and-dent defect is made under zero-pressure conditions, relatively deep denting is easily accomplished. While this simulates damage done when the line is being constructed or repaired, it does not reveal the nature of denting and gouging when the line is under pressure.

Thus, the second phase of external-damage studies was directed at evaluating the forces and energy required to dent and penetrate or puncture a pipe under pressure.

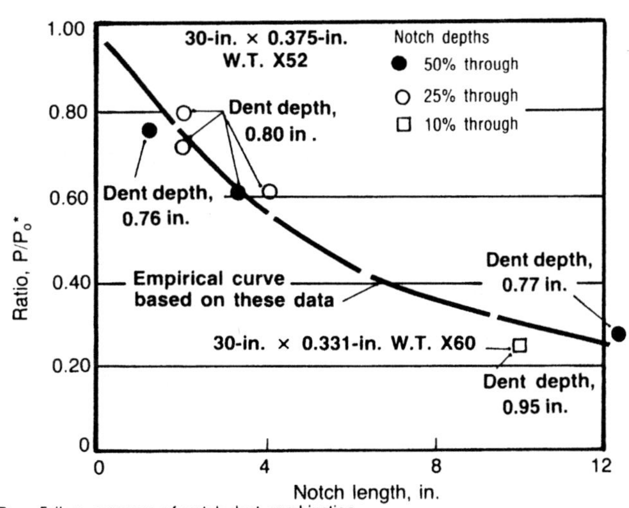

*P = Failure pressure of notch-dent combination

P_o = Failure pressure of notch (estimated)

FIG. 1. Gouge-in-dent experiments.

In penetration experiments, the specimen was pressurized to a predetermined level and then a wedge was slowly pressed into the specimen by means of a hydraulic jack. Both the wedge load and the depth of penetration were monitored continuously.

The apparatus used for the penetration experiments on pressurized pipe is the same equipment used to make gouge-in-dent defects. Instead of the round-bar indenter, however, a 1-in. long, 45° machined wedge oriented parallel to the axis of the pipe was pressed into the pressurized vessel.

The results of the four experiments are shown in Fig. 2; test conditions are given in Table 1. The variable "wedge travel" includes both indentation and penetration. Although the pipe were grades X52 and X60, all actual yield strengths were above 60 ksi (60,000 lb/in.^{2}) and the total range of yield strength was less than 10%.

As shown in Fig. 2, the heavy-wall pipes exhibited the greatest resistance to penetration and absorbed larger energies without penetration (total energy being equal to the area under each load vs travel curve). The permanent dents remaining after depressurizing of the 0.460-in. W.T. pipe were 0.17 and 0.30 in. for the pipes pressured to 1550 and 927 lb/in.^{2} gauge, respectively.

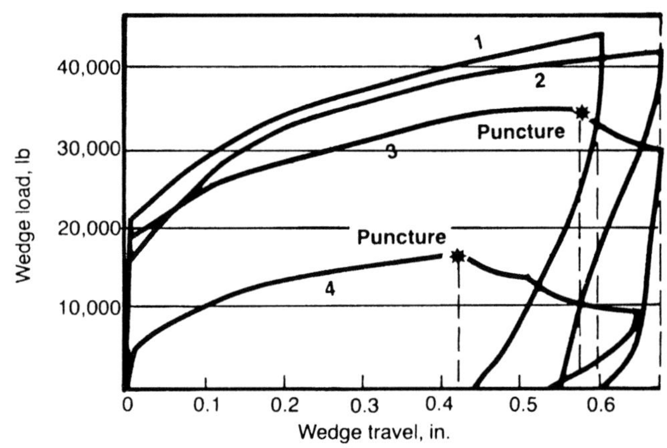

FIG. 2. Penetrating experiments: 20-in. pipe.

TABLE 1 Conditions for Penetration Tests

Test No.	Wall Thickness (in.)	Yield Strength (ksi)	Test Pressure (lb/in.^{2} gauge)	Hoop Stress (ksi)	Maximum Jack Load (in.)	Maximum Travel (in.)	Total Energy (in.-lb)
1	0.460	63.2	1,550	50.5	44,500	0.59	21,400[a]
2	0.460	63.2	927	30.2	42,500	0.67	25,700[a]
3	0.389	60.4	1,310	50.5	34,800	0.58	17,200
4	0.275	66.5	927	50.5	17,000	0.42	5,450

[a]Stopped before puncture occurred.

Fracture Toughness

This investigation examined the role of toughness in resisting loss of mechanical strength of the pipeline caused by external mechanical damage. After this investigation was initiated, Fearnehough [5] presented results that indicated toughness (based on Charpy shelf energy) influences the failure of externally damaged line pipe.

Two pipe sections (BH-1 and DQ-1) were selected for this investigation. Both sections were 30-in. X60 line pipe with specified 0.329-in. W.T. for DQ-1 and 0.312-in. W.T. for BH-1. These sections were evaluated to determine their mechanical properties in Table 2.

Four flaws were pressed into each specimen using a 5-in.-long indenter bar with a 0.250-in.-deep machined sharp edge (60° including angle) running its length.

The notch depths were variable along the length, deepest on the ends and shallowest in the center of the notch, and thus were different from previous notches.

Table 3 presents the pertinent failure details. The low-toughness specimen (BH-1) failed at the most severe flaw at a nominal hoop stress of 49.6 ksi, well below the ultimate stress of the material (101.7 ksi). In fact, this failure stress represents toughness-dependent behavior.

The high-toughness specimen (DQ-1) also failed at the most severe flaw but at a nominal hoop stress (90.9 ksi) nearly equivalent to the ultimate stress of the material (92.9 ksi).

TABLE 2 Toughness-Test Properties

Pipe number	BH-1	DQ-1
Pipe diameter, in.	30	30
Wall thickness, in.	0.355	0.333
Yield strength, ksi	72.1	73.8
Ultimate strength, ksi	101.7	92.9
2/3-Size Charpy shelf energy, ft-lb	18	58
DWTT[a] shelf energy, ft-lb	1000	1800

[a]Drop-weight tear test.

TABLE 3 Toughness-Test Failure Parameters

Specimen number	BH-1	DQ-1
Charpy shelf energy, ft-lb	18	58
Failure pressure, lb/in.^{2}	1175	2020
Failure stress, ksi	49.6	90.9
$\dfrac{\text{Failure stress } \sigma}{\text{Flow stress, } \bar{\bar{\sigma}}}$	0.604	1.084
Notch length, in.	5	5
$\dfrac{\text{Notch depth}}{\text{Wall thickness}} \left(\dfrac{d}{t}\right)$, %	≈10	≈10
Dent depth, in.	0.94	1.25

TABLE 4 Mechanical-Damage Flaw Initiation Prior to 1979

Test No.	Test Identification	Pipe No.	Pipe Size (in.)	Y (ksi)	⅔-CVN (ft-lb)	Flaw Length, $2c$ (in.)	Dent Depth (in.)	Crack Depth (in.)	P_f (lb/in.^{2} gauge)	σ (ksi)
1	78-1	BH-1	30×0.35	72.1	18	5.0	0.94	0.036	1175	49.6
2	78-2	DQ-1	30×0.33	73.8	58	5.0	1.25	0.033	2020	90.9
3	48	A-1862-2	24×0.38	54.1	25	2.6	0.73	0.150	1400	44.8
4	47	A-1862-1	24×0.38	54.1	25	2.7	0.81	0.188	1280	41.0
5	44	A-1861-1	24×0.38	61.3	23	2.7	0.78	0.188	1350	43.2
6	18-58	AD-3	30×0.38	56.1	29	3.3	0.80	0.190	980	38.3
7	18-60	AD-3	30×0.38	56.1	29	3.3	0.80	0.192	930	36.4
8	18-119	AC-3	30×0.38	53.8	38	2.0	0.80	0.096	1425	55.5
9	18-120	AC-3	30×0.38	53.8	38	2.0	0.80	0.095	1340	52.7
10	18-121	AC-3	30×0.38	53.8	38	4.0	0.80	0.095	1005	39.8
11	69-17	BA-2	30×0.33	65.3	20	10.0	0.95	0.033	360	16.3
12	70-1	BA-2	30×0.33	65.3	20	10.0	0.39	0.033	1375	62.4
13	70-2	BA-2	30×0.33	65.3	20	10.0	0.39	0.169	210	9.5
14	70-4	BA-1	30×0.33	71.2	22	10.0	0.56	0.033	725	33.0
15	70-5	BA-1	30×0.33	71.2	22	10.0	0.39	0.066	895	40.6
16	7-1	ZF-4	22×0.35	40.5	31	5.0	0.83	0.032	1390	43.2
17	7-2	ZF-4	22×0.35	40.5	31	10.0	0.83	0.035	880	27.3
18[a]	7-3	ZF-4	22×0.35	40.5	31	5.0	0.78	0.035	1170	36.6
19	7-4	ZF-4	22×0.35	40.5	31	5.0	0.86	0.035	1420	44.4
20	7-5	DQ-1	30×0.34	73.8	58	10.0	1.20	0.034	1625	72.5
21	7-6	DC-3	16×0.27	58.1	18	5.0	0.81	0.027	745	22.4
22	7-7	DC-3	16×0.27	58.1	18	10.0	0.80	0.028	200	5.9
23	7-8	DT-1	42×0.39	78.8	34	10.0	1.82	0.040	1040	54.2
24	7-9	DT-1	42×0.39	78.8	34	10.0	0.81	0.040	1320	69.1
25[a]	7-10	DT-2	42×0.39	78.8	21	5.0	0.80	0.040	940	50.1
26	7-11	DT-2	42×0.39	75.8	21	10.0	0.81	0.040	590	31.4
27	7-12	DC-3	16×0.27	58.1	18	5.0	0.91	0.030	780	23.3
28	7-13	DC-3	16×0.27	58.1	18	5.0	0.81	0.027	840	25.2
29	7-14	DT-2	42×0.39	75.8	21	10.0	0.87	0.020	1505	79.0
30	7-15	DT-2	42×0.39	75.8	21	10.0	0.82	0.082	365	19.6

[a]These tests exhibited brittle initiation and are excluded from the discussion.

The marked difference in burst strength of the two specimens is somewhat surprising, particularly since all indications are that the flaws in the high-toughness specimen (DQ-1) were more severe than in the low-toughness specimen (BH-1).

An interesting comparison arises from these data on the plot of Fig. 1. A review of defect-failure records indicates that the toughness of the earlier specimens was similar to the lower toughness of specimen BH-1.

Table 4 presents the pipe size, yield strength, toughness, and the dent-and-gouge data for these early experiments. The data for BH-1 and DQ-1 are included and presented in Fig. 3.

The tests results for both specimens BH-1 and DQ-1 are above the average curve reported earlier (in Fig. 1). However, the results for DQ-1 are substantially above the earlier results, while the BH-1 results are only marginally above the scatter of the original data.

This apparent inconsistency can be understood by considering the manner in which the flaws were created. The earlier experiments involved milling a notch and then forming a dent. This created a more severe flaw than those produced by this investigation.

The obvious conclusion to be drawn from these tests is that toughness plays a significant role in the failure pressure of gouge-in-dent defects.

Failure Pressure Parameters

Experiments were designed to build on those already conducted to examine crack length, crack depth, and toughness (Charpy shelf energy) independently.

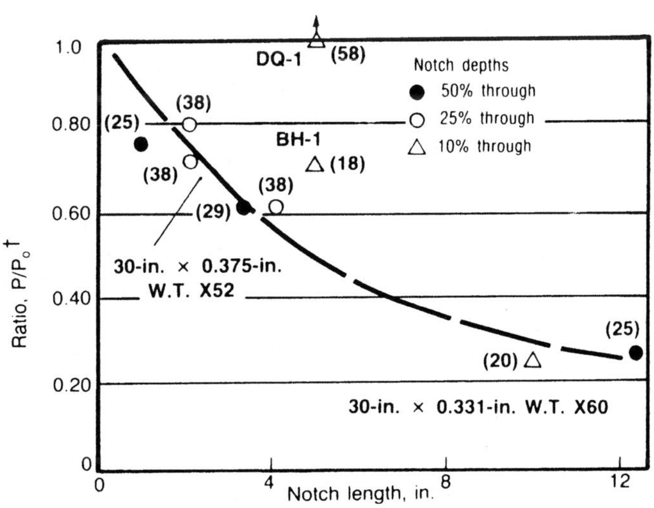

†P = Failure pressure of notch-dent combination; P_o = Failure pressure of notch (estimated).

FIG. 3. Gouge-in-dent: Two specimens. Numbers in parentheses indicate Charpy shelf energy in ft-lb. Nominal dent depths were 0.75 and 1 in. with actual depths ranging from 0.75 to 1.25 in.

A V-notch defect was machined on the outside surface to a certain depth and length and then indented from the outside surface. The defect was introduced at zero pipe pressure and the pipe was pressurized to failure.

Certain parameters were investigated becaused of past fracture-initiation research. The parameters investigated were the ratio of hoop stress at failure to flow stress ($\sigma/\bar{\sigma}$), crack depth to wall thickness (d/t), and dent depth to pipe diameter ($D/2R$).

The ratio of crack length to the pipe geometry factor $[2c/(Rt)^{1/2}]$ was examined independently because the local geometry of the dent was no longer Rt. In fact, the local curvature, because of the indentation, was inward, or just the reverse of the normal pipe curvature.

Fifteen additional experiments involving dent-and-gouge defects were conducted. This research involved experiments with pipe diameters of 16 in. through 42 in. The experimental results from these studies are shown in Table 4.

A synergistic effect of toughness level, flaw depth, flaw length, dent depth, and pipe geometry on the failure stress is believed to exist. To illustrate this point, consider Figs. 4 and 5.

Figure 4 is a plot of the ratio of failure stress to flow stress ($\sigma/\bar{\sigma}$) vs the ratio of gouge depth to wall thickness (d/t) for two different dent geometries [characterized by the ratio of dent depth to pipe radius (D/R)]. In Fig. 4, the toughness level and flaw length were fixed at $\approx$20 ft-lb and 10.0 in., respectively. Clearly, the deeper the gouge and dent, the lower the failure pressure. Figure 5 is a plot of the ratio $\sigma/\bar{\sigma}$ vs the ratio of dent depth to pipe radius (D/R) for varying toughness levels, where the flaw length ($2c$) and flaw depth-to-thickness ratio (d/t) have been fixed at 10.0 in. and $\approx$0.10, respectively.

This figure shows that deeper dents have a more detrimental effect on lower toughness materials.

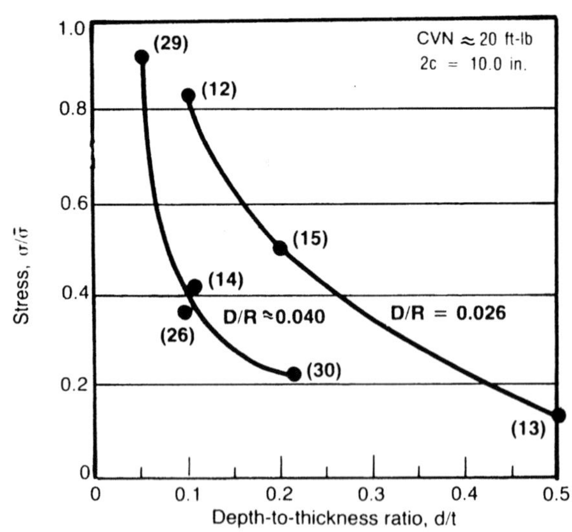

FIG. 4. Depth effect. Numbers in parentheses represent test numbers in Table 4.

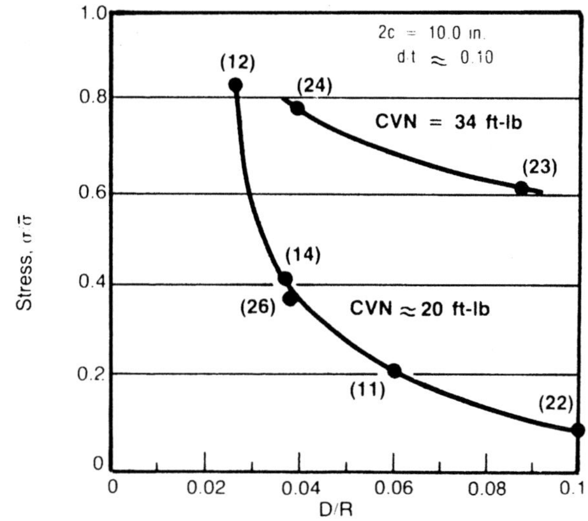

FIG. 5. Geometry, toughness effect. Numbers in parentheses represent test numbers in Table 4.

Figure 6 illustrates the synergistic effect of the variables, described above. A plot of $\sigma/\bar{\sigma}$ vs a Q parameter $CVN/[D/2R)(d/t)(2c)]$ is shown for experiments in Table 4. This plot is restricted to ranges of most interest to the pipeline industry by considering gouge depths less than 20% of the wall thickness and limiting the range on Q to less than 2000 ft-lb/in.

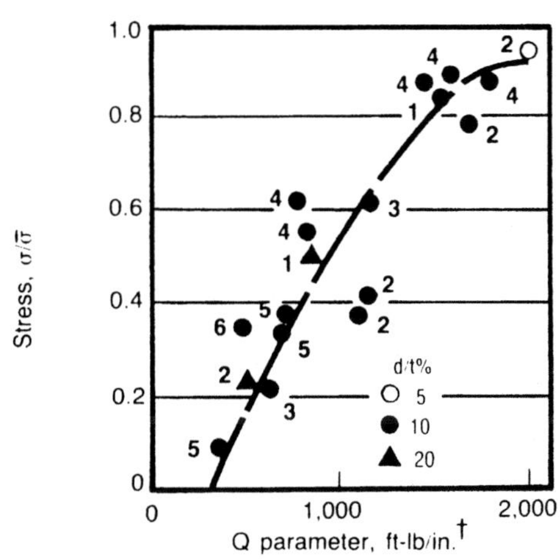

†2/3 size.

FIG. 6. Effect of variables. Numbers represent $D/2R$. Data from Table 4.

Part B. Effect of Temperature

Temperature and Failure

Another series of experiments was aimed at examining the effect of external mechanical damage (gouged and dented) on the fracture-initiation transition temperature and on the reduction of failure pressure during hold time.

Damaged pipes with known failure characteristics from prior experiments have been pressurized at various temperatures below and near the drop-weight-tear-test (DWTT) transition temperature to determine the effects of pipe temperature on failure pressure of the gouged-and-dented pipe.

Other pipes having defects with known failure characteristics were pressurized to various pressure levels below the defect failure pressure and monitored for flaw growth while held at constant pressure.

Experiments were conducted to examine the initiation transition temperature of gouged-then-dented pipe. The test temperatures ranged from +60 to −65°F, whereas the DWTT transition temperature for this steel was +45°F.

The four experiments all had nearly identical defects, a surface flaw 5 in. long with a gouge about 10% of the pipe wall thickness in depth. A 5-in.-long indenter was used to create a dent about 5% of the pipe diameter in depth. The range of failure pressures for these pipes was from 1225 to 130.5 lb/in.^{2} gauge. Data for these experiments are given in Table 5.

The coldest test, although not significantly lower in failure pressure, appears to have a slightly lower failure stress relative to the actual yield strength at test temperature (σ_H/Y). While no transition in failure pressure is evident, a transition in the failure type is evident between −29 and 10°F.

The experiments at and above 10°F failed as leaks, while the colder experiments failed as ruptures of brittle fractures. This transition in fracture type occurred 35 to 75°F below the DWTT transition temperature for this data set.

Two additional experiments that relate to the effects of holding the defects at constant pressure were conducted. The test conducted at +60°F (Table 4)

TABLE 5 Transition Temperature Failure Data

Test temperature, °F[a]	−65	−29	+10	+60
Test pressure, lb/in.^{2} gauge	1,235	1,305	1,250	1,215
Yield stress, ksi[b]	66.3	61.6	61.1	60.0
Average wall thickness, in.	0.462	0.468	0.462	0.450
Average notch depth (d), in.[c]	0.049	0.046	0.049	0.048
Average dent depth (D), in.[d]	1.85	1.80	1.80	1.80
σ_H/Y at test temperature	0.73	0.81	0.80	0.81
Failure type	Fracture	Fracture	Leak	Leak

[a]DWTT 85% SA transition temperature is +45°F.
[b]Yield stress at test temperature.
[c]Notch length was 5.0 in.
[d]Dent length was 5.0 in. using a 1-in. round bar indentor.

served as the control test for the hold experiments. (It was taken straight to failure with zero hold time.)

The first hold experiment was stopped at 960, 1130, and 1200 lb/in.^{2} gauge. At each of these hold levels, minimal flaw depth increase was observed for hold times of 30 min or less. The experiment was then depressurized to check instrumentation and repressurized to 1240 lb/in.^{2} gauge.

Flaw-depth increase was continuously observed during this hold; failure, as a leak, occurred after 120 min of continuous constant pressure.

A second experiment was conducted with improved instrumentation on pipe of the same heat and same flaw geometry. The first pressure hold period was at 1050 lb/in.^{2} gauge for 60 min with no flaw-depth increase observed.

The pipe was then pressured to 1100 lb/in.^{2} gauge and held for 120 min. Some flaw-depth activity on the crack-opening displacement gauge and electric potential was recorded during the first 30 min of the hold and then activity ceased.

The pipe was then pressured to 1175 lb/in.^{2} gauge and held for nearly 19 h. Considerable activity in the crack opening displacement (COD) and electric potential occurred during the first 100 min, but essentially no activity occurred for the remaining 17 h of this hold.

The pipe was then pressured to 1260 lb/in.^{2} gauge and failure occurred as a leak after a hold period of 130 min. Figure 7 shows the relationship between pipe pressure and the electric potential, a measure of crack growth through the wall thickness, for the hold times discussed. No significant crack growth was measured until the pressure reached 1000 lb/in.^{2} gauge or about 80% of the flaw failure pressure.

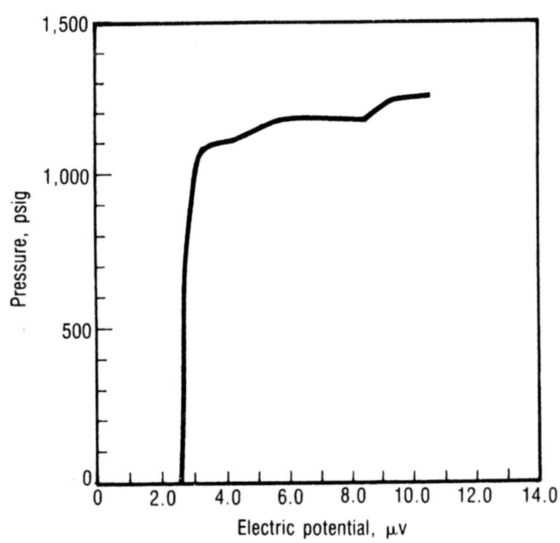

FIG. 7. Crack growth during hold for dent-and-gouge defect.

Rounding out of the dent during pressurization is shown in Fig. 8, where dent displacement (radially outward) is plotted relative to pipe pressure. This plot is nearly linear and shows that as soon as pressurization starts, the dent begins to move out. There was very minimal dent outward movement during the hold periods.

The final failure pressures of the two hold tests were essentially the same as the control test: 1240, 1260, and 1250 lb/in.^{2} gauge. The true minimum failure pressure under hold must lie somewhere between 1175 and 1260 lb/in.^{2} gauge. The last pressure increment was from 3 to 7% below the final failure pressure, approximately the reduction in failure pressure observed for undented surface flaws [10].

The hold-time experiments indicate stable growth from about 70 to 95% of the failure pressure for a similarly flawed pipe pressurized continuously to failure. This observation is also similar to past research results on flawed undented pipe [10].

Two-Step Damage

Earlier research studies into the failure pressures of dented-and-gouged pipe have been based upon results from two different methods of experimentally inducing damage. BCD conducted several mechanical-damage experimental programs in which damaging consisted of a two-step process.

A sharp notch was machined into the exterior of the pipe specimen surface and the entire area indented with a hydraulic ram. The root of the machined flaw was therefore coldworked in compression. All pipe specimens were at ambient pressure when the flaws were produced.

Investigations conducted at British Gas (BG) concerning mechanically damaged rings and pipe vessels induced the damaged sequence in the oppo-

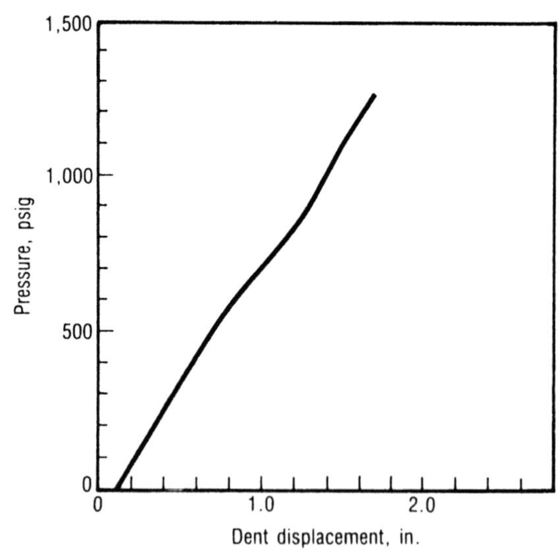

FIG. 8. Dent rounding.

site order. After a dent had been produced in a pipe at ambient pressure, a narrow flaw was machined in the base of the dent, thereby reducing residual stresses local to the flaw. The resulting defect proved to be less severe than the defect produced by the BCD method.

Mechanical-damage results have typically shown a significant amount of scatter resulting from lack of an appropriate fracture model for the dent-and-gouge configuration and inaccuracies associated with attempting to produce an identical gross deformation in a number of pipe specimens.

Based on the results of 132 burst tests that involve rings and some pipe vessels, BG developed a statistical analysis for predicting the failure stress based on Charpy energy and on critical gouge-and-dent depths [7]. The BG analysis typically considers only an infinite length flaw.

While this analysis gives good predictions for medium-diameter line pipe, it tends to deviate for both larger (>36 in.) and smaller (<24 in.) diameter pipe specimens.

BCD research shows similar trends in comparison with the BG results in that there is a synergistic effect of material toughness (CVN), gouge depth (d), gouge length ($2c$), and dent depth (D) on the failure stress of mechanically damaged pipe. To reduce experimental scatter, BCD developed the Q parameter, which attempts to illustrate the synergistic effects of the variables noted previously.

Figure 9 compares the results of past BCD pipe data with those from 19 BG experiments on 30-in. × 0.469-in. W.T. X52 pipe. The BG experiments all contained 15-in.-long axial gouges each having a depth of 25% of the pipe wall thickness but with variable dent depth and variable Charpy shelf energy. The gouges were machined into the bottom of an already indented pipe.

Figure 9 shows these data as "dented-then-notched pipes" on a plot of the Q parameter vs hoop stress at failure. The Battelle data ("notched-then-dented pipes") are the same as data shown previously (Fig. 6).

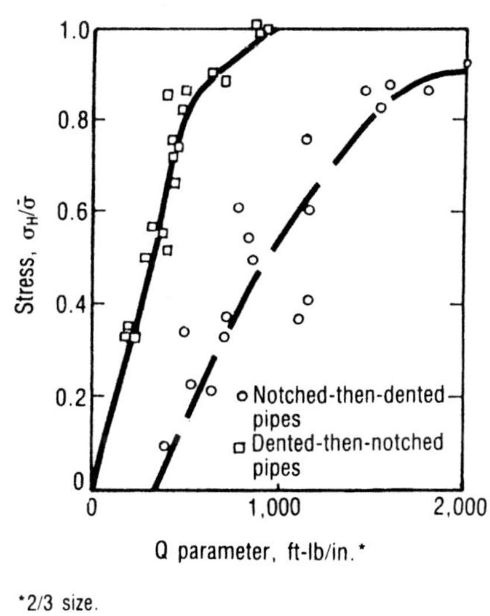

FIG. 9. Past experimental work.

As expected, the failure stress levels of the more severe BCD notched-then-pipe data are considerably lower than those of similar BG dented-then-notched pipe data. Although further modifications and refinements might be desirable for mechanical damage analysis, the Q factor provides an empirical starting point for mechanical-damage assessment.

Part C. Dynamic-Damage Research

This research on mechanical damage was aimed at damage done to pressurized pipe to stimulate in-service failures from mechanical damage.

A mechanical-damage machine was built to create mechanical-damage defects in test specimens up to half joints (20 ft) of 48-in.-diameter line pipe. The device for imparting damage to the pipe specimen is a blunt tool piece that is loaded by two hydraulic cylinders, one horizontal and one vertical.

The damaging process occurs dynamically, at a rate that can be controlled between 6 and 48 in./s. The specimens were completely water-filled and pressured to the desired pressure level with water.

Pipe specimens tested during this investigation consisted of 20- 30- and 42-in.-diameter longitudinally welded line pipe. Pipe grades ranged from API 5L X52 to X65. Wall thicknesses varied from 0.296 to 0.458 in., and Charpy V-notch (CVN) upper-shelf energies, based on ⅔-/thickness specimens, ranged from 15 to 45 ft-lb.

Tables 6 and 7 are a breakdown of chemical compositions and pertinent mechanical properties for each of the line-pipe steels used in this study.

Each complete test specimen consisted of a 7-ft length of the line-pipe test material, two heavy-wall pipe extensions welded onto each end of the center pipe, and two semiellipsoidal end caps welded onto each end. The total length of the test specimen approached 10 ft.

Test Procedure

The procedure used in this investigation consisted of characterization of each line-pipe material, fabrication of the test specimens, introduction of mechanical damage to each specimen under pressure, and analysis of the results.

Ten pipe joints were selected for this investigation, each with different mechanical properties. Each joint was evaluated to determine its specific chemical and mechanical properties.

The following tests were used in the evaluations: chemical analysis, strap tensile tests, and Charpy V-notch tests using ⅔-thickness specimens. The results of these tests are listed in Tables 6 and 7.

Once each material was properly characterized, individual test sections were removed from the pipe joints and final test vessels fabricated. A completed test specimen was then positioned into the loading framework and secured.

TABLE 6 Pipe Specimen Chemical Compositions

Pipe Identification	C	Mn	P	S	Si	V	Cb
CH-1	0.07	1.35	0.011	0.005	0.63	—	—
CZ-1	0.14	1.24	0.011	0.014	0.31	0.034	0.007
TT-5	0.25	1.27	0.010	0.025	0.03	0.070	—
LL-2	0.22	1.14	0.019	0.018	0.056	—	0.02
DE-1	0.16	0.81	0.012	0.024	0.028	—	0.012
EJ-1	0.11	1.35	0.007	0.009	0.28	—	0.024
EK-1	0.14	1.25	0.017	0.009	0.22	0.003	0.024

TABLE 7 Pipe Specimen Mechanical Properties

Pipe Identification	Size and Grade	Yield Strength (ksi)	Ultimate Strength (ksi)	CVN Upper Shelf ⅔-size (ft-lb)
CH-1	30 in. × 0.391-in. W.T. 5L X65	75.4	87.1	45
CZ-1	30 in. × 0.404-in. W.T. 5L X65	69.9	86.9	15
TT-5	30 in. × 0.300-in. W.T. 5L X60	61.8	87.5	16
LL-2	30 in. × 0.296-in. W.T. 5L X60	68.8	85.6	25
DE-1	20 in. × 0.300-in. W.T. 5L X52	56.4	72.0	23
EJ-1	20 in. × 0.381-in. W.T. 5L X60	64.6	77.7	42
EK-1	42 in. × 0.458-in. W.T. 5L X65	67.3	84.5	20
EK-2	42 in. × 0.458-in. W.T. 5L X65	67.3	85.1	21
TT-1	30 in. × 0.298-in. W.T. 5L X60	67.8	91.8	17
L-3	30 in. × 0.406-in. W.T. 5L X52	60.7	75.4	20

The tool for producing the gouge and dent was machined from a high-strength tool steel and air hardened to R_c52. The tool is essentially a blunt backhoe tooth with a rounded leading radius to prevent the tool from puncturing the pipe wall thickness. The tool was reground, as necessary, between tests to maintain a consistent shape.

After the initial setup has been completed, the pipe is pressurized with water to an appropriate stress level. A separate gas-filled accumulator is connected in series with the test vessel to help absorb any sudden pressure increases during the damage process. Since gas-transmission pipelines typically operate at 50, 60, or 72% of specified minimum yield strength (SMYS), most of these tests were conducted at one of these stress levels.

A light-beam oscillograph was used to record various dynamic data during each test. Displacement transducers measured the movement of the gouging tool along the pipe's longitudinal axis and also recorded the instantaneous vertical tool movement.

Pressure transducers mounted in each hydraulic ram monitored external forces applied to the pipe, while an additional pressure transducer monitored internal pipe pressure.

In the experiments, the first damage sequence was to produce a subcritical dent-and-gouge defect that did not cause pipe rupture. The pressure level in the pipe specimen was then lowered substantially below the test pressure, the pipe was rotated so that the damage would not occur in the same pipe location, and the specimen was repressurized.

Concurrently, the hydraulic loads were increased to produce a slightly more severe flaw on the next attempt. The second experiment was conducted. If failure did not occur, the process was repeated until a dynamic failure occurred.

Thus, a pipe specimen was produced that contained multiple flaws that gradually increased in severity. Although there was no appreciable hold time associated with this method of testing, the first flaws (the less severe) did experience a number of pressure cycles that the later flaws did not experience.

Post-test measurements included measurement of residual dents in the subcritical defects that did not fail (for comparison with the instantaneous dent depths measured during testing), measurement of gouge depths in the defect that failed, and a visual inspection of the fracture surface.

Dynamic Damage Results

The results of the experimental program are summarized in Tables 8 and 9. Table 8 presents the results of 17 experiments in which mechanical-damage flaws were of a critical size and caused immediate pipe rupture. Table 9 summarizes data collected on 60 subcritical dent-and-gouge defects that survived the damage operation.

Mechanical-damage defects introduced to pressurized pipe contained shallower dents and deeper gouges than damage to unpressurized pipe. Internal pressure in a pipe tends to "stiffen" the pipe wall to resist externally applied forces.

If the damaging tool piece is blunt, this stiffening effect will be beneficial because the line can absorb additional mechanical energy. A sharp tool, conversely, will not dent the pipe but will cut through the pipe wall thickness, causing a puncture. Internal pressure, in this case, accelerates the fracture-initiation event.

High-speed movies taken at Battelle have shown that the internal pressure of a pipe tends to return the pipe surface to a position intermediate to its original position and the dented position almost immediately after the pipe is indented by a passing tool.

If a dent-and-gouge defect is of a substantial length (more than 8 in.), then during pressurization the midlength of the gouge will attempt to return to the original contour of the pipe surface sooner than the ends of the gouge area. The ends of the gouge usually experience the greatest amount of residual indentation. It is in the more flexible center gouge region that the initial surface cracking occurs (Fig. 10).

TABLE 8 Mechanical-Damage Experiments Leading to Rupture

Test No.	BCL Pipe Identification	Pipe Grade (ksi)	Diameter (in.)	Wall Thickness, (in.)	CVN Upper Shelf ⅔-size, (ft-lb)	Test Pressure, (lb/in.^{2}gauge)	Stress Level % of SMYS (σ)	Gouge Gouge Length (in.)	Dent Tool Width (in.)	Depth Average (in.)	Depth Average (in.)	Q-Factor (ft-lb/in.)
1	CH-1-3	5L X65	30	0.391	45	1220	72(61)	21	0.50	0.095	0.80	330
2	CZ-1-1	5L X65	30	0.404	15	1260	72(61)	19.5	0.50	0.055	0.87	195
3	CZ-1-2	5L X65	30	0.404	15	1020	58(50)	21.75	0.50	0.032	1.01	259
4	CZ-1-3	5L X65	30	0.404	15	1750	100(86)	7.2	0.50	0.050	0.68	743
5	CZ-1-4	5L X65	30	0.404	15	1220	70(61)	15	0.50	0.032	0.78	486
6	TT-5-1	5L XV60	30	0.300	16	865	72(61)	8	0.50	0.050	0.96	375
7	LL-2-1	5L X60	30	0.296	25	710	60(50)	15.75	0.50	0.075	1.06	177
8	DE-1-1	5L X52	20	0.300	22	700	45(35)	9.8	0.38	0.035	1.05	370
9	DE-1-2	5L X52	20	0.300	22	1780	114(89)	7.2	0.38	0.060	0.70	440
10	EJ-1-1	5L X60	20	0.381	42	1000	44(35)	20.7	0.38	0.100	0.54	230
11	EJ-1-2	5L X60	20	0.381	42	2200	96(77)	9.9	0.38	0.098	0.34	760
12	DE-1-3	5L X52	20	0.300	22	1400	90(70)	7.0	0.56	0.070	0.44	610
13	EJ-1-3	5L X60	20	0.381	42	1710	75(60)	9.0	0.56	0.044	0.45	1800
14	DE-1-4	5L X52	20	0.300	22	1250	80(63)	8.4	0.56	0.060	0.76	340
15	EK-1-1	5L X65	42	0.458	20	1020	72(61)	15.5	0.75	0.048	0.77	670
16	EK-1-2	5L X65	42	0.458	20	1400	99(83)	15.5	0.75	0.027	0.60	1530
17	EK-2-2	5L X65	42	0.458	21	1200	85(71)	9.4	0.75	0.050	0.83	1040

TABLE 9 Mechanical Damage Experiments Leading to Subcritical Defects

Test No.	BCL Pipe Identification	Pipe Grade (ksi)	Diameter (in.)	Wall Thickness (in.)	CVN Upper Shelf 2/3 Size (ft-lb)	Test Pressure (lb/in.^{2}gauge)
1	CH-1-1	5L X65	30	0.391	45	1225
2	CH-1-1	5L X65	30	0.391	45	1220
3	CH-1-1	5L X65	30	0.391	45	1240
4	CH-1-2	5L X65	30	0.391	45	1020
5	CH-1-2	5L X65	30	0.391	45	1020
6	CH-1-2	5L X65	30	0.391	45	1025
7	CH-1-2	5L X65	30	0.391	45	1025
8	CH-1-2	5L X65	30	0.391	45	1025
9	CH-1-2	5L X65	30	0.391	45	1220
10	CZ-1-1	5L X65	30	0.404	15	1260
11	CZ-1-1	5L X65	30	0.404	15	1250
12	CZ-1-1	5L X65	30	0.404	15	1255
13	CZ-1-2	5L X65	30	0.404	15	1030
14	CZ-1-2	5L X65	30	0.404	15	1020
15	CZ-1-2	5L X65	30	0.404	15	1020
16	CZ-1-3	5L X65	30	0.404	15	1750
17	CZ-1-3	5L X65	30	0.404	15	1750
18	CZ-1-3	5L X65	30	0.404	15	1750
19	TT-5-1	5L XV60	30	0.300	16	865
20	LL-2-1	5L X60	30	0.296	25	710
21	LL-2-1	5L X60	30	0.296	25	710
22	LL-2-1	5L X60	30	0.296	25	710
23	DE-1-1	5L X52	20	0.300	22	700
24	DE-1-1	5L X52	20	0.300	22	700
25	DE-1-1	5L X52	20	0.300	22	700
26	DE-1-2	5L X52	20	0.300	22	1780
27	EJ-1-1	5L X60	20	0.381	42	1000
28	EJ-1-1	5L X60	20	0.381	42	1000
29	EJ-1-2	5L X60	20	0.381	42	2195
30	EJ-1-2	5L X60	20	0.381	42	2215
31	EJ-1-3	5L X60	20	0.381	42	1710
32	EJ-1-3	5L X60	20	0.381	42	1710

Stress Level % of SMYS ($\bar{\sigma}$)	Gouge Length (in.)	Tool Width (in.)	Gouge Depth Average (in.)	Dent Depth Average (in.)	Remaining Dent Depth Average (in.)	Q factor (ft-lb/in.)
72(55)	9.25	0.75	0.040	0.90		1585
72(55)	10.5	0.75	0.015	0.47		7131
73(56)	10.5	0.75	0.020	0.59		4260
60(46)	11	1.00	0.035	1.20		1143
60(46)	16.5	1.00		0.96		
60(46)	19	0.75	0.035	0.50		1588
60(46)	21	0.75	0.045	0.70		798
60(46)	21.5	0.50	0.095	0.93		278
72(55)	21	0.50	0.080	0.93		338
72(59)	21.5	0.50	0.010	0.24		3523
71(58)	23.5	0.50	0.020	0.22		1758
72(59)	24	0.50	0.045	0.45		374
59(48)	21.25	0.50	0.010	0.46	0.03	1860
58(47)	23.25	0.50	0.012	0.65	0.08	1002
58(47)	22.5	0.50	0.023	0.86	0.17	408
100(81)	24	0.50	0.010	0.25		3030
100(81)	23	0.50	0.017	0.48	0.02	969
100(81)	23	0.50	0.025	0.78	0.04	405
72(60)	24.25	0.50	0.018	0.60		550
60(45)	24.25	0.50	0.022	0.64	0.11	650
60(45)	24.50	0.50	0.025	0.84	0.16	431
60(45)	24.50	0.50	0.027	0.88	0.14	381
45(35)	24.0	0.38	0.025	0.367		599
45(35)	24.0	0.38	0.080	0.77		89
45(35)	24.0	0.38	0.100	0.77		71
114(89)	24.0	0.38	0.050	0.50		220
44(35)	24.0	0.38	0.060	0.39		570
44(35)	24.0	0.38	0.099	0.44		306
96(77)	24.0	0.38	0.050	0.28		953
97(78)	24.0	0.38	0.073	0.31		589
75(60)	22.9	0.56	0.082	0.14	0.10	1217
75(60)	22.4	0.56	0.066	0.39	0.09	555

(*continued*)

TABLE 9 *(continued)*

Test No.	BCL Pipe Identification	Pipe Grade (ksi)	Diameter (in.)	Wall Thickness (in.)	CVN Upper Shelf 2/3 Size (ft-lb)	Test Pressure (lb/in.^{2}gauge)
33	DE-1-4	5L X52	20	0.300	22	1250
34	EK-1-1	5L X65	42	0.458	20	1020
35	EK-1-1	5L X65	42	0.458	20	1020
36	EK-1-1	5L X65	42	0.458	20	1020
37	EK-1-1	5L X65	42	0.458	20	1020
38	EK-1-1	5L X65	42	0.458	20	1020
39	ED-1-1	5L X65	42	0.458	20	1020
40	EK-1-2	5L X65	42	0.458	20	1400
41	EK-1-2	5L X65	42	0.458	20	1400
42	EK-1-2	5L X65	42	0.458	20	1400
43	EK-2-1	5L X65	42	0.458	21	1200
44	EK-2-1	5L X65	42	0.458	21	1200
45	EK-2-1	5L X65	42	0.458	21	1200
46	EK-2-1	5L X65	42	0.458	21	1200
47	EK-2-1	5L X65	42	0.458	21	1200
48	EK-2-1	5L X65	42	0.458	21	1200
49	EK-2-1	5L X65	42	0.458	21	1200
50[a]	EK-2-1	5L X65	42	0.458	21	1200
51	EK-2-2	5L X65	42	0.458	21	1200
52	EK-2-2	5L X65	42	0.458	21	1200
53	EK-2-2	5L X65	42	0.458	21	1200
54	EK-2-2	5L X65	42	0.458	21	1200
55	TT-1-1	5L XV60	30	0.298	17	970
56	TT-1-1	5L XV60	30	0.298	17	970
57	TT-1-1	5L XV60	30	0.298	17	970
58	L-3-1	5L X52	30	0.406	20	970
59	L-3-1	5L X52	30	0.406	20	970
60	L-3-1	5L X52	30	0.406	20	970

[a]Test No. 50 pressured to failure at 1400 lb/in^2 gauge.

Stress Level % of SMYS ($\bar{\sigma}$)	Gouge Length (in.)	Tool Width (in.)	Gouge Depth Average (in.)	Dent Depth Average (in.)	Remaining Dent Depth Average (in.)	Q factor (ft-lb/in.)
80(63)	19.5	0.56	0.017	0.36	0.05	1106
72(61)	23.0	0.75	0.010	0.21	0.02	7965
72(61)	23.0	0.75	0.011	0.40		3802
72(61)	23.0	0.75	0.012	0.41	0.06	3400
72(61)	22.0	0.75	0.016	0.50	0.08	2186
72(61)	22.5	0.75	0.017	0.55	0.07	1829
72(61)	23.0	0.75	0.031	0.64	0.06	849
99(83)	23.0	0.75	0.008	0.32	0.04	6534
99(83)	23.0	0.75	0.022	0.37	0.02	2055
99(83)	23.4	0.75	0.034	0.46	0.04	1051
85(71)	23.6	0.75	0.025	0.29	0.02	2361
85(71)	23.3	0.75	0.027	0.51	0.04	1259
85(71)	23.0	0.75	0.030	0.52	0.07	1126
85(71)	23.5	0.75	0.042	0.49	0.03	816
85(71)	23.8	0.75	0.060	0.68	0.06	416
85(71)	2.0	0.75	0.080	0.56		4508
85(71)	19.0	0.75	0.025	0.56	0.04	1518
85(71)	22.0	0.75	0.028	0.62	0.07	1058
85(71)	22.8	0.75	0.013	0.28	0.04	4863
85(71)	23.5	0.75	0.019	0.50	0.09	1809
85(71)	23.5	0.75	0.033	0.53	0.07	983
85(71)	23.8	0.75	0.042	0.76	0.07	532
81(62)	24.0	0.50	0.025	0.48	0.06	520
81(62)	23.2	0.50	0.030	0.55	0.08	390
81(62)	23.8	0.50	0.050	0.76	0.07	165
69(51)	23.6	0.38	0.040	0.53	0.07	490
69(51)	24.0	0.38	0.050	0.63	0.07	320
69(51)	23.5	0.38	0.055	0.56	0.07	355

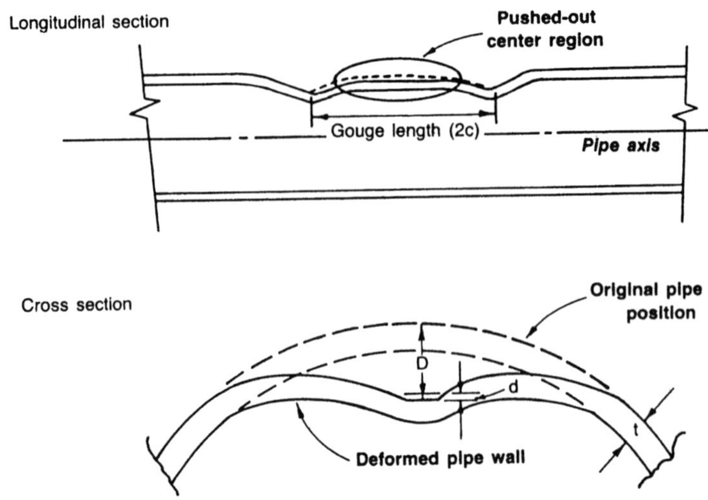

FIG. 10. Typical dent-and-gouge defect.

Many of the subcritical flaws produced in this study were sectioned and several but not all were found to contain some surface cracking.

It is evident that surface cracking plays a significant role in the failure characteristics of line pipe and must be adequately determined for assessing the severity of mechanical damage found in service.

Q Analysis

Raw data from the pipe experiments were reduced by use of the Q parameter analysis derived from previous work.

A Q factor was determined for each flaw configuration, and the value of hoop stress/flow stress was plotted as a function of the Q parameter, where flow stress for line-pipe steels is approximated by $\bar{\sigma} = \sigma_{yield} + 10$ ksi. Figure 11 presents the results of this exercise for those mechanical-damage flaws that failed during the time the damage was being done (Table 8).

A check of the data from Table 9 shows several experiments that contained surviving flaws with Q values lower than those which failed, suggesting they also should have failed.

Metallographic sections across these flaws and the failed flaws disclosed no consistent pattern. The cracking within the damage area was generally absent in the experiments with surviving flaws.

The amount of cold work at the gouged surface is always highly variable and is believed to be a significant factor in the lower failure-pressure experiments. Another mechanism believed to be significant is akin to friction welding.

If both the damaging tool and the pipe are dry and free of any kind of lubricant (oil, grease, coating), then the moving tool generates significant heat and tries to weld itself to the pipe. When this happens, small pieces of metal ride on the front edge of the tool and cause local gouging.

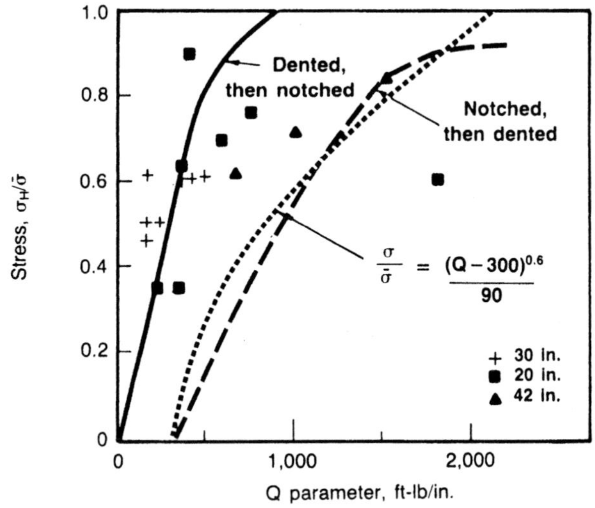

FIG. 11. Failure during introduction.

The Q parameter is based on the maximum dent depth. The dynamic damage-machine instrumentation measures the maximum dent depth also, although the pipe is internally pressurized.

As soon as the damage tool passes a location during the denting, the internal pressure starts removing the dent. The high-tensile stresses on the outside surface combined with the stress-raising effect of a gouge in this region of high local bending stresses cause the dynamic failure (failure at the time of occurrence).

If a defect is not severe enough to cause dynamic failure, then the indentation remaining has been partially removed as a result of the internal pressure.

If a defect such as this were found in the field, the only observable and measurable dent depth would be that of the partially removed dent.

Measurements of this remaining dent depth were recorded for several of the subcritical defects listed in Table 9.

These data were used to calculate the remaining dent depth to maximum dynamic-dent depth ratio.

The calculated values for each pipe were averaged, and the average data points for each pipe are plotted in Fig. 12 against the hoop stress ratioed to the pipe flow stress. An empirical curve has been fitted to the experimental data and an equation for this curve is shown in Fig. 12.

Additional unpublished data were examined and are shown in Fig. 13. The lower bound of the shaded area in Fig. 13 is defined by the equation shown in Fig. 12. The two midrange data points are from the penetration experiments discussed earlier in this article.

The upper bound of Fig. 13 was from deep, small-diameter spherical indentations with no gouge. These were only removed by about 50% of the original indention even at stress levels equal to the tensile yield of the pipe.

These data should be useful in estimating the margin of safety of a surviving mechanical-damage defect and will be discussed later.

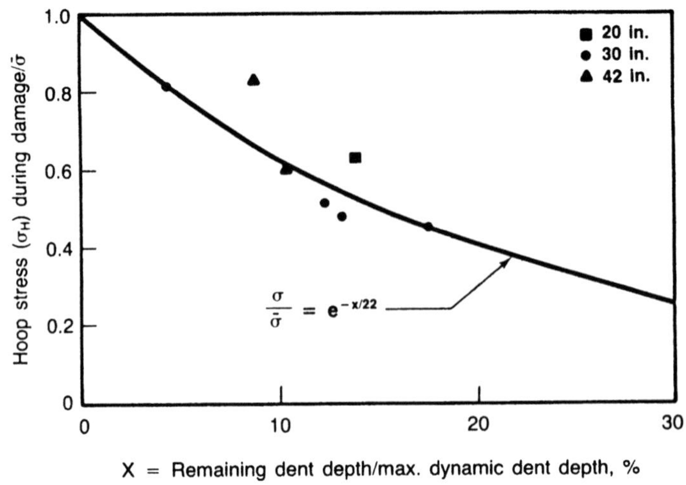

FIG. 12. Dent removal by pipe pressure.

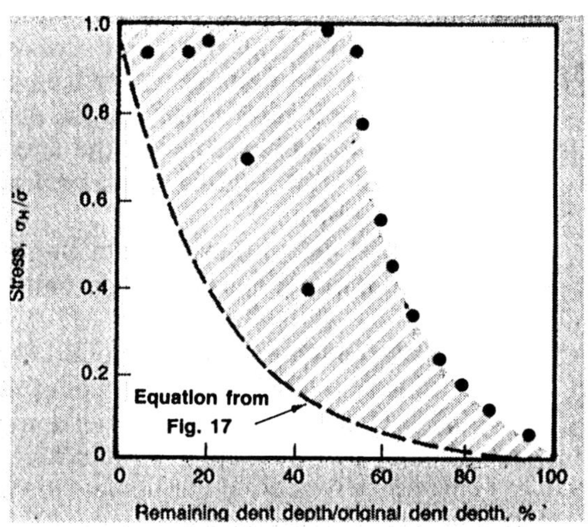

FIG. 13. Pipe pressure on dents.

Sustained Load

Two experiments examining sustained load were conducted on 30-in.-diameter pipe. Experimental procedures used for the two sustained load tests were similar to procedures described earlier, with the addition of four acoustic emission (AE) transducers that were mounted on the test pipe.

Acoustic emission events were used to monitor qualitatively crack activity during hold periods and to warn of impending pipe failure. Since all instrumentation had to be mounted prior to the pressuring and damaging of the test section, it was not possible to measure crack mouth opening with clip gauges or crack extension using the dc EP technique.

The AE transducers, however, could be mounted away from the gouge and dent to prevent being damaged and still retain sensitivity.

In an experiment the pipe was pressurized, mechanical damage was introduced to produce a subcritical dent-and-gouge defect, and the pipe was held at a constant pressure while acoustic emission (AE) data were monitored.

If flaw activity was detected and recorded in terms of AE events (i.e., acoustic energy released as a result of microscopic crack extension), the load was maintained.

If AE activity was seen to stop or substantially decrease, the internal pipe pressure was increased slightly and held again at constant pressure. Figure 14 illustrates a small portion of the typical AE data. The largest number of AE events occurred when the pressure was raised from one level to the next but ceased as pressurization was stopped. Only limited flaw activity was detected during most hold periods.

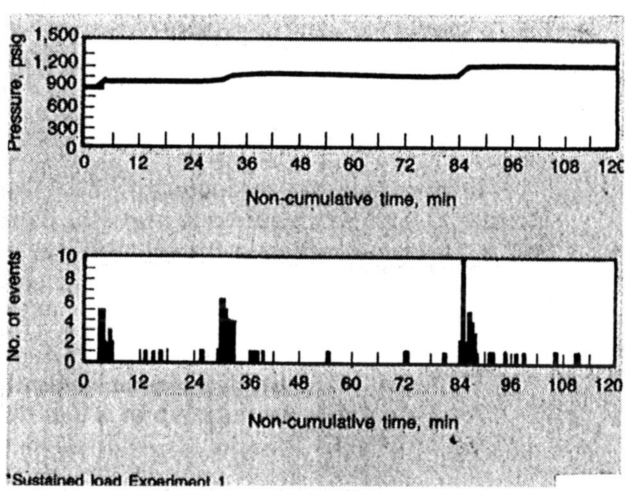

FIG. 14. AE events with pressure. Sustained load Experiment 1.

The failure stress for a specific flaw geometry in a sustained load experiment should be a reduced percentage of the continuous pressurization-to-failure stress, the reduced percentage being related to the length of time held.

In the two experiments conducted, the failure stress was much higher than that predicted for a mechanical-damage defect that fails with no appreciable hold time.

Both pipes reached a point of general yielding before rupture occurred.

At the point of general yielding, the validity of AE techniques for predicting crack extension is questionable. In the first experiment, the AE data showed a general increase in the number of events detected as the pipe approached failure pressure but was unable to predict the time of imminent pipe rupture. These two sustained-load experiments are shown in Fig. 15 in relation to the dented-then-notched failure curve developed earlier. At this time, this defect behavior is not understood because it contradicts earlier sustained-load studies in which defects were produced in line pipe at ambient pressures.

Data Analysis

These data have been analyzed by their measurable quantities. A correlation has been attempted with the Q parameter that includes pipe material toughness (CVN), gouge length ($2c$), dent-depth-to-pipe-diameter ratio ($D/2R$), and gouge-depth-to-pipe-wall-thickness ratio (d/t).

The Q parameter is compared with the pipe hoop stress as a ratio of flow stress ($\sigma_H/\bar{\sigma}$). The dent depth used in Q is the depth of the dent during introduction, which is the maximum dent depth.

If a surviving dent is found, its remaining depth can be measured and the maximum depth determined from Fig. 12.

The failures produced in this investigation suggest that the lower curve of Fig. 11 should be used to evaluate conservatively the total damage. A simple equation to fit the lower curve is shown on Fig. 11. A single data point failed below the lower curve (Test Number 13 of Table 7), while an earlier more severe (lower Q value) defect on the same pipe survived (Test Number 32 of Table 8).

A difference in the production of these two defects showed a continued increase in force was required to move the damage tool axially on the failure defect. This might indicate a frictional change that could have produced more heat and possibly changed the surface material within the gouge causing a low-failure pressure.

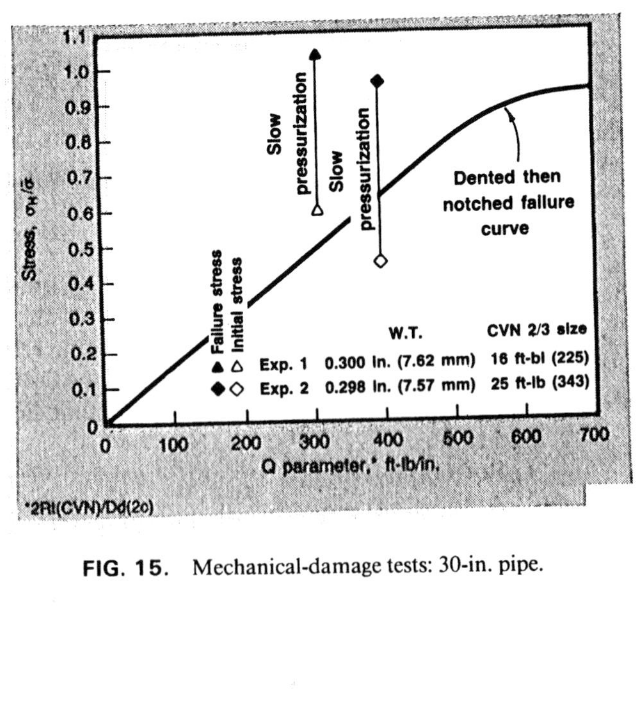

FIG. 15. Mechanical-damage tests: 30-in. pipe.

TABLE 10 Calculated Failure Pressure of Surviving Defects

Test No.[a]	Applied Stress Level,[b] % of (SMYS + 10 ksi)	Calculated[c] Max Dent Depth, (in.)	Calculated[d] Q-Factor (ft-lb/in.)	Calculated[e] $\sigma/\bar{\sigma}$ (SMYS + 10 ksi)	Calculated[f] Failure Pressure (lb/in.^{2} gauge)
13	51.0	0.20	8,560	>1	>2,020
14	50.5	0.53	2,460	>1	>2,020
15	50.5	1.13	620	0.355	717
17	86.6	0.63	1,475	0.773	1,560
18	86.6	1.26	500	0.268	542
20	51.4	0.75	665	0.383	530
21	51.4	1.09	400	0.175	242
22	51.4	0.96	420	0.196	271
31	64.1	1.02	120	0	0
32	64.1	0.92	170	0	0
33	67.2	0.57	795	0.459	854
34	62.4	0.19	13,205	>1	>1,636
36	62.4	0.58	3,605	>1	>1,636
37	62.4	0.77	2,130	>1	>1,636
38	62.4	0.67	2,250	>1	>1,636
39	62.4	0.58	1,395	0.740	1,211
40	85.6	1.17	2,680	>1	>1,636
41	85.6	0.58	1,965	0.952	1,557
42	85.6	1.17	620	0.354	579
43	73.4	0.29	3,372	>1	>1,636
44	73.4	0.59	1,554	0.803	1,314
45	73.4	1.03	810	0.469	767
46	73.4	0.44	1,330	0.713	1,166
47	73.4	0.88	460	0.233	381
49	73.4	0.59	2,060	0.984	1,609
50	73.4	1.03	910	0.521	852
51	73.4	0.59	3,300	1	1,636
52	73.4	1.32	980	0.556	909
53	73.4	1.03	720	0.418	684
54	73.4	1.03	560	0.313	512
55	69.8	0.76	590	0.332	462
56	69.8	1.01	380	0.156	217
57	69.8	0.88	255	0	0
58	57.8	0.58	555	0.310	520
59	57.8	0.58	440	0.213	358
60	57.8	0.58	405	0.183	306

[a]Data from Table 9.
[b](Test pressure) × R/t = applied stress. SMYS = specified minimum yield strength.
[c]Using equation on Fig. 12.
[d]Using equation on Fig. 11 for Q.
[e]Using $\sigma/\bar{\sigma} = (Q - 300)^{0.6}/90$ (simple equation to fit "notched, then dented" curve, Fig. 11).
[f]$P = \sigma/\bar{\sigma}$ from Column 5 times (SMYS + 10 ksi) times t/R.

Exploration of other explanations of the difference in failure pressure (mostly metallographic) revealed no additional information.

Those defects in Table 9 that did not fail can be examined as if they were surviving defects found in service to assess the conservativeness of the mechanical-damage failure criterion. (Only those with measured remaining dent depth can be used in this exercise.)

To define Q, the dent depth (D) was determined by the equation in Fig. 12 and the column "Remaining Dent Depth" in Table 9 to estimate the original dent depth. The "Test Pressure" from Table 9 was used to calculate hoop stress during damage which is the "operating pressure" for these defects

Pipe grade from Table 9 was used to calculate $\bar{\sigma}$. Since Charpy energy is generally unknown, a full-size CVN value of 25 ft-lb was selected for all 5L X52 pipe and 30 ft-lb for all 5L X60 and 5L X65 pipes for the Q in calculations.

Gouge depth and gouge length from Table 9 were used for d and $2c$ in Q as they are directly measurable in the field. Table 10 shows the resulting calculations.

The right-hand column of Table 10 is the calculated failure pressure for these surviving defects.

Comparing this calculated pressure with the column "Test Pressure" (a simulated operation pressure) from Table 9 indicates whether failure should have occurred. This comparison is made in Fig. 16 which shows that about 66% of these should have been failures.

All of the data points above the 1:1 line should have been failures according to the prediction. Because of the conservative nature of the prediction, they did not fail.

For those points below the 1:1 line, the failure pressure is above the pressure they were tested to; therefore no judgment can be made on prediction accuracy.

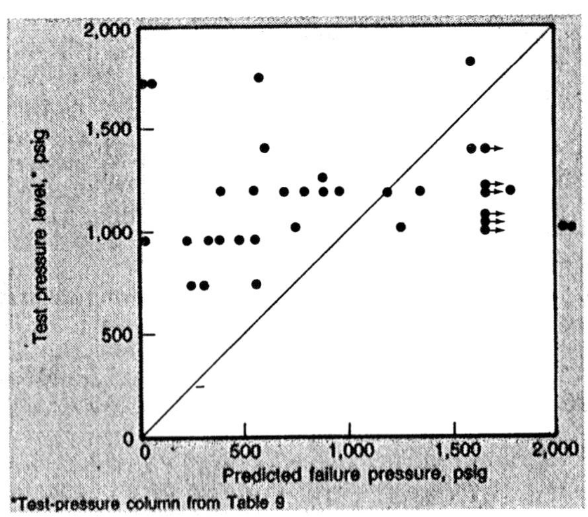

FIG. 16. Pressures of surviving defect.

The criterion being used is obviously quite conservative as it should be for these defect types.

The experiments described in this series have been conducted and analyzed to allow an evaluation of the serviceability of the line pipe with outside-force defects.

Methods to make this evaluation have been developed.

Results of this experimentation suggest the following: If mechanical-damage defects are found that are gouges within an indented area, pipeline pressure should be lowered to 2/3 of pressure level in the pipeline at the time the defect was discovered before conducting a detailed inspection. The gouge should be examined for cracking by grinding smooth a small area of the gouge making it suitable for crack inspection.

If a crack is found, a small circumferential grind should be made until the bottom of the crack is found and this crack depth should be added to the gouge depth to arrive at "d" in the analysis.

If the defect appears safe according to the criteria herein, it is probably safe to operate for a short time period, but it is recommended that it be replaced or repaired

An encircling sleeve constitutes the most appropriate repair [11], with the spaces between the dent and sleeve filled with polyester or epoxy molding compound.

References

1. D. J. Jones, G. S. Kramer, D. N. Gideon, and R. J. Eiber, *An Analysis of Reportable Incidents for Natural Gas Transmission and Gathering Lines 1970 through June 1984*, NG-18 Report No. 158, American Gas Association Catalog No. L51499, March 1986; *Oil Gas J.*, pp. 52–57 (March 16, 1987).
2. G. M. McClure and R. Eiber, *Research on Conditions Affecting Crack Initiation in Line Pipe*, Third Phase Report on Project NG-18 to the American Gas Association, July 11, 1960.
3. G. M. McClure et al., *Research on the Properties of Line Pipe*, Battelle Columbus Division's Report to the American Gas Association, AGA Catalog No. L00290, 1962.
4. J. F. Kiefner, *Fracture Initiation*, Fourth Symposium on Line Pipe Research, American Gas Association Catalog No. L30075, Dallas, Texas, November 1969.
5. G. D. Fearnehough and D. G. Jones, *An Approach to Defect Tolerance in Pipelines*, Conference on Defect Tolerance of Pressurized Vessels, Institute of Mechanical Engineers, May 1978.
6. R. J. Eiber et al., *The Effects of Dents on Failure Characteristics of Line Pipe*, Battelle Columbus Division's Report to the American Gas Association, AGA Catalog No. L51403, May 1981.

7. D. G. Jones, *The Significance of Mechanical Damage in Pipelines*, 3R International, July 1982.
8. M. E. Mayfield, W. A. Maxey, and G. M. Wilkowski, *Fracture Initiation Tolerance of Line Pipe*, Sixth Symposium on Line Pipe Research, AFA Catalog No. L30175, November 1979.
9. D. G. Jones and P. Hopkins, *The Influence of Mechanical Damage on Transmission Pipeline Integrity*, 1983 International Gas Research Conference, London, June 1983.
10. A. R. Duffy, *Hydrostatic Testing*, Fourth Symposium on Line Pipe Research, AGA Catalog No. L30075, Dallas, Texas, November 1969.
11. J. F. Kiefner, "Repair of Line Pipe Defects by Full-Encirclement Sleeves," *Weld. J.*, *56*(6), 26–34 (1977).

WILLARD A. MAXEY

Subsidence Strains

Adequate monitoring and proper intervention can significantly increase a pipeline's chances of surviving the strains of soil subsidence in an area of longwall mining.

The first part of this article on the effects of longwall mining on underground pipelines presents a technique for monitoring those effects. The concluding part examines intervention options and discusses the benefits of exposing pipelines in longwall mining areas.

Longwall mining can constitute a threat to the integrity of a pipeline by way of surface subsidence and soil strains.

The usual effects on a pipeline of mining-induced subsidence are increased axial and flexural strains affecting its longitudinal strength. In the presence of severe circumferentially oriented defects and added tensile strain, a rupture is possible. In the presence of added compressive strain, buckling of the pipe may occur.

Pipeline operators can use available predictive methods and geophysical data to estimate the potential effects of longwall mining, and they can monitor their pipelines and intervene if necessary to prevent a pipeline failure due to subsidence.

Monitoring Technique

Strain monitoring via the placement of strain gauges at critical locations is a satisfactory method of monitoring a pipeline during subsidence. A suitable strain monitoring technique is described here.

While it is theoretically possible to develop a rational fracture mechanics-based limit on tensile strain, such a limit cannot be realized in most practical subsidence situations. Compressive strain can be limited on a rational basis to prevent buckling.

When an operator's preselected limit on stress or strain is reached during subsidence, intervention may be required to prevent the limit from being exceeded.

To reduce the axial stress, the pressure can be lowered. Beyond that, further relief of axial strain may require taking the pipeline out of service and cutting it.

Flexural strains can only be reliably altered by raising or lowering portions of the pipeline to reduce curvature. This can be done without taking the pipeline out of service.

The concluding section discusses two examples of monitoring and intervention during longwall mining subsidence which reveal significant advantages to exposing the affected portion of a pipeline prior to and during subsidence rather than leaving it buried.

Material Loss

Land surface subsidence occurs when material is removed from the underlying strata as during longwall mining. The mining subsidence phenomenon is illustrated in Fig. 1, a vertical cross section through and perpendicular to the face advance direction of a coal seam undergoing longwall mining.

The draw angle is the angle between the vertical line at the edge of the mine face and the line connecting the edge of the face with the edge of the affected area on the ground surface. The draw angle defines the extent of the subsidence-affected area.

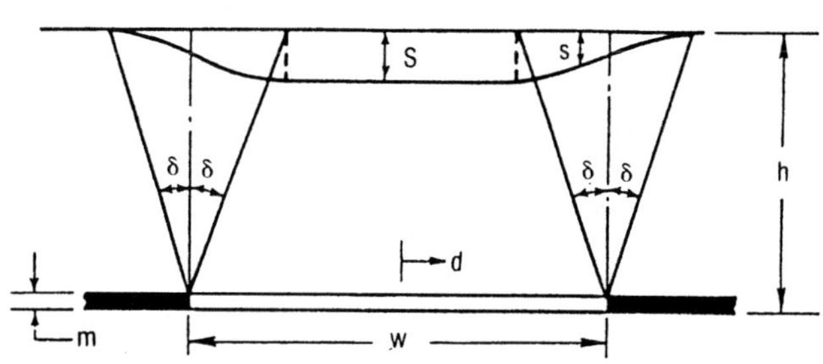

S = Maximum subsidence; here exaggerated for purposes of illustration. It is always less than m.
s = Subsidence at any given point in the affected area
m = Mined height of the coal seam
h = Distance from the floor of the seam to the ground surface
w = Width of the mined face
δ = Draw angle
d = Distance of any point from the center of the panel

FIG. 1. Longwall-mining subsidence.

The maximum subsidence is always less than the aimed seam height because of the overlying rock fracturing as it subsides to create voids that reduce the volume of material required to replace the coal and because of the "draw" of material from outside of the edges of the mined area.

The subsidence factor is defined as S/m and is always less than 1. The draw angle and subsidence factor for any given situation depend on the geophysical characteristics of the overlying strata.

Mining experience has shown that the nature of subsidence and the surface-soil strains created thereby are as illustrated in Fig. 2.

If the ratio of seam face width (w) to seam depth (h) is relatively small, the fan-shaped regions described by twice the draw angle overlap (Fig. 2a). In this case the width of the face (w) is said to be subcritical.

On the other hand, if the fan-shaped regions just barely meet or are separated, the width is said to be critical or supercritical (Figs. 2b and 2c, respectively).

In practical terms, this means that a subcritical area will have a lower subsidence factor but higher compressive soil strain above the center of the panel than a critical or a supercritical panel. An area of critical or supercritical width has zero strain at the center.

The surface-soil strain patterns for each type of panel configuration are shown in Fig. 2. Note that areas above the edges of the panel undergo tensile strain, while regions above the central portion of the panel undergo zero or compressive strain.

Pipeline strains due to subsidence arise because the curvature of the pipeline changes, because the pipeline must lengthen to accommodate the subsided profile, and because the soil–pipe interaction induces some of the soil strain into the pipeline if it remains buried during subsidence.

Note that a buried pipeline is not necessarily subjected to precisely these soil strains. It may slip relative to the soil, a situation which would cause some redistribution of the pipeline strain.

Exposed pipelines are probably not subjected at all to this pattern of axial strains. Instead, an exposed pipeline, if allowed to conform to the subsidence profile, would probably be subjected only to increased flexural strain and uniformity increased axial tension (unless it has preexisting slack).

Calculation methods are available for predicting subsidence [1–4]. The best known has been developed by the National Coal Board (NCB) of the United Kingdom [2].

Subsidence Calculation

Because the effects of subsidence on a pipeline are directly related to the nature and amount of subsidence, it is useful to be able to calculate the amount of subsidence associated with a given mining operation.

Without respect to any particular pipeline geometry, consider the subsidence effects resulting from a longwall mine of 600-ft face width, a seam with a 6-ft mined height, and four different cases of depths of the seam: 300, 400, 800, and 1200 ft below the surface.

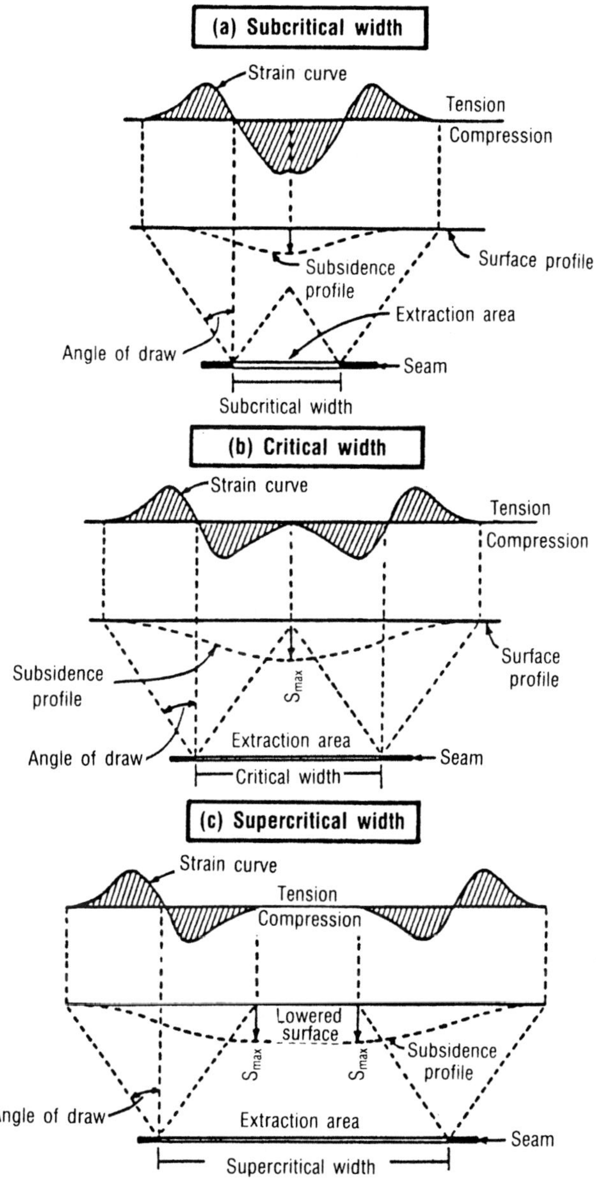

FIG. 2. Subsidence, soil strain curves. (From Ref. 3.)

The assumptions are made that the seam is horizontal, the ground surface is level, and the subsidence profile being calculated is perpendicular to the direction of face advance. The NCB technique is used in all four cases [2]. The actual calculations are too detailed for presentation here; therefore, only the results will be discussed.

The detailed calculations are contained in Kiefner [5]. The subsidence parameters of importance to these cases are presented in Table 1.

The subsidence profiles for the four cases based on the NCB method are compared in Fig. 3. Clearly, Cases 1 and 2 with the higher w/h ratios involve the more abrupt changes in curvature of the ground surface as well as the most subsidence.

In contrast, the curvatures of Cases 3 and 4 are less abrupt, but their limits of subsidence are also much further from the center than those of Cases 1 and 2. As will be shown, the curvature change becomes a very important factor for the pipeline operator to consider.

Strain Ranges

To some extent, soil strains due to subsidence can be induced in a buried pipeline. Therefore, it is useful to know the ranges of strains that can exist in the surface soil as the result of subsidence.

TABLE 1 Subsidence Factors: NCB Method[a]

Case Number	Seam Width (w) ft	Seam Width (w) m	Seam Depth (h) ft	Seam Depth (h) m	w/h	Subsidence Factor, S/m	Maximum Subsidence, s (ft)
1	600	183	300	91	2.00	0.89	5.34
2	600	183	400	122	1.50	0.87	5.22
3	600	183	800	244	0.75	0.74	4.44
4	600	183	1,200	366	0.50	0.49	2.94

[a] m, seam height = 6.0 ft.

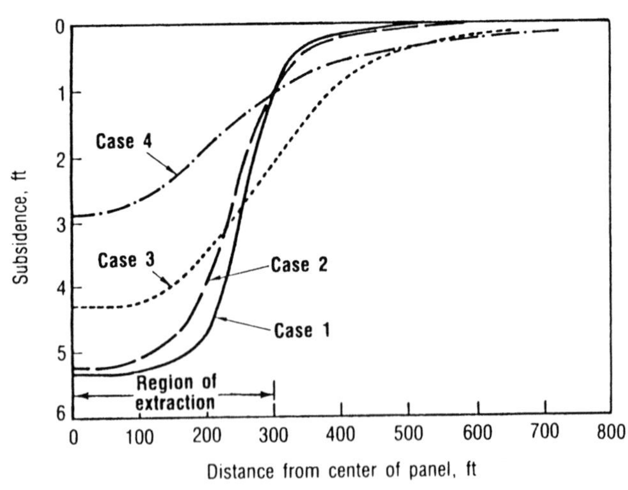

FIG. 3. Subsidence profiles: Four cases (NCB method).

The qualitative distribution of soil strains resulting from subsidence is displayed in Fig. 2. The NCB *Subsidence Engineers' Handbook* provides an empirically based procedure for calculating strain levels at various points along a subsidence profile [2]. The maximum tensile and compressive strains are proportional to the maximum subsidence (S) and inversely proportional to the depth of the seam (h).

For our four previous example cases, the peak strains are presented in Table 2. The complete strain profile for Case 1 is plotted in Fig. 4 along with the subsidence profile.

The potential risk to a pipeline crossing a mining subsidence region is that both the change in profile and the soil strain caused by subsidence (Fig. 4) will induce high strains in the pipe.

First, consider a buried line which is permitted to conform to the soil movement without any intervention. The pipeline may slip somewhat with respect to the soil, thus relieving some of the strain, but substantial axial strain may be induced in the pipeline by the soil.

TABLE 2 Peak Strains

Case Number	w/h	S (ft)	h (ft)	S/h	S/h^a Multiplier +	S/h^a Multiplier −	Maximum Strain +	Maximum Strain −
1	2.00	5.34	300	0.01780	0.65	0.51	0.0116	0.0091
2	1.50	5.22	400	0.01305	0.65	0.51	0.0085	0.0067
3	0.75	4.44	800	0.00555	0.65	0.80	0.0036	0.0044
4	0.50	2.94	1,200	0.00245	0.80	1.35	0.0020	0.0030

[a]From Ref. 5.

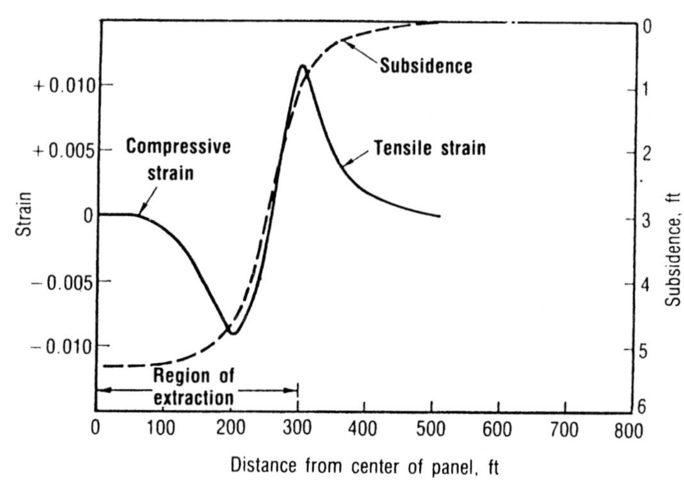

FIG. 4. Strain, subsidence profile: Case 1.

In addition, flexural strain is induced by the change in the pipeline profile, and even if the pipeline were not buried, a change in length (i.e., added axial strain) would accompany the change in profile. Even if the pipeline axial strain is only half the soil strain, Fig. 4 shows that it would be in the neighborhood of 0.5% (compressive at 200 ft from the panel center and tensile at 300 ft from the panel center).

Such strains could cause either a tensile failure in the presence of a significant girth-weld defect or a buckling failure even in sound pipe because that curvature as well as compressive strain is induced at one location.

Experience shows that pipelines can indeed buckle and fail from mining subsidence. The question then becomes: How can the risk of a subsidence failure be reduced? The answer is: By proper monitoring and, if necessary, by intervention and correction.

Subsidence is usually monitored by conventional surveying techniques which use benchmarks outside the subsidence area. In the case of a pipeline crossing a subsidence area, such surveying is extremely useful if not vital with respect to making intervention decisions.

In addition, it is useful to monitor pipeline strains directly. The possible monitoring and intervention responses for a pipeline operator range from doing nothing to shutting down the pipeline and relaying it after the subsidence has taken place.

The most reasonable and practical approaches involve leaving the line in service and either monitoring the pipeline without exposing it or exposing the pipeline throughout the subsidence area and monitoring it. In either case, the pipeline operator will probably establish a limit on strain and blowdown and cut the pipeline or move it to a less strained position if that limit is reached.

There are some practical ways for monitoring a pipeline during mining subsidence.

Strain Monitoring

The longitudinal strains in a pipeline can be monitored by means of electric resistance or vibrating wire strain gauges. The characterize completely the longitudinal strains at one station or cross section along a pipeline, a minimum of three strain gauges is needed.

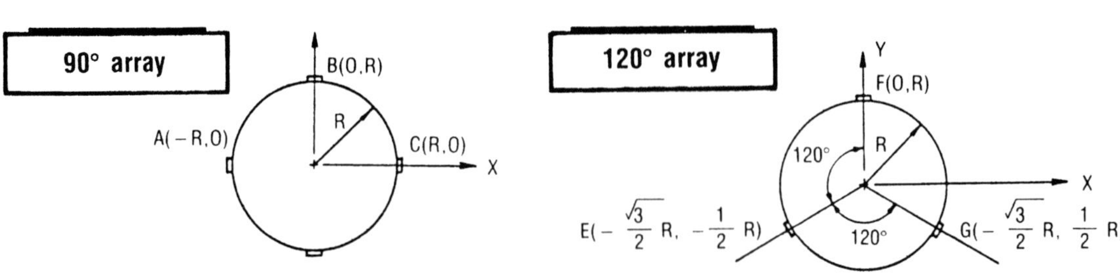

FIG. 5. Common arrays for strain gauges (longitudinally oriented).

While in principle the three longitudinally oriented gauges may be located in a wide variety of positions around the circumference, one of two types of arrays is generally used. Both types are illustrated in Fig. 5.

For the 90° array shown in Fig. 5, the longitudinal strain at any point (X, Y) around the pipe of radius R is Z_{90}, as defined in Eq. (1) (see Table 3 for all the numbered equations).

Z_{90} is a maximum or minimum at points defined by Eqs. (2) and (3). Similarly, for the 120° array shown in Fig. 5, the longitudinal strain at any point (X, Y) around a pipe of radius R is Z_{120} as defined in Eq. (4).

TABLE 3 Equations

$$Z_{90} = \frac{A + C}{2} + \left(\frac{C - A}{2}\right)\left(\frac{X}{R}\right) - \left(\frac{A + C - 2B}{2}\right)\left(\frac{Y}{R}\right) \tag{1}$$

$$\left(\frac{Y}{R}\right) = \pm \frac{C - A}{2a^2 + 2C^2 - 4AB - 4BC + 4B^2} \tag{2}$$

$$\left(\frac{Y}{R}\right) = \pm 1 - \left(\frac{X}{R}\right)^2 \tag{3}$$

$$Z_{120} = \frac{E + F + G}{3} - \frac{\sqrt{3}(E - G)}{3}\left(\frac{X}{R}\right) - \left(\frac{E + G - 2F}{3}\right)\left(\frac{Y}{R}\right) \tag{4}$$

$$\left(\frac{X}{R}\right) = + \frac{\sqrt{3}(E - G)}{2E^2 + F^2 + G^2 - EF - FG - EG} \tag{5}$$

$$\left(\frac{Y}{R}\right) = \pm 1 - \left(\frac{X}{R}\right)^2 \tag{6}$$

$$Z_{90\,\text{avg}} = \frac{A + C}{2} \tag{7}$$

$$Z_{120\,\text{avg}} = \frac{E + F + G}{3} \tag{8}$$

$$\sigma_p = \frac{\nu PR}{t} \tag{9}$$

$$\sigma_T = \alpha E(T_1 - T_2) \tag{10}$$

where ν = Poisson's ratio (0.3 for steel)
E = elastic modulas (30×10^6 lb/in.^{2})
α = coefficient of thermal expansion (6.5×10^{-6} in./in./°F)
P = internal pressure, lb/in.^{2} gauge
R = pipe radius, in.
t = pipe thickness, in.
T_2 = operating temperature of the pipe
T_1 = installation temperature of the pipe

$$\epsilon = kr \tag{11}$$

where ϵ = maximum flexural strain
r = pipe radius
k = curvature

Z_{120} is a maximum or minimum as defined by Eqs. (5) and (6).

The derivations of Eqs. (1)–(6) are presented in Kiefner [5]. The average longitudinal strain at the gauge station or cross section is shown in Eq. (7) for the 90° array and Eq. (8) for the 120° array.

It is important to note that either of these strain-gauge arrays can be used to completely characterize strains at one and only one location along the pipeline. To properly monitor a pipeline, one must instrument enough such stations to cover all of the possible points of critical strains along the pipeline in the subsidence area.

Fortunately, because subsidence is reasonably predictable, one can select the likely critical locations ahead of time. Note in Fig. 4, for example, that for the particular case shown, the maximum compressive soil strain (important if the line is not to be exposed) would be expected to occur at 200 ft from the center of the panel.

The point of maximum concave curvature due to subsidence is also located at about 200 ft from the center of the panel based on the subsidence profile shown in Fig. 4. Thus, it would make sense to concentrate monitoring stations around that region.

Similarly, as can be seen in Fig. 4, one would want to concentrate monitoring stations around the location 300 ft from the center of the panel where both the peak tensile soil strains and the maximum convex curvature occur. Adequate monitoring could be done with as few as four stations (two on each side of the center of the panel), but only if the pipeline operator were fortunate enough to pick the exact locations of maximum strains.

Because the subsidence phenomena are not as predictable as implied in Fig. 4, it is prudent to choose multiple gauge locations to "bracket" the potential areas of maximum strains. In choosing strain-gauge locations, one should keep in mind the steepness of the strain and curvature gradients as illustrated in Fig. 4.

Acceptable Strain

The question of how much subsidence-induced stress/strain is acceptable depends on several factors, some of which are difficult to evaluate. These include the initial stress in the pipeline before subsidence, the quality of the girth welds from the standpoint of defects, and the buckling strength of the pipeline.

In practice, the existing stress, the weld properties, the girth-weld quality, and the buckling strength of the pipe may not be known and, if not, they usually cannot be easily determined. Nevertheless, limits on added stress can and must be set even if they are arbitrary.

According to the ANSI/ASME B31.8 code, the longitudinal stress in a pipeline must not exceed 75% of the maximum allowable hoop stress. Thus, for a Class 1 location the limit is 54% of the specified minimum yield strength (SMYS) of the pipe.

Most pipelines are not stressed to that level and can tolerate some added stress. A pipeline operator may decide that he will not allow the total stress level to exceed this amount during a subsidence episode.

As will be shown, this is a very conservative approach to intervention in a subsidence area. In view of the magnitudes of the soil strains shown in Fig. 4, it would appear that this approach could be used only for cases where the pipeline is exposed.

As we examine the 54% of SMYS stress-limit approach, it is recalled that a long, straight, buried pipeline is fully restrained by the soil. Therefore, the longitudinal stress consists only of that due to the biaxial effect of pressure and to temperature change.

The pressure component of longitudinal stress (σ_p) is defined by Eq. (9) and that due to restraint of thermal expansion (σ_T) by Eq. (10).

For a 30-in. o.d., 0.375-in. W.T. X52 pipeline operating at 72% of SMYS which was installed at a temperature of 100°F and operates at a temperature of 60°F, $\sigma_p = 11.2$ ksi (11,200 psi) and $\sigma_T = 7.8$ ksi. Thus, the total existing axial stress is 19 ksi.

The allowable longitudinal stress is 28.1 ksi. Thus, a stress of 9.1 ksi can be added to this pipeline without the total stress exceeding 54% of SMYS.

Kiefner et al. present a method for calculating the optimum or at least an acceptable profile for a pipeline when it is lowered from an initially straight position [6]. The calculated profiles are based upon an assumed initial axial stress and the premise that the total stress cannot exceed 54% of SMYS.

Two additional points must be considered, however. First, the calculation is valid only if the pipeline is exposed throughout the subsidence area; the method of Kiefner [6] is designed for an exposed pipeline only because it includes no soil friction stresses which affect the axial stress.

As noted previously, the use of the 54% SMYS stress-limit approach is probably applicable only when the pipeline is exposed throughout the subsidence area.

Second, the curvature profile in an actual subsidence area may be more severe than the calculated profiles.

Limits to Prevent Buckling

Buckling or wrinkling of a pipeline is one form of failure which can arise from excessive subsidence. A pipeline with sound girth welds is much more likely to undergo wrinkling or buckling from excessive curvature than failure from excessive tensile stress.

Therefore, limiting strain to prevent buckling constitutes a sound approach for preventing damage to a pipeline in a subsidence area.

Buckling of a pressurized pipe due to excessive curvature is controlled by the amount and distribution of the imposed deformation, by the pipe geometry, and by the stress-strain behavior of the material. Because the buckling phenomenon in pipes of practical R/t ratios depends on the inelastic behavior of material, the calculation of the buckling load is usually a complex problem requiring a computer-based iterative solution.

One such solution method is a proprietary Battelle computer program COLAPS. To show how such a program might be used to set strain limits, the results of COLAPS runs for a number of practical cases are summarized in Table 4.

TABLE 4 Strain Limits to Prevent Pipe Wrinkling

o.d. (in.)	W.T. (in.)	Grade	Curvature at Wrinkling (microradians/ft)	Strain at Wrinkling (microstrain)	Ratio of Collapse to Wrinkling Curvature
8⅝	0.156	X52	7350	2641	3.6
12¾	0.250	GrB	4811	2556	4.5
12¾	0.250	X52	5123	2721	3.7
16	0.188	X60	4209	2806	2.6
16	0.250	X52	3963	2642	3.2
20	0.219	X52	2792	2327	2.8
20	0.312	GrB	Does not wrinkle[a]		—
24	0.250	X52	2312	2312	2.8
26	0.281	X52	2163	2344	2.8
30	0.375	X52	1935	2419	3.1
36	0.390	X60	1896	2844	2.4
42	0.469	X60	1645	2879	2.4

[a]Collapse occurs at 10,123 microradians/ft without intermediate wrinkling. The corresponding strain at collapse is 8436 microstrain.

Note that the curvatures are those at which these pipes begin to wrinkle. The ultimate collapse curvatures of the various pipe geometries in Table 4 are two to five times as large as the wrinkling curvatures. Also presented in Table 4 are the strains at the onset of wrinkling.

The relationship between curvature and strain is shown by Eq. (11).

In Table 4 it is noted that over this wide range of pipe geometries, the strain at wrinkling varies only from 2312 to 2879 microstrain. This is because wrinkling begins soon after the pipe begins to yield.

The pipes with the higher R/t ratios wrinkle at the lowest strains, as one would expect, but the effect of pipe geometry on the wrinkling strain is not very great. Also, note that total collapse by buckling occurs at strain levels two to five times the wrinkling strain levels.

Thus, limiting the strains during subsidence on the basis of wrinkling provides a significant factor of safety against collapse.

While Table 4 provides values of strain at wrinkling, it is desirable to avoid having the strains approach the wrinkling strain during a subsidence event. Therefore, a safety factor should be applied to the values in Table 4.

The choice of a safety factor is best left up to the individual pipeline operator. One can see, however, that the use of a strain limit such as 1000 microstrain would be adequately conservative.

Comparison of Limits

It has been stated that the idea of limiting the total stress to 54% of SMYS is very conservative. To compare it to a limit of 1000 microstrain, it is necessary to consider the 54% of SMYS limit in terms of pipe grade and the corresponding strain level. For other grades the 54% of SMYS limit corresponds to strain limits as shown in Table 5.

TABLE 5 Strain Limits

Pipe Grade	Strain at 54% of SMYS (microstrain)
B	630
X52	939
X60	1080
X65	1170

Thus the 54% of SMYS limit is more conservative for Grade B and X52 pipe, but the 1000 microstrain limit is more conservative for Grades X60 and up.

Note that in both cases the limits are on total strain, not added strain due to subsidence.

Limits based only on added strain may be more or less conservative depending on the initial strains in the pipeline.

For example, a limit of 1000 microstrain on added strain would be more conservative with respect to buckling if the pipeline has an existing tensile stress, whereas it is less conservative with respect to tensile rupture at a girth-weld defect.

In any case, the individual pipeline operator is in the best position to judge the quality and initial status of his pipeline. Therefore, he should have the flexibility to set a reasonable limit on the basis of his particular situation.

Intervention Options

Timely and proper intervention can save a pipeline from substantial damage due to soil-subsidence effects of longwall mining.

This part covers the range of intervention responses in such a situation and examines the benefits of exposing pipelines to subsidence in longwall-mining areas.

The readily apparent responses available to an operator whose pipeline is about to undergo subsidence due to mining are as mentioned briefly earlier:

Do nothing (acceptable if the anticipated added strains are very small).

Monitor subsidence and strain without exposing the pipeline and without taking it out of service.

Expose the pipeline and monitor subsidence and strain while leaving the pipeline in service.

Take the pipeline out of service, expose, and reposition it after the subsidence has taken place.

Since this article deals with responses involving monitoring, the first and the last cases will be omitted here.

Planning Activities

Prior to the subsidence event, the pipeline operator should learn as much as possible about the proposed longwall-mining activities. He should establish the depth of the coal, the mined height, the width of the seam, and the location and angle at which it will pass under the pipeline.

He should establish the proposed progress of the mining and continue to maintain contact with the mine officials to be advised of the actual progress. The operator should attempt to obtain subsidence profiles of previous mines in similar strata.

The mining company can often provide the necessary information, or the pipeline operator can retain a mining engineer or geologist to assist him in predicting the amount and distribution of the subsidence.

If no other information is available, the operator can make subsidence calculations based on NCB or other available methods. With this information, the operator must decide where and how to monitor the pipeline.

Establishing Gauge Locations

Given a predicted profile adequate for this particular geological region, the pipeline operator can establish appropriate survey stations and strain-gauge locations to monitor the subsidence.

As noted earlier, the locations of maximum curvature are of greatest importance with respect to strain monitoring. From the standpoint of settlement, the operator should plan to monitor elevations about every 25 ft along the pipeline throughout the subsidence area.

Excavation

The operator must choose whether to excavate the entire pipeline or only bellholes for installing strain gauges. As demonstrated, excavating the entire pipeline in the subsidence area prevents soil-imposed strains and thereby reduces the maximum strain level that the pipeline will experience.

Drawbacks to excavating the entire pipeline are the cost and the increased potential for excavation damage. Also, if the pipeline is exposed, adequate drainage of the ditch must be maintained to prevent flotation which could do more damage than the expected subsidence.

Additional advantages to complete excavation are that the entire pipeline becomes visible for inspection and the girth welds can be inspected and repaired if necessary. Also, the excavation of 500 to 1000 ft of a pipeline generally relieves any built-in axial compressive strains, thereby reducing the chances of buckling.

Strain, Curvature Limits

The pipeline operator needs to establish limits on the strain and/or curvature to protect the pipeline. The previously described bases for limiting strain should be considered.

Also, the subsidence profile can be compared to the optimum profile calculated as shown in Kiefner [6]. If these limits are reached or exceeded, the pipeline operator can consider several options as described below.

The following corrective actions can be taken to limit or reduce stresses or strains when the limiting values are reached.

Reducing pressure lowers the axial stress. The results may not show up on strain measurements, but the axial stress is nevertheless reduced by a calculable amount (Eq. 9).

Taking the pipeline out of service can eliminate the escape of the pipeline fluid if a failure is imminent or if the line is to be cut.

Cutting the pipeline will relieve accumulated axial strain, especially where the pipeline is completely exposed. It is less effective where the pipeline remains buried unless cuts are made at several locations. Cutting does not necessarily have a favorable effect on flexural strains.

Repositioning the pipeline may be accomplished with or without taking it out of service. The pipeline may be repositioned vertically up or down by means of additional excavation or supporting soil to change the profile to reduce flexural strains.

Two examples of monitoring of pipelines in subsidence areas are described briefly and in greater detail in Kiefner [5]. One involved monitoring a single pipeline which remained backfilled. The other involved monitoring three parallel, completely exposed pipelines. These examples illustrate the value of monitoring and provide a comparison between exposing the pipeline and leaving it backfilled.

Backfilled during Subsidence

A complete description of the monitoring pipeline that remained backfilled during subsidence is presented in Kiefner [5]. This 12¾-in. by 0.250-in. W.T. X42 gas pipeline had a maximum allowable operating pressure (MAOP) of 617 lb/in.^{2} gauge. The longwall panel was 650 ft wide, 800 to 1000 ft below the surface, and had a mining height of 5.3 to 5.7 ft.

The changes in strain in nine locations along the pipeline which occurred over the 43-day subsidence period are presented in Table 6. The strains at Station 5 are omitted because they were not presented by Hayes [7]. The individually measured strains (A, B, C in Table 6) were obtained from each of three longitudinally oriented strain gauges at each station arrayed as shown in Fig. 5 (the 90° array).

Other information presented in Table 6 includes the average axial strain at each station, $(A + C)/2$, the maximum and minimum strains, and the angle from the pipe to the location of the maximum strain.

Comparing this monitoring exercise with the one which follows will make clear that axial strain levels are significantly influenced by whether or not the pipeline is backfilled.

TABLE 6 Longitudinal Strains Due to Subsidence[a]

Gauge Station	Pipeline Distance (ft)	Change in Strain Oct. 11 to Nov. 22			$\frac{A+C}{2}$	$\frac{C-A}{2}$	$\frac{A+C-2B}{2}$	Location Maximum Strain		Clockwise Angle from Top to Point of Maximum Strain (degrees)	Maximum Tensile Strain or Minimum Compressive Strain	Maximum Compressive Strain or Minimum Tensile Strain
		A	B	C				$\frac{X}{R}$	$\frac{Y}{R}$			
1	0	430	460	400	415	−15	−45	−0.316	0.999	−18	461	369
2	97	740	870	700	720	−20	−150	−0.132	0.991	−6	871	569
3	199	510	560	490	500	−10	−60	−0.164	0.986	−9	561	439
4	299	440	240	590	515	75	285	0.263	−0.965	165	800	230
6	521	−70	−200	−120	−95	−25	105	0.232	0.973	13	13	−203
7	624	−340	−450	−420	−380	−40	70	0.496	0.868	30	−299	−461
8	722	−460	−340	−350	−405	−55	−65	−0.647	−0.762	−140	−320	−490
9	820	−620	−670	−570	−595	−25	75	−0.316	0.949	−18	−516	−674
10	920	−600	−490	−410	−505	−95	−15	−0.988	−0.516	−99	−409	−601

[a]In a 12¾-in. o.d. pipeline.

Exposed during Subsidence

Another significant monitoring exercise was carried out during longwall mining under three parallel pipelines. These pipelines were a 30-in o.d. by 0.375-in. W.T. X52, a 30-in. o.d. by 0.375-in. W.T. X52, and a 36-in. o.d. by 0.390-in. W.T. X60.

In this case the panel was 632 ft wide, 460 ft below the surface, and had a 6-ft mining height. Prior to the occurrence of subsidence, the three pipelines were exposed for a distance of more than 10,000 ft, well beyond the edges of the panel. All backfill was removed and at least the top half—and in most cases the top two-thirds—of the pipe was exposed.

Since the subsidence profiles of the three pipelines were similar, only that of Line 3 is discussed here. The subsidence occurred over a 29-day period; the maximum subsidence on Line 3 was 3.4 ft.

The changes in strain in Line 3 over the 29-day subsidence period are presented in Table 7. The individually measured strains (A, B, and C) in the table were obtained from each of three longitudinally oriented strain gauges at each station arrayed as shown in Fig. 5 (90°).

Other information in these tables includes the average axial strain at each station, $(A + C)/2$, the maximum and minimum strains, and the angle from the top of the pipe to the location of the maximum strain.

The strain changes due to subsidence in these three exposed pipelines followed a relatively predictable pattern. Because the pipelines were exposed, there was no instance of net axial compressive strain. The average change in axial strain at every station on all three lines was in the direction of increasing tensile strain.

Table 7 shows quantity $(A + C)/2$, which represents average axial strain and stayed within a fairly narrow range. The average for Line 3 was 136 microstrain. This level is much lower than the axial strain in the previously discussed 12¾-in. pipeline which was allowed to subside while remaining buried.

The main source of strain due to subsidence was from the induced curvature. This is evident from comparisons of maximum and minimum strains in Table 7.

Subsidence Effects

A significant difference in subsidence effects is indicated by a comparison of subsidence episodes.

First, it is clear that the buried pipeline underwent much higher axial strains than did the exposed pipelines. This is readily apparent where the axial strain levels for the two situations are compared.

The buried pipeline in Case 1 was subjected to axial strains $(A + C)/2$, ranging from −595 to +720 microstrain, whereas the maximum and minimum axial strains observed on the exposed Line 3 in Case 2 ranged from 108 to 159 microstrain.

Next, consider which of the two cases actually had the most severe curvatures. The answer is not readily apparent because the diameters of the

TABLE 7 Maximum Strains Due to Subsidence of Line 3

Gauge Station	Distance from Panel Center (ft)	Change in Strain Sept. 9 to Oct. 7			$\frac{A+C}{2}$	Clockwise Angle from Top To Point of Maximum Tensile Strain (degrees)	Maximum Tensile Strain	Maximum Compressive Strain or Minimum Tensile Strain
		A	*B*	*C*				
3–1	296	121	162	120	120.5	−1	162	79
3–2	261	124	467	108	116.0	−1	467	−235
3–3	221	119	500	118	118.5	0	500	−263
3–4	−23	125	−41	150	137.5	176	316	−41
3–5	−102	136	−144	159	147.5	178	439	−144
3–6	−143	196	−157	118	157.0	−172	473	−159
3–7	−488	148	78	147	147.5	180	217	78
					Av 135.0			

pipelines in the two cases were significantly different.

Flexural strains in Case 1 were lower than those in Case 2, but a valid comparison of the relative severities depends on comparisons of curvatures.

Recalling that curvature and strain are proportional (Eq. 11) and that pipe radius is the constant of proportionality, one can compare the maximum measured curvatures in the two cases. For examples, the maximum measured flexural strain due to subsidence in Case 1, observed at Station 4, was 295 microstrain.

This is calculable from the data of Table 6. It is one-half of the difference between the maximum and minimum tensile strains. This converts with Eq. (11) to a curvature of

$$295 \times \left(\frac{2 \times 12}{12.75}\right) = 555 \text{ microradians/ft}$$

Similarly, the maximum measured flexural strain in Case 2 was at Station 3-3, 382 microstrain. This is calculable from the data of Table 7 and converts to a curvature of

$$382 \times \left(\frac{2 \times 12}{36}\right) = 255 \text{ microradians/ft}$$

Hence the curvature of the 12¾-in. pipeline due to subsidence was actually greater than the curve of the 36-in. pipeline. This means that if the 12¾-in. pipeline in Case 1 had been a 36-in. pipeline, it would have exhibited much higher flexural strain than the 36-in. pipeline in Case 2.

These comparisons show that the pipeline in Case 1 was subjected to axial strains and curvatures much more than those of Case 2, even though the amount of subsidence in both cases was about the same.

It is strongly suggested that the reason for this situation is that the Case 1 pipeline remained backfilled whereas the pipelines in Case 2 were exposed during the subsidence.

Clearly, then, exposing a pipeline prior to subsidence has a substantial advantage with respect to pipeline integrity over leaving it buried.

Other advantages that accompany exposing the pipeline include the abilities visually to inspect the pipe before, during, and after subsidence, and to radiograph the girth welds. Another advantage is the relief of initial axial compressive strains (this occurs automatically once several hundred feet have been exposed).

The disadvantages that accompany exposing the pipeline are the added cost, the increased risk of mechanical damage, and the need to provide drainage of the ditch to avoid floatation.

Pipeline Survival

Review of subsidence phenomena and models and of two previous pipeline monitoring efforts during subsidence episodes here suggest the following: It is possible for a pipeline to survive safely a longwall-mining episode without being taken out of service. The chances of the pipeline surviving can be greatly enhanced by proper monitoring and intervention responses.

Further, longwall-mining subsidence for a given set of mining parameters can be predicted reasonably well by available empirical models. Subsidence adds to the longitudinal strain in a pipeline, but the added strain can be limited to an acceptable amount by means of appropriate responses.

One key response to preserving pipeline integrity is the monitoring of strains in critical areas. The available models of subsidence phenomena can be used to select the critical areas for strain-gauge locations.

A second key response to preserving pipeline integrity is timely intervention when indicated by the strain data. Intervention in terms of repositioning the pipeline without taking it out of service can be used to keep the strains within acceptable limits.

Having the pipeline exposed prior to subsidence leads to significantly lower and more uniformly distributed added strains. It also permits inspection prior to subsidence, repositioning to reduce strains or curvatures, and verification of no damage after subsidence.

References

1. B. M. Hall, "Subsidence Production Methods and Instrumentation for Caved Longwall Coal Mines," M.S. Thesis, Northwestern University, August 1980.
2. *Subsidence Engineers' Handbook*, National Coal Board Mining Department (UK), 1975.
3. T. Z. Jones and K. K. Kohli, Subsidence Engineering Workshop, Division of Mining, West Virginia Institute of Technology, Montgomery, West Virginia, July 18–19, 1985.
4. S. S. Peng and H. S. Chiange, *Longwall Mining*, 1984.
5. J. F. Kiefner, *Monitoring and Intervention on Pipelines in Mining Subsidence Areas*, NG-18 Report 155 to the Pipe Line Research Supervisory Committee of the Pipeline Research Committee, American Gas Association, August 8, 1986.
6. J. F. Kiefner, T. A. Wall, N. D. Ghadiali, K. Prabhat, and E. C. Rodabaugh, *Guidelines for Lowering Pipelines While in Service*, Report to the U.S. Department of Transportation, DOT-RSPA-DMT-30/84/8, February 25, 1985, available through the National Technical Information Service, Springfield, Virginia.
7. G. R. Hayes Jr., *Stresses Imposed on a Pipeline by Longwall Mining*, American Gas Association Operating Section Proceedings, No. 80-T-65, 1980.

JOHN F. KIEFNER

Rupture

Laboratory analyses of the failed section of the Moomba-to-Sydney, Australia, natural-gas pipeline predicted the probability of more failures due to stress-corrosion cracking (SCC) outside the initially replaced first 10 km of line.

As a result, another 20 km of pipeline was replaced, and changes were made in the line's operating procedures to reduce the risk of further failures due to SSC.

Part 1 of this article discusses the analyses following the failure, focusing on the pipeline's design and the subsequent tests of steel specimens from original and replacement joints.

Part 2 deals with further laboratory tests of crack initiation and growth under conditions that mirror the line's actual operating conditions.

Part 3 explains how the laboratory test results were able to predict future line failures and how line operating conditions were changed to avert such failures.

Part 1: Investigations and Analyses

The pipeline ruptured on July 25, 1982, at a point 4.2 km (2.6 miles) downstream from the CO_2 removal plant at Moomba (Fig. 1).

The failure was initiated at a stress-corrosion crack some 210-mm long and up to 7.3-mm deep. It ran as a longitudinal ductile fracture for about 13 m before the fractured section tore away from the main body of the pipeline, indicative of the good fracture toughness of the pipe steel.

Although accompanied by a fire, the failure caused no damage to property or injury to people. The failed section was replaced with new pipe and transportation of gas restored in about 50 h [1].

A detailed review of the original design of the pipeline confirmed that the design, construction methods, and quality were consistent with those being used throughout North America when the pipeline was designed and constructed.

At the time of its design, the Australian Pipeline Code was not in existence. An operating stress level of 80% of specified minimum yield strength (SMYS) in conjunction with high-level testing as provided for in the Canadian pipeline code and as recommended by the Battelle Memorial Institute, Columbus, Ohio (following an extensive research program in the late 1960s), was adopted.

Pipeline features

Length	1,300 km (807 miles)
Diameter	864 mm (34 in.)
Pipe wall thickness	8.3 mm (0.327 in.)
Steel grade	API 5LX - Grade X65
Coating	Over-the-ditch coal tar enamel with fibreglass reinforcing and coaltar impregnated outer wrap
Operating pressure	6,895 kPa (1,000 psi)
Hydrostatic test pressure	100% of specified minimum yield strength (SMYS) minimum
Operating stress	80% of SMYS 65% of actual yield stress
Specified maximum temperature	71° C. (160° F.)

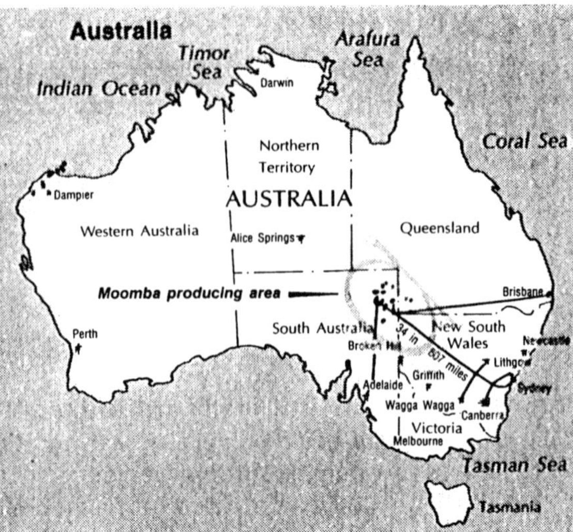

Source: 1986 International Petroleum Encyclopedia, Vol. 19, Tulsa, Okla: PennWell Publishing Co., 1986.

FIG. 1 The Moomba-Sydney pipeline. (Source: *1986 International Petroleum Encyclopedia,* Vol. 19, Pennwell Publishing Co., Tulsa, Oklahoma, 1986.)

A lower operating stress would have increased the size of the critical defect at which failure would occur and, hence, would have extended the time required for a failure to occur. The increase in the time to failure, however, would not have been significant nor would a lower operating stress of 72% of SMYS as specified in the current Australian pipeline code have prevented the failure. Although the nominal operating stress level of 80% of SMYS was relatively high, the actual operating stress level of 65% of yield strength was well within acceptable limits. A review of the original steel mill and pipe-mill records and the laboratory examination of steel samples from the failed pipe revealed no unusual characteristics in the steel.

The selection and quality of the external coating was, of course, subject to examination. Although the field examination of the over-the-ditch-applied coal-tar enamel indicated that some disbonded coating was present, the low cathodic-protection (CP) current of 37 $\mu A/m^2$ suggested to the Authority's advisers that the coating quality was of the standard that might reasonably be expected, given the age of the pipeline.

The first stage of the field investigation consisted of a flame ionization leak-detection survey over the first 30 km of the pipeline. This survey was carried out to determine whether any through-wall cracks existed in the pipeline. No leaks were located.

A Pearson survey was also carried out over the first 30 km of pipeline to establish if any particularly poor coating existed.

It was recognized that a Pearson survey would not locate disbonded coating unless associated with coating holidays, but it was considered that a Pearson survey was an essential part of the field data-gathering program. Despite subsequent work, no correlation between Pearson-survey results and stress-corrosion cracking (SCC) was established.

However, as indicated by the Pearson survey and an analysis of the cathodic-protection current records, subsequent excavation of the pipeline in a number of locations confirmed that no pitting or general corrosion of any consequence was present.

Soil Resistivity and SCC

The western end of the pipeline traverses alternate sand dunes and clay pans approximately at right angles. A soil-resistivity survey along the first 30 km of the pipeline showed low soil resistivities of the order of 160 Ω·cm in some locations. These results indicated wet and aggressive soil conditions in the clay pans, and high soil resistivities, indicating dry conditions, in the sand dunes.

Because the failure had occurred in the middle of the clay pan in an area of low soil resistivity, it was concluded that there might be a correlation between soil resistivity and the presence of SCC. Subsequent work showed this to be the case.

Gas temperature records indicated that the temperatures in the section of the line that included the failure were generally 45–60°C and approximately 50°C at the time of failure. The line had also been subjected to considerable pressure variations, a point returned to later.

Late in August 1982, the data which had been accumulated to that time were carefully examined. The pipe-wall temperature profile and the soil-resistivity survey results in particular indicated a high probability that other stress-corrosion cracks existed in the first 10 km of the pipeline, with a much lower probability of cracking downstream of 10 km.

In the interest of safety and security of supply of gas to New South Wales, the decision was made to build a 660-mm diameter bypass 10-km long starting from Moomba. The bypass was completed and put into service on October 30, 1982, and the first 10 km of the original pipeline was taken out of service.

The 10-km section of the original pipeline was then examined from inside with the ultrasonic technique. At the same time, the operating pipeline downstream of 10 km was examined at some 25 locations by excavation of the pipeline and removal of the coating and magnetic particle detection of the external pipe surface, particularly in areas of disbonded coating.

The excavations were carried out at locations of lowest soil resistivity which were considered to have the highest probability of SCC being present.

The excavation work was carried out working upstream from the 65-km point (kp).

Although disbonded coating was located in a number of excavations, no cracking was found between the 15- and 65-km locations on the pipeline. Some minor cracking was located between 10 and 15 km.

In addition, during a planned shutdown, an internal ultrasonic examination was carried out from kp 20.7 to kp 22.7. No cracking was detected.

AE Test

With examination of the operating pipeline downstream of the 10-km point having been by sample excavation, it was possible that a major undetected SCC defect existed in that section of the pipeline.

An acoustic emission test on the section of the operating pipeline most susceptible to SCC was therefore conducted. The test was successful with results indicating that no severe (near-failure) defects existed in that section of the pipeline. The maximum emission levels were about one-fifth of those which would be classified as serious.

Correlation and examination of all of the field data led to the following conclusions:

Significant SCC existed in the first 10 km of pipeline.
The size and frequency of occurrence of cracking in the section of the pipeline diminished in both directions with distance from the site of the failure.
SCC existed only in locations where the coating was disbonded, although not all disbonded coating was accompanied by cracking.
There was no cracking in locations having a high soil resistivity (above about 3000 Ω·cm) even where disbonded coating was present.
No cracking was located and probably no cracking exists downstream of the 15-kp probably because pipe-wall temperatures were substantially lower.
Minor cracking existed between 10 and 15 kp.

The decision to bypass the first 10 km and subsquently to replace the first 30 km was based upon the high probability that joints other than those associated with the failure contained stress-corrosion cracks.

Subsequent examinations confirmed this condition with typical groups of cracks found in various locations. The maximum crack lengths found in these various locations are indicated in Fig. 2.

The reduction in the maximum detected crack length with distance along the line was appreciably greater than the reduction in temperature with distance, as is apparent from a comparison of the two sets of data shown in Fig. 2.

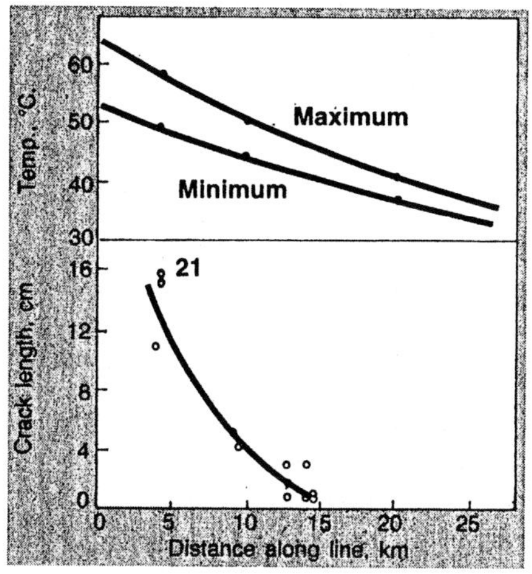

FIG. 2 Maximum crack lengths along pipeline.

Moreover, it is interesting to note that no cracks were detected closer to the inlet than 3.7 km, despite the higher temperatures in such locations.

Section Examinations

It is well known from experiences of stress-corrosion failures on pipelines in the United States that the cracks have a propensity for extending in the longitudinal direction along the pipe rather than in the depth or wall-thickness direction [2]. Such behavior was apparent at the region of the failure of the Moomba-Sydney line.

But in the hope of obtaining greater insight into this phenomenon, sections from other parts of this line where multiple cracking was detected have been examined in some detail. The length of each crack in a sample was measured after the successive removal of 0.125-mm cuts from the outer surface, the samples being previously flattened to facilitate this removal by grinding.

The crack measurements were further facilitated by the spraying of white paint onto the surface and then application of magnetic-particle inspection. These processes, together with the alternating metal removal, were continued until no cracks could be detected.

Figure 3 shows a plot of the maximum lengths of each crack from a number of samples against their depths, with a derived plot of the same data in Fig. 4. (Depth here refers to a dimension measured normal to the steel surface, whereas the cracks may have followed an oblique path, as indicated later.)

These figures illustrate that shallow cracks are initially relatively long (length-to-depth ratio, 10–20), after which the depth tends to increase with

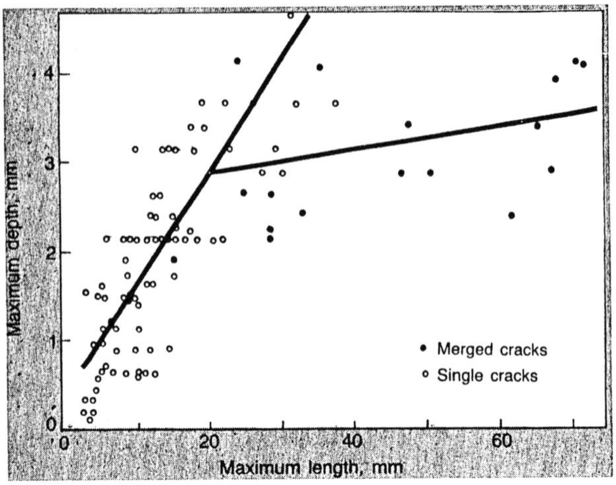

FIG. 3 Changes in crack length and depth.

less increase in length (length-to-depth ratio decreases from 10 to 5), until the depth reaches about 1.5 mm when the length increases again for small changes in depth.

It is possible that these changes in length-to-depth ratio initially reflect the easier deformation at a free surface, so that crack extension in the longitudinal direction occurs in preference to depthwise growth.

Stress concentration after this initial stage then results in depthwise growth, while the third stage, involving an increase in the length-to-depth

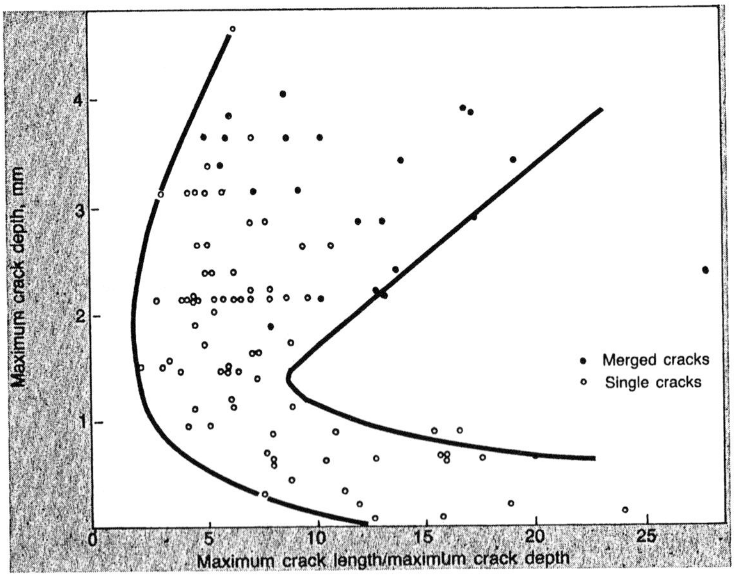

FIG. 4 Relationship of length/depth ratio.

ratio, relates to the joining of adjacent cracks. That the longer and deeper cracks were frequently formed by the joining of ajdacent cracks is shown by the crack profiles indicated in Fig. 5.

There the three longest cracks show the profiles to be expected from the joining of cracks, while the two shorter cracks show no such evidence. The plots of Figs. 3 and 4 indicate which cracks showed such evidence of joining.

Tapered-Specimen Tests

With the objectives of estimating threshold stresses and crack velocities, tapered tensile-specimen stress-corrosion tests were conducted on various steels that were incorporated in the original line or were available as candidate materials for the sections to be replaced [3].

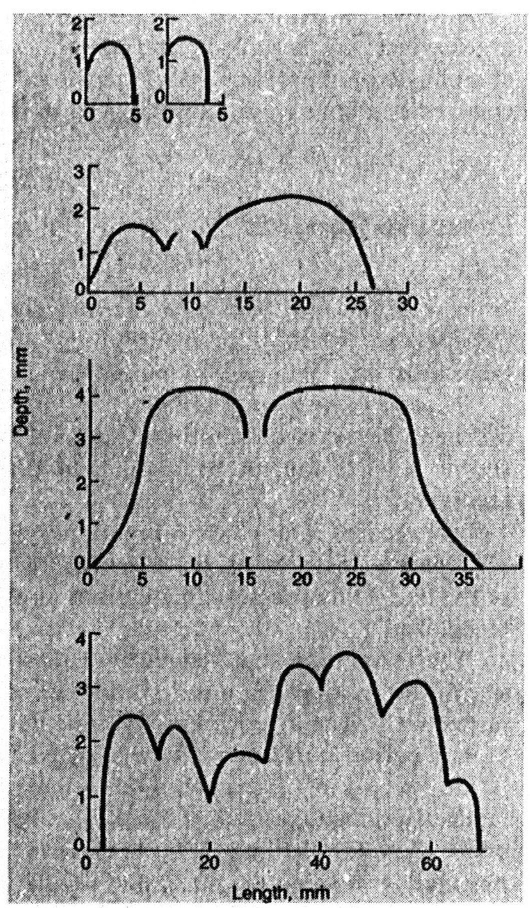

FIG. 5 Crack profiles: effect of joining for typical cracks.

The steels were tested with their surfaces in the conditions in which they were received, except for coating removal, where necessary, and degreasing. Where the specimens were cut from flat plate, their tensile axis was normal to the rolling direction. The stress-corrosion cracks then propagated in the direction that would obtain in practice.

The pipe sections were almost invariably tested with the axis parallel to the rolling direction to avoid the need to straighten the material and thereby possibly altering its cracking resistance.

The specimens were subjected to various cyclic loading conditions at a frequency of either 4.3×10^{-3} or 1.5×10^{-4} Hz. No differences were discernible in the results from specimens tested at these two different frequencies. The specimens were immersed in 1 *N* Na_2CO_3 + 1 *N* $NaHCO_3$, usually at 75°C and −650 mV (saturated calomel electrode; SCE), although some tests were conducted at lower temperatures.

Tests on material from the failed joint suggested that the sample available had suffered property changes during service or as a consequence of the failure, since tensile tests on substandard-size specimens indicated appreciably higher yield strengths than those given on the original mill certificates.

However, a feature of the stress-corrosion cracks produced in the laboratory tests on the failed joint material was that some cracks propagated at an angle of approximately 45° to the major tensile-stress direction. Such crack orientations were also observed in the region of the service failure.

Pressure Records

In view of the possibility that the material from the failed joint had been modified as a result of the failure, there is doubt surrounding the use of the laboratory data obtained on that material in considering the service failure.

Data from tests on a sample from a joint that had been in service but not near the service failure may be more reliable in this respect and are shown in Fig. 6 together with some data taken from the Moomba pressure records.

The records refer to the 6-month period immediately before the failure. That period is typical of the pressure fluctuations for most of the lifetime of the line. Only pressure fluctuations in excess of 20 kPa (2.9 lb/in.^{2}) were recognized.

The magnitude and frequency of these larger pressure fluctuations are shown in Fig. 6 by the location of the arrows and the associated numbers, respectively. The average pressure fluctuation in excess of 20 kPa in that 6-month period corresponded to a stress fluctuation of about 1.4 ksi (10 N/m^2; ksi = kips/in.^{2}; 1 kip = 1000 lb/in.^{2}), for which the laboratory data indicate a threshold stress of 52 ksi, that happens to coincide with the design hoop stress for the line. (The laboratory data were obtained at 75°C, appreciably exceeding the maximum temperature at the point of failure. But data presented later suggest that temperature variations in the relevant range have only a small effect upon threshold stress.)

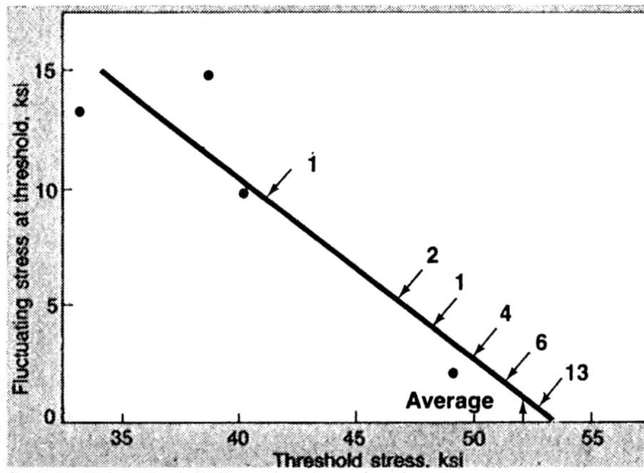

FIG. 6 SCC in steel MW 22.7. In 1 *N* Na_2CO_3 + 1 *N* $NaHCO_3$ at 75°C and −650 mV (SCE). Arrows indicate magnitude and frequency of pressure fluctuations on the Moomba-Sydney pipeline in the 6-month period before the 1982 failure.

While the average pressure variation suggests a threshold stress coincident with the design stress, there were 14 pressure fluctuations in the 6-month period prior to the failure which exceeded the average sufficiently to suggest that they could reduce the effective threshold stress below the value of 52 ksi, a point that will be discussed in Part 2.

Tapered-specimen tests on a range of steels available or used for the sections that had been or were to be replaced were conducted.

Figure 7 shows the data for one of these steels, while Fig. 8 shows the regression lines for other steels tested. The data points have been omitted for clarity, but their scatter about the regression lines was essentially similar to that shown in Fig. 7. The threshold stresses shown in Figs. 7 and 8 were calculated from the minimum cyclic load on the specimen at the position

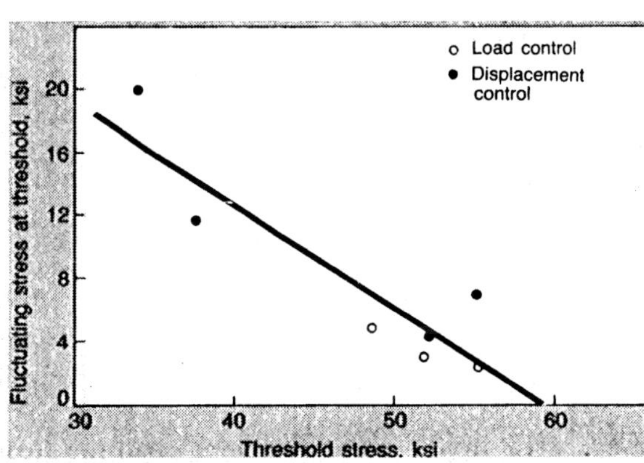

FIG. 7 Cracking of steel TPA 3. As an effect of the magnitude of the fluctuating stress component upon the steel's threshold stress; steel subjected to tests in 1 *N* Na_2CO_3 + 1 *N* $NaHCO_3$ at 75°C and − 650 mV (SCE).

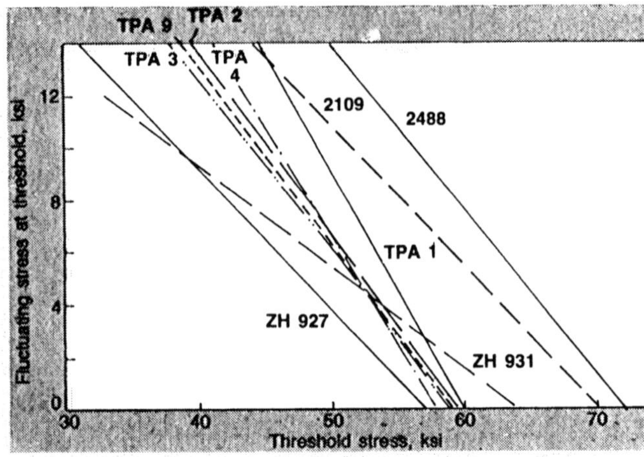

FIG. 8 Regression lines for various steels. For the effect of the magnitude of the fluctuating stress component upon the steel's threshold stress; steels exposed to 1 *N* Na_2CO_3 + 1 *N* $NaHCO_3$ at 75°C and 650 mV (SCE).

where cracking effectively ceased. To obtain the thresholds corresponding to the maximum cyclic load, the fluctuating stress at the threshold should simply be added to the thresholds calculated for the minimum loads.

Clearly this increases the steepness of the regression lines, and the thresholds corresponding to the mean cyclic loads would be midway between the regression lines for the minimum and maximum loads.

These various steels, which were all of the pearlite-reduced, controlled-rolled variety, had yield stresses ranging from 69.8 to 82.0 ksi, but there appeared to be no general relationship between SCC susceptibility, as measured by the threshold stress at zero fluctuating load, and yield strength for the various steels.

Nevertheless, for the average yield stress of about 75 ksi for these steels, the average threshold stress at zero fluctuating load is about 80% of the yield stress.

For a fluctuating stress of 10 ksi, the average threshold is reduced to about 60% of the yield stress.

Mill Scale

Two of the materials having the same compositions (ZH 927 and ZH 931) had mill scale on one side and were grit blasted to white metal on the opposite face.

The linear regression lines resulting from the measurements on the opposite faces of the specimens are shown in Fig. 9 and indicate a small shift to higher thresholds, at a given fluctuating stress, for the grit blasted as compared to the mill-scaled surfaces.

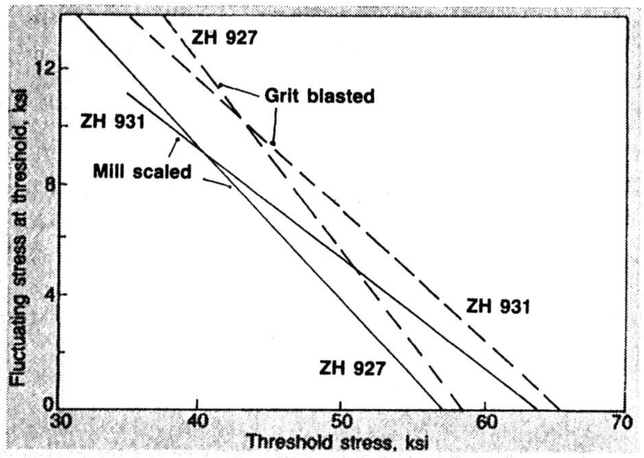

FIG. 9 Regression lines for different surfaces. For the effect of the magnitude of the fluctuating stress component upon the threshold stress for two steels exposed to 1 *N* Na_2CO_3 + 1 *N* $NaHCO_3$ at 75°C and −650 mV (SCE).

The grit-blasted surfaces displayed fewer cracks than those with mill scale present, although the maximum crack velocities showed no beneficial effect arising from the grit blasting.

In addition to these tests at 75°C, some tapered-specimen stress-corrosion tests were conducted on three of the steels at 50 and 40°C as more representative of maximum pipeline operating temperatures.

The cracking susceptibility may be regarded as being reflected in the threshold stress for cracking and the crack velocity above the threshold.

Earlier tests [4] on a pipeline steel in a carbonate–bicarbonate solution indicated an activation energy of about 10 kcal/mol (42 kJ/mol) for the crack velocity, but there are no data available in relation to the effect of temperature upon threshold stress, if any.

The results from slow strain-rate tests on steel ZH 927 at 40, 50, and 75°C gave an activation energy of 10.2 kcal/mol, in reasonable agreement with earlier determinations.

The effects of various fluctuating stresses upon the threshold stress at different temperatures for steels AL 395 and ZH 927 for a fluctuating stress of 10 ksi are shown in Fig. 10. Although only three results are available for each steel, it appears that there is an increase of about 3 ksi for each 10°C reduction in temperature.

Obviously, this is not a large change and is within reproducibility limits for threshold stress determination with tapered specimens. Such a change in threshold stress with temperature does not correlate with changes in yield stress with temperature.

The maximum crack velocities in those tapered specimen tests for these two steels showed an increase in velocity with incresing temperature, the regression line indicating an activation energy of 7.5 kcal/mol. Since only three temperatures were involved, it seems reasonable to regard that figure as approximating to the values mentioned earlier.

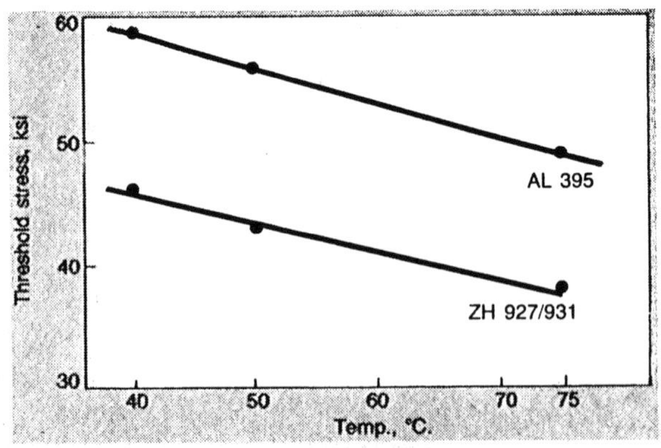

FIG. 10 Temperature effects on two steels. Effect upon the threshold stress for a fluctuating stress of 10 ksi for steels AL 395 (TPA 1) and ZH 927.

References for Part 1

1. T. Baker, Paper Presented at Australian Pipeline Industry Association Conference, October 1983.
2. R. L. Wenk, Paper T-1 Presented at the 5th Symposium on Line Pipe Research, American Gas Association, 1974, Catalogue No. L30174.
3. J. A. Beavers and R. N. Parkins, "Standard Test Procedure for Stress Corrosion Cracking Threshold Stress Determination," in *Corrosion '86,* NACE, Houston, Texas, 1986, Paper 321.
4. R. N. Parkins, Paper U-1 Presented at the 5th Symposium on Line Pipe Research, American Gas Association, 1974, Catalogue No. L30174.

T. N. BAKER
G. G. ROCHFORT
R. N. PARKINS

Part 2: Laboratory Tests on Crack Initiation

Laboratory tests of steel specimens from the original failed sections of the Moomba-to-Sydney, Australia, natural-gas pipeline and from replacement sections' steels focused on conditions promoting crack growth.

Line Integrity

At various stages following the failure on the Moomba-Sydney line, its repair and refurbishment, questions were raised about the integrity of the remaining parts of the line during its continued operation. Thus, during the building of the initial 10-km bypass and subsequently the 30-km bypass, it appeared likely that stress-corrosion cracks other than those which caused the failure would be present and indeed were shown to be so after the 10-km bypass was completed (Fig. 2).

The extent to which such existing cracks could propagate, and new cracks initiate, was difficult to judge. This was because of the changed operating conditions on the line (the temperatures, mean and fluctuating stresses being all lowered compared to the conditions prior to the failure) and because the laboratory data then available related to a higher temperature and higher stresses. Consequently, various laboratory test programs were undertaken involving temperatures and stressing the conditions similar to existing conditions on the line.

Simulation

Since the tests were designed to simulate operating conditions so far as possible, it was decided that they should be conducted in bending so that straightening of the specimens, with attendant changes in mechanical properties and possible stress-corrosion resistance, was not involved.

The material should be representative of the pipe still operating over part of the first 30 km and that had been in service at kp 22.7. The steel there was designated MW 22.7. Tests were conducted on specimens with and without precracks introduced to simulate the initiation and propagation of new and existing cracks at a temperature of 45°C and a potential of −610 mV (SCE), corresponding to the potential for maximum crack velocity at that temperature in 1 N Na_2CO_3 + 1 N $NaHCO_3$.

The loading conditions corresponded to a maximum stress of 50 ksi with load fluctuations of ±1.5% over intervals of 1 day following a triangular waveform. After the completion of each test on the initially plain specimens, the latter were sectioned along the centerline and prepared for microscopic examination whereby the maximum crack depth was measured.

The fatigue precracked specimens were immersed in liquid nitrogen and broken for examination by scanning electron microscopy to determine the extent of intergranular stress-corrosion crack growth following the completion of the test.

Decreasing Velocity

The results from the tests on the initially plain specimens are shown in Fig. 11, which indicates a decreasing crack velocity with increasing test time or numbers of stress cycles.

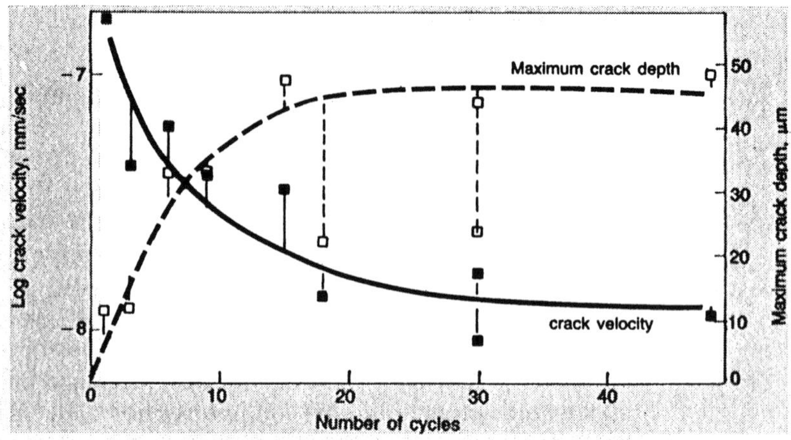

FIG. 11 Effect of stress cycles on crack depth. From tests on bend specimens of steel MW 22.7; initially plain mill-scaled surfaces.

In fact, the maximum crack depth probably reached a limiting size after about 15 cycles, as is apparent from Fig. 11; that is, the cracks stopped propagating. In two of the tests, one at 30 and the other at 18 cycles, the maximum crack depths were only about half the depth observed in other tests involving 15 or more cycles. This also probably indicates the cracks had stopped propagating.

In all tests, a relatively large number of stress-corrosion cracks were observed. While these were mostly confined to the central 5 cm of the specimen length over which the fiber stress was constant and maximal, some short cracks were also detected in the lower stressed regions.

Cracks ceasing to propagate after some period have been observed previously for pipeline steels in CO_3–HCO_3 solutions and have been shown to be due to the plastic strain rate or creep following initial loading, eventually falling to a value that cannot sustain cracking [1].

Of course, for higher mean stresses or larger fluctuating stresses the cracks either cease propagating at greater depths or after longer times. Or they continue to propagate until total failure occurs.

Nevertheless, for the particular loading conditions involved in these tests, and applicable to the then-operating conditions on the Moomba-to-Sydney line, it appeared unlikely that new stress-corrosion cracks were likely to initiate and propagate to a size that would result in a service failure.

Precracked Specimens

However, since it was possible that stress-corrosion cracks already existed in the line from the previous operating conditions, laboratory tests using precracked specimens were used to assess the likelihood of these propagating to give a service failure.

After completion of the first two tests involving precracked specimens, these were sectioned at the midpoint of the precrack with a view to measuring the extension of the precrack as an intergranular stress-corrosion crack. It was assumed that the crack growth would occur most readily in the region where the stress intensity factor was highest, that is, at the bottom of the 3.5-mm deep precrack.

No crack growth was detected in this region, despite evidence of intergranular SCC having occurred from the ends of the precracks where they emerged at the outer surfaces of the specimens.

At the outer surfaces, in fact, the cracks did not only extend in approximately straight lines from the tips of the precracks but also showed evidence of branching and following the lines of maximum shear stress, as was observed in much earlier work on the pipeline-cracking problem involving precracked specimens.

Since it appeared likely that stress-corrosion-crack growth would have extended for some distance down the precrack edges, the specimens were cooled in liquid nitrogen to facilitate fracture and the specimens broken open for subsequent examination by scanning electron microscopy.

In fact, all of the subsequent test specimens were so treated without any prior attempt at metallographic examination. The fracture surfaces had been cleaned of corrosion products with cathodic polarization to prevent attack upon the exposed metal before examination.

In all of the precracked specimens, it was found that extension of the precrack by stress corrosion was greatest near the surface where the precrack emerged, and the extent of intergranular stress corrosion diminished along the edge of the precrack as the depth from the outer surface increased.

The extents of intergranular crack growth from the two ends of the precrack in each specimen were very similar at the outer surface, but the depth to which such crack growth extended from the subsurface parts of the precrack sometimes showed marked variation between the two edges. Consequently, the data from these tests are presented as crack extension at the outer surface and the depth to which cracking occurred along the precrack edge, with two points for each measurement to show the differences between the observations for the two ends of the precrack.

Figure 12 shows the stress-corrosion-crack length from the precrack tips at the outer surfaces for various numbers of load cycles, together with the average crack velocity computed from the longer surface crack.

The length of the intergranular cracking at the surface increases with number of cycles, although the test involving only 1 cycle acquired a larger crack than expected from the other results and that tested for 34 cycles did not crack as much as had been expected.

The crack velocities from all of these tests average 1.9×10^{-7} mm/s and that may be compared with the average of all of the results on steel ZH 927 at 40 and 50°C in tapered specimen tests of 1.8×10^{-7} mm/s. As already mentioned, not only did the stress-corrosion crack extend from the precrack along the outer surface of the specimen, but the cracking also extended down into the specimen along the periphery of the precrack.

That extension also increased with number of cycles, as is apparent from

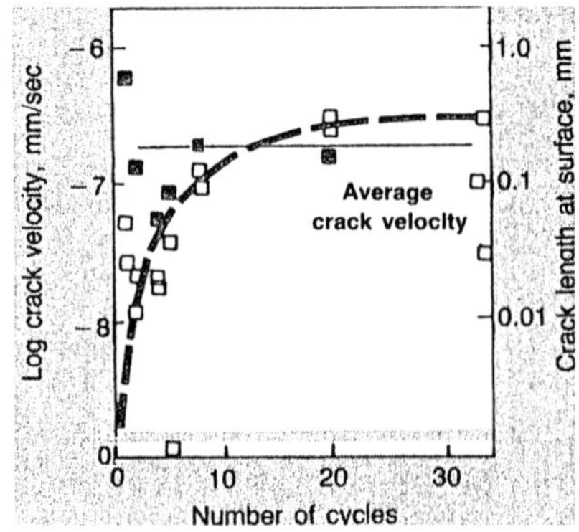

FIG. 12 Effect of stress cycles on crack length. From tests on bend specimens of steel MW 22.7.

Fig. 13, where again the test at 1 cycle gave more extensive, and that for 34 cycles less extensive, cracking than expected from the general trend.

Large Pressure Variations

These tests were all conducted under identical stressing conditions and involved fatigue precracks of essentially the same size, both related to conditions existing toward the Moomba end of the line.

Toward the Sydney end of the line, however, the pressure variations continued to be relatively large, although it appeared likely that the chances of the existence of stress-corrosion cracks in such regions would be appreciably lower than at the Moomba end in view of the markedly different temperatures and in the light of the data shown in Fig. 2.

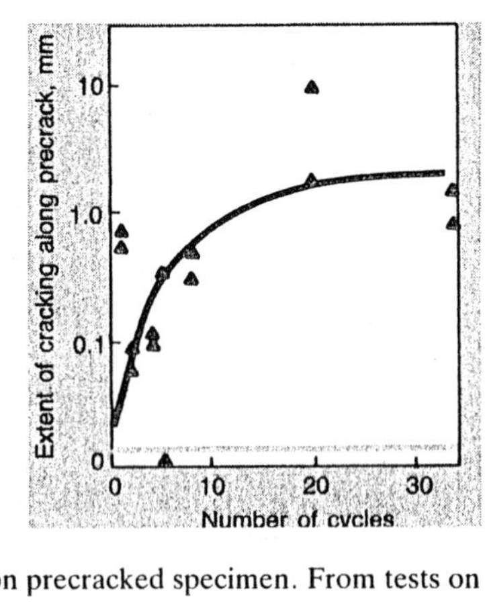

FIG. 13 Effect of cycles on precracked specimen. From tests on bend specimens of steel MW 22.7.

Nevertheless, it appeared prudent to conduct experiments on specimens containing precracks of various sizes and with stressing conditions simulating those toward the Sydney end of the line to determine the conditions under which the precracks would grow by stress corrosion.

The tests were conducted on a steel designated TPA 7. This steel had been stored since the building of the line and was representative of the pipe in service at the Sydney end of the line. The tests were also extended to include the material designated MW 22.7, as representative of the Moomba end.

The TPA 7 steel had been grit-blasted and coated with a zinc-rich paint on its outer surface. The latter was removed by further grit-blasting because it created problems in controlling the potentials of specimens during the test if it were left in place. The specimens were again tested in four-point symmetrical bending, to avoid straightening and possible modification of the SCC resistance, and were exposed to 1 N Na_2CO_3 + 1 N $NaHCO_3$ at 45°C. The frequency of loading was 1 cycle/d.

After completion of testing, the specimens were immersed in liquid nitrogen to facilitate their fracture and subsequently examined by scanning electron microscopy. With higher stresses, cracks invariably propagated from both ends of the precrack, but at lower stresses SCC was sometimes observed only at one end of the precrack.

The lengths of the crack extension at the outer surface of the specimen and the length of its progression along the precrack edge were measured, but for the crack velocities subsequently quoted, the longer of the extensions at the specimen surface was used. The appearance and shapes of the stress-corrosion cracks from such specimens were as indicated earlier.

Figure 14 shows the effect of precrack length upon the crack velocity (expressed as the crack extension at the surface of the bend specimen divided by the total test time) for both the TPA 7 and MW 22.7 samples loaded to various maximum stresses and fluctuating components of stress.

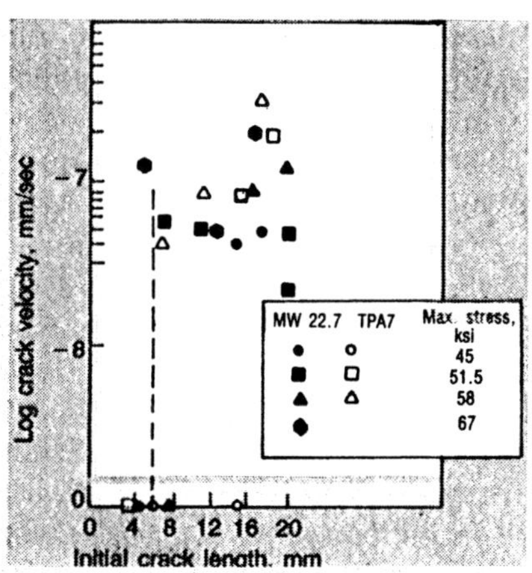

FIG. 14 Effect of precrack length on velocity. For various maximum stresses applied to bend specimens of TPA 7 and MW 22.7 steels.

For maximum stresses in the range 45 to 58 ksi (310 N/mm^2; ksi = kips/in.^{2}; 1 kip = 1000 lb/in.^{2}), it appears that precracks less than about 6 mm long are not likely to propagate, with some indications that at 45 ksi the minimum precrack size is appreciably greater and at 67 ksi probably less than that value.

The two different steels do not appear to behave very differently under the conditions of loading to which Fig. 14 refers. Similar tests were conducted involving smaller components of fluctuating stress, but with the same maximum stresses. More of these involved the MW 22.7 steel than the TPA 7 material because of the relative scarcity of the latter.

Plots similar to that of Fig. 14 gave limiting precrack sizes, and there was a general trend for the critical precrack size to increase with decreasing magnitude of fluctuating stress, as may be expected.

Data Treatment

There are a number of ways in which the data obtained in the tests upon precracked specimens may be treated.

One approach is to use stress intensity factors:

$$K = M\sigma\sqrt{a/Q}$$

where $Q = 1 + 1.464(a/c)^{1.65}$
σ = applied stress
a = maximum depth of the elliptical crack of length $2c$ at the surface
M = function of the geometry of the cracked specimen that can be obtained from tables in the paper from which the above expressions are taken [2]

The fluctuating component can then be expressed as ΔK, relating to the change in K between the maximum and minimum values for the corresponding applied loads.

The data from all of the tests upon precracked specimens of both steels, involving different K_{max} and ΔK values, are shown in Fig. 15 as a plot of K_{max} against crack velocity.

Although there is appreciable scatter, the broad trend is clear in that for the MW 22.7 steel there is a threshold value of K_{max} in the region of 19 ksi$\sqrt{\text{in.}}$ below which cracking was not observed, irrespective of the ΔK level, and above which cracking did occur with a trend toward higher crack velocities as K_{max} was increased. There is also a trend toward higher crack velocities the higher the ΔK level, as already mentioned.

With the exception of the one test at K_{max} = 18 ksi$\sqrt{\text{in.}}$, the results for the TPA 7 steel suggest that this may have a higher threshold for cracking, possibly in the region of about 21 ksi$\sqrt{\text{in.}}$

There is a single result shown in Fig. 15 which lies well beyond the broad

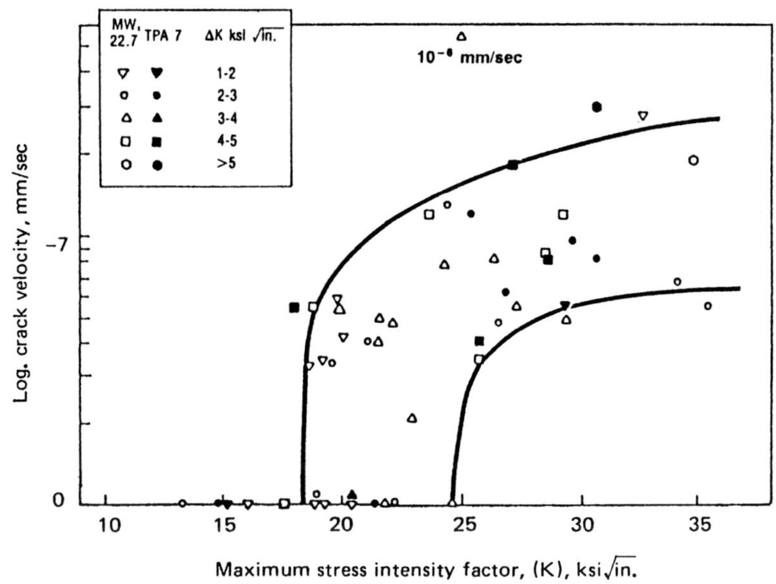

FIG. 15 Effect of stress intensity factor on crack velocity for various ΔK values for steels TPA 7 and MW 22.7.

trend of the results. This result refers to the test on the MW 22.7 steel at $K_{max} = 25\ \text{ksi}\sqrt{\text{in.}}$ which gave a crack velocity of 10^{-6} mm/s, i.e., about an order of magnitude higher than expected from the general trend of the results.

This specimen was odd in that the extension of the stress-corrosion crack along the surface was by about 600 μm, yet the progression of the crack in the depth direction along the precrack edge was only about 100 μm. Moreover, stress corrosion occurred at only one end of the precrack.

Almost invariably, for all other specimens, the crack extension at the surface was less than the depthwise progression of cracking along the precrack edge, so that this particular specimen was unusual, possibly because of some local metallurgical feature at one end of the precrack.

Nevertheless, it is a somewhat disturbing result, the equivalents of which have occasionally been seen in other tests on pipeline steels, since it represents a problem in predictability that could be of significance in the context of service failures.

Critical Crack Size

The broader implications of the above results are considered later, but what emerges from the data is that the critical crack size for stress-corrosion crack propagation under the stressing conditions that can exist toward the Sydney end of the line is in the region of 6 mm and that this dimension increases as the severity of the stressing conditions is diminished.

If crack growth does occur, it appears likely that the velocity will not be much higher than about 10^{-8} mm/s for much of the time, although that velocity would increase as the crack lengthened appreciably and the local stress intensity increased.

Large Stress Cycles

All of the laboratory tests involving cyclic loading of specimens involved repetition of the particular loading conditions chosen for a particular test. Consequently, nothing is known about the effects of superimposing, say, infrequent large stress cycles upon smaller repetitive cycles, as may happen in the service of a pipeline due to outages.

A program of experiments was therefore conducted which subjected specimens to small daily stress fluctuations for 6 days followed by one cycle in which the magnitude of the stress cycle was increased for 1 day. That weekly sequence was then repeated for a further 3 weeks in each test. The variable from test to test was the magnitude of the large stress fluctuation.

The repetitive stress cycles for 6 successive days involved changes from 46 to 47 to 46 ksi, followed by a single cycle from 46 ksi to some lower stress and back to 46 ksi in one day.

The minimum stress on the seventh day was varied from test to test, being 45, 44.1, 40, 38.4, 37.4, 35.5, or 34 ksi. In addition, two tests were conducted in which the stress cycle each day throughout the 4-week test was from 46 to 47 to 46 ksi, i.e., with no larger stress fluctuation to provide some baseline data.

The quantities used to assess the results from these tests were those of the maximum crack depth and the total number of cracks observed, including both faces of the specimen. Since the tests were related to the conditions that may occur in the new 30-km bypass, they were conducted on material from that location (designated steel 2109). The material was provided in the form of sections from pipe, and specimens for loading in tension were cut with the tensile axis parallel to the longitudinal axis of the pipe.

The specimens were parallel sided, with 5-cm gauge length and of full plate thickness, and were tested with the surfaces in the condition in which the material was received, except for cleaning in hot acetone just prior to testing.

The test conditions involved the usual carbonate–bicarbonate solution at 45°C and −610 mV (SCE). At the conclusion of each test, the specimen was cut longitudinally along the center line and prepared for metallographic examination.

In this examination the depth of each crack in the exposed part of the gauge length was measured on both surfaces of the specimen subjected to the crack environment.

The maximum crack depth and the total number of cracks were used for assessing the cracking response as a function of the stressing conditions.

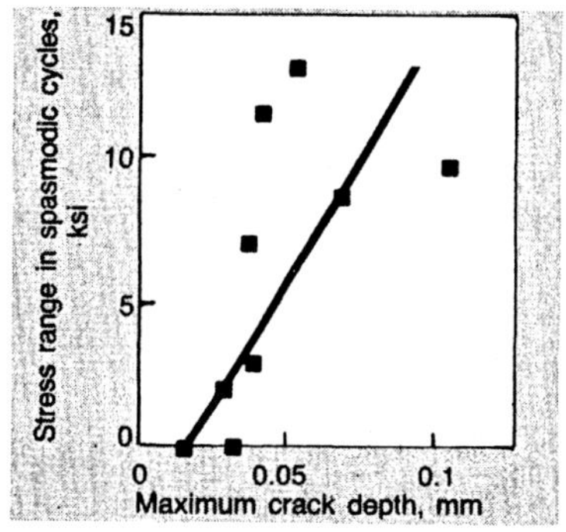

FIG. 16 The magnitude of spasmodic stress cycles between smaller stress cycles and the effect on crack depth detected in initially plain specimens of steel 2109.

Figures 16 and 17 show the results obtained for the maximum crack depth and crack numbers as a function of the magnitude of the large stress fluctuation applied each seventh day. The crack numbers show the more systematic trend, increasing markedly with increasing magnitude of stress fluctuation. The scatter in the data relating to the maximum crack depth is appreciably greater than for the crack numbers. Again, however, there appears to be a broad trend for deeper cracks to form the greater the magnitude of the stress fluctuation, depending upon the weight given to the results for the two highest fluctuating stresses or those for the two next highest stresses, which produced appreciably deeper cracks.

Perhaps the more conservative data are the more appropriate in the circumstances of dealing with a pressure vessel, but in any case there is the clear indication that the numbers of cracks increase markedly as the stress fluctuations increase in magnitude.

It is now reasonably clear that, even with the carefully controlled conditions associated with laboratory tests, stress-corrosion cracks continue to

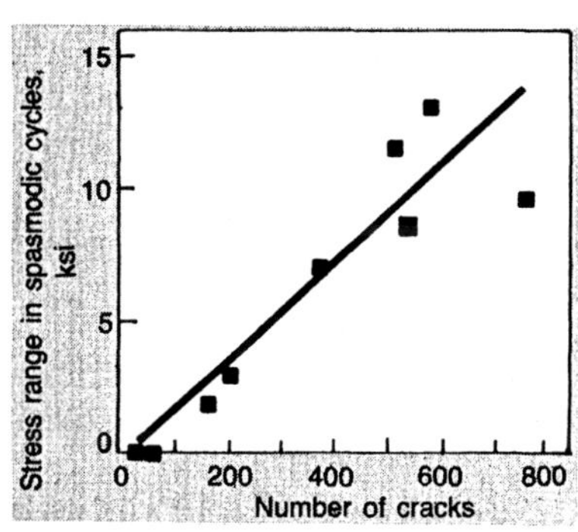

FIG. 17 The magnitude of spasmodic stress cycles between smaller stress cycles and the effect on the number of cracks detected in initially plain specimens of steel 2109.

initiate with the passage of time and that probably the vast majority of these cracks cease to propagate after extending for relatively short distances.

It appears likely that the more cracks that are initiated, the greater will be the chances of some continuing to propagate over markedly greater distances than the remainder. The results in Figs. 16 and 17 may simply be a reflection of such, since it is clear from the two figures that there is an approximate relationship between the maximum crack depth and the number of cracks observed.

Clearly, it would be preferable to have more data relating to this matter of spasmodic, larger stress fluctuations, but tests each lasting 4 weeks are not conducive to collecting data rapidly.

Nevertheless, the avaiable data point to the deleterious effects of large pressure excursions; the more these can be reduced, the better.

References for Part 2

1. R. N. Parkins, Paper U-1 Presented at the 5th Symposium on Line Pipe Research, American Gas Association, 1974, Catalogue No. L30174.
2. X. R. Wu, *Eng. Fract. Mech., 19,* 387 (1984).

T. N. BAKER
G. G. ROCHFORT
R. N. PARKINS

Part 3: Failure Prediction

Much of the recorded data concerned with realistic temperature and stressing conditions were obtained with a view to assessing the risk of another failure of the Moomba-Sydney pipeline. A test of those data would be to determine to what extent they could predict the original failure, about which much is known.

It has already been indicated in the context of Fig. 6:

1. That the relatively large spasmodic pressure variations to which the line was subject probably reduced the threshold below what it would have obtained in their absence.
2. But that the threshold for cracking was virtually coincident with the operating conditions when the latter did not involve large pressure excursions.

In any case, it is obvious from the fact of the failure that the stressing conditions exceeded the threshold, and predictability is therefore concerned with the time to failure, i.e., slightly less than 6 years, making use of relevant crack-velocity data.

Based on various laboratory studies, it now appears that stress-corrosion-crack velocity varies with time in something of the manner shown in Fig. 18. Unlike other models, it recognizes a period (Stage 1) in which the conditions for cracking are established, involving coating deterioration, generation of the cracking environment, and establishment of a cracking potential.

Once these conditions are met and if the stressing conditions exceed the relevant threshold, multiple cracks are initiated, but their growth rate rapidly diminishes (Stage 2), as indicated by the data of Fig. 11. This stage is probably of short duration, perhaps 2 or 3 weeks, and can be neglected by comparison with Stage 3. During that stage, the cracks will grow to much greater lengths despite the lower crack velocity, likely to be of the order of 10^{-8} mm/s.

In any case, experimental values for the crack velocity in Stage 3 will incorporate the higher values in Stage 2. When the cracks have reached about 6 mm in length (Fig. 14), the velocity will increase in Stage 4 to a value of about 2×10^{-7} mm/s (Fig. 15).

During Stage 4, however, or the latter parts of Stage 3, it is likely that cracks begin to join up (Figs. 3, 4, and 5) and account needs to be taken of this phenomenon.

In addition, it needs to be remembered that cracks are likely to extend simultaneously in both directions along the pipe so that, say, a 6-mm-long crack involves 3 mm of growth in each direction.

Finally, for the Moomba-Sydney line failure, it is known that the stress corrosion crack reached critical size for fast fracture at a length of about 210 mm.

The time for crack initiation (Stage 1) is not known and will be treated

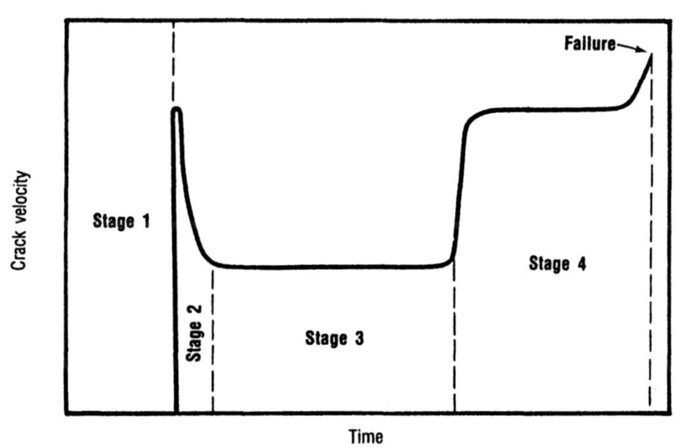

FIG. 18 Crack velocity variations during the lifetime of pipeline by stress-corrosion cracking.

as a variable. Stage 3 is the time for cracks to grow 3 mm, or to reach a total length of 6 mm, and the crack velocity in this stage will be allowed to vary from 3×10^{-8} mm/s to a maximum value of 2×10^{-7} mm/s. The latter value corresponds to the crack velocity in Stage 4.

In this last stage it will be assumed that cracks join up, the number of cracks so involved being treated as a variable, the crack velocity being 2×10^{-7} mm/s, and the final total crack length being 210 mm.

Thus,

$$\text{Time to failure} = \text{Stage 1} + \frac{3}{\mathrm{CV}_3} + \left(\frac{210}{n} - 3\right)\frac{1}{\mathrm{CV}_4}$$

where CV_3 and CV_4 = the crack velocities in Stages 3 and 4, respectively
n = the number of cracks that join

Critical Role of Coalescence

The times occupied in initiation (Stage 1) for various crack velocities in Stage 3 and number of cracks joining in Stage 4 are shown in Fig. 19. This indicates that the joining of cracks plays a critical role in the failure because when the number of cracks joining is less than about 7, the crack velocities in Stage 3 would need to approach those in Stage 4, and the initiation time would be extremely small.

Indeed, it is likely that if only one or two cracks were involved, the wall thickness would be penetrated before the crack length had reached critical size.

In other words, a leak rather than a rupture would be the result because there are various indications that the crack velocity through the wall is about

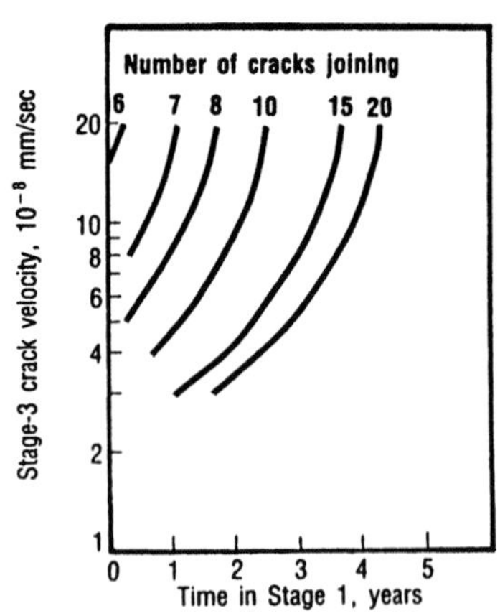

FIG. 19 The effect of the number of cracks joining to produce final failure and of various crack velocities during Stage 3 (Fig. 18) propagation upon the time for crack initiation when the final fracture is at 6 years.

four times less than the velocity in the longitudinal direction (Figs. 3, 4, and 5).

For a crack velocity of 2×10^{-7} mm/s throughout, therefore, the 8.3-mm wall would be penetrated in about 5.3 years when the crack length would only be about 16.6 mm, well below the critical size for fast fracture.

In fact, somewhat similar arguments lead to the conclusion that failure is unlikely to result from the joining of relatively large numbers of cracks (10–20), since these would only be associated with cracks some 5.5 to 2.7 mm deep when their total lengths had reached 210 mm and fast fracture would again be unlikely. Clearly this only leaves something like seven or eight joining cracks as realistic, which would be associated with crack depths of the correct order and would suggest an initiation time of about 0.5 years for probable crack velocities in Stage 3.

To obtain a total failure time of 6 years indicates that Stage 1 occupied about 0.5 years, Stage 3 about 1.5 years, and Stage 4 about 4 years, which seem reasonable.

It would appear that this would remain so even if an attempt were made to allow for the fact that the cracks frequently propagated obliquely rather than normally to the pipe surface, since the crack depth measurements, ignored this aspect of crack orientation.

Reasonable Success of SCC Tests

The various laboratory stress-corrosion tests performed were aimed at assessing the integrity of the line following the service failure. The reliability of the laboratory data in this respect was tested by determining the extent to which the data could predict the service failure. The general conclusion is that it is reasonably successful in this respect.

Thus, Fig. 6 indicates that the threshold stress for SCC, based upon the average pressure fluctuations, is coincident with the design hoop stress for the line, i.e., leaving no margin for safety.

Moreover, the spasmodic larger pressure cycles to which the line was subject would almost certainly result in the threshold stress being reduced below the design stress, as indicated by the results in Fig. 6 and supported by the laboratory tests involving spasmodic load cycles and depicted in Figs. 16 and 17.

While the stressing conditions to which the line was subject were demonstrated as likely to be conducive to SCC, the question remained as to whether the crack velocities involved would predict the failure of the line in a little under 6 years from its being brought into service.

The major difficulty with regard to predictability is that the time at which stress-corrosion cracks were initiated is not known.

However, for realistic average crack velocities of about 6×10^{-8} mm/s until cracks about 6 mm long were present and a crack velocity of about 2×10^{-7} mm/s (Figs. 12 and 15), thereafter the latter stage would occupy about 4 years and the former 1.5 years. These times would leave about 0.5 year as the time for cracks to initiate in a total lifetime of about 6 years.

These times appear reasonable and relate to the merging or coalescence of eight cracks during the final stage of crack growth.

This number is known to be realistic from examination of the fracture surfaces associated with the service failure.

The simple analysis performed in arriving at these numbers is of general significance in demonstrating that crack coalescence plays a critical role in producing a pipe rupture. Without such coalescence, a leak would have occurred before the conditions for rupture were achieved.

Further Failures

The prognosis from this analysis is that further failures were probably inevitable if the repair of the line had been restricted to the damaged joints or even to the first 10 km that were initially replaced. This conclusion was substantiated by the subsequent discovery of stress-corrosion cracks beyond 10 km (Fig. 2).

It is possible that stress corrosion cracks exist beyond the first 30 km of line now replaced, but until an inspection procedure is available for detecting them it appears reasonable to assume that any cracks will be small and that their continued propagation would be diminished, if not arrested, by modification of the operating procedure.

Changes in operating procedures to reduce temperature, mean and fluctuating processes, resulted in laboratory tests being conducted under conditions simulating these conditions. Indications are (Fig. 11) that if stress-corrosion cracks are initiated, they are not likely to propagate far (perhaps about 1 mm for a depth of 0.05 mm, Figs. 11 and 4) before ceasing to propagate.

For existing cracks to propagate under the reduced operating conditions, their lengths would need to exceed about 6 mm (Fig. 14).

Figure 2 suggests that such cracks are unlikely to exist beyond the 30 km of replaced pipe.

For the new pipe, all reasonable precautions were taken to reduce the risk of SCC by ensuring that the steel used was not markedly susceptible to stress corrosion, grit blasting its surface before applying a modern coating system, and subjecting it to lower temperatures and lesser mean and fluctuating pressures than obtained with the line that failed in service.

T. N. BAKER
G. G. ROCHFORT
R. N. PARKINS

Replacement Projects, Risk Analysis

A method has been developed for determining replacement priority among aging petroleum and natural-gas pipeline sections. The method, based on risk-analysis principles, aims at optimizing capital investment in the replacement project and ensuring public safety.

Favorable Record

The pipeline industry can boast of a favorable safety record compared with other transportation modes such as trucking, railways, and airlines. Nevertheless, several factors in recent years have contributed to increased emphasis on safety.

Public awareness of the potential consequences of accidents in the energy and chemical industries has put increased pressure on companies to maintain even better safety records.

Recent publicity has highlighted accidents in gas-transmission facilities that have resulted in significant damage and even loss of life. Such events prompt people to check "in their own backyards" for potential dangers to their communities.

Along with public awareness of safety issues comes public willingness to use the legal system to recover losses and assess punitive damages against operating companies involved in industrial accidents. Increasing the level of safety in a pipeline system can therefore decrease potential company liability.

The safety of a pipeline system depends upon the integrity of the components that make up the system. Developments in materials and construction technology have made new pipelines safer than in the past. Consequently, many existing pipeline systems fail to meet today's standards because the original codes have been updated to reflect technological advances.

By systematically upgrading higher-risk components of an existing pipeline system to meet current codes, operating companies not only increase system safety but can decrease operating and maintenance costs and reduce potential liabilities.

Pipeline Replacement

Many pipeline systems in use today, especially gas-distribution systems in large cities, utilize pipe which has already served well beyond its design life.

Although old sections of pipeline may be operating satisfactorily, those that do not meet current codes should be replaced. In addition, pipeline sections in areas that have undergone development and experienced changes in population densities (changes in construction zone classification) should be considered for replacement.

To bring a pipeline system up to code, large investments of capital and manpower are necessary [1]. In most cases these large expenditures require an operating company to schedule pipeline replacement systematically over an extended period of time.

Many companies have already taken the first step and begun to replace pipe according to schedules which call for complete replacement of older pipeline sections within 15 or 20 years [2, 3]. These companies can expect to increase the safety of their pipeline systems while decreasing long-term costs by replacing the most troublesome pipeline sections first.

A responsible analysis of a pipeline system requires information on each pipeline section of concern. In most cases these data are readily available.

However, in some cases the data necessary to conduct a thorough pipeline replacement analysis are distributed over several sources such as operating maps, alignment drawings, company contract documents, and operating reports.

The collection of data for older pipeline sections creates an immediate problem for many companies.

Operating company personnel are responsible for the day-to-day operation of their pipeline system and do not generally have the time to devote to labor-intensive work such as data collection. In many cases, bringing in an outside engineering contractor to do the work represents the most cost-effective solution.

Depending on the size of the company's system, a research program can take anywhere from 3 months to 2 years to complete.

Database

Current capabilities of microcomputer hardware and software enable pipeline replacement analyses for most companies to be conducted with personal computers. For most pipeline companies the investment in such computer equipment would be minimal because the required hardware and software are readily available.

One of several database management programs now on the market could be used to maintain a database on a pipeline system. The flexibility inherent in these programs allows for many forms of data analysis to be conducted.

Indeed, though the primary purpose of this data acquisition is for pipeline replacement analysis, the fact that pipeline information is already in a database allows a company to use it for many other analytical and economic purposes as well.

For the purposes of data collection and analysis, the database would be comprised of one entry for each pipeline segment in the system. A segment is defined as a section of pipeline for which all characteristics to be considered in the replacement analysis are constant for the length of the segment.

For a distribution system, a line segment may range from a few feet to several thousand feet in length. For a transmission system, a line segment can often be several miles in length.

In either case, each segment is identified as having a certain set of characteristics and will be considered as a unit in the analysis.

For each line segment in the database, information is collected in three categories.

The first category is devoted to information regarding the physical characteristics of the pipe itself. These data include pipe age, operating and design pressures, diameter and wall thickness, material of construction, and specification, joint, weld and longitudinal seam types, cathodic protection and soil data, strength test data, coating type and condition, and leak history.

The second category includes information regarding the relationship between the line and nearby population centers. These data include population density, types and proximities of nearby structures, and the type of product being transported in the line.

The final category contains pipeline location references such as operating map coordinates, street and intersection names, and the like.

More items can be easily added to the database system as required. The database should be a dynamic source of information on a company's pipeline system and as such can serve many purposes besides that of pipeline replacement analysis.

Priority

Once data have been collected into a database system on a microcomputer, they can be analyzed with the principles of risk analysis to determine optimum pipeline replacement sequence.

Pipeline-system risk analysis deals with the determination of probability of pipeline leakage and related effects to the public. Replacement-priority value may be defined as the product of the probability of a pipeline leak and its associated effects.

For example, a moderate-risk value may result from relatively high probability of leakage and relatively low impact or vice versa.

Determining the leak probability of a given pipeline segment depends upon several characteristics of the pipe as described earlier. The overall leak probability is actually the sum of several factors, each of which depends upon a particular characteristic of the pipe in question and is given a certain weight within the overall calculation.

Although the term "probability" is used, the value for leak probability is only a relative number to be used in comparing pipeline segments with one another.

For each factor of the leak-probability calculation, a certain characteristic of the pipe is examined and the factor assigned a value depending upon the result of the examination. Equation (1) illustrates this relationship:

$$P_k = f(D_k) \tag{1}$$

where P_k = leak probability factor for pipe characteristic D_k

For example, the coating factor would be given one of several predetermined values depending upon the type of coating used on the pipeline segment.

Once the value is determined, the factor is multiplied by its respective weighting factor and added into the overall leak probability:

$$P = \sum_{k=1}^{m} P_k W_{pk} \tag{2}$$

where P = overall leak probability
m = number of pipe characteristics examined in the analysis
W_{pk} = weighting factor for each characteristic

Figure 1 illustrates this procedure. Figure 2 illustrates the distribution of typical weighting factors for the leak-probability variables.

Calculation of a leak impact for a pipeline segment is analogous to that for leak probability. The leak impact, however, is derived from characteristics related to the location of the pipeline segment with regard to the proximity of structures and population centers and to the type of product being transported in the pipeline. Equation (3) illustrates this relationship:

$$I_k = f(D_k) \tag{3}$$

where I_k = leak impact factor for location or product characteristic D_k

For example, a given structure at a given distance from the pipeline segment is assigned a certain structure factor. In turn, the structure factor is multiplied by the structure weighting factor and added into the overall impact value:

$$I = \sum_{k=1}^{n} I_k W_{ik} \tag{4}$$

where I = overall leak impact
n = number of location or number of product characteristics examined in the analysis
W_{ik} = weighting factor for each characteristic

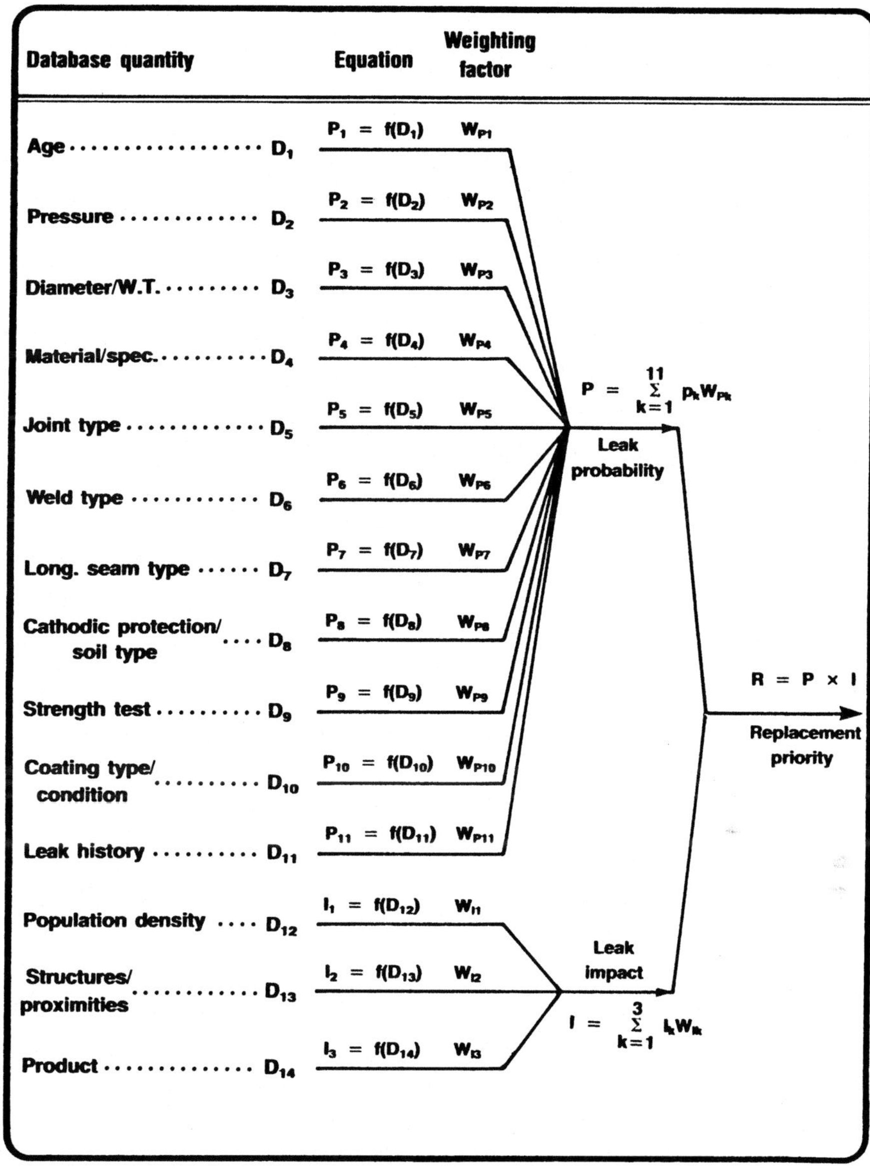

FIG. 1 Analysis procedure.

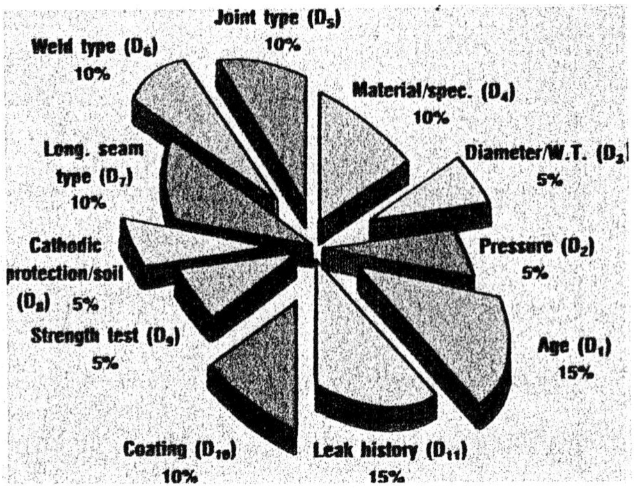

FIG. 2 Weighting factors for probability.

Figure 3 illustrates typical leak-impact weighting factors.

Leak probability and impact functions, represented by Eqs. (1) and (3), and all weighting factors are determined by a combination of statistical analysis and experienced engineering judgment. Several studies have been conducted which indicate which types of pipeline failure are most prevalent [4–6].

Once the leak probability and leak impact values are determined for a pipeline segment, they are multiplied together to produce the pipeline replacement value.

The overall replacement priority value for the pipeline is calculated as follows:

$$R = P \times l \tag{5}$$

where R = overall pipe replacement priority value

This value is a measure fo the perceived importance of replacing a given line segment relative to other segments being considered for replacement. Although many factors can be included in a systematic pipeline replacement analysis, a final determination of which replacement candidates actually to replace should include other appropriate criteria as well.

With prudent planning and budgeting as well as close coordination with regulatory bodies, other operating utilities, and right-of-way users, workable replacement programs can be set up which over a period of years will replace old or troublesome pipeline segments.

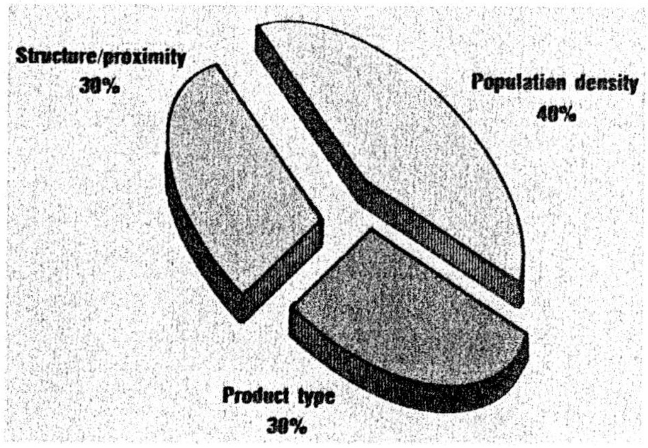

References

1. J. Watts, "1987 Construction Forecast," *Pipeline Gas J.*, pp. 20–22 (January 1987).
2. J. Watts, "Gas Utilities Focusing on 3Rs," *Pipeline Gas J.*, pp. 14–16 (December 1985).
3. J. Watts, "3Rs Keeping Utilities Busy," *Pipeline Gas J.*, pp. 12–14 (December 1986).
4. R. Eiber, D. Jones, and G. Kramer, "Outside Forces Cause Most Natural Gas Pipeline Failures," *Oil Gas J.*, pp. 52–57 (March 16, 1987).
5. Battelle's Columbus Laboratories, *An Analysis of Reportable Incidents for Natural Gas Transmission and Gathering Lines, 1970 through 1981,* Report to Pipeline Research Committee of the American Gas Association, NG-18 Report No. 139, February 1984.
6. *Pipeline Statistics,* U.S. Department of Transportation, Washington, D.C., 1983.

C. E. MYRICK
H. J. VAN DYKE
G. R. MAYER

10

Pipeline Economics and Costs

Pipeline Economics and Costs*

Operations improved in 1988 for United States regulated petroleum pipeline companies.

In almost all categories tracked by the Annual *Oil & Gas Journal* Pipeline Economics Reports, data from annual reports filed with the U.S. Federal Energy Regulatory Commission (FERC) showed improvements or were little worse than flat when compared with earlier data.

Most noticeable was the better than 9% increase from 1987 in net income for all (major and nonmajor) natural-gas transmission companies on operating revenues that rose for the first time since 1983 (Table 1).

Underlying this growth was a surge of more than 21% in volume of gas transported for others along with a slight increase in sales, the latter the first of the 1980s. This growth is reflected in the reversal in operating revenues which have been falling steeply since 1983 (Fig. 1).

TABLE 1 Pipeline Company Revenues, Incomes[a]

	Gas		Liquids	
	Operating revenue, $1000	Net income, $1000	Operating revenue, $1000	Net income, $1000
1980	48,128,791	3,098,018	6,356,037	1,912,411
1981	57,792,315	3,810,566	6,678,039	2,030,667
1982	64,850,548	3,566,041	7,140,469	2,162,447
1983	66,261,738	3,565,776	7,472,286	2,352,748
1984	64,006,035[b]	4,055,700[b]	7,823,582	2,544,764
1985	56,252,921[b]	5,190,817[b]	7,460,847	2,430,845
1986	40,664,635[b]	1,653,010[b]	7,287,062	2,051,372
1987	31,596,171[b,c]	1,438,203[b,c]	7,057,264[c]	2,475,411[c]
1988	32,736,380[b]	1,572,609[b]	6,861,938	2,504,986

[a]Source: U.S. FERC annual reports by regulated interstate natural-gas and liquids pipeline companies.

[b]Major and nonmajor companies under the FERC's classification system for natural-gas pipeline companies effective beginning with the 1984 reporting year. See Table 5 for definition of major and nonmajor companies.

[c]Revised from initially published data.

Editor's Note*: These data are updated each November by the editor of *Oil & Gas Journal*. The reader is encouraged to see the latest tabulation. Also, the cost data in this article should be updated by using the current cost indexes. The latest cost indexes available at the time of the printing of this volume are listed on page **xvi in the front matter.

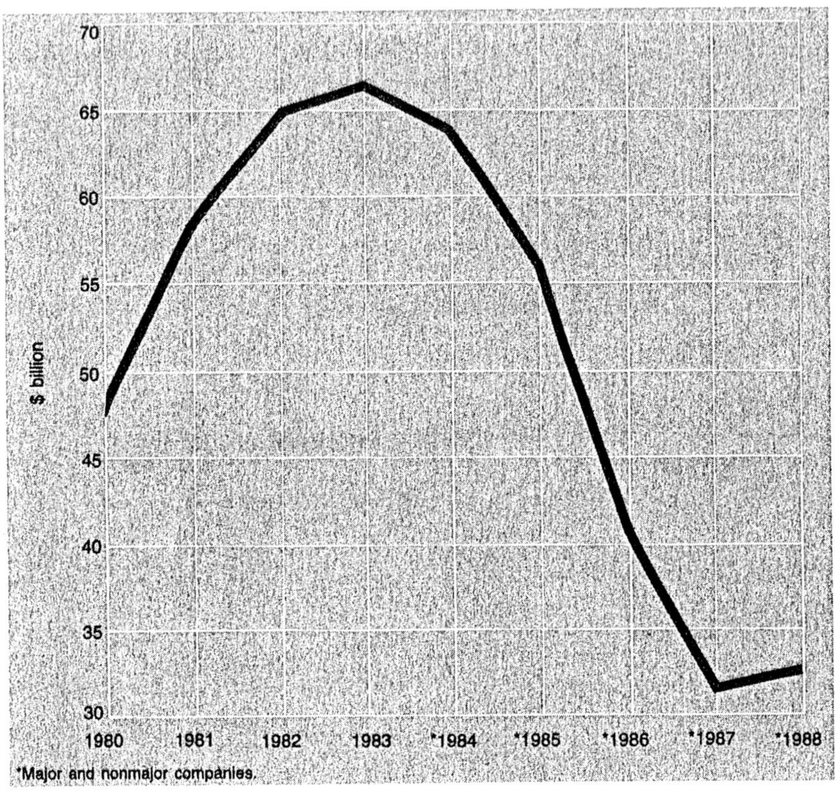

FIG. 1. Operating revenues: Natural-gas pipeline companies. (Source: U.S. FERC annual reports for natural-gas pipeline companies, 1980–1988.)

Although regulated common-carrier petroleum liquids (crude oil and product) pipelines showed less growth than natural-gas companies, they nonetheless held their ground in all categories. The strongest growth was better than a 3% increase in product-pipeline trunkline traffic. Overall traffic grew at slightly less than a 3% clip.

On the construction front, United States projects proposed for the 12-month period ending June 30, 1988, although off from that for the previous 12-month period, was nevertheless strong with more than 2400 miles of line proposed, 386 of which was targeted for offshore.

Bucking the Trends

Data from the 1988s FERC annual reports indicate that declines evident throughout this decade may be slowing, if not ending.

Net incomes for regulated major interstate natural-gas pipeline companies improved by 26.6% in 1988; operating revenues were up 1.45%. Net incomes for combined major and nonmajor gas-pipeline companies rose 9.35%; operating revenues were up 3.61% over 1987.

Operating revenues for major and nonmajor natural-gas transmission companies had been falling since 1983. Similarly, net incomes fell in 1983 and for 3 years beginning with 1985.

Petroleum liquids pipeline companies' net incomes in 1988 were up 1.2% on operating revenues that declined 2.8%. That decline is part of a trend begun in 1985. Net incomes' rise continues a rise begun in 1987.

Total pipeline mileage reported by major natural-gas companies was flat from 1987 to 1988 despite a 3.3% increase in reported transmission mileage. The same pattern held for combined major and nonmajor pipeline companies; total mileage growth was essentially flat at 0.5% despite a 2.83% increase in transmission mileage for both categories of companies.

The addition of transmission mileage nonetheless represented an increase in the nation's major gas-transmission system of almost 6000 miles.

Liquid pipeline companies showed virtually no change (+1.5%) in total pipeline mileage utilized during 1988 compared with 1987. Reflecting greater United States product demand but declining crude-oil production, mileage for product pipelines showed the largest gain at 2322 miles (+3%).

Total mileage reported for 1988 was 451,838 with 281,381 devoted to gas transmission, field, and storage, while 170,457 was for crude oil and product pipelines (Table 2).

Construction Pace

For the 12-month period ending June 30, 2410.37 miles of pipeline construction were proposed to the FERC, of which 386.59 miles were planned for offshore Gulf of Mexico (Table 3).

This total mileage was proposed in 105 construction projects, 21 of which were for offshore.

For the 12-month period July 1, 1987, to June 30, 1988, 5519.75 miles of pipeline were proposed of which 401.3 miles were planned for offshore.

Oil & Gas Journal's Worldwide Construction Report (p. 78, October 16, 1988) lists 70 individual pipeline projects totaling more than 9000 miles of United States pipeline construction reported by surveyed companies as either proposed, planned, or under way.

By comparison, *Oil & Gas Journal's* October 24, 1988, Construction Report listed 83 United States pipeline projects proposed, planned, or under way, totaling more than 11,400 miles.

Natural-gas pipeline projects reported in 1989's Construction Report total 8340 miles of pipeline. The report of 1988 listed gas-pipeline projects totaling more than 8700 miles.

Prospects for some of these lines being built are better now than at any time in the last 5 years.

Nonetheless, many projects are competing for the same markets, as Table 3 shows, or may still become uneconomical in the currently shifting petroleum markets.

This seems especially true for pipelines proposed to bring natural gas from Canadian producers into the United States, the myriad of major system additions and expansions into the United States Northeast, and projects competing to supply natural gas to EOR projects and power-generation markets in California.

Crude-oil projects in the latest *Oil & Gas Journal* Worldwide Construction Report total eight comprising 985 miles. Here, also, the construction report of 1988 lists six crude-oil pipeline projects totaling 1820 miles.

Finally, the 1989 *Journal* report shows 5 product-pipeline projects with 75 miles of pipeline. In 1988, 11 projects totaling 874 miles were reported.

TABLE 2 United States Interstate Pipeline Mileage[a]

	Miles		
Year	Gas	Liquid	Total
1979	264,134	169,794	433,928
1980	274,248	172,673	446,921
1981	274,634	172,815	447,449
1982	286,999	172,549	459,548
1983	285,204	167,819	453,023
1984	258,379[b]	173,922	432,301[b]
1985	279,395[b]	171,401	450,796[b]
1986	281,881[b]	170,014	451,895[b]
1987	280,085[b]	167,865	447,950[b]
1988	281,381[b]	170,457	451,838[b]

[a]Source: U.S. FERC Form 6 Annual Reports—Oil Pipelines; Forms 2 and 2A—Gas Pipelines.

[b]Reflects operating mileage as reported under the FERC's classification system for natural-gas pipeline companies effective beginning with the 1984 reporting year. Only major gas-pipeline companies are required to file mileage. See Table 5 for definition of major and nonmajor companies and details of companies reporting mileage for 1987.

TABLE 3 Current Pipeline Costs[a]

Size, in.	Location[b]	Length (miles)	Cost ($) Material	Labor	Miscellaneous[c]	R.O.W. Damages	Total	$/mile
			Land Pipelines					
2	Nebraska (R)	1.30	22,260	18,060	14,574	6,426	61,320	47,169
2	Nebraska (R)	2.00	24,000	21,000	19,000	7,000	71,000	35,500
2	Nebraska (R)	6.80	82,000	70,000	57,000	23,000	232,000	34,118
4	West Virginia-Pennsylvania	4.00	433,566	1,191,474	262,297	191,808	2,079,145	519,786
4	Nebraska (R)	30.00	508,005	412,155	332,600	146,651	1,399,411	46,647
6	Tennessee (L)	1.05	70,000	234,600	152,854	70,000	527,454	502,816
6	West Virginia-Pennsylvania	2.00	217,434	597,526	131,523	96,192	1,042,675	521,338
6	Oklahoma	5.02	174,672	171,748	173,580	72,000	592,000	117,928
8	Pennsylvania	1.41	212,000	534,000	103,695	45,000	894,695	634,535
8	Alabama	2.40	420,000	920,000	322,168	—	1,662,168	692,570
8	West Virginia-Pennsylvania	6.00	651,000	1,789,000	393,840	288,000	3,121,840	520,307
10	Connecticut (R)	0.90	112,400	575,700	445,473	135,000	1,268,573	1,409,526
10	Nevada (L)	1.70	191,000	87,000	161,996	20,000	459,996	262,855
12	Connecticut-Rhode Island	3.80	775,000	3,980,000	516,792	340,000	5,611,792	1,476,787
12	Connecticut (L)	4.90	1,000,000	4,035,000	541,625	555,000	6,131,625	1,251,352
12	Oklahoma	6.82	509,832	437,748	394,275	98,145	1,440,000	211,144
12	Oklahoma	12.40	977,000	715,500	294,500	386,700	2,373,700	191,427
12	New York	13.90	1,214,000	3,016,000	2,296,865	310,000	6,836,865	491,861
12	New York	26.00	2,291,000	3,685,000	1,526,717	1,375,000	8,877,717	341,451
16	Connecticut-Rhode Island (L)	2.10	545,000	2,170,000	313,112	365,000	3,393,112	1,615,768
16	North Carolina	7.50	776,700	2,159,600	758,111	185,700	3,880,111	317,348
16	Nevada	15.50	1,554,000	1,489,000	1,691,911	70,000	4,804,911	309,994
16	Louisiana	45.00	4,074,100	4,409,900	1,028,700	1,152,000	10,664,700	236,993
20	Alabama	0.25	343,300	229,700	117,266	4,400	694,666	2,778,664
20	New York	0.30	111,200	134,100	85,300	12,500	343,100	1,143,667
20	Pennsylvania	0.38	184,400	41,600	55,500	—	281,500	740,789

(*continued*)

TABLE 3 *(continued)*

Size, in.	Location[b]	Length (miles)	Cost ($) Material	Labor	Miscellaneous[c]	R.O.W. Damages	Total	$/mile
			Land Pipelines—continued					
20	Tennessee (R)	1.07	255,000	557,000	189,503	10,000	1,011,503	942,687
20	Pennsylvania	2.48	660,000	674,000	615,640	—	1,949,640	786,145
20	New Jersey (L)	3.81	748,700	1,507,587	602,575	187,500	3,046,362	799,570
20	New Jersey (L)	4.56	861,270	2,031,700	857,309	218,850	3,969,129	870,423
20	Massachusetts (R)	5.20	1,440,000	5,745,000	724,531	310,000	8,219,531	1,580,679
20	Ohio	5.61	1,413,000	3,276,000	568,380	196,000	5,453,380	972,082
20	Arkansas (L)	8.70	1,062,800	1,055,200	989,622	84,000	3,191,622	366,853
20	Pennsylvania-New Jersey	57.50	11,099,800	24,874,000	7,388,615	6,422,700	49,785,115	865,828
22	Tennessee (C)	0.48	325,000	857,600	375,680	8,000	1,566,280	3,263,083
24	Rhode Island (L)	1.60	665,000	2,225,000	323,296	285,000	3,498,296	2,186,435
24	Alabama	3.50	1,250,915	635,519	503,226	156,800	2,546,460	727,560
24	Alabama	4.17	1,288,399	567,935	516,747	186,900	2,559,981	613,904
24	Alabama	4.20	1,471,200	1,137,600	479,937	—	3,088,737	735,414
24	Pennsylvania (L)	5.00	1,222,000	3,154,000	518,000	216,000	5,110,000	1,022,000
24	Massachusetts (L)	5.50	2,010,000	5,125,000	780,694	920,000	8,835,694	1,606,490
24	Virginia	27.00	6,110,700	8,797,300	2,915,429	4,368,300	22,191,729	821,916
24	Pennsylvania	83.00	18,173,000	26,126,300	14,223,106	3,000,000	61,522,406	741,234
24	Alabama	86.00	14,372,100	12,218,500	4,642,560	2,752,000	33,985,160	395,176
24	Oklahoma-Arkansas	191.00	35,632,000	31,588,00	10,705,500	1,687,000	79,612,500	416,819
30	Alabama	1.00	757,800	395,400	251,630	17,600	1,422,430	1,422,430
30	Iowa-Illinois (C)	2.00	1,087,200	3,516,100	1,296,700	100,000	6,000,000	3,000,000
30	Pennsylvania	7.60	2,644,700	3,089,600	787,300	255,500	6,777,100	891,724
30	Arkansas (L)	12.00	2,646,800	2,182,900	2,115,062	115,900	7,060,662	558,389
30	New York (L)	12.30	4,150,000	5,221,800	2,167,800	492,000	12,031,600	978,179
30	California (L)	12.96	3,328,000	2,986,000	1,253,921	144,000	7,711,921	595,056
30	Pennsylvania (L)	15.15	1,917,490	2,870,686	1,208,779	82,400	6,079,355	1,180,457
30	Arkansas (L)	16.30	4,092,400	3,390,700	3,318,620	180,200	10,981,920	673,737
30	Oklahoma	22.61	5,265,100	4,112,900	4,183,065	255,300	13,816,365	611,073
30	Alabama	24.50	8,877,800	5,056,000	3,761,004	601,100	18,295,904	746,772
30	New Jersey-New York	38.70	22,151,000	49,449,000	9,519,927	3,072,00	84,191,927	2,175,502

30	Oklahoma-Arkansas (L)	50.20	11,151,200	9,131,700	8,964,918	484,800	29,732,618	592,283
30	Alabama	52.50	14,812,200	11,108,800	5,046,746	2,500,500	33,468,246	637,490
30	Texas	94.30	25,892,000	13,974,000	4,753,000	1,886,000	46,505,000	493,160
30	Oklahoma	97.50	26,469,000	14,608,000	4,881,000	1,804,000	47,762,000	489,867
36	New Jersey (L)	0.86	559,038	746,734	350,633	126,500	1,782,905	2,073,145
36	Connecticut (L)	2.00	1,400,000	2,600,000	425,114	395,000	4,820,114	2,410,057
36	Connecticut (L)	2.00	1,085,000	2,585,000	405,632	345,000	4,420,632	2,210,316
36	New Jersey (L)	4.13	2,066,180	3,363,092	1,484,816	608,120	7,522,208	1,821,358
36	Pennsylvania (R)	7.50	4,511,000	8,433,000	1,945,047	556,000	15,445,047	2,059,340
36	Pennsylvania (L)	8.00	4,209,042	4,664,288	2,562,896	256,000	11,692,226	1,461,528
36	New York (L)	8.20	5,245,000	10,430,000	1,711,735	2,060,000	19,446,735	2,371,553
36	Pennsylvania (L)	12.48	6,519,119	7,177,413	3,473,397	399,400	17,569,329	1,407,799
36	Pennsylvania (L)	12.59	6,621,560	7,235,685	3,909,853	402,800	18,169,898	1,443,201
36	Ohio-Pennsylvania (L)	17.53	8,454,000	12,520,000	2,840,433	611,000	24,425,433	1,393,350
36	Pennsylvania (L)	29.89	12,426,000	22,405,000	4,759,659	2,209,000	41,799,659	1,398,450
36	Pennsylvania-New Jersey	34.84	17,907,000	37,005,000	7,148,934	1,790,000	63,850,934	1,832,690
36	West Virginia-Pennsylvania	36.18	15,609,000	24,690,000	5,173,342	1,262,000	46,734,342	1,291,718
36	Idaho-Washington (L)	95.60	37,240,000	32,607,000	18,098,890	1,059,155	89,005,045	931,015
36	Idaho-Washington (L)	96.40	36,922,000	24,984,000	17,334,570	1,003,515	80,244,085	832,407
36	Pennsylvania	107.43	40,678,000	64,691,000	14,941,014	5,407,000	125,717,014	1,170,223
36	Pennsylvania	107.58	47,245,000	66,718,000	16,148,566	5,426,000	135,537,566	1,259,877
36	Idaho-Washington (L)	134.30	51,093,000	47,122,000	26,877,020	1,489,290	126,581,310	942,527
42	New Jersey	2.56	2,057,000	4,088,000	818,382	279,000	7,242,382	2,829,055
42	New Jersey (L)	3.86	2,109,756	5,020,357	2,106,567	568,100	9,804,780	2,540,098
42	New Jersey	5.50	4,276,000	6,849,000	1,490,507	320,000	12,935,507	2,351,910
42	Georgia (L)	12.05	5,462,610	3,995,160	2,593,514	327,760	12,379,044	1,027,307
42	Idaho-Washington (L)	103.70	52,649,000	29,986,000	22,275,090	1,201,555	106,111,645	1,023,256
48	Alabama (L)	15.17	8,743,954	5,975,585	3,978,801	461,130	19,159,470	1,262,984
Total—land projects		2,023.78	$627,865,702	$732,263,552	$276,475,508	$67,519,197	$1,704,123,959	$842,050
1988-report total/land		5,118.45	$1,111,105,674	$1,441,542,887	$606,328,053	$205,829,349	$3,364,805,963	$657,388
	Offshore Pipelines							
6	Louisiana	4.90	404,600	1,700,000	708,928	—	2,813,528	574,189
8	Alabama	2.40	420,000	920,000	322,168	—	1,662,168	692,570

(continued)

TABLE 3 *(continued)*

Size, in.	Location[b]	Length (miles)	Cost ($) Material	Labor	Miscellaneous[c]	R.O.W. Damages	Total	$/mile
			Offshore Pipelines—continued					
10	Louisiana	4.02	741,339	1,654,580	677,683	100	3,073,702	654,042
12	Alabama	9.00	1,807,000	2,673,000	623,900	2,000	5,105,900	567,322
12	Louisiana	34.30	5,467,100	4,000,000	3,123,000	—	12,591,000	367,085
12	Alabama	7.80	1,640,000	1,630,000	901,293	—	4,171,293	534,781
16	Alabama	8.70	2,460,000	3,510,000	1,222,825	—	7,192,825	826,762
16	Louisiana	10.22	2,636,090	3,364,930	1,744,192	246	7,745,458	757,873
16	Alabama	17.20	3,780,000	3,720,000	1,626,924	—	9,126,924	530,635
16	Louisiana	26.57	6,734,150	6,067,876	3,927,727	499	16,730,252	629,667
20	Mississippi	7.40	3,661,250	3,203,750	4,149,868	5,000	11,019,868	1,489,171
20	Alabama	10.10	2,620,000	2,710,000	1,141,397	—	6,471,397	640,732
20	Louisiana	25.39	7,484,096	8,940,018	4,811,499	476	21,236,089	836,396
20	Mississippi-Alabama	27.07	7,780,298	6,135,901	4,298,561	510	18,215,270	672,895
20	Alabama	29.60	10,000,000	7,560,000	2,394,880	10,000	19,964,880	674,489
20	Louisiana-Mississippi	30.74	10,027,720	7,085,757	5,176,936	565	22,290,978	725,146
24	Louisiana (L)	11.15	4,582,000	5,991,000	1,429,600	3,000	12,005,600	1,076,735
24	Alabama	13.80	6,504,810	9,788,504	4,169,284	105,061	20,567,659	1,490,410
24	Louisiana	20.03	7,809,000	5,200,000	2,237,830	—	15,246,830	761,200
24	Alabama	31.20	10,788,800	8,342,400	3,519,540	—	22,650,740	725,985
24	Louisiana	55.00	21,786,000	17,676,000	5,352,832	403,000	45,217,832	822,142
Total—Offshore Projects		386.59	$119,134,253	$111,873,716	53,561,767	$530,457	$285,100,193	737,474
1988 report total/offshore projects		401.30	$151,556,000	$84,126,000	$67,432,755	$5,000	$303,119,755	$755,345
Total—All Projects		2,410.37	$746,999,555	$844,137,268	$330,037,275	$68,049,654	$1,989,224,152	$825,278
1988—Report Total		5,519.75	$1,262,661,674	$1,525,688,887	$673,760,808	$205,829,349	$3,667,940,718	$664,509

[a]Source: U.S. FERC construction permit applications, July 1, 1988, to June 30, 1989.

[b]L = loop; R = replacement; c = river or channel crossing.

[c]Generally includes engineering, supervision, surveying, interest, administration and overheads, contingencies, afudc, and FERC fees.

United States Interstate Network

Revenue, income, and mileage changes cited earlier are evident on the pipeline-company tables (Tables 4 and 5).

These data are based on annual reports of the regulated interstate pipeline companies and provide a variety of detail on each of them, including pipeline mileage, crude-oil and refined-products deliveries, natural-gas sales and deliveries of gas for others, operating revenues, and net incomes.

In 1988's Pipeline Economics Report (*Oil & Gas J.*, p. 33, November 28, 1988), the *Oil & Gas Journal* began tracking volumes of gas transported for others by major interstate pipeline systems as a method of keeping track of the changing nature of the United States gas-transmission industry.

System Utilization

Comparisons between any years of United States petroleum and natural-gas pipeline mileages must be done with care for two reasons: The number of companies required to file reports with the FERC varies each year, and the FERC's system for classifying interstate natural-gas pipeline companies changed with the 1984 reporting year (*Oil & Gas J.*, p. 55, November 25, 1985).

Companies are now classified as "major" or "nonmajor" based on total natural-gas transmissions for each of the 3 previous reporting years. (See FERC Accounting and Reporting Requirements for Natural Gas Companies, para. 20-011.)

Major pipeline companies are those whose combined gas sold for resale and gas transported or stored for a fee exceeded 50 BCF at 14.73 lb/in.^{2} (60°F) in each of the 3 previous calendar years.

Nonmajors are companies not classified as majors and having had total gas sales of volume transactions exceeding 200 MMCF at 14.73 lb/in.^{2} (60°F) in each of the 3 previous calendar years.

One effect of this change has been that companies classified as nonmajor are exempt from filing certain data, chief among the excluded figures being mileage and gas-sales or transportation figures. Many nonmajor companies nonetheless file such data voluntarily, but consistency exists only among major companies.

In comparison, figures for nonmajor companies should be excluded from a calculation of total United States interstate pipeline utilization for any given year because of their unreliability year to year.

Therefore, combining 1988 mileage data for all regulated liquids-pipeline companies (170,457) with mileage reported by the major natural-gas pipeline companies (43 companies reporting 254,065 miles) yields a total figure of 424,522. This represents a 1% increase of 4210 miles over the comparable figures reported for 1987 (167,865 + 252,447 = 420,312).

This compares with a similarly small change for 1987 over 1986, a drop of 3343 miles for all liquids and major natural-gas pipelines.

TABLE 4 Liquid Pipelines[a]

Company	Miles of Pipeline				Deliveries (1000 bbl)
		Trunk			
	Gathering	Crude	Products	Total	Crude
Acorn Pipeline Co.	—	—	—	—	—
Airforce Pipeline Inc.	—	—	5	5	—
Algonquin Pipe Line Co.	33	51	—	84	650
All American Pipeline Co.	—	1,108	—	1,108	26,256
Allegheny Pipeline Co.	—	—	603	603	—
Amerada Hess Pipeline Co.	—	819[b]	—	819[b]	11,510
American Petrofina Pipeline Co.	—	1,104	—	1,104	27,539
Amoco Pipeline Co.	2,227	7,703	2,367	12,297	343,593
Antelope Pipe Line Co.	—	—	—	—	1330
ARCO Pipe Line Co.	1,610	2,398[b]	2,831	6,839[b]	403,694
Ashland Pipe Line Co.	501	787	—	1,288	95,845
Atlantic Pipeline Co.	—	11	929	940	48,043
ATTCO NGL Pipeline Co.	—	—	27	27	—
Badger Pipe Line Co.	—	—	331	331	—
Belle Fourche Pipeline Co.	1,253	728	—	1,981	26,240
Black Lake Pipe Line Co.	—	255	—	255	2,745
Buccaneer Pipe Line Co.	—	2	—	2	289
Buckeye Pipe Line Co. L. P.	—	—	3,390	3,390	—
Buckeye Pipe Line Co. of Mich L. P.	—	160	4	164	11,143
Butte Pipe Line Co.	—	373	—	373	33,490
C & T Pipeline Inc.	—	—	62	62	—
Calnev Pipe Line Co.	—	—	583	583	—
Chaparral Pipeline (NGL) Co.	—	—	940	940	—
Chase Transportation Co.	—	—	716	716	—
Chevron Pipe Line Co.	2,006	2,739	3,385	8,130	338,132
Chicap Pipe Line Co.	—	235	—	235	76,384
Chisholm Pipeline Co.	—	—	203	203	—
Ciniza Pipe Line Inc.	233	135	11	379	6,823
Citgo Pipeline Co.	602	847	3	1,452	106,731
CKB Petroleum Inc.	—	—	—	—	1,051
Clarco Pipe Line Co.	90	—	—	90	1,235
CNG Pipeline Co.	—	—	—	—	1,407
Coastal Pipeline Co.	—	—	8	8	—
Cochin Pipeline System	—	—	1,197	1,197	—
Collins Pipeline Co.	—	—	124	124	—
Colonial Pipeline Co.	—	—	5,274	5,274	—
Conoco Pipe Line Co.	4,103	1,654	2,138	7,895	202,159
Conquest Exploration Corp.	—	—	—	—	4
Cook Inlet Pipe Line Co.	—	55	—	55	10,869
Crown-Rancho Pipe Line Corp.	—	—	—	—	7,816
Diamond Shamrock Refg & Markt Co.	—	—	622	622	—
Dixie Pipeline Co.	—	—	1,303	1,303	—
Dome Pipeline Corp.	1	8	134	143	—
Emerald Pipe Line Corp.	—	—	113	113	—
Endicott Pipeline Co.	—	25	—	25	37,929
Enron Liquids Pipeline Co.	—	—	1,581	1,581	—

Deliveries (1000 bbl)		Total Trunkline Traffic (million bbl-miles)			Fiscal Data ($1000)			
Products	Total	Crude	Products	Total	Carrier Property	Change	Operating Revenue	Income
—	—	—	—	—	−3,388	—	—	−173
1,572	1,572	—	7	7	353	—	275	76
—	650	22	—	22	2,360	15	539	102
—	26,256	29,091	—	29,091	1,443,503	226,162	—	—
10,491	10,491	—	3,547	3,547	29,051	653	6,035	1,673
—	11,510	8,919	—	8,919	149,608	63	35,889	13,938
—	27,539	1,302	—	1,302	26,294	5,192	8,856	1,597
160,535	504,128	113,055	10,309	123,364	584,840	17,147	207,360	102,377
—	1330	115	—	115	—	—	365	49
142,222	545,916	155,744	10,371	166,115	2,412,291	−1,667	612,864	217,588
—	95,845	66,394	—	66,394	101,710	2,265	43,384	15,709
51,141	99,184	—	8,274	8,274	52,996	4,327	36,324	8,088
142	142	—	3	3	225	—	36	−40
59,610	59,610	—	—	—	25,685	306	11,848	3,299
—	26,240	2,483	—	2,483	57,700	1,570	14,680	5,217
—	2,745	469	—	469	10,563	174	1,541	69
—	289	—	—	—	176	—	43	−190
284,536	284,536	—	35,911	35,911	468,814	2,937	134,450	26,073
7,436	18,579	1,210	183	1,393	41,358	557	3,748	−2,097
—	33,490	7,918	—	7,918	19,039	331	9,591	2,211
89	89	—	59	59	350	—	145	−668
29,104	29,104	—	5,707	5,707	42,298	2,695	20,293	10,041
39,951	39,951	—	19,569	19,569	97,423	4,222	22,604	7,542
14,060	14,060	—	4,733	4,733	38,604	22	8,299	4,699
247,955	586,087	23,217	74,384	97,601	608,665	89,228	197,151	81,950
—	76,384	5,499	—	5,499	33,300	12	6,201	2,178
11,288	11,288	—	2,291	2,291	19,727	26	4,150	1,086
1,590	8,413	439	18	457	6,858	148	3,013	−266
3,231	109,962	3,071	3	3,074	42,479	−2,929	16,125	2,643
—	1,051	54	—	54	4,757	—	2,892	493
—	1,235	—	—	—	3,624	6	340	−51
—	1,407	—	—	—	4,144	—	1,652	708
14,321	14,321	—	1,065	1,065	6,608	—	2,657	696
22,959	22,959	—	20,370	20,370	240,592	485	54,033	32,140
41,629	41,629	—	—	5,162	15,344	—	3,330	252
635,620	635,620	—	658,014	658,014	1,367,578	20,971	500,464	157,057
59,371	261,530	15,316	9,950	25,266	234,005	37,518	84,080	38,222
—	4	—	—	—	300	—	29	−300
—	10,869	349	—	349	42,491	45	12,369	3,783
—	7,816	2,772	—	2,772	2,129	−100	2,510	1,871
15,100	15,100	—	7,391	7,391	34,314	−24	12,697	5,387
34,761	34,761	—	19,211	19,211	77,084	1,510	45,478	13,794
9,685	9,685	—	196	196	22,746	31	1,677	14,796
2,161	2,161	—	184	184	1,430	—	585	153
—	37,929	946	—	946	55,585	959	26,929	12,984
51,587	51,587	—	16,291	16,291	162,248	973	40,977	8,388

(continued)

TABLE 4 *(continued)*

Company	Miles of Pipeline				Deliveries (1000 bbl)
		Trunk			
	Gathering	Crude	Products	Total	Crude
Enterprise Pipeline Co.	—	—	98	98	—
Enterprise Products Co. of Miss.	—	—	125	125	—
EPC Partners Ltd.	—	—	268	268	—
Eureka Pipe Line Co., The	771	3	3	777	2,369
Explorer Pipeline Co.	—	—	1,397	1,397	—
Exxon Pipeline Co.	1,884	4,936[b]	2,558	9,378[b]	428,040
Farmland Industries Inc.	627	239	—	866	23,211
Four Corners Pipe Line Co.	281	1,183	14	1,478	111,710
Frontier Pipeline Co.	—	290	—	290	452
G & T Pipeline Co.	—	30	—	30	98
Gulf Central Pipeline Co.	—	—	1,975	1,975	—
Hess Pipeline Co.	—	442	—	442	29,651
Howell Crude Oil Co.[c]	—	—	—	—	—
Interstate Storage & Pipeline Co.	—	—	23	23	—
Jayhawk Pipeline Corp.	821	657	—	1,478	42,272
Kaneb Pipe Line Co.	—	—	2,080	2,080	—
Kerr-McGee Pipeline Corp.	8	56	—	64	1,439
Kiantone Pipeline Corp.	—	78	—	78	20,540
Koch Pipelines Inc.	1,016	1,125	1,185	3,326	48,060
Kuparak Transportation Co.	—	37	—	37	110,464
Lake Charles Pipe Line Co.	—	—	13	13	—
Lakehead Pipe Line Co. Inc.	—	2,615	—	2,615	473,074
Largo Co., The	—	—	—	—	961
Laurel Pipe Line Co.	—	—	351	351	—
Locap Inc.	—	59	—	59	228,708
Marathon Pipe Line Co.	1,018	1,726	1,466	4,210	356,132
Mark Oil Pipeline Co.	1	—	—	1	272
McMoRan Pipeline Co.	—	—	—	—	126
Meridan Oil Frontera Inc.	—	—	—	—	354
Mesa Transmission Co.	—	11	—	11	880
Mid-America Pipeline CO.	—	19	8,081	8,100	31,467
Mid-Valley Pipe Line Co.	—	95	—	95	126,588
Midland-Lea Inc.	—	101	275	376	—
Milne Point Pipe Line Co.	11	—	—	11	—
Minnesota Pipe Line Co.	—	509	—	509	80,265
Mitco Pipeline Co.	9	—	—	9	—
Mobil Alaska Pipeline Co.	—	818	—	818	29,966
Mobil Eugene Island Pipeline Co.	—	134	—	134	20,551
Mobil Pipe Line Co.	3,347	4,597	1,739	9,683	195,651
National Transit Co.	186	—	—	186	413
Navajo Pipeline Co.	—	70	323	393	—
Northern Rockies Pipe Line Co.	—	77	—	77	612
NW Pipeline Co.	—	—	—	—	2,552
Ohio Oil Gathering Corp.	731	—	—	731	3,980
Ohio River Pipe Line Co.	—	—	400	400	—
Oiltanking Pipeline Inc.	—	13	15	28	50,997

Deliveries (1000 bbl)		Total Trunkline Traffic (million bbl-miles)			Fiscal Data ($1000)			
Products	Total	Crude	Products	Total	Carrier Property	Change	Operating Revenue	Income
6,048	6,048	—	592	592	20,435	44	4,686	1,775
2,126	2,126	—	265	265	12,136	—	642	2,768
7,080	7,080	—	1,897	1,897	36,964	140	3,358	3,509
—	2,369	1	—	1	16,194	281	7,159	1,739
174,142	174,142	—	129,220	129,220	256,560	1,037	107,639	36,365
99,888	527,928	162,185	8,206	170,391	2,185,522	5,922	602,167	304,700
—	23,211	666	—	666	11,582	563	3,576	—
—	111,170	14,651	—	14,651	159,603	3,862	39,636	7,557
—	452	131	—	131	48,849	—	1,208	—
—	98	2	—	2	5,142	—	573	126
10,436	10,436	—	7,179	7,179	69,791	589	20,162	7,893
—	29,651	1,842	—	1,842	45,454	163	12,741	3,371
—	—	—	—	—	—	—	—	—
1,950	1,950	—	—	—	3,167	—	1,310	399
—	42,272	4,229	—	4,229	27,179	2,022	10,878	2,519
54,602	54,602	—	13,393	13,393	107,351	1,846	37,923	11,456
—	1,439	34	—	34	5,572	244	1,922	2,330
—	20,540	20	—	20	10,407	1,641	4,313	−170
55,382	103,442	2,391	25,344	27,735	300,059	10,311	57,945	19,190
—	110,464	3,624	—	3,624	126,148	—	68,411	54,950
90,560	90,560	—	454	454	10,938	43	4,528	1,872
—	473,074	366,739	—	366,739	537,718	27,449	216,492	59,077
—	961	38	—	38	12,654	14	1,974	712
33,694	33,694	—	5,220	5,220	46,979	825	14,372	3,514
—	228,708	13	—	13	120,511	155	20,545	2,368
238,129	594,261	53,855	18,166	72,021	361,501	10,601	144,706	46,999
—	272	—	—	—	922	—	190	−60
—	126	11	—	11	1,760	—	314	22
—	354	38	—	38	1,378	—	674	266
—	880	47	—	47	10,569	—	1,612	313
162,909	194,376	4,163	55,971	60,134	474,083	−1,289	152,294	45,475
—	126,588	68,366	—	68,366	100,342	805	44,297	13,294
—	—	—	—	—	20,325	—	—	—
—	—	—	—	—	26,930	45	—	−956
—	80,265	20,467	—	20,467	76,162	16,144	27,290	6,888
—	—	—	—	—	2,104	—	46	−146
—	29,966	23,849	—	23,849	398,440	140	91,641	50,016
—	20,551	1,629	—	1,629	46,965	2,848	18,305	9,415
107,655	303,306	66,505	12,134	78,639	307,869	4,012	160,536	82,961
—	413	—	—	—	3,619	57	1,059	605
5,788	5,788	—	1,221	1,221	8,556	27	6,077	2,946
—	612	47	—	47	1,048	—	152	1,423
—	2,552	260	—	260	6,386	—	2,603	1,865
—	3,980	—	—	—	9,904	685	7,837	240
16,113	16,113	—	2,012	2,012	25,957	2,853	8,256	2,553
52	51,049	623	1	624	16,016	4,709	4,961	2,042

(continued)

TABLE 4 *(continued)*

Company	Miles of Pipeline				Deliveries (1000 bbl)
		Trunk			
	Gathering	Crude	Products	Total	Crude
Olympic Pipe Line Co.	—	—	391	391	—
Osage Pipe Line Co.	—	136	—	136	30,492
Owensboro-Ashland Co.	—	266	—	266	71,501
Oxy NGL Pipeline Co.	—	—	106	106	—
Oxy Pipeline Inc.	—	—	—	—	605
Paloma Pipe Line Co.	—	—	—	—	21,034
Pennzoil Offshore Pipeline Co.	—	4	—	4	515
Permian Operating L.P.	—	—	—	—	—
Plantex Pipeline Co.	—	—	27	27	—
Phillips Alaska Pipeline Corp.	—	818[b]	—	818[b]	11,788
Phillips Pipe Line Co.	1,095	855	4,100	6,050	75,727
Pioneer Pipe Line Co.	—	—	299	299	—
PL Pipeline Co.	—	—	—	—	60
Plantation Pipe Line Co.	—	—	3146	3146	—
Platte Pipe Line Co.	—	1283	—	1,283	39,296
Pogo Offshore Pipeline Co.	—	—	—	—	1,571
Point Arguello Pipeline Co.	—	—	—	—	—
Point Pedernales Pipeline Co.	—	46	—	46	6,219
Polysar Hydrocarbons Inc.					
Portal Pipe Line Co.	294	614	—	908	20,810
Portland Pipe Line Corp.	—	177	—	177	23,813
Samedan Pipeline Corp.					
Seminole Pipeline Co.	106	—	634	740	—
Shamrock Pipe Line Corp., The	649	318	580	1,547	43,840
Shell Pipe Line Corp.	2,059	4,745	431	7,235	414,567
Sohio Alaska Pipeline Co.	—	818[b]	—	818[b]	380,878
Sohio Pipe Line Co.	1,101	547	277	1,925	182,648
Sonat Oil Transmission Inc.	—	42	—	42	950
Southcap Pipe Line Co.	—	—	—	—	43,658
Southern Pacific Pipe Line L.P.	—	—	3,174	3,174	—
Sun Oil Line Co. of Michigan	—	201	—	201	13,754
Sun Pipe Line Co.	379	441	1,758	2,578	85,241
Tecumseh Pipe Line Co.	—	205	—	205	19,539
Texaco Pipeline Inc.	2,075	3,216	984	6,275	360,598
Texas Eastern Products Pipeline Co.	—	—	3,373	3,373	—
Texas-New Mexico Pipe Line Co.	1735	2184	—	3,919	119,168
Total Pipeline Corp.	193	605	97	895	73,668
Trans Mountain Oil Pipeline Corp.	—	64	—	64	3,519
Trans-Ohio Pipeline Co.	—	—	133	133	—
Transco Terminal Co.	—	—	—	—	458
Unocal Pipeline Co.	693	1,528[b]	16	2,237[b]	85,725
West Emerald Pipeline Corp.	—	—	296	296	—
West Shore Pipe Line Co.	—	—	321	321	—
West Texas Gulf Pipe Line Co.	—	579	—	579	115,447
Western Oil Transportation	548	—	—	548	11,786
Whiskey Springs Oil Pipeline Co.	7	—	—	7	189

Deliveries (1000 bbl)		Total Trunkline Traffic (million bbl-miles)			Fiscal Data ($1000)			
Products	Total	Crude	Products	Total	Carrier Property	Change	Operating Revenue	Income
93,049	93,049	—	16,971	16,971	61,640	1,144	28,289	7,822
—	30,492	4,073	—	4,073	19,531	18	5,814	1,768
—	71,501	18,977	—	18,977	44,692	178	11,973	2,468
5,847	5,847	—	623	623	3,249	19	1,531	671
—	605	38	—	38	5,125	292	947	−279
—	21,034	1,309	—	1,309	5,783	20	2,727	882
—	515	14	—	14	4,025	560	412	46
—	—	—	—	—	—	—	—	—
125	125	—	3	3	2,521	2,521	169	70
—	11,788	8,032	—	8,032	126,915	68	32,493	15,579
222,775	298,502	14,232	48,743	62,975	380,103	32,422	160,365	54,764
12,185	12,185	—	2,826	2,826	13,725	318	6,779	1,695
—	60	7	—	7	1,039	—	112	−32
189,167	189,167	—	118,068	118,068	303,506	406	116,931	13,670
—	39,296	27,993	—	27,993	80,009	470	26,552	10,138
—	1,571	206	—	206	1,330	−7,942	1,065	7,091
—	—	—	—	—	—	—	—	—
—	6,219	143	—	143	50,216	−106	16,854	4,685
—	20,810	7,567	—	7,567	44,527	180	13,733	3,185
—	23,813	4,074	—	4,074	42,064	67	14,579	4,712
49,276	49,276	—	23,443	23,443	204,784	1,990	31,114	12,897
24,978	68,818	3,187	3,115	6,302	35,187	261	17,789	1,636
117,785	532,352	108,789	5,025	113,814	475,988	9,579	163,625	70,062
—	380,878	294,321	—	294,321	4,902,519	2,081	1,175,716	482,913
98,099	280,747	48,478	11,284	59,762	108,306	−79	83,216	16,975
—	950	9	—	9	5553	—	745	622
—	43,658	25,490	—	25,490	51,961	50	11,037	3,040
317,733	317,733	—	40,324	40,324	446,678	84,667	139,920	61,199
—	13,754	2,764	—	2,764	668	—	5,387	1,613
124,722	209,963	3,087	11,746	14,833	150,193	5,842	65,954	7,810
—	19,539	701	—	701	15,270	—	4,085	1,658
121,551	482,149	66,700	4,655	71,355	446,998	19,942	123,220	26,758
156,926	156,926	—	83,981	83,981	449,841	8,675	125,996	32,151
—	119,168	15,558	—	15,558	92,808	2,579	32,207	11,652
17,356	91,024	9,100	1,470	10,570	54,211	3,716	13,845	−209
—	3,519	184	—	184	6,756	33	1,185	−2,050
10,685	10,685	—	441	441	13,935	92	3,049	1,362
—	458	—	—	—	3,271	—	366	39
17,459	103,184	20,497	206	20,703	199,174	−19,790	56,818	22,711
2,578	2,578	—	562	562	1,947	26	1,180	152
71,622	71,622	—	8,537	8,537	35,257	339	18,845	5,790
—	115,447	47,336	—	47,336	47,167	647	28,937	12,055
—	11,786	—	—	—	29,735	98	6,040	−14,665
—	189	—	—	—	605	605	45	30

(continued)

TABLE 4 *(continued)*

Company	Miles of Pipeline				Deliveries (1000 bbl)
		Trunk			
	Gathering	Crude	Products	Total	Crude
White Shoal Pipeline Corp.	7	—	—	7	2,823
Williams Pipe Line Co.	—	—	6,780	6,780	2,718
Wolverine Pipe Line Co.	—	—	621	621	—
Wyco Pipe Line Co.	—	—	677	677	—
Yellowstone Pipe Line Co.	—	—	752	752	—
1988 Totals	34,311	55,900	80,246	170,457	6,509,367
1987 Totals	35,055	54,886	77,924	167,865	6,277,670

[a]Source: U.S. FERC Form No. 6's: Annual Report of Oil Pipelines, December 31, 1988.
[b]Mileage represents Trans-Alaska pipeline and is included in total only one to avoid duplication.
[c]Includes 1097 miles of ammonia pipeline.

TABLE 5 Gas Pipelines[a]

Company	Miles of Pipeline				Total Compressor Stations
	Transmission	Field	Storage	Total	Transmission
Alabama-Tennessee Natural Gas Co.	316	—	—	316	2
Algonquin Gas Transmission Co.[b]	983	—	—	983	5
Algonquin LNG Inc.		—	—	—	—
American Distribution Co.	38	—	—	38	—
ANR Pipeline Co.[b]	8,697	2,641	196	11,534	45
ANR Storage Co.	—	—	24	24	—
Arkansas Oklahoma Gas Corp.	360	174	—	534	5
Arkansas Western Gas Co.	1,296	360	—	1,656	5
Arkla Inc.[b]	6,292	3,598	66	9,956	23
Bayou Interstate Pipeline System	2	—	—	2	—
Bear Creek Storage Co.	—	—	25	25	—
Black Marlin Pipeline Co.	66	—	—	66	—
Blue Dolphin Pipe Line Co.	49	—	—	49	—
Bluefield Gas Co.	32	—	—	32	—
Boundary Gas Inc.	—	—	—	—	—
Canyon Creek Compression Co.	—	—	—	—	1
Caprock Pipeline Co.	39	14	—	53	2
Carnegie Natural Gas Co.	216	875	—	1,091	7
Chandeleur Pipe Line Co.	174	—	—	174	—
Cimarron Transmission Co.	—	41	—	41	—
CNG Transmission Corp.[b]	3,641	4,109	517	8,267	44
Colorado Interstate Gas Co.[b]	2,998	3,043	57	6,098	26
Columbia Gas Transmission Corp.[b]	10,950	6,647	1,207	18,804	111
Columbia Gulf Transmission Co.[b]	4,269	—	—	4,269	13
Columbia LNG Corp.	88	—	—	88	—

Deliveries (1000 bbl)		Total Trunkline Traffic (million bbl-miles)			Fiscal Data ($1000)			
Products	Total	Crude	Products	Total	Carrier Property	Change	Operating Revenue	Income
—	2,823	125	—	125	2,619	1	289	—
137,576	176,294	1,665	49,804	51,469	559,997	14,073	137,803	9,864
84,632	84,632	—	11,383	11,383	72,691	148	26,735	10,944
19,935	19,935	—	3,404	3,404	28,789	1,642	14,844	5,547
23,517	23,517	—	8,131	8,131	29,585	327	15,103	3,997
4,974,279	11,483,646	1,969,467	1,649,223	3,618,690	24,331,493	678,817	6,861,938	2,504,986
4,916,665	11,194,335	1,931,768	1,591,770	3,523,539	21,352,565	415,253	7,057,264	2,475,411

Total Compressor Stations		Volumes (MMCF)		Fiscal Data ($1000)			
Other	Total	Sold	Transmitted for Others	Gas Plant	Additions	Operating Revenues	Net Income
—	2	22,332	11,398	21,471	252	66,431	3,098
—	5	155,225	35,968	363,760	22,840	531,917	16,264
—	—	—	—	21,974	71	4,165	625
—	—	—	767	4,765	2	1,959	577
45	90	188,087	922,822	2,528,573	38,522	1,282,294	165,116
3	3	—	—	121,376	215	33,795	13,528
—	5	12,639	9,826	61,047	5,424	42,933	3,374
—	5	21,631	3,167	133,202		88,360	702
143	166	242,026	347,113	2,119,698	778,081	1,339,299	104,224
—	—	894	—	42	—	1,709	−29
1	1	—	—	143,746	−641	47,915	13,860
—	—	—	52,217	22,221	388	3,025	995
—	—	—	15,813	7,669	—	1,092	−484
—	—	959	—	2,370	135	4,374	143
—	—	32,317	—	—	—	74,391	—
—	1	—	40,013	22,232	11	7,889	1,078
2	4	99	14,094	4,379	584	1,046	181
7	14	14,213	2,162	75,141	4,441	47,346	640
—	—	—	28,285	20,865	232	930	−371
—	—	2,686	289	1,531	80	4,219	14
23	67	220,707	271,664	1,286,643	125,643	1,042,293	61,557
36	62	113,893	311,537	726,204	15,869	430,543	67,125
71	182	390,563	610,726	2,002,128	57,340	1,906,141	−146,407
22	35	—	833,060	1,080,801	23,895	129,161	12,592
—	—	—	1,611	201,957	—	12,229	856

(continued)

TABLE 5 *(continued)*

Company	Miles of Pipeline				Total Compressor Stations
	Transmission	Field	Storage	Total	Transmission
Consolidated System LNG Co.	—	—	—	—	—
Distrigas of Mass. Corp.	—	—	—	—	—
East Tennessee Natural Gas Co.[b]	1,091	—	—	1,091	15
Eastern Shore Natural Gas Co.	—	—	—	—	—
El Paso Natural Gas Co.[b]	10,765	10,926	18	21,709	88
Enserch Corp.	8,283	2,300	—	10,583	45
Equitable Gas Co.	—	—	—	—	—
Equitrans Inc.	545	1,485	135	2,165	6
Florida Gas Transmission Co.[b]	4,426	—	—	4,426	19
Freeport Interstate Pipeline Co.	9	—	—	9	—
Gas Gathering Corp.	—	31	—	31	—
Gas Transport Inc.	41	68	—	109	—
Gasdel Pipeline System Inc.	159	—	—	159	—
Glacier Gas Co.	34	3	—	37	1
Granite State Gas Transmission Inc.	105	—	—	105	1
Great Lakes Gas Transmission Co.[b]	1,321	—	—	1,321	14
Great Plains Natural Co.	65	—	—	65	—
Hampshire Gas Co.	—	18	—	18	—
High Island Offshore System[b]	203	—	—	203	1
Honeoye Storage Co.	11	19	—	30	—
Indiana Utilities Corp.	21	—	—	21	—
Inland Gas Co., The	128	394	—	522	—
Inter-City Minnesota Pipelines Ltd.	66	—	—	66	—
Interstate Power Co.	—	—	—	—	—
Iowa-Illinois Gas and Electric Co.	—	—	—	—	—
Jupiter Energy Corp.	27	—	—	27	—
K N Energy Inc.[b]	9,296	2,913	24	12,233	26
Kentucky West Virginia Gas Co.	108	2,078	—	2,186	9
Lawrenceburg Gas Transmission Corp.	6	—	—	6	—
Louisiana-Nevada Transit Co.	78	—	—	78	—
Lone Star Gathering Co.	39	7	—	46	—
Michigan Consolidated Gas Co.[b]	2,259	122	112	2,493	67
Michigan Gas Storage Co.[b]	563	—	148	711	1
Mid Louisiana Gas Co.	413	464	—	877	—
Midwestern Gas Transmission Co.[b]	903	—	—	903	15
MIGC Inc.	171	10	—	181	3
Miss. River Transmission Corp.[b]	2,018	145	74	2,237	—
Mitco Pipeline Co.	—	9	—	9	—
National Fuel Gas Distribution Corp.	641	335	—	976	—
National Fuel Gas Supply Corp.[b]	1,374	1,493	473	3,340	16
Natural Gas Pipeline of America[b]	10,724	1,753	294	12,771	56
North Penn Gas Co.	573	259	—	832	—
Northern Border Pipeline Co.[b]	824	—	—	824	2
Northern Natural, Div. of Enron[b]	16,318	7,398	208	23,924	123
Northwest Alaskan Pipeline Co.[b]	—	—	—	—	—
Northwest Pipeline Corp.[b]	3,543	3,476	7	7,026	29

Total Compressor Stations		Volumes (MMCF)		Fiscal Data ($1000)			
Other	Total	Sold	Transmitted for Others	Gas Plant	Additions	Operating Revenues	Net Income
—	—	—	—	—	—	15,427	4,408
—	—	15,252	—	49,113	3	50,727	15,744
1	16	92,276	154	143,470	8,332	261,594	4,250
—	—	7,638	3,496	14,965	687	24,481	921
12	100	186,401	1,121,964	2,557,466	78,988	1,041,722	9,399
—	45	206,371	275,920	1,214,537	62,565	1,055,712	-166,443
—	—	53,358	3,979	397,223	-194,318	276,021	-48,960
20	26	31,901	11,742	197,114	2,604	128,432	4,688
25	44	236,165	25,781	564,559	35,554	555,153	131
—	—	—	911	—	—	25	2
—	—	—	2,750	1,589	3	249	22
—	—	1,629	—	3,825	—	4,928	-56
—	—	—	13,225	14,774	-97	3,394	735
—	1	294	—	2,334	—	355	27
—	1	37,323	435	9,955	620	112,353	696
—	14	69,821	466,864	544,711	7,336	280,325	32,305
—	—	4,641	—	15,767	622	16,587	714
1	1	—	—	8,831	57	2,568	638
—	1	—	519,978	363,720	414	63,237	13,828
1	1	49	—	8,766	17	2,099	339
—	—	377	—	2,308	420	1,780	155
5	5	5,201	1	23,179	16	18,111	-330
—	—	8,074	3,160	3,494	15	18,351	-46
—	—	17,838	10,109	40,760	2,469	274,321	23,010
—	—	48,922	15,123	189,784	6,964	496,374	58,426
—	—	—	9,333	2,441	16	298	127
73	99	82,293	15,653	444,728	21,413	307,623	15,549
3	12	15,676	10,779	77,190	7,874	64,553	8,459
—	—	532	2,640	434	-41	2,962	58
—	—	3,221	23	2,669	145	4,894	611
—	—	—	376	1,656	7	393	88
8	75	207,784	125,044	1,514,257	76,521	1,197,639	49,406
1	2	3,177	102,810	54,353	860	108,980	1,984
—	—	19,566	1,503	89,029	6,154	52,576	9,050
—	15	115,514	2,174	171,086	2,965	386,796	937
2	5	2,676	14,196	13,811	684	9,373	-1,498
20	20	159,787	1,039	432,184	15,533	446,591	19,635
—	—	—	320	2,104	—	126	-133
2	2	135,126	22,860	705,280	53,106	687,980	21,408
3	19	149,742	58,204	303,047	22,544	476,770	22,535
21	77	393,822	1,042,406	2,447,468	52,916	1,603,450	-107,046
—	—	6,762	1,830	47,167	4,250	36,788	4,026
—	2	—	367,874	1,295,269	563	235,948	57,499
5	128	387,434	1,208,742	2,179,139	30,924	1,447,272	108,969
—	—	211,374	—	—	—	430,229	-389
26	55	100,140	450,495	1,351,983	15,980	563,066	5,964

(continued)

TABLE 5 *(continued)*

Company	Miles of Pipeline				Total Compressor Stations
	Transmission	Field	Storage	Total	Transmission
Ohio River Pipeline Corp.	24	—	—	24	—
Orange & Rockand Utilities Inc.	—	—	—	—	—
Overthrust Pipeline Co.[b]	89	—	—	89	—
Ozark Gas Transmission System	—	—	—	—	—
Pacific Gas Transmission Co.[b]	799	42	—	841	12
Pacific Interstate Offshore Co.	—	—	—	—	—
Pacific Interstate Transmission Co.	351	—	—	351	2
Pacific Offshore Pipeline Co.	—	—	—	—	—
Paiute Pipeline Co.	768	—	—	768	4
Panhandle Eastern Pipe Line Co.[b]	6,663	5,870	127	12,660	68
Pelican Interstate Gas System	85	—	—	85	—
Penn-Jersey Pipe Line	4	—	—	4	—
Penn-York Energy Corp.	—	—	39	39	—
Pennsylvania & Southern Gas Co.	—	—	—	—	—
Phillips Gas Pipeline Co.	153	—	—	153	—
Point Arguello Natural Gas Line	—	—	—	—	—
Questar Pipeline Co.[b]	1,558	745	50	2,353	13
Raton Gas Transmission Co. Inc.	21	—	—	21	—
Ringwood Gathering Co.	43	273	—	316	—
Sabine Pipe Line Co.	193	—	—	193	—
Sea Robin Pipeline Co.[b]	420	—	—	420	2
Seagull Interstate Corp.	7	5	—	12	—
Shenandoah Gas Co.	82	—	—	82	—
South County Gas Co.	—	—	—	—	—
South Georgia Natural Gas Co.	767	—	—	767	2
South Penn Gas Co.	17	—	—	17	—
Southeastern Natural Gas Co. of Ohio	—	—	—	—	—
Southern Energy Co. (LNG)	—	—	—	—	—
Southern Natural Co.[b]	7,656	108	48	7,812	48
Southwest Gas Storage Co.	—	—	31	31	—
Southwest Gas Transmission Co.	—	8	—	8	—
Stingray Pipeline Co.[b]	354	—	—	354	2
Superior Offshore Pipeline Co.	112	—	—	112	—
Tarpon Transmission Co.	40	—	—	40	—
T C P Gathering Co.	—	20	—	20	—
Tennessee Gas Pipeline Co.[b]	14,138	31	2	14,171	68
Texas Eastern Transmission Corp.[b]	9,859	48	196	10,103	77
Texas Gas Pipe Line Corp.	6	—	—	6	—
Texas Gas Transmission Corp.[b]	5,745	45	219	6,009	27
Texas Sea Rim Pipeline Inc.					
Trailblazer Pipeline Co.[b]	436	—	—	436	—
Transco Gas Supply Co.[b]	—	—	—	—	—
Transcontinental Gas Pipe Line Corp.[b]	10,458	149	121	10,728	40
Transwestern Pipeline Co.[b]	4,457	—	—	4,457	22
Trunkline Gas Co.[b]	4,044	498	27	4,569	21
Trunkline LNG Co.	—	—	—	—	—

Total Compressor Stations		Volumes (MMCF)		Fiscal Data ($1000)			
Other	Total	Sold	Transmitted for Others	Gas Plant	Additions	Operating Revenues	Net Income
—	—	7,573	—	3,684	15	19,124	161
—	—	21,996	—	131,550	7,937	443,409	44,238
—	—	—	63,460	63,041	—	9,978	2,244
—	—	—	37,470	146,300	418	25,255	2,787
1	13	371,350	120,037	415,515	8,114	793,401	9,518
—	—	16,476	—	12,014	—	42,358	552
—	2	93,733	—	63,513	313	305,762	1,528
—	—	9,701	—	148,804	3,812	72,922	6,023
—	4	9,377	41	55,682	4,762	28,923	982
67	135	54,106	536,066	1,583,659	65,059	597,144	-179,260
—	—	—	13,378	11,081	—	1,806	-27
—	—	—	755	234	—	62	-1
2	2	—	—	59,771	-433	25,788	6,325
—	—	4,691	1,347	19,033	1,343	23,263	1,306
—	—	619	42,362	48,603	494	3,630	-3,170
—	—	—	—	—	—	—	—
30	43	54,752	103,828	369,432	11,464	216,529	19,460
—	—	1,158	—	534	1	3,997	60
2	2	1,835	6,357	10,543	236	5,986	138
—	—	—	51,674	39,033	1,671	3,360	-748
1	3	3,244	184,256	252,176	1,696	51,908	5,693
—	—	—	10,815	6,234	—	673	-66
—	—	4,257	—	17,695	2,080	16,495	454
—	—	418	147	3,391	183	3,046	200
—	2	6,297	16,460	23,455	311	35,102	515
—	—	267	—	679	10	1,200	42
—	—	8	—	5	—	753	-108
—	—	—	—	155,228	51	11,092	467
13	61	153,048	52,052	1,313,248	52,603	781,757	79,379
1	1	—	—	68,967	1	20,924	6,253
—	—	—	17,098	1,539	—	355	58
—	2	—	240,169	247,798	522	32,156	5,849
—	—	—	50,360	12,894	13	652	90
—	—	—	24,225	21,088	—	1,456	192
—	—	—	1,596	1,163	21	178	63
1	69	380,964	1,103,741	3,669,048	66,672	1,593,414	854,388
8	85	679,106	302,781	3,096,396	197,129	2,054,952	61,947
—	—	1,995	261	433	108	3,860	87
9	36	210,931	634,590	950,659	28,762	816,137	31,356
—	—	—	58,036	269,206	-11	30,537	-3723
—	—	29,483	—	10	—	83,144	2,035
33	73	234,017	1,524,175	3,351,127	178,352	1,225,574	-3,192
97	119	90,472	261,206	696,462	9,421	391,473	22,973
14	35	156,674	341,001	1,008,259	27,101	628,491	-93,728
—	—	—	—	579,036	68	9,348	-48,920

(continued)

TABLE 5 *(continued)*

Company	Miles of Pipeline				Total Compressor Stations
	Transmission	Field	Storage	Total	Transmission
U-T Offshore System[b]	30	—	—	30	—
Union Light, Heat & Power Co.	—	—	—	—	—
United Cities Gas Co.	—	—	—	—	—
United Gas Pipe Line Co.[b]	7,362	2,383	28	9,773	29
Valero Interstate Transmission Co.	—	—	—	—	—
Valley Gas Transmission Inc.	—	200	—	200	—
Washington Gas Light Co.	379	—	—	379	—
West Texas Gas Inc.	—	—	—	—	—
West Texas Gathering Co.	—	—	—	—	—
Western Gas Interstate Co.	191	50	—	241	9
Western Transmission Corp.	—	61	—	61	—
Wheeler Gas, Inc.	—	—	—	—	—
Williams Natural Gas Co.[b]	5,857	3,770	175	9,802	40
Williston Basin Interstate Pipeline Co.[b]	3,071	847	148	4,066	23
Wyoming Interstate Co. Ltd.[b]	269	—	—	269	1
Zenith Natural Gas Co.	27	32	—	59	—
1988 Totals—majors (43)	186,723	62,800	4,542	254,065	1,232
1987 Totals—majors (44)	180,737	63,356	8,354	252,447	1,205
1988 Totals—all	204,192	72,393	4,796	281,381	1,336
1987 Totals—all	198,578	72,981	8,526	280,085	1,331

[a]Source: FERC Forms 2's and 2-A's for major and nonmajor natural-gas pipeline companies. Under criteria established for the 1984 reporting year (*Oil Gas J.*, p. 80, November 25, 1985). Major pipeline companies are those whose combined gas sold for resale and gas transported for a fee exceeded 50 BCF at 14.73 lb/in.^{2} (60°F) in each of the 3 previous calendar years. Nonmajors are natural-gas pipeline companies not classified as majors and having had total gas sales of volume transactions exceeding 200 MMCF at 14.73 lb/in.^{2} (60°F) in each of the 3 previous calendar years.
[b]Major natural-gas pipeline companies under FERC Accounting and Reporting Requirements for Natural Gas Companies, para. 20-011, effective February 2, 1985, beginning with 1984 reporting year.

At the end of 1988, reports from liquids-pipeline companies show a total of 34,311 miles of gathering lines, 55,900 miles of crude-oil trunk lines, and 80,246 miles of products pipelines.

For liquids pipeline companies, the criterion determining which company must file an annual report (Form 6) is whether the FERC determines the company to be an interstate common-carrier pipeline.

These reports for 1988 show that gathering lines decreased by 744 miles (2.1%). Crude lines increased by 1014 miles (1.9%) and products lines by 2322 miles (2.98%) over totals reported for 1987.

These figures are in line with the erratic pattern of liquids-pipeline utilization for the past 8 years (Table 2). Natural-gas transmission lines were up over 1987 by 5614 miles (2.83%) for all companies and by 5986 miles (3.3%) for major companies.

Field lines for all companies showed a fractional decrease of 588 miles and by 556 among major companies.

Total Compressor Stations		Volumes (MMCF)		Fiscal Data ($1000)			
Other	Total	Sold	Transmitted for Others	Gas Plant	Additions	Operating Revenues	Net Income
—	—	—	304,516	63,055	31	6,214	1,076
—	—	11,328	7,839	79,964	7,472	185,789	6,199
—	—	35,780	241	186,646	22,129	168,723	5,701
13	42	76,139	729,852	1,142,703	21,097	713,816	45,498
—	—	213	51,085	13,942	488	5,783	179
—	—	2,943	2,664	2,980	8	6,378	1,111
—	—	117,517	2	926,255	90,318	668,301	45,069
—	—	707	—	19,287	235	24,341	6,025
—	—	4,006	3,304	4,368	22	8,869	−1,005
—	9	3,395	—	8,197	87	7,476	339
—	—	—	5,221	600	20	528	276
—	—	21	—	—	−424	72	349
55	95	151,665	154,715	669,601	36,234	502,912	32,680
7	30	31,508	38,430	205,592	2,994	119,776	3,937
—	1	—	70,988	176,570	−198	23,196	1,668
2	2	1,127	38	784	157	2,743	67
884	2,116	6,383,690	15,665,971	44,018,804	2,140,075	26,740,552	1,415,225
827	2,032	6,266,593	12,880,681	42,040,766	1,208,123	26,358,565	1,117,717
938	2,274	7,523,542	16,542,847	50,910,100	2,250,038	32,736,380	1,572,609
897	2,228	7,301,292	13,505,501	49,969,424	1,596,481	31,596,171	1,438,203

For all companies as well as majors, storage lines decreased in 1988 over 1987: down 3730 miles (43.8%) for all companies, 3812 miles (45.7%) for majors.

Deliveries Rise

Throughput for liquids-pipeline companies in 1988 showed healthy gains over 1987.

Natural-gas pipelines sales rose slightly in 1988, by 117.1 BCF (+1.9%) for majors and by almost 222.2 BCF (3%) for all companies.

But the major story continues to be the shift of natural-gas pipeline companies from their historic role as buyers-for-resale of gas to transporters. This shift, going on now for almost 5 years, was evident in 1988 in the volumes of gas regulated interstate pipeline companies transported for others.

For 1988, all regulated United States interstate gas pipelines carried 16.5 TCF of gas for others; of those volumes, majors carried nearly 15.6 TCF. For all companies, volumes of gas transported for others rose 3.2 TCF (+24%); for majors the rise was almost 2.8 TCF (+21.6%).

Volumes of gas transported for others in 1988 by all interstate gas-pipeline companies (16.5 TCF) comprised 68% of total volumes (24.07 TCF) moved through the United States system, up from almost 65% for 1987. For major companies (whose total volumes slightly exceeded 22 TCF), the share was a bit larger: more than 71%. This share also represented an increase over the 1987 figure, up from 67%.

Crude-oil and product deliveries in 1988 again approached 11.5 billion bbl, an increase over 1987 of 289 million bbl, or only about 2.6%.

Crude-oil deliveries through the United States interstate system rose in 1988 by 231.7 million bbl, or an increase of 3.7%.

Product deliveries were up by 57.6 million bbl (1.2%).

Trunkline traffic for United States crude-oil and product pipelines showed a bit more life in 1988 than in 1987, advancing by 95.1 billion bbl-miles (2.7%) over that for 1987.

Crude-oil trunkline traffic increased by 37.7 billion bbl-miles (2%). Product traffic also increased by 57.4 billion bbl-miles (3.6%).

Rankings

Oil & Gas Journal ranks the top 10 natural-gas and liquids pipeline companies in three categories: mileage, natural-gas sales or liquids barrel-mile throughput, and operating income (Tables 6 and 7). These rankings are broken out from the pipeline-company tables (Tables 4 and 5).

For all natural-gas pipeline companies, net income as a portion of gas-plant investment reversed a slide evident in recent years. This portion stood at 3.1% for 1988, up from 2.9% in 1987. It was 3.4% in 1986, 4.5% in 1985, and 8.7% in 1984.

For major gas pipelines, net income as a portion of gas-plant investment mirrored the change, rising to 3.2% for 1988 after falling to 2.6% in 1987 from 3.2% in 1986. It was 4.3% in 1985, and 7.6% in 1984.

For 1988, all gas companies reported an industry gas-plant investment totaling almost $51 billion compared to $49.9 billion in 1987, $49.4 billion for 1986, and $48.3 billion for 1985.

Majors' gas-plant investment in 1988 increased to $44 billion from slightly less than $42 billion for 1987 and $42.2 billion for 1986.

For interstate common-carrier liquids pipeline companies, net income as a percentage of investment in carrier property fell to 10.3% in 1988 from 11.6% in 1987. It was 9.2% in 1986, 11.3% in 1985, and 13.1% in 1984.

Liquids pipelines' investment in carrier property for 1988 increased by more than $2.9 billion (+14%) over 1987 which had fallen to slightly less than $22.3 billion compared with $22.4 billion for 1986, $21.6 billion in 1985, and $19.4 billion in 1984.

Another measure of the profitability of oil and natural-gas pipeline companies in recent years is the percentage net income represents of opera-

TABLE 6 Top 10 Interstate Liquids Pipelines—1988[a]

Company	Mileage[b]	Company	Trunkline traffic (million bbl-miles)	Company	Income $1000
1. Amoco Pipeline Co.	12,297C	Colonial Pipeline Co.	658,014P	Sohio Alaska Pipeline Co.	482,913
2. Mobil Pipe Line Co.	9,683C	Lakehead Pipe Line Co. Inc.	366,739P	Exxon Pipeline Co.	304,700
3. Exxon Pipeline Co.	9,378C[c]	Sohio Alaska Pipeline Co.	294,321C	ARCO Pipe Line Co.	217,588
4. Chevron Pipe Line Co.	8,130P	Exxon Pipeline Co.	170,391C	Colonial Pipeline Co.	157,057
5. Mid-America Pipeline Co.	8,100P[d]	ARCO Pipe Line Co.	166,115C	Amoco Pipeline Co.	102,377
6. Conoco Pipeline Co.	7,895C	Explorer Pipeline Co.	129,220P	Mobil Pipe Line Co.	82,961
7. Shell Pipe Line Corp.	7,235C	Amoco Pipeline Co.	123,362C	Chevron Pipe Line Co.	81,950
8. ARCO Pipe Line Co.	6,839C[c]	Plantation Pipe Line Co.	118,068P	Shell Pipe Line Corp.	70,062
9. Williams Pipe Line Co.	6,780P	Shell Pipe Line Corp.	113,814C	Southern Pacific Pipe Line L.P.	61,199
10. Texaco Pipeline Co.	6,275C	Chevron Pipe Line Co.	97,601P	Lakehead Pipe Line Co. Inc.	59,077
Total	81,794		2,237,645		$1,619,884
Part of all companies, %	48.73		61.84		64.67
1987 totals	80,744		2,104,875		$1,636,671

[a]Source: U.S. FERC Form 6: Annual Report of Oil Pipeline Companies, December 31, 1988,
[b]C = mostly or exclusively crude oil; P = mostly or exclusively product.
[c]Includes 818 miles for Trans-Alaska Pipeline; to avoid duplication, the total includes 818 miles only once.
[d]Includes 1097 miles of ammonia pipelines.

TABLE 7 Top 10 United States Gas-Pipeline Companies—1988[a]

Company	Mileage	Company	Sales (MMCF)	Company	Net Income ($1000)
1. Northern Natural, Div. of Enron	23,924	Texas Eastern Transmission Corp.	679,106	Tennessee Gas Pipeline Co.	854,388
2. El Paso Natural Gas Co.	21,709	Natural Gas Pipeline of America	393,822	ANR Pipeline Co.	165,116
3. Columbia Gas Transmission Corp.	18,804	Columbia Gas Transmission Corp.	390,563	Northern Natural, Div. of Enron	108,969
4. Tennessee Gas Pipeline Co.	14,171	Northern Natural, Div. of Enron	387,434	Arkla Inc.	104,224
5. Natural Gas Pipeline of America	12,771	Tennessee Gas Pipeline Co.	380,964	Southern Natural Co.	79,379
6. Panhandle Eastern Pipe Line Co.	12,660	Pacific Gas Transmission Co.	371,350	Colorado Interstate Gas Co.	67,125
7. K N Energy Inc.	12,233	Arkla Inc.	242,026	Texas Eastern Transmission Corp.	61,947
8. ANR Pipeline Co.	11,534	Florida Gas Transmission Co.	236,197	CNG Transmission Corp.	61,557
9. Transcontinental Gas Pipe Line Corp.	10,728	Transcontinental Gas Pipe Line Corp.	234,017C	Iowa-Illinois Gas and Electric Co.	58,426
10. Ensearch Corp.	10,583	CNG Transmission Corp.	220,707	Northern Border Pipeline Co.	57,499
Totals	149,117		3,536,186		$1,618,630
Part of majors, %	58.69		55.39		114.37
Part of all companies, %	52.99		47.00		102.93
1987 totals	151,514		3,608,745		$1,173,625

[a]Source: U.S. FERC Form 2: Annual Report for Major Natural Gas Companies, December 31, 1988.

ting income. Through 1987, trends for liquids-pipeline companies and for natural-gas pipeline companies had been heading in opposite directions for 10 years.

For liquids-pipeline companies in 1988, this figure rose to 36.5% from 35% for 1987, 28% in 1986, and 21.9% in 1978. The figure for 1986 represents the only year the percentage has declined (from 32.6% for 1985).

By contrast for all natural-gas pipeline companies, income as a percentage of operating revenues has been steadily falling in the past 10 years.

At the end of 1988, however, it increased to 4.8% from 4.5% for 1987, the latter portion fractionally higher than for 1986. The 10-year decline is evident when the portion for 1978 is noted: 8.3%.

For majors in 1988, income as a percentage of operating revenues was 5.3%.

Tracked Companies

In this annual Pipeline Economics Report, *Oil & Gas Journal* has for several years been tracking investment by five crude-oil pipeline companies and five products-pipeline companies.

Table 8 indicates that investment by the five crude-oil pipelines stood at $1711.2 million at the end of 1988 compared with $1661.8 million at the end of 1987, an increase of $49.4 million (2.9%).

Investment by the five products pipeline companies was $2934.5 million at the end of 1988 compared with $2890.7 million at the end of 1987, a slight increase of $43.8 million (1.5%).

Figure 2 illustrates the investment split in the crude-oil and products pipeline companies.

Construction Activity

As Fig. 2 shows, the costs of line pipe and fittings and of laying pipelines make up roughly 70% of the cost of a pipeline system.

These elements of pipeline construction costs have risen very little for the past few years, after rapidly increasing before 1982.

Cost Trends

Table 9 lists a 10-year land-construction cost trend for natural-gas pipelines with diameters ranging from 8 to 36 in. The table's data are based on a mix of actual costs of completed pipeline construction and projected costs filed under CP dockets with the FERC.

As the table reflects, the average cost per mile for any given diameter may fluctuate from one year to the next as projects' costs are affected by geographic location, terrain, population density, or other factors.

TABLE 8 Investment in Liquids Pipelines[a]

	Company and Investment ($)						
	A	B	C	D	E	Total ($)	%
Crude Pipelines Investment by Five Companies							
Land	1,787,499	814,497	292,759	1,001,563	1,194,975	5,091,293	0.30
Right of way	12,288,388	1,006,210	316,392	16,428,203	15,627,321	45,666,514	2.67
Line pipe	138,917,243	25,705,045	7,797,471	97,880,236	154,588,584	424,888,579	24.83
Line pipe fittings	9,888,276	922,085	1,386,293	17,591,909	28,249,661	58,038,224	3.39
Pipeline construction	203,189,839	27,810,895	10,931,879	222,373,192	228,611,206	692,917,011	40.49
Buildings	22,411,241	4,967,781	2,473,633	13,752,261	10,738,992	54,343,908	3.18
Boilers	—	—	—	—	—	—	—
Pumping equipment	25,374,049	5,524,747	2,414,882	17,381,420	48,587,557	99,282,655	5.80
Machine tools and machinery	—	—	—	33,140	—	33,140	0.00
Other station equipment	93,848,747	21,881,549	4,601,583	45,653,071	33,947,215	199,932,165	11.68
Oil tanks	14,531,122	5,991,716	5,678,530	17,074,367	26,633,026	69,908,761	4.09
Delivery facilities	—	—	5,357,432	1,569,207	6,820,018	13,746,657	0.80
Communication systems	886,153	3,196,680	70,566	11,373,089	3,692,707	19,219,195	1.12
Office furniture and equipment	1,973,669	250,029	230,421	2,257,128	899,485	5,610,732	0.33
Vehicles and other work equip.	8,509,180	846,444	512,646	6,250,181	245,786	16,364,237	0.96
Other property	—	1,309,285	—	—	4,767,315	6,076,600	0.36
Total	$533,605,406	$100,226,963	$42,064,487	$564,603,848	$470,618,967	$1,711,119,671	100.00[b]
Product Pipelines Investment by Five Companies							
Land	4,422,773	1,763,961	574,646	936,246	3,250,670	10,948,296	0.37
Right of way	29,644,947	10,829,964	19,697,239	8,777,747	6,834,819	75,784,716	2.58
Line pipe	345,795,587	67,590,791	107,841,055	87,382,990	98,196,581	706,807,004	24.09
Line pipe fittings	72,357,196	12,889,746	15,970,328	1,732,982	21,615,003	124,565,255	4.24
Pipeline construction	582,101,592	109,237,361	203,656,797	01,968,825	143,747,696	1,140,712,271	38.87
Buildings	23,951,907	4,067,919	5,267,726	10,625,847	16,747,663	60,661,062	2.07
Boilers	—	—	—	—	—	—	—
Pumping equipment	45,685,261	6,182,370	30,499,386	28,200,014	20,059,120	130,626,151	4.45
Machine tools and machinery	—	—	—	—	407,435	407,435	0.01
Other station equipment	120,823,235	21,212,683	37,825,620	23,395,650	116,266,346	319,523,534	10.89
Oil tanks	48,834,579	17,463,369	8,964,158	17,465,781	61,094,267	153,822,154	5.24

Delivery facilities	—	—	9,509,419	10,821,471	47,518,440	67,849,330	2.31
Communication systems	1,353,992	558,106	2,371,291	8,878,183	—	13,161,572	0.45
Office furniture and equipment	10,277,081	2,426,975	2,638,126	640,018	958,420	16,940,620	0.58
Vehicles and other work equip.	5,641,235	1,853,004	7,957,202	2,449,576	2,332,710	2,332,727	0.69
Other property	73,357,374	—	19,089,144	—	—	92,446,518	3.15
Total	$1,364,246,459	$256,076,249	$471,862,137	$303,275,330	$539,029,170	$2,934,489,345	100.00[b]

[a]Source: U.S. FERC Form 6, Annual Report of Oil Pipeline Companies, December 31, 1988.
[b]Actual total may be inexact due to rounding.

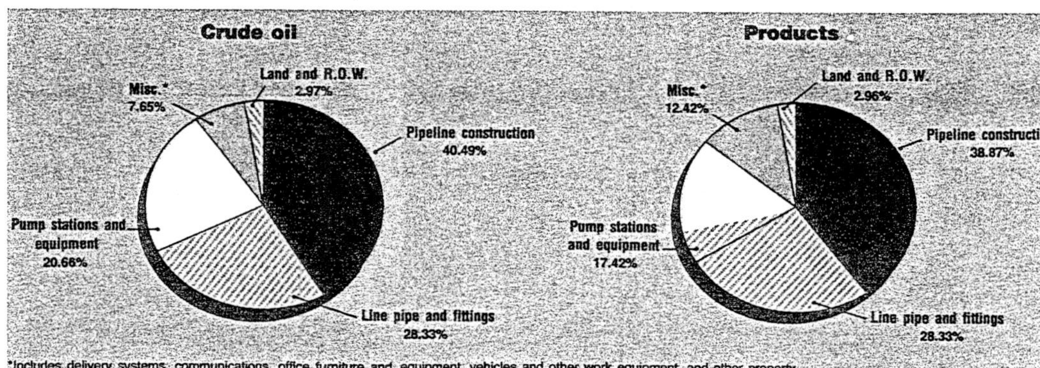

FIG. 2. Liquids-pipeline investment. (Source: U.S. FERC Form 6, oil-pipeline company annual reports, 1988.)

TABLE 9 10-Year Land Construction Cost Trends

Size	Year[a]	Average Cost ($/mile)					Range ($/mile)	
		R.O.W.	Material	Labor	Miscellaneous	Total	Low	High
8 in.	1989	33,945	130,785	330,581	883,558	578,869	520,307	692,570
	1988	7,619	58,695	56,381	37,363	160,058	116,856	220,000
	1987	20,032	55,430	75,568	30,581	181,611	151,193	652,036
	1986	14,763	44,703	63,102	32,958	155,527	137,033	193,846
	1985	3,141	28,144	47,317	16,283	94,884	64,203	228,866
	1984	22,339	49,974	175,733	74,086	322,132	223,767	540,655
	1983	6,244	48,392	89,600	23,088	167,324	159,940	178,020
	1982	8,996	46,002	67,950	31,769	154,717	73,596	572,926
	1981	8,119	34,860	23,528	20,884	87,391	55,981	109,078
	1980	8,247	39,040	102,839	13,036	163,162	149,926	200,554
12 in.	1989	43,896	96,918	227,288	79,788	447,890	191,427	1,476,787
	1988	52,772	85,487	268,169	89,656	496,085	261,697	1,873,852
	1987	26,018	81,675	161,790	67,404	336,826	166,473	669,332
	1986	43,656	94,786	243,533	101,683	483,657	186,180	1,201,356
	1985	26,582	74,620	106,447	46,164	253,813	131,765	767,548
	1984	23,035	147,750	211,783	93,481	476,050	338,872	608,685
	1983	5,135	91,492	73,042	20,754	190,423	177,050	218,265
	1982	10,536	84,972	102,390	32,173	230,071	130,252	669,227
	1981	3,855	71,444	65,613	11,861	152,773	55,981	369,018
	1980	2,610	54,235	41,089	14,111	112,045	78,428	901,282
16 in.	1989	25,288	99,141	145,913	54,092	324,434	236,993	1,615,768
	1988	34,217	114,232	185,798	56,558	390,805	203,181	2,256,422
	1987	34,775	109,042	187,216	68,857	399,890	190,946	3,334,886[b]
	1986	21,474	104,368	183,134	53,546	362,523	302,102	1,071,477
	1985	8,511	80,529	47,708	20,091	156,839	154,746	167,876
	1984	15,106	131,139	111,234	82,747	340,226	231,790	363,718
	1983	31,542	337,088	929,969	278,159	1,576,758	447,890	2,370,748
	1982	6,278	104,319	113,380	21,936	245,913	175,715	430,841
	1981	9,203	110,684	133,968	40,424	294,279	140,674	590,112
	1980	7,404	74,571	57,681	18,050	157,706	118,085	471,972

20 in.	1989	82,862	202,309	446,538	135,703	867,411	366,853	2,778,664
	1988	27,225	139,302	170,120	58,787	345,225	246,260	2,776,416
	1987	35,655	139,346	172,071	68,265	415,337	274,034	904,487
	1986	29,227	134,332	182,623	82,657	428,839	175,000	1,077,419
	1985	108,870	187,379	190,886	144,974	632,109	353,088	1,277,454
	1984	16,735	175,406	190,088	63,302	445,531	226,415	616,250
	1983	12,682	147,354	154,811	46,073	360,934	237,600	476,104
	1982	12,284	124,148	128,148	30,649	295,229	210,378	482,464
	1981	8,556	122,659	114,147	30,370	275,732	227,679	1,741,597
	1980	11,470	130,438	73,751	48,213	263,872	157,947	533,265
24 in.	1989	26,328	165,032	193,096	75,348	459,804	446,819	2,186,435
	1988	49,494	231,264	360,603	145,358	786,719	431,093	4,289,167[c]
	1987	34,785	254,677	271,319	110,959	671,741	405,987	2,043,000
	1986	43,746	245,807	378,917	170,094	838,565	355,428	1,607,660
	1985	26,648	296,486	305,437	151,657	780,228	469,318	4,893,740
	1984	29,133	289,487	298,521	128,390	745,531	395,676	1,772,049
	1983	31,416	310,583	501,401	148,993	995,393	608,781	1,649,454
	1982	28,537	508,226	807,847	120,895	1,465,505	[b]	[b]
	1981	30,078	269,414	395,204	127,790	822,486	350,898	826,088
	1980	14,220	162,019	152,406	52,114	380,759	230,122	820,295
30 in.	1989	26,090	294,249	285,222	116,421	721,981	489,867	3,000,000
	1988	47,502	260,549	322,973	137,019	768,043	526,637	4,858,168[c]
	1987	25,506	257,158	308,383	161,008	752,054	477,195	1,584,483
	1986	34,035	268,922	253,838	120,603	677,397	188,835	1,571,948
	1985	60,408	320,219	314,878	152,035	847,541	668,640	5,778,947
	1984	20,885	336,043	376,631	183,034	916,592	637,546	1,413,826
	1983	23,872	253,071	348,478	116,084	741,506	660,086	1,683,143
	1982	20,756	296,737	362,787	93,103	773,383	548,396	1,179,760
	1981	14,135	309,364	201,037	73,281	597,817	561,358	4,684,745
	1980	13,206	227,229	134,491	46,018	420,944	319,482	921,772

(continued)

TABLE 9 *(continued)*

Size	Year[a]	Average Cost ($/mile)					Range ($/mile)	
		R.O.W.	Material	Labor	Miscellaneous	Total	Low	High
36 in.	1989	35,408	417,820	529,578	180,613	1,163,419	832,407	2,410,057
	1988	59,395	385,007	449,650	204,591	1,098,643	758,869	2,625,105
	1987	45,461	415,118	637,294	144,784	1,242,657	1,170,963	1,923,266
	1986	360,582	428,572	468,960	151,769	1,085,359	714,286	5,264,368
	1985	21,967	389,751	296,742	157,536	865,997	629,179	6,443,590
	1984	36,063	358,876	362,667	152,787	910,393	665,480	2,143,680
	1983	51,436	498,047	490,568	306,666	1,346,717	703,112	2,724,493
	1982	13,429	375,281	176,287	58,733	623,730	615,160	833,546
	1981	5,948	344,374	358,072	74,415	782,809	693,945	1,573,759
	1980	20,677	344,664	268,753	87,850	721,944	528,710	1,105,515

[a] 1980–1982 by calendar year. 1983–1988 based on FERC construction-permit applications, by fiscal year.
[b] One report only.
[c] Involves a river crossing.

The cost-per-mile trends from 1987 to 1988 are up, with four of seven diameter classifications showing increases. These are for 8, 16, 20, and 36-in. categories.

Yearly fluctuations in these figures are illustrated in construction figures for a 12-in. pipeline. These had leaped between 1984 and 1985 by 90% but fell in 1988 by more than 30%.

For the period reported on in this year's report, however, the total average cost-per-mile for 12-in. pipeline fell again, this time by $48,195/mile, or 9.7%.

Plans and Miles

Table 3 lists 84 land-pipeline construction projects and 21 offshore for a total of 105, compared to 129 land and 3 offshore reported in last year's Pipeline Economics Report.

Land projects proposed during the period surveyed represent 2025.78 miles of pipeline at an estimated price tag of more than $1.7 billion, and the offshore pipeline projects represent 386.59 miles of pipeline with an estimated price of just more than $285 million.

Total estimated costs for these projects, land and offshore, reach almost $2 billion.

Total costs of projects surveyed in 1988's Pipeline Economics Report ran more than $3.6 billion divided between $3.3 billion for land and $303 million for offshore.

All projects proposed during the period covered (July 1, 1988, to June 30, 1989) represent 2410.37 miles, a decrease over mileage proposed for the period immediately preceding of 5519.75 miles, or −56.3%.

Examination of proposed projects for the latest period reveals the extent to which United States natural-gas pipeline companies want to go after the anticipated heavy gas demand in the United States Northeast. Many projects reflect looping in Pennsylvania, New Jersey, and Connecticut.

By way of comparison, this year's report includes data on a major Canadian gas-transmission system expansion project (Fig. 3). By the same token, anticipated heavy gas demand in California EOR and electric-generating markets shows up in the Idaho-Washington projects.

And development of the large gas fields immediately offshore Alabama is reflected in the mileage proposed for that area.

Cost Trends, Components

Cost-per-mile figures may reveal more about cost trends than aggregate costs.

For proposed United States gas-pipeline projects in the 1988–89 period surveyed, the average land cost per mile was $842,050/mile compared with $657,388/mile for the 1987–88 period, an increase of 28%.

For offshore projects, the 1988–89 figure was $737,474 compared with $755,345/mile for 1987–88, a decrease of 2.4%.

Combined land and offshore projects show an average cost of $825,278/mile for this year's coverage; for the 1987–88 period, the figure was $664,509/mile.

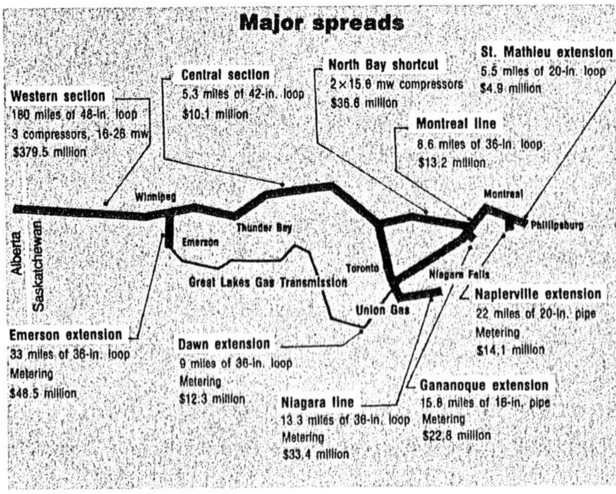

Cost elements

Type	Costs, $1,000				
	Pipeline	Compression	Metering	Other	Total
Materials	262,708	67,275	7,935	7,700	345,618
Installation	220,146	44,019	5,140	0	269,305
Land	10,095	0	450	0	10,545
Engineering	4,701	2,040	250	0	6,991
Contingency	24,883	5,667	689	385	31,623
Overhead	6,270	1,428	174	97	7,969
Afudc	29,084	6,624	805	450	36,963
Total	**$557,887**	**$127,052**	**$15,442**	**$8,632**	**$709,014**

FIG. 3. TransCanada Pipeline's proposed expansion. (Source: Canadian National Energy Board. All costs are direct costs and estimated in $Can as of June 1989.)

Analyses of the four major categories of pipeline construction cost—material, labor, right-of-way and damages, and miscellaneous—can also indicate trends within each group.

"Miscellaneous" generally includes engineering, supervision, contingencies, allowances for funds used during construction (afudc), administration and overheads, and FERC filing fees.

For the 105 projects surveyed for the 1988–89 period covered in this report, cost-per-mile data for the four categories are as follows:

Material—Land, $309,938/mile; offshore, $308,167/mile; total, $309,654/mile.
Labor—Land, $361,472/mile; offshore, $289,386/mile; total, $349,920/mile.
R.O.W. and damages—Land, $33,330/mile; offshore, $1372/mile; total, $28,209/mile.
Miscellaneous—Land, $136,479/mile; offshore, $138,549/mile; total, $136,810/mile.

Table 3 lists proposed pipelines in order of increasing diameter and increasing lengths within each pipeline diameter.

The average cost per mile for the projects shows no clear-cut trend related to either length or geographic area.

In general, however, the cost per mile within a given diameter indicates that the longer the pipeline, the lower the incremental cost for construction.

Nonetheless, road, highway, and river crossing and marshy or rocky terrain each strongly affects pipeline construction costs.

Figure 4, derived from Table 3 for land and offshore pipelines, shows the major cost-component split for pipeline construction cost.

Material and labor are shown to make up almost 80% of the cost of

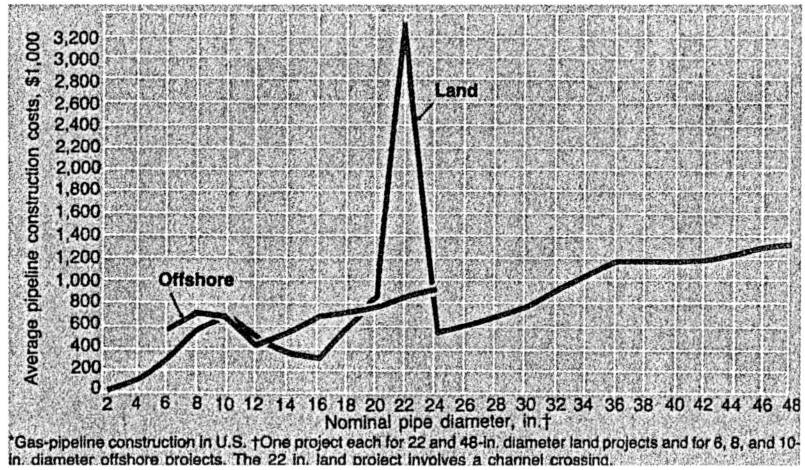

FIG. 4. Gas-pipeline construction costs. (Source: U.S. FERC construction-permit applications, July 1, 1988 to June 30, 1989.)

constructing land pipelines and almost 81% of the cost for offshore pipelines.

Figure 5 plots the average pipeline construction cost for land and offshore gas-pipeline construction projects listed in Table 3.

Compressor Stations

Compressor-station costs make up another major cost element of natural-gas pipelines. Table 10 lists 44 land and 2 offshore compressor-station projects. FERC applications for these projects cover the same period as for pipelines: July 1, 1988, to June 30, 1989.

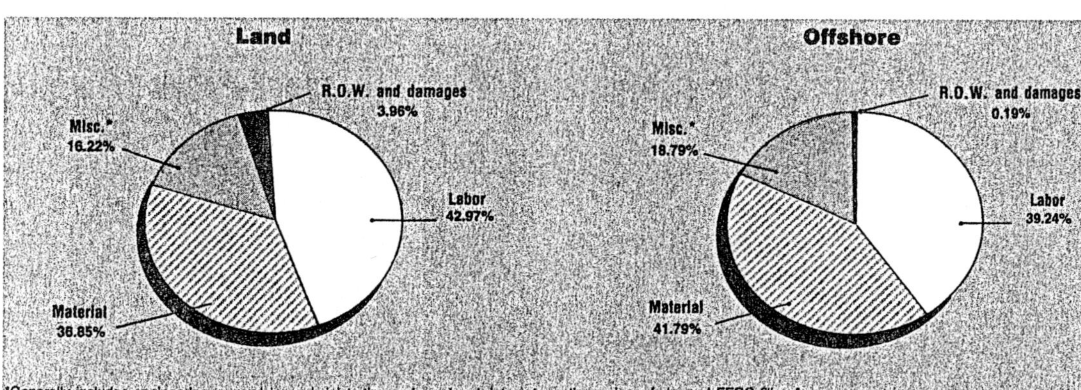

FIG. 5. Average costs of gas-pipeline construction in the United States. (Source: U.S. FERC construction-permit applications, July 1, 1988 to June 30, 1989.)

TABLE 10 Current Compressor-Station Costs[a]

Location[b]	Horsepower	$					
		Equipment and material	Labor	Land	Miscellaneous	Total	$/hp
Tennessee	1,100	720,000	575,000	100,000	294,172	1,689,172	1,536
Tennessee	1,100	870,000	600,000	—	305,981	1,775,981	1,615
North Carolina	1,200	1,501,000	894,000	—	357,088	2,752,088	2,293
Tennessee	1,280	955,000	600,000	—	317,791	1,872,791	1,463
Tennessee	1,280	1,180,000	775,000	84,000	408,650	2,447,650	1,912
Tennessee	1,280	1,445,000	900,000	550,000	516,748	3,411,748	2,665
Pennsylvania	2,200	2,425,400	575,500	—	479,100	3,480,000	1,582
New York	2,700	4,157,000	3,221,000	3,866,185	46,000	11,290,185	4,181
Pennsylvania	3,000	4,250,000	1,181,000	—	815,120	6,246,120	2,082
Pennsylvania	3,000	1,030,000	582,000	—	340,000	1,952,000	651
Pennsylvania	4,000	5,294,000	5,665,000	400,000	1,341,132	12,700,132	3,175
Alabama	4,000	6,200,000	1,500,000	100,000	2,576,069	10,376,069	2,594
Alabama	4,000	4,402,000	3,219,000	450,000	1,037,151	9,108,151	2,277
Arkansas	4,390	3,451,000	854,000	—	1,672,468	5,977,468	1,362
Virginia	4,390	2,450,000	930,000	—	719,556	4,099,556	934
Virginia	4,390	2,450,000	930,000	—	719,556	4,099,556	934
W. Virginia	4,390	2,356,000	894,000	—	390,000	3,940,000	897
Arkansas	4,500	4,807,000	1,200,000	100,000	2,606,288	8,713,288	1,936
Mississippi	5,500	2,088,700	700,300	—	933,055	3,722,055	677
California	5,300	2,798,000	1,330,000	35,000	701,271	4,864,271	918
Pennsylvania	5,800	3,260,141	1,345,376	—	1,539,935	6,145,452	1,060
Pennsylvania	5,900	3,450,700	1,215,300	—	894,000	5,560,000	942
Alabama	6,000	4,510,000	3,210,000	—	2,421,156	10,141,156	1,690
Oklahoma	6,750	7,429,500	1,635,000	—	3,365,962	12,430,462	1,842
Pennsylvania	8,000	7,116,500	2,029,500	—	2,437,737	11,583,737	1,448

Arkansas	8,000	5,714,000	1,000,000	—	2,549,113	9,263,113	1,158
Alabama	8,500	3,230,500	873,500	—	1,298,370	5,402,370	636
Oklahoma	8,700	8,626,000	3,366,000	80,000	1,374,000	13,446,000	1,546
Pennsylvania	11,000	4,817,000	2,673,000	25,000	1,114,094	8,629,094	784
Pennsylvania	11,000	7,611,000	4,887,000	30,000	1,596,975	14,124,975	1,284
Pennsylvania	11,000	8,547,000	5,502,000	30,000	1,796,077	15,875,077	1,443
Pennsylvania	11,000	8,012,000	4,538,000	—	1,509,035	14,059,035	1,278
Ohio	11,000	8,510,000	6,042,000	40,000	1,723,493	16,315,493	1,483
New Jersey	12,000	8,849,600	4,147,700	375,000	4,342,814	17,715,114	1,476
Alabama	12,600	4,429,500	1,210,500	—	1,832,207	7,472,207	593
Georgia	12,600	4,115,500	1,137,500	—	1,710,158	6,963,158	553
Pennsylvania	12,600	5,376,000	1,423,500	—	1,904,475	8,703,975	691
Pennsylvania	12,600	5,376,000	1,423,500	—	1,691,646	8,491,146	674
Pennsylvania	12,600	5,133,700	1,748,900	—	1,619,400	8,502,000	675
South Carolina	12,600	4,106,500	1,128,500	—	1,705,060	6,940,060	551
Pennsylvania	13,500	11,004,667	4,301,000	—	3,043,113	18,348,780	1,359
South Dakota	16,000	8,984,000	2,644,000	—	4,455,900	16,083,900	1,005
Pennsylvania	22,000	12,776,000	8,599,000	50,000	2,731,822	24,156,822	1,098
Texas	31,050	9,704,000	3,740,000	100,000	2,772,450	16,316,450	450
Total—Land projects	345,800	215,519,908	96,946,576	6,415,185	68,306,188	387,187,857	1,120
1988—Report total land projects	660,565	338,483,025	258,899,573	15,517,911	164,573,742	777,474,251	1,177
Offshore Stations							
Alabama	1,200	1,738,000	833,000	—	778,910	3,349,910	2,792
Mississippi	11,000	4,355,500	1,149,500	—	1,650,198	7,155,198	650
Total—Offshore projects	12,200	6,093,500	1,982,500	—	2,429,108	10,505,108	861
1988—Report total offshore projects	—	—	—	—	—	—	—
Total—All projects	358,000	221,613,408	98,929,076	6,415,185	70,735,296	397,692,965	1,111
1988—Report total all projects	660,565	338,483,025	258,899,573	15,517,911	164,573,742	777,474,251	1,177

[a]Source: U.S. FERC construction-permit applications, July 1, 1988, to June 30, 1989.

[b]Generally includes engineering, surveying, interest, administration and overheads, contingencies, afudc, and FERC fees.

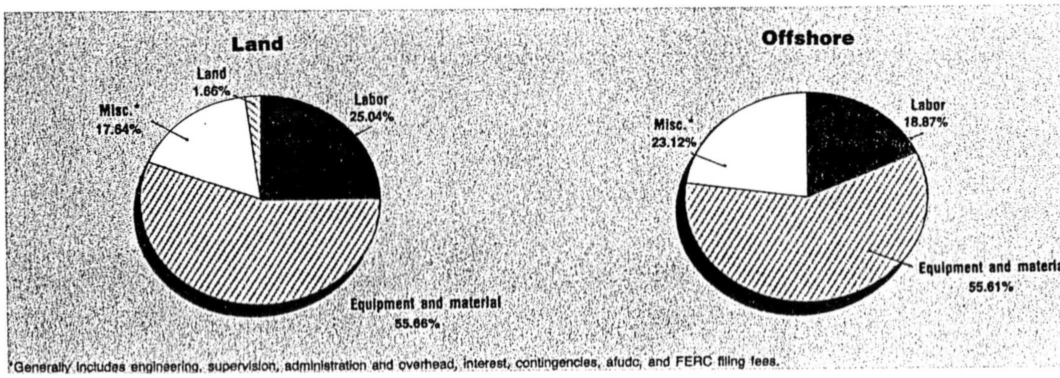

FIG. 6. Compressor-station cost. (Source: U.S. FERC construction-permit applications, July 1, 1988 to June 30, 1989.)

Costs for new land compressor stations range from a low of \$551/hp for a 12,600-hp station in Connecticut to a high of \$4181/hp for a 2700-hp station in New York. Cost-per-horsepower figures show no particular correlation with compressor-station size or location.

Figure 6 shows the cost split for land compressor stations based on data in Table 10.

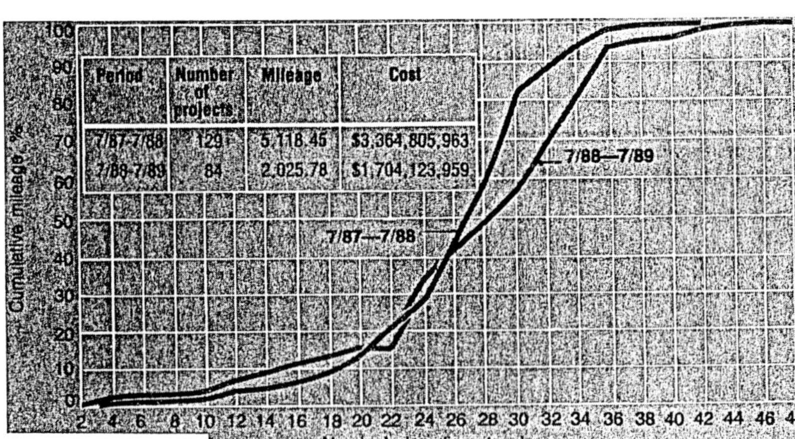

FIG. 7. Diameter distribution for pipelines (U.S. land, gas pipelines). (Source: U.S. FERC construction-permit applications, July 1, 1987 to June 30, 1988, and July 1, 1988 to June 30, 1989.)

Effect of Size

Another element affecting the capital expenditures for pipeline construction is the distribution of the diameter and mileage.

Figure 7's curves reflect the cumulative percentage of mileage for pipeline diameters of 3 to 42 in. for July 1, 1987, to June 30, 1988, and for 2 in. to 48 in. for July 1, 1988, to June 30, 1989.

For the 1987–88 period, 71% of the construction proposed was for diameters of 24, 26, 30, 36, and 42 in. A similar distribution pattern was evident for the more recent period in which diameters of 24, 30, and 36 in. occupied more than 65% of the proposed construction.

As an aid in determining the detailed costs for each major element in pipeline and compressor-station construction for the pipelines and compressor stations, Table 11–31, covering specific projects, are included in this report. These give detailed cost breakdowns for each major component in a construction-cost estimate.

They cover cost information from the FERC applications for the 1988–89 period surveyed and listed in Table 3. These detailed tables represent both a variety of land geographic locations in the United States and pipeline and compressor stations for both land and offshore facilities.

Operating Costs

Once a natural-gas pipeline is laid and compressor stations constructed, operating costs become the next consideration.

As an aid in estimating this element, transmission expenses for interstate gas pipelines for 1986 and 1987 have been included (Table 32). Its data are based on 253,656 miles of pipeline operated and 7.8 TCF of gas sold for 1986 majors and on 248,458 miles of pipeline operated and 6.5 TCF of gas sold for 1987 majors.

TABLE 11 Onshore Pipeline Project: 12.4 miles of 12 in., Oklahoma

Item	Cost ($)
Material	977,000
Installation	715,500
Rights-of-way	386,700
Engineering and supervision	208,000
Interest during construction	61,550
Insurance and warehousing	5,500
FERC filing fee	19,450
Total	2,373,700
$/mile	191,427

TABLE 12 Onshore Pipeline Project: 103.7 miles of 42-in. Loop, Idaho-Washington

Item	Cost ($)
Mainline coated pipe	42,257,000
Valves	592,000
Misc. pipe and fittings	737,000
Casing, 48-in.	123,000
Valve shelters, pipeline markers, miscellaneous	99,000
Mainline pipe inland freight	6,459,000
Other pipe inland freight	12,000
Use tax	2,737,000
Pipeline construction	28,584,000
Other contracts	310,000
Construction use tax in Washington	1,092,000
Survey contract	104,000
Radiographic contract	430,000
Property damage payments	548,000
Right-of-way (est.)	653,555
Restoration	1,612,000
Line pack	554,000
Miscellaneous (est.)	19,575,090
Total	106,111,645
$/mile	1,023,256

TABLE 13 Onshore Pipeline Project: 134.3 miles of 36-in. Loop, Oregon

Item	Cost ($)
Mainline coated pipe	46,395,000
Valves	1,516,000
Misc. pipe and fittings	449,000
Casing, 42-in.	49,000
Valve shelters, pipeline markers, miscellaneous	137,000
Mainline pipe inland freight	2,540,000
Other pipe inland freight	7,000
Pipeline construction	45,524,000
Other contracts	1,598,000
Survey contract	135,000
Radiographic contract	551,000
Property damage payments	709,000
Right-of-way (est.)	780,290
Restoration	2,087,000
Line pack	515,000
Blow gas	218,000
Miscellaneous (est.)	23,371,020
Total	126,581,310
$/mile	924,527

TABLE 14 Onshore Pipeline Project: 34.3 miles of 12.75 in., Louisiana

Item	Cost ($)
Line pipe:	
12.75 in. × 0.375-in. W.T. × 42	4,165,400
Riser/fab pipe:	
12.75 in. × 0.500-in. W.T. × 42	36,000
Corrosion/weight coating	1,002,700
Cathodic protection	143,000
Valves and fittings	120,000
Install 34.3 miles of 12-in. pipe, 2 risers, 3 subsea tap assemblies	4,000,000
Survey	150,000
Engineering	200,000
Diving inspection	360,000
Inspection	230,000
Bond	40,000
Insurance	240,000
Miscellaneous (est.)	1,903,900
Total	12,591,000
$/mile	367,085

TABLE 15 Onshore Pipeline Project: 4.9 miles of 6 in., Louisiana

Item	Cost ($)
Line pipe:	
6.625 in. × 0.375-in. W.T. × 42	289,800
Riser/fab pipe:	
6.625 in. × 0.432-in. W.T. × 42	7,700
Corrosion/weight coating	37,100
Cathodic protection	20,000
Valves and fittings	50,000
Install 4.9 miles of 6-in. pipe, 1 riser, 1 subsea tie-in	1,700,000
Survey	16,500
Engineering	22,000
Diving inspection	75,000
Inspection	100,000
Bond	17,000
Insurance	53,000
Miscellaneous (est.)	425,428
Total	2,813,528
$/mile	574,189

TABLE 16 Onshore Pipeline Project: 0.25 miles of 20 in., Alabama ("Gateway")

Item	Cost ($)
Right-of-way and cost of acquisition	4,400
Pipe:	
20 in. OD × 0.375-in. W.T. 78.60# API 5L, × 60	26,500
20 in. OD × 0.500-in. W.T. 104.13# API 5L, × 42	23,400
Coating material—enamel	6,700
Meter and regulator facilities	192,600
Miscellaneous material	94,100
Construction, equipment and labor	229,700
Engineering—surveying and mapping	700
Administrative and field engineering and inspection	67,000
General overhead	41,900
Interest	6,000
FERC fees (est.)	1,666
Total	694,666
$/mile	2,778,664

TABLE 17 Onshore Pipeline Project: 57.5 miles of 20 in. Replacement, Pennsylvania-New Jersey

Item	Cost ($)
Materials	11,099,800
Installation	24,874,000
Right-of-way and damages	6,422,700
Other costs	3,291,600
Allowance for funds used during construction (afudc)	949,200
General administration	2,463,600
Construction overheads	607,100
FERC fees (est.)	77,115
Total	49,785,115
$/mile	865,828

TABLE 18 Onshore Pipeline Project: 24.5 miles of 30 in., Alabama ("Gateway")

Item	Cost ($)
Right-of-way and cost of acquisition	601,100
Pipe:	
30 in. OD × 0.328-in. W.T.	
103.94# API 5L, × 65	5,219,900
30 in. OD × 0.406-in. W.T.	
128.32# API 5L, × 65	426,200
30 in. OD × 0.500-in. W.T.	
157.53# API 5L, × 60	226,200
Coating material—enamel	770,200
Meter and regulator facilities	647,900
Miscellaneous material	1,587,400
Construction equipment	
and labor installation	5,056,000
Engineering: surveying and mapping	70,200
Administrative and field	
engineering and inspection	1,677,300
General overhead	1,058,300
Interest	911,300
FERC fee (est.)	43,904
Total	18,295,904
$/mile	746,772

TABLE 19 Onshore Pipeline Project: 8 miles of 36-in. Loop, Pennsylvania

Item	Cost ($)
Rights-of-way and damages	256,000
Pipe:	
36 in. OD × 0.462-in. W.T. × 65	997,170
36 in. OD × 0.554-in. W.T. × 65	1,109,418
36 in. OD × 0.750-in. W.T. × 65	479,797
Internal coating	117,886
External coating	367,533
Valves and fittings	524,750
Rock shield	115,000
Freight	199,640
Miscellaneous	297,848
Installation:	
36-in. line	3,286,229
Bored crossings	204,000
Other installations	1,174,059
Surveys	
Field engineering and supervision	416,670
Miscellaneous (est.)	2,026,226
Total	11,692,226
$/mile	1,461,528

TABLE 20 Onshore Pipeline Project: 0.48 miles of 22-in. Embayment Crossing Loop, Tennessee

Item	Cost ($)
Right-of-way and damages	8,000
Material:	
22 in. OD × 0.375-in. W.T. × 52 DSAW w/thin film epoxy and concrete coating	86,400
22 in. OD × 0.375-in. W.T. × 52 DSAW wt/thin film epoxy coating	14,720
22 in. OD × 0.312-in. W.T. × 52 DSAW w/thin film epoxy coating	33,440
Valve, 22-in. ball	112,000
Valve, 6-in. plug	21,000
Miscellaneous material and sales tax	57,440
Installation:	
Lay and bury	61,600
River crossing	720,000
Fabrication, road bore and miscellaneous construction	64,000
Nondestructive testing	12,000
General supervision, engineering	50,000
Field inspection	62,400
Contingencies	65,000
Miscellaneous (est.)	198,280
Total	1,566,280
$/mile	3,263,083

TABLE 21 Onshore Pipeline Project: 7.5 miles of 16 in., North Carolina

Item	Cost ($)
Materials	776,700
Installation costs	2,159,600
Rights-of-way, damages	185,700
Other costs	419,350
Allowance for funds used during construction (afudc)	49,700
General adminstration	195,900
Construction overheads	82,600
FERC fee (est.)	10,561
Total	3,880,111
$/mile	517,348

TABLE 22 Onshore Pipeline Project: 191 miles of 24 in., Oklahoma-Arkansas

Item	Cost ($)
Mainline pipe:	
24 in. × 0.312-in. W.T. × 60	31,713,000
24 in. × 0.375-in. W.T. × 60	717,000
Valve assemblies	455,000
Trap assemblies	946,000
Other	1,801,000
Pipeline construction	31,588,000
Pipeline radiographic inspection	430,000
Right-of-way and damages	1,687,000
Surveys	230,000
Engineering, construction management	3,000,000
Contingency	3,754,000
Allowance for funds used during construction	3,272,000
FERC filing fee	19,500
Total	79,612,500
$/mile	416,819

TABLE 23 Onshore Pipeline Project: 2 miles of 30-in. River Crossing, Iowa-Illinois

Item	Cost ($)
Company material	1,087,200
Contract label, materials	3,516,100
Supervision	250,000
Miscellaneous company costs including FERC fee ($1,470)	275,000
Land and right-of-way	100,000
Contingency	248,900
Interest and overheads	522,800
Total	6,000,000
$/mile	3,000,000

TABLE 24 Offshore Pipeline Project: 55.0 miles of 24 in., Alabama-Louisiana

Item	Cost ($)
Route and environmental survey	206,000
Right-of-way, damages, and permits	403,000
Material:	
Pipe:	
24 in. OD × 0.500-in. W.T. × 52	13,433,000
24 in. OD × 0.688-in. W.T. × 65	414,000
Coating	6,485,000
Other material	1,454,000
Installation:	
Contract portion	16,952,000
Company portion	724,000
Contingency	2,003,000
Miscellaneous	3,143,832
Total	45,217,832
$/mile	822,142

TABLE 25 Offshore Pipeline Project: 9 miles of 12 in., Alabama

Item	Cost ($)
Route and environmental survey	42,000
Right-of-way, damages, and permits	2,000
Material:	
Pipe:	
12 in. OD × 0.406-in. W.T. × 42	1,016,000
12 in. OD × 0.500-in. W.T. × 42	30,000
Coating	568,000
Other material	193,000
Installation	
Contract portion	2,528,000
Company portion	145,000
Contingency	227,000
Miscellaneous (est.)	354,900
Total	5,105,900
$/mile	567,322

TABLE 26 Onshore Compressor Project: Install 4000 hp, Alabama

Item	Cost ($)
Land	450,000
Site improvements	50,000
Other structures	217,000
Compressor station equipment:	
Main compressor	1,850,000
Spare parts	450,000
Major gas piping	253,000
Station electrical system	582,000
Instrumentation	170,000
Other equipment	875,000
Installation:	
Structures	974,000
Main compressor	653,000
Major gas piping	653,000
Station electrical system	155,000
Instrumentation	58,000
Other equipment	676,000
Contingency	404,000
Miscellanous (est.)	633,151
Total	9,108,151
$/hp	2,277

TABLE 27 Onshore Compressor Project: Add 12,600 hp, Pennsylvania

Item	Cost ($)
Material:	
Structures	150,000
Yard improvements, structures	29,000
Station piping	748,000
Main compressor units	4,095,000
Electrical equipment	56,000
Other equipment	262,000
Concrete	36,000
Contractor's overhead	3,000
Field engineering, supervision	87,500
Installation:	
Structures	50,000
Yard improvements, structures	81,000
Station piping	748,000
Main compressor units	315,000
Electrical equipment	45,000
Other equipment	112,500
Concrete	72,000
Contractor's overhead	6,000
Field engineering, supervision	300,000
Other miscellaneous	1,507,975
Total	8,703,975
$/hp	691

TABLE 28 Onshore Compressor Project: Install 16,000 hp, South Dakota

Item	Cost ($)
Material	8,984,000
Labor	2,644,000
Engineering	891,000
Company supervision	1,256,000
Contingencies	1,252,000
Interest and overheads (est.)	1,038,000
Filing fees (est.)	18,900
Total	16,083,900
$/hp	1,005

TABLE 29 Onshore Compressor Project: Add 12,600 hp, Pennsylvania

Item	Cost ($)
Material:	
Structures	150,000
Yard improvements, structures	29,000
Station piping	748,000
Main compressor units	4,095,000
Electrical equipment	56,000
Other equipment	262,000
Concrete	36,000
Contractor's overhead	3,000
Field engineering, supervision	87,500
Installation:	
Structures	50,000
Yard improvements, structures	81,000
Station piping	748,000
Main compressor units	315,000
Electrical equipment	45,000
Other equipment	112,500
Concrete	72,000
Contractor's overhead	6,000
Field engineering, supervision	300,000
Miscellaneous (est.)	1,295,146
Total	8,491,146
$/hp	674

TABLE 30 Offshore Compressor Project: Install 11,000 hp, Mississippi

Item	Cost ($)
Material:	
Structures	84,000
Yard improvements, structures	12,000
Station piping	526,000
Main compressor units	3,350,000
Electrical equipment	62,000
Other equipment	281,500
Concrete	40,000
Contractor's overhead	10,000
Field engineering, supervision	85,000
Installation:	
Structures	56,000
Yard improvements, structures	31,000
Station piping	551,000
Main compressor units	335,000
Electrical equipment	30,000
Other equipment	86,500
Concrete	60,000
Contractor's overhead	2,000
Field engineering, supervision	425,000
Miscellaneous (est.)	1,128,198
Total	7,155,198
$/hp	650

TABLE 31 Offshore Compressor Project: Install 1200 hp, Alabama

Item	Cost ($)
Material:	
Structures	125,500
Yard improvements, structures	53,000
Station piping	426,000
Main compressor units	700,000
Electrical equipment	46,000
Other equipment	367,500
Concrete	20,000
Contractor's overhead	10,000
Field engineering, supervision	49,000
Installation:	
Structures	72,500
Yard improvements, structures	79,000
Station piping	451,000
Main compressor units	70,000
Electrical equipment	32,000
Other equipment	93,500
Concrete	35,000
Contractor's overhead	2,000
Field engineering, supervision	190,000
Miscellaneous (est.)	527,910
Total	3,349,910
$/hp	2,792

TABLE 32 Natural-Gas Pipeline Transmission Expenses[a,b]

	Expenses ($)		Cost/mile ($)		Cost/MMCF sold ($)	
	1986 Majors[c]	1987 Majors[d]	1986 Majors[c]	1987 Majors[d]	1986 Majors[c]	1987 Majors[d]
Operation expenses:						
Supervision and engineering	159,699,942	151,679,238	629.59	610.48	20.43	23.45
System control and load dispatching	28,679,686	26,858,420	113.07	108.10	3.67	4.15
Communication system expenses	28,265,422	29,310,758	111.43	117.97	3.62	4.53
Compressor station labor and expenses	239,180,275	251,651,867	942.93	1,012.85	30.60	38.90
Gas for compressor station fuel	532,572,604	367,406,730	2,099.59	1.478.75	68.14	56.80
Other fuel and power for compressor stations	46,066,937	63,842,107	181.61	256.95	5.89	9.87
Mains	182,900,300	168,358,688	721.06	677.61	23.40	26.03
Measuring and regulating station expenses	62,517,614	61,623,174	246.47	248.02	8.00	9.53
Transmission and compression of gas by others	1,267,266,607	1,049,122,569	4,996.01	4,222.53	162.13	162.18
Other transmission expenses	36,206,154	38,278,436	142.74	154.06	4.63	5.92
Rents	17,142,978	18,421,763	67.68	74.14	2.18	2.85
Total operation expenses	2,597,098,519	2,226,553,750	10,238.66	8,961.49	332.27	344.19
Maintenance expenses:						
Supervision and engineering	35,545,963	34,596,228	140.14	139.24	4.55	5.35
Structures and improvements	30,638,479	21,181,592	120.79	85.25	3.92	3.27
Mains	93,717,959	91,003,249	369.47	366.27	11.99	14.07
Compressor station equipment	197,872,245	197,326,728	780.08	794.21	25.32	30.50
Measuring and regulating station equipment	12,539,338	12,117,702	49.43	48.77	1.60	1.87
Communication equipment	13,913,234	12,077,950	54.85	48.61	1.78	1.87
Other equipment	3,138,461	3,259,046	12.37	13.12	0.40	0.50
Total maintenance expenses	387,365,679	371,562,495	1,527.13	1,495.47	49.56	57.44
Total transmission expenses	$2,984,464,198	$2,598,116,245	$11,765.79	$10,456.96	$381.83	$401.63
Total miles of transmission pipeline	253,656	248,458				
Total natural gas sold, MMCF	7,816,259	6,468,964				

[a]Source: Statistics of Interstate Natural Gas Pipeline Companies—1986 and 1987. U.S. Department of Energy.

[b]U.S. interstate pipelines for calendar years 1986 and 1987.

[c]44 of 135 companies filing Forms 2's or 2A's with the U.S. Federal Energy Regulatory Commission; majors and companies whose combined gas sold for resale and gas transported or stored for a fee exceed 50 BCF (at 14.7 lb/in.^{2}; 60°F) in each of the 3 previous calendar years.

[d]43 of 126 companies filing Form 2's and 2A's with the U.S. FERC for 1987; majors defined in previous note and in U.S. FERC Accounting and Reporting Requirements for Natural Gas Companies, para. 20-011, effective February 2, 1985, beginning with the 1984 reporting year (*Oil Gas J.*, p. 80, November 25, 1985).

The highest component of transmission-operating expenses is transmission and compression of gas by others: for majors in 1986, this component was $4996/mile or $162/MMCF; for majors in 1987, this component was $4222/mile or $162/MMCF.

The second highest cost component of transmission-operating expense is gas for compressor-station fuel: for majors in 1986, $2100/mile or $68/MMCF; for majors in 1987, $1479/mile or $57/MMCF.

The highest cost component of maintenance expense covers compressor-station equipment: for majors in 1986, $780/mile or $25/MMCF; for majors in 1987, $794/mile or $30/MMCF.

The second highest cost component for maintenance is the maintenance of pipelines: for 1986 majors, $369/mile or $12/MMCF; for majors in 1987, $366/mile or $14/MMCF.

Table 32 indicates that the highest component of transmission expenses for natural-gas pipelines is the operation expense: for 1986 majors, $10,239/mile or $332/MMCF; for 1987 majors, $8961/mile or $344/MMCF.

Unit-maintenance expenses for 1986 majors were $1527/mile or $59/MMCF; for 1987 majors, $1495/mile or $402/MMCF.

During 1986, major natural-gas pipeline companies spent almost $2.6 billion for operation expenses and nearly $390 million on maintenance. Total expenditures by 1986 majors for operation and maintenance expenses reached almost $3 billion. Total operation expenses for majors in 1987 fell to slightly more than $2.2 billion, a drop of more than $370 million (off a sharp 14.3%). This follows a similarly sharp 14.8% drop from 1985 to 1986. Operation costs for the 3-year period 1985 through 1987 dropped by 27%. Total maintenance expenses in 1987 fell to slightly more than $371 million from a 1986 cost of $386 million, a decrease of almost $16 million (4%). Total transmission expenses showed a decline between 1986 and 1987. In the latter year they reached slightly less than $2.6 billion, a drop from 1986 of a little more than $386 million (down a noticeable 13%).

On a unit basis, the total 1986 transmission expense for majors was $11,766/mile or $382/MMCF of gas sold; for 1987 majors, $10,457/mile or $402/MMCF.

WARREN R. TRUE

Construction, Worldwide

Tables 1–4 present factual data which pipeline operating companies require when planning new construction projects. Included are basic data on some of the world's leading pipeline construction and engineering firms—their physical and fiscal capabilities, equipment, personnel, and experience.

This huge study is made annually by the editors of *Pipe Line Industry* and is updated in each January issue. The reader is urged to see the latest report. This study for 1989 is used here by special permission from *Pipe Line Industry*, copyright © 1990 by Gulf Publishing Co., Houston, Texas 77252.

The Worldwide Construction Scoreboard (Table 4) is of particular value to those who supply, service, or construct the facilities that the pipeline industry requires. Information is furnished on the more than 85,042 miles of pipeline construction underway, planned, or actively under study throughout the world. These figures include 16,596 miles of potential construction of lines in the United States and 68,446 miles in other countries. These figures do not include worldwide gas distribution facilities. Capital requirements through the end of the century are estimated to exceed $200 billion.

The new Scoreboard lists 20,005 miles of pipelines scheduled to start construction in 1990, including 6,456 miles in the United States and 13,549 miles in other countries.

Pipeline construction in the United States showed a downward trend in 1989, with 6026 mi of all types of lines installed during the year.

In 1990, United States companies plan to install 6456 mi of gas and liquids facilities. Again, this total does not include any mileage from the proposed gas lines into the Northeast Corridor or the lines to transport natural gas to the heavy oil fields of California for steam oil recovery facilities.

During 1989, 964,716 hp in compression facilities was installed in new or existing natural gas transmission line compressor stations. A total of 240,978 hp of pumping facilities was installed in either new pump stations or existing facilities on crude and products pipelines.

In 1990, companies are expected to install 1,553,870 hp in compressor station facilities and 574,961 hp in liquid line pump stations.

This forecast for 1990 construction and the review of 1989 actual construction does not reflect activities in the USSR or China.

TABLE 1 United States/Outside United States Construction to Exceed 20,005 Miles[a]

	United States		Outside U.S.		Total World	
	1989 (actual)	1990 (forecast)	1989 (actual)	1990 (forecast)	1989 (actual)	1990 (forecast)
Gas lines:						
Transmission[b]	3,364	4,062	6,965	5,657	10,329	9,719
Gathering	262	378	532	856	794	1,234
Total	3,626	4,440	7,497	6,513	11,123	10,953
Crude lines:						
Trunk lines	490	524	3,524	2,843	4,014	3,367
Gathering	254	314	618	432	872	746
Total	744	838	4,142	3,275	4,886	4,113
Product lines:						
Refined products	1,216	742	1,394	2,022	2,610	2,764
CO_2	39	34	—	—	39	34
Other	67	106	512	401	579	507
Total (all products)	1,322	882	1,906	2,423	3,228	3,305
Offshore:						
Gas	232	178	507	1,181	739	1,359
Oil	102	118	229	157	331	275
Total	334	296	736	1,338	1,070	1,634
Grand total	6,026	6,456	14,281	13,549	20,307	20,005

[a]Excludes Russia and China
[b]Includes interstate transmission company reports, FERC, and intrastate reports in *Pipe Line Industry* survey (mainlines and laterals).

TABLE 2 Diameter Breakout on Pipe Usage (miles) (gas distribution not included)

Nominal diameter[a] (inches)	United States 1989 (actual)	United States 1990 (forecast)	Outside U.S. 1989 (actual)	Outside U.S. 1990 (forecast)	Total World 1989 (actual)	Total World 1990 (forecast)
2–4	331	417	967	1473	1298	1890
6	227	328	625	317	852	645
8	224	375	993	614	1217	989
10	475	287	589	1818	1064	2105
12	264	335	1617	1055	1881	1390
14	111	84	287	444	398	528
16	1619	977	1310	1157	2929	2134
18	86	180	228	270	314	450
20	544	415	480	935	1024	1350
22	15	—	60	—	75	—
24	170	354	1075	1657	1245	2011
26	70	76	319	102	389	178
28	—	—	32	42	32	42
30	623	796	687	752	1310	1548
32	—	—	25	60	25	60
34	—	12	396	169	396	181
36	539	676	658	416	1197	1092
40	—	—	—	440	—	440
42	5	316	1284	341	1289	657
44	—	—	106	—	106	—
48	—	18	295	365	295	383
56–60	6	4	489	243	495	247

[a]Total of diameters does not equal industry total as some companies did not indicate diameters.

TABLE 3 Station Additions to Exceed 2,128,861 Horsepower

	United States[a] 1989 (actual)	United States[a] 1990 (forecast)	Outside U.S. 1989 (actual)	Outside U.S. 1990 (forecast)	Total World 1989 (actual)	Total World 1990 (forecast)
Compressor	212,637	504,230	757,079	1,049,640	969,716	1,553,870
Pump	72,840	142,810	168,138	432,181	240,978	574,991
Total	285,477	647,040	925,217	1,481,821	1,210,694	2,128,861

[a]United States figures include interstate transmission company reports to FERC and intrastate reports in *Pipe Line Industry* surveys.

TABLE 4 Worldwide Construction Scoreboard

United States

Company	Est. Cost US$ (mm)	Project Location	Length (Miles)	Pipe Diam. in Inches	Type of Service	Compressor or Pump Station hp and Location	Project Status *	Contractor and/or Engineering Firm	Completion Date
State of Alaska		Prudhoe Bay	33	6	Trunkline, NGL		Planned		
Algonquin Gas Transmission Co.	10	Medway, Holliston Hopkinton, Mass.	8	12	Transmission, Gas		Planned		
	51.6	Maryland	13.5	24	Transmission, Gas	12,000	Before FERC		
	87	N.Y., Md., Conn.	36.2		Transmission, Gas	18,000	Before FERC		
	113	N.J., N.Y., Md., Conn., & R.I.	44.7 12.7	36 16	Transmission, Gas		Before FERC		
	74	New York	25.3 2.4	36, 42		22,000	Before FERC		
Amoco Oil Corp.	•	Offshore High Island, Texas	50	6	Trunkline, Crude		Await Start	John Brown Engrs. & Constr. (E)	1990
	•	Minnesota	1	8	Trunkline, Products (Relocation)				1990
	•	Missouri	2	8	Trunkline, Products (Relocation)				1990
Amoco Pipeline Co.	•	Louisiana	0.4	16	Gathering, Crude (Replacement)		Planned		1990
	•	Missouri	20.4	12	Trunkline, Products (Replacement)		Planned		1990
	•	North Dakota	11.5 8	4,6 8	Gathering, Crude Trunkline, Crude (Replacement)		Planned Planned		1990 1990
	•	Oklahoma	19	8	Trunkline, Crude (Replacement)		Planned		1990
	•	Texas	10.4	4,8	Gathering, Crude (Replacement)		Planned		1990
	•	Texas	14.5	8,12, 26	Trunkline, Crude (Replacement)		Planned		1990
	•	Texas	2	18	Trunkline, Products (Replacement)		Planned		1990
ANR Pipeline Co.		Ill., Iowa, Kans., Mo., Ohio & Wis.	160.7		Trans., Loops, Gas	21,000	Before FERC		
ANR Pipeline Co./Arkla, Inc.	120	Wilburton, Okla. to Sherrill, Ark.	200	30	Transmission, Gas		Before FERC		
Apex Pipeline System	1.2 Bil.	Alberta, Canada to California	1,400		Transmission, Gas		Under Study		
Appalachian Oil & Gas Co.		Kentucky	60	12	Transmission, Gas		Under Study		
Arco Alaska, Inc.	•	Prudhoe Bay, Alaska	6	60	Transmission, Gas		Working	H.C. Price Constr. Co. (C)	Aug. '90
Arkansas Western Gas Co.	• 1.3	Washington County, Ark.	5.2	16	Transmission, Gas		Planned		Fall 1990
Arkla Energy Services	• •	Louisiana Louisiana	31 11	30 24	Transmission, Gas Transmission, Gas		Before FERC		
Ashland Pipe Line Co.	• 5.5	Morrow & Richland counties, Ohio	17	16	Trunkline, Crude		Under Study		Oct. '90
	• 1.4	Warrick County, Ind.	2.3	20	Trunkline, Crude		Await Start	Co. Crews	July '90
	• 0.2	Various locations Ohio	1	8	Trunkline, Products		Under Study	Co. Crews	Oct. '90
Baltimore Gas & Electric Co.	• 20	Baltimore area	45	1 to 26	Distribution, Gas		Working	Co. Crews & Contractors	Dec. '90
	• 1	Baltimore County	1	20	Distribution, Gas		Planned		June '90
	• 1.5	Baltimore County	2	12	Distribution, Gas		Planned		Oct. '90
	• 8	Baltimore County	9	24	Distribution, Gas		Planned	Fluor Daniel (E)	Oct. '90

*LEGEND

UNDER STUDY: Includes all pipe line projects which are being considered or for which a feasibility study is being made or has been made but no decision has been reached regarding its being built.

PLANNED: Includes all pipe lines planned or proposed for construction. The proposal may have been made to the operating company, government authorities, etc.; but, excludes U.S. gas pipe lines presented to FERC for approval.

BEFORE FERC: Gas pipe line projects within the U.S. where planned construction has been submitted to FERC for approval, but not yet granted.

FERC APPROVED: Projects authorized for construction.

APPROVED: Projects outside the U.S. approved for construction by a government or other agency.

AWAITING START: Projects out for bid or awaiting start of construction.

WORKING: Projects underway.

(E) Engineering Firm

(C) Contractor

• New listing since September 1989 Scoreboard.

(continued)

TABLE 4 *(continued)*

Company	Est. Cost US$ (mm)	Project Location	Length (Miles)	Pipe Diam. in Inches	Type of Service	Compressor or Pump Station hp and Location	Project Status •	Contractor and/or Engineering Firm	Completion Date
Baltimore Gas & Electric Co. (*Cont.*)	• 7	Baltimore County	7	20	Distribution, Gas		Planned		Oct. '91
	• 12	Harford County	14	24	Distribution, Gas		Planned	Northeastern Land Services (E)	Dec. '92
Battle Creek Gas Co.	• 0.115	Battle Creek, Mich.	3	2,4	Distribution, Gas (New Mains)		Planned		Dec. '90
	• 0.739	Battle Creek, Mich.	14	2,4,	Distribution, Gas (Replacement Mains)		Planned		Dec. '90
	• 0.125	Battle Creek, Mich.	4	3/4, 1, 1 1/4	Distribution, Gas (New Mains)		Planned		Dec. '90
	• 0.597	Battle Creek, Mich.	18	3/4 & 1 1/4	Distribution, Gas (Replacement Mains)		Planned		Dec. '90
Big Rock Oil & Gas Co.		San Antonio, Texas	44	4 to 8	Trunkline, Crude		Under Study		
Cascade Natural Gas Corp.	0.5	Bend, Ore.	4.9	4	Distribution, Gas		Planned		
	0.3	Mt. Vernon, Wash.	2.5	6	Distribution, Gas		Planned		
Champlain Pipeline Co.	268.6	Highgate Springs, Vt. to West Medway, Maine to Pelham, N.H.	253		Transmission, Gas		Before FERC		
Chandeleur Pipe Line Co. (Chevron Pipeline Co.)		Offshore Mobile Bay	53.6	20	Transmission, Gas		Before FERC		
Chevron Pipe Line Co.	4.5	Central California	20	18	Trunkline, Crude		Planned		1990
Chevron Pipe Line, Sun Pipeline Co.—joint venture	13.5	Taft to Emidio, Calif.	34	18	Trunkline, Crude		Under Study		
CITGO Pipeline Co.	• 2.5	East Texas, Beaumont to Longview	150	12	Trunkline, Crude		Under Study		1990
Citizens Gas & Coke Utility	7.6	Marion County, Indiana	48.8	1 to 20	Distribution, Gas		Working	Co. Crews & Contractors	Oct. '90
Citrus Interstate Pipeline Co.	32.4	Bayou La Batre to Mobile, Ala.	50	30	Transmission, Gas		Before FERC		
City Public Service Co.	2.4	San Antonio, Texas	5	16	Distribution, Gas		Planned		
			5	12	Distribution, Gas		Planned		
Colonial Pipeline Co.	•	Jefferson County, Fla.	45	12	Trunkline, Products		Planned		Late 1990
Colorado Interstate Gas Co.	2.8	Colo.,Kans., Okla., Texas, Utah, Wyo.	13.4	2 to 6	Gathering, Gas		Planned		Dec. '90
	• 7.9	Adams County, Colo.			Transmission, Gas	8,000	Before FERC		1991
	• 7.4	Pueblo County Colo.			Transmission, Gas	6,000	Before FERC		1991
Columbia Gas Transmission Corp.	•	Pa., W. Va. & Va.	51.5	24 & 36	Trans., Loop, Gas	17,600	Before FERC		1990
	• 25.2	Md., N.J., Va., Pa., & Ohio	17.3	20 & 24	Transmission, Gas	7,590	Before FERC		Nov. '90
	• 34.3	Penn. & N.J.	48.9	20	Transmission, Gas		Working		Nov. '90
	• 33.5	Ky., N.Y., Ohio, Pa. & W.Va.	112.3	3 to 20	Trans. & Gathering, Gas		Planned & Approved		1990-1991
Columbia Gulf Transmission Co.	• 5.6	Jefferson Davis & Cameron parishes, La.	8	24	Transmission, Loop, Gas		Await Start		Oct. '90
	• 10.1	West Cameron & East Cameron	20.3	16	Transmission, Gas		Await Start		July '90
Conoco Pipe Line Co. (West Texas line)	•	Dallas to Abilene, Texas	180	6	Trunkline, Crude		Planned		Late 1990
Consolidated Natural Gas Transmission Corp.	• 22.2	W. Va.	46	20	Transmission, Gas	2,160 @ Jackson	Before FERC		1990
	• 13.0	New York	12.3	30	Transmission, Gas		Before FERC		1990
	• 23	Pennsylvania	25	30	Transmission, Gas		Before FERC		1990
	• 11	N.Y., Pa.	2.3	30	Transmission, Gas	1,350 & 2,250	Before FERC		1990
	• 9.6	W. Va.	18	20	Transmission, Gas		Before FERC		1990
CNG Trans. Corp. and Texas Eastern Transmission Corp.	• 54	Va. to W. Va.	25.19	24	Transmission, Gas	11,000 @ Sta. 21A 6,500 @ Sta. 23	Before FERC		
	• 39	Va. & W. Va.				6,000 @ Leesburg, Va. and Lambert, W.Va.	Before FERC		
Delhi Gas Pipeline Corp.		North Texas	30	4 to 12	Transmission, Gas		Planned		
Delta Pipeline Co.	• 82	Wilburton, Okla. to Fort Smith, Arkansas	190	24	Transmission, Gas		Before FERC		
Diamond Shamrock Oil Co.		Amarillo, Texas to Albuquerque, N. Mex.	150	10	Trunkline, Crude		Planned		
Eastern American States Transmission Co. (ANR Pipeline Co.)	288	Ohio & Pa.	76	24	Transmission, Gas		Before FERC		
			265	30					
Elizabethtown Gas Co.	0.70	Washington Twp., N.J.	3.5	8	Distribution, Gas		Under Study		Dec. '90
El Paso Natural (Mojave Companion) (Coal Seam Project)	10.8	Arizona	14.8	30	Trans., Loop, Gas		Before FERC		1990
	22.5	San Juan County, N Mex.	5.4	34	Trans., Loop, Gas		Before FERC		Mid 1990
			6.5	34	Trans., Loop, Gas				
			14.2	30	Trans., Loop, Gas				

TABLE 4 *(continued)*

Company	Est. Cost US$ (mm)	Project Location	Length (Miles)	Pipe Diam. in Inches	Type of Service	Compressor or Pump Station hp and Location	Project Status *	Contractor and/or Engineering Firm	Completion Date
	13.2	La Paz County, Ariz.	29.96	30	Trans., Loop, Gas		Before FERC		Mid 1990
(Wenden to Ehrenberg)		La Paz County, Ariz.				5,300 @ Wenden	Before FERC		Mid 1990
	•	Various locations N.Mex., Ariz., & Colo.	241.6	30,34, 36	Trans., Loop, Gas	69,300	Before FERC		
Enron Corp. (Black Marlin Pipeline Co.)	• 28	Texas and Oklahoma	39	30	Transmission, Gas		Working		1991
Enron Corp./ Channel (joint venture)	• 8.0	Port Lavaca, Texas	12	20	Transmission, Gas		Working		1990
Enron Corp. (Florida Gas Trans. Corp.)	90.5	Louisiana to Florida	43.8	4 to 20	Transmission, Gas	36,200	Before FERC		1991
Enron Corp. (HPL)	• 3.5	Nueces, Texas			Transmission, Gas	2,800	Working		1990
Enron Corp. (Intratex)	• 8.0	LaGrange, Texas			Transmission, Gas	15,600	Working		1990
Enron Corp. (Northern Natural Gas)	• 57	Iowa, Illinois & Wisconsin	45 50	30 16	Transmission, Gas	4,000 @ Waterloo, Ia.	Before FERC		1990
	• 38	Oklahoma & Texas	89	20	Transmission, Gas	1,500 @ Roger Mills County	Before FERC		1991
	• 11	Offshore Texas	13	16	Transmission, Gas	3,000	Planned		1990
Enron Corp. (Texoma Pipeline)	• 35	Texas, Oklahoma	39	30	Transmission, Gas	6,000 @ Texas			1990
Enron Corp. (Transwestern Pipeline)	• 94	San Juan & McKinley counties, N.M.	121	30	Transmission, Gas	6,500	Before FERC		1990
	• 153	New Mexico & Arizona	210	30	Transmission, Gas		Before FERC		1992
	• 2.5	Ward County, Texas	4	30	Transmission, Gas		Before FERC		1990
Enstar Natural Gas Co.	185	Wasilla to Fairbanks, Alaska	285	20	Transmission, Gas		Planned		
	35	Fairbanks, Alaska	375		Distribution, Gas		Planned		
Equitable Gas	1.65	Braxton County, W. Va.	1.3	20	Transmission, Gas	350 @ Harrison County & 250 @ Lewis County	Planned		
	1.02	Taylor County, W. Va.	5.3	10	Distribution, Gas		Planned		
	40	Braxton County, W. Va.	1.3	16	Gathering, Gas		Planned		
Explorer Pipeline Co.	20	Port Arthur to Dallas area, Texas			Trunkline, Products	28,000	Under Study		July '90
Falcon Seaboard Pipeline Co.		Plattsburg area, N.Y.	26	12	Transmission, Gas		Planned		
Florida Gas Transmission Co., Tennessee Gas Pipeline Co. & Southern Natural Gas Co.	• 48	Offshore, inshore Mobile area, Ala.	53	36	Transmission, Gas		Before FERC		
Gateway Pipe Line Co. (United Gas Pipe Line Co.)	• 20	Mobile County	25	30	Transmission, Gas		Before FERC		1991
Great Lakes Gas Transmission Co.	•	Mich., Wis. & Minn.	68.5 465.1	36 42	Trans., Loop, Gas	2 @ 27,000 ea.	Before FERC		
	•	Minn., Mich. & Wis.	178	36	Trans., Loop, Gas		Working	Murphy Bros. (C) Associated Pipe Line Constr. (C) Welded Constr. Co. (C)	Mar. '90
Green Canyon Pipe Line Co.	48.8	Green Canyon to South Marsh Island Blk. 106	66.2		Transmission, Gas		Before FERC		
Indiana Ohio Pipeline Co. (Panhandle Eastern Corp.)	• 70	Grant County, Ind., to Warren County, Ohio	110.7	30	Transmission, Gas		Await Start	H.C. Price Co., Murphy Bros. (C)	Summer 1990
Iowa-Illinois Gas & Electric Co.	•	Moline & Rock Island counties, Ill	10.5		Transmission, Gas		Before FERC		
Iroquois Gas Transmission System (TransCanada PipeLines, Algonquin Gas Transmission & Tennessee Gas Pipeline Co.)		N.Y., Conn., Maine, Pa.	460	30	Transmission, Gas		Before FERC		
Jubilee Pipeline Co.	67	Mobile Bay, Ala.	81.6	8 to 24	Transmission, Gas	6,000	Before FERC		1991
Kentucky West Virginia Gas Co.	6.6	Perry County to Knott County, Ky.	25.22	12	Transmission, Gas		Before FERC		
KN Energy, Inc.	• 1.46	Buffalo & Dawson counties, Nebr.	30.8	4 & 16	Transmission, Gas	600	Before FERC		1990
	2	Colorado	15	6	Lateral, Gas		Before FERC		
Long Beach Gas Dept.	4.2	Long Beach	4.8	3 to 6	Distribution, Gas		Planned	Co. Crews	June '90
Magnolia Pipeline Co. (Energen Corp. & Transco Energy)	• 55	Tuscaloosa County to Chilton County, Ala.	55 40	24 16 to 24	Transmission, Gas Laterals, Gas		Working	Harcro, Inc. (C)	Spring 1990

TABLE 4 *(continued)*

Company	Est. Cost US$ (mm)	Project Location	Length (Miles)	Pipe Diam. in Inches	Type of Service	Compressor or Pump Station hp and Location	Project Status *	Contractor and/or Engineering Firm	Completion Date
Meridian Oil Gathering Co	•	San Juan Basin, N Mex.	120	10 to 20	Gathering & Laterals, Gas		Planned		1990
Mex/US Interstate Pipeline Co. (Wagners Brown Partnership)		South Texas	30	30	Transmission, Gas		Before FERC		
Minnegasco, Inc		Minnesota Minnesota	32 200	 1 to 20	Transmission, Gas Distribution, Gas		Planned Planned		
Missouri Pipeline Co. (Omega Pipeline Co. and Sun Pipe Line Co.)		St. Louis area, Mo.	85	12	Transmission, Gas		Before Mo. Public Service Comm.		
Mobile Bay Pipe Line Co.		Offshore Mobile Bay, Ala.	97	4 to 16	Gathering & Laterals, Gas		Planned		
Mobil Corp.	•	Los Angeles, Calif. area	75	16	Trunkline, Crude		Planned		Fall 1990
Mobil Exploration & Production, USA, Inc.		Hockley	39	12	Trunkline, CO_2		Planned		
Mohave Pipeline System (Enron/ El Paso)	214	Arizona to California	121.5 12.5 233.5 17	42 36 30 24	Transmission, Gas	14,080 @ Topock, Ariz.	Before FERC		1990
Mountain Fuel Supply Co.	• 2.6 • 4.4	Utah and Southwest, Wyo. Utah and Southwest, Wyo.	20 203	6 to 10 1¼ to 6	Transmission, Gas Distribution, Gas		Planned Planned		Dec. '90 Dec. '90
Natural Gas Pipeline Co. of America (Amarillo Inter-connect Project)		So. Okla. to NE Texas	191	30	Transmission, Gas		Working	H.B. Zachry (C) & South-west Pipeline Constr. Co. (C)	Mar.'90
Natural Gas Pipeline Co. of America	• 5 • 2.3 • 4.4	Cameron Parish, La Fayette County, Ill. Harrison County, Texas	 93 1.03 2.28 .25	 8 10 8 6	 Lateral, Gas Water Line Lateral, Gas Lateral, Gas	3,600	Before FERC Before FERC Before FERC		Aug. '90 Oct. '90 Oct. '90
National Fuel Gas Supply Co./ Penn-York Energy Co.	46.5	N.Y. & Pa.	40	24	Transmission, Gas	18,000	Before FERC		
New Jersey Natural Gas Co.	• 4.0 • 4.0 • 3.0	Ocean County, N.J. Ocean County, N.J. Monmouth County,	4.5 3.4 3	16 & 20 20 16	Transmission, Gas Transmission, Gas Distribution, Gas		Working Working Planned	J.F. Kiely Constr. Co. (C) Henkels & McCoy (C) Elgate (E)	June '90 Apr. '90 June '90
Niagara Interstate Pipeline System	417	N.Y. & Pa.	160	42	Transmission, Gas	11,500 @ Liedy	Before FERC		
Niagara Mohawk Power Co.	25	Clayton to Watertown, N.Y.	27	16	Transmission, Gas		Before FERC		
Noark Pipeline System (Intercon Gas, Inc., Grubb Industries, Inc. & Southwestern Energy Co.)	• 73	Northern Arkansas	285 44	10 & 16 6	Transmission, Gas		Before Arkansas Public Service Commission		1991
Northeast Gas Transmission Co.	86	Maryland	45.9	36	Transmission, Gas	2,000	Before FERC		
Northern Natural Gas Co.	11.0	North of Janesville to SW corner of Waukesha County, Wis.	30	24	Transmission, Gas		Under Study		
Northwest Pipeline Corp.	•	Ignacio, Colo. to Blanco, N.M.	33	30	Transmission, Gas	10,000 @ Ignacio & 8,000 @ Blanco	Planned	Co. Crews	Jan. '91
Oklahoma-Arkansas Pipeline Co. (International Paper Co., Reliance Pipeline Co. and Texas Eastern-Arkoma)	• 273	Okla., Ark., Miss.	240 112	36 30	Transmission, Gas		Before FERC		
Pacific Alaska LNG Association	1 Bil.	Offshore Cook Inlet to Nikiski, Alaska	300	2 to 24	Transmission, Gas		Under Study		1992
Pacific Gas & Electric Co.	565	Malin, Ore. to Fresno County, Calif.	120 294	36 42	Transmission, Gas		Before CPUC		1993
Pacific Gas Transmission Co.	635	B.C.-Idaho line to California-Oregon line	430	42	Trans., Loop, Gas		Before FERC		1993
Puget-Columbia Industrial Pipeline Co.	200	Semac, British Columbia to Oregon	398		Transmission, Gas		Under Study	Morrison-Knudsen Engrg (E)	
Questar Pipeline Co.		Daggett County, to Unitah County, Utah	81	20	Transmission, Gas		Before FERC		Summer 1990
Rocky Mountain Pipeline Co.	515	Sage, Wyo. to Searchlight, Nev.	583	36	Transmission, Gas	21,000 @ Sage	Planned		
Saginaw Bay Pipeline Co.	65	Bay City to Kalkaska, Mich.	125	10 to 20	Transmission, Gas		Before Mich. Public Serv. Comm.		Late 1990

TABLE 4 *(continued)*

Company	Est. Cost US$ (mm)	Project Location	Length (Miles)	Pipe Diam. in Inches	Type of Service	Compressor or Pump Station hp and Location	Project Status *	Contractor and/or Engineering Firm	Completion Date
Seagull Energy Corp./Quantum Chemical Corp.	•	Deer Park to Mont Belvieu, Texas	73	8 & 10	Trunkline, Products		Planned		Fall 1990
Seagull Shoreline System	• 5	Offshore Texas			Transmission, Gas	6,000	Planned		1990
Southcoast Transmission Corp. (Inland Development Co.)	1 Bil.	Montana, Idaho, Utah, Nevada to Calif.	995	42	Transmission, Gas		Under Study		
Southeastern Michigan Gas Co.	.5	Southeastern Michigan	17	2 to 6	Distribution, Gas		Planned		Feb. '91
Southern Indiana Gas & Electric Co	1.75	Posey County, Ind.	6.5	16	Transmission, Gas		Planned		1990
Southern Interstate Support System	641	Wyoming to California	928	20 & 28	Trunkline, CO_2		Under Study		
Southern Natural Gas Co.	• 2.8	Offshore La.	3.8	8	Transmission, Gas		Before FERC		Oct. '90
	• 1.0	Orleans Parish., La.	1.5	8	Transmission, Gas		Await Start		Mar. '90
	• 2.2	Offshore La.	4.3	12	Transmission, Gas		Working		June '90
Southwestern Public Service Co.		Raton, N. Mex to Lubbock, Texas	300		CO_2 & Coal Slurry Lines		Under Study		1990
Standard Gas Marketing Co.		Bolivan County, Miss., to Desha County, Ark.	30	8	Lateral, Gas		Before FERC		
St. Lawrence System (PennEast CNG Transmission & Texas Eastern Co.)	322	New York	38 83	36 24	Transmission, Gas	54,000	Before FERC		
Sunshine Natural Gas System (Coastal Corp.)	410	Mobile County, Ala. to Miami area	608	24 & 30	Transmission, Gas	33,500 @ 2 Sta.	Before FERC		
Tarpon Offshore Pipeline System (Corpus Christi Oil & Gas Co.)		Offshore Texas	250	24	Transmission, Gas		Planned		
Tennessee Gas Pipeline Co.	62	Mobile County to Forrest County, Ala.	88	30	Transmission, Gas		Before FERC		Mid 1991
	38.7	N.Y. & Pa.	20.1 8.6	30 24	Transmission, Gas	3,165	Before FERC		
	131.6	Conn., Md., R.I. & N.Y.	82.2 6.5 11.5	30 12 20	Transmission, Gas	18,800	Before FERC		
	149.2	Ky., Md., N.Y. & Pa.	55.2	30 & 36	Transmission, Gas	29,500	Before FERC		
Texas Eastern Transmission Corp.	72.5	Onshore Mobile Bay to Offshore Vioscaknoll	55	24	Transmission, Gas		Before FERC		
	44.11	Staten Island, N.Y.				11,000	Before FERC		
(Penn-Jersey System)		Pennsylvania	28.8	36	Trans., Loop, Gas		Before FERC		
Texas Eastern Transmission Corp./ ANR Pipeline Co.	153.3	Onshore Mobile Bay to Kosciusco, Miss.	204	30	Transmission, Gas		Before FERC		
	28	Alabama				22,000	Before FERC		
T.P.C. Services, Inc.	•	Offshore Texas	18.24	16	Dual Phase Transmission, Gas & NGL		Before FERC		
Transcontinental Gas Pipe Line Corp.	19.95	Alabama	15.17	48	Trans., Loop, Gas		Before FERC		Nov. '90
	12.60	Georgia	12.05	42	Trans., Loop, Gas		Before FERC		Nov. '90
	16.56	New Jersey				16,000	Working		Nov. '90
	7.22	Pennsylvania	7.37	16, 30	Trans., Loop, Gas		Before FERC		Nov. '90
	4.16	Pennsylvania	4	30	Trans., Loop, Gas		Before FERC		Oct. '90
	7.83	Georgia				1 each @	Before		Nov. '90
	7.85	South Carolina				12,600	FERC		
	7.83	Alabama				12,600	Before		Nov. '90
	4.07	Mississippi				5,500	FERC		Nov. '90
	• 9.10	Pennsylvania				12,600	Working		Dec. '90
	• 2.32	Louisiana	2.9	10	Transmission, Gas		Under Study		Oct. '90
	• 3.49	Texas	4.2	8	Transmission, Gas		Under Study		Oct. '90
	• 10.37	Pennsylvania	8.22	36	Trans., Loop, Gas		Working		Oct. '90
	• 14.06	Pennsylvania	10.6	36	Trans., Loop, Gas		Working		Oct. '90
	• 7.25	New Jersey	3.86	42	Trans., Loop, Gas		Working		Oct. '90
	• 5.83	New Jersey	4.69	20	Trans., Loop, Gas		Working		July '90
Transok, Inc.	•	Wilburton & Atoka Counties, Okla.	62	24	Transmission, Gas		Planned		
Transwestern Pipeline Co.	•	New Mexico to California	210	30	Trans., Loop, Gas		Planned		
Trunkline Gas Co.	•	Offshore S. Timbalier area, La.	20	24	Transmission, Gas		Before FERC		
United Gas Pipe Line Co.	• 22.8	Marchand to St. Rose, La.	34.6	24	Transmission, Gas		Planned		June '90
Virginia Natural Gas Co.		Dulles Airport to Mechanicsville, Va.	144	24	Transmission, Gas		Planned		
UNOCAL Corp.	30	Union Island to Rodeo, Calif.	65	12	Trunkline, Gas		Planned	Spec Services (E)	Dec. '90
Wesco Pipe Line Co.	2.3	Oklahoma	32	6	Trunkline, Crude & Products		Under Study		

TABLE 4 *(continued)*

Company	Est. Cost US$ (mm)	Project Location	Length (Miles)	Pipe Diam. in Inches	Type of Service	Compressor or Pump Station hp and Location	Project Status *	Contractor and/or Engineering Firm	Completion Date
Wisconsin Natural Gas Co.	• 2.3	Waukesha to New Berlin, Wis.	7	20	Transmission, Gas		Planned		Fall 1990
	0.5	Appleton, Wis.	2	8,12	Transmission, Gas		Planned		Summer 1990
Yukon Pacific Pipeline Co.		North Slope to Anderson Bay, Alaska (near Valdez)	820	36	Transmission, Gas		Planned		

Canada

Outside U.S.A.

Company	Est. Cost US$ (mm)	Project Location	Length (Miles)	Pipe Diam. in Inches	Type of Service	Compressor or Pump Station hp and Location	Project Status *	Contractor and/or Engineering Firm	Completion Date
Altamont Pipeline Project (Amoco Canada Ltd., Shell Canada Ltd., Mobil Canada Ltd.)	• 580	Alberta to Opal, Wyo.	615		Transmission, Gas		Planned		
Alberta Natural Gas Co. Ltd.	•	Between Crowsnest and Kingsgate, B.C.	50	42	Trans., Loop, Gas		Planned		Early 1992
BP Resources Ltd. & Esso Canada Resources	146	N. Saskatchewan River to Cold Lake area	53		Trunkline, Water		Under Study		1991
Canadian Western Natural Gas Co.	• .51	Southern Alberta	7.5	4	Trans., Loop, Gas				Feb. '90
	• .41	Southern Alberta	3	4	Trans., Loop, Gas				Aug. '90
	• 2.60	Western Alberta	13	10	Trans., Loop, Gas				Sept. '90
Coastal Corp. & Union Enterprises (Niagara Pipeline System)	62	Hamilton to Niagara Falls, Ont.	57	24	Transmission, Gas		Before NEB		
Consumers Gas		Ontario & Quebec	108	1¼ to 12	Distribution, Gas (Mains)		Planned		May '90
Foothills Pipe Lines Ltd.	• 3.43 Bil.	MacKenzie Delta to Boundry Lake on Alberta-B.C. border	820	42	Transmission, Gas		Before NEB		1997
Gas Inter-Cite Quebec, Inc.	500	80 Communities in Quebec	4,000		Distribution, Gas		Approved		1991
ICG Scotia Gas Ltd.	400	Nova Scotia	1,500	1¼ to 18	Distribution, Gas		Planned		Dec. '92
ICG Utilities (Alberta) Ltd.	• 2.0	Alberta rural areas	180	¾ to 3	Distribution, Gas		Await Start		Nov. '90
	• 0.4	Alberta rural areas	10	3, 4	Transmission, Gas		Planned		Dec. '90
	0.3	Bonnyville, Alta.	20	¾ to 3	Distribution, Gas		Await Start		Nov. '90
ICG Utilities (Manitoba) Ltd.	•	Winnipeg	5.0	2	Distribution, Gas (Mains)		Working	Harris Holdings (C)	Mar. '90
	•	Winnipeg, Brandon Steinbach & other rural areas	41.8	2,4, & 8	Distribution, Gas (Mains)		Planned		Mar. '91
Interprovincial	• 225 (Can.)	Sarnia, Ontario to Montreal, Quebec	520	10	Trunkline, Gas		Under Study		
		Alberta & Manitoba				3 @ 2,500	Planned		July '90
New Brunswick Government	• 600	Quebec City, Que. to Northern New Brunswick	373		Transmission, Gas		Under Study		
Northwestern Utilities Ltd.	• 2.7	Lodgepole, Alta.			Transmission, Gas	2 @ 1,100	Planned		1990
	• .75	Grande Prairie, Alta.	4	8	Transmission, Gas (Relocation)		Planned		Aug. '90
	• .8	Edmonton, Alberta	2	16	Transmission, Gas (Relocation)		Planned		July '90
	• 1.7	Auburndale, Alta.			Gathering, Gas	1 @ 500	Planned		Feb. '90
NOVA Corporation of Alberta	•	Various locations Alberta	14	4 to 8	Transmission, Gas	117,205	Approved		Summer 1990
	•	Various locations Alberta	224	4 to 20	Transmission, Gas		Under Study		Apr. '90
	•	Various locations Alberta	306	4 to 42	Transmission, Gas	43,583	Under Study		Summer 1990
Pacific Coast Energy Corp.	•	Southern British Columbia	330	10	Transmission, Gas		NEB Approved		Sept. '91
Pacific Coast Energy Corp. (Westcoast Energy, Inc. & Alberta Energy Corp.)		Vancouver Island, B.C.			Transmission, Gas		Working	Majestic Pipelines (C)	End 1990
Peace Pipe Line Ltd.	• 2.0	Laglace	12.5	8	Trunkline, Loop, Crude		Planned		Mar. '90
	• 3.25	Glenevis	12.5	12	Trunkline, Loop, Products		Planned		Apr. '90
Polar Gas Project		Mackenzie Delta to Edson, Alta.	1,330	36	Transmission, Gas		Before NEB		Nov. '96
Sable Gas Systems Ltd.	780	Sable Island to Nova Scotia to Maine	164	30	Transmission, Gas		Before NEB	J.P. Penney Offshore Engrg. (E)	
Shell Canada Ltd. Husky Oil Operations Ltd. & Gulf Canada Resources Ltd.		Between Ram River and Strachan, Alta.	400	3,6,8 10,12	Gathering, Gas Trunkline, Water		Before ERCB		Summer 1991

TABLE 4 *(continued)*

Company	Est. Cost US$ (mm)	Project Location	Length (Miles)	Pipe Diam. in Inches	Type of Service	Compressor or Pump Station hp and Location	Project Status *	Contractor and/or Engineering Firm	Completion Date
Soligaz Pipelines Project (Petromont, Inc., Soquip, Noverco, SNG Group)	•	Sarnia, Ont. to Montreal	497		Trunkline, LPG		Planned		Late 1990
TransCanada PipeLines Ltd.	354	Saskatchewan & Manitoba	189	48	Transmission, Gas	3 @ 35,000	Before NEB		Fall 1990
	61	Southwestern & Southern Ontario	22 20	36 30	Transmission, Gas		Before NEB		Fall 1990
	66	Eastern Ontario	9 16	36 16	Transmission, Gas	2 @ 21,000	Before NEB		Fall 1990
	23	Quebec	15 22	20 12	Transmission, Gas		Before NEB		Fall 1990
	345	Saskatchewan and Manitoba	186	48	Transmission, Gas	1 @ 18,000 2 @ 30,000	Before NEB		Fall 1991
	226	Eastern Ontario	87 3	36 30	Transmission, Gas	1 @ 21,000 7 @ 8,500 1 @ 5,500	Before NEB		Fall 1991
	103	Quebec	6 69	36 24	Transmission, Gas	3 @ 5,500	Before NEB		Fall 1991
	14	Southwestern Ontario	10	36	Transmission, Gas		Before NEB		Fall 1991
	• 145	Saskatchewan and Manitoba	91	48	Transmission, Gas		Before NEB		Fall 1992
	• 104	Eastern Ontario	70	36	Transmission, Gas	1 @ 5,500	Before NEB		Fall 1992
Trans Mountain Pipe Line Co. Ltd.		Alberta to British Columbia	725		Trunkline, Crude		Under Study		
Union Gas Ltd.		Dawn, Ontario				35,000	Planned		Fall 1990
		Lobo Township and Blandford Township, Ontario				1 ea. @ 33,200	Planned		Fall 1990
	•	Southwestern Ontario	5.04	10	Gathering and Transmission, Gas		Planned		Summer 1990
	•	Southwestern Ontario	21	16	Transmission, Gas		Planned		Summer 1990
	•	Southwestern Ontario	2.36	26 & 34	Transmission, Gas		Planned		Summer 1990
	•	Southwestern Ontario	41.88	48	Transmission, Gas		Planned		Summer 1990
Westcoast Energy, Inc.	•	North British Columbia	16.7	8	Gathering, Gas		Await Start		Mar. '90
	250	South British Columbia	330	10	Transmission, Gas		Planned		Sept. '90

Europe/USSR

Company	Est. Cost US$ (mm)	Project Location	Length (Miles)	Pipe Diam. in Inches	Type of Service	Compressor or Pump Station hp and Location	Project Status *	Contractor and/or Engineering Firm	Completion Date
Aethylen-Rohrleitungs-Ges. Nat. Mij der Pijpleidingen (Belgium-Germany)		Antwerp to Tessenderlo, Belgium	39	10, 12	Trunkline, Products		Working	NMP/Haecon (E)	1990
Distrigaz (Belgium)	21.3	Brussels to Ronquieres	19	24	Trans., Loop, Gas		Planned		1990
	168	Zeebrugge to Blaregnies	91	40	Transmission, Gas		Planned		End 1992
Dansk Underground Consortium (Denmark)	• 77	Denmark	50		Transmission, Gas		Planned		
Statoil, Dansk Olie & Naturgas (DONG) (Joint Venture) (Denmark)	270	Ekofisk field to Tyra field	118		Transmission, Gas		Under Study		
Ministry of Trade of Finland (Finland)	300	Rihimaki, Finland to Gaevle, Sweden (inshore & offshore)	250	20	Transmission, Gas		Under Study		
Greek Government (Greece)		USSR to Northern Greece	746		Transmission, Gas		Under Study		Early 1990
Gasunie (Holland)	• 14	Rysbergen to Nispen	12.5	36	Transmission, Gas		Planned		Dec. '90
	• 8	Monster to Gaag	12.5	12	Transmission, Gas		Planned		Oct. '90
	• 48	Balgzand to Oud.tocht	19	48	Transmission, Gas		Planned		Oct. '91
	• 22	Ten Boer to Borger Compagnie	11.5	42	Transmission, Gas		Under Study		Oct. '93
	• 40	Velzen to Oud.tocht	34.5	24	Transmission, Gas		Planned		Oct. '93
	• 50	Ommel.w to Witteveen	34.5	24	Transmission, Gas		Planned		Oct. '92
	• 10	Waalwyk to Kedichem	12.5	12	Transmission, Gas		Planned		Oct. '91
Hungarian Oil & Gas Trust (Hungary)		Pilisvorosvar to Dorog	22	30	Transmission, Gas		Planned		1990
Bord Gais Eireann (Ireland)	31.5	Limerick, Waterford, Progheda & Dundalk			Distribution, Gas		Planned		
City of Genoa (Italy)	•	Genoa	30		Distribution, Gas		Planned	British Gas (E)	1993
Consorzio Valle Agno (Italy)		Valdagno to Montecchio Maggiore	16.25	16,28	Trunkline, Water		Working	Saipem (C)	Oct. '90
Snam S.p.A. (Italy)	9	Recco to Ge-Staglieno	15.5	12,16, 18	Lateral, Gas		Working	SICIM (C) Snamprogetti (E)	Apr. '90
		USSR to Italy (Italian section)	112 (Exp.)	42	Transmission, Gas		Working (E)	Snamprogetti (C)	1990
Cons. Valle Agno (Italy)	•	Valli Agno to Chiampov., Italy	26		Water line		Working	Saipem S.p.A. (C)	Oct. '90

TABLE 4 *(continued)*

Company	Est. Cost US$ (mm)	Project Location	Length (Miles)	Pipe Diam. in Inches	Type of Service	Compressor or Pump Station hp and Location	Project Status *	Contractor and/or Engineering Firm	Completion Date
NAM, Elf Petroland Energie, Belneer Nederland (Nogat System) (Netherlands)	•	Blk.F3 offshore to inshore terminal at Den Helder	93	36	Transmission, Gas		Await Start	McDermott-ETPM J.V. (C)	1991
			155	10	Trunkline, NGL		Await Start	John Brown Engineers & Constr. (E)	1991
Scanpipe (Norway)		Norway & Sweden	300		Transmission, Gas		Under Study		
Statoil (Norway)	1.61	North Sea (Zeepipe) (Phase I)	816	40	Trunkline, Gas		Await Start	European Contractors (C)	1993
		North Sea (Sleipner-Statpipe)	40	30	Transmission, Gas		Approved		1993
		North Sea (Sleipner) Kaarstoe, Norway	245	20	Trunkline, NGL		Approved		1993
Petroquimica e Gas de Portugal (Portugal)	•	Various locations Portugal	2,175		Distribution, Gas		Under Study	Services Energeticos SA (E)	2010
CAMPSA (Spain)	56	Tarragona to Barcelona to Gerona	134	8 & 12	Trunkline, Products and Terminals	2,000	Working	A.F.E./ O.S.H.S.A. (C)	1990
	19	Valladolid to Salamanca	70	8	Trunkline, Products and Terminals	800	Planned	Initec (E)	1990
	19	Palencia to Leon	70	8	Trunkline, Products and Terminals	800	Await Start	Diseprosa (E)	1990
	• 17	Castellon to Valencia	37	10	Trunkline, Products and Terminals	2,400	Planned	Intecsa (E)	1991
ENAGAS S.A. and Gaz de France (Spain)		From France through Spain to Portugal border	900		Transmission, Gas		Under Study		
Swedegas (Sweden)		Gothenburg to Oreboro to Gavle to Stockholm	248	24	Transmission, Gas		Planned		1990
Vattenfall/Swedegas A/B Statoil (Sweden)		Norway to Sweden to Germany	1,000	56	Transmission, Gas		Under Study		
Gasverbund Mittelland AG (Switzerland)	10	Northwest Switzerland	15	8	Transmission, Gas		Planned	Helbling Engrg. (E)	1991
BP Exploration (UK)		Miller field to St. Fergus	150.4	30	Transmission, Gas		Await Start	European Marine Contractors Ltd. (C) British Pipe Coaters Ltd. (C)	Sept. '90
	•	Miller platform to Brae A platform	4.9	18	Trunkline, Crude				
	•	Forties field North Sea	105	36	Trunkline, Crude		Await Start	ETPM Services (C)	July '90
British Gas plc (UK)		Easington to Paull	15.6	36	Transmission, Gas		Planned		Oct. '91
		Cambridge to Epping	37	36	Transmission, Gas		Planned		Oct. '90
		Steppingley to Whitwell	12.5	36	Transmission, Gas		Planned		Oct. '90
		Easington to Killingholme	25	8	Condensate, Products		Planned		Oct. '90
	•	Epping to Horndon	16	36	Transmission, Gas		Planned		Oct. '92
	•	Beeford to Hornsea	2.5	18	Transmission, Gas		Planned		Oct. '91
British Petroleum plc (Forties Pipeline) (UK)	•	Forties field to Cruden Bay terminal, UK	105	36	Trunkline, Crude		Await Start	ETPM Services (C) British Pipe Coaters (C)	1990
Gas Transmission UK Ltd. (British Dept. of Energy) (UK)	•	Barton terminal to Thames Estuary, London	180	30 to 36	Transmission, Gas		Planned		1992
Invermoray Hydrocarbon Utility Co. Ltd. (Costain Group) (UK)	606	Bruce field to Nigg Bay	275	24	Trunkline, NGL		Under Study		
Mobil North Sea Ltd. (UK)	•	Offshore Beryl field to St. Fergus	210	30	Transmission, Gas		Await Start	J.P. Kenny (E) Ralph M. Parsons Co. Ltd. (E)	1992
Petrofina (UK) Ltd. (UK)		South Humberside to Hertfordshire	139		Trunkline, Products		Planned	John Brown (E)	
Phillips UK (UK)		Audrey field	70	30	Transmission, Gas		Planned		
Shell/Esso (UK)		Offshore Clipper field to Bacton Terminal	46	24	Transmission, Gas & Condensate		Planned		Oct. '90
		Gannet field	35	3,4, 10,12	Gathering, Gas & Oil & Gas lift line		Planned		1990
Russia-Japan (USSR)	3.8 Bil.	Off Sakhalin Island to Sekasti, USSR	143		Transmission, Gas & Liquids		Under Study		
USSR		Yanburg to Zakavkazye to Gorky	2,814	56	Transmission, Gas		Planned		1990
		Gorky to Talla to Kiev	1,960	56	Transmission, Gas		Planned		1990
		Kiev to Novolzhyr	1,713	56	Transmission, Gas		Planned		1990
		Various locations	4,944		Transmission, Gas		Planned		1990
East German and Czechoslovakia governments (East Germany)	• 700	Postak to Berlin, East Germany to Dresden to Prague, Czech.	1,100		Transmission, Gas		Under Study		
Bayerngas GmbH (West Germany)	•	Amerdingen to Anwalting	32	32	Transmission, Gas		Planned		1991

TABLE 4 *(continued)*

Company	Est. Cost US$ (mm)	Project Location	Length (Miles)	Pipe Diam. in Inches	Type of Service	Compressor or Pump Station hp and Location	Project Status *	Contractor and/or Engineering Firm	Completion Date
Erdgas-Verkaufs-GmbH, Munster (West Germany)	40	Rehden to Frenswegen	67	18	Transmission, Gas		Working	Preussag/ Winter Rohrbau (C)	Oct. '90
Gasversorgung Suddeutschland GmbH (West Germany)		Amerdingen to Scharenstetten	37	28	Transmission, Gas		Planned		1990
Government of the Federal Republic of Germany (West Germany)	10.5	Lechfield to Leipheim	44	10	Trunkline, Products		Working	Mannesmann Anlagenbau AG/ Sudrohrbau GmbH (C)	End 1990
Ruhrgas AG (West Germany)		Obermichelbach -Amerdingen	65	36	Distribution, Gas		Working	Pipeline Engrg. GmbH (E)	End 1990
		GieBen	3	8	Loop, Gas		Working	Pipeline Engrg. GmbH (E)	End 1990
		Bonn-Euskirchen	3	12	Loop, Gas		Working	Pipeline Engrg. GmbH (E)	End 1990
	•	Weidenhousen to Asslar	9	12	Loop, Gas		Working	Pipeline Engrg. GmbH (E)	End 1990
	•	Wunstorf	3	8	Loop, Gas		Working & Planned	Pipeline Engrg. GmbH (E)	1990-91
	•	Amerdingen to Anwalting	32	32	Distribution, Gas		Working	Pipeline Engrg. GmbH (E)	1990-91
Wintershall A.G. (Middle German connecting line) (West Germany)	• 408	Emden to Lugwigshafen	349		Transmission, Gas		Planned		1993
INA Naftaplin (Yugoslavia)		Invana IKA offshore field to Pula	29	18	Transmission, Gas		Planned	Brown & Root (UK) (E)	

Middle East

Company	Est. Cost US$ (mm)	Project Location	Length (Miles)	Pipe Diam. in Inches	Type of Service	Compressor or Pump Station hp and Location	Project Status *	Contractor and/or Engineering Firm	Completion Date
Iran-Turkey governments	10	Iran, Turkey, Greece to Italy	4,000		Transmission, Gas		Under Study		
National Iranian Oil Co. (Iran)	1 Bil.	Dalan field to Khuzestan field	212		Transmission, Gas		Planned		
North Oil Co. Kirkuk (Iraq)	190	Kirkuk, Iraq	300	14	Gathering, Water Trunkline, Crude		Working	Mannesmann Anlagenbau AG, Dusseldorf/FRG (C)	Aug.'90
SCOP (Iraq)	•	Khor al-Zubair Port to Basra refinery	50		Trunkline, Products		Planned		1992
		North to South oil fields	400	42	Trunkline, Crude		Planned		1996
		West Qarma field	186		Gathering, Trunkline, Crude		Working	TSMPC (C)	1990
Israel-Egypt governments (Israel)		Gulf of Suez Ashdon, Israel	323		Transmission, Gas		Under Study		
Ministry of Energy and Mineral Resources (Jordan)		Al-Risha field to Azzarqa	217		Transmission, Gas		Under Study		
Water Authority of Jordan (South Amman water project) (Jordan)	44	Al Qaim, Iraq to Amman, Jordan	750		Trunkline, Water		Under Study		
Kuwait National Petroleum Co. (Kuwait)		Kuwait to Yanbu, Saudi Arabia	700	Dual 36	Trunkline, Crude		Under Study	Santa Fe Braun (E)	
Ministry of Electricity & Water (Kuwait)		Basrah, Iraq to Kuwait			Trunkline, Water		Under Study	Braun Trans-world Corp. (E)	
Qatar General Petroleum Co. (Qatar)		North field offshore	387	12,20, 34	Transmission, Gas Trunkline, NGL		Working	Saipem (C) Bechtel/ Technip (C)	Nov.'90
Aramco (Saudi Arabia)		Petroline	62		Trunkline, Loops, Oil		Await Start	Mannesmann Anlagenbau A.G. (C)	1990
Ankara Electric & Gas Authority (Turkey)	6.1	Ankara	10		Transmission, Gas		Working	Tekfen (C) Sofregaz (E)	1990
Ankara Natural Gas Project (Ankara Metropolitan Municipality-Directorate of Electricity Gas Omnibus) (Turkey)	140.39	Ankara	621	1/4 to 24	Distribution, Gas		Working	Amec Intn'l. Constr./Kutlutas Constr. & Trade Industry Ltd. J.V. (E&C)	Jan.'92
Botas (Turkey)	•	Tekirdag to Eskisehir	155	4 to 30	Transmission, Gas		Working	Altila Dogan (C)	Mid 1990
Istanbul Electricity, Transway & Tunnel Authority (Turkey)	400	Istanbul, Turkey	1,299		Distribution, Gas		Planned	Sofregaz (E)	1991
Istanbul Gaz Digitma AS (Turkey)		Istanbul	1,243		Distribution, Gas		Working	Societe Auxiliare d'Enterprises (E&C)	1991
Turkey Government (Turkey)		Seyhan and Ceyhan Rivers to the Arabian Peninsula	1,267		Trunkline, Water		Under Study	Brown & Root (UK) (E)	
Turkish Ministry of Defense (Nato Infrastructure Dept.) (Turkey)	•	Aliaga to Cigili	23	8	Trunkline, Products		Planned		Mid 1990

TABLE 4 *(continued)*

Company	Est. Cost US$ (mm)	Project Location	Length (Miles)	Pipe Diam. in Inches	Type of Service	Compressor or Pump Station hp and Location	Project Status *	Contractor and/or Engineering Firm	Completion Date
Abu Dhabi National Oil Co. (ADNOC) (U.A.E.)		Maqta to Al Ain	93		Transmission, Gas		Approved		
		Habshan to Fujairah	186	48	Trunkline, Crude		Under Study		
		Mugata field	37	24	Transmission, Gas			Arab Engrg. Co. (E)	
		Ruwais to Umm Al-Nar	140		Trunkline, Products		Under Study		
		Umm Al-Nar to Jabel Ali	59		Trunkline, Products		Under Study		
		Jabel Ali to Layyah	37		Trunkline, Products		Under Study		
		Layyah to Ras al-Khalimah	56		Trunkline, Products		Under Study		
		Layyah to Fulairah	81		Trunkline, Products		Under Study		
Emirates General Petroleum Co. (UAE)		Sharjah to Dubai	50		Transmission, Gas		Planned	Pipe Line Engrg. GmbH (E)	
		Offshore Bukha field to Sohar	217		Transmission, Gas & NGL		Under Study	Consolidated International Petr. Co. (E) Brown & Root (UK) (E)	
North Yemen Government (North Yemen)		Alif field to coast	372		Transmission, Gas		Under Study		1991
South Yemen Government (South Yemen)		Shawba field to Bir' Ali	124		Trunkline, Products		Planned		

Africa

Company	Est. Cost US$ (mm)	Project Location	Length (Miles)	Pipe Diam. in Inches	Type of Service	Compressor or Pump Station hp and Location	Project Status *	Contractor and/or Engineering Firm	Completion Date
Maghreb Arab Co. for Transportation of Natural Gas (Algeria)		Algeria to Tunisia to Libya	249		Transmission, Gas		Under Study		
Badreddin Petroleum Co. (Shell Mining & Egyptian General Petroleum Co. (Egypt)		Badr-al-din field to El Amiriya	160	20	Transmission, Gas		Working	Petrojet (C) EPPI (E)	Mid 1990
Egypt & Jordan (Trans-Sinai Line) (Egypt)		Nev'ot to Ras Misalla, Sinai Peninsula	280		Trunkline, Crude		Under Study		
Egyptian General Petroleum/Shell Winning (Egypt)		Amiriyah region	186		Transmission, Gas		Planned		1996
Petroleum Pipeline Co. (Egypt)		Egypt	180	20	Distribution, Gas		Planned	Snamprogetti (E)	
Kenya Pipeline Co. Ltd. (Kenya)	•	Nairobi to Kisumu/Eldoret	277	6 & 8	Trunkline, Products	670	Await Start	NKK/Lavalin/ SOGEA (C)	Early 1992
	•	Kisumu, Kenya to Kampala, Uganda to Kagali, Rwanda to Bujumburu, Burundi	304	8	Trunkline, Products		Under Study		
Libya Agriculture Secretariat (Great Man-made) (Libya)		Sarir Qattush wells to Tripoli (Phase II GMR)		90	Trunkline, Water		Await Start	Brown & Root (UK) (E) Dong Ah (C)	
Trans-Maghrebian Natural Gas Co. (Sonatrach, ETAP, National Oil Co. of Libya) (Libya)		Trans-Med line at Tunisian border to Tripoli tos Misirata	670		Transmission, Gas		Under Study		
Madagascar Government (Madagascar)		Offshore West Manambolo field to Antonanarivo	220		Transmission, Gas		Under Study		
Soc. Cherifienne des Petroles (Morocco)		Mohammedia to Sidi Kacem	110	14	Trunkline, Oil		Under Study	O.T.P. (E)	1991
Economic Community of West Africa (Nigeria)	2.8 Bil.	Nigeria to Ghana to Benin to Toto, Nigeria & Escravos to Niger	1,100		Transmission, Gas		Under Study		
Nigerian National Petroleum Co. (Nigeria)		Throughout Nigeria	1,200	12 & 16	Trunkline, Crude		Under Study		
Soekor (South Africa)		Offshore to Mossel Bay to Capetown	248		Transmission, Gas		Under Study		
Mining and Energy Authority (Zaire)		Matdi to Muanda	93		Trunkline, Products		Under Study	Technip (E)	

Western Pacific

Company	Est. Cost US$ (mm)	Project Location	Length (Miles)	Pipe Diam. in Inches	Type of Service	Compressor or Pump Station hp and Location	Project Status *	Contractor and/or Engineering Firm	Completion Date
Australian Gas Light Co. and Oilco of Australia NL (Australia)	•	Denison Trough to Gladstone	125		Gathering, Gas		Planned		1992

TABLE 4 *(continued)*

Company	Est. Cost US$ (mm)	Project Location	Length (Miles)	Pipe Diam. in Inches	Type of Service	Compressor or Pump Station hp and Location	Project Status •	Contractor and/or Engineering Firm	Completion Date
Bridge Oil Ltd. (Australia)		Silver Springs	74	4	Trunkline, LPG		Planned		
Broome Oil Pipeline Co. (Australia)		Canning Basin to Broome	375	6	Trunkline, Crude		Planned	William Bros.-CMPS Engrs. (E)	
CSR Petroleum (Australia)	50	Central Queensland, Australia	150	1 to 6	Gathering and Transmission, Gas		Planned		July '90
Doral Resources NL (Australia)	• 15	Dampier to Perth			Transmission, Gas		Under Study		Mid 1991
Elf Aquitaine & Partners (Australia)	2.80 Bil.	Petrel & Tern fields to shore at Bonaporte Gulf to Darwin, Australia	234		Transmission, Gas & LNG Plant		Under Study		
Esso Expro Australia & BHP Petroleum (Australia)		Bass Strait fields to Longford, Victoria	Dual 29		Transmission, Gas & Dual phase line		Planned		
The Pipeline Authority (Australia)	16	Wagga Wagga or Junee to Narrandera Leeton & Griffith	95	6	Transmission, Gas		Under Study		
	12	Marsden to Forbes & Parkes	59	6	Transmission, Gas		Under Study		
	41	Albury to Wagga Wagga	82	12	Transmission, Gas		Under Study		
Queensland Water Resources Commission (Australia)		Fitzroy River to Stanwell Power Station, Queensland	17	36	Trunkline, Water		Planned		Fall 1992
Woodside Petroleum Development Ltd. (Australia)		North Rankin field and Dampier-Pibara areas	22 118	30 10 & 14	Transmission, Gas Laterals, Gas		Planned Planned		
Pertamina & Perun Gas Negara (Indonesia)	7.5 Bil.	Offshore Natuna Java & Sumatra	1,864		Transmission and Distribution, Gas		Under Study	Gasunie Engrg. Co. (E)	1990
Pertamina-KODECO (Indonesia)		Madura shore to East Java shore	39.4	24	Gathering, Gas		Under Study		
Perusahaan Umum Gas Negara (Indonesia)	•	Indonesia	187		Distribution, Gas		Planned		End 1992
Petronas Carigali SDN BHD (Malaysia)	•	Dulang field	1.9 1.6 1.6	12,18 10,16 6	Gathering, Oil & Gas Trunkline, Oil Transmission, Gas		Working	Intec/Technip (E) McDermott (C)	July '90
Petronas (Peninsula Gas Project) (Malaysia)		Kerteh to Port Kelang	294 128 86	36 30 6 to 20	Transmission, Gas		Working	MMC Gas Co. (Malaysia Mining Corp.), Entrepose Intn'l. & IMEG (C)	1992
Bougainville Copper Co. (Papua, New Guinea)		Bougainville Island	20	34	Slurry, Copper		Planned	Minenco-Bechtel (E)	
Niagini Gulf Oil Pty. Ltd. (Papua, New Guinea)		Iagitu to Lae	373		Trunkline, Crude		Under Study		
New Zealand Energy Ministry (New Zealand)		North Island	145	14 & 20	Transmission, Gas		Under Study		
Southgas Resources (New Zealand)		Invercargil area Dunedin area	62 112		Transmission, Gas		Under Study		

Southern Asia

Company	Est. Cost US$ (mm)	Project Location	Length (Miles)	Pipe Diam. in Inches	Type of Service	Compressor or Pump Station hp and Location	Project Status •	Contractor and/or Engineering Firm	Completion Date
Bangladesh Oil & Gas Corp. (Bangladesh)	110	Kailashtilla field to Ashuganj	110	24	Transmission, Gas		Planned		
		Beano to Kailashtilla	110 110	20 8	Transmission, Gas Trunkline, NGL		Under Study		
		Kailashtilla to Ashugani	107	5	Trunkline, NGL		Under Study		
		Sylhef field to Ashiganj	100		Transmission, Gas		Under Study		
Titus Gas Co. (Bangladesh)	75	Ashiganj to Elega Bangladesh	219 205 2,000		Transmission, Gas Distribution, Gas Distribution, Gas		Planned Planned		1990 1996
Burma Government (Burma)		Payagon gas field to Ranyoon	50	18	Transmission, Gas		Planned		
Beijing Municipality (China)		Beijing area	150		Transmission, Coal Gas		Working	Lummus Crest, Kaiser Engrg. (E) Compagnia Tecnica Internazionale Progetti (C)	
China National Oil Corp.— Pipeline Bureau (China)	•	Zhongyuan field to Luoyang, Henan Province	180	16	Trunkline		Planned		
Gas Utilization of China National Offshore (China)		South China Sea to Hainan Island and to Guangdong (Qiang-Shen Project)	683	24			Planned	Fluor Daniel (PRC) (E) Pipeline Engrg. (E)	

TABLE 4 *(continued)*

Company	Est. Cost US$ (mm)	Project Location	Length (Miles)	Pipe Diam. in Inches	Type of Service	Compressor or Pump Station hp and Location	Project Status *	Contractor and/or Engineering Firm	Completion Date
Ministry of Energy (China)		Zhongyuan fields to Henan	174		Trunkline, Crude		Planned		
		Dagang and Shengli oil fields	200		Trunkline and Laterals, Crude		Planned	Novacorp (E)	
		Zhongyuan to Cangzhouand Jilu	239		Transmission, Gas		Planned	Novacorp (E)	
National Coal Development Corp. (China)		Shanxi Province to Jiangsu Province to Qinhuangdao	530	28	Slurry, Coal (Dual Lines)		Under Study	Bechtel (E)	
		Shanxi Province to Hamtong	600				Under Study	Bechtel (E)	
Oil & Natural Gas Commission (ONGC) (India)	22	Hazira, India	15	24	Trunkline, Products		Working	Land & Marine (India) (C) EIL (E)	1990
		Trivandrum to Jammu	2,600		Transmission, Gas		Under Study		
		Calcutta to Kandla	3,880		Transmission, Gas		Under Study		
	700	Offshore & inshore India	560	26,30, 36	Transmission, Gas		Planned		
		Naraspurt Kovvur	46	8	Transmission, Gas		Planned		
		Duliajou to New Boafaigaan, Assam	400		Transmission, Gas		Under Study		
Oil India Ltd. (India)		Kandla to Bhatind, Punjab	600		Trunkline, Products		Under Study		
		Hazira to Kogali to Viramgam	200	24	Trunkline, Crude		Planned		
		Haldia to Barauni	325	12	Trunkline, Crude		Planned		
Japan National Oil Corp., Japan Natural Gas Asso. and Federation of Electric Power Companies (Japan)		Honshu Island	1,864		Transmission and Distribution, Gas		Under Study		
Tokyo Gas Co. Ltd. (Japan)		Tokyo to Kanagawa	25	30	Transmission, Gas		Working	Nippon Kokan, Nippon Steel (C)	1994
Pakistan Oil & Gas Corp. (Pakistan)		Badinfield to Karachi to Jamshoro	186		Trunkline, Crude		Approved		
PARCO (Pakistan)	•	Karachi to Multan	540	16	Trunkline, Products		Under Study	O.T.P. (E)	1990
	•	Multan to Lahore	225	16	Trunkline, Products		Under Study	O.T.P. (E)	1991
Sui Gas Transmission Co. (Pakistan)		Indus/Right Bank	311		Transmission, Gas		Under Study	Fluor (Great Britain) Ltd./Enar, Petrotech (E)	
Chinese Petroleum Corp. (Taiwan)	26.4	Keelung to Taoyuan	111	12	Trunkline, Oil		Planned		
	10	Keelung to Taoyuan	37	12	Transmission, Gas		Planned		
		Taipei to Khaesiung	217	26	Transmission, Gas		Working	CTC Corp. (C) M.W. Kellogg (E)	Mid 1990
Bangkok Aviation Services (Thailand)		Refinery to Don Muang Airport	25		Trunkline, Products		Under Study	Dorsch Consulting (E)	
Petroleum Authority of Thailand (Thailand)		Backchak Refinery to Ayutthaya	50	14 to 16	Trunkline, Products		Under Study	Bechtel, Inc. (E)	
		Namphung field to Central Plains District	248		Transmission, Gas		Under Study	Bechtel, Inc. (E)	
Public Works Dept., Bangkok (Thailand)		Khorat Water System	61		Trunkline, Water		Planned		

Mexico/Central America

Company	Est. Cost US$ (mm)	Project Location	Length (Miles)	Pipe Diam. in Inches	Type of Service	Compressor or Pump Station hp and Location	Project Status *	Contractor and/or Engineering Firm	Completion Date
Government of Costa Rica (Costa Rica)	450	Rincon to Gandoca	94		Trunkline, Crude		Planned	Oleoducto de Costa Rica/ Mogos-Moin, S.A. (E&C)	
Petroleos Mexicanos (PEMEX) (Mexico)		Matapionche to Paso Del Toro	26.7	20	Transmission, Gas		Planned		
		Sen to Pijije	9.3	12	Gathering, Gas & Crude		Planned		
		Pijije to	24.8	16	Trunkline, Crude		Planned		
		Sen to Oxiacaque	21.7	10	Trunkline, Products		Planned		
Panama Government (Panama)	4.5 Bil.	Puerto Armuelles to Chiriqui Grande	81		Trunkline, Methanol		Under Study	Ebasco Services, Inc. (E)	1990

South America

Company	Est. Cost US$ (mm)	Project Location	Length (Miles)	Pipe Diam. in Inches	Type of Service	Compressor or Pump Station hp and Location	Project Status *	Contractor and/or Engineering Firm	Completion Date
Gaz del Estado (Argentina)		Rosario, Argentina to Porto Alegre, Brazil	342	16	Transmission, Gas		Under Study		1990
		Santa Rosa to Gral. Pico	69	8	Transmission, Gas		Under Study		
		Briloche to Esquel	231	4 & 8	Transmission, Gas		Under Study		
		Pergamino to Bragado	110	6 to 12	Transmission, Gas		Under Study		

TABLE 4 *(continued)*

Company	Est. Cost US$ (mm)	Project Location	Length (Miles)	Pipe Diam. in Inches	Type of Service	Compressor or Pump Station hp and Location	Project Status •	Contractor and/or Engineering Firm	Completion Date
Bolivia Government (Bolivia)	• 1 Bil.	Santa Cruz, Bolivia to Matto Grasso State, Brazil	373		Transmission, Gas		Under Study		
Comgas (Brazil)		São Paulo State	9 20 10	10 12 8	Transmission, Gas		Planned		
Ministry of Mines and Energy (Brazil)	5 Bil.	Carauan to São Paulo	1,864		Transmission, Gas		Under Study		
National Petroleum Council (Brazil)	200	Rio de Janeiro and São Paulo	232		Laterals & Distribution, Gas		Under Study		
Petroleo Brasileiro S.A.—PETROBRAS (Brazil)	70	Guarerema to Lorena to Japeri	160	16	Trunkline, Products		Working	Techint (C) Mendes Junior (C)	1991
		Duque de Caxias to Ilha D'Agua	12 12	14 16	Trunkline, Products		Working		1990
		Cabiunas to Rio de Janeiro	40	20	Transmission, Gas		Planned		1991
		Barra do Furado to Cabiunas I	43	42	Trunkline, Crude		Planned		1990
		Guarerema to REPLAN II	94	16	Trunkline, Products		Planned		1991
		RLAM-TEMADRE	6	16, 18	Trunkline, Products		Planned		1990
		RLAM to COPENE	23	14	Trunkline, Products		Planned		1990
		REDUC to REVAP	174	32	Trunkline, Crude		Planned		1992
		Urucu to P.Sao	124	12	Transmission, Crude		Planned		1993
		TORGUA to Sao Goncalo	6	6	Trunkline, Products		Under Study		1991
		REPAR to Florianopolis	174	8 & 10	Trunkline, Products		Planned		1991
		TEMADRE to Itabuna	227	8 & 10	Trunkline, Products		Planned		1991
		REPLAN to Brasilia	690	10 & 18	Trunkline, Products		Planned		1992
ENAP-Magallanes (Chile)	• 0.6	Posesion 5 Plat. to Inshore	2.8	8	Transmission, Gas		Await Start	Enap-Magallanes (E&C)	Sept. '90
	• 0.2	Daniel field	1.5	6	Transmission, Gas		Await Start	Enap-Magallanes (E&C)	Apr. '90
	• 0.3	Daniel field	1.5	8	Trunkline, Crude		Await Start	Enap-Magallanes (E&C)	Apr. '90
Soc. National del Oleoductos (SONACOL) (Chile)		San Vincente to Temuco	168	6	Trunkline, Products		Under Study		
Ecopetrol (Colombia)	•	Neiva to Natagaina	75	6	Trunkline, Products		Working	Prodegas-Mora Mora/JV (C)	June '90
	•	Oleoducto Toldado to Gualanday	61	10	Trunkline, Crude		Await Start		June '90
	•	Galan to Vizcaina	15	20	Trunkline, Crude		Planned		Dec. '90
	•	Girardota to Bello	18	12	Trunkline, Products		Planned		June '90
	•	Pozos Colorados to Ayacucho	132	14	Trunkline, Products		Await Start		Dec. '90
	•	Variante Tulua	13	8	Trunkline, Products		Planned		Mar. '90
	•	Guacari to Paso La Torre	30	8	Trunkline, Products		Planned		Oct. '90
	•	Buenaventura to Dagua	56	8	Trunkline, Products		Planned		Oct. '91
Hocol, S.A., Esso Colombia S.A. & Total (Colombia)	•	Tenay to Vasconia	60 340	12 20	Trunkline, Crude		Working	Willbros Colombia S.A. (C)	July '90
Oficinas de Distrito de Oleoductos (Colombia)		Casanara field to Velasquez	167		Trunkline, Crude		Planned		
Oleoducto de Colombia, Ecopetrol, Hocol, S.A. Esso Colombia S.A., Total-Elf-LL&B (joint venture) (Colombia)		Vasconia to Covenas	476	24	Trunkline, Crude		Working	Spie Capag (C) Techint (C)	Feb. '91
Promigas S.A. (Colombia)	•	Pajaro to Bogota	600 400	20 24	Transmission, Gas		Under Study		
Petroperu (Peru)	•	Chambra 1284 field to Blk. 1-AB Northern Peru	18		Gathering, Crude		Planned		
Shell Exploratora y Productora de Peru (Peru)	750	Ucayali distr. of the Amazon to coast	329 329		Transmission, Gas Trunkline, NGL		Under Study		1990
Corpoven S.A. (Venezuela)	800	Anaco to Edo. Anzoategui	500	12 & 36	Transmission, Gas		Working		1992
	14	Yaritagua to Edo. Yaracuy/ Acarigua Edo. Portuguesa	44	10	Distribution, Gas		Under Study		1992
Maraven S.A. (Venezuela)		San Lorenzo to El Vigia	172	12	Trunkline, Products		Planned		Mid 1991
		El Vigia to Tachira State	84	12					

United States Gas Lines

Some of the larger natural gas transmission and gathering systems scheduled to be installed in 1990 are listed below. The caveat is, as usual, the Federal Energy Regulatory Commission's (FERC) timely actions on pending applications. The lines are:

Columbia Gas Transmission Co., 112 mi of 3-in. to 30-in. transmission and gathering lines and 20 mi of 16-in. offshore lines
El Paso Natural Gas Co., 30 mi of 30-in.
Enron's Transwestern Pipeline Co., 121 mi of 30-in.
Enron's Northern Natural Gas Co., 45 mi of 30-in. and 50 mi of 16-in.
Enron's Texoma Pipeline, 39 mi of 30-in.
Indiana Ohio Pipeline Co., 111 mi of 30-in.
Magnolia Pipeline Co., 554 mi of 24-in. transmission lines and 40 mi of 16-in. to 24-in. laterals
Natural Gas Pipeline Company of America, 191 mi of 30-in.

Other liquid lines to be installed in 1990 include:

Amoco Pipeline Co., 50 mi of 6-in. offshore lines
Amoco Pipe Co., 87 mi of 4-in. to 18-in. crude and products lines throughout its system
Meridian Oil Gathering Co., 120 mi of 10 to 20-in.
Mobil Oil Co., 75 mi of 16-in. in California
Seagull Energy Co., 73 mi of 8-in. and 10-in. products lines
UNOCAL Corp., 65 mi of 12-in. crude lines in California

International Areas

Construction of natural gas lines in Canada, Europe, the Middle East, and Southern Asia will increase in 1990. Some of the gas and liquids line projects in 1990 include:

Nova Corp., 544 mi of various size lines in Canada
TransCanada PipeLine Ltd., 271 mi of gas lines in Canada
Pacific Coast Energy Corp., 116 mi of 10-in. gas lines in British Colombia
Oleoducto de Colombia, 476 mi of 24-in. crude lines in Colombia
Petrobras, 160 mi of 16-in. products lines in Brazil

Offshore Lines

Pipeline construction in the North Sea shows a large increase, with contracts awarded to install, over a 2 to 3-year period, the following lines:

Statoil (Norway), 816 mi of 40-in. gas lines
NAM (Netherlands), 93 mi of 36-in. gas lines
BP Expro (UK), 150 mi of 30-in. gas lines from the Miller field
British Petroleum (UK), 105 mi of 36-in. crude lines (Forties Field to Cruden Bay)
Mobil North Sea Ltd. (UK), 210 mi of 30-in. gas lines from the Beryl field

Edited by JOHN J. McKETTA

Index